Handbook of High-Temperature Superconductivity

J. Robert Schrieffer

Handbook of High-Temperature Superconductivity

Theory and Experiment

J. Robert Schrieffer
Editor

James S. Brooks
Associate Editor

J. Robert Schrieffer
National High Magnetic Field Laboratory
Florida State University
1800 East Paul Dirac Drive
Tallahassee Florida 32310

James S. Brooks
Department of Physics and
National High Magnetic Field Laboratory
1800 East Paul Dirac Drive
Florida State University
Tallahassee Florida 32310 USA
brooks@magnet.fsu.edu

Front Cover Image: *Angle resolved phase sensitive determination of the in-plane superconducting gap in* $YBa_2Cu_3O_{7-\delta}$. Combined SQUID microscope images of a series of 2-junction YBCO/Nb rings, with one junction angle fixed at 167.5 degrees relative to the majority twin a-axis direction of the YBCO, and the other junction angle varying in 5 degree intervals. The images, each of a square area 150 microns on a side and taken after the rings were cooled in zero field, are arranged in a polar plot. They show that the rings were either in the n = 0 or the n = 1/2 flux quantum states. The transitions from the n = 0 to n = 1/2 flux quantum states occur at angles slightly different from (2m + 1) 45 degrees, m an integer, because of a small s-wave component in addition to the predominant d-wave component to the in-plane superconducting gap in this high temperature cuprate perovskite superconductor. Image appears courtesy of J.R. Kirtley. Data were originally published in J.R. Kirtley, C.C. Tsuei, Ariando, C.J.M. Verwijs, S. Harkema, and H. Hilgenkamp, *Nature Physics* **2**, 190 (2006).

Library of Congress Control Number: 2006926925

ISBN-10: 0-387-35071-3   eISBN-10: 0-387-68734-3
ISBN-13: 978-0-387-35071-4 eISBN-13: 978-0-387-68734-6

Printed on acid-free paper.

© 2007 Springer Science + Business Media, LLC.

All rights reserved. This work may not be translated or copied in whole or in part without the written permission of the publisher (Springer Science+Business Media, LLC. 233 Spring Street, New York, NY 10013, USA), except for brief excerpts in connection with reviews or scholarly analysis. Use in connection with any form of information storage and retrieval, electronic adaptation, computer software, or by similar or dissimilar methodology now known or hereafter developed is forbidden. The use in this publication of trade names, trademarks, service marks and similar terms, even if they are not identified as such, is not to be taken as an expression of opinion as to whether or not they are subject to proprietary rights.

10 9 8 7 6 5 4 3 2 1

springer.com

Preface

Low temperature superconductivity was discovered by H. Kammerlingh-Onnes in 1911, at the University of Leiden. He was awarded the 1913 Nobel Prize in Physics, partly for this discovery, i.e., that at low enough temperatures, certain metals become perfect conductors of electricity. In 1933, Meissner and Oschenfeld discovered that a superconductor (SC) is also a perfect diamagnet, i.e., that the magnetic field vanishes in the bulk of a SC. In 1957, J. Bardeen, L.N. Cooper and J.R. Schrieffer (BCS) advanced the pairing theory of superconductivity which gives a quantitative account of many properties of low temperature SCs, and makes a number of predictions of novel phenomena which have been confirmed in a large variety of experiments. BCS were awarded the Nobel Prize in 1972 for the pairing theory. Through intensive experimental research, the maximum T_c was raised to 21° K in an alloy NbGeAl. In 1986, G. Bednorz and K.A. Müller discovered "high temperature superconductivity" in the layered cuprate $La_{2-x}Ba_xCuO_4$ at 30° K, for which they were awarded the 1987 Nobel Prize in Physics. $T_c \sim 93°$ K was discovered by P. Chu in the ternary compound of YBaCuO soon there after.

The maximum T_c found to date is in a mercury based cuprate, which has $T_c = 133°$ K at ambient pressure ($\sim 160°$ K under pressure). Through concerted experimental and theoretical efforts, strong evidence has been adduced that the attractive electron pairing interaction in HTS cuprates is magnetic in origin.

A lot has happened since 1986. The problem of high temperature superconductivity, and more generally that of metallic strongly correlated systems, remains a major open problem in condensed matter physics, and it is the focus of intensive research. As the reader will see from the many chapters to follow, the authors are meeting these challenges. There have been incredible advances in materials, in sample quality and in single crystals, in hole and electron doping, and in the development of sister compounds with lower T_c's that allow access to the normal state with available high magnetic fields. Probes for structure and dynamics such as scanning-tunneling probe spectroscopy, angle resolved photoemission, and neutron scattering have greatly advanced. High precision resonance and thermodynamic methods, low energy optical probes, and high pressures have likewise been brought to bear on the problems. The authors' statement in the introductory section of Chapter 3 articulates a broad central theme of this treatise: "This revolution.." (in this case in reference to ARPES) "..and its scientific impact result from dramatic advances in four essential components: instrumental resolution and efficiency, sample manipulation, high quality samples and well-matched scientific issues." On the theoretical front, the deceptively simple problem of a "doped Mott Insulator," when applied to the cuprates, turns out to be only the starting point of what rapidly becomes a huge and complex problem. To go beyond BCS, new phenomena need new theories: not only high T_c, but pairing, interactions, symmetry, pseudogaps, inhomogeneity and stripes, the proximity of magnetism and superconductivity, sensitivity to impurities, and non-Fermi liquid normal state properties must all be addressed.

We have selected the title "Handbook of High Temperature Superconductivity" to describe this treatise since many of the articles go into considerable depth in both experimental and theoretical methodologies.

The treatise begins in Chapter 1 with Müller's review of hole-doped cuprates where he argues that the dynamical coexistence of bipolarons and fermions are essential features of both the normal and superconducting states. In Chapter 2 Kirtley and Tafuri briefly review the information obtained from tunneling into conventional superconductors and describe why the situation is more complicated and interesting in the cuprates. They then describe experimental methods for making tunneling contacts, the evidence for and implications of d-wave symmetry, the superconducting gap, the pseudogap, quasiparticle interactions, and other aspects of high temperature superconductors. In Chapter 3, the technique of angle resolved photoemission spectroscopy (ARPES) is described in some detail by Zhou, Cuk, Devereaux, Nagaosa, and Shen, and the impact of ARPES on our understanding of the electronic structure, such as Fermi surface, gap anisotropy and d-wave character, and pseudogap behavior is reviewed. Of special importance is their presentation of the latest results on the electron-phonon interaction in the cuprates. In Chapter 4 Bonn and Hardy review microwave studies of high temperature superconductors, where considerable background and detail is given to the methods employed. Results on the penetration depth leading to the "superfluid stiffness" parameter, the surface resistance that yields the microwave conductivity, and a discussion of the role of superconducting fluctuations are presented. In Chapter 5 Slichter reviews the area of magnetic resonance (predominantly NMR, but also briefly ESR) in high temperature superconductors. The spin lattice relaxation time, transverse relaxation time, and the Knight shift are discussed for both YBCO, LSCO in terms of information gained on the electron spin susceptibility, and on the pairing state. In Sr doped and undoped LCO, analysis of line widths and shapes yield information about local (spatial) spin modulations, and spin glass behavior.

Neutron scattering in the cuprates is presented in Chapter 6 by Tranquada in the context of magnetic excitations and antiferromagnetic correlations for both hole and (briefly) electron doped systems. The evolution of the spin dynamics with doping, from the antiferromagnetism of the parent insulators through the universal magnetic excitation spectrum found near optimal doping, is discussed. The nature of stripe order and its possible relevance are also covered. In the summary, the nature of magnetic excitations revealed by neutron scattering is discussed in the context of current theoretical work. In Chapter 7 Orenstein treats optical conductivity and spatial inhomogeneity in the cuprates, first in an overview of the field. An additional spectral feature seen in the so-called "terahertz gap" in many cuprates is discussed, and is assigned to the spatial variation of the superfluid density. It is shown that optical conductivity can provide critical information about inhomogeneity in the cuprates. In Chapter 8 Geballe and Koster consider the wide range of superconducting transition temperature (T_c) values in the cuprates and re-visit the notion that interactions are confined to the CuO_2 layers. They provide evidence that T_c enhancements found in the cuprates that contain charge reservoir layers can be understood in terms of pairing interactions in the charge reservoir layers, and also propose linear quasiparticles to account for superconductivity in the one dimensional double chain cuprates. In Chapter 9, Fisher, Gordon, and Phillips review the thermodynamic properties of high temperature superconductors. More recent results (mostly specific heat) based on better samples and new interpretations are featured, and are reported for the energy gap, fluctuation effects, vortices, flux-lattice melting, the pseudogap, stripes, and chemical substitutions. Some attention is also given to experimental methodology.

The various anomalies in the normal state transport properties of cuprates are reviewed by Hussey in Chapter 10. Experimental work on in-plane and inter-plane electrical transport, Hall effect and Kohler's rule, thermal transport, and the Nernst-Ettingshausen effect, are reviewed for materials over a wide range of doping. Despite the wide-range of crystallographic structures in the different cuprate families, a remarkably generic picture emerges, suggesting the transport behavior is largely associated with a single CuO_4 unit. Theoretical attempts at explaining this mysterious behavior are also summarized. A comprehensive review of high pressure effects on elemental, binary, and high T_c superconductors is given by Schilling in Chapter 11. Hydrostatic, non-hydrostatic, and uniaxial pressure effects are discussed. One conclusion is that pressure effects seem to point to the structure of the CuO_2 planes as the most important parameter that determines T_c, where "the closer the planes are to being square and flat, and the smaller their area A, the higher the value of T_c". The result $T_c \sim A^{-2}$ is considered to be one of the most important results that pressure has yet given us for high temperature superconductors. Future prospects for combining pressure with other simultaneous measurements to resolve other aspects of the high T_c problem are also discussed. In Chapter 12 Brooks reviews in parallel quasi-one and quasi-two dimensional organic superconductors, and their close relationship to the Mott Hubbard model. Both conventional and unconventional (p-wave and d-wave) superconducting properties are discussed, and similarities and differences between organic and cuprate and perovskite systems are described.

In the next three chapters theoretical aspects of high temperature superconductivity are treated. Scalapino, in Chapter 13, reviews numerical studies of the two-dimensional one-band Hubbard model which show that this model exhibits the basic phenomena seen in the cuprates. These show that, at half-filling, the ground state of the system is a Mott-Hubbard antiferromagnetic insulator. Then, upon doping the system away from half filling a pseudo-gap can appear and at low temperatures evidence for d-wave pairing and striped phases are found. The near degeneracy of these phases is also reminiscent of the behavior of the actual cuprate materials. This chapter concludes with a discussion of what numerical methods tell us about the momentum, frequency and spin structure of the pairing interaction in this model. In Chapter 14 Lee reviews previous theoretical work on high temperature superconductivity, and argues that the one-band Hubbard model in the strong coupling limit (t–J model with t′) can capture the physics. To make further progress, the treatment involves the constraint of no-double occupancy and thereby gauge theories. The predicted pseudogap and vortex structure lead to a description of the phase diagram and the onset of T_c. A number of other fundamental theoretical issues including RVB, spin liquids, fractionalization and emergent phenomena are also discussed. Kivelson and Fradkin, in Chapter 15, consider the role of inhomogeneity for the mechanism of high temperature superconductivity. In reviewing the field, the authors observe that superconductivity is common, but high temperature superconductivity is rare and confined to a small subset of materials. They analyze a class of model inhomogeneous doped Mott insulators, which are shown conclusively to exhibit high temperature superconductivity. Generalizing from this, they propose that an optimal degree (and form) of inhomogeneity (probably self-organized) is an essential feature of the mechanism. The relation of this notion to the occurrence of competing orders is clarified. The chapter contains an interesting appendix on "what defines high temperature superconductivity?".

We depart from the cuprates in Chapter 16 where Pugh, Saxena and Lonzarich consider novel quantum states and unconventional forms of superconductivity which may occur on the border of long range magnetic order in heavy-fermion and related itinerant electron magnetic materials. The chapter begins by considering the simplest deviations from the standard low temperature theory of metals that are observed on the border of long-range ferromagnetic order in metals where no superconductivity arises. It then describes cases on the border of antiferromagnetism where superconducting instabilities are prevalent. The effective dimensionality and proximity of density instabilities in some heavy-fermion superconductors are considered in light of Cooper pair formation. The case of superconductivity on the border of ferromagnetism is also described. Open questions to our current understanding are highlighted and possible future advances are discussed. Some of the materials described in the chapter have some similarities with high temperature superconductors and these are considered. An important aspect of this chapter is the description of the next generation of high pressure and low temperature instrumentation to further advance research in the important area of magnetic metals, quantum phase transitions and superconductivity.

We think you will find this treatise essential to obtain a global view of high temperature superconductivity, including the experimental and theoretical methods involved, the materials, the relationships with heavy-fermion and organic systems, and the many formidable remaining problems and challenges.

J.R. Schrieffer
J.S. Brooks

Acknowledgments

The contributors would like to acknowledge that the origin of this treatise arose from the insight, enthusiasm and persuasive influence of J.R. Schrieffer. We have all greatly benefited from his kind and personal manner, and his fundamental advances in the field of condensed matter physics.

Contents

Preface v

Acknowledgments ix

List of Contributors xxi

Credit Lines xxiii

1 From Single- to Bipolarons with Jahn–Teller Character and Metallic Cluster-Stripes in Hole-Doped Cuprates

K. A. Müller

1.1.	The Original Jahn–Teller Polaron Concept and Its Shortcomings	1
1.2.	Recent Experiments Probing Delocalized Properties	2
1.3.	Probing of Local Properties	4
1.4.	The Intersite JT-Bipolaron Concept Derived from EXAFS, EPR, and Neutron Scattering	5
1.5.	Two-Component Scenario	7
1.6.	JT-Bipolarons as the Elementary Quasiparticles to Understand the Phase Diagram and Metallic Clusters or Stripes	9
1.7.	Substantial Oxygen Isotope Effects	12
1.8.	Concluding Remarks	17
Bibliography		17

2 Tunneling Measurements of the Cuprate Superconductors

J. R. Kirtley and F. Tafuri

2.1.	Introduction		19
2.2.	General Concepts		20
	2.2.1.	Types of Junction Structures	20
	2.2.2.	Generalized Junction Conductance	22
	2.2.3.	The Tunnel and Proximity Effects	22
	2.2.4.	Andreev Reflection and Bound States	25
	2.2.5.	The Josephson Effect: General Features	27
		Andreev Reflection in SNS Junctions	28
2.3.	Means of Preparing Tunnel Junctions		32
	2.3.1.	Junctions with Single Crystals	32

		2.3.2.	Grain Boundary Junctions	32
			Bicrystal Junctions	32
			Biepitaxial Junctions	33
			Step-Edge Junctions	34
			Electron Beam Junctions	34
		2.3.3.	Junctions with Artificial Barriers	35
			Noble Metal Barriers	35
			Perovskite and Layered Materials Barriers	36
		2.3.4.	Interface-Engineered Junctions	37
		2.3.5.	Junctions with HTS Rather than YBCO	37
			$La_{1.85}Sr_{0.15}CuO_4$-Based Trilayer with One-Unit-Cell-Thick Barrier	37
			Electron Doped HTS	38
			Ca and Co Doped YBCO: Insights into the Overdoped Regime	38
			Ultra-Thin Films and Superlattices	38
			Intrinsic Stacked Junctions	38
2.4.		π-Rings and $0 - \pi$-Junctions		39
2.5.		Tunneling Spectroscopy		44
		2.5.1.	Superconducting Gap	44
			General Features	44
			Temperature Dependence	50
			Momentum Dependence	53
			Doping Dependence	57
			Macroscopic Quantum Effects	59
		2.5.2.	Pseudogap	60
			Temperature Dependence	60
			Magnetic Field Dependence	62
		2.5.3.	Linear Conduction Background	64
		2.5.4.	Zero-Bias Anomalies	65
		2.5.5.	Atomically Resolved Conductivity Modulation Effects	69
		2.5.6.	Strong Coupling Effects	72
			Electron–Phonon	73
			Electron–Magnon	74
2.6.		Conclusions		75
Bibliography				75

3 Angle-Resolved Photoemission Spectroscopy on Electronic Structure and Electron–Phonon Coupling in Cuprate Superconductors

X. J. Zhou, T. Cuk, T. Devereaux, N. Nagaosa, and Z.-X. Shen

3.1.	Introduction		87
3.2.	Angle-Resolved Photoemission Spectroscopy		88
	3.2.1.	Principle	88
	3.2.2.	Technique	90
3.3.	Electronic Structures of High Temperature Superconductors		95
	3.3.1.	Basic Crystal Structure and Electronic Structure	95
	3.3.2.	Brief Summary of Some Latest ARPES Results	98
3.4.	Electron–Phonon Coupling in High Temperature Superconductors		98
	3.4.1.	Brief Survey of Electron–Phonon Coupling in High-Temperature Superconductors	99

		3.4.2.	Electron–Phonon Coupling: Theory	102
			General	102
			Weak Coupling—Perturbative and Self-Energy Description	106
			Strong Coupling—Polaron	110
		3.4.3.	Band Renormalization and Quasiparticle Lifetime Effects	111
			El–Ph Coupling Along the $(0,0)$–(π,π) Nodal Direction	111
			Multiple Modes in the Electron Self-Energy	116
			El–Ph Coupling Near the $(\pi,0)$ Antinodal Region	118
			Anisotropic El–Ph Coupling	122
		3.4.4.	Polaronic Behavior	124
			Polaronic Behavior in Parent Compounds	124
			Doping Dependence: From Z~0 Polaron to Finite Z Quasiparticles	128
			Doping Evolution of Fermi Surface: Nodal–Antinodal Dichotomy	130
		3.4.5.	Electron–Phonon Coupling and High Temperature Superconductivity	135
3.5.	Summary			137
Bibliography				138

4 Microwave Electrodynamics of High Temperature Superconductors

D. A. Bonn and W. N. Hardy

4.1.	Introduction			145
4.2.	Electrodynamics of Superconductors			146
	4.2.1.	London Theory		146
	4.2.2.	Surface Impedance Approximation		147
	4.2.3.	Non-local Electrodynamics		151
	4.2.4.	Excitation Spectrum of a d-Wave Superconductor		151
		Phenomenological Pairing Model		152
		Effect of Impurities		154
4.3.	Experimental Techniques			156
	4.3.1.	Penetration Depth Techniques—Single Crystals		158
		Excluded Volume Techniques		158
		Far Infrared Reflectivity: $|R|e^{i\theta}$		159
		Measurement of Internal Field Distribution in Mixed State		160
		Zero-Field Gadolinium ESR		161
	4.3.2.	Penetration Depth Techniques—Thin Films		161
		Low Frequency Mutual Inductance Techniques		161
		Thin Film Resonator Techniques		161
		Millimetre Wave Transmission		162
		Far-Infrared Reflection		162
		Slow Muon Beam Method		162
	4.3.3.	Penetration Depth Techniques—Powders		162
4.4.	Measurement of Surface Resistance R_s			163
	4.4.1.	Single Crystals		163
		Cavity Perturbation		163
		Broadband Bolometric Spectroscopy		165
		Thin Film Methods		165

4.5.	Penetration Depth		166
	4.5.1.	Complementary Roles of λ and R_s	166
	4.5.2.	$YBa_2Cu_3O_{6+x}$	167
	4.5.3.	Penetration Depth Anisotropy in $YBa_2Cu_3O_{6+x}$	170
	4.5.4.	Oxygen Doping Effects	171
	4.5.5.	Other Materials	174
		$Bi_2Sr_2CaCu_2O_{8+\delta}$	174
		$Tl_2Ba_2CaCu_2O_8$	174
		$Tl_2Ba_2CuO_{6+\delta}$	174
		$La_{1-x}Sr_xCuO_4$	175
		$HgBa_2Ca_2Cu_3O_{8+\delta}$	175
		Electron Doped Thin Films and Single Crystals	175
	4.5.6.	$\hat{c}$-Axis Penetration Depth	177
4.6.	Surface Resistance		179
	4.6.1.	$YBa_2Cu_3O_{6+x}$ $\hat{ab}$-Plane	180
	4.6.2.	Disorder and Quasiparticle Damping	185
	4.6.3.	Other Materials—ab-Plane	187
	4.6.4.	Low Temperature Limit	193
	4.6.5.	Anisotropy	200
4.7.	Fluctuations		202
Bibliography			209

5 Magnetic Resonance Studies of High Temperature Superconductors

Charles P. Slichter

5.1.	Introduction		215
5.2.	Basic NMR Theory and Experiment		216
	5.2.1.	The Resonance Spectrum	216
	5.2.2.	Exciting a Resonance	217
	5.2.3.	Spin–Lattice Relaxation	219
	5.2.4.	Double Resonance	220
	5.2.5.	NMR in Superconductors	221
5.3.	NMR in Normal State Metals		221
5.4.	NMR in Conventional BCS Superconductors		223
5.5.	The Cuprate Spin Hamiltonian		224
5.6.	YBCO above T_C		226
	5.6.1.	One or Two Components?	226
	5.6.2.	The Spin Pseudogap	227
	5.6.3.	The Spin–Lattice Relaxation Time	227
	5.6.4.	Transverse Relaxation and T_{2G}	232
	5.6.5.	Scaling Relationships	234
5.7.	YBCO Below T_C: NMR Evidence About the Pairing State		236
	5.7.1.	The Knight Shift	236
	5.7.2.	Spin–Lattice Relaxation	239
5.8.	LSCO		240
	5.8.1.	The Spectrum	240
	5.8.2.	One or Two Components	243
	5.8.3.	The Incommensurate State	244
	5.8.4.	Spatial Modulation	245

	5.8.5.	The High Temperature Properties	248
	5.8.6.	The Low Temperature Properties: Wipeout	248
5.9.	Brief Review of EPR		252
Bibliography			254

6 Neutron Scattering Studies of Antiferromagnetic Correlations in Cuprates

John M. Tranquada

6.1.	Introduction		257
6.2.	Magnetic Excitations in Hole-Doped Superconductors		259
	6.2.1.	Dispersion	259
	6.2.2.	Spin Gap and "Resonance" Peak	262
	6.2.3.	Discussion	263
6.3.	Antiferromagnetism in the Parent Insulators		264
	6.3.1.	Antiferromagnetic Order	264
	6.3.2.	Spin Waves	267
	6.3.3.	Spin Dynamics at $T > T_N$	271
6.4.	Destruction of Antiferromagnetic Order by Hole Doping		272
6.5.	Stripe Order and Other Competing States		274
	6.5.1.	Charge and Spin Stripe Order in Nickelates	274
	6.5.2.	Stripes in Cuprates	276
	6.5.3.	Spin-Density-Wave Order in Chromium	279
	6.5.4.	Other Proposed Types of Competing Order	280
6.6.	Variation of Magnetic Correlations with Doping and Temperature in Cuprates		280
	6.6.1.	Magnetic Incommensurability vs. Hole Doping	280
	6.6.2.	Doping Dependence of Energy Scales	282
	6.6.3.	Temperature-Dependent Effects	283
6.7.	Effects of Perturbations on Magnetic Correlations		284
	6.7.1.	Magnetic Field	284
	6.7.2.	Zn Substitution	286
	6.7.3.	Li-Doping	286
6.8.	Electron-Doped Cuprates		286
6.9.	Discussion		288
	6.9.1.	Summary of Experimental Trends in Hole-Doped Cuprates	288
	6.9.2.	Theoretical Interpretations	289
Bibliography			290

7 Optical Conductivity and Spatial Inhomogeneity in Cuprate Superconductors

J. Orenstein

7.1.	Introduction		299
	7.1.1.	Optical Conductivity of Superconductors	299
	7.1.2.	Optical Conductivity and the Cuprates	300
7.2.	Low Frequency Optical Conductivity in the Cuprates		301
	7.2.1.	YBCO Single Crystals: Success of the Two-Fluid Model	301
	7.2.2.	The BSCCO System: Failure of the Two-Fluid Description	303
	7.2.3.	Additional Examples	307

7.3. Optical Conductivity vs. Hole Concentration in BSCCO 309
    7.3.1. Systematics of the Conductivity Anomaly 309
    7.3.2. Quantitative Modeling of $\sigma(\omega, T)$ 312
7.4. Collective Mode Contribution to Optical Conductivity 314
    7.4.1. Origin of the Collective Contribution 314
    7.4.2. Optical Conductivity in the Presence of Inhomogeneity 316
    7.4.3. Extended Two-Fluid Model 316
    7.4.4. Comparison of Model and Experiment 320
7.5. Summary and Outlook 321
    7.5.1. Summary 321
    7.5.2. Outlook and Directions of Future Research 321
Bibliography 323

8 What T_c can Teach About Superconductivity
T. H. Geballe and G. Koster

8.1. Introduction 325
8.2. Cuprate Superconductivity 326
    8.2.1. Pairing and T_cs in the Cuprates 327
            The Cu Ion 327
8.3. Interactions Beyond the CuO_2 Layers 328
    8.3.1. Pairing Centers in the Charge Reservoir Layer Cuprates 329
    8.3.2. Negative-U Center Electronic Pairing in a Model System 330
    8.3.3. The Chain-Layer Cuprates 334
    8.3.4. Other Chain Layer Compounds 338
8.4. Superconductivity Originating in the CuO_2 Layers 339
8.5. Summary 341
Bibliography 341

9 High-T_c Superconductors: Thermodynamic Properties
R. A. Fisher, J. E. Gordon, and N. E. Phillips

9.1. Introduction 345
    9.1.1. Scope and Organization of the Review 345
    9.1.2. Cuprate Superconductors: Occurrence; Structures; Nomenclature;
            Phase Diagram; Characteristic Parameters 346
    9.1.3. Magnetic Properties; Critical-Field Measurements 349
    9.1.4. Specific-Heat Measurements 350
            Specific Heat: Component Contributions; Field and Temperature
                Dependences; Nomenclature 350
            Specific Heat: Experimental Techniques 352
            Specific Heat: Problems and Uncertainties in Analysis of Data 353
9.2. Low-Temperature Specific Heat 353
    9.2.1. Zero-Field "Linear" Term 354
    9.2.2. Evidence for Line Nodes in the Energy Gap 357
9.3. Chemical Substitutions 360
    9.3.1. Rare-Earth Substitutions on the Y and La Sites 361
    9.3.2. General Effects of Substitutions on the Cu Sites 362
    9.3.3. Effects of Zn Substitution on the Cu Sites 364

9.4.	Stripes		367
9.5.	Specific-Heat Anomaly at T_c: Fluctuations; BCS Transition, BEC		372
	9.5.1.	Gaussian and Critical Fluctuations:	372
		Fluctuations: Optimally-Doped Samples in Zero Field	373
		Fluctuations: Optimally Doped Samples in Field	375
		Fluctuations: Under- and Over-Doped Samples	376
	9.5.2.	BCS to BEC	376
9.6.	Vortex-Lattice Melting		380
	9.6.1.	Introduction; Early Measurements on YBCO	380
	9.6.2.	Other Measurements on YBCO	381
	9.6.3.	Measurements on Other HTS	386
9.7.	Calorimetric Evidence for the Pseudogap		386
	9.7.1.	Determination of the Electron Specific Heat of $YBa_2Cu_3O_{6.97}$	387
	9.7.2.	Use of the Differential Method to Obtain the Conduction-Electron Specific Heat of $YBa_2Cu_3O_{6+x}$—A Simplified Discussion	388
	9.7.3.	Other Specific-Heat Results and Their Interpretation	390
Bibliography			390

10 Normal State Transport Properties
N. E. Hussey

10.1.	Introduction	399
10.2.	Evolution of the In-Plane Resistivity with Doping	400
	10.2.1. Introduction	400
	10.2.2. Optimally Doped Cuprates	401
	10.2.3. Underdoped Cuprates	404
	10.2.4. Overdoped Cuprates	406
10.3.	The Out-of-Plane Transport	406
	10.3.1. Introduction	406
	10.3.2. Optimal Doped Cuprates	407
	10.3.3. Underdoped Cuprates	408
	10.3.4. Overdoped Cuprates	409
10.4.	The Anomalous Hall Coefficient and Violation of Kohler's Rule	410
	10.4.1. Introduction	410
	10.4.2. Magnitude of R_H	410
	10.4.3. The Inverse Hall Angle $\cot \vartheta_H(T)$	411
	10.4.4. Theoretical Modeling of $\rho_{ab}T$ and $R_H(T)$ in Cuprates	412
	10.4.5. In-Plane Magnetoresistance	414
10.5.	Impurity Studies	416
10.6.	Thermal Transport	417
	10.6.1. Introduction	417
	10.6.2. Thermoelectric Power	418
	10.6.3. Thermal Conductivity	418
	10.6.4. Nernst–Ettinghausen Effect	419
10.7.	Discussion and Summary	419
Bibliography		422

11 High-Pressure Effects

J. S. Schilling

11.1.	Introduction	427
11.2.	Elemental Superconductors	430
	11.2.1. Simple Metals	430
	Nonalkali Metals	430
	Alkali Metals	433
	11.2.2. Transition Metals	436
11.3.	Binary Superconductors	437
	11.3.1. A-15 Compounds	437
	11.3.2. A Special Case: MgB_2	438
	11.3.3. Doped Fullerenes A_3C_{60}	439
11.4.	Multiatom Superconductors: High-T_c Oxides	442
	11.4.1. Nonhydrostatic Pressure Media	446
	11.4.2. Structural Phase Transitions	446
	11.4.3. Oxygen Ordering Effects	447
	11.4.4. Intrinsic Pressure Dependence $T_c^{\text{intr}}(P)$	451
	11.4.5. Uniaxial Pressure Results	453
11.5.	Conclusions and Outlook	455
Bibliography		457

12 Superconductivity in Organic Conductors

J. S. Brooks

12.1.	Introduction	463
12.2.	Organic Building Blocks and Electronic Structure	464
12.3.	"Conventional" Properties of Organic Superconductors	466
12.4.	The "Standard Model" for Metallic, Insulating, and Antiferromagnetic Ground States	475
	12.4.1. Band Filling and Its Consequences	475
	12.4.2. Can Superconductivity Emerge From the "Standard Model"?	479
	12.4.3. But What if it is Really Just Phonons?	481
12.5.	"Unconventional" Properties of Organic Superconductors	481
	12.5.1. Q1D Materials and p-Wave Pairing	481
	12.5.2. Q2D Materials and d-Wave Pairing	482
	12.5.3. Magnetic Field Induced Superconductivity and Possible FFLO States	483
12.6.	Comparison of High T_c Superconductors with Organic Conductors	486
12.7.	Summary and Future Prospects	488
Bibliography		490

13 Numerical Studies of the 2D Hubbard Model

D. J. Scalapino

13.1.	Introduction	495
13.2.	Numerical Techniques	496
	13.2.1. Determinantal Quantum Monte Carlo	497
	13.2.2. The Dynamic Cluster Approximation	499
	13.2.3. The Density Matrix Renormalization Group	501

13.3.	Properties of the 2D Hubbard Model	503
	13.3.1. The Antiferromagnetic Phase	504
	13.3.2. $d_{x^2-y^2}$ Pairing	506
	13.3.3. Stripes	510
	13.3.4. The Pseudogap	512
13.4.	The Structure of the Effective Pairing Interaction	516
13.5.	Conclusions	522
Bibliography		524

14 t–J Model and the Gauge Theory Description of Underdoped Cuprates

Patrick A. Lee

14.1.	Introduction	527
14.2.	Basic Electronic Structure of the Cuprates	528
14.3.	Phenomenology of the Underdoped Cuprates	531
14.4.	Introduction to RVB and a Simple Explanation of the Pseudogap	534
14.5.	Slave-Boson Formulation of t–J Model and Mean Field Theory	536
14.6.	$U(1)$ Gauge Theory of the URVB State	541
14.7.	$SU(2)$ Slave-Boson Theory of Doped Mott Insulators	546
	14.7.1. $SU(2)$ Slave-Boson Mean-Field Theory at Finite Doping	547
	14.7.2. Effect of Gauge Fluctuations: Enhanced (π, π) spin Fluctuations in Pseudogap Phase	550
	14.7.3. σ-Model Effective Theory and New Collective Modes in the Superconducting State	551
	14.7.4. Vortex Structure	554
	14.7.5. Phase Diagram	555
14.8.	Spin Liquids, Deconfinement, and the Emergence of Gauge Fields and Fractionalized Particles	557
14.9.	Application of Gauge Theory to the High T_c Superconductivity Problem	559
	14.9.1. Spin Liquid, Quantum Critical Point, and the Pseudogap	560
	14.9.2. Signature of the Spin Liquid	562
14.10.	Summary and Outlook	563
Bibliography		565

15 How Optimal Inhomogeneity Produces High Temperature Superconductivity

Steven A. Kivelson and Eduardo Fradkin

15.1.	Why High Temperature Superconductivity is Difficult	570
15.2.	Dynamic Inhomogeneity-Induced Pairing Mechanism of HTC	572
	15.2.1. Pairing in Hubbard Clusters	573
	15.2.2. Spin-Gap Proximity Effect	574
15.3.	Superconductivity in a Striped Hubbard Model: A Case Study	576
	15.3.1. Zeroth-Order Solution: Isolated two-Leg Ladders	578
	15.3.2. Weak Inter-Ladder Interactions	579
	15.3.3. Renormalization-Group Analysis and Inter-Ladder Mean Field Theory	580
	15.3.4. The $x \to 0$ Limit	581
	15.3.5. Relation to Superconductivity in the Cuprates	582

15.4.	Why There is Mesoscale Structure in Doped Mott Insulators	582
15.5.	Weak Coupling Vs. Strong Coupling Perspectives	584
15.6.	What is so Special About the Cuprates?	585
	15.6.1. Is Charge Order, Or Fluctuating Charge Order, Ubiquitous?	585
	15.6.2. Does the "Stuff" Between the Cu–O Planes Matter?	586
	15.6.3. What About Phonons?	588
	15.6.4. What About Magnetism?	588
	15.6.5. Must We Consider Cu–O Chemistry and the Three-Band Model?	589
	15.6.6. Is d-Wave Crucial?	589
	15.6.7. Is Electron Fractionalization Relevant?	590
15.7.	Coda: High Temperature Superconductivity is Delicate But Robust	590
Bibliography		592

16 Superconducting States on the Border of Itinerant Electron Magnetism

Emma Pugh, Siddharth Saxena, and Gilbert Lonzarich

16.1.	Introduction	597
16.2.	Uncharted Territory: The New Frontier	597
16.3.	Logarithmic Fermi Liquid	598
16.4.	The Puzzle of MnSi	599
16.5.	Superconductivity on the Border of Magnetism	600
16.6.	Three Dimensional vs. Quasi-Two-Dimensional Structures	600
16.7.	Density Mediated Superconductivity	601
16.8.	The Search for Superconductivity on the Border of Itinerant Ferromagnetism	602
16.9.	Why Don't All Nearly Magnetic Materials Show Superconductivity?	605
16.10.	From Weak to Strong Coupling	607
16.11.	Superconductivity Without Inversion Symmetry	608
16.12.	Quantum Tuning	608
16.13.	Concluding Remarks	611
Bibliography		611

Index ... 615

List of Contributors

D.A. Bonn, Department of Physics and Astronomy, University of British Columbia, 6224 Agricultural Rd., Vancouver, BC, Canada V6T 1Z1.

J.S. Brooks, Physics/NHMFL Florida State University, 1800 East Paul Dirac Drive Tallahassee, FL 23210 USA.

T. Cuk, Department of Physics, Applied Physics and Stanford Synchrotron Radiation Laboratory, Stanford University, Stanford, CA 94305, USA.

T. Devereaux, Department of Physics, University of Waterloo, Ontario, Canada N2L 3GI.

R.A. Fisher, Department of Chemistry, University of California at Berkeley and Lawrence Berkeley National Laboratory, Berkeley, CA 94720, USA.

E. Fradkin, Department of Physics, University of Illinois at Urbana-Champaign, 1110 West Green Street, Urbana, IL 61801-3080, USA.

T.H. Geballe, Department of Applied Physics and Department of Materials Science, Stanford University, Stanford, CA 94305, USA.

J.E. Gordon, Physics Department, Amherst College, Amherst, MA 01002, USA.

W.N. Hardy, Department of Physics and Astronomy, University of British Columbia, 6224 Agricultural Rd., Vancouver, BC, Canada V6T 1Z1.

N.E. Hussey, H. H. Wills Physics Laboratory, University of Bristol, Tyndall Avenue, Bristol, BS8 1TL, UK.

J.R. Kirtley, IBM, T.J. Watson Research Center, Yorktown Heights, NY 10598, USA.

S.A. Kivelson, Department of Physics, Stanford University, Stanford CA 93105, USA; Department of Physics and Astronomy, University of California Los Angeles, Los Angeles, CA 90095 1547, USA.

G. Koster, Geballe Laboratory for Advanced Materials, Stanford University, Stanford, CA 94305, USA.

P.A. Lee, Department of Physics, Massachusetts Institute of Technology, Cambridge, MA 02139, USA.

G. Lonzarich, Cavendish Laboratory, University of Cambridge, J.J. Thomson Avenue, Cambridge CB3 0HE, UK.

K.A. Müller, University of Zürich, Winterthurerstr. 190, Ch-8057 Zürich, Switzerland.

N. Nagaosa, CREST, Department of Applied Physics, University of Tokyo, Bunkyo-ku, Tokyo 113-8656, Japan.

J. Orenstein, Physics Department, University of California, Berkeley, CA, 94720, USA Materials Science Division, Lawrence Berkeley National Laboratory, Berkeley, CA 94720, USA.

N.E. Phillips, Department of Chemistry, University of California at Berkeley and Lawrence Berkeley National Laboratory, Berkeley, CA 94720, USA.

E. Pugh, Cavendish Laboratory, University of Cambridge, J.J. Thomson Avenue, Cambridge CB3 0HE, UK.

S. Saxena, Cavendish Laboratory, University of Cambridge, J.J. Thomson Avenue, Cambridge CB3 0HE, UK.

D.J. Scalapino, Department of Physics, University of California, Santa Barbara, CA 93106-9530, USA.

J.S. Schilling, Department of Physics, Washington University, CB 1105, One Brookings Dr., St. Louis, MO 63130, USA.

Z.-X. Shen, Department of Physics, Applied Physics and Stanford Synchrotron Radiation Laboratory, Stanford University, Stanford, CA 94305, USA.

C.P. Slichter, Research Professor of Physics, Department of Physics, University of Illinois Urbana/Champaign, Urbana, IL 61801, USA.

F. Tafuri, Dip. Ingegneria dell'Informazione, Seconda Università di Napoli, 29-81031 Aversa (CE), Italy.

J.M. Tranquada, Condensed Matter Physics & Materials Science Department, Brookhaven National Laboratory, Upton, NY 11973, USA.

X.J. Zhou, Department of Physics, Applied Physics and Stanford Synchrotron Radiation Laboratory, Stanford University, Stanford, CA 94305, USA; Advanced Light Source, Lawrence Berkeley National Lab, Berkeley, CA 94720, USA; National Laboratory for Superconductivity, Institute of Physics & Beijing National Laboratory for Condensed Matter Physics, Chinese Academy of Sciences, Beijing 100080, China.

Credit Lines

The Contributors are grateful to the Authors and Publishers for permission to reproduce figures that appear in the following chapters:

Chapter 1

Fig. 1.1. Reprinted with permission from [10]. Copyright (2002) by the Taylor & Francis Group.
Fig. 1.2. Reprinted with permission from [12]. Copyright (2001) by the American Institute of Physics.
Fig. 1.3. Reprinted with permission from [15]. Copyright (1996) by the American Physical Society.
Fig. 1.4. Reprinted with permission from [18]. Copyright (1997) by the American Physical Society.
Fig. 1.5. Reprinted with permission from [20]. Copyright (2001) by the American Physical Society.
Fig. 1.8. Reprinted from [28], with permission from Springer Science+Business Media.
Figs. 1.9, 1.10. Reprinted with permission from [29]. Copyright (2004) by the American Physical Society.
Fig. 1.11. Reprinted with permission from [31]. Copyright (2005) by the American Physical Society.
Figs. 1.12, 1.13. Reprinted from [32], with permission from IOP Publishing Limited.
Fig. 1.14. Reprinted from [35], with permission from Springer Science+Business Media.

Chapter 2

Fig. 2.7. Reprinted with permission from [44]. Copyright (1997) by the American Physical Society.
Fig. 2.8. Reprinted from [20], with permission from IOP Publishing Limited.
Fig. 2.13. Reprinted with permission from [177]. Copyright (2002) by the American Institute of Physics.
Fig. 2.14. Reprinted from [114], with permission from the Nature Publishing Group.

Fig. 2.15. Reprinted with permission from [57]. Copyright (1989) by the American Physical Society.
Fig. 2.16. Reprinted with permission from [204]. Copyright (1991) by the American Physical Society.
Fig. 2.17. Reprinted with permission from [206]. Copyright (1999) by Elsevier.
Fig. 2.18. Reprinted with permission from [208]. Copyright (2003) by the American Physical Society.
Fig. 2.19. Reprinted with permission from [215]. Copyright (1998) by the American Physical Society.
Fig. 2.20. Reprinted with permission from [202]. Copyright (1998) by the American Physical Society.
Fig. 2.21. Reprinted with permission from [220]. Copyright (1998) by the American Physical Society.
Fig. 2.22. Reprinted with permission from [219]. Copyright (1999) by the American Physical Society.
Fig. 2.23. Reprinted with permission from [223]. Copyright (2000) by the American Physical Society.
Fig. 2.24. Reprinted from [229], with permission from the Nature Publishing Group.
Fig. 2.25. Reprinted with permission from [76]. Copyright (2002) by the American Physical Society.
Fig. 2.26. Reprinted with permission from [232]. Copyright (2005) by the American Physical Society.
Fig. 2.27. Reprinted with permission from [235], with permission from the Nature Publishing Group.
Fig. 2.28. Reprinted with permission from [219]. Copyright (1999) by the American Physical Society.
Fig. 2.29. Reprinted with permission from [78]. Copyright (2005) by the American Physical Society.
Fig. 2.30. Reprinted with permission from [218]. Copyright (1998) by the American Physical Society.
Fig. 2.31. Reprinted with permission from [230]. Copyright (2001) by the American Physical Society.
Fig. 2.32a. Reprinted with permission from [301]. Copyright (2004) by the American Physical Society.
Fig. 2.32b. Reprinted with permission from [300]. Copyright (2001) by the American Physical Society.
Fig. 2.33. Reprinted with permission from [328]. Copyright (1990) by the American Physical Society.
Fig. 2.34. Reprinted with permission from [336]. Copyright (1993) by the American Physical Society.
Fig. 2.35. Reprinted with permission from [357]. Copyright (2004) by the American Physical Society.
Fig. 2.36. Reprinted with permission from [354]. Copyright (1997) by the American Physical Society.
Fig. 2.37. Reprinted with permission from [373]. Copyright (1998) by the American Physical Society.
Fig. 2.38. Reprinted from [404], with permission from the Nature Publishing Group.

Credit Lines

Fig. 2.39. Reprinted from [405], with permission from the Nature Publishing Group.
Fig. 2.40. Reprinted from [412], with permission from the Nature Publishing Group.
Fig. 2.41. Reprinted with permission from [426]. Copyright (1991) by the American Physical Society.
Fig. 2.42. Reprinted with permission from [439]. Copyright (2001) by the American Physical Society.

Chapter 3

Fig. 3.4. Reprinted with permission from [25]. Copyright (2005) by the American Physical Society.
Fig. 3.5. Reprinted with permission from [27]. Copyright (2005) by Elsevier.
Fig. 3.6a Reprinted with permission from [31]. Copyright (2002) by the American Physical Society.
Fig. 3.6b. Reprinted with permission from [24]. Copyright (1999) by the American Physical Society.
Fig. 3.7. Reprinted with permission from [35]. Copyright (2005) by the Taylor & Francis Group.
Fig. 3.9. Reprinted from [40], Courtesy of International Business Machines Corporation copyright 1989 © International Business Machines Corporation.
Fig. 3.10a Reprinted with permission from [41]. Copyright (1987) by the American Physical Society.
Fig. 3.10b Reprinted with permission from [49]. Copyright (2000) by the American Physical Society.
Fig. 3.10c. Reprinted with permission from [42]. Copyright (1991) by the American Physical Society.
Fig. 3.13. Reprinted with permission from [100]. Copyright (1993) by the American Physical Society.
Fig. 3.14. Reprinted with permission from [101]. Copyright (1998) by the American Physical Society.
Fig. 3.15. Reprinted with permission from [103]. Copyright (2003) by the American Physical Society.
Fig. 3.16a. Reprinted from [104], with permission from Wiley-VCH Verlag GmBH & Co.
Fig. 3.16b. Reprinted with permission from [108]. Copyright (1999) by the American Physical Society.
Fig. 3.17. Reprinted with permission from [106] Copyright (1995) by the American Physical Society.
Fig. 3.18 Reprinted with permission from [109]. Copyright (2001) by the American Physical Society.
Fig. 3.19. Reprinted from [114], with permission from Wiley-VCH Verlag GmBH & Co.
Fig. 3.21. Reprinted from [128], with permission from the Nature Publishing Group.
Fig. 3.22. Reprinted from [132], with permission from the Nature Publishing Group.
Fig. 3.23. Reprinted with permission from [135]. Copyright (2006) by Elsevier.
Fig. 3.24. Reprinted from [133], with permission from the Nature Publishing Group.
Fig. 3.26. Reprinted with permission from [142]. Copyright (2005) by the American Physical Society.

Fig. 3.28. Reprinted with permission from [54]. Copyright (2003) by the American Physical Society.
Fig. 3.29. Reprinted with permission from [147]. Copyright (2004) by the American Physical Society.
Fig. 3.30. Reprinted with permission from [129]. Copyright (2001) by the American Physical Society.
Figs. 3.31-3.33. Reprinted with permission from [148]. Copyright (2004) by the American Physical Society.
Fig. 3.34a. Reprinted with permission from [158]. Copyright (1998) by the American Physical Society.
Fig. 3.34b. Reprinted with permission from [113]. Copyright (2004) by the American Physical Society.
Fig. 3.34c. Reprinted with permission from [167]. Copyright (2005) by the American Physical Society.
Fig. 3.34d. Reprinted with permission from [164]. Copyright (2002) by the American Physical Society.
Fig. 3.35a. Reprinted with permission from [162]. Copyright (2005) by the American Physical Society.
Fig. 3.35b. Reprinted from [51], with permission from IOP Publishing Limited.
Fig. 3.36. Reprinted with permission from [113]. Copyright (2004) by the American Physical Society.
Fig. 3.37. Reprinted with permission from [175]. Copyright (2004) by the American Physical Society.
Fig. 3.38. Reprinted with permission from [167]. Copyright (2005) by the American Physical Society.
Figs. 3.39, 3.40. Reprinted with permission from [166]. Copyright (2003) by the American Physical Society.
Figs. 3.41, 3.42. Reprinted with permission from [113]. Copyright (2004) by the American Physical Society.
Fig. 3.43. Reprinted with permission from [134]. Copyright (2004) by the American Physical Society.
Fig. 3.44. Reprinted with permission from [180] and [181]. Copyrights (2001) and (2001) by the American Physical Society.
Fig. 3.45, 3.46. Reprinted from [163], with permission from the American Association for the Advancement of Science.

Chapter 4

Figs. 4.3, 4.4. Reprinted with permission from [86]. Copyright (2002) by the Taylor & Francis Group
Fig. 4.5. Reprinted with permission from [76]. Copyright (1994) by the American Physical Society.
Fig. 4.7. Reprinted from [65], with permission from Springer Science+Business Media.
Fig. 4.8. Reprinted from [38], with permission from Springer Science+Business Media.
Fig. 4.9. Reprinted from [65], with permission from Springer Science+Business Media.

Fig. 4.10. Reprinted with permission from [181]. Copyright (1997) by the American Physical Society.
Fig. 4.12. Reprinted from [65], with permission from Springer Science+Business Media.
Fig. 4.15. Reprinted with permission from [120]. Copyright (1996) by the American Physical Society.
Fig. 4.16. Reprinted with permission from [27]. Copyright (1996) by the American Physical Society.
Fig. 4.17. Reprinted with permission from [149]. Copyright (1996) by the American Physical Society.
Fig. 4.19. Reprinted with permission from [84]. Copyright (2004) by the American Physical Society.
Fig. 4.21a. Reprinted with permission from [15]. Copyright (1992) by the American Physical Society.
Fig. 4.21b. Reprinted with permission from [94]. Copyright (1998) by the American Physical Society.
Fig. 4.23. Reprinted with permission from [141]. Copyright (1991) by the American Physical Society.
Fig. 4.24. Reprinted with permission from [67]. Copyright (2006) by the American Physical Society.
Figs. 4.25-4.28. Reprinted from [16], with permission from the World Scientific Publishing Co.
Fig. 4.29. Reprinted with permission from [145]. Copyright (2001) by the American Physical Society.
Fig. 4.30. Reprinted with permission from [90]. Copyright (1995) by the American Physical Society.
Figs. 4.31, 4.32. Reprinted with permission from [120]. Copyright (1996) by the American Physical Society.
Figs. 4.33, 4.34. Reprinted with permission from [27]. Copyright (1997) by Elsevier.
Fig. 4.35. Reprinted with permission from [175]. Copyright (1994) by the American Physical Society.
Figs. 4.36, 4.37. Reprinted with permission from [106]. Copyright (2000) by the American Physical Society.
Fig. 4.38. Reprinted with permission from [71]. Copyright (2004) by the American Physical Society.
Fig. 4.44. Reprinted with permission from [99]. Copyright (1998) by the American Physical Society.
Figs. 4.45, 4.46. Reprinted with permission from [92]. Copyright (1994) by the American Physical Society.
Fig. 4.49. Reprinted with author's permission from [226].

Chapter 5

Fig. 5.1. Reprinted with permission from [26]. Copyright (1991) by the American Physical Society.
Fig. 5.2. Reprinted with permission from [27]. Copyright (1989) by the American Physical Society.

Fig. 5.3.	Reprinted with permission from [22]. Copyright (1989) by the American Physical Society.
Fig. 5.4.	Reprinted with permission from [29]. Copyright (1997) by the American Physical Society.
Fig. 5.5.	Reprinted with permission from [31]. Copyright (1989) by the American Physical Society.
Fig. 5.6.	Reprinted with permission from [32]. Copyright (1991) by the American Physical Society.
Fig. 5.7.	Reprinted with permission from [36]. Copyright (1994) by the American Physical Society.
Figs. 5.8, 5.9.	Reprinted with permission from [40]. Copyright (1991) by the American Physical Society.
Figs. 5.10, 5.11.	Reprinted with permission from [28]. Copyright (1990) by the American Physical Society.
Fig. 5.13.	Reprinted with permission from [45]. Copyright (1990) by the American Physical Society.
Fig. 5.14.	Reprinted from [47], with permission from the Institute of Pure and Applied Physics.
Figs. 5.15, 5.16.	Reprinted with permission from [49]. Copyright (1990) by the American Physical Society.
Fig. 5.17.	Reprinted with permission from [51]. Copyright (1992) by the American Physical Society.
Fig. 5.18.	Reprinted with permission from [52]. Copyright (1993) by the American Physical Society.
Fig. 5.19.	Reprinted with permission from [53], with permission from the Institute of Pure and Applied Physics.
Fig. 5.20.	Reprinted with permission from [55]. Copyright (1998) by the American Physical Society.
Fig. 5.22.	Reprinted with permission from [36]. Copyright (1994) by the American Physical Society.
Fig. 5.23.	Reprinted from [70], with permission from Springer Science+Business Media.
Fig. 5.24.	Reprinted from [76], with permission from Springer Science+Business Media.
Figs. 5.25, 5.26.	Reprinted with permission from [77]. Copyright (2002) by the American Physical Society.
Figs. 5.27, 5.28.	Reprinted from [72], with permission from Springer Science+Business Media.
Fig. 5.29.	Reprinted with permission from [78]. Copyright (1993) by the American Physical Society.
Figs. 5.30, 5.31.	Reprinted with permission from [89]. Copyright (2001) by the American Physical Society.

Chapter 6

Fig. 6.5.	Reprinted with permission from [40]. Copyright (2004) by the American Physical Society.
Fig. 6.7a.	Reprinted with permission from [55]. Copyright (1999) by the American Physical Society.

Fig. 6.9.	Reprinted with permission from [68]. Copyright (2001) by the American Physical Society.
Fig. 6.11.	Reprinted with permission from [32]. Copyright (1996) by the American Physical Society.
Fig. 6.12.	Reprinted with permission from [127]. Copyright (2002) by the American Physical Society.
Fig. 6.13b.	Reprinted with permission from [150]. Copyright (2000) by the American Physical Society.
Fig. 6.15.	Reprinted with permission from [41]. Copyright (2004) by the American Physical Society.
Fig. 6.16a.	Reprinted with permission from [131]. Copyright (2002) by the American Physical Society.
Fig. 6.18a.	Reprinted from [18], with permission from Wiley-VCH Verlag GmBH & Co.
Fig. 6.19.	Reprinted with permission from [17]. Copyright (2001) by the American Physical Society.

Chapter 7

Fig. 7.1.	Reprinted with permission from [3]. Copyright (1999) by the American Physical Society.
Fig. 7.2.	Reprinted with permission from [17]. Copyright (2003) by the American Physical Society.
Figs. 7.3, 7.4.	Reprinted with permission from [6]. Copyright (1996) by the American Physical Society.
Fig. 7.6.	Reprinted with permission from [22]. Copyright (2002) by the American Physical Society.
Fig. 7.9.	Reprinted from [8], with permission from Springer Science+Business Media.
Fig. 7.10.	Reprinted with permission from [25]. Copyright (1999) by the American Physical Society.

Chapter 8

Fig. 8.2.	Reprinted with permission from [43]. Copyright (2004) by the American Physical Society
Fig. 8.4.	Reprinted with permission from [107]. Copyright (1994) by the American Physical Society.
Fig. 8.5.	Reprinted with permission from [25]. Copyright (2003) by the American Physical Society.
Figs. 8.6, 8.7.	Reprinted with author's permission from [62].
Fig. 8.8.	Reprinted with permission from [92]. Copyright (2004) by the American Physical Society.

Chapter 9

Figs. 9.1, 9.2. We are grateful to F. Hardy for preparing these figures.
Figs. 9.17-9.19. Reprinted with permission from [160]. Copyright (1999) by Elsevier.
Figs. 9.20, 9.21. Reprinted with permission from [147]. Copyright (1999) by Elsevier.
Fig. 9.25. Reprinted with permission from [16]. Copyright (1993) by the American Physical Society.
Fig. 9.26. Reprinted with permission from [258]. Copyright (1994) by Elsevier.

Chapter 10

Fig. 10.1a. Reprinted with permission from [11]. Copyright (1996) by the American Physical Society.
Fig. 10.1b. Reprinted with permission from [12]. Copyright (2004) by the American Physical Society.
Fig. 10.3a. Reprinted with permission from [23]. Copyright (2000) by the American Physical Society.
Fig. 10.3b. Reprinted with permission from [25]. Copyright (1998) by the American Physical Society.
Fig. 10.4a. Reprinted with permission from [35]. Copyright (1994) by Elsevier.
Fig. 10.4b. Reprinted with permission from [45]. Copyright (2001) by the American Physical Society.
Fig. 10.5a. Reprinted with permission from [55]. Copyright (1998) by the American Physical Society.
Fig. 10.5a. Reprinted with permission from [13]. Copyright (2000) by the American Physical Society.
Fig. 10.6. Reprinted from [73], with permission from the Nature Publishing Group.
Fig. 10.7. Reprinted with permission from [79]. Copyright (1994) by the American Physical Society.
Fig. 10.8. Reprinted with permission from [17]. Copyright (1997) by Elsevier.
Fig. 10.9a. Reprinted with permission from [90]. Copyright (2000) by the American Physical Society.
Fig. 10.9b. Reprinted with permission from [88]. Copyright (1991) by the American Physical Society.
Fig. 10.10. Reprinted with permission from [93]. Copyright (1995) by the American Physical Society.
Fig. 10.11. Reprinted with permission from [43]. Copyright (1996) by the American Physical Society.
Fig. 10.12. Reprinted from [129], with permission from Springer Science+Business Media.

Chapter 11

Figs. 11.3, 11.4.	Reprinted with permission from [46]. Copyright (2003) by the American Physical Society.
Fig. 11.6.	Reprinted with permission from [93]. Copyright (2003) by Elsevier.
Fig. 11.7.	Reprinted with permission from [110]. Copyright (1996) by the American Physical Society.
Fig. 11.8 (left).	Reprinted with permission from [9]. Copyright (1994) by the American Physical Society.
Fig. 11.8 (right).	Reprinted with permission from [8]. Copyright (1994) by Elsevier.
Fig. 11.9.	Reprinted with permission from [122]. Copyright (1991) by Elsevier.
Fig. 11.11.	Reprinted with permission from [128]. Copyright (2000) by the American Physical Society.
Fig. 11.12.	Reprinted with permission from [137]. Copyright (1991) by Elsevier.
Fig. 11.13.	Reprinted with permission from [139]. Copyright (2005) by the American Physical Society.
Fig. 11.14.	Reprinted with permission from [147]. Copyright (1997) by Elsevier.
Fig. 11.15.	Reprinted from [134], with permission from IOP Publishing Limited.
Fig. 11.16.	Reprinted from [127], with permission from Springer Science+Business Media.
Fig. 11.17.	Reprinted from [174], with permission from Springer Science+Business Media.

Chapter 12

Figs. 12.7, 12.16.	The author is grateful to J. Wosnitza and M. Lang for aid in the preparation of these two figures.

Chapter 13

Fig. 13.1.	Reprinted from [19], with permission from the World Scientific Publishing Co.
Fig. 13.2.	Reprinted with permission from [6]. Copyright (2005) by the American Physical Society.
Figs. 13.3, 13.4.	Reprinted with permission from [45]. Copyright (1993) by the American Physical Society.
Fig. 13.5.	Reprinted with permission from [15] and [16]. Copyrights (1985) and (1989) by the American Physical Society.
Fig. 13.6.	Reprinted with permission from [15] and[17]. Copyrights (1985) and (1990) by the American Physical Society.
Fig. 13.7.	Reprinted with permission from [15] and [16]. Copyrights (1985) and (1989) by the American Physical Society.
Fig. 13.8.	Reprinted with permission from [15] and [16]. Copyrights (1985) and (1989) by the American Physical Society.
Fig. 13.9.	Reprinted with permission from [47]. Copyright (1990) by the American Physical Society.
Fig. 13.10.	Reprinted with permission from [16]. Copyright (1989) by the American Physical Society.

Figs. 13.11, 13.12.	Reprinted with permission from [27]. Copyright (2005) by the American Physical Society.
Fig. 13.13.	Reprinted with permission from [54]. Copyright (1997) by the American Physical Society.
Figs. 13.14, 13.15.	Reprinted with permission from [14]. Copyright (2005) by the American Physical Society.
Fig. 13.16.	Reprinted with permission from [20]. Copyright (1995) by the American Physical Society.
Fig. 13.17.	Reprinted with permission from [68]. Copyright (1996) by the Scientific and Technological Research Council of Turkey.
Fig. 13.18.	Reprinted with permission from [20]. Copyright (1995) by the American Physical Society.
Fig. 13.19.	Reprinted with permission from [28]. Copyright (2001) by the American Physical Society.
Fig. 13.20.	Reprinted with permission from [42]. Copyright (2006) by the American Physical Society.
Figs. 13.21, 13.22	Reprinted with permission from [41]. Copyright (1994) by the American Physical Society.
Figs. 13.23-13.25.	Reprinted with permission from [42]. Copyright (2006) by the American Physical Society.
Fig. 13.26.	Reprinted with permission from [70]. Copyright (2006) by the American Physical Society.

Chapter 14

Fig. 14.2.	Reprinted with permission from [15]. Copyright (1997) by the American Physical Society.
Fig. 14.3.	Reprinted with permission from [23]. Copyright (1997) by Elsevier.
Fig. 14.5.	Reprinted with permission from [67]. Copyright (1992) by the American Physical Society.
Fig. 14.6.	Reprinted with permission from [81]. Copyright (1996) by the American Physical Society.

Chapter 16

Fig. 16.1.	Reprinted from [100], with permission from the Nature Publishing Group.
Figs. 16.4-16.6.	Reprinted with author's permission from [22].
Fig. 16.7.	Reprinted from [94], with permission from the Nature Publishing Group.

From Single- to Bipolarons with Jahn–Teller Character and Metallic Cluster-Stripes in Hole-Doped Cuprates

K. A. Müller

Experiments, published in the past dozen years, are reviewed which are considered as relevant in hole-doped cuprates in understanding the microscopic pairing mechanism. They range from those which are wavevector dependent, such as photoemission and inelastic neutron scattering, to those which probe local properties as EXAFS, XANES, muon rotation, and EPR. Of importance is the time scale which the different techniques probe, also including optical picosecond excitations. All of them point in a consistent way to the presence of two kinds of quasiparticles of fermionic and vibronic character. The latter are theoretically derived from symmetry considerations to be of intersite Jahn–Teller type. Of central importance are also the substantial oxygen isotope effects observed at the pseudogap temperature T^*, at the superconducting transition temperature T_c and on the London penetration depth λ_L, all being a function of hole doping. The former are ascribed to real space bipolaron formation whereas the latter are quantitatively reproduced by the momentum space analogue, i.e., a two-component model. From the latter it follows necessarily that the lattice distortions in the vibronic ground state are of the local Q_2 type Jahn–Teller conformation. Finally, the most recent findings are reviewed, regarding the agglomeration of bipolarons in forming clusters or stripes with metallic character, even at very low dopings and temperatures.

1.1. The Original Jahn–Teller Polaron Concept and Its Shortcomings

The concept which led to the discovery of high temperature superconductivity (HTS) in hole-doped La_2CuO_4 [1] is Jahn–Teller (JT) polaron. Thomas and his group in Basilea used the famous Holstein Hamiltonian for a linear molecular chain and calculated with a variational method the effective mass of the polaron as a function of arbitrary Jahn–Teller coupling [2]. Holstein had only deduced the extreme limiting cases of either an entirely localized and or a completely extended molecular polaron. The result of the Basilea paper was that the JT polaron had a very large effective mass and had experimentally not been observed at the time.

The La_2CuO_4 and the subsequently discovered cuprate superconductors have, when undoped, all aniferromagnetic (AF) ground states. Consequently, a displaced JT-polaron will leave behind it a trail of reversed Cu spins in the corresponding CuO_2 plane of the cuprates,

K. A. Müller • University of Zürich, Winterthurerstr. 190, Ch-8057 Zürich, Switzerland

which increases the immobility of the polaron in question. This fact enhanced the skepticism of the community with respect to the original concept of the author. However, very early Hirsch pointed out [3] that in the case of bipolaron formation the Cu spins would remain in the AFM ground state after a bipolaron of any kind had passed by. In the following, the more recent experiments will be reviewed and commented, which all indicate the presence of Jahn–Teller bipolarons whose binding energy is substantially reduced upon reaching optimum doping from the low doping side. At this point, one should also note that in many theories the electronic repulsion, i.e., the Hubbard U, on the Cu^{2+} is of the order of 10 eV further adding to the skepticism regarding the original concept. However, quite early Schrieffer and others [4] pointed out that in the overdoped regime U is substantially reduced from this value.

1.2. Recent Experiments Probing Delocalized Properties

Three years after the paper of Hirsch [3] appeared, the important one of Alexandrov, Kabanov and Mott [5] was published in which they introduced the basic equations for the bulk properties starting from the bipolaron concept. However, the temperature dependence for the susceptibility they obtained deviated substantially from the measured ones for certain hole dopings. It was shown by Müller et al. [6] that the addition of a Pauli temperature-independent term, due to Fermions, yielded good agreement with the data. This meant that two types of carriers, bipolarons and Fermions, were present simultaneously. A review by Mihailovic and the author [7] on the occasion of the 10th anniversary of the discovery of the HTS emphasized the existence of two types of quasiparticles as established by the experimental techniques known at that time: magnetic susceptibility, EXAFS, the Mössbauer effect, pulsed photoexcitation, NMR/NQR, and far-infrared response. At this point, it is important to note that right after the HTS discovery, Gor'kov and Sokol [8] supported the view of the existence of two types of particles, namely those of fermionic and polaronic character. Later and independently, Enz and Galasiewicz proposed theoretically that only the simultaneous presence of light and heavy quasiparticles coupled to each other could lead to the observed high values of T_c [9].

In this paragraph I summarize two more recent experiments carried out with techniques that yield information on wavevector-dependent properties which support the viewpoint of the existence of two quasiparticles: photoemission and inelastic neutron scattering.

Angle resolved photoemission (ARPES) data by Lanzara et al. [10] clearly showed a common feature in different high temperature superconductors, that is the signature of two types of carriers: The quasiparticle energies vs. (rescaled) wavevector plots in the $\Gamma - \pi, \pi$ direction of the Brillouin zone for the hole doped Bi2212, Bi2201, and LSCO show a kink, while NCCO, the electron doped cuprate, does not show any such behavior (Figure 1.1).

The kink appears near 70 meV at a characteristic wavevector in the center of the Brillouin zone and separates two different group velocities. They are due to two different quasiparticles, one of fermionic character, near the Fermi energy E_F, and the other of more bosonic character at larger binding energies. Most recent data of this group [11] agree with this interpretation, since the dispersion near E_F does not show any oxygen isotope effect ($^{16}O \rightarrow ^{18}O$), whereas the dispersive part below the kink shows a substantial one, as expected for a polaronic particle.

Probing the vibronic excitations, inelastic neutron scattering is a sensitive tool in providing a deeper understanding of the particles present. Egami and collaborators showed that

From Single- to Bipolarons with Jahn–Teller Character and Metallic Cluster-Stripes 3

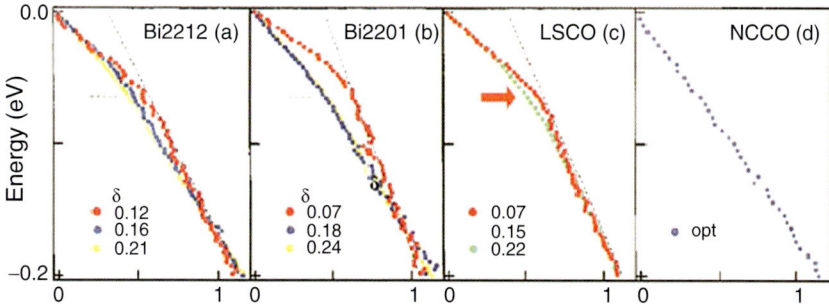

Figure 1.1. The quasiparticle dispersions vs. the rescaled momentum k' for three p-type material systems (nodal direction): (a) Bi2212; (b) Bi2201; (c) LSCO. The arrow indicates the frequency values obtained by inelastic neutron diffraction data. The dispersions are compared with n-type superconductor NCCO (panel d) along ΓY. The dotted lines are guide to the eye obtained fitting, the linear part with a linear function [10].

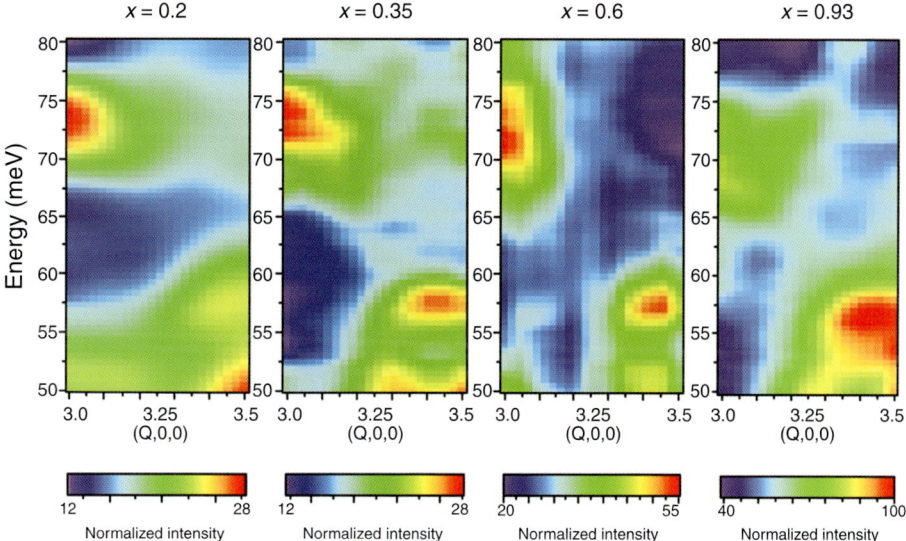

Figure 1.2. Composition dependence of the inelastic neutron scattering intensity from YBCO single crystals with $x = 0.2, 0.35, 0.6,$ and 0.93, at $T = 10$ K [12].

the LO phonon spectra in YBCO and LSCO change significantly with oxygen doping concentrations, for example in $YBa_2Cu_3O_{6+x}$ from $x = 0.2$ to 0.93 (Figure 1.2) [12].

There is a distinct feature in the dispersion at 60–80 meV that occurs in the Brillouin zone along the (100) wavevector, as indicated in the figure. The intensity redistribution of the excitation reflects the change of the ratio of the two types of quasiparticles present. The same features have also been observed in LSCO [13]. Their determined symmetry will be referred to in theoretical Section 1.4.

To end this paragraph, it is important to emphasize that the Fermi surface in the Brillouin zone as detected by photoemission in the bismutates [14] evidences the presence of two kinds of carriers near E_F, one along the Γ-M direction, with polaronic, and one along the Γ-X direction with fermionic character.

1.3. Probing of Local Properties

Due to the polaronic character of the involved quasiparticles, the results from local probes testing real space properties are of considerable relevance. Typical experiments sensitive to the local structure are EXAFS, XANES, certain inelastic neutron scattering data, EPR, and NMR/NQR. Important are the time windows which these techniques offer. They range in decreasing order from 10^{-13} to 10^{-6} s from left to right. The one with the narrowest time window, i.e., the shortest interaction time, yields a nearly "frozen" configuration of the polaron involved.

Therefore, let us start with the EXAFS experiments performed by the group of Bianconi in Rome [15]: With EXAFS the local environment, for example around the Cu ion in the CuO_2 plane, can be determined on a time scale of 10^{-13} s. The results are shown in the lower part of Figure 1.3 for LSCO doped with 0.15% holes.

Their analysis suggested the existence of two types of configurations, one being "LTT" distorted octahedra and simultaneously tilted by $\sim 16°$ most likely arising sterically due to the Cu–O instantaneous elongation. The distortions are reminiscent of a "Q_2"-type local mode,

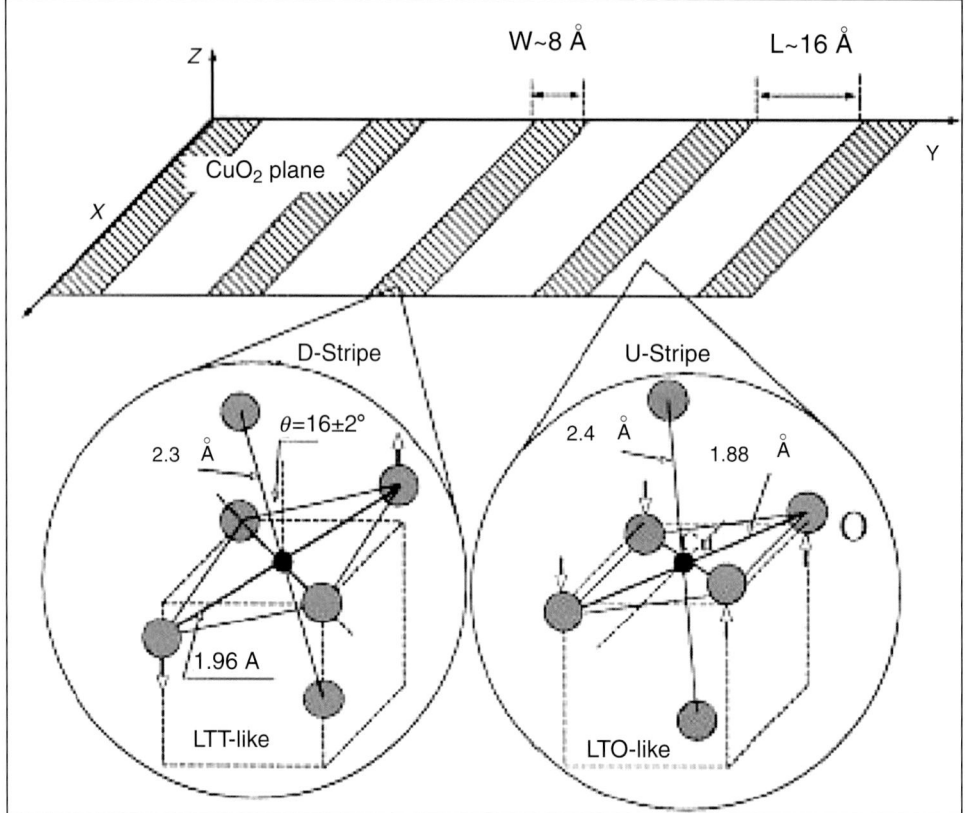

Figure 1.3. Stripe formation at T^* for $La_{2-x}Sr_xCuO_4$, $x = 0.15$. Pictorial view of the distorted CuO_6 octahedra (left side) of the "LTT type" assigned to the distorted (D stripes) of width ≈ 8 Å and of the undistorted octahedra (U stripes) of width $L \approx 16$ Å. The superlattice of quantum stripes of wavelength $\lambda = L + W$ is shown in the upper part [15].

familiar in the Jahn–Teller effect, a point we will return to later. The other type of configuration was essentially an undistorted octahedron shown on the right hand side in the lower part of Figure 1.3. From earlier x-ray data, the group proposed that nanodomains are formed. Different from their interpretation, however, we assign the alternating bands of "stripes" to charge rich (D) and charge-poor (U) regions. The stripes consist of distorted unit cell D-bands of width ~ 8 Å and undistorted unit cell U-bands of width ~ 16 Å, as shown in the upper part of Figure 1.3. The U regions are locally LTT like, while the D regions have LTO like CuO_6 octahedra.

NQR is another important technique to probe the local structure, as used by Imai and collaborators which yielded comparable tilt angles of the octahedra [16].

Instantaneous lattice inhomogeneities, as discussed above, are associated with the local octahedral tilts, or more generally, with a pair distribution function (PDFs) of such tilts as found by the Billinge group [17]. They inferred from neutron diffuse scattering, that for underdoped LSCO, the $x = 0.1$ data, corresponding to "average $3°$ tilts," can best be reproduced by a superposition of heavily ($5°$) tilted and untilted ($0°$) octahedra. They support qualitatively the existence of two different types of lattice conformations in terms of stripes as indicated in Figure 1.3.

Electron paramagnetic resonance (EPR) is a further powerful technique to probe local properties in condensed matter. An intrinsic EPR line observed on quasilocalized holes in $La_{2-x}Sr_xCuO_4$ by the group of Elschner in Darmstadt was analyzed by Kochelaev [18]. The recorded signal was typical for a paramagnetic center with spin $S = 1/2$ having axial symmetry, i.e., gyromagnetic ratios $g_\perp$ and $g_\parallel$. The parallel axis was directed perpendicular to the CuO_2 plane.

The model for the analysis of the experimental results was based on the so called three spin-polaron (TSP), earlier proposed by Emery and Reiter [19]. This polaron is created by the p-hole on the oxygen atom in the CuO_2 plane and two d-holes on the adjacent Cu atoms. Since these holes are coupled to the isotropic antiferromagnetic exchange interaction, the ground state of the TSP has spin $S = 1/2$ in agreement with the observations. At the same time the temperature dependence of the EPR line width was similar to that found in LSCO doped by Mn^{2+} impurities (see Section 1.7). Another experimental evidence for this model was the temperature dependence of the g-factors: $g_\parallel$ decreases with decreasing temperature to a rather unusual value $g_\parallel < 2$, and a crossover with $g_\perp$ takes place (see the left panel of Figure 1.4). Such a behavior was consistent with *dynamical* Q_2-type Jahn–Teller distortions of the TSP (see Figure 1.4), and its anisotropic effective exchange coupling with the surrounding Cu^{2+} ions. Later on, the model was found to apply also in the interpretation of the phase separation observed by EPR (see Section 1.7).

1.4. The Intersite JT-Bipolaron Concept Derived from EXAFS, EPR, and Neutron Scattering

On the basis of the three experiments mentioned in the above title Kabanov and Mihailovic [20] proposed the formation of small bipolarons due to the Jahn–Teller distortions created by two holes, which occupy the same orbitals separated from each other by a distance of the order of a lattice constant.

They suggested a phenomenological interaction with a coupling constant of the form:

$$g(q) = g_0[(q - q_c)^2 + \Gamma]^{-1/2}, \qquad (1.1)$$

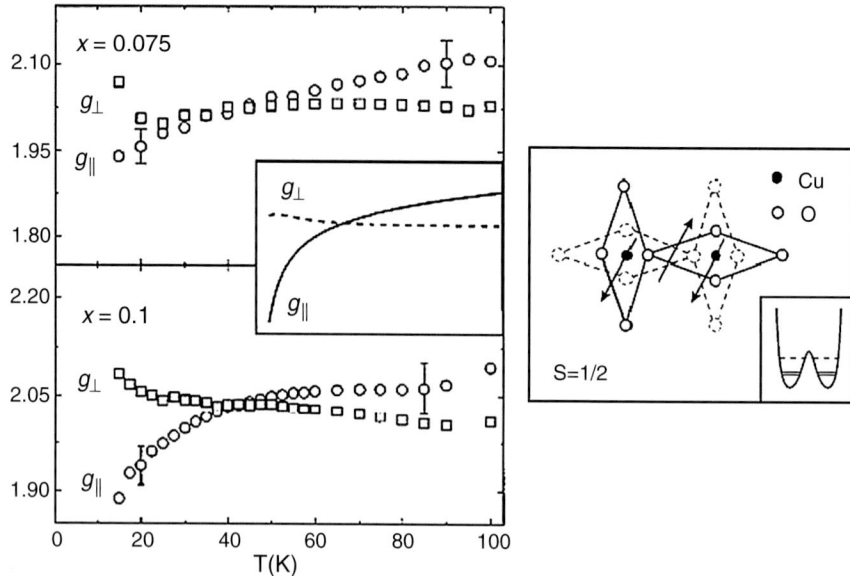

Figure 1.4. Right panel: Three-spin magnetic polaron which is regarded as the EPR active center in the CuO_2 plane. The Jahn–Teller distorted polaron has two degenerate configurations as indicated by the dashed lines. The inset shows the corresponding double-well potential with the excited vibronic states (dashed lines) and the ground state split by tunneling (solid lines) [18]. Left panel: Temperature dependence of the g-factors for two different doping concentrations. The inset shows the results obtained form model calculations based on the TSP of the right panel.

which is resonant at the wavevector q_c. Their group theoretical analysis showed that couplings between $q \neq 0$ phonons and the twofold degenerate electron states including spin takes place (Panel 1) all with the resonant coupling structure of Eq. (1.1).

$$\begin{aligned}
H_{\text{int}} = & \sum_{l,s} \sigma_{0,l} \sum_{k_0=1}^{4} \sum_{\vec{k}} g_0(k_0,\vec{k}) \exp[i\vec{k}l] \left(b_{\vec{k}}^+ + b_{-\vec{k}}\right) \\
& + \sum_{l,s} \sigma_{3,l} \sum_{k_0=1}^{4} \sum_{\vec{k}} g_1(k_0,\vec{k})(k_x^2 - k_y^2)\exp[i\vec{k}l] \left(b_{\vec{k}}^+ + b_{-\vec{k}}\right) \\
& + \sum_{l,s} \sigma_{1,l} \sum_{k_0=1}^{4} \sum_{\vec{k}} g_2(k_0,\vec{k}) k_x k_y \exp[i\vec{k}l] \left(b_{\vec{k}}^+ + b_{-\vec{k}}\right) \\
& + \sum_{l,s} \sigma_{2,l} S_{z,l} \sum_{k_0=1}^{4} \sum_{\vec{k}} g_3(k_0,\vec{k}) k_x k_y \exp[i\vec{k}l] \left(b_{\vec{k}}^+ + b_{-\vec{k}}\right).
\end{aligned}$$

Panel 1 Interaction between phonons $k \neq 0$ and twofold degenerate electronic states (a $k \neq 0$ Jahn–Teller effect).

By symmetry, there are four coupling terms as shown in Panel 1. In front of each is a Pauli matrix $\sigma_{i,l}$ reflecting the twofold degeneracy of the state. The first term stems from the coupling to the breathing mode. The second and third terms are due to the interactions with the x^2-y^2 and xy JT modes, and the fourth proportional to the $\sigma_{2,l}$ matrix is a consequence of the magnetic interaction. Measurements of the ratios of the g_1 and g_2 JT coupling constants vs. the magnetic coupling g_3 would clarify the long standing discussions on the importance of lattice distortions as compared to the magnetic origin of HTS in cuprates (Figure 1.5).

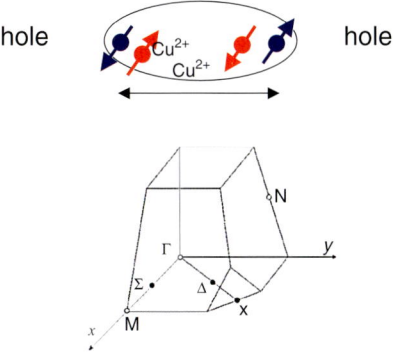

Figure 1.5. The intersite Jahn–Teller pairing interaction. The Brillouin zone (BZ) of $La_{2-x}Sr_xCuO_4$ corresponding to the tetragonal phase with point group D_{4h} is shown below [20].

1.5. Two-Component Scenario

In quantum theory there are always two complementary descriptions of matter, one in terms of particles, as above, and, on the other hand, one in terms of waves. The latter has recently been proposed by Bussmann-Holder and Keller [21].

Since the above outlined results and theoretical concepts suggest that the physics of cuprates are dominated by real space local properties rather than by a momentum space representation, as is true for conventional superconductors, a correspondence between both is difficult to achieve. It can, however, approximately be obtained by considering *subspaces in k-space* as the relevant ones, to which charge rich distorted and charge poor undistorted regimes are ascribed. The distorted charge rich regimes are characterized by strong electron lattice interactions which form charge-lattice-bound states, i.e., polarons. These are randomly distributed over the lattice at high temperatures, but the accompanying huge strain fields force them into a dynamically ordered phase of self-organized stripe segments (Figure 1.6) in an AF

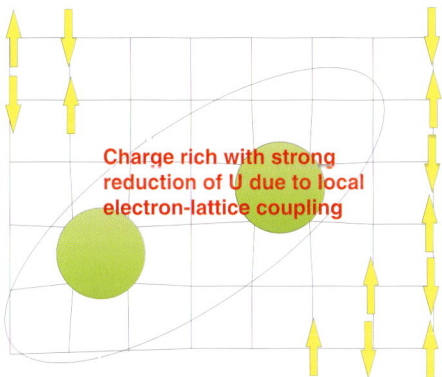

Figure 1.6. Schematic view of the doping effect and polaron formation within an AF matrix. Inside the polaron distorted regimes the system exhibits metallic properties since *the electron–lattice coupling compensates the U term*. These regions correspond to the D stripes in Figure 1.3, whereas the remaining AF matrix represents the U stripes.

background. Within the charge rich areas the electron lattice interaction compensates the on-site Coulomb repulsion [3, 22] to create metallic droplets. The considered system is analogous to a two-component scenario [8, 9] which has been described in detail previously [21, 23]. Here, only the important effects arising from polaron formation, are described. The energy band dispersions [24] within the two components are assumed to be of the same form, however, as outlined above, with different k-space weight, namely:

$$E_{sp,ch} = -2t_1(\cos k_x a + \cos k_y b) + 4t_2 \cos k_x a \cos k_y b \\ + 2t_3(\cos 2k_x a + \cos 2k_y b) \mp t_4(\cos k_x a - \cos k_y b)^2/4 - \mu, \quad (1.2)$$

where t_1, t_2, t_3, t_4 are nearest, next nearest, third nearest neighbor, and interplanar hopping integrals, respectively, and a, b are the in-plane lattice constants with $a \neq b$ to account for the orthorhombic distortion. μ is the chemical potential which controls the number of particles and is directly proportional to doping [25]. Applying a standard Lang–Firsov decoupling scheme [26], important renormalization effects on the band energies appear which are: a band shift proportional to Δ^*, and an exponential band narrowing by means of which all hopping integrals are renormalized like:

$$t_i \to \tilde{t}_i = t_i \exp\left[-\gamma^2 \coth\frac{\hbar\omega}{2kT}\right],$$

where ω is the relevant lattice mode frequency. The effects of these renormalizations are that isotope effects appear due to the isotope dependence of the polaronic coupling constant γ. Numerical investigations of the two-component system show that the average superconducting gap E_g is linearly dependent on the superconducting transition temperature T_c in accordance with experimental results obtained from Andreev reflection spectroscopy [27] (Figure 1.7).

In addition, T_c is enhanced due to the polaronic coupling by more than 30% as compared to the bare case, and the bell-shaped dependence of T_c on doping is realized as well [23]. The coupling constants are of intermediate strength, and a collapse of T_c is observed for too large

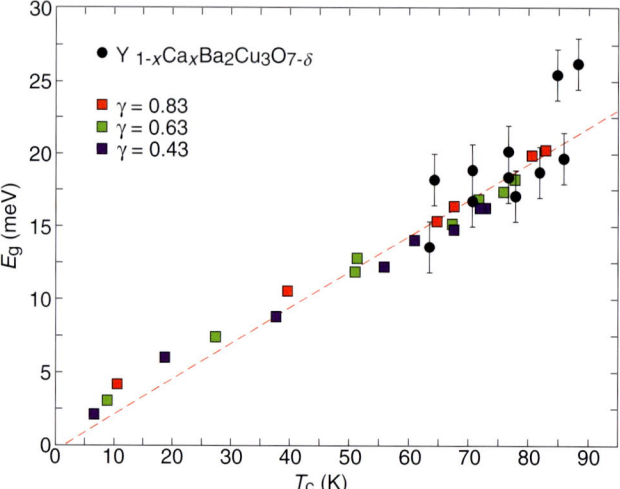

Figure 1.7. The average superconducting energy gap E_g as a function of the superconducting transition temperature T_c. Squares are calculated values with $\gamma = 0.43, 0.63, 0.83$ (blue, green, red), respectively, whereas black symbols are experimental data points for $Y_{1-x}Ca_xBa_2Cu_3O_{7-\delta}$ taken from [27]. The ratio of $t_2/t_1 = 0.3$.

couplings since in this limit localization sets in. Isotope effects are another consequence and will be discussed in a subsequent section.

1.6. JT-Bipolarons as the Elementary Quasiparticles to Understand the Phase Diagram and Metallic Clusters or Stripes

We note, that Weisskopf had shown for classical superconductors, that

$$\Delta \propto E_F \lambda / \xi_0,$$

where λ is the screening length. This implies that the superconducting gap/temperature $\Delta \approx T_c$ is large when the coherence length ξ_0 is small. Thus a model based on small-pairs can be one which captures the relevant physics of HTS.

In fact, if pairs are small, then superconductivity can take place through a kind of percolation, with pair size l_p smaller than the coherence length, and larger than the lattice scale a, as described in the first part of Section 1.4 for the intersite polaron.

$$a < \xi_0 < l_p.$$

The picture that emerges is that of Jahn–Teller induced mesoscopic pairs, which fluctuate and percolate coherently [28]. A quantitative development of this picture (Figure 1.8) yields:

1. An understanding of the minimum coherence length observed experimentally.
2. The correct percentage of holes for which the onset of cuprate superconductivity (6%) is observed.
3. The correct percentage of holes to achieve the maximum value of T_c, i.e., T_c^m (15%) and T_c^m itself.

From Figure 1.8 one can also visualize that there are extended (multi)-bipolaronic percolated regions which have metallic character. This, indeed, was confirmed by recent EPR experiments by Shengelaya et al. [29] in the very low doping regime of LSCO, i.e., for less than 6%.

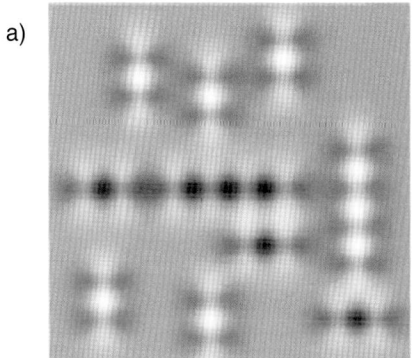
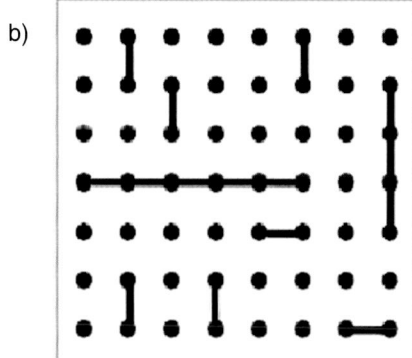

Figure 1.8. (a, left panel). The amplitude of the lattice deformation caused by pairs described by the mesosocopic Jahn–Teller model. The picture corresponds to a "snapshot" at 6% doping at $T = 0$ K. (b, right panel) The bond percolation model describing the situation in (a) [28].

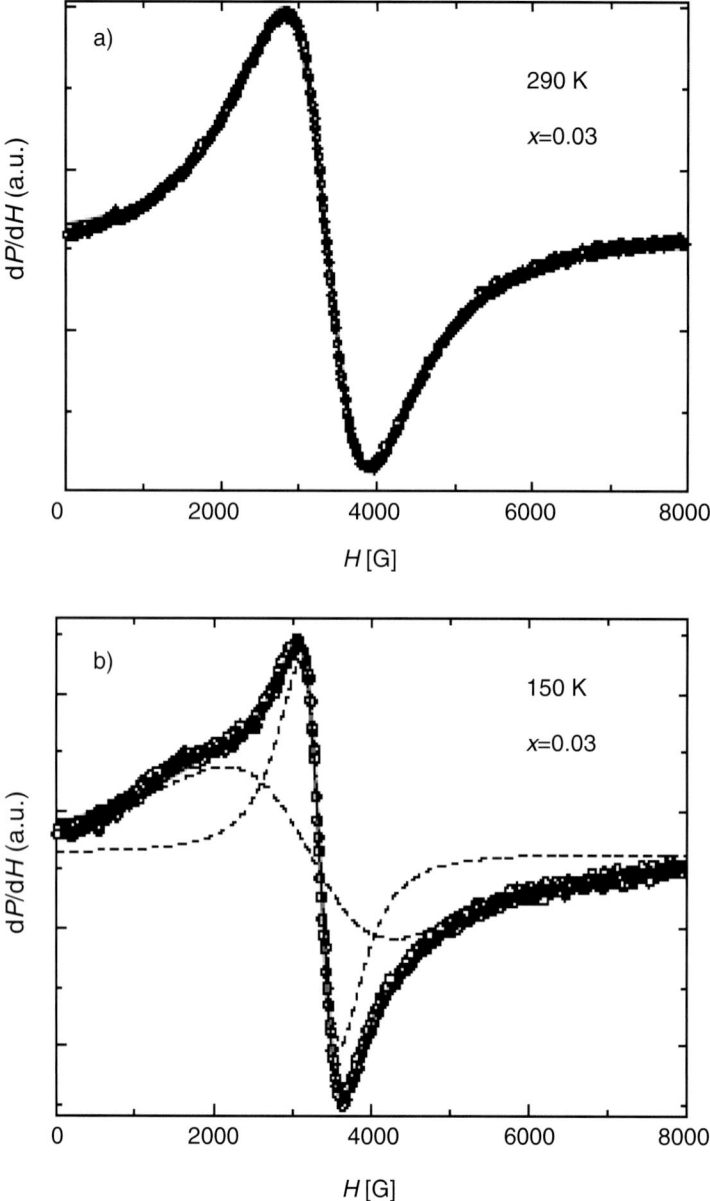

Figure 1.9. (a) EPR signal of $La_{1.97}Sr_{0.03}Cu_{0.98}Mn_{0.02}O_4$ measured at $T = 290$ K. The fit to a Lorentzian line shape is represented by the solid line. (b) EPR signal at $T = 150$ K. The solid line is a fit with a sum of two Lorentzians represented by dashed lines [29].

Using Mn^{2+} as probe, two EPR lines with the same resonance frequency are detected, a narrow and a broad one. The width of the narrow one is oxygen isotope independent, whereas the broad one is isotope dependent (see Figure 1.9). We recall that in EPR the *derivatives* of the lines are recorded. Furthermore, it was shown that the resonances occur in the bottleneck

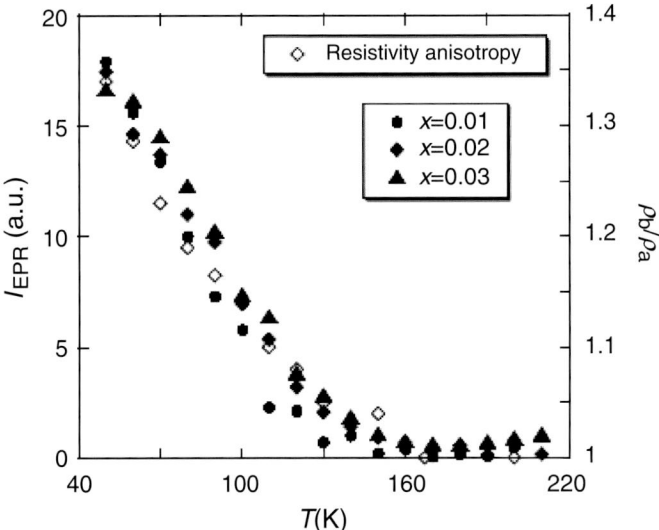

Figure 1.10. Temperature dependence of the narrow EPR line intensities in $La_{2-x}Sr_xCu_{0.98}Mn_{0.02}O_4$ and the resistivity anisotropic ratio in $La_{1.97}Sr_{0.03}CuO_4$ [29].

region, where the absorbed microwave energy of the Mn^{2+} is transmitted to the lattice via the Cu^{2+} ions.

The narrow line is assigned to Mn^{2+} ions located in metallic regions of the sample, and the broad one to those near single polarons. Upon cooling, the narrow EPR line grows exponentially in intensity whereas the broad one nearly disappears. The activation energy Δ deduced from the exponential behavior of the narrow EPR line is 460 ($\pm$50) K, *independent* of hole doping concentrations between 1% and 6%, the experimental range. The activation energy is, within experimental error, the same as the one derived from Raman and inelastic neutron scattering experiments for bipolarons. Therefore it is suggestive that bipolaron formation is the elementary process for the formation of metallic segments. This finding has a macroscopic consequence as well, since the EPR intensity follows the same temperature dependence as the in-plane resistivity anisotropy in LSCO for the same doping range (Figure 1.10). This same temperature behavior of the microscopic EPR and macroscopic resistivity anisotropy as shown in Figure 1.9 is astounding. This result suggests that bipolarons are the microscopic entities responsible for the formation of metallic clusters or stripes, respectively, to which the observed resistivity anisotropy has been attributed [30]. The bipolaron formation can then be the origin for the formation of hole-rich regions by attracting additional holes via elastic coupling forces. Because of the high anisotropy of the elastic forces, these regions are expected to self-organize into dynamical stripe patterns. Therefore, the bipolaron formation energy Δ can also be regarded as an energy scale for the onset of stripe formation associated with the pseudogap.

The existence of an essential heterogeneity in cuprate superconductors due to the coexistence of two types of quasiparticles became also apparent early via femtosecond experiments of the Ljubljana group [7]. In these experiments an excitation pulse is followed by a probing pulse, and the change in reflectivity R is measured as a function of time delay. From the exponential decay the lifetime τ_R of a quasiparticle (QP) is obtained. Within the bipolaron pairing picture two QPs recombine to form a bipolaron of size l_0. As a consequence, τ_R is determined

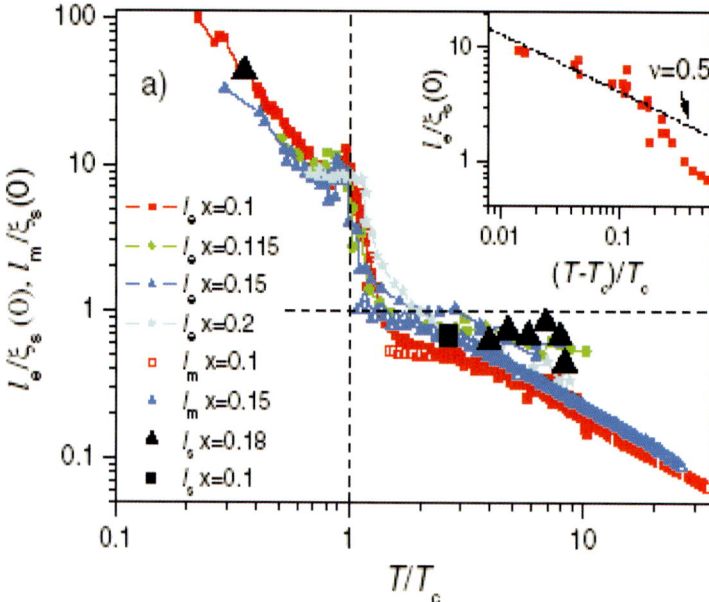

Figure 1.11. A plot of the escape lengths l_o (solid color symbols), l_m (open symbols), and l_s (black symbols), as a function of T/T_c for LSCO at different doping levels. In all cases, the length scale is normalized to $\xi_s(0)$ [31].

by the time which acoustic phonons admit for the bipolaron volume, corresponding in first approximation to $\tau_R = l_o v_s$, where v_s is the sound velocity. In Figure 1.11 l_o normalized by the coherence length ξ_o, is plotted vs. T/T_c for various doping levels [31].

From Figure 1.11 one sees that l_o in the vicinity of T_c is closely the same as the coherence length ξ_o, a quite remarkable finding. Furthermore, also the renormalized mean free path l_m is plotted in that figure. One observes an astounding agreement between l_o and l_m over the entire temperature range ($T_c < T < 300$ K) where the data sets overlap. The conclusion of Mihailovic [31] is: "Upon cooling, bipolarons are formed at $kT^* = 2\Delta$. They lead to a charge-inhomogeneous state. These objects form and dissociate according to thermal fluctuations, leading to a state which is *dynamically* inhomogeneous, in agreement with what has been outlined at the beginning of this section (Figure 1.8). The dimensions of these objects are determined by the balance of Coulomb repulsion and lattice attraction as discussed in [19], and are of the order of: $\xi_o \sim 1-2$ nm above T_c. As the temperature is reduced, the density of pairs starts to coalesce into larger segments, which is reflected by the increasing length scales observed at low temperatures." From the EPR data this is even true in the very underdoped regime where superconductivity is absent. However, for doping concentrations larger than 6% a phase percolation threshold for the metallic regions is reached, and a macroscopically phase-coherent state occurs at T_c [20]. There the characteristic length scale becomes comparable to the superconducting coherence length ξ_o.

1.7. Substantial Oxygen Isotope Effects

In the past years the onset of a pseudogap at a temperature T^* has been reported by various experiments where the interpretations are quite different from the one just outlined. They are, in part, based on theories which presuppose a rigid lattice and with a homogeneous

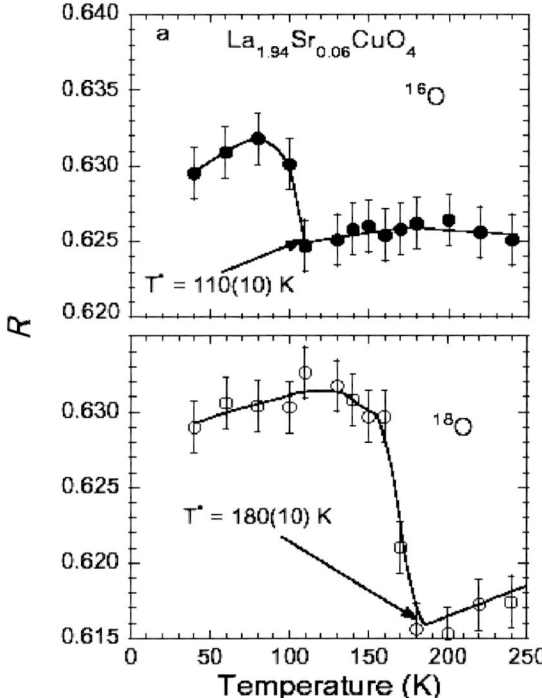

Figure 1.12. The oxygen isotope effect on the charge-stripe ordering temperature T^* in $La_{1.94}Sr_{0.06}CuO_4$ and the doping dependence of T^*. The temperature dependence of the XANES peak intensity ratio R for the ^{16}O and ^{18}O samples of $La_{1.94}Sr_{0.06}CuO_4$ [32].

charge distributions. On the other hand, if one associates the observed temperature T^* as the one where local distortions begin to form then isotope effects are expected which should be observable, and lattice sensitive experiments are warranted.

Changes in the local structure are probed by XANES, where an x-ray photon ejects an electron from a Cu^{2+} ion, and the electron waves interact with the O^{2-} neighbors. Thus, this technique is fast, and entirely nonmagnetic. Data plots as presented in a 1999 paper Lanzara et al. [32] show fluorescence counts vs. photon energy, with peaks related to in plane neighboring oxygen ions of Cu^{2+} and others corresponding to the La/Sr out of plane ions. The latter ones can be considered as structurally nearly immobile at T^*, thus are used as reference compared to the peaks stemming from the Cu–O interferences.

The temperature dependence of the XANES peak intensity ratios, Cu–O to Cu–La/Sr, shows a dip at $T^* \sim 110$ K which is associated with stripe formation. Since this technique probes only oxygen neighbors of the Cu ions, it is a site specific way to investigate effects of isotopic substitutions. In fact, there is a giant isotope effect, with $^{16}O \rightarrow {}^{18}O$ substitution causing a rise of T^* to ~ 180 K (Figure 1.12). This is, to our knowledge, the largest oxygen isotope effect ever reported in the literature. The dynamics of the bipolarons forming the stripes (see Section 1.5) are exponentially dependent on the oxygen ion mass due to the polaronic character of the quasiparticles. The compound with the heavier mass, i.e., ^{18}O, then requires a larger thermal energy to dissociate the stripes at T^* into single polarons. Several techniques can be used to investigate T^* vs. doping La by Sr, and experiments like XANES, EPR, and NQR probe widely different time scales (10^{-13} s, 10^{-9} s, 10^{-7} s). Nonetheless,

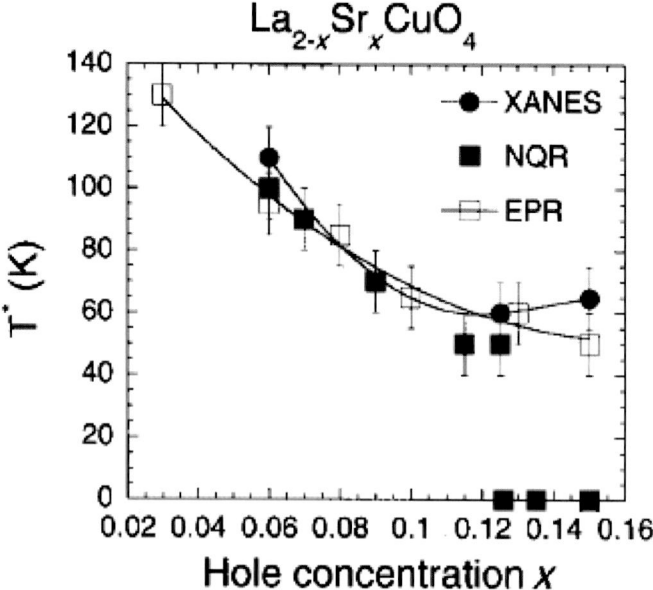

Figure 1.13. The doping dependence of the charge-stripe ordering temperature in $La_{2-x}(BaSr)_xCuO_4$ system from NQR, XANES, and recent EPR experiments [32].

the data fall on the same curve below optimum doping at x_m, corresponding to T_c^m, see Figure 1.13 [33]. Above this doping level, T^* is zero in the NQR data since this technique is the slowest one and the dynamic distortions average to zero above. However, T^* as derived from the EPR data, with a two orders of magnitude shorter time scale, remained finite above x_m exhibiting nearly the same value as at optimum doping. This excludes the existence of a quantum critical end point near optimum doping, predicted by a number of theories based on magnetic interactions, since the existence of such a point requires that T^* has to vanish.

A quantitative confirmation of the XANES results in Figure 1.12 are the inelastic time of flight, neutron scattering measurements by Rubio Temprano et al. [34], in which the line width of a Ho^{3+} transition that substitutes for Y^{3+} was recorded. This work on the isotopic series $HoBa_2{}^nCu_4{}^pO_8$ where $n = 63, 65$ and $p = 16, 18$, reveals a large T^* isotope effect for oxygen ($\alpha_O^* = -2.2$) and an even larger value for copper ($\alpha_{Cu}^* = -4.9$) [34] (Figure 1.14). Thus *both* oxygen and copper dynamics play a role in stripe formation of YBCO, on a time scale of $\sim 10^{-12}$ s, that cannot be accounted for by magnetic interactions.

The oxygen isotope shift reflects the JT coupling, whereas the Cu isotope shift is ascribed to the so-called "umbrella" mode in which the Cu motion is present due to the lack of inversion symmetry at the planar Cu site. This is a consequence of the pyramidal oxygen coordination of Cu in YBCO. Indeed, subsequent results for Ho in LSCO yielded a comparable oxygen isotope shift for T^* as found in YBCO but none for exchanging ^{63}Cu by ^{65}Cu [35]. In LSCO the Cu is octahedrally coordinated and only the quadrupolar JT modes can be active, the inversion symmetry precludes an asymmetric "umbrella" motion. In conjunction the magnitude of T_c of LSCO is about half of the one of YBCO with $T_c = 92$ K at optimum doping. Therefore, it is reasonable to assume that the coupling to the umbrella mode in YBCO is responsible for the T_c enhancement as compared to LSCO, whereas the coupling to the JT mode accounts for a similar order of magnitude T_c as in LSCO. The former may be assigned to another bipolaron coupling first proposed by Alexandrov [36].

From Single- to Bipolarons with Jahn–Teller Character and Metallic Cluster-Stripes 15

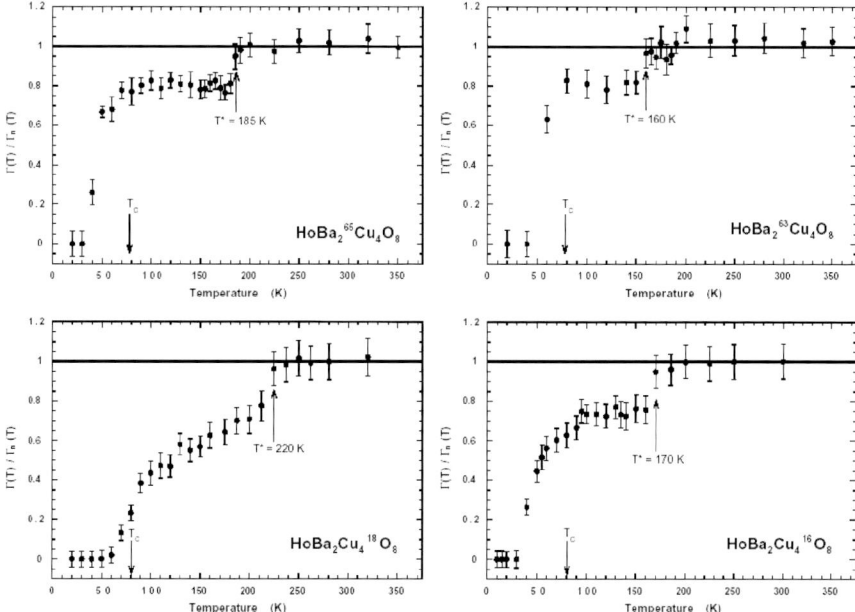

Figure 1.14. Temperature dependence of the reduced linewidth observed for both copper (upper panels) and oxygen (lower panels) isotope substituted $HoBa_2Cu_4O_8$ [34,35]. The solid line represents the normal state linewidth expected by the Korringa law.

In order to explore the role of the lattice for superconductivity isotope experiments have been the standard probe. In HTS, early on, isotope effects on T_c have been reported which are vanishingly small at optimum doping but increase substantially in the underdoped regime to even exceed considerably the BCS value in the immediate vicinity to the AF regime [37]. This finding is inconsistent with conventional phonon mediated superconductivity where the isotope effect should be constant and independent of doping, but suggestive of unconventional electron lattice interactions. This latter view point has been substantiated by recent pioneering low-energy muon SR techniques performed by the group of Keller in Zürich [38]. They measured the muon relaxation time from which they obtained the London penetration depth λ_L. The important result of this technique is that the penetration depth carries an isotope effect, which is neither expected within conventional BCS theory nor within purely electronic models.

An understanding of both effects can be achieved within the above described two-component scenario where polaronic band renormalization effects are its primary cause. Within this scenario, the average superconducting energy gap E_g is isotope dependent and a linear correlation between the gap isotope effect and the one on T_c is observed. Furthermore, a similar relation is obtained for the isotope effect on the penetration depth, and both results are compared to each other in Figure 1.15a. The overall agreement is remarkable, even though in the underdoped regime the isotope effect on the penetration depth seems to saturate whereas the theoretical results remain linear. This discrepancy is attributed to the fact that the polaronic coupling is taken to be doping independent and a mean-field approximation is used. The isotope effect on T_c is calculated within the two-component scenario. Comparison between experimental data and theory is made in Figure 1.15b. Principally, all four hopping

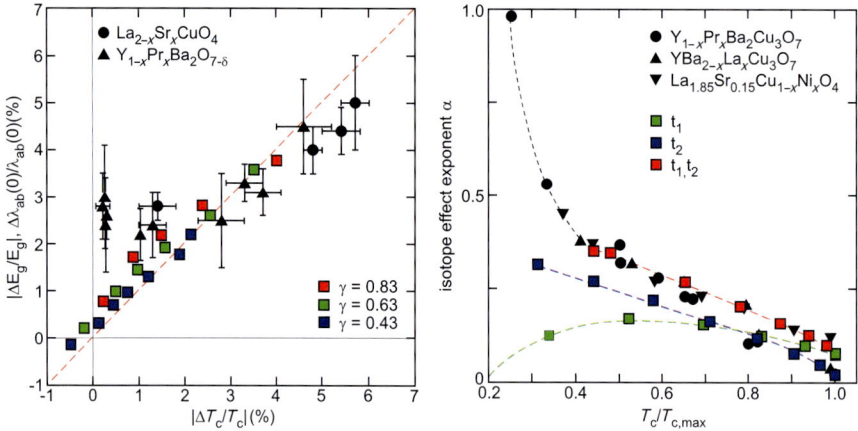

Figure 1.15. (a) Relative isotope shift of the average gap ($|\Delta E_g/E_g|$) as a function of the relative isotope shift of the superconducting transition temperature T_c ($|\Delta T_c/T_c|$). Squares are calculated values with $\gamma = 0.43, 0.63, 0.83$ (blue, green, red), respectively. The dashed red line is a guide to the eye. Black symbols refer to experimental oxygen-isotope effect data of the zero-temperature in-plane magnetic penetration depth $\lambda_{ab}(0)$ and T_c taken from [38]. The ratio of $t_2/t_1 = 0.3$. (b) Oxygen isotope exponent α as a function of T_c/T_c^m, where T_c^m is the maximum T_c observed in a given cuprate family. The open black symbols stand for experimental data points taken from [37], whereas the green full symbols are theoretical ones.

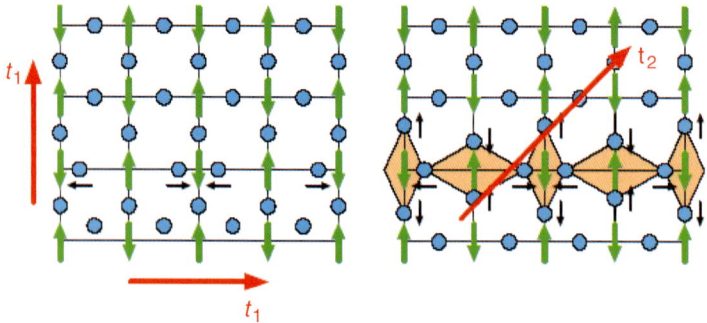

Figure 1.16. The relevant lattice modes which renormalize the hopping integrals t_1, t_2. The black arrows indicate the ionic displacements, the small blue circles denote the oxygen ions, and the green arrows stand for the copper spins.

integrals could contribute equally to an isotope effect. However, the numerical results show so far that only two hopping integrals, namely t_1, t_2, are of relevance where t_1 yields the wrong doping dependence of the isotope exponent α, and t_2 gives the correct trend (Figure 1.15b).

Most importantly, one can conclude from these results on the relevant lattice displacements which are responsible for the isotope effect. It has been suggested that the half-breathing mode (Figure 1.16, left panel), which shows substantial anomalies in the superconducting compounds only [12], is the important mode for the pairing mechanism. This mode is certainly sensitive to doping and to T_c, but, from the above results, can be ruled out as the origin of the isotope effects, since the half-breathing mode involves only the nearest neighbor hopping element t_1 along $<10>$, $<01>$, respectively. The same holds for the full-breathing mode by the same arguments. Consequently, it can be concluded that the most likely candidate to cause the observed effects is the Q_2-*type Jahn–Teller* mode (Figure 1.16, right panel), which has also been shown to be associated with paramagnetic polaron signals observed by EPR [18].

The two modes discussed above are shown in Figure 1.16 to illustrate their role on both hopping integrals t_1 and t_2. As can be seen, only the Q_2-type mode can renormalize t_2 pronouncedly, whereas both the half- and full-breathing modes do not.

Finally, it is worth noting that also the isotope effect on the London penetration depth can consistently be explained within this polaronic picture as has been shown in [21, 23].

1.8. Concluding Remarks

Many theories for cuprate superconductors focus on the strong electronic correlations, present in the undoped systems. They ignore from the beginning, effects stemming from the lattice which is taken as a rigid framework. The many experiments which are of quite distinctive character outlined here, tell a different and new story, especially if the time scale of a particular experiment is sufficiently short. Furthermore, the observations of substantial and unconventional oxygen isotope effects have been reviewed. For all of them the vibronic character of the ground state is manifested in a clear manner. Especially, EPR (the technique used by the author for more than half a century) was able to contribute at the forefront in high temperature superconductivity research. It points to the Jahn–Teller effect as being of outstanding relevance for the vibronic character of the ground state. This, however, was the concept which led to the discovery of the HTS in cuprates [1]. But, instead of single JT polarons, JT intersite bipolarons have been identified as the relevant quasiparticles responsible for the formation of metallic clusters and superconductivity in hole-doped cuprates. Both quasiparticles, manifest the heterogeneity which is present in charge, spin, and lattice distortion. This physics is neither that of a pure bipolaronic superconductor nor a pure fermionic system. The dynamical coexistence of bipolarons and fermions is a prerequisite where their interplay results in a two-component scenario [8].

Acknowledgment

The author thanks indeed A. Bussmann-Holder for her help in editing and complementing the present review.

This paper has adopted parts of an earlier publication which appeared with a different title [39].

Bibliography

1. J. G. Bednorz and K. A. Müller, Z. Phys. B **64**, 189 (1986); Adv. Chem. **100**, 757 (1988), Nobel Lecture.
2. K. H. Höck, H. Nickisch, and H. Thomas, Helv. Phys. Acta **56**, 237, (1983).
3. J. E. Hirsch, Phys. Rev. B **47**, 5351 (1993).
4. J. R. Schrieffer, X. G. Wen, and S. C. Zhang, Phys. Rev. B **39**, 11663 (1989).
5. A. S. Alexandrov, V. V. Kabanov, and N. F. Mott, Phys. Rev. Lett. **77**, 4796 (1996).
6. K. A. Müller, G. M. Zhao, K. Conder, and H. Keller, J. Phys.: Condens. Matter **10**, L291 (1998).
7. D. Mihailovic and K. A. Müller, *High-T_c Superconductivity 1996: Ten Years after the Discovery*, NATO ASI. Ser. E., vol. 343 (Kluwer, Dordrecht, 1997).

8. L. P. Gor'kov and A. V. Sokol, JETP Lett. **46**, 420 (1987).
9. C. P. Enz and Z. M. Galasiewicz, Solid State Commun. **66**, 49 (1988).
10. A. Lanzara, P. V. Bogdanov, X. Zhou, S. A. Kellar, D. I. Feng, E. D. Liu, T. Yoshida, H. Eisaki, A. Fujimori, K. Kishio, J. L. Shimoyama, T. Noda, S. Uchida, Z. Hussain, and Z. X. Shen, Nature **412**, 510 (2001); Z. X. Shen, A. Lanzara, S. Ishiara, N. Nagosa, Philos. Mag. B **82**, 1349 (2002).
11. G. H. Gweon, T. Sasagawa, S. Y. Zhou, J. Graf, H. Tagaki, D. H. Lee, and A. Lanzara, Nature **430**, 187 (2004).
12. R. J. McQueeny, Y. Petrov, T. Egami, M. Yethiraj, M. Shirane, Y. Endoh, Phys. Rev. Lett. **82**, 628 (1999); Y. Petrov, T. Egami, R. J. McQueeny, M. Yethiraj, H. A. Mook, and F. Dogan, cond-mat/0003414 (2000).
13. J. H. Chung, T. Egami, R. J. McQueeny, M. Yethiraj, M. Arai, T. Yokoo, Y. Petrov, H. A. Mook, Y. Endoh, Y. Tajima, S. Forst, and F. Dogan, Phys. Rev. B **67**, 0145 (2003).
14. D. S. Marshall, D. S. Dessau, A. G. Loeser, C. -H. Park, A. Y. Matsuura, J. N. Eckstein, I. Bozovic, P. Fournier, A. Kapitulnik, W. E. Spicer, and Z. -X. Shen, Phys. Rev. Lett. **76**, 4841 (1996).
15. A. Bianconi, N. L. Saini, A. Lanzara, M. Missori, T. Rosetti, H. Oyanagi, H. Yamaguchi, K. Oka, and T. Itoh, Phys. Rev. Lett. **76**, 3412 (1996).
16. T. Imai, C. P. Slichter, K. Yoshimura, and K. Kosuge, Phys. Rev. Lett. **70**, 1002 (1993).
17. E. S. Bozin, S. J. L. Billinge, G. H. Kwei, and H. Tagaki, Phys. Rev. B **59**, 4445 (1999).
18. B. J. Kochelaev, J. Sichelschmidt, B. Elschner, W. Lemor, and A. Loidl, Phys. Rev. Lett. **79**, 4274 (1997).
19. V. J. Emery and G. Reiter, Phys. Rev. B **38**, 4547 (1988).
20. V. V. Kabanov and D. Mihailovic, J. Supercond. **13**, 959 (2000); D. Mihailovic, and V. V. Kabanov, Phys. Rev B **63**, 054505 (2001).
21. A. Bussmann-Holder and H. Keller, Eur. Phys. J. B **44**, 487 (2005).
22. J. B. Goodenough and J. -S. Zhou, Phys. Rev. B **42**, 4276 (1990).
23. A. Bussmann-Holder, H. Keller, and K. A. Müller, *Structure and Bonding*, vol. 114 (Springer, Berlin Heidelberg New York, 2005), p. 367.
24. E. Pavarini, I. Dasgupta, T. Saha-Dasgupta, O. Jepsen, and O. K. Andersen, Phys. Rev. Lett. **87**, 047003 (2001).
25. K. M. Shen, F. Ronning, D. H. Lu, W. S. Lee, N. J. C. Ingle, W. Meevasana, F. Baumberger, A. Damascelli, N. P. Armitage, L. L. Miller, Y. Kohsaka, M. Azuma, M. Takano, H. Takagi, and Z. -X. Shen, Phys. Rev. Lett. **93**, 267002 (2004).
26. S. G. Lang and Yu. A. Firsov, Sov. Phys. JETP **16**, 1302 (1963).
27. A. Kohen, G. Leibowitch, and G. Deutscher, Phys. Rev. Lett. **90**, 207005 (2003).
28. D. Mihailovic, V. V. Kabanov, and K. A. Müller, Europhys. Lett. **52**, 254 (2002).
29. A. Shengelaya, M. Brun, B. J. Kochelaev, A. Safina, K. Conder, and K. A. Müller, Phys. Rev. Lett. **93**, 017001 (2004).
30. Y. Ando, K. Segawa, S. Komiya, and A. N. Lavrov, Phys. Rev. Lett. **88**, 137005 (2002).
31. D. Mihailovic, Phys. Rev. Lett. **94**, 207001 (2005).
32. A. Lanzara, G. M. Zhao, N. L. Saini, A. Bianconi, K. Conder, H. Keller, and K. A. Müller, J. Phys.: Condens. Matter. **11**, L 54 (1999).
33. K. A. Müller, J. Supercond. **13**, 863 (2000).
34. D. Rubio Temprano, J. Mesot, S. Jansen, K. Conder, A. Furrer, A. Mutka, and K. A. Müller, Phys. Rev. Lett. **84**, 1990 (2000).
35. D. Rubio Temprano, J. Mesot, S. Jansen, K. Conder, A. Furrer, A. Sokolov, V. Trounov, M. Kazanov, J. Karpinski, and K. A. Müller, Eur. Phys. J. **B 19**, R5 (2001).
36. A. S. Alexandrov, *Theory of Superconductivity*, IOP Series Cond. Mat., T. Spicer (ed.) (2003), pp. 174–183.
37. R. Khasanov, A. Shengelaya, E. Morenzoni, K. Conder, I. M. Savic, and H. Keller, J. Phys.: Condens. Matter. **16**, S4439 (2004), and refs. therein.
38. R. Khasanov, D. G. Eshchenko, H. Luetkens, E. Morenzoni, T. Prokscha, A. Suter, N. Garifianov, M. Mali, J. Roos, K. Conder, and H. Keller, Phys. Rev. Lett. **92**, 057602 (2004).
39. K. A. Müller, Essential heterogeneities in hole-doped cuprate superconductors, in *Structure and Bonding*, vol. 114 (Springer, Berlin Heidelberg New York, 2005), p. 1.

2

Tunneling Measurements of the Cuprate Superconductors

John Robert Kirtley and Francesco Tafuri

After a very brief description of what has been learned from tunneling measurements in conventional superconductors, we provide an overview of general concepts relevant to the cuprates. These include the types of junction structures used, effects due to variable junction transparency from the point contact to the tunneling regimes, proximity effects, Andreev scattering, unconventional pairing symmetry, and possible broken time reversal symmetry. We describe the various methods used for obtaining tunneling junctions in the high-temperature cuprate superconductors. We describe how the unconventional pairing symmetry of the cuprate superconductors leads to π-rings and 0–π-junctions, and how these effects have been used to determine that the gap in the cuprates has predominantly $d_{x^2-y^2}$ pairing symmetry. We then turn to tunneling spectroscopy. The superconducting gap, the pseudogap, and zero bias conductance peaks are closely interrelated. The superconducting gap and zero bias conductance peaks can be understood in terms of transport between electrodes with $d_{x^2-y^2}$ pairing symmetry through low and high transmissivity barriers. It is controversial whether the pseudogap represents an order competing with superconductivity or preformed Cooper pairs. Similarly, there are many indications of broken time reversal symmetry in tunneling spectroscopy measurements, but not in measurements of π-ring and 0–π-junctions. Conductivity modulations in atomically resolved scanning tunneling spectroscopy certainly can arise from quasiparticle interference effects, but there is also evidence for nondispersive conductivity modulations, expected from stripe models. We describe tunneling evidence for strong coupling effects involving phonon and magnon interactions with the quasiparticles in the superconducting state.

2.1. Introduction

Tunneling measurements played an important role in the development of our understanding of conventional superconductors, providing direct evidence for a gap in the density of states of a superconductor, [1] high precision measurements of the size, shape, temperature, and field dependence of this gap, [2] values for the electron–phonon spectral density $\alpha^2 F(\omega)$, as well as the renormalized coulomb pseudopotential μ^*. [3–5] These measurements and calculations provided strong evidence for the electron–phonon mechanism for superconductivity in conventional superconductors. The tunneling of Cooper pairs between conventional superconductors [6, 7] demonstrated the macroscopic quantum coherence of the superconducting state, as well as providing a wealth of fundamental phenomena and applications [8].

J. R. Kirtley • IBM T.J. Watson Research Center, Yorktown Heights, NY, USA
F. Tafuri • Dip. Ingegneria dell'Informazione, Seconda Università di Napoli, Aversa (CE), Italy

However, tunneling into the cuprate superconductors is much more complex, and arguably much more interesting, than tunneling into conventional superconductors. Some of the properties of the cuprate superconductors that make this such a rich topic include:

(1) *Unconventional pairing symmetry:* There is now overwhelming evidence that the cuprate superconductors have predominantly $d_{x^2-y^2}$ pairing symmetry [9–12]. Some of the evidence for this symmetry from tunneling measurements will be reviewed here. The momentum dependent sign changes in the superconducting gap function associated with this pairing symmetry open the way for many interesting π-SQUID and 0–π-junction devices. These sign changes also cause zero energy bound states at tunnel junction interfaces in certain geometries. It is predicted that subdominant pairing symmetries can appear at surfaces and interfaces of unconventional superconductors. The presence of these subdominant components can break time reversal symmetry, leading to interesting tunneling effects.

(2) *Pseudogap behavior:* A reduction of the density of states near the Fermi surface develops well above the superconducting critical temperature in many of the cuprate superconductors. It is controversial whether this pseudogap is due to preformed pairs or some competing order, and how the pseudogap is related to the superconducting gap.

(3) *Spatial inhomogeneities:* Scanning tunneling measurements show that at least some cuprate samples have substantial spatial inhomogeneities in their tunneling density of states. Part of this inhomogeneity may be due to structural inhomogeneities, and part can be attributed to quasiparticle interferences resulting from scattering from the normal core of vortices, or impurities. However, there have also been reports of periodic inhomogeneities that do not have the dispersive properties expected for quasiparticle interference, and instead may be due to a pinned intrinsic modulation in the electronic density of states.

The task of describing the field of tunneling in the high temperature superconductors is daunting. It is impossible to provide a complete survey of this extraordinarily rich topic here. We therefore intend this chapter to be representative, rather than exhaustive. Just as for conventional superconductors, it is rare in the high temperature superconductors to have definitive measurements of both the Cooper pair and quasi-particle tunneling in the same experiment because of their different energy scales. These energy scale differences are even more dramatic for HTS than for conventional superconductors. Several excellent reviews of various aspects of tunneling in the cuprate superconductors have appeared previously [13–23]. We hope to build upon this previous work, and apologize in advance for work that we neglect.

2.2. General Concepts

2.2.1. Types of Junction Structures

A superconducting junction is traditionally thought of as a thin insulating layer (I) separating a superconductor (S) from a normal metal (N) or another superconductor (S′) (Figure 2.1a). Fifteen years of activity have clearly demonstrated that high critical temperature superconductors (HTS) represent a formidable materials science challenge, especially when dealing with junctions. This is due to the structural complexity of HTS, the ease of oxygen desorption, the extreme difficulty of growing good barriers, etc. These problems have direct

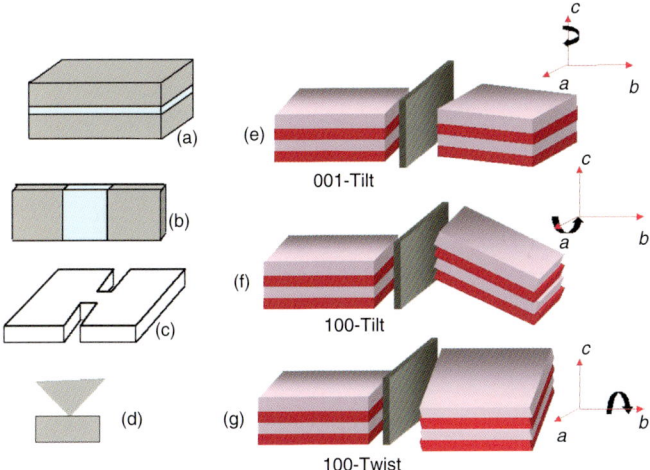

Figure 2.1. Different configurations of weak-links: junction with (a) insulating or (b) normal metal barrier, (c) microbridge, (d) point contact, (e) 001 tilt, (f) 100 tilt, and (g) 100 twist grain boundary junction.

consequences on the fabrication and physics of Josephson junctions: for instance, the equivalent of the classical trilayer junction structure, which is commonly fabricated with low T_c superconductors, has not yet been reproduced with HTS.

We extend our definition of a junction by considering more transmissive barriers, which are often more appropriate for HTS. In doing so we gain in our understanding of the Josephson effect and of other subgap spectroscopic effects. The barrier transparency can be changed by substituting a N′ layer for the I layer. The resulting S–N′–S′ structure will exhibit the Josephson effect for thicknesses of the normal layer N′ up to a few microns (Figure 2.1b). The proximity effect, the mutual influence of a superconductor layer in contact with a normal metal layer; [24] and Andreev reflection, the microscopic process in which a dissipative electrical current is converted at a S/N interface into dissipationless supercurrent, [25] enter the phenomenology of the Josephson effect and can dominate over tunnel effects. In some regimes, roughly defined through the barrier transparency and characteristic scaling lengths such as the coherence length ξ, all these effects may coexist. These concepts can be reasonably extended to barriers composed of semiconductors.

Another way to form a junction is by creating a microrestriction or point contact in a superconducting thin film (Figure 2.1c, d). For widths of the order of a few times the coherence length, the microbridge will behave as a Josephson weak-link, i.e., a system characterized by weak superconductivity [8]. This type of junction depends very critically on the dimensions of the microbridge and its typical scaling lengths. In the limit of long microbridges, Josephson behavior disappears. The difficulties encountered in dealing with HTS thin films and interfaces with HTS motivated intense research toward alternative junction designs which could exploit the intrinsic properties of HTS. One such property is that boundaries between grains with different orientations are Josephson weak links. This has lead to the development of a wide family of grain boundary Josephson junctions [19].

Each HTS grain boundary can be considered as the composition of the three fundamental operations of tilt around the c-axis (001 tilt) (Figure 2.1e), tilt of the c-axis around the a- or b-axis (100 tilt) (Figure 2.1f) and twist around the b-axis (100 twist) (Figure 2.1g). In Figure 2.1e–g the orientation of the left electrode has been fixed, but it can also change. We label D–D junctions using the notation $d_{\theta_1}/d_{\theta_2}$, where θ_1 and θ_2 are the angles of the antinode directions of the $d_{x^2-y^2}$ pairing wavefunctions with respect to the junction normal on the two sides of the junction, respectively. Grain boundaries influence the Josephson phenomenology in a manner which is still not completely clear and also subject to microstructural barrier imperfections [19, 22]. Nevertheless beautiful and clean experiments have been realized using grain boundary junctions, demonstrating for instance a prevailing d-wave order parameter (OP) symmetry [11] and the feasibility of some simple device concepts. [19]

Finally, successful metallic-like barriers have been reported in the literature. The barrier can for instance be a Au layer or damaged HTS material; best results have been obtained by using a LTS as a counterelectrode.

2.2.2. Generalized Junction Conductance

We outline the formulation of the general problem of conductance in a junction. The classical ohmic scaling law for the conductance of a conductor connected to two reservoirs (the contacts) is $G = \sigma W/L$, where W, L, and σ are the width, the length, and the conductivity of the conductor, respectively. This expression becomes invalid as the dimensions become smaller. In this case the conductor is called mesoscopic, i.e., at the borderline between the microscopic and the macroscopic world. It is modeled by a phase-coherent disordered region connected by ideal leads (without disorder) to two electron reservoirs [26–30].

Two factors have to be taken into account to evaluate conductance in the mesoscopic regime (1) the interface resistance between the conductor and each electrode, independent of the length L of the sample and (2) the number of conducting channels (transverse modes) in the conductor, which are discrete and do not scale with W for small dimensions. The zero temperature Landauer formula incorporates both factors [26, 27]:

$$G = (2e^2/h)M\mathcal{T}, \tag{2.1}$$

where $\mathcal{T}$ represents the average probability that an electron injected from one end of the conductor will transmit to the other end and M is the number of transverse modes in the conductor. The net current flowing at any point of the device

$$I = (2e/h)(\mu_1 - \mu_2)M\mathcal{T}. \tag{2.2}$$

At finite temperatures, transport takes place through multiple energy channels in the energy range $\mu_1 +$ (a few $k_B T$) $\geq E \geq \mu_2 -$ (a few $k_B T$) weighted by the energy distributions of the two leads. Here μ_1, μ_2, and T are the electrochemical potentials in the electrodes (1) and (2) and the temperature, respectively.

2.2.3. The Tunnel and Proximity Effects

Keeping in mind the great variety of behaviors of HTS tunnel junctions and weak links, it is of interest to discuss the transition from the scattering formalism to the tunneling formalism, where most of the original formulation on the Josephson effect and on superconducting junctions has been developed. The tunneling transfer Hamiltonian formalism can be considered as, in some respects, a weak coupling version of the scattering formalism. Consider for the moment a nonsuperconducting contact. In the tunneling limit, Fermi's golden rule gives the current I between two bulk electrodes with voltage difference V [28]:

$$I = \frac{4\pi e}{h} \sum_{\mathbf{k},\mathbf{q}} |T_{\mathbf{kq}}|^2 [f_k(1-f_q) - f_q(1-f_k)]\delta(\varepsilon_{\mathbf{k}} - \varepsilon_{\mathbf{q}} - eV), \tag{2.3}$$

where $T_{\mathbf{k},\mathbf{q}}$ are the tunneling matrix elements, which depend on the initial and final state momenta $\mathbf{k}$, $\mathbf{q}$, f_1 and f_2 are Fermi functions, and $\varepsilon_{\mathbf{k}}$ and $\varepsilon_{\mathbf{q}}$ are the initial and final state energies. If we assume that the matrix elements depend only on energy: $|T_{kq}| = M(E)$ (these simplifications will be removed in the discussion of "General Features" in Section 2.5.1), and integrate over momenta at constant energy this equation becomes

$$I = \frac{4\pi e}{h} \int [f_1(E) - f_2(E)] M(E)^2 \rho_1(E) \rho_2(E) dE, \quad (2.4)$$

where $\rho_1(E)$ ($\rho_2(E)$) is the density of states in the electrode 1 (2), respectively [28]. The change in the current in response to a change in the potential μ_2 (keeping μ_1 constant) can be written as:

$$\frac{\partial I}{\partial \mu_2} = \frac{4\pi e}{h} [M(E)^2 \rho_1(E) \rho_2(E)]_{E=\mu_2}. \quad (2.5)$$

In these expressions we have assumed low temperature and neglected any change in $M(E)$, ρ_1, and ρ_2 due to the applied bias. This equation relates the slope of the current–voltage curve to the density of states in the leads. As is widely known, this allows one to use current–voltage measurements to deduce the density of states ρ_2 in one lead, if the density of states ρ_1 in the other lead is known.

The equivalent expression in the scattering formalism is:

$$I = \frac{2e}{\hbar} \int [f_1(E) - f_2(E)] \bar{T}(E) dE, \quad (2.6)$$

where the contact transparency $\bar{T} = M\mathcal{T}$. The expressions (2.4) and (2.6) are consistent if: $\bar{T} = 2\pi M(E)^2 \rho_1(E) \rho_2(E)$ [29, 30]. Independently of the details on the structures where the two expressions can be applied, the similarity between the expressions above qualitatively suggests the contiguity of the scattering and tunneling formalisms. The scattering formalism describes all types of interfaces, including highly transmissive situations and the tunneling limit. When the electrodes are superconductors, new phenomena will occur but the similarities and analogies between scattering and tunnel approaches will still be relevant.

It is straightforward to understand from the expressions above that in a N–I–S junction, the tunnel effect allows a direct measurement, through the conductance of the junction, of the density of states in a S, and therefore of the energy gap [2]. This is shown in Figure 2.2a, where the conductance $G(V)$ usually observed in the limit $T \to 0$ in a LTS tunnel junction is shown schematically. Distinctive features are the peak in $G(V)$ close to the gap value Δ/e and zero conductance up to voltages of about Δ. The density of states of more complicated electrodes, such as the same S in the presence of impurities or backed by a normal metal (S/N electrode), are still reflected in the $G(V)$ of the tunnel junction. $G(V)$ will obviously be quite different depending on the nature and morphology of the electrodes and of the barrier. We are particularly interested in pointing out the smearing of the peak structure corresponding to the gap value Δ/e, and the reduction of Δ, as indicated schematically in Figure 2.2b. The presence of impurities or of an N layer generally makes lower energy states available. The density of states in a S/N bilayer is particularly instructive for our aims, since it is tightly connected to the physics of the proximity effect (PE). This effect plays a crucial role in non-homogeneous systems as well as in Josephson systems and weak links. The proximity effect and its microscopic elementary mechanism (Andreev reflection) will subtly enter into many sections of this review. Here we limit the discussion to an operative definition and give a few examples of induced effects.

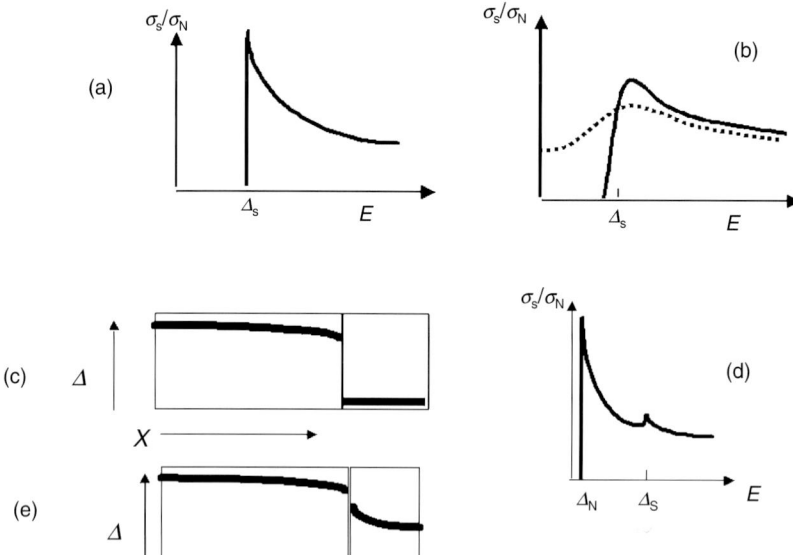

Figure 2.2. (a) Conductance in an ideal S–I–N junction; in the limit of $T \to 0$ it gives the density of states in S; (b) conductance in S–I–N in presence of pair-breaking; (c) order parameter spatial profile in a weakly coupled S/N bilayer $\gamma_B \ll 1$; (d) conductance for a S/N–I–N junction for a weakly coupled S/N bilayer; (e) order parameter spatial profile in a strongly coupled S/N bilayer.

The proximity effect describes the mutual influence of a S and a N layer in contact with each other. N can be replaced by a semiconductor or a ferromagnet with consequent implications. The S will induce some superconducting properties in N within a distance of the coherence length $\xi_N = (\hbar D/k_B T)^{1/2}$ in N (T is the temperature and D is the normal metal diffusion constant). In addition, the superconducting properties will be weaker in the S within a length of the order of its coherence length from the interface. This mutual influence is also controlled by the nature of the interface (barrier transparency,...) and by the boundary conditions, which again involve the respective coherence lengths and the thickness of the N and S layers [24, 29, 31].

In the bi-layer considered above, if the S and N layers are weakly coupled (low transmission barrier interface) the order parameter induced in N will be very poorly correlated to the one in S [31–33]. The profile of the order parameter will have a sharp jump at the S/N boundary, and the gap induced in N (Δ_N) will be much lower than the one in S ($\Delta_N \ll \Delta_S$) (Figure 2.2c). In N the density of states will be substantially rescaled to the induced gap (Figure 2.2d). For strongly coupled S and N layers, the more intense mutual influence is clearly visible in the OP profile along the bilayer (Figure 2.2e). The Δ_N value is much closer to Δ_S [24, 31]. The density of states can be also probed at distinct locations [34]. The density of states shows a depression at the Fermi energy over a characteristic energy of the order of the Thouless energy. When the normal metal is disconnected from any electron reservoir a true energy gap is expected provided that the system is disordered or chaotic [35].

The properties of S/N bilayers can be properly studied in the "dirty" limit on the basis of the Usadel equations [36, 37]. This approach has the advantage of taking into account a varying boundary resistance and describing, in some detail, the nature of the prefactor of the exponential dependence of I_C on the ratio L/ξ_N (L being the barrier thickness). The order parameter in a S/N bilayer can be expressed through two dimensionless parameters: $\gamma = \rho_S \xi_S / (\rho_N \xi_N)$ and $\gamma_B = R_B/(\rho_N \xi_N)$, where $\rho_{N,S}$ and $\xi_{N,S}$ are the normal state resistivities of the junction materials and their coherence lengths in N and S, respectively, while R_B is the specific resistance of the S/N boundary. γ_B is a measure of the coupling between the two slabs of the bilayer: the higher the value of the resistance, the weaker the coupling between N and S is. The tunnel regime at the S/N interface is obtained for $\gamma_B \gg 1$. Direct information on the spatial variation of the order parameter is given by γ. For $\gamma \ll 1$ (rigid boundary conditions) the effects on the superconductor due to the proximity of the normal metal are small, in contrast with the limit $\gamma \gg 1$ (soft boundary conditions). In this case many quasiparticles diffuse from N to S. Both the parameters γ and γ_B are related to the carrier concentration N_S, due to the dependence of $\rho_{N,S}$ and $\xi_{N,S}$.

The proximity effect also allows a simple understanding of how a supercurrent can flow in a normal conductor of appropriate dimensions when placed between two superconductors. We have a supercurrent as long as the two order parameters of each electrode overlap in the barrier region. This supercurrent has the special attributes of the Josephson effect since it is related to the phase difference of the electrodes.

This intuitive picture is complementary to the description in terms of Andreev reflection (AR), extensively discussed in Section 2.2.4. In the latter scenario the multiple Andreev reflections give rise to discrete energy levels or resonances in the energy gap. These current carrying states are localized near the junction and decay exponentially into the bulk. AR is the key mechanism for the superconducting PE. It provides phase correlations in a system of noninteracting electrons over distances much longer than the microscopic lengths.

2.2.4. Andreev Reflection and Bound States

At the interface between a normal metal and a superconductor, dissipative electrical current is converted into dissipation-less supercurrent. An electron excitation slightly above the Fermi level in the normal metal is reflected at the interface as a hole excitation slightly below the Fermi level (see Figure 2.3). The missing charge of $2e$ is removed as a Cooper pair. This scattering mechanism is called Andreev reflection (AR) or retroreflection [25]. This is a branch-crossing process which converts electrons into holes and vice versa, and therefore changes the net charge in the excitation distribution. The reflected hole (or electron) has a shift in phase compared to the incoming electron (or hole) wave-function:

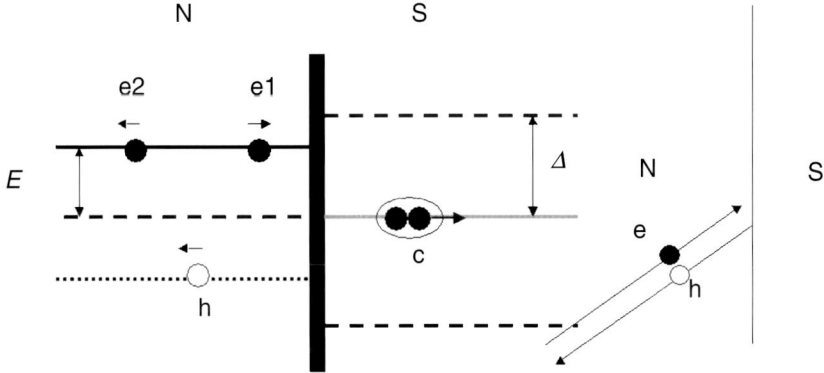

Figure 2.3. Andreev reflection: energy and spatial representation.

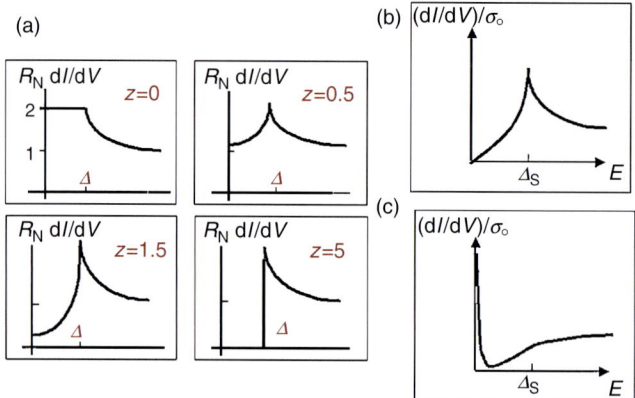

Figure 2.4. (a) Conductance as a function of transmission probability for a S/N junction in the BTK model. (b) Density of states in a d-wave superconductor. (c) Density of states in a D–I–N junction with N facing the node of D.

$\phi_{\text{hole}} = \phi_{\text{elect}} + \phi_{\text{superc}} + \arccos(E/\Delta)$ ($\phi_{\text{elect}} = \phi_{\text{hole}} - \phi_{\text{superc}} + \arccos(E/\Delta)$), where Δ and ϕ_{superc} are the gap value and the superconducting phase of the S. The macroscopic phase of the S and the microscopic phase of the quasiparticles are therefore mixed through Andreev reflection. The Andreev-reflected holes act as a parallel conduction channel to the initial electron current, doubling the normal state conductance of the S/N interface for applied voltages less than the superconducting gap $eV < \Delta$ [38, 39]. Blonder, Tinkham, and Klapwijk [38] (BTK) introduced the dimensionless parameter Z, proportional to the potential barrier at the interface, to describe the barrier transparency $\overline{T} = 1/(1+z^2)$, allowing a continuous description from a highly transmissive barrier to the tunnel limit. Conductance for a S/N junction is displayed in Figure 2.4a for different values of the parameter Z.

The Landauer conductance expression (Eq. 2.1) has been extended to the case of an S–N interface by Beenakker [40] through scattering matrix theory:

$$G_{\text{NS}} = \frac{2e^2}{\pi \hbar} \sum_{n=1}^{N} \frac{T_n^2}{(2-T_n)^2}. \qquad (2.7)$$

Here the T_ns are the transmission eigenvalues of the disordered normal part. The difference in the behavior of the transmission eigenvalues T_n will lead to different mesoscopic behaviors of tunnel junctions and metallic weak links. While in tunnel junctions many small T_ns are relevant, in weak links most T_ns are close to zero or unity.

An important characteristic of the subgap bound states in d-wave junctions is the existence of a finite density of states at zero energy (Figure 2.4c). A delta function peak is found at zero energy, as first predicted by Hu [41]. These zero energy states (ZES), which can be revealed in differential conductance of a N–I–d-wave junction when N is facing the node of the d-wave electrode, can be visualized by considering an Andreev bound state created at a S–I interface.

The basic process is illustrated in Figure 2.5. An electron traveling toward a surface is reflected back into the d-wave superconductor and is subsequently Andreev reflected into a hole by the positive pair potential. Then the hole follows the same path backward, reflected at the surface, and finally Andreev reflected into another electron by the negative pair potential. Due to the d-wave symmetry, the Andreev reflections connect superconducting gaps with a phase difference of π, since at one end of the trajectory the positive and on the other the

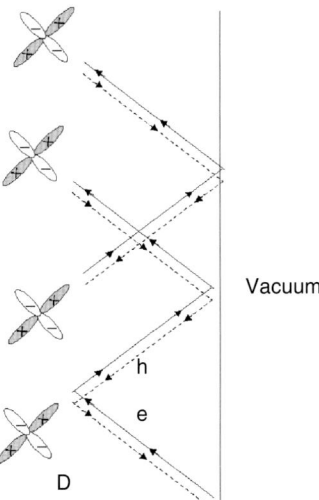

Figure 2.5. Andreev bound states along the surface of a d-wave superconductor rotated by 45° relative to the surface.

negative lobe of the d-wave function is involved in the reflection process [41]. The analogy to the Josephson junction is quite direct. The spontaneous currents are also carried by the quasiparticles, similar to the SNS case [42, 43]. The surface of the d-wave superconductor plays the role of the junction interface with barrier transparency $\overline{T} = 1$, and the sign change in the pair potential corresponds to the phase difference $\phi = \pi$.

The dependence of the ZES has been calculated as a function of interface orientation, barrier transparency and temperature, and a comprehensive description of effects related to them has been given by Tanaka and Kashiwaya [44]. As expected, ZES are more relevant for higher misorientation angles and lower temperatures. The theory can be considered an extension to S/N contacts involving d-wave superconductors of the original phenomenological approach by Blonder, Tinkham, and Klapwijk [38] (BTK) and of the subsequent microscopic advanced version by Arnold [45].

In superconducting quantum point contacts each transport mode contributes one quantized conductance unit to the total conductance and one quantized supercurrent unit to the critical current, in analogy with normal quantum point contacts. Backscattering processes play no role in the conductance through quantum point contacts, since the transport through them is ballistic. The energy separation between the modes becomes very small and the quantum effects are smeared with increasing point contact width. In the short junction limit ($L \ll \xi_0$) a stepwise change in the supercurrent and conductance was observed in a mechanically controllable break junction [46]. Similar behavior was also observed in a long junction ($L \gg \xi_0$) by varying the gate voltage of the split gate in the two-dimensional electron gas (2DEG) of a S–2DEG–S Josephson junction [47]. The correlation between current and conductance steps has also been analyzed recently [48].

2.2.5. The Josephson Effect: General Features

The two basic Josephson equations, originally derived for a S_1–I–S_2 junction with an insulating barrier, are:

$$I = I_c \sin(\phi), \qquad (2.8)$$

$$\dot{\phi} = 2eV/\hbar, \quad (2.9)$$

where $\phi = \phi_1 - \phi_2$ (1 and 2 refer to the left and right electrode, respectively). The microscopic derivation can be found in [6,8,49]. We have the Josephson effect as long as the macroscopic wave functions of the two electrodes overlap in the barrier region. Coulomb $E_C = e^2/(2C)$ and Josephson $E_J = I_c \Phi_0/(2\pi c)$ energies will be associated with each junction. The behavior of quantum Josephson junctions, with either a well-defined charge or phase variable, will depend on the relative magnitude of E_C and E_J (phase for $E_J \gg E_C$, charge for $E_J \ll E_C$).

The critical current for two generalized d-wave superconductors can be written, on the basis of general symmetry arguments, in the case of time reversal symmetry, and to lowest order as [51]:

$$I_c = (C_{2,2}[\cos(2\theta_1)\cos(2\theta_2)] + S_{2,2}[\sin(2\theta_1)\sin(2\theta_2)] + ...). \quad (2.10)$$

The first term of expression Eq.(2.10) corresponds to the well known Sigrist–Rice clean limit formula [50,52]:

$$I_c = A_S[\cos(2\theta_1)\cos(2\theta_2)]. \quad (2.11)$$

In the case that $S_{2,2} = -C_{2,2}$, the sum of the first two terms leads to the dirty limit expression:

$$I_c = A_S[\cos(2\theta_1 + 2\theta_2)], \quad (2.12)$$

when disorder effects and faceting are taken into account. In these expressions a negative supercurrent can be translated as a phase shift of π at the junction, but since an arbitrary phase shift can be added, these π shifts are only meaningful when considered in closed superconducting loops.

Particular choices of θ_1 and θ_2 (for instance in 45° GB junctions) can also make the $\sin(\phi)$ component negligible. Higher order corrections (in particular the second harmonic) in the current–phase relation may play a more relevant role in these limits: $I_{C0} = I_1 \max_\varphi \{\sin\varphi - a\sin 2\varphi\}$. A wide range of issues related to the dependence of the junction supercurrent on phase have been extensively discussed in the review by Golubov, Kupryanov, and Ilichev [53].

We complete our discussion of the Josephson effect by presenting the main concepts behind d-wave induced effects, the presence of bound states, second harmonic in the current–phase relation, and time reversal symmetry breaking. All these affect the tunneling spectra. Further details on unusual properties of the Josephson effect associated with unconventional superconductivity are given in [22].

Andreev Reflection in SNS Junctions

In an S_1–N–S_2 structure the electron obtains an extra phase of $\phi_1 - \phi_2 + \pi$ (see Figure 2.6) in each Andreev reflection. The Josephson effect can be reformulated in terms of this property and of quasiparticle bound states. The spectrum of the elementary excitations of a N layer in contact with S on both sides is quantized if $E < \Delta$. In particular, the expression of the bound state energy in a S–N–S one-dimensional system, in the short junction limit $L \ll \xi_0$, where L is the N thickness and ξ_0 the coherence length, is [42,43]

$$E = \pm \Delta_0 \sqrt{1 - \overline{T}\sin^2(\phi/2)}. \quad (2.13)$$

This quantization implies the presence of a coherent connection between the phases of the order parameter symmetry in both superconductors. The energy of the junction will depend on the relative phase ϕ, and this dependence remains in force also when the width of the normal metal layer greatly exceeds the dimension of the Cooper pair [42, 43]. Therefore in multiple Andreev reflection processes, the electrons/holes cannot escape from the normal metal and they also do not gain energy, thus generating bound states and the consequent supercurrent.

Andreev levels in Josephson junctions are shown in Figure 2.6. The current can be obtained via the derivative of the free energy with respect to ϕ (or similarly through the phase dispersion of the energy of the Andreev state $(dE/d\phi)$):

$$I_J = (2\pi/\Phi_o)(\partial F/\partial \phi), \qquad (2.14)$$

where F is the free energy determined by the Andreev bound states. The Andreev bands $E(\phi)$ have width (dispersion) proportional to the junction transparency $\vec{T}$.

The current contribution can also be separated for each $\vec{k}_F$:

$$\vec{J}_{k_F} = -\frac{2e}{h} L_{k_F} \frac{dF}{d\phi}\bigg|_{k_F} \vec{k}_F \qquad (2.15)$$

and, once integrated over all directions, can be decomposed into $I_\perp$, the perpendicular component passing through the junction (Josephson current), and $I_\parallel$, the component parallel to the metal layer. The detailed general expressions can be found in [54, 55]. Higher order contributions (components carried by the multiple reflection process at the interface) in the current–phase relation are taken into account. The lowest energy state is given by $I_\perp(\phi = \phi_o) = 0$.

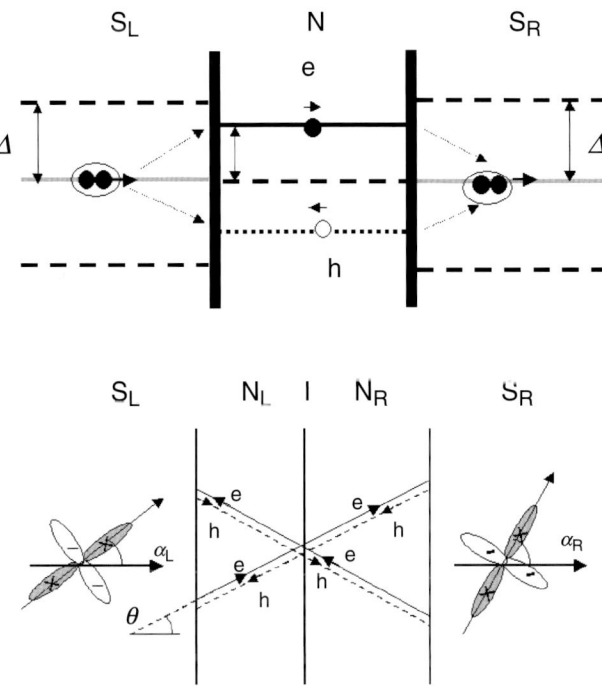

Figure 2.6. Andreev reflections in S–N–S junctions (a) supercurrent; (b) supercurrent in d-wave junctions.

We first consider a S–N–D (D is a d-wave superconductor) junction. This simpler configuration allows the introduction of some basic concepts that will be extended to D–D junctions. Surprisingly, the component of the current parallel to the surface is not zero for $\phi_0 \neq 0$ or π. In the total current perpendicular to the interface, all odd harmonics of the general expression $I(\phi) = I_1(\theta)\sin(\phi) + I_2(\theta)\sin(2\phi) + \cdots$ cancel, and the Josephson coupling is reduced. This is because for junctions with angle β between the interface normal and the d-wave antinode orientation such that $0 < \beta < \pi/4$, for each bound state that sees the "+" lobe with phase ϕ_d, there is a mirror bound state with orientation $-\beta$ that sees the "−" lobe of the OP symmetry with phase $\phi_d + \pi$. The leading term in $I_\perp$ is of the order $\sin(2\phi)$, and the stable ground state with $I_\perp = 0$ is at $\phi = \pm\pi/2$. The Josephson current parallel to the interface, however, has contributions from the odd harmonics and the leading order is $\sin(\phi)$. The presence of a finite parallel current component in the ground state constitutes a spontaneous current and is a manifestation of broken time reversal symmetry, since there is a degenerate state with reversed current (Figure 2.7) [54].

The nature of the Andreev levels changes with the incidence angle θ. For $22.5° < |\theta| < 67.5°$, for instance, mid-gap states are formed at ϕ=0. Elsewhere no mid-gap states are formed at $\phi = 0$, and the Andreev levels resemble those formed in a classical Josephson junction composed of s-wave superconductors ($E_{conv} = \Delta(0)\sqrt{1 - \overline{T}(0)\sin^2(\phi/2)}$). Δ and $\overline{T}$ will be a function of the specific orientation, which will be not indicated in the formulas but can be inferred from the type of structure [44, 54].

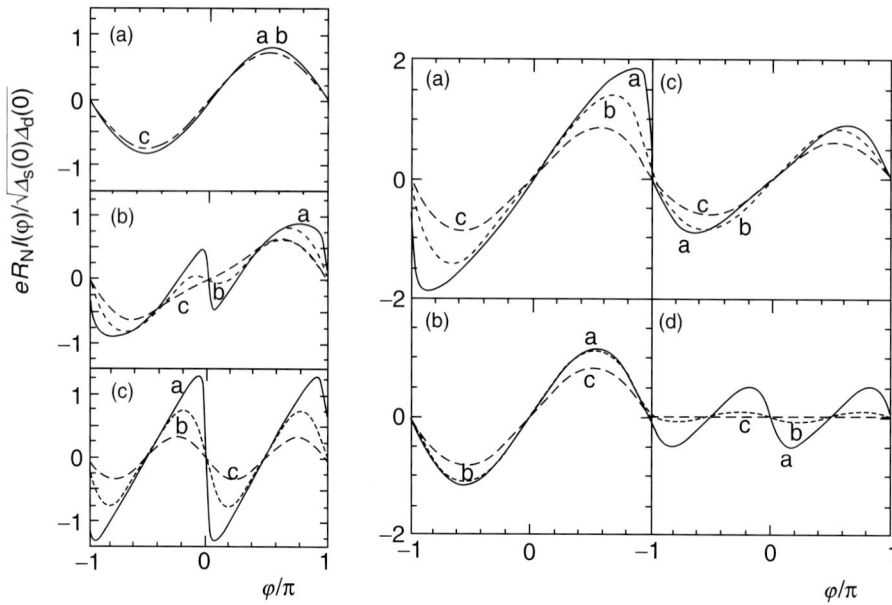

Figure 2.7. Normalized supercurrent for different angles θ_1 and θ_2 between the interface normal and the d-wave antinodes on the two sides of the junction, for temperatures $T/T_c = 0.05$, a; $T/T_c = 0.3$, b; and $T/T_c = 0.6$, c. Left panel: S–I–D junction; $\lambda_0 d_i = 1$, where $\lambda_0 = \sqrt{2mU_0/\hbar^2}$, U_0 and d_i are the barrier potential and thickness, respectively; $\kappa = k_F/\lambda_0 = 0.5$; with (a) $\theta_2 = 0$, (b) $\theta_2 = \pi/8$, (c) $\theta_2 = \pi/4$. Right panel: D–I–D junction with $\kappa = 0.5$; (a) $\theta_2 = 0$, $\lambda_0 d_i = 0$; (b) $\theta_2 = 0$, $\lambda_0 d_i = 1$, $\theta_2 = \pi/4$; (c) $\theta_2 = \pi/8$, $\lambda_0 d_i = 1$, (d) $\theta_2 = \pi/4$, $\lambda_0 d_i = 1$. For a misorientation angle of $\theta_2 = \pi/4$ the contribution of the second harmonic becomes dominant (from Tanaka and Kashiwaya [44]).

In D–D junctions the dispersion relation for the Andreev bound states (and the corresponding angle integrated current–phase relation) is particularly significant for some specific misorientations, which encompass all practical experimental situations. A comprehensive derivation of these expressions with detailed references can be found in the review by Löfwander, Shumeiko, and Wendin [20]. Here we limit ourselves to some aspects relevant to experimental results. For a D–D junction with $\theta_1 = \pi/4$, $\theta_2 = \pi/4$ (0-junction with large I_C) and $\theta_1 = \pi/4$, $\theta_2 = -\pi/4$ (π-junction with large I_C) orientations the solutions of the corresponding spectral equations are $E_\pm = \pm |\Delta|\sqrt{\bar{T}} \cos(\phi/2)$ and $E_\pm = \pm |\Delta|\sqrt{\bar{T}} \sin(\phi/2)$, respectively (Figure 2.8).

At zero temperature, only the level below zero energy is populated, while the level above is empty, and the currents will be $j_{MGS} = \frac{ek_F}{\hbar}(|\Delta|\sqrt{\bar{T}}) \sin(\phi/2) \mathrm{sgn}(\cos(\phi/2))$ in the first case and π-shifted with respect to the first configuration in the second. If the surface states at the two sides of the junctions have equal energies, the coupling becomes resonant. In this resonant case, the splitting of the levels, and as a consequence the width of the Andreev band, will be particularly large, proportional to $\bar{T}^{1/2}$, and causes Josephson coupling.

In conclusion this last section summarizes how surfaces may hybridize and form bound states in superconducting junctions, and Andreev reflection may lead to the formation of zero energy quasiparticle bound states in d-wave superconductors. The existence of midgap states enhances the Josephson current at low temperatures.

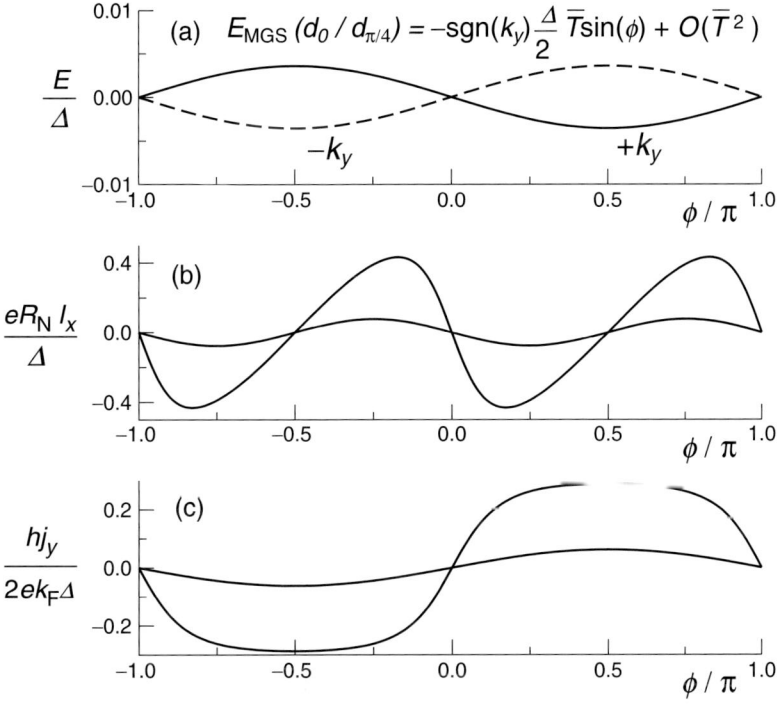

Figure 2.8. (a) Andreev band for a D–D junction with $\theta_1 = 0$, $\theta_2 = \pi/4$. (b) The angle integrated current–phase relation and (c) the phase dependence of the surface current density along the junction interfaces calculated for an injection angle $= \pi/9$ and junction transparency $\bar{T} = 0.01$ for temperatures $T = 0.01T_c$ and $0.001T_c$, respectively (adapted from Löfwander et al. [20]).

2.3. Means of Preparing Tunnel Junctions

In contrast with low-temperature superconductors, a clear distinction between tunnel junctions and structures with direct conduction does not hold for HTS junctions. Except for a few established cases, such as in scanning tunneling microscopy measurements described below, the barrier is often intermediate between the tunnel and direct contact limits. We therefore will present most of the means for fabricating junctions in high-temperature superconductors in one section.

2.3.1. Junctions with Single Crystals

The first generation of HTS junctions could not rely on good quality films, and employed bulk materials as electrodes, either in break junctions, where a bulk piece is broken in two pieces, possibly at low temperatures; or in point contact junctions, where one electrode is a sharp tip. This latter technique has been widely employed in the past also for LTS JJs [8,56]. Performances of the very first break and point contact junctions were mostly controlled by the quality of the single crystals employed, and were always limited by the poor reproducibility of a natural barrier in intrinsic complex systems such as the HTS. Significant subsequent achievements were made possible by surface treatment techniques followed by the deposition of an artificial barrier and a counterelectrode (normal metal or LTS), which avoided the critical step of the deposition of an artificial barrier. These approaches were successfully applied to measurements of the energy gap [57] and order parameter symmetry [58–62] in YBCO.

2.3.2. Grain Boundary Junctions

Grain boundary (GB) junctions [19] (a general classification has been given in Figure 2.1e–g) take advantage of a significant reduction of the critical current between two grains with different orientations, which generates weak coupling and Josephson-like behavior between the two electrodes. The natural intrinsic barrier avoids problems related to an artificial barrier. These junctions, despite the limits discussed below, can be considered of good quality and made several significant experiments possible. Grain boundary critical currents decay exponentially with increasing misorientation angle, which can be roughly interpreted as due to an increase of the thickness of the GB barrier with increasing misorientation angle θ. Most existing data is on YBCO junctions, but similar angular dependencies of the grain boundary J_C have been reported for all other high-T_c materials, [19] including electron doped materials.

The nomenclature used for grain boundary junctions typically distinguishes between the asymmetric case, where one grain is crystallographically aligned with the boundary (Figure 2.9a), and the symmetric case, in which the crystallographic misorientations of the two grains relative to the grain boundary are the same (Figure 2.9b). For example, a 45° symmetric boundary corresponds to 22.5°–22.5° misorientations, while an asymmetric boundary corresponds to 0°–45°. Ninety degree boundaries are particularly relevant limit cases, which commonly occur in a-axis oriented thin films [63–65].

Bicrystal Junctions

The bicrystal technique, based on the union of two substrates with different crystal orientations (see Figure 2.9a,b), is the most direct way to create a grain boundary junction [66–68]. Epitaxial HTS films reproduce the relative orientations of the two substrates. This

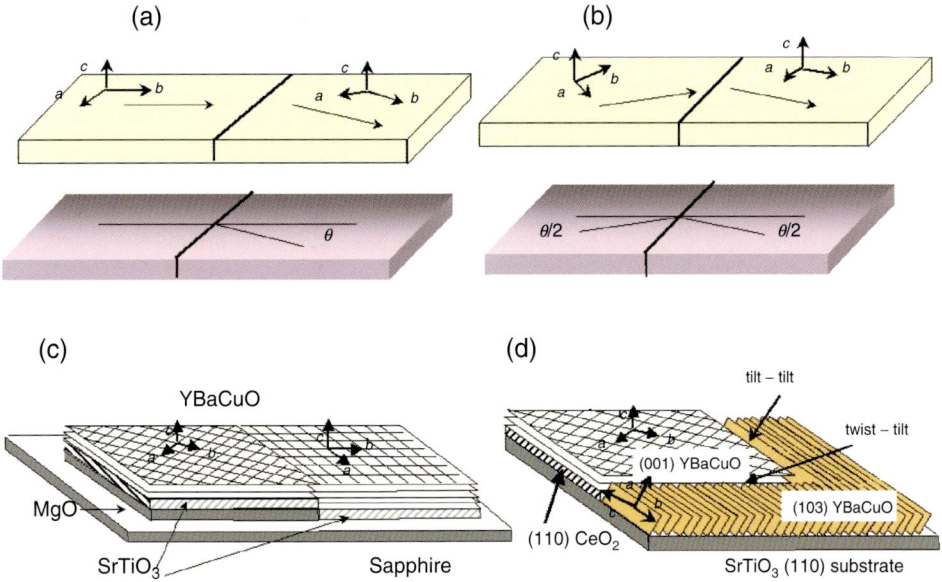

Figure 2.9. Sketch of bicrystal junction in asymmetric (a) and symmetric (b) configuration, and of the grain boundary structures in classical biepitaxial (c) and CeO_2 (d)-based out-of-plane biepitaxial junctions. In (d) the two limit configurations, tilt and twist, are indicated. In (d) the presence of the CeO_2 produces an additional 45° in-plane rotation of the YBCO axes with respect to the in-plane directions of the substrate.

is the only way to vary the relative orientations of the electrodes in all possible combinations [19, 67]. High values of the $I_C R_N$ parameter have been observed in the less frequently used c-axis tilt GBs [69].

Bicrystalline substrates of many compounds, including $SrTiO_3$, doped $SrTiO_3$, MgO, yttria-stabilized zirconia (YSZ), $NdGaO_3$, $LaAlO_3$, silicon, and sapphire have been used [19, 70, 71]. The other techniques (step, step-edge, and biepitaxial) can overcome the constraints imposed by the underlying bi-, tri-, or tetra-crystal substrate, such as where the junctions are placed. Other techniques all employ photolithographic means to define the grain boundary interfaces. Although they are more flexible in the placement of the junctions for fundamental studies and more efficient for circuit design, the GBs produced by these other techniques are limited to particular misorientation angles, and the performances of some of these junctions may be less reliable.

Biepitaxial Junctions

The biepitaxial technique uses changes of the orientation of HTS films induced by epitaxial growth on structured template layers. In the original technique [72], a MgO template layer on a r-plane sapphire produces an in-plane rotation by 45° of a $SrTiO_3$/YBCO bilayer compared to an identical bilayer grown directly on the sapphire (the grain boundary has a 45° tilt around the [001] direction) (Figure 2.9c). This junction makes use of the epitaxial relationships: $SrTiO_3$ [110] || Al_2O_3 [11$\bar{2}$0] and $SrTiO_3$ [100] || MgO [100] || Al_2O_3 [11$\bar{1}$0] [73]. Subsequent works employed various materials combinations to induce variations of the in-plane orientation [74].

More recently the biepitaxial technique has been extended to novel configurations, in which one of the electrodes does not grow along the c-axis orientation [75–77]. A specific feature of these structures is the use of a (110)-oriented MgO or CeO_2 (Figure 2.9d) buffer layer, deposited on (110) $SrTiO_3$ substrates. YBCO grows along the [001] direction on the MgO and on the CeO_2 seed layers, while it grows along the [103]/[013] direction on $SrTiO_3$ substrates. The presence of the CeO_2 produces an additional 45° in-plane rotation of the YBCO axes with respect to the in-plane directions of the substrate (Figure 2.9d). As a consequence, the grain boundaries are the product of two 45° rotations, a first one around the c-axis, and a second one around the b-axis. This configuration produces the desired 45° misorientation between the two electrodes to enhance d-wave order parameter effects.

The biepitaxial grain boundaries have a lower transmission probability than other types of grain boundaries, and are closer to the tunnel-limit. This is probably the key feature which lead to the first successful observation of the angular dependence of I_C in all HTS junctions [76] (see Figure 2.25) and later to the first observation of macroscopic quantum tunneling (MQT) (see Figure 2.29) and energy level quantization in high-T_c Josephson junctions [78].

Step-Edge Junctions

Grain boundaries are also nucleated by growing a HTS film over a suitable step patterned into the substrate [79–84]. The resulting structure will strongly depend on the morphology of the step [83, 85]. However, one grain boundary is typically nucleated at the bottom of the step, and another at its top, and these are in series electrically. The step-edge junctions can be positioned anywhere on the substrate, being defined by photolithography (in the most advanced by using an amorphous carbon mask and ion-beam etching or reactive ion-etching). The step height (200–300 nm) is usually larger than the film thickness; both the step angle and the material substrate play crucial roles. Better performances are achieved for high step angles. Various detailed studies on the correlation between YBCO step-edge junction characteristics with microstructure have been carried out [83, 84]. Problems of reproducibility are severe constraints for the use of these junctions for applications.

Electron Beam Junctions

Josephson junctions are also produced by weakening superconducting properties in narrow microbridges. Different sources of irradiation (electron beam in particular) [86–89] have been used. In HTS irradiation causes displacement defects, which act as strong scattering centers in the Cu–O planes and are primarily oxygen defects for energies of the order of 100 keV. At higher irradiation energy (>300 keV) Cu defects may be created as well. Decreasing the carrier concentration (done by removing chain-oxygens for instance at lower irradiation energies) decreases the doping level, which lowers T_c. The barrier region, which is the irradiated region, is never exposed to air or broken in this technique.

Focused electron beam irradiation [90, 91] is also used to modify the properties of the GB Josephson junctions. In this case the situation is further complicated by the presence of a grain boundary, which acts as a sink for the migration of defects, affecting the kinetics of their accumulation.

Electron irradiation changes the current–voltage $(I-V)$ characteristics, and presumably the barrier as well as the microstructure of the grain boundary by modifying the oxygen content in the vicinity of such interfaces. These changes can be controlled by varying the electron dose and partially restored by isothermal annealing of the junctions.

Other examples of a barrier that can be controllably adjusted "a posteriori" come from proton and light irradiation [92–95], causing a decrease of I_C and increase in R_N similarly to e-beam irradiation. Results have been interpreted in terms of a tunneling barrier height increasing with fluence. Ion irradiation (1 MeV H$^+$) has been used to remove Andreev bound states in tunnel YBCO/Pb junctions [96].

2.3.3. Junctions with Artificial Barriers

While grain boundary junctions are only based on thin films, junctions with artificial barriers have been realized both on thin films and single crystals, as mentioned previously. It is obvious that junctions based on single crystals have a scientific value only for fundamental experiments. For instance, the first YBCO (single crystal)–insulator–Pb (Nb) junctions [57–59] have been promptly replaced by a second generation based on thin-films [97, 98]. Barriers have been fabricated through different methods.

Noble metals and oxide-like materials are the most commonly used barriers. Different geometries and counterelectrodes are used to take advantage of the various features (longer coherence length, anisotropy,...). Ramp-type junctions, for instance, allow for the use of well-established c-axis HTS thin film technology while allowing the main current to flow in the a–b planes. Optimizing interface resistance has basically driven the research activities on oxide-barriers. Interface resistance might be due to mismatches in carrier density, lattice constant, thermal expansion, and dimensionality [99]. This was the impetus for strategies to reduce lattice mismatch, including matching expansion coefficients in the c-direction (PrBaCuO and Pr-doped YBCO), increasing the carrier density and driving YBCO into the over-doped region (Ca-doped YBCO), reducing the carrier concentration by cation substitution on lattice sites far from the CuO$_2$ planes (YBa$_2$(Cu$_{1-x}$Co$_x$)$_3$O$_{7-x}$), and replacing Cu atoms directly on the CuO$_2$ plane layers using for example Zn or Ni [99].

Noble Metal Barriers

Au [98, 101–105] and Ag [106] have been used as a barrier for junctions based both on single crystals and thin films because of their good compatibility with HTS. Various counterelectrodes (both LTS and HTS) and configurations have been used with the high-T_c superconducting material as a base electrode. YBCO S–N–S junctions in a step-edge geometry, for instance, have been fabricated by special inhibiting layers introduced to ensure proper separation of the superconducting electrodes, with the final junction conductance through a gold barrier. Focused ion beam (FIB) has also been used to define narrow trenches where YBCO films break naturally [107].

Junctions employing a LTS counterelectrode often perform better than those employing HTS thin films for both electrodes, but have a limited working temperature range. Examples are YBCO–Au–Nb (ramp-type) [98], YBCO–Ag–Pb [108, 109], YBCO–Au–Pb [110], and YBCO–Ag–PbIn [111, 112] junctions.

A ramp-edge junction technology has been introduced [98, 100] (Figure 2.10e) that allows the photolithographic patterning of high quality junctions. In this technique, ramp edges are produced in [001] oriented, pulse laser deposited YBCO films using photolithography and Ar ion etching. The devices are returned to the deposition chamber, etched and cleaned, and then thin layers of YBCO and Au are deposited by pulsed laser deposition in situ. The junctions are completed with Nb. This process eliminates a degraded layer of YBCO next to the Au, improving the junction characteristics. This technology has been used to make

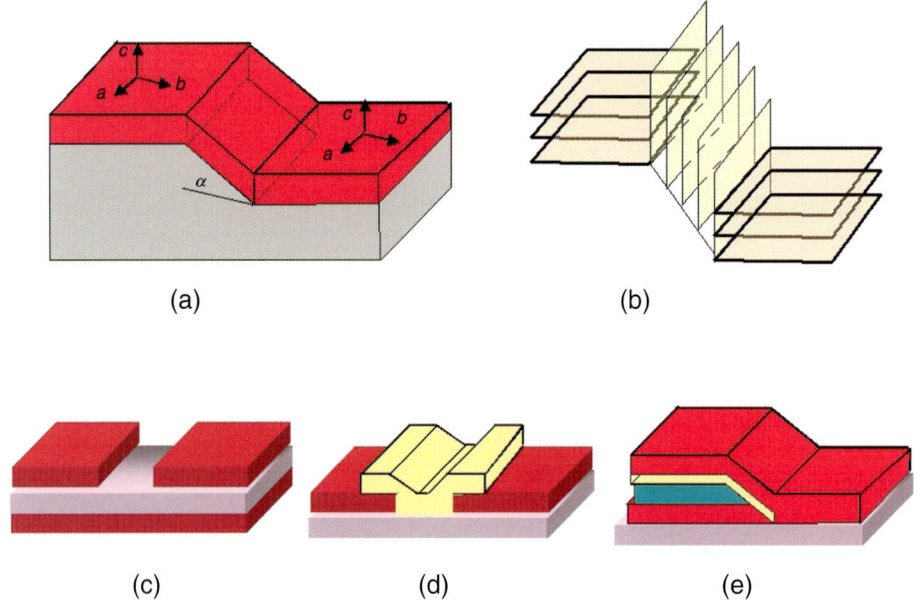

Figure 2.10. (a) and (b) Step junctions for high α angles, (c) and (d) SNS coplanar junctions; in (c) the barrier is predeposited or occurs through a suitable substrate while in (d) the normal metal barrier is deposited in a narrow trench; (e) SNS ramp-edge junction: in improved versions a degraded layer of YBCO next to the Au is eliminated through suitable surface treatment before in situ deposition of the Au barrier and Nb counterelectrode [98, 100].

junctions with intentional facets that reproduce the unusual magnetic interference patterns seen in asymmetric 0–45° grain boundary junctions, and to make very large arrays of π-rings [98, 113, 114]. The excellent properties of these junctions will be discussed in various sections below.

Perovskite and Layered Materials Barriers

Comparative studies between cubic perovskite barrier materials $CaRuO_3$, $SrRuO_3$ [115], and $La_{0.5}Sr_{0.5}CoO_3$ [99], and layered materials such as $Y_{0.7}Ca_{0.3}Ba_2Cu_3O_{7-x}$, $YBa_2Cu_{2.79}Co_{0.21}O_{7-x}$, and $La_{1.4}Sr_{0.6}CuO_4$ [116] have been carried out in step-edge geometry junctions. Oxygen deficiency/disorder has also been considered as the source of the interface resistance [117]. As a matter of fact for the cubic perovskite barriers, characterized by a large difference in thermal expansion coefficients with respect to YBCO, higher values of the normal state resistance have been measured ($R_N A$ of the order of $10^{-8} \Omega$ cm^2), as compared with the layered materials ($R_N A$ of the order of $10^{-10} \Omega$ cm^2). Co-doped and Ca-doped YBCO are significant terms of comparison, being an overdoped (underdoped) version of YBCO with larger (smaller) carrier density, lower T_c, and smaller (larger) anisotropy than YBCO, respectively. Proximity effects have been shown to occur for both these barriers [16, 99].

PrBaCuO and Pr-doped YBCO based oxides ($Y_{0.3}Pr_{0.7}Ba_2Cu_3O_{7-x}$, $Y_{0.6}Pr_{0.4}Ba_2Cu_3O_{7-x}$) have also been widely employed as a barrier from the early stages in different geometries [118–122], also as a function of Ga doping [123], with $R_N A$ of the order of 10^{-7}–$10^{-8} \Omega$ cm^2 [124]. Barriers are typically varied between 6 and 30 nm producing different values of J_C. The $I_C R_N$ has been found to scale nearly linearly with barrier thickness, ranging from 0.8 to 5 mV. Ga-doped junctions appear to be less sensitive to variations in the barrier thickness [125].

Additional transport issues can be investigated by exploiting barrier properties. For instance, bulk PBCO is reported to behave like a variable-range hopping conductor, caused by the relatively high density of localized states, and PBCO barriers may allow the study of effects of two localized states in an inelastic tunneling process [126, 127]. A-axis YBCO–PBCO–YBCO junctions (meant to exploit the longer in-plane coherence lengths) have been also realized on (100) LaSrGaO$_4$ [128] and vicinal (001) LaAlO$_3$–SrAl$_{0.5}$Ta$_{0.5}$O$_3$ substrates [129], with spreads in I_C and R_N of 11% and 8.8%, respectively. Josephson behavior has been claimed to occur for barriers 80 nm thick, and coherence lengths of the order of 20 nm have been found [128].

2.3.4. Interface-Engineered Junctions

Interface-engineered junctions have a thin barrier layer, typically on the ramp edges, made by damaging the YBCO base electrode surface using ion bombardment [130, 132]. During the counterelectrode deposition process the surface is then recrystallized. This technique has also been applied to produce an all YBCO c-axis trilayer [130–133].

A recent comprehensive study on good quality interface-engineered junctions, with magnetic modulation of the critical current above 80% and critical current density ranging from 10^2 to 10^6 A/cm^2 at $T = 4.2$ K, suggests that they should be regarded as an array of microscopic SNS contacts embedded in an insulating barrier with random orientation [133]. This filamentary structure prefers special orientations, inhibiting effects particular to d-wave pairing symmetry.

2.3.5. Junctions with HTS Rather than YBCO

Although most of the results presented above refer to junctions made with YBCO, other superconductors have also been used because of their special properties. For instance, Bi and Ta based compounds, with their large anisotropy, are preferred for intrinsic junctions. Ca-doped YBCO junctions exploit the over-doping property of this compound [134], while electron-doped compounds could illuminate novel physical aspects. Most HTS compounds have dominant d-wave order parameter symmetry; the presence of additional subdominant components may depend on the material and the interface geometry.

La$_{1.85}$Sr$_{0.15}$CuO$_4$-Based Trilayer with One-Unit-Cell-Thick Barrier

The most significant step toward the goal of an all-HTS trilayer with an insulating barrier is the structure composed of La$_{1.85}$Sr$_{0.15}$CuO$_4$ (LSCO) electrodes separated by a one-unit-cell-thick La$_2$CuO$_4$ (LCO) barrier [135]. This achievement can be considered the follow-up of an intense research activity started years ago on BiSrCaCuO (2212) [136, 137].

Bozovic et al. [138] claimed a "giant" proximity effect in LSCO junctions from observations of Josephson current for LCO barrier thicknesses ranging from 1 to 15 unit cells (up to 20 nm), much thicker than the coherence and mean free path lengths. This cannot be understood using conventional theory. They suggested that the supercurrent was mediated by resonant tunneling through a series of energy-aligned states within the barrier layer [138]. These experiments used the conversion of the junction from S–N–S to S–I–S through annealing at low temperature in vacuum (which drives LCO insulating leaving LSCO almost intact) [138].

A long-range or anomalous proximity effect has also been discussed in other types of junctions (see for instance the references in [138]) and within the context of a quantum phase transition between the low carrier-concentration insulating antiferromagnetic phase and the high carrier concentration metallic and superconducting phase [139].

Electron Doped HTS

Bicrystal junctions are the most common type of junction involving electron-doped cuprate superconductors. In the electron doped $La_{1.85}Sr_{0.14}CuO_4$ compound [140], barriers are on average less transmissive than those on YBCO bicrystals, with $J_C = 6 \times 10^3$ A/cm^2 at $T = 4.2$ K, $R_N A = 5 \times 10^{-8}$ A/cm^2, for a misorientation angle of 24°; $J_C = 3$ A/cm^2 at $T = 4.2$ K, $R_N A = 10^{-6}$ A/cm^2, for a misorientation angle of 36°.

The same technology as for YBCO has been used to produce $Nd_{2-x}Ce_xCuO_{4-y}$ (NCCO) zig-zag ramp-type junctions. Both optimally doped ($x = 0.15$) and overdoped ($x = 0.165$) samples were prepared with a bilayer of 150 nm (001)-oriented NCCO and 35 SrTiO$_3$ and with 160 nm Nb top electrode. A 12-nm NCCO interlayer and a 12-nm Au barrier were used as a barrier. In the optimally doped case, J_C and $I_C R_N$ values of 30 A/cm^2 and 30 µV at $T = 4.2$ K, respectively, were achieved. As a consequence the Josephson penetration depth λ_J (see section 2.4) was about 65 µm, comparable with the zig-zag facet length. $R_N A$ was about 10^{-6} Ω/cm^2. Anomalous controllable magnetic patterns were observed, giving evidence of a predominant d-wave OP symmetry, without any change into s-wave at low temperatures [141].

Ca and Co Doped YBCO: Insights into the Overdoped Regime

An enhancement of the critical current in bicrystal junctions has been achieved by overdoping the superconductor [142] and in particular through Ca and Co doped YBCO [143,144]. Ca and Co doped YBCO junctions have given the best results in enhancing J_C. J_C has been studied as a function of Ca concentration, giving evidence of optimum doping of $x = 0.3$ (for instance for a grain boundary angle of 24° $J_C = 7 \times 10^6$ A/cm^2 at $T = 4.2$ K, about one order of magnitude higher than the nondoped YBCO case) [143].

Ultra-Thin Films and Superlattices

Bicrystal junctions based on ultra-thin films have been realized. Josephson junctions composed of only a few superconducting CuO$_2$ planes (six layers in particular) have been realized by exploiting ultrathin $[Ba_{0.9}Nd_{0.1}CuO_{2+x}]_5/[CaCuO_2]_2/[Ba_{0.9}Nd_{0.1}CuO_{2+x}]_5/[CaCuO_2]_2/[Ba_{0.9}Nd_{0.1}CuO_{2+x}]_5$ (CBCO) structures (5/2/5/2/5). The CBCO film is only 8 nm thick. The Josephson effect was measured even in junctions 5 mm wide [145].

Intrinsic Stacked Junctions

The strongly anisotropic, layered crystal structure of the cuprates allows intrinsic stacked junctions, which are fabricated from bulk single crystals (Figure 2.11). The supercurrent in these junctions is mostly along the c-axis. Most successful results have been achieved in $Bi_2Sr_2CaCu_2O_8$ [146] and $Tl_2Ba_2Ca_2Cu_3O_{10}$ single crystals and thin films [147, 148]. In a-axis oriented YBCO thin films only flux flow behavior has been observed [147].

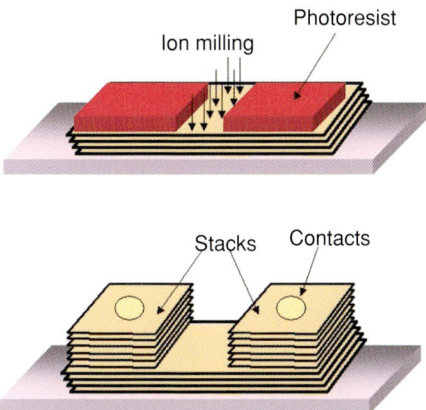

Figure 2.11. An example of intrinsic stacked junctions. The mesas with c-axis transport are defined through mechanical etching.

Josephson coupling between CuO_2 double layers has been proved, and most of the materials behaved like stacks of S–I–S JJs with effective barriers of the order of the separation of the CuO_2 double layers (1.5 nm) (J_C typically 10^3 a/cm^2). The I–V curves exhibited large hysteresis and multiple branches, indicative of a series connection of highly capacitive junctions. Practical realizations (see Figure 2.11) of intrinsic stacked junctions have been designed to avoid heating effects in principle [17]. However, at high voltages caution is required when extracting information from the current–voltage characteristics because of possible unavoidable heating problems. Recently macroscopic quantum tunneling has been claimed to occur in BiSCCO intrinsic junctions [149].

2.4. π-Rings and $0 - \pi$-Junctions

Bulaevskii et al. [150] proposed in 1977 that a superconducting loop including a Josephson junction could have an intrinsic π-phase shift in the absence of an externally applied field or current. Such a ring is now termed a π-ring. They speculated that such a π-phase shift could result from spin-flip assisted tunneling within the tunnel junction itself. This process has not yet been observed experimentally. However, three other mechanisms for introducing a π-phase shift into a superconducting ring have been demonstrated (1) by taking advantage of the momentum dependence of the pairing wavefunction in unconventional superconductors [50, 52, 151, 152], (2) by introducing a π-phase shift by tunneling through ferromagnetic layers [153–155], and (3) by running supercurrent through two closely spaced electrodes along the ring [156–158]. In this section we will focus on the first route, using unconventional superconductors, for producing π-rings. Geshkenbein, Larkin, and Barone suggested using the properties of π-rings to test for unconventional pairing symmetry in the heavy fermion superconductors [151, 152]. Sigrist and Rice [50, 52, 159] proposed that the paramagnetic Meissner effect [160–165] which occurred in ceramic samples of $Bi_2Sr_2CaCu_2O_8$ was due to naturally occurring π-rings due to Josephson contacts between the grains, and suggested using a controlled geometry as a test of d-wave superconductivity in the high-T_c cuprate superconductors.

The controlled geometries that have been used for observing the effects of an intrinsic π-shift in a superconducting ring can be divided into two classes. In the first, the $0-\pi$ junction illustrated in Figure 2.12a, one section of a Josephson junction has an intrinsic π phase shift relative to the other. For the purposes of discussion we take a geometry in which the junction normal is in the z-direction, the junction has a width W in the x-direction, and the junction depth in the y-direction is small compared with the Josephson penetration depth $\lambda_J = \sqrt{\hbar/2e\mu_0 d j_c}$, where d is the spacing between the superconducting faces making up the junction, and j_c is the Josephson critical current per area of the junction. If the x-dependent intrinsic phase drop is $\theta(x)$, which can take the values 0 or π, the local supercurrent density across the junction follows the relation $j_s = j_c \sin(\phi + \theta)$, and the quantum mechanical phase difference across the junction $\phi(x)$ follows the Sine–Gordon relation

$$\frac{\partial^2 \phi}{\partial x^2} = \frac{1}{\lambda_J^2} \sin(\phi(x) + \theta(x)). \qquad (2.16)$$

Analytical [166–169] and numerical [170–174] solutions of Eq. (2.16) have been published. In the "short-junction" limit $W \ll \lambda_J$, $\partial^2 \phi / \partial x^2 \to 0$, and $\phi(x) = \phi_0 + 2\pi \Phi x/(\Phi_0 W)$, where ϕ_0 is a constant and Φ is the total magnetic flux threading the junction in the y-direction. Then for equal lengths of 0- and π intrinsic phase shifts in the junction the critical current becomes

$$I_c(\Phi) = I_0 |\sin^2(\pi \Phi/2\Phi_0)/(\pi \Phi/2\Phi_0)|, \qquad 0-\pi \text{ junction} \qquad (2.17)$$

with a minimum at zero applied flux (solid line in Figure 2.12b). This is to be compared with the expression for a conventional junction in the short junction limit:

$$I_c(\Phi) = I_0 |\sin(\pi \Phi/\Phi_0)/(\pi \Phi/\Phi_0)|, \qquad 0-\text{junction} \qquad (2.18)$$

which has a maximum at zero applied flux (dashed line in Figure 2.12b). As the width W of the junction becomes comparable to λ_J, the amplitude of the oscillations in the critical current with applied field becomes smaller [170, 171].

In the opposite limit, in which $W \gg \lambda_J$, a $0-\pi$ junction spontaneously generates a Josephson vortex at the intersection between the regions with 0 and π intrinsic phase shift. This vortex generates $\Phi_0/2 = h/4e$ ($N = 1/2$) total flux threading through the junction in the y-direction. Junctions of intermediate length $W \sim \lambda_J$ spontaneously generate "semifluxons" with total flux less than $\Phi_0/2$ (Figure 2.12) [168, 172–174]. Josephson semifluxons with higher quantum number (3/2, 5/2, etc.) are in principle allowed, but are energetically unstable to the formation of an $N = 1/2$ Josephson vortex at the $0-\pi$ intersection, plus integer Josephson vortices elsewhere in the junction [167, 175].

A second geometry for phase sensitive tests of the pairing symmetry in unconventional superconductors is the superconducting quantum interference device (SQUID), a superconducting ring with at least one Josephson junction, with in general an intrinsic phase shift ε upon circling the ring. Consider a symmetric two-junction SQUID with junction critical currents I_c, phase drops ϕ_1, ϕ_2 across the two junctions, and total inductance L, with an intrinsic phase shift of π (Figure 2.12d). The total current I_B through the SQUID is

$$I_B = I_c(\sin(\phi_1) + \sin(\phi_2)). \qquad (2.19)$$

The requirement of a single valued wave function leads to the condition

$$2\pi N = \pi + \phi_2 - \phi_1 + \beta(\sin\phi_1 - \sin\phi_2) + 2\pi \Phi_e/\Phi_0, \qquad (2.20)$$

Tunneling Measurements of the Cuprate Superconductors

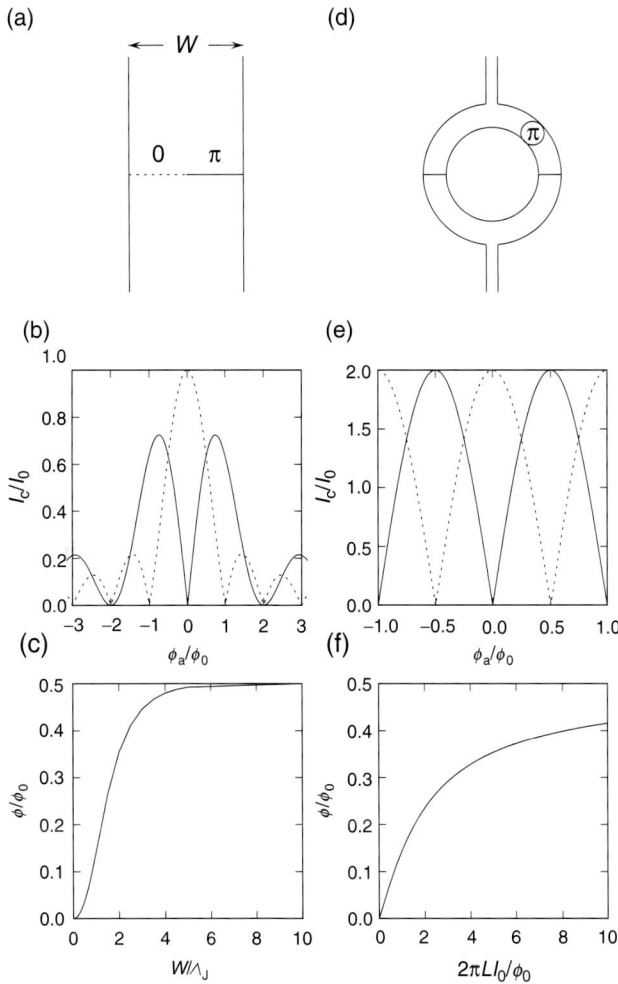

Figure 2.12. Two basic geometries and techniques for phase sensitive tests of pairing symmetry in unconventional superconductors. (a) Depicts a 0–π junction. The junction critical current as a function of field for a symmetric 0–π junction in the short junction limit (b, solid line) has a minimum at zero field. Also shown is the analogous curve for a conventional junction (b, dashed line). The spontaneously generated magnetic flux in a symmetric 0–π junction is plotted in (c) as a function of the total width of the junction W divided by the Josephson penetration depth λ_J. (d) Depicts a two-junction SQUID with an intrinsic π phase shift. The critical current of a symmetric, two-junction π-SQUID in the limit $2\pi L I_0 \ll \Phi_0$ (e, solid line), where L is the total SQUID inductance and I_0 is the single junction critical current, is shifted by $\Phi_0/2$ relative to that for a 0-SQUID in the same limit (e, dashed line). The spontaneously generated flux in a symmetric, two-junction π-SQUID is plotted in (f) as a function of $\beta = 2\pi L I_0/\Phi_0$.

where N is an integer, $\beta = 2\pi L I_c/\Phi_0$ and Φ_e is the externally applied flux through the SQUID. Plotted as the solid line in Figure 2.12e is the critical current, the maximum allowed supercurrent, through such a symmetric π-SQUID under the conditions of Eqs. (2.19) and (2.20). The dashed line in Figure 2.12e is the critical current for a conventional 0-SQUID. The dependence of the critical current on applied field is shifted by one half-period $\Phi_0/2$ for the π-SQUID relative to that for the 0-SQUID. Asymmetric π-SQUIDs have been considered in [10, 176].

Pairing symmetry experiments on the cuprate superconductors using 0–π junctions and π-SQUIDs have been summarized in [10, 11]. The first such experiments were by Wollman et al. [58, 59]. They made π-rings and 0–π-junctions between single crystals of YBa$_2$CuO$_{7-\delta}$ (YBCO) and Pb, a conventional superconductor. They observed a phase shift of π in the dependence of the critical currents of their π-SQUIDs on applied flux, relative to that expected for a conventional SQUID, and they observed a minimum in the critical current of their 0–π junctions at zero applied field. Brawner and Ott [60] formed π-SQUIDs using Nb point contacts to single crystals of Nb and also saw phase shifts in the magnetic interference patterns. Both the Wollman et al. and the Brawner and Ott experiments were in the limits $\beta \ll 1$ or $W \gg \lambda_J$: the spontaneous currents were small.

These early phase sensitive pairing symmetry experiments had characteristics that differed from the ideal behaviors displayed in Figure 2.12 because of asymmetries in the junction critical currents, inhomogeneities in the junction critical current densities, and problems of trapped flux. However, it is now possible to make 0–π junctions and π-rings using the cuprate superconductors without these complications. An example [177] is shown in Figure 2.13. Here optimally doped YBCO was epitaxially deposited by laser deposition on bi- and quad-crystals of SrTiO$_3$ in geometries chosen to form 0- (Figure 2.13a–c) and π-SQUIDs (Figure 2.13d–f) for a predominantly d$_{x^2-y^2}$ superconductor. The SQUIDs were designed to have small β factors, so that asymmetries in the junction critical currents would not result in

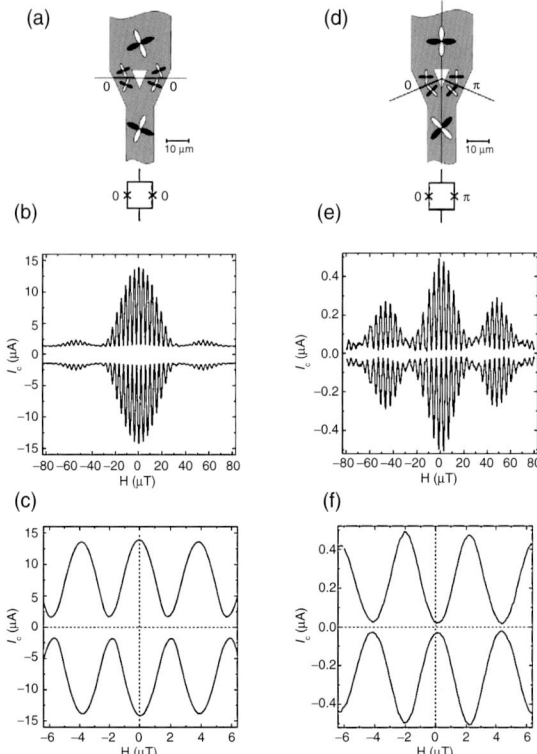

Figure 2.13. Geometry (a,d) and measured critical currents (b,c,e,f) for all high-T_c YBCO 0- (a–c) and π-SQUIDs (d–f). The 0-SQUID shows a maximum, while the π-SQUID shows a minimum, in the critical current at zero applied field, as expected for predominantly d$_{x^2-y^2}$ pairing symmetry (adapted from Schulz et al. [177]).

unintentional shifts in the magnetic interference patterns, and relatively large junction areas, so that stray magnetic fields could be detected from the single junction interference characteristics, which should be symmetric in field in the absence of a stray field. The junction interference characteristics, the "envelopes" in Figure 2.13 b,e, are quite symmetric, indicating small stray fields, and the SQUID critical currents have a minimum for the π-SQUID (Figure 2.13c), and a maximum for the 0-SQUID (Figure 2.13f), as expected if YBCO is a predominantly $d_{x^2-y^2}$ superconductor.

The first experiments to detect the spontaneous flux predicted for π-SQUIDs and 0–π junctions fabricated from cuprate superconductors were by Tsuei et al. [11]. They formed π-rings [178] and 0–π junctions [179] using a tricrystal geometry with grain boundary weak links. This allowed very high junction critical currents, with high β factors and short λ_Js, so that the spontaneous magnetizations were very close to $\Phi_0/2$. These spontaneous currents were imaged with a scanning SQUID microscope [180–183]. Tsuei and Kirtley used tricrystal pairing symmetry tests to infer that the optimally hole doped cuprates YBCO [178, 184], $Tl_2Ba_2CuO_{6+\delta}$ [185, 186], and $Bi_2Sr_2CaCu_2O_{8+\delta}$ [187], and the optimally electron doped cuprates $Nd_{1.85}Ce_{0.15}CuO_{4-\delta}$ and $Pr_{1.85}Ce_{0.15}CuO_{4-\delta}$ [188] have predominantly $d_{x^2-y^2}$ pairing symmetry, and that this symmetry persists in the hole-doped cuprates over a broad doping range [175]. They also used a variable sample temperature scanning SQUID microscope [189–191] to image the Josephson vortex at the tricrystal point in optimally doped YBCO as a function of temperature, and concluded that, within experimental error, it had half of the superconducting flux quantum of flux from 0.5 K to within a few degrees of T_c [192]. The conclusion of predominantly $d_{x^2-y^2}$ pairing symmetry in optimally electron doped superconductors was confirmed by Chesca et al. using grain boundary π-SQUID interferometers fabricated from $La_{2-x}Ce_xCuO_{4-y}$ [193] and Ariando et al., using $Nd_{1.85}Ce_{0.15}CuO_{4-y}$/Nb ramp edge zigzag junctions [141]. Ariando et al. also demonstrated predominantly $d_{x^2-y^2}$ pairing symmetry in overdoped $Nd_{1.835}Ce_{0.165}CuO_{4-y}$/Nb zigzag junctions [141]. The question of the pairing symmetry of the electron-doped superconductors will be discussed further in Section 2.5.4. Mathai et al. [194, 195] made π-SQUIDs between thin films of YBCO and Pb, also imaging the resultant spontaneous currents with a SQUID microscope. The Mathai et al., SQUIDs however, had β factors close to 1, so that the spontaneous currents were small. They used a sensor SQUID bias reversing scheme to distinguish between 0- and π-SQUIDs. Tricrystal geometries were also used for interferometry phase sensitive pairing symmetry experiments [196], and to reproduce the magnetometry experiments of Tsuei and Kirtley [197]. Predominantly d-wave pairing symmetry could also be inferred from the characteristics of asymmetric 0–45° c-axis grain boundary junctions, which have rapidly alternating 0- and π-junctions due to facetting. Such facetting results in unusual magnetic interference patterns [198, 199] and spontaneous flux generation in the grain boundaries [200]. The rapid alternation in sign of the local Josephson critical current due to facetting can result in "splinter" Josephson fluxons with flux a fraction of the conventional flux quantum [201].

Early π-rings and 0–π junctions were made with techniques that would be difficult to use to place several devices on the same substrate. However, recently a ramp-edge junction technology has been introduced [98, 100] that allows the photolithographic patterning of high quality junctions.

This technology has been used to make junctions with intentional facets [98] that reproduce the unusual magnetic interference patterns seen in asymmetric 0–45° grain boundary junctions [198–200], and to make very large arrays of π-rings [113, 114]. A particularly striking example of the spontaneous generation of half-flux quantum vortices in 0–π junctions is displayed in Figure 2.14, which shows scanning SQUID microscope images of such

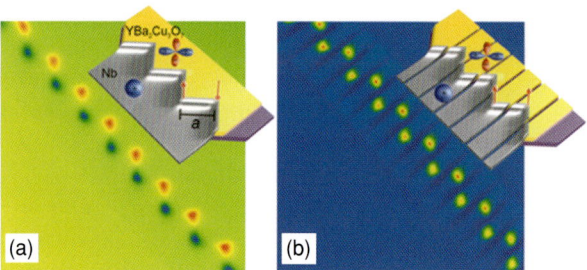

Figure 2.14. Schematics (insets) and scanning SQUID microscope images of facetted YBCO–Nb junctions. There is an intrinsic π-shift in the superconducting phase normal to the junction interface at each facet corner, which causes the spontaneous generation of a half-flux quantum vortex. The half-fluxons order strongly antiferromagnetically for the electrically connected junction (a), but weakly for electrically disconnected junctions (b) (from Hilgenkamp et al. [114]).

facetted YBCO-Nb junctions [113, 114]. A half-flux quantum vortex is generated at each facet corner as the sample is cooled through the Nb superconducting transition temperature. The directions of circulation of the spontaneous supercurrents order strongly antiferromagnetically when the facet corners are electrically connected (Figure 2.14a), but only weakly when the facet corners are electrically disconnected (Figure 2.14b). Two-dimensional arrays of electrically disconnected, photolithographically patterned π-rings show short range antiferromagnetic correlations when cooled in zero field, but do not show ordering beyond a few lattice spacings [113, 114].

2.5. Tunneling Spectroscopy

2.5.1. Superconducting Gap

General Features

Several earlier reviews of tunneling measurements of the superconducting gap in the cuprates exist [13–15, 21]. There is now a consensus that the superconducting gap in many optimally doped high-T_c cuprates at low temperatures and high tunneling resistances is consistent with predominantly $d_{x^2-y^2}$ pairing symmetry. However, the interpretation of tunneling spectra in the high-T_c superconductors is more complex than for conventional superconductors. A general expression for quasiparticle tunneling across a normal metal–insulator–superconductor (NIS) junction at zero temperature in the tunneling (low interface transmission) limit can be written as [202]:

$$I = \pm 2\pi e \sum_{\mathbf{k},\mathbf{q}} |T_{\mathbf{kq}}|^2 [1 \mp \xi_{\mathbf{k}}/E_{\mathbf{k}}]\delta(\xi_{\mathbf{q}} - eV \pm E_{\mathbf{k}})\theta(|eV| - E_{\mathbf{k}}), \qquad (2.21)$$

where I is the current per junction area, V is the voltage, the $\pm$ sign indicates the polarity of the S relative to the N, and $\mathbf{k}$ and $\mathbf{q}$ label the wave vectors for S and N, respectively. The step function θ represents the Fermi function at zero temperature and the δ function reflects energy conservation (elastic tunneling). $E_{\mathbf{k}}$ and $\xi_{\mathbf{k}}$ are the quasiparticle and normal state dispersions in the S electrode, respectively. A free-electron dispersion $\xi_{\mathbf{q}} = \hbar^2 q^2/2m$ can be assumed for the normal metal. If we set the matrix elements $|T_{\mathbf{kq}}|^2 =$ a constant, set the coherence factor $1 \mp \xi_\mathbf{k}/E_\mathbf{k}=1$ (electron–hole symmetry), use the standard BCS model relating the superconductor gap function $\Delta_\mathbf{k}$ to the S electrode dispersions, $E_\mathbf{k}^2 = \xi_\mathbf{k}^2 + \Delta_\mathbf{k}^2$, and take $\Delta_\mathbf{k} = \Delta$, Eq. (2.21) reduces to the Giaever expression for the normalized conductance [2]

$$\frac{(dI/dV)_s}{(dI/dV)_n} = N_s(eV), \qquad (2.22)$$

with the BCS quasiparticle density of states given by

$$N_s(E) = |E|/\sqrt{E^2 - \Delta^2}. \qquad (2.23)$$

However, modeling of the tunneling conductance in the gap region of the cuprate superconductors must in general include the effects of unconventional pairing symmetry, band structure, and energy and momentum dependent matrix elements, and the full expression Eq. (2.21) should be used.

An example of an early tunneling measurement of the energy gap in YBCO is shown in Figure 2.15. In these experiments, single crystals of optimally doped YBCO were lightly etched, after which elemental metals were deposited to complete tunnel junctions. Reproducible results were obtained, with an appreciable apparent density of states at zero bias, a linear background conductance at high voltages, and a complicated gap structure. Although the linear conductance background was originally interpreted as a density of states effect [57, 203], it often does not appear (see for example the STM measurements in Figure 2.18) and may in fact be representative of the tunneling process itself. We will discuss the linear conductance background in Section 2.5.3.

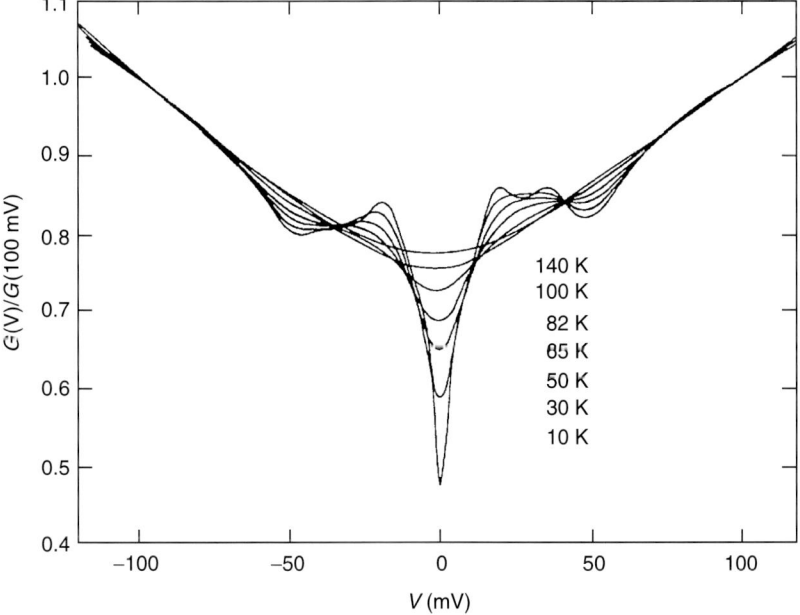

Figure 2.15. Voltage dependence of G (V)/G (100 mV) for a etched single crystal $YBa_2Cu_3O_7$/Pb junction. The lowest temperature curve has the lowest zero-bias conductance. The polarity refers to the $YBa_2Cu_3O_7$ electrode (from Gurvitch et al. [57]).

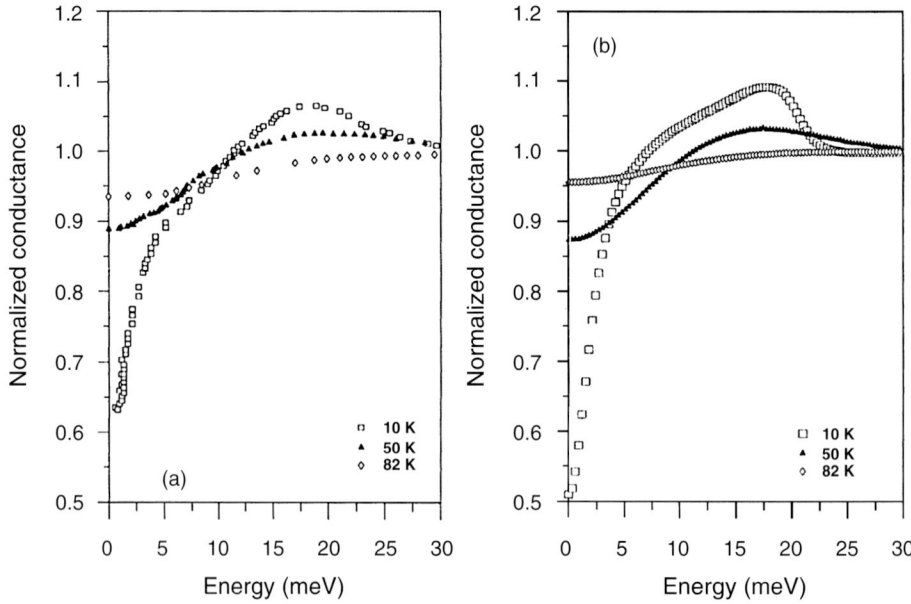

Figure 2.16. (a) Experimental values of the normalized conductance vs voltages of a planar YBCO–Pb tunnel junction at different temperatures (10, 50, 82 K) [57]. (b) Computed values of the normalized conductance vs. voltage at the same temperatures as (a), modeling the cuprate electrode as a weakly coupled superconducting-normal metal bilayer (from Di Chiara et al. [204]).

Some of the properties of the early tunneling measurements, such as excess conductance at zero bias, and smeared out gap features, can be explained by the influence of proximity effects in the transport properties [204, 205]. An example is shown in Figure 2.16, which compares experimental data from [57] with calculations modeling the high critical temperature electrode as a weakly coupled superconducting–normal metal bilayer.

However, nearly ideal characteristics have been obtained from tunneling into the cuprates using a number of different techniques, if the properties expected for tunneling into a $d_{x^2-y^2}$ superconductor are taken into account. For example, Figure 2.17 shows data for point-contact SIN tunneling into a Tl-2201 single crystal in the c-axis direction [206]. The experimental data has been normalized by dividing out an estimated normal state conductance background. The solid line is a fit to the data using an empirical expression for the tunneling density of states

$$N_s(E) = \int f(\theta) \frac{E - i\Gamma}{\sqrt{(E - i\Gamma)^2 - \Delta(\theta)^2}} d\theta, \qquad (2.24)$$

where Γ is a lifetime broadening function, $\Delta(\theta) = \Delta_0 \cos(2\theta)$ is the $d_{x^2-y^2}$ gap symmetry, and $f(\theta) = 1 + 0.4\cos(4\theta)$ is an empirical function to account for the momentum dependence of the matrix elements in the c-axis direction. In this fit $\Delta_0 = 25$ meV and $\Gamma = 1$ meV. Although this very simple functional form provides a good fit to the data, the same workers also fit their tunneling data, without normalizing out the normal state conductance, using models also including the effects of band structure, group velocity, and tunneling direction [206, 207].

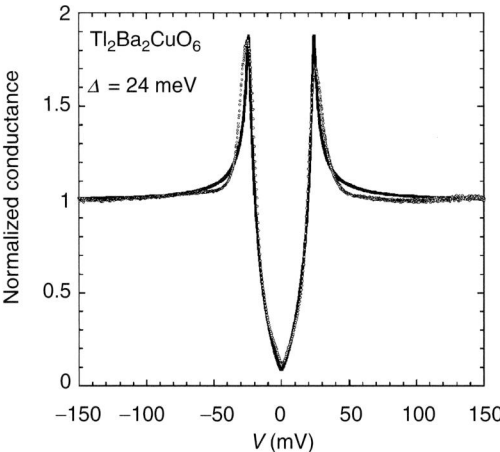

Figure 2.17. Point-contact SIN tunneling conductance for an optimally doped Tl-2201 single crystal at 4.2 K, normalized to an estimated normal state conductance line (open circles). The solid line is a fit assuming $d_{x^2-y^2}$ pairing symmetry (from Ozyuzer et al. [206]).

A third example (Figure 2.18) is from Hoogenboom et al. [208]. The experimental data is taken from STM measurements of single crystals of Bi-2212 with various doping levels and T_cs. The modeling is as follows: Hoogenboom et al. write the tunneling conductance as

$$\frac{dI}{dV} \propto -\int d\omega \sum_{\mathbf{k},n} |T_\mathbf{k}|^2 A_n(\mathbf{k}, w) f'(\omega - eV), \quad (2.25)$$

where f is the Fermi function and A_n is the spectral function in the sample, with n labeling the electronic bands. The spectral function (assumed n independent) is given by

$$A(\mathbf{k}, w) = -\frac{1}{\pi} \text{Im} \left[\frac{1}{\omega + i\Gamma - \xi_\mathbf{k} - \Sigma(\mathbf{k}, \omega)} \right]. \quad (2.26)$$

For the modeling of Figure 2.18, the lifetime broadening $\Gamma = 1$ meV, and the self-energy $\Sigma(\mathbf{k}, \omega)$ is that of the conventional $d_{x^2-y^2}$ BCS model:

$$\Sigma(\mathbf{k}, \omega) = \frac{|\Delta_\mathbf{k}|^2}{\omega + i\Gamma + \xi_\mathbf{k}}, \quad (2.27)$$

with $\Delta_\mathbf{k} = \Delta_0(\cos k_x - \cos k_y)/2$. The doping dependent band bonding (B) and antibonding (A) dispersions are given by

$$\xi_\mathbf{k}^{A,B} = -2t(\cos k_x + \cos k_y) + 4t' \cos k_x \cos k_y - 2t''(\cos 2k_x + \cos 2k_y)$$
$$\pm \frac{1}{4} t_\perp (\cos k_x - \cos k_y)^2 + \Delta\varepsilon. \quad (2.28)$$

Here the interlayer coupling is set by $t_\perp$. The doping dependent parameters $t, t', t'', t_\perp$, and $\Delta\varepsilon$ were derived from ARPES measurements. For the modeling of Figure 2.18 the tunneling matrix elements were taken to be momentum independent ($T_\mathbf{k} = T_0$). There is remarkably

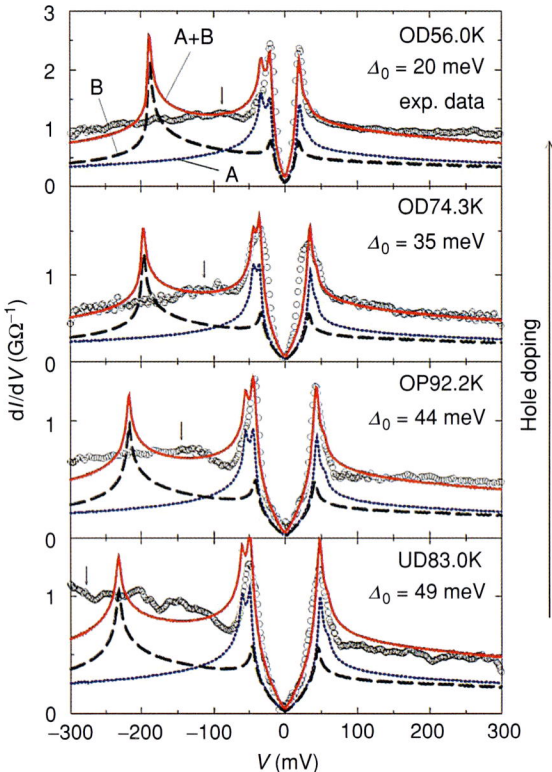

Figure 2.18. STM tunneling data from a series of single crystal Bi-2212 samples with varying doping levels. The lines are modeling using the conventional BCS model, a d-wave superconducting gap, and an isotropic matrix element, but with band structure parameters derived from photoemission measurements. The contributions from two bands (labeled A and B) are shown separately, and their sum should be compared to the experimental data (circles) (from Hoogenboom et al. [208]).

good agreement between this model and experiment, aside from the sharp peak predicted by the model at negative voltages derived from the B band, and the lack of an undershoot just below the gap edge at negative voltages. This second feature, which was modeled by Hoogenboom et al. including interaction with a bosonic mode at wavevector (π, π) and energy $\Omega = 5.4 k_b T_c$, will be discussed in more detail in Section 2.5.6.

A number of workers have extended the Blonder, Tinkham, and Klapwijk (BTK) model [38] (see Section 2.2.4) treatment of transport through superconducting contacts to the case of unconventional pairing symmetry [41, 209–214]. Hasselbach and Kirtley simply averaged the BTK expressions over a gap distribution appropriate for the unconventional pairing symmetry of URu$_2$Si$_2$ [209]. Hu [41] and Tanaka and Kashiwaya [211, 212] extended the BTK model to account explicitly for the phases of the propagating charges. Wei et al. [215] further extended the modeling to sum over a realistic band structure, and to account for the effects of a directional transmission by including a Gaussian "tunneling cone" factor. They write the modified BTK expression for the tunneling current as

$$I_{NS} = G_{NN} \int\int e^{-k_t^2/\beta^2} d^2k_t \int_{-\infty}^{\infty} [1 + A(E_\mathbf{k}, \Delta_\mathbf{k}, Z) - B(E_\mathbf{k}, \Delta_\mathbf{k}, Z)][f(E_\mathbf{k} - eV) - f(E_\mathbf{k})] dE, \quad (2.29)$$

Tunneling Measurements of the Cuprate Superconductors

where A and B are the Andreev-reflection and normal-reflection coefficients, G_{NN} is the normal-state junction conductance, and β is the tunneling cone width. The generalized BTK kernel is given by [211, 212]

$$1 + A - B = \frac{16(1 + |\Gamma_+|^2)\cos^4\theta_s + 4Z^2(1 - |\Gamma_+\Gamma_-|^2)\cos^2\theta_s}{|4\cos^2\theta_s + Z^2[1 - \Gamma_+\Gamma_-\exp(i\phi_- - i\phi_+)]|^2}, \qquad (2.30)$$

where

$$\Gamma_\pm = \frac{E}{|\Delta_\pm|} - \sqrt{\frac{E^2}{|\Delta_\pm|^2} - 1}, \qquad (2.31)$$

and $\exp(i\phi_\pm) = \Delta_\pm/|\Delta_\pm|$ represents the phase of the pair potential $\Delta_\pm = \Delta(\theta_{s,\pm})$ experienced by an Andreev-reflected electron (or hole) propagating at an angle $\theta_{s,+}$ or $\theta_{s,-} = \pi - \theta_{s,+}$ relative to the junction normal.

Figure 2.19 shows STM tunneling conductance data into the {110} and {001} faces of an optimally doped YBCO single crystal, divided by an estimated normal state conductance

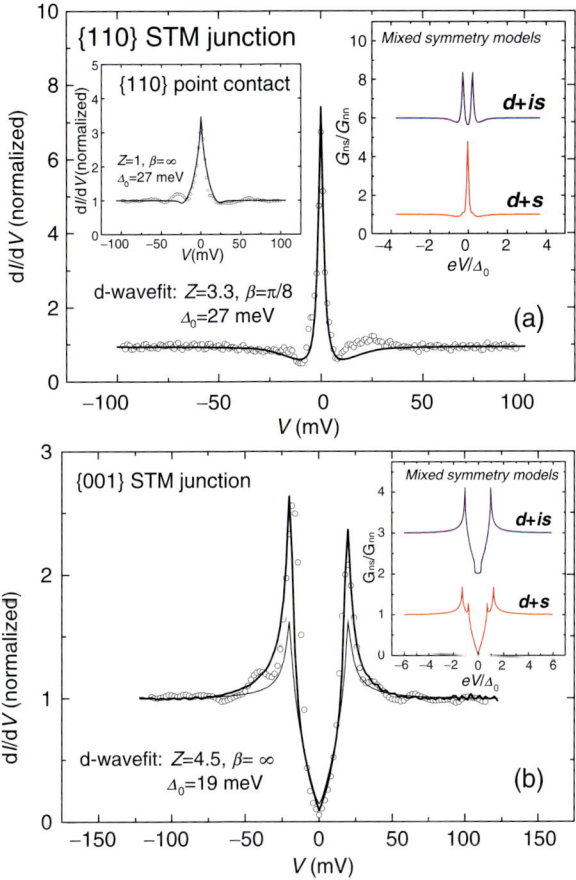

Figure 2.19. Conductance data for an STM tunnel junction on the {110} (a) and {001} crystal faces (b) of YBCO, using a Pt/Ir tip, normalized to an estimated normal state background (open symbols). The solid lines are fits assuming a $d_{x^2-y^2}$ pairing symmetry. In (b) the modeling is with (thick line) and without (thin line) a correction for band structure effects (from Wei et al. [215]).

background. To model these data Wei et al. [215] used the same form for the gap as Hoogenboom et al. [208], with a somewhat simpler one-band tight-binding structure for the electron dispersion. They showed that their data could be consistently modelled using a pure $d_{x^2-y^2}$ pairing symmetry, for tunneling into various crystalline faces, in both the point-contact and tunneling limits. They further found that the enhanced conductance overshoots at the gap edges, and the asymmetry of these overshoots, were the result of the proximity of the Fermi level to the 2D van Hove singularity [202]. More details on the zero bias conductance peak often seen in tunneling into the cuprate superconductors will appear in Section 2.5.4.

Temperature Dependence

The BCS weak coupling limit for the ratio $2\Delta/k_B T_c$ is 3.54 for s-wave [25] and 4.3 for $d_{x^2-y^2}$ [216] pairing symmetries. The high-T_c superconductors have a value for this ratio which is typically larger than the weak coupling limits [13–15, 21]. For example, in the work described above, the tunneling measurements of YBCO by Gurvitch et al. was interpreted in terms of two gaps of 19 meV and 4–5 meV [57, 203], possibly associated with the ab plane and c-axis directions, respectively. The larger value gives $2\Delta/k_B T_c = 5$, although a geometric mean of the two values gives $2\Delta/k_B T_c = 4.1$. The measurements of the $T_c = 86$ K single Tl-2201 crystals of Ozyuzer et al. [206] gave fit gaps of $\Delta = 25$ meV, implying $2\Delta_0/k_B T_c = 6.7$. The optimally doped Bi-2212 crystal of Hoogenboom et al. [208] had $2\Delta/k_B T_c = 11.1$, while the optimally doped YBCO of Wei et al. had $2\Delta/k_B T_c = 7.2$–7.5 within the ab planes, and $2\Delta/k_B T_c = 4.9$ in the c-axis direction. An assessment of the systematic variation of gap size with critical temperature is complicated by the strong variation of gap size with doping. An early review [14] found that $2\Delta/k_B T_c$ was roughly proportional to $5.4 T_c$. However, sample quality, experimental techniques, and modeling sophistication have all improved with time. The results of a more recent review by Wei et al. [202] are shown in Figure 2.20. It shows

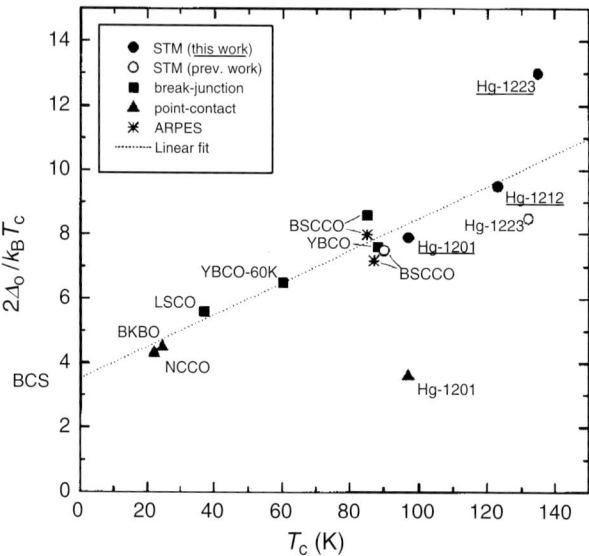

Figure 2.20. Plot of $2\Delta_0/k_B T_c$ vs. T_c for various (assumed optimally doped) high-T_c cuprates, including the non-cuprate BKBO for comparison (from Wei et al. [202]).

that the ratio $2\Delta/k_B T_c$ is close to the BCS value for low T_cs, but increases as T_c gets larger. Panagopoulos and Xiang [217] argue that the density of states in the node regions of the d-wave gap function follow a momentum dependence $\Delta_{\mathbf{k}} = \Delta_0 \cos(2\theta)$ that scales with T_c like $2\Delta_0/k_B T_c = 4.3$ as expected for weak-coupling d-wave superconductivity, but that the maximum gap value does not follow this scaling, especially for underdoped cuprates. In this argument thermodynamic measures of the gap, such as penetration depth, and specific heat, would be sensitive to the low-energy quasiparticle states in the gap region, while spectroscopy measures, such as angle resolved photoemission and tunneling, would be more sensitive to the high-gap regions in momentum.

Determination of the temperature dependence (as opposed to the T_c dependence described above) of the energy gap is complicated in the cuprate superconductors by the presence of the pseudogap (see Section 2.5.2). A number of workers report that the superconducting gap has a temperature dependence that is much weaker than BCS, and that the gap closes by "filling in" states, rather than by a narrowing in energy [21,218,219]. An example is shown in Figure 2.21. The experimental data is scanning tunneling spectroscopy of a slightly underdoped Bi-2212 single crystal by Renner et al. [218]. The fits are by Franz and Millis, [220] using Eq. (2.25) with a tunneling matrix element $M_{\mathbf{k}}(\omega) \propto |\cos(2\theta)|$, $\Delta_{\mathbf{k}} = \Delta_d \cos(2\theta)$, and a spectral function given by Eq. (2.26) (with $\Gamma_1 \equiv \Gamma$) but a self-energy given by Eq. (3.7) with $\Gamma = 0$. Franz and Millis find that the gap is relatively insensitive to temperature, converting smoothly from a superconducting gap to the pseudogap as the temperature is

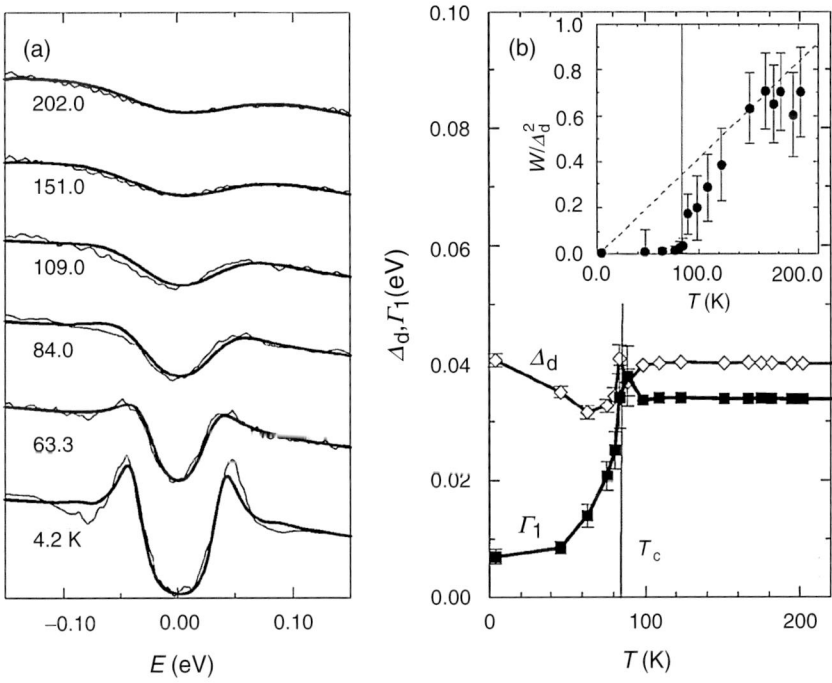

Figure 2.21. (a) Fit to scanning tunneling spectroscopy data of Renner et al. [218] for slightly underdoped single crystal Bi-2212 at selected temperatures. (b) Parameters of the fit as a function of T. W is the width of a Gaussian distribution of phase fluctuation in their model. Above T_c both Δ_d and Γ_1 are fixed to their values at 84 K (from Franz and Millis [220]).

increased through T_c. However, the single particle scattering rate Γ_1 increases dramatically as the temperature is increased [221]. Franz and Millis attribute this increase to transverse phase fluctuations of a Kosterlitz–Thouless vortex–antivortex unbinding transition.

The interpretation of a superconducting gap that continuously evolves with increasing temperature into the pseudogap is supported by the break tunnel junction measurements of Miyakawa et al. [219] summarized in Figure 2.22.

However, measurements using intrinsic junctions [17, 146–148, 222] indicate that the superconducting gap and the pseudogap can coexist [223, 224], and that the superconducting gap closes by narrowing in energy as the temperature approaches T_c, while the pseudogap persists above T_c, with a value for the gap energy which is very similar to the low-temperature superconducting gap. An example is shown in Figure 2.23. This figure shows values for the superconducting gap energy determined from the peak voltage $v_s = 2\Delta_s/e$ in the conductance–voltage characteristic, as well as from the spacing δv_s between quasiparticle branches, for both optimally doped and overdoped Bi-2212. Also shown are the peak values of the pseudogap hump voltage v_{pg} as a function of temperature. Zasadzinski [21] cautions that the intrinsic junctions have very high current densities, and nonequilibrium phenomena [225] could complicate the interpretation of such data.

The picture of coexisting superconducting gaps and pseudogaps at the same temperature is supported by scanning tunneling spectroscopy measurements on $Bi_2Sr_2CaCu_2O_{8+\delta}$ [226–229] and $Bi_2Sr_2CuO_{6+\delta}$ [230] single-crystal samples. An example is in Figure 2.24, which displays a series of tunneling spectra along the line in (b), with regions with low gap and high gap edge peaks alternating with regions of high gap but broad gap edge peaks. These can be

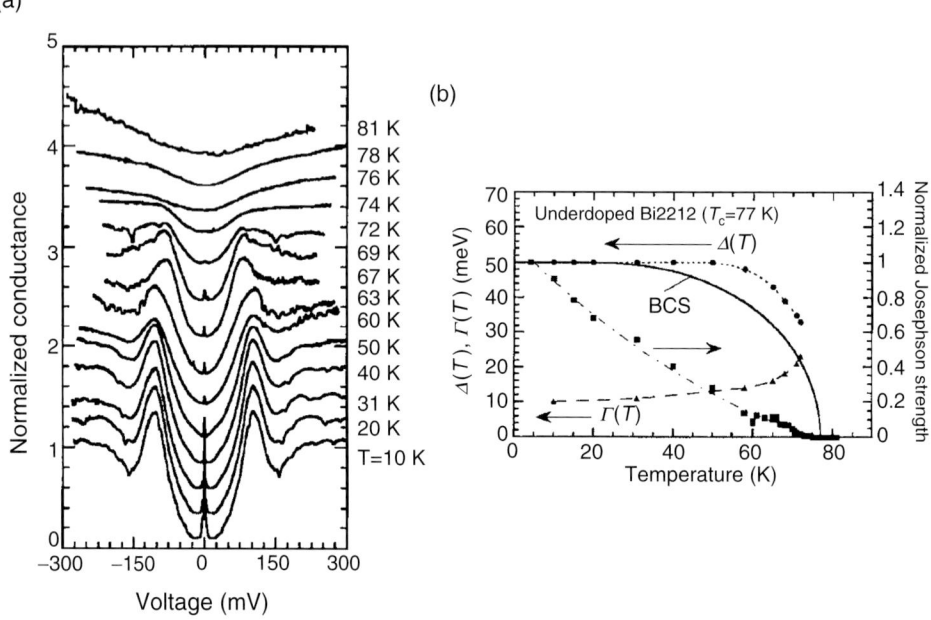

Figure 2.22. (a) Temperature dependence of SIS tunneling conductance on an underdoped Bi-2212 ($T_c = 77$ K) break junction. For clarity each conductance curve has been normalized by its value at 200 mV and (except for the 10 K curve) has been offset vertically. (b) Temperature dependence of superconducting gap $\Delta(T)$ (circle), quasiparticle scattering rate $\Gamma(T)$ (triangle), and Josephson strength $I_c^* R_n$ (square) normalized by $I_c^* R_n$ (4.2 K). Here, the Josephson current I_c^* is estimated from the peak in conductance at zero bias. The full curve represents the BCS superconducting gap $\Delta(T)$ (from Miyakawa et al. [219]).

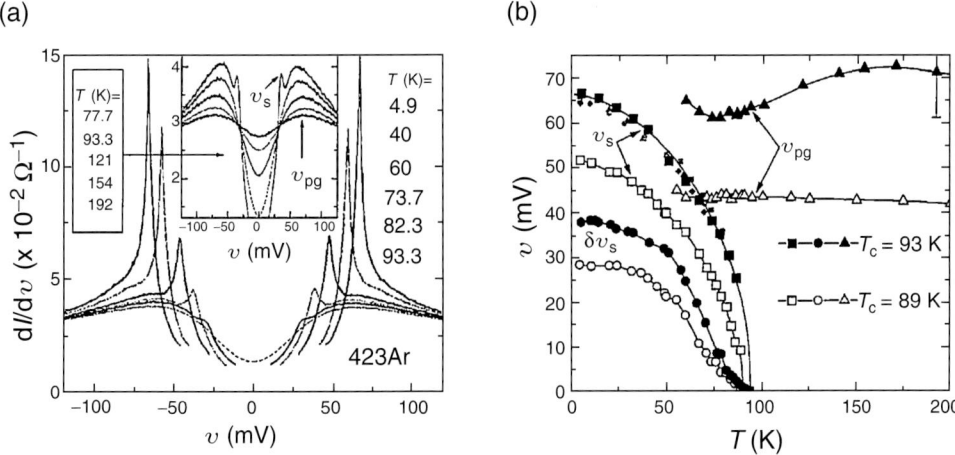

Figure 2.23. (a) Dynamic conductance $\sigma(v)$, at different temperatures for an optimally doped Bi-2212 single crystal mesa intrinsic junction. The insert shows structure due to the superconducting gap and the pseudogap coexisting at $T = 77.7$ K. (b) Temperature dependence of parameters of optimally doped (solid symbols) and overdoped (open symbols) Bi-2212 intrinsic junctions: the superconducting peak voltage $v_s = 2\Delta_s/e$, the spacing between quasiparticle branches δv_s, and the pseudogap hump voltage v_{pg} (from Krasnov et al. [223]).

associated with superconducting gap and pseudogap regions, respectively, coexisting in the sample.

Momentum Dependence

Early attempts to measure the dependence of the critical current of high-T_c-conventional superconductor junctions on junction angle relative to the cuprate crystalline axes [195, 231] were hindered by the uncontrolled nature of the junction interfaces. However, recently success in this area has been reported using two very different junction technologies. Lombardi et al. [76] have measured the angular dependence of the Josephson critical currents of c-axis tilt biepitaxial grain boundary YBCO junctions (Figure 2.25). These junctions, which are formed by the grain boundary between a (001) and a (103) oriented film, have crystalline rotations about two axes and relatively low interface transmission probabilities. The solid symbols in Figure 2.25 are normalized critical current densities from junctions with widths of 10 μm (triangles) and 4 μm (stars), as a function of the angle θ of the junction normal relative to the a or b axis of the (001) film. In the Sigrist–Rice phenomenological approach, [50] in which the Josephson current is proportional to the projection of the momentum-dependent energy gap onto the junction normals, the critical current density is given by

$$J_c = J_0(n_x^2 - n_y^2)_L(n_x^2 - n_y^2)_R \sin(\phi), \qquad (2.32)$$

where J_0 is the maximum Josephson current density, and n_x, n_y are the x, y components of the junction normals on the two sides of the junction, and a pure $d_{x^2-y^2}$ pairing symmetry has been assumed. In the biepitaxial grain boundary structure of Figure 2.25a, this reduces to

$$J_c \sim \sin 2\theta (2 - \cos^2\theta)(1 - 3\sin^2\theta)/(1 + \sin^2\theta), \qquad (2.33)$$

which is plotted as the dashed line in Figure 2.25b. The experimental critical current has minima at approximately 0°, 35°, and 90°, as expected for a $d_{x^2-y^2}$ superconductor with this

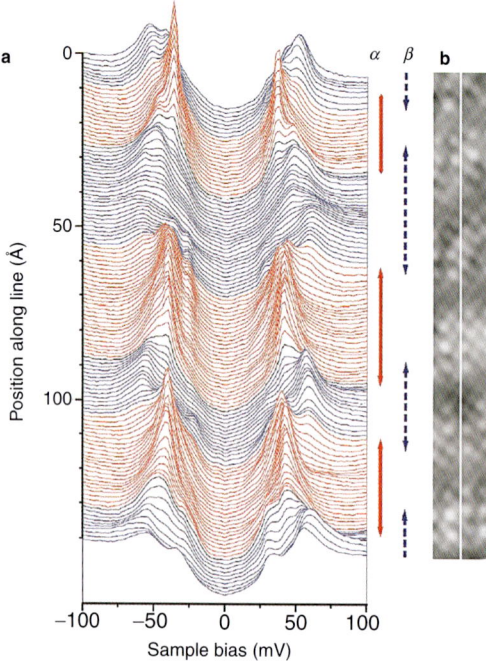

Figure 2.24. (a) Series of scanning tunneling dI/dV measurements from an underdoped ($T_c = 79$ K) Bi-2212 single crystal sample illustrating two distinct types of regions, labeled α and β. The α-domain spectra have low gap magnitudes and sharp gap edge peaks whose amplitude is low at the edges of the domain and rises to a maximum at the center. The β-domain spectra have high gap magnitude and very broad gap-edge peaks whose amplitude is relatively low and constant. (b) Surface topography along the trajectory along which the spectra in (a) were measured, demonstrating atomic resolution (from Lang et al. [229]).

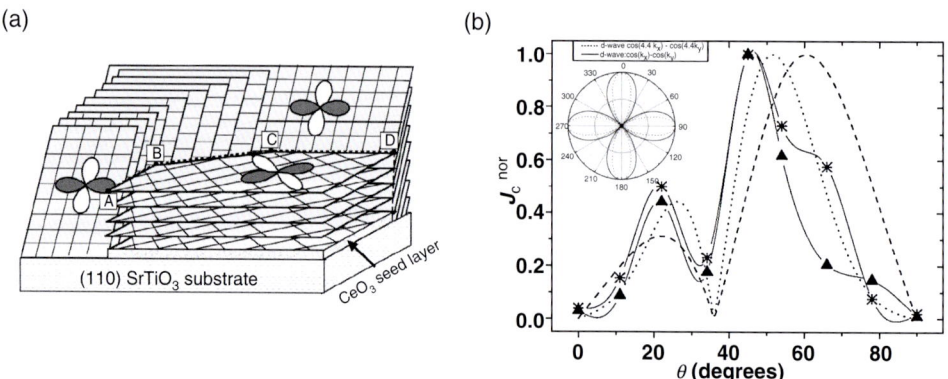

Figure 2.25. Normalized critical current density J_c vs. angle θ for two sets of c-axis tilt biepitaxial YBCO junctions, with width 10 μm (triangles) and 4 μm (stars). The solid lines connecting the symbols are guides to the eye. The dashed line is the Sigrist–Rice formula assuming pure $d_{x^2-y^2}$ pairing symmetry in this geometry; the dotted line is for a wave function with slightly narrower lobes, as shown in the inset (from Lombardi et al. [76]).

geometry. Note that the minimum angle of 35° is different from 45° because of the compound angle formed by the crystalline axes in this type of grain boundary.

Smilde et al. [232] have used the YBCO–Nb ramp-edge technology of Figure 2.14 to produce a series of junctions with varying junction normals relative to the a-axis. In this case the junctions were made with either twinned or untwinned YBCO films [233]. Junctions made with twinned YBCO films showed the fourfold symmetry, with nodes at 45°, expected for a $d_{x^2-y^2}$ symmetry (Figure 2.26). Those made with untwinned YBCO showed nodes offset by about 5°, consistent with a small s-wave component to the gap, as expected for this orthorhombic supercoductor [61, 62, 234]. The measured angular dependence of the critical currents for YBCO using this ramp-edge junction technology was well fit using an in-plane gap with 83% $d_{x^2-y^2}$, 13% isotropic s-wave, and 5% anisotropic s-wave pairing symmetry, resulting in a gap amplitude 50% higher in the b (Cu–O chain) direction than in the a-direction.

Measurements of the junction critical currents are insensitive to the orbital component of the phase of the pairing wavefunction. However, recently two-junction YBCO–Nb rings were made with the ramp-edge junction technology in the geometry illustrated in the inset to Figure 2.27 [235]. In these samples one junction angle relative to the YBCO crystalline axes was held fixed, while the other was changed in 5° increments from ring to ring. Therefore the rings alternated between having, and not having, an intrinsic sign change in the pairing wavefunctions normal to the two junction interfaces, and these rings alternated between spontaneously generating a half-flux quantum worth of flux when cooled in zero field and having no spontaneous flux. Note that the transition between the presence and absence of spontaneous flux occurs at angles slightly different from multiples of 45°. This reflects the orthorhombic symmetry of YBCO, and is the result of the gap being slightly larger in the b-axis direction, parallel to the chains, than in the a-axis direction. These results are consistent with those of Smilde et al., [232] but also show that the pairing wavefunction has sign changes, and therefore has predominantly $d_{x^2-y^2}$, as opposed to anisotropic s-wave symmetry. The presence of spontaneous magnetization of the rings with, to within the precision of the measurements, either 0 or $\Phi_0/2 = h/4e$ integrated total flux, confirms that the in-plane pairing wavefunction in optimally doped YBCO has momentum dependent sign changes, with little, if any, imaginary component to the gap in any crystalline direction [236, 237].

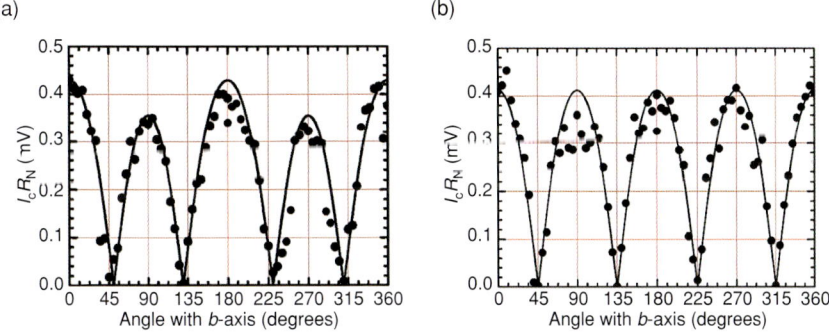

Figure 2.26. $I_c R_n$-product varying the junction orientation with respect to the crystalline axes for monocrystalline (a) and twinned (b) thin-film $YBa_2Cu_3O_7$/Au/Nb ramp-type junctions at $T = 4.2$ K and in zero magnetic field. The solid lines are fits to the data using a linear combination of isotropic s, $d_{x^2-y^2}$, and $s_{x^2-y^2}$ symmetry functions with the corresponding coefficients 0.12, 0.83, and 0.05, respectively (from Smilde et al. [232]).

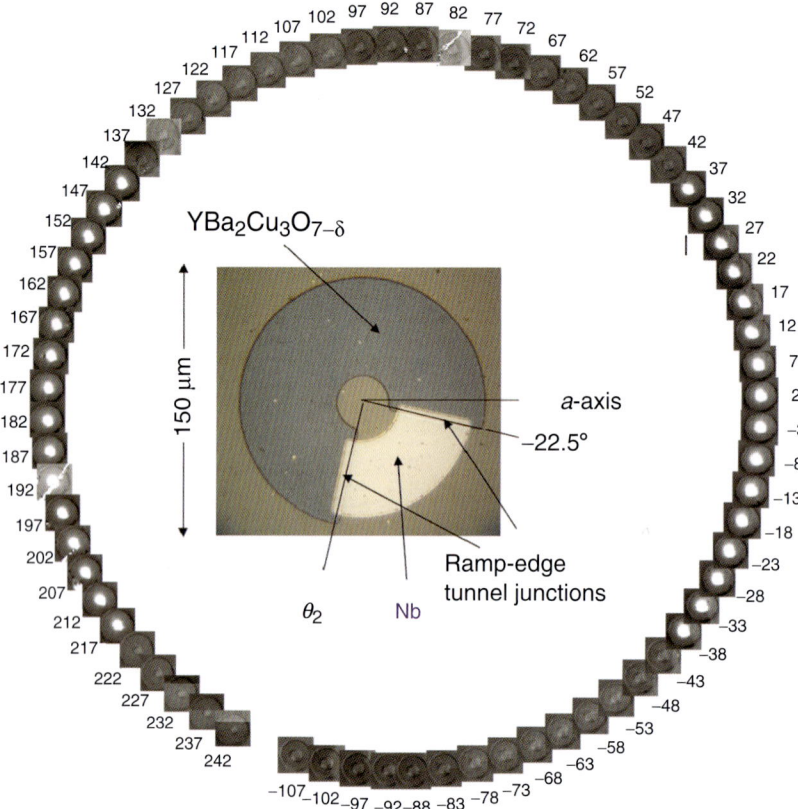

Figure 2.27. SQUID microscope images of a series of two-junction SQUID rings made with YBa$_2$Cu$_3$O$_{7-\delta}$/Au/Nb ramp-type junctions. The inset is a photograph of one of the rings. The YBCO sections had outer diameters and inner diameters of 130 μm and 30 μm, while the Nb section diameters were 120 μm and 40 μm, respectively. One junction angle was held fixed at $-22.5°$ relative to the YBCO a-axis. The second junction angle was varied in 5° intervals from ring to ring. The SQUID microscope images are labeled with $\theta = \theta_2 - 90°$, the angle of the second junction normal relative to the a-axis. The YBCO films were estimated from x-ray scattering measurements to be 85% untwinned (from Kirtley et al. [235]).

One would expect the Josephson coupling in the c-axis direction between a two-dimensional, pure d$_{x^2-y^2}$ superconductor and an s-wave superconductor to vanish because of cancellation of the contributions from the positive and negative antinodes. A finite Josephson supercurrent is expected if there is some three-dimensional character to the cuprate Fermi surface [44], or if there is some s-wave admixture. The Josephson tunneling from a superconductor with a real s-wave component, as expected for an orthorhombic superconductor such as YBCO, is complicated by the presence of twinning. It has been suggested theoretically [234, 238] and demonstrated experimentally [239] that the d$_{x^2-y^2}$ component is phase-locked, while the s-wave component changes sign, across the twin boundary in a predominantly d$_{x^2-y^2}$ superconductor. This may explain why the Josephson $I_c R_n$ products for tunneling between a cuprate and a conventional superconductor in the c-axis direction are often quite low [61, 62, 97, 109].

There have been several measurements of the Josephson critical current density in Bi$_2$Sr$_2$CaCu$_2$O$_{8+\delta}$ bicrystal junctions with varying misorientation (twist) angles ϕ about the c-axis direction. Some of these experiments show a strong dependence of the critical current on ϕ [240–244], while others do not [245–248]. The lack of a strong ϕ dependence has been taken as evidence for at least a small s-wave component in this material [247–249]. It has been proposed that strong, local fluctuations could break the fourfold symmetry of an otherwise tetragonal superconductor [250]. However, both the observation of the half-flux quantum effect in the appropriate tricrystal geometry [187] and the observation of $I_c(H)$ characteristics indicative of rapid sign changes in the critical currents along the grain boundaries of asymmetric 0–45° bicrystal grain boundaries [251] indicate that $Bi_2Sr_2CaCu_2O_{8+\delta}$ has predominantly $d_{x^2-y^2}$ pairing symmetry in the CuO_2 planes. Many photoemission experiments indicate that $Bi_2Sr_2CaCu_2O_{8+\delta}$ has a highly anisotropic gap consistent with $d_{x^2-y^2}$ pairing symmetry [252–255]. It has been suggested [251] that the lack of a strong ϕ dependence in some c-axis twist bicrystal junctions could be the result of the presence at the interface of a layer with relatively large s-wave components, either due to intrinsic causes [256] or due to impurity effects. Such impurities would not be visible in transmission electron microscopy, which only images columns of atoms.

Further, it is believed [257, 258] that the one-electron interplane hopping in BSCCO is proportional to $(\cos k_x - \cos k_y)^2$, where k_x, k_y are the momentum parallel to the in-plane a, b axes, respectively. Support for this view comes from STM measurements [256]. Scalapino [259] has suggested that the one-electron interplane hopping in a c-axis twist junction could take the form

$$\sum_{kk's}(a + b(\cos k_x - \cos k_y))(a + b(\cos k'_x - \cos k'_y))c^\dagger_{1ks}c_{2k's}, \qquad (2.34)$$

where $c^\dagger_{ks}$ creates an electron on layer 1 with planar momentum k, with similar primed operators and momenta referring to electrons on layer 2. Then the pair tunneling Hamiltonian would have a term proportional to

$$\sum_{kk'} a^2b^2(\cos k_x - \cos k_y)(\cos k'_x - \cos k'_y)c^\dagger_{1k\uparrow}c^\dagger_{1-k\downarrow}c_{2-k'\downarrow}c_{2k'\uparrow}. \qquad (2.35)$$

This equation has the physical interpretation that the 4s orbitals of the Bi and Cu act as a ball-and-socket connection between the CuO_2 planes which is invariant under a twist rotation. Equation (2.35) would result in a finite and nearly angle independent Josephson coupling across c-axis twist junctions if the coefficients a, b were angle independent. A similar microscopic picture was used to derive the separable form required to understand c-axis infrared conductivity by Hirschfeld, Quinlan, and Scalapino [260].

While it is interesting that some of the c-axis twist experiments show little dependence of the Josephson current density on twist angle, they cannot carry the same weight as the in-plane tests described above as tests of pairing symmetry, since the in-plane experiments vary with macroscopic geometry in the way predicted by simple theory, while the c-axis twist experiments show no dependence on the parameter of interest: The lack of a twist angle dependence in some experiments could be the result of factors having nothing to do with the Cooper pairing symmetry.

Doping Dependence

It is generally agreed that the superconducting gap measured by tunneling spectroscopy in Bi-2212 increases with decreasing doping below optimal doping [208, 218, 219, 261–263],

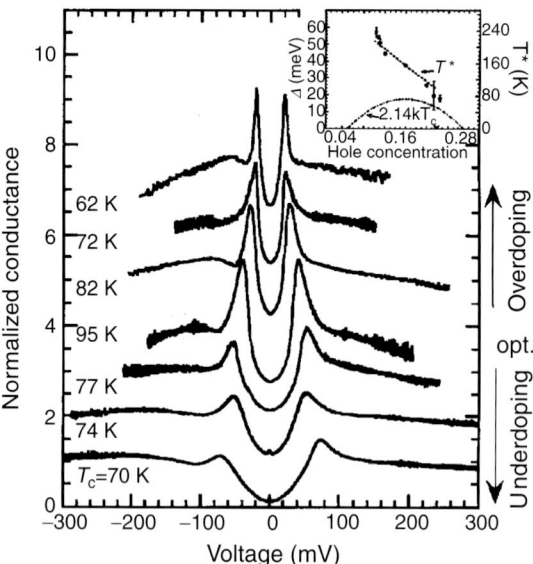

Figure 2.28. Point contact SIN tunneling conductances in $Bi_2Sr_2CaCu_2O_{8+\delta}$ for various hole doping levels from underdoped to overdoped. Data are normalized by a constant value and offset for clarity. The inset shows energy gap vs. hole doping from SIS and SIN junctions (dots) along with a linear fit (dashed line) to values for T^* derived from Ref. [264] using $\Delta = 2.14 k_B T^*$ (from Miyakawa et al. [219]).

and that the maximum gap scales with the pseudogap temperature T^* rather than the superconducting critical temperature T_c [218, 219, 262, 263]. Figure 2.28 shows the results from point-contact tunneling spectroscopy on Bi-2212 for various dopings. For overdoped cuprates the pseudo- and superconducting gaps converge, and it is believed that the gap scales with T_c. The gaps become larger, and the gap peaks become smaller and more rounded, as the samples become more underdoped. The inset in Figure 2.28 plots the measured gap values as a function of doping, which agree well with the dashed line $\Delta = 2.14 k_B T^*$, where T^* is the pseudogap derived from transport measurements [264]. Yeh et al. [265] report in a study of the doping dependence of YBCO that the gap Δ peaks at optimal doping, but the ratio $2\Delta/k_B T_c$ increases with decreasing doping, although much less strongly than in Bi-2212, from a value of about 4.3 for strongly overdoped YBCO. Deutscher [263, 266] reports that while the gap measured for a number of cuprate superconductors from tunneling measurements (Δ_p) scales with T^* in the underdoped regime, the gap measured from Andreev scattering measurements (Δ_c) scales with T_c, and argues that the former is a measure of the single-particle excitation energy out of the condensed pairs, while the latter is the coherence energy of the macroscopic quantum condensate of the paired charges. Mourachkine draws similar conclusions for an under-electron-doped $Nd_{1.85}Ce_{0.15}CuO_{4+\delta}$ sample [267].

Although most of the recent tunneling spectroscopy studies indicate predominantly $d_{x^2-y^2}$ pairing symmetry in the optimally doped cuprates, there have been reports of a doping-induced change in the pairing wavefunction in some cuprates. For example, tunneling spectroscopy suggests a significant s-wave component in the pairing wavefunction in overdoped $Y_{1-x}Ca_xBa_2Cu_3O_{7-\delta}$ [265]. Analysis of the zero magnetic field splitting of the zero bias peak in YBCO as a function of doping indicates a change in symmetry from $d_{x^2-y^2}$ to $d_{x^2-y^2}+id_{xy}$ or $d_{x^2-y^2}+is$ in overdoped $YBa_2Cu_3O_7$ [268] and $Y_{1-x}Ca_xBa_2Cu_3O_{7-\delta}$ [269] thin films.

A drop in the scattering rate of uncondensed carriers below 15 K inferred from penetration depth measurements of Ca-doped YBCO is reported as supporting evidence for a complex order parameter in strongly overdoped YBCO [270]. This has been cited as evidence for the existence of a quantum critical point near optimal doping in YBCO. Analysis of Andreev scattering data from $La_{2-x}Sr_xCuO_4$ single crystals are best fitted by an anisotropic (extended) s-wave gap of the form $\Delta(\theta) = \Delta_0 + \Delta_1 \cos(4\theta)$, with a maximum gap value of 15 meV, and a minimum gap of 5 meV, although the authors state that it is hard to distinguish between the extended s-wave gap functional form, and an s + d form [271, 272]. Further, Kohen et al. report an is component to the pairing potential inferred from Andreev reflection measurements on $Y_{1-x}Ca_xBa_2Cu_3O_{7-\delta}$ which is enhanced as the contract transparency is increased [273]. They interpret this as a proximity effect between the cuprate superconductor and the normal metal tip.

However, all of the π-SQUID or 0–π-junction experiments described above indicate predominantly $d_{x^2-y^2}$ pairing symmetry with little, if any, imaginary component to the order parameter for optimally doped superconductors [11], while tricrystal pairing symmetry tests provide the same conclusion for several of the cuprates as a function of doping over a wide doping range [175].

Macroscopic Quantum Effects

The high temperature superconductors, having d-wave pairing symmetry, have many low energy quasiparticle states. These states could be expected to produce large dissipation in tunneling measurements, making the high-temperature superconductors unsuitable for the observation of macroscopic quantum effects. High temperature superconductors also have the interesting features of unconventional order parameter symmetry and a possible high $\sin(2\phi)$ component in the Josephson current–phase relationship. One test of the role of dissipation in high-T_c superconductors is through the study of macroscopic quantum tunneling (MQT) [274–277] in the escape from the zero-voltage state to the voltage state in the current–voltage characteristics of Josephson junctions. Very low-dissipation junctions were required to observe macroscopic quantum tunneling and energy level quantization through resonant activation in junctions incorporating low-temperature superconductors [278–280].

Bauch et al. reported the first observation of MQT in a d-wave superconductor [78]. They chose for these studies CeO_2 based biepitaxial GB junctions in the tilt limit, which had low average barrier transparency, leading to 90% hysteretic behavior in the current–voltage characteristic. The escape rate of the superconducting phase ϕ from a local minimum in the washboard potential into the running state has been studied as a function of temperature in analogy with experiments on low-T_c junctions. The dependence of the distribution width σ of switching current probability distributions on temperature is reported in Figure 2.29. The measured σ saturates below 50 mK, indicating a crossover from the thermal to the MQT regime. To rule out the possibility that the saturation of σ is due to any spurious noise or heating in the measurement setup the switching current the probability distributions were measured for a reduced critical current ($I_{C0} = 0.78\,\mu A$) by applying an external magnetic field $B = 2$ mT. The width σ for $B = 2$ mT and the data for $B = 0$ mT are shown in the inset of Figure 2.29. The data in the presence of a magnetic field clearly show a smaller width σ, which does not saturate down to the base temperature.

The low average barrier transparency for these junctions strongly reduces dissipation from nodal quasiparticles, explaining why dissipation mechanisms related to a d-wave junction do not prevent the observation of MQT. The low dissipation argues that the presence of

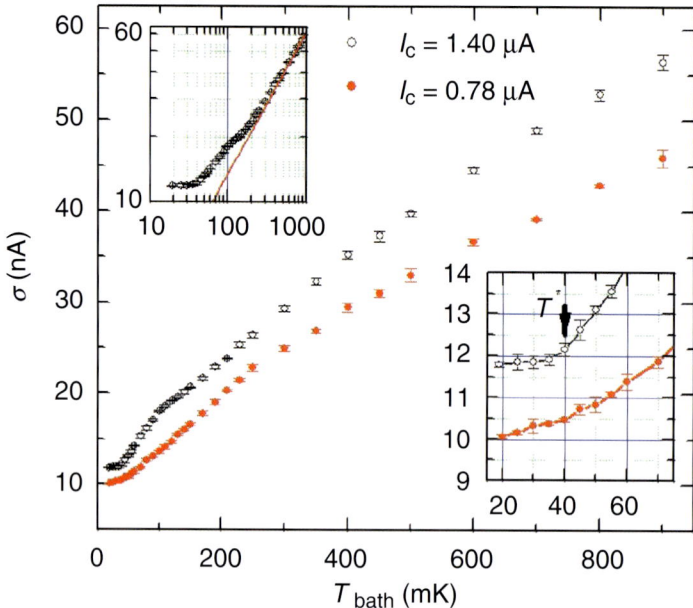

Figure 2.29. Temperature dependence of the width σ of the escape probability distribution from the zero-voltage to the voltage state for a YBCO biepitaxial Josephson junction at $B = 0\,\text{T}$ (open circles) and $B = 2\,\text{mT}$ (full points). The solid line in the upper left insert shows the calculation for the thermally activated widths on a log-linear scale, using $I_{C0} = 1.40\,\mu\text{A}$, $C = 1\,\text{pF}$, and $R = 80\,\Omega$. The lower right insert displays the low temperature data on an expanded scale (adapted from Bauch et al. [78]).

low energy quasiparticles does not prevent macroscopic quantum behavior, such as required for solid-state quantum computers, and may open the way for experiments aimed at demonstrating coherence in systems taking advantage of the "quiet" configuration offered by the d-wave symmetry [281].

2.5.2. Pseudogap

There is often in the cuprate high-T_c superconductors a "pseudogap," a reduction in the density of states near the Fermi surface, which has properties that distinguish it from the superconducting gap [282]. The pseudogap does not reduce the tunneling density of states to zero at the Fermi surface, it tends to be more pronounced for underdoped than for overdoped cuprates, and it persists in temperature well above the superconducting T_c for strongly underdoped cuprates. There have been many proposals for the source of the pseudogap behavior, including spin fluctuations [283, 284], condensation of preformed pairs [218], SO(5) symmetry [285], spin–charge separation [286], and phase fluctuations of a superconducting state with finite local pairing amplitude [220, 287, 288]. Angle resolved photoemission spectroscopy has shown that the pseudogap in Bi-2212, just like the superconducting gap [252, 253, 255], has an in-plane momentum dependence consistent with $d_{x^2-y^2}$ symmetry [289].

Temperature Dependence

As described in Section 2.5.1, in tunneling measurements, especially in underdoped cuprates, the superconducting gap evolves continuously into the pseudogap as the temperature increases [218,219,230,262,290,291], although there is some evidence for coexistence of the superconducting gap and the pseudogap, possibly associated with spatial inhomogeneities, [227,229] at temperatures close to T_c [223,224].

It has been reported that the pseudogap closing temperature T^* is below the superconducting T_c over much of the doping range in the electron doped superconductors, arguing against it being an indicator of precursor superconductivity [292,293].

The pseudogap energy scale does not appear to be strongly temperature dependent. Instead the psuedogap seems to close by a gradual filling in of states [218,219,262]. An example is displayed in Figure 2.30, which shows STM spectra for tunneling into an underdoped single crystal of Bi-2212. The peaks at the gap edges in the superconducting state disappear at T_c, but the superconducting dip in the density of states continuously evolves into the pseudogap, remnants of which persists up to room temperature.

Since the energy scales of the pseudogap and superconducting gap track each other closely, the pseudogap, like the superconducting gap, increases with doping below optimal doping in Bi-2212. Kugler et al. [230] report that the pseudogap energy Δ_p scales like $2\Delta/k_B T^* = 4.3$, where it is the pseudogap temperature T^*, rather than T_c, which reflects the mean-field critical temperature of the superconductor (see Figure 2.31).

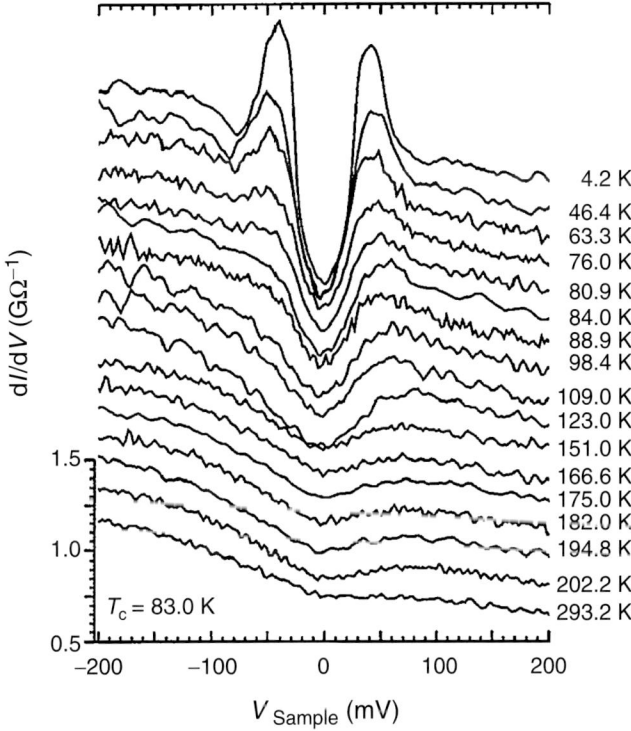

Figure 2.30. Scanning tunneling microscope spectra measured as a function of temperature on underdoped ($T_c = 83$ K) Bi-2212. The conductance scale corresponds to the 293 K spectrum, and the other spectra are offset vertically for clarity (from Renner et al. [218]).

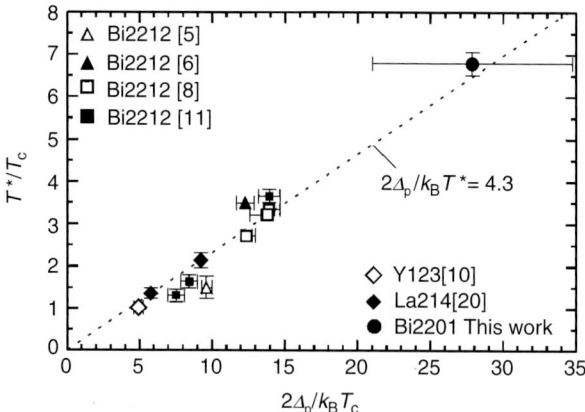

Figure 2.31. T^*/T_c vs. $2\Delta_p/k_B T_c$ for various cuprates compared to the mean-field relation $2\Delta_p/k_B T^* = 4.3$, where T^* replaces T_c. The numbers in square brackets refer to reference numbers from the original text (from Kugler et al. [230]).

Magnetic Field Dependence

The high-T_c cuprates have coherence lengths ξ much shorter than their penetration depths λ. For example, the in-plane coherence length $\xi_{ab} \sim 3.5$ nm [294] for optimally doped $EuBa_2Cu_3O_{7-\delta}$, and the in-plane penetration depth $\lambda_{ab} \sim 120$–160 nm for optimally doped $YBa_2Cu_3O_{7-\delta}$ [295]. A superconductor with $\kappa \equiv \lambda/\xi > 1/\sqrt{2}$ is termed type II. In a type II superconductor the magnetic field penetrates in a range of magnetic field $H_{c2} > H > H_{c1}$ in the form of superconducting vortices with quantized flux $\Phi_0 = h/2e$ [296]. In the extreme type II limit $\kappa \gg 1$ appropriate for the high-T_c cuprates, $H_{c1} = \Phi_0 \ln \kappa / 4\pi\lambda^2$ and $H_{c2} = \Phi_0^2 / 2\pi\xi^2$ [297]. For the examples given above this corresponds to $H_{c1} \sim 0.025$–0.04 T and $H_{c2} \sim 27$ T. Krasnov et al [298]. report that the peak at the superconducting gap edge in Bi-2212 disappears in fields parallel to the c-axis direction of $H_{c2} = 10$ T at $T = 80$ K and $H_{c2} = 14$ T at $T = 72$ K, in agreement with previous transport measurements [299], but that fields of this magnitude have little effect on the pseudogap.

Shibauchi and Krusin-Elbaum [300, 301] have inferred from c-axis interlayer tunneling measurements in Bi2212 that the pseudogap closes at a critical magnetic field H_{pg} which has a temperature dependence quite different from that of the superconducting critical field H_{sc} (Figure 2.32b). The low-temperature pseudogap closing field H_{pg} has Zeeman scaling: it varies with the pseudogap temperature T^* as $g\mu_B H_{pg} \approx k_B T^*$ (Figure 2.32a), where $g=2.0$ for fields parallel to the c-axis. Further, the anisotropy of the pseudogap closing fields $H_{pg}^{||ab}/H_{pg}^{||c} \approx 1.35$ corresponds to the anisotropy of the g factor in the cuprates. This implies that there is little orbital frustration. These results would seem to favor the competing order explanation, as opposed to the preformed pairs explanation, for the origin of the pseudogap. However, Shibauchi and Krusin-Elbaum conclude that their results mean that, in a preformed pair scenario, there is little orbital motion of the pairs. This may be the case if the charges self-organize into microstripes below T^* [250, 304]. We will discuss tunneling evidence for spatial inhomogeneities in the cuprates in Section 2.5.5.

Superconducting vortices have been imaged with scanning tunneling microscopy in YBCO [305, 306] and Bi-2212 [218, 307–310]. The imaging experiments are complicated by the very strong vortex pinning in these materials [307], making it difficult to image regular

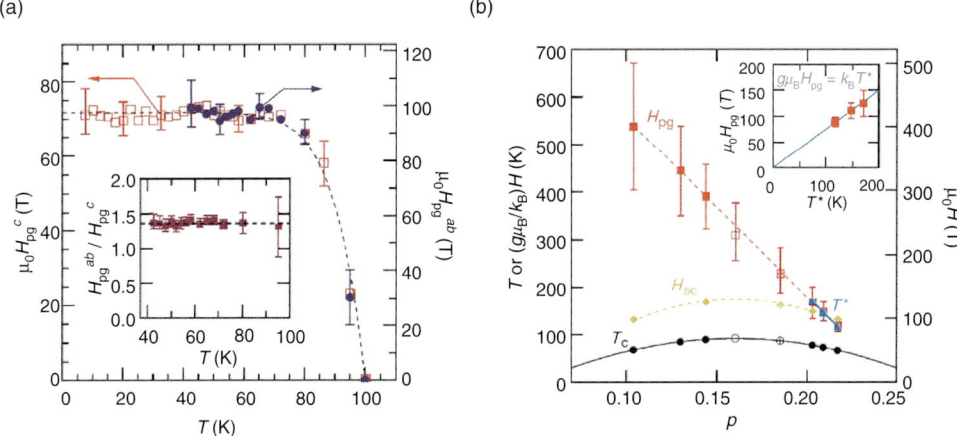

Figure 2.32. (a) H–T diagram showing the pseudogap closing field $H_{pg}(T)$ for $H \parallel c$ (left-hand side) and $H \parallel ab$ (right hand side), inferred from interlayer tunneling measurements in overdoped ($T_c \sim 60$ K) Bi-2212. In contrast to the characteristic fields in the superconducting state, which are highly anisotropic, the small ratio $\gamma_{pg} = H_{pg}^{\parallel ab}(T)/H_{pg}^{\parallel c}(T) \sim 1.35$ is temperature independent (inset) and corresponds to the anisotropy of the g factor. (b) Doping dependences of the low-temperature pseudogap closing field H_{pg}, the low temperature superconducting critical field $H_{sc}(T \to 0)$, the pseudogap characteristic temperature T^*, and the superconducting critical field T_c. The hole doping concentration p was obtained from the empirical formula $T_c/T_c^{max} = 1 - 82.6(p - 0.16)$ [2, 302, 303], with $T_c^{max} = 92$ K. The right-hand-side field scale directly translates onto the Zeeman energy scale on the left-hand side as $(g\mu_B/k_B)H.H_{pg}$ (squares) and T^* (triangles), obtained separately in the same crystals in the overdoped regime, give a scaling $g\mu_B H_{pg} \approx k_B T^*$ with $g = 2.0$ (inset) (from Shibauchi et al. [300] and Krusin-Elbaum et al. [301]).

vortex lattices or the fourfold vortex core symmetry expected for a superconductor with predominantly $d_{x^2-y^2}$ symmetry [311, 312]. Hoogenboom et al. have found evidence in their STM images for quantum tunneling of vortices between pinning sites [307]. The pseudogap spectra measured by STM in the core of superconducting vortices at low temperatures and high magnetic fields appears very similar to that measured at zero field above T_c, for both underdoped ($T_c = 83.0$ K) and overdoped ($T_c = 74.3$ K) Bi-2212 [305].

The superconducting vortex core acts as a potential well for quasiparticle states, leading to the formation of localized states [313–315]. Low temperature scanning tunneling spectroscopy in the vortex normal cores shows two low voltage resonances in both YBCO [306] and BSCO [308, 310]. The vortex states decay away from the vortex core center with a decay length of 2.2 ± 0.3 nm [310]. The discovery of discrete states split away from zero voltage in the vortex cores is surprising, since one expects the low-energy quasiparticle states in a vortex to be extended along the nodes of the gap in a d-wave superconductor, resulting in a broad peak centered at zero voltage [316–318]. Hoogenboom et al. have shown that the vortex core state energies in Bi-2212 depend linearly on the gap and are independent of magnetic field [308]. This is in contrast to the vortex bound states in a conventional superconductor, which have energies that are proportional to the square of the gap ($E \propto \Delta^2/E_F$) [313]. There has been a great deal of work attempting to explain the STM spectroscopy measurements of vortex core states in d-wave superconductors. Franz and Tešanović [319, 320] have concentrated on the properties of Block waves, as opposed to Landau levels, associated with an array of vortices. Han et al. [321] have used the t–J model, and Balatsky [322] has invoked the influence of the magnetic field, to induce an out-of-phase id_{xy} component to the gap, which would give a splitting of the core states away from zero energy. Kishine et al. [323] have used SU(2) slave-boson theory. Berthod et al. [324] have argued that strong correlation effects, calculated using a Cooperon-propagator description of HTS, can produce bound state zero bias conduction peaks in agreement with experiment. Although there is not a fourfold symmetric structure to the vortex core observed in scanning tunneling images, there is a fourfold symmetry to quasiparticle interference patterns generated by vortices as scattering centers [309]. Such quasiparticle interference effects will be discussed further in Section 2.5.5.

2.5.3. Linear Conduction Background

Often tunneling data on the high-T_c cuprate superconductors show a linearly increasing conductance background for high resistance contacts, while for low resistance contacts there is a roughly linearly decreasing background [325–327]. Both have been interpreted as density of states effects, with the inferred density of states either increasing [203, 215], or decreasing [325] away from the Fermi energy. A strong linear conductance background is observed in a number of tunneling systems [328], and is not necessarily observed in tunneling into the high-T_c superconductors. See, for example the STM data in Figure 2.18. A number of other mechanisms for this effect have been proposed, including space–charge limited currents [329], charging effects [330], resonance tunneling [331], voltage dependent tunneling matrix elements [14], or tunneling into impurity states [332]. Several proposals share in common the view that the linear conduction background reflects the dynamics of the tunneling process itself: Anderson and Zou [333] proposed that the tunneling process excited both holons and spinons. Integration over one of the degrees of freedom leads to a predicted linear conductance background. Varma et al. proposed that the linear conductance background arises from tunneling into the very short lifetime states of a marginal Fermi liquid [334]. Kirtley and Scalapino [328] have proposed that it is due to inelastic tunneling with a broadly distributed spectral weight $F(\omega)$ of inelastic scattering energy losses. In that case the inelastic contribution to the total tunneling current can be written as:

$$I_i(V) \sim \int_{-\infty}^{\infty} d\omega F(\omega)[n(\omega) + 1] \int_{-\infty}^{\infty} dE\{f(E)[1 - f(E + eV - \hbar\omega)]$$
$$- f(E + eV + \hbar\omega)[1 - f(E)]\} + \int_{0}^{\infty} d\omega F(\omega)n(\omega)$$
$$\int_{-\infty}^{\infty} dE\{f(E)[1 - f(E + eV + \hbar\omega)] - f(E + eV - \hbar\omega)[1 - f(E)]\}, \quad (2.36)$$

where $n(\omega)$ and $f(E)$ are the usual Bose and Fermi factors. The first term in Eq. (2.36) corresponds to emissions of excitations in the tunneling process, and the second term corresponds to absorption. Equation (2.36) reduces at zero temperature to

$$\frac{dI_i(V)}{d(eV)} \sim \int_{0}^{eV} d\omega F(\omega), \quad (2.37)$$

so that a constant spectral weight $F(\omega)$ would give rise to a conductance which rises linearly with V. Figure 2.33(a) shows tunneling data from a $La_{2-x}Sr_xCuO_4$–In junction as a function of temperature. There is no sign of a tunneling gap in this data, presumably because of a nonsuperconducting surface layer, but there is a large linear conductance background. The theoretical curves in Figure 2.33(b) are fits to the data using a spectral weight function derived from fits to NMR spin-relaxation data for the cuprate superconductors. The contribution to

Tunneling Measurements of the Cuprate Superconductors

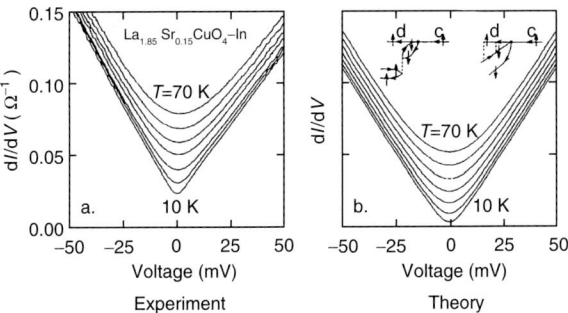

Figure 2.33. (a) Experimental conductance–voltage characteristic for a $La_{2-x}Sr_xCuO_4$–In junction as a function of temperature. (b) Theory prediction, including the effects of inelastic tunneling from a broad distribution of inelastic excitations. Inset: Diagrams associated with the inelastic spin-fluctuation tunneling (from Kirtley and Scalapino [328]).

the total conductance from inelastic tunneling from a broad spectral distribution has a thermal broadening of $5.4k_BT$ of the discontinuity in slope at zero voltage, compared with the $3.5k_BT$ expected for thermal broadening of elastic tunneling features. Experimental data from several types of tunnel junctions shows this characteristic $5.4k_BT$ thermal broadening [14, 335]. Analysis of Al–Al oxide–Cr–Pb tunnel junctions, which have a very large linear conductance background, show that taking the ratio of the tunneling data in the superconducting state, divided by that in the normal state, reduces the amplitudes of the strong coupling Pb phonon peaks, and that a better procedure is to subtract out a large linear conductance background due to inelastic tunneling [335].

Kirtley has extended the BTK analysis for the current–voltage characteristics of NS contacts to the case of inelastic tunneling with a broad distribution of energy losses [336]. This analysis assumed an isotropic s-wave superconducting gap. It would be very interesting to extend the theory of Wu [41] and Tanaka and Kashiwaya [211, 212] to inelastic scattering processes. Kirtley finds (see Figure 2.34) that when the elastic and inelastic contributions are summed, the small barrier height Z behavior of a plateau of width Δ near zero voltage, followed by a linear decrease in the conductance, evolves smoothly into a gap of width Δ at zero voltage, followed by a linear increase in conductance for large Z [336]. Grajcar [337] has extended this analysis to include a finite quasiparticle scattering rate Γ, and find good agreement with experiment on $Bi_2Sr_2CaCu_2/Sr_2TiO_3$–Au point contacts with varying thicknesses of $SrTiO_3$.

2.5.4. Zero-Bias Anomalies

Tunneling measurements of the high-T_c superconductors often show peaks or dips in conductance centered on zero voltage. Reviews of this topic appear in [18, 20, 23, 338–344]. Many different mechanisms for zero-bias anomalies in tunneling measurements of the high-T_c cuprates have been proposed, including charging effects in metallic inclusions in the tunneling barrier [345], electron–electron Coulomb interactions [346], and phase diffusion [347]. Early investigations in the cuprates [339] focused most often on an exchange-scattering interaction between tunneling electrons and isolated magnetic spins in the tunneling barrier [348–351], first described theoretically by Applebaum [352] and Anderson [353]. There are several reasons for doubting this interpretation as the explanation for the zero bias conductance peaks

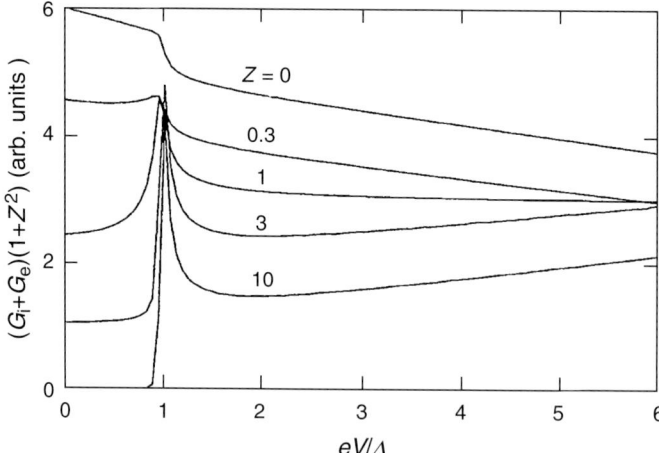

Figure 2.34. Calculated sum of elastic and inelastic tunneling components, with the ratio of the two contributions scaled so that they are equal for the dimensionless barrier height $Z \gg 1$ at $eV = 5\Delta$ (assumed isotropic s-wave). Each successive curve is shifted up by 1 unit for clarity. As Z increases, the total conductance curves evolve from a conductance plateau of width Δ superposed on a linearly decreasing background, to a conductance gap with width Δ on a linearly increasing background (from Kirtley [336]).

(ZBCPs) in many of the tunneling measurements in the cuprates. First, in the Applebaum–Anderson mechanism the splitting in the energies of the ZBCPs away from zero should be linear with magnetic field, whereas in tunneling in the ab plane directions into YBCO, the splitting is nonlinear in field, with an anomalously large g factor [339]. Further, often the appearance of the ZBCP is correlated with the onset of superconductivity [340], and the temperature dependence is not that expected for the Applebaum–Anderson mechanism [354]. Perhaps the most convincing evidence against the Applebaum–Anderson mechanism is the anisotropy of the field dependent splitting of the ZBCP: when the magnetic field is applied parallel to the c-axis there is a strong field dependence, but a much weaker dependence is observed if the field is applied perpendicular to the c-axis [355, 356].

Figure 2.35 shows the splitting of the ZBCP in magnetic fields applied parallel (a,b) and perpendicular (c) to the c-axis in (110) oriented films of YBCO [357]. This behavior is consistent with the interpretation of Hu [41] and Tanaka and Kashiwaya [18, 211, 212], that the ZBCP in the cuprates is due to a zero-energy bound state arising from Andreev scattering and the sign changes accompanying the $d_{x^2-y^2}$ pairing symmetry. (see the discussion in "General Features" in Section 2.5.1.) We will adopt this interpretation for the rest of the present section.

The magnetic field splitting of the ZBCP has been attributed to a Doppler shift in energy equal to $\mathbf{v_s} \cdot \mathbf{p_F} \cos \Theta$, where $\mathbf{v_s}$ is the superfluid velocity associated with the Meissner screening currents, $\mathbf{p_F}$ is the Fermi momentum, and Θ is the angle between the tunneling quasiparticle and the sample surface [354, 358, 359]. This provides a natural explanation for the magnetic field anisotropy, since the Meissner screening currents are much weaker for fields applied perpendicular to the c-axis than parallel to it. Within this picture, since $\mathbf{v_s}$ is proportional to applied field, as long as there is little flux penetration into the films, the ZBCP splitting should be linear in field, saturating at fields of order the thermodynamic critical field H_c, as observed by Covington et al. [354]. However, sometimes nonzero splittings are seen even at very small fields [354].

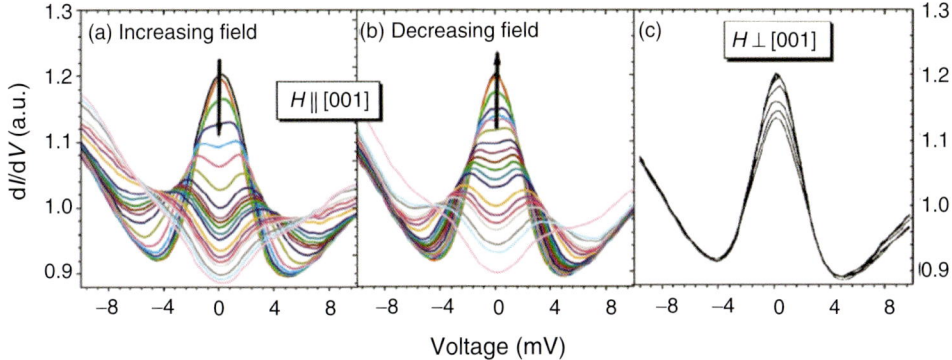

Figure 2.35. Normalized dynamical conductance dI/dV vs. bias V for increasing (a) and decreasing (b) applied fields for an YBCO (110)-oriented film (T_c = 88 K, film thickness d = 60 nm) at 4.2 K. Applied fields parallel to the (001) axis in Tesla: 0, 0.1, 0.3, 0.5, 0.7, 0.9, 1.2, 1.5, 1.8, 2.1, 3.0, 3.5, 5, 6, 7, 11, 13, and 15. (c) Behavior of the same junction for magnetic field applied perpendicular to the c-axis at fields (in Tesla): 0, 0.5, 1, 2, 4, 8, 12, and 15.5 (from Beck et al. [357]).

An example is shown in Figure 2.36, in which a splitting of the ZBCP is seen even at zero applied field. Such zero field splitting has been taken as evidence for states with broken time reversal symmetry (BTRS) in the cuprates [354]. Although there has been a report of bulk BTRS in the pseudogap state of an underdoped cuprate [360, 361], it is generally accepted that there is little if any bulk BTRS in the superconducting state in optimally doped cuprates [362, 363]. However, there have been many predictions of broken time reversal symmetry at surfaces and interfaces of cuprate superconductors [55, 364–371]. Fogelström et al [358]. have calculated the phase diagram of a surface induced state, in which the dominant $d_{x^2-y^2}$ pairing symmetry is suppressed by the presence of the surface, allowing a subdominant pairing interaction to coexist with a $\pi/2$ relative phase difference at low temperatures. This phase difference leads to a spontaneous supercurrent, and a surface BTRS state is achieved at low temperatures. The solid line in Figure 2.36 is the prediction of Fogelström et al. for the magnetic field dependence of the splitting of the ZBCP, assuming that the subdominant order parameter has is symmetry.

Some caution should be used in interpreting zero field splitting of the zero bias conductance peak as evidence for the presence or absence of TRSB. For example, Asano et al. have reported that impurity scattering near the interface also causes splitting of the ZBCP [372]. In addition, Flatté and Byers [373] have shown that good agreement with a number of different tunneling spectra, including ones with an apparently split ZBCP, can be made using self-consistent calculations of the electronic structure near strongly scattering impurities (Figure 2.37).

Further, Tanuma et al. have shown that the ZBCP will not be split in the presence of TRSB unless the transmission coefficients of the junctions are sufficiently small [376]. Note that neither zero field splitting nor field splitting of the ZBCP is observed in grain boundary junctions [343]. April et al. have shown that the amplitude of the ZBCP decreases in YBa$_2$Cu$_3$O$_{7-\delta}$/Pb junctions upon ion irradiation, without reducing the junction quality [96].

Caution should also be used in interpreting the absence of a ZBCP as evidence for s-wave pairing symmetry. Several groups have argued that the presence of the ZBCP in grain boundary junctions of the hole-doped cuprate superconductors, and the absence of such a peak in similar junctions in the electron-doped cuprates (although exceptions to this absence have

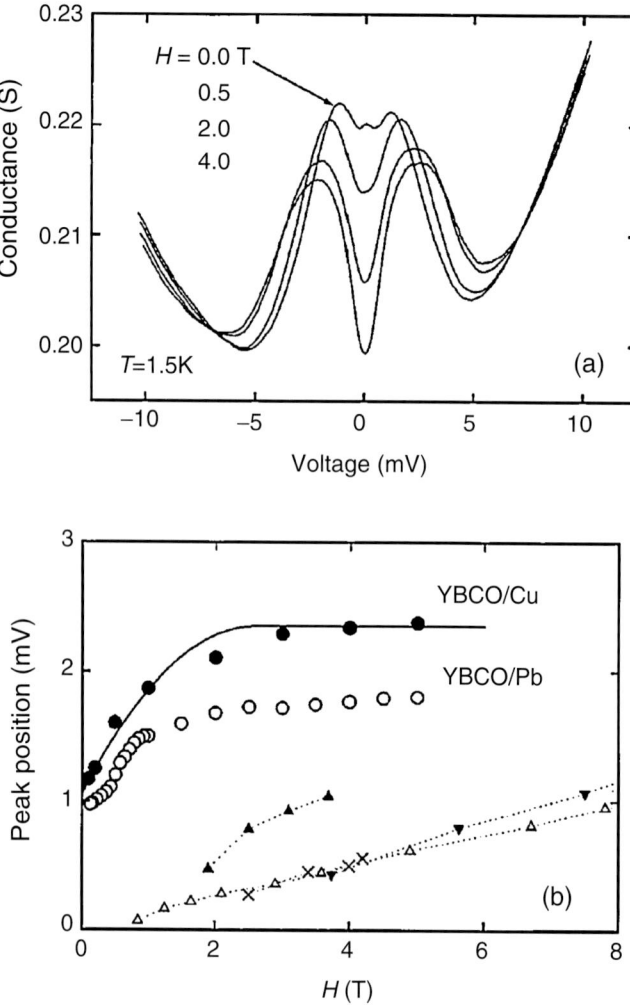

Figure 2.36. (a) Magnetic field dependence of the ZBCP from a YBCO/Cu tunnel junction. (b) A compendium of data on the magnetic field-induced splitting of ZBCP's. Data from YBCO/Cu and YBCO/Pb junctions are indicated by closed and open circles, respectively. The theoretical curve for the subdominant order parameter being A_{1g} (s-wave) is shown as a full line [358]. The remaining symbols represent data from junctions with magnetic scattering centers in the tunneling barrier (from Covington et al. [354]).

been reported [341]), implies that the latter have s-wave pairing symmetry [342, 343, 377]. Further Biswas et al. [378] and Qazilbash et al. [379] argue that the presence of a ZBCP in point contact junctions involving underdoped $Pr_{2-x}Ce_xCuO_4$ and its absence in optimally and overdoped samples of the same material imply a change in pairing symmetry from d- to s- as a function of doping. Although the results from various phase insensitive pairing symmetry tests are mixed, with some indicating s-wave pairing at some doping levels [380–384], and others indicating d-wave pairing [385–390], the tricrystal experiments of Tsuei et al [188]. in optimally doped samples of both $Nd_{2-x}Ce_xCuO_4$ and $Pr_{2-x}Ce_xCuO_4$, the SQUID interference measurements of Chesca et al. [193] in the electron doped cuprate $La_{2-x}Ce_xCuO_{4-y}$, and

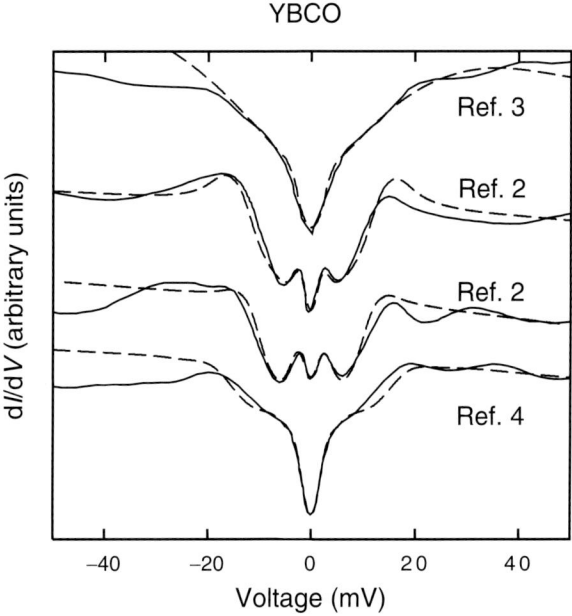

Figure 2.37. Comparison of theoretical results (dashed lines) with measurements (solid lines), using a self-consistent calculation of the YBCO electronic structure, including a realistic band structure and strong impurity scattering. The impurity parameters used were, from top down, $V_0 = 2t$, $n_i = 1.7\%$; $V_0 = 2.5t$, $n_i = 0.5\%$; $V_0 = 2.5t$, $n_i = 1.1\%$; $V_0 = 2t$, $n_i = 1.4\%$; where V_0 is the strength of the impurity scattering, t is the nearest neighbor hopping element, and n_i is the impurity concentration. Ref. 2 refers to Geerk et al. [374], Ref. 3 to Takeuchi et al. [375], and Ref. 4 to Gurvitch et al. [57] (from Flatté et al. [373]).

the zig-zap ramp type junction experiments of Ariando et al. [141] in both optimally doped and overdoped $Nd_{2-x}Ce_xCuO_{4-y}$, all show strong evidence for $d_{x^2-y^2}$ pairing symmetry. It has been proposed that some of the features in the transport measurements interpreted to be consistent with s-wave symmetry could in fact be due to band structure effects [391].

Beck et al. [23, 357] have suggested that the large splitting of the ZBCP for decreasing fields (Figure 2.35), must be due to some mechanism other than Doppler shifts, because the barrier to vortex exit is small, and therefore the Meissner screening currents are small for decreasing fields. They suggest instead that there is a field induced id_{xy} component of the order parameter. This view is supported by the $H^{1/2}$ field dependence of this second contribution to the splitting, in agreement with theoretical predictions by Laughlin [392].

2.5.5. Atomically Resolved Conductivity Modulation Effects

The cuprate superconductors are derived from parent compounds which are antiferromagnetic insulators. Proximity to the Mott insulating state may therefore be important for many physical properties in these materials [393]. One consequence of this proximity is possible short-range spatial inhomogeneities. Tunneling measurements have been used to study three types of inhomogeneities: random fluctuations, inhomogeneities due to quasiparticle scattering and interference, and stripes.

The doping concentration in the cuprate superconductors is so small, and the coherence length is so short, that even statistical fluctuations in the doping concentration may have

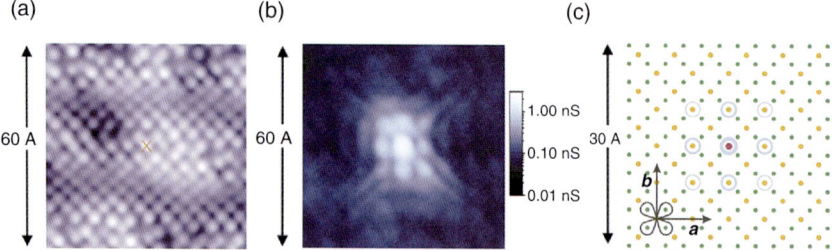

Figure 2.38. Relationship between the position of the Bi atoms on the crystal surface of Bi-2212, the resonant DOS structure of an impurity Zn atom, and the positions of the Cu and O atoms in the superconducting plane two layers below. (a) and (b) are simultaneously acquired 60×60 Å, high-spatial-resolution STM images of the topography and differential conductance at $V_{\text{sample}} = -1.5$ mV. The bright center of the scattering resonance in (b) coincides with the position of the Bi atom marked by an X in (a). (c) A 30×30 Å schematic representation of the square CuO_2 lattice, showing its relative orientation to the exposed BiO surface two layers above in (a) (from Pan et al. [404]).

important physical consequences [394]. Although some atomically resolved STM measurements show high spatial homogeneity of the superconducting gap [395], others report very large fluctuations in the local density of states and gap size, over length scales of a few nm, in STM measurements of optimally doped $Bi_2Sr_2CaCu_2O_{8+x}$ (Bi-2212) [226–228], as well as in underdoped Bi-2212 [229] and Bi-2201 [230] (see for example Figure 2.24). These experiments show strong correlations between the integrated local density of states and the energy gap size, with larger energy gaps being associated with smaller densities of states [227]. In addition, strong gap inhomogeneities and one-dimensional scattering resonances were reported from tunneling into the Cu–O chains in Y-123 [396–399].

Strong zero energy quasiparticle scattering resonances, consistent with scattering from atomic-scale nonmagnetic impurities, have been reported in low temperature STM measurements of Bi-2212 [400, 401]. These resonances arise from virtual or virtual-bound quasiparticle states inside the gap of a d-wave superconductor [402, 403], and their observation for nonmagnetic scatterers is inconsistent with s-wave superconductivity in the high-T_c cuprates. Images of individual nonmagnetic zinc impurity atoms in Bi-2212 (Figure 2.38) reveal a fourfold symmetric quasiparticle cloud, aligned with the nodes of the $d_{x^2-y^2}$ superconducting gap [404]. Scanning tunneling spectroscopy measurements of magnetic Ni atoms in Bi-2212 (Figure 2.39) show two impurity states, one above and one below the Fermi level [405]. These can be explained as previously spin-degenerate impurity states split by their interaction with the Ni moment [406]. Surprisingly, for the Ni impurity overall particle–hole symmetry is observed in the impurity-state spectrum, and the superconducting gap magnitude does not change as the impurity site is approached, indicating that the Ni atom does not significantly perturb the superconductivity. This is in contrast to nonmagnetic Zn, which locally destroys superconductivity [404].

In addition, effects due to weak quasiparticle scattering are also seen. In an isotropic s-wave superconductor with a circular normal-state Fermi surface, a weak scattering center will form circular ripples in the local tunneling density of states $N(x, \omega)$ due to quasiparticle interference scattering. However, $d_{x^2-y^2}$-wave superconductors form ripples from a scattering center which appear as a set of rays whose wavelength and amplitude vary with angular direction and bias voltage [407–409]. These interference patterns were first noted surrounding

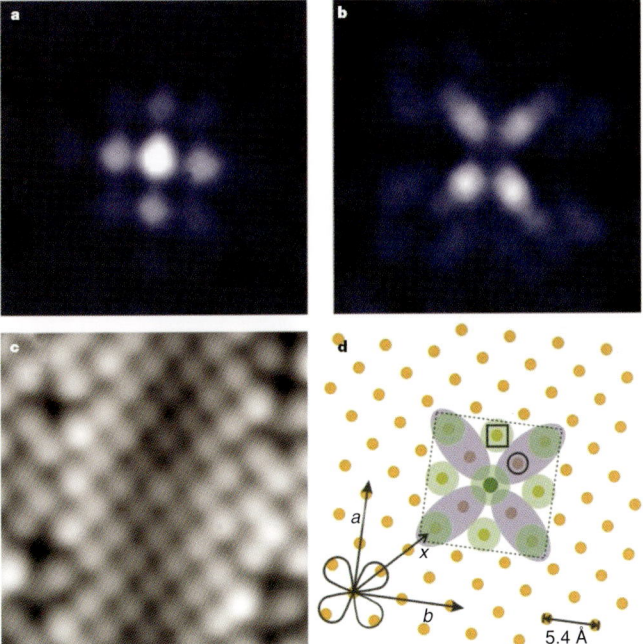

Figure 2.39. STM data on the impurity state at a single Ni atom in Bi-2212. (a), (b), 35 Å square differential conductance maps above an Ni atom at (a) +9 mV and (b) −9 mV. (c) A 35 Å square atomic resolution topograph of the BiO surface obtained simultaneously with the maps. (d) Schematic of the relative position of the Ni atom relative to the Cu atoms in the invisible CuO_2 plane (from Hudson et al. [405]).

vortex cores in Bi-2212 [309], but have also been seen (with smaller amplitudes) using Fourier transform techniques in zero field [410, 411].

The dispersion in these interference patterns can be well understood considering only the band structure of the quasiparticle states, and the $d_{x^2-y^2}$ symmetry of the superconducting gap [412]. This is illustrated in Figure 2.40, which compares the scattering wavevectors of the local maxima in the Fourier transformed local density of states with theory. The derived angular dependence of the gap is fit by the form $\Delta(\theta_k) = \Delta_0[A\cos(2\theta_k) + B\cos(6\theta_k)]$, with $\Delta_0 = 39.3$ meV, $A = 0.818$, and $B = 0.182$ (solid line in Figure 2.40). There is some systematic difference between the gap functional form derived from FT-STS and that from ARPES measurements, but the general agreement is quite good, providing yet another confirmation of $d_{x^2-y^2}$ pairing symmetry in the cuprate superconductors.

It has been proposed that when holes are doped into an antiferromagnetic insulator, they can form a slowly fluctuating array of metallic, quasi-one-dimensional stripes [250, 413–415]. In this view the mechanism of pairing is the generation of a spin gap in spatially confined Mott-insulating regions near the metallic stripes. In underdoped and optimally doped cuprates phase coherence occurs at a temperature well below the pairing temperature, while in overdoped materials pairing and phase coherence occur at the same temperature, as in conventional superconductors. It is well established that static stripe order occurs in $La_{2-x}Sr_xNiO_{4+\delta}$ and $La_{1.6-x}Nd_{0.4}Sr_xCuO_4$ [416], but it is more difficult to establish fluctuating stripe order in the cuprate superconductors [417]. Scanning tunneling microscopy has insufficient time

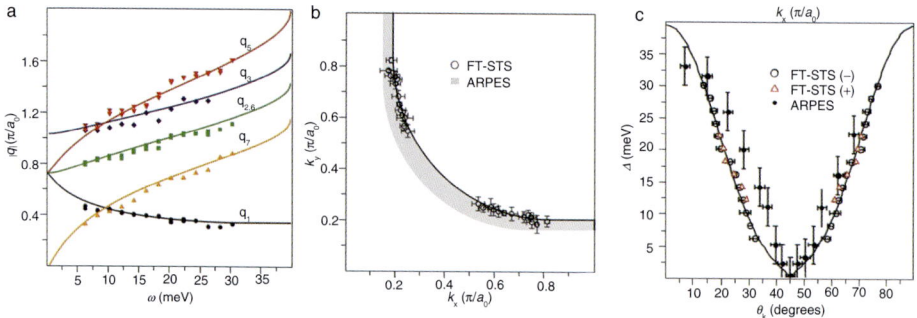

Figure 2.40. (a) The measured dispersion $\mathbf{q}(\omega)$ of a number of quasiparticle interference local maxima in the Fourier transform scanning tunneling spectroscopy (FT-STS) of Bi-2212. The solid lines represent theoretical predictions based on fits to the band dispersion $\mathbf{k}_s(\omega)$ and momentum dependent superconducting gap $\Delta(\mathbf{k}_s)$ as determined from FT-STS data. (b) The locus of scattering momenta $\mathbf{k}_s(\omega)$ extracted using the measured positions of the scattering wavevectors $\mathbf{q}(\omega)$. The solid line is a fit to the data, assuming the Fermi surface is a combination of a circular arc joined with two straight lines. The gray band represents ARPES measurements [255]. (c) A plot of the energy gap $\Delta(\theta_k)$ determined from the filled-state measurements, shown as open circles, and empty-state measurements, shown as open triangles. The solid line is a fit to the data. The filled circles represent ARPES measurements [255] (from McElroy et al. [412]).

resolution to image fluctuating stripes, but it is possible that these stripes may be pinned by defects. A distinction between quasiparticle interference and pinned stripe effects can be made by measuring the dispersion of the observed features: quasiparticle interference should have a strong dependence of wavelength on energy, but stripes should be nondispersive [417]. Several reports of energy-independent modulations of the tunneling density of states in Bi-2212 have been made [411, 418–420]. There is disagreement on the exact nature of these modulations, but the peak component is approximately $2\pi/4a_0 \pm 25\%$ along the Cu–O chain directions. It has also been reported that there is a correlation between modulations in the coherence peak heights and modulations in the low-energy density of states, indicating that the latter are charge density modulations that interact with superconductivity [419]. Recently McElroy et al. reported charge modulations in underdoped Bi-2212 that were spatially localized in regions with exceptionally large (pseudo) gaps but no coherence peaks [421, 422]. It should be noted that the strong breaking of the fourfold rotation symmetry predicted in the stripe scenario is not observed in any of the STM studies, but this may be due to the presence of strong disorder [417].

2.5.6. Strong Coupling Effects

One of the early triumphs of superconductive tunneling was the derivation of the effective electron–phonon coupling function times the phonon density of states $\alpha^2 F(\omega)$ (phonon spectral density function) and the Coulomb pseudopotential term $\mu^\star$ from tunneling data, and their use to correctly infer transition temperatures in conventional superconductors [3, 5, 423, 424]. There has been much effort to do the same thing with the high-T_c cuprate superconductors. This effort has concentrated on two types of features: small oscillations in the conductance–voltage characteristics due to electron–phonon interactions, and a "dip" feature somewhat above the gap energy commonly attributed to electron–magnon interactions.

Electron–Phonon

Schrieffer, Wilkins, and Scalapino [424] showed that the tunneling electronic density of states in a superconductor is given by

$$\frac{N_s(\omega)}{N(0)} = \mathrm{Re}\left\{\frac{|\omega|}{[\omega^2 - \Delta^2(\omega)]^{1/2}}\right\}, \quad (2.38)$$

where $N(0)$ is the electronic density of states at the Fermi surface unrenormalized by the electron–phonon interaction. In conventional superconductors at temperatures $T \ll T_c$, it is assumed that $N_s(\omega)/N(0) \approx (\mathrm{d}I/\mathrm{d}V)_s/(\mathrm{d}I/\mathrm{d}V)_n|_{eV=\hbar\omega}$. The gap $\Delta(\omega)$, which is energy independent in the BCS formulation, has inflections in strong-coupling conventional superconductors at energies corresponding to peaks in $\alpha^2 F(\omega)$. The most commonly used method for inferring the phonon spectral function from tunneling data was devised by McMillan and Rowell [3]. It involves inserting an educated guess for the phonon spectral function and the Coulomb pseudopotential into the normal and pairing self-energy integral equations, calculating corrections, and iterating to convergence. The procedure produces remarkable agreement between tunneling, neutron scattering, and first-principles calculations for the phonon spectral densities for a number of conventional superconductors [5].

Workers have inferred strong coupling between the electrons and phonons from renormalization of the tunneling density of states in a number of cuprate superconductors, including $Nd_{1.85}Ce_{0.15}CuO_{4-y}$ [425–427], $La_{1.85}Sr_xCuO_4$ [428], Bi-2212 [429–435], and a number of other layered Bi-cuprates [436]. An example is shown in Figure 2.41, in the electron-doped superconductor $Nd_{1.85}Ce_{0.15}CuO_{4-y}$. The solid line is the electron–phonon spectral function $\alpha^2 F(\omega)$ inferred from point contact tunnel junctions [425] using a McMillan–Rowell inversion procedure modified to allow for the effects of a proximity induced layer of reduced superconductivity on the surface [437]. The inversion procedure gave reasonable values of $\mu^\star \sim 0.1$ and $\lambda \sim 1$, and predicted values for T_c (including only an electron–phonon interaction) in good agreement with experiment. The dot-dashed line in Figure 2.41 is the phonon density of states determined from neutron scattering measurements [438]. Figure 2.41 illustrates the difficulty in analyzing electron–phonon strong coupling effects in the cuprate

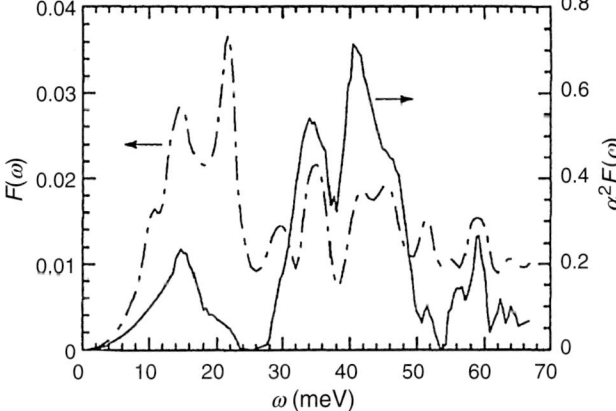

Figure 2.41. The electron–phonon spectral function $\alpha^2 F(\omega)$ for $Nd_{1.85}Ce_{0.15}CuO_{4-y}$ (solid line), and the phonon density of states $F(\omega)$ from neutron scattering on single crystal Nd_2CuO_4 (dot-dashed line) (from Tralshawala et al. [426]).

superconductors: the phonon densities of states are complex and cover a broad frequency range, and the tunneling data to date is not very reproducible. Further, Trashawala et al. used the phonon spectral function $\alpha^2 F(\omega)$ to calculate the temperature dependence of the normal state resistivity of $Nd_{1.85}Ce_{0.15}CuO_{4-y}$, and found that there must be an additional nonphonon contribution [426, 427].

Electron–Magnon

It was observed quite early [325] that there is a reproducible "dip" above the coherence peak in the tunneling conductance–voltage characteristic in several of the cuprate superconductors. This dip is more pronounced for a bias direction corresponding to removal of quasiparticles from the cuprate superconductor. For optimally doped superconductors the position of the dip corresponds to $eV \sim 2\Delta$, and roughly scales with the gap as a function of doping [440]. This lead to the belief that the dip was due to an electron–electron pairing interaction [440]. However, later observations showed that the dip occurs at an energy Ω relative to the superconducting gap energy that scales as $4.9 k_B T_c$ with doping, and that the amplitude of the dip is largest for optimal doping (Figure 2.42) [439]. This lead Zasadzinski et al. [439] to infer that this dip is due to a resonance spin excitation [441, 442]. The peak/dip/hump features in angle-resolved photoemission spectroscopy near $(0, \pi)$ [443–445], and the resonance peak

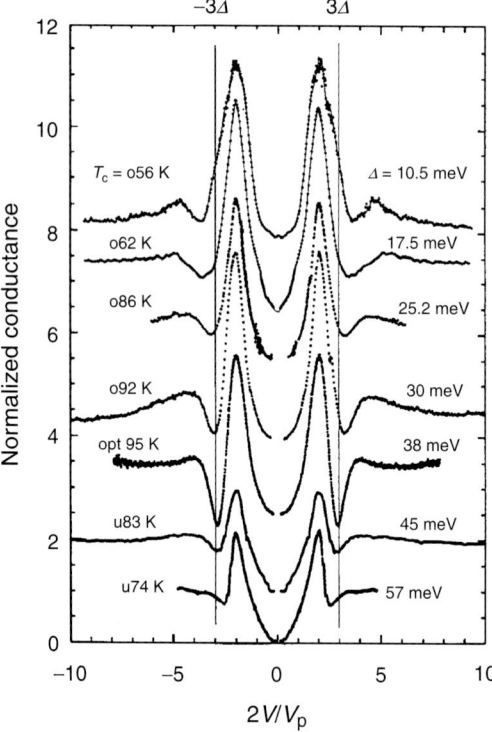

Figure 2.42. SIS conductances of Bi-2212 for various hole dopings. Notation is o=overdoped, opt=optimally doped, and u=underdoped. The voltage axis has been rescaled in units of Δ. Each curve has been rescaled and shifted for clarity. The Josephson current has been removed from each curve, and the inferred conductance at zero bias for each curve is close to zero (from Zasadzinski et al. [439]).

below 2Δ in inelastic neutron-scattering data [446, 447] have also been attributed to a spin-fluctuation mechanism. Hoogenboom et al. [208] have modeled scanning tunneling spectra of BSCCO including a collective bosonic mode with energy $\Omega = 5.4 k_B T_c$ at wave vector (π, π), consistent with the neutron scattering data, and find good agreement with experiment. These observations have lent support to the proposal that spin-fluctuations play an important role in superconductivity in the cuprates [448].

2.6. Conclusions

Tunneling measurements tell us much about the nature of superconductivity in the cuprate high-T_c superconductors. The pairing wavefunction has predominantly $d_{x^2-y^2}$ pairing symmetry. The sign changes in the pairing wavefunction associated with this pairing symmetry lead to spontaneous, persistent supercurrents, even in the absence of externally applied magnetic fields, in a number of different ring and junction geometries. These sign changes also lead to zero energy bound states at surfaces and interfaces. The superconducting gap and the pseudogap coexist in some cuprates at some doping levels and temperatures, but the pseudogap can persist to temperatures well above the superconducting gap. The pseudogap magnetic closing field varies with the pseudogap temperature with Zeeman scaling $g \mu_B H_{pg} \approx k_B T^*$. Superconducting quasiparticles couple strongly to both phonons and spin-fluctuations. There are strong inhomogeneities in the superconducting gap and in the local tunneling density of states in many cuprates, which become more pronounced with increased underdoping.

However, there are still many unresolved issues that tunneling can address. The most important is: What is the mechanism for Cooper pairing in the cuprates? Some authors have argued that analysis of tunneling measurements provide electron–phonon coupling strengths sufficiently large to explain the high critical temperatures observed. Others argue that the dip structure observed in tunneling indicates that spin fluctuations must play an important role. Tunneling spectroscopy observations of zero field splitting of the zero bias conductance peak indicates the presence of broken time reversal symmetry, but measurements of π-SQUIDs and 0–π junctions shows no evidence for such effects. The evidence for spatially inhomogeneous tunneling densities of states from impurity resonances and quasiparticle interference effects is quite strong, but there is also evidence of nondispersive density of states modulations, indicative of stripes.

Much work remains.

Bibliography

1. J. Bardeen, L. N. Cooper, and J. R. Schrieffer, Phys. Rev. **108**, 1175 (1957).
2. I. Giaever, Phys. Rev. Lett. **5**, 147 (1960).
3. W. L. McMillan and J. M. Rowell, Phys. Rev. Lett. **14**, 108 (1965).
4. W. L. McMillan, Phys. Rev. **167**, 331 (1968).
5. J. P. Carbotte, Rev. Mod. Phys. **62**, 1027 (1990).
6. B. D. Josephson, Phys. Lett. **1**, 251 (1962).
7. P. W. Anderson and J. M. Rowell, Phys. Rev. Lett. **10**, 230 (1963).
8. A. Barone and G. Paterno (Wiley, New York, 1982); K. K. Likharev, Dynamics of Josephson Junctions and Circuits (Gordon and Breach, New York, 1986).
9. D. J. Scalapino, Phys. Rep. **250**, 329 (1995).
10. D. J. van Harlingen, Rev. Mod. Phys. **67**, 515 (1995).
11. C. C. Tsuei and J. R. Kirtley, Rev. Mod. Phys. **72**, 969 (2000).

12. C. C. Tsuei and J. R. Kirtley, in *The Physics of Superconductors*, vol. I, K. H. Bennemann and J. B. Ketterson (eds.) (Springer Heidelberg New York Berlin 2003), p. 647.
13. K. E. Gray, M. E. Hawley, and E. R. Moog, Tunneling spectroscopy of novel superconductors, in *Novel Superconductivity*, S. A. Wolf and V. Z. Krezin (eds.) (Plenum, New York, 1987), p. 611.
14. J. R. Kirtley, Int. J. Mod. Phys. **B4**, 201 (1990).
15. T. Hasegawa, H. Ikuta, and K. Kitizawa, Tunneling Spectroscopy of Oxide Superconductors, in *Physical Properties of High Temperature Superconductors III*, D. M. Ginsberg (ed.) (World Scientific, Singapore, 1992).
16. K. A. Delin and A. W. Kleinsasser, Supercond. Sci. Technol. **9**, 227 (1996).
17. A. A. Yurgens, Supercond. Sci. Technol. **13**, R85 (2000).
18. S. Kashiwaya and Y. Tanaka, Rep. Prog. Phys. **63**, 1641 (2000).
19. H. Hilgenkamp and J. Mannhart, Rev. Mod. Phys. **74**, 485 (2002).
20. T. Löfwander, V. S. Shumeiko, and G. Wendin, Supercond. Sci. Technol. **14**, R53 (2001).
21. J. Zasadzinski, in *The Physics of Superconductors*, vol. I, K. H. Bennemann and J. B. Ketterson (eds.) (Springer, Berlin Heidelberg New York, 2003), p. 591.
22. F. Tafuri and J. R. Kirtley, Rep. Prog. Phys. **68**, 2573 (2005).
23. G. Deutscher, Physics **77**, 109 (2005).
24. P. G. de Gennes, Rev. Mod. Phys. **36**, 225 (1964).
25. A. F. Andreev, Zh. Eksp. Teor. Fiz. **46**, 1823 (1964). (Sov. Phys. JETP **19**, 1228).
26. R. Landauer, IBM J. Res. Dev. **1**, 223 (1957).
27. R. Landauer, IBM J. Res. Dev. **32**, 306 (1988).
28. E. Wolf, *Principles of Electron Tunneling Spectroscopy* (Oxford University Press, Oxford, 1985).
29. Y. Imry, *Introduction to Mesoscopic Physics* (Oxford University Press, 1997).
30. S. Datta, *Electronic Transport in Mesoscopic Systems* (Cambridge University Press, Cambridge, 1995).
31. A. A. Golubov and M. Yu. Kupryanov, Sov. Phys. JETP **69**, 805 (1989).
32. W. L. McMillan, Phys. Rev. **175**, 537 (1968).
33. V. Z. Kresin, Phys. Rev. B **25**, 157 (1982).
34. S. Gueron, H. Pothier, N. O. Birge, D. Esteve, and M. H. Devoret. Phys. Rev. Lett. **77**, 3025 (1996).
35. K. M. Frahm, P. W. Brouwer, J. A. Melsen and C. W. J. Beenakker, Phys. Rev. Lett. **76**, 2981 (1996).
36. M. Yu. Kupryanov and V. F. Lukichev, Sov. J. Low Temp. Phys. **8**, 526 (1982).
37. A. A. Golubov, M. Yu. Kupryanov, and V. F. Lukichev, Sov. J. Low Temp. Phys. **10**, 418 (Fiz. Nizk. Temp. **10**, 799) (1984).
38. G. E. Blonder, M. Tinkham, and T. M. Klapwijk, Phys. Rev. B **25**, 4515 (1982).
39. S. N. Artemenko, A. F. Volkov, and A.V. Zaitsev, Zh. Eksp. Teor. Fiz. **76**, 1816 (Sov. Phys. JETP) (1979).
40. C. W. J. Beenakker, Phys. Rev. B **46**, 12841 (1992).
41. C. -R. Hu, Phys. Rev. Lett. **72**, 1526 (1994).
42. I. O. Kulik, Sov. Phys. JETP **30**, 944 (1969).
43. I. O. Kulik, JETP Lett. **11**, 275 (1970).
44. Y. Tanaka and S. Kashiwaya, Phys. Rev. B **56**, 892 (1997).
45. G. B. Arnold, J. Low Temp. Phys. **59**, 143 (1985).
46. C. J. Muller, J. M. van Ruitenbeek, and L. J. de Jongh, Phys. Rev. Lett. **69**, 140 (1992).
47. H. Takayanagi, T. Akazaki, and J. Nitta, Phys. Rev. Lett. **75**, 3533 (1995).
48. T. Bauch, E. Hurfeld, V. M. Krasnov, P. Delsing, H. Takayanagi, and T. Akazaki, Phys. Rev. B **71**, 174502 (2005).
49. K. K. Likharev, *Dynamics of Josephson Junctions and Circuits* (Gordon and Breach, New York, 1986).
50. M. Sigrist and T. M. Rice, J. Phys. Soc. Jpn **61**, 4283 (1992).
51. M. B. Walker and J. Luettmer-Strathmann, Phys. Rev. B **54**, 588 (1996).
52. M. Sigrist and T. M. Rice, Rev. Mod. Phys. **67**, 503 (1995).
53. A. A. Golubov, M. Yu. Kupryanov and E. Ilichev, Rev. Mod. Phys. **76**, 411 (2004).
54. A. Huck, A. van Otterlo, and M. Sigrist, Phys. Rev. B **56**, 14163 (1997).
55. M. Sigrist, Prog. Theor. Phys. **99**, 899 (1998).
56. T. Van Duzer and C. W. Turner, *Principles of Superconductive Devices and Circuits* (Elsevier, New York, 1981).
57. M. Gurvitch, J. M. Valles, Jr., A. M. Cucolo, R. C. Dynes, J. P. Garno, L. F. Schneemeyer and J. V. Wasczak, Phys. Rev. Lett. **63**, 1008 (1989).
58. D. A. Wollman, D. J. Van Harlingen, W. C. Lee, D. M. Ginsberg, and A. J. Leggett, Phys. Rev. Lett. **71**, 2134 (1993).
59. D. A. Wollman, D. J. Van Harlingen, J. Giapintzakis, and D. M. Ginsberg, Phys. Rev. Lett. **74**, 797 (1995).
60. D. A. Brawner and H. R. Ott, Phys. Rev. B **50**, 6530 (1994).

61. A. G. Sun, D. A. Gajewski, M. B. Maple, and R. C. Dynes, Phys. Rev. Lett. **72**, 2267 (1994).
62. R. Kleiner, A. S. Katz, A. G. Sun, R. Summer, D. A. Gajewski, S. H. Han, S. I. Woods, E. Dantsker, B. Chen, K. Char, M. B. Maple, R. C. Dynes, and J. Clarke, Phys. Rev. Lett. **76**, 2161 (1996).
63. C. B. Eom, A. F. Marshall, Y. Suzuki, B. Boyer, R. F. Pease, and T. H. Geballe, Nature **353**, 544 (1991).
64. C. B. Eom, A. F. Marshall, Y. Suzuki, T. H. Geballe, B. Boyer, R. F. W. Pease, R. B. van Dover, and J. M. Phillips, Phys. Rev. B **46**, 11902 (1992).
65. B. H. Moeckly, D. K. Lathrop and R. A. Buhrman, Phys. Rev. B **47**, 400 (1993).
66. J. Mannhart, P. Chaudhari, D. Dimos, C. C. Tsuei, and T. R. McGuire, Phys. Rev. Lett. **61**, 2476 (1988).
67. P. Chaudhari, J. Mannhart, D. Dimos, R. Tsuei, J. Chi, M. M. Oprysko, and M. Scheuermann, Phys. Rev. Lett. **60**, 1653 (1988).
68. D. Dimos, P. Chaudhari, and J. Mannhart, Phys. Rev. B **41**, 4038 (1990).
69. U. Poppe, Y. Y. Divin, M. I. Faley, J. S. Wu, C. L. Jia, P. Shadrin, and K. Urban, IEEE Trans. Appl. Supercond. **11**, 3768 (2001).
70. Z. G. Ivanov, P. A. Nilsson, D. Winkler, J. A. Alarco, T. Claeson, E. A. Stepansov, and A. Ya. Tsalenchuk, Appl. Phys. Lett. **59**, 3030 (1991).
71. R. Gross, P. Chaudhari, M. Kawasaki, and A. Gupta, Phys. Rev. B **42**, 10735 (1992).
72. K. Char, M. S. Colclough, S. M. Garrison, N. Newman, and G. Zaharchuk, Appl. Phys. Lett. **59**, 733 (1991).
73. S. J. Rosner, K. Char, and G. Zaharchuk, Appl. Phys. Lett. **60**, 1010 (1991).
74. K. Char, M. S. Colclough, L. P. Lee, and G. Zaharchuk, Appl. Phys. Lett. **59**, 2177 (1991).
75. F. Tafuri, F. Miletto Granozio, F. Carillo, A. Di Chiara, K. Verbist, and G. Van Tendeloo, Phys. Rev. B **59**, 11523 (1999).
76. F. Lombardi, F. Tafuri, F. Ricci, F. Miletto Granozio, A. Barone, G. Testa, E. Sarnelli, J. R. Kirtley, and C. C. Tsuei, Phys. Rev. Lett. **89**, 207001 (2002).
77. F. Granozio Miletto, U. Scotti di Uccio, F. Lombardi, F. Ricci, F. Bevilacqua, G. Ausanio, F. Carillo, and F. Tafuri, Phys. Rev. B **67**, 184506 (2003).
78. T. Bauch, F. Lombardi, F. Tafuri, A. Barone, G. Rotoli, P. Delsing, and T. Claeson, Phys. Rev. Lett. **94**, 087003 (2005); T. Bauch, T. Lindström, F. Tafuri, G. Rotoli, P. Delsing, T. Claeson, F. Lombardi, Science **311**, 57 (2006).
79. R. W. Simon, J. F. Burch, K. P. Daly, W. D. Dozier, R. Hu, A. E. Lee, J. A. Luine, H. M. Manasevit, C. E. Platt, S. M. Scharzberg, D. St. John, M. S. Wire, and M. J. Zani, in *Science and Technology of Thin Film Superconductors 2*, R. D. McConnel and R. Noufi (eds.) (Plenum, New York, 1990), p. 549.
80. J. Luine, J. Bulman, J. Burch, K. Daly, A. Lee, C. Pettiette-Hall, S. Schwarzbek, and D. Miller, Appl. Phys. Lett. **61**, 1128 (1992).
81. C. L. Jia, B. Kabius, K. Urban, G. J. Cui, J. Schubert, W. Zander, A. I. Braginsky, and C. Heiden, Physica C **196**, 211 (1991).
82. H. R. Yi, Z. G. Ivanov, D. Winkler, Y. M. Zhang, H. Olin, and P. Larsson, Appl. Phys. Lett. **65**, 1177 (1994).
83. F. Lombardi, Z. G. Ivanov, G. M. Fischer, E. Olsson, and T. Claeson, Appl. Phys. Lett. **72**, 249 (1998).
84. K. Hermann, G. Kunkel, M. Siegel, J. Schubert, W. Zander, A. I. Braginski, C. L. Jia, B. Kabius, and K. Urban, J. Appl. Phys. **78**, 1132 (1995).
85. C. L. Jia, B. Kabius, K. Urban, K. Hermann, G. J. Cui, J. Schubert, W. Zander, A. I. Braginski, and C. Heiden, Physica C **175**, 545 (1991).
86. S. K. Tolpygo, S. Shokhor, B. Nadgorny, A. Bourdillon, J.-Y. Lin, S. H. Hou, J. M. Phillips, and M. Gurvitch, Appl. Phys. Lett. **63**, 1696 (1993).
87. B. A. Davidson, J. E. Nordman, B. M. Hinaus, M. S. Rzchowski, K. Siangchaew, and M. Libera, Appl. Phys. Lett. **68**, 3811 (1996).
88. A. J. Pauza, D. F. Moore, A. M. Campbell, A. N. Broers, and K. Char, IEEE Trans. Appl. Supercond. **5**, 3410 (1995).
89. A. J. Pauza, W. E. Booij, K. Herrmann, D. F. Moore, M. G. Blamire, D. A. Rudman, and L. R. Vale, J. Appl. Phys. **82**, 5612 (1997).
90. F. Tafuri, S. Shokhor, B. Nadgorny, M. Gurvitch, F. Lombardi, and A. Di Chiara, Appl. Phys. Lett. **71**, 125 (1997).
91. F. Tafuri, B. Nadgorny, S. Shokhor, M. Gurvitch, F. Lombardi, F. Carillo, A. Di Chiara, and E. Sarnelli, Phys. Rev. B **57**, R14076 (1998).
92. E. M. Jackson, B. D. Weaver, and G. R. Summers, Appl. Phys. Lett. **64**, 511 (1994).
93. E. M. Jackson, B. D. Weaver, G. P. Summers, P. Shapiro, and E. A. Burke, Phys. Rev. Lett. **74**, 3033 (1995).
94. J. Elly, M. G. Medici, A. Gilabert, F. Schmidl, P. Siedel, A. Hoffmann, and I. K. Schuller, Phys. Rev. B **56**, R8507 (1997).

95. R. Adam, R. Kula, R. Sobolewski, J. M. Murduck, and C. Pettiette-Hall, Appl. Phys. Lett. **67**, 3801 (1995).
96. M. Aprili, M. Covington, E. Paraoanu, B. Niedermeier, and L. H. Greene, Phys. Rev. B **57**, R8319 (1998).
97. A. S. Katz, A. G. Sun, R. C. Dynes, K. Char, Appl. Phys. Lett. **66**, 105 (1995).
98. H. J. H. Smilde, Ariando, D. H. A. Blank, G.J. Gerritsma, H. Hilgenkamp, and H. Rogalla, Phys. Rev. Lett. **88**, 057004 (2002).
99. L. Antognazza, S. J. Berkowitz, T. H. Geballe, and K. Char, Phys. Rev. B **51**, 8560 (1995).
100. H. -J. H. Smilde, H. Hilgenkamp, G. Rijnders, H. Rogalla, and D. H. A. Blank, Appl. Phys. Lett. **80**, 4579 (2000).
101. R. H. Ono, J. A. Beall, M. W. Cromar, T. E. Harvey, M. E. Johansson, C. D. Reintsema, and D. A. Rudman, Appl. Phys. Lett. **59**, 1126 (1991).
102. C. D. Reintsema, R. H. Ono, G. Barnes, L. Borcherdt, T. E. Harvey, G. Kunkel, D. A. Rudman, L. R. Vale, N. Missert, and P. A. Rosenthal, IEEE Trans. Appl. Supercond. **5**, 3405 (1995).
103. B. D. Hunt, M. C. Foote, and L. J. Bajuk, Appl. Phys. Lett. **59**, 982 (1991).
104. P. A. Rosenthal, E. N. Grossman, R. H. Ono, and L. R. Vale, Appl. Phys. Lett. **63**, 1984 (1993).
105. M. Bode, M. Grove, M. Siegel, and A.I. Braginski, J. Appl. Phys. **80**, 6378 (1996).
106. M. S. DiIorio, S. Yoshizumi, K. -Y. Yang, J. Zhang, and M. Maung, Appl. Phys. Lett. **58**, 2552 (1991).
107. C. H. Chen, I. Jin, S. P. Pai, Z. W. Dong, C. J. Lobb, T. Vankatesan, K. Edinger, J. Orloff, and J. Melngilis, Appl. Phys. Lett. **73**, 1730 (1998).
108. M. Lee and M. R. Beasley, Appl. Phys. Lett. **59**, 591 (1991).
109. J. Lesueur, M. Aprili, A. Goulon, T. J. Horton, and L. Dumoulin, Phys. Rev. B **55**, R3398 (1997).
110. A. G. Sun, A. Truscott, A. S. Katz, R. C. Dynes, B. W. Veal, and C. Gu, Phys. Rev. B **54**, 6734 (1996).
111. I. Takeuchi, C. J. Lobb, Z. Trajanovic, P. A. Warburton, and T. Venkatesan, Appl. Phys. Lett. **68**, 1564 (1996).
112. I. Takeuchi, Y. Gim, F. C. Wellstood, C. J. Lobb, Z. Trajanovic, and T. Venkatesan, Phys. Rev. B **59**, 7205 (1999).
113. J. R. Kirtley, C. C. Tsuei, Ariando, H. -J. Smilde, and H. Hilgenkamp, Phys. Rev. B **72**, 214521 (2005).
114. H. Hilgenkamp, Ariando, H. -J. Smilde, D. H. A. Blank, H. Rogalla, J. R. Kirtley, and C. C. Tsuei, Nature **422**, 50 (2003).
115. K. Char, M. S. Colclough, T. H. Geballe, and K. E. Myers, Appl. Phys. Lett. **62**, 196 (1993).
116. L. Antognazza, B. H. Moeckly, T. H. Geballe, and K. Char, Phys. Rev. B **52**, 4559 (1995).
117. K. Char, L. Antognazza, and T. H. Geballe, Appl. Phys. Lett. **63**, 2420 (1993).
118. J. B. Barner, Appl. Phys. Lett. **59**, 982 (1991).
119. C. Stolzel, M. Siegel, G. Adrian, C. Krimmer, J. Sollner, W. Wilkens, G. Schuelz, and H. Adrian, Appl. Phys. Lett. **63**, 2970 (1993).
120. M. I. Faley, U. Poppe, H. Soltner, C. L. Jia, M. Siegel, and K. Urban, Appl. Phys. Lett. **63**, 15 (1993).
121. E. Polturak, G. Koren, D. Cohen, E. Aharoni, and G. Deutscher, Phys. Rev. Lett. **67**, 3038 (1991).
122. J. Gao, Yu. Boguslavskij, B. B. G. Klopman, D. Terpstra, G. J. Gerritsma, and H. Rogalla, Appl. Phys. Lett. **59**, 2754 (1991).
123. M. A. J. Verhoeven, G. J. Gerritsma, H. Rogalla, and A. A. Golubov, Appl. Phys. Lett. **69**, 848 (1996).
124. M. A. J. Verhoeven, G. J. Gerritsma, H. Rogalla, and A. A. Golubov, IEEE Trans. Appl. Supercond. **5**, 2095 (1995).
125. I. H. Song, E. H. Lee, B. M. Kim, I. Song, and G. Park, Appl. Phys. Lett. **74**, 2053 (1999).
126. L. I. Glazman and K. A. Matveev, Sov. Phys. JETP **67**, 1276 (1988).
127. Y. Xu, A. Matsuda, and M. R. Beasley, Phys. Rev. B **42**, 1492 (1990).
128. I. Takeuchi, P. A. Warburton, Z. Trajanovic, C. J. Lobb, Z. W. Dong, T. Venkatesan, M. A. Bari, W. E. Booij, E. J. Tarte, and M. G. Blamire, Appl. Phys. Lett. **69**, 112 (1996).
129. K. Kuroda, T. Takami, K. Nishi, Y. Wada, and T. Ozeki, Jpn. J. Appl. Phys. **40**, L609 (2000).
130. B. H. Moeckly and K. Char, Appl. Phys. Lett. **71**, 2526 (1997).
131. B. H. Moeckly, Appl. Phys. Lett. **78**, 790 (2001).
132. B. H. Moeckly and K. Char, Phys. Rev. B **65**, 174504 (2002).
133. J. Yoshida, H. Katsuno, K. Nakayama, and T. Nagano, Phys. Rev. B **70**, 54511 (2004).
134. C. W. Schneider, R. R. Schulz, B. Goetz, A. Schmehl, H. Bielenfeldt, H. Hilgenkamp, and J. Mannhart, Appl. Phys. Lett. **75**, 850 (1999)
135. I. Bozovic, G. Logvenov, M. A. J. Verhoeven, P. Caputo, E. Goldobin, and T. H. Geballe, Nature **422**, 873 (2000).
136. G. F. Virshup, M. E. Klusmeier-Brown, I. Bozovic, and J. N. Eckstein, Appl. Phys. Lett. **60**, 2288 (1992).
137. M. E. Klusmeier-Brown, I. Bozovic, J. N. Eckstein, and G. F. Virshup, Appl. Phys. Lett. **61**, 2806 (1992).

138. I. Bozovic, G. Logvenov, M. A. J. Verhoeven, P. Caputo, E. Goldobin, and M. R. Beasley, Phys. Rev. Lett. **93**, 157002 (2004).
139. R. S. Decca, H. D. Drew, E. Osquiguil, B. Maiorov, and J. Guimpel, Phys. Rev. Lett. **85**, 3708 (2000).
140. A. Beck, O. M. Froelich, D. Koelle, R. Gross, H. Sato, and M. Naito Appl. Phys. Lett. **68**, 3341 (1996).
141. Ariando, D. Darminto, H. -J. H. Smilde, V. Leca, D. H. A. Blank, H. Rogalla, and H. Hilgenkamp, Phys. Rev. Lett. **94**, 167001 (2005).
142. G. Hammerl, A. Schmehl, R. R. Schulz, B. Goetz, H. Bielefeldt, C. W. Schneider, H. Hilgenkamp, and J. Mannhart, Nature **407**, 162 (2000).
143. A. Schmehl, B. Goetz, R. R. Schulz, C. W. Schneider, H. Bielefeldt, H. Hilgenkamp, and J. Mannhart, Europhys. Lett. **47**, 110 (1999).
144. G. A. Daniels, A. Gurevich, and D.C. Larbalestier, Appl. Phys. Lett. **77**, 3251 (2000).
145. F. Tafuri, J. R. Kirtley, F. Lombardi, P. G. Medaglia, P. Orgiani, and G. Balestrino, Sov. J. Low Temp. Phys. **30**, 785 (2004).
146. R. Kleiner, F. Steinmeyer, G. Kunkel, and P. Müller, Phys. Rev. Lett. **68**, 2394 (1992).
147. R. Kleiner and P. Müller, Phys. Rev. B **49**, 1327 (1994).
148. A. Yurgens, D. Winkler, T. Claeson, and N. V. Savaritsky, Appl. Phys. Lett. **70**, 1769 (1997).
149. K. Inomata, S. Sato, K. Nakajima, A. Tanaka, Y. Takano, H. B. Wang, M. Nagao, H. Hatano, and S. Kawabata, Phys. Rev. Lett. **95**, 107005 (2005); X. Y. Jin, J. Lisenfeld, Y. Koval, A. Lukashenko, A. V. Ustinov, and P. Müller, Phys. Rev. Lett. **96**, 177003 (2006).
150. L. N. Bulaevskii, V. V. Kuzii, and A. A. Sobyanin, JEPT Lett. **25**, 290 (1977).
151. V. B. Geshkenbein and A. I. Larkin, JETP Lett. **43**, 395 (1986).
152. V. B. Geshkenbein, A. I. Larkin, and A. Barone, Phys. Rev. B **36**, 235 (1987).
153. A. V. Andreev, A. I. Buzdin, and R. M. Osgood, III, Phys. Rev. B **43**, 10124 (1991).
154. V. V. Ryazanov, V. A. Oboznov, A. V. Veretennikov, and A. Yu. Rusanov, Phys. Rev. B **65**, 020501(R) (2001).
155. A. Bauer, J. Bentner, M. Aprili, M. L. Della-Rocca, M. Reinwald, W. Wegscheider, and C. Strunk, Phys. Rev. Lett. **92**, 217001 (2004).
156. E. Goldobin, A. Sterck, T. Gaber, D. Koelle, and R. Kleiner, Phys. Rev. Lett. **92**, 057005 (2004).
157. E. Goldobin, D. Koelle, and R. Kleiner, Phys. Rev. B **70**, 174519 (2004).
158. T. Gaber, E. Goldobin, A. Sterck, R. Kleiner, D. Koelle, M. Siegel, and M. Neuhaus, Phys. Rev. B **72**, 054522 (2005).
159. T. M. Rice and M. Sigrist, Phys. Rev. B **55**, 14647 (1997).
160. P. Svedlindh, K. Niskanen, P. Norling, P. Nordblad, L. Lundgren, B. Lönnberg, and T. Lundström, Physica C **162–164**, 1365 (1989).
161. W. Braunisch, N. Knauf, V. Kataev, S. Neuhausen, A. Grütz, A. Kock, B. Roden, D. Khomskii, and D. Wohlleben, Phys. Rev. Lett. **68**, 1908 (1992).
162. W. Braunisch, N. Knauf, G. Bauer, A. Kock, A. Becker, B. Freitag, A. Grütz, V. Kataev, S. Neuhausen, B. Roden, D. Khomskii, D. Wohlleben, J. Bock, and E. Preisler, Phys. Rev. B **48**, 4030 (1993).
163. Ch. Heinzel, Th. Theilig, and P. Ziemann, Phys. Rev. B **48**, 3445 (1993).
164. K. Niskanen, J. Magnusson, P. Nordblad, P. Svedlindh, A. -S. Ullström, and T. Lundström, Physica B **194–196**, 1549 (1994).
165. K. N. Shrivastava, Solid State Commun. **90**, 589 (1994).
166. J. H. Xu, J. H. Miller, Jr., and C. S. Ting, Phys. Rev. B **51**, 11958 (1995).
167. V. G. Kogan, J. R. Clem, and J. R. Kirtley, Phys. Rev. B **61**, 9122 (2000).
168. E. Goldobin, D. Koelle, and R. Kleiner, Phys. Rev. B **66**, 100508(R) (2002).
169. H. Susanto, S. A. van Gils, T. P. P. Visser, Ariando, H J. H. Smilde, H. Hilgenkamp, Phys. Rev. B **68**, 104501 (2003).
170. C. S. Owen and D. J. Scalapino, Phys. Rev. **164**, 538 (1967).
171. J. R. Kirtley, K. A. Moler, and D. J. Scalapino, Phys. Rev. B **56**, 886 (1997).
172. E. Goldobin, D. Koelle, and R. Kleiner, Phys. Rev. B **67**, 224515 (2003).
173. A. Zenchuk and E. Goldobin, Phys. Rev. B **69**, 024515 (2004).
174. E. Goldobin, N. Stefanakis, D. Koelle, and R. Kleiner, Phys. Rev. B **70**, 094520 (2004).
175. C. C. Tsuei, J. R. Kirtley, G. Hammerl, J. Mannhart, H. Raffy, and Z. Z. Li, Phys. Rev. Lett. **93**, 187004 (2004).
176. H. J. H. Smilde, Ariando, D. H. A. Blank, H. Hilgenkamp, and H. Rogalla, Phys. Rev. B **70**, 024519 (2004).
177. R. R. Schulz, B. Chesca, B. Goetz, C. W. Schneider, A. Schmehl, H. Bielefeldt, H. Hilgenkamp, and J. Mannhart, Appl. Phys. Lett. **76**, 912 (2002).
178. C. C. Tsuei, J. R. Kirtley, C. C. Chi, Lock See Yu-Jahnes, A. Gupta, T. Shaw, J. Z. Sun, and M. B. Ketchen, Phys. Rev. Lett. **73**, 593 (1994).

179. J. R. Kirtley, C. C. Tsuei, M. Rupp, J. Z. Sun, L. S. Yu-Jahnes, A. Gupta, M. B. Ketchen, K. A. Moler, and M. Bhushan, Phys. Rev. Lett. **76**, 1336 (1996).
180. F. P. Rogers, Master's Thesis, MIT, Boston (1983).
181. L. N. Vu and D. J. van Harlingen, IEEE Trans. Appl. Supercond. **3**, 1918 (1993).
182. R. C. Black, A. Mathai, F. C. Wellstood, E. Dantsker, A. H. Miklich, D. T. Nemeth, J. J. Kingston, and J. Clarke, Appl. Phys. Lett. **62**, 2128 (1993).
183. J. R. Kirtley, M. B. Ketchen, K. G. Stawiasz, J. Z. Sun, W. J. Gallagher, S. H. Blanton, and S. J. Wind, Appl. Phys. Lett. **66**, 1138 (1995).
184. J. R. Kirtley, C. C. Tsuei, J. Z. Sun, C. C. Chi, L. S. Yu-Jahnes, A. Gupta, M. Rupp, and M. B. Ketchen, Nature **373**, 225 (1995).
185. C. C. Tsuei, J. R. Kirtley, M. Rupp, J. Z. Sun, A. Gupta, and M. B. Ketchen, Science **272**, 329 (1996).
186. C. C. Tsuei, J. R. Kirtley, Z. F. Ren, J. H. Wang, H. Raffy, and Z. Z. Li, Nature **387**, 481 (1997).
187. J. R. Kirtley, C. C. Tsuei, H. Raffy, Z. Z. Li, A. Gupta, J. Z. Sun, and S. Megert, Europhys. Lett. **36**, 707 (1996).
188. C. C. Tsuei and J. R. Kirtley, Phys. Rev. Lett. **85**, 182 (2000).
189. J. R. Kirtley, C. C. Tsuei, K. A. Moler, V. G. Kogan, J. R. Clem, and A. J. Turberfield, Appl. Phys. Lett. **74**, 4011 (1999).
190. T. Morooka, S. Nakayama, A. Odowara, M. Ikeda, S. Tanaka, and K. Chinone, Appl. Supercond. **9**, 3491 (1999).
191. A. Ya. Tzalenchuk, Z. G. Ivanov, S. Pehrson, T. Claeson, A. Lohmus, IEEE Trans. Appl. Supercond. **9**, 4115 (1999).
192. J. R. Kirtley, C. C. Tsuei, and K. A. Moler, Science **285**, 1373 (1999).
193. B. Chesca, K. Ehrhardt, M. Mößle, R. Straub, D. Koelle, R. Kleiner, and A. Tsukada, Phys. Rev. Lett. **90**, 057004 (2003).
194. A. Mathai, Y. Gim, R. C. Black, A. Amar, and F. C. Wellstood, Phys. Rev. Lett. **74**, 4523 (1995).
195. Y. Gim, A. Mathai, R. Black, A. Amar, and F. C. Wellstood, IEEE Trans. Appl. Supercond. **7**, 2331 (1997).
196. J. H. Miller, Jr., Q. Y. Ying, Z. G. Zou, N. Q. Fan, J. H. Xu, M. F. Davis, and J. C. Wolfe, Phys. Rev. Lett. **74**, 2347 (1995).
197. A. Sugimoto, T. Yamaguchi, and I. Iguchi, Physica C **367**, 28 (2002).
198. C. A. Copetti, F. Rüders, B. Oelze, Ch. Buchal, B. Kabius, and J. W. Seo, Physica C **253**, 63 (1995).
199. J. Mannhart, B. Mayer, and H. Hilgenkamp, Z. Phys. B: Condens. Matter **101**, 175 (1996).
200. J. Mannhart, H. Hilgenkamp, B. Mayer, Ch. Gerber, J. R. Kirtley, K. A. Moler and M. Sigrist, Phys. Rev. Lett. **77**, 2782 (1996).
201. R. G. Mints, I. Papiashvili, J. R. Kirtley, H. Hilgenkamp, G. Hammerl, and J. Mannhart, Phys. Rev. Lett. **89**, 067004 (2002).
202. J. Y. T. Wei, C. C. Tsuei, P. J. M. van Bentum, Q. Xiong, C. W. Chu, and M. K. Wu, Phys. Rev. B **57**, 3650 (1998).
203. J. M. Valles, Jr., R. C. Dynes, A. M. Cucolo, M. Gurvitch, L. F. Schneemeyer, J. P. Garno, and J. V. Waszczak, Phys. Rev. B **44**, 11986 (1991).
204. A. Di Chiara, F. Fontana, G. Peluso, and F. Tafuri, Phys. Rev. B **44**, 12026 (1991).
205. A. Di Chiara, F. Fontana, G. Peluso, and F. Tafuri, Phys. Rev. B **48**, 6695 (1993).
206. L. Ozyuzer, Z. Yosuf, J. F. Zasadzinski, T. -W. Li, D. G. Hings, and K. E. Gray, Physica C **320**, 9 (1999).
207. Z. Yusof, J. F. Zasadzinski, L. Coffey, and N. Miyakawa, Phys. Rev. B **58**, 514 (1998).
208. B. W. Hoogenboom, C. Berthod, M. Peter, Ø. Fischer and A. A. Kordyuk, Phys. Rev. B **67**, 224502 (2003).
209. K. Hasselbach, J. R. Kirtley and P. Lejay, Phys. Rev. B **46**, R5826 (1992).
210. J. H. Xu, J. H. Miller, and C. S. Ting, Phys. Rev. B **53**, 3604 (1996).
211. Y. Tanaka and S. Kashiwaya, Phys. Rev. Lett. **74**, 3451 (1995).
212. S. Kashiwaya, Y. Tanaka, M. Koyanagi, and K. Kajimura, Phys. Rev. B **53**, 2667 (1996).
213. M. B. Walker, P. Pairor, and M. E. Zhitomirsky, Phys. Rev. B **56**, 9015 (1997).
214. Ch. Bruder, Phys. Rev. B **41**, 4017 (1990).
215. J. Y. T. Wei, N. -C. Yeh, D. F. Garrigus, and M. Strasik, Phys. Rev. Lett. **81**, 2542 (1998).
216. H. Won and K. Maki, Phys. Rev. B **49**, 1397 (1994).
217. C. Panagopoulos and T. Xiang, Phys. Rev. Lett. **81**, 2336 (1998).
218. Ch. Renner, B. Revaz, J. -Y. Genoud, K. Kadowaki, and Ø. Fischer, Phys. Rev. Lett. **80**, 149 (1998).
219. N. Miyakawa, J. F. Zasadzinski, L. Ozyuzer, P. Guptasarma, D. G. Hinks, C. Kendziora, and K. E. Gray, Phys. Rev. Lett. **83**, 1018 (1999).
220. M. Franz and A. J. Millis, Phys. Rev. B **58**, 14572 (1998).
221. D. A. Bonn, P. Dosanjh, R. Liang, and W. N. Hardy, Phys. Rev. Lett. **68**, 2390 (1992).

222. K. Schlenga, W. Biberacher, G. Hectfischer, R. Kleiner, B. Schey, O. Waldmann, W. Walkenhorst, P. Müller, F. X. Reéi, H. Savary, J. Schneck, M. Brinkmann, H. Bach, K. Westerholt, Physica C **235–240**, 3273 (1994).
223. V. M. Krasnov, A. Yurgens, D. Winkler, P. Delsing, and T. Claeson, Phys. Rev. Lett. **84**, 5860 (2000).
224. M. Suzuki and T. Watanabe, Phys. Rev. Lett. **85**, 4787 (2000).
225. M. Suzuki and K. Tanabe, Jpn. J. Appl. Phys. **35**, L482 (1996).
226. T. Cren, D. Roditchev, W. Sacks, J. Klein, J. -B. Moussy, C. Deville-Cavellin, and M. Laguës, Phys. Rev. Lett. **84**, 147 (2000).
227. S. H. Pan, J. P. O'Neal, R. L. Badzey, C. Chamon, H. Ding, J. R. Engelbrecht, Z. Wang, H. Eisaki, S. Uchida, A. K. Gupta, K. -W. Ng, E. W. Hudson, K. M. Lang, and J. C. Davis, Nature **413**, 282 (2001).
228. C. Howald, P. Fournier, and A. Kapitulnik, Phys. Rev. B **64**, 100504R (2001).
229. K. M. Lang, V. Madhaven, J. E. Hoffman, E. W. Hudson, H. Eisaki, S. Uchida, and J. C. Davis, Nature **415**, 412 (2002).
230. M. Kugler, Ø. Fischer, Ch. Renner, S. Ono, and Y. Ando, Phys. Rev. Lett. **86**, 4911 (2001).
231. D. J. van Harlingen, J. E. Hilliard, B. L. T. Plourde, and B. D. Yanoff, Physica C **317–318**, 410 (1999).
232. H. J. H. Smilde, A. A. Golubov, Ariando, G. Rijnders, J. M. Dekkers, S. Harkema, D. H. A. Blank, H. Rogalla, and H. Hilgenkamp, Phys. Rev. Lett. **95**, 257001 (2005).
233. J. M. Dekkers, G. Rijnders, S. Harkema, H. J. H. Smilde, H. Hilgenkamp, H. Rogalla, and D. H. A. Blank, Phys. Lett. **83**, 5199 (2003).
234. M. B. Walker, Phys. Rev. B **53**, 5835 (1996).
235. J. R. Kirtley, C. C. Tsuei, Ariando, C. J. M. Verwijs, S. Harkema, and H. Hilgenkamp, Nat. Phys. **2**, 190 (2006).
236. M. R. Beasley, D. Lew, and R. B. Laughlin, Phys. Rev. B **49**, 12330 (1994).
237. T. -K. Ng and C. M. Varma, Phys. Rev. B **70**, 054514 (2004).
238. M. B. Walker and J. Luettmer-Strathmann, J. Low. Temp. Phys. **105**, 483 (1996).
239. K. A. Kouznetsov, A. G. Sun, B. Chen, A. S. Katz, S. R. Bahcall, J. Clarke, R. C. Dynes, D. A. Gajewski, S. H. Han, M. B. Maple, J. Giapintzakis, J. -T. Kim, and D. M. Ginsberg, Phys. Rev. Lett. **79**, 3050 (1997).
240. N. Tomita, Y. Takahashi, M. Mori, and Y. Uchida, Jpn. J. Appl. Phys., Part 2 **31**, L942 (1992).
241. J. -L. Wang, X. Y. Cai, R. J. Kelley, M. D. Vaudin, S. E. Babcock, and D. C. Larbalestier, Physica C **230**, 189 (1994).
242. Y. Takano, T. Hatano, A. Fukuyo, A. Ishii, M. Ohmori, S. Arisawa, K. Togano, and M. Tachiki, Phys. Rev. B **65**, 140513R (2002).
243. M. Tachiki, Y. Takano, and T. Hatano, Physica C **367**, 343 (2002).
244. K. Maki and S. Haas, Phys. Rev. B **67**, 020510(R) (2003).
245. Q. Li, Y. N. Tsay, M. Suenaga, G. D. Gu, and N. Koshizuka, Physica C **282–287**, 1495 (1997).
246. Y. Zhu, Q. Li, Y. N. Tsay, M. Suenaga, G. D. Gu, and N. Koshizuka Phys. Rev. B **57**, 8601 (1998).
247. Q. Li, Y. N. Tsay, M. Suenaga, R. A. Klemm, G. D. Gu, and N. Koshizuka, Phys. Rev. Lett. **83**, 4160 (1999).
248. Yu. I. Latyshev, A. P. Orlov, A. M. Nikitina, P. Monceau, and R. A. Klemm, Phys. Rev. B **70**, 094517 (2004).
249. R. A. Klemm, Philos. Mag. **85**, 801 (2005).
250. V. J. Emery, S. A. Kivelson, and O. Zachar, Phys. Rev. B **56**, 6120 (1997).
251. C. W. Schneider, W. K. Neils, H. Bielefeldt, G. Hammerl, A. Schmehl, H. Raffy, Z. Z. Li, S. Oh, J. N. Eckstein, D. J. van Harlingen, and J. Mannhart, Europhys. Lett. **64**, 489 (2003).
252. Z. -X. Shen, D. S. Dessau, B. O. Wells, D. M. King, W. E. Spicer, A. J. Arko, D. Marshall, L. W. Lombardo, A. Kapitulnik, P. Dickinson, S. Doniach, J. DiCarlo, A. G. Loeser, and C. H. Park, Phys. Rev. Lett. **70**, 1553 (1993).
253. Z. X. Shen, W. E. Spicer, D. M. King. D. S. Dessau, and B. O. Wells, Science **267**, 343 (1995).
254. J. Ma, C. Quitmann, R. K. Kelley, H. Berger, G. Margaritondo, and M. Onellion, Science **267**, 862 (1995)
255. H. Ding, M. R. Norman, J. C. Campuzano, M. Randeria, A. F. Bellman, T. Yokoya, T. Takahashi, T. Mochiku, and K. Kadowaki, Phys. Rev. B **54**, R9678 (1996).
256. S. Misra, S. Oh, D. J. Hornbaker, T. DiLuccio, J. N. Eckstein, and A. Yazdani, Phys. Rev. Lett. **89**, 087002 (2002).
257. S. Chakravarty, A. Sudbo, P. W. Anderson, and S. Strong, Science **261**, 337 (1993).
258. O. K. Andersen, A. I. Lichtenstein, O. Jepsen, and F. Paulsen, J. Phys. Chem. Solids **56**, 1573 (1995).
259. D. J. Scalapino, private communication.
260. P. J. Hirschfeld, S. M. Quinlan, and D. J. Scalapino, Phys. Rev. B **55**, 12742 (1997).
261. J. Liu, Y. Li, and C. M. Lieber, Phys. Rev. B. **49**, 6234 (1994).
262. N. Miyakawa, P. Guptasarma, J. F. Zasadzinski, D. G. Hinks, and K. E. Gray, Phys. Rev. Lett. **80**, 157 (1998).
263. G. Deutscher, Nature **397**, 410 (1999).
264. T. Nakano, N. Momono, M. Oda, and M. Ido, J. Phys. Soc. Jpn **67**, 2622 (1998).

265. N. -C. Yeh, C. -T. Chen, G. Hammerl, J. Mannhart, A. Schmehl, C. W. Schneider, R. R. Schulz, S. Tajima, K. Yoshida, D. Garrigus, and M. Strasik, Phys. Rev. Lett. **87**, 87003 (2001).
266. A. Auerbach and E. Altman, Phys. Rev. Lett. **85**, 3480 (2000).
267. A. Mourachkine, Europhys. Lett. **50**, 663 (2000).
268. Y. Dagan and G. Deutscher, Phys. Rev. Lett. **87**, 177004 (2001).
269. A. Sharoni, O. Millo, A. Kohen, Y. Dagan, R. Beck, G. Deutscher, and G. Koren, Phys. Rev. B **65**, 134526 (2002).
270. E. Farber, G. Deutscher, B. Goshunov, and M. Dressel, Europhys. Lett. **67**, 834 (2004).
271. N. S. Achsaf, D. Goldschmidt, G. Deutscher, A. Revcolevschi, and A. Vietkine, J. Low Temp. Phys. **105**, 329 (1996).
272. G. Deutscher, N. Achsaf, D. Goldschmidt, A. Revsolevschi, and A. Vietkine, Physica C **282–287**, 140 (1997).
273. A. Kohen, G. Leibovitch, and G. Deutscher, Phys. Rev. Lett. **90**, 207005 (2003).
274. A. J. Leggett, J. Phys. (Paris) Colloq. **39**, C6–1264 (1980).
275. A. J. Leggett, Prog. Theor. Phys. (Suppl.) **69**, 80 (1980).
276. A. O. Caldeira and A. J. Leggett, Phys. Rev. Lett. **46**, 211 (1981).
277. A. O. Caldeira and A. J. Leggett, Ann. Phys. (N.Y) **149**, 374 (1983).
278. R. F. Voss and R. A. Webb, Phys. Rev. Lett. **47**, 265 (1981).
279. M. H. Devoret, J. M. Martinis, and J. Clarke, Phys. Rev. Lett. **55**, 1908 (1985).
280. J. M. Martinis, M. H. Devoret, and J. Clarke, Phys. Rev. B **35**, 4682 (1987).
281. L. B. Ioffe, V. B. Geshkenbein, M. V. Feigel'man, A. L. Fauchere, and G. Blatter, Nature **398**, 679 (1999).
282. T. Timusk and B. Statt, Rep. Prog. Phys. **62**, 61 (1999).
283. D. Pines, Physica C **282**, 273 (1997)
284. A. V. Chubukov and J. Schmalian, Phys. Rev. B **57**, R11085 (1998).
285. S. -C. Zhang, Science **275**, 1089 (1997).
286. P. A. Lee and X.-G. Wen, Phys. Rev. Lett. **78**, 4111 (1997).
287. Y. J. Uemura, G. M. Luke, B. J. Strenlieb, J. H. Brewer, J. F. Carolan, W. N. Hardy, R. Kadono, J. R. Kempton, R. F. Kiefl, S. R. Kreitzman, P. Mulhern, T. M. Riseman, D. Ll. Williams, B. X. Yang, S. Uchida, H. Takagi, J. Gopalakrishnan, A. W. Sleight, M. A. Subramanian, C. L. Chien, M. Z. Cieplak, G. Xiao, V. Y. Lee, B. W. Statt, C. E. Stronach, W. J. Kossler, and X. H. Yu, Phys. Rev. Lett. **62**, 2317 (1989).
288. V. J. Emery and S. A. Kivelson, Phys. Rev. Lett. **74**, 3253 (1995).
289. H. Ding, T. Yokoya, J. C. Campuzano, T. Takahashi, M. Randeria, M. R. Norman, T. Mochiku, K. Kadowaki, and J. Giapintzakis, Nature **382**, 51 (1996).
290. J. H. Tao, F. Lu, and E. L. Wolf, Physica C **282**, 1507 (1997).
291. A. Biswas, P. Fournier, V. N. Smolyaninova, R. C. Budhani, J. S. Higgins, and R. L. Greene, Phys. Rev. B **64**, 104519 (2001).
292. S. Kleefisch, B. Welter, A. Marx, L. Alff, R. Gross, and M. Naito, Phys. Rev. B **63**, 100507(R) (2001).
293. L. Alff, Y. Krockenberger, B. Welter, M. Schonecke, R. Gross, D. Manske, and M. Naito, Nature **422**, 698 (2003).
294. Y. Tajima, M. Hikita, T. Ishii, H. Fuke, K. Sugiyama, M. Date, A. Yamagishi, A. Katsui, Y. Hidaka, T. Iwata, and S. Tsurumi, Phys. Rev. B **37**, 7956 (1988).
295. D. N. Basov, R. Liang, D. A. Bonn, W. N. Hardy, B. Dabrowski, M. Quijada, D. B. Tanner, J. P. Rice, D. M. Ginsberg, and T. Timusk, Phys. Rev. Lett. **74**, 598 (1995).
296. A. A. Abrikosov, Sov. Phys. JETP-USSR **5**, 1174 (1957).
297. M. Tinkham, *Introduction to Superconductivity* (McGraw-Hill, New York, 1975).
298. V. M. Krasnov, A. E. Kovalev, A. Yurgens, and D. Winkler, Phys. Rev. Lett. **86**, 2657 (2001).
299. N. Morozov, L. Krusin-Elbaum, T. Shibauchi, L. N. Bulaevskii, M. P. Maley, Yu. I. Latyshev, and T. Yamashita, Phys. Rev. Lett. **84**, 1784 (2000).
300. T. Shibauchi, L. Krusin-Elbaum, M. Li, M. P. Maley, and P. H. Kes, Phys. Rev. Lett. **86**, 5763 (2001).
301. L. Krusin-Elbaum, T. Shibauchi, and C. H. Mielke, Phys. Rev. Lett. **92**, 097005 (2004).
302. J. W. Loram, J. L. Luo, J. R. Cooper, W. Y. Liang, and J. L. Tallon, Physica **341–348C**, 831 (2000).
303. J. L. Tallon and J. W. Loram, Physica **349C**, 53 (2001).
304. C. C. Tsuei and T. Doderer, Eur. Phys. J. B **10**, 257 (1999).
305. Ch. Renner, B. Revaz, K. Kadowaki, I. Maggio-Aprile, and Ø. Fischer, Phys. Rev. Lett. **80**, 3606 (1998).
306. I. Maggio-Aprile, Ch. Renner, A. Erb, E. Walker, and Ø. Fischer, Phys. Rev. Lett. **75**, 2754 (1995).
307. B. W. Hoogenboom, M. Kugler, B. Revaz, I. Maggio-Aprile, Ø. Fischer and C. Renner, Phys. Rev. B **62**, 9179 (2000).
308. B. W. Hoogenboom, K. Kadowaki, B. Revaz, M. Li, Ch. Renner, and Ø. Fischer, Phys. Rev. Lett. **87**, 267001 (2001).

309. J. E. Hoffman, E. W. Hudson, K. M. Lang, V. Madhaven, H. Eisaki, S. Uchida, and J. C. Davis, Science **295**, 466 (2002).
310. S. H. Pan, E. W. Hudson, A. K. Gupta, K. -W. Ng, H. Eisaki, S. Uchida, and J. C. Davis, Phys. Rev. Lett. **85**, 1536 (2000).
311. A. J. Berlinsky, A. L. Fetter, M. Franz, C. Kallin, and P. I. Soininen, Phys. Rev. Lett. **75**, 2200 (1995).
312. K. Maki, H. Won, M. Kohmoto, J. Shiraishi, Y. Morita, and G. -F. Wang, Physica C **317–318**, 353 (1999).
313. C. Caroli, P. G. de Gennes, and J. Matricon, Phys. Lett. **9**, 307 (1964).
314. H. F. Hess, R. B. Robinson, R. C. Dynes, J. M. Valles, Jr., and J. V. Waszczak, Phys. Rev. Lett. **62**, 214 (1989).
315. Ch. Renner, A. D. Kent, Ph. Niedermann, Ø. Fischer, and F. Lévy, Phys. Rev. Lett. **67**, 1650 (1991).
316. Y. Wang and A. H. MacDonald, Phys. Rev. B **52**, R3876 (1995).
317. Y. Morita, M. Kohmoto, and K. Maki, Phys. Rev. Lett. **78**, 4841 (1997).
318. K. Yasui and T. Kita, Phys. Rev. Lett. **83**, 4168 (1999).
319. M. Franz and Z. Tešanović, Phys. Rev. Lett. **80**, 4763 (1998).
320. M. Franz and Z. Tešanović, Phys. Rev. Lett. **84**, 554 (2000).
321. J. H. Han and D.-H. Lee, Phys. Rev. Lett. **85**, 1100 (2000).
322. A. V. Balatsky, Phys. Rev. B **61**, 6940 (2000).
323. J. -I. Kishine, P. A. Lee, and X. -G. Wen, Phys. Rev. Lett. **86**, 5365 (2001).
324. C. Berthod and B. Giovannini, Phys. Rev. Lett. **87**, 277002 (2001).
325. Q. Huang, J. F. Zasadzinski, and K. E. Gray, Physica C **161**, 141 (1989).
326. T. Hasegawa, M. Nantoh, and K. Kitizawa, Jpn. J. Appl. Phys. **30**, L276 (1991).
327. H. Srikanth and A. K. Raychaudhuri, Phys. Rev. B **45**, 383 (1992).
328. J. R. Kirtley and D. J. Scalapino, Phys. Rev. Lett. **65**, 798 (1990).
329. I. K. Yanson and A. A. Vlasenko, JETP Lett.-USSR **13**, 292 (1971).
330. H. R. Zeller and I. Giaever, Phys. Rev. **181**, 789 (1969).
331. J. Halbritter, Surf. Sci. **122**, 80 (1982).
332. J. C. Phillips, Phys. Rev. B **41**, 8968 (1990).
333. P. W. Anderson and Z. Zou, Phys. Rev. Lett. **60**, 132 (1988).
334. C. M. Varma, P. B. Littlewood, S. Schmitt-Rink, E. Abrahams, and A. E. Ruckenstein, Phys. Rev. Lett. **63**, 1996 (1989).
335. J. R. Kirtley, S. Washburn, and D. J. Scalapino, Phys. Rev. B **45**, 336 (1992).
336. J. R. Kirtley, Phys. Rev. B **47**, 11379 (1993).
337. M. Grajcar, A. Plecenik, P. Seidel, and A. Pfuch, Phys. Rev. B **51**, 16185 (1995).
338. T. Walsh, Int. J. Mod. Phys. B **6**, 125 (1991).
339. J. Lesueur, L. H. Greene, W. L. Feldmann, and A. Inam, Physica C **191**, 325 (1992).
340. A. M. Cucolo, Physica C **305**, 85 (1998).
341. F. Hayashi, E. Ueda, M. Sato, K. Kurahashi, and K. Yamada, J. Phys. Soc. Jpn **67**, 3234 (1998).
342. L. Alff, S. Kleefisch, U. Schoop, M. Zittartz, T. Kemen, T. Bauch, A. Marx, and R. Gross, Eur. Phys. J. B **5**, 423 (1998).
343. L. Alff and R. Gross, Superlatt. Microstruct. **25**, 1041 (1999).
344. L. H. Greene, M. Covington, M. Aprili, E. Badica, and D. E. Pugel, Physica B **280**, 159 (2000).
345. I. Giaever and H. R. Zeller, Phys. Rev. Lett. **20**, 1504 (1968).
346. Y. Imry and Z. Ovadyahu, Phys. Rev. Lett. **49**, 841 (1982).
347. T. Walsh, J. Moreland, R. H. Ono, and T. S. Kalkur, Phys. Rev. Lett. **66**, 516 (1991).
348. R. N. Hall, J. H. Racette, and H. Ehrenreich, Phys. Rev. Lett. **4**, 456 (1960).
349. A. F. G. Wyatt, Phys. Rev. Lett. **13**, 401 (1964).
350. R. A. Logan and J. M. Rowell, Phys. Rev. Lett. **13**, 404 (1964).
351. J. M. Rowell and L. Y. L. Shen, Phys. Rev. Lett. **17**, 15 (1966).
352. J. Applebaum, Phys. Rev. Lett. **17**, 91 (1966).
353. P. W. Anderson, Phys. Rev. Lett. **17**, 95 (1966).
354. M. Covington, M. Aprili, E. Paraoanu, L. H. Greene, F. Xu, J. Zhu, and C. A. Mirkin, Phys. Rev. Lett. **79**, 277 (1997).
355. R. Krupke and G. Deutscher, Phys. Rev. Lett. **83**, 4634 (1999).
356. M. Aprili, E. Badica, and L. H. Greene, Phys. Rev. Lett. **83**, 4630 (1999).
357. R. Beck, Y. Dagan, A. Milner, A. Gerber, and G. Deutscher, Phys. Rev. B **69**, 144506 (2004).
358. M. Fogelstrom, D. Rainer, and J. A. Sauls, Phys. Rev. Lett. **79**, 281 (1997).
359. M. Fogelstrom, D. Rainer, and J. A. Sauls, Phys. Rev. B **70**, 012503 (2004).

360. A. Kaminski, S. Rosenkranz, H. M. Fretwell, J. C. Campuzano, Z. Li, H. Raffy, W. G. Cullen, H. You, C. G. Olson, C. M. Varma, and H. Höchst, Nature **416**, 610 (2002).
361. M. E. Simon and C. M. Varma, Phys. Rev. Lett. **89**, 247003 (2002).
362. S. Spielman, J. S. Dodge, L. W. Lombardo, C. B. Eom, M. M. Fejer, T. H. Geballe, and A. Kapitulnik, Phys. Rev. Lett. **68**, 3472 (1992).
363. T. W. Lawrence, A. Szöke and R. B. Laughlin, Phys. Rev. Lett. **69**, 1439 (1992).
364. M. Sigrist, D. B. Bailey, and R. B. Laughlin, Phys. Rev. Lett. **74**, 3249 (1995).
365. M. Matsumoto and H. Shiba, J. Phys. Soc. Jpn **64**, 3384 (1995).
366. M. Matsumoto and H. Shiba, J. Phys. Soc. Jpn **64**, 4867 (1995).
367. L. J. Buchholtz, M. Palumbo, D. Rainer, and J. A. Sauls, J. Low Temp. Phys. **101**, 1079 (1995).
368. M. Matsumoto and H. Shiba, J. Phys. Soc. Jpn **65**, 2194 (1996).
369. M. Sigrist, K. Kuboki, P. A. Lee, A. J. Millis, and T. M. Rice, Phys. Rev. B **53**, 2835 (1996).
370. W. Belzig, C. Bruder, and M. Sigrist, Phys. Rev. Lett. **80**, 4285 (1998).
371. M. Fogelström and S. -K. Yip, Phys. Rev. B **57**, R14060 (1998).
372. Y. Asano, Y. Tanaka, and S. Kashiwaya, Phys. Rev. B **69**, 214509 (2004).
373. M. E. Flatté and J. M. Byers, Phys. Rev. Lett. **80**, 4546 (1998).
374. J. Geerk, X. X. Xi, and G. Linker, Z. Phys. B **73**, 329 (1988).
375. I. Takeuchi, J. S. Tsai, Y. Shimakawa, T. Manako, and Y. Kubo, Physica (Amsterdam) **158C**, 83 (1989).
376. Y. Tanuma, Y. Tanaka, and S. Kashiwaya, Phys. Rev. B **64**, 214519 (2001).
377. S. Kashiwaya, T. Ito, K. Oka, S. Ueno, H. Takashima, M. Koyanagi, Y. Tanaka, and K. Kajimura, Phys. Rev. B **57**, 8680 (1998).
378. A. Biswas, P. Fournier, M. M. Qazilbash, V. N. Smolyaninova, H. Balci, and R. L. Greene, Phys. Rev. Lett. **88**, 207004 (2002).
379. M. M. Qazilbash, A. Biswas, Y. Dagan, R. A. Ott and R. L. Greene, Phys. Rev. B **68**, 024502 (2003).
380. D. H. Wu, J. Mao, S. N. Mao, J. L. Peng, X. X. Xi, T. Venkatesan, R. L. Greene, and S. M. Anlage, Phys. Rev. Lett. **70**, 85 (1993).
381. A. Andreone, A. Cassinese, A. Di Chiara, R. Vaglio, A. Gupta, and E. Sarnelli, Phys. Rev. B **49**, 6392 (1994).
382. B. Stadlober, G. Krug, R. Nemetschek, R. Hackl, J. L. Cobb, and J. T. Markert, Phys. Rev. Lett. **74**, 4911 (1995).
383. L. Alff, S. Meyer, S. Kleefisch, U. Schoop, A. Marx, H. Sato, M. Naito, and R. Gross, Phys. Rev. Lett. **83**, 2644 (1999).
384. J. A. Skinta, T. R. Lemberger, T. Greibe, and M. Naito, Phys. Rev. Lett. **88**, 207003 (2002).
385. J. R. Cooper, Phys. Rev. B **54**, R3753 (1996).
386. J. D. Kokales, P. Fournier, L. V. Mercaldo, V. V. Talanov, R. L. Greene, S. M. Anlage, Phys. Rev. Lett. **85**, 3696 (2000).
387. R. Prozorov, R. W. Giannetta, P. Fournier, and R. L. Greene, Phys. Rev. Lett. **85**, 3700 (2000).
388. N. P. Armitage, D. H. Lu, D. L. Feng, C. Kim, A. Damascelli, K. M. Shen, F. Ronning, Z. X. Shen, Y. Onose, Y. Taguchi, and Y. Tokura, Phys. Rev. Lett. **86**, 1126 (2001).
389. T. Sato, T. Kamiayama, T. Takahashi, K. Kurahashi, and K. Yamada, Science **291**, 1517 (2001).
390. H. Balci, V. N. Smolyaninova, P. Fournier, A. Biswas, and R. L Greene, Phys. Rev. B **66**, 174510 (2002).
391. H. G. Luo and T. Xiang, Phys. Rev. Lett. **94**, 027001 (2005).
392. R. B. Laughlin, Phys. Rev. Lett. **80**, 5188 (1988).
393. S. Sachdev, Rev. Mod. Phys. **75**, 913 (2003).
394. I. Martin and A. V. Balatsky, Physica C **357–360**, 46 (2001).
395. Ch. Renner and Ø. Fischer, Phys. Rev. B **51**, 9208 (1995).
396. H. L. Edwards, J. T. Markert and A. L. de Lozanne, Phys. Rev. Lett. **69**, 2967 (1992).
397. H. L. Edwards, D. J. Derro, A. L. Barr, J. T. Markert and A. L. de Lozanne, Phys. Rev. Lett. **75**, 1387 (1995).
398. H. L. Edwards, D. J. Derro, A. L. Barr, J. T. Markert, and A. L. de Lozanne, Phys. Rev. Lett. **75**, 1387 (2002).
399. D. J. Derro, E. W. Hudson, K. M. Lang, S. H. Pan, J. C. Davis, J. T. Markert, and A. L. de Lozanne, Phys. Rev. Lett. **88**, 097002 (2002).
400. E. W. Hudson, S. H. Pan, A. K. Gupta, K. -W. Ng, and J. C. Davis, Science **285**, 88 (1999).
401. A. Yazdani, C. M. Howald, C. P. Lutz, A. Kapitulnik, and D. M. Eigler, Phys. Rev. Lett. **83**, 176 (1999).
402. A. V. Balatsky, M. I. Salkola, and A. Rosengren, Phys. Rev. B **51**, 15547 (1995).
403. M. I. Salkola, A. V. Balatsky, and D. J. Scalapino, Phys. Rev. Lett. **77**, 1841 (1996).
404. S. H. Pan, E. W. Hudson, K. M. Lang, H. Eisaki, S. Uchida, and J. C. Davis, Nature **403**, 746 (2000).
405. E. W. Hudson, K. M. Lang, V. Madhavan, S. H. Pan, H. Eisaki, S. Uchida, and J. C. Davis, Nature **411**, 920 (2001).

406. M. I. Salkola, A. V. Balatsky, and J. R. Schrieffer, Phys. Rev. B **55**, 12648 (1997).
407. Q. -H. Wang and D. -H. Lee, Phys. Rev. B **67**, 020511(R) (2003).
408. L. Capriotti, D. J. Scalapino, and R. D. Sedgewick, Phys. Rev. B **68**, 014508 (2003).
409. D. Zhang and C. S. Ting, Phys. Rev. B **69**, 012501 (2004).
410. J. E. Hoffman, K. McElroy, D.-H. Lee, K. M. Lang, H. Eisaki, S. Uchida, and J. C. Davis, Science **297**, 1148 (2002).
411. C. Howald, H. Eisaki, N. Kaneko, M. Greven, and A. Kapitulnik, Phys. Rev. B **67**, 014533 (2003).
412. K. McElroy, R. W. Simmonds, J. E. Hoffman, D. -H. Lee, J. Orenstein, H. Eisaki, S. Uchida, and J. C. Davis, Nature **422**, 592 (2003).
413. J. Zaanen and O. Gunnarsson, Phys. Rev. B **40**, 7391 (1989).
414. V. J. Emery and S. A. Kivelson, Physica C **209**, 597 (1993).
415. V. J. Emery, S. A. Kivelson, and J. M. Tranquada, Proc. Natl Acad. Sci. USA **96**, 8814 (1999).
416. J. M. Tranquada, B. J. Sternlieb, J. D. Axe, Y. Nakamura, and S. Uchida, Nature (London) **375**, 561 (1995).
417. S. A. Kivelson, I. P. Bindloss, E. Fradkin, V. Oganesyan, J. M. Tranquada, A. Kapitulnik, and C. Howard, Rev. Mod. Phys. **75**, 1201 (2003).
418. M. Vershinen, S. Misra, S. Ono, Y. Abe, Y. Ando, and A. Yazdani, Science **303**, 1995 (2004).
419. A. Fang, C. Howald, N. Kaneko, M. Greven, and A. Kapitulnik, Phys. Rev. B **70**, 214514 (2004).
420. S. Misra, M. Vershinin, P. Phillips, and A. Yazdani, Phys. Rev. B **70**, 220503(R) (2004).
421. K. McElroy, D.-H. Lee, J. E. Hoffman, K. M. Lang, E. W. Hudson, H. Eisaki, S. Uchida, J. Lee, and J. C. Davis, cond-mat/0404005.
422. K. McElroy, D.-H. Lee, J. E. Hoffman, K. M. Lang, J. Lee, E. W. Hudson, H. Eisaki, S. Uchida, and J. C. Davis, Phys. Rev. Lett. **94**, 197005 (2005).
423. J. M. Rowell and L. Kopf, Phys. Rev. **137**, A907 (1965).
424. J. R. Schreiffer, D. J. Scalapino, and J. W. Wilkins, Phys. Rev. Lett. **10**, 336 (1963).
425. Q. Huang, J. F. Zasadzinski, N. Trashwala, K. E. Gray, D. G. Hinks, J. L. Peng, and R. L. Greene, Nature **347**, 369 (1990).
426. N. Tralshawala, J. F. Zasadzinski, L. Coffey, and Q. Huang, Phys. Rev. B **44**, 12102 (1991).
427. N. Tralshawala, J. F. Zasadzinski, L. Coffey, W. Gai, M. Romalis, Q. Huang, R. Vaglio, and K. E. Gray, Phys. Rev. B **51**, 3812 (1995).
428. G. Deutscher, N. Hass, Y. Yagil, A. Revcolevschi, and G. Dhalenne, J. Supercond. **7**, 371 (1994).
429. S. I. Vedeneev, A. A. Tsvetkov, A. G. M. Jansen, and P. Wyder, Physica C **235–240**, 1851 (1994).
430. R. S. Gonnelli, S. I. Vedeneev, O. V. Dolgov, and G. A. Ummarino, Physica C **235–240**, 1861 (1994).
431. B. A. Aminov, M. A. Hein, G. Müller, H. Piel, Ya. G. Ponomarev, D. Wehler, M. Boeckholt, L. Buschmann, and G. Güntherodt, Physica C **235–240**, 1863 (1994).
432. R. S. Gonnelli, D. Andreone, and G. A. Ummarino, Physica C **235–240**, 1865 (1994).
433. D. Shimada, Y. Shiina, A. Mottate, Y. Ohyagi, and N. Tsuda, Phys. Rev. B **51**, 16495 (1995).
434. Y. Shiina, D. Shimada, A. Mottate, Y. Ohyagi, and N. Tsuda, J. Phys. Soc. Jpn **64**, 2577 (1995).
435. R. Aoki and H. Murakami, J. Low Temp. Phys. **105**, 1231 (1996).
436. N. Tsuda, E. Arai, A. Mottate, T.. Kogawa, N. Numasaki, and D. Shimada, Physica C **235–240**, 1889 (1994).
437. E. L. Wolf, J. Zasadzinski, G. B. Arnold, D. F. Moore, J. M. Rowell, and M. R. Beasley, Phys. Rev. B **22**, 1214 (1980).
438. W. Reichardt, F. Gompf, M. Pintschovius, N. Pyka, B. Renker, P. Bourges, G. Collin, A. S. Ivanov, N. L. Mirofanov, and A. Yu. Rumiantsev, in *Electron–Phonon Interaction in Oxide Superconductors*, R. Baquero (ed.) (World Scientific, Singapore, 1991).
439. J. F. Zasadzinski, L. Ozyuzer, N. Miyakawa, K. E. Gray, D. G. Hinks, and C. Kendziora, Phys. Rev. Lett. **87**, 067005 (2001).
440. Y. DeWilde, N. Miyakawa, P. Guptasarma, M. Iavarone, L. Ozyuzer, J. F. Zasadzinski, P. Romano, D. G. Hinks, C. Kendziora, G. W. Crabtree, and K. E. Gray, Phys. Rev. Lett. **80**, 153 (1998).
441. Ar. Abanov and A. V. Chubukov, Phys. Rev. B **61**, R9241 (2000).
442. M. Eschrig and M. R. Norman, Phys. Rev. Lett. **85**, 3261 (2000).
443. J. C. Campuzano, H. Ding, M. R. Norman, H. M. Fretwell, M. Randeria, A. Kaminski, J. Mesot, T. Takeuchi, T. Sato, T. Yokoya, T. Takahashi, T. Mochiku, K. Kadowaki, P. Guptasarma, D. G. Hinks, Z. Konstantinovic, Z. Z. Li, and H. Raffy, Phys. Rev. Lett. **83**, 3709 (1999).
444. M. R. Norman and H. Ding, Phys. Rev. B **57**, R11089 (1998).
445. Z.-X. Shen and J. R. Schrieffer, Phys. Rev. Lett. **78**, 1771 (1997).

446. H. F. Fong, B. Keimer, D. Reznik, D. L. Milius, and I. A. Aksay, Phys. Rev. B **54**, 6708 (1996).
447. P. Dai, H. A. Mook, S. M. Hayden, G. Aeppli, T. G. Perring, R. D. Hunt, and F. Doğan, Science **284**, 1344 (1999).
448. A. V. Chubukov, D. Pines, and J. Schmalian, in *The Physics of Superconductors*, vol. I, K. H. Bennemann and J. B. Ketterson (eds.) (Springer, Berlin Heidelberg New York, 2003), p. 495.

3

Angle-Resolved Photoemission Spectroscopy on Electronic Structure and Electron–Phonon Coupling in Cuprate Superconductors

X. J. Zhou, T. Cuk, T. Devereaux, N. Nagaosa, and Z. -X. Shen

3.1. Introduction

In addition to the record high superconducting transition temperature (T_c), high temperature cuprate superconductors [1, 2] are characterized by their unusual superconducting properties below T_c, and anomalous normal state properties above T_c. In the superconducting state, although it has long been realized that superconductivity still involves Cooper pairs [3], as in the traditional BCS theory [4–6], the experimentally determined d-wave pairing [7] is different from the usual s-wave pairing found in conventional superconductors [8,9]. The identification of the pairing mechanism in cuprate superconductors remains an outstanding issue [10]. The normal state properties, particularly in the underdoped region, have been found to be at odd with conventional metals which is usually described by Fermi liquid theory; instead, the normal state at optimal doping fits better with the marginal Fermi liquid phenomenology [11]. Most notable is the observation of the pseudogap state in the underdoped region above T_c [12]. As in other strongly correlated electron systems, these unusual properties stem from the interplay between electronic, magnetic, lattice, and orbital degrees of freedom. Understanding the microscopic process involved in these materials and the interaction of electrons with other entities is essential to understand the mechanism of high temperature superconductivity.

Since the discovery of high-T_c superconductivity in cuprates [1], angle-resolved photoemission spectroscopy (ARPES) has provided key experimental insights in revealing the electronic structure of high temperature superconductors [13–15]. These include, among others, the earliest identification of dispersion and a large Fermi surface [16], an anisotropic superconducting gap suggestive of a d-wave order parameter [17], and an observation of the pseudogap in underdoped samples [18]. In the mean time, this technique itself has experienced a dramatic improvement in its energy and momentum resolutions, leading to a series of new discoveries not thought possible only a decade ago. This revolution of the ARPES technique and its scientific impact result from dramatic advances in four essential components: instrumental resolution and efficiency, sample manipulation, high quality samples, and well-matched scientific issues.

X. J. Zhou • Dept. of Physics, Applied Physics and Stanford Synchrotron Radiation Laboratory, Stanford University, Stanford, CA 94305; Advanced Light Source, Lawrence Berkeley National Lab, Berkeley, CA 94720; National Laboratory for Superconductivity, Institute of Physics & Beijing National Laboratory for Condensed Matter Physics, Chinese Academy of Sciences, Beijing 100080, China
T. Cuk • Dept. of Physics, Applied Physics and Stanford Synchrotron Radiation Laboratory, Stanford University, Stanford, CA 94305
T. Devereaux • Department of Physics, University of Waterloo, Ontario, Canada N2L 3GI
N. Nagaosa • CREST, Department of Applied Physics, University of Tokyo, Bunkyo-ku, Tokyo 113-8656, Japan
Z. -X. Shen • Dept. of Physics, Applied Physics and Stanford Synchrotron Radiation Laboratory, Stanford University, Stanford, CA 94305

The purpose of this treatise is to go through the prominent results obtained from ARPES on cuprate superconductors. Because there have been a number of recent reviews on the electronic structures of high-T_c materials [13–15], we will mainly present the latest results not covered previously, with a special attention given on the electron–phonon interaction in cuprate superconductors. What has emerged is rich information about the anomalous electron–phonon interaction well beyond the traditional views of the subject. It exhibits strong doping, momentum and phonon symmetry dependence, and shows complex interplay with the strong electron-electron interaction in these materials.

3.2. Angle-Resolved Photoemission Spectroscopy

3.2.1. Principle

Angle-resolved photoemission spectroscopy is a powerful technique for studying the electronic structure of materials (Figure 3.1) [19]. The information of interest, i.e., the energy and momentum of electrons in the material, can be inferred from that of the photoemitted electrons. This conversion is made possible through two conservation laws involved in the photoemission process:

(1) Energy conservation: $E_B = h\nu - E_{kin} - \Phi$
(2) Momentum conservation: $K_{||} = k_{||} + G$

where E_B represents the binding energy of electrons in the material; $h\nu$ the photon energy of incident light; E_{kin} the kinetic energy of photoemitted electrons; Φ work function; $k_{||}$ momentum of electrons in the material parallel to sample surface; $K_{||}$ projected component of momentum of photoemitted electrons on the sample surface which can be calculated from the kinetic energy by $\hbar K_{||} = \sqrt{2m E_{kin}} \sin\theta$ with $\hbar$ being Planck constant; G reciprocal lattice vector. Therefore, by measuring the intensity of the photoemitted electrons as a function of the kinetic energy at different emission angles, the electronic structure of the material under study, i.e., energy and momentum of electrons, can be probed directly [19].

For three-dimensional materials, the electronic structure also relies on $k_\perp$, the momentum perpendicular to the sample surface. Because of the symmetry breaking near the sample surface, the momentum perpendicular to the sample surface is not conserved. In order to obtain $k_\perp$, one has to consider the inner potential which can be obtained in various ways [19]. For strictly two-dimensional materials or quasi-two-dimensional materials such as the cuprate

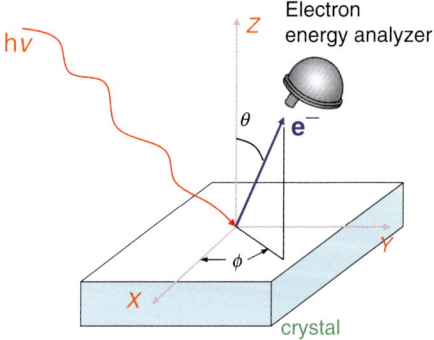

Figure 3.1. Schematic of angle-resolved photoemission spectroscopy.

superconductors discussed in this treatise, to the first approximation, one may treat $k_\perp$ as a secondary effect. However, one should always be wary about the residual three-dimensionality in these materials and its effect on photoemission data [20].

The photoemission process can be understood intuitively in terms of a "three step model" [21]: (1) excitation of the electrons in the bulk by photons, (2) transport of the excited electrons to the surface, and (3) emission of the photoelectrons into vacuum. Under the "sudden approximation" (described below), photoemission measures the single-particle spectral function $A(k, \omega)$, weighted by the matrix element M and Fermi function $f(\omega)$: $I \sim A(k,\omega)|M|^2 f(\omega)$ [22,23]. The matrix element $|M|^2$ term indicates that, besides the energy and momentum of the initial state and the final state, the measured photoemission intensity is closely related to some experimental details, such as energy and polarization of incident light, measurement geometry, and instrumental resolution. The inclusion of the Fermi function accounts for the fact that the direct photoemission measures only the occupied electronic states.

The single-particle spectral function $A(k,\omega)$ can be written in the following way using the Nambu–Gorkov formalism

$$A(k, \omega) = -(1/\pi) \mathrm{Im} G_{11}(k, \omega), \tag{3.1}$$

$$\widehat{G}(k, \omega) = \frac{Z(k, \omega) \omega \tau_0 + (\varepsilon(k) + \chi(k, \omega)) \tau_2 + \phi(k, \omega) \tau_1}{(Z(k, \omega) \omega)^2 - (\varepsilon(k) + \chi(k, \omega))^2 - \phi(k, \omega)^2}, \tag{3.2}$$

where Z, χ, and ϕ represent a renormalization due to either electron–electron or electron–phonon interactions and $\varepsilon(k)$ is the bare-band energy. τ_0, τ_1, τ_2 are the matrices and G_{11} represents the Pauli electronic charge density channel measured in photoemission. In the weak coupling case, $Z = 1$, $\chi = 0$, and $\phi = \Delta$, the superconducting gap. The same formalism can be extended to the normal state by setting $\phi = 0$. In the normal state, the spectral function can be written in a more compact way [22, 23], in terms of the real and imaginary parts of the electron self-energies $\mathrm{Re}\,\Sigma$ and $\mathrm{Im}\,\Sigma$

$$A(k, \omega) = \frac{1}{\pi} \frac{|\mathrm{Im}\,\Sigma(k, \omega)|}{(\omega - \varepsilon(k) - \mathrm{Re}\,\Sigma(k, \omega))^2 + (\mathrm{Im}\,\Sigma(k, \omega))^2}, \tag{3.3}$$

where $\mathrm{Re}\,\Sigma$ describes the renormalization of the dispersion and $\mathrm{Im}\,\Sigma$ describes the lifetime.

In relating the photoemission process in terms of single particle spectral function $A(k, \omega)$, it is helpful to recognize some prominent assumptions involved:

(1) The excited state of the sample (created by the ejection of the photoelectron) does not relax in the time it takes for the photoelectron to reach the detector. This so-called "sudden-approximation" allows one to write the final state wave-function in a separable form, $\Psi_f^N = \Phi_f^k \Psi_f^{N-1}$, where Φ_f^k denotes the photoelectron and Ψ_f^{N-1} denotes the final state of the material with $N - 1$ electrons. If the system is noninteracting, then the final state overlaps with a single eigenstate of the Hamiltonian describing the $N - 1$ electrons, revealing the band structure of the single electron. In the interacting case, the final state can overlap with all possible eigenstates of the $N - 1$ system.

(2) In the interacting case, $A(k, \omega)$ describes a "quasiparticle" picture in which the interactions of the electrons with lattice motions as well as other electrons can be treated as a perturbation to the bare band dispersion, $\varepsilon(k)$, in the form of a self-energy, $\Sigma(k, \omega)$. The validity of this picture as well as (1) rests on whether or not the spectra can be understood in terms of well-defined peaks representing poles in the spectral function.

(3) The surface is treated no differently from the bulk in this $A(k, \omega)$. In reality surface states are expected and are observed and can lead to confusion in the data interpretation [14]. Surface termination also affects photoemission process [24].

In addition to the matrix element M, there are other extrinsic effects which contribute to measured photoemission spectrum, e.g., the contribution from inelastic electron scattering. On the way to get out from inside the sample, the photoemitted electrons will experience scattering from other electrons, giving rise to a relatively smooth background in the photoemission spectrum.

3.2.2. Technique

As shown in Figure 3.1, an ARPES system consists of a light source, chamber and sample manipulation and characterization systems, and an electron energy analyzer. Figure 3.2 is an example of a modern ARPES setup with the following primary components:

(1) *Light source.* Possible light sources for angle-resolved photoemission are X-ray tubes, gas-discharge lamps, synchrotron radiation source, and VUV lasers. Among them, the synchrotron radiation source is the most versatile in that it can provide photons with continuously tunable energy, fixed or variable photon polarization, high energy resolution, and high photon flux. The latest development of the VUV laser is significant as a result of its super-high energy resolution and super-high photon flux. In addition, the lower photon energy achievable by the VUV lasers makes the measured electronic structure more bulk-sensitive in certain materials [25]. However, the strong final state effect may limit its application to certain material systems.

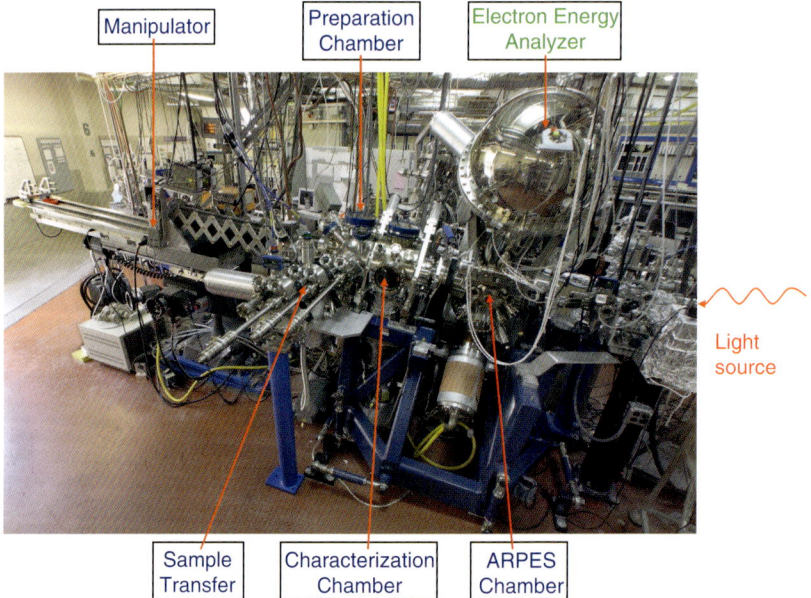

Figure 3.2. A representative ARPES system on Beamline 10.0.1 at the Advanced Light Source, Lawrence Berkeley National Lab.

Electronic Structure and Electron–Phonon Coupling in Cuprate Superconductors

(2) *Chambers and sample manipulation and characterization systems.* In most of the photon energy range commonly used (20–100 eV), the escape depth of photoemitted electrons is on the order of 5–20 Å, as shown in Figure 3.3 [26]. This means that photoemission is a surface-sensitive technique. Therefore, obtaining and retaining a clean surface during measurement is essential to probe the intrinsic electronic properties of the sample. To achieve this, the ARPES measurement chamber has to be in ultra-high vacuum, typically better than 5×10^{-11} Torr. A clean surface is usually obtained either by cleaving samples in situ in the chamber if the samples are cleavable or by sputtering and annealing process if the sample is hard to cleave. The quality of the surface can be characterized by low energy electron diffraction (LEED) or other techniques such as scanning tunneling microscopy (STM). The sample transfer system is responsible for quickly transferring samples from air to UHV chambers while not damaging the ultra-high vacuum. The manipulator is responsible for controlling the sample position and orientation, it also holds a cryostat that can change the sample temperature during the measurement. An advanced low temperature cryostat which can control the sample temperature precisely and has multiple degrees of translation and rotation freedoms is critical to an ARPES measurement.

(3) *Electron energy analyzer.* An analyzer measures the intensity of photoemitted electrons as a function of their kinetic energy, i.e., energy distribution curve (EDC), at a given angle relative to the sample orientation. The dramatic improvement of the ARPES technique in the last decade is in large part due to the advent of modern electron energy analyzer, in particular, the Scienta series hemisphere analyzers. The enhancement of the performance lies in mainly three aspects:

(a) *Energy resolution improvement.*
The energy resolution of the electron energy analyzer improves steadily over time. The upgrade of the one-dimensional multichannel detection scheme of the VSW analyzer allows efficient measurement with ∼20 meV energy resolution. Among others, it enabled the discovery of the d-wave superconducting gap structure [17]. The first introduction of the Scienta

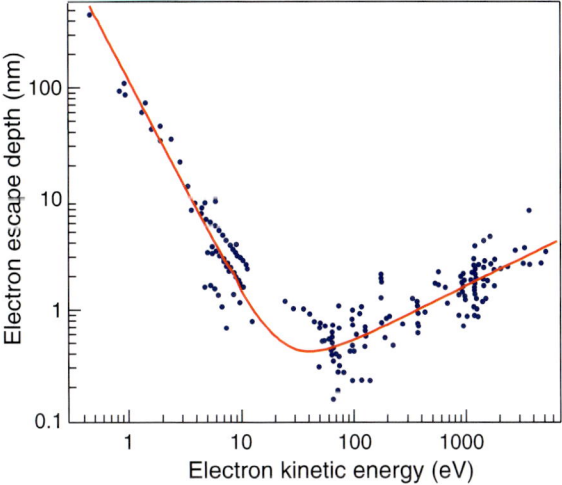

Figure 3.3. Escape depth of photoemitted electron as a function of kinetic energy [26]. For elements and inorganic compounds, the escape depth is found to follow the "universal curve" (red solid line).

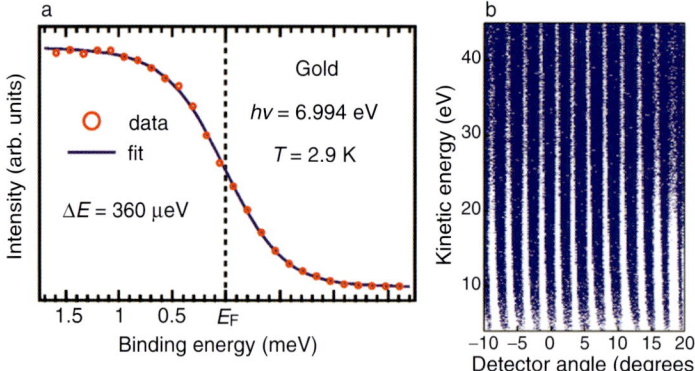

Figure 3.4. (a) Ultrahigh-resolution photoemission spectrum of an evaporated gold film measured using Scienta R4000 analyzer at a temperature of 2.9 K (red circles), together with the Fermi–Dirac function at 2.9 K convolved by a Gaussian with full width at half maximum of 360 μeV (a blue line). Total energy resolution of 360 μeV was confirmed from the very good match between the experimental and calculated spectra [25]. The energy resolution from the VUV laser is estimated to be 260 μeV. (b) Angle mode testing image of Scienta R4000 electron analyzer. The test was performed using "wire-and-slit" setup, with the angle interval between adjacent slits being 2.5°. In this particular angular mode, the analyzer collects emission angle within 30° simultaneously.

200 analyzer in the middle 1990s dramatically improved the energy resolution to better than 5 meV. The latest Scienta R4000 analyzer has improved the energy resolution further to better than 1 meV, as shown in Figure 3.4 [25].

We note that the total experimental energy resolution relies on both the analyzer resolution and the light source resolution. Sample temperature can also cause thermal broadening which is a limitation in some cases. The necessity of multiple degrees of rotation controls as well as the exposure of the surface during an ARPES measurement often puts a lower limit on the sample temperature. In addition, one should be aware of some intrinsic effects associated with the photoemission process, i.e., space charge effect and mirror charge effect [27]. When pulsed light is incident on a sample, the photoemitted electrons experience energy redistribution after escaping from the surface because of the Coulomb interaction between them (space charge effect) and between photoemitted electrons and the distribution of mirror charges in the sample (mirror charge effect). These combined Coulomb interaction effects give rise to an energy shift and a broadening whose magnitude depends on the photon energy, photon flux, beam spot size, emission angles, etc. For a typical third-generation synchrotron light source, the energy shift and broadening can be on the order of 10 meV (Figure 3.5) [27]. This value is comparable to many fundamental physical parameters actively studied by photoemission spectroscopy and should be taken seriously in interpreting photoemission data and in designing next generation experiments.

(b) *Momentum resolution.*
The introduction of the angular mode operation in the new Scienta analyzers has also greatly improved the angular resolution, from a previous ∼ 2° to 0.1°–0.3°. This improvement of the momentum resolution allows one to observe detailed structures in the band structure and Fermi surface, as well as subtle but important many-body effects. As an example, recent identification of two Fermi surface sheets (so-called "bilayer splitting") in $Bi_2Sr_2CaCu_2O_8$ (Bi2212) (Figure 3.6) is largely due to such an improvement of momentum resolution [28–30], combined with the advancement of theoretical calculations [24].

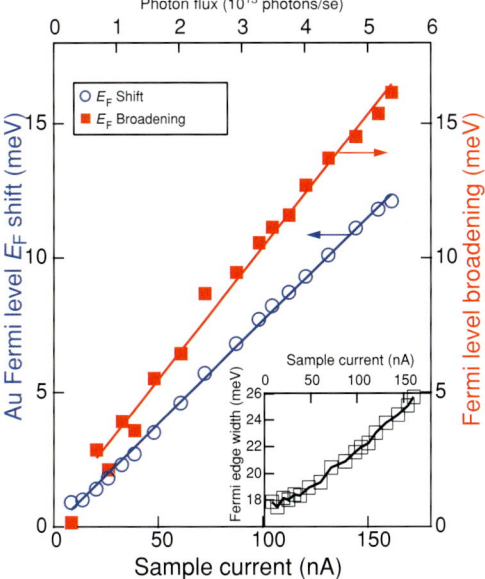

Figure 3.5. Space charge and mirror charge effects in photoemission [27]. Fermi edge broadening (solid square) and the Fermi edge shift (open circle) as a function of sample current. The beam spot size is ∼0.43×0.30 mm. The inset shows the measured overall Fermi edge width as a function of the sample current, which includes all contributions including the beamline, the analyzer, and the temperature broadening. The net broadening resulting from pulsed photons is obtained by deconvolution of the measured data, taking the width at low photon flux as from all the other contributions.

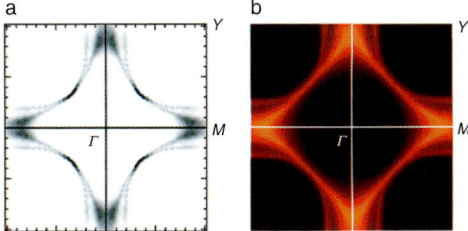

Figure 3.6. (a) Experimentally measured Fermi surface in Pb-doped Bi2212 [31]. (b) Calculated Fermi surface of Bi2212 [24].

(c) *Two-dimensional multiple angle detection.*
Traditionally, the electron energy analyzer collects one photoemission spectrum, i.e., energy distribution curve (EDC), at one measurement for each emission angle. Modern electron energy analyzers collect multiple angles simultaneously. As shown in Figure 3.4b, the latest Scienta R4000 analyzer can collect photoemitted electrons in the angle range of 30° simultaneously. Therefore, at one measurement, the raw data thus obtained, shown in Figure 3.7a, is a two-dimensional image of the photoelectron intensity (represented by false color) as a function electron kinetic energy and emission angle (and hence momentum). This two-dimensionality greatly enhances data collection efficiency and provides a convenient way of analyzing the photoemission data.

As shown in Figure 3.7, the traditional way to visualize the photoemission data is by means of so-called energy distribution curves (EDCs), which represent photoelectron intensity as a function of energy for a given momentum. The 2D image comprising the raw data is then equivalent to a number of EDCs at different momenta (Figure 3.7b). The peak position at different momenta will give the energy–momentum dispersion relation determining the real part of electron self-energy ReΣ. The EDC linewidth determines the quasiparticle lifetime, or the imaginary part of electron self-energy ImΣ. However, the EDC lineshape is usually complicated by a background at higher binding energy, the Fermi function cutoff near the Fermi level, and an undetermined bare band energy which make it difficult to extract the electron self-energy precisely.

An alternative way to visualize the 2D data is to analyze photoelectron intensity as a function of momentum for a given electron kinetic energy [32] by means of momentum distribution curves (MDCs) [33,34]. This approach provides a different way of extracting the electron self-energy. As shown in Figure 3.7c, the MDCs exhibit well-defined peaks with flat

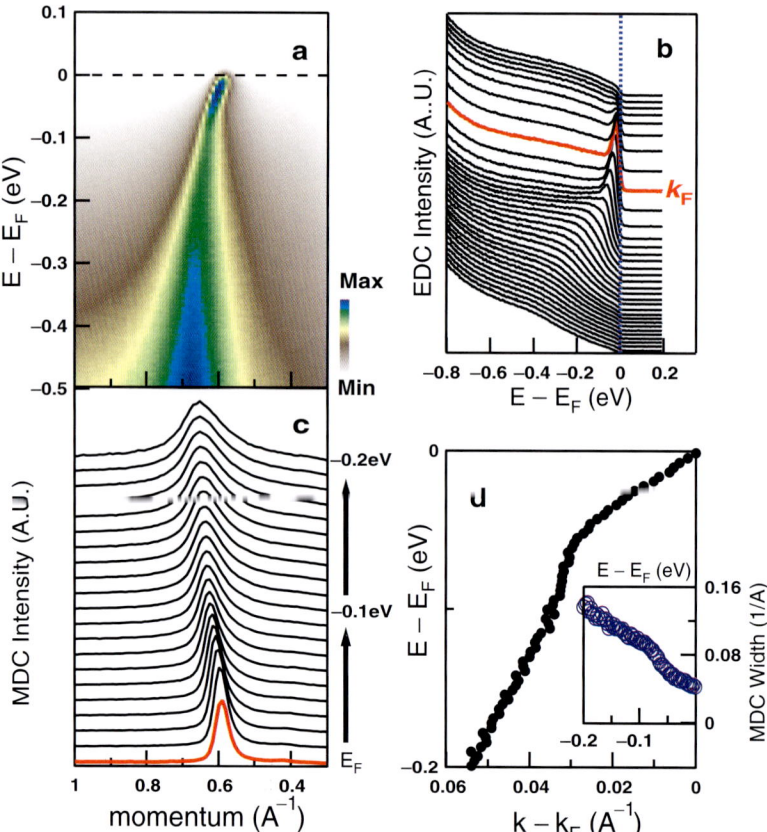

Figure 3.7. Illustration of the MDC method for extracting the electron self-energy. (a) Raw photoemission data for LSCO with x =0.063 (T_c ∼12 K) along the (0,0)–(π,π) nodal direction at 20 K [35]. The two-dimensional data represent the photoelectron intensity (denoted by false color) as a function of energy and momentum. (b) Energy distribution curves (EDCs) at different momenta. The EDC colored red corresponds to the Fermi momentum k_F. (c) Momentum distribution curves (MDCs) at different binding energies. The MDC colored red corresponds to the Fermi level. (d) Energy–momentum dispersion relation extracted by the MDC method. The inset shows the MDC width as a function of energy.

backgrounds; moreover, they can be fitted by a Lorentzian lineshape. When the bandwidth is large, the band dispersion ϵ_k can be approximated as $\epsilon_k = v_0 k$ in the vicinity of the Fermi level. Under the condition that the electron self-energy shows weak momentum dependence, $A(k,\omega)$ indeed exhibits a Lorentzian lineshape as a function of k for a given binding energy. By fitting a series of MDCs at different binding energies to obtain the MDC position $\tilde{k}$ and width Γ (full-width at half maximum, FWHM) (Figure 3.7d) [35], one can extract the electron self-energy directly as: Re $\Sigma = \hbar\omega - \tilde{k}v_0$ and Im $\Sigma = \Gamma v_0/2$.

It is worthwhile to point out the latest effort in attempting to overcome the surface sensitivity issue related with photoemission. As seen from Figure 3.3, in the usual photon energy range used for valence band photoemission, the photoemitted electron escape depth is on the order on 5–10 Å. Therefore, it is always an issue whether the photoemission results obtained in this energy range represents the bulk properties. To overcome such a problem, there have been two approaches by employing either high photon energy or lower photon energy. As seen from Figure 3.3, when the photon energy is on the order of 1 keV, the electron escape depth can be increased to ~20 Å [36]. However, this modest enhancement of the bulk sensitivity comes at a price of sacrificing both the energy resolution and momentum resolution. On the other hand, when the photon energy is low, one can see that the electron escape depth increases dramatically. Note that this "universal" curve is obtained from metals, whether the same curve can be applied to oxide materials remains unclear yet. In addition to the potential enhancement of the bulk sensitivity, one may further improve the energy and momentum resolution by going to lower photon energy.

3.3. Electronic Structures of High Temperature Superconductors

3.3.1. Basic Crystal Structure and Electronic Structure

A common structural feature of all cuprate superconductors is the CuO_2 plane (Figure 3.8a) which is responsible for the low lying electronic structure; the CuO_2 planes are sandwiched between various block layers which serve as charge reservoirs to dope CuO_2 planes [37,38]. For the undoped parent compound, such as La_2CuO_4, the valence of Cu is 2+, corresponding to $3d^9$ electronic configuration. Since the Cu^{2+} is surrounded by four oxygens in the CuO_2 plane and apical oxygen(s) or halogen(s) perpendicular to the plane, the crystal field splits the otherwise degenerate five d-orbitals, as schematically shown in Figure 3.9 [39]. The four lower energy orbitals, including xy, xz, yz, and $3z^2 - r^2$, are fully occupied, while the

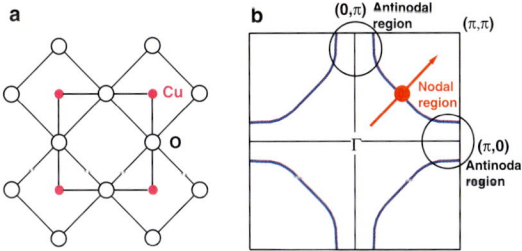

Figure 3.8. (a) Schematic of the real-space CuO_2 plane. The CuO_2 plane consists of copper (pink solid circles) and oxygen (black open circles). (b) The corresponding Brillouin zone in a reciprocal space. In the first Brillouin zone, the area near ($\pi/2$, $\pi/2$) (denoted as red circle) is referred to as nodal region, and the $(0,0)-(\pi,\pi)$ direction is the nodal direction (red arrow). The area near (π,0) and (0,π) is referred to as the antinodal region (shaded circles). The blue solid line shows a schematic Fermi surface.

orbital with highest energy, $x^2 - y^2$, is half-filled. Since the energies of the Cu d-orbitals and O 2p-orbitals are close, there is a strong hybridization between them. As a result, the topmost energy level has both Cu $d_{x^2-y^2}$ and O $2p_{x,y}$ character.

The same conclusion is also drawn from band structure calculations (Figure 3.10a) [39]. According to both simple valence counting (Figure 3.9) and band structure calculation (Figure 3.10a), the undoped parent compound is supposed to be a metal. However, strong Coulomb interactions between electrons on the same Cu site makes it an antiferromagnetic insulator with an energy gap of 2 eV [42, 43]. The basic theoretical model for the electronic structure most relevant to our discussion is the multiband Hubbard Hamiltonian [44, 45] containing d states on Cu sites, p states on O sites, hybridization between Cu–O states, hybridization between O–O states, and Coulomb repulsion terms. In terms of hole notation, i.e., starting from the filled-shell configuration ($3d^{10}$, $2p^6$) corresponding to a formal valence of Cu^{1+} and O^{2-}, the general form of the model can be written as [46]

$$H = \sum_{i\sigma} \varepsilon_d d^+_{i\sigma} d_{i\sigma} + \sum_{l\sigma} \varepsilon_p p^+_{l\sigma} p_{l\sigma} + \sum_{<li>\sigma} t_{pd} p^+_{l\sigma} d_{i\sigma} + \text{h.c.}$$
$$+ \sum_i U_d n_{i\uparrow} n_{i\downarrow} + \sum_{<ll'>\sigma} t_{O-O} p^+_{l\sigma} p_{l'\sigma} + \text{h.c.}$$
$$+ \sum_{<il>\sigma\sigma'} U_{pd} n_{l\sigma} n_{i\sigma'} + \sum_l U_p n_{l\uparrow} n_{l\downarrow}, \quad (3.4)$$

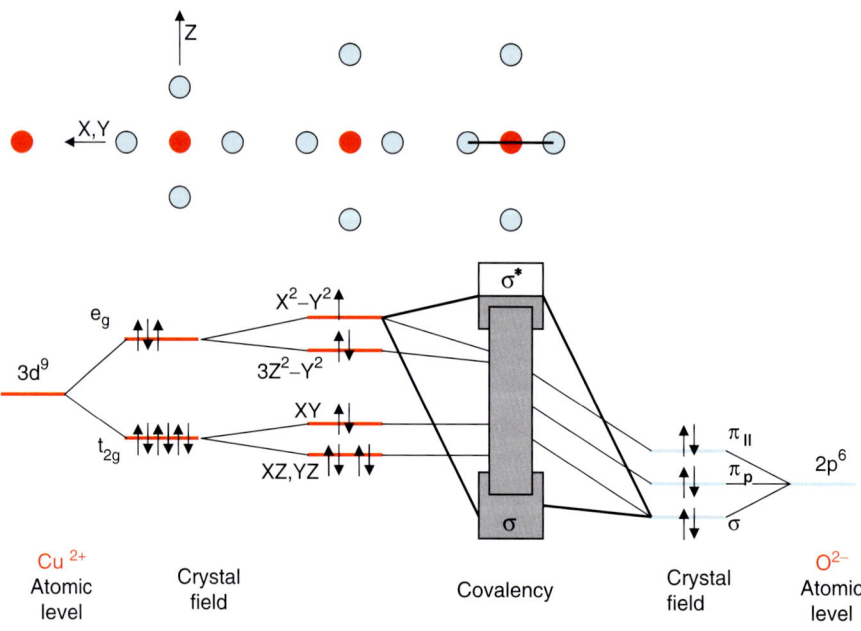

Figure 3.9. Bonding in CuO_2 plane [40]. The atomic Cu 3d level is split due to the cubic crystal field into e_g and t_{2g} states. There is a further splitting due to an octahedral crystal field into $x^2 - y^2$, $3z^2 - r^2$, xy, and xz, yz states. For divalent Cu which has nine 3d electrons, the uppermost $x^2 - y^2$ level is half filled, while all other levels are completely filled. There is a strong hybridization of the Cu states, particularly the $x^2 - y^2$ states, with the O 2p states thus forming a half-filled two-dimensional Cu $3d_{x^2-y^2}$–O $2p_{x,y}$ antibonding $dp\sigma$ band. The hybridization of the other 3d levels is smaller and is indicated in figure only by a broadening.

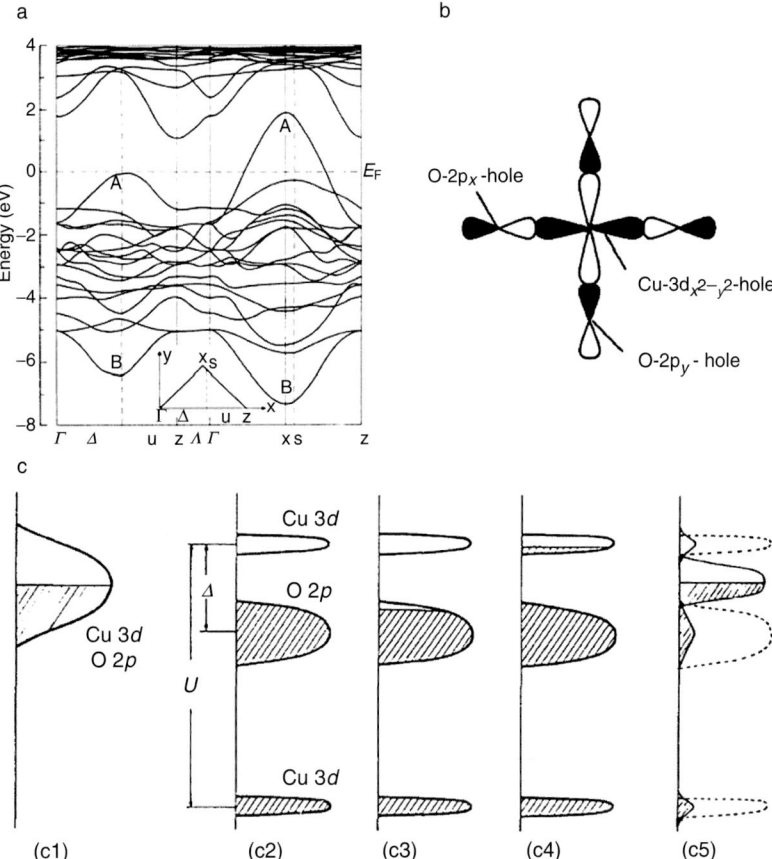

Figure 3.10. (a) LDA calculated band structure of La_2CuO_4 [41]. The band-labeled B is bonding band between Cu $3d_{x^2-y^2}$ and O 2p states while the band-labeled A is the corresponding antibonding band that is half-filled. (b) Schematic of Zhang-Rice singlet state [48, 49]. (c) Schematic energy diagrams for undoped and doped CuO_2 planes [42]. (c1) Band picture for a half-filled (undoped) CuO_2 plane (Fermi liquid). (c2) Charge-transfer insulating state of the CuO_2 plane with split Cu 3d bands due to on-site Coulomb repulsive interaction U. The O 2p band is separated by a charge transfer energy Δ from the upper Cu 3d band. (c3) and (c4) show rigid charge transfer energy bands doped with holes and electrons, respectively. (c5) Formation of mid-gap states inside the charge transfer gap.

where the operator $d_{i\sigma}^+$ creates Cu ($3d_{x^2-y^2}$) holes at site i, and $p_{l\sigma}^+$ creates O(2p) holes at the site l. U_d is the on-site Coulomb repulsion between two holes on a Cu site. The third term accounts for the direct overlap between Cu–O orbitals. The fifth terms describes direct hopping between nearest-neighbor oxygens, and U_{pd} in the sixth term is the nearest-neighbor Coulomb repulsion between holes on Cu and O atoms. Qualitatively, this model gives the energy diagram in Figure 3.10c.

Simplified versions of model Hamiltonians have also been proposed. Notably among them are the single-band Hubbard model [47] and $t-J$ model [48]. The $t-J$ Hamiltonian can be written in the following form [46, 50]

$$H_{t-J} = -t \sum_{<ij>,\sigma} (\tilde{c}_{i\sigma}^\dagger \tilde{c}_{j\sigma} + \text{h.c.}) + J \sum_{<ij>} (\mathbf{S}_i \cdot \mathbf{S}_j - \hat{n}_{i\uparrow}\hat{n}_{j\downarrow}/4), \quad (3.5)$$

where the operator $\tilde{c}_{i\sigma}^{\dagger} = c_{i\sigma}^{\dagger}(1-\hat{n}_{i-\sigma})$ excludes double occupancy, $J = 4t^2/U$ is the antiferromagnetic exchange coupling constant, and $\mathbf{S}_i$ is the spin operator. Since the hopping process may also involve the second (t') and third (t'') nearest neighbors, an extended t−J model, the $t-t'-t''-J$ model, has also been proposed [51].

3.3.2. Brief Summary of Some Latest ARPES Results

ARPES has provided key information on the electronic structure of high temperature superconductors, including the band structure, Fermi surface, superconducting gap, and pseudogap. These topics are well covered in recent reviews [14, 15] that we will not repeat here. Instead, we briefly summarize some of the latest developments not included before.

Band structure and Fermi surface: The bilayer splitting of the Fermi surface is well established in the overdoped Bi2212 [28–30], as shown in Figure 3.6 and also suggested to exist in underdoped and optimally doped Bi2212 [52–55]. Recent measurements also show that there is a slight splitting along the $(0,0)-(\pi,\pi)$ nodal direction [56]. The measurement on four-layered $Ba_2Ca_3Cu_4O_8F_2$ has identified at least two clear Fermi surface sheets [57].

Superconducting gap and pseudogap: Since the first identification of an anisotropic superconducting gap in Bi2212 [17], subsequent measurements on the superconductors such as Bi2212 [58–61], Bi2201 [62, 63], Bi2223 [64–66], $YBa_2Cu_3O_{7-\delta}$ [67], LSCO [68] have established a universal behavior of the anisotropic superconducting gap in these hole-doped superconductors which is consistent with d-wave pairing symmetry (although it is still an open question whether the gap form is a simple d-wave-like $\Delta(k) = \Delta_0[\cos(k_x a) - \cos(k_y a)]$ or higher harmonics of the expansion should be included). The measurements on electron-doped superconductors also reveal an anisotropic superconducting gap [69, 70].

One interesting issue is, if a material has multiple Fermi surface sheets, whether the superconducting gap on different Fermi surface sheets is the same. This issue traces back to superconducting $SrTiO_3$ where it was shown from tunneling measurements that different Fermi surface sheets may show different superconducting gaps [71]. With the dramatic advancement of the ARPES technique, different superconducting gaps on different Fermi surface sheets have been observed in 2H-$NbSe_2$ [72] and MgB_2 [73]. For high-T_c materials, Bi2212 shows two clear FS sheets, but no obvious difference of the superconducting gas has been resolved [61]. In $Ba_2Ca_3Cu_4O_8F_2$, it has been clearly observed that the two Fermi surface sheets have different superconducting gaps [57].

Time reversal symmetry breaking: It has been proposed theoretically that, by utilizing circularly polarized light for ARPES, it is possible to probe time-reversal symmetry breaking that may be associated with the pseudogap state in the underdoped samples [74,75]. Kaminski et al. first reported the observation of such an effect [76]. However, this observation is not reproduced by another group [77] and the subject remain controversial [78].

3.4. Electron−Phonon Coupling in High Temperature Superconductors

The many-body effect refers to interactions of electrons with other entities, such as other electrons, or collective excitations like phonons, magnons, and so on. It has been recognized from the very beginning that many-body effects are key to understanding cuprate physics.

Due to its proximity to the antiferromagnetic Mott insulating state, electron−electron interactions are extensively discussed in the literature [14, 15]. In this treatise, we will mostly review the recent progress in our understanding of electrons interacting with bosonic modes, such as phonons. This progress stems from improved sample quality, instrumental resolution, as well as theoretical development. In a complex system like the cuprates, it is not possible to isolate various degrees of freedom as the interactions mix them together. We will discuss the electron−boson interactions in this spirit, and will comment on the interplay between electron−phonon and electron−electron interactions whenever appropriate. Here by bosonic modes, we are referring to collective modes with sharp collective energy scale such as the optical phonons and the famous magnetic resonance mode seen in some cuprates [79–81], but not the broad excitation spectra such as those from the broad electron/spin excitations as these issues have been discussed in previous reviews. Furthermore, we believe the effects due to sharp mode coupling seen in cuprates are caused by phonons rather than the magnetic resonance. Our reason for not attributing the observed effect to magnetic resonance will become apparent from the rest of the manuscript. With more limited data, other groups have taken the view that the magnetic resonance is the origin of the boson coupling effect. For this reason, we will focus more on our own results in reviewing the issues of electron−phonon interaction in cuprates.

The electron−phonon interactions can be characterized into two categories (1) weak coupling where one can still use the perturbative self-energy approach to describe the quasiparticle and its lifetime and mass and (2) Strong coupling and polaron regime where this picture breaks down.

3.4.1. Brief Survey of Electron−Phonon Coupling in High-Temperature Superconductors

It is well-known that, in conventional superconductors, electron−phonon (el−ph) coupling is responsible for the formation of Cooper pairs [4]. The discovery of high temperature superconductivity in cuprates was actually inspired by possible strong electron−phonon interaction in oxides owing to polaron formation or in mixed-valence systems [1]. However, shortly after the discovery, a number of experiments lead some people to believe that electron−phonon coupling may not be relevant to high temperature superconductivity. Among them are [82]:

(1) *High critical transition temperature T_c*

So far, the highest T_c achieved is 135 K in $HgBa_2Ca_2Cu_3O_8$ [83] at ambient pressure and ∼160 K under high pressure [84]. Such a high T_c was not expected in simple materials using the strongly coupled version of BCS theory or the McMillan equations.

(2) *Small isotope effect on T_c.*

It was found that the isotope effect in optimally doped samples is rather small, much less than that expected for strongly coupled phonon-mediated superconductivity [85].

(3) *Transport measurement.*

The linear resistivity−temperature dependence in optimally doped samples and the lack of a saturation in resistivity over a wide temperature range have been taken as an evidence of weak electron−phonon coupling in the cuprate superconductors [86].

(4) *d-wave symmetry of the superconducting gap.*

It is generally believed that electron−phonon coupling is favorable to s-wave coupling.

(5) *Structural instability*.

It is generally believed that sufficiently strong electron–phonon coupling to yield high T_c will result in structural instability [87].

Although none of these observations can decisively rule out the electron–phonon coupling mechanism in high-T_c superconductors, overall they suggest looking elsewhere. Instead, strong electron–electron correlation has been proposed to be the mechanism of high-T_c superconductivity [88]. This approach is attractive since d-wave pairing is a natural consequence. Furthermore, the high temperature superconductors evolve from antiferromagnetic insulating compounds where the electron–electron interactions are strong [8, 9].

However, there is a large body of experimental evidence also showing strong electron–phonon coupling in high-temperature superconductors [89–91]. Among them are:
(1) *Isotope effect*.

As seen in Figure 3.11, although at the optimal doping, the oxygen isotope effect on T_c is indeed small, it gets larger and becomes significant with reduced doping [93]. In particular, near the "1/8" doping level, the isotope effect in $(La_{2-x}Sr_x)CuO_4$ and $(La_{2-x}Ba_x)CuO_4$ is anomalously strong, which is related to the structural instability [94]. Furthermore, the measurement of an oxygen isotope effect on the in-plane penetration depth also suggests the importance of lattice vibration for high-T_c superconductivity [95].
(2) *Optical spectroscopy and Raman scattering*.

Raman scattering [96] and infrared spectroscopy [97] reveal strong electron–phonon interaction for certain phonon modes. Some typical vibrations related to the in-plane and apical oxygens are depicted in Figure 3.12. In $YBa_2Cu_3O_{7-\delta}$, it has been found that, the B_{1g} phonon, which is related to the out-of-plane, out-of-phase, in-plane oxygen vibrations

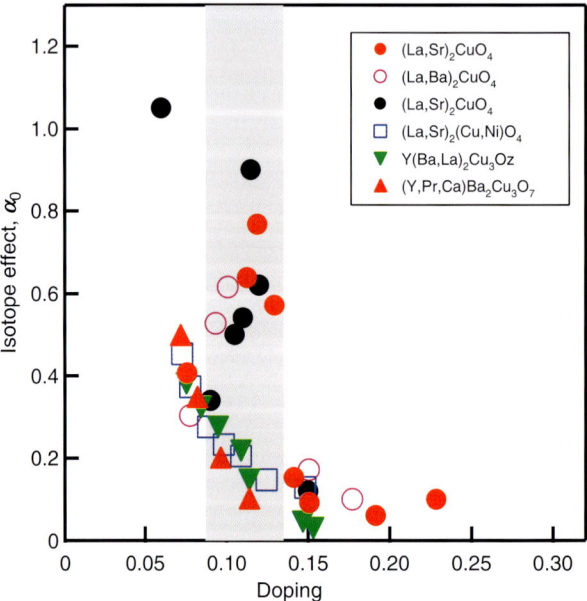

Figure 3.11. Doping dependence of the oxygen isotope effect α_0 on T_c in several classes of cuprates [92–94]. The "1/8 anomaly" data found in LSCO system is highlighted in the shaded region.

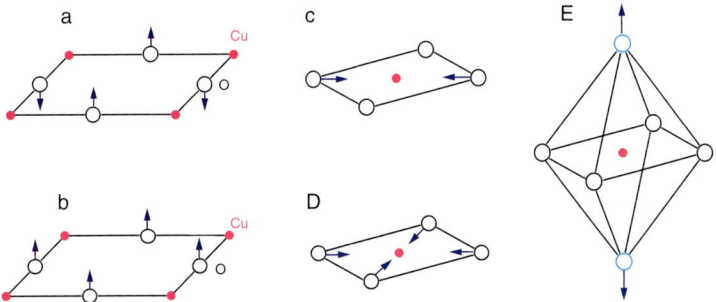

Figure 3.12. Schematic of B_{1g} mode (a), A_{1g} mode (b), half-breathing mode (c), full-breathing mode (d) and apical oxygen mode (e).

(see Figure 3.12), exhibits a Fano-like lineshape (Figure 3.13) and shows an abrupt softening upon entering the superconducting state [98–100]. The A_{1g} modes, as found in HgBa$_2$Ca$_3$Cu$_4$O$_{10}$ (Hg1234) [101] and in HgBa$_2$Ca$_2$Cu$_3$O$_8$ (Hg1223) [102], exhibit especially strong superconductivity-induced phonon softening (Figure 3.14). Infrared reflectance measurements on various cuprates found that the frequency of the Cu–O stretching mode in the CuO$_2$ plane is very sensitive to the distance between copper and oxygen [97].

Figure 3.15 shows Raman data as a function of doping in LSCO [103]. The sharp structures at high frequency are signals from multiphonon processes, which can only occur if the electron–phonon interaction is very strong. One can see that this effect is very strong in undoped and deeply underdoped regime, and gets weaker with doping increase.

(3) *Neutron scattering*.

Neutron scattering measurements have provided rich information about electron–phonon coupling in high temperature superconductors [104–106]. As seen from Figure 3.16a, the in-plane "half-breathing" mode exhibits strong frequency renormalizations upon doping along (001) direction [104, 107]. In (La$_{1.85}$Sr$_{0.15}$)CuO$_4$, it is reported that, at low temperature, the half-breathing mode shows a discontinuity in dispersion (Figure 3.16b) [108]. In YBCO, neutron scattering indicates that the softening of the B_{1g} mode upon entering the superconducting state is not just restricted near $q = 0$, as indicated by Raman scattering (Figure 3.13), but can be observed in a large part of the Brillouin zone (Figure 3.17) [106].

(4) *Material and structural dependence*.

There is a strong material and structural dependence to the high-T_c superconductivity, as exemplified in Figure 3.18 [109, 110]. Empirically it is found that, for a given homologous series of materials, the optimal T_c varies with the number of adjacent CuO$_2$ planes, n, in a unit cell: T_c goes up first with n, reaching a maximum at $n = 3$, and goes down as n further increases. For the cuprates with the same number of CuO$_2$ layers, T_c also varies significantly among different classes. For example, the optimal T_c for one-layered (La$_{2-x}$Sr$_x$)CuO$_4$ is 40 K while it is 95 K for one-layered HgBa$_2$CuO$_4$. These behaviors are clearly beyond simplified models that consider CuO$_2$ planes only, such as the t–J model. In fact, such effects were taken as evidence against theoretical models based on such simple models and in favor of the interlayer tunneling model [111]. Although the interlayer tunneling model has inconsistencies with some experiments, the issue that the material dependence cannot be explained by single band Hubbard and t–J model remains to be true.

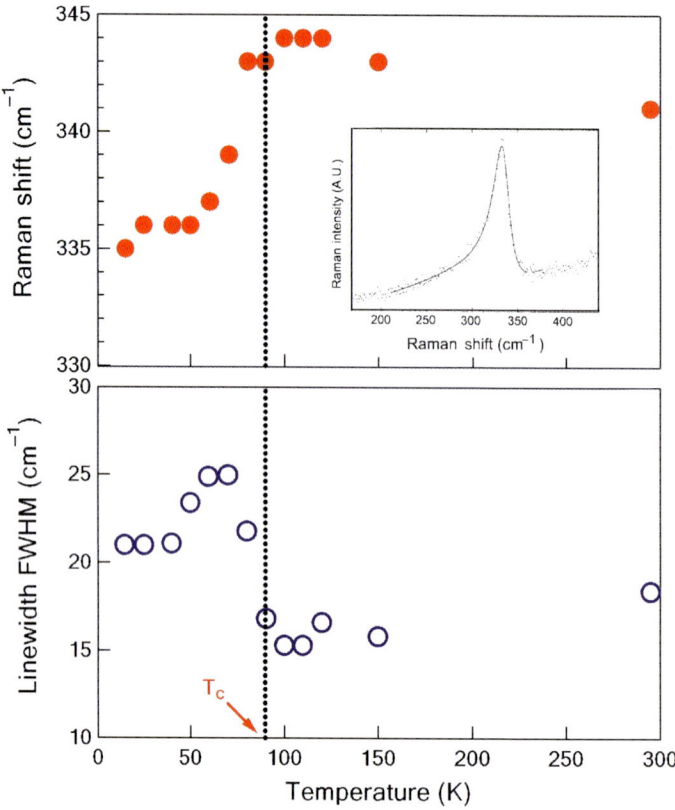

Figure 3.13. Anomalous softening of the B_{1g} phonon when YBCO is cooled below T_c [100]. The inset shows the fit of a Fano function to the phonon peak at $T = 72$ K [98].

The above results suggest that the lattice degree of freedom plays an essential role. However, the role of phonons has not been scrutinized as much, in particular in regard to the intriguing question of whether high-T_c superconductivity involves a special type of electron–phonon coupling. In other words, the complexity of electron–phonon interaction has not been as carefully examined as some of the electronic models. As a result, many naive arguments are used to argue against electron–phonon coupling as if the conclusions based on simple metals are applicable here. Recently, a large body of experimental results from angle-resolved photoemission, as we review below, suggest that electron–phonon coupling in cuprates is not only strong but shows behaviors distinct from conventional electron–phonon coupling. In particular, the momentum dependence and the interaction between electron–phonon interaction and electron–electron interaction are very important.

3.4.2. Electron–Phonon Coupling: Theory

General

Theory of electron–phonon interaction in the presence of strong electron correlation has not been developed. Given both interactions are important in cuprates, it is difficult a priori to have a good way to address these issues. In fact, we believe that an important outcome of

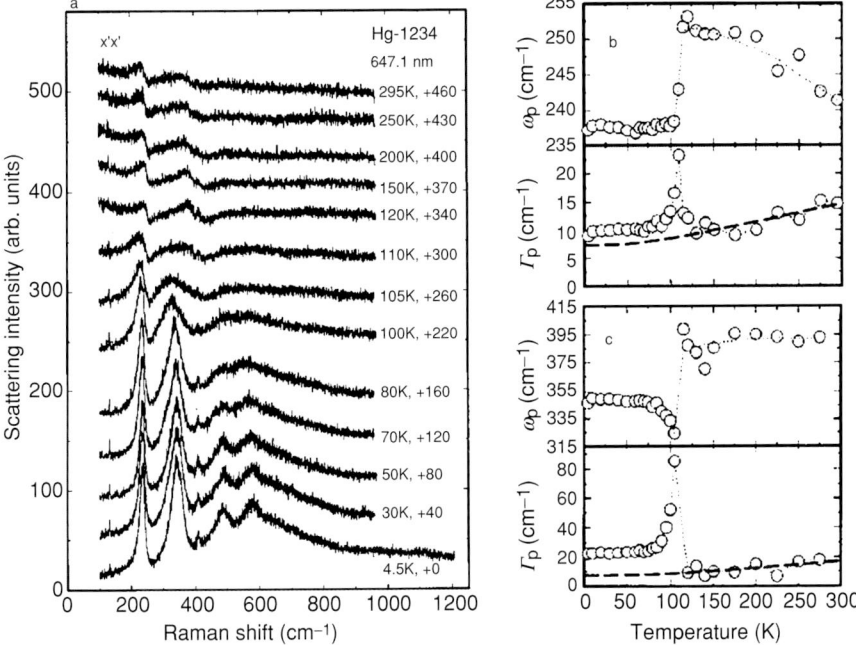

Figure 3.14. Raman spectra of Hg1234 showing a giant superconductivity-induced mode softening across T_c = 123 K [101]. The modes at 240 and 390 cm^{-1} correspond to A_{1g} out-of-plane, in-phase vibration of oxygens in the CuO$_2$ planes. Upon cooling from room temperature to 4.5 K, the 240 cm^{-1} A_{1g} mode shows an abrupt drop in frequency at T_c from 253 to 237 cm^{-1} and the 390 cm^{-1} mode drops from 395 to 317 cm^{-1} [101].

our research is the stimulus to develop such a theory. In the mean time, our strategy is to separate the problem in different regimes and see to what extent we can develop a heuristic understanding of the experimental data. Such empirical findings can serve as a guide for comprehensive theory. We now start our discussion with an overview of existing theories of electron–phonon physics.

The theories of electron–phonon coupling in condensed matter have been developed rather separately for metals and insulators. In the former case, the dominant energy scale is the kinetic energy or the Fermi energy ε_F on order of 1–10 eV, and the phonon frequency $\Omega \sim$ 1–100 meV is much smaller. The Fermi degeneracy protects the many-body fermion system from perturbations and only the small energy window near the Fermi surface responds. Therefore even if the lattice relaxation energy $E_{LR} = g^2/\omega$ for the localized electron is comparable to the kinetic energy ε_F the el–ph coupling is essentially weak and the perturbative treatment is justified. The dimensionless coupling constant λ is basically the ratio of E_{LR}/ε_F, which ranges $\lambda \cong$ 0.1–2 in the usual metals. In the diagrammatic language, the physics described above is formulated within the framework of the Fermi liquid theory [112]. The el–el interaction is taken care of by the formation of the quasiparticle, which is well defined near the Fermi surface, and the el–ph vertex correction is shown to be smaller by the factor of Ω/ε_F and can be neglected. Therefore the multiphonon excitations are reduced and the single-loop approximation or at most the self-consistent Born approximation is enough to capture the physics well, i.e., Migdal–Eliashberg formalism.

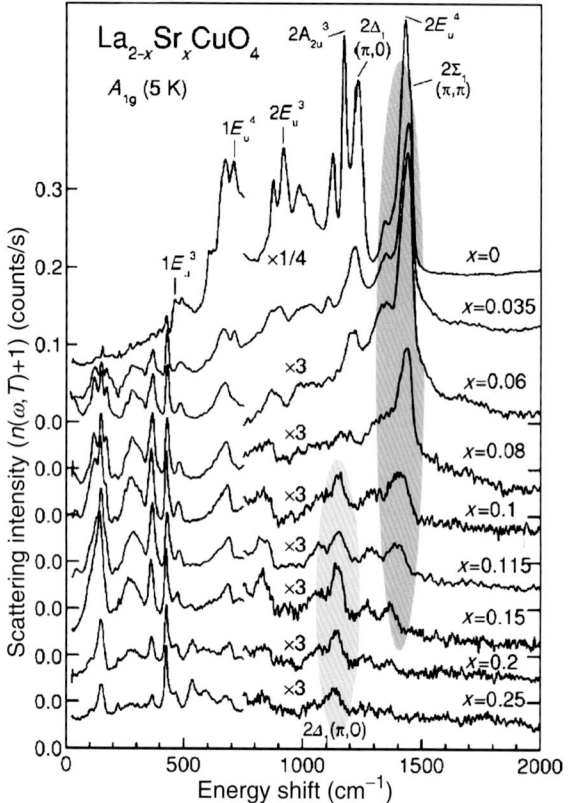

Figure 3.15. A_{1g} two-phonon Raman spectra in LSCO at different dopings. The dark gray area indicates that the two-phonon peak of the (π,π) LO mode is strong and the light gray area indicates that the two-phonon peak of the $(\pi,0)$ LO mode is strong [103].

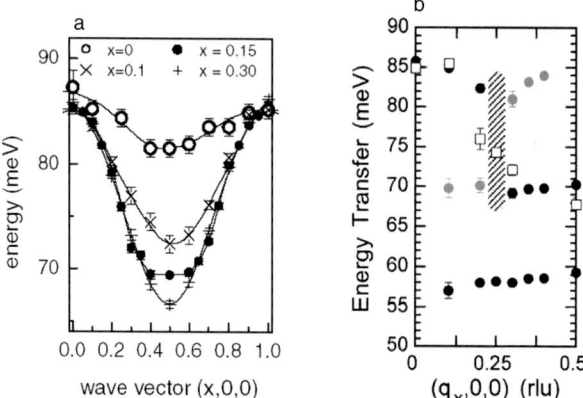

Figure 3.16. (a) Dispersion of the Cu–O bond-stretching vibrations in the (100)-direction in $(La_{2-x}Sr_x)CuO_4$ [104]. (b) Anomalous dispersion of LO phonons in $La_{1.85}Sr_{0.15}CuO_4$. A 10 K data are filled circles and room temperature data are empty squares. Gray shaded circles indicate the frequency of the weak extra branch seen at 10 K [108].

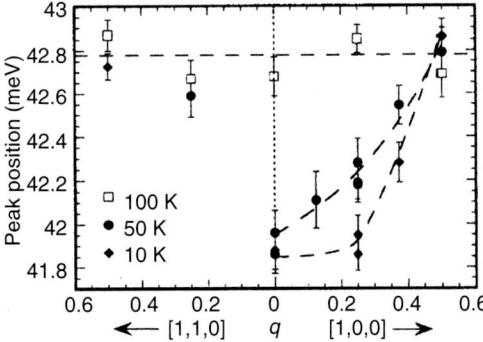

Figure 3.17. q dependence of B_{1g} mode peak position at different temperatures in YBCO. Dashed lines are guides to the eye [106].

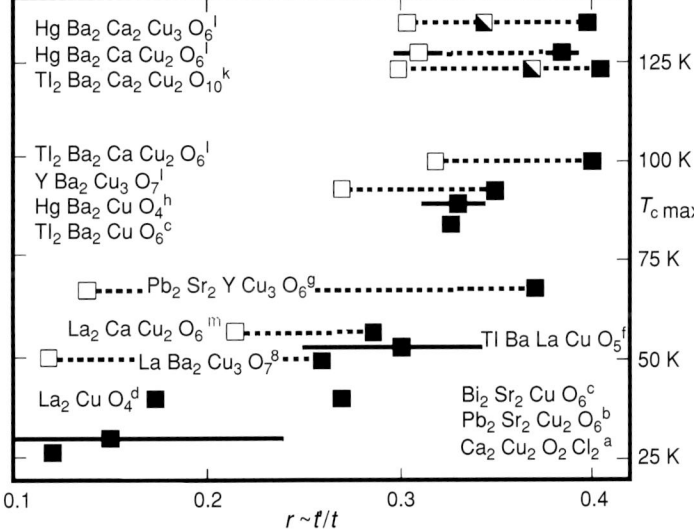

Figure 3.18. Correlation between calculated range parameter r and observed T_{cmax} where r is controlled by the energy of the axial orbital, a hybrid between Cu 4s, apical-oxygen $2p_z$, and farther orbitals [109]. Filled squares: single-layer materials and most bonding subband for multilayers. Empty squares: most antibonding subband. Half-filled squares: nonbonding subband. Dotted lines connect subband values. Bars give k_z dispersion of r in primitive tetragonal materials. For reference for a–m, refer to [109].

When a carrier is put into an insulator, on the other hand, it stays near the bottom of the quadratic dispersion and its velocity is very small. The kinetic energy is much smaller than the phonon energy, and the carrier can be dressed by a thick phonon cloud and its effective mass can be very large. This is called the phonon polaron. Historically the single carrier problem coupled to the optical phonon through the long range Coulomb interaction, i.e., Fröhlich polaron, is the first studied model, which is defined in the continuum. When one considers the tight-binding models, which is more relevant to the Bloch electron, the bandwidth W plays the role of ε_F in the above metallic case. Then again we have three energy scales, W, E_{RL}, and Ω. Compared with the metallic case, the dominance of the kinetic energy is not trivial, and the competition between the itinerancy and the localization is the key issue in the polaron problem, which is controlled by the dimensionless coupling constant $\lambda = E_{RL}/W$. Another dimensionless coupling constant is $S = E_{RL}/\Omega$, which counts the number of phonon quanta in the phonon cloud around the localized electron. This appears in the overlap integral of the two-phonon wavefunctions with and without the phonon cloud as:

$$\langle \text{phonon vacuum}|\text{phonon cloud}\rangle \propto e^{-S}. \quad (3.6)$$

This factor appears in the weight of the zero-phonon line of the spectral function of the localized electron and S can be regarded as the maximum value for the number of phonons N_{ph} near the electron. In a generic situation, N_{ph} is controlled by λ, and there are cases where N_{ph} shows an (almost) discontinuous change from the itinerant undressed large polaron to the heavily dressed small polaron as λ increases. This is called the self-trapping transition. Here a remark on the terminology "self-trapping" is in order. Even for the heavy mass polaron, the ground state is the extended Bloch state over the whole sample and there is no localization. However, a small amount of disorder can cause the localization. Therefore in the usual situation, the formation of the small polaron implies the self-trapping, and we use this language to represent the formation of the thick phonon clouds and huge mass enhancement. In cuprates, it is still a mystery why the transport properties of the heavily underdoped samples do not show the strong localization behavior even though the ARPES shows the small polaron formation as will be discussed in "Polaronic Behavior in Parent Compounds" in Section 3.4.4.

Now the most serious question is what is the picture for the el–ph coupling in cuprates? The answer seems not so simple, and depends both on the hole-doping concentration, momentum, and energy. The half-filled undoped cuprate is a Mott insulator with antiferromagnetic ordering, and a single hole doped into it can be regarded as the polaron subjected to the hole–magnon and hole–phonon interactions. At finite doping, but still in the antiferromagnetic (AF) order, the small hole pockets are formed and the hole kinetic energy can be still smaller than the phonon energy. In this case the polaron picture still persists. The main issue is to what range this continues. One scenario is that once the antiferromagnetic order disappears the metallic Fermi surface is formed and the system enters the Migdal–Eliashberg regime. However, there are several physical quantities such as the resistivity, Hall coefficient, optical conductivity, which strongly suggest that the physics still bears a strong characteristics of doped holes in an insulator rather than a simple metal with large Fermi surface. Therefore the crossover hole concentration x_c between the polaron picture and the Migdal–Eliashberg picture remains an open issue. Probably, it depends on the momentum/energy of the spectrum. For example, the electrons have smaller velocity and are more strongly coupled to the phonons in the antinodal region near $(\pm\pi, 0)$, $(0, \pm\pi)$, remaining polaronic up to higher doping, while in the nodal region, the electrons behave more like the conventional metallic ones since the velocity is large along this direction. Furthermore, the low energy states near the Fermi energy are well described by Landau's quasiparticle and Migdal–Eliashberg theory, while the higher energy states do not change much with doping even at $x \cong 0.1$ [113] suggestive of polaronic behavior. In any event, the dichotomy between the hole doping picture and the metallic (large) Fermi surface picture is the key issue in the research of high-T_c superconductors.

Weak Coupling—Perturbative and Self-Energy Description

We review first the Migdal–Eliashberg regime, in which the electron–phonon interaction results in single-phonon excitations and can be considered as a perturbation to the bare band dispersion. In this case, dominant features of the mode-coupling behavior can be

Electronic Structure and Electron–Phonon Coupling in Cuprate Superconductors

captured using the following form for the self-energy:

$$\widehat{\Sigma}(k,\omega) = T/N \sum_{q,\nu} g^2(k,q) D(q, i\nu) \tau_3 \widehat{G}(k-q, i\omega - i\nu) \tau_3, \quad (3.7)$$

where $D(q,\omega) = 2\Omega_q/(\omega^2 - \Omega_q^2)$ is the phonon propagator, Ω_q is the phonon energy, T is temperature, N is the number of particles, and τ_3 is the Pauli matrix.

In this form of the self-energy, corrections to the electron–phonon vertex, g, are neglected as mentioned above [115]. Furthermore, we assume only one-iteration of the coupled self-energy and Green's function equations. In other words, in the equation for the self-energy, Σ, we assume bare electron and phonon propagators, G_0 and D_0. With these assumptions, the imaginary parts of the functions Z, χ, and ϕ, denoted as Z_2, χ_2, and ϕ_2, are:

$$Z_2(k,\omega)\omega = \sum_q g^2(k,q)(\pi/2)\{[\delta(\omega - \Omega_q - E_{k-q})$$
$$+ \delta(\omega - \Omega_q + E_{k-q})][f(\Omega_q - \omega) + n(\Omega_q)]$$
$$+ [-\delta(\omega + \Omega_q - E_{k-q}) - \delta(\omega + \Omega_q + E_{k-q})][f(\Omega_q + \omega) + n(\Omega_q)]\}, \quad (3.8)$$

$$\chi_2(k,\omega) = \sum_q g^2(k,q)(\pi \varepsilon_{k-q}/2E_{k-q})\{[-\delta(\omega - \Omega_q - E_{k-q})$$
$$+ \delta(\omega - \Omega_q + E_{k-q})][f(\Omega_q - \omega) + n(\Omega_q)]$$
$$+ [\delta(\omega + \Omega_q - E_{k-q}) - \delta(\omega + \Omega_q + E_{k-q})][f(\Omega_q + \omega) + n(\Omega_q)]\}, \quad (3.9)$$

$$\phi_2(k,\omega) = \sum_q g^2(k,q)(\pi \Delta_{k-q}/2E_{k-q})\{[\delta(\omega - \Omega_q - E_{k-q})$$
$$- \delta(\omega - \Omega_q + E_{k-q})][f(\Omega_q - \omega) + n(\Omega_q)]$$
$$+ [-\delta(\omega + \Omega_q - E_{k-q}) + \delta(\omega + \Omega_q + E_{k-q})][f(\Omega_q + \omega) + n(\Omega_q)]\}, \quad (3.10)$$

where $f(x)$, $n(x)$, are the Fermi, Bose distribution functions and E_k is the superconducting state dispersion, $E_k^2 = \varepsilon_k^2 + \Delta_k^2$.

The above equations are essentially those of Eliashberg theory for strongly coupled superconductors. Although λ can be large (> 1), i.e., "strongly coupled," the vertex corrections and multiphonon processes are still negligible due to the Fermi degeneracy and small Ω/E_F [116]. To illustrate the essential features of mode coupling, we consider an Einstein phonon coupled isotropically to a parabolic band. We present this calculation in the spirit of Engelsberg and Schrieffer, who first calculated the spectral function for an electron–phonon coupled system [117] and which provided the foundation for the later work by Scalapino, Schrieffer, and Wilkins [118] in the superconducting state. Figure 3.19 plots $-Z_2\omega + \chi_2$, the imaginary part of the phonon self-energy, Im Σ, that represents the renormalization to the diagonal channel of the electron propagator, or the one in which the charge number density is subjected to electron–phonon interactions. This part of the self-energy gives a finite lifetime to the electron, and consequently broadens the peak in the spectra (Im Σ in $A(k,\omega)$ (Eq. 3.3) is the half-width-at-half-maximum, HWHM of the peak). In the normal state, $-Z_2\omega + \chi_2$

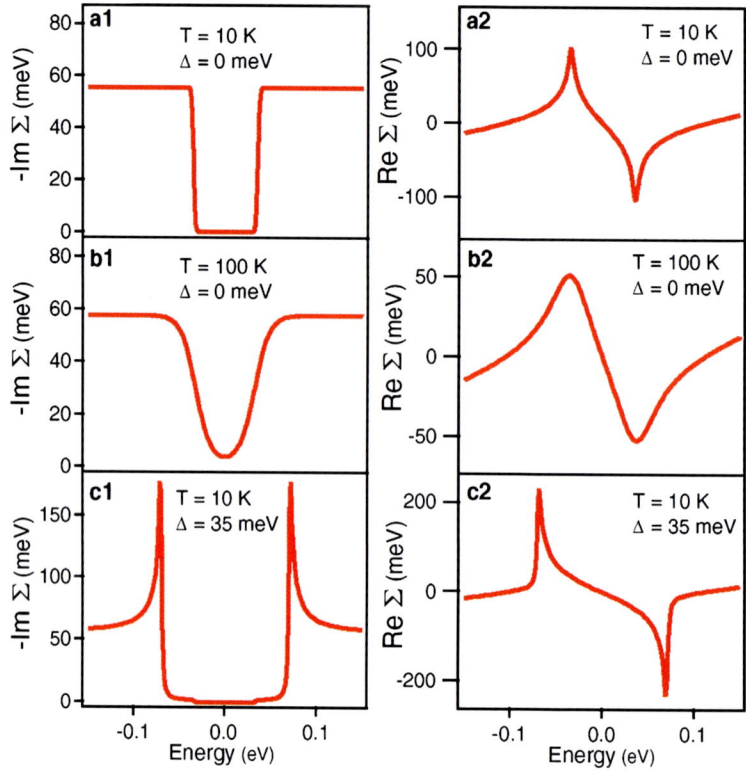

Figure 3.19. Self-energy for electrons coupled to an Einstein mode with $\Omega = 35$ meV and electron–phonon vertex $g = 0.15$ eV [114]. (a1), (b1), and (c1) plots $\mathrm{Im}\,\Sigma = -Z_2\omega + \chi_2$ for a normal state electron at 10 K, for a normal state electron at 100 K, and for an electron in an s-wave superconducting state at 10 K, respectively. (a2), (b2), and (c2) plots the corresponding real parts, $\mathrm{Re}\,\Sigma$, obtained using the Kramers–Kronig relation.

takes the familiar form:

$$\mathrm{Im}\,\Sigma(k,\omega) = \Sigma_q - \pi g^2(k,q)[2n(\Omega_q) + f(\Omega_q + \omega) + f(\Omega_q - \omega)]\delta(\omega - E_{k-q}), \qquad (3.11)$$

which when integrated over q becomes:

$$\mathrm{Im}\,\Sigma(k,\omega) = \int d\Omega\, \alpha_k^2 F(\Omega)[2n(\Omega) + f(\Omega + \omega) + f(\Omega - \omega)], \qquad (3.12)$$

where $\alpha_k^2 F(\Omega)$, the Eliashberg function, represents the coupling of the electron with Fermi surface momentum k, to all Ω phonons connecting that electron to other points on the Fermi surface.

For the normal state electron at 10 K (Figure 3.19a1), there is a sharp onset of the self-energy that broadens the spectra beyond the mode energy; for the normal state electron at 100 K (Figure 3.19b1), the onset of the self-energy is much smoother and occurs over ~50 meV; for the superconducting state electron (Figure 3.19c1), there is a singularity that causes a much more abrupt broadening of the spectra at the energy $\Omega + \Delta$. The superconducting state singularity is due to the density of states pile-up at the gap energy; the energy at which the decay onsets shift by Δ, since below the gap energy there are no states to which a hole created by photoemission can decay. For each of these imaginary parts of the self-energy, one can use the Kramers–Kronig transform to obtain the real part of the self-energy, which renormalizes the peak position (Re Σ in $A(k, \omega)$ (Eq. 3.3) changes the position of the peak in the spectral function). The real self-energies thereby obtained are also plotted in Figure 3.19a2–c2. In the superconducting state, again there is a singularity that causes a more abrupt break from the bare-band dispersion at the energy $\Omega + \Delta$.

For most metals, where the electrons are weakly interacting, and therefore the poles of the spectral function are well defined, one would expect such a treatment to hold and indeed it does, as evidenced by several cases *including* Beryllium [119, 120] and Molybdenum [121]. A priori, one might not expect the same to hold in ceramic materials such as the copper-oxides, where the copper d-wave electrons are localized and subject to strong electron–electron and electron–phonon interactions. Nonetheless, in the superconducting state of the copper-oxides at optimal and overdoped regime, one recovers narrow peaks (20–30 meV) of the spectral function. The above self-energy, then, is able to describe the phenomenology of the mode-coupling behavior for the superconducting state. The difference between the self-energy induced for a particular mode and coupling constant in the normal state at $T = 100$ K (Figure 3.19) and the superconducting state at $T = 10$ K (Figure 3.19) also shows the extent to which one can expect a temperature-dependent mode coupling in the high-T_c cuprates.

To illustrate the salient features of mode coupling on the dispersion, we consider a linear bare band coupled to an Einstein phonon in the normal state at $T = 10$ K. The effect of electron–phonon interaction on the one-electron spectral weight $A(k, \omega)$ of a $d_{x^2-y^2}$ superconductor has been simulated by Sandvik et al. [122]. In Figure 3.20, we show image plots, EDCs, MDC-derived dispersions, and the MDC-extracted widths for two different coupling constants (the case of stronger coupling is a factor of five increase in the vertex, g^2).

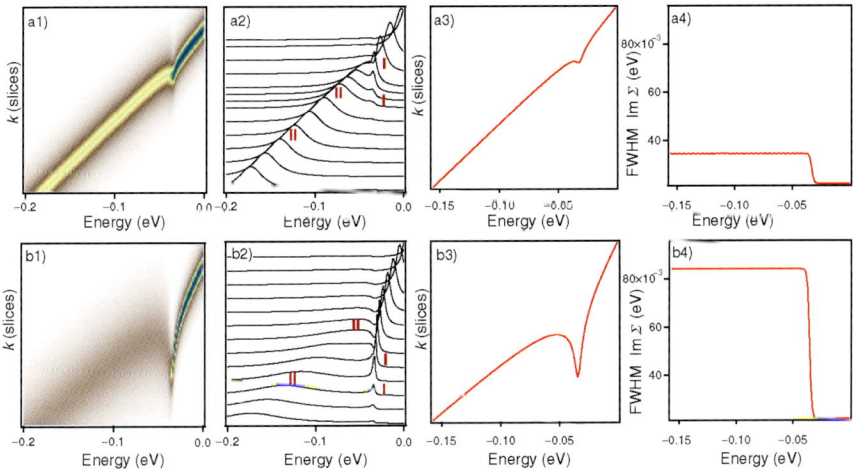

Figure 3.20. Simulated electron–phonon coupling using Einstein model. Spectral function (a1, b1), EDCs (a2, b2), MDC-derived dispersion (a3, b3), and the MDC-derived width (a4, b4) (imaginary part of self-energy) for two different couplings (a weak, b five times stronger) to a linear bare band.

There are three characteristic signatures of mode-coupling behavior evident:

(1) A break up of a single dispersing peak into two branches (Figure 3.20a1, b1)—a peak that decays as it asymptotically approaches the mode energy (I in Figure 3.20a2, b2), and a hump that traces out a dispersing band (II in Figure 3.20a2, b2).

(2) In the image plots (Figure 3.20a1, b1), a significant broadening of the spectra beyond the mode energy is readily apparent. This is also the origin of the broad hump of the dispersing band seen in the EDCs (Figure 3.20a2, b2) and the step in the extracted widths (or lifetime) (Figure 3.20a4, b4).

(3) At the mode energy itself, there is a "dip" between the peak and the hump in the EDCs (Figure 3.20a2, b2) leading to the "peak-dip-hump" structure often discussed in the literature.

From these generic features of electron–phonon coupling, one could ascertain the mode energy and coupling strength. Theoretically, the mode energy should be the energy to which the peak in the EDC curve decays. If there is a well-defined peak that has enough phase space range to decay, the last point at which it can be measured is the best indication of the mode energy. Otherwise, estimates can be made from the EDC, MDC-derived dispersions, and the position of the step in the MDC widths. The coupling strength is indicated by the extent of the break up of the spectra into a peak and a hump, the sharpness of the "kink" in the MDC-derived dispersion, and the magnitude of the step in the MDC-derived widths. Quantitative assessments of the coupling strength, however, require either a full model calculation or an extraction procedure to invert the phonon density of states coupled to the electronic spectra.

Strong Coupling—Polaron

When the kinetic energy of the particles is less than the phonon energy, the dressing of the phonon cloud could be large and the el–ph coupling enters into the polaron regime. A single particle coupled to the phonon is the typical case, on which extensive theoretical studies have been done. Let $g(q)$ be the coupling constant of the phonon with wavenumber q to the electrons, and the lattice relaxation energy E_{LR} is estimated as $E_{LR} \cong \langle |g(q)|^2 \rangle / \Omega$. When this E_{LR} is smaller than the bandwidth, the effective reduction of the el–ph coupling due to the rapid motion of the electron, i.e., the motional narrowing, occurs and the weight of the one-phonon side-band is of the order of $g(q)/W$ with the number of the phonon quanta N_{ph} being estimated as $N_{ph} \sim \langle |g(q)|^2 \rangle / W^2 \sim S(\Omega/W)^2$ where $S = E_{LR}/\Omega$. As the el–ph coupling constant increases, the polaron state evolves from this weak-coupling large polaron to the strong-coupling small polaron. This behavior is nonperturbative in nature, and the theoretical analysis is rather difficult. One useful method is the adiabatic approximation where the frequency of the phonon is set to be zero while E_{LR} remains finite. In this limit, one can regard the phonon as a classical lattice displacement, whose Fourier component is denoted by Q_q. Then one can investigate the stability of the weak-coupling large polaron state, i.e., zero distortion state in the present approximation, by the perturbative way. Namely the energy gain second order in $g(q)$ reads

$$\delta E = -\frac{1}{N} \sum_{q,\Omega} \frac{g(q)^2}{E(q) - E(0)} Q_q Q_{-q} \qquad (3.13)$$

with the energy dispersion $E(k)$ of the electron. Here the electron is at the ground state with the energy $E(0)$ in the unperturbed state. Introducing the index ℓ characterizing the range of the coupling as $g(q) \propto q^{-\ell}$, and considering the smallest possible wavenumber $q_{\min} \propto N^{-1/d}$ for the linear sample size $L = N^{1/d}$ in spatial dimension d, one can see that the index

$$s = d - 2(1 + \ell) \tag{3.14}$$

separates the two different behavior of δE. For $s > 0$, δE for $q = q_{\min}$ goes to zero as $N \to \infty$, which suggests that the weak-coupling state is always locally stable, separated by an energy barrier from the strong-coupled small polaron state. This means that a discontinuous change from the weak- to strong-coupling polaron states occurs where the mass becomes so heavy that the carrier is easily localized by impurities. Namely, the self-trapping transition occurs. For $s < 0$, on the other hand, the zero distortion state is always unstable for infinitesimal $g(q)$ and hence the lowest energy state continues smoothly as the coupling increases, i.e., no self-trapping transition. The most relevant case of the short range el–ph coupling in two dimensions, i.e., $d = 2$, $\ell = 0$, corresponds to $s = 0$, and hence is the marginal class. Therefore whether the self-trapping transition occurs or not is determined by the model of interest, and is nontrivial.

For the study of the polaron in the intermediate to strong-coupling region, one needs to invent a reliable theoretical method to calculate the energy, phonon cloud, effective mass, and the spectral function. Up to very recently, it has been missing but the diagrammatic quantum Monte Carlo method [123] combined with the stochastic analytic continuation [124] enabled the "numerically exact" solution to this difficult problem. By this method, the crossover from the weak- to strong-coupling regions have been analyzed accurately for various models [125, 126]. With this method, the polaron problem in the t–J model has been studied, and detailed information on the spectral function is now available which can be directly compared with experimental results. It is found that the self-trapping transition occurs in the two-dimensional t–J polaron model, and in comparison with experiment, the realistic coupling constant for the undoped case corresponds to the strong-coupling region. Namely the single hole doped into the undoped cuprates is self-trapped. Section 3.4.4 gives more details of how the polaron model relates to such experimentally determinable quantities as the lineshape, dispersion, and the chemical potential shift with doping.

Now we turn to the ARPES measurements that can be related to the two regimes of electron–phonon coupling. We will first review the band renormalization effects along the $(0,0)$–(π,π) nodal direction and near the $(\pi,0)$ antinodal region. The weak electron–phonon coupling picture is useful in accounting for many observations. However, there are experimental indications that defy the conventional electron–phonon coupling picture. Then we will move on to review the polaron issue which manifests in undoped and heavily underdoped samples.

3.4.3. Band Renormalization and Quasiparticle Lifetime Effects
El–Ph Coupling Along the $(0,0)$–(π,π) Nodal Direction

The nodal direction denotes the $(0,0)$–(π,π) direction in the Brillouin zone (Figure 3.8b). The d-wave superconducting gap is zero along this particular direction. As shown in Figures 3.21 and 3.22a, the energy–momentum dispersion curves from MDC method exhibit an abrupt slope change ("kink") near 70 meV. The kink is accompanied by an accelerated drop in the MDC width at a similar energy scale (Figure 3.22b). The existence of the kink has been well established as ubiquitous in hole-doped cuprate materials [127–133]:

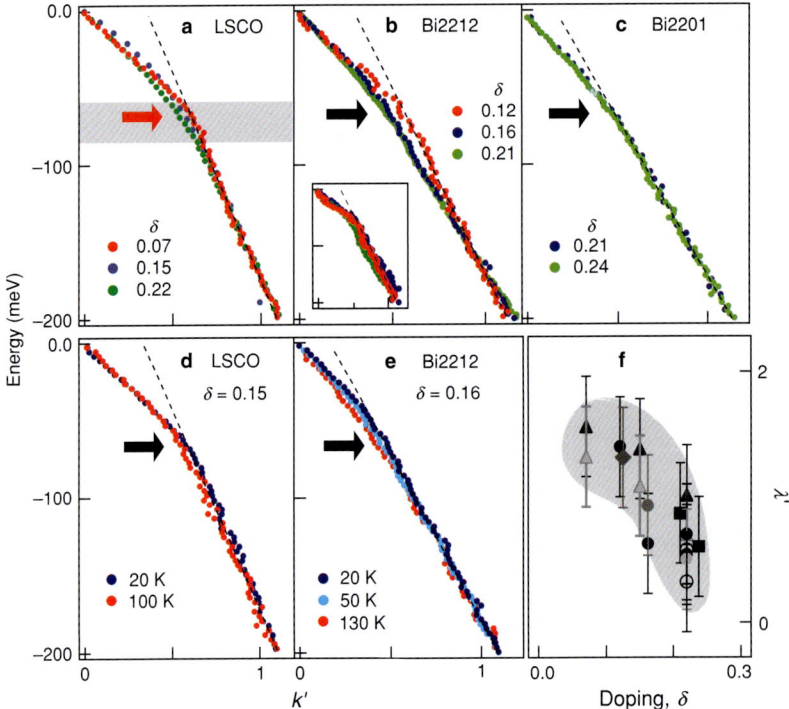

Figure 3.21. Ubiquitous existence of a kink in the nodal dispersion of various cuprate materials [128]. Top panels (a, b, c) plot dispersions along $(0, 0)$–(π,π) direction (except for panel b inset, which is off this direction) as a function of the rescaled momentum k' for different samples and at different doping levels (δ): (a) LSCO at 20 K, (b) Bi2212 in superconducting state at 20 K, and (c) Bi2201 in normal state at 30 K. Dotted lines are guides to the eye. Inset in (b) shows that the kinks in the dispersions off the $(0, 0)$–(π,π) direction sharpen upon moving away from the nodal direction. The black arrows indicate the position of the kink in the dispersions. Panels (d) and (e) show the temperature dependence of the dispersions for (d) optimally doped LSCO ($x = 0.15$) and (e) optimally-doped Bi2212, respectively. Panel (f) shows the doping dependence of the effective electron–phonon coupling strength λ' along the $(0, 0)$–(π,π) direction. Data are shown for LSCO (filled triangles), Nd-doped LSCO (1/8 doping; filled diamonds), Bi2201 (filled squares), and Bi2212 (filled circles in the first Brillouin zone and unfilled circles in the second zone). The different shadings represent data obtained in different experimental runs. Shaded area is a guide to the eye.

1. It is present in various hole-doped cuprate materials, including $Bi_2Sr_2CaCu_2O_8$ (Bi2212), $Bi_2Sr_2CuO_6$ (Bi2201), $(La_{2-x}Sr_x)CuO_4$ (LSCO) and others. The energy scale (in the range of 50–70 meV) at which the kink occurs is similar for various systems.

2. It is present both below T_c and above T_c.

3. It is present over an entire doping range (Figure 3.22a). The kink effect is stronger in the underdoped region and gets weaker with increasing doping.

While there is a consensus on the data, the exact meaning of the data is still under discussion. The first issue concerns whether the kink in the normal state is related to an energy scale. Valla et al. argued that the system is quantum critical and thus has no energy scale, even though a band renormalization is present in the data [33]. Since their data do not show a sudden change in the scattering rate at the corresponding energy, they attributed the kink in Bi2212 above T_c to the marginal Fermi liquid (MFL) behavior without an energy scale [130]. Others believe the existence of energy scale in the normal and superconducting states has a common origin, i.e., coupling of quasiparticles with low-energy collective excitations

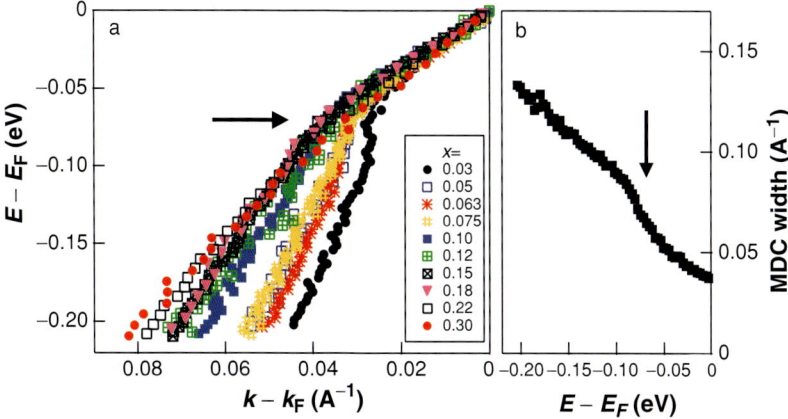

Figure 3.22. Doping dependence of the nodal electron dynamics in LSCO and universal nodal Fermi velocity [132]. (a) Dispersion of LSCO with various doping levels ($x = 0.03 - 0.30$) measured at 20 K along the $(0,0)$–(π,π) nodal direction. The arrow indicates the position of kink that separates the dispersion into high-energy and low-energy parts with different slopes. (b) Scattering rate as measured by MDC width (full-width-at-half-maximum, FWHM) of the LSCO ($x = 0.063$) sample measured at 20 K. The arrow indicates a drop at an energy ∼70 meV.

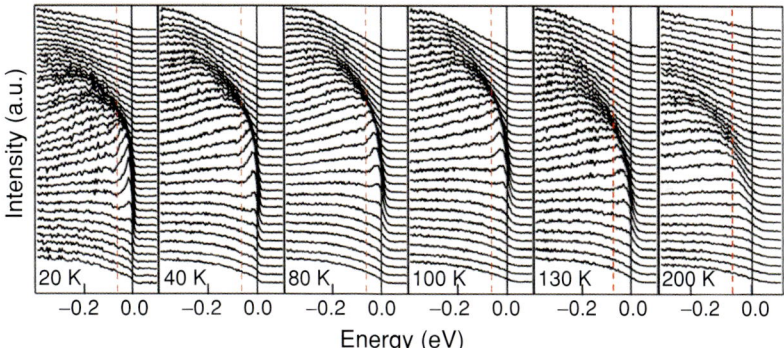

Figure 3.23. Energy distribution curves (EDC) in the normal state of underdoped Bi2201 ($T_c = 10$ K) at several temperatures (from 20 to 200 K) [135]. A dip in the EDCs can be clearly observed almost for all the temperatures. The dip position (dotted line) is 60 meV and is roughly temperature independent.

(bosons) [127–129]. The sharp kink structure in dispersion and concomitant existence of a drop in the scattering rate which is becoming increasingly clear with the improvement of signal to noise in the data, as exemplified in underdoped LSCO ($x = 0.063$) in the normal state (Figure 3.22b) [134], are apparently hard to reconcile with the MFL behavior.

The existence of a well-defined energy scale over an extended temperature range is best seen in Bi2201 compound [135]. As shown in Figure 3.23, the spectra reveal a "peak-dip-hump" structure up to temperatures near 130 K, almost ten times the superconducting critical temperature T_c. Such a "peak-dip-hump" structure is very natural in an electron–phonon coupled system, but will not be there if there is no energy scale in the problem as argued by Valla et al. [33].

A further issue concerns the origin of the bosons involved in the coupling, with a magnetic resonance mode [129, 130] and optical phonons [128] being possible candidates

considered. The phonon interpretation is based on the fact that the sudden band renormalization (or "kink") effect is seen for different cuprate materials, at different temperatures, and at a similar energy scale over the entire doping range [128]. For the nodal kink, the phonon considered in the early work was the half-breathing mode, which shows an anomaly in neutron experiments [107, 108]. Unlike the phonons, which are similar in all cuprates, the magnetic resonance (at correct energy) is observed only in certain materials and only below T_c. The absence of the magnetic mode in LSCO and the appearance of magnetic mode only below T_c in some cuprate materials are not consistent with its being the cause of the universal presence of the kink effect. Whether the magnetic resonance can cause any additional effect is still an active research topic [136, 137].

To test the idea of electron–phonon coupling, an isotope exchange experiment has been carried out [133]. When exchanging ^{18}O and ^{16}O in Bi2212, a strong isotope effect has been reported in the nodal dispersions (Figure 3.24). Surprisingly, however, the isotope effect mainly appears in the high binding energy region above the kink energy; at the lower binding energy near the Fermi level, the effect is minimal. This is quite different from the conventional electron–phonon coupling where isotope substitution will result in a small shift of phonon energy while keeping most of the dispersion intact. The origin of this behavior is still being investigated.

It is interesting to note in Figure 3.22a that the energy scale of the kink also serves as a dividing point where the high and low energy dispersions display different doping dependence [132]. The dispersion in this Figure were obtained by the MDC method. In Figure 3.25a, we reproduce some of these MDC-extracted dispersions, but we also plot the dispersion extracted using EDCs by following the EDC peak position. Since the first derivative of the dispersion, $\partial E/\partial k$, corresponds to velocity, the dispersions at high binding energy (-0.1 to -0.25 eV)

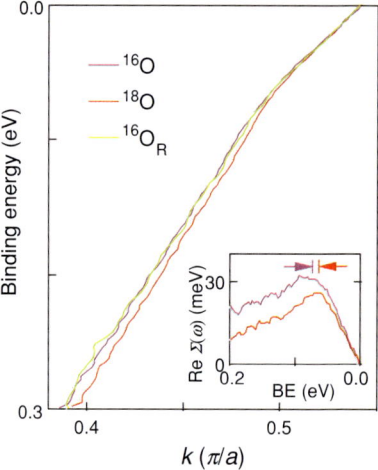

Figure 3.24. Isotope-induced changes of the nodal dispersion [133]. The data were taken on optimally doped Bi$_2$Sr$_2$CaCu$_2$O$_8$ samples ($T_c \sim 91$–92 K) with different oxygen isotopes ^{16}O and ^{18}O at $T \sim 25$ K along the nodal direction. The low energy dispersion is nearly isotope-independent, while the high energy dispersion is isotope-dependent. The effect is reversible by isotope resubstitution (green). Inset shows the real part of the electron self-energy, ReΣ, obtained from the dispersion by subtracting a line approximation for the one-electron band E_k, connecting two points (one at E_F and the other at a 300-meV binding energy) of the ^{18}O dispersion.

and low binding energy (0 to −0.05 eV) are fitted by straight lines to quantitatively extract velocities, as plotted in Figure 3.25b [138].

While nodal data clearly reveal the presence of coupling to collective modes with well-defined energy scale, there are a couple of peculiar behaviors associated with the doping evolution of the nodal dispersion. One obvious anomaly is the difference of low energy velocity obtained from MDC and EDC methods (Figure 3.25b). As seen from Figures 3.22a and 3.25b, the low-energy dispersion and velocity from the MDC method is insensitive to doping over the entire doping range, giving the so-called "universal nodal Fermi velocity" behavior [132]. Similar behavior was also reported in Bi2212 [130]. However, improved LSCO data where we can resolve a well-defined quasiparticle peak to extract dispersion using EDC method reveal a dichotomy in EDC and MDC derived dispersions, particularly for low doping (Figure 3.25), like $x = 0.01$ [139]. This discrepancy between EDC and MDC cannot be reconciled within the conventional el−ph interaction picture, as simulations considering experimental resolutions show.

In terms of conventional electron−phonon coupling, if one considers that the "bare band" does not change with doping but the electron−phonon coupling strength increases with decreasing doping, as it is probably the case for LSCO, one would expect that the low energy dispersion and velocity show strong doping dependence, while the high-energy ones

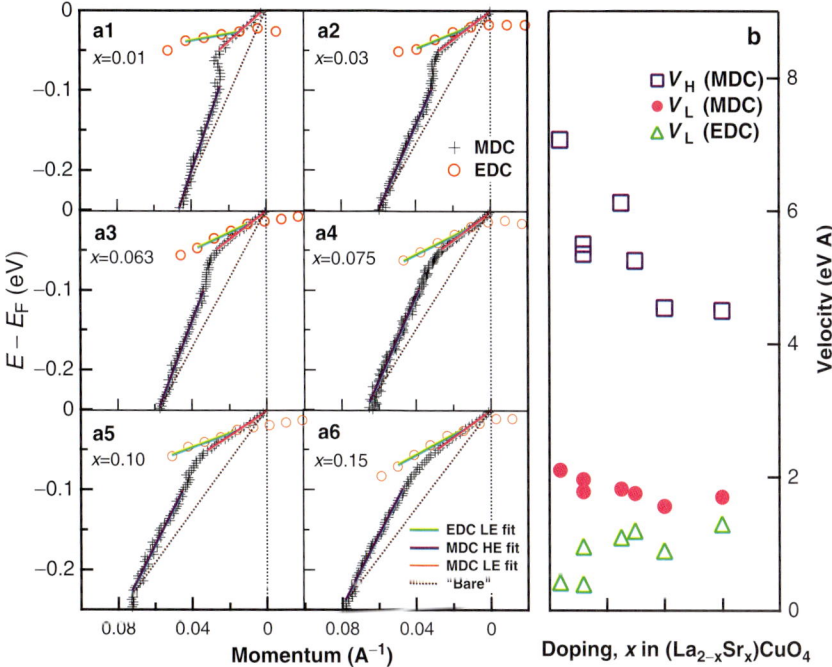

Figure 3.25. (a). Energy−momentum dispersions for LSCO with different dopings, using both EDC and MDC methods [138]. The EDC low-energy velocity is obtained by fitting the EDC dispersion linearly in the intermediate energy range because the data points very close to Fermi level is affected by the Fermi cutoff while the data at higher energy have large uncertainty because the EDCs are broader. The MDC low (high) energy velocity v_H is obtained by fitting MDC dispersion at binding energy 0–50 meV (100–250 meV) using a linear line. (b). Low and high-energy velocities as a function of doping obtained from MDC and EDC dispersions.

converge. However, one sees that the high-energy velocity is highly doping dependent. Moreover, its trend is anomalous if one takes electron–electron interaction into account. It is known in cuprates that, with decreasing doping, the electron–electron interaction gets stronger. According to conventional wisdom, this would result in a larger effective mass and smaller velocity. However, the doping dependence of the high-energy velocity is just opposite to this expectation, as seen from Figure 3.25b.

Therefore, these anomalies indicate a potential deviation from the standard Migdal–Eliashberg theory and the possibility of a complex interplay between electron–electron and electron–phonon interactions. As we discuss later, this phenomenon is a hint of polaronic effect where the traditional analysis fails. Such a polaron effect gets stronger in deeply underdoped system even along the nodal direction. Under such a condition, one needs to use EDC derived dispersion when the peaks are resolved.

Multiple Modes in the Electron Self-Energy

In conventional superconductors, the successful extraction of the phonon spectral function, or the Eliashberg function, $\alpha^2 F(\omega)$, from electron tunneling data played a decisive role in cementing the consensus on the phonon-mediated mechanism of superconductivity [140]. For high temperature superconductors, the extraction of the bosonic spectral function can provide fingerprints for more definitive identification of the nature of bosons involved in the coupling.

In principle, the ability to directly measure the dispersion, and therefore, the electron self-energy, would make ARPES the most direct way of extracting the bosonic spectral function. This is because, in metals, the real part of the electron self-energy Re Σ is related to the Eliashberg function $\alpha^2 F(\Omega; \varepsilon, \hat{\mathbf{k}})$ by:

$$\mathrm{Re}\,\Sigma(\hat{\mathbf{k}}, \varepsilon; T) = \int_0^\infty d\Omega\, \alpha^2 F(\Omega; \varepsilon, \hat{\mathbf{k}}) K\left(\frac{\varepsilon}{kT}, \frac{\hbar\Omega}{kT}\right), \qquad (3.15)$$

where

$$K(y, z) = \int_{-\infty}^{\infty} dx \frac{2z}{x^2 - z^2} f(x + y), \qquad (3.16)$$

with $f(x)$ being the Fermi distribution function. Such a relation can be extended to any electron–boson coupling system and the function $\alpha^2 F(\omega)$ then describes the underlying bosonic spectral function. We note that the form of Re $\Sigma(\hat{\mathbf{k}}, \varepsilon; T)$ (Eq. 3.15) can be derived by taking the Kramers–Kronig transformation of Im Σ for the normal state as shown above (Eq. 3.12). Unfortunately, given that the experimental data inevitably have noise, the traditional least-square method to invert an integral problem is mathematically unstable.

Very recently, Shi et al. have made an important advance in extracting the Eliashberg function from ARPES data by employing the maximum entropy method (MEM) and successfully applied the method to Be surface states [141]. The MEM approach [141] is advantageous over the least squares method in that (i) It treats the bosonic spectral function to be extracted as a probability function and tries to obtain the most probable one, (ii) More importantly, it is a natural way to incorporate the priori knowledge as a constraint into the fitting process. In practice, to achieve an unbiased interpretation of data, only a few basic physical constraints to the bosonic spectral function are imposed: (a) It is positive. (b) It vanishes at the limit $\omega \to 0$.

(c) It vanishes above the maximum energy of the self-energy features. As shown in the case of Be surface state, this method is robust in extracting the Eliasberg function [141].

Initial efforts have been made to extend this approach to underdoped LSCO and evidence for electron coupling to several phonon modes has been revealed [142]. As seen from Figure 3.26, from both the electron self-energy (Figure 3.26a), and the derivative of their fitted curves (Figure 3.26a), one can identify two dominant features near ∼ 40 and ∼60 meV. In addition, two addition modes may also be present near ∼25 and ∼75 meV [142]. The multiple features in Figure 3.26b show marked difference from the magnetic excitation spectra measured in LSCO which is mostly featureless and doping dependent [143]. In comparison, they show more resemblance to the phonon density-of-states (DOS), measured from neutron scattering on LSCO (Figure 3.26c) [144], in the sense of the number of modes and their positions.

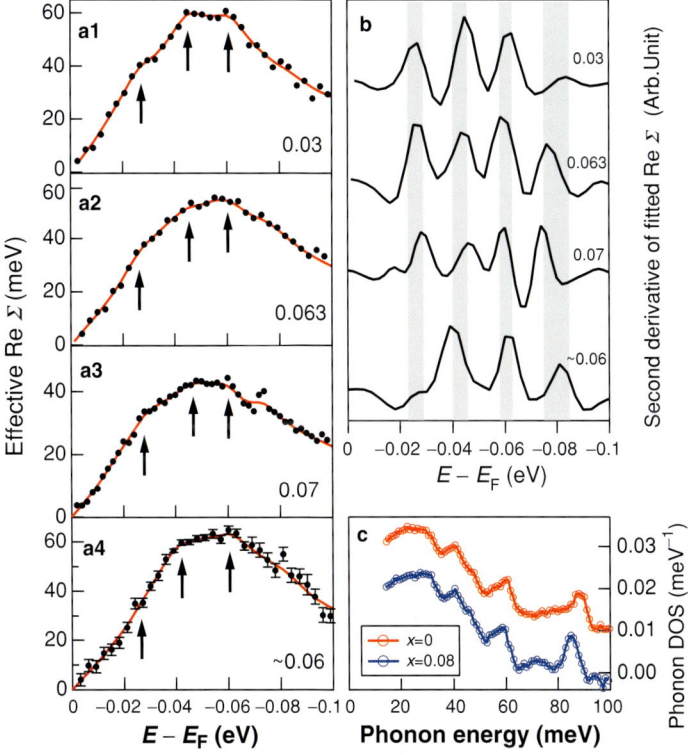

Figure 3.26. Multiple modes coupling in electron self-energy in LSCO [142]. (a) The effective real part of the electron self-energy for LSCO $x = 0.03$ (a1), 0.063 (a2), 0.07 (a3) and ∼0.06 (a4) samples. Data (a1–a3) were taken using Scienta 2002 analyzer while data (a4) were taken using Scienta R4000 analyzer. Data (a1–a3) were taking using 10 eV pass energy at an energy resolution of ∼18 meV. Data (a4) were taken a $x \sim 0.06$ sample using 5 eV pass energy with an energy resolution of ∼12 meV. For clarity, the error bar is only shown for data (a4) which becomes larger with increasing binding energy. The arrows in the figure mark possible fine structures in the self-energy. The data are fitted using the maximum entropy method (solid red lines). (b) The second-order derivative of the calculated ReΣ. The four shaded areas correspond to energies of (23–29), (40–46), (58–63), and (75–85) meV where the fine features fall in. (c) The phonon density of state $F(\omega)$ for LSCO $x = 0$ (red) and $x = 0.08$ (blue) measured from neutron scattering [144].

This similarity between the extracted fine structure and the measured phonon features favors phonons as the bosons coupling to the electrons. In this case, in addition to the half-breathing mode at 70–80 meV that we previously considered strongly coupled to electrons [128], the present results suggest that several lower energy optical phonons of oxygens are also actively involved. Particularly we note that the mode at ∼60 meV corresponds to the vibration of apical oxygens.

We note that, in order to be able to identify fine structure in the electron self-energy, it is imperative to have both high energy resolution and high statistics [145]. These requirements have made the experiment highly challenging because of the necessity to compromise between two conflicting requirements for the synchrotron light source: high energy resolution and high photon flux. Further improvements in photoemission experiments will likely enable a detailed understanding of the boson modes coupled to electrons, and provide critical information for the pairing mechanism.

One would like to extend this method to the superconducting state, in momentum around the BZ, and to higher temperatures. The superconducting state could, in principle, be achieved by using the BCS dispersion of the quasiparticles rather than the normal state dispersion and is currently under study. However, considering the anisotropy of the el–ph coupling detailed below, the anisotropy of the underlying band structure, and the d-wave superconducting gap, extending the procedure in momentum may be somewhat more difficult. The $\alpha^2 F(\omega, \varepsilon, \vec{k})$ used for the above form of the real part of the self-energy is assumed to be only weakly dependent on the initial energy ε and momentum k of the electron. But again, one in principle could begin to consider a different form of the calculated Re Σ and then apply the MEM method with it instead. Extending the method to higher temperatures, for example ∼100 K for normal state Bi2212 data, may be, however, a limitation that cannot be overcome. The method's strength is in resolving fine structures due to the phonon density of states. Those fine structures occur predominantly at lower temperatures. At higher temperatures of ∼100 K, the imaginary and real parts of the self energy get broadened on the order of the phonon energy itself. In that case, two or more neighboring phonons would contribute to the electronic renormalization at a given energy, both broadening the fine structures in the data and weakening the resolving power of the method itself. So, while the MEM method can directly extract fine features from ARPES data in agreement with neutron scattering without implicitly assuming a phonon model, it does not have the freedom to incorporate the temperature and momentum dependence needed to describe the ARPES data in both superconducting and normal states, near the vHS and near the node. Both modeling of the data and direct extraction, then, are needed, to gain a complete picture of the mode-coupling features in the data.

El–Ph Coupling Near the (π,0) Antinodal Region

The antinodal region refers to the (π,0) region in the Brillouin zone where the d-wave superconducting gap has a maximum (Figure 3.8b). Recently, a low-energy kink was also identified near the (π,0) antinodal region in Bi2212 [54, 129, 146, 147]. This observation was made possible thanks to the clear resolution of the bilayer splitting [28–30]. As there are two adjacent CuO_2 planes in a unit cell of Bi2212, these give rise to two Fermi surface sheets from the higher-binding-energy bonding band (B) (thick red curves in Figure 3.27c) and the lower-binding- energy antibonding band (A) (thick black curves in Figure 3.27c).

Consider a cut along (π,π) − (π,−π) across (π,0) in Bi2212, both above T_c (Figure 3.27a) and below T_c (Figure 3.27b) [54]. Superimposed are the dispersion of the bonding

Electronic Structure and Electron–Phonon Coupling in Cuprate Superconductors

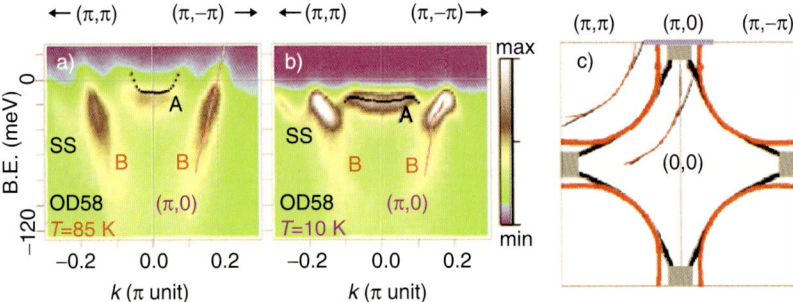

Figure 3.27. Antinodal kink near $(\pi,0)$ in a heavily overdoped Bi2212 sample ($T_c \sim 58$ K). (a) Normal-state data (T = 85 K) near the antinodal region. (b) Superconducting-state data from the same sample at 10 K, showing the emergence of a dispersion kink in the bilayer split-B band. The B band dispersions (red curves) were determined by fitting MDC peak positions. The black dots represent A band EDC peak positions. (c) Brillouin zone with bonding band B (thick red) and antibonding band A (thick black) Fermi surfaces, as well as momentum-cut locations for panels (a) and (b) (blue bars). The two sets of thin curves are replicas of Fermi surface originating from the superstructure in Bi2212.

band determined from the MDC (red lines) and antibonding band from the EDC (black dots). When the bandwidth is narrow, the applicability of the MDC method in obtaining dispersion becomes questionable so one has to resort to the traditional EDC method. In the normal state, the bonding-band dispersion (Figure 3.27a) is nearly linear and featureless in the energy range of interest. Upon cooling to 10 K (Figure 3.27b), the dispersion, as well as the near-E_F spectral weight, is radically changed. In addition to the opening of a superconducting gap, there is a clear kink in the dispersion around 40 meV.

Gromko et al. [54] reported that the antinodal kink effect appears only in the superconducting state and gets stronger with decreasing temperature. Their momentum-dependence measurements show that the kink effect is strong near $(\pi,0)$ and weakens dramatically when the momentum moves away from the $(\pi,0)$ point. Excluding the possibility that this is a by-product of a superconducting-gap opening, they attributed the antinodal kink to the coupling of electrons to a bosonic excitation, such as a phonon or a collective magnetic excitation. The prime candidate they considered is the magnetic-resonance mode observed in inelastic neutron scattering experiments.

The temperature and momentum dependence identified for a range of doping levels has also led others to attribute the effect to the magnetic resonance [129, 146]. However, there are some inconsistencies with this interpretation (1) the magnetic resonance has not yet been observed by neutron scattering in such a heavily doped cuprate and (2) the magnetic resonance has little spectral weight, and may be too weak to cause the effect seen by ARPES. Furthermore, the electron–phonon coupling in the early tunneling spectra, such as Pb, appeared prominently only in the superconducting state. The linear MDC-derived dispersion in the normal state of Bi2212 at $(\pi,0)$ that Gromko et al. reports [54] is not conclusive enough proof that the same mode does not couple to the electrons in the normal state. On the other hand, the clear determination of mode-coupling by Gromko et al. in the antinodal region, where the gap is maximum, without the complication of bilayer splitting or superstructure, suggests that the renormalization effects seen by ARPES in the cuprates may indeed by related to the microscopic mechanism of superconductivity.

Cuk et al. [147] and Devereaux et al. [148] have recently proposed a new interpretation of the renormalization effects near antinodal region seen in Bi2212. Specifically, the key observation that prompted them to rule out the magnetic resonance interpretation is the unraveling of the existence of the antinodal kink even in the normal state. This observation was made possible by utilizing the EDC method because the MDC method is not appropriate when the assumed linear approximation of the bare band fails near $(\pi,0)$ where the band bottom is close to Fermi level E_F. Figure 3.28a1, b1 and c show dispersions in the normal state of an optimally doped sample which consistently reveal a 40 meV energy scale that has eluded detection previously. Upon entering the superconducting state, the energy scale shifts to 70 meV consistent with a gap opening of 35–40 meV. This coupling is also found to be more extended in a Brillouin zone than previously reported [54]. In Figure 3.29, we show data from the optimally doped Bi2212 sample for a large portion of the BZ in the superconducting state [147]. The renormalization occurs at 70 meV throughout the BZ and increases in strength from the nodal to antinodal points. Similar behaviors are also noted by others [129] (Figure 3.30). The increase in coupling strength can be seen in the following ways: Near $(\pi,0)$, the band breaks up into two bands (peak and hump) as seen in Figure 3.29a2, a3. For cuts taken in the $(0,0) - (\pi,\pi)$ direction, the band dispersion is steeper and the effects of mode-coupling, though significant, are less pronounced.

It is quite natural that phonon modes of different origin and energy preferentially couple to electrons in certain k-space regions. While the detection of multiple modes in the normal

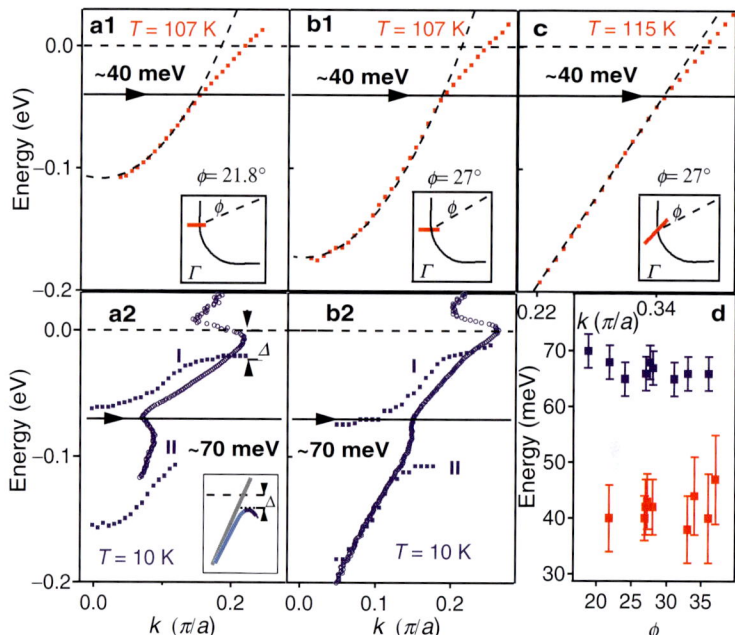

Figure 3.28. Antinodal kink near $(\pi,0)$ in the normal state (a1,b1,c) and superconducting state (a2,b2) in an optimally-doped Bi2212 [147]. The dispersions in a1, b1, and c were derived by the EDC method; the position of the momentum cuts is labeled in the insets. The red dots are the data; the fit to the curve (black dashed line) below the 40-meV line is a guide to the eye. The dispersions at the same location in the superconducting state (10 K) are shown in (a2) and (b2), which were derived by the MDC method (blue circles). Also plotted in (a2) and (b2) are the peak (blue squares, I) and hump positions (blue squares, II) of the EDCs for comparison. The inset of (a2) shows the expected behavior of a Bogoliubov-type gap opening. The s-like shape below the gap energy is an artifact of the way the MDC method handles the backbend of the Bogoliubov quasiparticle. (d) Kink positions as a function of momentum cuts in the antinodal region.

Electronic Structure and Electron–Phonon Coupling in Cuprate Superconductors

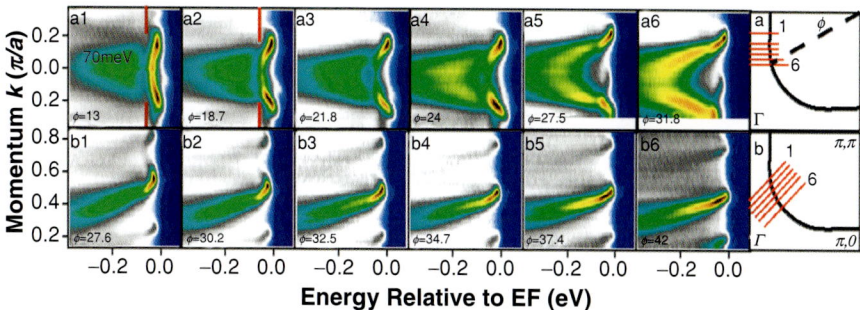

Figure 3.29. Anisotropic electron–boson coupling in Bi2212 [147]. Image plots in (a1–a6) and (b1–b6) are cuts taken parallel to $(0, \pi)$–(π, π) and $(0, 0)$–(π, π) respectively, at the locations indicated in the Brillouin zone ((a) and (b)) at 15 K for an optimally doped sample ($T_c = 94$ K).

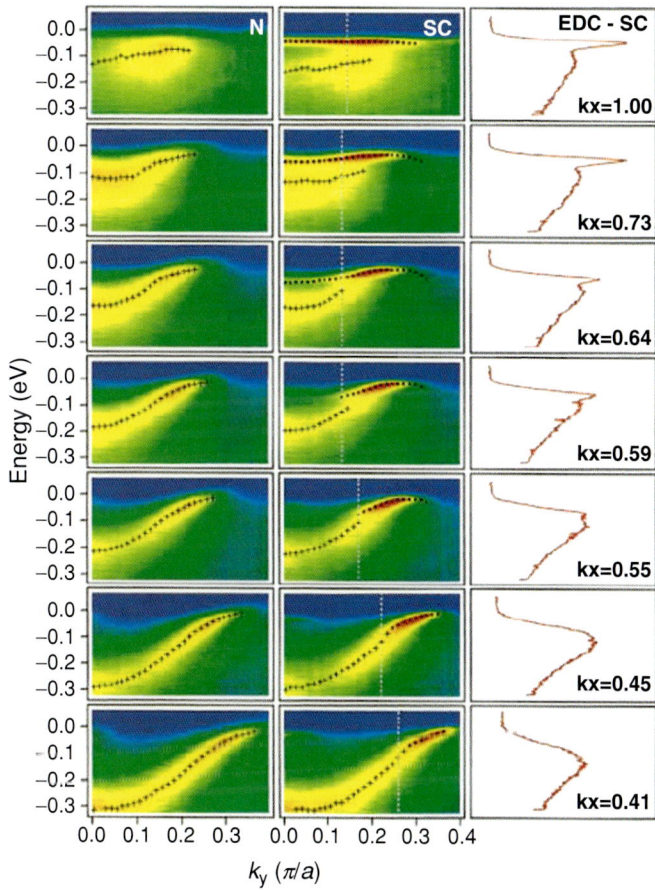

Figure 3.30. Momentum dependence of photoemission data in optimally doped Bi2212 [129]. Left panels: Photoemission data in the normal state ($T = 140$ K) along selected cuts parallel to $M(\pi, 0) - Y(\pi, \pi)$. EDC peak positions are indicated by crosses. Middle panels: Photoemission data in the superconducting state ($T = 40$ K) at the same cuts as for left panels. Crosses indicate positions of broad high energy peaks, dots sharp low energy peaks. Right panels: EDCs at locations marked by the vertical lines in the middle panels.

state of LSCO [142] suggests that several phonons may be involved, only one has the correct energy and momentum dependence to understand the prominent signature seen in the superconducting state. This new interpretation [147] attributes the renormalization seen in the superconducting state to the "bond-buckling" B_{1g} phonon mode involving the out-of-plane, out-of-phase motion of the in-plane oxygens. The bond-buckling phonon is observed at 35 meV in the B_{1g} polarization of Raman scattering in an optimally doped sample, the same channel in which the ~35–40 meV d-wave superconducting gap shows up [99, 149, 150]. Applying simple symmetry considerations and kinematic constraints, it is found that this B_{1g} buckling mode involves small momentum transfers and couples strongly to electronic states near the antinode [148]. In contrast, the in-plane Cu–O breathing modes involve large momentum transfers and couple strongly to nodal electronic states. Band renormalization effects are also found to be strongest in the superconducting state near the antinode, in full agreement with angle-resolved photoemission spectroscopy data (Figure 3.31). The dramatic temperature dependence stems from a substantial change in the electronic occupation distribution and the opening of the superconducting gap [147, 148]. It is important to note that the electron–phonon coupling, especially the one with B_{1g} phonon, explains the temperature and momentum dependence of the self-energy effects that were taken as key evidence to support the magnetic resonance interpretation of the data. Compounded with the findings that cannot be explained by the magnetic resonance as discussed earlier, this development makes the phonon interpretation of the kink effect self-contained.

Anisotropic El–Ph Coupling

The full Migdal–Eliashberg-based calculation consists of a tight-binding band structure and el–ph coupling to the breathing mode as well as the B_{1g} bond-buckling mode and is based on an earlier calculation [151]. The electron–phonon coupling vertex $g(k, q)$, where k represents the initial momentum of the electron and q the momentum of the phonon is determined on the basis of the oxygen displacements for each mode in the presence of the underlying band-structure. In the case of the breathing mode, the in-plane displacements of the oxygen modulate the CuO_2 nearest neighbor hopping integral as well as the site energies. In the case of the bond-buckling mode, one must suppose that the mirror plane symmetry across the CuO_2 plane is broken in order for electrons to couple linearly to phonons. The mirror

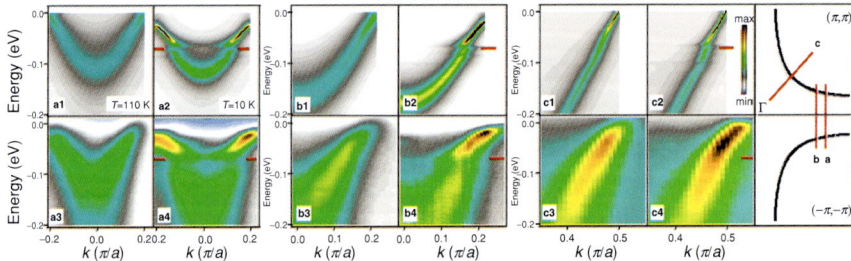

Figure 3.31. Comparison between the calculated and measured spectral function in Bi2212 including electron–phonon coupling for three different momentum cuts (a, b, c) through the Brillouin zone. (a1,b1,c1) and (a2,b2,c2) show the calculated spectral functions in the normal and superconducting states, respectively [148]. The measured spectral functions are shown in (a3,b3,c3) for the normal state and in (a4,b4,c4) for the superconducting state. The corresponding momentum cuts a, b, and c are shown in the rightmost panel. The red markers in the superconducting state indicate 70 meV. The simulation includes B_{1g} oxygen buckling mode and half-breathing mode.

plane symmetry can be broken by the presence of a crystal field perpendicular to the plane, tilting of the Cu–O octahedral, static in-plane buckling, or may be dynamically generated.

The $g_{B_{1g}}(k,q)$ form factor leads to preferential $q \sim 2k_F$ scattering between the parallel pieces of Fermi surface in the antinodal region, as shown in Figure 3.32 depicting $g(k,k')$ for the buckling mode (where $k' = k - q$) for an electron initially at the antinode (k_{AN}; upper-left) and for an electron initially at the node (k_N; bottom-left). This coupling anisotropy partially accounts for the strong manifestation of electron–phonon coupling in the antinodal region where one sees a break up into two bands. The breathing mode, in contrast, modulates the hopping integral and has a form factor, $g_{br}(k,q)$, that leads to preferential scattering for large q and couples opposing nodal states. This coupling anisotropy then accounts for the 70 meV energy scale seen most prominently in a narrow k-space region near the nodal direction in the normal state of LSCO. Figure 3.32 also shows that the magnitude of the electron–phonon vertex is largest for an electron initially sitting at the node, k_N, that scatters to the opposing nodal state. For more details on this calculation, see Devereaux et al. [148].

The anisotropy of the mode-coupling in both the superconducting state data and the calculation is peculiar to the cuprates. Such a strong anisotropy in the electron–phonon coupling is not traditionally expected. In cuprates, the sources of the anisotropy are (1) an electron–phonon vertex for the B_{1g} bond-buckling mode and the breathing mode that depends both on the electron momentum k as well as the phonon momentum q. This comes from a preferential scattering in the Brillouin zone, in which nodal states couple to other antinodal states and antinodal states to other antinodal states, (2) a strongly anisotropic electronic band structure characterized by a van Hove Singularity (vHS) at $(\pi, 0)$. In the antinodal region and along the $(\pi, 0)$–(π, π) direction in which $2k_F$ scattering is preferred, the bands are narrow, giving rise to a larger electronic density of states near the phonon energy and therefore a stronger manifestation of electron–phonon coupling, (3) a d-wave superconducting gap, and (4) a collision of energy scales in the antinodal region that resonate to enhance the above effects—the vHS at ~ 35 meV in the tight-binding model that best fits the data, the

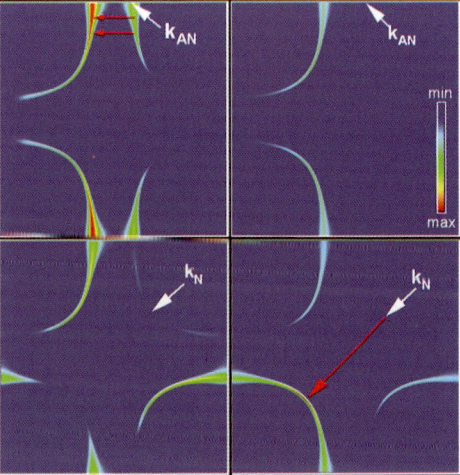

Figure 3.32. Plots of the electron–phonon coupling $|g(\mathbf{k},\mathbf{q})|^2$ for initial $\mathbf{k}$ and scattered $\mathbf{k}' = \mathbf{k} - \mathbf{q}$ states on the Fermi surface for the buckling mode (left panels) and breathing mode (right panels) for initial fermion $\mathbf{k}$ at an antinodal (top panels) and nodal (bottom panels) point on the Fermi surface, as indicated by the arrows. The red/blue color indicates the maximum/minimum of the el–ph coupling vertex in the BZ for each phonon [148].

maximum d-wave gap at ~35 meV, and the bond-buckling phonon energy at ~35 meV. All these three factors collide to give the anisotropy of the mode-coupling behavior in the superconducting and normal states. For a detailed look at how each plays a role in the agreement with the data, please see Cuk et al. [147]. The coincidence of energy scales, along with the dominance of the renormalization near the antinode, indicates the potential importance of the B_{1g} phonon to the pairing mechanism, which is consistent with some theories based on the B_{1g} phonon [89, 152–154] but remains to be investigated.

The cuprates provide an excellent platform to study anisotropic electron– phonon interaction. In one material, such as optimally doped Bi2212, the effective coupling can span λ of order ~1 at the node to 3 at the antinode (Figure 3.33) [147, 148]. In addition to the large variation of coupling strength, there is a strong variation in the kinematic considerations for electron–phonon coupling. In the nodal direction, the band bottom is far from the relevant phonon energy scales. However, at the antinode, the relevant phonon frequencies approach the bandwidth. Indeed the approximation of Migdal, in which higher order vertex corrections to the el–ph coupling are neglected due to the smallness of $(\lambda * \Omega_{ph}/E_F)$, may be breaking down in the antinodal region. Non-adiabatic effect beyond the Migdal approximation have been considered and are under continuing study [155].

3.4.4. Polaronic Behavior

Polaronic Behavior in Parent Compounds

The parent compounds of the cuprate superconductors, being antiferromagnetic Mott insulators, provide an ideal testing ground for investigating the dynamics of one hole in an antiferromagnetic background. Indeed, many theories have been formed and tested by ARPES on a number of compounds, among them are $Sr_2CuO_2Cl_2$ (SCOC) [156–160], $Ca_2CuO_2Cl_2$ (CCOC) [113, 160–163], Nd_2CuO_4 [164], and La_2CuO_4 [165–167]. However, several aspects of the data can only be explained by invoking polaron physics, as we will now discuss.

The ARPES measurements on SCOC [156, 158] and CCOC [113, 162] give essentially similar results. As seen in Figure 3.34a, b, along the $(0,0)–(\pi,\pi)$ direction, the lowest energy feature disperses toward lower binding energy with increasing momentum, reaches its lowest binding energy position near $(\pi/2,\pi/2)$ where it becomes sharpest in its lineshape, and

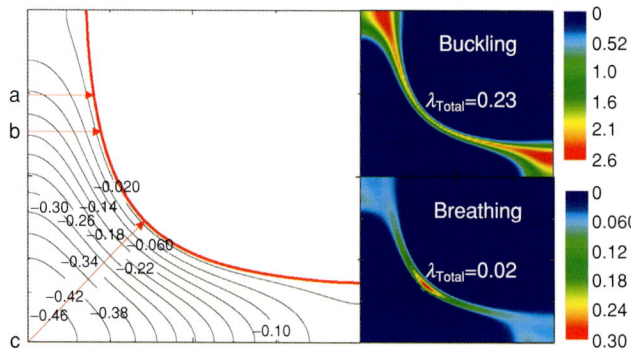

Figure 3.33. Plots of the electron–phonon coupling λ_k in the first quadrant of the BZ for the buckling mode (right top panel) and breathing mode (right bottom panel). The color scale is shown on the right for each phonon. The left panel shows energy contours for the band structure used [148].

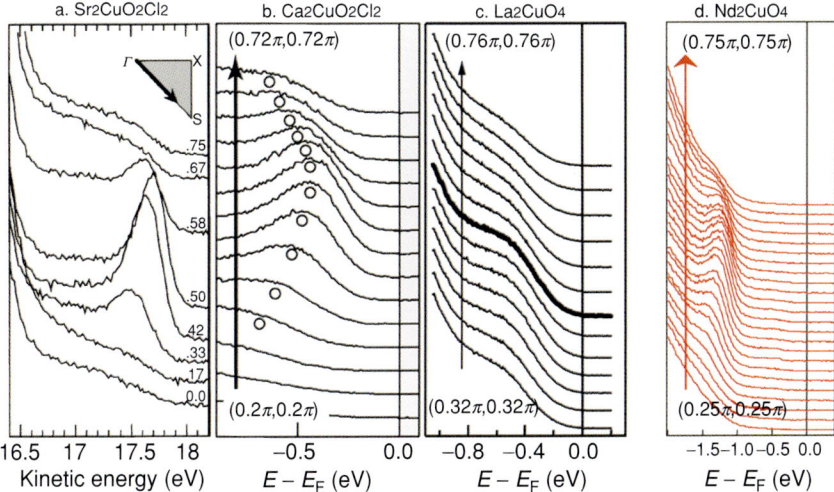

Figure 3.34. (a). Photoemission spectra along the $(0,0)$ and (π,π) direction in $Sr_2CuO_2Cl_2$ [156, 158], $Ca_2CuO_2Cl_2$ [113, 162], La_2CuO_4 [166, 167] and Nd_2CuO_4 [164].

then suddenly loses intensity after passing $(\pi/2,\pi/2)$ and bends back to high binding energy. This behavior can be more clearly seen in the image plot of Figure 3.35a [162] where the "band" breaks into two branches. The lowest binding energy feature shows a dispersion of ~ 0.35 eV while an additional band at high binding energy (Figure 3.35a) is very close to the unrenormalized band predicted by band theory [162]. The dispersion of low binding-energy band along the $(0,0)-(\pi,\pi)$ direction and other high symmetry directions are shown in Figure 3.35b by keeping track of the EDC peak position [51]. The total dispersion of the peak is ~ 0.35 eV. This is in contrast to the predictions of one-electron band calculations which gives an occupied band width of ~ 1.5 eV and total bandwidth of ~ 3.5 eV [168]. Nevertheless, it is consistent with the calculations from the t–J model where the predicted occupied bandwidth is $\sim 2.2J$ [50, 169]. This indicates that the dynamics of one-hole in an antiferromagnetic background is renormalized from scale t to scale J.

While the t–J model and experiments show agreement along the $(0,0)-(\pi,\pi)$ direction, there are discrepancies along other directions, such as the $(0,0)-(\pi,0)$ and $(0,\pi)-(\pi/2,\pi/2)-(\pi,0)$ directions [156]. The later intensive theoretical effort resolved this issue by incorporating the hopping to the second (t') and third (t'') nearest-neighbors [170]. More precise calculations of the dispersion in the $t-t'-t''-J$ model are performed by using a self-consistent Born approximation (SCBA) [171]. These calculations show a satisfactory agreement with experimentally derived dispersion, as shown in Figure 3.35b [51].

However, there remain a few prominent puzzles related to the interpretation of the photoemission data in undoped parent compounds [113]. The first prominent issue is the linewidth of the peak near $(\pi/2,\pi/2)$. As highlighted in Figure 3.36a, the width of the sharpest peak near $(\pi/2,\pi/2)$ is ~ 300 meV which is comparable with the entire occupied bandwidth $2J \approx 350$ meV [113]. This is much broader than that from t–J model calculations and too broad to be considered as a coherent quasiparticle peak for which the quasiparticle peak is basically resolution limited, as exemplified by the data on Sr_2RuO_4 in Figure 3.36a. An early attempt interpreted this anomalously large linewidth to additional interaction with a nonmagnetic boson bath of excitations, such as phonons [159]. But this interpretation meets with difficulty in

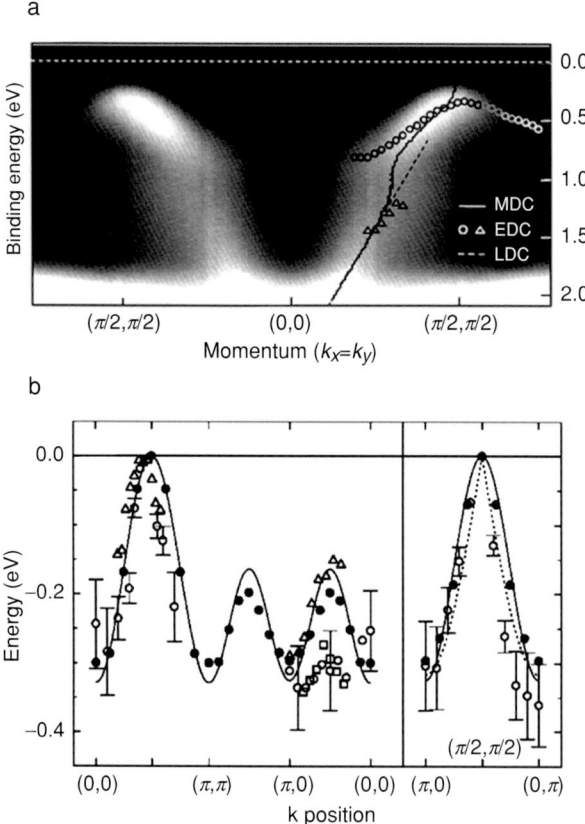

Figure 3.35. (a). Intensity plot of ARPES data as functions of the binding energy and momentum for $Ca_2CuO_2Cl_2$ along the $\Gamma(0,0)$–(π,π) direction [162]. The data was symmetrized around the Γ point. Also shown on the plot are the dispersions obtained by following the peak positions of the MDCs (solid line) and the EDCs (circle and triangles). The results are compared with the shifted dispersion from the LDA calculation (dashed line). (b) Energy dispersion of quasiparticle for insulating $Sr_2CuO_2Cl_2$ measured from the top of the band. Experimental data are taken from [156] (open circles), [157] (open triangles) and [158] (open squares). Solid circles: the results of the self-consistent Born approximation (SCBA) for the $t-t'-t''-J$ model with $t = 0.35$ eV, $t' = -0.12$ eV, $t'' = 0.08$ eV and $J = 0.14$ eV. The solid lines are obtained by fitting the SCBA data to a dispersion relation given by $E_0(k) + E_1(k)$, being $t'_{eff} = -0.038$ eV and $t''_{eff} = 0.022$ eV. The dashed line along the $(\pi, 0)$–$(0,\pi)$ direction represents the spinon dispersion from [172].

explaining little renormalization in the dispersion from this "extra interaction" because dispersion and linewidth are closely related. A diagrammatic quantum Monte Carlo study [175] showed that this problem can be resolved by considering the polaron effect in the t–J model. Namely the dispersion for the center of mass of the spectral function obeys that of the pure t–J model, while the lineshape is strongly modified. The details of this will be given below.

Another unresolved issue is the chemical potential μ. For an insulator, μ is not well defined, and may be pinned by surface defects or impurities and will vary between different samples. If one considers that the peak A in Figure 3.36a represents a quasiparticle peak, one would expect the chemical potential to vary anywhere above the top of this valence band. However, the experimental chemical potential clearly sets a lower bound that is ~ 0.45 eV apart from the peak A (Figure 3.36b) [113]. Shen et al. [113] invented a new method to

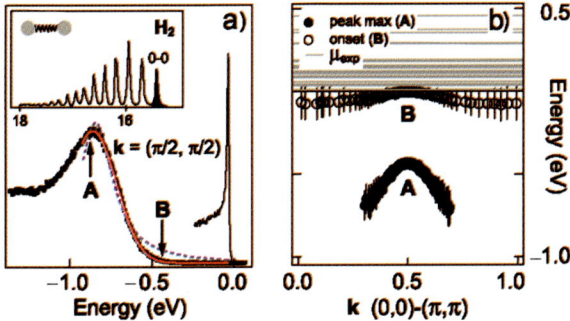

Figure 3.36. (a) Photoemission spectrum of $Ca_2CuO_2Cl_2$ at $k = (\pi/2, \pi/2)$ with fits to a Lorentzian spectral function (dashed) and Gaussian (red or gray) [113]. A and B denote the peak maximum and the onset of spectral weight, respectively. Comparison with Fermi-liquid system Sr_2RuO_4 is shown (thin black). Upper inset shows photoemission spectra from H_2 [173]. (b) Dispersion of A and B along $(0,0)$–(π,π), along with experimental values for the chemical potential μ (lines).

determine the chemical potential using both the energy of the nonhybridized oxygen orbital and the detailed line-shape of $(Ca_{2-x}Na_x)CuO_2Cl_2$ (Na–CCOC).

The resolution of these discrepancies between experiment and expectation leads to identifying polaron physics as responsible for the bulk of the lineshape in underdoped cuprates. In fact, the photoemission spectra in the underdoped cuprates resemble the Frank–Condon effect seen in photoemission spectra of molecules such as H_2 [173] (inset of Figure 3.36), where only the "0–0" peak (filled black) represents the H final state with no excited vibrations and comprises only $\sim 10\%$ of the total intensity. In the solid state, this "0–0" would correspond to the quasiparticle or the coherent part of the spectral function, A_{coh}, whereas the excited states comprise the incoherent part, A_{inc}. This behavior is reminiscent of polarons, and such models have been invoked in systems where strong couplings are present [174]. In this picture, in the undoped compound, the true QP (B) is hidden within the tail of spectral intensity, with a quasiparticle residue Z vanishingly small, while feature A is simply incoherent weight associated with shake-off excitations.

From the viewpoint of polaron physics, the cuprates offer a unique and first opportunity to compare experimental spectra with theory in detail. The single hole interacts both with magnons and phonons. The hole–magnon interaction has been successfully analyzed in terms of the self-consistent Born approximation [171]. The success of the Born approximation results from a "saturation" effect; namely the single spin 1/2 can flip only once, and hence magnon clouds do not become large enough to induce the self-trapping transition to the small polaron. On the other hand, phonon clouds can be larger and larger as the coupling constant g increases and can lead to a self-trapping transition. The t–J model coupled to phonons in the polaronic regime has illuminated one-hole dynamics in the parent compound in the following way [175]. (1) With increasing electron–phonon coupling strength, the spectral function experiences a transition from weak-coupling, to intermediate coupling, and to strong-coupling regimes. (2) In the strong coupling regime, the spectral function consists a ground state resonance (as indicated by vertical arrows) with vanishing intensity and a broad peak denoted as "coherent C," as shown in Figure 3.37a. (3) The broad peak C shows strong momentum dependence while the lowest state is dispersionless. These results are in good correspondence to

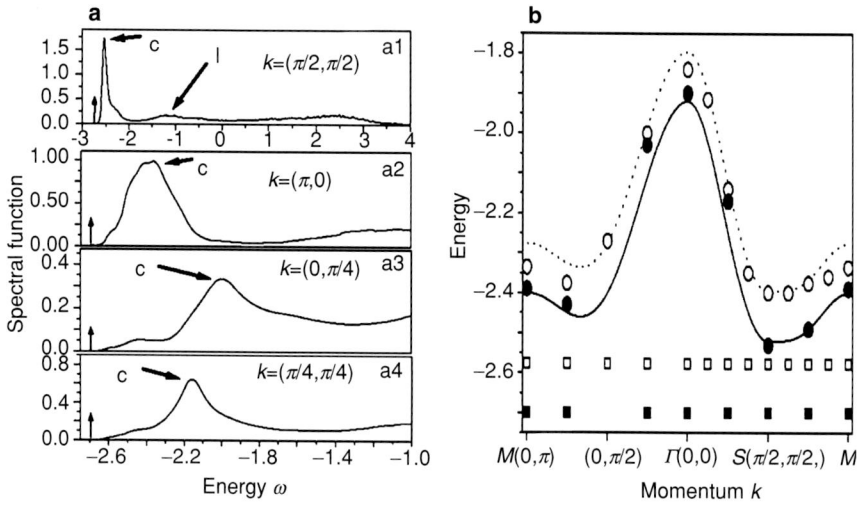

Figure 3.37. (a) Calculated hole spectral function in ground state at $J/t = 0.3$ for different momenta [175]. (a1) Full energy range for $k = (\pi/2, \pi/2)$. (a2)–(a4) Low energy part for different momenta. Slanted arrows show broad peaks which can be interpreted in ARPES spectra as "coherent" (C) and incoherent (I) part. Vertical arrows indicate position of ground state resonance which is not seen in the vertical scale of the figure. (b) Dispersion of resonance energies at $J/t = 0.3$. Broad resonance (filled circles) and lowest polaron resonance (filled squares) at $g = 0.231125$; third broad resonance (open circles) and lowest polaron resonance (open squares) at $g = 0.2$. The solid curves are dispersions of a hole in the pure t–J model at $J/t = 0.3$.

experimental observations. The most surprising result is that the broad resonance has the momentum dependence of the t–J model without coupling to phonons (shown in Figure 3.37b). In the Franck–Condon effect for molecules a similar result occurs. The center of the shake-off band corresponds to the hole motion in the background of the frozen lattice configuration, i.e, the dispersion of the hole remains that of the noninteracting limit, while the line-width broadens. A more elaborated analytic treatment of the t–J polaron model in the Franck–Condon approximation [176] successfully reproduced this Monte Carlo results. The calculated spectral function line-shape most consistent with experiment has a $\lambda \cong 0.9$–1.3, well within the strong-coupling, small-polaron regime. Recent realistic shell model calculation [167] also concluded $\lambda = 1.2$ for La_2CuO_4.

In La_2CuO_4, a broad feature near –0.5eV (Figure 3.38) was identified as the lower Hubbard band [165, 166]. The electron–phonon coupling strength, calculated using a shell model, puts La_2CuO_4 in the polaron regime, similar to $Ca_2CuO_2Cl_2$. In this picture, the –0.5eV feature corresponds to the phonon side-band while the real quasiparticle residue is very weak. As shown in Figure 3.38, the calculated spectral function agrees well with the measured data [167].

Doping Dependence: From Z~0 Polaron to Finite Z Quasiparticles

We next turn to the question of how the small polaron state evolves as a function of doping, connecting to the Migdal–Eliashberg regime discussed in Section C. There are two possible ways to dope the Mott insulator, schematically shown in Figure 3.10c [42, 177] (1) Upon doping, the chemical potential shifts to the top of the valence band for hole doping (Figure 3.10c3) or to the bottom of the conduction band for electron doping (Figure 3.10c4).

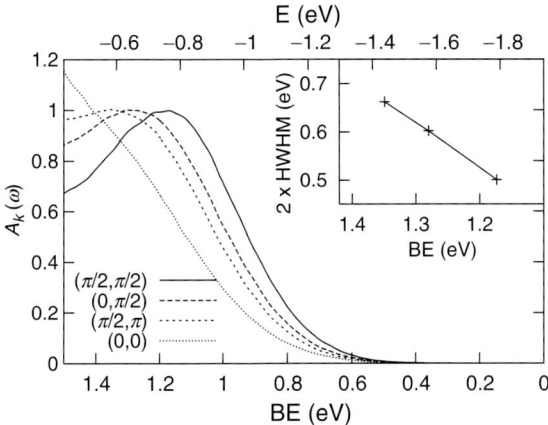

Figure 3.38. Polarons in La_2CuO_4 [167]. Calculated ARPES spectra for the undoped La_2CuO_4 system at $T = 0$ for different **k** normalized to the height of the phonon side band. The lower abscissa shows binding energies (BE) and the upper abscissa the energies of the final states corresponding to the spectral features. The inset shows the dependence of the width of the phonon side band on its binding energy. The width of the (0, 0) spectrum is poorly defined and not shown.

(2) The chemical potential is pinned inside the charge transfer gap. Upon doping, new states will form inside the gap (Figure 3.10c5).

Recent ARPES measurements on lightly doped $(La_{2-x}Sr_x)CuO_4$ compounds provide a good window to look into this issue. As shown in Figure 3.39a, e for undoped La_2CuO_4, the main feature is the broad peak near –0.5 eV which exhibits weak dispersion [166]. There is also little spectral weight present near the Fermi level. However, upon only a doping of $x = 0.03$, the electronic structure undergoes a dramatic change. A new dispersive band near the Fermi level develops along the $(0,0)-(\pi,\pi)$ nodal direction (Figure 3.39e, right panel), while along the $(0,0)-(\pi,0)$ direction a saddle band residing –0.2eV below the Fermi level develops. Even for more lightly doped samples, such as $x = 0.01$, new states near the Fermi level are created [139]. Note that, for these lightly doped samples, the original –0.5 eV remains, although with weakened spectral weight (Figure 3.39d). So, the –0.5 eV peak and the new dispersive band coexist at doping levels close to the parent compounds.

The systematic evolution of the photoemission spectra near the nodal and antinodal regions with doping in LSCO is shown in Figure 3.40a,b [166]. The nodal quasiparticle weight, Z_{QP}, integrated over a small energy window near the Fermi level, is shown in Figure 3.40c. In the underdoped region, it increases with increasing doping nearly linearly, and no abrupt change occurs near the nonsuperconductor−superconductor transition at $x \sim 0.05$.

$(Ca_{2-x}Na_x)CuO_2Cl_2$ (Na−CCOC) is another ideal system to address the doping evolution of the electronic structure. The precise measurement of the chemical potential (Figure 3.41a), in conjunction with the identification of polaron physics in the underdoped compounds, provides a globally consistent picture of the doping evolution of the cuprates [113]. Instead of measuring the chemical potential with deep core level spectroscopy (the usual method) [179], one utilizes orbitals in the valence band at lower energies (Figure 3.41a,b). The measured chemical shift, $\Delta \mu$, exhibits a strong doping dependence, $\partial \mu / \partial x = -1.8 \pm 0.5$ eV/hole, comparable to the band structure estimation ($\sim$–1.3 eV/hole) (Figure 3.41c).

Figure 3.42(a-d) show the doping evolution of the near-E_F EDCs plotted relative to μ_0 of the undoped sample (determined in Figure 3.41c). With doping, feature A evolves smoothly

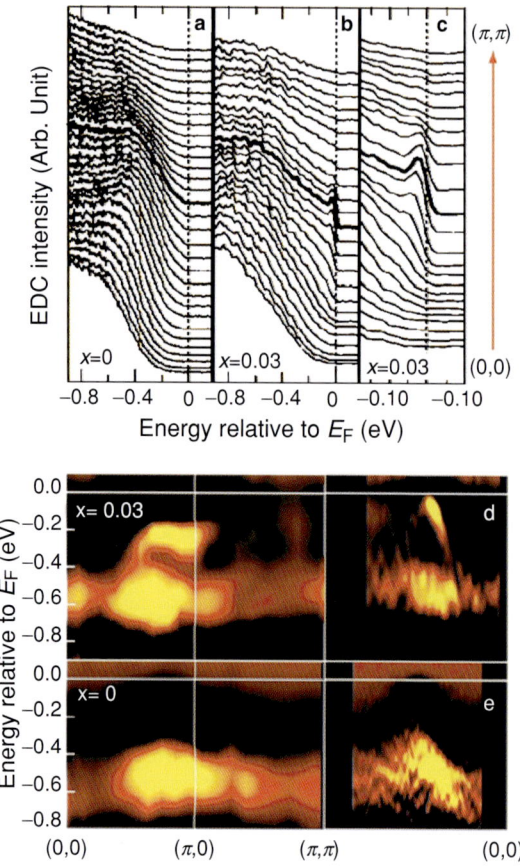

Figure 3.39. Creation of nodal quasiparticles in lightly doped LSCO [166]. ARPES spectra for LSCO with $x = 0$ and $x = 0.03$. Panels a and b are EDC's along the nodal direction $(0,0)$–(π, π) in the second Brillouin zone (BZ). The spectra for $x = 0.03$ are plotted on an expanded scale in panel c. Panels d and e represent energy dispersions deduced from the second derivative of the EDC's.

into a broad, high energy hump with a backfolded dispersion similar to the parent insulator (symbols), while B shifts downward relative to its position in the undoped compound. Spectral weight increases with doping at B, and a well-defined peak emerges for the $x = 0.10$ and 0.12 samples, resulting in a coherent, low-energy band. The dispersion of the high-energy hump (A), tracked using the local maxima or second derivative of the EDCs, shows little change as a function of doping (Figure 3.41e). The lowest energy excitations (feature B, $-0.05\,\text{eV} < E < E_F$), tracked using MDC analysis, evolve with doping in such a way that the quasiparticle dispersion (v_F) and Fermi wave vectors (k_F) virtually collapse onto a single straight line.

Doping Evolution of Fermi Surface: Nodal–Antinodal Dichotomy

So far, we have discussed the doping evolution along the nodal direction, and seen that a sharply defined quasiparticle peak develops out of the small weight near the chemical potential in the undoped samples. We now discuss the doping evolution in other directions of

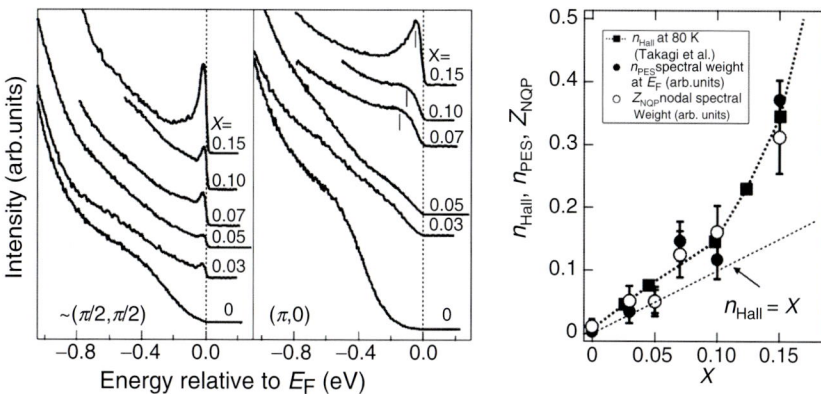

Figure 3.40. ARPES spectra at $\mathbf{k} = \mathbf{k}_F$ in the nodal direction in the second BZ (a) and those at $(\pi, 0)$ (b) for various doping levels in $(La_{2-x}Sr_x)CuO_4$ [166]. (c) Doping dependence of the nodal QP spectral weight, Z_{NQP}, and the spectral weight integrated at E_F over the entire second Brillouin zone, n_{PES} [166]. They show similar doping dependence to the hole concentration evaluated from Hall coefficient (n_{Hall}) [178].

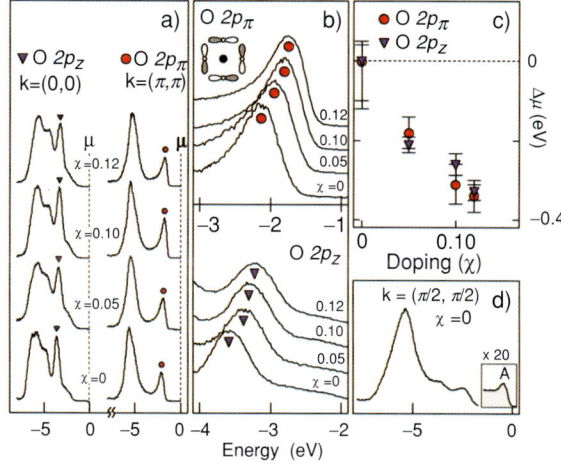

Figure 3.41. Chemical potential shift in Na–CCOC [113]. (a) Valence band spectra for $x = 0, 0.05, 0.10$, and 0.12 compositions at $k = (0, 0)$ and (π, π). O $2p_z$ and O $2p_\pi$ states are marked by triangles and circles, respectively. (b) Shifts of the O $2p_z$ and O $2p_\pi$ peaks shown on an expanded scale. (c) Doping dependence of chemical potential $\Delta\mu$ determined from (b). (d) Valence band at $k = (\pi/2, \pi/2)$, showing the lower Hubbard band (A) on an expanded scale.

the Brillouin zone. Surprisingly, one finds that the coherent peak near the Fermi level in the lightly doped samples is confined to the nodal region, and quickly disappears with momentum around the Brillouin zone. The spectral weight near the Fermi level, confined to the $(\pi/2, \pi/2)$ nodal region, forms a so-called "Fermi arc." This dichotomy between nodal and antinodal excitations is shown in Figure 3.43 [134]. For the $x = 0.063$ sample, which is close to the nonsuperconductor–superconductor transition and therefore heavily underdoped, the spectral weight near Fermi level is mainly concentrated near the nodal region (Figure 3.43a). The coherent peaks in the EDCs (Figure 3.43c1) near the nodal region disappear as one approaches the antinodal region, where the EDCs exhibit a step rather than a peak. The LSCO $x = 0.09$

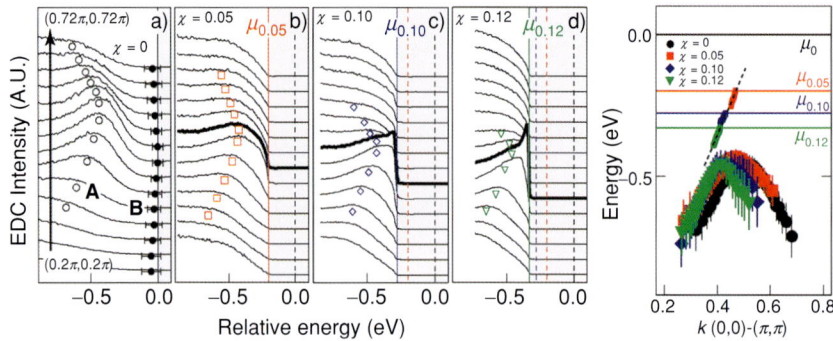

Figure 3.42. (a–d) EDC spectra of Na–CCOC $x = 0$ (a), 0.05 (b), 0.10 (c), and 0.12 (d) from $(0.2\pi, 0.2\pi)$ to $(0.72\pi, 0.72\pi)$ with hump positions marked by open symbols and the EDC at k_F shown in bold [113]. Data are plotted on a relative energy scale referenced to the shift in μ shown in Figure 3.41c (e) Summary of hump (symbols) from Figure (a–d) and MDC dispersions (lines).

sample exhibits similar behavior (Figure 3.43c2). In contrast, for overdoped LSCO such as $x = 0.22$ (Figure 3.43c3), sharp peaks are observable along the entire Fermi surface. These observations indicate that the electrons near the antinodal region experience additional scattering in underdoped samples. Therefore, as shown in Figure 3.44, the "Fermi surface" in LSCO evolves from the "Fermi arc" in lightly doped samples, to a hole-like Fermi surface in underdoped samples, and to an electron-like Fermi surface in overdoped samples ($x > 0.15$).

The evolution of electronic structure with doping in $(Ca_{2-x}Na_x)CuO_2Cl_2$ exhibits marked resemblance to that in $(La_{2-x}Sr_x)CuO_4$ [163]. As summarized in Figure 3.45, at low doping, the quasiparticle weight is again confined to the nodal region and the weight in the quasiparticle peak, Z_{qp}, increases with increasing doping, consistent with LSCO. In the Na−CCOC system, recent scanning tunneling microscopy (STM) work has revealed a real space pattern of $4a_0 \times 4a_0$ two-dimensional charge ordering [182]. In momentum space, as seen from Figure 3.46, strong Fermi surface nesting exists in Na−CCOC with a nesting vector insensitive to doping close to $2 \times \pi/4$ that may account for the broad near-E_F spectra in the antinodal region. In LSCO, neutron scattering has also indicated the existence of dynamic stripes [183]. These similarities suggest an intrinsic commonality between the low-lying excitations across different cuprate families and may imply a generic microscopic origin for these essential nodal states irrespective of other ordering tendencies. At very low doping levels, the nodal excitations should entirely dominate the transport properties, consistent with the high-temperature metallic tendencies observed in very lightly doped cuprates [184]. Thus any microscopic models of charge ordering must simultaneously explain and incorporate the existence of coherent nodal states and broad antinodal excitations.

The nodal−antinodal dichotomy of quasiparticle dynamics in the normal state also exists in Bi2212 [185]. A number of possible mechanisms have been proposed to account for the antinodal spectral broadening in the normal state. A prime candidate is the (π, π) magnetic excitations observed in various cuprates [79–81]. As schematically shown in Figure 3.43b, this excitation will give rise to "hot spots" on the Fermi surface that can be connected by (π, π) momentum transfer. Electrons around these hot spots experience additional scattering from the (π, π) magnetic scattering. The same mechanism has also been proposed for $(Nd_{2-x}Ce_x)CuO_4$ for which the spectral broadening is localized to the expected "hot spot" [186]. However, in LSCO, the same magnetic response, magnetic resonance mode, is

Electronic Structure and Electron–Phonon Coupling in Cuprate Superconductors

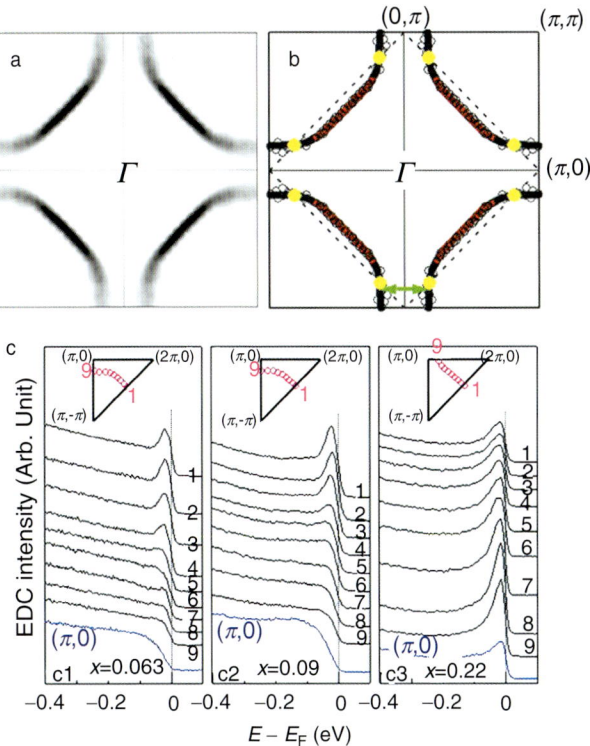

Figure 3.43. Dichotomy between nodal and antinodal excitations in LSCO [134]. (a) Spectral weight near a small energy window of Fermi level as a function of k_x and k_y for LSCO $x = 0.063$ sample measured at ∼20 K. The original data was taken in the second Brillouin zone and converted into the first Brillouin zone and symmetrized under fourfold symmetry. (b) Experimental Fermi surface for LSCO $x = 0.063$ sample. The black open circles are obtained from the MDC peak position at E_F. The solid lines are guides to the eye for the measured Fermi surface. The red lines represent the portion of Fermi surface where one can see quasiparticle peaks. The dotted black line represents the antiferromagnetic Brillouin zone boundary; its intersection with the Fermi surface gives eight hot spots (solid yellow circles) from (π,π) magnetic excitations. The double-arrow-ended green line represents a nesting vector between the antinodal part of the Fermi surface. (c) EDCs on Fermi surface for LSCO $x = 0.063$ (c1), 0.09 (c2), and 0.22 (c3) samples. All samples are measured at ∼20 K. The corresponding momentum position is marked in the upper inset of each panel. Also included are the spectra at $(\pi,0)$ points, colored as blue.

not observed. Instead, incommensurate magnetic peaks are observed at low energy (below 15 meV) [183], which broaden rapidly with increasing energy although the magnetic fluctuation can persist up to 280 meV [187]. Intrigued by the fact that the extra broadening sets in when the Fermi surface turns from the $(\pi,0)-(0,\pi)$ diagonal direction to the $(0,0)-(\pi,0)$ or the $(0,0)-(0,\pi)$ direction in LSCO sample with x = 0.063 (Figure 3.43b, c), an alternative mechanism was proposed [134] in which the scattering in question causes a pair of electrons on two parallel antinodal segments to be scattered to the opposite ones (Figure 3.43b). In the normal state, this scattering can cause a quasiparticle to decay into two quasiparticles and one quasihole. The antinodal spectral broadening occurs as a result of the frequent occurrence of such a decay which renders the normal state quasiparticle ill defined.

Another potential explanation for the broad antinodal features may come from models based on the polaron picture discussed before [113, 167, 175]. In such a scenario, the strong

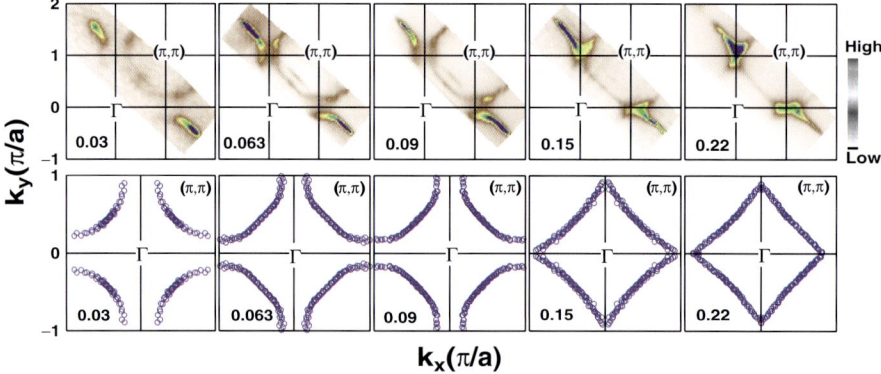

Figure 3.44. Doping evolution of underlying "Fermi surface" in $(La_{2-x}Sr_x)CuO_4$ [134, 166, 180, 181]. The data were measured at a temperature of ∼20 K.

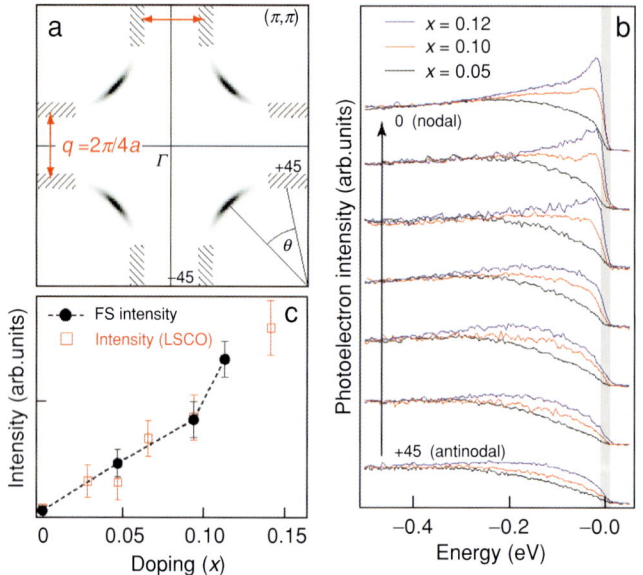

Figure 3.45. Dichotomy between nodal and antinodal excitations in Na–CCOC [163]. (a) Schematic of the low-lying spectral intensity for $(Ca_{2-x}Na_x)CuO_2Cl_2$ ($x = 0.10$). The hatched regions show the nested portions of Fermi surface, and the Fermi surface angle is defined in the lower right quadrant. (b) EDCs taken at equal increments along the FS contour from the nodal direction (top) to the antinodal region (bottom) for $x = 0.05$, 0.10, and 0.12 at a temperature of 15 K. (c) The doping evolution of the low-lying spectral weight (circles), along with corresponding data from $La_{2-x}Sr_xCuO_4$ (squares), with the error bars representing the uncertainty in integrated weight as well as sample-to-sample variations.

coupling of the electrons to any bosonic excitations would result in $Z \ll 1$, and spectral weight is transferred to incoherent, multiboson excitations. An effective anisotropic coupling could lead to a larger Z (weaker coupling) along the nodal direction and a much smaller, yet still finite Z, at the antinodes (strong coupling). In this picture, the antinodal polaron effect in LSCO (Figure 3.43c) [134] is much weaker than Na−CCOC (Figure 3.45b) [163] if one compares the spectral weight near Fermi level around the antinodal region. Regardless of the

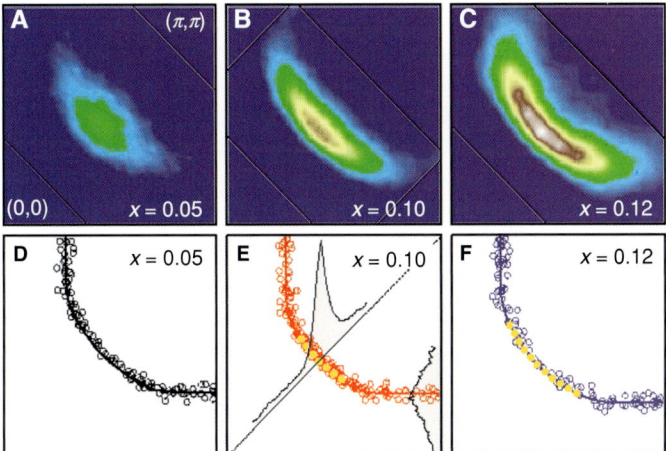

Figure 3.46. Doping evolution of "Fermi surface" in Na-CCOC [163]. (A – C) The momentum distribution of spectral weight within a ±10-meV window around E_F for $x = 0.05$, 0.10, and 0.12 in one quadrant of the first Brillouin zone. Data were taken at 15 K and symmetrized along the (0,0)–(π,π) line. The data acquisition range is shown within the black lines. The FS contours shown in (D – F) were compiled from more than four samples for each composition with different photon energies and photon polarizations. Data from these samples constitute the individual points; the best fit is shown as a solid line. The region in which a low-energy peak was typically observed is marked by gold circles. The gray shaded areas in (E) represent the momentum distribution of intensity at $E_F \pm 10$ meV along the (0,0)–(π,π) and ($\pi,0$)–(π,π) high-symmetry directions.

microscopic explanation, the broad and nested antinodal FS segments observed by ARPES are consistent with the propensity for two-dimensional charge ordering in the lightly doped cuprates seen in STM experiments on Na−CCOC [182] and Bi2212 [188–190]. Furthermore, an explanation based on an anisotropic coupling (coming from either polaron physics or the magnetic resonance) may not be sufficient to cause the two-dimensional charge order; it may be a combination of strong coupling and Fermi surface nesting which ultimately stabilizes the antinodal charge-ordered state.

3.4.5. Electron–Phonon Coupling and High Temperature Superconductivity

Much of the physics discussed in this review has attributed essential features of the ARPES data to electron−phonon coupling, and if not to electron−phonon coupling alone, to electron−phonon coupling in an antiferromagnetic background. The question remains as to how this electron−phonon coupling can account for high-temperature superconductivity with d-wave pairing seen in the cuprates. It is often assumed that el−ph coupling leads to s-wave pairing, and that therefore such a mechanism contradicts with the d-wave symmetry of the Cooper pairing in the cuprates. Instead, electronic correlations have been thought to be consistent with d-wave pairing. However, while strong electronic correlations will suppress the Cooper pair amplitude on the same orbital, and hence induce a d-wave like symmetry, they do not tell us much about the explicit pairing mechanism. One of the early studies on possible phononic mechanisms of high T_c superconductivity [154] pointed out that the out-of-plane displacement of the oxygen, i.e., the buckling mode, combined with antiferromagnetic correlations, leads to $d_{x^2-y^2}$ pairing. Bulut and Scalapino [191] studied the various phonon modes from the viewpoint of the possible pairing force. They found that the interaction which becomes more positive as the momentum transfer increases helps $d_{x^2-y^2}$ pairing (the case for buckling mode, but not the case for the apical oxygen mode or the in-plane breathing mode).

One can understand the nature of the q momentum dependence by considering how the phonon couples to the electron density. For deformation phonons, the coupling is dipolar driven and thus small for small q, the case for the breathing modes. This also includes infrared active phonons. Yet for Raman active modes, which couple via the creation of isotropic and quadrupolar moments, the coupling is generally strongest for small q. Specifically for the cuprates, such strong k, q dependencies occur explicitly for c-axis phonons, which include the Raman active in-phase buckling A_{1g}, out-of-phase buckling B_{1g} and modes involving the apical oxygen A_{1g}. The k momentum dependence comes from the phonon eigenvectors as well as the direction of charge-transfer induced by the phonon. For example, for the B_{1g} phonon the eigenvectors enforce a change of sign when k_x and k_y are interchanged, a factor $\sim \cos(k_x a) - \cos(k_y a)$, while for the apical charge transfer coupling between Cu and the three oxygen orbitals, a factor $\sim [\cos(k_x a) - \cos(k_y a)]^2$ emerges.

As discussed by Bulut and Scalapino [191] among others, the q dependence of phonons can be important to give $d_{x^2-y^2}$ pairing. In particular, if the attractive electron–phonon interaction falls off for momentum transfers q along the diagonal, then conceptually the interaction is of the same structure as the magnetic pairing from antiferromagnetic spin fluctuations. This type of structure occurs for both B_{1g} and A_{1g} c-axis Raman-active phonons, and thus they contribute to the pairing interaction in the d-wave channel, parameterized by λ_d

$$\lambda_d = \frac{2 \sum_{k,k'} d_k d_{k'} \mid g(k, k-k') \mid^2 \delta(\varepsilon_k) \delta(\varepsilon_{k'})}{\Omega_{ph} \sum_k \delta(\varepsilon_k) d_k^2} \quad (3.17)$$

with the d-wave form factor $d_k = [\cos(k_x a) - \cos(k_y a)]/2$. However, the A_{1g} phonons predominantly contribute to the s-pairing channel (replacing d_k by 1 in the above equation) in the absence of any Coulomb interaction, leaving the B_{1g} phonon as the largest contributor to d-wave pairing, as found in LDA studies [192].

However, Coulomb interactions change this picture. They cannot be neglected since they are necessary to screen the long-wavelength nature of isotropic charge fluctuations. The screened electron–phonon interaction $\bar{g}$ is of the form

$$\bar{g}(k, q, \Omega) = g(k, q) + \frac{V(q) \Pi_{g,1}(q, \Omega)}{1 - V(q) \Pi_{1,1}(q, \Omega)}, \quad (3.18)$$

where $V(q) = 4\pi e^2/q^2$ is the 3D Coulomb interaction and $\Pi_{a,b}(q, \Omega)$ is the frequency-dependent polarizability calculated with vertices a, b, respectively. Note if g were independent of momentum, then the effective electron–phonon coupling would be screened by the dielectric function $\varepsilon(q, \Omega) = 1 - V(q) \Pi_{1,1}(q, \Omega)$. Particularly in the limit $q \to 0$ we recover complete screening and $\bar{g} = 0$ for $\Omega = 0$, restating particle number conservation, while for $\Omega = \Omega_{ph}$ the renormalized coupling is of order Ω_{ph}/Ω_{pl}. However, any fermion k-dependence of the electron–phonon coupling survives screening *even at $q = 0$* as shown by Abrikosov and Genkin [193], and the effective charge vertex in this limit is $\bar{g}(k, q \to 0) = g(k, q \to 0) - \delta g$, with $\delta g = \langle g(k, q \to 0) \rangle$, and $\langle \cdots \rangle$ denotes an average over the Fermi surface, defined as

$$\langle A \rangle = \frac{\sum_k A(k) \delta(\varepsilon(k))}{\sum_k \delta(\varepsilon(k))}. \quad (3.19)$$

Thus screening removes the constant part of the electron–phonon interaction and can highlight the d-wave channel. This is important if the bare coupling is highly anisotropic with the Fermionic momentum k, the case of the apical oxygen coupling.

Moreover, the issue of strong local correlations on electron–phonon interactions has been recently readdressed by the Hubbard X operator method [90, 194] and quantum Monte Carlo simulations [195]. Assuming no specific phonon and that phonons couple to the on-site charge density, i.e., diagonal coupling, these works found enhanced forward scattering (i.e., small momentum transfer), while large momentum transfer process were suppressed. Therefore, $d_{x^2-y^2}$ pairing can occur by el–ph coupling. Furthermore, the vertex correction explains the absence of phonon features in the resistivity, since the transport relaxation rate contains the factor $1-\cos\theta$ (θ: the angle between the initial and final state momenta) which reduces the contribution for forward scatterings. There has been controversy as to whether the vertex correction for the off-diagonal el–ph coupling, which modulates the bond, also enhances forward scattering and suppresses large momentum transfers [110]. The in-plane half-breathing mode, which modulates the bond, exhibits a sharp softening with doping in neutron scattering [107] and has been studied in particular. The Zhang–Rice singlet couples to the half-breathing mode much stronger than estimated in LDA calculations, and the vertex correction leads to an effective attractive interaction for $d_{x^2-y^2}$ pairing [196]. On the other hand, later analysis [197] has shown that the cancellation of terms reduces the off-diagonal coupling, and the diagonal coupling dominates even after the vertex correction has been taken into account. In understanding the effects of this vertex correction on experimental spectra, one should note that the correction works differently for phononic and electronic self-energies. The sum rules [198] conclude that the phononic self-energy is reduced by an additional factor of x (hole concentration) as compared to the electronic self-energy. Intuitively, the difference between phononic and electronic self-energies arises because a small number of holes cannot influence phonons as much as phonons, in which atoms vibrate at every site, can influence a single hole.

In summary, local coulomb repulsion suppresses charge density modulations, which in turn decreases the strength of the electron–phonon interaction at large momentum transfers. This has two effects: first, as a consequence the contribution of *all* phonons to the resistivity will be reduced by the correlation effect. Second, and more relevant to pairing, small q phonons will have an accentuating λ for d-wave pairing since the coupling will decrease faster for large q than without correlations. Thus it appears that Coulomb interactions in general can have a dramatic impact on electron–phonon driven $d_{x^2-y^2}$ pairing. However, theoretical developments are still needed in order to treat the simultaneous importance of strong correlations and electron–phonon coupling. This is a promising direction for future research.

3.5. Summary

ARPES experiments have been instrumental in identifying the electronic structure, observing and detailing the electron–phonon mode coupling behavior, and mapping the doping evolution of the high-T_c cuprates. The spectra evolve from the strongly coupled, polaronic spectra seen in underdoped cuprates to the Migdal-Eliashberg like spectra seen in the optimally and overdoped cuprates. In addition to the marked doping dependence, the cuprates exhibit pronounced anisotropy with direction in the Brillouin zone: sharp quasiparticles along the nodal direction that broaden significantly in the antinodal region of the underdoped cuprates, an anisotropic electron–phonon coupling vertex for particular modes identified in the optimal and overdoped compounds, and preferential scattering across the two parallel pieces of Fermi surface in the antinodal region for all doping levels. This also contributes to the pseudogap effect. To the extent that the Migdal–Eliashberg picture applies, the spectra of the cuprates bear resemblance to that seen in established strongly coupled electron–phonon superconductors such as Pb. On the other hand, the cuprates deviate from this conventional picture. In the underdoped regime, the carriers are best understood as small polarons in an antiferromagnetic, highly electron correlated background, while the doped compounds require an anisotropic electron–phonon vertex to detail the prominent mode coupling signatures in the superconducting state. Electronic vertex corrections to the electron–phonon coupling furthermore may enhance, and for certain phonons, determine, the anisotropy of the electron–phonon coupling. A consistent picture emerges of the cuprates, combining strong, anisotropic electron–phonon coupling, particular phonon modes that could give rise to such a coupling, and an electron–electron interaction modifying the el–ph vertex. Such a combination, albeit with further experimental and theoretical effort, may indeed lead to an understanding of the high-critical transition temperature with d-wave pairing in the cuprate superconductors.

Acknowledgments

We would like to first thank our colleagues and collaborators on the photoemission work: N. P. Armitage, Felix Bauberger, Pavel V. Bogdanov, Veronique Brouet, Yulin Chen, Andrea Damascelli, D. Dessau, J. F. Douglas, Donglai Feng, Atsushi Fujimori, M. Z. Hasan, Zahid Hussain, Nik J. C. Ingle, Scot A. Kellar, C. Kim, Alessandra Lanzara, Weisheng Lee, Donghui Lu, Norman Mannella, W. Meevasana, T. Mizokawa, Jin Nakamura, Filip Ronning, Kyle M. Shen, Z. Sun, K. Tanaka, B. Wannberg, Wanli Yang and Teppei Yoshida. We are grateful for our collaborators on high-quality samples that are critical to the success of our photoemission experiment: Yoichi Ando, M. Azuma, Hiroshi Eisaki, M. Greven, Genda Gu, A. Iyo, T. Kakeshita, N. Kaneko, K. Kishio, Y. Kohsaka, Seiki Komiya, Chengtian Lin, L. L. Miller, Takuya Noda, Takao Sasagawa, J.-I. Shimoyama, H. Takagi, M. Takano, J. W. X. Ti, Shinichi Uchida, Jiwu Xiong, Fang Zhou and Zhongxian Zhao.

We would also like to thank our collaborators on the theoretical effort and discussions: O. K. Andersen, A. V. Balatsky, Arun Bansil, T. Egami, O. Gunnarsson, W. Hanke, S. Ishihara, A. Kapitulnik, S. Kivelson, Bob Laughlin, Dung-Hai Lee, T. K. Lee, M. Lindroos, S. Maekawa, R. S. Markiewicz, A. S. Mishchenko, Ward Plummer, O. Rosch, Seppo Sahrakorpi, G. Sawatzky, D. J. Scalapino, Junren Shi, T. Tohyama, J. Zaanen, and Zhengyu Zhang. The work at the ALS and SSRL is supported by the DOE's Office of BES, Division of Material Science, with contract DE-FG03-01ER45929-A001 and DE-AC03-765F00515. The work at Stanford was also supported by NSF grant DMR-0304981 and ONR grant N00014-04-1-0048-P00002. NN is supported by NAREGI project and Grant-in-Aids (Grant No. 15104006, No. 16076205, and No. 17105002) from the Ministry of Education, Culture, Sports, Science, and Technology. TPD acknowledges support from NSERC, PREA, ONR grant N00014-05-1-0127 and the A. von Humboldt Foundation. XJZ is also supported by "100-Talent" Project of CAS and the Outstanding Young Researcher Foundation.

Bibliography

1. J. G. Bednorz and K. A. Müller, Z. Phys. B **64**, 189 (1986).
2. M. K. Wu, J. R. Ashburn, C. J. Torng, P. H. Hor, R. L. Meng, L. Gao, Z. J. Huang, Y. Q. Wang, and C. W. Chu, Phys. Rev. Lett. **58**, 908 (1987).

3. C. E. Gough et al., Nature **326**, 855 (1987).
4. J. Bardeen, L. N. Cooper, and J. R. Schrieffer, Phys. Rev. B **108**, 1175 (1957).
5. W. L. McMillan and J. M. Rowell, Phys. Rev. Lett. **14**, 108 (1965).
6. For a recent review, see F. Marsiglio and J. P. Carbotte, in *The Physics of Conventional and Unconventional Superconductors,* K. H. Bennemann and J. B. Ketterson (eds.) (Springer, Berlin Heidelberg, New York).
7. See, e.g., C. C. Tsuei and J. R. Kirtley, Rev. Mod. Phys. **72**, 969 (2000) and references therein.
8. D. J. Scalapino, Phys. Rep. **250**, 329 (1995).
9. D. Pines and P. Monthoux, J. Phys. Chem. Solid **56**, 1651 (1995).
10. D. J. Scalapino, Science **284**, 1282 (1999); J. Orenstein, Nature **72**, 333 (1999), and references therein.
11. C. M. Varma, P. B. Littlewood, and S. Schmitt-Rink, Phys. Rev. Lett. **63**, 1996 (1989).
12. For a review, see T. Timusk and B. Statt, Rep. Prog. Phys. **62**, 61 (1999).
13. Z.-X. Shen and D. S. Dessau, Phys. Rep. **253**, 1 (1995).
14. A. Damascelli, Z. Hussain, and Z.-X. Shen, Rev. Mod. Phys. **75**, 473 (2003).
15. J. C. Campuzano, M. R. Norman, and M. Randeria, in *Physics of Superconductors*, vol. II, K. H. Bennemann and J. B. Ketterson (eds.) (Springer, Berlin Heidelberg, New York, 2004), pp. 167–273.
16. C. G. Olson, R. Liu, A.-B. Yang, D. W. Lynch, A. J. Arko, R. S. List, B. W. Veal, Y. C. Chang, P. Z. Jiang, and A. P. Paulikas, Science **245**, 731 (1989).
17. Z.-X. Shen, D. S. Dessau, B. O. Wells, D. M. King, W. E. Spicer, A. J. Arko, D. Marshall, L. W. Lombardo, A. Kapitulnik, P. Dickinson, S. Doniach, J. DiCarlo, T. Loeser, and C. H. Park, Phys. Rev. Lett. **70**, 1553 (1993).
18. D. S. Marshall, D. S. Dessau, A. G. Loeser, C.-H. Park, A. Y. Matsuura, J. N. Eckstein, I. Bozovic, P. Fournier, A. Kapitulnik, W. E. Spicer, and Z.-X. Shen, Phys. Rev. Lett. **76**, 4841 (1996); A. G. Loeser, Z.-X. Shen, D. S. Dessau, D. S. Marshall, C. H. Park, P. Fournier, and A. Kapitulnik, Science **273**, 325 (1996); H. Ding, T. Yokoya, J. C. Campuzano, T. Takahashi, M. Randeria, M. R. Norman, T. Mochiku, K. Hadowaki, and J. Giapintzakis, Nature **382**, 51 (1996).
19. S. Hufner, *Photoelectron Spectroscopy: Principles and Applications*, (Springer, Berlin Heidelberg, New York, 1995).
20. A. Bansil, M. Lindroos, S. Sahrakorpi, and R. S. Markiewicz, Phys. Rev. B **71**, 012503 (2005); S. Sahrakorpi, M. Lindroos, R. S. Markiewicz and A. Bansil, Phys. Rev. Lett. **95**, 157601 (2005).
21. C. N. Berglund and W. E. Spicer, Phys. Rev. **136**, A1030 (1964).
22. L. Hedin and S. Lundqvist, in *Solid State Physics*, F. Seitz, D. Turnbull, and H. Ehrenreich (eds.) (Academic, New York, 1969).
23. M. Randeria et al., Phys. Rev. Lett. **74**, 4951 (1995).
24. A. Bansil and M. Lindroos, Phys. Rev. Lett. **83**, 5154 (1999).
25. T. Kiss, F. Kanetaka, T. Yokoya, T. Shimojima, K. Kanai, S. Shin, Y. Onuki, T. Togashi, C. Zhang, C. T. Chen, and S. Watanabe, Phys. Rev. Lett. **94**, 057001 (2005).
26. M. P. Seah and W. A. Dench, Surf. Interf. Anal. **1**, 2 (1979).
27. X. J. Zhou, B. Wannberg, W. L. Yang, V. Brouet, Z. Sun, J. F. Douglas, D. Dessau, Z. Hussain, and Z.-X. Shen, J. Electron. Spectrosc. Relat. Phenom. **142**, 27 (2005).
28. D. L. Feng et al., Phys. Rev. Lett. **86**, 5550 (2001).
29. Y. D. Chuang et al., Phys. Rev. Lett. **87**, 117002 (2001).
30. P. V. Bogdanov et al., Phys. Rev. B **64**, 180505 (2001).
31. P.V. Bogdanov, A. Lanzara, X. J. Zhou, W. L. Yang, H. Eisaki, Z. Hussain, and Z. X. Shen, Phys. Rev. Lett. **89**, 167002 (2002).
32. P. Aebi et al., Phys. Rev. Lett. **72**, 2757 (1994).
33. T. Valla, A. V. Fedorov, P. D. Johnson, B. O. Wells, S. L. Hulbert, Q. Li, G. D. Gu, and N. Koshizuka, Science **285**, 2110 (1999).
34. S. LaShell, E. Jensen, and T. Balasubramanian, Phys. Rev. B **61**, 2371 (2000)
35. X. J. Zhou, Z. Hussain, and Z.-X. Shen, Synchrotron Radiat. News **18**, 15 (2005).
36. A. Sekiyama, T. Iwasaki, K. Matsuda, Y. Saitoh, Y. Onuki, and S. Suga, Nature (London) **403**, 396 (2000).
37. Y. Tokura and T. Arima, Jpn. J. Appl. Phys., Part 1, **29**, 2388 (1990).
38. R. J. Cava, J. Am. Ceram. Soc. **83**, 5 (2000).
39. W. E. Pickett, Rev. Mod. Phys. **61**, 433 (1989).
40. J. Fink, N. Nucker, H. A. Romberg, and J. C. Fuggle, IBM J. Res. Dev. **33**, 372 (1989).
41. L. F. Mattheiss, Phys. Rev. Lett. **58**, 1028 (1987).
42. S. Uchida, T. Ido, H. Takagi, T. Arima, Y. Tokura, and S. Tajima, Phys. Rev. B **43**, 7942 (1991).
43. D. Vaknin, S. K. Sinha, D. E. Moncton, D. C. Johnston, J. M. Newsam, C. R. Safinya, and H. E. King, Phys. Rev. Lett. **58**, 2802 (1987).

44. V. J. Emery, Phys. Rev. Lett. **58**, 2794 (1987).
45. C. M. Varma, S. Schmitt-Rink and E. Abrahams, Solid State Commun. **62**, 681 (1987).
46. T. M. Rice, in *Leshouches 1991 Session LVI, Strongly Interacting Fermions and High-T_c Superconductivity*, B. Doucot, and J. Zinn-Justin (eds.) (North-Holland, Amsterdan 1991).
47. P. W. Anderson, Science **235**, 1196 (1987).
48. F. C. Zhang, and T. M. Rice, Phys. Rev. B **37**, R3759 (1988); **41**, 7243 (1991).
49. C. Durr, S. Legner, R. Hayn, S. V. Borisenko, Z. Hu, A. Theresiak, M. Knupfer, M. S. Golden, J. Fink, F. Ronning, Z.-X. Shen, H. Eisaki, and S. Uchida, C. Janowitz, R. Müller, R. L. Johnson II., K. Rossnagel, L. Kipp, and G. Reichardt, Phys. Rev. B **63**, 14505 (2000).
50. E. Dagotto, Rev. Modern Phys. **66**, 763 (1994).
51. T. Tohyama and S. Maekawa, Supercond. Sci. Technol. **13**, R17 (2000).
52. Y.-D. Chuang, A. D. Gromko, A. V. Fedorov, Y. Aiura, K. Oka, Yoichi Ando, M. Lindroos, R. S. Markiewicz, A. Bansil, and D. S. Dessau, Phys. Rev. B **69**, 094515 (2004).
53. D. L. Feng, C. Kim, H. Eisaki, D. H. Lu, A. Damascelli, K. M. Shen, F. Ronning, N. P. Armitage, N. Kaneko, M. Greven, J.-I. Shimoyama, K. Kishio, R. Yoshizaki, G. D. Gu, and Z.-X. Shen, Phys. Rev. B **65**, 220501 (2002).
54. A. D. Gromko *et al.*, Phys. Rev. B **68**, 174520 (2003).
55. A. A. Kordyuk, S. V. Borisenko, M. S. Golden, S. Legner, K. A. Nenkov, M. Knupfer, J. Fink, H. Berger, L. Forro, and R. Follath, Phys. Rev. B **66**, 014502 (2002).
56. A. A. Kordyuk, S. V. Borisenko, A. N. Yaresko, S.-L. Drechsler, H. Rosner, T. K. Kim, A. Koitzsch, K. A. Nenkov, M. Knupfer, J. Fink, R. Follath, H. Berger, B. Keimer, S. Ono, and Yoichi Ando, Phys. Rev. B **70**, 214525 (2004).
57. Y. L. Chen, A. Iyo, W. L. Yang, X. J. Zhou, D. H. Lu, H. Eisaki, Z. Hussain, and Z.-X. Shen, unpublished work.
58. H. Ding, J. C. Campuzano, A. F. Bellman, T. Yokoya, M. R. Norman, M. Randeria, T. Takahashi, H. Katayama-Yoshida, T. Mochiku, K. Kadowaki, and G. Jennings, Phys. Rev. Lett. **74**, 2784 (1995); H. Ding, J. C. Campuzano, A. F. Bellman, T. Yokoya, M. R. Norman, M. Randeria, T. Takahashi, H. Katayama-Yoshida, T. Mochiku, K. Kadowaki, and G. Jennings, Phys. Rev. Lett. **75**, 1425 (1995).
59. H. Ding, M. R. Norman, J. C. Campuzano, M. Randeria, A. F. Bellman, T. Yokoya, T. Takahashi, T. Mochiku, and K. Kadowaki, Phys. Rev. B **54**, 9678 (1996).
60. J. Mesot, M. R. Norman, H. Ding, M. Randeria, J. C. Campuzano, A. Paramekanti, H. M. Fretwell, A. Kaminski, T. Takeuchi, T. Yokoya, T. Sato, T. Takahashi, T. Mochiku, and K. Kadowaki, Phys. Rev. Lett. **83**, 840 (1999).
61. S. V. Borisenko, A. A. Kordyuk, T. K. Kim, S. Legner, K. A. Nenkov, M. Knupfer, M. S. Golden, J. Fink, H. Berger, and R. Follath, Phys. Rev. B **64**, 140509 (2002).
62. J. M. Harris, P. J. White, Z.-X. Shen, H. Ikeda, R. Yoshizaki, H. Eisaki, S. Uchida, W. D. Si, J. W. Xiong, Z.-X. Zhao, and D. S. Dessau, Phys. Rev. Lett. **79**, 143 (1997).
63. T. Sato, T. Kamiyama, Y. Naitoh, and T. Takahashi, I. Chong, T. Terashima, and M. Takano, Phys. Rev. B **63**, 132502 (2001).
64. D. L. Feng, A. Damascelli, K. M. Shen, N. Motoyama, D. H. Lu, H. Eisaki, K. Shimizu, J.-I. Shimoyama, K. Kishio, N. Kaneko, M. Greven, G. D. Gu, X. J. Zhou, C. Kim, F. Ronning, N. P. Armitage, and Z.-X Shen, Phys. Rev. Lett. **88**, 107001 (2002).
65. R. Müller, C. Janowitz, M. Schneider, R.-S. Unger, A. Krapf, H. Dwelk, A. Müller, L. Dudy, R. Manzke, and H. Hoechst, J. Supercond. **15**, 147 (2002).
66. T. Sato, H. Matsui, S. Nishina, T. Takahashi, T. Fujii, T. Watanabe, and A. Matsuda, Phys. Rev. Lett. **89**, 67005 (2002).
67. D. H. Lu, D. L. Feng, N. P. Armitage, K. M. Shen, A. Damascelli, C. Kim, F. Ronning, Z.-X. Shen, D. A. Bonn, R. Liang, W. N. Hardy, A. I. Rykov, and S. Tajima, Phys. Rev. Lett. **86**, 4370 (2001).
68. A. Ino, C. Kim, T. Mizokawa, Z.-X. Shen, A. Fujimori, M. Takaba, K. Tamasaku, H. Eisaki, and S. Uchida, J. Phys. Soc. Jpn **68**, 1496 (1999).
69. N. P. Armitage, D. H. Lu, D. L. Feng, C. Kim, A. Damascelli, K. M. Shen, F. Ronning, Z.-X. Shen, Y. Onose, Y. Taguchi, and Y. Tokura, Phys. Rev. Lett. **86**, 1126 (2001).
70. T. Sato, T. Kamiyama, T. Takahashi, K. Kurahashi, and K. Yamada, Science **291**, 1517 (2001).
71. G. Binnig, A. Baratoff, H. E. Hoenig, and J. G. Bednorz, Phys. Rev. Lett. **45**, 1352 (1980).
72. T. Yokoya, T. Kiss, A. Chainani, S. Shin, M. Nohara, and H. Takagi, Science **294**, 2518 (2001).
73. S. Souma, Y. Machida, T. Sato, T. Takahashi, H. Matsui, S.-C. Wang, H. Ding, A. Kaminski, J. C. Campuzano, S. Sasaki, and K. Kadowaki, Nature **423**, 65 (2003).
74. C. M. Varma, Phys. Rev. B **61**, R3804 (2000).

75. M. E. Simon and C.M. Varma, Phys. Rev. Lett. **89**, 247003 (2002).
76. A. Kaminski, S. Rosenkranz, H. M. Fretwell, J. C. Campuzano, Z. Li, H. Raffy, W. G. Cullen, H. You, C. G. Olsonk, C. M. Varma, and H. Hochst, Nature **416**, 610 (2002).
77. S.V. Borisenko, A. A. Kordyuk, A. Koitzsch, T. K. Kim, K. A. Nenkov, M. Knupfer, J. Fink, C. Grazioli, S. Turchini, and H. Berger, Phys. Rev. Lett. **92**, 207001 (2004).
78. S. V. Borisenko, A. A. Kordyuk, A. Koitzsch, M. Knupfer, J. Fink, H. Berger, and C. T. Lin, Nature **431**, 1 (2004); J. C. Campuzano, A. Kaminski, S. Rosenkranz, and H. M. Fretwell, Nature **431**, 2 (2004).
79. J. Rossat-Mignod, L. P. Regnault, C. Vettier, P. Bourges, P. Burlet, J. Bossy, J. Y. Henry, and G. Lapertot, Physica C **185–189**, 86 (1991); H. A. Mook, M. Yethiraj, G. Aeppli, T. E. Mason, and T. Armstrong, Phys. Rev. Lett. **70**, 3490 (1993); P. Bourges, in *The Gap Symmetry and Fluctuations in High Temperature Superconductors*, J. Bok, G. Deutscher, D. Pavuna, and S. A. Wolf (eds.), (Plenum, New York, 1998) p. 349; H. F. Fong, P. Bourges, Y. Sidis, L. P. Regnault, J Bossy, A. S. Ivanov, D. L. Milius, I. A. Aksay, and B. Keimer, Phys. Rev. B **61**, 14773 (2000); P. Dai, H. A. Mook, R. D. Hunt, and F. Dogan, Phys. Rev. B **63**, 054525 (2001).
80. H. F. Fong, P. Bourges, Y. Sidis, L. P. Regnault, A. S. Ivanov, G. D. Gu, N. Koshizuka, and B. Keimer, Nature **398**, 588 (1999).
81. H. He, P. Bourges, Y. Sidis, C. Ulrich, L. P. Regnault, S. Pailhs, N. S. Berzigiarova, N. N. Kolesnikov, and B. Keimer, Science **295**, 1045 (2002).
82. P. B. Allen, Nature **412**, 494 (2001).
83. A. Schilling, M. Cantoni, J. D. Guo, and H. R. Ott, Nature (London) **363**, 56 (1993).
84. C. W. Chu, L. Gao, F. Chen, Z. J. Huang, R. L. Meng, and Y. Y. Xue, Nature (London) **365**, 323 (1993).
85. B. Batlogg, R. J. Cava, A. Jayaraman, R. B. van Dover, G. A. Kourouklis, S. Sunshine, D. W. Murphy, L. W. Rupp, H. S. Chen, A. White, K. T. Short, A. M. Mujsce, and E. A. Rietman, Phys. Rev. Lett. **58**, 2333 (1987).
86. M. Gurvitch and A. T. Fiory, Phys. Rev. Lett. **59**, 1337 (1987); S. Martin, A. T. Fiory, R. M. Fleming, L. F. Schneemeyer, and J. V. Waszczak, Phys. Rev. B **41**, 846 (1990).
87. M. Cohen and P. W. Anderson, in *Superconductivity in d- and f-Band Metals*, D.H. Douglass (ed.) (AIP, New York, 1972), p. 17.
88. P. W. Anderson, *The Theory of Superconductivity in the High-T_c Cuprates* (Princeton University Press, Princeton, NJ, 1997).
89. K. A. Müller, Physics C **341–348**, 11 (2000); K. A. Müller, *Proceedings of the 10th Anniversary HTS Workshop, March 12–16.*, B. Batlogg et al. (ed.) (World Scientific, Houston, 1996).
90. For a review, see M. L. Kulic, Phys. Rep. **338**, 1 (2000).
91. A. S. Alexandrov, and N. F. Mott, Rep. Prog. Phys. **57**, 1197 (1994).
92. G.-M. Zhao, K. Conder, H. Keller and K. A. Müller, J. Phys.: Condens. Matter **10**, 9055 (1998).
93. T. Schneider and H. Keller, Phys. Rev. Lett. **86**, 4899 (2001), and references therein.
94. M. K. Crawford, W. E. Farneth, E. M. McCarron III, R. L. Harlow and A. H. Mouden, Science **250**, 1390 (1990); M. K. Crawford, M. N. Kunchur, W. E. Farneth, E. M. McCarron III, and S. J. Poon, Phys. Rev. B **41**, 282 (1990).
95. J. Hofer et al., Phys. Rev. Lett. **84**, 4192 (2000).
96. For reviews on electron–phonon coupling in Raman Scattering on high-T_c materials, see C. Thomson and M. Cardona, in *Physical Properties of High Temperature Superconductors I*, D. M. Ginzberg (ed.) (World Scientific, Singapore, 1989), p. 409; C. Thomson, in *Light Scattering in Solids VI*, M. Cardona and G. Guntherodt (eds.) (Springer, Berlin, Heidelberg, New York, 1991), p. 285; M. Cardona, Physica C **317–318**, 30 (1999).
97. S. Tajima, T. Ido, S. Ishibashi, T. Itoh, H. Eisaki, Y. Mizuo, T. Arima, H. Takagi, and S. Uchida Phys. Rev. B **43**, 10496 (1991).
98. C. Thomson, M. Cardona, B. Gegenheimer, R. Liu, and A. Simon, Phys. Rev. B **37**, 9860 (1988).
99. B. Friedl, C. Thomsen, and M. Cardona, Phys. Rev. Lett. **65**, 915 (1990).
100. E. Altendorf, X. K. Chen, J. C. Irwin, R. Liang, and W. N. Hardy, Phys. Rev. B **47**, 8140 (1993).
101. X. J. Zhou, V. G. Hadjiev, M. Cardona, Q. M. Lin, and C. W. Chu, Phys. Status Solidi (a) **202**, R7 (1997); V. G. Hadjiev, X. J. Zhou, T. Strohm, M. Cardona, Q. M. Lin, C. W. Chu, Phys. Rev. B **58**, 1043 (1998).
102. X. J. Zhou, M. Cardona, D. Colson, and V. Viallet, Phys. Rev. B **55**, 12770 (1997).
103. S. Sugai, H. Suzuki, Y. Takayanagi, T. Hosokawa, and N. Hayamizu, Phys. Rev. B **68**, 184504 (2003).
104. For a recent review, see L. Pintschovius, Phys. Status Solidi (b) **242**, 30 (2005).
105. T. Egami and S. J. L. Billinge, in *Physical Properties of High Temperature Superconductors V*, (ed.) D. Ginsberg (World Scientific, Singapore, 1996) p. 265.
106. D. Reznik, B. Keimer, F. Dogan, and I. A. Aksay, Phys. Rev. Lett. **75**, 2396 (1995).
107. L. Pintschovius and M. Braden, Phys. Rev. B **60**, R15039 (1999).
108. R. J. McQueeney et al., Phys. Rev. Lett. **82**, 628 (1999).

109. E. Pavarini, I. Dasgupta, T. Saha-Dasgupta, O. Jepsen, and O. K. Andersen, Phys. Rev. Lett. **87**, 47003 (2001).
110. Z.-X. Shen, A. Lanzara, S. Ishihara, and N. Nagaosa, Philos. Mag. B **82**, 1349 (2002).
111. P. W. Anderson, Science **268**, 1154 (1995).
112. A. A. Abrikosov, L. P. Gorkov, and I. E. Dzyaloshinski, *Methods of Quantum Field Theory in Statistical Physics* (Dover Publications, Inc., New York, 1963).
113. K. M. Shen, F. Ronning, D. H. Lu, W. S. Lee, N. J. C. Ingle, W. Meevasana, F. Baumberger, A. Damascelli, N. P. Armitage, L. L. Miller, Y. Kohsaka, M. Azuma, M. Takano, H. Takagi, and Z.-X. Shen, Phys. Rev. Lett. **93**, 267002 (2004).
114. T. Cuk, D. H. Lu, X. J. Zhou, Z.-X. Shen, T. P. Devereaux, and N. Nagaosa, Phys. Status Solidi. (b) **242**, 11 (2005).
115. A. B. Migdal, Zh. Eksperim. i Teor. Fiz. **34**, 1438; [translation: Soviet Phys.-JETP **7**, 996 (1958)]
116. G. M. Eliashberg, Zh. Eksperim. i Teor. Fiz. **38**, 966 (1960); [translation: Soviet Phys.-JETP **11**, 696 (1960)]
117. S. Engelsberg and J. R. Schrieffer, Phys. Rev. **131**, 993 (1963).
118. D. J. Scalapino, J. R. Schrieffer, J. W. Wilkins, Phys. Rev. **148**, 263 (1966).
119. S. LaShell, E. Jensen, and T. Balasubramanian, Phys. Rev. B **61**, 2371 (2000).
120. M. Hengsberger, D. Purdie, P. Segovia, M. Garnier, and Y. Baer, Phys. Rev. Lett. **83**, 592 (1999).
121. T. Valla, A. V. Fedorov, P. D. Johnson, and S. L. Hulbert, Phys. Rev. Lett. **83**, 2085 (1999).
122. A. W. Sandvik, D. J. Scalapino, and N. E. Bickers, Phys. Rev. B **69**, 094523 (2004).
123. N. V. Prokof'ev and B. V. Svistunov, Phys. Rev. Lett. **81**, 2514 (1998).
124. A. S. Mishchenko, N. V. Prokof'ev, A. Sakamoto, and B. V. Svistunov, Phys. Rev. B **62**, 6317 (2000).
125. A. S. Mishchenko, N. Nagaosa, N. V. Prokof'ev, A. Sakamoto, and B. V. Svistunov, Phys. Rev. B **66**, 020301 (2002).
126. A. S. Mishchenko, N. Nagaosa, N. V. Prokof'ev, A. Sakamoto, and B. V. Svistunov, Phys. Rev. Lett. **91**, 236401 (2003).
127. P. V. Bogdanov, A. Lanzara, S. A. Kellar, X. J. Zhou, E. D. Lu, W. J. Zheng, G. Gu, J.-I. Shimoyama, K. Kishio, H. Ikeda, R. Yoshizaki, Z. Hussain, and Z. X. Shenm, Phys. Rev. Lett. **85**, 2581 (2000).
128. A. Lanzara, P. V. Bogdanov, X. J. Zhou, S. A. Kellar, D. L. Feng, E. D. Lu, T. Yoshida, H. Eisaki, A. Fujimori, K. Kishio, J.-I. Shimoyama, T. Noda, S. Uchida, Z. Hussain, and Z.-X. Shen, Nature **412**, 510 (2001).
129. A. Kaminski, M. Randeria, J. C. Campuzano, M. R. Norman, H. Fretwell, J. Mesot, T. Sato, T. Takahashi, and K. Kadowaki, Phys. Rev. Lett. **86**, 1070 (2001).
130. P. D. Johnson, T. Valla, A. V. Fedorov, Z. Yusof, B. O. Wells, Q. Li, A. R. Moodenbaugh, G. D. Gu, N. Koshizuka, C. Kendziora, S. Jian, and D. G. Hinks, Phys. Rev. Lett. **87**, 177007 (2001).
131. S. V. Borisenko, A. A. Kordyuk, T. K. Kim, A. Koitzsch, M. Knupfer, J. Fink, M. S. Golden, M. Eschrig, H. Berger, and R. Follath, Phys. Rev. Lett. **90**, 207001 (2003).
132. X. J. Zhou, T. Yoshida, A. Lanzara, P. V. Bogdanov, S. A. Kellar, K. M. Shen, W. L. Yang, F. Ronning, T. Sasagawa, T. Kakeshita, T. Noda, H. Eisaki, S. Uchida, C. T. Lin, F. Zhou, J. W. Xiong, W. X. Ti, Z. X. Zhao, A. Fujimori, Z. Hussain, and Z.-X. Shen , Nature **423**, 398 (2003).
133. G.-H. Gweon, T. Sasagawa, S. Y. Zhou, J. Graf, H. Takagi, D.-H. Lee, and A. Lanzara, Nature **430**, 187 (2004).
134. X. J. Zhou, T. Yoshida, D.-H. Lee, W. L. Yang, V. Brouet, F. Zhou, W. X. Ti, J. W. Xiong, Z. X. Zhao, T. Sasagawa, T. Kakeshita, H. Eisaki, S. Uchida, A. Fujimori, Z. Hussain, and Z.-X. Shen, Phys. Rev. Lett. **92**, 187001 (2004).
135. A. Lanzara, P. V. Bogdanov, X. J. Zhou, N. Kaneko, H. Eisaki, M. Greven, Z. Hussain, and Z. -X. Shen, J. Phys. Chem. Solids **67**, 239 (2006).
136. H.-Y. Kee, S. Kivelson, and G. Aeppli, Phys. Rev. Lett. **88**, 257002 (2002).
137. Ar. Abanov, A. V. Chubukov, M. Eschrig, M. R. Norman, and J. Schmalian, Phys. Rev. Lett. **89**, 177002 (2002).
138. X. J. Zhou, T. Yoshida, W. L. Yang, S. Komiya, Y. Ando, F. Zhou, J. W. Xiong, W. X. Ti, Z. X. Zhao, T. Sasagawa, T. Kakeshita, H. Eisaki, S. Uchida, A. Fujimori, Z. Hussain, and Z.-X. Shen, unpublished work.
139. X. J. Zhou, T. Yoshida, W. L. Yang, V. Brouet, S. Komiya, Y. Ando, A. Fujimori, Z. Hussain, and Z.-X. Shen, unpublished work.
140. J. M. Rowell, P. W. Anderson, and D. E. Thomas, Phys. Rev. Lett. **10**, 334 (1963); D. J. Scalapino, J. R. Schrieffer, and J. W. Wilkins, Phys. Rev. **148**, 263 (1966).
141. J. R. Shi, S.-J Tang, B. Wu, P.T. Sprunger, W.L. Yang, V. Brouet, X. J. Zhou, Z. Hussain, Z.-X. Shen, Z. Y. Zhang, E. W. Plummer, Phys. Rev. Lett. **92**, 186401 (2004).
142. X. J. Zhou, J. Shi, T. Yoshida, T. Cuk, W. L. Yang, V. Brouet, J. Nakamura, N. Mannella, S. Komiya, Y. Ando, F. Zhou, W. X. Ti, J. W. Xiong, Z. X. Zhao, T. Sasagawa, T. Kakeshita, H. Eisaki, S. Uchida, A. Fujimori, Zhenyu Zhang, E. W. Plummer, R. B. Laughlin, Z. Hussain, and Z.-X. Shen, Phys. Rev. Lett. **95**, 117001 (2005).

143. S. M. Hayden, G. Aeppli, H. A. Mook, T. G. Perring, T. E. Mason, S.-W. Cheong, and Z. Fisk, Phys. Rev. Lett. **76**, 1344 (1996); H. Hiraka, Y. Endoh, M. Fujita, Y. S. Lee, J. Kulda, A. Ivanov, and R. J. Birgeneau, J. Phys. Soc. Jpn **70**, 853 (2001); H. Goka, S. Kuroshima, M. Fujita, K. Yamada, H. Hiraka, Y. Endoh, and C. D. Frost, Physica C **388–389**, 239 (2003); J. M. Tranquada, H. Woo, T. G. Perring, H. Goka, G. D. Gu, G. Xu, M. Fujita, and K. Yamada, Nature 429 **534** (2004).
144. R. J. McQueeney, J. L. Sarrao, P. G. Pagliuso, P. W. Stephens, and R. Osborn, Phys. Rev. Lett. **87**, 077001 (2001).
145. T. Valla, Phys. Rev. Lett. **96**, 119701 (2006); X. J. Zhou, Junren Shi, W. L. Yang, Seiki Komiya, Yoichi Ando, W. Plummer, Z. Hussain, Z.-X. Shen, Phys. Rev. Lett. **96**, 119702 (2006).
146. T. K. Kim, A. A. Kordyuk, S. V. Borisenko, A. Koitzsch, M. Knupfer, H. Berger, and J. Fink, Phys. Rev. Lett. **91**, 167002 (2003).
147. T. Cuk, F. Baumberger, D. H. Lu, N. Ingle, X. J. Zhou, H. Eisaki, N. Kaneko, Z. Hussain, T. P. Devereaux, N. Nagaosa, and Z.-X. Shen, Phys. Rev. Lett. **93**, 117003 (2004).
148. T. P. Devereaux, T. Cuk, Z.-X. Shen, and N. Nagaosa, Phys. Rev. Lett. **93**, 117004 (2004).
149. T. P. Devereaux, Phys. Rev. Lett. **72**, 396 (1994).
150. M. Opel, R. Hackl, T. P. Devereaux, A. Virosztek, A. Zawadowski, A. Erb, E. Walker, H. Berger, and L. Forro, Phys. Rev. B **60**, 9836 (1999).
151. T. P. Devereaux, A. Virosztek, and A. Zawadowski, Phys. Rev. B **59**, 14618 (1999).
152. O. Jepsen, O. K. Andersen, I. Dasgupta, and S. Savrasov, J. Phys. Chem. Solids **59**, 1718 (1998); O. K. Andersen, S. Y. Savrasov, O. Jepsen, and A. I. Liechtenstein, J. Low Temp. Phys. **105**, 285 (1996).
153. D. J. Scalapino, J. Phys. Chem. Solids **56**, 1669 (1995).
154. A. Nazarenko and E. Dagotto, Phys. Rev. B **53**, R2987 (1996).
155. M. Botti, E. Cappelluti, C. Grimaldi, and L. Pietronero, Phys. Rev. B **66**, 054532 (2002); E. Cappelluti and L. Pietronero, Phys. Rev. B **68**, 224511 (2003).
156. B. O. Wells, Z.-X. Shen, A. Matsuura, D. M. King, M. A. Kastner, M. Greven, and R. J. Birgeneau, Phys. Rev. Lett. **74**, 964 (1995).
157. S. LaRosa, I. Vobornik, F. Zwick, H. Berger, M. Grioni, G. Margaritondo, R. J. Kelley, M. Onellion, and A. Chubukov Phys. Rev. B **56**, R525 (1997).
158. C. Kim, P. J. White, Z.-X. Shen, T. Tohyama, Y. Shibata, S. Maekawa, B. O. Wells, Y. J. Kim, R. J. Birgeneau, and M. A. Kastner, Phys. Rev. Lett. **80** 4245 (1998).
159. J. J. M. Pothuizen, R. Eder, N. T. Hien, M. Matoba, A. A. Menovsky, and G. A. Sawatzky, Phys. Rev. Lett. **78**, 717 (1997).
160. F. Ronning, C. Kim, K. M. Shen, N. P. Armitage, A. Damascelli, D. H. Lu, D. L. Feng, Z.-X. Shen, L. L. Miller, Y.-J. Kim, F. Chou, and I. Terasaki, Phys. Rev. B **67**, 035113 (2003).
161. F. Ronning, C. Kim, D. L. Feng, D. S. Marshall, A. G. Loeser, L. L. Miller, J. N. Eckstein, I. Bozovic, and Z.-X. Shen, Science **282**, 2067 (1998).
162. F. Ronning, K. M. Shen, N. P. Armitage, A. Damascelli, D. H. Lu, Z.-X. Shen, L. L. Miller, and C. Kim, Phys. Rev. B **71**, 94518 (2005).
163. K. M. Shen, F. Ronning, D. H. Lu, F. Baumberger, N. J. C. Ingle, W. S. Lee, W. Meevasana, Y. Kohsaka, M. Azuma, M. Takano, H. Takagi, Z.-X. Shen, Science **307**, 901 (2005).
164. N. P. Armitage, F. Ronning, D. H. Lu, C. Kim, A. Damascelli, K. M. Shen, D. L. Feng, H. Eisaki, Z.-X. Shen, P. K. Mang, N. Kaneko, M. Greven, Y. Onose, Y. Taguchi, and Y. Tokura, Phys. Rev. Lett. **88**, 257001 (2002).
165. A. Ino, C. Kim, M. Nakamura, T. Yoshida, T. Mizokawa, Z.-X. Shen, A. Fujimori, T. Kakeshita, H. Eisaki, and S. Uchida, Phys. Rev. B **62**, 4137 (2000).
166. T. Yoshida, X. J. Zhou, T. Sasagawa, W. L. Yang, P. V. Bogdanov, A. Lanzara, Z. Hussain, T. Mizokawa, A. Fujimori, H. Eisaki, Z.-X. Shen, T. Kakeshita, and S. Uchida, Phys. Rev. Lett. **91**, 27001 (2003)
167. O. Rösch, O. Gunnarsson, X. J. Zhou, T. Yoshida, T. Sasagawa, A. Fujimori, Z. Hussain, Z.-X. Shen, and S. Uchida, Phys. Rev. Lett. **95**, 227002 (2005).
168. D. L. Novikov, A. J. Freeman, and A. D. Jorgenson, Phys. Rev. B **51**, 6675 (1995); L. F. Mattheiss, Phys. Rev. B **42**, 354 (1990).
169. Z. P. Liu and E. Manousakis, Phys. Rev. B **45**, 2425 (1992).
170. A. Nazarenko, K. J. E. Vos, S. Haas, E. Dagotto, and R. Gooding, Phys. Rev. B **51**, 8676 (1995); B. Kyung and R. A. Ferrell, Phys. Rev. B **54**, 10125 (1996); T. Xiang and J. M. Wheatley, Phys. Rev. B **54**, R12653 (1996); V. I. Belinicher, A. L. Chernyshev, and V. A. Shubin, Phys. Rev. B **54**, 14914(1996); R. Eder, Y. Ohta, and G. A. Sawatzky, Phys. Rev. B **55**, R3414(1997); T. K. Lee and C. T. Shih, Phys. Rev. B **55**, 5983(1997); F. Lema F and A. A. Aligia, Phys. Rev. B **55**, 14092 (1997); P. W. Leung, B. O. Wells, and R. J. Gooding, Phys. Rev. B **56**, 6320 (1997); O. P. Sushkov, G. A. Sawatzky, R. Eder, and H. Eskes, Phys. Rev. B **56**, 11769 (1997).

171. S. Schmitt-Rink, C, M. Varma, and A. E. Ruckenstein, Phys. Rev. Lett. **60**, 2793 (1988); C. L. Kane, P. A. Lee and N. Read, Phys. Rev. B **39**, 6880 (1989); G. Martinez and P. Horsch, Phys. Rev. B **44**, 317 (1991).
172. R. B. Laughlin, Phys. Rev. Lett. **79**, 1726 (1997).
173. D. W. Turner, *Molecular Photoelectron Spectroscopy* (Wiley, New York, 1970).
174. L. Perfetti, H. Berger, A. Reginelli, L. Degiorgi, H. Hohst, J. Voit, G. Margaritondo, and M. Grioni, Phys. Rev. Lett. **87**, 216404 (2001); D. S. Dessau, T. Saitoh, C.-H. Park, Z.-X. Shen, P. Villella, N. Hamada, Y. Moritomo, and Y. Tokura, Phys. Rev. Lett. **81**, 192 (1998); V. Perebeinos and P. B. Allen, Phys. Rev. Lett. **85**, 5178 (2000).
175. A. S. Mishchenko and N. Nagaosa, Phys. Rev. Lett. **93**, 036402 (2004).
176. O. Rösch, and O. Gunnarsson, Eur. Phys. J. B **43**, 11 (2005).
177. M. A. van Veenendaal, G. A. Sawatzky, and W. A. Groen, Phys. Rev. B **49**, 1407 (1994).
178. H. Takagi, T. Ido, S. Ishibashi, M. Uota, S. Uchida, and Y. Tokura, Phys. Rev. B **40**, 2254 (1989).
179. A. Ino, T. Mizokawa, A. Fujimori, K. Tamesaku, H. Eisaki, S. Uchida, T. Kimura, T. Sasagawa, and K. Kishio, Phys. Rev. Lett. **79**, 2101 (1997).
180. X. J. Zhou, T. Yoshida, S. A. Kellar, P.V. Bogdanov, E. D. Lu, A. Lanzara, M. Nakamura, T. Noda, T. Kakeshita, H. Eisaki, S. Uchida, A. Fujimori, Z. Hussain, and Z.-X. Shen, Phys. Rev. Lett. **86**, 5578 (2001).
181. T. Yoshida, X. J. Zhou, M. Nakamura, S. A. Kellar, P. V. Bogdanov, E. D. Lu, A. Lanzara, Z. Hussain, A. Ino, T. Mizokawa, A. Fujimori, H. Eisaki, C. Kim, Z.-X. Shen, T. Kakeshita, and S. Uchida, Phys. Rev. B **63**, 220501 (2001).
182. T. Hanaguri, C. Lupien, Y. Kohsaka, D.−H. Lee, M. Azuma, M. Takano, H. Takagi, and J. C. Davis, Nature **430**, 1001 (2004).
183. J. M. Tranquada, B. J. Sternlieb, J. D. Axe, Y. Nakamura, and S. Uchida, Nature (London) **375**, 561 (1995); K. Yamada, C. H. Lee, K. Kurahashi, J. Wada, S. Wakimoto, S. Ueki, H. Kimura, Y. Endoh, S. Hosoya, G. Shirane, R. J. Birgeneau, M. Greven, M. A. Kastner, and Y. J. Kim, Phys. Rev. B **57**, 6165 (1998).
184. Y. Ando, A. N. Lavrov, S. Komiya, K. Segawa, and X. F. Sun, Phys. Rev. Lett. **87**, 017001 (2001).
185. Z.-X. Shen and J. R. Schrieffer, Phys. Rev. Lett. **78**, 1771 (1997).
186. N. P. Armitage, D. H. Lu, C. Kim, A. Damascelli, K. M. Shen, F. Ronning, D. L. Feng, P. Bogdanov, Z.-X. Shen, Y. Onose, Y. Taguchi, Y. Tokura, P. K. Mang, N. Kaneko, and M. Greven, Phys. Rev. Lett. **87**, 147003 (2001).
187. S. M. Hayden, G. Aeppli, H. A. Mook, T. G. Perring, T. E. Mason, S.-W. Cheong, and Z. Fisk, Phys. Rev. Lett. **76**, 1344 (1996).
188. M. Vershinin, S. Misra, S. Ono, Y. Abe, Y. Ando, A. Yazdani1, Science **303**, 1995 (2004); The charge ordering here is under debate, see U. Chatterjee, M. Shi, A. Kaminski, A. Kanigel, H. M. Fretwell, K. Terashima, T. Takahashi, S. Rosenkranz, Z. Z. Li, H. Raffy, A. Santander-Syro, K. Kadowaki, M. R. Norman, M. Randeria, and J. C. Campuzano, Phys. Rev. Lett. **96**, 107006 (2006).
189. J. E. Hoffman, E. W. Hudson, K. M. Lang, V. Madhavan, H. Eisaki, S. Uchida, and J. C. Davis, Science **295**, 466 (2002).
190. K. McElroy, D.-H. Lee, J. E. Hoffman, K. M. Lang, E. W. Hudson, H. Eisaki, S. Uchida, J. Lee, and J. C. Davis, Phys. Rev. Lett. **94**, 197005 (2005).
191. N. Bulut and D. J. Scalapino, Phys. Rev. B **54**, 14971 (1996).
192. O. K. Andersen, O. Jepsen, A. I. Liechtenstein, and I. I. Mazin, Phys. Rev. B **49**, 4145 (1994).
193. A. A. Abrikov and V. M. Genkin, Zh. Eksp. Teor. Fiz. **65**, 842 (1973) [Sov. Phys. JETP **38**, 417 (1974)].
194. R. Zeyher and M. Kulic, Phys. Rev. B **53**, 2850 (1996).
195. Z. B. Huang, W. Hanke, E. Arrigoni, and D. J. Scalapino, Phys. Rev. B **68**, 220507 (2003).
196. S. Ishihara and N. Nagaosa, Phys. Rev. B **69**, 144520 (2004).
197. O. Rösch and O. Gunnarsson, Phys. Rev. Lett. **92**, 146403 (2004).
198. O. Rösch and O. Gunnarsson, Phys. Rev. Lett. **93**, 237001 (2004).

4

Microwave Electrodynamics of High Temperature Superconductors

D. A. Bonn and W. N. Hardy

4.1. Introduction

Measurements of electrodynamics at microwave frequencies play an important and varied role in the study of superconductivity because they give diverse information on both the superconducting groundstate and the excitations out of that state. In particular, the imaginary part of the microwave conductivity can be used to determine the penetration depth, a key length scale in the superconducting state that gives access to the superfluid density or, more correctly, the superfluid phase stiffness. Measurements of the real part of the conductivity give a wealth of information on dissipation associated with quasiparticle excitations out of the groundstate. Such measurements figured prominently throughout the history of the study of s-wave BCS superconductors, starting with the development of the London model of superconducting electrodynamics and the subsequent measurements of the penetration depth. Later, detailed work on the temperature dependence of the penetration depth and the temperature and frequency dependence of microwave absorption provided evidence of the energy gap predicted by BCS theory.

The wealth of information on s-wave BCS superconductors was hard-won over many years of theory and experiment. One reason for this is the multiple lengthscales in a superconducting sample; London penetration depth, coherence length, quasiparticle mean free path and sample dimensions can all come into play. For instance, in the early work on microwave penetration depth in the elements, the possibility that the coherence length can be comparable to, or larger than, the London penetration depth led to measured values of penetration depth that differed from the predictions of the London model. Subsequent careful studies as a function of purity and mean free path clarified this complex situation and ultimately provided access to the coherence length as well as the penetration depth. Impurities provide further complexity, especially for the microwave absorption which necessarily involves the mechanisms that scatter quasiparticles and lead to dissipation. A practical payoff for this effort has been the development of high Q superconducting resonators used in particle accelerators.

The high temperature superconductors have also provided a complex and lengthy challenge, though for quite different reasons. The foremost difference is that the cuprates have been found to be d-wave superconductors, with nodes in the superconducting energy gap that have far-reaching effects on the electrodynamic properties. The presence of nodes in the gap brings with it a serious complication—many physical properties are more sensitive

---

D.A. Bonn and W.N. Hardy • Department of Physics and Astronomy, University of British Columbia 6224 Agricultural Rd., Vancouver, BC, Canada V6T 1Z1

to impurities than is the case for s-wave superconductors where Anderson's theorem places serious constraints on the influence of nonmagnetic impurities. Sensitivity to defects is particularly problematic in these complex quaternary or even pentenary compounds. Nature has given us an easily disturbed superconducting state in materials that are extremely difficult to grow in a highly perfect crystalline form.

Another major difference in the cuprates is that the large energy gap associated with having a high T_c carries with it a very short coherence length. This has the simplifying effect that the microwave electrodynamics are local under most measurement conditions, but it also brings in new phenomena by making critical fluctuations much more important than they are in lower T_c materials. The issue of fluctuations is made even more important by the relatively low phase stiffness associated with the rather long London penetration depth in the cuprates. Added to all of these novel features is the extreme anisotropy of these layered materials plus structural complications such as coupled bilayers of CuO_2 in some materials, CuO chain layers in the $YBa_2Cu_3O_{6+x}$ family and in $YBa_2Cu_4O_8$. This leaves a vast parameter space to explore since a complete experimental picture would require measurements of the microwave properties in all three crystallographic directions, as a function of temperature, carrier doping, and purity, in several different compounds. In the following chapter we will review the progress on this topic achieved over nearly two decades. After an introduction to the pieces of theory most relevant to microwave measurements, we will introduce the experimental techniques used in the field. Because of the central role that superfluid phase stiffness plays in the cuprates, a detailed review of microwave measurements of the penetration depth will also include an overview of other types of penetration depth measurements. Following this is a review of the present state of understanding of the microwave conductivity of the cuprates and a separate section on the role of superconducting fluctuations. Throughout this there will be a bias towards discussion of work on single crystals where the best case has been made for sample quality. There will be regular reference to thin film results where they help complete the picture. Recent reviews by Maeda [127], Trunin and Golubov [194], as well as the earlier review by Trunin [193] should be consulted for complementary treatments of topics found in this review.

4.2. Electrodynamics of Superconductors

4.2.1. London Theory

At first glance, the fact that the coherence length ξ is typically very short and the penetration depth λ rather long in the high temperature superconductors guarantees that one is in the London limit $\lambda \gg \xi$ where the electrodynamics of the superfluid are local. There are, however, situations in the cuprates where non-local effects might be observable and these will be discussed near the end of this section. In most microwave measurement conditions these effects do not come into play, so a good starting point for discussing the electrodynamics of the cuprates is the London model.

In the local limit, the current density $\vec{J}$ is related to the local electric field $\vec{E}$ by a conductivity tensor:

$$\vec{J} = \sigma \vec{E} \qquad (4.1)$$

or equivalently to the local magnetic field $\vec{B}$

$$\nabla \times (\sigma^{-1} \vec{J}) = -i\omega \vec{B}. \qquad (4.2)$$

We will be concerned with the diagonal components of the conductivity tensor σ_{ii} and where needed will denote the three components for a typical orthorhombic cuprate as σ_a, σ_b and σ_c. For the convention $e^{i\omega t}$ for the time dependence of $\vec{E}$, each of the diagonal components of the conductivity tensor can be written in the form $\sigma = \sigma_1(\omega, T) - i\sigma_2(\omega, T)$.

4.2.2. Surface Impedance Approximation

For local electrodynamics and in situations where the sample surface is flat on the scale of the penetration depth, the surface impedance Z_s suffices to describe the response of the sample to an applied EM field. In the local limit where Eqs. (4.1) or (4.2) are valid, it follows directly from Maxwell's equations that $Z_s = R_s + iX_s$, defined by the ratio of the tangential electric and magnetic fields at the surface of the sample (e.g. E_x/H_y), is given by

$$Z_s(\omega) = \left(\frac{i\mu_0\omega}{\sigma_1 - i\sigma_2}\right)^{1/2}, \tag{4.3}$$

where R_s is referred to as the surface resistance and X_s the surface reactance. Within the same local approximation the propagation constant κ is given by

$$\kappa = (-i\mu_0\omega\sigma)^{1/2}. \tag{4.4}$$

For this discussion, it is useful to have in mind a simple picture of the T and ω dependence of the electrodynamics we are likely to encounter. Figure 4.1 gives a schematic view of the situation for (a) $T > T_c$, (b) $T = 0$ and (c) $0 < T < T_c$. The curve in (a) can be taken as that of a Drude metal where

$$\sigma_1 = \frac{ne^2}{m^*}\frac{\tau}{1+\omega^2\tau^2} \text{ and } \sigma_2 = \frac{ne^2}{m^*}\frac{\omega\tau^2}{1+\omega^2\tau^2}, \tag{4.5}$$

where n/m^* is the ratio of normal carrier density over effective mass and τ is the current relaxation time of the charge carriers. The DC conductivity is $\sigma_{DC} = ne^2\tau/m^*$. This model is just meant as an illustration, since for cuprates we know that σ_1 falls off more like $1/\omega$ than $1/\omega^2$. In the normal state, the quasiparticle scattering rate $1/\tau$ is generally much greater than the measuring frequency of microwave techniques (as shown).

At low temperatures, some fraction of the area under $\sigma_1(\omega, T > T_c)$ appears as a δ-function at $\omega = 0$, representing the response of the superfluid condensate. For the $\hat{ab}$-plane response, the remaining area under $\sigma_1(\omega, T = 0)$ at microwave to far-infrared frequencies is relatively small for good quality cuprate samples having low disorder. This corresponds to the so-called clean limit where the energy scale of the superconducting gap is larger than the quasiparticle (qp) scattering rate. Strictly speaking, the language of "clean limit" vs. "dirty limit" is not entirely appropriate for the high temperature superconductors since the scattering rate is dominated by inelastic scattering rather than elastic impurity scattering. For instance, there is a rapid drop in $1/\tau$ as T is lowered below T_c, as discussed in Section 4.4, which suggests that the materials are in the very clean limit. However, this scattering rate is necessarily strongly frequency dependent, so that the low scattering seen at microwave frequencies is not a good guide to the behaviour in the infrared. Still, the relatively small oscillator strength left at low frequency as $T \to 0$ in good samples, together with the strong screening by the superfluid, make measurements of $\sigma_1(\omega, T < T_c)$ a challenge at microwave, THz, and far infrared frequencies. Add to this the fact that a d-wave superconductor does not have the sharp onset of absorption seen in $\sigma_1(\omega)$ at the gap edge of an s-wave superconductor and one readily sees

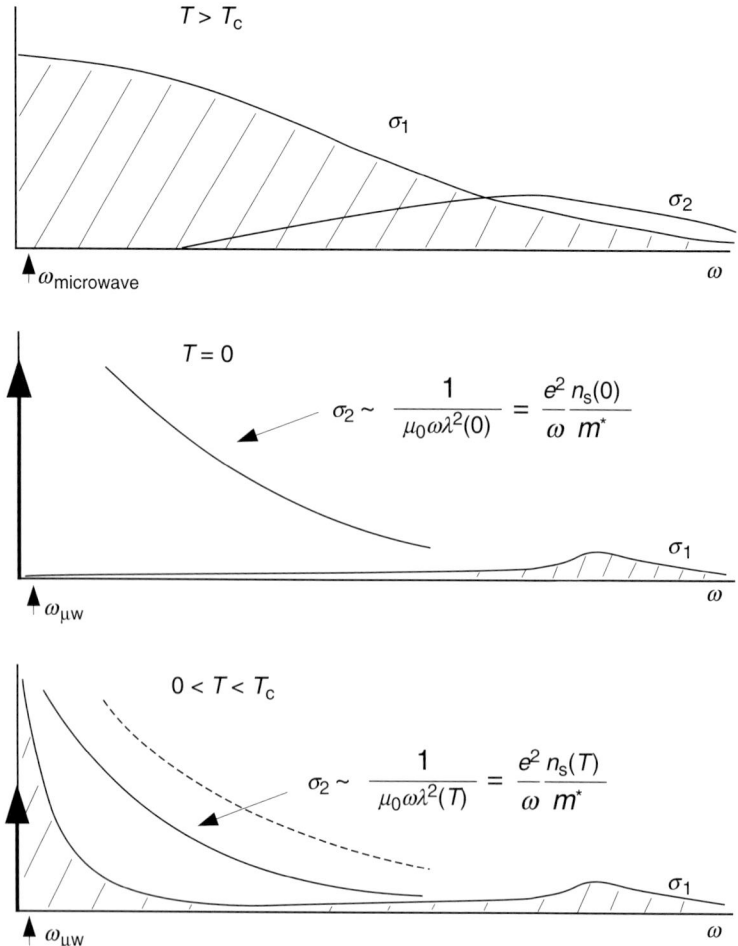

Figure 4.1. Schematic behaviour of $\sigma_1(\omega, T)$, $\sigma_2(\omega, T)$ for a superconductor.

why measurements of the conductivity proved difficult for spectroscopists trying to discern the superconducting gap.

As one raises the temperature, $\sigma_1(\omega > 0, T)$ increases as quasiparticles become thermally excited. At the same time, the strength of the δ-function in σ_1 at $\omega = 0$ has a corresponding decrease. (The accuracy of this implied sum rule is discussed at the beginning of Section 4.5.) Thus a measurement of the temperature dependence of the δ-function strength gives the increase in area under $\sigma_1(\omega > 0, T)$, which is determined by the spectrum of the quasiparticle excitations. In practice, one is usually measuring the contribution to $\sigma_2(\omega, T)$ from the δ-function in $\sigma_1(\omega, T)$. If we take the weight of the δ-function to be $(\pi/2)(e^2/m^*)n_s(T)$ then the contribution to the imaginary part of σ (by Kramers–Kronig) is

$$\sigma_2^{\text{SF}} = \frac{e^2}{m^*} n_s(T) \frac{1}{\omega}.$$

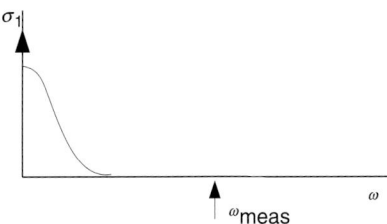

Figure 4.2. Situation where measurement frequency is too high for measurement of superfluid density.

Here we define the magnetic penetration depth λ via the imaginary part of σ,

$$\sigma_2 = \frac{1}{\mu_0 \omega \lambda^2(T)}$$

such that the electromagnetic propagation constant $\kappa = [-i\mu_0\omega\sigma]^{1/2}$ is equal to $1/\lambda$ in the case that $\sigma = \sigma_1 - i\sigma_2$ is dominated by σ_2. Furthermore, if the measuring frequency is low enough that σ_2 is mainly due to the superfluid response, then

$$\sigma_2 = \frac{e^2}{m^*} n_s(T) \frac{1}{\omega} = \frac{1}{\mu_0 \omega \lambda^2(T)}$$

and $\lambda(T)$ reduces to the usual London penetration depth $\lambda_L(T)$.

Below frequencies of a few GHz or so, σ_2 in the high T_c materials is completely dominated by the response of the superfluid condensate and contributions to σ_2 from the normal fluid are negligible except very close to T_c. However, the normal fluid may contribute at higher frequencies, depending on the quasiparticle scattering rate. When $1/\tau$ is lower than the measurement frequency (see Figure 4.2), $\sigma_2(\omega)$ approaches $(e^2/m^*)(n_{\text{total}}/\omega)$, where n_{total} includes both the superfluid and normal fluid (i.e. a narrow response of the normal fluid near $\omega = 0$, responds at higher frequencies as a superfluid). One therefore has to interpret Far IR (and even some mm wave) measurements that do not extend to $\omega \approx 0$ with care. If $1/\tau$ is strongly temperature dependent and falls below the minimum measurement frequency as one lowers the temperature, the superfluid condensate may appear to have very little temperature dependence, even at relatively high temperatures. Here, Far IR measurements are useful for obtaining n_{total}/m^* at $T = 0$ but generally not n_s/m^* and certainly not its temperature dependence. In conjunction with independent "low" frequency measurements, they also give valuable information on the initial fall of $1/\tau$.

Of course, there are wide variations in the temperature dependence of $1/\tau$: in materials doped with Zn or Ni or in most thin films, $1/\tau$ is extrinsic and has a relatively weaker temperature dependence. On the other hand, in high purity single crystals of $YBa_2Cu_3O_{6+x}$, $1/\tau$ falls to microwave frequencies ($\approx 30\,\text{GHz} = 1\,\text{cm}^{-1}$) for $T \leq 40\,\text{K}$. This is well below the minimum far IR measurement frequency achievable in single crystal work (20–$50\,\text{cm}^{-1}$), and even millimetre wave measurements have to be carefully scrutinized. Dähne et al. [38] for example were able to use the frequency dependence of λ in the millimetre wave region to extract τ. On the other hand, de Vaulchier et al. [40] saw no frequency dependence of λ in their films, up to $500\,\text{GHz}$, presumably due to higher extrinsic scattering rates.

Having reviewed the general phenomenology that one will encounter, we return to the actual task of extracting σ_1 and σ_2 from physical measurements. We begin with some simple limits. In the normal state and at low frequencies ($\omega \ll 1/\tau$), $\sigma_1(\omega) \gg \sigma_2(\omega)$ so that $R_s = X_s = \sqrt{\mu_0\omega/(2\sigma_{DC})}$ and $\kappa = (1-i)\sqrt{\mu_0\omega\sigma_{DC}/2} = (1-i)/\delta$ where δ is the classical skin depth, which is typically of order microns at microwave frequencies. For the high temperature superconductors this may be comparable to one of the sample dimensions and, if so, one cannot directly use the surface impedance approximation.

In the superconducting state below T_c, the DC resistivity is zero and is represented in the conductivity spectrum by a δ-function at $\omega = 0$ with an oscillator strength determined by the superfluid density or, equivalently, the penetration depth. This term in the conductivity is $\sigma_1(\omega, T) = \pi\delta(\omega)/\mu_0\lambda^2(T)$ and, through a Kramers–Kronig relation, gives rise to a dominant term in the imaginary part, $\sigma_2(\omega, T) = 1/\mu_0\omega\lambda^2(T)$. Below T_c, one can thus write a general expression for the conductivity away from $\omega = 0$

$$\sigma(\omega, T) = \sigma^*(\omega, T) - i\frac{1}{\mu_0\omega\lambda^2(T)}. \tag{4.6}$$

The term σ^* represents all contributions to the conductivity other than the superfluid contribution and is mainly real at low frequencies ($\omega\tau \ll 1$ where τ is the transport lifetime). So at low frequency $\sigma^*(\omega, T)$ can be replaced by a purely real $\sigma_1(\omega, T)$ and the imaginary part of the conductivity is determined by the superfluid term in Eq. (4.6). Except near T_c, $\sigma_2 \gg \sigma_1$ and Eqs. (4.3) and (4.6) yield simple approximations for the surface impedance:

$$R_s = \frac{\mu_0^2}{2}\omega^2\lambda^3(T)\sigma_1(\omega, T) \tag{4.7}$$

$$X_s = \mu_0\omega\lambda(T). \tag{4.8}$$

We see that a measurement of $X_s(T)$ allows a direct determination of $\lambda(T)$ and the associated quantities $n_s(T)e^2/m^* = \omega\sigma_2(\omega, T) = 1/(\mu_0\lambda^2(T))$, whereas $\sigma_1(\omega, T)$ can only be extracted from R_s if values of $\lambda(T)$ are also available. It turns out that $X_s(T)$ is rarely measured directly, especially on small single crystals, and the typical situation is that $\lambda(T)$ is measured by one or more of a variety of techniques, not necessarily involving microwaves. In fact, obtaining reliable values of $\lambda(T)$, a quantity of intrinsic importance on its own but also crucial to the analysis of microwave data, has turned out to be a very difficult task for the case of the cuprate superconductors. In the next section we briefly review the various methods that have been used. The reader is referred to Hardy et al. [65] for more details.

For completeness, we include a general expression for extracting $\sigma_1(\omega, t)$ in the superconducting state and right through T_c into the normal state:

$$\sigma_1 = \left[\left(\frac{\sigma_s}{2} \pm \left[\frac{\sigma_s^2}{4} - \sigma_2\sigma_s\right]^{\frac{1}{2}}\right)^2 - \sigma_2^2\right]^{1/2}, \tag{4.9}$$

where the $+(-)$ sign is used for the case $\sigma_1 > (<)\sqrt{3}\sigma_2$ and $\sigma_s = \mu_0\omega/2R_s^2$. This simplifies to Eq. (4.7) when the approximation $\sigma_2 \gg \sigma_1$ is made. While the full expression must be used for quantitative analysis very close to T_c, the approximate version, Eq. (4.7), is much more convenient and transparent.

4.2.3. Non-local Electrodynamics

Non-local effects occur when there exist physical correlation lengths that exceed the penetration lengths of the applied electromagnetic fields, which in the present case are of order 100–200 nm or more.

For the response of the superfluid, the relevant physical scale is the coherence length which is much smaller that λ for the cuprates, and in most cases one is safely in the local limit. However, Kosztin and Leggett [108] show that for B applied parallel to the $\hat{c}$-axis of a clean d-wave superconductor, non-local effects in $\lambda(T)$ may appear below a crossover temperature of about 1 K (an effect arising from the coherence length becoming large in the nodal directions). For the response of the "normal fluid", the situation is more complicated. For clean materials at low temperatures one can easily have in-plane quasiparticle mean-free-paths that are greater than the in-plane penetration depth. Nevertheless, since the transport is strongly two-dimensional, non-local effects are greatly suppressed for excitation fields applied parallel to the *ab* plane: the quasiparticles largely remain within the penetration depth between scattering events. However, there exist geometries where this condition does not hold. An example is the case where B is applied parallel to an *ac* or *bc* face of a crystal; here the quasiparticles can exit the field penetration region before a scattering event [166]. To the best of our knowledge, this has not been studied experimentally. In the discussion that follows, we assume local electrodynamics.

4.2.4. Excitation Spectrum of a d-Wave Superconductor

It is now generally, although not universally, agreed that *all* of the cuprates have an order parameter with predominately $d_{x^2-y^2}$ character, with four nodal lines along the *c*-direction of the crystals. The electronic structure is two-dimensional in character, although the degree of this two-dimensionality varies enormously, both between families and with doping; it can also be a strong function of the temperature. Furthermore, except for the occurrence of nodes (not seen in the "old" superconductors) the phenomenology of the superconductivity is conventional in many respects. This is in stark contrast to the normal state, which is anything but conventional and which has bedeviled the community for almost two decades. The pairing mechanism is undoubtedly closely tied up with the physics of a doped Mott insulator and the ensuing strong electron–electron correlations [116], but the details are still the subject of continuing debate.

One aspect of our own particular view of the superconductivity in the cuprates, which may be less universally agreed upon, concerns the issue of homogeneity. It is our opinion that over a wide region of the phase diagram, the superconductivity is intrinsically homogeneous, by which we mean that the more perfect the samples become, the stronger and more homogeneous the superconductivity becomes. Of course, one should not rule out the possibility of "intrinsic" phenomena such as fluctuating stripes or other competing order, the suppression of which might strengthen the superconductivity (the suppression of T_c near $1/8$ doping is a case in point [4, 121]).

A superconductor with nodes is particularly susceptible to many kinds of imperfections. Disorder, short coherence lengths, and the weak screening characteristic of the cuprates combine to place stringent constraints on the crystallinity of the materials, which must be considered if one wants to study the intrinsic properties of the superconducting state. The fact that superconductivity requires doping of the CuO_2 planes, generally by dopants that have some (or total) randomness associated with their positions, further complicates the situation.

Finally, many of the high temperature superconducting compounds, when grown under typical conditions tend to have cation disorder that may or may not be amenable to strategies for their reduction. The effect of cation disorder on the superconductivity has been studied in considerable detail recently, by Eisaki et al. [45] and Fujita et al. [52], where is it shown that the reduction of T_c is very sensitive to the details of the disorder, for example, the ionic radius of the substituted cation. However, in all cases, *more* disorder is correlated with a *larger* reduction in T_c.

In regions of the phase diagram, where T_c varies quickly with doping, it is clear that any randomness in the doping will lead to large spatial variations in the strength of the superconductivity. In an extreme case, near the AFM-SC border for example, some regions will be inherently antiferromagnetic and other regions superconducting. When proximity effects are added to the mix, one has a situation where it is extremely difficult to extract the "intrinsic" physics from measured properties. Microwave measurements typically have neither wavevector nor spatial resolution, so, as with most bulk probes, one cannot à priori distinguish intrinsic from extrinsic behaviour: one has to observe how the properties evolve with sample perfection. In this review much of the focus is on this evolution.

Phenomenological Pairing Model

At present, there is no "theory" of high temperature superconductivity that has the all-encompassing reach of the BCS theory for conventional superconductors in which the temperature and frequency dependence of the basic physical observables can be calculated from $T = 0$ to $T = T_c$. While some aspects of the superconducting state in the cuprates mirror those of "d-wave BCS" superconductivity (i.e. the BCS solution for a fermionic system to which some interaction favouring a d-wave ground state has been added), there remain important differences. For example, Lee and Wen [117] pointed out the essential disconnect between the zero temperature superfluid density $\rho_s(0)/m$ and the temperature dependence of the normal fluid density: $(\rho_n(T)/m) = (\rho_s(0)/m) - (\rho_s(T)/m)$. In BCS theory these are necessarily tied together and set by the properties of the gap function. For the cuprates as the hole doping is reduced and one approaches the Mott insulator $\rho_s(0)/m$ must, and does, go to zero, whereas $\rho_n(T)/m$ is less doping dependent. From the experimentalist's point of view it is still premature to try to fit data to strong- or weak-coupling BCS, given the lack of a global theory. However, at low temperature, where the excitation spectrum for the quasiparticles has settled down, it *is* fruitful to measure quantities such as normal fluid density, specific heat, thermal and electrical conductivity, etc. to empirically determine the qp parameters. Hussey [86] summarizes existing experiments relevant to this task, and in particular, critically examines the self-consistency of various results. In the present review, we will confine our attention to the superfluid density and the electrical conductivity σ_1.

The basic elements of the phenomenological pairing theory put forward by Lee and coworkers [116, 117, 210] are as follows. One assumes that the elementary excitations in the superconducting state are well defined quasiparticles (qp), with dispersion $E(\mathbf{k}) = \left[(\epsilon_\mathbf{k} - \mu)^2 + \Delta_\mathbf{k}^2\right]^{\frac{1}{2}}$, where $\Delta_\mathbf{k} = \frac{1}{2}\Delta_0 \left(\cos k_x a - \cos k_y a\right)$ is a d-wave gap for an assumed tetragonal lattice, with lattice parameter a. For $\epsilon_\mathbf{k}$, a tightbinding approximation is assumed with $\epsilon_\mathbf{k} = 2t \left(\cos k_x a + \cos k_y a\right)$. There is now considerable evidence that, for the generally orthorhombic crystals the predominant state is a "distorted" $d_{x^2-y^2}$ state where the effect of orthorhombicity is mainly to make one set of opposing lobes somewhat larger than the other.

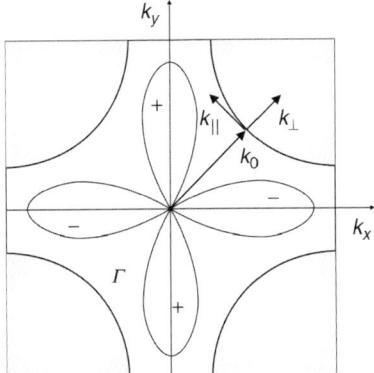

Figure 4.3. Schematic diagram of the single-band tight-binding Fermi surface for a high temperature superconductor, showing the shift of the co-ordinate system to the nodes. Γ is the centre of the Brillouin zone. The lobes represent the superconducting gap [86].

This will shift the nodal points ($\Delta(\mathbf{k}) = 0$) from the 45° positions, but there is no evidence to show this is a large effect in the high temperature superconductors. Also, there is no guarantee that the gap function has the pure $d_{x^2-y^2}$ form where $\Delta(\mathbf{k}) = \Delta_0 \cos 2\phi$ where ϕ is the azimuthal in-plane angle. For low temperature properties, this is inconsequential since only the slope of Δ near the node is of importance. The effect of orthorhombicity on the band structure can be more substantial, but here these complications are ignored.

For calculating low temperature properties, it is convenient and standard to rotate by 45° to a coordinate system $(k_\perp, k_\parallel)$ whose origin is centred at one of the nodes with $k_\perp$ and $k_\parallel$ the momentum normal and tangential to the Fermi surface, respectively (see Figure 4.3)

$$k_\perp = \frac{1}{\sqrt{2}}(k_x + k_y) - |\mathbf{k}_0|,$$

$$k_\parallel = \frac{1}{\sqrt{2}}(k_x - k_y).$$

Linearizing the spectrum around one of the four values of $\mathbf{k}_0$, one obtains:

$$\epsilon_\mathbf{k} = \hbar v_F k_\perp,$$

$$\Delta_\mathbf{k} = \hbar v_\Delta k_\parallel,$$

where v_F is the Fermi velocity associated with dispersion of the qp's *normal* to the Fermi surface (in the nodal region),

$$\mathbf{v}_F = \frac{\partial \epsilon_\mathbf{k}}{\partial \mathbf{k}} = v_F \hat{\mathbf{k}}_\perp \qquad v_F = 2\sqrt{2}\, t a\, \sin(k_{0x}a)$$

and v_Δ gives the slope of the gap in the nodal region and is associated with the dispersion of the qp's *along* the Fermi surface:

$$\mathbf{v}_\Delta = \frac{\partial \Delta_\mathbf{k}}{\partial \mathbf{k}} = v_\Delta \hat{\mathbf{k}}_\parallel \qquad v_\Delta = \frac{1}{\sqrt{2}} \Delta_0 a\, \sin(k_{0x}a).$$

In the region of the node, the excitation spectrum can thus be written,

$$E(\mathbf{k}) = \left(\epsilon_\mathbf{k}^2 + \Delta_\mathbf{k}^2\right)^{\frac{1}{2}} = \hbar\left(v_F^2\, k_\perp^2 + v_\Delta^2\, k_\parallel^2\right)^{\frac{1}{2}}.$$

Adding contributions from the four nodes, one obtains the angle-averaged density of states:

$$N_s(E) = \left(\frac{2}{\pi\hbar^2}\right)\left(\frac{1}{v_F v_\Delta}\right) E$$

which is linear in E (limit of low E, no impurities). From this, many of the low-energy properties of the system can be calculated. In particular, the normal fluid density for one CuO_2 plane is given by [210]:

$$\frac{\rho_n}{m} = \frac{2\ln 2}{\pi}\, \alpha_{FL}^2\, \frac{v_F}{v_\Delta}\, T.$$

In the first version of the model [117] the qp current was assumed to be given by:

$$\mathbf{j}(\mathbf{k}) = -e\,\frac{\partial \epsilon_\mathbf{k}}{\partial \mathbf{k}} = -e\, \mathbf{v}_\mathbf{k}.$$

Millis et al. [133] pointed out that a Fermi liquid correction should be applied and the qp current was modified to $j(\mathbf{k}) = -e\, \alpha_{FL}\, v_k$ [210]. This accounts for the α_{FL}^2 factor in the expression for ρ_n/m. Results for other quantities are listed by Hussey [86].

The ratio v_F/v_Δ, sometimes referred to as the Dirac anisotropy ratio α_D, is an important parameter for high temperature superconductors. It is possible to extract α_D from the low temperature universal thermal conductivity κ_0/T [188]. Combined with a value of $\alpha_{FL}^2\, \alpha_D$ from the temperature dependence of the superfluid density via $\lambda(T)$, one can find the Fermi liquid parameters. However, the value of α_D^2 obtained depends on the absolute magnitude of $[\Delta\lambda(T)/\Delta T]/\lambda^3(0)$. While $\Delta\lambda(T)/\Delta T$ is easily measured, the value of $\lambda(0)$ is not. This results in relatively large error bars for α_{FL}^2.

Lee and Wen [117] further assume that the full Drude weight of the doped holes appears in the zero temperature superfluid density so that $(\rho_s(0)/m) = (x/a^2 m)$, where x is the hole doping. In this phenomenological model, motivated by experiment, the wide divergence from standard BCS results is explicit: $\rho_s(0)/m$ depends on x while $\rho_n(T)/m$ does not. This is one of the central issues that must be addressed by microscopic theories.

Effect of Impurities

The review of Hussey [86] contains a very useful discussion of the self-consistent T-matrix approximation (SCTMA) that is widely used to treat the effect of impurities in metals, semiconductors, and superconductors. As pointed out by Hussey, impurity substitution has proven to be a powerful probe in the study of complex many-body systems in general. For the cuprates studies of T_c suppression, increase of residual in-plane resistivity ρ_0, change in temperature dependence of the penetration depth, impurity-induced effects in the low-T specific heat and thermal conductivity, electrical conductivity, and impurity related bound states seen by scanning tunnelling microscopy (STM) have all yielded important information. For unconventional superconductors, the SCTMA has been studied in

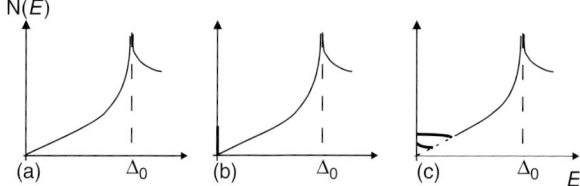

Figure 4.4. a.) The density of states of a clean d-wave superconductor. b.) The appearance of a bound state at the Fermi level due to a non-magnetic unitary scatterer. c.) The broadening of the bound state distribution for a dilute density of non-magnetic impurities, leading to a finite density of states at zero-energy that increases with increasing impurity concentration. The rounding of the coherence peak in the presence of impurities is not shown [86].

most detail by Hirschfeld and coworkers (see Nunner and Hirschfeld for references to earlier work [140]).

For the high temperature superconductors Hirschfeld et al. [75, 76] adopted a generalized BCS model with a $d_{x^2-y^2}$ state. The scattering from a single point impurity at low temperatures is characterized by c, the cotangent of the s-wave scattering phase shift. The finite density of scatterers n_i, defines a temperature independent elastic scattering rate, $\Gamma \equiv n_i n/(\pi N_0)$ where n is the electron density and N_0 the density of states (DOS) at the Fermi level. In the normal state limit ($\Delta_\mathbf{k} \to 0$) the electrical conductivity reduces to a Drude form: $\sigma = \sigma_0/[1 + (\Omega \tau_N)^2]$ where $\sigma_0 \equiv n e^2 \tau_N/m$ with $1/2\tau_N = \Gamma_N = \Gamma/(1+c^2)$ and Ω is the microwave frequency.

For $\Delta_\mathbf{k} \neq 0$, the impurities modify the d-wave density of states in such a way that there is a region of width γ near the Fermi level where the DOS is finite (Figure 4.4). For temperatures $T^* < \gamma/k \ll T_c$, the so-called "gapless" regime, the superconducting properties reflect the temperature dependence of their normal state analogues, but scaled according to the reduced DOS. Above T^* one enters the "pure" regime where properties approach those of a clean d-wave superconductor. In the resonant scattering (unitary) limit where $c \simeq 0$, γ is of order $(\Gamma \Delta_0)^{\frac{1}{2}}$ (to within logarithmic factors). However, in the Born limit where $c \gg 1$, $\gamma \sim \Delta_0 \exp[-\Delta_0/\Gamma_N]$, which can be extremely small in the weak scattering limit. Because of the exponential dependence on Δ_0/Γ_N, the physics of the gapless region may become inaccessible, even when Γ_N is not particularly small.

In the gapless regime, and for $\hbar \Omega \ll kT$, the conductivity becomes $\sigma_1 \simeq \sigma_{00} [1 + (\pi^2/12)(T/\gamma)^2]$ where $\sigma_{00} = ne^2/[m\pi \Delta_0(0)]$ is the universal limit first derived by Lee [118]. The lack of dependence of σ_{00} on impurity concentration is due to the fact that an increase in impurity density *increases* the density of quasiparticles, while *decreasing* the quasiparticle lifetime by the same factor.

Durst and Lee [44] later showed that both vertex and Fermi liquid corrections needed to be applied. (Vertex corrections account for differences in forward vs. backward scattering amplitudes, and Fermi liquid corrections account for the fact that the superconducting state emerged from a Fermi liquid with strong electron–electron correlations). They found that vertex corrections modify the electrical conductivity, and Fermi liquid corrections renormalize both electrical and spin conductivity, while the thermal conductivity maintains its universal value. If all three could be measured on the same sample, the Fermi liquid and vertex corrections could be independently determined. See [188] for measurements of the universal thermal conductivity.

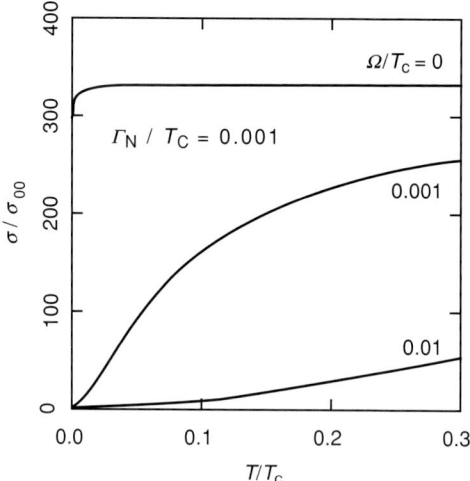

Figure 4.5. Normalized low-T conductivity σ/σ_{00} vs. the reduced temperature T/T_c in the Born limit [76].

Returning to the temperature and frequency dependence of σ_1, Hirschfeld et al. [76] give for the pure regime,

$$\sigma_1(\Omega) = \frac{ne^2}{m} \int_{-\infty}^{\infty} \left(-\frac{\partial f}{\partial \omega}\right) \left|\frac{\omega}{\Delta_0}\right| \frac{\tau(\omega)}{1 + \Omega^2 \tau^2(\omega)} d\omega \qquad (4.10)$$

with $1/\tau = -2\,\text{Im}\,\Sigma_0(\omega)$, valid for all scattering strengths. Here $\Sigma_0(\omega)$ is the averaged self-energy due to impurity scattering in the SCTM approximation. The conductivity thus takes the form of a sum of Drude lineshapes. However, in an unconventional superconductor $1/\tau(\omega)$ will *always* have a non-trivial dependence on energy, so that $\sigma_1(\Omega)$ should *never* have a simple Drude form deep in the superconducting state. The observation of approximate Drude shapes for σ_1 in high quality YBa$_2$Cu$_3$O$_{6+x}$ crystals [67] at low temperatures posed a challenge for theory, and a number of alternatives to simple point scattering were studied, including the cumulative effect of scattering by a variety of dilute and/or weak scatterers [14], scattering of qp's from order parameter "holes" [71] and extended scatterers [44, 140].

Very recently the experiments of Turner et al. [202] on YBa$_2$Cu$_3$O$_{6.5}$ with highly ordered oxygen chains clearly show a $\sigma_1(\omega, T)$ for the $\hat{a}$ crystal direction that exhibits most of the expected characteristics of Born scattering in a $d_{x^2-y^2}$ state at low temperatures. We note here that for weak (Born) scattering in a $d_{x^2-y^2}$ state, $1/\tau(\omega) \simeq \Gamma\,\omega/\Delta_0$ for, to within logarithmic corrections, $\omega \ll \Delta_0$. Thus in the pure regime, the low frequency value of the conductivity ($\Omega\tau \ll 1$) becomes $\sigma_1 = \sigma_0 \gg \sigma_{00}$. This is illustrated in Figure 4.5 for various values of Ω/T_c, for the choice $\Gamma_N/T_c = 0.01$ [76]. Here the low frequency limit of σ_1 will remain well above the universal limit until extremely low temperatures.

4.3. Experimental Techniques

In this section we will restrict our attention to the linear response of high temperature superconductors in the region below approximately 100 GHz. Typically (although not always) one is in a "skin-effect" region where the penetration depth of the microwave magnetic field is much shorter than any dimension of the crystal. For very practical reasons there is usually a distinct separation between the microwave region and the higher frequency regions (far infrared, mid-infrared, etc.). First, when the free space wavelength of the EM radiation is comparable to the sample size, serious and often insurmountable difficulties arise due to diffraction effects. This crossover occurs at approximately 90 GHz ($= 3 \text{ cm}^{-1}$) when the sample sizes are of order a few millimetres. Second, below about 30 cm^{-1} the reflectivity of samples in the superconducting state become extremely high, $>99\%$, so that the traditional single-reflection optical methods become problematic. Of course the reflectivity becomes even higher in the microwave region, but here the techniques of cavity perturbation, which are equivalent to multiple-reflection methods, more than make up for the increased sample reflectivity.

For the cuprates, where the best materials tend to be small single crystals, one is usually left with a spectral gap in the region $3-30 \text{ cm}^{-1}$. In cases where high quality large-area films can be produced with thicknesses much less than the magnetic penetration depth λ, then one has the option of transmission experiments, using for example femtosecond time-domain spectroscopy techniques [141]. Orenstein and co-workers [183], as well as others, have very successfully applied these methods to thin films. However, in general one cannot bridge the microwave/far-IR spectroscopy gap using thin films, since they almost always have significantly different properties from bulk single crystals. For example, it is possible to have $YBa_2Cu_3O_{6+x}$ thin films with penetration depths significantly larger than those for single crystals (factor 2) yet show a linear temperature dependence for $\Delta\lambda(T)$ at low T. This is not explainable in terms of impurity or defect scattering and one needs something analogous to the Swiss-cheese model proposed by Nachumi et al. [135] where parts of the sample are non-superconducting.

We note also that in principle, near-field techniques can also overcome diffraction limitations. Here, one uses an "antenna", much smaller than the free space wavelength, placed in close proximity to the sample surface and one gauges the properties of the sample by the effect on the antenna. The "antenna" could be a needle shaped probe [87,119] or a small hole in a waveguide [129] or a small hole in a resonant cavity [28]. These techniques are analogous to STM and allow high spatial resolution or, equivalently, allow the use of smaller samples for a given microwave frequency. Calibration and other issues tend to prevent these methods from becoming general spectroscopic techniques, but they are nevertheless extremely useful in special cases.

Classical optical reflection techniques measure the power reflectivity over a very wide range of frequencies, and then use various Kramers–Kronig transforms of the data to extract the desired properties, such as the real and imaginary parts of the conductivity tensor. For $\hat{a}\hat{b}$-plane properties of single crystals one typically measures the reflectivity of the natural growth face of the platelet-like crystals, along with gold or lead films evaporated in situ on the crystal as a reference reflectivity [12, 80, 81]. Here the requirement is a flat face, with dimensions substantially larger than the longest wavelengths of interest. One can also measure $\hat{c}$-axis properties in the case that the c-dimension is large enough. For microwave techniques, where the microwave wavelength is much larger than the crystal dimensions, the experimental issues are very different. For good conductors and superconductors, one generally places the crystal in regions where the applied electric fields are as small as possible and the applied ac magnetic field as uniform as possible.[1] Here the issues are (1) what are the actual fields at the surface of the crystal and (2) along what crystal directions do the induced currents flow? In this case the detailed shape of the crystal and its orientation with respect to the applied B field have to be taken into account. In these circumstances one should in principle work with ellipsoidal

---

[1] There are some situations where it is preferable to place the sample in the electric field region. For 1D conductors this is the *only* option. See Maeda et al. [127] and Peligrad et al. [156].

shaped samples in order to have reasonably well-controlled demagnetizing effects, but this is difficult to do with the cuprates and is rarely done. Instead one tends to use limiting forms of ellipsoids in low demagnetizing-effect geometries (such as thin platelets with the face parallel to B) or one tries to correct for non-ellipsoidal shapes [161]. Some of the issues will be discussed in more detail later, but suffice it to say that great care has to be taken to avoid errors.

4.3.1. Penetration Depth Techniques—Single Crystals

Single crystals, thin films, and powders all have their place in the quest to understand the cuprates. While properties of the best single crystals still seem to be the closest to intrinsic, such samples are often only available as small platelets. In addition, crystals are grown close to equilibrium conditions where certain doping regimes cannot be reached, and thin films or powders are necessary. Thin films are also central to many applications, so that direct measurements of their properties are essential. The measurement methods are conveniently classified according to the form of the samples.

Excluded Volume Techniques

In this method, one places the crystal in a small magnetic field and measures the "effective" volume of the crystal. This can be carried out all the way from DC (using SQUID magnetometry), to audio frequencies (AC susceptometry) to rf and microwaves (cavity perturbation). In the situation where the demagnetizing effects are negligible (i.e. the fields at the surface of the sample are almost everywhere equal to the applied field), the effective volume is just the geometrical volume of the crystal minus the volume penetrated by the magnetic field. As an example, for a thin platelet with the field applied parallel to the broad face, the effective volume is approximately $ab(c - 2\lambda)$ where ab is the area of the flat face and c is the thin dimension. Since single crystals typically have $c > 20\,\mu$m and λ is of order 100 nm, in order to extract the absolute value of λ to within 10% from a measurement of the effective volume, one would have to know the thickness of the crystal to better than 1/1,000. In addition there are demagnetizing corrections and calibration factors that have to be accurately determined. A direct attack in this direction is essentially impractical given the small and not perfectly regular shaped crystals one is dealing with. Therefore one generally has to be content with a measurement of the temperature dependence of λ: $\Delta\lambda = \lambda(T) - \lambda(T_0)$ where T_0 is some reference temperature (usually the base temperature of the apparatus). To do this, one needs to be able to change the temperature of the sample without affecting the calibration factors of the apparatus and at the same time avoiding (or correcting for) effects of sample holder materials that have temperature dependent magnetic properties. One arrangement that works extremely well is the use of a sapphire hot finger in vacuum, pioneered by Sridhar and Kennedy [184] and by Rubin et al. [168] and widely used. High purity sapphire has a very small magnetic susceptibility, is an excellent thermal conductor, and has very low microwave loss even up to mm wave frequencies.

We note here a very clever refinement of the excluded volume technique devised by Prozorov et al. [162], where the sample is coated with a thin film of a low T_c superconductor such as aluminium. The low T_c film excludes the magnetic flux until its T_c is reached, whereupon the film becomes transparent at the 10 MHz measurement frequency. To within an error of order ± 15 nm, determined by the accuracy to which λ and the film thickness are known, one can extract $\lambda(T \sim 0)$ for small crystals. The method requires good quality Al films and is not immune to the usual demagnetizing problems for $\vec{B} \parallel \hat{c}$ or $\hat{c}$-axis contamination for $\vec{B} \parallel \hat{ab}$-plane.

For excluded volume techniques applied to high temperature superconductors, one has to be extremely careful to ensure that thermal expansion of the sample is not affecting the measurements. The issue is addressed in [42, 197] and is described in some detail in Hardy et al. [65]. Trunin [193], also treats this problem, along with other important issues. As an example, using the thermal expansion data of Kraut et al. [109] one can show that for a crystal with $c = 100\,\mu$m, measured in the geometry of the preceding paragraph, errors of order 30% in $\Delta\lambda$ can arise in optimally doped $YBa_2Cu_3O_{6+x}$ at ~70K. The effects become proportionally larger for thicker crystals; for very thin crystals, the effects are smaller and corrections have adequate accuracy.

Now consider the situation where the magnetic field is applied parallel to the $\hat{c}$-axis (perpendicular to the broad face) a geometry used when one wants to restrict the currents to the ab plane. In this geometry there are large demagnetizing effects, and the fields at the edge of the sample can be one or two orders of magnitude larger than the applied field. If we approximate the usual square or rectangular ab plane shape by a circle, then one has a geometry that has been studied extensively in the literature (see Brandt [22] and references therein). There are no analytical solutions, and for the case where $\lambda \ll c \ll 2a$ the numerical solutions are also not available. Prozorov et al. [161] have proposed approximate solutions that appear to be useful for not too thin crystals. Nefyodov et al. treat the $b \gg a > c$ geometry (long slab) in detail and suggest corrections for finite b [137]. There are two issues that arise (1) what is the relationship between the physical dimensions of the crystal and the measured effective volume (which, due to demagnetizing effects, is much larger than the physical volume of the crystal) and (2) what are the thermal expansion effects. It is easy to see that thermal expansion effects are likely to dominate for this geometry, and given the lack of accurate solutions for the effective volume, it is generally not possible to correct for thermal effects. An exception is the Al marker film technique of Prozorov et al. [161] described above.

A number of groups [78, 130, 175, 186] have extracted values of $\lambda(T)$ from microwave cavity perturbation methods by assuming that in the normal metal, the real and imaginary parts of the surface impedance are equal; this is valid for $\sigma_1 \gg \sigma_2$, which is the case when the quasiparticle scattering rate is much greater that the observing microwave frequency. The method works as follows: the change in resonant frequency of the measurement cavity as one varies the sample temperature from $T = 0$ to $T > T_c$ is, to within a calibration constant, given by $\delta(T) - 2\lambda(0)$ where $\delta(T)$ is the sample skin depth. To within the same calibration constant, the change in width of the cavity response (change in $1/Q$ of the cavity), gives $\delta(T)$ directly (this assumes that in the superconducting state the sample losses are negligible in comparison). Comparing the two results gives $\lambda(0)$. The weak point in the scheme is that one must neglect or correct for thermal expansion contributions to the frequency shift. For the $\vec{B}_{rf} \perp c$ geometry, the expansion effects are relatively small for very thin crystals, but then one has to contend with contributions from λ_c or δ_c to the frequency shifts ($\hat{c}$-axis contributions). On the other hand, for the $\vec{B}_{rf} \parallel c$ geometry where currents flow only in the ab plane, thermal expansion effects will dominate the shifts. In either case, the method has to be applied with extreme care [65]. A further cautionary note: Kusko et al. [112] have found that $R_s \neq X_s$ for $T > T_c$ in some underdoped samples, a condition that violates the basic assumption of the method.

Far Infrared Reflectivity: $|R|e^{i\theta}$

If one can measure $|R(\omega)|^2 \equiv$ power reflectivity over a wide enough frequency range that one can perform a Kramers–Kronig on $|R(\omega)|$ to obtain $\theta(\omega)$, the Fresnel formula will yield σ_1 and σ_2. The superfluid contributes a δ-function in $\sigma_1(\omega, T)$ at $\omega = 0$ and a component to $\sigma_2 = (\mu_0 \omega \lambda^2)^{-1} = n_s e^2/m^* \omega$. If at low frequencies there is a region of frequencies where the quantity $(\mu_o \omega \sigma_2)^{-1/2}$ tends to a constant, then it is reasonable to assume one is dealing with a $\sigma_1(\omega)$ concentrated near $\omega = 0$. However, as discussed previously, the value of λ obtained may not be the London penetration depth λ_L if there are contributions from $\sigma_1(\omega)$ other than the superfluid δ-function. The method has the advantage that it gives absolute values for λ, and by changing the polarization and the faces one can measure λ_a, λ_b and λ_c independently.

Another approach is to look for the plasmon associated with the superfluid by which $1/|\epsilon(\omega)|$ peaks at the plasma frequency $\Omega_p = (Ne^2/\epsilon_0 m^*)^{1/2}$. Again, any narrow Drude component in $\sigma_1(\omega)$ will be included in Ω_p. The method is particularly useful when, by using grazing incidence with respect to the ab plane, one can pick up the $\hat{c}$-axis plasmon [97] in highly anisotropic materials where other methods to measure λ_c may be impractical.

Measurement of Internal Field Distribution in Mixed State

In Type II superconductors, with an applied magnetic field $H > H_{c_1}$ and weak pinning, a regular lattice of vortices with density B/Φ_0 is formed. Away from an isolated vortex the magnetic field falls to zero with scale length λ_L. For $H \gg H_{c_1}$ the density of vortices is high enough that the internal field B is relatively uniform, however, $\overline{\Delta B^2}$ is set by $1/\lambda^2$ to within a constant. More generally, the detailed field distribution contains considerable information beyond the value of λ.

Muon Spin Rotation (μSR) has been applied with great success to the cuprate superconductors, in many cases giving the first values of λ_L. The 100% spin-polarized positive muons are implanted one at a time into the sample, where they quickly thermalize and take up a preferred interstitial position in the crystalline lattice. The muons decay with an average lifetime of 2.2 μs with the emission of an energetic positron, emitted preferentially along the direction of the muon spin. Using a start counter and positional β^+ counters, a histogram can be built up which contains information on the precession of the muon spin. Something closely analogous to the "free induction decay" in NMR is so obtained, with a corresponding Fourier transform that gives the distribution of magnetic fields within the sample.

This method has the advantages that it is a bulk measurement (implant distance typically a few hundred microns), it gives absolute values for $\lambda(T)$ and, perhaps most importantly, it contains additional information. As an example, the shape of the high field part of the distribution is sensitive to the details of the vortex core and μSR is one of the very few methods that can measure the coherence length ξ.

It has the disadvantage that rather large (~ 0.5 cm^2), thick (~ 0.3 mm) samples are required and further, it is difficult to measure λ_a, λ_b separately, or to measure λ_c at all. On the other hand the method works with ceramics or powders, although it is now clear that some early measurements on unaligned powders gave misleading results. For example, in 1987 Harshman et al. [68] obtained a temperature dependence of $1/\lambda^2$ in a YBa$_2$Cu$_3$O$_{6+x}$ ceramic that was very flat at low temperatures. This was interpreted as evidence for s-wave superconductivity. More recent μSR results on single crystal YBa$_2$Cu$_3$O$_{6+x}$ now agree quite well with, for example, microwave methods. With further refinements, the μSR method can now be used to measure the *field dependence* of quantities such as λ and ξ [181].

In principle NMR can give more or less equivalent information to that from μSR. Unfortunately, the non-spin 1/2 species in the cuprates have NMR linewidths that are much too broad for this purpose, and the spin-1/2 species, such as Y, give signals that are generally very weak. The technique of β-NMR where nuclear spin-polarized radioactive species are implanted is showing great promise [60], and intense radioactive beams are becoming routinely available. This technique will allow smaller and thinner samples to be used.

Zero-Field Gadolinium ESR

This novel method is based on the zero-field electron spin resonance (ESR) of small amounts of Gd substituted for Y in $YBa_2Cu_3O_{6+x}$ (Pereg-Barnea et al. [157]). Using the broadband bolometric microwave technique developed by Turner et al. [201], it was possible to measure $\chi''(\omega)$ for the three zero-field transitions of the $S = 7/2$ $^{3+}$Gd ion. The integrated absorption strengths are directly proportional to the number of spins exposed to $B_1(t)$, which in turn is controlled by λ. The method can yield $\lambda(T=0)$ for a, b, and c.

4.3.2. Penetration Depth Techniques—Thin Films

Excluded volume techniques are very difficult to apply to films whose lateral dimensions are of order of millimetres but whose thickness is of order 1,000 Å. Here, large demagnetizing effects become unavoidable and are not usually under control. Even when the field is applied parallel to the film, a small misorientation of the film, or inhomogeneities in the applied field will strongly distort the applied field.

Low Frequency Mutual Inductance Techniques

This technique works extremely well for films that are thin enough. It has been used by several groups [199, 200] usually in a configuration with the primary and secondary coils on opposite sides of the film. The film starts to screen the applied ac field when $t = \lambda \cdot \lambda / R$ where R = radius of coil and t is the thickness of the film. Thus, substantial screening can occur for films that are much thinner than the penetration depth λ. Very roughly, $H_{sec}/H_{prim} = (1 + Rt/\lambda^2)^{-1}$, so that for $R = 0.5$ cm, $t = \lambda \simeq 150$ nm, the film attenuates the drive field by a factor of 40,000. It is essential to reduce unwanted direct pickup, and films with diameters as large as $4''$ have been used to solve this problem [51]. Because of the large attenuation, one must avoid macroscopic defects in the films. The method is restricted to probing in-plane currents and in its conventional form does not allow separation of λ_a and λ_b (typically, the films are micro-twinned). This method has the strong advantage of yielding absolute values of λ with fairly good precision.

Thin Film Resonator Techniques

Here the thin film is itself part of a resonant circuit. For the parallel plate method, one measures the transverse electromagnetic (TEM) resonance(s) of a face-to-face pair of films separated by a thin dielectric. This technique was pioneered by Taber et al. [189], and used by Anlage et al. [6], Ma et al. [126] and others. The fundamental mode corresponds to the lateral dimension being equal to a half wavelength in the dielectric medium. Very high resolution is possible, but the method does not normally yield absolute values for λ; thermal expansion effects may also be important. More generally, any microstrip resonator can be used to measure $\Delta\lambda(T)$. A variant developed by Andreone et al. [5] which avoids patterning of the film of interest uses a microstrip ring ($YBa_2Cu_3O_{6+x}$) with the film of interest (NCCO) as the ground plane. While these methods do not generally give absolute values of $\lambda(0)$, by fabricating coplanar waveguide resonators of different geometries out of a single thin film, Valenzuela et al. [207] were able to extract $\lambda(0)$ by comparing frequency shift data for the resonators. An accuracy of $\pm$ 30 nm was achieved.

Millimetre Wave Transmission

This technique is related to the mutual inductance technique, but can yield considerably more information such as R_s, the frequency dependence of λ, etc. Here too the transmitted signal is strongly attenuated by the films: the fields being reduced by the factor $\lambda_{EM} t / 2\pi \lambda^2$, where here the free space wavelength of the radiation takes the place of the coil dimension and we have assumed that $t \ll \lambda$. The reduction can be very large at low temperatures, and leakage *around* the film can be difficult to suppress. Also, the dielectric properties of the substrate have to be measured separately. Phase coherent detection permits direct measurement of σ_1 and σ_2 and has been used by Dähne et al. [38] and others. In a non-phase coherent setup, de Vaulchier [40] used light pipe optics with the film electrically sealed to an aperture to obtain $\lambda(T)$ at fixed mm wave frequencies. Feenstra et al. [46] used the frequency dependance of the transmission of a focussed mm wave beam to account for substrate effects and also to obtain $\lambda(T)$.

Time domain terahertz spectroscopy was first applied by Nuss et al. to superconducting films [141] and was then taken up by other groups [25, 29, 50, 212]. It expands the available frequency range to 1,000 GHz or more, and also greatly increases the available power level for probing non-linear effects. The work of Orenstein's group on the electrodynamics of the vortex state [152, 183] and on quasiparticle lifetimes [36, 144], is a striking example of how powerful the method can be.

Far-Infrared Reflection

This works best for films that are thick enough that reflections from the second surface can be neglected. With films, one can also work at grazing incidence where the reflectivity is sensitive to properties in the direction perpendicular to the substrate (usually but not always, the $\hat{c}$-axis). Thus one can detect the $\hat{c}$-axis plasmon $(1/|\epsilon| \to \infty)$ and so measure n_s/m^* for the $\hat{c}$-direction.

Slow Muon Beam Method

With the development of very slow muon beams, groups at Paul Scherrer Institute have succeeded in directly measuring the penetration of weak magnetic fields into a superconductor by varying the muon energy and thus the average muon penetration. For example, the method has had spectacular success in measuring isotopic effects in $\lambda(0)$, with stated accuracies of 1%. At present the method requires sample areas of order several cm^2, which rules out its use for small crystal platelets [95, 96].

4.3.3. Penetration Depth Techniques—Powders

Although powders are not the ideal form for precision measurement of the electrodynamics of the cuprates, there are situations where grain-aligned powders are particularly

Microwave Electrodynamics of High Temperature Superconductors

useful. One is where good quality single crystals or thin films are not available; the second is where one wants to study the effect of impurities over a wide range of concentrations. In this latter case, controlling the impurity concentrations is achieved more easily in powder samples.

Panagopoulos and coworkers have made extensive use of grain aligned powders. A particularly good example is their work on $HgBa_2Ca_2Cu_3O_{8+\delta}$ [149]. The grains were aligned by suspending them in epoxy cured in a 7 T magnetic field; the resulting angular distribution of the $\hat{c}$-axis was 1.7° FWHM. The penetration depth was extracted from the temperature dependent magnetization measured in a weak (1–10 G) probe field (DC SQUID or AC susceptometer). For this excluded volume technique, thermal expansion effects are negligible because the grains are small. The magnetization is related to the penetration depth via London's model [173], which requires knowledge of the grain size distribution and the assumption of spherical grains.

4.4. Measurement of Surface Resistance R_s

Here again the methods are conveniently classed according to the type of sample, small single crystals or thin films. We do not consider powders or ceramic samples since the measured R_s will be far from the intrinsic values.

4.4.1. Single Crystals
Cavity Perturbation

For small single crystals almost all of the methods amount to some form of cavity perturbation, shown schematically in Figure 4.6. Here a small conducting sample is placed in a microwave resonant cavity. For a good conductor one is usually in the extreme skin effect region, where the screening currents in the sample flow with a depth that is smaller than any dimension or radius of curvature of the sample. Except for extreme anisotropy, we

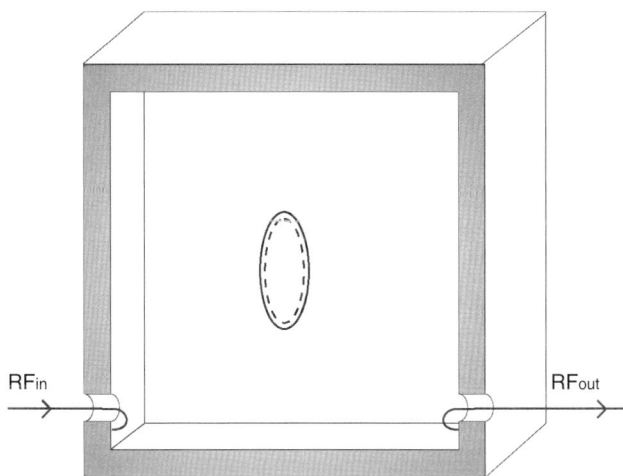

Figure 4.6. Ellipsoidal metallic or superconducting sample in a resonant cavity.

need only consider currents parallel to the surface. Now imagine that the conductivity of the sample is *infinite* so that the penetration depth is zero and there is no loss. A transmission curve yields the resonant frequency of the cavity (slightly perturbed by the presence of the sample), and its quality factor Q_0. We express this result as a complex frequency $\tilde{\omega}_0$ where $e^{i\tilde{\omega}_0 t} = e^{i\omega_0 t} e^{-(\omega_0/2Q)t}$ is the free solution for the cavity mode, so that $\tilde{\omega}_0 = \omega_0(1 + i/2Q)$. Now we set σ to its actual *finite* value and remeasure the complex mode frequency. Within the surface impedance approximation it can be shown that

$$\delta\tilde{\omega} = \tilde{\omega} - \tilde{\omega}_0 \propto i \int_{\text{sample}} Z_s J_s^2 ds, \tag{4.11}$$

where J_s is the surface current density (A/m), which is assumed to be unchanged from the $\sigma = \infty$ value i.e. the EM fields at the surface are unchanged. Thus

$$\frac{\delta\omega}{\omega_0} = \frac{\omega - \omega_0}{\omega_0} = -KX_s \tag{4.12}$$

and

$$\delta(1/Q) = 1/Q - 1/Q_0 = 2KR_s. \tag{4.13}$$

All of this requires that $\delta\omega/\omega \ll 1$. The constant K is set by the geometry of the cavity and sample; in simple geometries K can be calculated. More commonly, K is determined by using a sample of similar shape with known properties.

The key problem with the method is that both $\delta\omega/\omega_0$ and $\delta(1/Q)$ are referenced to the zero-skin-depth ($\sigma_1 = \infty$) case, which is not accessible experimentally, and this places limitations on the information that can be extracted. For R_s, this may not be a severe restriction, if either R_s drops to a very low value at low T, or the sample can be moved. However, for $\delta\omega/\omega_0$, neither the $T = 0$ limit or movement of the sample gets around the fact that there is a zero order shift in the cavity frequency set by the size and shape of the sample. For single crystals, one almost always uses the hot finger technique referred to earlier [168, 184], where the sample temperature can be varied independently, or *nearly* independently, of all other parts of the apparatus. In addition, the hot finger assembly may, or may not, be moveable. We consider the two situations in turn.

If the sample position is *moveable*, then the unperturbed cavity Q can be determined and, in principle, $R_s(T)$ obtained directly. The only caveat is that the presence of the sample slightly changes the current distribution in the cavity walls and therefore Q. The size of such second-order effects can be estimated using, for example, samples of a superconductor such as Pb with lower loss. As already discussed in "Excluded Volume Techniques" in Section 4.3.1, because of thermal expansion effects and calibration inaccuracies, one cannot use the value of the $T \sim 0$ frequency shift to obtain $\lambda(T_0)$.

In the case where the sample position is *fixed*, one can only obtain values of $\Delta\lambda(T)$ and $\Delta R_s(T) = R_s(T) - R_s(T_0)$. Usually this gives more accurate values of $\Delta\lambda(T)$, since the reproducibility of the sample position is not an issue, although one still has to ensure that the sample does not move as the temperature of the hot finger is changed.

In the case that Z_s is anisotropic, as it is for the high temperature superconductors, the appropriate integral of $Z_s \times J_s^2$ has to be carried out. For simple geometries this is readily done, and by making measurements with different sample orientations with respect to the cavity fields one can often separate out the various components of Z_s. Another possible complication is that the sample may not be in the simple surface impedance regime, for example in the normal state when the skin depth δ may be larger than one of the sample dimensions.

In this case more detailed solutions of Maxwell's equations are required, which are specific to particular situations and are beyond the scope of the present discussion (see, for example, Hardy et al. [64], Ning et al. [139], Gough et al. [57]).

We close this section by noting that time domain measurements of Q are generally more accurate than frequency domain transmission techniques. Here one applies a short pulse of microwaves at the cavity resonance and measures the exponential ring-down of the power emitted by the cavity. The advantage of this procedure is that any instability in the cavity frequency, caused for example by small motions of the hot finger, does not affect the envelope of the free decay. The UBC group has implemented this method for most of its cavity perturbation measurements, finding improvements by more than an order of magnitude [94].

Broadband Bolometric Spectroscopy

Recently, Turner et al. [201] have developed a bolometric technique that dispenses with resonant cavities and measures directly the power absorbed by the sample, namely

$$P_{abs} = R_s \int H_{rf}^2 dS. \tag{4.14}$$

Here the sample is placed near the end of a shorted transmission line, allowing R_s to be measured as a continuous function of frequency. The key to the success of the method is the in situ use of a normal metal reference sample that calibrates the absolute rf field strength. So far, the method has been implemented for the frequency region 0.1–20 GHz and temperature regime from about 1 to 10 K.

Thin Film Methods

The microwave techniques available for thin films have been reviewed by Klein [101] and fall into roughly four categories: resonator endplate replacement, dielectrically loaded resonator, planar resonator, and quasioptical free space resonator or transmission.

In the endplate replacement technique [41, 103], the high T_c sample forms one endplate of a cylindrical cavity operating in a TE_{0np} mode. This class of mode is chosen so that, by symmetry, no current is required to flow between the sample film and the body of the resonator. In the dielectric resonator technique, a dielectric cylinder (e.g. sapphire) is sandwiched between either two high T_c thin films (symmetric resonator [123]) or is placed on top of a single film (asymmetric version [102]). In the first version, the dielectric cylinder is chosen to have a diameter much less than that of the film, so that the evanescent fields are largely confined by the high T_c material itself. If the loss in the dielectric is small, a direct measure of the surface resistance of unpatterned films is obtained. The asymmetric version requires a shielding cavity; nevertheless, very low values of R_s can be measured [102].

For patterned films, one uses the stripline, microstrip, or coplanar transmission line resonator, geometries used for practical microwave devices. In the microstrip and stripline cases, one or two groundplanes are required, respectively, and their losses must be accounted for. The coplanar geometry [159] has the advantage that the ground planes are not an essential part of the transmission line, can be placed well away from the fields, and therefore can be made of normal material. The latter method seems capable of providing absolute values of $\lambda(T)$ as well as accurate values of $R_s(T)$. It should be noted that in general all of the patterned transmission line resonators have large currents at the patterned edges and are therefore rather sensitive to damage produced by the patterning process.

Free space methods where, for example, the high T_c thin film is part of a semi-confocal Fabry–Perot resonator [91, 131] are well suited to measurements of R_s at high frequencies (100 GHz range), and scanning versions have been particularly useful for measurement of large area thin films. [79]. Dähne et al. [38] used a quasioptical Mach–Zehnder interferometer to obtain both σ_1 and σ_2 in the range 100–350 GHz.

4.5. Penetration Depth

There exists a large body of data on the magnetic penetration depth, obtained by a wide variety of techniques and covering the dependence on cuprate families, crystallographic directions, and especially the doping systematics (see for example Uemura et al. [203, 204], Panagopoulos et al. [151], Tallon et al. [192], and references within). A comprehensive review of this data would take us too far afield from our focus on microwave properties, and in any case would be much too lengthy. We have chosen to concentrate on results that are of fairly direct relevance to microwave measurements, or results where microwave measurements have shed light on important issues. In addition, personal views of what is well established and what needs further work will be included.

4.5.1. Complementary Roles of λ and R_s

Before beginning the presentation of published results for high temperature superconductors, it is useful to clarify the complementary roles played by measurements of λ and of R_s. In the linear response regime, the real and imaginary conductivities, $\sigma_1(\omega, T)$ and $\sigma_2(\omega, T)$, are related by Kramers–Kronig transforms, so in principle it is only necessary to measure one of the quantities, for example $\sigma_1(\omega, T)$. As a practical matter, for ω in the microwave region, σ_1 is almost never known over a wide enough frequency range to carry out meaningful KK transforms. Typically, measurements of $\lambda(T)$ gives us the delta-function in σ_1 at $\omega = 0$, and we may have a few values of R_s from which we can extract σ_1 at the same few frequencies. This is very different from the situation in the Far IR and higher frequencies where data can be taken with as fine a frequency grid as desired and up to very high frequencies. It is helpful, therefore, that a restricted spectral-weight sum rule likely holds for the frequencies relevant to the microwave properties. Specifically, if ω' extends to a few hundred wavenumbers to include most of the Drude-like conductivity (roughly 10,000 GHz), then

$$\int_0^{\omega'} \sigma_1(\omega, T) \, d\omega \approx \text{const.} \qquad (4.15)$$

As one cools below T_c, two things happen: the approximate Drude-like $\sigma_1(\omega, T)$ narrows as the inelastic scattering weakens, and at the same time some of the spectral weight appears as a δ-function at $\omega = 0$ (the superfluid). The possibility also exists that there is exchange of spectral weight between this "Drude" region and much higher frequencies. This spectral weight shift is controlled by the high energy physics and is therefore of fundamental importance. The establishment of the systematics is currently the subject of major efforts by several groups [21, 134, 169]. Although at present unanimity on the details has yet to be achieved, the shift of spectral weight is small, so that for our purposes the low frequency restricted sum rule is a useful first approximation.

The expectation of an approximate sum rule helps us to fill in our picture of the lower-frequency electrodynamics where we typically have incomplete information. For example it is becoming clear that, depending on the quality of the samples and the particular cuprate family they belong to, a substantial portion of the "Drude" spectral weight does not condense into the δ-function at low T. Using the sum rule, one can estimate the effect on the superfluid density if we can measure R_s at a few strategic frequencies. Alternatively, if we know both $\lambda(0)$ and the uncondensed spectral weight then we can get an estimate of the fraction of sample that is non-conducting or otherwise inactive. It appears that some films, for example, have reduced superfluid density without the obvious presence of strong scatterers; establishing the magnitude and location of the missing spectral weight is a useful tool for sorting out this behaviour.

In a review of microwave properties, the formal significance of a measurement of $\lambda(T)$ is that (a) it gives the main part of $\sigma_2(\omega, T)$ that dominates the superconducting electrodynamics in the microwave region and (b) it is also necessary for extracting $\sigma_1(\omega)$. Given this key role in the electrodynamics, together with the fact that our knowledge of $\sigma_1(\omega, T)$ is generally rather incomplete, it is important to reliably establish the systematics of $\lambda(T)$.

A comprehensive review of the existing results would be much too lengthy. We concentrate on data that we consider to be representative of the best results available for the various compounds, and try to cover the main issues of interest. For more details, we refer the reader to the review by Bonn and Hardy [16].

4.5.2. $YBa_2Cu_3O_{6+x}$

The most studied of the cuprates is the least anisotropic (which simplifies measurements of the electrodynamics), and the most homogeneous electronically. On the other hand, it is orthorhombic with highly conducting CuO chains which, while incidental to the superconductivity, interfere with the interpretation of the electrodynamics of the CuO_2 planes, the quantity of central interest. In order to separate chain from non-chain effects one generally has to work with untwinned crystals.

In Figure 4.7 we show the data of Hardy et al. [64] on twinned $YBa_2Cu_3O_{6.95}$ ($T_c = 93$ K). A 1 GHz superconducting cavity perturbation method was used to obtain

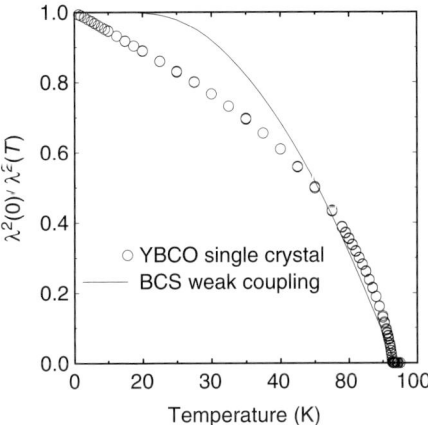

Figure 4.7. $\lambda_{ab}^2(0)/\lambda_{ab}^2(T)$ vs. T for twinned $YBa_2Cu_3O_{6.95}$ using microwave values for $\Delta\lambda_{ab}(T)$ and $\lambda_{ab}(0) \approx$ 1400 Å from μSR measurement on similar crystals. The solid line is the weak coupling BCS s-wave prediction (Hardy et al. [64], [65]).

$\Delta\lambda(T) = \lambda(T) - \lambda(1.3\ \text{K})$, and $\lambda_{ab} \simeq 1{,}400\ \text{Å}$ was taken from μSR measurements on similar crystals. The data deviates substantially from the weak-coupling BCS s-wave result (solid line) at both low and high temperatures. The linear dependence seen at low temperature is now generally agreed to be the intrinsic behaviour of virtually all high temperature superconductors, both for single crystals and thin films. Measurements in the region near T_c where $1/\lambda^2$ appears to approach the temperature axis with infinite slope, indicate non-mean-field behaviour. This latter behaviour is *not* universally observed for reasons that are not well understood—see Section 4.7. for further discussion.

Virtually all early measurements on all forms of $YBa_2Cu_3O_{6+x}$ (ceramics, thin films, single crystal) and by a variety of techniques, were interpreted in terms of a uniform single gap or two-gap BCS ground state. For instance measurements by Klein et al. of $\lambda(T)$ in $YBa_2Cu_3O_{6+x}$ films grown on $NdGaO_3$ and $LaAlO_3$ showed features that suggested the presence of two gaps, but also exhibited considerable sample dependence [101]. With improved samples and measurement techniques, a consensus emerged that $1 - \lambda^2(0)/\lambda^2(T)$ does not show activated behaviour at low T, being predominantly linear up to $T_c/3$, giving over to T^2 at lower temperatures, the cross-over temperature depending on the purity and perfection of the sample [73]. For crystals grown in yttria-stabilized zirconia crucibles the crossover is in the 1–5 K range, but small concentrations (0.3%) of Zn can raise this to 30 K.

Using 9.6 GHz cavity perturbation, Mao et al. [130] observed $1 - \lambda_{ab}^2(0)/\lambda_{ab}^2(T)$ in single crystals grown in zirconia crucibles to be linear from 4 K to more than 40 K. In contrast, most thin film data show $1 - \lambda^2(0)/\lambda^2(T)$ varying as T^2 over a fairly wide range of temperatures. For example, the high resolution microstrip resonator data of Anlage and Wu followed T^2 extremely well over the whole temperature range [8]. Lee et al. [115], using the low temperature mutual inductance technique on laser ablated films on $SrTiO_3$ ($T_c = 88$ K), saw a T^2 dependence below 25 K and generally good agreement with the single crystal results [64] above this temperature. In later work by Ma et al. [125], $\Delta\lambda(T)$ of films from a variety of sources was measured using the parallel plate resonator technique. They found that the higher the quality of the film (higher T_c), the narrower the T^2 region and the closer the match to the UBC single crystal data. Indeed, earlier measurements by Gao et al. [54] on high quality ($T_c = 90$ K) commercial film from Conductus, had shown a linear region between 6 and 30 K.

Similar results were obtained by De Vaulchier et al. [40], who used power transmission in the 120–500 GHz range to measure $\lambda(T)$ absolutely. For films with a relatively large value of $\lambda(0)$ (3,400 ± 200 Å; $T_c = 86$ K) $\Delta\lambda$ accurately followed T^2. For a much higher quality film ($\lambda_0 = 1{,}570$ Å; $T_c = 92$ K) they observed $\Delta\lambda(T) \propto T$ from 4 to 40 K, with $\Delta\lambda/\Delta T$ very close to that for the single twinned single crystal [64]. Collectively, the cited data establish that the intrinsic low T behaviour of $\Delta\lambda$ for optimally doped $YBa_2Cu_3O_{6+x}$ is close to linear and the T^2 dependence observed in many films is due to defects.

Dähne et al. [38] used a quasioptical Mach–Zehnder interferometer, to study $\lambda(T)$ at 300 GHz in very high quality films (dc-sputtered on (110) $NdGaO_3$) and compared these results to data at 18.9 GHz measurements on the same films. Below $0.6T_c$, they found $\lambda(T)$ to be much less temperature dependent at 300 GHz than at 18.9 GHz (Figure 4.8). This was interpreted as the effect of a quasiparticle scattering rate that was lower than the 300 GHz measuring frequency, and they extracted $1/2\pi\tau \approx 170$ GHz below 30 K (see discussion in Section 4.2.2 and, in particular, Figure 4.2). By contrast, in the work of de Vaulchier [40], no frequency dependence of $\lambda(T)$ was seen up to 500 GHz, presumably due to a higher scattering rate. We note that in the early 1990s Bonn et al. [18] inferred scattering rates in single crystals

Microwave Electrodynamics of High Temperature Superconductors

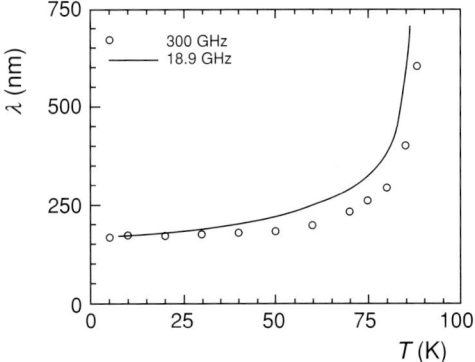

Figure 4.8. Data of Dähne et al. [38] which shows strong frequency dependence of $\lambda(T)$ for very clean YBCO films.

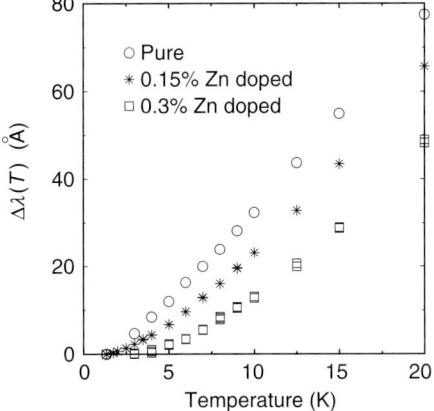

Figure 4.9. Low T penetration depth for $YBa_2(Cu_{1-x}Zn_x)_3O_{6.95}$ for $x = 0$ (nominally pure), 0.0015 and 0.003 (Bonn et al. [18], Hardy et al. [65]).

as low as 30 GHz and in recent work in very high quality samples, values as low as 3 GHz have been reported [202].

The question remains: if the scattering rates in de Vaulchier et al.'s films were so large, why did not they see $\Delta\lambda \propto T^2$ instead of T? In part, this may be explained by the fact that two defects producing the same quasiparticle scattering rate need not be equally effective in changing T to T^2 (i.e. in producing states at the fermi energy). For example, Bonn et al. [18] and Achkir et al. [2] found that Zn impurities were much more effective in producing a change over to the T^2 dependence than was Ni, although these impurities were about equally effective as scattering centres (Bonn et al. [18, 222]). Figure 4.9 shows the effect of $x = 0.15$ and 0.3% Zn on $\Delta\lambda(T)$ for $YBa_2(Cu_{1-x}Zn_x)_3O_{6.95}$. The addition of 0.7% Ni had less effect than 0.15% Zn. In contrast, Ulm et al. [206] studied the behaviour of both $\lambda(0)$ and the temperature dependence of $1/\lambda^2(T)$ in films for 2–6% Ni and Zn impurities, and found Ni and Zn to have about equal effects. They found that $n_s(0)$ decreases by a factor of 2 for each percentage of dopant. This discrepancy is yet to be explained. We also point out that many film results show a rather large $\lambda(0)$, (i.e. low superfluid density), yet have a linear temperature dependence of

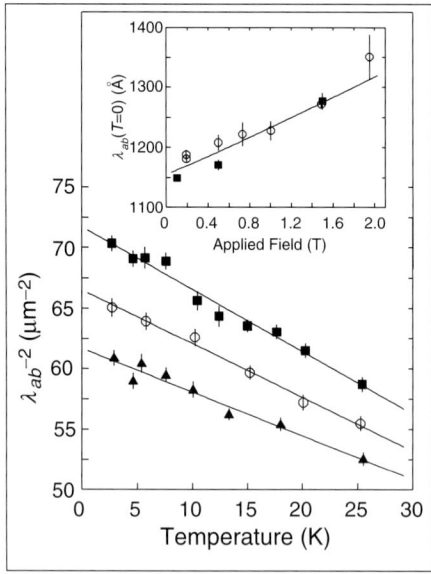

Figure 4.10. Temperature dependence of $1/\lambda_{ab}^2$ in YBa$_2$Cu$_3$O$_{6.95}$ for applied fields of 0.2 (solid square), 1.0 (circles), and 1.5 T (solid triangles) (Sonier et al. [181]).

λ down to quite low T. This also cannot be explained by simple defect scattering and seems to suggest inhomogeneous superconductivity: the overall low superfluid density results from parts of the sample being non- or weakly superconducting.

Numerous complementary measurements have confirmed the linear behaviour of the penetration depth and superfluid density at temperatures well below T_c. As an example, we show in Figure 4.10 the quality of μSR results that was achieved by the mid-1990s. The temperature dependence of $1/\lambda_{ab}^2$ from the work of Sonier et al. [181] on single crystals clearly sees the linear temperature dependence (which agrees quite well with microwave measurements), but in addition shows a non-negligible field dependence. In later work by Sonier et al. [182] the explicit dependence of $\lambda(0)$ on applied magnetic field was measured and compared to the theory of Amin et al. [3] which included both non-linear and non-local effects in the vortex state.

4.5.3. Penetration Depth Anisotropy in YBa$_2$Cu$_3$O$_{6+x}$

Substantial anisotropies in the normal state electrical conductivities and thermal conductivities have been observed in YBa$_2$Cu$_3$O$_{6+x}$ (see Gagnon et al. and references therein [53]). Zhang et al. [221] were the first to measure $\lambda_a(T)$ and $\lambda_b(T)$ separately using a combination of Far IR and microwave techniques. The original microwave measurements were made at 1 GHz with $H \perp \hat{c}$ on a crystal thin enough that the effect of currents in the $\hat{c}$ direction could be ignored. IR measurements on the same crystal gave $\lambda_a(0) = 1{,}600$ Å and $\lambda_b(0) = 1{,}030$ Å, a rather large anisotropy. However, this is not inconsistent with the n/m^* anisotropy observed in the normal state conductivities, nor the μSR results of Tallon et al. [191] for λ_a/λ_b inferred from the dependence of μSR relaxation on oxygen content. Later, the technique was refined to the point where $\lambda_a(T)$, $\lambda_b(T)$, and $\lambda_c(T)$ could be determined absolutely from 1.3 K to T_c (see Hardy et al. [63], Bonn et al. [17]). This was accomplished by measuring $\Delta\lambda_a(T)$

Microwave Electrodynamics of High Temperature Superconductors

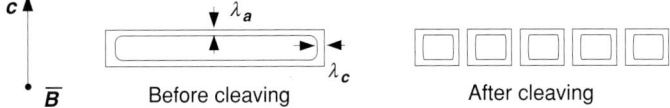

Figure 4.11. Arrangement to measure the $\hat{c}$-axis penetration depth, $\Delta\lambda_c(T)$.

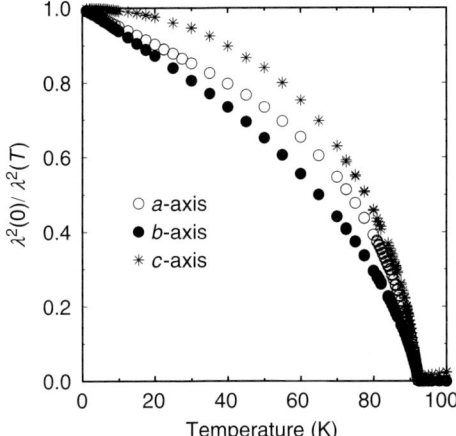

Figure 4.12. $\lambda_i^2(0)/\lambda_i^2(T)$ vs. T for $i = a, b, c$ in YBa$_2$Cu$_3$O$_{6.95}$ obtained from microwave $\Delta\lambda_i(T)$ and Far IR derived values of $\lambda_i(0)$ (see Table 2) (Zhang et al. [221], Hardy et al. [65]).

and $\Delta\lambda_b(T)$ before and after the (approximately square) crystal was cleaved into five or more bars (Figure 4.11). For $H \parallel$ to the long axis of the bars, the effect of λ_c is multiplied up by the number of pieces. A measurement perpendicular to the bars should not be affected to first order and serves as a control on the procedure.

Figure 4.12 shows the results of this approach, where the Far IR measurements of $\lambda_i(0)$ have been incorporated. Qualitatively, Figure 4.12 shows that $\lambda_a(T)$ and $\lambda_b(T)$ have rather similar temperature dependencies, which is very strong evidence that it is not the chains that are causing the linear low temperature dependence. The $\hat{c}$-axis behaviour is rather different, being much flatter, with a nearly quadratic temperature dependence at low temperatures that will be discussed in more detail in Section 4.5.6. The extent to which the results depend upon measurements of $\lambda(0)$ from other techniques will be discussed further in the next section.

4.5.4. Oxygen Doping Effects

A thorough testing of models of the superconducting state of the cuprates would require measurements of all three components of the penetration depth across as wide a range of doping as possible. For this purpose, YBa$_2$Cu$_3$O$_{7-\delta}$ is the "cleanest" material in many senses, however, it is complicated by its strongly conducting chains and, except for $\delta = 0$ or for oxygen ordered phases such as Ortho-II, it has considerable disorder in the chains. Therefore it is extremely important to see how the electrodynamics evolve with oxygen deficiency δ and chain disorder in order to disentangle the effects of chains, planes, and defects. One reassuring advantage of this system is that it is known that oxygen defects seem to have relatively little

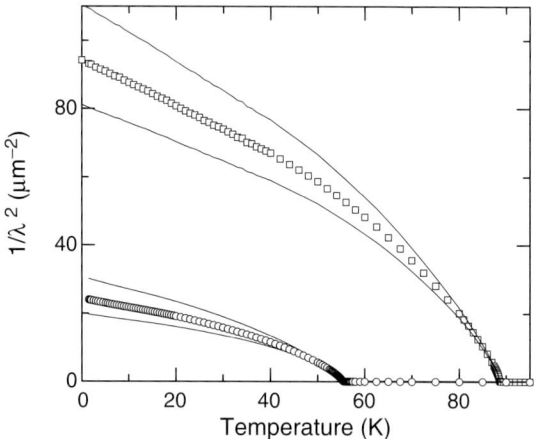

Figure 4.13. Superfluid density in the a-direction for two oxygen-ordered phases: the Ortho-I phase with nearly full chains at $YBa_2Cu_3O_{6.993}$ (open squares) and the Ortho-II phase with alternating full and empty chains at $YBa_2Cu_3O_{6.52}$ (open circles). The lines indicate the uncertainty in the absolute value stemming from uncertainty in the measurements of λ (1.2 K).

effect on the ab plane surface resistance (Bonn et al. [19], Fuchs et al. [51]) i.e. the O defects are weak electronic scatterers.

A serious problem standing in the way of a systematic study of the penetration depth turns out to be the issue of accurately determining the absolute value of $\lambda(0)$. It is the authors' opinion that each technique for measuring this quantity carries with it assumptions and potential systematic errors, compounded by a significant amount of sample dependence, as mentioned above in the discussion of thin film measurements vs. single crystal measurements. Thus, while microwave measurements are very good at accurately determining the temperature dependence relative to some base temperature $\Delta\lambda(T) = \lambda(T) - \lambda(T_0)$, the conversion of this quantity to a superfluid density or phase stiffness forces a reliance on other techniques. Figure 4.13 illustrates the extent of this problem by showing the superfluid density in the a-direction of YBCO at two different dopings [221]. The curves come from combining cavity perturbation measurements of $\Delta\lambda(T)$ with Gd-ESR measurements of the absolute value of λ_a at 1.2 K. The curves bracketing each of the two data sets indicate the serious impact of the 10% uncertainty estimated for the Gd-ESR measurements. The problem is that the 10% uncertainty is more than doubled when squaring λ to get $1/\lambda^2$ and if one is interested in the slope of the superfluid density, the uncertainty is more than tripled to $\pm 33\%$. Such uncertainties must be kept in mind when trying to draw conclusions that involve comparison from sample to sample, or from technique to technique. In particular, a quantitative comparison between the two dopings shown here is somewhat inconclusive due to the large relative uncertainty in the magnitudes of the curves. Bonn et al. [17] had previously suggested that the $\hat{a}$-axis superfluid density scales when normalized to $\lambda(0)$ and T_c and much has been made of the observation that the slope of the superfluid density's temperature dependence is doping independent [117]. In the face of the uncertainties shown in Figure 4.13, it is difficult to draw a strong conclusion as to whether or not this is really the case.

Figure 4.14 shows the most recent set of results for the in-plane superfluid density of the two best-ordered phases in the $YBa_2Cu_3O_{6+x}$ system: the Ortho-I phase with nearly every CuO chain filled at a composition of $YBa_2Cu_3O_{6.99}$ (slightly overdoped, $T_c = 89$), and the Ortho-II phase with every other CuO chain filled at a composition of $YBa_2Cu_3O_{6.52}$

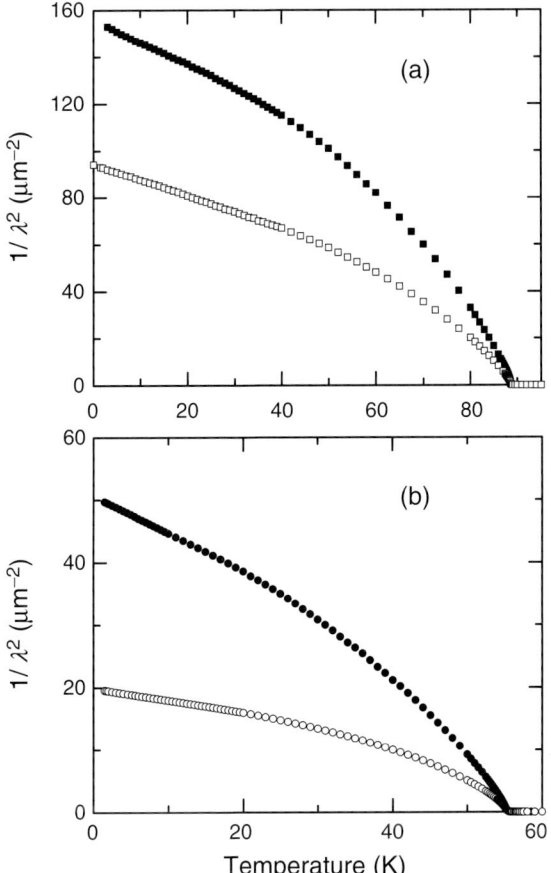

Figure 4.14. The anisotropic in-plane superfluid density for (a) slightly overdoped $YBa_2Cu_3O_{6.99}$ ($T_c = 89$ K) and (b) underdoped $YBa_2Cu_3O_{6.52}$ ($T_c = 59$ K). Solid symbols are for currents running in the $\hat{b}$-direction, open symbols are for $\hat{a}$-axis currents.

(underdoped, $T_c = 89$). As noted above, these figures make use of values for $\lambda(1.2\,\text{K})$ found using Gd-ESR and, as is the case with any technique, there is significant uncertainty in the overall magnitude each of these curves. With this caveat in mind, there are still many conclusions that can be drawn from the data. There is certainly large in-plane anisotropy in both samples due to the presence of the CuO chains. For the fully doped sample with $T_c = 89$ K, the results are in accord with the calculations made by Atkinson [10], who considered a realistic model for d-wave superconductivity in a bilayer material (i.e. two quasi-two-dimensional fermi cylinders) plus proximity effect induced superconductivity on an open fermi surface arising from quasi-one-dimensional chain band hybridized with the planes. Earlier models [11] predicted an upturn in the $\hat{b}$-axis superfluid density as the proximity-induced superconductivity comes into play, but Atkinson showed that this upturn might be suppressed by the remnant chain defects (slight oxygen deficiency) in this material. A similar story might hold true for the Ortho-II phase as well, although in that system one would also have to take into account the folded Brilloiun zone associated with the doubling of the unit cell, something that might help account for the very large anisotropy shown for in Figure 4.14b.

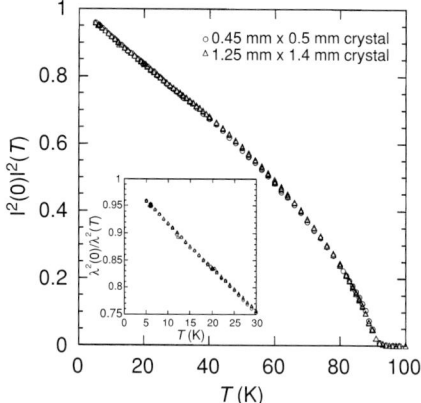

Figure 4.15. Temperature dependence of the superfluid fraction for $Bi_2Sr_2CaCu_2O_8$ single crystal, assuming $\lambda_{ab}(0) = 2,100$ Å. (Lee et al. [120]).

4.5.5. Other Materials

$Bi_2Sr_2CaCu_2O_{8+\delta}$

The Stanford group [126] used cleaved $Bi_2Sr_2CaCu_2O_{8+\delta}$ crystals in the parallel plate method (here $H \perp \hat{c}$) and obtained a low temperature T^2 dependence for $1/\lambda^2$. However, they concluded this was likely contaminated by $\hat{c}$-axis conduction due to the extreme anisotropy. This strongly 2D behaviour also complicates measurement of $\lambda(T)$ by μSR in the vortex state because the coherence of the vortices from layer to layer is strongly dependent on H and T [37]. The first study of λ_{ab} that showed a linear T dependence was that of Jacobs et al. [90] obtained by microwave cavity perturbation with $H \parallel \hat{c}$, for which $\lambda^2(0)/\lambda^2(T)$ vs. T/T_c was quite similar to the $\hat{a}$-axis data of Zhang et al. [221]. This was followed shortly after by the results of Lee et al. [120] which are shown in Figure 4.15. The low temperature dependence is very linear, the overall dependence being rather similar to $YBa_2Cu_3O_{6+x}$ data, but falling closer to the weak coupling d-wave result.

$Tl_2Ba_2CaCu_2O_8$

Ma et al. [125] measured $\Delta\lambda_{ab}(T)$ for two pairs of commercial films (STI, CA) and fitted the data to $bT^2/(T + T^*)$ with $T^* = 25$ and 40 K, respectively. For the first pair of films the overall dependence of $\lambda^2(0)/\lambda^2(T)$ vs. T/T_c matched the $YBa_2Cu_3O_{6+x}$ single crystal data [64] rather well above T^*. The agreement for the second pair (considered inferior quality) was less good.

$Tl_2Ba_2CuO_{6+\delta}$

$\lambda_{ab}(T)$ has been measured for the tetragonal, single layer compound $Tl_2Ba_2CuO_{6+\delta}$ by Broun et al. using a 35 GHz superconducting cavity technique with $H \parallel \hat{c}$ [26, 27]. The material is nearly optimally doped, with a T_c of 78 K. Below 20 K, $\Delta\lambda(T) \propto T$ with slope of 13 Å/K. This is larger than the corresponding slopes for $YBa_2Cu_3O_{6+x}$ and $Bi_2Sr_2CaCu_2O_{8+\delta}$ (4.8 and 10.2 Å/K, respectively). Figure 4.16 shows $\lambda^2(0)/\lambda^2(T)$ vs. T

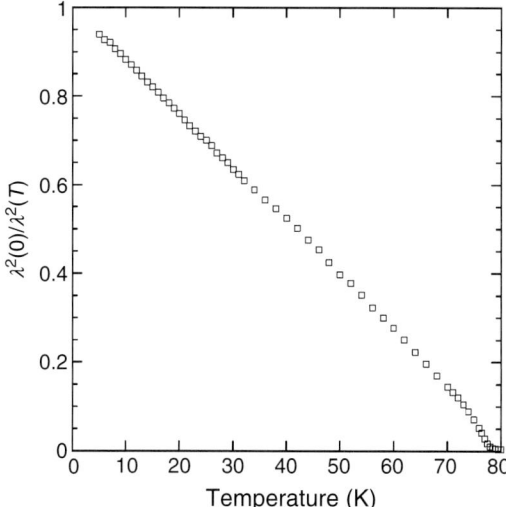

Figure 4.16. Temperature dependence of the superfluid fraction for $Tl_2Ba_2CuO_{6+\delta}$ [27] using $\lambda_{ab}(0) = 1,650$ Å from μSR measurements [26, 27].

with $\lambda(0) = 1,650$ Å taken from μSR measurements [205]. It varies linearly with T over almost the whole temperature range (i.e. very similar to weak coupling d-wave), with a sudden downturn within a few degrees of T_c that may be due to fluctuations.

$La_{1-x}Sr_xCuO_4$

Shibauchi et al. [175] measured both $\Delta\lambda_{ab}(T)$ and $\Delta\lambda_c(T)$ of single crystals of $La_{1-x}Sr_xCuO_4$ for $0.09 < x < 0.19$ using platelets with $\hat{c} \perp$ and $\parallel$ to the face. The large demagnetization factors were determined from Pb reference samples. They obtained $\lambda_{ab}(0) = 0.4$ μm and $\lambda_c(0) = 5$ μm (values consistent with optical and μSR measurements) by the method described earlier. The overall T dependence of $1/\lambda_{ab}^2(T)$ follows weak coupling s-wave BCS, but the resolution was insufficient to rule out a T^2 dependence at low T. $1/\lambda_c^2(T)$ is much flatter and the data was fit to a model involving Josephson coupled 2D superconducting layers (s-wave BCS).

$HgBa_2Ca_2Cu_3O_{8+\delta}$

Panagopoulos et al. [149] were able to measure both $\lambda_{ab}(T)$ and $\lambda_c(T)$ in grain aligned $HgBa_2Ca_2Cu_3O_{8+\delta}$ powders. They found $\lambda_{ab}(0)$ and $\lambda_c(0) = 2,100$ and $61,000$ Å, respectively, and a temperature dependence of $1/\lambda_{ab}^2(T)$ that fitted rather well to a d-wave model with $\Delta(0)/kT_c = 2.14$ (Figure 4.17). On the other hand $1/\lambda_c^2(T)$ followed the behaviour of a Josephson coupled d-wave superconductor [114]. Overall the results again look remarkably similar to the $YBa_2Cu_3O_{6+x}$ data.

Electron Doped Thin Films and Single Crystals

The general expectation of theories based on Hubbard-like models is that the pairing mechanism and ground state symmetry should be the same for both hole- and electron-doped

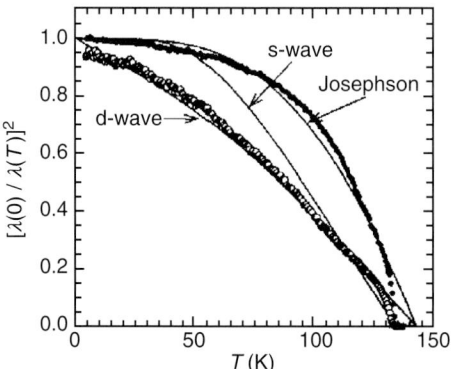

Figure 4.17. Plots of $\lambda_{ab}^2(0)/\lambda ab^2(T)$ (open circles) and $\lambda_c^2(0)/\lambda_c^2(T)$ (closed circles) vs. T for HgBa$_2$Ca$_2$Cu$_3$O$_{8+\delta}$. The solid lines are theoretical predictions from weak-coupling BCS theory for s-wave, weak coupling BCS theory for d-wave, and a Josephson coupled d-wave superconductor (Panagopoulos et al. [149]).

samples. It is now well established that the hole-doped cuprates have d-wave symmetry. The answer regarding the symmetry of the electron-doped cuprates has been slower in coming, although d-wave symmetry is favoured by the preponderance of recent data.

Early measurements of $\Delta\lambda_{ab}(T)$ in Nd$_{2-x}$Ce$_x$CuO$_{4-\delta}$ by Wu et al. [214], Anlage et al. [8], and Andreone et al. [5] exhibited no regions that follow T or T^2, and convincing fits to BCS s-wave were made. Cooper [34] subsequently pointed out that the results were contaminated by the paramagnetism of the rare-earth ions present in these materials. Subsequently, Kokales et al. [106] reported T^2 power law behaviour for $\Delta\lambda_{ab}(T)$ in single crystals and thin films of Pr$_{1.85}$Ce$_{0.15}$CuO$_{4-\delta}$, where the paramagnetism is much reduced. They also studied an Nd$_{2-x}$Ce$_x$CuO$_{4-\delta}$ single crystal in an improved microwave geometry to reduce the paramagnetic contamination. Here, above 4 K, a power law region was also observed. At the same time, Prozorov et al. [160] convincingly showed that Pr$_{1.85}$Ce$_{0.15}$CuO$_{4-\delta}$ gave a T^2 power law dependence down to 0.4 K; they were also able to quantitatively assess the interfering effect of the Nd magnetism in Nd$_{2-x}$Ce$_x$CuO$_{4-\delta}$. More recently, Snezhko et al. [180] obtained power law dependencies for Pr$_{1.85}$Ce$_{0.15}$CuO$_{4-\delta}$ films with $x=0.13$, 0.15, and 0.17, the optimally doped sample ($x = 0.15$, $T_c = 20.5$ K) giving the first indication of a linear regime at higher temperature. Given that half-integral flux indicative of d-wave pairing has been reported in tri-crystal experiments with both Nd$_{2-x}$Ce$_x$CuO$_{4-\delta}$ and Pr$_{1.85}$Ce$_{0.15}$CuO$_{4-\delta}$ thin films [198], the case for a d-wave ground state appears to be very strong. Recently however, results obtained by the Ohio state group on electron-doped MBE films suggest a crossover from a d-wave to an s-wave state as the doping falls below optimal [178]. In separate work they found gap-like behaviour with Δ/T_c that was reproduceable over a series of five optimally doped Pr$_{1.85}$Ce$_{0.15}$CuO$_{4-\delta}$ films [179]. In later work on Pr$_{1.85}$Ce$_{0.15}$CuO$_{4-\delta}$ films on buffered SrTiO$_3$ substrates where the measurements were taken to lower temperatures (0.5 K), they found a full gap with $\Delta/T_c = 0.3$–1.0 for all dopings [98]. In samples for which a gap is claimed, all the values of Δ/T_c are small, suggesting non-intrinsic properties. However, given the quality of the samples and the low temperatures reached, it seems difficult to come up with a convincing scenario for why this particular experiment does not fit the d-wave scenario.

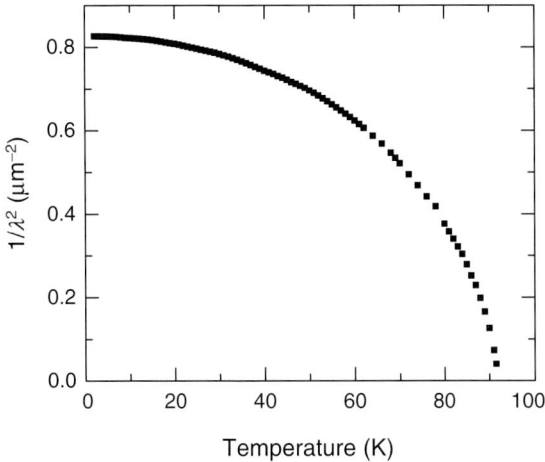

Figure 4.18. Superfluid density in the c-direction for the Ortho-I phase with nearly full chains at the composition $YBa_2Cu_3O_{6.993}$ [83].

4.5.6. $\hat{c}$-Axis Penetration Depth

The $\hat{c}$-axis results in all of the cuprate families contrast sharply with those found for the in-plane superfluid density. Although there have been reports of linear temperature dependence in the low temperature behaviour of the $\hat{c}$-axis penetration depth (Mao et al. [130], Nefyodov et al. [137]), these have most likely been artefacts due to the difficulties in separating in-plane and out-of-plane contributions. For instance, Mao et al. [130] found $\lambda_c^2(0)/\lambda_c^2(T)$ dropping much *faster* (factor 3 or more) than $\lambda_{ab}^2(0)/\lambda_{ab}^2(T)$, but the use of data from geometries with $H_{rf} \perp \hat{c}$ and $H_{rf} \parallel \hat{c}$ without adequate control over thermal and demagnetizing effects in a thin plate-like sample, likely produced this anomalous behaviour. The alternative technique of cleaving a thin plate, as illustrated in Figure 4.11 has now been complemented by a number of other geometries. In the $YBa_2Cu_3O_{6+x}$ system, Hosseini et al. polished a thick crystal into a thin plate with the $\hat{c}$-axis lying in the plane of the thin plate [83]. Measurement with $H_{rf} \perp \hat{c}$ on such a plate is dominated by the $\hat{c}$-axis penetration depth, with a small admixture of in-plane penetration depth. Further thinning of the crystal allows the contributions to be separated. The result of this procedure is shown in Figure 4.18 and is in good agreement with the earlier results obtained by cleaving a thin crystal. The measurements show both 3DXY critical behaviour near T_c, discussed further in Section 4.7, and a nearly quadratic temperature dependence at low temperatures that is ubiquitous across many cuprate families. (See the review of Maeda et al. [127] for a particularly good discussion of $\hat{c}$-axis properties.)

The difference between the T^2 dependence for $1/\lambda^2$ in the $\hat{c}$-direction, vs. T in the $\hat{a}\hat{b}$ plane, is an interesting and important question. If the superconductor is simply anisotropic with coherent hopping between the CuO_2 planes, then the power laws have to be the same for all directions [164]. Alternatively, if one has incoherent tunnelling, independent of the in plane wavevector $k_\parallel$, then cancellation from the sign change in the (pure tetragonal) d-wave order parameter would completely suppress any $\hat{c}$-axis transport. Thus some dependence of the hopping on $k_\parallel$ seems to be required.

Radtke et al. [164] used an incoherent hopping model with a particular choice of inelastic impurity scattering to obtain T^2 for the temperature-dependent part of $1/\lambda_c^2$. This Fermi liquid based approach also connects the normal state resistivity to $1/\lambda_c^2$. They point out that the "confinement" approach to $\hat{c}$-axis coupling, where transport is impeded by some form of spin–charge separation in the CuO$_2$ layers (see for example, Chakravarty and Anderson [33] and references therein) give somewhat similar results. However, in the confinement picture the inelastic scattering is an *intrinsic* part of the interlayer transport, as opposed to boson-assisted hopping in the Radtke et al. model. In fact, the requirement of impurities is a worrying feature of this latter model : $1/\lambda_c^2$ should be a strong function of sample perfection. We do not believe this is to be a generic feature of the cuprates, at least for YBCO. Hirschfeld et al. [74] included both elastic impurity and inelastic spin-fluctuation scattering and obtained $\Delta \lambda_c(T) \propto T^3$ at low T in the clean limit, crossing over to a T^2 dependence when the system becomes sufficiently dirty. Xiang and Wheatley [217, 218] (see also Xiang et al. [216]) made an important advance by incorporating the intrinsic $k_\parallel$ dependence of the interlayer hopping integral $\epsilon_\perp$. For tetragonal symmetry, LDA band structure calculations show that $\epsilon_\perp \propto (\cos k_x a - \cos k_y a)^2$. This is an intrinsic property of tetragonal high temperature superconductors and the fact that $\epsilon_\perp$ vanishes at the position of the d-wave nodes has a profound effect on the $\hat{c}$-axis transport. In fact in the clean limit, $1/\lambda_c^2 \propto T^5$. One power of T comes from the d-wave DOS and the other T^4 is from a $(\cos k_x a - \cos k_y a)^4$ factor in $\epsilon_\perp^2(k_\parallel)$ which varies as E^4 at low E. The T^5 dependence, the observation of which would give convincing support for the theory of Xiang and Wheatley, has only been seen in the tetragonal compound HgBa$_2$CuO$_{4+\delta}$ [150], and needs to be confirmed.

For compounds with orthorhombic symmetry, there is no requirement for the hopping integral to vanish, and in any case the node in the d-wave gap need not coincide with the minimum in $\epsilon_\perp$. Other T dependencies can then result. We believe that more attention needs to be paid to the issue of whether the strong $k_\parallel$ dependence of the hopping is an essential part of the $\hat{c}$-axis electrodynamics or not. Much of the theoretical work completely ignores this special property of the cuprates.

Most recently, the geometry with $H_{\rm rf} \perp \hat{c}$ has been used for a sample doped to an oxygen content where oxygen ordering at room temperature can be used to tune the hole doping of a single sample [84]. The advantage of this doping technique, peculiar to YBa$_2$Cu$_3$O$_{6+x}$, is that the doping can be tuned in a single sample, allowing one to avoid systematic errors associated with comparing measurements on samples with different dimensions. Figure 4.19 shows the $\hat{c}$-axis superfluid density of a very underdoped sample, close to where T_c falls to zero, with different T_cs corresponding to slight differences in oxygen content and oxygen ordering. As with $\hat{c}$-axis measurements on YBa$_2$Cu$_3$O$_{6+x}$ and other cuprates, the low temperature behaviour is not linear, but closer to a quadratic or even slightly higher power of T.

Sheehy and Franz [174] proposed a unified theory of the $\hat{ab}$-plane and $\hat{c}$-axis penetration depth based on a d-wave BCS model augmented by a phenomenological charge renormalization factor, suggested by Ioffe and Millis [89], that is close to unity near the nodes up to a cut-off energy E_c, but vanishingly small at higher energies. Sheehy and Franz are able to fit the $\hat{c}$-axis data of Hosseini et al. [84] very well, except near T_c where fluctuation effects start to become important. The values of E_c derived from the fits are found to vary linearly with T_c, reaching 10 meV for $T_c = 20$ K. A missing piece of experimental information is a direct measure of T_c vs. the doping x for these underdoped samples. Following the work of Liang et al. [121], where it is proposed that the $\hat{c}$-axis lattice parameter can be used as a convenient and accurate measure of x, the relationship between T_c and x at very low doping should be experimentally accessible.

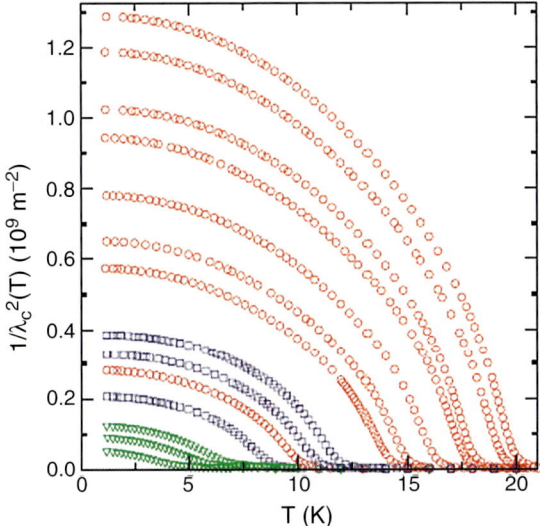

Figure 4.19. Superfluid density in the c-direction for an underdoped sample near $YBa_2Cu_3O_{6.35}$ with progressive alteration of the doping through small changes in oxygen content and oxygen ordering [84].

4.6. Surface Resistance

Surface resistance measurements provide information on low frequency energy dissipation, something complementary to the information on the superfluid that comes from penetration depth measurements (surface reactance). Dissipation at finite frequency in superconductors can come from a number of different sources: direct excitation of quasiparticles, absorption by thermally excited quasiparticles, the effects of grain boundaries, sample inhomogeneity, and other extrinsic loss mechanisms can all come into play and must be disentangled from one another. All of these sources of dissipation involve defects—even the intrinsic dissipation associated with quasiparticle excitations must invoke scattering mechanisms if there is to be real absorption from microwave fields. One could say that microwave conductivity is all about defects. This makes the task of understanding microwave surface resistance a significant challenge because one is not certain a priori which defects are important in a given sample. Added to this is the challenge that theoretical work on defects and scattering in d-wave superconductors is an ongoing pursuit occurring in parallel with the measurements and is still an area of active debate and calculation.

To simplify the task of studying $R_s(T)$ as much as possible, one must turn to measurements on single crystals. There was early work on microwave loss in sintered powders [185], but such measurements mix up the in-plane and out-of-plane properties of these anisotropic materials and are also likely to be strongly influenced by grain boundaries. There is also a wealth of microwave measurements on thin films, driven by the potential for microwave applications such as filters based on the low microwave dissipation that can be achieved at liquid nitrogen temperatures. These film measurements are likely to be influenced by higher defect density than in crystals and also have more chance of being affected by grain boundaries. In principle, the highest quality epitaxial films could exceed the perfection of crystals, but that is a possibility that has yet to be realized.

Among the different members of the cuprate family, the $YBa_2Cu_3O_{6+x}$ system holds a particularly important place because it is possible to grow crystals with relatively high purity and very good cation stoichiometry. This is a crucial point because if one is to disentangle the influence of different types of defects, it is important to start with a sample that has as small a density and variety of defects as possible. The majority of the cuprates have considerable cation disorder, either through cation doping as in the case of $La_{2-x}Sr_xCuO_{4+\delta}$ or through natural cation cross-substitution such as the presence of Cu on the Tl sites in $Tl_2Ba_2CuO_{8+\delta}$. The exceptions to this cation defect problem are $YBa_2Cu_3O_{6+x}$, $YBa_2Cu_4O_8$, and oxygen doped $La_2CuO_{4+\delta}$. Of these three, $YBa_2Cu_4O_8$ is potentially the most perfect because it has stoichiometric CuO chains as well as good cation stoichiometry. However, this system has been relatively little studied because of the need to grow crystals under high pressure. $La_2CuO_{4+\delta}$ has two staged phases with the extra oxygen δ ordered in the c-direction, but has not been studied with microwave techniques. The $YBa_2Cu_3O_{6+x}$ has been the most studied of these three systems. In addition to its potential for high purity and cation order, it has wide range of doping available by changing the oxygen content of the CuO_x chains and there are even phases where the chain oxygens form well-ordered superstructures. The negative aspect of this means of doping is that the chains and their coupling to the CuO_2 bilayers potentially have a complicated influence on the microwave properties.

For the reasons outlined above, the following sections will lean heavily on the extensive measurements available on $YBa_2Cu_3O_{6+x}$ and then, where there is data available, compare them to results on other systems.

4.6.1. $YBa_2Cu_3O_{6+x}$ $\hat{ab}$-Plane

Historically, the field of course did not start out with the cleanest samples possible, nor were the microwave measurement techniques fine-tuned to the properties of the cuprates. The first round of cavity perturbation measurements of crystals did not have sufficient sensitivity to measure the surface resistance much below T_c, but instead focussed attention on $R_s(T)$ near T_c. Rubin et al. [168] and Wu et al. [213] both reported surface resistances that dropped farther and more sharply below T_c than the early measurements on sintered samples. One noteworthy feature of these early measurements was Rubin et al.'s observation that $R_s(T)$ dropped very rapidly in the first few degrees below T_c as shown in Figure 4.20. The behaviour of $R_s(T)$ just below T_c in any superconductor is strongly affected by the divergence of $\lambda(T)$ as T_c is approached from below. This loss of screening by the superfluid causes $R_s(T) \propto \omega^2\lambda^3(T)$ to change very rapidly near T_c. However, the behaviour $R_s(T)$ in the cuprates near T_c is further influenced by two factors not seen in conventional superconductors. One is the critical fluctuations near T_c (discussed in Section 4.7), which lead to a faster temperature dependence of $\lambda(T)$ near T_c than occurs in a mean-field BCS superconductor.

The other factor influencing $R_s(T)$ is the absence of a coherence peak in $\sigma_1(T)$ below T_c in the cuprates. In an s-wave BCS superconductor, the superconducting gap opens up below T_c and as the density of states peaks at the gap edge develop they give rise to a peak in $\sigma_1(T)$ below T_c. It is referred to as a coherence peak because the BCS coherence factors determine which physical measurements will show evidence of the peaked density of states. However, the coherence peak in $\sigma_1(T)$ is greatly suppressed in a d-wave superconductor. Moreover, for optimally doped and underdoped cuprates, spectroscopic measurements such as tunnelling and photoemission indicate that the superconducting gap does not close as T_c is approached— the transition is different from that of even a d-wave BCS superconductor. The exact behaviour of $\sigma_1(T)$ has led to some controversy because it requires simultaneous measurements of

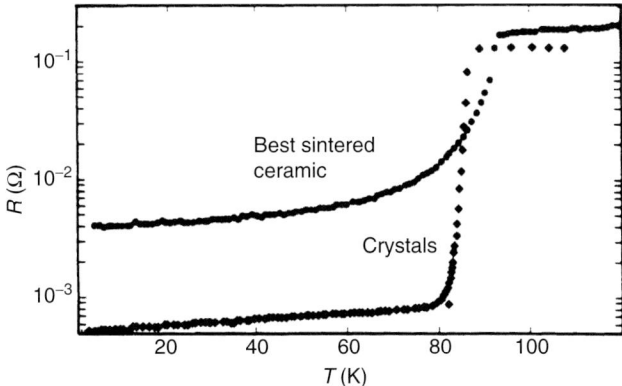

Figure 4.20. The in-plane microwave surface resistance R_s of a mosaic of crystals of $YBa_2Cu_3O_{6.95}$ measured at 5.95 GHz by Rubin et al. was found to be much lower than that of early sintered ceramics [168]. The rapid drop in microwave power absorption is caused by the quick onset of superfluid screening in this material.

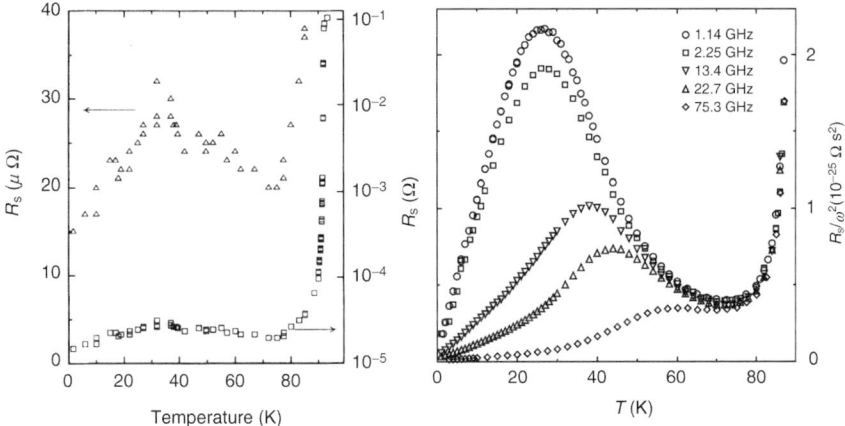

Figure 4.21. The in-plane microwave surface resistance at 2.95 GHZ of the an early generation of $YBa_2Cu_3O_{6+x}$ crystals (left [15]) and the recent generation of high purity crystals grown in $BaZrO_3$ (right [94]) different generations of $YBa_2Cu_3O_{6+x}$ crystals.

$R_s(T)$ and $X_s(T)$, both of which change very rapidly with T, and because the behaviour is highly susceptible to inhomogeneity broadening the transition. Holczer et al. [78] and Klein et al. [104] measured $R_s(T)$ and $X_s(T)$ in $YBa_2Cu_3O_{6+x}$ and $Bi_2Sr_2CaCu_2O_{8+\delta}$ in order to extract $\sigma_1(T)$. They observed a peak in $\sigma_1(T)$ near T_c, but it was sharper than a conventional BCS coherence peak. In addition to the unusual sharpness of this feature, the absence of a coherence peak in NMR measurements [62] led to the suggestion that this feature in $\sigma_1(T)$ had another origin. Olson and Koch [142] and Glass and Hall [56] attributed it to a broadened superconducting transition. As samples have improved, this feature has given way to a sharp peak in $\sigma_1(T)$ very close to T_c that can be attributed to superconducting fluctuations. Horbach et al. [82] first pointed out that fluctuations in the conductivity, which were already apparent in DC resistivity measurements, would give rise to a sharp peak in $\sigma_1(T)$ at T_c.

The more dramatic feature to arise as sample quality improved is illustrated in the sequence of measurements shown in Figure 4.21, beginning with the results from an early

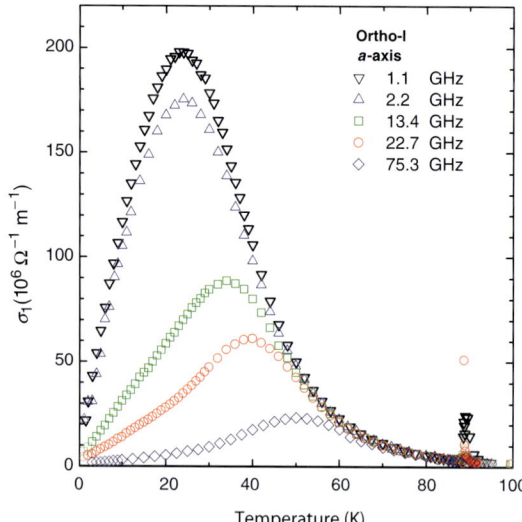

Figure 4.22. The $\hat{a}$-axis microwave conductivity of a high purity YBa$_2$Cu$_3$O$_{6.99}$ crystal shows a large, frequency dependent peak caused by the development of long-lived quasiparticles in the superconducting state.

generation of crystals grown rapidly in gold foils. This measurement by Bonn et al. [15] showed the first hint of a broad peak in $R_s(T)$, which was subsequently confirmed by measurements at different frequencies and with different crystals by Shibauchi et al. [176], Zhang et al. [222], Anlage et al. [9], Kitano et al. [100], Mao et al. [130], and Trunin et al. [196]. As sample purity was improved, first through a shift to growth in yttria-stabilized zirconia, then to growth in BaZrO$_3$ crucibles, the broad peak in $R_s(T)$ became more prominent. Paradoxically, as sample quality improves the microwave loss increases. Since $\lambda(T)$ has no such feature, the peak in $R_s(T)$ must come from a peak in $\sigma_1(T)$. As discussed in Section 4.2.2, $\sigma_1(T)$ can be extracted from measurements of $R_s(T)$ provided that one also has measurements of $X_s(T)$. Ideally, one wants both quantities measured on the same sample at the same frequency. However, it is often sufficient to know $\lambda(T)$ and then to make a small correction to account for any screening by the quasiparticle contribution to σ_2 [67]. Figure 4.22 shows the results of such an analysis of $\hat{a}$-axis measurements of a high purity crystal of YBa$_2$Cu$_3$O$_{6.99}$. The sharp spike in $\sigma_1(T)$ near 90 K is due to critical fluctuations and marks T_c for this sample. Below this, both the temperature dependence and frequency dependence can be explained by the development of very long quasiparticle lifetimes in the superconducting state.

The appearance of frequency dependence in $\sigma_1(\omega, T)$ below 60 K indicates that the width of the conductivity spectrum has moved to microwave frequencies, more than two orders of magnitude narrower than the conductivity spectrum in the normal state. The peak in the temperature dependence occurs because the quasiparticle conductivity spectrum initially narrows below T_c much faster than it loses oscillator strength as the carriers condense into the superfluid, causing the conductivity to initially rise with decreasing temperature. Below about 20 K the spectrum stops narrowing and the declining normal fluid oscillator strength dominates the temperature dependence, causing it to decline with decreasing temperature. As a crude phenomenological model, one can fit the conductivity spectra to a Drude model with two competing temperature dependences, a normal fluid density $n_n(T)$ that declines with temperature and a lifetime $\tau(T)$ that increases with decreasing temperature. Such a form was

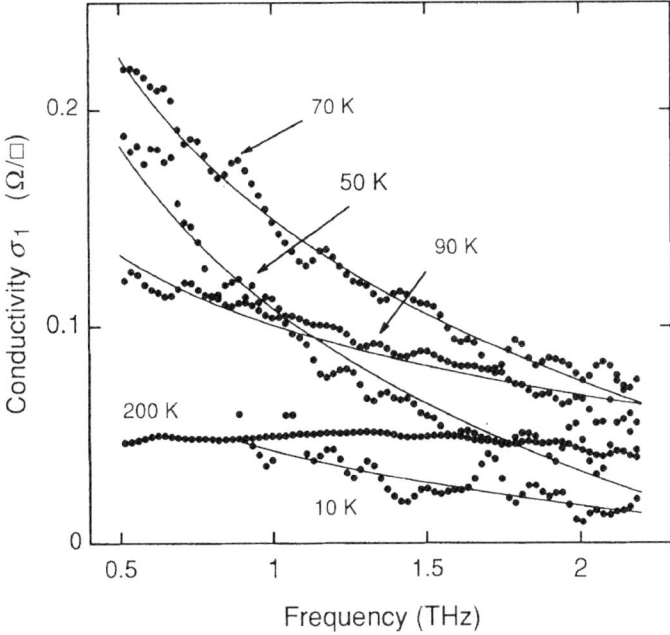

Figure 4.23. The evolution of the conductivity spectrum with temperature for a thin film of $YBa_2Cu_3O_{6+x}$ shows narrowing below T_c in the THz frequency range [141].

loosely justified by Hirschfeld et al. [75, 76] who showed that within a T-matrix model for point scattering, the conductivity spectrum of thermally excited quasiparticles in a d-wave superconductor is given by Eq. (4.10) which has a Drude-like form. If $\tau(\omega)$ were actually frequency independent, this expression would reduce to the Drude form. The frequency-independent lifetime of a Drude model is certainly not valid for nodal quasiparticles in a d-wave superconductor, but a parameterization in terms of quasiparticle density and lifetime is a useful start.

Figure 4.24 shows the scattering rate $1/\tau(T)$ derived from a Drude fit to the data of Figure 4.22. This dramatic temperature dependence has been attributed to a collapse in the inelastic scattering of quasiparticles below T_c. This picture is supported by several other measurements sensitive to quasiparticle lifetimes. An early experimental indication of this narrowing in $\sigma_1(\omega)$ was seen in far infrared measurements made just below T_c by Romero et al. [167]. Behaviour qualitatively similar to the microwave measurements was observed first in THz measurements on thin films by Nuss et al. [141] and later by Spielman et al. [183]. Figure 4.23 shows measurements on a thin film of $YBa_2Cu_3O_{6+x}$ that have the same trend of a peak in the spectrum $\sigma(\omega)$ that narrows rapidly below T_c. Measurements of the temperature dependence of thermal conductivity $\kappa(T)$ below T_c also show a pronounced peak that has been attributed to the development of long quasiparticle lifetimes [163]. There was some controversy over this since the thermal conductivity could also exhibit a peak if the phonon damping declines rapidly below T_c, an effect seen in neutron scattering of phonons in conventional superconductors. However, measurements of the thermal Hall conductivity [110, 111, 219] have provided a means of separating out the electronic contribution and have shown that the lifetime associated with thermal transport does indeed become

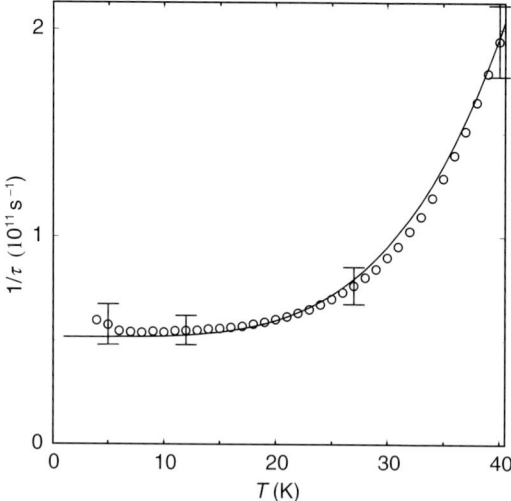

Figure 4.24. An estimate of the transport scattering rate of quasiparticles can be extracted from Drude fits to the $\hat{a}$-axis microwave conductivity spectrum of a high purity $YBa_2Cu_3O_{6.99}$ crystal. The rapid temperature dependence is either a very high power law or even an exponential temperature dependence [67].

very long below T_c. Studies as a function of Zn-doping in $YBa_2Cu_3O_{6.95}$ also support the view that a rise in quasiparticle lifetime makes a major contribution to thermal conductivity [74].

There has been some work on modelling the collapse of the scattering rate below T_c in terms of electron–phonon scattering that is strongly temperature-dependent [47, 195]. However, the large inelastic scattering rate seen in transport measurements, which is linear in both temperature and frequency over a very wide range for optimally doped cuprates, argues against an electron–phonon dominated transport scattering in the normal state. If it is some form of electron–electron scattering then it is possible that the development of the superconducting gap suppresses this scattering mechanism below T_c. Hirschfeld et al. [76] attributed the collapse of inelastic scattering to spin fluctuations gapped by the opening of a d-wave energy gap. This and related work by Schachinger et al. [171, 172] and Rieck et al. [165] build a model for the scattering rate that is motivated by measured normal state properties, particularly the spin susceptibility in the normal state and the normal state DC resistivity. The essential idea is to tie together the electron–electron interactions as the source of the high critical temperature, the d-wave pairing state, and the anomalous transport properties. In any such model where the transport is dominated by electron–electron, or electron-spin fluctuation scattering, the scattering is suppressed as the superconducting gap develops below T_c. Later, Walker and Smith pointed out that the damping measured in measurements of charge transport requires Umklapp processes that scatter nodal quasiparticles [209]. The shape of the Fermi surface and the position of the nodes in the energy gap mean that the nodal quasiparticles must scatter from excitations far from the nodes in the gap. This in turn means that the Umklapp processes are strongly suppressed because they involve scattering from parts of the Fermi surface that are gapped. In support of this notion, Figure 4.24 shows a scattering rate derived from Drude fits to $\sigma_1(\omega)$ like those shown in Figure 4.41. Hosseini et al. [67] fit the temperature dependence to a high power law, but the data is equally well fit by an exponential as suggested by Walker and Smith. Subsequently, Duffy et al. [43] separated Umklapp and normal scattering in the model of scattering by spin fluctuations and were able to reconcile the microwave scattering rate with the different temperature dependence seen in thermal transport [223], which is influenced by both Umklapp and normal scattering processes.

4.6.2. Disorder and Quasiparticle Damping

We have already discussed the strong influence that defects can have on the penetration depth, and this is doubly true for the microwave surface resistance. Here, defects not only affect the low temperature asymptotic behaviour, they also directly affect the broad peak in $\sigma_1(T)$ by limiting the rapid increase in the quasiparticle lifetime τ below T_c. The increase in the height of the peak in $R_s(T)$ and $\sigma_1(T)$ shown in Figure 4.21 and 4.22 is one indication that defects control this feature. When the quasiparticle damping $1/\tau(T)$ drops rapidly below T_c due to the suppression of inelastic scattering, it eventually runs into a limit controlled by scattering from defects. In optimally doped $YBa_2Cu_3O_{6+x}$ this elastic scattering limit causes $\sigma_1(T)$ to turn down as the temperature decreases further. Decreases in the defect density as the purity of $YBa_2Cu_3O_{6+x}$ improved made this turnaround occur at a lower temperature, making the peak in $\sigma_1(T)$ and $R_s(T)$ larger in size and lower in temperature [94]. The main point is that if τ really increases by orders of magnitude below T_c, then the conductivity should become extremely sensitive to low levels of point defects. The first hint of this was an early thin film study by Lippert et al. [122] where a sample was found to have a somewhat lower microwave loss after irradiation with oxygen ions.

Studying single crystals, which start from a point of relatively low defect density, provides a more systematic means of testing impurity effects. The 100-fold or more increase in τ below T_c suggests a mean free path for the charge carriers in the $\hat{ab}$-plane that is on the surprising scale of hundreds of nanometres for high purity crystals of $YBa_2Cu_3O_{7-\delta}$. It should only take impurity concentrations on the order of a few tenths of a percent to substantially affect this mean free path. For $YBa_2Cu_3O_{7-\delta}$ two of the most studied impurities are Ni and Zn, since they both can substitute for Cu on the CuO_2 planes [23, 24, 128], but offer the contrast that Ni is magnetic, but Zn is not. Figure 4.25 shows $R_s(T)$ for a sample of $YBa_2Cu_3O_{6.95}$ grown in yttria-stabilized zirconia (99.9% pure) and for a crystal with 0.75%

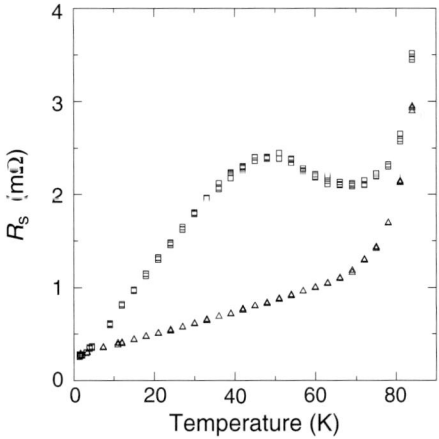

Figure 4.25. The broad peak in $R_s(T)$ of a nominally pure crystal of $YBa_2Cu_3O_{6.95}$ at 34.8 GHz (open boxes) can be completely eliminated by 0.75% substitution of Ni (open triangles) for Cu [16, 18, 220, 222].

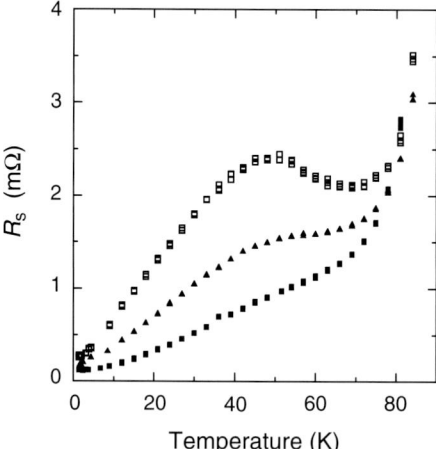

Figure 4.26. The influence of Zn substitution on $R_s(T)$ of YBa$_2$Cu$_3$O$_{6.95}$ at 34.8 GHz. The broad peak at 40 K in nominally pure samples (squares) is reduced to a plateau at 0.15% (solid triangles) and $R_s(T)$ becomes completely monotonic by 0.31% (solid boxes) substitution if Zn for Cu [16, 18, 220, 222].

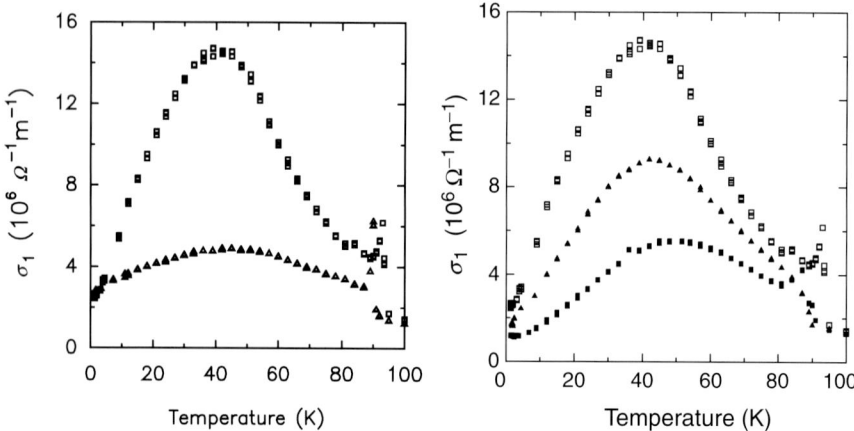

Figure 4.27. (a) Left: at 0.75% substitution of Ni for Cu (open triangles) in YBa$_2$Cu$_3$O$_{6.95}$ the broad peak seen $\sigma_1(T)$ for pure samples (open boxes) is almost completely eliminated [16, 18, 220]. (b) Right: low levels of Zn substitution for Cu in YBa$_2$Cu$_3$O$_{6.95}$ (nominally pure, open boxes; 0.15%, solid triangles; 0.31%, solid boxes) cause the peak in $\sigma_1(T)$ to shrink and shift to higher temperature [16, 18, 220].

substitution of Ni for Cu atoms [18, 222]. At this concentration the broad peak in $R_s(T)$ is completely eliminated. Figure 4.26 shows that at very low concentrations Zn substitution reduces the peak to a plateau (0.15% substitution) or eliminates it entirely (0.31% substitution). These low levels of impurities lead to surface resistances that are very similar to the lowest loss thin films (see reviews by Klein [101], Piel and Muller [158], Newman and Lyons [138]).

Figure 4.27 shows the conductivities extracted from the surface resistance measurements shown in Figures 4.25 and 4.26. Ni substitution at the 0.75% level almost completely eliminates the peak in $\sigma_1(T)$. The evolution of the suppression of the peak is seen clearly in a study of lower concentrations of Zn impurities. At 0.15% the peak is greatly diminished, though still clearly present in the conductivity. However, at this impurity level, the diminished

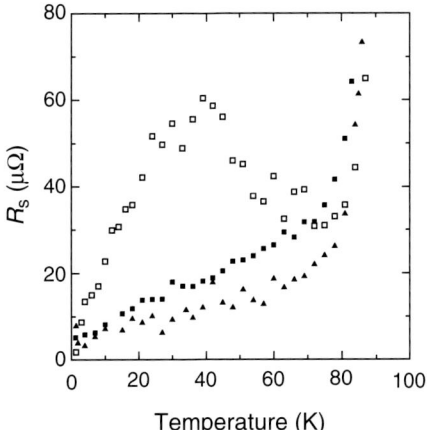

Figure 4.28. At 3.8 GHz, the peak in the surface resistance of a nominally pure crystals (open squares) is not present in a crystal doped with 0.75% Ni (solid triangles). At 1.4% Ni doping (solid boxes), the overall surface resistance starts to increase [16, 17].

peak no longer gives rise to a clear peak in $R_s(T)$. The influence of the $\lambda^3(T)$ screening term that comes into the surface resistance contributes a monotonically decreasing factor that reduces the peak seen in $\sigma_1(T)$ to a plateau in $R_s(T)$. At 0.31% Zn doping the peak in $\sigma_1(T)$ is so diminished that $R_s(T)$ decreases monotonically with temperature.

Qualitatively, the effects of Zn and Ni impurities support the explanation of the broad peak in $R_s(T)$ and $\sigma_1(T)$ in terms of a scattering rate that increases rapidly below T_c. The peak in $\sigma_1(T)$ is diminished and moves up in temperature because the rapidly changing $\tau(T)$ runs into a limit set by impurity scattering. When this limit is reached $\sigma_1(T)$ and $R_s(T)$ decrease with temperature, following the decreasing density of thermally activated quasiparticles. These results are for rather low impurity concentrations that affect τ without causing large changes in either T_c or $\lambda(0)$. Once sufficient impurities are added that $\hbar/\tau$ becomes comparable to $k_B T_c$, the materials are no longer in the clean limit and impurities significantly increase $\lambda(0)$. Ulm et al. [206] have demonstrated that quite large changes in $\lambda(0)$ can be achieved with Ni impurity doping without destroying superconductivity in $YBa_2Cu_3O_{7-\delta}$. Also, as noted in the discussion of penetration depth in Section 4.4.2, the penetration depth can be quite large in thin films, without substantial suppression of T_c, perhaps due to inhomogeneity. The effect of an increase in λ can start to influence the microwave surface resistance rather quickly because $R_s(T) \propto \lambda^3$. For the case of Ni impurities in $YBa_2Cu_3O_{6.95}$ the effect of increased λ is already encountered at 1.4% substitution. Figure 4.28 shows $R_s(T)$ at 3.8 GHz for a nominally pure sample, and for 0.75 and 1.4% Ni substitution. At 0.75% Ni impurities suppress the peak in $R_s(T)$, just as was seen at 34.8 GHz, but at 1.4% substitution the overall magnitude of $R_s(T)$ becomes higher again. If one is interested in the practical problem of minimizing microwave loss there appears to be an optimum impurity concentration for doing so.

4.6.3. Other Materials—*ab*-Plane

A number of measurements on relatives of $YBa_2Cu_3O_{6+x}$ with different rare earths on the Y site, such as $LuBa_2Cu_3O_{6+x}$ [30] exhibit a peak in $R_s(T)$ and $\sigma_1(T)$, similar to that shown above for $YBa_2Cu_3O_{6+x}$. An unusual example is measurements of $GdBa_2Cu_3O_{6+x}$ by

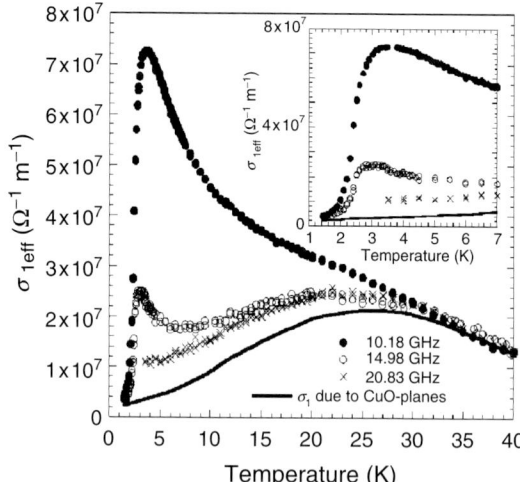

Figure 4.29. The effective conductivity of an optimally doped $GdBa_2Cu_3O_{6+x}$ crystal is determined by both quasi-particle conductivity and the susceptibility of the magnetic Gd ions. Despite these fluctuating spins, which also order at 2.2 K, the underlying quasiparticle conductivity still has a large peak due to long quasiparticle lifetimes [145].

Ormeno et al. [145]. Figure 4.29 shows the "effective conductivity" $\sigma_{1\text{eff}}(\omega, T)$ obtained from the standard treatment of surface impedance data to extract conductivities. Such a treatment does not extract the true quasiparticle conductivity because the Gd ions are magnetic and one would need to change the permittivity μ_0 in Eq. (4.3) to $\mu = \mu_0(1 + \chi' + i\chi'')$. As a consequence, the large peak seen near 3 K is associated not with conductivity, but with the susceptibility of the Gd ions, which go through a Neel transition at 2.2 K. As shown in [145], Ormeno et al. were able to separate out the magnetic effects and show that the quasiparticle conductivity is very similar to that of $YBa_2Cu_3O_{6+x}$, something quite remarkable since the Gd ions are sandwiched by the CuO_2 bilayer yet their fluctuations and ordering have no impact on the quasiparticles.

Microwave measurements on other cuprates have progressed more slowly, not only because of the concerns about sample quality but also because there are technical difficulties associated with microwave measurements on highly anisotropic materials. Since the majority of microwave measurements on superconductors are performed in microwave magnetic fields, any one measurement necessarily mixes together the surface impedance of two different directions in an orthorhombic material. $YBa_2Cu_3O_{6+x}$ is the least anisotropic of the cuprates, and multiple measurements on samples with different orientations and aspect ratios can be used to extract the R_s and X_s for currents running parallel to each of the three principle axes. This becomes more difficult as anisotropy increases and the properties of one direction dominate others. Some materials, especially $Bi_2Sr_2CaCu_2O_{8+\delta}$ have the additional problem of being highly susceptible to cleaving perpendicular to the $\hat{c}$-axis, making it difficult to cut and polish samples. Nevertheless, successful measurements of R_s of single crystals of $Bi_2Sr_2CaCu_2O_{8+\delta}$ were eventually achieved in cavity perturbation measurements that also simultaneously determined the temperature dependence of the penetration depth $\Delta\lambda(T)$.

Figure 4.30 shows $R_s(T)$ of $Bi_2Sr_2CaCu_2O_{8+\delta}$ at 10 GHz measured by Jacobs et al. [90]. The linear temperature dependence of $R_s(T)$ is qualitatively similar to that seen in Ni-doped $YBa_2Cu_3O_{6.95}$, suggesting that these crystals of $Bi_2Sr_2CaCu_2O_{8+\delta}$ have a

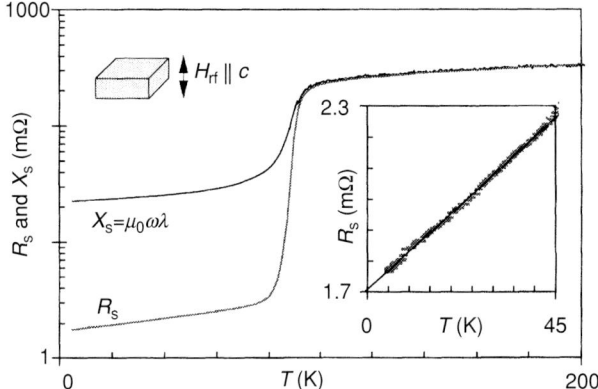

Figure 4.30. $R_s(T)$ of $Bi_2Sr_2CaCu_2O_{8+\delta}$ at 10 GHz measured by Jacobs et al. [90]. The monotonic, linear temperature dependence below about 60 K is similar to that seen in Ni-doped $YBa_2Cu_3O_{6.95}$ shown in Figure 4.25.

significant level of disorder. The linear temperature dependence can come from two sources, a linear temperature dependence of $\lambda(T)$, $\sigma_1(T)$, or perhaps both. To first order, the two contributions give

$$\frac{\partial R_s}{\partial T} = \frac{\mu_0^2}{2}\omega^2\left(\lambda_0^3\frac{\partial \sigma_1}{\partial T} + 3\lambda^2\sigma_0\frac{\partial \lambda}{\partial T}\right) \quad (4.16)$$

The technical difficulty with separating the contributions is that it depends strongly on the choice of $\lambda(T=0)$ used in the analysis. The uncertainty in $\lambda(T=0)$ for $Bi_2Sr_2CaCu_2O_{8+\delta}$ data reported by Jacobs et al. means that the linear temperature dependence of $R_s(T)$ could be attributed to a large extent to the linear temperature dependence of $\lambda(T)$.

As happened with $YBa_2Cu_3O_{6.99}$, frequency dependent measurements provide a less ambiguous picture. Figure 4.31 shows the surface resistance of $Bi_2Sr_2CaCu_2O_{8+\delta}$ measured at three microwave frequencies by Lee et al. [120]. Although $R_s(T)$ again does not have the prominent peak seen in high purity $YBa_2Cu_3O_{6.99}$, there is some concave downwards curvature in the lower two frequencies that is qualitatively similar to that seen at some defect densities in Figures 4.25 and 4.26 for $YBa_2Cu_3O_{6.95}$ with light impurity doping. Also, the slight frequency-dependence apparent in this curvature suggests that the conductivity spectrum $\sigma_1(\omega)$ is developing structure in the microwave regime. These effects are more apparent when $\sigma_1(\omega, T)$ is extracted, as shown in Figure 4.32. Like in $YBa_2Cu_3O_{6.99}$, $\sigma_1(T)$ rises substantially below T_c, suggesting that the quasiparticle scattering time also rises rapidly below T_c in this material. However, $\sigma(T)$ neither rises as high, nor does it develop as strong a frequency dependence as occurs in $YBa_2Cu_3O_{6.99}$. It would seem that the width of the peak in $\sigma_1(\omega)$ does not get nearly as narrow, possibly being limited by a higher density of defects.

Another noteworthy feature of the data shown in Figure 4.32 is the residual conductivity at low temperatures. Quantitatively, the value of σ_1 at low frequency and temperature is comparable to that seen in high purity $YBa_2Cu_3O_{6.99}$ (Figure 4.22). The very different look of the temperature dependence is that $\sigma_1(T)$ rises much higher in $YBa_2Cu_3O_{6.99}$ before falling to this residual value. The other significant difference is that because $\sigma_1(\omega)$ falls off much more slowly in $Bi_2Sr_2CaCu_2O_{8+\delta}$, there is considerably more oscillator strength remaining at low frequency and low temperature than is seen in $YBa_2Cu_3O_{6.99}$. We will return to this issue

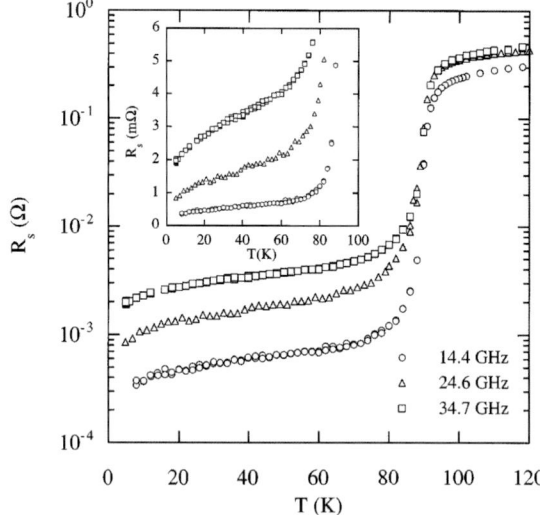

Figure 4.31. $R_s(T)$ of $Bi_2Sr_2CaCu_2O_{8+\delta}$ at three different microwave frequencies [120].

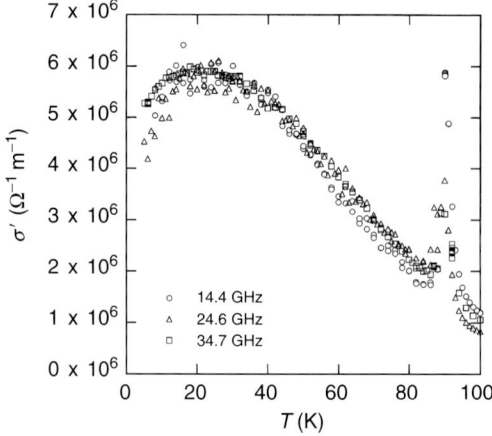

Figure 4.32. $\sigma_1(\omega, T)$ of $Bi_2Sr_2CaCu_2O_{8+\delta}$ extracted from the $R_s(T)$ shown in Figure 4.31 [120].

in Section 4.6.4, but note here that THz measurements by Corson et al. [36] also confirm that there is considerable uncondensed oscillator strength in $Bi_2Sr_2CaCu_2O_{8+\delta}$.

$Tl_2Ba_2CuO_{6+\delta}$ is one other material that possesses a high T_c at optimal doping and has yielded to microwave measurements. Figure 4.33 shows microwave measurements of $R_s(T)$ by Broun et al. [26, 27] that bear a striking quantitative similarity to the results for $Bi_2Sr_2CaCu_2O_{8+\delta}$. The conductivity extracted from these measurements is shown in Figure 4.34 and has the familiar rise below T_c suggesting a rapidly increasing quasiparticle scattering time. The rise is not as dramatic as in high purity $YBa_2Cu_3O_{6.99}$, but is more pronounced than it is in $Bi_2Sr_2CaCu_2O_{8+\delta}$. Supporting this comparison is the frequency dependence that develops at low temperatures, less pronounced than $YBa_2Cu_3O_{6.99}$, but more pronounced than $Bi_2Sr_2CaCu_2O_{8+\delta}$. Like $Bi_2Sr_2CaCu_2O_{8+\delta}$, the $Tl_2Ba_2CuO_{6+\delta}$ data has

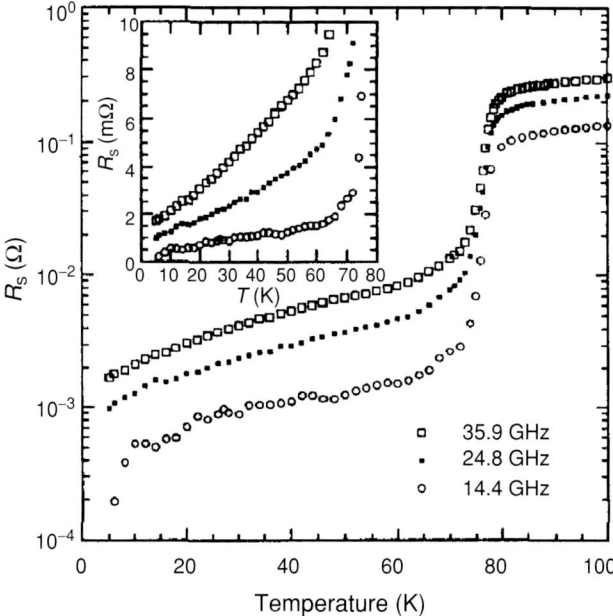

Figure 4.33. $R_s(T)$ of $Tl_2Ba_2CuO_{6+\delta}$ at three different microwave frequencies [27].

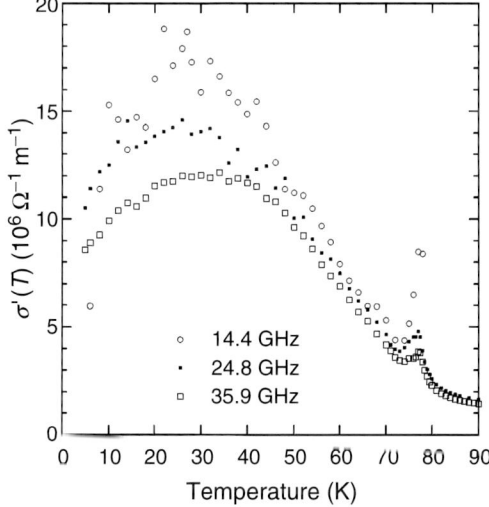

Figure 4.34. Conductivity of $Tl_2Ba_2CuO_{6+\delta}$ at three different microwave frequencies [27] is strikingly similar to that measured for $Bi_2Sr_2CaCu_2O_{8+\delta}$, though with some indication that it develops a somewhat longer quasiparticle lifetime at low temperatures.

substantial uncondensed oscillator strength, which is noteworthy because $Tl_2Ba_2CuO_{6+\delta}$ is a structurally simpler material, possessing single CuO_2 layers rather than bilayers, and having no CuO chains. It does, however, have interstitial oxygen dopants and cation non-stoichiometry comparable in magnitude to that of $Bi_2Sr_2CaCu_2O_{8+\delta}$, which might explain the similarity of the two systems. Another puzzling similarity of the two systems is that although the height of the peak in $\sigma_1(T)$ suggests more disorder than in YBa$_2$Cu$_3$O$_{6.99}$, the temperature at which the peak occurs is quite low, something that differs from the naive argument made above in Section 4.6.2 that defects should cause the peak to diminish *and* shift up in temperature. This puzzle has recently been addressed by Nunner et al. [140] and is discussed further in the following section.

There have also been R_s measurements of the single-layer cuprates with relatively low T_cs at optimal doping; the hole doped material La$_{1-x}$Sr$_x$CuO$_{4+\delta}$ and the electron-doped materials Pr$_{1-x}$Ce$_x$CuO$_{4+\delta}$ Nd$_{1-x}$Ce$_x$CuO$_{4+\delta}$. As noted in Section 4.5, disorder is a particularly serious problem in these systems, in part because the means of doping them is to add cation defects in close proximity to the CuO$_2$ planes. None of these materials exhibits the linear $\lambda(T)$ expected for a clean d-wave superconductor and there has been controversy over whether the behaviour is a power law associated with a dirty d-wave superconductor or exponential temperature dependence expected for an s-wave superconductor. Similarly, the surface resistance of these materials shows signs of substantial disorder. Figure 4.35 shows the only available measurements of $R_s(T)$ in La$_{1.85}$Sr$_{0.15}$CuO$_{4+\delta}$, performed by Shibauchi et al. [175]. The results are qualitatively similar to those for Bi$_2$Sr$_2$CaCu$_2$O$_{8+\delta}$, although the superconducting transition seems broader. The magnitude of R_s below T_c is somewhat higher, probably because λ is larger in this material, resulting in weaker superfluid screening.

Figure 4.36 shows the most recent R_s measurements on two of the electron-doped cuprates [106]. After a concerted effort to improve samples of these materials, one must still note the very high residual loss in these materials. $R_s(T)$ falls by less than a factor of 10 at 9.6 GHz, as compared to the drop by a couple of orders of magnitude shown in Figure 4.35 for La$_{1.85}$Sr$_{0.15}$CuO$_{4+\delta}$. Although Nd$_{1.85}$Ce$_{0.15}$CuO$_4$ has the additional problem that the Nd moments contribute to the measured surface impedance, the Pr$_{1.85}$Ce$_{0.15}$CuO$_4$ measurements actually have the higher residual loss. It is not at all clear whether or not this residual is intrinsic, which makes it nearly impossible to draw any conclusions about $\sigma_1(\omega, T)$. Kokales et al. [106] demonstrated this by extracting the conductivity with and without subtracting the residual loss first, as shown in Figure 4.37. The shape of $\sigma_1(T)$ depends greatly on the treatment of the residual loss—it either rises monotonically or exhibits a peak. If one treats the data in the same way as all of the $R_s(T)$ measurements discussed above, without doing any subtraction from the raw data, one would conclude that $\sigma_1(T)$ increases monotonically below T_c.

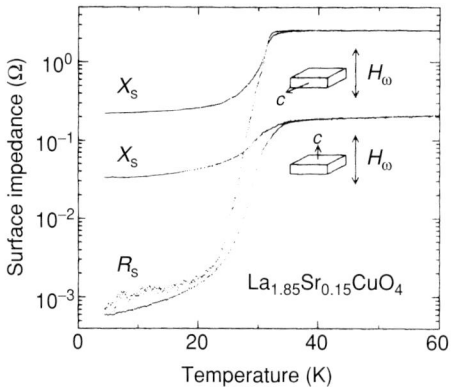

Figure 4.35. The surface resistance of La$_{1.85}$Sr$_{0.15}$CuO$_{4+\delta}$ at 10 GHz has some of the qualitative features seen in Bi$_2$Sr$_2$CaCu$_2$O$_{8+\delta}$ [175].

Microwave Electrodynamics of High Temperature Superconductors

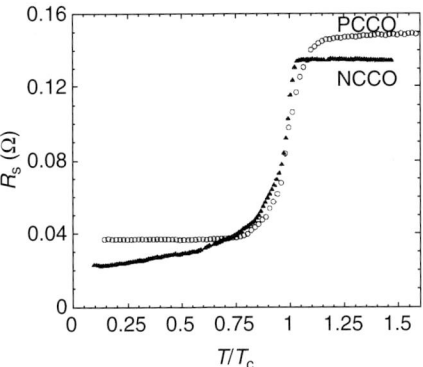

Figure 4.36. The surface resistance of crystals of Nd$_{1.85}$Ce$_{0.15}$CuO$_4$ and Pr$_{1.85}$Ce$_{0.15}$CuO$_4$ exhibit considerably more residual loss than the electron-doped materials.

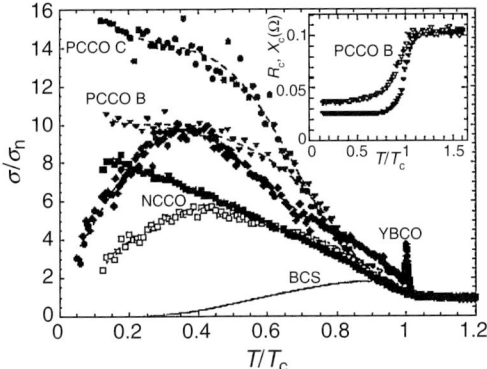

Figure 4.37. The large residual loss of Nd$_{1.85}$Ce$_{0.15}$CuO$_4$ and Pr$_{1.85}$Ce$_{0.15}$CuO$_4$ crystals leads to relatively high uncertainty in how to extract the conductivity. If one subtracts the residual from R_s a peak in $\sigma_1(T)$ can be generated, but there is a monotonic increase with decreasing temperature if one treats that data as all other data discussed above [106].

Qualitatively, this would be a step further down the path from the prominent peak seen in YBa$_2$Cu$_3$O$_{6.99}$, through the weaker peaks seen in Tl$_2$Ba$_2$CuO$_{6+\delta}$ and Bi$_2$Sr$_2$CaCu$_2$O$_{8+\delta}$, to a monotonic rise below T_c.

4.6.4. Low Temperature Limit

One of the most unsettled aspects of the microwave loss in the cuprates at the present time is the behaviour in the low temperature limit. For s-wave superconductors $R_s(T)$ falls to zero as $T \to 0$, following an Arrhenius law due to the presence of an energy gap everywhere on the Fermi surface. In practice there is always some residual loss attributed to extrinsic effects and in the best measurements the residual is small enough that the exponential temperature dependence is clear. It was in fact common practice to simply subtract the value of the residual loss from $R_s(T)$ in order to better show the exponential behaviour on a semi-log plot. Because of the technological importance of low microwave loss for high Q superconducting cavities, considerable experimental effort was expended in reducing it and for materials such as Nb and Pb, the sensitivity to actually do the measurements was achieved by constructing an entire resonant cavity out of the material of interest [55]. There is also a body of theoretical work on mechanisms for the residual losses, such as the impurity phases that plague Nb due to oxides and hydrides in the surface (see e.g. [61]).

Although it has been tempting to subtract residual loss from $R_s(T)$ in the cuprates in order to clarify the temperature dependence, this has turned out to be a questionable practice because of the possibility of non-zero $R_s(T)$ at T = 0 in a d-wave superconductor. The first stimulus on this issue was the work of Lee et al. [118] who showed that impurities in a d-wave superconductor could generate states at the Fermi energy that were delocalized, giving rise to a non-zero conductivity at zero temperature. For point scatterers, $\sigma_1(T \to 0)$ at low frequencies is expected to approach a universal limit,

$$\sigma_{00} = \frac{e^2}{\pi^2 \hbar d} \frac{v_F}{v_\Delta}, \tag{4.17}$$

where d is the average spacing of the CuO_2 [118]. Surprisingly, this residual σ_1 is independent of the impurity scattering time τ. A simple picture of this is that impurities in a d-wave superconductor generate low-lying states that give rise to absorption, even at zero temperature, but also affect the impurity scattering time τ_{imp}. This gives two dependencies on impurity concentration that cancel, leaving a residual conductivity that does not depend on τ_{imp}. The relatively simple prediction, the $T \to 0$ limit, was later shown to need vertex corrections [44] and Fermi liquid corrections [133]. Away from $T = 0$, $\sigma_1(T)$ of a superconductor that has nodes in its energy gap is expected to have power law temperature dependence at low T, rather than the exponentially activated behaviour of a superconductor without nodes in the gap [76, 77, 105]. The initial deviation from this limit is expected to be quadratic, in contrast to the linear temperature dependence of the superfluid density.

The challenge then is that one has both intrinsic and extrinsic mechanisms for generating residual loss at low temperatures and this ambiguity stands in the way of understanding the low temperature conductivity. A further complication is that the temperature dependent part of the low temperature $\sigma_1(T)$ is highly sensitive to the nature of the defects. If the scatterers are unitary, one expects the low temperature behaviour to be apparent below a temperature and energy scale $\gamma \sim \sqrt{\Gamma_n \Delta(0)}$, where Γ_n is the impurity scattering rate in the normal state. Since $\Delta(0)$ is large in the cuprates, the low temperature limit should be at easily accessible temperatures if the scatterers are unitary, making it possible to observe both σ_{00} and the quadratic temperature dependence. However, in the Born limit, the crossover to the $T \to 0$ limiting behaviour occurs at an exponentially small scale $\gamma \sim \Delta_0 \exp(-\Delta_0/\Gamma_n)$ and would then be below the lowest temperatures for experiments performed in helium baths (1.2 or 4.2 K).

In terms of materials parameters that are appropriate for $YBa_2Cu_3O_{6.95}$, where thermal conductivity gives $v_F/v_\Delta \sim 14$ [32] and the average plane spacing is $2/11.6$ Å, the universal conductivity limit is $\sigma_{00} \sim 3 \times 10^5 \, \Omega^{-1} m^{-1}$, which is about 1/3 of the normal state conductivity near T_c. Although this seems large, it is not easily measured at microwave frequencies, because the $\omega^2 \lambda^3$ screening term in $R_s(T)$ means that this conductivity corresponds to a rather low microwave loss. For $\lambda = 120$ nm $R_s(T \to 0) \approx 1.6 \times 10^{-8} f^2$, where R_s is in Ω and f is the measurement frequency in GHz. The dependence of $R_s(T)$ on the square of the frequency must be kept in mind when comparing loss measurements performed at different frequencies.

The first time that the low temperature R_s was measured in single crystals of $YBa_2Cu_3O_{7-\delta}$ was the early measurements of Rubin et al. [168]. Although their measurements were performed in a high Q, Nb resonator, the large size of their cylindrical resonator, with a

TE$_{011}$ mode at 5.95 GHz, meant that the resolution of $R_s(T)$ for small crystals was not high, only about 500 μΩ. However, at low temperature they directly measured microwave power absorption with a bolometric technique and achieved a resolution of 15 μΩ at 3 K. The crystals showed no measurable microwave loss on this scale, but the resolution is still a factor of 10 larger than the estimate of the d-wave limit. Cavity perturbation measurements since then still have difficulty achieving the resolution that would be needed to see the zero temperature limit σ_{00} proposed by Lee et al. Losses in the sample holder and non-perturbative effects tend to be comparable to, or larger than, the predicted value. Aside from this technical issue, it now seems to be the case that many defects are not in the unitary limit, with the consequence that typical measurements down to 1.2 or 4.2 K are not yet in the regime where $\sigma(T)$ reaches its low temperature behaviour. As noted in the discussion of the penetration depth, the one scatterer that does seem to be unitary is Zn substituted on planar Cu sites and Figure 4.27b shows the expected T^2 dependence of $\sigma_1(T)$ at 35 GHz. However, this measurement did not have the resolution to unambiguously determine the $T \to 0$ limit.

Although cavity perturbation measurements do not typically have the resolution to detect the predicted limit σ_{00}, they do nevertheless resolve a temperature-dependent $R_s(T)$ down to 1.2 K that is an order of magnitude larger than the expected loss for a d-wave superconductor with point scatterers in the low temperature limit. Explaining this loss has been the subject of intense study. It was suggested in Section 4.6.1 that once the inelastic scattering has finished its rapid decline below T_c, $\sigma_1(T)$ is controlled by defect scattering that varies substantially from material to material. The challenge has been to explain this conductivity, which is much larger than σ_{00} and has a temperature dependence different from the T^2 expected away from the $T \to 0$ limit. One clue in this regard has been a closely related universal thermal conductivity [59, 187]. Taillefer et al. reported observing such a *thermal* conductivity that was both independent of impurity concentration at mK temperatures and whose magnitude was consistent with the expectations for the universal conductivity [190]. It was subsequently pointed out that the universal charge conductivity might deviate from the thermal conductivity because of Fermi liquid corrections and vertex corrections [44]. Perhaps more importantly, the microwave conductivity measurements have not yet been done at the sub-1-Kelvin temperatures used in the thermal conductivity studies. Unless the point scatterers are in the unitary limit, the temperature range where one ought to see the universal limit is likely below the range of the existing microwave measurements. This has been partly confirmed by Hill et al.'s measurement of the electronic contribution to the thermal conductivity, shown in Figure 4.38. The zero field electronic thermal conductivity starts from the universal limit at very low temperatures, but quickly rises with the expected T^2 temperature dependence well below 1.2 K.

Although the universal conductivity limit has yet to be properly addressed by microwave measurements, there is now a wealth of information on the temperature- and frequency-dependent conductivity which can be used to test models of defect scattering in a d-wave superconductor. As far as the measured quantity $R_s(T)$ is concerned, the most common temperature dependence seen at microwave frequencies is close to linear in T. For example, this can be seen in both early and recent YBa$_2$Cu$_3$O$_{6+x}$ (Figure 4.21), in Ni-doped YBa$_2$Cu$_3$O$_{6.95}$ (Figures 4.25 and 4.28), in Bi$_2$Sr$_2$CaCu$_2$O$_{8+\delta}$ (Figure 4.30) and in many thin film experiments [101, 138, 158]. As was pointed out in Eq. (4.16), $R_s(T)$ can have a linear temperature dependence coming from the linear temperature dependence of $\lambda(T)$, a linear temperature dependence of $\sigma_1(T)$, or both. For typical values of the low T conductivity $\sigma_0 \sim 5 \times 10^6 \, \Omega^{-1} \, m^{-1}$, the low T penetration depth $\lambda_0 \sim 150$ nm, and a penetration depth slope $\partial \lambda / \partial T \sim 0.5$ nm/K^{-1} the conductivity would have to have a linear term of order

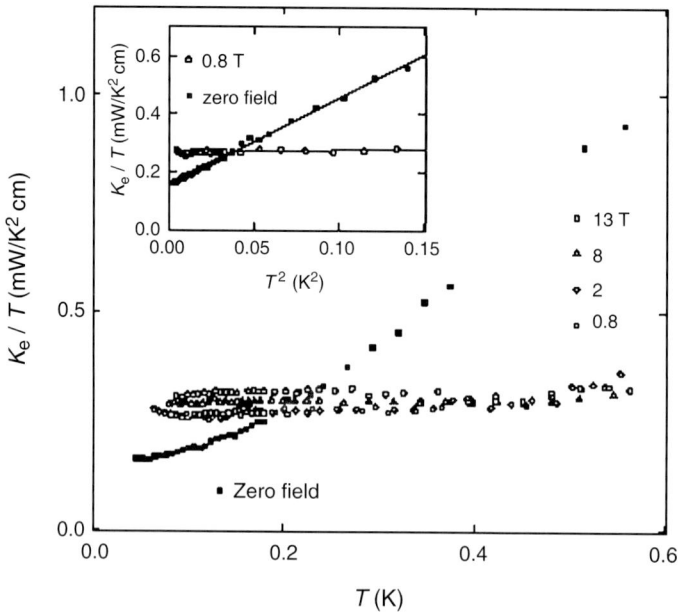

Figure 4.38. The electronic contribution to the thermal conductivity of YBa$_2$Cu$_3$O$_{6.99}$ in zero field (solid squares) rises from the universal limit with the expected T^2 temperature dependence. This rise below 1 K suggests scatterers that are neither in the Born or the unitary limit and is also one reason why microwave measurements above 1.2 K fail to see the universal limit [72].

$\partial \sigma_1 / \partial T \sim 10^5 \, \omega^{-1} \, \text{m}^{-1} \text{K}^{-1}$ to make a competitive contribution to a linear term in $R_s(T)$. It appears that Tl$_2$Ba$_2$CuO$_{6+\delta}$ and Bi$_2$Sr$_2$CaCu$_2$O$_{8+\delta}$ and Ni-doped YBa$_2$Cu$_3$O$_{6.95}$ are just entering the regime where $\sigma_1(T)$ has an impact on the temperature dependence of $R_s(T)$. In this situation, Lee et al. [120] pointed out that the exact shape of $\sigma_1(T)$ becomes rather sensitive to the choice of λ_0 used in the analysis of surface impedance measurements. For instance, the weak peak in $\sigma_1(T)$ shown in Figure 4.32 can change in height and position if λ_0 is altered, which will have some effect on the low T behaviour inferred for $\sigma_1(T)$. However, one can still draw the conclusion that for many materials $\sigma_1(T)$ rises from a residual value at 1.2 K that is an order of magnitude larger than the predicted σ_{00} and that the temperature dependence is weak and either linear or sublinear in T.

The low temperature behaviour of $\sigma_1(T)$ is much less ambiguous in YBa$_2$Cu$_3$O$_{6.99}$ because the shape of $R_s(T)$ is controlled much more by $\sigma_1(T)$ than by $\lambda(T)$. One still makes two observations related to those above—$\sigma_1(T)$ has a residual value at 1.2 K that is an order of magnitude higher than σ_{00} and the temperature dependence is either linear or slightly sublinear. There has been considerable theoretical progress over several years in understanding why the typical behaviour of $\sigma_1(T)$ seems so different from the initial expectation of a d-wave superconductor. The problem is that $\sigma_1(T) \propto T$ shown in Figure 4.22 and a Drude lineshape for $\sigma_1(\omega)$ suggests a temperature and frequency independent $1/\tau_{\text{imp}}$, which is the expectation in a normal metal, but not the expectation for the reduced phase space available for scattering nodal quasiparticles in a d-wave superconductor. Figure 4.39 shows the frequency dependence of $1/\tau(\omega)$ expected from a T-matrix treatment of point scattering of nodal quasiparticles, calculated by Hirschfeld et al. [76]. For unitary scattering a resonance develops at $\omega = 0$ which has been seen experimentally by scanning tunnelling spectroscopy (STS) measurements on

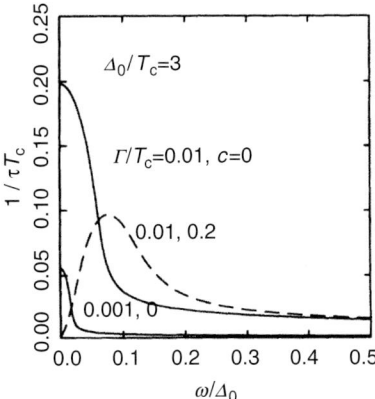

Figure 4.39. The impurity relaxation rate calculated in a T-matrix approximation for point scatterers. In the unitary limit ($c = 0$) there is a peak at $\omega = 0$ in $1/\tau(\omega)$ and away from unitary scattering the peak moves out to finite energy [76].

Zn-doped $Bi_2Sr_2CaCu_2O_{8+\delta}$ [39] and which gives rise to a $1/\tau(\omega)$ peaked at $\omega = 0$. Away from unitary scattering, the resonance moves out to finite frequency, as seen in STS on Ni impurities, and this moves the peak in $1/\tau(\omega)$ to finite frequency. Whatever the phase shift, one does not expect the ω-independent $1/\tau(\omega)$ implied by $\sigma_1(\omega, T)$ in $YBa_2Cu_3O_{6.99}$. Berlinsky et al. [14] quantified this problem by extracting a frequency-dependent quasiparticle self-energy from the data of Figure 4.22. They found that $\sigma(\omega, T)$ could be modelled with a $1/\tau(\omega)$ that was linear in ω as expected for non-unitary scatterers, but with a large, additive constant term. The constant term gives the nearly linear behaviour of $\sigma_1(T)$ and the linear term gives a slight frequency-dependent change in curvature seen in the data.

We will return to recent improvements on the scattering models, but first will show data a little closer to the expectations for weak point scatterers. Figure 4.40 shows a detailed view of the low temperature conductivity spectrum for high purity $YBa_2Cu_3O_{6.52}$ in the Ortho-II ordered state where every other CuO chain is empty. The data have the shape expected for Born-limit scattering: the peak is cusp-shaped, with a temperature-independent value of σ_0, a temperature dependent width Γ, and a power law frequency dependence slower than ω^2. These features are consequences of measuring in a regime where the temperature is above the range where one would see the universal conductivity ($k_B T \gg \gamma$), but at low enough temperature that one is only sampling the range where $1/\tau(\omega) \propto \omega$ (Figure 4.39). The damping Γ has a linear temperature dependence as expected for Born scattering, but deviates from this by tending towards a finite width as $T \to 0$. Put another way, the curves scale as $\omega/(T + T_0)$ rather than the ω/T scaling expected for Born scattering. This suggests some residual oscillator strength, of order a few percent of the superfluid oscillator strength, that does not condense, reminiscent of the uncondensed oscillator strength in other materials, though it is much smaller in magnitude.

Figure 4.41 shows similar broadband data for the $\hat{a}$-axis of $YBa_2Cu_3O_{6.99}$. Here one sees a similar cusp-shaped spectrum developing only at the lowest temperatures and $\sigma_1(\omega)$ evolving into more Drude like curves at higher temperature, as was suggested by the fixed frequency data. The $YBa_2Cu_3O_{6.52}$ broadband data seems to be in the regime where $1/\tau(\omega) \propto \omega$ for the measured temperatures 1.2–7 K, whereas the $YBa_2Cu_3O_{6.99}$ data leaves this regime above 3 K. This could occur if the scattering lies neither in the Born limit nor the unitary

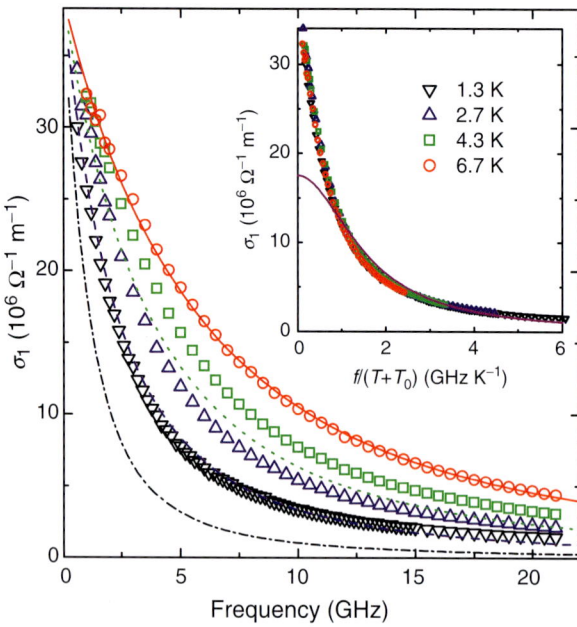

Figure 4.40. $\sigma_1(\omega, T)$ in the a-direction of the Ortho-II phase of $YBa_2Cu_3O_{6.52}$ with every other CuO chain filled. The red solid line is a fit to the 6.7 K of a model for Born scattering in a d-wave superconductor. The fit captures the cusp-like shape, but fails to predict the correct evolution as a function of temperature. The data has a slower narrowing with temperature than the Born model which approaches zero width as $T \to 0$ ($f/(T+T_0)$) scaling rather than the f/T scaling of the Born model [202]. The inset shows the complete failure of a Drude lineshape to fit these spectra.

limit, in which case the position of the resonance in the density of states and $1/\tau(\omega)$ would depend upon the scattering phase shift and the size of the energy gap Δ_0, both of which could differ between these samples. So, the most obvious step to take in explaining the microwave conductivity is to move away from the simple limits of Born and unitary scattering. As mentioned above and indicated in Figure 4.38, this move has also been suggested by measurements of thermal conductivity at low temperatures. Hill et al. [72] resolved the T^2 deviation away from the universal thermal conductivity limit in samples similar to those used in the microwave measurements above. The crossover temperature dependence suggests that the crossover energy scale is $\gamma \sim 0.25$ K for these high purity crystals. This crossover is too low in temperature to be generated by a sensible number of unitary scatterers (it would imply a mean free path of millimetres) and too high in temperature to be the exponentially small γ expected for Born scattering.

There have been a number of attempts to explain the microwave conductivity in terms of scatterers with a phase shift between the Born and unitary limits [58, 69, 172]. As a recent example, Schachinger et al. [170, 172] focussed in particular on the $YBa_2Cu_3O_{6+x}$ spectroscopic data shown in Figures 4.40 and 4.41 and were able to produce the different spectral shapes observed. However, they were not able to produce the extra residual spectral weight seen at 1.2 K and thus are also unable to completely reproduce the evolution of the conductivity spectra with temperature. Dwelling on the residual spectral weight at 1.2 K and the temperature dependence is more than just a matter of sorting out fine details; in most of the cuprates other than the $YBa_2Cu_3O_{6+x}$ system there tends to be much larger uncondensed oscillator strength

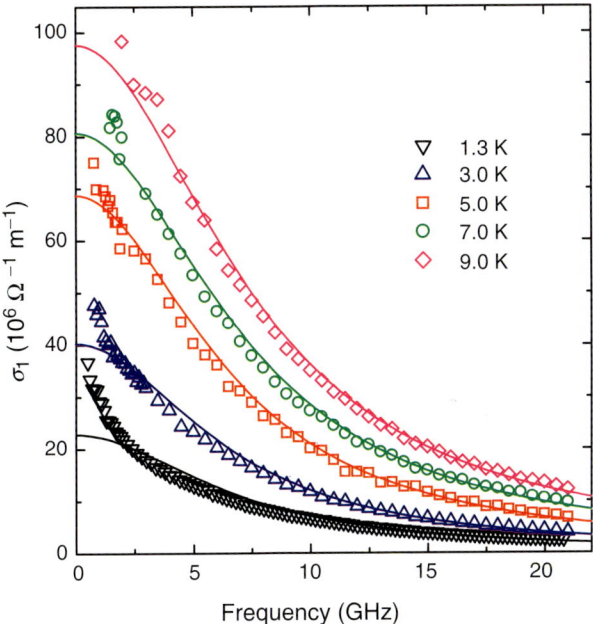

Figure 4.41. The conductivity spectrum of a fully doped sample of $YBa_2Cu_3O_{6.99}$ in the $\hat{a}$-direction. The Drude fits to the spectra highlight the evolution from a cusp-like shape to a more Lorentzian line shape with increasing temperature [202].

at 1.2 K, dominating the microwave and THz properties and presumably other physical properties as well. This oscillator strength is apparent in Figures 4.32 and 4.34 for the $Tl_2Ba_2CuO_{6+\delta}$ and $Bi_2Sr_2CaCu_2O_{8+\delta}$ systems. In $Bi_2Sr_2CaCu_2O_{8+\delta}$ thin films it has been studied in some detail at THz frequencies [36] and Orenstein has attributed it to a collective mode associated with inhomogeneity [143], a natural direction to take in light of the mesoscale inhomogeneity observed in scanning tunnelling spectroscopy measurements on this material.

A related direction taken to explain the microwave conductivity spectra has been to consider more complex scattering models than the T-matrix treatment of point scatterers. For instance, a number of authors have taken the direction of considering the local suppression of the superconducting order parameter around an impurity site, rather than treating impurity effects as spatially homogeneous [49,70,224]. Most recently, there has been considerable attention paid to off plane disorder, rather than point scatterers lying within a CuO_2 plane [1,225]. This direction is given experimental backing by the recent correlation found between the location of interstitial oxygen dopant atoms and the patchy electronic spectra seen in scanning tunnelling spectroscopy by McElroy et al. [132]. Nunner et al. [140] have taken an approach that includes both point scatterers and extended scatterers due to off-plane disorder, in order to try to explain the puzzles and diverse behaviour of the microwave conductivity of the cuprates. This approach is remarkably successful on several fronts; it comes very close to the correct temperature and frequency dependence of σ_1 in clean, fully-doped $YBa_2Cu_3O_{6.99}$ yet is also able to explain the quite different behaviour of $Bi_2Sr_2CaCu_2O_{8+\delta}$. The success in modelling the $Bi_2Sr_2CaCu_2O_{8+\delta}$ microwave data suggests that this inclusion of off-plane disorder can account for the apparent large oscillator strength observed at 1.2 K in that material. While not the same physics as the normal mode in an inhomogeneous

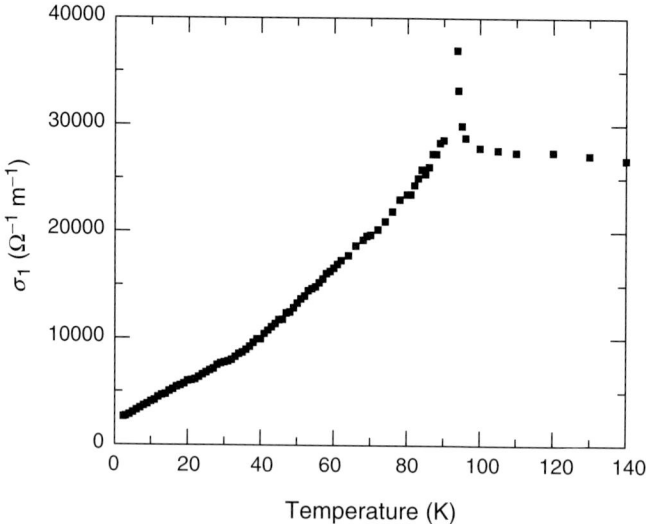

Figure 4.42. The $\hat{c}$-axis microwave conductivity of a sample of $YBa_2Cu_3O_{6.95}$ measured at 18 GHz [83].

superconductor suggested by Orenstein, the direction is related. Namely, one needs to consider the real materials and their defects if one is to make sense of comparisons across the entire cuprate family. This is particularly important in light of the fact that so much of our understanding of the cuprates in recent years has come from surface sensitive probes (ARPES and STS) performed on $Bi_2Sr_2CaCu_2O_{8+\delta}$, a material that seems far away from high purity samples of the $YBa_2Cu_3O_{6+x}$ system.

4.6.5. Anisotropy

As with the out-of-plane superfluid response, the out-of-plane conductivity has been studied much less than the in-plane and there are still unresolved issues of differences between samples and families of compounds. The technical difficulty of isolating the out of plane response has been tackled in several different ways: by changing the orientation of a single sample to compare measurements with $\vec{H}_{rf} \parallel \hat{c}$ and $\vec{H}_{rf} \perp \hat{c}$, by cleaving or polishing a sample to vary the $\hat{c}$-axis contribution (Figure 4.11), and by measuring in a microwave electric field with $\vec{E}_{rf} \parallel \hat{c}$. As discussed in Section 4.5.6, changing the orientation of a thin platelet sample runs into difficulties with severe changes in demagnetizing factors, leading to difficulty in separating in-plane and out-of-plane contributions. Mao et al. [130] ran into this difficulty in $\hat{c}$-axis measurements of superfluid density, which showed the linear temperature dependence seen in-plane, rather than the nearly T^2 seen in most subsequent measurements. Similarly, the peak they observed in $\sigma_1(T)$ for $\hat{c}$-axis currents might be dismissed as a contribution from the peak in the in-plane conductivity. However, when Kitano et al. [100] sought to avoid this by using thick samples with similar demagnetizing factors for different sample orientations, they also observed a broad peak in $\sigma_1(T)$ for a sample of $YBa_2Cu_3O_{6+x}$ near optimal doping. Two underdoped samples showed conflicting behaviour, one had a peak and one had $\sigma_1(T)$ falling rapidly and monotonically below T_c.

Hosseini et al. sought to extract $\sigma_1(T)$ in the $\hat{c}$-direction by cleaving a thin plate, as illustrated in Figure 4.11. The optimally doped sample in that case showed a precipitous drop

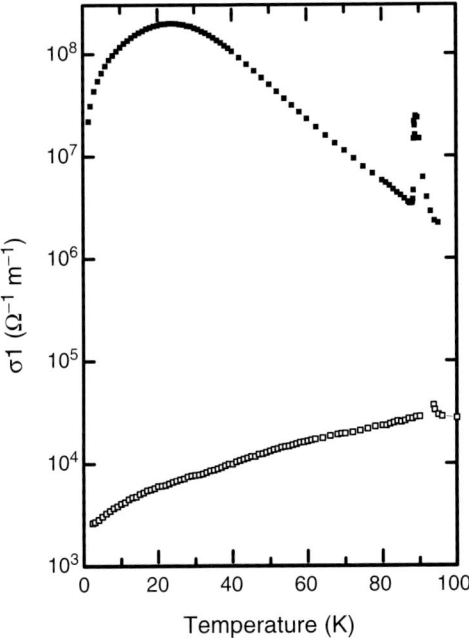

Figure 4.43. The anisotropy of the microwave conductivity of $YBa_2Cu_3O_{6.95}$ is illustrated by plotting the $\hat{c}$-axis conductivity taken from Figure 4.42 (open squares) along with the $\hat{a}$-axis conductivity (filled squares) at 1 GHz taken from Figure 4.22 [83]. Note that while the $\hat{a}$-axis conductivity is strongly frequency dependent and the 1 GHz data is close to the zero frequency limit, *no* dependence on the microwave frequency is expected for the $\hat{c}$-axis conductivity.

and then a weak rise at low temperatures [85]. Because of concerns that the process of cleaving a thin crystal might introduce problems, Hosseini later tried polishing a relatively thick sample into a thin blade in order to make the $\hat{c}$-axis contribution dominate a measurement with $\vec{H}_{rf} \parallel \hat{a}$ [83]. Further thinning of the sample and remeasurement allowed an unambiguous extraction of the $\hat{c}$-axis conductivity, shown in Figure 4.42. The more recent measurement also shows a rapid drop below T_c, but without the weak rise again at low temperatures, suggesting the latter is an artefact of the cleaving. The cause of the sample dependence observed by Kitano et al. and the disagreement with Hosseini et al. is not clear at the present time. One possibility is that the tunnelling process responsible for $\hat{c}$-axis transport is highly sensitive to the details of the intervening layers, including impurity effects, oxygen content and oxygen order, but more systematic measurements would be needed to resolve this. Notwithstanding the fine details, the microwave measurements all show that the conductivity continues to be very anisotropic in the superconducting state, as illustrated in Figure 4.43. If the hopping matrix element, with its strong dependence on the in-plane momentum, is indeed a controlling factor in $\hat{c}$-axis transport properties, then this anisotropy to some extent reflects the difference between nodal quasiparticles and excitations away from the vicinity of the nodes. The in-plane conductivity is dominated by nodal quasiparticles that can develop exceptionally long transport lifetimes. However, the anisotropic hopping suppresses the $\hat{c}$-axis transport of these nodal quasiparticles.

Xiang and Hardy [215] applied the cold spot scattering model of Ioffe and Millis [88] to a calculation of σ_1 for the $\hat{c}$-axis, taking account of the $k_\parallel$ dependence of the interlayer hopping integral $\epsilon_\perp$ (cf. section 4.5.6). They obtained a T^3 dependence for σ_{1c} at intermediate temperatures, which agreed with the $Bi_2S_2CaCu_2O_{8+\delta}$ data of Latyshev et al. [113] and the

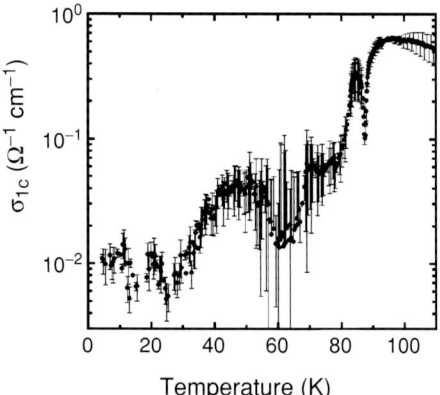

Figure 4.44. The microwave conductivity of $Bi_2S_2CaCu_2O_{8+\delta}$ is much lower than that found in the $YBa_2Cu_3O_{6+x}$ system, requiring measurements in microwave electric field on small samples to separate the $\hat{c}$-axis conductivity [99].

early data of Hosseini et al. [85] on $YBa_2Cu_3O_{6.95}$. However, the improved experimental data on $YBa_2Cu_3O_{6+x}$ (see Figure 4.42) has a weaker T dependence, so that the comparison with experiment needs to be revisited.

The rest of the cuprates are even more anisotropic than the $YBa_2Cu_3O_{6+x}$ samples discussed above, both in their normal state transport properties and in the anisotropy of λ. $Bi_2S_2CaCu_2O_{8+\delta}$ is so anisotropic that quite different microwave techniques must be used, taking into account that the $\hat{c}$-axis properties are close to being those of a dielectric rather than a conductor, and that the $\hat{c}$-axis penetration depth is of order the typical sample size. Kitano et al. [99] tackled this problem by working with small samples in microwave electric fields aligned along the $\hat{c}$-axis. The behaviour of one such sample is shown in Figure 4.44, where one notes not only the drop in the microwave conductivity below T_c, but also the very low value of the conductivity overall.

4.7. Fluctuations

Shortly after the discovery of the high temperature superconductors, it was pointed out [124] that the region of temperature around T_c where mean-field theories (BCS; Ginzberg–Landau) break down and fluctuation effects become noticeable, would be substantial. In the case of conventional superconductors where the coherence lengths are of order hundreds of Å, the number of Cooper pairs within a coherence volume is enormous and mean field behaviour is both expected and observed. For the high temperature superconductors ξ_{ab} and ξ_c are about 15 and 3 Å, respectively, giving much smaller coherence volumes and correspondingly wider critical regions.

The universality class for a 3D superconductor with a complex order parameter $\Psi = |\Psi|e^{i\phi}$ is 3D-XY, the same as for liquid ^{4}He, except extremely close to T_c where the charged nature of the superfluid enters [48]. Fisher et al. estimated the location of the crossover from mean-field to critical behaviour by setting the GL expression for λ^2/ξ equal to the 3D-XY value. In the critical region with reduced temperature $t = (T_c - T)/T_c$, for $T < T_c$, the crossover value of t, t_x is given by

$$t_x = \left[\frac{\lambda_{\perp 0}^2}{\xi_{\perp 0}} \frac{\pi c_s}{\gamma \Lambda_{T_c}} \right]^2,$$

where $\lambda_{\perp 0}$ is the zero temperature in-plane magnetic penetration depth, $\xi_{\perp 0}$ is the in-plane coherence length, c_s is a universal parameter estimated to be 0.4, $\gamma = (m_\perp/m_z)^{\frac{1}{2}}$ and $\Lambda_T = \frac{\Phi_0^2}{16\pi^2 T}$. Taking $\gamma \simeq 0.2$, $\kappa = \frac{\lambda_{\perp 0}}{\xi_{\perp 0}} \simeq 100$, $\lambda_{\perp 0} \simeq 1,600\,\text{Å}$ one obtains $t_x \simeq 0.2$ which translates to 20 K for $T_c = 100$ K. Allowing the choice of parameters to vary somewhat, values of order 10^{-2} to 10^{-1} can result. This form of the Ginzberg criterion suggests a rather wide fluctuation regime. Fisher et al. [48] acknowledge that alternative criteria exist for the crossover which yield much narrower critical regions. Lobb [124], for example, gave estimates that varied from 0.1 K for his best estimates of H_{c2} and κ, up to 1 K for more speculative values of these parameters. Experimentally, it appears that the fluctuation regime is wide, perhaps as much as 10 K for optimally doped YBa$_2$Cu$_3$O$_{6+x}$, with critical exponents close to the 3D-XY values. However, this expected behaviour has not been universally observed, and discrepancies between thin films and single crystals seem to be common.

Early transport and thermodynamic properties in zero applied field were initially treated in terms of Gaussian fluctuations, but later analyses suggested critical scaling in the 3D-XY universality class. Later, even clearer indications were observed in specific heat measurements made in applied magnetic fields. (See for example Overend et al. and references therein [147].) In these types of experiments the fluctuation signal is usually superposed on a large background, the substraction of which tends to be problematic. A notable exception was the thermal expansivity study of Pasler et al. [153] where by using the difference in the $\hat{a}$- and $\hat{b}$-axis expansivity, the background was largely suppressed. For optimally doped YBa$_2$Cu$_3$O$_{6+x}$ single crystals, critical fluctuations consistent with 3D-XY scaling were observed over the temperature range $|T - T_c| = 10$ K , with nearly equal amplitudes above and below T_c .

For the microwave properties, critical fluctuations will contribute to both the real and imaginary components of σ. We first note that σ_1, which is controlled by the non-universal dynamical critical exponent z, is the more problematic quantity to study. First of all, σ_1 quickly becomes much smaller than σ_2 below T_c and increasingly difficult to extract from R_s. We also note that the fluctuation contribution to σ_1 appears above a background conductivity that has to be subtracted. On the other hand $\sigma_2 \propto 1/\lambda^2$ has the advantage that $1/\lambda^2$ tends to zero as $T \to T_c$ in a manner determined solely by critical fluctuations, without any background subtractions required to be made. Note that for most of this discussion we are ignoring the finite measurement frequency as far as $1/\lambda^2$ is concerned, since the data is fit far enough away from T_c that finite frequency effects are negligible. Later in this section, where the data of Kamal et al. [93] is examined very close to T_c (Figure 4.47c), the effect of the finite measuring frequency becomes apparent. For the superfluid density or equivalently $1/\lambda^2$ in the critical region one has

$$\frac{1}{\lambda^2} \propto t^\nu, \tag{4.18}$$

where $\nu/2 = 0.333$ for 3D-XY vs. 0.5 for mean field. Thus, for mean field behaviour $1/\lambda^2$ should approach zero linearly as $T - T_c$, with a finite slope, whereas for 3D-XY behaviour the slope should become infinite. In particular $1/\lambda^3$, not $1/\lambda^2$, should be linear with $T - T_c$.

Kamal et al. [92] made careful measurement of $\Delta\lambda(T)$ in very high quality optimally doped YBa$_2$Cu$_3$O$_{6+x}$ crystals using cavity perturbation at 900 MHz. Figure 4.45 compares $1/\lambda^3$ vs. $1/\lambda^2$ and one sees that 3D-XY behaviour seems to be followed over a rather large temperature interval, as much as 10 K. Figure 4.46 is a log–log plot of $1/\lambda(t)$ vs. reduced

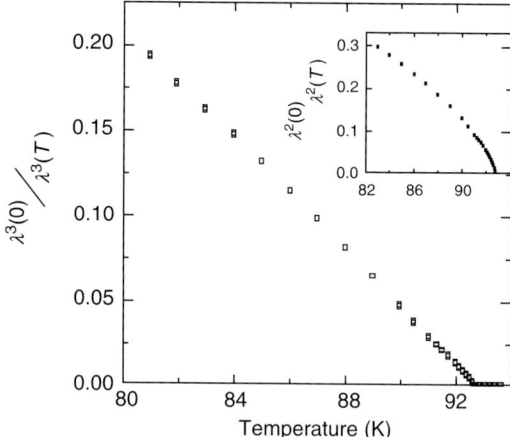

Figure 4.45. $1/\lambda^2(T)$ of a single crystal of $YBa_2Cu_3O_{6.95}$ [92] shows a power law different from the $(T_c - T)/T_c$ behaviour expected for a mean field phase transition.

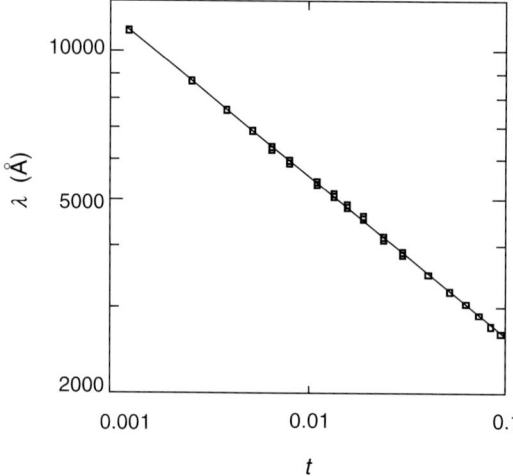

Figure 4.46. A log–log plot near T_c of $\lambda(T)$ for a single crystal of $YBa_2Cu_3O_{6.95}$ [92] displays a power law behaviour consistent with the 3D-XY critical exponent $\nu/2=0.33$.

temperature. A fit to the data gave $\nu/2 = 0.33 \pm 0.01$ consistent with 3D-XY scaling. The uncertainty in the exponent represents the range of values obtained for the range of choices of T_c. For these types of measurements, one needs to obtain a value of $\lambda(T = 0)$ from other measurements, and a representative value of 1,400 Å was taken from μSR and Far IR experiments. Later analysis, allowing $\lambda(T = 0)$ to vary from 1,300 to 1,500 Å, but fitting over the same temperature interval, yielded values of $\nu/2$ from 0.343 to 0.355. Data from cavity perturbation measurements at 22.7 GHz gave values from 0.328 to 0.351.

The recent measurements of λ using zero field ESR of Gd-doped $YBa_2Cu_3O_{6+x}$ of Pereg-Barnea et al. [157] suggest that the literature values of $\lambda(T = 0)$ used by the UBC group to convert their $\Delta\lambda(T)$ measurements to absolute $\lambda(T)$, were probably too high. For the optimally doped $YBa_2Cu_3O_{6+x}$ twinned crystal used by Kamal et al., we estimate that

$\lambda_{ab}(0)$ may have been as low as 1,200 Å. This has the effect of decreasing the apparent width of the fluctuation regime somewhat, but does not shift the values of $\nu/2$ very much. More recent measurements on highly ordered Ortho-II crystals with $T_c = 56$ K also showed a wide fluctuation regime with critical exponents close to 0.33 for $1/\lambda(T)$ [93].

Anlage et al. [7] presented microwave results for both σ_1 and σ_2 in optimally doped $YBa_2Cu_3O_{6+x}$, reporting a large asymmetry in the width of the fluctuation regime, being up to 5 K below T_c, but less than 0.6 K above T_c. Their results for λ were consistent with 3D-XY scaling and thus in agreement with the results of Kamal et al. [92]. In contrast, Paget et al. [148] found 3D-XY behaviour to be absent in their measurements on $YBa_2Cu_3O_{6+x}$ thin films, made by a variety of methods; sputtering co-evaporation and pulsed-laser deposition. They concluded that the critical region, if it existed, was less than 0.5 K and perhaps less than 0.2 K.

This small sampling of the many published results for λ^{-2} delineates the problem that existed until the late 1990s with the understanding of fluctuations: specific heat and thermal expansivity seemed to show a wide region of 3D-XY behaviour on both sides of T_c, some microwave measurements on single crystals agreed on a wide fluctuation region below T_c, whereas most low frequency thin film measurements showed no discernable critical fluctuation effects. The situation has evolved substantially in the past few years. For example Osborne et al. [146] report two-coil inductance measurements (10–100 kHz) on epitaxially grown Bi2212 films, finding static 3D-XY critical exponents in the superfluid density, and a dynamical critical exponent of $z = 2$. Most recently, the Stuttgart group has made extensive 9.5 GHz microwave conductivity measurements on $Bi_2Sr_2CaCu_2O_{8+\delta}$, $Bi_2Sr_2Ca_2Cu_3O_{10+\delta}$, and $YBa_2Cu_3O_{6+x}$ thin films, emphasizing the importance of short-wavelength cutoff effects [154, 177] in the theoretical modelling of the fluctuation conductivity [155]. The short-wavelength cutoff yields strong effects at higher temperatures above T_c, and a small but experimentally detectable feature at T_c. Previous microwave studies reporting both Gaussian and critical behaviour [7, 136, 208] did not include such effects in their analyses, and would have to be revisited.

The most ambitious study of fluctuations in the microwave region is that of Booth et al. [20] where the rf conductivity of a $YBa_2Cu_3O_{6+x}$ thin film was measured for $T > T_c$ over the frequency range 45 MHz–45 GHz. By a scaling analysis they obtained a dynamical critical exponent z of 2.3–3, larger than the value 2 expected for the "relaxational" version of the 3D-XY model. The quantity z is non-universal, so this result is of particular interest. However, their data yielded a static scaling exponent of about 1.0, at odds with the 3D-XY scaling value of 2/3, and therefore at odds with previous results of the same group on single crystals where 3D-XY scaling was observed [7].

It is the opinion of the present authors that there remain many unresolved issues, both theoretical and experimental. First of all, there remains the unsettling variation in the actual measurements: more attention needs to be paid to the quality and nature of the samples. In particular, the reasons for the fluctuation behaviour being so sensitive to sample defects needs to be properly elucidated. As a concrete example of these unresolved issues, we show in Figures 4.47 and 4.48, data of Kamal et al. [93] for σ_1 and σ_2 on an optimally doped $YBa_2Cu_3O_{6+x}$ crystal of particularly high quality. In Figure 4.47b, c we zoom in on the region near T_c by first expanding the vertical scale of the data (Figure 4.47b) and then the horizontal scale (Figure 4.47c).

In Figure 4.48 we compare the fluctuation peak, for the same crystal, but at two different frequencies (1.14 and 22.7 GHz). One expects the two peaks to occur at the same temperature and we presume the shift is due to a difference in the thermometer calibration for the two

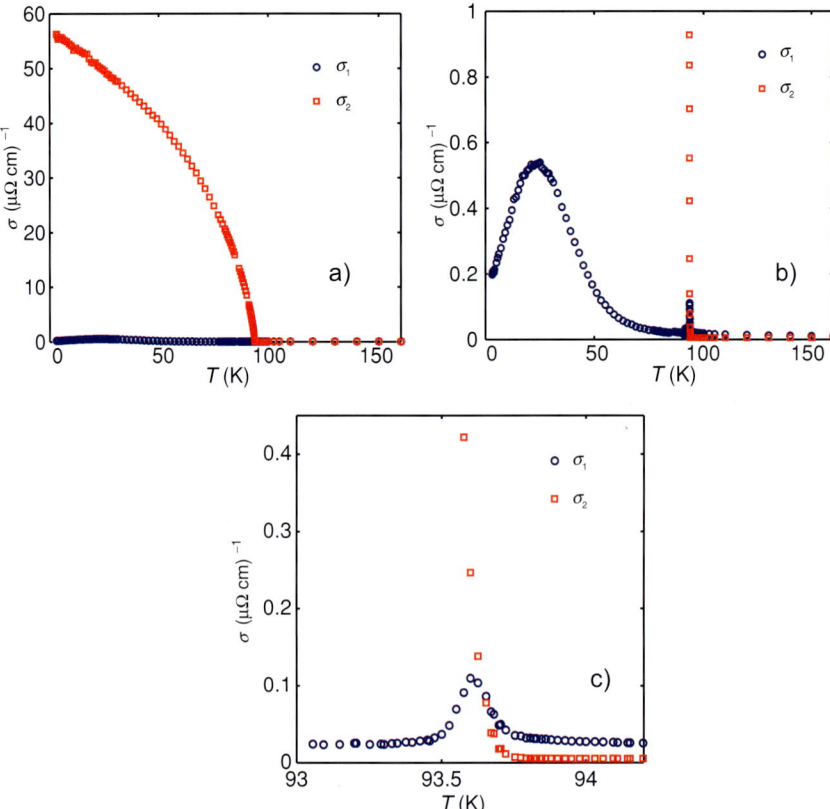

Figure 4.47. The real and imaginary parts of σ at 1.1 GHZ shown on three different scales to display the narrowness of the peak in $\sigma_1(T)$ near T_c in a single crystal of YBa$_2$Cu$_3$O$_{6.95}$ [93].

experimental setups. We call attention to the widths of the fluctuation peaks: about 0.07 K for $f = 1.14$ GHz and 0.3 K for $f = 22.7$ GHz. This is to be compared to approximately 10 K for the 9.5 GHz data of Peligrad et al. [155] on Bi$_2$Sr$_2$Ca$_2$Cu$_3$O$_{10+\delta}$, about 0.8 K for the 100 kHz data of Osborne et al. [146] on MBE grown Bi$_2$Sr$_2$CaCu$_2$O$_{8+\delta}$ and about 0.9 K for the 9.6 GHz data of Anlage et al. [7] on an optimally doped YBa$_2$Cu$_3$O$_{6+x}$ crystal. Clearly, there is a serious lack of universality: the width of the conductivity peaks at 9.5 GHz should be somewhere intermediate between the 1.14 and 22.7 GHz data rather than being much wider; the peak for the 100 kHz data should be much narrower than it is.

One might argue that Bi$_2$Sr$_2$CaCu$_2$O$_{8+\delta}$, Bi$_2$Sr$_2$Ca$_2$Cu$_3$O$_{10+\delta}$, and YBa$_2$Cu$_3$O$_{6+x}$ are different compounds and the non-universal quantities need not be the same. However, one expects that near T_c the underlying relaxation mechanism for the carriers should be dominated by the inelastic processes, and therefore similar in magnitude, since the dc resistivities above T_c for all of these materials are about the same. It is difficult to escape the conclusion that the observed widths of the conductivity fluctuation peaks in the published literature either have substantial contributions from a distribution of T_cs or are in fact dominated by them. One cannot expect the analysis of such data to accurately reflect intrinsic properties. The Harris criterion [66] by which disorder should not affect critical exponents in a system with heat capacity exponent $\alpha < 0$ does not apply to a macroscopic distribution of T_cs.

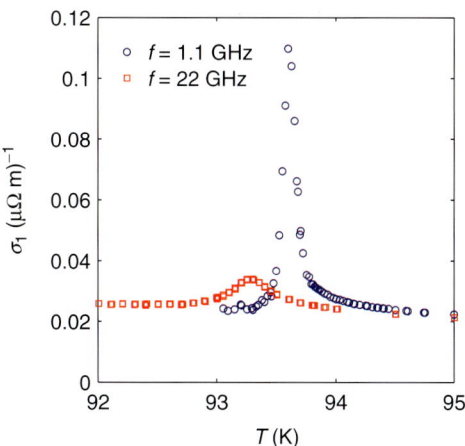

Figure 4.48. The peak in $\sigma_1(T)$ near T_c in a single crystal of $YBa_2Cu_3O_{6.95}$ at two different frequencies [93].

While it is not our intention to present a detailed analysis of the Kamal et al. data, there are some useful comments that can be made, under the reasonable assumption that lower frequency data will be more affected by a distribution of T_cs (because the intrinsic width of the fluctuation peak must be smaller). Under this assumption, one can conclude that the intrinsic width of the fluctuation peak measured at 22.7 GHz is close to the observed width of 0.3 K (because the 1.14 GHz data gives a substantially narrower peak of 0.07 K). If one evaluates the ratio $\sigma_1(1.14\,\text{GHz})/\sigma_1(22.7\,\text{GHz})$ at T_c one obtains a value of about 8.0. From the scaling result [211] $\sigma_1(T_c) \propto 1/\omega^{(z-1)/z}$, the data give a high value of 3.3 for the dynamical critical exponent z. A value closer to the "expected" value of 2 would require the fluctuation peak at 1.14 GHz to be almost a factor 2 *smaller* than observed. An observed fluctuation peak that is too high could *not* be the result of a distribution of T_cs and we conclude that there is fairly solid evidence that z is not 2, but something greater than 3 and perhaps greater than 3.3. Booth et al. [20] also reported values greater than 2 (2.35–3).

One surprise in these diverse results on different samples is that there has been no sign of 2D-XY critical behaviour in these highly anisotropic layered materials. The expectation in 2D is that when the phase stiffness determined from $1/\lambda^2(T)$ falls below a critical value, a Berezinski–Kosterlitz–Thouless (BKT) transition occurs [13, 107], a transition driven by a proliferation of vortices and exhibiting a discontinuous drop in phase stiffness at T_{2D}. The critical phase stiffness at which a superconducting sheet is expected to undergo such a transition is

$$d/\lambda^2 = (8\pi\mu_0/\Phi_0^2)T_{2D}, \tag{4.19}$$

where d is the thickness of the superconducting sheet. In the cuprates, evidence of such a transition has been seen by Zuev et al. [226] in very thin films of $YBa_2Cu_3O_{6+x}$ as shown in Figure 4.49. In the thin films they studied, a marked downturn in $\lambda^2(T)$ occurs in accordance with Eq. (4.19), but only if one takes the thickness of the superconducting layer to be the entire thickness of the film, d. For microwave and lower frequency probes, there seems to be no obvious evidence in the cuprates of such a downturn associated with the layer-spacing of the CuO_2 layers, which would be the expectation if the superconducting transition of bulk samples was governed by a Berezinski–Kosterlitz–Thouless (BKT) transition due to the nearly 2D nature of the materials. We should note, however, that Corson et al. [35], while looking

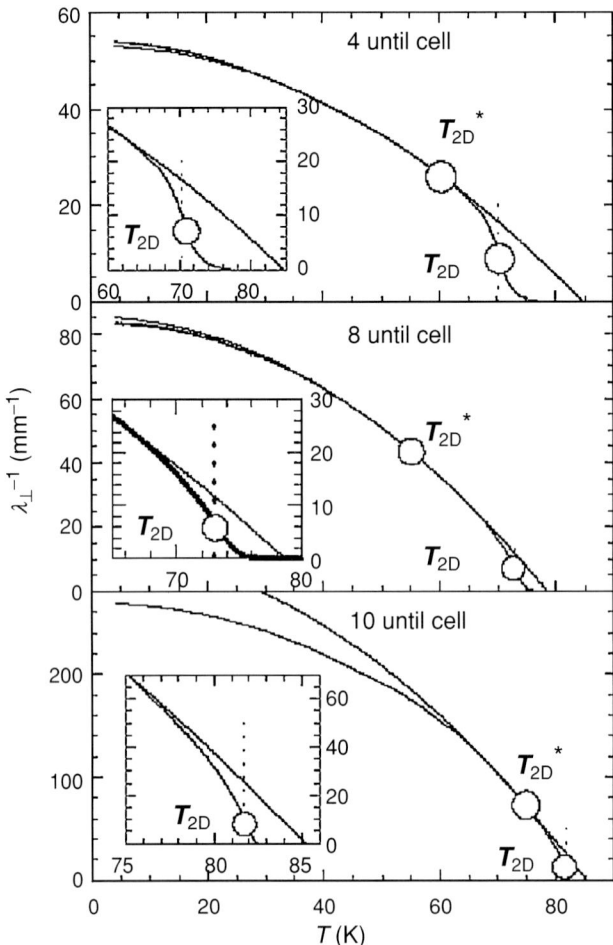

Figure 4.49. $1/\lambda_\perp = d/\lambda^2(T)$ measured at 50 kHz for ultra-thin, optimally doped YBa$_2$Cu$_3$O$_{6+x}$ films of varying thickness. The downturn associated with a BKT transition is indicated by the open circles labelled T_{2D} [226].

for evidence of preformed pairs and partial phase coherence above T_c, measured σ_1 and σ_2 at millimetre wave frequencies (100–600 GHz) and uncovered evidence for BKT behaviour in underdoped Bi$_2$Sr$_2$CaCu$_2$O$_{8+\delta}$. Their analysis relies on tracking the phase coherence time as a function of temperature and doping, and is not presented in a way that allows comparison with fluctuation measurements presented earlier in this section.

These results near T_c raise questions of what level of theory is required to handle the superconducting transition in the high temperature superconductors. They are never going to be as good a system as liquid ^{4}He in which to study fluctuation effects, because sample inhomogeneities will limit the range of reduced temperatures that can be reached and one will likely have to deal with non-universal behaviour. However, the importance of tackling the problem extends beyond simply understanding the superconducting transition in the cuprates, since superconducting fluctuations are thought to play a substantial role throughout the underdoped side of the cuprate phase diagram [31, 35].

Acknowledgments

The authors are indebted to the many people who have contributed to the research discussed in this chapter. We are particularly grateful to C. Kasier for assistance in preparing the manuscript. The work on high temperature superconductors at the University of British Columbia is the product of the combined efforts of our colleagues and students; particularly, Ruixing Liang, D. Broun, R. Harris, P. Turner, A. Hosseini, J. Bobowski, D. Peets, J. DeBenedictis, Kuan Zhang, D. C. Morgan, S. Kamal, B. Gowe, J. F. Carolan, D. J. Baar, P. Dosanjh, P. Schleger, T. Pereg-Barnea, G. Mullins, C. Bidinosti, K. Mossman, A. Wong, M. Hayden, and R. Knobel. We are also indebted to many others working in the field who have contributed to our understanding of the electrodynamics of superconductors; particularly, P.J. Hirschfeld, J. Orenstein, T. Timusk, C. C. Homes, D. Basov, A. J. Berlinsky, C. Kallin, N. Goldenfeld, D. J. Scalapino, P. A. Lee, A. Mills, S. Kivelson, J. Preston, T. Lemberger, T. Xiang, I. Affleck, M. Franz, J. Halbritter, R. Klemm, K. Scharnberg, E. Nicol, W. Atkinson, and J. P. Carbotte. Many complementary experimental studies by our collaborators have also informed this work, including the research led by K. A. Moler, N. P. Ong, L. Taillefer, R. Kiefl, J. Brewer, J. Sonier, B. Statt, B. Buyers, Z. X. Shen, G. A. Sawatzky, and A. Damascelli. Work at UBC has been supported by the Natural Science and Engineering Research Council of Canada and by the Canadian Institute for Advanced Research.

Bibliography

1. E. Abrahams and C. M. Varma, *Proc. Natl Acad. Sci.*, 97:5714, 2000.
2. D. Achkir, M. Poirier, D. A. Bonn, Ruixing Liang, and W. N. Hardy, *Phys. Rev. B*, 48:13184, 1993.
3. M. H. S. Amin, I. Affleck, and M. Franz, *Phys. Rev. B*, 58:5848, 1998.
4. Y. Ando and K. Segawa, *Phys. Rev. Lett.*, 88:167005, 2002.
5. A. Andreone, A. Cassinese, A. Di Chiara, R. Vaglio, A. Gupta, and E. Sarnelli, *Phys. Rev. B*, 49:6392, 1994.
6. S. M. Anlage, B. W. Langley, G. Deutscher, J. Halbritter, and M. R. Beasley, *Phys. Rev. B*, 44:9764, 1991.
7. S. M. Anlage, J. Mao, J. C. Booth, D. H. Wu, and J. L. Peng, *Phys. Rev. B*, 53:2792, 1996.
8. S. M. Anlage and D.-H. Wu, *J. Supercond.*, 5:395, 1992.
9. S. M. Anlage, D.-H. Wu, J. Mao, X. X. Xi, T. Venkatesan, J. L. Peng, and R.L. Greene, *Phys. Rev. B*, 50:523, 1994.
10. W. A. Atkinson, *Phys. Rev. B*, 59:3377, 1999.
11. W. A. Atkinson and J. P. Carbotte, *Phys. Rev. B*, 52:10601, 1995.
12. D. N. Basov, R. Liang, D. A. Bonn, W. N. Hardy, B. Dabrowski, M. Quijada, D. B. Tanner, J. P. Rice, D. M. Ginsberg, and T. Timusk, *Phys. Rev. Lett.*, 74:598, 1995.
13. V. L. Berezinskii, *Sov. Phys. JETP*, 32:493, 1971.
14. A. J. Berlinsky, D. A. Bonn, R. Harris, and C. Kallin, *Phys. Rev. B.*, 61:9088, 2000.
15. D. A. Bonn, P. Dosanjh, R. Liang, and W. N. Hardy, *Phys. Rev. Lett.*, 68:2390, 1992.
16. D. A. Bonn and W. N. Hardy, in *Physical Properties of High Temperature Superconductors*, edited by D. M. Ginsberg, Chapter 2, page 7, World Scientific, Singapore, 1996.
17. D. A. Bonn, S. Kamal, A. Bonakdarpour, R. Liang, W. N. Hardy, C. C. Homes, D. N. Basov, and T. Timusk, *Czechoslovak J. Phys.*, 46:3195, 1996.
18. D. A. Bonn, S. Kamal, K. Zhang, R. Liang, D. J. Baar, E. Klein, and W. N. Hardy, *Phys. Rev. B*, 50, 1994.
19. D. A. Bonn, Kuan Zhang, R. Liang, D. J. Baar, D. C. Morgan, and W. N. Hardy, *J. Supercond.*, 6:219, 1993.
20. J. M. Booth, D. H. Wu, S. B. Quadri, E. F. Skelton, M. S. Osofsky, A. Pique, and S. M. Anlage, *Phys. Rev. Lett.*, 77:4438, 1996.
21. A. V. Boris, N. N. Kovaleva, O. V. Dolgov, T. Holden, C. T. Lin, B. Keimer, and C. Bernhard, *Science*, 304:708, 2004.
22. E. H. Brandt, *Phys. Rev. B*, 54:4246, 1996.
23. F. Bridges, J. B. Boyce, T. Claeson, T. H. Geballe, and J. M. Tarascon, *Phys. Rev. B*, 42:2137, 1990.
24. F. Bridges, G. Li, J. B. Boyce, and T. Claeson, *Phys. Rev. B*, 48:1266, 1993.

25. S. D. Brorson, R. Buhleier, I. E. Trofimov, J. O. White, C. Ludwig, F. Balakirev, H. U. Habermeier, and J. Kuhl, *J. Opt. Soc. Am.*, 13:1979, 1996.
26. D. M. Broun, D. C. Morgan, R. J. Ormeno, S. F. Lee, A. P. Mackenzie, A. W. Tyler, and J. R. Waldram, *Physica C*, 282–287:1467, 1997.
27. D. M. Broun, D. C. Morgan, R. J. Ormeno, S. F. Lee, A. W. Tyler, A. P. Mackenzie, and J. R. Waldram, *Phys. Rev. B*, 56:11443, 1997.
28. D. M. Broun, P. J. Turner, W. Huttema, S. Ozcan, B. Morgan, Ruixing Liang, W. N. Hardy, and D. A. Bonn, *cond-mat/0509223*, 2005.
29. M. Brucherseifer, A. Meltzow-Altmeyer, P. H. Bolivar, H. Kurz, and P. Seidel, *Physica C*, 399:53, 2003.
30. J. Buan, B. P. Stojkovic, N. I. Sraeloff, A. M. Goldman, C. C. Huang, O. T. Valls, T. Jacobs, S. Sridhar, C. R. Shih, H. D. Yang, J-Z. Liu, and R. Shelton, *Phys. Rev. B*, 54:7462, 1996.
31. E. W. Carlson, S. A. Kivelson, V. J. Emery, and E. Manousakis, *Phys. Rev. Lett.*, 83:612, 1999.
32. M. Chaio, R. W. Hill, C. Lupien, L. Taillefer, P. Lambert, R. Gagnon, and P. Fournier, *Phys. Rev. B*, 62:3554–3558, 2000.
33. S. Chakravarty and P. W. Anderson, *Phys. Rev. Lett.*, 72:3859, 1994.
34. J. R. Cooper, *Phys. Rev. B*, 54:R3753, 1996.
35. J. Corson, R. Mallozzi, J. Orenstein, J. N. Eckstein, and I. Bozovic, *Nature*, 398:221, 1999.
36. J. Corson, J. Orenstein, Seongshik Oh, J. O'Donnell, and J. N. Eckstein, *Phys. Rev. Lett.*, 85, 2000.
37. R. Cubitt, E. M. Forgan, G. Yang, S. L. Lee, D. M. Paul, H. A. Mook, M. Yethiraj, P. Kes, T. Li, A. Menovsky, Z. Tarnawski, and K. Mortensen, *Nature*, 365:407, 1993.
38. U. Dähne, Y. Goncharov, N. Klein, N. Tellmann, G. Kozlov, and K. Urban, *J. Supercond.*, 8:129, 1995.
39. J. C. Davis, *Mater. Today*, 5:24, 2002.
40. L. A. de Vaulchier, J. P. Vieren, Y. Guldner, N. Bontemps, R. Combescot, Y. Lemaitre, and J. C. Mage, *Europhys. Lett.*, 33:153, 1996.
41. L. Drabeck, G. Gruner, J. J. Chang, A. Inam, X. D. Wu, L. Nazer, T. Venkatesan, and D. J. Scalapino, *Phys. Rev. B*, 40:7350, 1989.
42. M. Dressel, O. Klein, S. Donovan, and G. Gruner, *Int. J. Infrared Millim. Wave*, 14:2489, 1993.
43. D. Duffy, P. J. Hirschfeld, and D. J. Scalapino, *Phys. Rev. B*, 64:224522, 2001.
44. A. C. Durst and P. A. Lee, *Phys. Rev. B*, 62:1270, 2000.
45. H. Eisaki, N. Kaneko, D. L. Feng, A. Damascelli, P. K. Mang, K. M. Shen, Z. X. Shen, and M. Greven, *Phys. Rev. B*, 69:064512, 2004.
46. B. J. Feenstra, F. C. Klaassen, D. van der Marel, Z. H. Barber, R. P. Pinaya, and M. Decroux, *Physica C*, 278:213, 1997.
47. H. J. Fink, *Phys. Rev. B*, 58:9415, 1998.
48. D. S. Fisher, M. P. A. Fisher, and D. A. Huse, *Phys. Rev. B*, 43:130, 1991.
49. M. Franz, C. Kallin, A. J. Berlinsky, and M. I. Salkola, *Phys. Rev. B*, 56:7882, 1997.
50. J. Frenkel, F. Gao, Y. Liu, J. F. Whitaker, C. Uher, S. Y. Hou, and J. M. Phillips, *Phys. Rev. B*, 54:1355, 1996.
51. A. Fuchs, W. Prusseit, P. Berberich, and H. Kinder, *Phys. Rev. B*, 53:14745, 1996.
52. K. Fujita, T. Noda, K. M. Kojima, H. Eisaki, and S. Uchida, *Phys. Rev. Lett.*, 95:097006, 2005.
53. R. Gagnon, S. Pu, B. Ellman, and L. Taillefer, *Phys. Rev. Lett.*, 78:1976, 1997.
54. F. Gao, J. W. Kruse, C. E. Platt, M. Feng, and M. V. Klein, *Appl. Phys. Lett.*, 63:2274, 1993.
55. S. Giordano, H. Hahn, H. J. Halama, C. Varmazis, and L. Rinderer, *J. Appl. Phys.*, 44:4185, 1973.
56. N. E. Glass and W. F. Hall, *Phys. Rev. B*, 44:4495, 1991.
57. C. E. Gough and N. J. Xon, *Phys. Rev. B*, 50:488, 1994.
58. M. Graf, M. Palumbo, and D. Rainer, *Phys. Rev. B*, 52:10588, 1995.
59. M. J. Graf, S. K. Yip, J. A. Sauls, and D. Rainer, *Phys. Rev. B*, 53:15147, 1996.
60. P. H. H. Ackermann and H. J. Stockmann, in *HF Ints. of Radioact. Nuclei*, volume 31 of *Topics in Current Physics*, Chapter 5, page 291, Springer, Berlin, Heidelberg, New York, 1983.
61. J. Halbritter, *Z. Phys.*, 266:209, 1974.
62. P. C. Hammel, M. Takigawa, R. H. Heffner, Z. Fisk, and K. C. Ott, *Phys. Rev. Lett.*, 63:1992, 1989.
63. W. N. Hardy, in Proceedings of the 10th anniversary HTS workshop, edited by B.Batlogg et al., page 223, World Scientific, Singapore, 1996.
64. W. N. Hardy, D. A. Bonn, D. Morgan, R. Liang, and K. Zhang, *Phys. Rev. Lett.*, 70:3999, 1993.
65. W. N. Hardy, S. Kamal, and D. Bonn, in *The Gap Symmetry and Fluctuations in High-T_c Superconductors*, page 373, Plenum New York, 1998.
66. A. B. Harris, *J. Phys. C*, 7:1671, 1974.

67. R. Harris, P. J. Turner, Saeid Kamal, A. R. Hosseini, P. Dosanjh, G. K. Mullins, J. S. Bobowski, C. P. Bidinosti, D. M. Broun, Ruixing Liang, W. N. Hardy, and D. A. Bonn, *Phys. Rev. B*, 74:104508, 2006.
68. D. R. Harshman, G. Aeppli, E. J. Ansaldo, B. Batlogg, J. H. Brewer, J. F. Carolan, R. J. Cava, M. Celio, A. C. D. Chaklader, W. N. Hardy, S. R. Kreitzman, G. M. Luke, D. R. Noakes, and M. Senba, *Phys. Rev. B*, 36:2386, 1987.
69. S. Hensen, G. Muller, C. Rieck, and K. Scharnberg, *Phys. Rev. B*, 56:6237, 1997.
70. M. Hettler and P. Hirschfeld, *Phys. Rev. B*, 59:9606, 1999.
71. M. Hettler and P. Hirschfeld, *Phys. Rev. B*, 61:11313, 2000.
72. R. W. Hill, C. Lupien, M. Sutherland, E. Boaknin, D. G. Hawthorn, C. Proust, F. Ronning, L. Taillefer, R. Liang, D. A. Bonn, and W. N. Hardy, *Phys. Rev. Lett.*, 92:027001, 2004.
73. P. J. Hirschfeld and N. Goldenfeld, *Phys. Rev. B*, 48:4219, 1993.
74. P. J. Hirschfeld and W. O. Putikka, *Phys. Rev. Lett.*, 77:3909, 1996.
75. P. J. Hirschfeld, W. O. Puttika, and D. J. Scalapino, *Phys. Rev. Lett.*, 71:3705, 1993.
76. P. J. Hirschfeld, W. O. Puttika, and D. J. Scalapino, *Phys. Rev. B*, 50:10250, 1994.
77. P. J. Hirschfeld, P. Wolfle, J. A. Sauls, D. Einzel, and W. O. Puttika, *Phys. Rev. B*, 40:6695, 1989.
78. Holczer, L. Forro, L. Mihaly, and G. Gruner, *Phys. Rev. Lett.*, 67:152, 1991.
79. W. L. Holstein, L. A. Parisi, Z. Y. Shen, C. Wilker, M. S. Brenner, and J. S. Martens, *J. Supercond.*, 6:191, 1993.
80. C. C. Homes, D. A. Bonn, R. Liang, W. N. Hardy, D. N. Basov, T. Timusk, and B. P. Clayman, *Phys. Rev. B*, 60:9782, 1960.
81. C. C. Homes, T. Timusk, R. Liang, D. A. Bonn, and W. N. Hardy, *Phys. Rev. Lett.*, 71:1645, 1993.
82. M. L. Horbach, W. van Saarloos, and D. A. Huse, *Phys. Rev. Lett.*, 67:3464, 1991.
83. A. Hosseini, Ph.D., University of British Columbia, 2000.
84. A. Hosseini, D. M. Broun, D. E. Sheehy, T. P. Davis, M. Franz, W. N. Hardy, R. Liang, and D. A. Bonn, *Phys. Rev. Lett.*, 93:107003, 2004.
85. A. Hosseini, S. Kamal, D. A. Bonn, Ruixing Liang, and W. N. Hardy, *Phys. Rev. Lett.*, 81:1298, 1998.
86. N. E. Hussey, *Adv. Phys.*, 51:1685–1771, 2002.
87. A. Imtiaz and S. M. Anlage, *Ultramicroscopy*, 94:209, 2003.
88. L. B. Ioffe and A. J. Millis, *Phys. Rev. B*, 58:11631, 1998.
89. L. B. Ioffe and A. J. Millis, *J. Phys. Chem. Solids*, 63:2259, 2002.
90. T. Jacobs, S. Sridhar, Q. Li, G. D. Gu, and N. Koshizuka, *Phys. Rev. Lett.*, 75:4516, 1995.
91. B. Jia-shan, Z. Shi-ping, W. Ke-qing, L. Wei-gen, D. Ai-lei, and W. Shu-hong, *J. Supercond.*, 4:253, 1991.
92. S. Kamal, D. A. Bonn, N. Goldenfeld, P. J. Hirschfeld, Ruixing Liang, and W. N. Hardy, *Phys. Rev. Lett.*, 73:1845, 1994.
93. S. Kamal, D. A. Bonn, R. Liang, and W. N. Hardy, *unpublished*.
94. S. Kamal, R. Liang, A. Hosseini, D. A. Bonn, and W. N. Hardy, *Phys. Rev. B*, 58:8933, 1998.
95. R. Khasanov, D. G. Eshchenko, H. Luetkens, E. Morenzoni, T. Prokscha, A. Suter, N. Garifianov, M. Mali, J. Roos, K. Conder, and H. Keller, *Phys. Rev. Lett.*, 92:057602, 2004.
96. R. Khasanov, A. Shengelaya, E. Morenzoni, M. Angst, K. Conder, I. M. Savi, D. Lampakis, E. Liarokapis, A. Tatsi, and H. Keller, *Phys. Rev. B*, 68:220506, 2003.
97. J. H. Kim, H. S. Somal, M. T. Czyzyk, D. van der Marel, A. Wittlin, A. M. Gerrits, V. H. M. Duijn, N. T. Hien, and A. A. Menovsky, *Physica C*, 247:297, 1995.
98. M.-S. Kim, J. A. Skinta, T. R. Lemberger, A. Tsukada, and M. Naito, *Phys. Rev. Lett.*, 91:087001, 2003.
99. H. Kitano, T. Hanaguri, and A. Maeda, *Phys. Rev. B*, 57:10946, 1998.
100. H. Kitano, T. Shibauchi, K. Uchinokura, A. Maeda, H. Asaoka, and H. Takei, *Phys. Rev. B*, 51:1401, 1995.
101. N. Klein, in *Materials Science Forum*, volume 130–132 of *Materials Science Forum*, page 373, Trans Tech Publications, 1993.
102. N. Klein, U. Dahne, U. Poppe, N. Tellmann, K. Urban, S. Orbach, S. Hensen, G. Muller, and H. Piel, *J. Supercond.*, 5:195, 1992.
103. N. Klein, G. Muller, H. Piel, B. Roas, L. Schultz, U. Klein, and M. Peiniger, *Appl. Phys. Lett.*, 54:757, 1989.
104. O. Klein, K. Holczer, and G. Gruner, *Phys. Rev. Lett.*, 68:2407, 1992.
105. R. A. Klemm, K. Scharnberg, D. Walker, and C. T. Rieck, *Z. Phys. B*, 72:139, 1988.
106. J. D. Kokales, P. Fournier, L. V. Mercaldo, V. V. Talanov, R. L. Greene, and S. M. Anlage, *Physica C*, 341–348:1655, 2000; J. D. Kokales, P. Fournier, L. V. Mercaldo, V. V. Talanov, R. L. Greene, and S. M. Anlage, *Phys. Rev. Lett.*, 85:3696, 2000.
107. J. M. Kosterlitz and D. J. Thouless, *J. Phys. C*, 6:1181, 1973.
108. I. Kosztin and A. J. Leggett, *Phys. Rev. Lett.*, 79:135, 1997.

109. O. Kraut, C. Meingast, G. Brauchle, H. Claus, A. Erb, G. Muller-Vogt, and H. Wuhl, *Physica C*, 205:139, 1993.
110. K. Krishana, J. M. Harris, and N. P. Ong, *Phys. Rev. Lett.*, 75:3529, 1995.
111. K. Krishana, N. P. Ong, Y. Zhang, Z. A. Xu, R. Gagnon, and L. Taillefer, *Phys. Rev. Lett.*, 82:5108, 1999.
112. C. Kusko, Z. Zhai, N. Hakim, R. S. Markiewicz, S. Sridhar, D. Colson, V. Viallet-Guillen, A. Forget, Yu. A. Nefyodov, M. R. Trunin, N. N. Kolesnikov, A. Maignan, A. Daignere, and A. Erb, *Phys. Rev. B*, 65:132501, 2002.
113. Y. I. Latyshev, T. Yamashita, L. N. Bulaevskii, M. J. Graf, A. V. Balatsky, and M. P. Maley, *Phys. Rev. Lett.*, 82, 1999.
114. W. E. Lawrence and S. Doniach, *Proceedings of the 12th International Conference on L.T. Physics*, edited by B. Batlogg et al., page 223, Academic, New York, 1996.
115. J. Y. Lee, K. M. Paget, T. R. Lemberger, S. R. Foltyn, and X. Wu, *Phys. Rev. B*, 50:3337, 1994.
116. P.A. Lee, N. Nagaosa, and X.-G. Wen, *Rev. Mod. Phys.*, 78:17, 2006.
117. P.A. Lee and X. G. Wen, *Phys. Rev. Lett.*, 78:4111, 1997.
118. P. A. Lee, *Phys. Rev. Lett.*, 71:1887, 1993.
119. S.-C. Lee and S. M. Anlage, *IEEE Trans. Appl. Supercond.*, 13:3594, 2003.
120. S. F. Lee, D. Morgan, R.J. Ormeno, D. M. Broun, R. A. Doyle, J. R. Waldram, and K. Kadowakil, *Phys. Rev. Lett.*, 77:735, 1996.
121. R. Liang, D. A. Bonn, and W. N. Hardy, *Phys. Rev. B* 73:180505, 2006.
122. M. Lippert, J. P. Strobel, G. Saemann-Ischenko, S. Orbach, S. Hensen, G. Muller, H. Piel, J. Schutzmann, K. Renk, B. Roas, and W. Gieres, *Physica C*, 185–189:1041, 1991.
123. O. Llopis and J. Graffeuil, *J. Less Common Metals*, 164–165:1248, 1990.
124. C. J. Lobb, *Phys. Rev. B*, 36:3930, 1987.
125. Z. Ma, PhD Thesis.
126. Z. Ma, R. C. Taber, L. W. Lombardo, A. Kapitulnik, M. R. Beasley, P. Merchant, C. B. Eom, S. Y. Hou, and J. M. Phillips, *Phys. Rev. Lett.*, 71:781, 1993.
127. A. Maeda, H. Kitano, and R. Inoue, *J. Phys.: Condens. Matter*, 17:R143, 2005.
128. H. Maeda, A. Koizumi, N. Bamba, E. Takayama-Muromachi, F. Izumi, H. Asano, K. Shimizu, H. Moriwaki, H. Maruyama, Y. Kuroda, and H. Yamazaki, *Physica C*, 157:483, 1989.
129. S. Mair, B. Gompf, and M. Dressel, *Appl. Phys. Lett.*, 84:1219, 2004.
130. J. Mao, D. H. Wu, J. L. Peng, R. L. Greene, and S. M. Anlage, *Phys. Rev. B*, 51:3316, 1995.
131. J. S. Martens, V. M. Hietela, D. S. Ginley, T. E. Zipperian, and G. K. G. Hohenwarter, *Appl. Phys. Lett.*, 58:2543, 1991.
132. K. McElroy, J. Lee, J. A. Slezak, D. H. Lee, H. Eisaki, S. Uchida, and J. C. Davis, *Science*, 309:1048, 2005.
133. A. J. Millis, S. M. Girvin, L. B. Ioffe, and A. I. Larkin, *J. Phys. Chem. Solids*, 59:1742, 1998.
134. H. J. A. Molegraaf, C. Presura, D. van der Marel, P. H. Kes, and M. Li, *Science*, 295:2239, 2002.
135. B. Nachumi, A. Karen, K. Kojima, M. Larkin, G. M. Luke, J. Merrin, O. Tchernyshov, Y. J. Uemura, N. Ichikawa, M. Goto, and S. Uchida, *Phys. Rev. Lett.*, 77:5421, 1996.
136. G. Nakielski, D. Gorlitz, C. Stodte, M. Welters, A. Kramer, and J. Kotzler, *Phys. Rev. B*, 55:6077, 1997.
137. Y. A. Nefyodov, M. R. Trunin, A. A. Zhohov, I. G. Naumenko, G. A. Emelchenko, D. Y. Vodolazov, and I. L. Maksimov, *Phys. Rev. B*, 67:144504, 2003.
138. N. Newman and W. G. Lyons, *J. Supercond.*, 6:119, 1993.
139. H. Ning, H. Duan, P. D. Kirven, A. M. Hermann, , and T. Datta, *J. Supercond.*, 5:503, 1992.
140. T. S. Nunner and P. J. Hirschfeld, *Phys. Rev. B*, 72:014514, 2005.
141. M. C. Nuss, P. M. Mankiewich, M. L. O'Malley, E. H. Westerwick, and P. B. Littlewood, *Phys. Rev. Lett.*, 66:3305, 1991.
142. H. K. Olson and R. H. Koch, *Phys. Rev. Lett.*, 68:2406, 1991.
143. J. Orenstein, *Physica C*, 390:243, 2003.
144. J. Orenstein, J. Bokor, E. Budiarto, J. Corson, R. Mallozzi, I. Bozovic, and J. N. Eckstein, *Physica C*, 282–287:252, 1997.
145. R. J. Ormeno, C. E. Gough, and G. Yang, *Phys. Rev. B*, 63:104517, 2001.
146. K. D. Osborne, D. J. V. Harlingen, V. Aji, N. Goldenfeld, S. Oh, and J. N. Eckstein, *Phys. Rev. B*, 68:144516, 2003.
147. N. Overend, M. A. Howson, and I. D. Lawrie, *Phys. Rev. Lett.*, 72:3238, 1994.
148. K. M. Paget, B. R. Boyce, and T. R. Lemberger, *Phys. Rev. B*, 59:6545, 1999.
149. C. Panagopoulos, J. R. Cooper, G. B. Peacock, I. Gameson, P. P. Edwards, W. Schmidbauer, and J. W. Hodby, *Phys. Rev. B*, 53:2999, 1996.
150. C. Panagopoulos, J. R. Cooper, and T. Xiang, *Phys. Rev. Lett.*, 79:2320, 1997.

151. C. Panagopoulos, T. Xiang, W. Anukool, J. Cooper, Y. S. Wang, and C. W. Chu. *Phys. Rev B*, 67:220502, 2003.
152. B. Parks, S. Spielman, J. Orenstein, D. T. Nemeth, F. Ludwig, J. Clarke, P. Merchant, and D. J. Lew, *Phys. Rev. Lett.*, 74:3265, 1995.
153. V. Pasler, P. Schweiss, C. Meingast, B. Obst, H. Wuhl, A. I. Rykov, and S. Tajima, *Phys. Rev. Lett.*, 81:1094, 1999.
154. D.-N. Peligrad, M. Merhring, and A. Dulcic, *Phys. Rev. B*, 67:174515, 2003.
155. D.-N. Peligrad, M. Merhring, and A. Dulcic, *Phys. Rev. B*, 69:144516, 2004.
156. D.-N. Peligrad, B. Nebendahl, M. Merhring, A. Dulcic, M. Pozek, and D. Paar, *Phys. Rev. B*, 64:224504, 2001.
157. T. Pereg-Barnea, P. J. Turner, R. Harris, G. K. Mullins, J. S. Bobowski, M. Raudsepp, R. Liang, D. A. Bonn, and W. N. Hardy, *Phys. Rev. B*, 69:184513, 2004.
158. H. Piel and G. Muller, *IEEE Trans. Mag.*, 27:854, 1991.
159. A. Porch, M. J. Lancaster, and R.G. Humphreys, *IEEE Trans. MTT*, 43:306, 1995.
160. R. Prozorov and R. W. Gianetta, *Phys. Rev. Lett.*, 85:3700, 2000.
161. R. Prozorov, R. W. Gianetta, A. Carrington, and F. M. Araujo—Moreira, *Phys. Rev. B*, 62:115, 2000.
162. R. Prozorov, R. W. Gianetta, A. Carrington, P. Fournier, R. L. Greene, P. Guptasarma, D. G. Hinks, and A. R. Banks, *Appl. Phys. Lett.*, 77:4202, 2000.
163. R. C. Yu, M. B. Salamon and W. C. Lee, *Phys. Rev. Lett.*, 69:1431, 1992.
164. R. J. Radtke, V. N. Kostur, and K. Levin, *Phys. Rev. B*, 53:R522, 1996.
165. C. T. Rieck, K. Scharnberg, and J. Ruvalds, *Phys. Rev. B*, 60:12432, 1999.
166. C. T. Rieck, D. Straub, and K. Scharnberg, *J. Low Temp. Phys.*, 117:1295, 1999.
167. D. B. Romero, C. Porter, D. Tanner, L. Forro, D. Mandrus, L. Mihaly, G. L. Carr, and G. P. Williams, *Phys. Rev. Lett.*, 68:1590, 1992.
168. D. L. Rubin, K. Green, J. Gruschus, J. Kirchgessner, D. Moffat, H. Padamsee, J. Sears, Q. S. Shu, L. F. Schneemeyer, and J. V. Waszczak, *Phys. Rev. B*, 38:6538, 1988.
169. A. F. Santander-Syro, R. P. S. M. Lobo, N. Bontemps, Z. Konstatinovic, Z. Z. Li, and H. Raffy, *Europhys. Lett. B*, 62:568, 2003.
170. E. Schachinger and J. Carbotte, *Phys. Rev. B*, 67:134509, 2003.
171. E. Schachinger and J. P. Carbotte, *Phys. Rev. B*, 57:7970, 1998.
172. E. Schachinger and J. P. Carbotte, *Phys. Rev. B*, 65:064514, 2002.
173. D. Schoenberg, *Superconductivity*, Cambridge University Press, Cambridge, 1954.
174. D. E. Sheehy, T. P. Davis, and M. Franz, *Phys. Rev. B*, 70:054510, 2004.
175. T. Shibauchi, H. Kitano, K. Uchinokura, A. Maeda, T. Kimura, and K. Kishio, *Phys. Rev. Lett.*, 72:2263, 1994.
176. T. Shibauchi, A. Maeda, H. Kitano, T. Honda, and K. Uchinokura, *Physica C*, 203:315, 1992.
177. E. Silva, R. Marcon, S. Sarti, R. Fastampa, M. Giura, M. Boffa, and A.M. Cucolo, *Eur. Phys. J. B*, 37:277, 2004.
178. J. A. Skinta, M.-S. Kim, T. R. Lemberger, T. Greibe, and M. Naito, *Phys. Rev. Lett*, 88:207005, 2002.
179. J. A. Skinta, T. R. Lemberger, T. Greibe, and M. Naito, *Phys. Rev. Lett*, 88:207003, 2002.
180. A. Snezhko, R. Prozorov, D. D. Lawrie, R. W. Giannetta, J. Gauthier, J. Renaud, and P. Fournier, *Phys. Rev. Lett.*, 92:157005, 2004.
181. J. E. Sonier, J. H. Brewer, R. F. Kiefl, D. A. Bonn, S. R. Dunsiger, W. N. Hardy, Ruixing Liang, W.A. MacFarlane, R. I. Miller, T. M. Riseman, D. R. Noakes, C. E. Stronach, and M. F. White Jr., *Phys. Rev. Lett.*, 79:2875, 1997.
182. J. E. Sonier, R. F. Kiefl, J. H. Brewer, D. A. Bonn, S. R. Drensiger, W. N. Hardy, Ruixing Liang, W. A. MacFarlane, and T. M. Riseman, D. R. Noakes and C. E. Stronach, *Phys. Rev. B* 55:11789, 1997.
183. S. Spielman, B. Parks, J. Orenstein, D. T. Nemeth, J. Clarke, P. Merchant, and D. J. Lew, *Phys. Rev. Lett.*, 73:1537, 1994.
184. S. Sridhar and W. L. Kennedy, *Rev. Sci. Instrum.*, 59:531, 1988.
185. S. Sridhar, C. A. Shiffman, and H. Hamdeh, *Phys. Rev. B*, 36:2301, 1987.
186. H. Srikanth, B. A. Willemsen, T. Jacobs, S. Sridhar, A. Erb, E. Walker, and R. Flkiger, *Phys. Rev. B*, 55:14733, 1997.
187. Y. Sun and K. Maki, *Europhys. Lett.*, 32:355, 1995.
188. M. Sutherland, D. G. Hawthorn, R. W. Hill, F. Ronning, S. Wakimoto, H. Zhang, C. Proust, E. Boaknin, C. L. L. Taillefer, R. Liang, D. A. B. W. N. Hardy, N. E. Hussey, T. Kimura, and M. N. H. Takagi, *Phys. Rev. B*, 67:174520, 2003.
189. R. C. Taber, *Rev. Sci. Instrum.*, 61:2200, 1990.
190. L. Taillefer, B. Lussier, R. Gagnon, K. Behnia, and H. Aubin, *Phys. Rev. Lett.*, 79:483, 1997.

191. J. L. Tallon, C. Bernhard, U. Binninger, A. Hofer, G. V. M. Williams, E. J. Ansaldo, J. I. Budnick, and C. Niedermayer, *Phys. Rev. Lett.*, 74:1008, 1995.
192. J. L. Tallon, J. W. Loram, J. R. Cooper, C. Panangopoulos, and C. Bernhard, *Phys. Rev. B*, 68:180501, 2003.
193. M. R. Trunin, *J. Supercond.*, 11:381, 1998.
194. M. R. Trunin and A. A. Golubov, in *Spectroscopy of High-T_c Superconductors. A Theoretical View*, page 159, Taylor and Francis, London, 2003.
195. M. R. Trunin, Y. A. Nefyodov, and H. J. Fink, *JETP Lett.*, 91:801, 2000.
196. M. R. Trunin, A. A. Zhukov, G. A. Emelchenko, and I. G. Naumenko, *JETP Lett.*, 65:938, 1997.
197. Y. Tsuchiya, K. Iwaya, K. Kinoshita, T. Hanaguri, H. Kitano, A. Maeda, K. Shibata, T. Nishizaki, and N. Kobayashi, *Phys. Rev. B*, 63:184517, 2001.
198. C. C. Tsuei and J. R. Kirtley, *Phys. Rev. Lett.*, 85:182, 2000.
199. S. J. Tureaure, A. A. Pesetskim, and T. R. Lemberger, *J. Appl. Phys.*, 83:4334, 1998.
200. S. J. Tureaure, E. R. Ulm, and T. Lemberger, *J. Appl. Phys.*, 79:4221, 1996.
201. P. J. Turner, D. M. Broun, S. Kamal, M. E. Hayden, J. S. Bobowski, R. Harris, D. Morgan, J. S. Preston, D. A. Bonn, and W. N. Hardy, *Rev. Sci. Instrum.*, 75:8124, 2004.
202. P. J. Turner, R. Harris, S. Kamal, M. E. Hayden, D. M. Broun, D. C. Morgan, A. Hosseini, P. Dosanjh, G. K. Mullins, J. S. Preston, R. Liang, D. A. Bonn, and W. N. Hardy, *Phys. Rev. Lett.*, 90:237005, 2003: P. J. Turner, Ph,D., university of British Columbia, 2004.
203. Y. J. Uemura, *Solid State Commun.*, 126:23, 2003.
204. Y. J. Uemura, *J. Phys.: Condens. Matter*, 16:S4515, 2004.
205. Y. J. Uemura, L. P. Le, A. Keren, G. Luke, W. D. Wu, Y. Kubo, Y. Shimakawa, M. Subramanian, and J. T. Markert, J. L. Cobb, *Nature*, 364:605, 1993.
206. E. R. Ulm, J. T.Kim, T. R. Lemberger, S. R. Foltyn, and X. Wu, *Phys. Rev. B*, 51:9193, 1995.
207. A. A. Valenzuela, G. Solkner, J. Kessler, and P. Russer, *Mater. Sci. Forum*, 130–132:349, 1993.
208. J. R. Waldram, D. M. Broun, D. C. Morgan, R. Ormeno, and A. Porch, *Phys. Rev. B*, 59:1528, 1999.
209. M. B. Walker and M. F. Smith, *Phys. Rev. B*, 61:11285, 2000.
210. X.-G. Wen and P. A. Lee, *Phys. Rev. Lett*, 80:2193, 1998.
211. R. A. Wickham and A. T. Dorsey, *Phys. Rev. B*, 61:6945, 2000.
212. I. Wilke, M. Khazan, C. Rieck, P. Kuzel, T. Kiser, C. Jaekel, and H. Kurz, *J. Appl. Phys.*, 87:2984, 2000.
213. D.-H. Wu, W. L. Kennedy, C. Zohopoulos, and S. Sridhar, *Appl. Phys. Lett.*, 55:696, 1989.
214. D.-H. Wu, J. Mao, S. N. Mao, J. L. Peng, T. Venkatesan, X. X. Xi, R. L. Greene, and S. M. Anlage, *Phys. Rev. Lett.*, 70:85, 1993.
215. T. Xiang and W. N. Hardy, *Phys. Rev. B*, 63:024506, 2000.
216. T. Xiang, C. Panagopoulos, and J. R Cooper, *Int. J. Mod. Phys.*, B12:1007, 1998.
217. T. Xiang and J. M. Wheatley, *Phys. Rev. Lett*, 76:134, 1996.
218. T. Xiang and J. M. Wheatley, *Phys. Rev. Lett*, 77:4632, 1996.
219. B. Zeini, A. Freimuth, B. Buchner, R. Gross, A. P. Kampf, M. Klaser, and G. Muller-Vogt, *Phys. Rev. Lett.*, 82:2175, 1999.
220. K. Zhang, Ph.D., University of British Columbia, 1995.
221. K. Zhang, D. A. Bonn, S. Kamal, R. Liang, D. J. Baar, W. N. Hardy, D. Basov, and T. Timusk, *Phys. Rev. Lett.*, 73:2484, 1994.
222. K. Zhang, D. A. Bonn, R. Liang, D. J. Baar, and W. N. Hardy, *Appl. Phys. Lett.*, 62:3019, 1993.
223. Y. Zhang, N. P. Ong, P. W. Anderson, D. A. Bonn, R. Liang, and W. N. Hardy, *Phys. Rev. Lett.*, 86:890, 2001.
224. M. E. Zhitomirsky and M. B. Walker, *Phys. Rev. Lett.*, 80:5413, 1998.
225. L. Zhu, P. J. Hirschfeld, and D. J. Scalapino, *Phys. Rev. B*, 70:214503, 2004.
226. Y. Zuev, J. A. Skinta, M.-S. Kim, T. R. Lemberger, E. Wertz, K. Wu, and Q. Li, *cond-mat 0407113*, 2004.

5

Magnetic Resonance Studies of High Temperature Superconductors

Charles P. Slichter

Since most magnetic resonance studies of high T_C materials have been made by nuclear magnetic resonance (NMR), the article focuses primarily on NMR, with only brief reference to electron-spin resonance (ESR).

Most atoms in the cuprates have isotopes that can be studied. We review basic NMR theory of the resonance spectrum, how signals are observed, the theory of spin–lattice relaxation time, T_1, problems of observing NMR in superconductors, the theory of T_1 and magnetic shifts in normal state metals, the theory of NMR in conventional BCS superconductors. The spin Hamiltonian in cuprates is discussed together with its relationship to one vs. two-component theories of superconductivity.

For YBCO, the arguments are given for a one-component picture and for the existence of a spin gap. Utilizing a relationship due Moriya, relating NMR T_1 to the imaginary part of the electron-spin susceptibility, the Millis, Monien, and Pines (MMP) theory of T_1 is described, as well as its use to evaluate the parameters of the spin Hamiltonian. The phenomenological form of the temperature dependence of T_1 at the Cu, O, and Y sites is given in terms of the spin susceptibility and several parameters. The transverse relaxation time, T_{2G}, is defined and shown to give information about the real part of the electron-spin susceptibility. Certain scaling relationships are described and NMR tests of them are given.

Measurements of the Cu Knight shift in the superconducting state are described, as is their use to conclude that the spin pairing is singlet. T_1 data show that the orbital pairing cannot be s-state but is well described by d-state.

For LSCO, the situation is more complex. The spectrum is described and explained. Arguments are presented that the system may require a two-component description. The incommensurate nature of the peaks reported in neutron diffraction poses problems understanding the T_1 data using the conventional MMP analysis. The NMR line shapes and widths give compelling evidence that the charge density and the local spin susceptibility are spatially modulated over length scales of the order of a few lattice constants. Studies at high temperatures of Sr-doped and undoped LCO show that in this temperature regime the systems all act much like Heisenberg antiferromagnets. Nuclear quadrupole resonance studies at low temperatures (room temperature and below) of systems that are thought, from neutron diffraction, to have stripe ordering, show that they have great similarity to spin glasses.

5.1. Introduction

Nuclear magnetic resonance (NMR) and electron-spin resonance (ESR) have been powerful techniques for the study of condensed matter. For the cuprates, the literature involving

---

Charles P. Slichter • Research Professor of Physics, Department of Physics, University of Illinois Urbana/Champaign, Urbana, IL, USA

NMR is more extensive since special techniques are needed to find an ESR signal. The technique that works is to dope the cuprate with a low concentration of Gd or Mn ions. The spin-7/2 electrons on Gd give an ESR signal whose properties give indirect information about the electron spins of the CuO_2 planes. Thus, most of this article is focused on NMR. We start by listing some of the strengths of NMR for study of the cuprates.

Most of the atoms in the cuprates have isotopes that can in principle give rise to NMR signals: ^{89}Y, ^{135}Ba, ^{137}Ba, ^{63}Cu, ^{65}Cu, ^{17}O, ^{139}La, ^{207}Pb, ^{203}Tl, ^{205}Tl, ^{209}Bi, ^{151}Eu, ^{153}Eu. Unfortunately, the abundant isotope of O, ^{16}O, does not have a nuclear moment, so studies of O require enriching the sample with ^{17}O. NMR nuclei are sensitive to magnetic effects and, if their spins are greater than 1/2, to electrical or charge effects. As a result, NMR can distinguish between the signals of a given nuclear isotope present at more than one crystal site. For example, in YBCO, NMR can probe the Cu (or O) atoms in the CuO planes separately from the Cu (or O) atoms in the chains, and can distinguish between oxygen atoms in the plane that form Cu–O bonds parallel vs. perpendicular to the direction of the CuO chains.

Moreover, NMR can study both static and dynamic effects. It can follow how they change with temperature as well as with how they change with magnetic field strength. Although NMR is a point probe, it is also able to obtain wavevector dependence in some cases, as we shall see.

Since one needs many nuclei to detect an NMR signal, NMR essentially probes bulk properties. Since one can easily calibrate the intensity of an NMR signal using a reference compound, one can tell whether one is observing signals from all the nuclei in the sample or only a fraction. Such intensity studies have revealed in some cases the probable onset of a spin-glass state as the temperature is lowered.

Magnetic resonance is a rather specialized topic yet is especially powerful. Many scientists therefore lack the background to read much of the magnetic resonance literature. Accordingly, this paper has a twofold goal (1) to provide a background in magnetic resonance useful for reading the magnetic resonance literature on high temperature superconductors and (2) to summarize some of the principal findings about high temperature superconductors obtained by magnetic resonance. The paper is not an effort to record all of the important work. There just is not enough space given these two goals. Consequently, much beautiful and important magnetic resonance work alas will not be found in this review.

5.2. Basic NMR Theory and Experiment

5.2.1. The Resonance Spectrum

The general physical and chemical environment of a nucleus is contained in the parameters of the nuclear spin Hamiltonian [1] H.

$$H = -\gamma_n \hbar \sum_{\alpha=x,y,z} I_\alpha(1 + K_{\alpha\alpha})B_{0\alpha} + \frac{h}{2I(2I-1)}\left[\nu_\alpha(3I_z^2 - I^2) + (\nu_{xx} - \nu_{yy})(I_x^2 - I_y^2)\right]. \tag{5.1}$$

The first term is the Zeeman energy of the nuclear magnetic moment in the applied magnetic field B and the second term is the interaction of the nuclear electric quadrupole moment with the electric field gradient. K is the magnetic shift tensor. In this expression we have assumed, for simplicity, that the principal axes of the shift and field gradient tensors coincide. The presence of a magnetic shift tensor expresses the displacement in frequency of the NMR resonant frequency from the frequency it has in some other substance that is used as a frequency reference, so values quoted for K will vary with the reference. Likewise, theoretical interpretations of experimental results also require a theory of the reference. A thorough discussion of such matters for the cuprates has been made by Renold, Heine, Weber, and Meier [2].

K is customarily decomposed into two terms, K^L, called the orbital or chemical shift, and K^S, the spin or Knight shift.

$$K = K^L + K^S. \tag{5.2}$$

The chemical shift arises from electric currents induced in the electron cloud by application of the applied magnetic field. It is often in the range of several percent for Cu in the cuprates. Ordinarily one expects the chemical shift to be independent of temperature. The Knight shift arises from polarization of the electron spins. As we shall see, it is closely related to the electron-spin susceptibility. Since the electron-spin susceptibility in the high T_C materials is often temperature dependent, so is the Knight shift.

If a strong magnetic field is applied parallel to the z-axis, the energy levels become to a first approximation:

$$E_m = -\gamma_n \hbar (1 + K_{zz}) m B_0 + \frac{h}{2I(2I-1)} v_{zz}(3m^2 - I(I+1)), \quad m = I, I-1, \ldots, -I. \tag{5.3}$$

This result is exact if the electric quadrupole interaction is axially symmetric about the z-direction.

The application of an alternating magnetic field induces transitions in which the m quantum number changes by one, corresponding to angular frequencies

$$\omega = \gamma_n (1 + K_{zz}) B_0 + \frac{6\omega_{zz}}{2I(2I-1)} k. \tag{5.4}$$

k changes by integers from $-(2I-1)/2$ to $+(2I-1)/2$. Thus for nuclei with spin-3/2 such as ^{63}Cu or ^{65}Cu, k is $(-1, 0, 1)$, whereas for spin-5/2 nucleus such as ^{17}O, k is $(-2, -1, 0, 1, 2)$.

In the absence of an applied magnetic field, one has the case known as pure quadrupole resonance. Then, if the electric field gradient is axially symmetric, the frequencies become:

$$v = \frac{6v_{zz}}{4I(2I-1)} |k|. \tag{5.5}$$

For a system with spin-3/2, the pure quadrupole frequency is $v_{zz}/2$.

5.2.2. Exciting a Resonance

Observation of a magnetic resonance signal arises by inducing transitions among the various energy levels by applying alternating magnetic fields whose frequency matches the corresponding energy difference. Most NMR studies of high T_C materials have employed pulse methods. The simplest in principle is to apply a single, strong, short radio frequency (*rf*) pulse of alternating magnetic field to the sample and then observe the NMR signal that follows the turnoff of the pulse. That signal is called the *free induction* signal. Its Fourier transform in time gives one the absorption spectrum in frequency space. This result is a rigorous result of theory. This method is commonly used for NMR study of liquids, for example to determine molecular structures.

As practical matter, the radio frequency pulse disturbs the amplifiers of the detection system, so the signal amplifiers need a short time to recover before the NMR signal can be observed. For liquids, the NMR signals however persist for a long time, so one can wait for the amplifier recovery before observing the signal.

For solids, the NMR signals have much shorter duration, so amplifier recovery poses a problem. Signals are usually observed by a two-pulse method that generates so-called *spin echoes*. If the first pulse is at time zero, the second at time τ, the NMR echo signal appears at time 2τ. This method has the advantage that the signal is separated in time from the time of the strong radio frequency pulses so the amplifiers have recovered by the time of the signal. The disadvantage of this method is that NMR theory of spin echoes has been solved rigorously only for special cases.

As one varies the time τ, the amplitude of the echo signal varies, in general falling off with time. The time dependence may be quite complex, perhaps even exhibiting oscillations. If the decay form is exponential, the time constant is conventionally denoted as T_2. The time dependence of the free induction signal following a single rf pulse is often determined partly by magnet inhomogeneity. The tradition is to call this decay time T_2^*, a term that only has precise meaning if one also gives the mathematical form of the decay.

On occasion, the echo envelope is a Gaussian function of time. Then the custom is to characterize the Gaussian time scale by the symbol T_{2G}. This situation is of importance for high T_C materials.

In general one starts with a system in thermal equilibrium under the action of a static Hamiltonian H_0. The various energy levels, n, are then populated according to the equation

$$p_n = \frac{e^{-E_n/k_B T}}{Z}, \tag{5.6}$$

where p_n is the probability of occupation of level n and Z is the partition function

$$Z = \sum_k e^{-E_k/k_B T}. \tag{5.7}$$

Application of the rf pulses disturbs the system from thermal equilibrium, but during the times that the pulses are off, the system still obeys the Hamiltonian H_0. Description of the system is then conveniently given in terms of the time dependent density matrix $\rho(t)$ obeying

$$\rho(t) = e^{(-iH_0 t/\hbar)} \rho(0) e^{(+iH_0 t/\hbar)}, \tag{5.8}$$

where $\rho(0)$ is the value after the end of the most recent pulse.

Experimentally, we observe the magnetization in the absence of the rf magnetic fields. For example, the transverse magnetization in the x-direction is given by

$$\langle M_x(t) \rangle = \text{Tr}(\gamma I_x \rho(t)). \tag{5.9}$$

In thermal equilibrium, the density matrix is diagonal in the representation that diagonalizes H_0. Only those components of magnetization that are independent of time are nonzero. Application of the rf pulses creates off diagonal elements of the density matrix. They decay with time giving rise to the T_2 effects described above.

The diagonal elements of ρ give the populations of the energy levels, p_n. They are disturbed from thermal equilibrium by the pulses. The recovery to thermal equilibrium requires an energy transfer between the spin system and the outside world (conventionally called *the lattice*). The time needed to achieve thermal equilibrium is conventionally called T_1, the *spin–lattice relaxation time*.

In discussing magnetic resonance, there are two viewpoints that are often convenient. The first is to consider a single spin. Then the situation is described by a Hamiltonian such as that of Eq. (5.1). There are $2I + 1$ energy levels. This viewpoint is useful for determining the frequencies at which resonance occurs or the behavior of the system while the rf fields are on. It is used for consideration of T_1 processes. The second viewpoint is to include many spin sites. This approach is needed if one wants to consider the effects of spin–spin coupling. It is needed if one wants to consider T_2 or T_{2G} processes.

5.2.3. Spin–Lattice Relaxation

Turning now to spin–lattice relaxation, we adopt the single spin viewpoint. The recovery to thermal equilibrium following application of one or more rf pulses is often described by a set of coupled rate equations among the $2I + 1$ energy levels

$$\frac{dp_n}{dt} = \sum_k (W_{nk} p_k - W_{kn} p_n), \tag{5.10}$$

where W_{nk} is the probability per second of a thermally induced transition from state k to state n. These have normal mode solutions that are decaying real exponentials solutions. One of the normal modes is the total population (the sum of the p_ns). It does not decay. So there are $N-1$ exponentials if there are N energy levels. For a spin-1/2 system, there is a single exponential response. It is called the spin–lattice relaxation time T_1. For a spin-3/2 system, there are three exponentials. One still speaks of a spin–lattice relaxation process but there is no obvious single time to call T_1.

In general, a single pulse applied at a single transition frequency will excite more than one normal mode of recovery. The magnitude of the rate coefficients W_{nk} is determined by the relaxation mechanism. For nuclei, applying the principle of detailed balance, one finds that to achieve thermal equilibrium

$$\frac{W_{nk}}{W_{kn}} = \frac{e^{-E_n/k_B T}}{e^{-E_k/k_B T}}. \tag{5.11}$$

For nuclei, the small size of the energy level splittings compared to $k_B T$ will make the ratio essentially unity. In fact, if one defines p_n to be the deviation of the population from the thermal equilibrium value, one can simply set the ratio of Eq. (5.11) to be unity.

As an example, for spin-3/2 one then has

$$\begin{aligned}W_{3/2,1/2} = W_{1/2,3/2} = W_{-3/2,-1/2} = W_{-1/2,-3/2} &\equiv W_1 \\ W_{1/2,-1/2} = W_{-1/2,1/2} &\equiv W_2.\end{aligned} \tag{5.12}$$

These equations help one determine the relaxation mechanism. For example, if it arises from fluctuating magnetic fields

$$W_2 = \frac{4}{3} W_1. \tag{5.13}$$

In this circumstance, a single parameter W_1 will describe all three exponential time constants, providing a better description than the term T_1. When there are quadrupole splittings, the rf pulse excites only a single transition. A 90° pulse, for example, applied to the central transition produces a nonzero off-diagonal element of density matrix ($\rho_{1/2,-1/2}$), equalizes the populations of the $+1/2$ and $-1/2$ levels and leaves the populations of the $= 3/2$ and $-3/2$ states unchanged. The resultant distribution is then not one of thermal equilibrium. The off-diagonal element gives rise to an NMR signal according to Eq. (5.9). It dies out from the T_2 processes.

Thermal relaxation (T_1) processes will then change all of the populations to reestablish thermal equilibrium. We can follow these processes in time by exciting the various transitions at later times to produce a spin echo signal.

For example, for the case of a spin-3/2 system with relaxation by fluctuating magnetic fields, a spin echo signal, $S(t)$, following a 90° pulse applied to the central transition at time zero will grow at later times according to the law

$$\frac{S(t)}{S(0)} = 1 - \left[0.1 \exp\left(\frac{2}{3}W_1(t)\right) + 0.9 \exp(4W_1 t)\right]. \tag{5.14}$$

So the data can be examined to see whether a single parameter W_1 provides a fit. If it does, the relaxation mechanism is magnetic fluctuations.

Since the populations are coupled, application of a pulse to any one transition will cause the population difference between any other pair of levels to be disturbed. Thus a pulse applied to the 3/2, 1/2 transition will initially change the population of the 1/2 level. It will therefore immediately affect the signal produced by exciting the 1/2 to $-1/2$ transition. Although it will initially leave the populations of the $-1/2$ and $-3/2$ levels undisturbed, as time goes on, these populations will change from thermal equilibrium values before eventually returning to thermal equilibrium.

5.2.4. Double Resonance

Haase [3] has invented a powerful and useful double resonance method based on these principles. For example, in optimally doped LSCO, the transition frequencies of the central transitions of the planar and apical oxygen atoms are close, and thus the resonance lines overlap and are hard to distinguish from one another. Their quadrupole splittings differ greatly, however. By applying a pulse to the 3/2 to 1/2 transition of the planar O, he can change the intensity of its 1/2 to $-1/2$ spin echo signal, while leaving unchanged that contributed by the central transition of the apical O. He then employs a so-called add–subtract sequence in which he first pulses the planar 3/2 to1/2 transition and records the signal from an echo from the central transition, then allows the system to equilibrate, then records an echo from the central transition without first pulsing the planar 3/2 to 1/2 transition. He subtracts the second echo signal from the first echo signal. In this process, the apical oxygen signal vanishes since it is the same in both cases, whereas the planar O central transition signal remains since it is different between the two echoes. Thus, he could make precise measurements of the central transition of the planar oxygen in this manner. He utilized this method to distinguish signals arising from the two isotopes of Cu.

Another useful technique for studying the spatial dependence of the NMR environment is *spin echo double resonance (SEDOR)*. In this technique we consider two species that we label I and S that interact either directly through their nuclear magnetic moments, or indirectly via their coupling to the valence electrons as in the so-called J coupling of liquid NMR or the RKKY coupling in solids. We observe the spin echoes of the I-spins. As a result of the coupling, the precession frequency of the I-spins depends on the orientation of the S-spins. Thus, a 180° flip of the S-spins changes the precession frequency of the I-spins. In one particularly useful embodiment of SEDOR, one uses a 90–180° pulse sequence to produce the echoes of the I-spins. One records the echo amplitude $A_I(\tau_S)$ of the I-spins as a function of the time τ_S at which one applies a 180° pulse to the S-spins. If there were no I–S coupling, A_I would be independent of τ_S. If one has a spin–spin interaction, H_{IS} of the form

$$H_{IS} = a\hbar I_z S_z. \tag{5.15}$$

then,

$$A_I(\tau_S) = A_I(0)\cos(a\tau_S). \tag{5.16}$$

The variation of A_I with τ_S tells one the coupling strength, and the frequency at which one must excite the S spins tells one the resonance frequency of the S spin that is producing the SEDOR effect. Note that the S spins can be of the same nuclear species (e.g., both ^{63}Cu), of the same chemical species but different isotope (as with ^{63}Cu and ^{65}Cu), or different chemical species (^{63}Cu and ^{17}O). (For the case of two nuclei of the same species, if the strength of the rf fields is strong enough to cover both their absorption lines, one does not need two separate rf driving signals since the 180° pulse that flips the I spins will also flip the S spins by 180°.)

The importance of the SEDOR result is that, since it depends on the existence of nuclear–nuclear spin–spin coupling, it measures a correlation of NMR environments between two sites that are near neighbors.

5.2.5. NMR in Superconductors

Observation of NMR signals from samples that are in the superconducting state has certain special features. Type I superconductors exclude the magnetic field except for a thin layer at the surface. How then can one do NMR in a superconductor? This problem was first solved by Hebel and Slichter [4, 5] and independently by Redfield [6] by using a magnetic field cycling method. The concept is to start with the sample below the zero field superconducting transition temperature. One then turns on a magnet of sufficient strength to suppress the superconductivity. The nuclei become polarized in this initial magnetic field. One then turns off the magnet, cooling the nuclear spins by adiabatic demagnetization while rendering the sample superconducting. The cold nuclei then warm toward the lattice temperature. After a time t_{off}, one turns the magnet back on, rendering the sample normal and quickly inspects the size of the NMR signal. By studying how this signal size varies with t_{off}, one can deduce the nuclear spin–lattice relaxation time in the superconducting state.

If there is a quadrupole coupling, one could still do a pure quadrupole resonance in the superconducting state thus avoiding problems with penetration of the static magnetic field into the sample. However, the rf magnetic fields are still excluded except for the superconducting skin depth. Such signals have been reported by Hammond and Knight [7] and by Simmons [8, 9], however the inhomogeneity of the rf field complicates data interpretation.

The discovery of Type II superconductors made possible direct observation of NMR in the superconducting state. The static magnetic field penetrates the sample by vortices. In practice, the static field is quite homogeneous, but it does vary spatially, being largest in the vortex cores. Redfield [10] did the first experiments mapping the field distribution in the mixed state. The variation in magnetic field as one moves away from the vortex provides a nearly one-to-one correspondence between NMR frequency and distance from the vortex that has been useful for studying effects of the supercurrent flow on the density of states [11–13].

5.3. NMR in Normal State Metals

In normal metals, the nucleus interacts via its magnetic moment with the magnetic moments of the conduction electrons. The polarization of the electron spins in the applied field gives rise to a magnetic field at the nucleus that produces the Knight shift. In addition, the interaction gives rise to scattering of the conduction electrons in which the nucleus may exchange energy with the conduction electrons, providing the mechanism for the nuclear spin–lattice relaxation process. The form of the interaction consists of the conventional interaction between distant dipoles, plus the Fermi contact interaction, H_Fermi, that describes the coupling when the electron orbit penetrates the nucleus:

$$H_\text{Fermi} = -\gamma_e \gamma_n \hbar^2 I \cdot S \delta(r_e - R_n). \tag{5.17}$$

When present, the Fermi interaction frequently dominates the conventional dipolar coupling, and gives rise to a magnetic field at the nucleus, ΔH_{0e},

$$\Delta H_{0e} = \frac{8\pi}{3} \gamma_e \hbar \sum_{k,m} |u_k(0)|^2 m_k f(E_k - m_k \gamma_e \hbar H_0), \tag{5.18}$$

where f is the Fermi function, k the electron wave vector, and m_k the eigenvalue of the component of electron spin, S_{zk}, along the direction of the applied static field H_0. This equation can be evaluated to give the Knight shift

$$\begin{aligned}
\frac{\Delta H_{0e}}{H_0} &= \frac{8\pi}{3} (\gamma_e \hbar)^2 \sum_{m_k} \int m_k^2 |u_k(0)|^2 \rho(E_k) \frac{\partial f(E_k)}{\partial E_k} dE_k \\
&= \frac{8\pi}{3} (\gamma_e \hbar)^2 \left\langle |u_k(0)|^2 \right\rangle_{E_F} \rho(E_F)/2 \\
&= \frac{8\pi}{3} \left\langle |u_k(0)|^2 \right\rangle_{E_F} \chi_S,
\end{aligned} \tag{5.19}$$

where χ_S is the conduction electron-spin susceptibility and where the bracket indicates averaging over the values at the Fermi energy.

To visualize the nuclear spin–lattice relaxation process, one may note that the magnetic coupling (Eq. (5.17)) produces scattering of the system initially in a state $m_I m_S k$ to a final state $m_I + 1, m_S - 1 k'$. The resulting T_1 then obeys the equation

$$\frac{1}{T_1} = C \left\{ \left\langle |u_k(0)|^2 \right\rangle_{E_F} \right\}^2 \int f(E) \rho(E) [1 - f(E - \Delta E)] \rho(E - \Delta E) dE. \tag{5.20}$$

C is a numerical factor that depends on some spin sums, and where ΔE is the difference in energy between the initial and final nuclear spin states. More generally

$$\frac{1}{T_1} = C \int \left\langle |\langle i|V|f\rangle|^2 \right\rangle_E f(E) \rho(E) [1 - f(E - \Delta E)] \rho(E - \Delta E) dE, \tag{5.21}$$

where V is the electro–nuclear interaction and i and f are initial and final electron states. For a normal metal, this expression reduces to

$$\frac{1}{T_1} = C \left\langle |\langle i|V|f\rangle|^2 \right\rangle_{E_F} (\rho(E_F))^2 k_B T. \tag{5.22}$$

The linear dependence of $1/T_1$ on T was first enunciated by *Heitler and Teller* before the discovery of magnetic resonance [14]. It expresses the fact that only electrons in the tail of the Fermi distribution can scatter since the energy change on scattering is only derived from the change in the nuclear Zeeman energy, a quantity much less than $k_B T$.

Korringa [15] discovered that by combining Eqs. (5.20) and (5.21), one gets the *Korringa relation* relating Knight shift to spin–lattice relaxation time:

$$T_1 T K_S^2 = \frac{\hbar}{4\pi k_B} \left(\frac{\gamma_e}{\gamma_n}\right)^2. \tag{5.23}$$

Equation (5.20) is based on the one electron theory of metals. It includes band structure effects but not electron–electron coupling. Pines [16] has generalized the formula to include electron–electron coupling

$$T_1 T K_S^2 = \frac{\hbar}{4\pi k_B} \left(\frac{\gamma_e}{\gamma_n}\right)^2 \frac{\chi_e^S}{\chi_0^S} \frac{\rho_0(E_F)}{\rho(E_F)}, \tag{5.24}$$

where the subscript "0" refers to the noninteracting electron values, and other values include the electron–electron coupling.

We call the quantity $T_1 T K_S^2$ the *Korringa product*. Note that in the absence of electron–electron interactions, the Korringa product depends on universal constants that are independent of the particular system. Thus it is independent of band structure or other such effects.

If we express the interaction of the nuclear spin I with the electron spins S_i as

$$H = \sum_i I \cdot A_i \cdot S_i \tag{5.25}$$

an alternate form of the expression for T_1 is useful for discussing high T_C superconductors. We discuss it in Section 5.6.3.

5.4. NMR in Conventional BCS Superconductors

NMR measurements of spin–lattice relaxation and Knight shift provided some of the first verifications of the BCS theory of superconductivity. They are described in Cooper's Nobel Lecture [17]. MacLaughlin [18] has published a comprehensive review of NMR studies prior to high T_C.

The first measurements of T_1 were made by Hebel and Slichter a few months before the creation of the BCS theory. They studied ^{27}Al. Thinking in terms of a two fluid picture, a popular model at that time, they had expected that the relaxation rate would be slower in the superconducting state. Instead they found that the relaxation rate increased by a factor of two within about a 15% drop in temperature below T_C.

In the classical BCS theory of superconductivity, the electrons form pairs of opposite spin and momentum, so that the spin pairing is into a spin singlet ($S=0$), while the orbital pairing is also into an $L=0$ (orbital s-wave) state. There is a temperature dependent gap, $\Delta(T)$, in the density of states that goes to zero as T approaches T_C and levels off near $T=0$ at a Δ_0 approximately equal to $1.75 T_C$.

Using the BCS theory, with the generous help of its authors, Hebel and Slichter found that one gets the expression for spin–lattice relaxation in a superconductor by simple modification of Eq. (5.21). We introduce the symbol ε to specify the energy of an electron state in a metal in the normal state, measured with respect to the Fermi energy. Then, below T_C, the energy, E, to occupy such a state becomes

$$E = \sqrt{\varepsilon^2 + \Delta(T)^2}. \tag{5.26}$$

The density of states $\rho_S(E)$ is zero for $|E| \leq \Delta(T)$ and

$$\rho_S(E) = \rho_N(0) \frac{E}{\sqrt{E^2 - \Delta(T)^2}} \tag{5.27}$$

for $|E| \geq \Delta(T)$. The scattering matrix V_{if}^2 is replaced by

$$V_{Sif}^2 = \frac{V_1^2 + V_2^2}{2}, \tag{5.28}$$

where

$$\begin{aligned} V_1^2 &= V_{if}^2 \left(1 + \frac{\Delta^2}{EE'}\right) \\ V_2^2 &= V_{if}^2 \frac{\Delta^2}{EE'}. \end{aligned} \tag{5.29}$$

The prime distinguishes the final state from the initial state. BCS found that for ultrasonic absorption, Eq. (5.21) describes things if instead one uses

$$V_{Sif}^2 = \frac{V_1^2 - V_2^2}{2}. \tag{5.30}$$

The change in sign causes the two phenomena to have very different temperature behavior for temperatures just below T_C. As one cools from T_C, the ultrasonic absorption rate drops precipitously whereas the NMR relaxation rate rises rapidly. The existence of the two terms that either add or subtract signs arises from the pair nature of the BCS wave function. In a conventional one-electron theory, there is only one term so all low energy scattering processes should have the same T dependence. BCS point out that the contrast between the two temperature dependences is strong proof of the pairing condition.

At low temperatures, the BCS theory shows that

$$\frac{1}{T_1} \propto \exp(-\Delta_0/k_B T). \tag{5.31}$$

Thus, measurement of the T dependence of $1/T_1$ near absolute zero gives one the value of low temperature energy gap.

An expression for the Knight shift was derived by Yosida [19]. Using the relationship, from Eq. (5.20)

$$K_S = \frac{\Delta H}{H_0} \propto \int \rho(E) \frac{\partial f(E)}{\partial E} dE$$

with the BCS expressions for density of states and E, one finds the *Yosida function*, $Y_0(T/T_C)$.

The most complete test of the BCS predictions have been obtained for T_1 by Masuda and Redfield [20] for Al and for Knight shift by Knight [21].

5.5. The Cuprate Spin Hamiltonian

The key elements found for all the cuprates appear to be the CuO_2 planes. We might think of them as composed of two systems of electrons. The first system is a set of Cu^{2+} ions, much like $(3d)^9$ atoms with a hole in the $x^2 - y^2$ orbitals, where we take the x and y directions to be along the crystallographic a and b axes, respectively. The second system is the holes in the oxygen orbitals.

We might then have Hamiltonians

$$^{63}H = {}^{63}H_{\text{Zeeman}} + {}^{63}I \cdot A \cdot S_d(0) + {}^{63}H_Q \qquad (5.32)$$

and

$$^{17}H = {}^{17}H_{\text{Zeeman}} + {}^{17}I \cdot C_p \cdot S_p(0) + {}^{17}H_Q, \qquad (5.33)$$

where $S_d(0)$ and $S_p(0)$ are the on-site electron spins. Making use of the fact that an applied field H_0 produces thermal average magnetizations

$$\begin{aligned} \langle M_d \rangle &= \chi_d H_0 \\ \langle M_p \rangle &= \chi_p H_0 \end{aligned} \qquad (5.34)$$

we get for the Knight shifts

$$^{63}K_S = \frac{A}{\gamma_{63}\gamma_e \hbar^2} \chi_d, \qquad (5.35)$$

$$^{17}K_S = \frac{C_p}{\gamma_{17}\gamma_e \hbar^2} \chi_p. \qquad (5.36)$$

As we shall see, the above Hamiltonians omit important contributions arising from transferred hyperfine couplings that give rise to a shift contribution by the electron spin on one atom to the magnetic field acting on nuclei of neighboring atoms.

The discovery by Takigawa, Hammel et al. [22] that the powder average Knight shift is positive, suggested that A alone would not work since for an isolated atom the powder average shift comes from the presence of core polarization and is expected to be negative. Mila and Rice [23] proposed the addition of the term $B_d(k)$. From studies of the effect of O doping on the ^{89}Y NMR shift, Alloul et al. found that the ^{89}Y shift arose from transferred hyperfine coupling to the Cu magnetization.

Zhang and Rice [24], however, recognized that there would be a strong exchange coupling between the Cu electron spin and those on its four neighboring O atoms.

They proposed that one O electron spin might pair with the Cu spin to form a spin singlet state (*the Zhang–Rice singlet*) and that this might be the mobile charge carrier produced by O doping. Such considerations led to the proposal that the system posses only a single component, a concept that can be expressed in the nuclear Hamiltonians as follows:

$$\begin{aligned} ^{63}H &= {}^{63}H_{\text{Zeeman}} + {}^{63}I \cdot \left(A S_d(0) + B \cdot \sum_{k=nn} S_d(k) \right) + {}^{63}H_Q \\ ^{17}H &= {}^{17}H_{\text{Zeeman}} + {}^{17}I \cdot C \cdot \sum_{k=nn} S_d(k) + {}^{17}H_Q. \end{aligned} \qquad (5.37)$$

For these equations, the Knight shift relations are

$$\begin{aligned} ^{63}K_{Szz} &= \frac{(A_{zz} + 4B)}{\gamma_e \gamma_{63} \hbar^2} \chi_d \\ ^{17}K_{Szz} &= \frac{2C}{\gamma_e \gamma_{17} \hbar^2} \chi_d, \end{aligned} \qquad (5.38)$$

where we have defined the z-direction to correspond to the direction of the applied magnetic field, assumed to lie along one of the principal axes of the hyperfine tensors. We have also assumed that the two tensors B and C are isotropic.

More general Hamiltonians have been proposed by Curro et al. [25] in their work on heavy fermion systems:

$$^{63}H = {}^{63}H_{\text{Zeeman}} + {}^{63}I \cdot A \cdot S_d(0) + {}^{63}I \cdot \left(B_d \cdot \sum_{k=nn} S_d(k) + B_p \cdot \sum_{l=nn} S_p(l) \right) + {}^{63}H_Q, \quad (5.39)$$

$$^{17}H = {}^{17}H_{\text{Zeeman}} + {}^{17}I \cdot \left(C_p \cdot S_p(0) + C_d \cdot \sum_{k=nn} S_d(k) \right) + {}^{17}H_Q, \quad (5.40)$$

where $S_d(0)$ is the on-site Cu electron spin and $S_d(k)$ is the Cu electron spin at neighbor site k, etc.

For this set of equations, the Knight shift formulas are more complicated. Defining

$$\begin{aligned}
\chi_{dd}(r,t) &= (\gamma_e \hbar)^2 \langle S_d(r,t) S_d(0,0) \rangle \\
\chi_{pp}(r,t) &= (\gamma_e \hbar)^2 \langle S_p(r,t) S_p(0,0) \rangle \\
\chi_{pd}(r,t) &= (\gamma_e \hbar)^2 \langle S_p(r,t) S_d(0,0) \rangle = \chi_{dp}(r,t)
\end{aligned} \quad (5.41)$$

with

$$\begin{aligned}
\chi &= \chi_{dd} + 2\chi_{pd} + \chi_{pp} \\
\langle S_d \rangle &= (\chi_{dd} + \chi_{dp}) H_0, \langle S_p \rangle = (\chi_{pd} + \chi_{pp}) H_0
\end{aligned} \quad (5.42)$$

for the Knight shifts

$$\begin{aligned}
^{63}K &= \frac{(A+4B_d)}{\gamma_e \gamma_{63} \hbar^2}(\chi_{dd} + \chi_{pd}) + \frac{2B_p}{\gamma_e \gamma_{63} \hbar^2}(\chi_{pd} + \chi_{pp}) \\
^{17}K &= \frac{2C_d}{\gamma_e \gamma_{17} \hbar^2}(\chi_{dd} + \chi_{pd}) + \frac{C_p}{\gamma_e \gamma_{17} \hbar^2}(\chi_{pd} + \chi_{pp}).
\end{aligned} \quad (5.43)$$

Comparing Eq. (5.38) with Eqs. (5.35), (5.36), and (5.43) we see that a simple test for the one-component system is that the ratio of the O to the Cu Knight shift be independent of temperature.

5.6. YBCO above T_c

By *YBCO* we mean the family of materials such as $YBa_2Cu_3O_{6+y}$ (*123O_{6+y}*) and $YBa_2Cu_4O_8$ (*1248*).

5.6.1. One or Two Components?

It has been generally agreed in the NMR community that the YBCO materials obey the "one component" Hamiltonians of Eq. (5.40). The evidence arises from measurements of the temperature dependence of the Knight shifts. For all the YBCO family $^{63}K_{cc}$, where the c-axis is normal to the CuO_2 planes, is nearly independent of temperature. The explanation is believed to be the accidental relationship

$$A_{cc} + 4B \approx 0 \quad (5.44)$$

As a result, measurements of $^{63}K_{cc}$ do not distinguish between the various models. The most clear-cut test has been given by Takigawa et al. [26] using $123O_{6.63}$.

Figure 5.1 shows their data for three ^{17}Ks and $^{63}K_{ab}$ (the Cu shift for magnetic field lying in the plane of the *ab* axes). As can be seen, by proper normalization at a single temperature, all the data lie on top of one another over the entire temperature range from 300 K to temperatures well below T_C. Thus, these data fit the single-component equations.

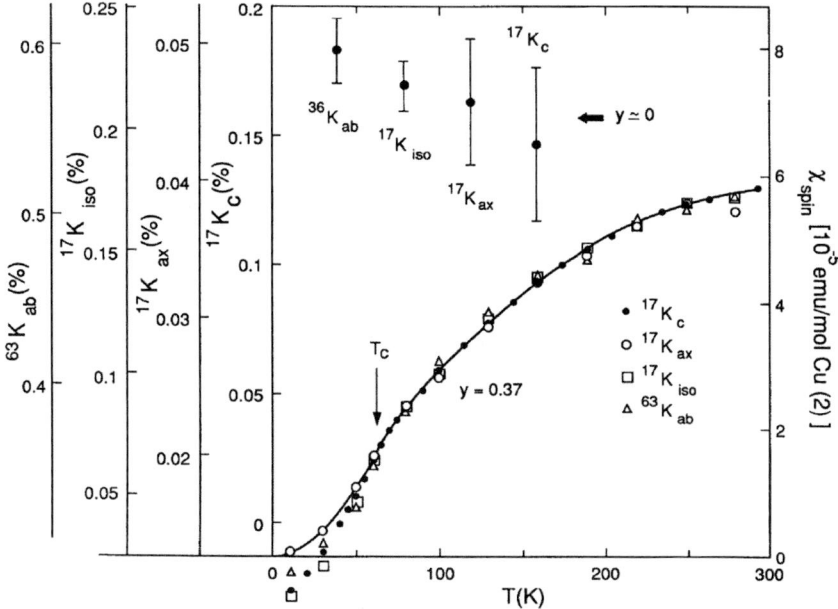

Figure 5.1. The temperature dependence of the Cu and O Knight shifts, showing that the three ^{17}O shifts and the ^{63}Cu shift all have the same temperature dependence [26].

5.6.2. The Spin Pseudogap

Figure 5.2 shows data from Alloul et al. [27] for 123O$_x$ showing how the temperature dependence of the Knight shift of ^{89}Y resonance depends on the amount, x, of O doping. Figure 5.3 shows ^{63}K data from Takigawa, Hammel, et al. [22] and Fig. 5.12 shows data from Barrett et al. [28] for 123O$_7$. We discuss the data below T_C in Section 5.7. Above T_C, $^{63}K_{ab}$ is nearly independent of temperature. Figure 5.4 shows $^{63}K_{ab}$ data of Curro et al. for 1248 [29]. These data are much like that of 123O$_{6.63}$, except they show a maximum at about 500 K. In conventional metals, the spin susceptibility is independent of temperature. The data for the O$_{6.63}$ and the 1248 samples show that the susceptibility falls off at lower temperatures. Several authors have tried to fit such data with formulas such as

$$\chi(T) = \chi_0 \frac{1}{1 + e^{\Delta/2T}} \qquad (5.45)$$

suggested by Tranquada [30] based on neutron scattering experiments. For O$_{6.63}$, Takigawa found a good fit with $\Delta = 150 K$. Formulas such as Eq. (5.43) suggest that the magnetic susceptibility arises from excited states whose occupancy falls with falling temperatures. The quantity Δ then appears to represent something like an energy gap to a family of excited states. The phenomena are given the name *spin pseudogap*. It also shows up for NMR in measurements of the T dependence of T_1.

5.6.3. The Spin–Lattice Relaxation Time

The mechanism of spin–lattice relaxation is magnetic fluctuations. This result is well demonstrated for Cu by data from Pennington et al. [31]. Figure 5.5 shows the recovery of the

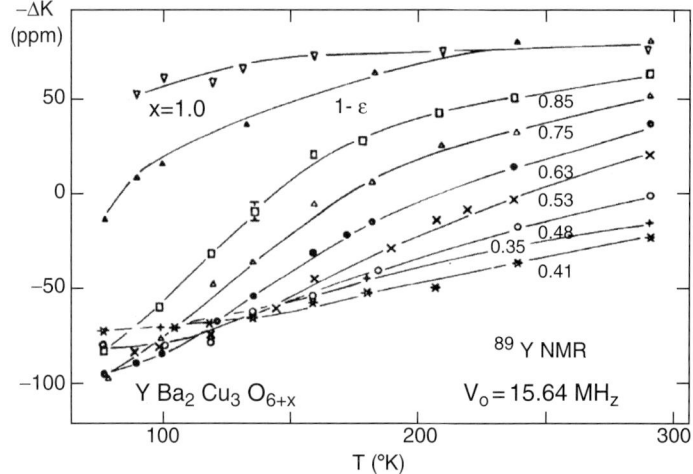

Figure 5.2. The doping dependence of the ^{89}Y Knight shift in YBCO$_x$. The drop-off at low temperatures arises from the spin gap. These data represent the discovery of the spin gap and show how the gap varies with doping [27].

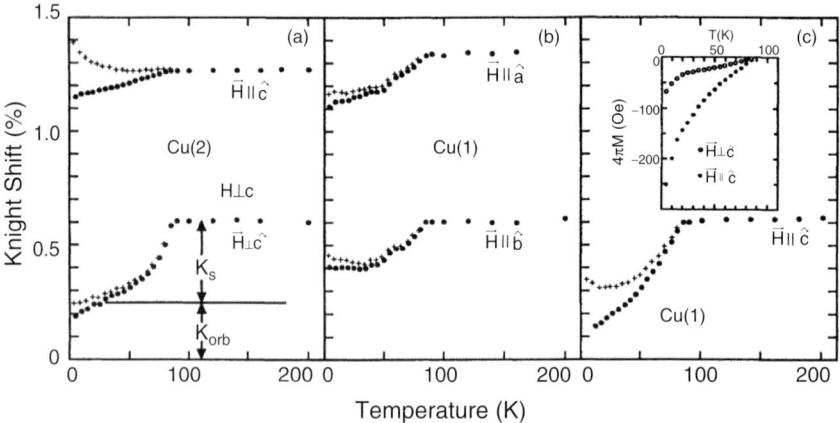

Figure 5.3. The ^{63}Cu magnetic shift ($K_L + K_S(T)$) in optimally doped YBCO for the chains (Cu(1)) and planes (Cu(2)). The intercept at $T = 0$ gives K_L. The crosses are raw data, the filled squares are data corrected for Meissner screening [22].

^{63}Cu NMR echo signal following an initial destruction of the population difference between the various energy levels of the three transitions. A single parameter W_1 makes possible the fit for the time dependence of the three NMR transitions, as described by Eq. (5.14).

Initially, a major mystery was the fact that the Cu and O T_1s had very different temperature dependences. For example, for 123O$_7$, one finds that T_1T for O is independent of temperature as in a conventional metal whereas for the planar Cu it is nearly a linear function of T [32] (Figure 5.6). The explanation for the different temperature dependence at these neighboring sites in the crystal was explained by Shastry [33] and independently by Hammel et al. [34] as being due to the fact that the wavelength dependence of spin fluctuations differs between the Cu and O sites. Thus, fluctuations in the electron-spin magnetization near the antiferromagnetic wave vector cancel at the O site but not at the Cu site.

Magnetic Resonance Studies of High Temperature Superconductors

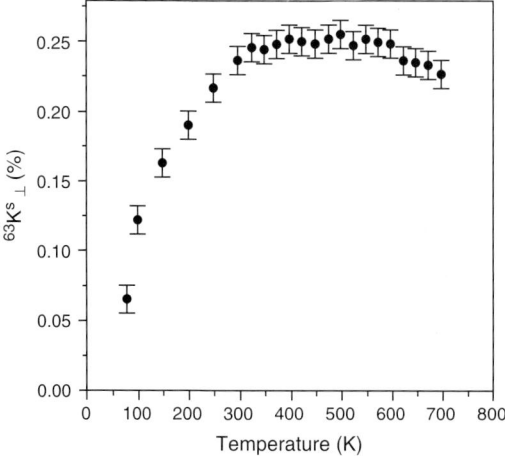

Figure 5.4. $^{63}K_{ab}$ for the stoichiometric material 1248 showing the maximum in Knight shift at about 500 K [29].

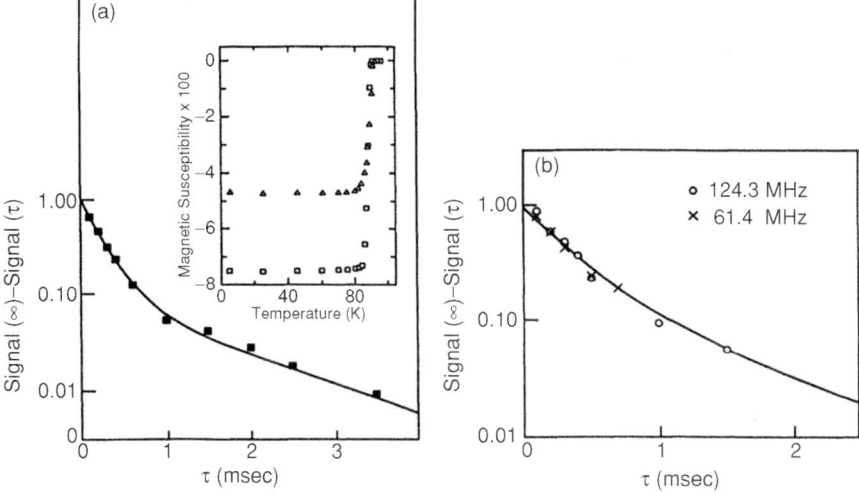

Figure 5.5. Recovery of the planar ^{63}Cu signal for the central transition (a) and the upper satellite (b). The data are fit with a single parameter W_1 using the formulas for relaxation from fluctuating magnetic fields [31].

Millis, Monien, and Pines (*MMP*) proposed a simple theory based on the relationship due to Moriya [35],

$$\left(\frac{1}{iT_1}\right)_\alpha = \frac{2\gamma_i^2 k_B T}{(\gamma_e \hbar)^2} \sum_{q,\alpha' \neq \alpha} \left| {}^i A_{\alpha'\alpha'} \right|^2 \frac{\chi''_{\alpha'\alpha'}(q,\omega_0)}{\omega_0}, \tag{5.46}$$

where ω_0 is the nuclear precession frequency, $A_{\alpha'\alpha'}$ the electron–nuclear hyperfine coupling, and $\chi''_{\alpha'\alpha'}$ the imaginary part of the electron-spin susceptibility. The concept behind this formula is that the nucleus precessing at angular frequency ω_0 produces a magnetic field at the

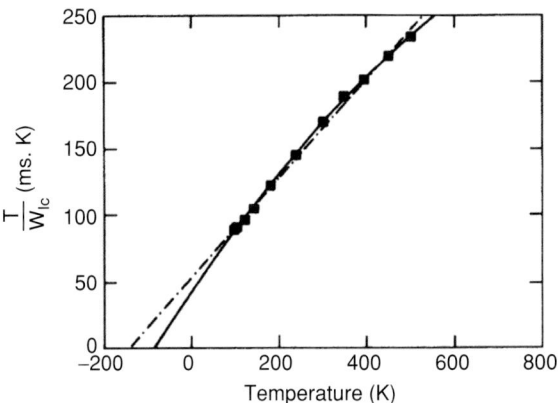

Figure 5.6. $^{63}T_1T(=T/^{63}W_1)$ vs. temperature for optimally doped YBCO showing its nearly linear temperature dependence [32].

electrons at this frequency through the hyperfine coupling, leading to energy absorption by the electron via $\chi''_{a'a'}$.

Using Eq. (5.37), MMP have

$$^{63}A_{zz}(q) = A_c + 2B(\cos(q_x a) + \cos(q_y a)) \qquad (5.47)$$
$$^{63}A_{xx}(q) = A_{ab} + 2B(\cos(q_x a) + \cos(q_y a)).$$

From Eq. (5.33) we get

$$\left|^{17}A_{\alpha\alpha}(q)\right|^2 = 2C^2\left(1 + \frac{1}{2}[\cos(q_x a) + \cos(q_y a)]\right). \qquad (5.48)$$

These expressions show that the hyperfine coupling acts as a filter in q-space. In particular, they show that for antiferromagnetic spin fluctuations for which the arguments of the cosine functions are π, the coupling to the O vanishes, whereas that to the Cu does not.

Barzykin, Pines, and Thelen [36] define the quantities $^{63}F_\|$ and $^{63}F_\perp$ to simplify Eq. (5.44).

$$^{63}F_\| = |A_x|^2 + |A_y|^2 \qquad (5.49)$$

and

$$^{63}F_\perp = (|A_x|^2 + |A_y|^2)/2 + |A_z|^2. \qquad (5.50)$$

Figure 5.7 shows how the F's vary in q-space.

MMP express the complex susceptibility arising from antiferromagnetic fluctuations peaking at a point Q in q space as

$$\chi_{AF} = \frac{\alpha\xi^2}{1 + |Q - q|^2 \xi^2 - i(\omega/\omega_{SF})}, \qquad (5.51)$$

where ξ represents a spatial correlation length for spin fluctuations and ω_{SF} a characteristic frequency or the inverse of a relaxation time. Making use of the fact that the NMR frequency is low, they find

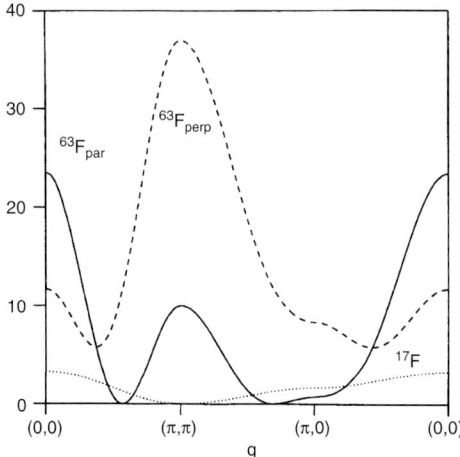

Figure 5.7. The variation in wave vector space of the Cu and O form factors, F, showing the peaking of the Cu response at the (π, π) point, and the suppression of the O form factor there [36].

$$\chi'_{AF} = \left[\frac{\alpha \xi^2}{(1+|q-Q|^2 \xi^2)} \right]$$
$$\chi''_{AF} = \left[\frac{\alpha \xi^2}{(1+|q-Q|^2 \xi^2)^2} \right] \omega / \omega_{SF}. \qquad (5.52)$$

Imai et al. [37] include in addition a correction to correspond to the region near $q = 0$.

$$\chi'(q) = \chi'_{AF}(q) + \chi'_B$$
$$\chi''(q) = \chi''_{AF}(q) + \chi'_B \omega / \Gamma_B. \qquad (5.53)$$

There are thus four hyperfine coupling constants A_{ab}, A_c, B and C. To determine them we need four experimental conditions. The first is Eq. (5.44)

$$A_c + 4B = 0.$$

For the second, we assume that the correlation length is long so that the spin fluctuations peak strongly at the antiferromagnetic wave vector. Then, using Eq. (5.46) and (5.47), we find that the ratio of the Cu T_1's for two orientations of applied magnetic fields, is given by

$$^{63}R \equiv \frac{T_{1c}}{T_{1ab}} = \left[1 + \frac{[A_c - 2B]^2}{[A_{ab} - 2B]^2} \right]. \qquad (5.54)$$

The third equation used [38] expresses the zero field NMR frequency of the antiferromagnetic parent compound.

$$\frac{\mu_{\text{eff}}}{\mu_e} \left| {}^{63}A_{ab} - 4{}^{63}B \right| = \hbar {}^{63}\omega_{AF}, \qquad (5.55)$$

where μ_e is the free electron-spin magnetic moment and μ_{eff} is the effective value for the antiferromagnetic ground state. Manousakis [39] finds the ratio μ_{eff}/μ_e to be 0.62. Using $^{63}R = 3.73$ Barzykin, Pines, and Thelen find

$$B = 3.82 \times 10^{-7} \text{ eV}, \ A_c = -4B, \ A_{ab} = 0.84B.$$

From the temperature dependence of the Knight shift, they find $C = 0.91B$.

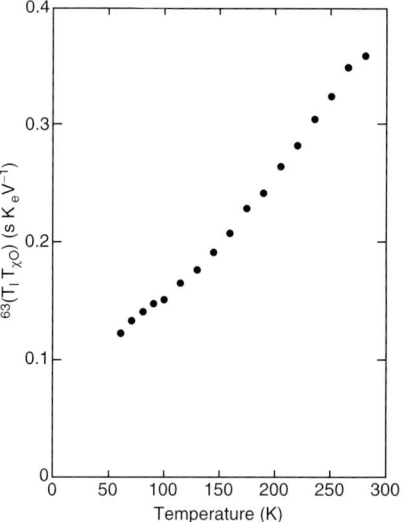

Figure 5.8. The temperature variation of $^{63}T_1 T \chi_0^S(T)$ for YBCO$_{6.63}$ showing the nearly linear dependence on T [40].

We can rather satisfactorily describe the data of various investigators by saying that for YBCO$_7$

$$^{89}W_1/T = P$$
$$^{17}W_1/T = Q \qquad (5.56)$$
$$^{63}W_1/T = R/(T + T_\theta),$$

where P, Q, R, and T_θ are constants independent of temperature.

For the underdoped materials, T_1 (or $1/W_1$) differs from that of YBCO-7. Figures 5.8 and 5.9 display data of Takigawa [40] showing plots of $^{63}T_1 T \chi_0^S(T)$ and $^{17}T_1 T \chi_0^S(T)$ for the 6.63 material. From these figures, we see that for the underdoped materials, we can replace these equations with

$$^{89}W_1/T = P'\chi_0^S$$
$$^{17}W_1/T = Q'\chi_0^S \qquad (5.57)$$
$$^{63}W_1/T = R'\chi_0^S/(T + T_\theta).$$

Thus, the spin gap effect noted in Knight shift or spin susceptibility appears also in spin–lattice relaxation. Note that the first set is a special case of the second since the susceptibility of YBCO-7 is essentially independent of temperature.

From these equations we see that the quantity $^iT_1 T \chi_0^S$ for nuclear species "i" is the quantity that has a very simple temperature dependence.

5.6.4. Transverse Relaxation and T_{2G}

In conventional metals, the line width δH of the NMR lines are determined by the magnetic fields produced by the neighboring nuclear spins ($\delta H \approx \gamma_n \hbar I \sqrt{Z}/a^3$ where a is

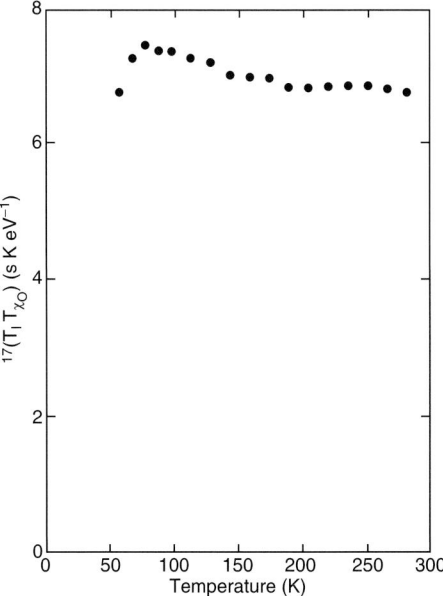

Figure 5.9. The temperature dependence of $^{17}T_1T\chi_0^S(T)$ for YBCO$_{6.63}$ showing that it is nearly independent of T [40].

the nearest neighbor distance and Z the number of nearest neighbors). This coupling limits the extent of the spin echo decay envelope. Pennington et al. [41, 42] discovered that for the 123O$_7$, the Cu echo decay rate is much faster. They found that the Cu spin echo envelope was Gaussian. The echo signal $S(t)$ at time t obeys the relationship.

$$S(t) = S(0)\exp[-(t/T_{2G})^2/2] \tag{5.58}$$

Pennington and Slichter showed that T_{2G} arose from an indirect nuclear spin–nuclear spin coupling via the hyperfine coupling to the Cu electron spins, and could be expressed in terms of the real part of the static electron-spin susceptibility $\chi'(q)$.

Defining the hyperfine coupling

$$G_z(r, r_1) = A_C \delta_{r,r_1} + B \sum_{\rho=nn} \delta_{r,r_1+\rho} \tag{5.59}$$

then the hyperfine coupling to spin I_{1z} is

$$H_{zz} = \sum S_z(r) G_z(r, r_1) I_z(r_1) \tag{5.60}$$

giving us an effective coupling to a second spin at position r_2

$$H_{12zz} = -I_z(r_2) \left[\sum_{r,r'} G_z(r_2, r') \frac{\chi'(r', r)}{(\gamma_e \hbar)^2} G_z(r, r_2) \right] I_z(r_1)$$
$$= I_z(r_2) a_{12}^z I_z(r_1). \tag{5.61}$$

Since

$$\chi'(r', r) = \sum_q \frac{\chi'(q)}{N} \exp[iq \cdot (r' - r)], \quad (5.62)$$

where N is the number of Cu atoms per unit volume, a_{12} can be expressed in terms of $\chi'(q)$.
The result is

$$(1/T_{2G})^2 = \left(\frac{1}{8}\right) \sum_j \left|a_{1j}^z\right|^2. \quad (5.63)$$

Thelen and Pines have put these results in a convenient form. For the isotope ^{63}Cu, the result is

$$(1/T_{2G})^2 = \frac{0.69}{8\hbar^2} \left[\frac{1}{N} \sum_q f_C(q)^4 \chi'(q)^2 - \left\{ \frac{1}{N} \sum_q f_C(q)^2 \chi'(q) \right\}^2 \right] \quad (5.64)$$

$$f_C(q) \equiv A_c + 2B \left[\cos(q_x a) + \cos(q_y a)\right],$$

where $f_C(q)$ is the form factor. The 0.69 factor arises from consideration of the isotopic abundance. The couplings of the x and y components are much smaller. As a result, only zz couplings of the same isotope contribute, and the form of the echo envelope is rather accurately a Gaussian.

An important conclusion is that measurements of T_1 give information about the imaginary part of the electron-spin susceptibility, and measurements of T_{2G} give information about the real part.

A description of the dependence of the coupling a_{1j}^z on the relative positions of $r_j - r_1$ has been given by Haase et al. [43]. They also discuss the dependence of T_{2G} on the correlation length ζ.

5.6.5. Scaling Relationships

A number of authors [44] have considered the role of scaling in analyzing the NMR data, expressed in a relationship between the energy scale parameter ω_{SF} and the magnetic correlation length, ζ.

The scaling is expressed by a relationship between them

$$\omega_{SF} \propto \zeta^{-z}. \quad (5.65)$$

Assuming that the correlation length is large compared to the lattice constant and assuming the MMP form for the electron-spin susceptibility, Barzykin and Pines conclude that

$$\begin{aligned} ^{63}T_1 T &= C_1 \omega_{SF}/\alpha \\ 1/T_{2G} &= C_2 \alpha \zeta, \end{aligned} \quad (5.66)$$

where C_1 and C_2 are two constants. Using $B = 3.82 \times 10^{-7}$ eV, $A_c = -4B$, $A_{ab} = 0.84B$, they find $C_1 = 132 (\text{s K/eV}^2)$ and $C_2 = 295 (\text{eV/s})$.
Then

$$\begin{aligned} \frac{T_1 T}{T_{2G}} &= C_1 C_2 \omega_{SF} \zeta \\ \frac{T_1 T}{T_{2G}^2} &= C_1 C_2^2 \alpha \omega_{SF} \zeta^2. \end{aligned} \quad (5.67)$$

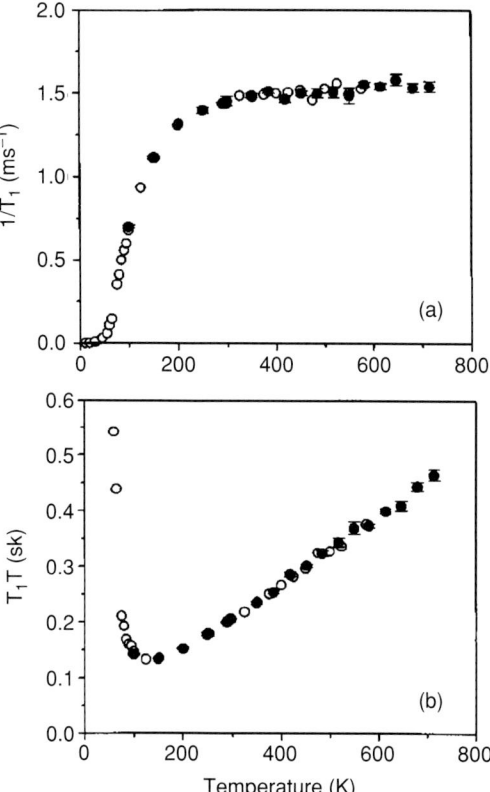

Figure 5.10. $^{63}T_1$ vs. temperature for 1248 showing the spin gap around 100 K and $^{63}T_1T$ vs. temperature showing the linear temperature dependence at higher temperatures predicted by scaling theories [28].

Barzykin and Pines have argued that there are two crossover temperatures T^* and $T_{cr}(T^* < T_{cr})$ that vary with doping. Below T^* one is in the pseudogap region. They propose that between these temperatures one should find $z = 1$ characterized by

$$\omega_{SF} \propto \xi^{-1}. \tag{5.68}$$

Above T_{cr} the system exhibits nonuniversal $z = 2$ mean field behavior, so that

$$\omega_{SF} \propto 1/\xi^2. \tag{5.69}$$

They argue that T_{cr} occurs at the temperature, T_{max}, where the static susceptibility (and thus Knight shift) has its maximum. From Figure 5.4, that is 500 K in 1248.

Therefore, for $z = 1$, T_1T/T_{2G} is independent of temperature, while for $z = 2$, $T_1T/(T_{2G})^2$ is proportional to α, which they believe is independent of temperature.

Figures 5.10 and 5.11 show data of Curro et al. [28] for 1248 ($T_C = 81$ K). In Figure 5.11, the upturn in T_1T below 160 K is evidence of the spin gap. Note at higher temperatures, the plot is linear in T, also a prediction from scaling theory.

Figure 5.11 shows the tests of the data using these scaling laws. It appears that the data fit $z = 1$ from 160 K to 500 K, then above 500 K have a better fit to $z = 2$.

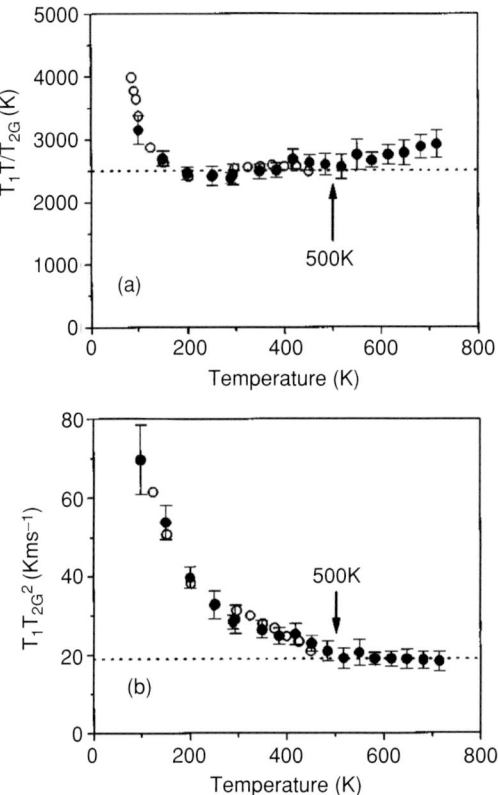

Figure 5.11. Tests for the scaling parameter z in 1248 using Eq. (5.67), indicating that $z = 1$ from 160 K to 500 K (upper figure) and $z = 2$ above 500 K [28].

5.7. YBCO Below T_C: NMR Evidence About the Pairing State

5.7.1. The Knight Shift

Knight shift measurements are the best evidence that the superconducting state is a spin singlet. The first Knight shift measurements in the superconducting state were made by Takigawa, Hammel et al. for $123O_7$ (Figure 5.3). As we have remarked, they found that the powder average Knight shift at T_C was positive relative to its value at $T = 0$. The experiments are complicated by the fact that the Meissner effect screens the metal, making a diamagnetic correction necessary. This they did from measurements of the magnetization and use of formulas for demagnetizing factors. Barrett et al. [45] (reference 27) avoided these difficulties by using the ^{89}Y NMR as a probe of the internal field. Their results are shown in Figure 5.12.

Before discussing their method in detail, we remark on some conclusions from their data. The first is that the shift for applied field along the c-direction (normal to the CuO_2 planes) is independent of T over the whole temperature range. The same result is found for the other YBCO materials. Thus, this component of shift shows no signs of the temperature dependence of the magnetic susceptibility, leading to the conclusion that it arises solely from orbital effects, i.e., is a chemical shift only. The explanation requires that $A_c + 4B = 0$.

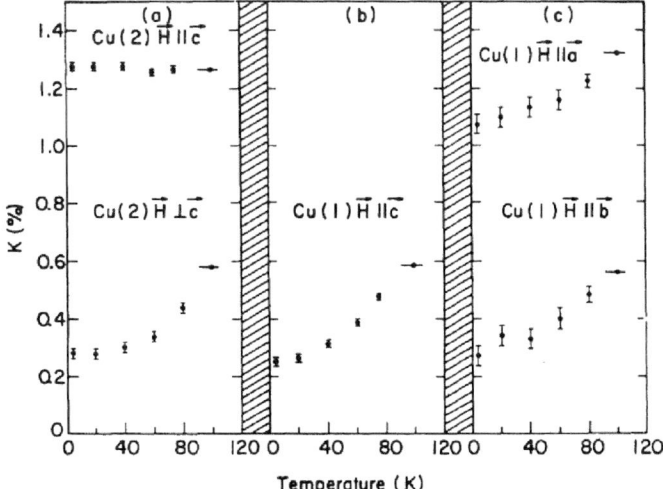

Figure 5.12. ^{63}Cu shift, $K_L + K_S(T)$, for optimally doped YBCO for chain(Cu(1)) and plane(Cu(2)) using the method of Barrett et al. [45].

As pointed out by Pennington et al., the chemical shift can be expressed to a good approximation as

$$K_{\alpha\alpha}^L = 2\beta^2 \left[\sum_n \frac{\langle 0| L_\alpha |n\rangle \langle n| L_\alpha/r^3 |0\rangle}{E_n - E_0} + cc \right]. \tag{5.70}$$

Using reasonable values for the energy levels and the matrix elements one can explain the zero temperature shifts as being orbital i.e., chemical shifts. If this is the case, the spin susceptibility at $T = 0$ vanishes, and the system is in a spin singlet state.

If the system were in a spin triplet state, some components of the Knight shift will not vanish at $T = 0$. In particular, if the trace of the spin susceptibility at T_C is $3\chi_0$, then at $T = 0$, it is $2\chi_0$ [46].

Meier and colleagues have carried out extensive many-electron calculations of the orbital shifts. They conclude that the orbital shifts need to be corrected for the fact that the chemical shift reference material, CuO, has a nonnegligible orbital shift, hence recommend adding 0.15% to all the literature values of the orbital shifts. They also find that so doing brings theory and experiment into agreement for the *aa* and *bb* orbital shifts. However, their theoretical *cc* shift is only about 50% of the experimental result. There is as yet no explanation of this discrepancy.

We now return to an explanation of the experiment of Barrett et al. If H_{int} is the internal field, that is the applied field corrected for the demagnetizing effects,

$$\begin{aligned} \nu_{63}(T) &= \gamma_{63} \left(1 + {}^{63}K_L + {}^{63}K_S(T)\right) H_{int} \\ \nu_{89}(T) &= \gamma_{89} \left(1 + {}^{89}K_L + {}^{89}K_S(T)\right) H_{int}. \end{aligned} \tag{5.71}$$

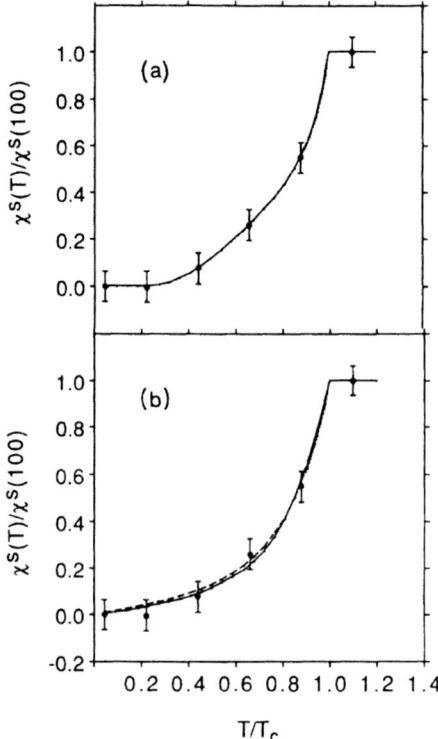

Figure 5.13. Normalized spin susceptibility in the superconducting state determined from ^{63}Cu(2) Knight shift compared theoretical results of Monien (solid and dashed lines), using strong coupling Yosida functions, based on (a) isotropic and anisotropic s-wave models and (b) and two d-wave models [45].

Taking the ratio, and utilizing the smallness of the K_S we get

$$\frac{\nu_{63}(T)}{\nu_{89}(T)} = \frac{\gamma_{63}}{\gamma_{89}} \left(1 + {}^{63}K_L + {}^{63}K_S(T) - {}^{89}K_L + {}^{89}K_S(T) \right). \tag{5.72}$$

The K_Ls are found from the value of the frequencies at $T = 0$, leaving then an equation with the one unknown $^{63}K_S(T) - {}^{89}K_S(T)$.

There are now two approaches that can be made. The first is to assume $^{89}K_S(T)$ is so small that it can be neglected. The second is to assume that both shifts are proportional to the spin susceptibility. Then a plot of $^{63}K_S(T) - {}^{89}K_S(T)$ vs. T gives one the correct temperature dependence of either nuclear Knight shift. Barrett et al. made the latter assumption in their data plots.

A spin singlet can go with various orbital pairing states ($L = 0, 2, 4$, etc.). In principle, the temperature dependence of the spin susceptibility and thus the Knight shift can make the distinction. For $L = 0$, the slope of $K_S(T)$ at $T = 0$ is zero, whereas for a d-wave, there is linear dependence. Figure 5.13 shows theoretical calculations by Monien to fit the Barrett data for the $L = 0$ and $L = 2$ cases. Unfortunately, the experimental precision is inadequate for these data to indicate a choice between the two values of L.

5.7.2. Spin–Lattice Relaxation

The experience with conventional superconductors led experimenters to realize that measurements of T_1 in the superconducting state would be of interest. The first measurements were by Imai et al. [47] and by Kitaoka et al. [48]. Figure 5.14 shows the data of Imai for 123O$_7$. There are two striking features. There is no sign of a coherence peak just below T_C, and the temperature dependence at lower temperatures obeys a T^3 power law rather than an exponential in reciprocal temperature. The power law is what one would expect if there were nodal lines in the energy gap in k-space. Such behavior rules out orbital s-wave pairing. It suggests orbital d-wave or higher.

Figure 5.15 shows data of Martindale et al. [49] of ^{63}Cu and ^{17}O in 123O$_7$, again illustrating the T^3 dependence. Figure 5.16 shows the Cu data plotted in a manner [ln $(1/T_1)$ vs. $1/T$] that would yield a straight line at low temperature for an orbital s-state. On this plot, the slope is a measure of the energy gap needed to excite quasiparticles. Thus, the plot shows that the apparent gap at low temperatures is less than it is a high temperatures.

In the normal state, the spin–lattice relaxation rate is much faster for applied field parallel to the c-axis than for applied field perpendicular to the c-axis. The ratio of these rates is independent of temperature. Barrett et al. [50] discovered that in the superconducting state the anisotropy in relaxation rate is strongly temperature dependent. The explanation for the change in behavior was found by Bulut and Scalapino [51] and later also by Thelen, Pines, and Lu [52]. The physical cause of the T dependence is the fact that for a $d_{x^2-y^2}$ orbital state, there are nodes along the $|k_x| = |k_y|$ lines. Figure 5.17 shows a comparison of experiment with theory.

Figure 5.18 shows the Fermi surface and several spanning vectors. At high temperatures, the spin fluctuations peak at the antiferromagnetic wave vector $(\pi/a, \pi/a)$. At low temperatures, only wave vectors that join points near the nodes can play a role in spin–lattice

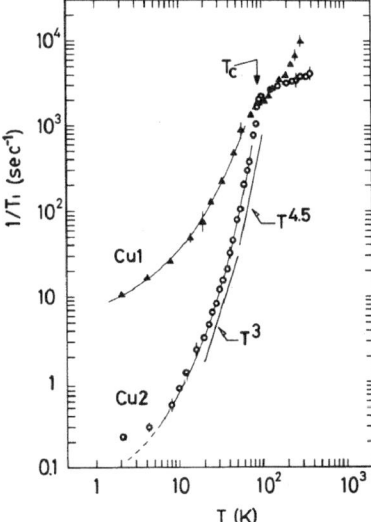

Figure 5.14. Spin–lattice relaxation rate, $1/T_1$, for optimally doped YBCO for planar Cu(2) and chain Cu(1) showing a low temperature power law. The absence of a coherence peak just below T_C and the power law rather than an exponential $1/T$ dependence rule out the conventional s-wave pairing [47].

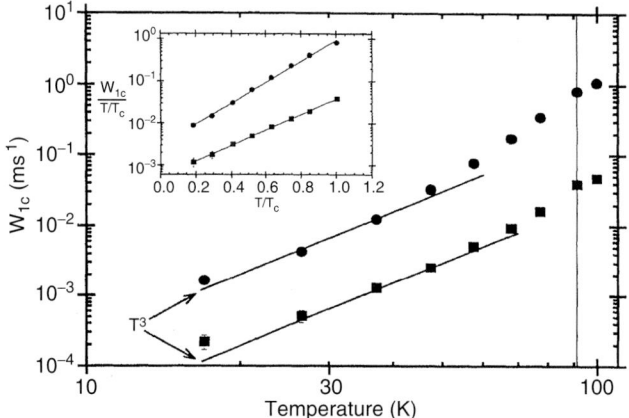

Figure 5.15. NMR T_1s of ^{63}Cu(2) and ^{17}O in optimally doped YBCO, measured a low static magnetic field, showing that both nuclear relaxation rates obey the T^3 law expected for d-wave orbital pairing [49].

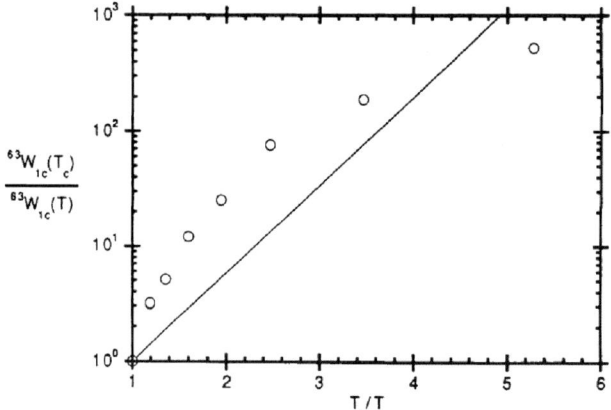

Figure 5.16. $\ln(1/^{63}T_1)$ vs. T_C/T for planar Cu in YBCO$_7$ compared (solid line) to the temperature dependence for a conventional BCS weak coupling, spin-singlet, orbital s-wave pairing with $\Delta(0)/k_B T_C = 1.75$ [49].

relaxation. Thus, the anisotropy measurements show that the nodes lie along the $|k_x| = |k_y|$ lines.

A similar effect is found for T_{2G} (see Figure 5.19). The data of Itoh et al. [53] are well fit by Bulut and Scalapino for an orbital d-wave but not by an orbital s-wave.

5.8. LSCO

5.8.1. The Spectrum

Yoshimura et al. [54] studied Cu NQR in LSCO for $x = 0.15, 0.20$, and 0.30. For a single nuclear species of spin-3/2, one expects a single NQR line corresponding to the transition between the degenerate $\pm 1/2$ levels and the degenerate $\pm 3/2$ levels Since there

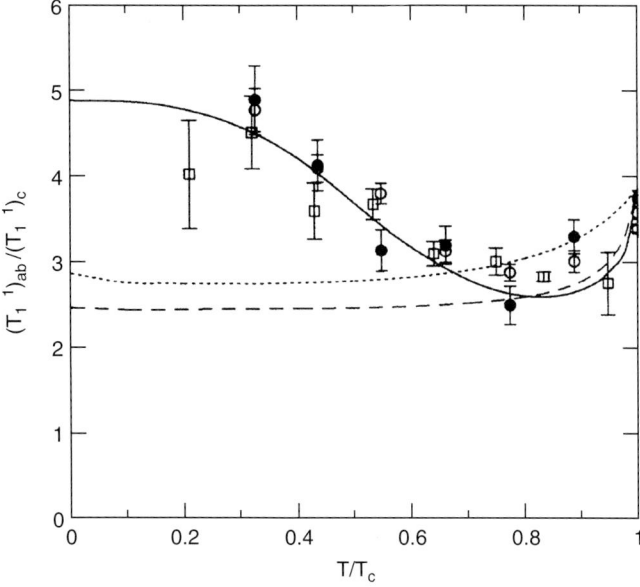

Figure 5.17. Anisotropy in the ^{63}Cu spin–lattice relaxation rate, $(1/T_1)_{ab}/(1/T_1)_c$, below T_C [51].

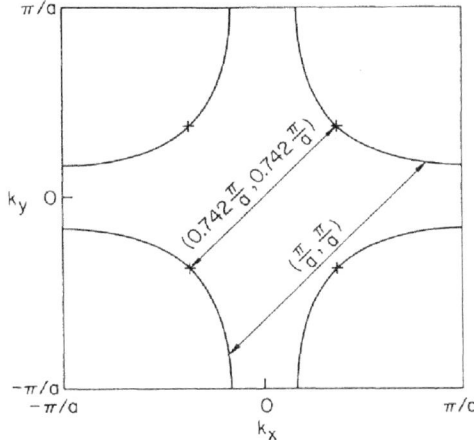

Figure 5.18. Fermi surface for YBCO and several vectors [52].

are two isotopes, ^{63}Cu and ^{65}Cu with very similar properties, we expect to find two lines, one for each isotope. Yoshimura et al. found that the spectrum consisted of four resonance lines. For each isotope, one line is much stronger than the other line. They labeled these the A and B lines, respectively. The relative strength of the B line grew with doping. In pure La_2CuO_4, there is only a single line whose frequency is very close to that of the A line. It was hypothesized that the B line arose from Cu nuclei in which the nearest La site was occupied by a Sr.

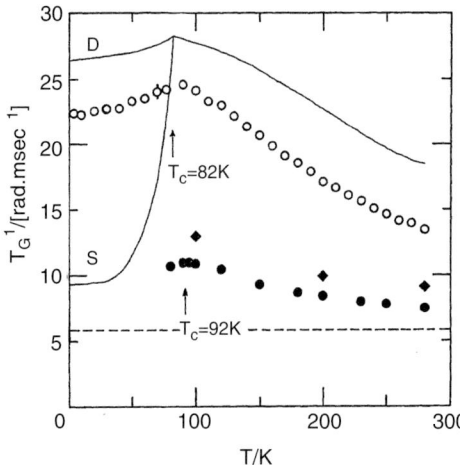

Figure 5.19. T_{2G} vs. temperature compared to theoretical predictions for s-wave and d-wave orbital pairing [53].

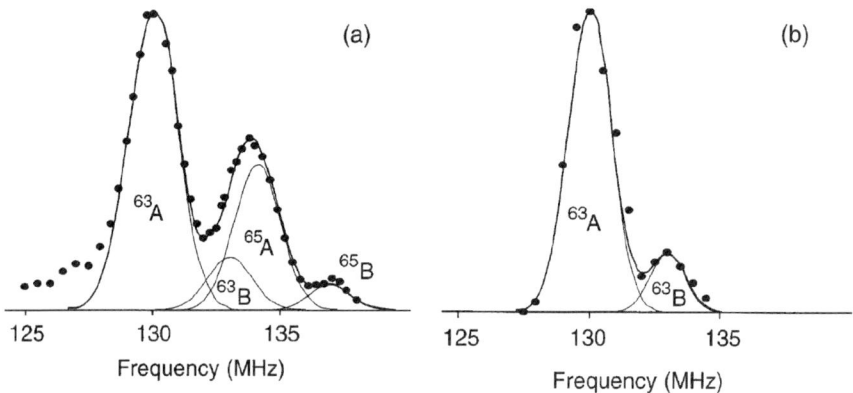

Figure 5.20. NMR resonances of the upper frequency Cu satellite: (a) conventional NMR showing A and B lines of both isotopes; (b) double resonance signal that selectively eliminates the ^{65}Cu signal, revealing the A and B lines of ^{63}Cu [55].

Haase et al. [55, 56] utilized a double resonance method to suppress the signals from the ^{65}Cu nuclei, making it possible to see the ^{63}Cu spectrum clearly. Their data are shown in Figure 5.20. The A and B lines are clearly visible. They showed that from the intensity and the axial symmetry of the B site spectrum that it arose from Cu sites in which a Sr atom replaces a La atom in the nearest La site as proposed by Yoshimura et al. Hunt et al. [57] grew a crystal of $La_{1.875}Ba_{.125}CuO_4$ containing only ^{63}Cu and demonstrated the same thing. We show their figures later in this article in the discussion of low temperature spectra.

The NQR lines of the undoped sample are narrow (a width of 60 kHz), but for a doping of only $x = 0.06$, the NQR lines become 2 MHz wide. The width changes little for higher concentrations. Haase found that the for $x = 0.15$ the A line has a breadth of 2.3 MHz, the B line 1.8 MHz. Thus the line broadening is not a simple result of broadening from the Sr sites, but evidently arises because the presence of the Sr atoms or the doping level they produce, initiate a new phenomena that broaden the lines. The A and B sites appear to be very similar in all NMR properties such as T_1 and T_{2G}.

The doping changes the average electric field gradient at the Cu sites as well as the average electric field gradient. Both increase linearly with doping. Haase and Shushkov have utilized results from atomic spectroscopy to analyze the NMR quadrupole splitting in terms of the hole content of the orbitals that are involved. They are able to determine the hole contents n_d and n_p on the Cu and O and thus to determine the doping $n_d + 2n_p - 1$ as a function of Sr concentration x. They find that for LSCO almost all the doped holes reside on the O.

5.8.2. One or Two Components

There are several issues with respect to $La_{2-x}Sr_xCuO_4$ (LSCO) that at this writing are not resolved. The first is whether this system is a one- or two-component system.

The evidence in favor of a two-component system comes from several sources. Johnston [58] and Nakano et al. [59] have shown that the spin magnetic susceptibility of LSCO consists of two terms. The first, $\chi_1(T)$, is temperature dependent and reaches a maximum value, χ_{1MAX}, at a temperature T_{MAX}. They show that $\chi_1(T/T_{MAX})/\chi_{1MAX}$ obeys a scaling law with temperature as the doping level is varied. The other term in the susceptibility, $\chi_2(T)$, is temperature independent in the normal state, but varies with doping.

Figure 5.21 shows a plot of ^{63}K vs. ^{17}K from Haase [60] for optimally doped LSCO. For a one-component system this should be a single straight line through the origin according to Eq. (5.41). The linear region for large shifts is in the normal state. It is linear but, if extended, does not go through the origin. Equation (5.39) describes the Knight shifts for a two-component system. There are three susceptibilities, but they occur in only two ways:

$$\chi_a = \chi_{dd} + \chi_{pd}$$
$$\chi_b = \chi_{pd} + \chi_{pp}. \quad (5.73)$$

Thus if we consider χ_1 and χ_2 to be linear combinations of χ_a and χ_b we can account for Haase's data, however at this time we are not able to determine the composition of the linear combination.

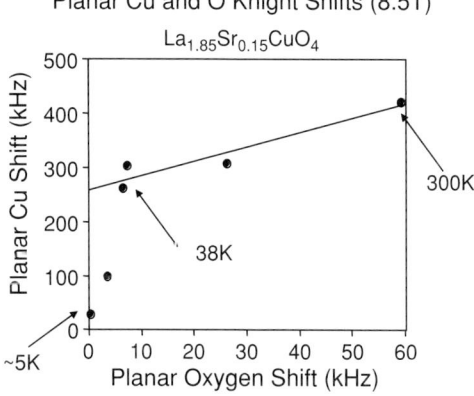

Figure 5.21. $^{63}K_{ab}$ vs. for optimally doped LSCO indicating the possible need for a two-component theory. For a single-component theory, the data should fall on a single line through the origin [60].

Walstedt, Shastry, and Cheong [61] have also challenged the one-band model based on studies of the spin–lattice relaxation time of optimally doped LSCO. They find that $1/^{17}T_1T$ of the planar oxygens increases with temperature, whereas calculations based on neutron data predict that it should fall with temperature. On the other hand, $1/^{17}T_1T$ for the planar oxygen is rather well fit, though not perfectly, by theoretical curves based on neutron data. In another paper, Walstedt, Cheong, Sundstrom, and Greenblatt [62] show that the transfer hyperfine coupling constant, C, of the MMP theory is much smaller in the paramagnetic state of the parent material, the antiferromagnet La_2CuO_4, than in LSCO although the usual MMP assumption is that the hyperfine coupling coefficients are nearly the same in all materials. Indeed, a fundamental number used in determining the size of the hyperfine coupling constants is the size to the Cu hyperfine coupling in the antiferromagnetic state of the parent compounds $YBa_2Cu_3O_6$ and La_2CuO_4.

Gorkov [63] has analyzed the ^{63}Cu T_1 data from a number of cuprates. He finds that after an appropriate vertical offset, all the $1/^{63}T_1$ curves collapse onto the same T dependence above their T_C and below 300 K. In this region, he argues that the nuclear spin–lattice relaxation is a sum of two parallel processes.

$$1\big/^{63}T_1 = 1\big/^{63}\overline{T_1}(x) + 1\big/^{63}\tilde{T}_1(T). \tag{5.74}$$

He conjectures that the doping dependent term is associated with the incommensurate nature of the neutron scattering peaks. He attributes the temperature dependent term to the crossover from the local regime to the dynamical regime. Thus one process arises from "stripe-like" excitations and the other from a "universal" temperature dependent term. So he has a picture of dynamical phase separation into coexisting metallic and incommensurate magnetic phases.

5.8.3. The Incommensurate State

Indeed the second difficulty with understanding LSCO is related to the fact that the neutron scattering peaks of LSCO are slightly displaced from the $(\pi/a, \pi/a)$ point, whereas the peaks in YBCO$_7$ occur at $(\pi/a, \pi/a)$. As can be seen from Figure 5.7, the form factor for the planar oxygens vanishes at the $(\pi/a, \pi/a)$ point. If instead the fluctuations peak at a different point, they make a much bigger contribution to the O relaxation rate. Barzykin, Pines, and Thelen [64] and Zha, Barzykin, and Pines [65] discuss this problem.

For LSCO, the peaks occur at $[(\pi \pm \delta)/a, \pi/a]$ and $[\pi/a, (\pi \pm \delta)/a]$. For optimally doped LSCO, $\delta\delta = 0.245\pi$.

^{63}R changes to

$$^{63}R = \frac{1}{2}\left[1 + \frac{[A_c - 2B(1 + \cos\delta)]^2}{[A_{ab} - 2B(1 + \cos\delta)]^2}\right]. \tag{5.75}$$

The resulting relaxation rate ratio then should change from 3.7 to 4.5. In fact, it then changes to 2.3 according to Haase et al. [66]. The ratio of the Cu to O relaxation rates also changes since the antiferromagnetic fluctuations now become prominent for the O. In the long correlation length limit, the Cu/OT_1 ratio becomes

$$\frac{^{63}(1/T_1)}{^{17}(1/T_1)} = \frac{[A_{ab} - 2B(1 + \cos\delta)]^2}{2C^2(1 - \cos\delta)}. \tag{5.76}$$

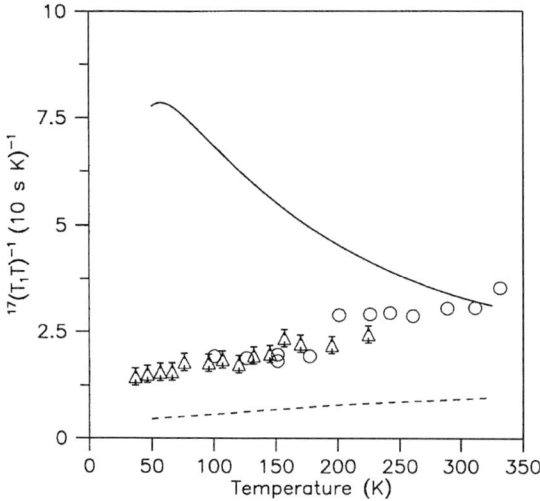

Figure 5.22. O spin–lattice relaxation ($1/T_1T$) vs. temperature for optimally doped LSCO, compared to two theoretical models: solid line assuming the spin fluctuation peak occurs at the wave vector given by the neutron diffraction incommensurate peak, the dashed curve assuming the fluctuations peak at the commensurate, (π, π), point [36].

Figure 5.22 from Barzykin, Pines, and Thelen [36] compares various theoretical and experimental results for O spin–lattice relaxation. It is evident that predictions based on the known degree of incommensurability from neutrons are far from the experimental results.

One possible explanation [67] is that the fluctuations are commensurate spatially except for regions of phase slip, much like McMillan's concept of discommensurations [68] for charge density waves. Independently, Phillips [69] has proposed a similar idea.

5.8.4. Spatial Modulation

Haase et al. [70–72] have discovered that the NMR properties of LSCO are spatially modulated. The evidence is very clear that the modulation is large and short range, but to date the exact form has not been established from NMR.

In simple materials such as copper metal or the alkali halides, the width of the central transition ($+1/2 \to -1/2$) is determined the dipolar interaction between neighboring nuclei, perhaps augmented by the indirect coupling when it is sufficiently large. For optimally doped LSCO with a strong applied field (8.3 T) parallel to the c-axis, the ^{63}Cu central transition width is 1.9 MHz for the A transition and 1.5 MHz for the B transition whereas the nuclear spin–spin coupling, which can be characterized by $1/T_{2G}$, is of the order of 10 kHz [73–75].

Thus nuclear spin–spin coupling cannot explain this width. If the electron-spin susceptibility were nonuniform owing to long range doping variations, the effect would not show up in the linewidth since $A_c + 4B = 0$. However, if the wavelength of Knight shift variation were comparable to the lattice constant, such a variation would contribute to line width. This result led Haase et al. to carry out a spin echo double resonance experiment in which they looked at the echo at one site while probing the effect of tuning the frequency of a second pulse. Figure 5.23 shows the data. The solid line is the normal ^{63}Cu spin echo line shape. The filled circles show the destruction of ^{63}Cu intensity, inspected at various line positions, produced by an 8 μs ^{65}Cu π pulse (corresponding to a 125 kHz resolving power) applied at the ^{65}Cu line

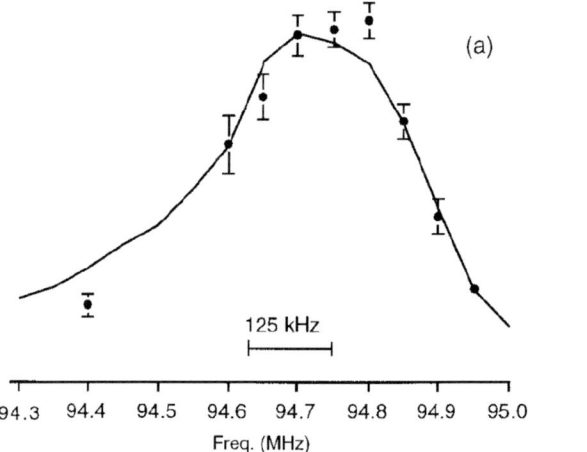

Figure 5.23. Comparison of normal ^{63}Cu spin echo line shape (solid curve) with ^{63}Cu line shape obtained at different ^{63}Cu frequencies by spin echo double resonance in which π-pulses are applied at the center of the ^{65}Cu line. The fact that the ^{63}Cu line shape is the same in both cases shows that at neighboring sites the Cu spins may have very different frequencies, showing that the modulation is short wavelength. Data taken at 300 K [70].

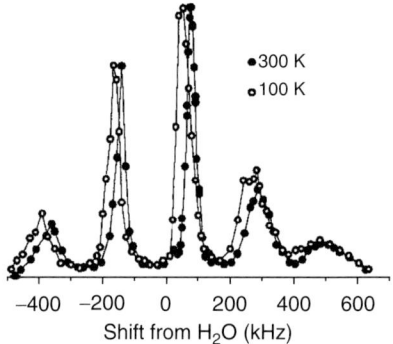

Figure 5.24. ^{17}O lineshapes for optimally doped LSCO, showing the asymmetry about the central transition that show that Knight shift and electric field gradient are modulated spatially coherently [76].

center, showing that neighboring Cu nuclei are typically far apart in frequency. Therefore, the modulation is on the scale of the lattice constant.

Oxygen NMR data show that both the electric field gradient and the Knight shift are modulated and that they are modulated in step (larger Knight shift goes with larger quadrupolar coupling). This result is immediately evident from the ^{17}O spectrum. Figure 5.24 shows the data of Haase et al. [76] for optimally doped LSCO. The striking feature of the spectrum is that the resonance lines to the left of the central transition are narrower and higher than those on the right. If there were only a modulation of the quadrupolar coupling the spectral pattern would be symmetric about the central transition, the central transition would not be broadened, the outer two lines would be twice as broad as the inner pair. If only the Knight shift were broadened, the pattern would be symmetric about the central transition, and all five lines would have the same width. Haase et al. show that the spectrum can be completely explained by assuming that for transition $n(= 2, 1, 0, -1, -2)$ the resonance frequency obeys

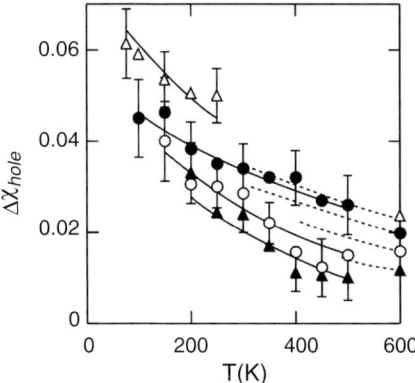

Figure 5.25. Temperature dependence of the hole density modulation vs. temperature deduced from ^{63}Cu T_1 measurements. Data, from top to bottom, for $x = 0.16, 0.115, 0.07, 0.04$ [77].

an equation

$$v_n = (1 + K_0) + n v_{q,0} + (M + nR)h, \qquad (5.77)$$

where h is a parameter that characterizes the modulation and M and R parameters that describe the effect of modulation of h on the Knight shift and the electric field gradient. If one assumes that h is distributed with some function $f(h)$, then f gives one the line shape. In this model, if one knows the shape of any one transition, one can calculate the shape of the other four. Indeed the data satisfy this condition.

Since the average Knight shift and the average electric field gradient change with doping, it is logical to interpret these data as saying that there is a spatial variation of the doping. If one uses this formula to describe the average frequency of the NMR lines (the first moment of the spectrum), in a field of 8.3 T, $R/M = 1$. But to fit the line shapes, one finds that $R/M = 2$, a result probably of the fact that the modulation is not of infinite wavelength.

Singer et al. [77] have studied ^{63}Cu NQR in LSCO with doping, x, from 0.06 to 0.16. These samples were grown using isotopically pure ^{63}Cu. They find that the spin–lattice relaxation rate varies across the NQR line, but that at a given NQR frequency, the T_1 is roughly the same for all dopings that have absorption at that frequency. They therefore interpret the data as saying that there is a charge variation within the sample, the size of which they can estimate from the average change in $1/T_1$ with doping. Figure 5.25 shows the variation in hole concentration deduced in this manner.

Figure 5.26 shows their T_1 data. Both the A and B lines are evident, as well as the variation of the B line with doping that gives solid proof that the B site arises from Cu sites in which the nearest La is replaced by a Sr atom. They describe the situation as arising from charge patches. In a complex calculation of electric field gradients and NQR line widths, they argue that they can determine the approximate size of the patches and the charge excess associated with the patch. They estimate the patch size as being greater than 3 nm.

A quantitative measure of the modulation can be found by calculating the second moment of the NMR line broadening. We introduce the spin operator σ_i (for Cu atoms at position i) that measures the difference between S_{zi} and its spatial average $< S_{zi} >$. Then, utilizing Eq. (5.40), we find for the second moments of the apical and planar Os and the planar Cu

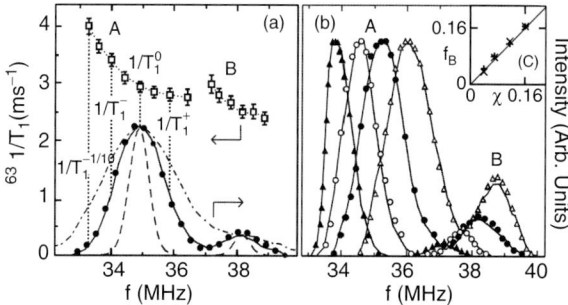

Figure 5.26. (a) Frequency dependence of $(1/^{63}T_1)$ and NMR signal strength for $x = 0.015$ sample showing A and B lines. Dashed, solid, and dot-dashed lines for patch size infinite, 3, and 1.5 nm. (b) A and B lines for, left to right, $x = 0.04, 0.07, 0.115, 0.016$ [77].

$$\left\langle \Delta\nu^2_{\text{apical},\alpha} \right\rangle_{17} = E_\alpha^2 \langle \sigma_0^2 \rangle$$
$$\left\langle \Delta\nu^2_{\text{planar},\alpha} \right\rangle_{17} = 2C_\alpha^2 \langle \sigma_0^2 + \sigma_0\sigma_1 \rangle \tag{5.78}$$
$$\left\langle \Delta\nu^2_c \right\rangle_{63} = A_c^2 \langle \sigma_0^2 \rangle + 2A_c B \langle \sigma_0^2 + \sigma_0\sigma_1 \rangle + 4B^2 [\langle \sigma_1\sigma_2 \rangle + 2\langle \sigma_1\sigma_3 \rangle],$$

where the central Cu is at position labeled 0 and the neighbors on the $+x, -y, -x, -y$ axes are labeled 1, 2, 3, 4, respectively.

The ratio of C to A are obtained from comparing the temperature variation of the apical to the planar ^{17}O knight shifts. Thus, from the apical and planar oxygen second moments, one can obtain $\langle \sigma_0\sigma_1 \rangle / \langle \sigma_0^2 \rangle$. Combining these results with the second moment of the Cu central transition, one can obtain $[\langle \sigma_1\sigma_2/\sigma_0^2 \rangle + 2\langle \sigma_1\sigma_3/\sigma_0^2 \rangle]$.

Figure 5.27 shows results from Haase et al. for $\langle \sigma_0\sigma_1 \rangle / \langle \sigma_0^2 \rangle$ for three separate dopings. The value -1 corresponds to perfect antiferromagnetic order to the shift variations. We see that even for the optimally doped material ($x = 0.15$), we approach this limit at low temperatures. The $x = 0.1$ doping has nearly antiferromagnetic lattice modulations even at room temperatures, whereas the $x = 0.2$ doping remains rather far from this limit over the entire temperature range.

5.8.5. The High Temperature Properties

Imai et al have studied $^{63}1/T_1$ [78] and $^{63}1/T_{2G}$ [79] for undoped (Figure 5.28) and Sr doped $La_{2-x}Sr_xCuO_4$ (Figure 5.29). Analysis of the results for the undoped parent material La_2CuO_4 shows that it is well described by theories for the two-dimensional quantum Heisenberg antiferromagnet. The striking result of Figure 5.29 is that at high temperature, the T_1s for all dopings become nearly identical, showing that at high temperature the excitations in these materials are basically those of the parent antiferromagnet.

5.8.6. The Low Temperature Properties: Wipeout

In 1995, Tranquada et al. [80] showed by neutron scattering that $La_{1.6-x}Nd_{0.4}Sr_xCuO_4$ for x near to 1/8 exhibits charge stripe order for temperatures below 65 K, followed by spin stripe order below 50 K. The Nd atoms stabilize the crystal structure into the low temperature tetragonal (LTT) form and the doping near $x = 1/8$ was chosen to pin the stripes in a static

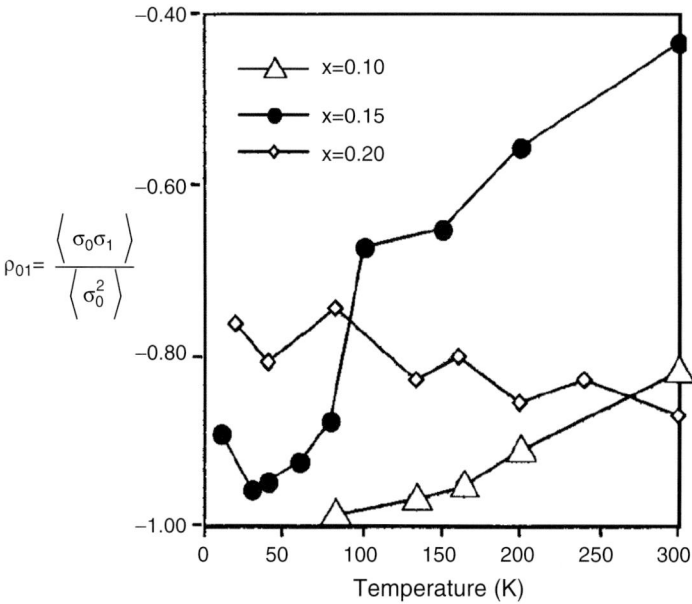

Figure 5.27. Temperature dependence of the normalized Cu electron spin–spin correlation function between first neighbors for LSCO with $x = 0.1$, 0.15, and 0.20. -1 corresponds to a 100% antiferromagnetic correlation of the Knight shift modulation [72].

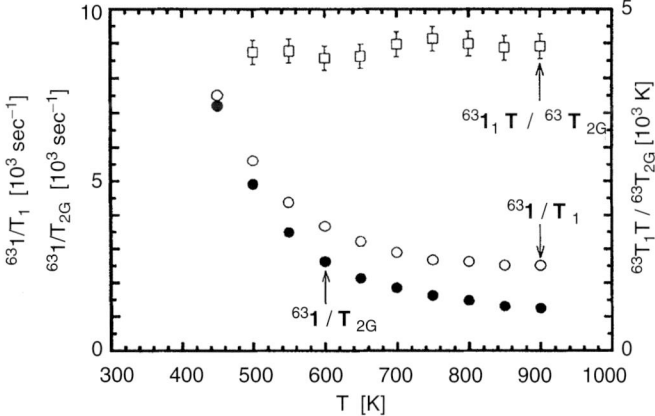

Figure 5.28. $1/(^{63}T_1)$, $1/(^{63}T_{2G})$, and $(^{63}T_1T)/(^{63}T_{2G})$ vs. temperature for La_2CuO_4. The data for $(^{63}T_1T)/(^{63}T_{2G})$ show that it obeys $z = 1$ scaling. The data agree with predictions of the two dimensional Heisenberg model [79].

configuration. It is therefore natural to ask whether the existence of such stripes shows up in NMR studies. Several groups have carried out investigations.

Hunt et al. [81] studied ^{63}Cu NMR in $La_{1.875}Ba_{0.125}CuO_4$, $La_{1.48}Nd_{0.4}Sr_{0.12}CuO_4$, $La_{1.68}Eu_{0.2}Sr_{0.12}CuO_4$ by NQR. (The Eu atoms differ from the Nd atoms in lacking a permanent magnetic moment.) Hunt et al. find dramatic intensity loss (termed "wipeout") in the

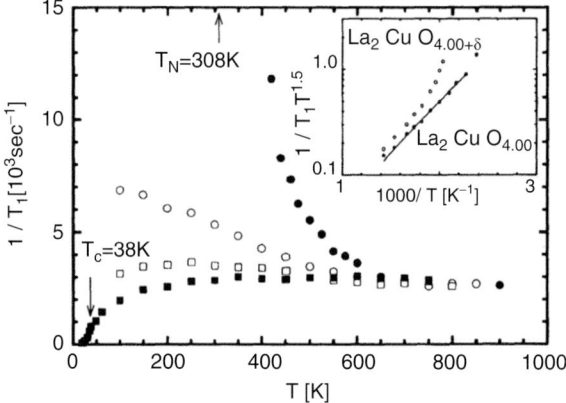

Figure 5.29. $1/(^{63}T_1)$ for LSCO for $x = 0, 0.04, 0.075, 0.15$ (top to bottom curves) showing that at high temperatures the relaxation behavior is very similar to that of the undoped material, hence is dominated by spin fluctuations [78].

NQR signal on cooling that they associate with the formation of charge stripe because its onset corresponding to the charge ordering temperature observed by neutrons. They report similar effects for $La_{2-x}Sr_xCuO_4$ for x between 1/16th and 1/8th. Evidently, they attribute the signal loss to electric quadrupole effects although they do not present a detailed mechanism. In a second paper [82], they point out that fluctuations in the charge stripe order will lead to loss of the NQR resonance. They also mention that the charge order turns on slow electron-spin fluctuations that may also reduce the NQR signal intensity.

Curro et al. [83] studied the ^{63}Cu and ^{139}La resonances in $La_{1.8-x}Eu_{0.2}Sr_xCuO_4$. Since the La atoms are outside of the CuO_2 planes, the coupling of the ^{139}La nuclei is much weaker than that of the ^{63}Cu nuclei. Consequently, Curro et al. were able to observe the ^{139}La resonance even at temperatures where the ^{63}Cu signal was wiped out. From the ^{139}La T_1, they showed that the Cu electron spins were fluctuating with a correlation time τ_C that obeyed a law

$$\tau_C = \tau_0 \exp(E_a/k_B T). \tag{5.79}$$

where E_a was distributed over a range comparable to its average value of the order of 80 K. They found that this picture described data for the Eu doped samples with $x = 0.15$ and 0.015, and for pure LSCO with $x = 0.014$.

Further insight is obtained by electron-spin resonance (ESR). Kataev et al. [84, 85] have studied LSCO and Eu-doped LSCO by ESR. The technique consists of replacing about 1% of the La atoms with Gd. The resulting Gd^{3+} ions have an electron spin of 7/2 and a g-value of 2. There is an axial electric field splitting giving rise to a Hamiltonian, H, in the presence of an applied magnetic field H_0

$$H = D\left[S_z^2 - S(S+1)/3\right] + g\mu_B H_0 \cdot S,$$

where D turns out to be about $0.28\,\mathrm{cm}^{-1}$. They observe a strong broadening of the ESR line at low temperatures (below about 60 K for a sample with $x = 0.1$) that they attribute to effects on the Gd T_1 of "extremely slow antiferromagnetic dynamics."

Spin glasses have been studied previously by NMR or NQR. For example, Chen and Slichter [86] studied the ^{93}Nb NQR resonance in iron-doped 2H-niobium diselenide. The

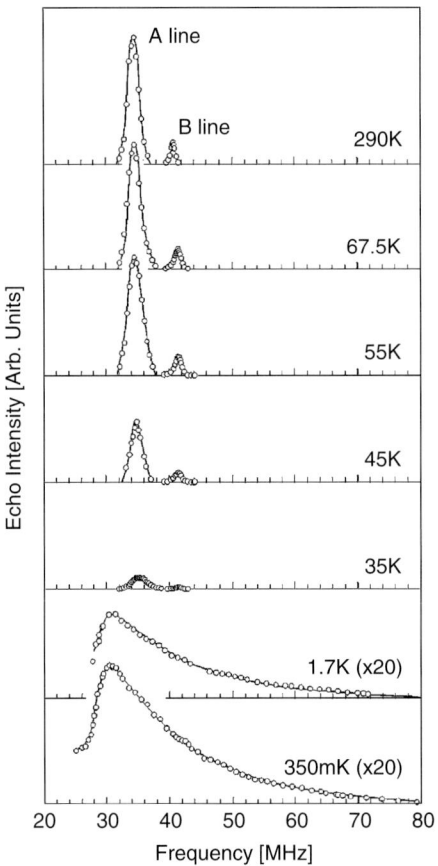

Figure 5.30. ^{63}Cu NQR lines at various temperatures for La$_{1.85}$Ba$_{0.125}$CuO$_4$. The sample was isotopically pure. The signal is smaller in the temperature range near 35 K owing to the wipeout phenomena, but is recovered at the two lowest temperatures where the spin fluctuations to a rate less than the NQR linewidth [89].

^{93}Nb spin of 9/2 gives several transitions for the NQR, the highest frequency one being the 9/2 to 7/2 transition at 10 MHz. They found a well-defined minimum in NQR intensity at the spin glass temperature that they attribute to slowing down of the spin fluctuations.

The classic spin glass system is Cu doped with Mn. From studies at high temperatures and high magnetic fields, the magnetic resonance of Cu atoms at the sites near to the magnetic atom [87] for systems of low Mn content, were found. Utilizing these results, Alloul [88] could predict the resonance frequency of the first neighbor Cu to an Mn atom at zero applied field and at temperatures well below the spin glass temperature where the orientations of the Mn magnetic moments were frozen. He successfully observed both the first and fourth neighbors to the Mn in zero applied magnetic field at 1.25 K. Hunt et al. [89] performed a similar experiment on La$_{1.875}$Ba$_{0.125}$CuO$_4$ in which only the isotope ^{63}Cu was present. Figure 5.30 shows the Cu NQR signal as a function of temperature (The weak signal labeled "B" comes from Cu atoms where the nearest La atom has been replaced by a Ba atom). The signal nearly disappears at 35 K, but returns as one cools near 1 K. Hunt et al. made efforts to fit the low temperature line shape using several stripe models, but find the best fit comes from

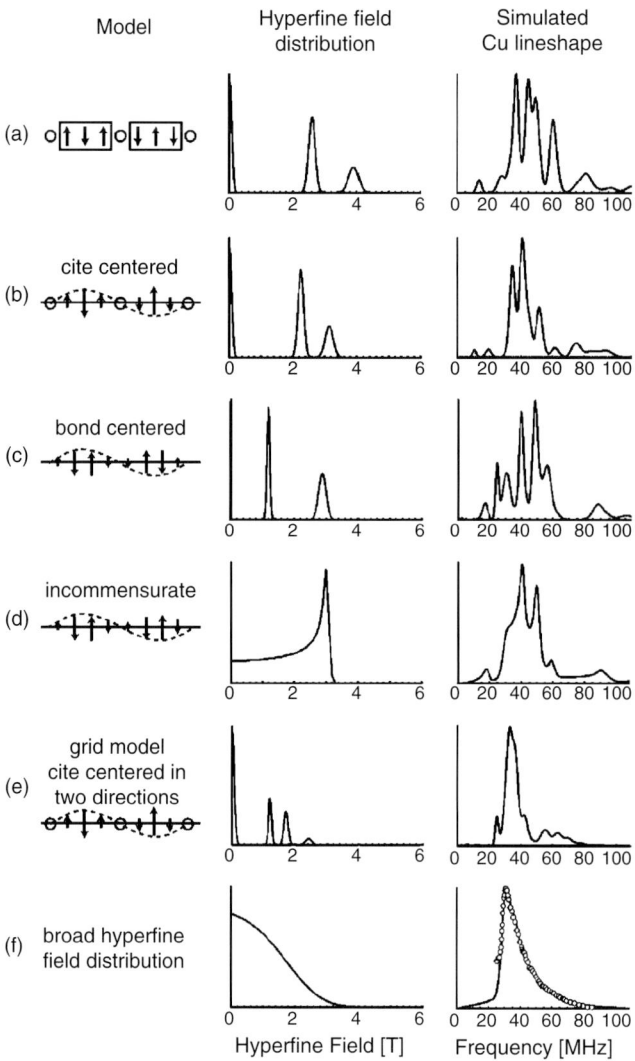

Figure 5.31. Comparison with several theoretical models to account for the low temperature lineshapes of Figure 5.30. The best fit comes from a broad hyperfine distribution such as one might get from some forms of two-dimensional spatial modulations [89].

a simple broad distribution of hyperfine fields. The results of several such model calculations are shown in Figure 5.31.

Thus, NMR and ESR show that "stripe" phases have many of the characteristics of a spin glass with spin fluctuations that are slow even on the NMR time scale at low temperatures.

5.9. Brief Review of EPR

Although the Cu^{2+} ion is a well-known EPR atom (a $3d^9$ configuration), and was one of the first studied in the early days of EPR, for example in copper sulfate, initially no EPR signal was seen in the cuprate high T_C materials. The reason was probably because of the wide line width. So the alternative approach has been to study the cuprates by doping with magnetic atoms such as Fe^{3+} or Mn^{2+} that might go in the CuO_2 planes, or with atoms such as Gd^{3+}, Fe^{3+}, or Mn^{2+} that dope between the planes. Of these Gd^{3+}, Fe^{3+}, and Mn^{2+} are S-state ions ($L = 0$), and thus particularly good EPR candidates.

Alekseevski et al. [90] made one of the first successful studies of EPR in a cuprate, studying the Gd EPR of $GdBa_2Cu_3O_{6+x}$ for a range of x. Although the Gd is present to 100%, the resonance can be seen although it is 1,000 gauss wide.

Jánossy et al. [91] studied $YBa_2Cu_3O_{6+x}$ with 0.001 of the Y atoms replaced by Gd. At such a low concentration, mutual electron spin flips between Gd atoms are unlikely, making the Gd an effective local probe. Various crystal field lines are resolved and exhibit structure that reveals the spectra for various local orders of the chain O atoms as x was varied. In another publication [92] with $x = 0.76$ (putting the sample in the pseudogap regime) they showed that the Gd^{3+} frequency had a shift that displayed the pseudogap, having a temperature dependence identical to that of the ^{89}Y NMR Knight shift. In another set of experiments, Kataev et al. [93] studied stripe phenomena in $La_{1.84-x}Gd_{0.01}Eu_{0.15}Sr_xCuO_4$, as we have discussed.

Kochelaev et al. [94] used Mn^{2+} EPR to study the spin dynamics of Mn^{2+}-doped LSCO. They found bottle-necked behavior arising evidently because the Mn and Cu electron systems are strongly coupled by exchange. Shengelaya et al. [95] also studied Mn-doped LSCO. For lightly doped material ($x < 0.06$), they found two lines at low temperatures ($T < 50$ K), attributing the narrow line to hole-rich, metallic regions and the broad line to hole-poor antiferromagnetic regions. The narrow line grows exponentially in intensity as the temperature is lowered with an activation energy of about 460 K. In another paper [96], they studied the effect of the O isotopic mass in the same material (Mn doped LSCO), changing the O doping from ^{16}O to ^{18}O. For Sr doping of 0.06, they find that the ^{18}O sample has a somewhat broader line width at lower temperature than the sample made with ^{16}O. The effect is largest in the underdoped region and disappears in the overdoped region. They present an elegant and sophisticated theory involving coupling of lattice distortion modes to the Dzyaloshinky–Moriya type spin terms. They point out that the relevant modes are Jahn–Teller active and are considered relevant to possible bipolaron formation. Thus, these experiments may support an explanation of superconductivity based on the electron–lattice interaction through the Jahn–Teller effect.

Acknowledgment

The author is indebted to a number of people for stimulating discussions about many of the topics discussed in this review. Among them are David Pines, Jörg Schmalian, Dirk Morr, Hartmut Monien, Takashi Imai, Jürgen Haase, Nicholas Curro, Dylan Smith, Sean Barrett, Charles Pennington, Dale Durand, Joe Martindale, Craig Milling, Raivo Stern, Donald Ginsberg, Doug Scalapino, Rachel Wortis, David Johnston, Lev Gorkov, and Peter Meier. Special thanks go to Dylan Smith for help with preparation of the manuscript.

This research has been supported in part by the US Department of Energy, Division of Material Sciences, through the Frederick Seitz Materials Research Laboratory at the University of Illinois at Urbana-Champaign under Grant No. DEFG02-91ER45439.

Bibliography

1. A more complete discussion of magnetic resonance theory can be found in *Principles of Magnetic Resonance* by C. P. Slichter, 3rd edition (Springer, Berlin Heidelberg New York, reprinted 2005).
2. S. Renold, T. Heine, J. Weber, and P. F. Meier, "Nuclear magnetic resonance chemical and paramagnetic field modifications in La_2CuO_4", Phys. Rev. B **67**, 024501 (2003).
3. J. Haase, N. J. Curro, R. Stern, and C. P. Slichter, "New methods for NMR of superconductor", Phys. Rev. Lett. **81**, 1489 (1998).
4. L. C. Hebel and C. P. Slichter, "Nuclear relaxation in superconducting aluminum", Phys. Rev. **107**, 901 (1957).
5. L. C. Hebel and C. P. Slichter "Nuclear spin relaxation in normal and superconducting aluminum", Phys. Rev. **113**, 1504 (1959).
6. Y. Masuda and A. G. Redfield, "Nuclear spin–lattice relaxation in superconducting aluminum", Phys. Rev. **125**, 159 (1962).
7. R. H. Hammond and W. D. Knight, "Nuclear quadrupole resonance in superconducting gallium", Phys. Rev. **120**, 762 (1959).
8. W. W. Simmons, *Pure Quadrupole Resonance in Indium Metal*, Ph.D. Thesis, University of Illinois, Urbana, Illinois (1960).
9. W. W. Simmons and C. P. Slichter "Nuclear quadrupole resonance in indium metal", Phys. Rev. **121**, 1580 (1961).
10. G. Redfield "Local field mapping in mixed-state superconducting vanadium by nuclear magnetic resonance", Phys. Rev. **162**, 367 (1967).
11. G. E. Volovik, JETP Lett. **58**, 469 (1993).
12. J. Curro, C. Milling, J. Haase, and C. P. Slichter, Phys. Rev. B **62**, 3473 (2000).
13. D. K. Morr and R. Wortis, Phys. Rev. B **61**, 882, (2000).
14. W. Heitler and E. Teller, Proc. R. Soc. **A155**, 629 (1936).
15. J. Korringa, Physica **16**, 601, (1950).
16. D. Pines, *Solid State Physics*, vol. 1, F. Seitz and D. Turnbull (eds.) (Academic, New York, 1955).
17. L. Cooper, *Nobel Lectures, 1971-1980*, S. Lundquist (ed.) (World Scientific, Singapore, 1992).
18. D. E. MacLaughlin, Magnetic resonance in the superconducting state, in *Solid State Physics*, vol. 31, H. Ehrenreich, F. Seitz, and D. Turnbull (eds.) (Academic, New York, 1976), pp. 1–69.
19. K. Yosida, Phys. Rev. **110**, 769 (1958).
20. Y. Masuda and A. G. Redfield, Phys. Rev. **125**, 159 (1962).
21. F. Wright, Jr., W. A. Hines, and W. D. Knight, Phys. Rev. Lett. **18**, 115 (1967).
22. M. Takigawa, P. C. Hammel, R. H. Heffner, and Z. Fisk, Phys. Rev. B **39**,7371 (1989).
23. F. Mila and T. M. Rice, "Analysis of Magnetic Resonance Experiments in $YBa_2Cu_3O_7$", Physica C **157**, 561 (1989).
24. F. C. Zhang and T. M. Rice, "Effective Hamiltonian for the superconducting Cu oxides", Phys. Rev. B **37**, 3759 (1988).
25. N. J. Curro, B. -L, Young, J. Schmalian, and D. Pines, "Scaling in the emergent behavior of heavy-electron materials", Phys. Rev. B **70**, 235117 (2004).
26. M. Takigawa, P. C. Hammel, R. H. Heffner, Z. Fisk, K. C. Ott, and J. D. Thompson, Phys. Rev. B **43**, 247 (1991).
27. H. Alloul, T. Ohno, and O. Mendels, Phys. Rev. Lett. **63**, 1700 (1989).
28. S. E. Barrett, D. J. Durand, C. H. Pennington, C. P. Slichter, T. A. Friedmann, J. P. Rice, and D. M. Ginsberg, Phys. Rev. B **44**, 6283 (1990).
29. N. J. Curro, T. Imai, C. P. Slichter, and B. Dabrowski, Phys. Rev. B **56**, 877 (1997).
30. J. M. Tranquada, J. D. Axe, N. Ichikawa, Y. Nakamura, S. Uchida, and B. Nachumi, "Neutron-scattering study of stripe-phase order of holes and points in $La_{1.48}Nd_{0.4}Sr_{0.12}CuO_4$", Phys. Rev. B **54**, 7489 (1996).
31. C. H. Pennington, D. J. Durand, C. P. Slichter, J. P. Rice, E. D. Bukowski, and D. M. Ginsberg, "Static and dynamic NMR tensors of $YBa2Cu_3O_{7-\delta}$", Phys. Rev. B **39**, 2902 (1989).
32. S. Barrett, J. A. Martindale, D. J. Durand, C. H. Pennington, C. P. Slichter, J. P. Rice, E. D. Bukowski, and D. M. Ginsberg, Phys. Rev. Lett. **66**, 108 (1991).
33. B. S. Shastry, Phys. Rev. Lett. **63**, 1288 (1989).
34. P. C. Hammel, M. Takigawa, R. H. Heffner, Z. Fisk, and K. C. Ott, Phys. Rev. Lett. **63**, 1992 (1989).
35. T. Moriya, J. Phys. Soc. Jpn **18**, 516 (1963).
36. V. Barzykin, D. Pines, and D. Thelen, Phys. Rev. B **50**, 16052 (1994).
37. T. Imai, C. P. Slichter, A. P. Paulikas, and B. Veal, Phys. Rev. B **47**, 9158 (1993).
38. H. Monien, D. Pines, and C. P. Slichter, Phys. Rev. B **41**, 11120 (1990).

39. E. Manousakis, Rev. Mod. Phys. **63**, 1 (1991).
40. H. Monien, D. Pines, and M. Takigawa, Phys. Rev. B **43**, 258 (1991).
41. C. H. Penninigton, D. J. Durand, C. P. Slichter, J. P. Rice, E. D. Bukowski, and D. M. Ginsberg, Phys. Rev. B **39**, 274 (1989).
42. C. H. Pennington and C. P. Slichter, Phys. Rev. Lett. **66**, 381 (1991).
43. J. Haase, D. K. Morr, and C. P. Slichter, Phys. Rev. B **59**, 7191 (1999).
44. A. Chubukov and S. Sachdev, Phys. Rev. Lett. **71**, 169 (1993); V. Barzykin, D. Pines, A. Sokol, and D. Thelen, Phys. Rev. B **49**, 1544 (1994); A. Sokol and D. Pines, Phys. Rev. Lett. **71**, 2813 (1994).
45. S. E. Barrett, D. J. Durand, C. H. Pennington, C. P. Slichter, T. A. Friedmann, J. P. Rice, and D. M. Ginsberg, Phys. Rev. B **44**, 6283 (1990).
46. A. J. Leggett, Rev. Mod. Phys. **47**, 331 (1975).
47. T. Imai, t. Shimizu, H. Yasuoka, Y. Ueda, and K. Kosuge, Phys. Soc. Jpn **57**, 2280 (1988).
48. Y. Kitaoka, s. Hiramatsu, T. Kondo, and K. Asayama J. Phys. Soc. Jpn **57**, 30 (1988).
49. J. A. Martindale, S. E. Barrett, K. E. O'Hara, C. P. Slichter, W. C. Lee, and D. M. Ginsberg Phys. Rev. B **47**, 9155 (1993).
50. S. E. Barrett, J. A. Martindale, D. J. Durand, C. H. Pennington, C. P. Slichter, T. A. Friedman, and D. M. Ginsberg, Phys. Rev. Lett. **66**, 108 (1991); see also M. Takigawa, J. L. Smith, and W. L. Hults, Phys. Rev. B **44**, 7764 (1991).
51. N. Bulut and D. J. Scalapino, Phys. Rev. Lett. **68**, 706 (1992).
52. D. Thelen, D. Pines, and J. P. Lu, Phys. Rev. B **47**, 9151 (1993).
53. Y. Itoh, H. Yasuoka, Y. Fujiwara, Y. Ueda, T. Machi, I. Tomeno, K. Tai, N. Koshizuka, and S. Tanaka, J. Phys. Soc. Jpn **61**, 1287 (1992).
54. K. Yoshimura, T, Imai, T. Shimizu, Y. Ueda, K. Kosuge, and H. Yasuoaka, J. Phys. Soc. Jpn **58**, 3057 (1989).
55. J. Haase, N. J. Curro, R. Stern, and C. P. Slichter, Phys. Rev. Lett. **81**, 1489 (1998).
56. J. Haase, R. Stern, D. G. Hinks, and C. P. Slichter, *Stripes and Related Phenomena*, A. Bianconi, and N. L. Saini (eds.) (Plenum, New York, 2000), p. 303.
57. A. W. Hunt, P. M. Singer, A. F. Cederström, and T. Imai, Phys. Rev. B **64**, 134525 (2001).
58. D. C. Johnston, Phys. Rev. Lett. **62**, 957 (1989).
59. T. Nakano, M. Oda, C. Manabe, N. Momomo, Y. Miura, and M. Ido, Phys. Rev. B **49**, 16000 (1994).
60. J. Haase and C. P. Slichter (private communication).
61. R. E. Walstedt, B. S. Shastry, and S. -W. Cheong, Phys. Rev. Lett. **72**, 3610 (1994).
62. R. E. Walstedt, S. -W. Cheong, J. Sundstrom, and M. Greenblatt, Phys. Rev. Lett. **80**, 2457 (1998).
63. L. P. Gorkov and G. B. Teitel'baum, JETP Lett. **80**, 195 (2004).
64. V. Barzykin, D. Pines, and D. Thelen, Phys. Rev. B **50**, 16052 (1994).
65. Y. Zha, V. Barzykin, and D. Pines, Phys. Rev. B **54**, 7561 (1996).
66. J. Haase and C. P. Slichter
67. C. P. Slichter, private communication
68. W. L. McMillan, Phys. Rev. B **12**, 1187 (1975) and **14**, 1496 (1976).
69. J. C. Phillips, Solid State Commun. **84**, 189 (1992).
70. J. Haase, C. P. Slichter, R. Stern, C. T. Milling, and D. G. Hinks, J. Supercond. **13**, 723 (2000).
71. J. Haase, C. P. Slichter, R. Stern, C. T. Milling, and D. G. Hinks, Physica C **341** (2000).
72. J. Haase, C. P. Slichter, C. T. Milling J. Supercond. **15**, 339 (2002).
73. T. Imai, C. P. Slichter, K. Yoshimura, and K. Kosuge, Phys. Rev. Lett. **70**, 1002 (1993).
74. T. Imai, and C. P. Slichter, Phys. Rev. Lett. **71**, 1254 (1993).
75. T. Imai, C. P. Slichter, K. Yoshimura, M. Katoh, and K. Kosuge, Physica B **197**, 601 (1994).
76. J. Haase, R. Stern, C. T. Milling, C. P. Slichter, and D. G. Hinks, "NMR evidence for spatial modulations in the cuprates", J. Supercond. **13**, 723–726 (2000).
77. P. M. Singer, A. W. Hunt, and T. Imai, Phys. Rev. Lett. **88**, 047602 (2002).
78. T. Imai, C. P. Slichter, K. Yoshimura, and K. Kosuge, Phys. Rev. Lett. **70**, 1002 (1993).
79. T. Imai, C. P. Slichter, K. Yoshimura, M. Katoh, and K. Kosuge, Phys. Rev. Lett. **71**, 1254 (1993).
80. J. M. Tranquada, B. J. Stemlieb, J. D. Axe, Y. Nahamara, S. Uchida, Nature (London) **375**, 561 (1995).
81. A. W. Hunt, P. M. Singer, K. R. Thurber, and T. Imai, Phys. Rev. Lett. **82**, 4300 (1999).
82. P. M. Singer, A. W. Hunt, A. F. Cederström, and T. Imai, Phys. Rev. B **60**, 15345 (1999).
83. N. J. Curro, P. C. Hammel, B. J. Suh, M. Hücker, B. Büchner, U. Ammerahl, and A. Revcolevschi, Phys. Rev. Lett. **85**, 642 (2000).
84. V. Kataev, Yu. Greznev, G. Teitel'baum, M. Breuer, and N. Knauf, Phys. Rev. B **48**, 13042 (1993).
85. V. Kataev, B. Rameev, B. Büchner, M. Hücher, and R. Borowski, Phys. Rev. B **55**, R3394 (1997).

86. M. C. Chen and C. P. Slichter, Phys. Rev. B **27**, 278 (1983).
87. J. B. Boyce and C. P. Slichter, Phys. Rev. Lett. **32**, 61 (1974); D. M. Follstaedt, D. Abbas, T. S. Stakelon, and C. P. Slichter, Phys. Rev. B **47** (1976); T. J. Aton, T. S. Stakelon, and C. P. Slichter, Phys. Rev. B **18**, 3337 (1978).
88. H. Alloul, Phys. Rev. Lett. **42**, 603 (1979).
89. A. W. Hunt, P. M. Singer, A. F. Cederström, and T. Imai, Phys. Rev. B **64**, 134525 (2001).
90. A. Alekseevski et al., J. Low Temp. Phys. **77**, 87 (1989).
91. A. Jánossy et al., Physica C **171**, 457 (1990) and **181**, 11 (1991).
92. A. Jánossy, L. -C. Brunel, J. R. Cooper, Phys. Rev. B **54**, 10186 (1995).
93. Reference 85.
94. B. I. Kochelaev, L. Kan, Elschner, and S. Elschner, Phys. Rev. B **49**, 13106 (1994).
95. A. Shengelaya, M. Bruun, B. I.Kochalaev, A. Safinia, K. Conder, and K. A. Müller, Phys. Rev. Lett. **93**, 017001 (2004).
96. A. Shengelaya, H. Keller, K. A. Müller, B. I. Kochalaev, and K. Conder, Phys. Rev. B **63**, 144513 (2001).

6

Neutron Scattering Studies of Antiferromagnetic Correlations in Cuprates

John M. Tranquada

Neutron scattering studies have provided important information about the momentum and energy dependence of magnetic excitations in cuprate superconductors. Of particular interest are the recent indications of a universal magnetic excitation spectrum in hole-doped cuprates. That starting point provides motivation for reviewing the antiferromagnetic state of the parent insulators, and the destruction of the ordered state by hole doping. The nature of spin correlations in stripe-ordered phases is discussed, followed by a description of the doping and temperature dependence of magnetic correlations in superconducting cuprates. After describing the impact on the magnetic correlations of perturbations such as an applied magnetic field or impurity substitution, a brief summary of work on electron-doped cuprates is given. The chapter concludes with a summary of experimental trends and a discussion of theoretical perspectives.

6.1. Introduction

Neutron scattering has played a major role in characterizing the nature and strength of antiferromagnetic interactions and correlations in the cuprates. Following Anderson's observation [1] that La_2CuO_4, the parent compound of the first high-temperature superconductor, should be a correlated insulator, with moments of neighboring Cu^{2+} ions antialigned due to the superexchange interaction, antiferromagnetic order was discovered in a neutron diffraction study of a polycrystalline sample [2]. When single-crystal samples became available, inelastic studies of the spin waves determined the strength of the superexchange, J, as well as weaker interactions, such as the coupling between CuO_2 layers. The existence of strong antiferromagnetic spin correlations above the Néel temperature, T_N, has been demonstrated and explained. Over time, the quality of such characterizations has improved considerably with gradual evolution in the size and quality of samples and in experimental techniques.

Of course, what we are really interested in understanding are the optimally doped cuprate superconductors. It took much longer to get a clear picture of the magnetic excitations in these compounds, which should not be surprising given that there is no static magnetic order, the magnetic moments are small, and the bandwidth characterizing the magnetic excitations is quite large. Nevertheless, we are finally at a point where a picture of universal behavior, for at least two families of cuprates, is beginning to emerge. Thus, it seems

---

John M. Tranquada • Condensed Matter Physics & Materials Science Department, Brookhaven National Laboratory, Upton, NY 11973, USA

reasonable to start our story with recent results on the excitation spectrum in superconducting $YBa_2Cu_3O_{6+x}$ and $La_{2-x}Sr_xCuO_4$, and the nature of the spin gap that appears below the superconducting transition temperature, T_c. (Note that these are hole-doped superconductors, which is where most of the emphasis will be placed in this chapter.) An important result is that this spectrum looks quite similar to that measured for $La_{1.875}Ba_{0.125}CuO_4$, a compound in which T_c is depressed toward zero and ordered charge and spin stripes are observed. The nature of stripe order and its relevance will be discussed later.

Following the initial discussion of results for the superconductors in Section 6.2, one can have a better appreciation for the antiferromagnetism of the parent insulators, presented in Section 6.3. The destruction of antiferromagnetic order by hole doping is discussed in Section 6.4. In Section 6.5, evidence for stripe order, and for other possible ordered states competing with superconductivity, is considered. Section 6.6. discusses how the magnetic correlations in superconducting cuprates evolve with hole-doping and with temperature. While doping tends to destroy antiferromagnetic order, perturbations of the doped state can induce static order, or modify the dynamics, and these effects are discussed in Section 6.7. A brief description of work on electron-doped cuprates is given in Section 6.8. The chapter concludes with a discussion, in Section 6.9, of experimental trends and theoretical perspectives on the magnetic correlations in the cuprates. It should be noted that there is not space here for a complete review of neutron studies of cuprates; some earlier reviews and different perspectives appear in [3–8].

Before getting started, it is useful to first establish some notation. A general wave vector $\mathbf{Q} = (h, k, l)$ is specified in units of the reciprocal lattice, $(a^*, b^*, c^*) = (2\pi/a, 2\pi/b, 2\pi/c)$. The CuO_2 planes are approximately square, with a Cu–Cu distance of $a \approx b \sim 3.8$ Å. Antiferromagnetic order of Cu moments ($S = \frac{1}{2}$) in a single plane causes a doubling of the unit cell and is characterized by the wave vector $\mathbf{Q}_{AF} = (\frac{1}{2}, \frac{1}{2}, 0)$, as indicate in Figure 6.1; however, the relative ordering of the spins along the c-axis can cause the intensities of $(\frac{1}{2}, \frac{1}{2}, L)$ superlattice peaks to have a strongly modulated structure factor as a function of L. For the magnetic excitations, we will generally be interested in their dependence on $\mathbf{Q}_{2D} = \mathbf{Q} - (\frac{1}{2}, \frac{1}{2}, L)$ associated with an individual CuO_2 plane and ignoring the L dependence.

The magnetic scattering function measured with neutrons can be written as [9, 10]

$$\mathcal{S}(\mathbf{Q}, \omega) = \sum_{\alpha,\beta} \left(\delta_{\alpha,\beta} - Q_\alpha Q_\beta / Q^2 \right) \mathcal{S}^{\alpha\beta}(\mathbf{Q}, \omega), \tag{6.1}$$

where

$$\mathcal{S}^{\alpha\beta}(\mathbf{Q}, \omega) = \frac{1}{2\pi} \int_{-\infty}^{\infty} dt\, e^{-i\omega t} \sum_{\mathbf{r}} e^{i\mathbf{Q}\cdot\mathbf{r}} \langle S_{\mathbf{0}}^\alpha(0) S_{\mathbf{r}}^\beta(t) \rangle. \tag{6.2}$$

Here $S_{\mathbf{r}}^\beta(t)$ is the $\beta\, (= x, y, z)$ component of the atomic spin at lattice site $\mathbf{r}$ and time t and the angle brackets $\langle \ldots \rangle$ denote an average over configurations. For inelastic scattering, it is possible to relate $\mathcal{S}(\mathbf{Q}, \omega)$ to the imaginary part of the dynamical spin susceptibility, $\chi''(\mathbf{Q}, \omega)$ via the fluctuation-dissipation theorem,

$$\mathcal{S}(\mathbf{Q}, \omega) = \frac{\chi''(\mathbf{Q}, \omega)}{1 - e^{-\hbar\omega/k_B T}}. \tag{6.3}$$

Another useful quantity is the "local" susceptibility $\tilde{\chi}''(\omega)$, defined as

$$\tilde{\chi}''(\omega) = \int d\mathbf{Q}_{2D}\, \chi''(\mathbf{Q}, \omega). \tag{6.4}$$

Neutron Scattering Studies

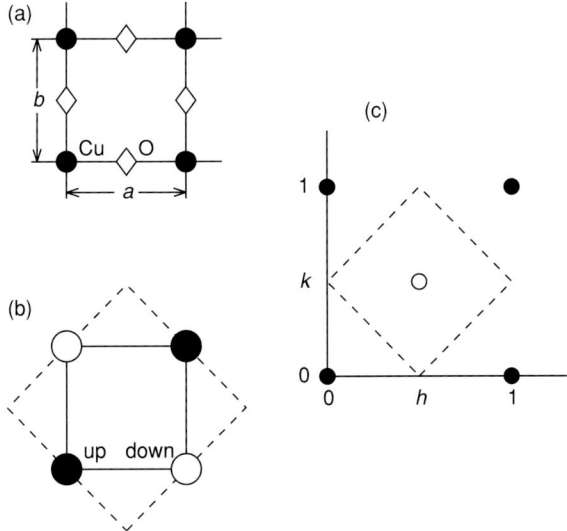

Figure 6.1. (a) CuO_2 plane, indicating positions of the Cu and O atoms and identifying the lattice parameters, a and b. (b) Sketch of antiferromagnetic order of Cu moments, with filled (empty) circles representing up (down) spins. Solid line indicates the chemical unit cell; dashed line denotes the magnetic unit cell. (c) Filled circles: Bragg peak positions in reciprocal space corresponding to the chemical lattice. Empty circle: magnetic Bragg peak due to antiferromagnetic order. Dashed line indicates the antiferromagnetic Brillouin zone.

Experimentally, the integral is generally not performed over the entire first Brillouin zone, but rather over the measured region about $\mathbf{Q}_{AF}$.

6.2. Magnetic Excitations in Hole-Doped Superconductors

6.2.1. Dispersion

Most of the neutron scattering studies of cuprate superconductors have focused on two families: $La_{2-x}Sr_xCuO_4$ and $YBa_2Cu_3O_{6+x}$. The simple reason for this is that these are the only compounds for which large crystals have been available. For quite some time it appeared that the magnetic spectra of these two families were distinct. In $La_{2-x}Sr_xCuO_4$, the distinctive feature was incommensurate scattering, studied at low energies (<20 meV) [11–13], whereas for $YBa_2Cu_3O_{6+x}$ the attention was focused on the commensurate scattering ("41-meV" or "resonance" peak [14–18]) that grows in intensity (and shifts in energy [19]) as the temperature is cooled below T_c. A resonance peak was also detected in $Bi_2Sr_2CaCu_2O_{8+\delta}$ [20–22] and in $Tl_2Ba_2CuO_{6+\delta}$ [23]. Considerable theoretical attention has been directed towards the resonance peak and its significance (e.g., see [24–26]). The fact that no strongly temperature-dependent excitation at $\mathbf{Q}_{AF}$ was ever observed in $La_{2-x}Sr_xCuO_4$ raised questions about the role of magnetic excitations in cuprate superconductivity.

While considerable emphasis has been placed on the resonance peak, it has been clear for quite some time that normal-state magnetic excitations in under-doped $YBa_2Cu_3O_{6+x}$ extend over a large energy range [30, 31], comparable to that in the antiferromagnetic parent compound [32, 33]. The first clear signature that the excitations below the resonance are incommensurate, similar to the low-energy excitations in $La_{2-x}Sr_xCuO_4$ [11, 13], was obtained by Mook et al. [70] for $YBa_2Cu_3O_{6.6}$. That these incommensurate excitations disperse inwards toward the resonance energy was demonstrated in $YBa_2Cu_3O_{6.7}$ by Arai et al. [35] and in $YBa_2Cu_3O_{6.85}$ by Bourges et al. [36]. More recent measurements have established a common picture of the dispersion [27, 28, 37–39].

A schematic of the measured dispersion is shown in Figure 6.2, with the energy normalized to that of the commensurate excitations, E_r. (Note that the distribution of intensity is not intended to accurately reflect experiment, especially in the superconducting state.) The figure also indicates the **Q** dependence of magnetic scattering at fixed excitation energies. For $E < E_r$, measurements on crystallographically twinned crystals indicate a fourfold intensity pattern, with maxima at incommensurate wave vectors displaced from $\mathbf{Q}_{AF}$ along [100] and [010] directions. For $E > E_r$, Hayden et al. [27] infer for their $YBa_2Cu_3O_{6.6}$ sample a fourfold structure that is rotated by 45° compared to low energies, whereas Stock et al. [28] find an isotropic ring of scattering for $YBa_2Cu_3O_{6.5}$. (These differences are minor compared to the overall level of agreement.) The spectrum with a finite spin gap is applicable to measurements below T_c; the gap fills in above T_c, where it also becomes difficult to resolve any incommensurate features.

Figure 6.3 shows a direct comparison of measurements on $La_{2-x}Sr_xCuO_4$ [40] and under-doped $YBa_2Cu_3O_{6+x}$ [27, 28], with energy scaled by the superexchange energy, J (see Table 6.1 in Section 6.3). Also included in the figure are results for $La_{2-x}Ba_xCuO_4$ with

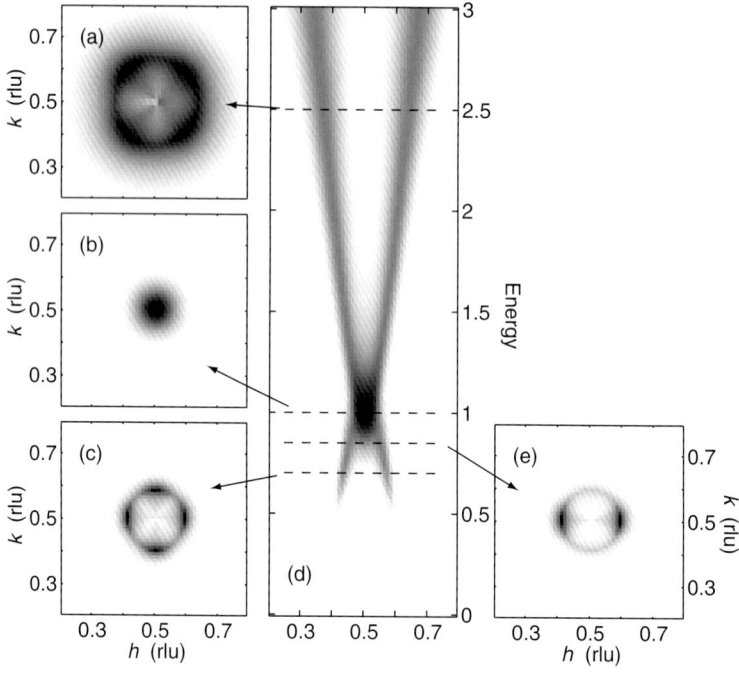

Figure 6.2. Schematic plots intended to represent neutron scattering measurements of $\chi''(\mathbf{Q}, \omega)$ in superconducting $YBa_2Cu_3O_{6+x}$ at $T \ll T_c$. Panels (a), (b), and (c) represent the distribution of scattering in reciprocal space about $\mathbf{Q}_{AF}$ at relative energies indicated by the dashed lines in (d), for a twinned sample. (d) χ'' along $\mathbf{Q} = (h, \frac{1}{2}, L)$ as a function of energy (normalized to the saddle-point energy, which is doping dependent). (a-d) modeled after [27, 28]. (e) Anisotropic distribution of scattering inferred for a detwinned, single-domain sample of $YBa_2Cu_3O_{6.85}$, after [29].

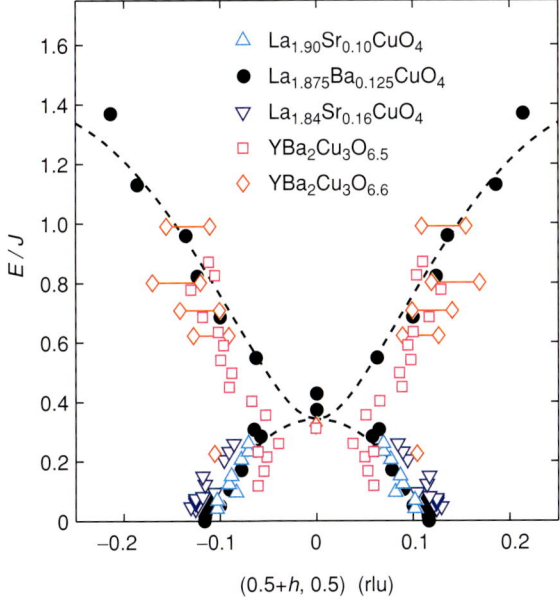

Figure 6.3. Comparison of measured dispersions along $\mathbf{Q}_{2D} = (0.5+h, 0.5)$ in $La_{2-x}Sr_xCuO_4$ with $x = 0.10$ (up triangles) and 0.16 (down triangles) from Christensen et al. [40], in $La_{1.875}Ba_{0.125}CuO_4$ (filled circles) from [42], and in $YBa_2Cu_3O_{6+x}$ with $x = 0.5$ (squares) from Stock et al. [28] and 0.6 (diamonds) from Hayden et al. [27]. The energy has been scaled by the superexchange energy J for the appropriate parent insulator as given in Table 6.1. For $YBa_2Cu_3O_{6.6}$, the data at higher energies were fit along the [1,1] direction; the doubled symbols with bars indicate two different ways of interpolating the results for the [1,0] direction. The upwardly dispersing dashed curve corresponds to the result of Barnes and Riera [45] for a two-leg spin ladder, with an effective superexchange of $\sim \frac{2}{3}J$; the downward curve is a guide to the eye.

$x = \frac{1}{8}$ [41, 42], a compound of interest because of the occurrence of charge and spin stripe order [43] (to be discussed later) and a strongly suppressed T_c. At the lowest energies, the spin excitations rise out of incommensurate magnetic (two-dimensional) Bragg peaks. Besides the presence of Bragg peaks, the magnetic scattering differs from that of $YBa_2Cu_3O_{6+x}$ in the absence of a spin gap. The results for optimally doped $La_{2-x}Sr_xCuO_4$, with $x = 0.16$, interpolate between these cases, exhibiting the same inward dispersion of the excitations towards $\mathbf{Q}_{AF}$ (measured up to 40 meV) and a spin gap of intermediate magnitude in the superconducting state [40]. The degree of similarity among the results shown in Figure 6.3 is striking, and suggests that the magnetic excitation spectrum may be universal in the cuprates [40, 44].

For optimally doped $YBa_2Cu_3O_{6+x}$, the measured dispersive excitations are restricted to a narrower energy window, as shown in Figure 6.4. Nevertheless, excitations are observed to disperse both downward and upward from E_r, and the qualitative similarity with dispersions at lower doping is obvious.

Anisotropy of the magnetic scattering as a function of $\mathbf{Q}_{2D}$ can be measured in specially detwinned samples of $YBa_2Cu_3O_{6+x}$, as the crystal structure has an anisotropy associated with the orientation of the CuO chains. (Note that it is a major experimental challenge to detwin samples of sufficient volume to allow a successful inelastic neutron scattering study.) An initial study of a partially detwinned sample of $YBa_2Cu_3O_{6.6}$ by Mook et al. [46] demonstrated that, for the incommensurate scattering at an energy corresponding to 70% of the saddle point, the intensity is quite anisotropic, with maxima along the $\mathbf{a}^*$ direction (perpendicular

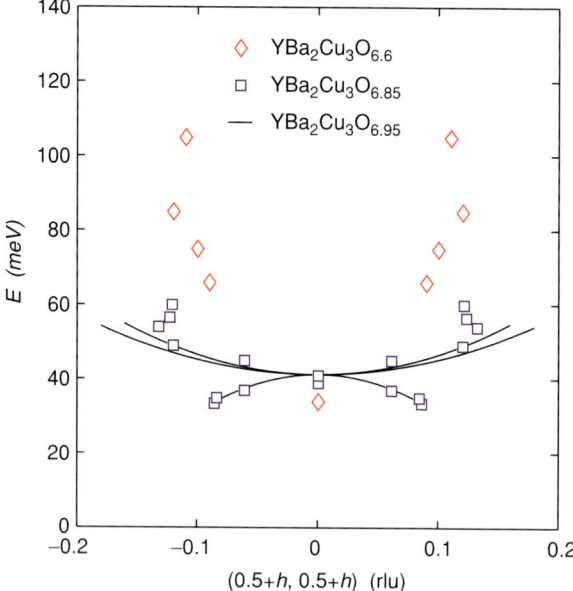

Figure 6.4. Comparison of measured dispersions along $Q_{2D} = (0.5 + h, 0.5 + h)$ in the superconducting state of $YBa_2Cu_3O_{6+x}$ with $x = 0.6$ (diamonds) from Hayden et al. [27] and 0.85 (squares) from Pailhès et al. [37]. The solid lines represent the model dispersion (and variation in dispersion) compatible with measurements on $YBa_2Cu_3O_{6.95}$ from Reznik et al. [38].

to the orientation of the CuO chains). A recent study of an array of highly detwinned crystals of $YBa_2Cu_3O_{6.85}$ by Hinkov et al. [29] found substantial anisotropy in the peak scattered intensity for an energy of 85% of the saddle point, but also demonstrated that scattered intensity at that energy forms a circle about Q_{AF} [see Figure 6.2(e)]. Measurements on a partially detwinned sample of $YBa_2Cu_3O_{6.5}$ by Stock et al. [28,47] suggest a strong anisotropy in the scattered intensity at $0.36E_r$, but essentially perfect isotropy for $E > E_r$.

6.2.2. Spin Gap and "Resonance" Peak

For optimally doped cuprates, the most dramatic change in the magnetic scattering with temperature is the opening of a spin gap, with redistribution of spectral weight from below to above the gap. A clear example of this has been presented recently by Christensen et al. [40] for $La_{2-x}Sr_xCuO_4$ with $x = 0.16$; their results are shown in Figure 6.5. For the energy range shown, the scattering is incommensurate in Q, with the dispersion indicated in Figure 6.3. In the normal state, the amplitude of χ'' heads to zero only at $\omega = 0$; in the superconducting state, weight is removed from below a spin gap energy of $\Delta_s \approx 8$ meV, and shifted to energies just above Δ_s. This is apparent both for the plot of the peak amplitude of χ'' in Figure 6.5(a), and for the Q-integrated χ'' in (b); within the experimental uncertainty, the spectral weight below 40 meV is conserved on cooling through T_c [40]. Another important feature of the spin gap is that its magnitude is independent of Q [48]. This is of particular interest because it is inconsistent with a weak-coupling prediction of χ'' for a d-wave superconductor, assuming that the spin response comes from quasiparticles [49].

The behavior is similar near optimal doping in $YBa_2Cu_3O_{6+x}$ [17, 36–38], with the difference being that the spin gap energy of ~ 33 meV is much closer to $E_r = 41$ meV. The

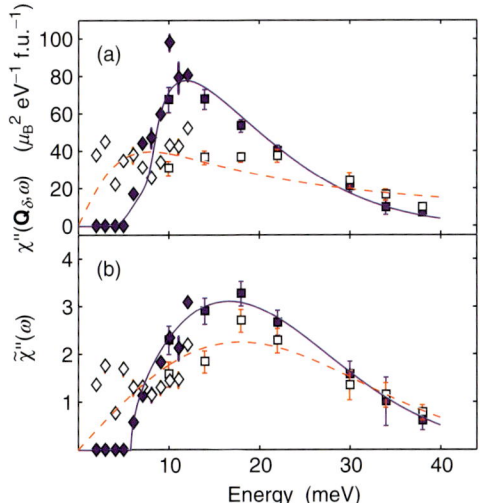

Figure 6.5. (a) The fitted peak intensity of $\chi''(\mathbf{Q}_\delta, \omega)$ (where $\mathbf{Q}_\delta$ is the peak position) for $La_{2-x}Sr_xCuO_4$ with $x = 0.16$. (b) The local susceptibility, $\tilde{\chi}''(\omega)$. Filled symbols: $T < T_c$; open symbols: $T > T_c$. Results are a combination of data from time-of-flight measurements (squares) and triple-axis measurements (diamonds). Lines are guides to the eye. From Christensen et al. [40].

strongest intensity enhancement below T_c occurs at E_r, where χ'' is peaked at $\mathbf{Q}_{AF}$; however, there is also enhanced intensity at energies a bit below and above E_r, where χ'' is incommensurate [36, 37]. The spin gap Δ_s decreases and broadens with underdoping, so that the region over which $\chi'' \approx 0$ is no more than a few meV for $YBa_2Cu_3O_{6.5}$ [30, 39, 47].

Besides the temperature dependence, there is also a similar behavior of the enhanced intensity for these two cuprate families in response to an applied magnetic field. As the cuprates are type-II superconductors with a very small lower critical field, an applied magnetic field can enter a sample as an array of vortices. Dai et al. [50] showed that application of a magnetic field of 6.8 T along the c-axis of $YBa_2Cu_3O_{6.6}$ at $T \ll T_c$ caused a reduction of the intensity at E_r by $\sim 30\%$. A study of $La_{2-x}Sr_xCuO_4$ with $x = 0.18$ found that application of a 10-T field along the c-axis caused a reduction of the intensity maximum at 9 meV of about 25% (with an increase in Q width) and a shift of some weight into the spin gap [51]. The field-induced increase of weight within the spin gap of $La_{2-x}Sr_xCuO_4$ ($x = 0.163$) was first observed by Lake et al. [52].

By focusing on Δ_s rather than E_r, it is possible to identify a correlation between magnetic excitations and T_c that applies to a variety of cuprates. Fig. 6.6 shows a plot of T_c as a function of the spin-gap energy for several different cuprates near optimal doping. This trend makes clear that the magnetic excitations are quite sensitive to the superconductivity, but, by itself, it does not resolve the issues of whether or how magnetic correlations may be involved in the pairing mechanism.

6.2.3. Discussion

From the results presented above, it now appears that there may be a universal magnetic excitation spectrum for the cuprates. On entering the superconducting state, a gap in the magnetic spectrum develops, with spectral weight redistributed from below to above the gap. The magnitude of the spin gap is correlated with T_c.

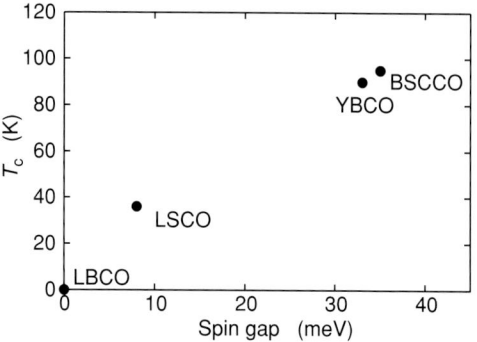

Figure 6.6. Plot of T_c vs. spin-gap energy, Δ_s, in cuprates near optimal doping. LSCO: $La_{2-x}Sr_xCuO_4$ ($x = 0.16$) from [40]; YBCO: $YBa_2Cu_3O_{6.85}$ from [36]; BSCCO: $Bi_2Sr_2CaCu_2O_{8+\delta}$ with Δ_s estimated by scaling E_r with respect to $YBa_2Cu_3O_{6+x}$, from [20]; LBCO: $La_{2-x}Ba_xCuO_4$ ($x = \frac{1}{8}$) from [41].

A long-standing question concerns the role of magnetic excitations in the mechanism of high-temperature superconductivity, and some varying perspectives are presented in later chapters of this book. An underlying issue concerns the nature of the magnetic excitations themselves. Given that $La_{2-x}Sr_xCuO_4$ and $YBa_2Cu_3O_{6+x}$ exhibit antiferromagnetically ordered phases when the hole-doping of the CuO_2 planes goes to zero, one approach is to look for a connection between the magnetic correlations in the superconducting and in the correlated-insulator phases. On the other hand, the magnetic response of common metallic systems (such as chromium) is tied to the low-energy excitations of electrons from filled to empty states, across the Fermi surface. This motivates attempts to interpret the magnetic excitations in terms of electron–hole excitations. It is not clear that these contrasting approaches can be reconciled with one another [53], but, in any case, there are presently no consensual criteria for selecting one approach over another.

An experimentalist's approach is to consider the correlations in the superconducting cuprates in the context of related systems. Thus, in the following sections we consider experimental results for antiferromagnetic cuprates, other doped transition-metal-oxide systems, perturbations to the superconducting phase, and the doping dependence of the magnetic correlations in the superconductors. A comparison of theoretical approaches is better discussed within the full context of experimental results.

6.3. Antiferromagnetism in the Parent Insulators

6.3.1. Antiferromagnetic Order

In the parent insulators, the magnetic moments of the copper atoms order in a 3D Néel structure. Powder neutron diffraction studies first demonstrated this for La_2CuO_4 [2], and later for $YBa_2Cu_3O_{6+x}$ [54]. The magnetic moments tend to lie nearly parallel to the CuO_2 planes. The details of the magnetic structures are tied to the crystal structures, so we will have to consider these briefly.

The crystal structure of La_2CuO_4 is presented in Figure 6.7. The CuO_2 planes are stacked in a body-centered fashion, so that the unit cell contains two layers. Below 550 K each CuO_6 octahedron rotates about a [110] axis of the high-temperature tetragonal cell.

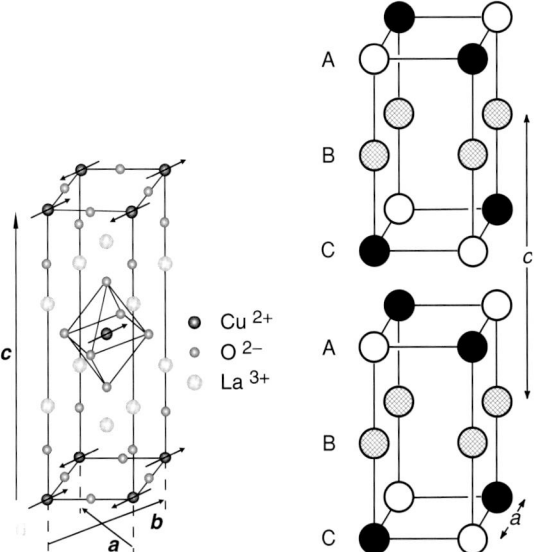

Figure 6.7. Left: crystal structure of La_2CuO_4. Arrows indicate orientation of magnetic moments on Cu sites in the antiferromagnetic state. From Lee et al. [55]. Right: Magnetic structure of $YBa_2Cu_3O_6$. Circles: Cu atoms; lines: paths bridged by oxygen. Filled and empty circles represent Cu^{2+} sites with opposite spin orientations; hatched circles denote nonmagnetic Cu^{1+} sites. After Tranquada et al. [56].

Neighboring octahedra within a plane rotate in opposite directions, causing a doubling of the unit cell volume and a change to orthorhombic symmetry, with the $\mathbf{a}_O$ and $\mathbf{b}_O$ axes rotated by 45° with respect to the Cu–O bond directions. In the orthorhombic coordinates, the octahedral tilts are along the $\mathbf{b}_O$ direction (but $b_O > a_O$, contrary to naive expectations).

In the antiferromagnetic phase of La_2CuO_4, the spins point along the b_O-axis, and they have the stacking sequence shown in Figure 6.7 [2,57]. As the octahedral tilts break the tetragonal symmetry of the planes, they allow spin–orbit effects, in the form of Dzyalozhinsky–Moriya (DM) antisymmetric exchange, to cause a slight canting of the spins along the c-axis. This canting is in opposite directions for neighboring planes, resulting in no bulk moment, but a modest magnetic field can flip the spins in half of the planes, yielding a weakly ferromagnetic state [58]. The tendency to cant in the paramagnetic state above T_N leads to a ferromagnetic-like susceptibility at high temperatures and a cusp at T_N [59]. Studies of quasi-1D cuprates have made it clear that the DM (and additional) interactions are quite common [60]; however, the tetragonal CuO_2 planes of other layered cuprate antiferromagnets cause the effects of the DM interaction to cancel out, so that there is no canting [61,62].

In the early diffraction studies, the La_2CuO_4 powder samples contained some excess oxygen and the first crystals had contamination from flux or the crucible, thus resulting in a reduced ordering temperature. (It is now known that excess oxygen, in sufficient quantity, can segregate to form superconducting phases [63].) It was eventually found that by properly annealing a crystal one can obtain a sample with $T_N = 325$ K [64]. The ordered magnetic moment is also sensitive to impurity effects. In a study of single crystals with different annealing treatments, Yamada et al. [65] found that the ordered moment is correlated with T_N, with a maximum Cu moment of $0.60(5)$ μ_B [apparently determined from the intensity of the (100) magnetic reflection alone].

Table 6.1. Compilation of some neutron scattering results for a number of layered cuprate antiferromagnets. m_{Cu} is the average ordered moment per Cu atom at $T \ll T_N$. The superexchange energy J corresponds to the value obtained from the spin wave velocity after correction for the quantum-renormalization factor $Z_c = 1.18$. For crystal symmetry, O = orthorhombic, T = tetragonal.

Compound	T_N (K)	m_{Cu} (μ_B)	J (meV)	Crystal symmetry	Layers per cell	Refs.
La_2CuO_4	325(2)	0.60(5)	146(4)	O	1	[64, 65, 68]
$Sr_2CuO_2Cl_2$	256(2)	0.34(4)	125(6)	T	1	[69–71]
$Ca_2CuO_2Cl_2$	247(5)	0.25(10)		T	1	[72]
Nd_2CuO_4	276(1)	0.46(5)	155(3)	T	1	[73–76]
Pr_2CuO_4	284(1)	0.40(2)	130(13)	T	1	[73, 77]
$YBa_2Cu_3O_{6.1}$	410(1)	0.55(3)	106(7)	T	2	[32, 78]
$TlBa_2YCu_2O_7$	>350	0.52(8)		T	2	[79]
$Ca_{0.85}Sr_{0.15}CuO_2$	537(5)	0.51(5)		T	∞	[80]

The magnetic coupling between layers in La_2CuO_4 is quite weak because each Cu sees two up spins and two down spins at nearly the same distance in a neighboring layer. The small orthorhombic distortion of the lattice removes any true frustration, resulting in a small but finite coupling. There are, however, several other cuprate antiferromagnets with a similar centered stacking of layers, but with tetragonal symmetry (see Table 6.1). Yildirim et al. [66] showed that the long-range order (including spin directions) can be understood when one takes into account zero-point spin fluctuations, together with the proper exchange anisotropies [67].

The parent compounds of the electron-doped superconductors, Nd_2CuO_4 and Pr_2CuO_4, have somewhat more complicated magnetic structures. Nd moments and induced moments on Pr couple to the order in the CuO_2 planes, resulting in noncollinear magnetic structures and spin reorientation transitions as a function of temperature; these are described in the review by Lynn and Skanthakumar [75]. The magnetic structures and transitions have been explained by Sachidanandam et al. [81] by taking account of the single-ion anisotropy and crystal-field effects for the rare-earth ions. Further discussion is given by Petitgrand et al. [82].

The crystal structure of $YBa_2Cu_3O_{6+x}$ contains pairs of CuO_2 layers (bilayers). There is also a third layer of Cu atoms, but in $YBa_2Cu_3O_6$ these are nonmagnetic Cu^{1+} ions. (Added oxygen goes into this layer, forming the CuO chains of $YBa_2Cu_3O_7$.) The magnetic structure of $YBa_2Cu_3O_6$ is indicated in Figure 6.7. Because of the relative antiferromagnetic ordering of the bilayers, together with a spacing that is not determined by symmetry, there is a structure factor for the magnetic Bragg peaks that depends on Q_z. This structure factor also affects the spin-wave intensities, as will be discussed.

It is not possible to determine the spin direction from zero-field diffraction measurements due to the tetragonal symmetry of the lattice and inevitable twinning of the magnetic domains. Nevertheless, Burlet et al. [83] were able to determine the spin direction by studying the impact of a magnetic field applied along a $[1, -1, 0]$ direction of a $YBa_2Cu_3O_{6.05}$ single crystal. They found that in zero field, the spins must lie along [100] or [010] directions (parallel to the Cu–O bonds), and that the applied field rotates them toward [110]. This result has been confirmed by electron-spin resonance studies of $YBa_2Cu_3O_{6+x}$ with a small amount of Gd substituted for Y [84].

A complication in studies of magnetic order involving some of the first crystals of $YBa_2Cu_3O_{6+x}$ arose from inadvertent partial substitution of Al ions onto the Cu(1) ("chain") site. The Al presumably came from the use of crucibles made of Al_2O_3. Kadowaki et al [85], performing one of the first single-crystal diffraction studies on a $YBa_2Cu_3O_{6+x}$ sample with T_N of 405 K, found that below 40 K a new set of superlattice peaks appeared, indicating a doubling of the magnetic unit cell along the c-axis. It was later demonstrated convincingly, by comparing pure and Al-doped crystals, that the low-temperature doubling of the magnetic period only occurs in crystals with Al [78, 86]. A model explaining how the presence of Al on Cu chain sites can change the magnetic order was developed by Andersen and Uimin [87].

To evaluate the ordered magnetic moment, it is necessary to have knowledge of the magnetic form factor. In all of the early studies of antiferromagnetic order in cuprates it was assumed that the spin-density on a Cu ion is spherical; however, this assumption is far from being correct. The magnetic moment results from the half-filled $3d_{x^2-y^2}$ orbital, which deviates substantially from sphericity. The proper, anisotropic form factor was identified by Shamoto et al. [88] and shown to give an improved description of magnetic Bragg intensities for $YBa_2Cu_3O_{6.15}$. An even better measurement of magnetic Bragg peaks was done on a small crystal of $YBa_2Cu_3O_{6.10}$ by Casalta et al. [78]. They obtained a Cu moment of $0.55(3)\mu_B$. Use of the proper form factor is important for properly evaluating the magnetic moment, as there is always a gap between $Q = 0$ (where the magnitude of the form factor is defined to be 1) and the Q value of the first magnetic Bragg peak. It does not appear that anyone has gone back to reevaluate the magnetic diffraction data on other cuprates, such as La_2CuO_4 or $Sr_2CuO_2Cl_2$ using the anisotropic form factor.

The maximum observed Cu moments are consistent with a large reduction due to zero-point spin fluctuations as predicted by spin-wave theory. The moment m is equal to $g\langle S\rangle\mu_B$, where a typical value of the gyromagnetic ratio g is 2.1. Without zero-point fluctuations, one would expect $m \approx 1.1\mu_B$. Linear spin-wave theory predicts $\langle S\rangle = 0.303$ [89], giving $m \approx 0.64\mu_B$, a bit more than the largest observed moments. Further reductions can occur due to hybridization effects [90, 91].

The ordered moments of the oxy-chlorides listed in Table 6.1 seem surprisingly small. While this might be due to hybridization effects, it is interesting to note that there is a correlation between m_{Cu} and T_N for the first five antiferromagnets in the table, which all share a similar body-centered stacking of the CuO_2 planes. The correlation is illustrated in Figure 6.8. The ratio T_N/J is expected to be sensitive to the interlayer exchange J' [92], and J' varies substantially among these compounds; however, I am not aware of any predicted dependence of m_{Cu} on J'. A correlation between m_{Cu} and T_N/J has been reported for quasi-1D antiferromagnets, but such a correlation is expected in that case [93].

6.3.2. Spin Waves

The starting point for considering magnetic interactions in the cuprates is the Heisenberg hamiltonian

$$H = J \sum_{\langle i,j\rangle} \mathbf{S}_i \cdot \mathbf{S}_j, \tag{6.5}$$

where $\langle i, j\rangle$ denotes all nearest-neighbor pairs, each included once. This hamiltonian can be derived in second-order perturbation from a Hubbard model for a single, half-filled band of electrons. Such a model includes a nearest-neighbor hopping energy t and the Coulomb repulsion energy U for two electrons on the same site; in terms of these parameters, $J = 4t^2/U$ [94]. Spin-wave theory can be applied to the Heisenberg hamiltonian to calculate the dispersion of spin fluctuations about $\mathbf{Q}_{AF}$ [95]. At low energies the spin waves disperse linearly with $\mathbf{q} = \mathbf{Q} - \mathbf{Q}_{AF}$ (see Figure 6.9), having a velocity $c = \sqrt{8}SZ_c Ja/\hbar$, where $Z_c \approx 1.18$ [96]

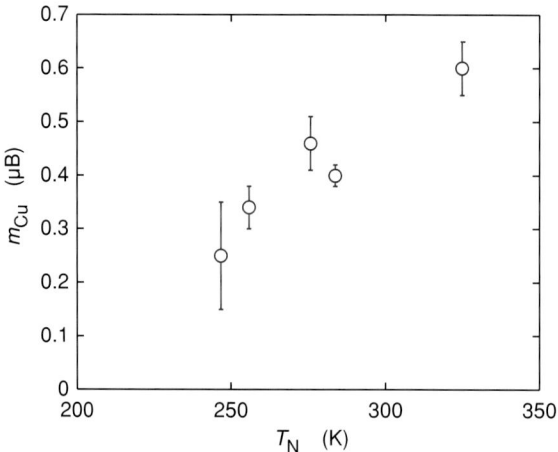

Figure 6.8. Ordered magnetic moment per Cu atom vs. T_N for the first five compounds in Table 6.1, all of which have a similar body-centered stacking of CuO_2 layers.

is a quantum-renormalization factor. Thus, by measuring the spin-wave velocity, one can determine J.

Spin-wave measurements have been performed for a number of cuprates, and some results for J are listed in Table 6.1. (Complementary measurements of J can be obtained by two-magnon Raman scattering [97].) To calculate the values of J from spectroscopically determined parameters, one must consider at least a three-band Hubbard model [98]. Recent ab initio cluster calculations [99, 100] have been able to achieve reasonable agreement with experiment. While the magnitude of J in layered cuprates is rather large, it is not extreme; a value of $J = 226(12)$ meV has been measured for Cu–O chains in $SrCuO_2$ [101].

To describe the experimental dispersion curves in greater detail, one must add more terms to the spin hamiltonian. For example, in a *tour de force* experiment, Coldea et al. [68] have measured the dispersion of spin waves in La_2CuO_4 along high-symmetry directions of the 2D Brillouin zone, as shown in Figure 6.9. The observed dispersion along the zone boundary, between $(\frac{1}{2}, 0)$ and $(\frac{3}{4}, \frac{1}{4})$, is not predicted by the simple Heisenberg model. To describe this, they consider the additional terms that appear when the perturbation expansion for the single-band Hubbard model is extended to fourth order. The most important new term involves four-spin cyclic exchange about a plaquette of four Cu sites [102–104]. Coldea and coworkers were able to fit the data quite well with the added parameters (see lines through data points in Figure 6.9a), obtaining, at 10 K, $J = 146(4)$ meV and a cyclic exchange energy $J_c = 61(8)$ meV [68]. (Superexchange terms coupling sites separated by two hops are finite but negligible.)

If, instead of expanding to higher order, one extends the Hubbard model to include hopping between next-nearest-neighbor Cu sites, one can calculate a superexchange term J' between next-nearest neighbors that is on the order of 10% of J [105, 106]. It turns out, however, that fitting the measured dispersion with only J and J' requires that J' correspond to a ferromagnetic interaction [68], which is inconsistent with the model predictions.

In $YBa_2Cu_3O_{6+x}$, the effective exchange coupling between Cu moments in nearest-neighbor layers is substantial. Its effect is to split the spin waves into acoustic and optic

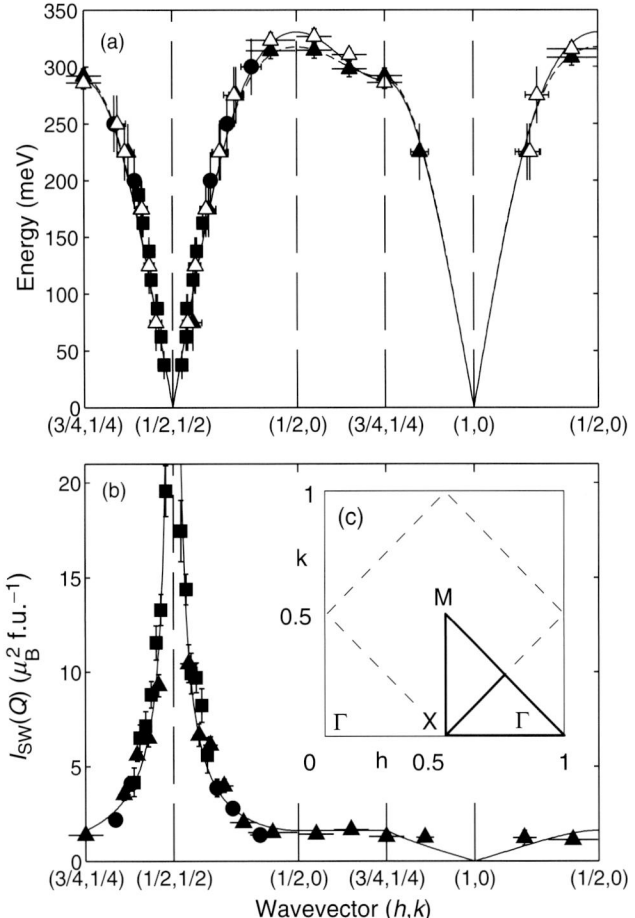

Figure 6.9. (a) Spin-wave dispersion in La_2CuO_4 along high-symmetry directions in the 2D Brillouin zone, as indicated in (c); $T = 10$ K (295 K): open (filled) symbols. Solid (dashed) line is a fit to the 10-K (295-K) data. (b) Spin-wave intensity vs. wave vector. Line is prediction of linear spin-wave theory. From Coldea et al. [68].

branches, having odd and even symmetry, respectively, with respect to the bilayers. The structure factors for these excitations are [107]

$$g_{ac} = \sin(\pi z l), \tag{6.6}$$
$$g_{op} = \cos(\pi z l), \tag{6.7}$$

where $z = d_{Cu-Cu}/c$ is the relative spacing between Cu moments along the c-axis within a bilayer ($d_{Cu-Cu} \approx 3.285$ Å [108]); the intensity of the spin-wave scattering is proportional to g^2. An example of the intensity modulation due to the acoustic-mode structure factor in the antiferromagnetic state is indicated by the filled circles in Figure 6.10.

The energy gap for the optical magnons has been measured to be approximately 70 meV [32, 33]. Experimental results for the spin wave dispersion and the spectral weight are shown in Figure 6.11. The magnitude of the gap indicates that the intrabilayer exchange is 11(2) meV [32, 33].

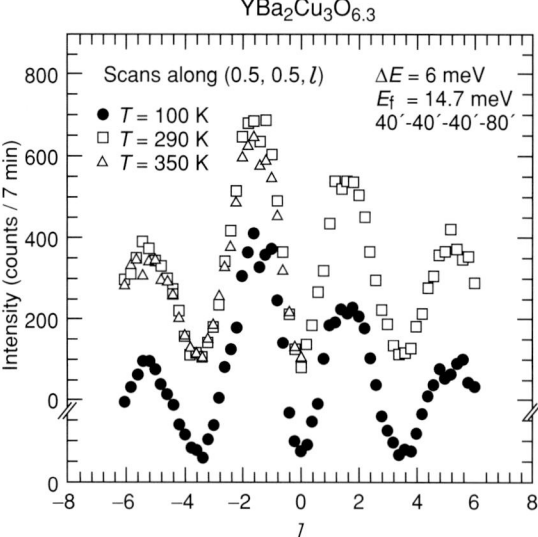

Figure 6.10. Scans along the quasi-2D antiferromagnetic scattering rod $\mathbf{Q} = (\frac{1}{2}, \frac{1}{2}, l)$ at a fixed energy transfer of 6 meV for a crystal of $YBa_2Cu_3O_{6.3}$ with $T_N = 260(5)$ K. The sinusoidal modulation is due to the inelastic structure factor. The asymmetry in the scattering as a function of l is an effect due to the experimental resolution function [9]; the decrease in intensity at large $|l|$ is due to the magnetic form factor. After Tranquada et al. [107].

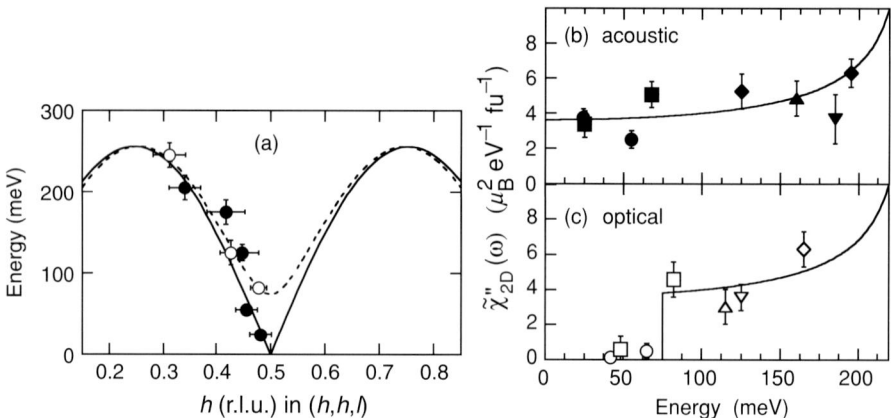

Figure 6.11. Experimental results for $YBa_2Cu_3O_{6.15}$. (a) Dispersion of acoustic (filled symbols, solid line) and optic (open symbols, dashed line) spin-wave modes. (b) **Q**-integrated dynamic susceptibility, $\tilde{\chi}''(\omega)$ for the acoustic, and (c) optic modes. From Hayden et al. [32].

At low energies, there are other terms that need to be considered. There need to be anisotropies, with associated spin-wave gaps, in order to fix the spin direction; however, an atom with $S = \frac{1}{2}$ cannot have single-ion anisotropy. Instead, the anisotropy is associated with the nearest-neighbor superexchange interaction. Consider a pair of nearest-neighbor spins, $\mathbf{S}_i$ and $\mathbf{S}_j$, within a CuO_2 plane, with each site having tetragonal symmetry. The Heisenberg Hamiltonian for this pair can be written

$$H_{\text{pair}} = J_\| S_i^\| S_j^\| + J_\perp S_i^\perp S_j^\perp + J_z S_i^z S_j^z, \tag{6.8}$$

where $\parallel$ and $\perp$ denote directions parallel and perpendicular to the bond within the plane, and z is the out-of-plane direction. Yildirim et al. [67] showed that the anisotropy can be explained by taking into account both the spin–orbit and Coulomb exchange interactions. To discuss the anisotropies, it is convenient to define the quantities $\Delta J \equiv J_{\rm av} - J_z$, where $J_{\rm av} \equiv (J_\parallel + J_\perp)/2$, and $\delta J_{\rm in} \equiv J_\parallel - J_\perp$ [66]. For the cuprates, $J_{\rm av} \gg \Delta J > \delta J_{\rm in} > 0$. The out-of-plane anisotropy, $\alpha_{\rm XY} = \Delta J / J_{\rm av}$, causes the spins to lie, on average, in the x–y plane, and results in a spin-wave gap for out-of-plane fluctuations. The in-plane anisotropy $\delta J_{\rm in}/J_{\rm av}$, contributing through the quantum zero-point energy [66, 109], tends to favor alignment of the spins parallel to a bond direction, and causes the in-plane spin-wave mode to have a gap. The effective coupling between planes (which can involve contributions from several interactions [66]) leads to (weak) dispersion along Q_z.

For stoichiometric La_2CuO_4, the out-of-plane spin gap is 5.5(5) meV, corresponding to $\alpha_{\rm XY} = 1.5 \times 10^{-4}$ [110]. The in-plane gap of 2.8(5) meV has a contribution from anisotropic exchange of the Dzyaloshinsky–Moriya type [111, 112], as well as from $\delta J_{\rm in}$. No dispersion along Q_z has been reported.

For antiferromagnetic $YBa_2Cu_3O_{6+x}$, an out-of-plane gap of about 5 meV has been observed [107, 113, 114], indicating an easy-plane anisotropy similar to that in La_2CuO_4. No in-plane gap has been resolved; however, the in-plane mode shows a dispersion of about 3 meV along Q_z [107, 113, 114]. The latter dispersion is controlled by the effective exchange between Cu moments in neighboring bilayers through the nonmagnetic Cu(1) sites, which is on the order of $10^{-4}J$.

6.3.3. Spin Dynamics at $T > T_N$

That strong 2D spin correlations survive in the CuO_2 planes for $T > T_N$ initially came as a surprise [115]. Such behavior was certainly uncommon at that point. Detailed studies have since been performed measuring the instantaneous spin correlation length ξ as a function of temperature in La_2CuO_4 [116] and in $Sr_2CuO_2Cl_2$ [70, 117]. The correlation length is obtained using an experimental trick to integrate the inelastic scattering over excitation energy, and using the formula

$$\mathcal{S}(\mathbf{q}_{\rm 2D}) = \int d\omega\, \mathcal{S}(\mathbf{q}_{\rm 2D}, \omega) = \frac{\mathcal{S}(0)}{1 + q_{\rm 2D}^2 \xi^2}. \tag{6.9}$$

Here, $\mathbf{q}_{\rm 2D}$ is the momentum-transfer component parallel to the planes, and the scattering is assumed to be independent of momentum transfer perpendicular to the planes. (The experimental energy integration is imperfect, but, by proper choice of incident neutron energy, does capture most of the critical scattering.)

To theoretically analyze the behavior of the correlation length, Chakravarty, Halperin, and Nelson [118] evaluated the 2D quantum nonlinear σ model using renormalization group techniques; their results were later extended to a higher-order approximation by Hasenfratz and Niedermayer [119]. The essential result is that

$$\xi/a \sim e^{2\pi\rho_s/T}, \tag{6.10}$$

where the spin stiffness ρ_s is proportional to J (see [116] for a thorough discussion). The experimental results are in excellent agreement with theory, with essentially no adjustable parameters. The unusual feature of $\xi(T)$ is the exponential, rather than algebraic, dependence on temperature; nevertheless, note that it is consistent with achieving long-range order at

$T = 0$. The robustness of the experimentally observed spin correlations is due to the large value of J, comparable to 1,500 K, and the weak interlayer exchange, J'. The 3D ordering temperature can be estimated as [120]

$$kT_N \approx J' \left(\frac{m}{m_0}\right)^2 \left(\frac{\xi}{a}\right)^2, \qquad (6.11)$$

where $m/m_0 = 0.6$ is the reduction of the ordered moment due to quantum fluctuations. Because of the small J', the correlation length can reach the order of $100a$ before ordering occurs [116].

Although $Sr_2CuO_2Cl_2$ has essentially the same structure as La_2CuO_4, its tetragonal symmetry leads one to expect, classically, that the net interlayer exchange should be zero; however, an analysis by Yildirim et al. [66] has shown that a finite interaction of appropriate size results when quantum zero-point energy is taken into account. Because of its relatively low T_N of 257 K, it has been possible to detect in $Sr_2CuO_2Cl_2$ a crossover to XY-like behavior about 30 K above T_N, as reported in a ^{35}Cl NMR study [121]. This behavior results from the small easy-plane exchange anisotropy common to the layered cuprates [122]. Using neutrons to study the same material, it was possible to show that the characteristic fluctuation rate in the paramagnetic state follows the behavior $\Gamma = \omega_0 \sim \xi^{-z}$ with $z = 1.0(1)$ [123], consistent with dynamic scaling theory for the 2D Heisenberg antiferromagnet [124].

There has been less work done on the paramagnetic phase of $YBa_2Cu_3O_{6+x}$, as the inelastic structure factor, Eq. (6.6), complicates the experimental trick for energy integration. There are also complications to studying $YBa_2Cu_3O_{6+x}$ samples at elevated temperatures, as oxygen can easily diffuse into and out of crystals as one heats much above room temperature. In any case, Figure 6.10 shows that the bilayers remain correlated at $T > T_N$ [107].

6.4. Destruction of Antiferromagnetic Order by Hole Doping

The long-range antiferromagnetic order of La_2CuO_4 is completely destroyed when 2% of Sr (measured relative to Cu concentration) is doped into the system [125]. Adding holes effectively reduces the number of Cu spins, so one might consider whether the reduction of order is due to dilution of the magnetic lattice. For comparison, an extensive study of magnetic dilution has been performed by Vajk et al. [126] on $La_2Cu_{1-z}(Zn,Mg)_zO_4$, where Cu is substituted by nonmagnetic Zn and/or Mg. They found that long-range antiferromagnetic order was lost at the classical 2D percolation limit of $z \approx 41\%$. Thus, holes destroy magnetic order an order of magnitude more rapidly than does simple magnetic dilution.

The reduction of the Néel temperature at small but finite doping is accompanied by a strong depression of the antiferromagnetic Bragg intensities, together with an anomalous loss of intensity at $T < 30$ K [127]. Matsuda et al. [127] showed recently that the latter behavior is correlated with the onset of incommensurate magnetic diffuse scattering below 30 K. In tetragonal coordinates, this scattering is peaked at $(\frac{1}{2}, \frac{1}{2}, 0) \pm \frac{1}{\sqrt{2}}(\epsilon, \epsilon, 0)$. To be more accurate, it is necessary to note that the crystal structure is actually orthorhombic, with the unit-cell axes rotated by 45°; the magnetic modulation is uniquely oriented along the b_O^* direction (see inset of Figure 6.12c).

The doping dependence of the transition temperature, ordered moments, and incommensurability are shown in Figure 6.12. The facts that (a) the volume fraction of the incommensurate phase grows with x for $x \leq 0.02$ (inset of Figure 6.12b) and (b) the

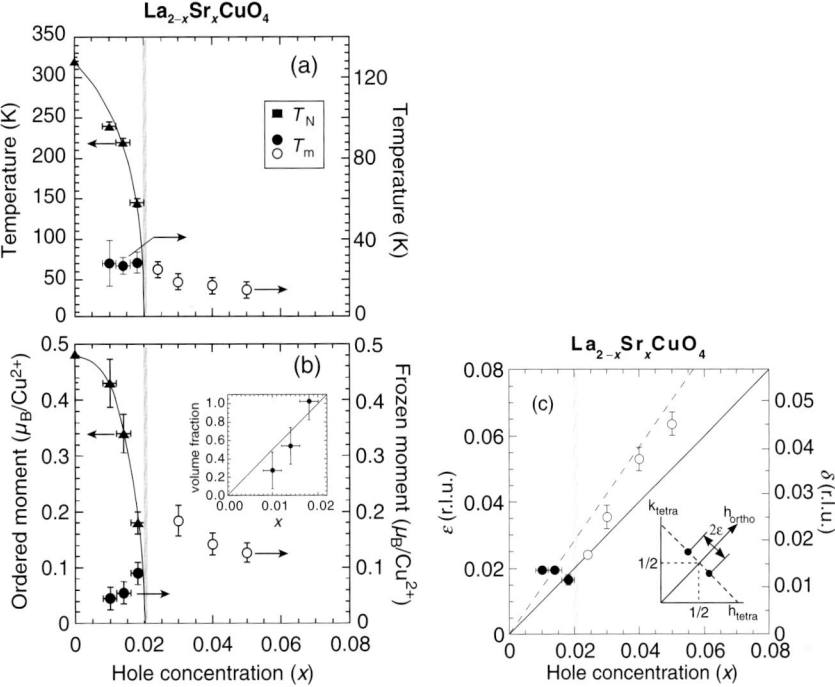

Figure 6.12. Results for lightly doped $La_{2-x}Sr_xCuO_4$. (a) Magnetic transition temperatures for commensurate (triangles) and incommensurate (circles) order vs. hole concentration. (b) Commensurate ordered moment at $T = 30\,K$ and incommensurate frozen moment at $T = 4\,K$ vs. hole concentration. Inset shows estimated volume fraction of incommensurate phase. (c) Variation of the incommensurability ϵ vs. hole concentration; $\delta = \epsilon/\sqrt{2}$. Solid and broken lines correspond to $\epsilon = x$ and $\delta = x$, respectively. Inset shows the positions of the incommensurate superlattice peaks in reciprocal space. From Matsuda et al. [127], including results from [128–131].

incommensurability does not change for $x \leq 0.02$ strongly suggest that an electronic phase separation occurs [127]. Thus, it appears that commensurate antiferromagnetic order is not compatible with hole doping. The disordered potential due to the Sr dopants may be responsible for the finite range of doping over which the Néel state appears to survive. The diagonally modulated, incommensurate spin-density-wave phase induced by doping survives up to the onset of superconductivity at $x \approx 0.06$ [131], and it corresponds to what was originally characterized as the "spin-glass" phase, based on bulk susceptibility studies [132, 133].

Further evidence for electronic phase separation comes from studies of oxygen-doped $La_2CuO_{4+\delta}$ (for a review, see [134]). The oxygen interstitials are mobile, in contrast to the quenched disorder of Sr substitution, and so they can move to screen discrete electronic phases. For $\delta < 0.06$, a temperature-dependent phase separation is observed between an oxygen-poor antiferromagnetic phase and an oxygen-rich superconducting phase [135, 136]; further miscibility gaps are observed between superconducting phases at higher oxygen content [63]. A sample with $\delta \approx 0.05$ and quenched disorder (due to electrochemical oxygenation in molten salt) exhibited reduced Néel and superconducting transition temperatures [137]. The interesting feature in this case was the observation of a decrease in the antiferromagnetic order with the onset of the superconductivity, suggesting a competition between the two phases.

In $YBa_2Cu_3O_{6+x}$ the situation is somewhat more complicated, as the doping of the planes is coupled to the tetragonal–orthorhombic (T–O) transition [138–140] that occurs in the vicinity of $x = 0.3$–0.4, depending on the degree of annealing. In the tetragonal phase, an isolated oxygen atom entering the "chain" layer simply converts neighboring Cu(1) sites from Cu^{1+} to Cu^{2+}; holes are created when chain segments form [54, 141]. The transfer of holes from the chains to the planes must be screened by displacements in the Ba–O layer that sits between, and a large jump in this screening occurs at the T–O transition [138–140]. Thus, one tends to find a discontinuous jump from a very small planar hole density in the antiferromagnetic phase just below the T–O transition to a significant density ($\sim$0.05 holes/Cu) just above.

Antiferromagnetic order has been observed throughout the tetragonal phase of $YBa_2Cu_3O_{6+x}$, with T_N decreasing rapidly as the T–O transition (at $x \approx 0.4$) is approached [54, 142]. A study of a set of carefully annealed powder samples, for which the T–O transition occurred at $x \approx 0.2$, indicated antiferromagnetic order in the orthorhombic phase at $x = 0.22$ and 0.24 with $T_N = 50(15)$ and 20(10) K, respectively. For tetragonal crystals with $x \sim 0.3$, a drop in the antiferromagnetic Bragg intensity has been observed below ≈ 30 K [107, 142]; as the Bragg intensity decreased, an increase in diffuse intensity along the 2D antiferromagnetic rod (with an acoustic bilayer structure factor) was found. This latter behavior might be related to the apparent phase separation in $La_{2-x}Sr_xCuO_4$ with $x < 0.02$ [127] discussed above.

The best study of a single-crystal sample just on the orthorhombic side of the T–O boundary is on $YBa_2Cu_3O_{6.35}$, a sample with $T_c = 18$ K [143]. Quasielastic diffuse scattering is observed at the antiferromagnetic superlattice positions. The peak intensity of this central mode grows on cooling below $\sim$30 K, but the energy width decreases below T_c. These results indicate there is no coexistence of long-range antiferromagnetic order in the superconducting phase. The spin–spin correlation length is short ($\sim$8 unit cells), suggesting segregation of hole-poor and hole-rich regions [143].

A possibly related response to doping is observed in the bilayer system $La_{2-x}(Sr,Ca)_xCaCu_2O_{6+\delta}$. Studies of crystals with $x = 0.1$–0.2 reveal commensurate short-range antiferromagnetic order that survives to temperatures > 100 K [144, 145], despite evidence from optical conductivity measurements for a substantial hole density in the CuO_2 planes [146]. Thus, there seems to be a local phase separation between hole-rich regions and antiferromagnetic clusters having an in-plane diameter on the order of 10 lattice spacings.

6.5. Stripe Order and Other Competing States

6.5.1. Charge and Spin Stripe Order in Nickelates

To understand cuprates, it seems sensible to consider the behavior of closely related model systems. One such system is $La_{2-x}Sr_xNiO_{4+\delta}$, a material that is isostructural with $La_{2-x}Sr_xCuO_4$. It is obtained by replacing $S = \frac{1}{2}$ Cu^{2+} ions ($Z = 29$) with $S = 1$ Ni^{2+} ions ($Z = 28$). One might consider the nickelates to be uninteresting as they are neither superconducting nor metallic (for $x < 0.9$) [147, 148]; however, the insulating behavior is inconsistent with the predictions of band theory, and it is important to understand why.

Pure La_2NiO_4 is an antiferromagnetic insulator [155] that is easily doped with excess oxygen, as well as by Sr substitution for La [134]. Doping the NiO_2 planes with holes reduces T_N more gradually than in the cuprates [156]. It is necessary to dope to a hole concentration of $n_h = x + 2\delta \sim 0.2$ before the commensurate antiferromagnetic order is replaced by stripe order [151, 152, 156]. Figure 6.13a shows a schematic of diagonal stripe order appropriate for $n_h \approx 1/4$. The charge stripes, with a hole filling of one per Ni site, act as antiphase domain

Neutron Scattering Studies

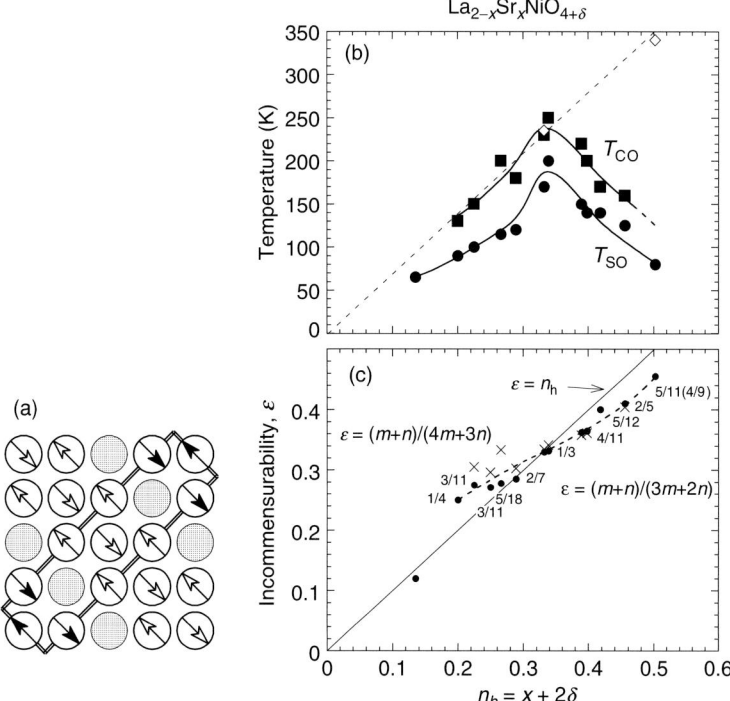

Figure 6.13. (a) Cartoon of diagonal stripe order in an NiO$_2$ plane (only Ni sites indicated) for $n_h \approx 1/4$. Magnetic unit cell is indicated by double lines, shaded circles indicate charge stripes with a hole density of one per Ni site. (b) Transition temperatures for charge order, T_{co} (squares), and spin order, T_{so} (circles), measured by neutron diffraction. Open diamonds: transition temperatures from transport measurements [149]. (c) Incommensurability vs. n_h. Circles (crosses) results at low temperature (high temperature, near T_{so}). Fraction labels are approximate long-period commensurabilities. (b) and (c) from Yoshizawa et al. [150], including results from [151–154].

walls for the magnetic order, so that the magnetic period is twice that of the charge order. The nature of the stripe order has been deduced from the positions and intensities of the superlattice peaks [134, 157]. The characteristic wave vector for spin order is $\mathbf{q}_{so} = \mathbf{Q}_{AF} \pm \frac{1}{\sqrt{2}}(\epsilon, \epsilon, 0)$ and that for charge order is $\mathbf{q}_{co} = \frac{1}{\sqrt{2}}(2\epsilon, 2\epsilon, 0) + (0, 0, 1)$. When the symmetry of the average lattice does not pick a unique orientation, modulations rotated by 90° will also be present in separate domains. The fact that diagonal stripe order has a unique modulation direction within each domain has been confirmed by electron diffraction [158]. Evidence for significant charge modulation has also been provided by nuclear magnetic resonance studies [159, 160]. The charge-ordering transition is always observed to occur at a higher temperature than the spin ordering, as shown in Figure 6.13b, with the highest ordering temperatures occurring for $x = 1/3$ [149, 161].

The magnetic incommensurability ϵ is inversely proportional to the period of the magnetic modulation. It increases steadily with doping, as shown in Figure 6.13c, staying close to the line $\epsilon = n_h$, indicating that the hole-density within the charge stripes remains roughly constant but the stripe spacing decreases with doping. For a given sample, the incommensurability changes with temperature, tending toward $\epsilon = 1/3$ as $T \to T_{co}$ [162]. In a sample with ordered oxygen interstitials, ϵ has been observed to pass through lock-in plateaus on warming, indicating a significant coupling to the lattice [154].

The impact of hole-doping on the magnetic interactions has been determined from measurements of the spin-wave dispersions for crystals with $x \approx 1/3$ [163–165]. Analysis shows that the superexchange J within an antiferromagnetic region is 27.5(4) meV [165], which is only a modest reduction compared to $J = 31(1)$ meV in undoped La_2NiO_4 [166]. The effective coupling across a charge stripe is found to be $\approx 0.5J$, a surprisingly large value. In the spin-wave modeling, it was assumed that there is no magnetic contribution from the charge stripes; however, it is not obvious that this is a correct assumption. Combining an $S = \frac{1}{2}$ hole with an $S = 1$ Ni ion should leave at least an $S = \frac{1}{2}$ per Ni site in a domain wall. Recently, Boothroyd et al. [167] have discovered quasi-1D magnetic scattering that disperses up to about 10 meV and becomes very weak above 100 K. This appears to correspond to the spin excitations of the charge stripes.

Inelastic neutron scattering measurements at $T > T_{co}$ indicate that incommensurate spin fluctuations survive in the disordered state [163, 168], implying the existence of fluctuating stripes. This result is consistent with optical conductivity studies [169, 170] which show that while the dc conductivity approaches that of a metal above room temperature, the dynamic conductivity in the disordered state never shows the response of a conventional metal.

The overall message here is that a system very close to the cuprates shows a strong tendency for charge and spin to order in a manner that preserves the strong superexchange interaction of the undoped parent compound. It is certainly true that Ni^{2+} has $S = 1$ while Cu^{2+} has $S = \frac{1}{2}$, and this can have a significant impact on the strength of the charge localization in the stripe-ordered nickelates [171]; however, the size of the spin cannot, on its own, explain why conventional band theory breaks down for the nickelates. The electronic inhomogeneity observed in the nickelates suggests that similarly unusual behavior might be expected in the cuprates.

6.5.2. Stripes in Cuprates

Static charge and spin stripe orders have only been observed in a couple of cuprates, $La_{2-x}Ba_xCuO_4$ [41, 172] and $La_{1.6-x}Nd_{0.4}Sr_xCuO_4$ [43, 173] to be specific. The characteristic 2D wave vector for spin order is $\mathbf{q}_{so} = \mathbf{Q}_{AF} \pm (\epsilon, 0)$ and that for charge order is $\mathbf{q}_{co} = (2\epsilon, 0)$. A cartoon of stripe order consistent with these wave vectors is shown in Figure 6.14(a); the inferred charge density within the charge stripes is approximately one hole for every two sites along the length of a stripe. The magnetic unit cell is twice as long as that for charge order. It should be noted that the phase of the stripe order with respect to the lattice has not been determined experimentally, so that it could be either bond-centered, as shown, or site-centered.

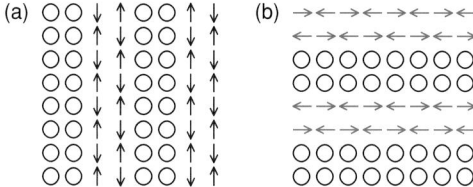

Figure 6.14. Cartoons of equivalent domains of (a) vertical and (b) horizontal bond-centered stripe order within a CuO_2 plane (only Cu sites shown). Note that the magnetic period is twice that of the charge period. The charge density along a stripe is one hole for every two sites in length.

In a square lattice, the domains of vertical and horizontal stripes shown in Figure 6.14 are equivalent; however, each breaks the rotational symmetry of the square lattice. In fact, static stripe order has only been observed in compounds in which the average crystal structure for each CuO_2 plane exhibits a compatible reduction to rectangular symmetry. This is the case for the low-temperature-tetragonal (LTT) symmetry (space group $P4_2/ncm$) of $La_{2-x}Ba_xCuO_4$ and $La_{1.6-x}Nd_{0.4}Sr_xCuO_4$ [174], where orthogonal Cu–O bonds are inequivalent within each plane, but the special direction rotates by 90° from one layer to the next. Because planes of each orientation are equally represented in the LTT phase, both stripe domains are equally represented. The correlation between lattice symmetry and stripe ordering is especially clear in studies of the system $La_{1.875}Ba_{0.125-x}Sr_xCuO_4$ by Fujita and coworkers [175–177].

When diffraction peaks from orthogonal stripe domains are present simultaneously, one might ask whether the diffraction pattern is also consistent with a checkerboard structure (a superposition of orthogonal stripe domains in the same layer) [178]. There are a number of arguments against a checkerboard interpretation. (1) The observed sensitivity of charge and spin ordering to lattice symmetry would have no obvious explanation for a checkerboard structure, with its four-fold symmetry. (2) For a pattern of crossed stripes, the positions of the magnetic peaks should rotate by 45° with respect to the charge-order peaks. One would also expect to see additional charge-order peaks in the [110] direction. Tests for both of these possibilities have come out negative [179]. It is possible to imagine other two-dimensional patterns that are consistent with the observed diffraction peaks [178]; however, the physical justification for the relationship between the spin and charge modulation becomes unclear in such models. (3) The intensity of the charge-order scattering is strongly modulated along Q_z, with maxima at $l = n \pm \frac{1}{2}$, where n is an integer. This behavior is straightforwardly explained in terms of unidirectional stripe order tied to local lattice symmetry, with Coulomb repulsion between stripes in equivalent (next-nearest-neighbor) layers [180]. For a checkerboard pattern, one would expect correlations between nearest-neighbor layers, which would give a Q_z dependence inconsistent with experiment.

There has also been a report of stripe-like charge order and incommensurate spin fluctuations in a $YBa_2Cu_3O_{6+x}$ sample with a nominal $x = 0.35$ [181]. The superconducting transition for this sample, having a midpoint at 39 K and a width of 10 K, is a bit high to be consistent with the nominal oxygen content [143]; this may indicate some inhomogeneity of oxygen content in the very large melt-grown crystal that was studied. Weak superlattice peaks attributed to charge order corresponding to vertical stripes with $2\epsilon = 0.127$ retain finite intensity at room temperature. The difference in magnetic scattering at 10 K relative to 100 K shows a spectrum very similar to that in Figure 6.3, with $E_r \approx 23$ meV and $\epsilon \approx 0.06$. While these experimental results are quite intriguing, it would be desirable to confirm them on another sample. In any case, it is interesting to note that a recent muon spin rotation (µSR) study by Sanna et al. [182] identified local magnetic order at low temperatures in $YBa_2Cu_3O_{6+x}$ with $x \leq 0.39$, and coexistence with superconductivity for $x \geq 0.37$.

Elastic incommensurate scattering consistent with stripe order has been observed in stage-4 $La_2CuO_{4+\delta}$ with $T_c = 42$ K, although charge order has not been detected [55]. An interesting question is whether static stripe order coexists homogeneously with high-temperature superconductivity in this sample. The fact that the four-layer staging of the oxygen interstitials creates two inequivalent types of CuO_2 layers suggests the possibility that the order parameters for stripe order and superconductivity might have their maxima in different sets of layers. A µSR study of the same material found that only about 40% of the muons detected a local, static magnetic field [183]. While it was argued that the best fit to the time dependence of the zero-field muon spectra was obtained with an inhomogeneous island model, the data may also be compatible with a model of inhomogeneity perpendicular to the planes.

Beyond static order, we can consider the excitations of a stripe-ordered system. It has already been noted in Section 6.2.1 that the magnetic excitations of $La_{1.875}Ba_{0.125}CuO_4$ at low temperature exhibit a similar dispersion to good superconductors without stripe order. The overall spectrum is only partially consistent with initial predictions of linear spin-wave theory for a stripe-ordered system [184–186]; however, it is reasonably reproduced by calculations that consider weakly coupled two-leg spin ladders (of the type suggested by Figure 6.14) [187–189] or that treat both spin– and electron–hole excitations of a stripe-ordered ground state [190].

The temperature dependence of the magnetic scattering at low energies (≤ 10 meV) has been reported by Fujita et al. [41]; Figure 6.15 shows some of the results. On the left, one can see that, in the stripe-ordered state ($T = 8$ and 30 K), the **Q**-integrated dynamic susceptibility is independent of frequency and temperature. Such behavior is consistent with expectations for spin waves. In the disordered phase (65 K and above), $\tilde{\chi}''(\omega)$ heads linearly to zero at zero frequency; however, at 10 meV the decrease with temperature is relatively slow. The temperature dependence of $\tilde{\chi}''(\omega)$ at 3 and 6 meV is shown in more detail on the right side, in panel (a). There is a rapid drop above T_{co} at 3 meV, but the change at 6 meV is more gradual. There is a substantial increase in **Q** width of the incommensurate peaks at the transition, as shown in (b). Interestingly, there is also a significant drop in incommensurability at the

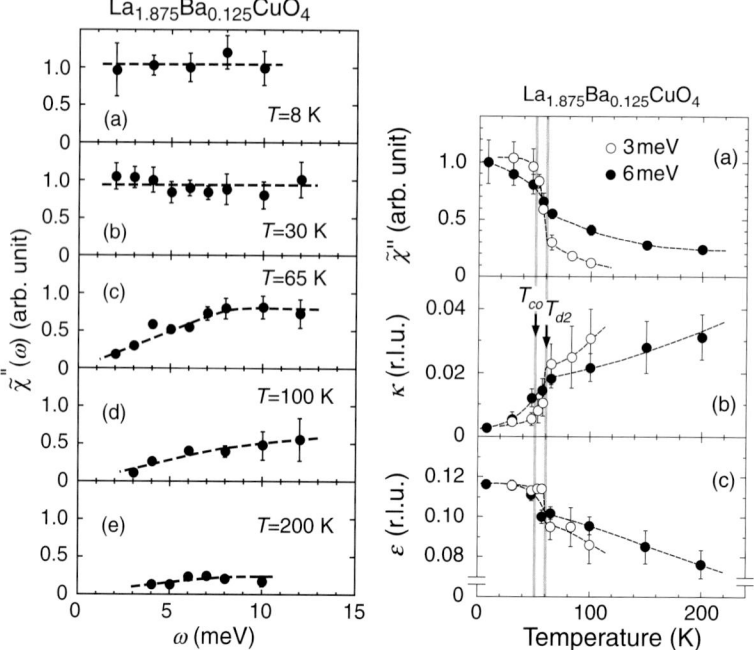

Figure 6.15. Results for low-energy inelastic magnetic scattering in $La_{1.875}Ba_{0.125}CuO_4$. Left: local susceptibility, $\tilde{\chi}''(\omega)$, as a function of $\hbar\omega$ for temperatures below (a) and (b) and above (c)–(e) the charge-ordering temperature, T_{co}. Right: temperature dependence, for $\hbar\omega = 3$ and 6 meV, of (a) local susceptibility, (b) κ, half-width in **Q** of the incommensurate peaks, (c) incommensurability ϵ. Vertical lines denote T_{co} and T_{d2}, the transition to the LTT phase. The dashed lines are guides to the eye. From Fujita et al. [41].

transition, shown in (c), with a continuing decrease at higher temperatures. Similar results for $La_{1.6-x}Nd_{0.4}Sr_xCuO_4$ with $x = 0.12$ were obtained by Ito et al. [191]. The jump in ϵ on cooling through T_{co} may be related to commensurability effects in the stripe-ordered state.

The results in the disordered state ($T > 60$ K) of $La_{1.875}Ba_{0.125}CuO_4$ look similar to those found in the normal state of $La_{2-x}Sr_xCuO_4$ [12]. The continuous variation of the magnetic scattering through the transition suggests that the nature of the underlying electronic correlations is the same on both sides of the transition. The simplest conclusion seems to be that dynamic stripes are present in the disordered state of $La_{1.875}Ba_{0.125}CuO_4$ and in the normal state of $La_{2-x}Sr_xCuO_4$. The similarity of the magnetic spectrum to that in $YBa_2Cu_3O_{6+x}$ (see Figure 6.3) then suggests that dynamic stripes may be common to under- and optimally doped cuprates.

6.5.3. Spin-Density-Wave Order in Chromium

Chromium and its alloys represent another system that has been proposed as a model for understanding the magnetic excitations in superconducting cuprates [192]. Pure Cr has a body-centered-cubic structure and exhibits antiferromagnetic order that is slightly incommensurate [193]. Overhauser and Arrott [194] first proposed that the order was due to a spin-density-wave instability of the conduction electrons. Lomer [195] later showed that the amplitude of the SDW order could be understood in terms of approximate nesting of separate electron and hole Fermi surfaces. The ordering can be modified by adjusting the Fermi energy through small substitutions of neighboring 3d elements. For example, adding electrons through substitution of less than a percent of Mn is enough to drive the ordering wave vector to commensurability, whereas reducing the electron density with V causes the incommensurate ordering temperature to head to zero at a concentration of about 3.5% [192].

The magnetic excitations in pure Cr have a very large spin-wave velocity [196, 197], similar to the situation for cuprates. The results seem to be qualitatively consistent with calculations based on Fermi-surface nesting [198, 199]. A study of paramagnetic $Cr_{0.95}V_{0.05}$ at low temperature [200] has revealed incommensurate excitations at low energy that broaden with increasing energy. χ'' has a peak at about 100 meV, but remains substantial up to at least 400 meV. A comparison of the magnitude of the experimental χ'' with ab initio calculations [201] indicates a substantial exchange enhancement over the bare Lindhard susceptibility [200].

Given that Cr is cubic, there are three equivalent and orthogonal nesting wave vectors. Within an ordered domain, the ordering wave vector consists of just one of these three possibilities. Along with the SDW order, there is also a weak CDW order that appears. A neutron diffraction study showed that the intensity of the CDW peaks scales as the square of the intensity of the SDW peaks, indicating that the CDW is a secondary order parameter and that the ordering transition is driven by the magnetic ordering [202]. It is natural to compare this behavior with that found in stripe ordered cuprates. The behavior in the latter is different, with the charge order peaks appearing at a higher temperature than those for spin order in $La_{1.6-x}Nd_{0.4}Sr_xCuO_4$ with $x = 0.12$ [203]. That result indicates that either charge ordering alone, or a combination of charge and spin energies, drive the initial ordering [204], so that stripe order is distinct from SDW order.

There are certainly some similarities between the magnetic excitations of Cr alloys and those of optimally doped cuprates. The fact that the magnetism in Cr and its alloys is caused by Fermi-surface nesting has led many people to assume that a similar mechanism might explain the excitations of superconducting cuprates, as discussed elsewhere in this book. Some arguments against such an approach have been presented in Section VI of [205].

6.5.4. Other Proposed Types of Competing Order

New types of order beyond spin-density waves or stripes has been proposed for cuprates. One is d-density-wave (DDW) order, which has been introduced by Chakravarty et al. [206] to explain the d-wave-like pseudogap seen by photoemission experiments on underdoped cuprates. (A related model of a staggered-flux phase was proposed by Wen and Lee [207] with a similar motivation; however, their model does not have static order.) The model of DDW order involves local currents that rotate in opposite directions about neighboring plaquettes within the CuO_2 planes. The orbital currents should induce weak, staggered magnetic moments oriented along the c-axis. Because of the large size of the current path in real space, the magnetic form factor should fall off very rapidly with $\mathbf{Q}_{2D}$ in reciprocal space. Mook et al. [208] have done extensive measurements in search of the proposed signal in $YBa_2Cu_3O_{6+x}$ with several values of x. The measurements are complicated by the fact that large crystals are required to achieve the necessary sensitivity, while the largest crystals available are contaminated by a significant amount of Y_2BaCuO_5. Stock et al. [209] studied a crystal of $YBa_2Cu_3O_{6.5}$ with unpolarized neutrons, and concluded that no ordered moment could be seen to a sensitivity of $\sim 0.003 \mu_B$. Using polarized neutrons, Mook et al. [210] have seen, in the spin-flip channel, a weak peak at $\mathbf{Q}_{AF}$ on top of a large background. Without giving an error bar, they suggest that the associated moment might be $0.0025 \mu_B$. They concluded that "the present results provide indications that orbital current phases are not ruled out" [210].

Varma [211] has proposed a different model of ordered orbital currents, in which the currents flow between a single Cu ion and its four coplanar O neighbors. This state breaks time-reversal and rotational symmetry but not translational symmetry. Thus, magnetic scattering from the c-axis-oriented orbital moments should be superimposed on nuclear Bragg scattering from the crystal lattice. Information on the nature of the orbital currents is contained in a strongly $\mathbf{Q}$-dependent structure factor. The only practical way to detect such a small magnetic signal on top of the strong nuclear peaks would be with polarized neutrons. Lee et al. [212] performed extensive polarized-beam studies on $La_{2-x}Sr_xCuO_4$ and $YBa_2Cu_3O_{6+x}$ single crystals. They found no positive evidence for the proposed magnetic moments, with a sensitivity of $0.01 \mu_B$ in the case of 3D order, and $0.1 \mu_B$ in the case of quasi-2D order. Simon and Varma [213] have since proposed a second pattern of orbital currents that would have a different magnetic structure factor from the original version. Positive results in $YBa_2Cu_3O_{6+x}$ corresponding to this second pattern have recently been reported by Fauqué et al. [214].

6.6. Variation of Magnetic Correlations with Doping and Temperature in Cuprates

6.6.1. Magnetic Incommensurability vs. Hole Doping

The doping dependence of the low-energy magnetic excitations in superconducting $La_{2-x}Sr_xCuO_4$ has been studied in considerable detail [13, 131]. In particular, the $\mathbf{Q}$ dependence has been characterized. We already saw in Section 6.4 that the destruction of antiferromagnetic order by hole doping leads to diagonal spin stripes. As shown in Figure 6.16(a), the magnetic incommensurability ϵ grows roughly linearly with x across the "spin-glass" regime. (The results in this region are from elastic scattering.) At $x \approx 0.055$, there is an insulator to superconductor transition, and along with that is a rotation of the incommensurate

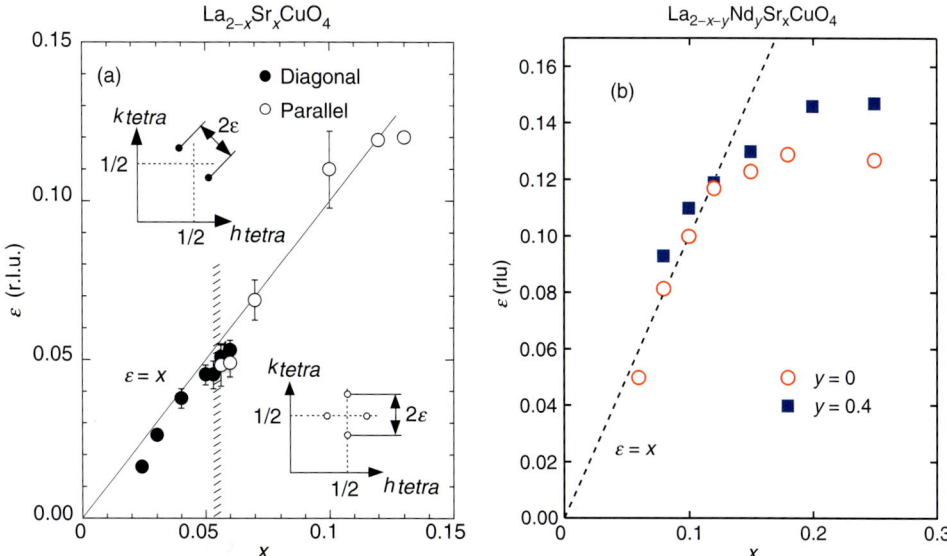

Figure 6.16. Variation of the magnetic incommensurability ϵ [as defined in the insets of (a)] for (a) lightly doped $La_{2-x}Sr_xCuO_4$, and (b) $La_{2-x}Sr_xCuO_4$ with and without Nd-codoping. In (a) the filled (open) symbols correspond to diagonal (bond-parallel, or vertical) spin stripes. Adapted From Fujita et al. [131]. In (b), open circles are from measurements of excitations at $E \sim 3$ meV and $T \approx T_c$ in $La_{2-x}Sr_xCuO_4$ from Yamada et al. [13]; filled squares are from elastic scattering on $La_{1.6-x}Nd_{0.4}Sr_xCuO_4$ from Ichikawa et al. [173].

peaks (as shown in the insets), indicating a shift from diagonal to vertical (or bond-parallel) stripes [131]. The rotation of the stripes is not as sharply defined as is the onset of the superconductivity—there is a more gradual evolution of the distribution of stripe orientations as indicated by the measured peak widths, especially around the circle of radius ϵ centered on Q_{AF}. Interestingly, the magnitude of ϵ continues its linear x dependence through the onset of superconductivity.

In the superconducting phase, ϵ continues to grow with doping up to $x \sim 1/8$, beyond which it seems to saturate, as indicated by the circles [13] in Figure 6.16(b). Interestingly, the same trend in incommensurability is found for static stripe order in Nd-doped $La_{2-x}Sr_xCuO_4$ [173], as indicated by the filled squares in the same figure. The differences in wave vector for a given x may reflect a change in the hole density of the charge stripes when they become statically ordered in the anisotropic lattice potential of the LTT phase.

While low-energy incommensurate scattering is also observed in overdoped $La_{2-x}Sr_xCuO_4$, Wakimoto et al. [215] have found that the magnitude of χ'', measured at $E \sim 6$ meV, drops rapidly for $x > 0.25$, becoming negligible by $x = 0.30$. The decrease in the magnitude of χ'' is correlated with the fall off in T_c. Interestingly, these results suggest that the superconductivity weakens as magnetic correlations disappear.

In $YBa_2Cu_3O_{6+x}$, the incommensurability of the magnetic excitations at $E < E_r$ is resolvable only for $T \ll T_c$. The presence of a substantial spin gap in the superconducting state, together with the dispersion of the magnetic excitations, makes it difficult to compare directly the results for $YBa_2Cu_3O_{6+x}$ with the behavior of $La_{2-x}Sr_xCuO_4$ shown in Figure 6.16(b). Dai et al. [17] have determined ϵ at energies just above the spin gap; the results

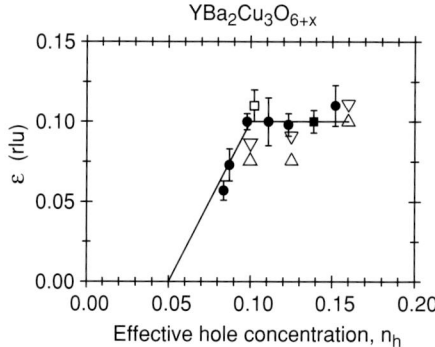

Figure 6.17. Magnetic incommensurability in $YBa_2Cu_3O_{6+x}$ (circles and squares) measured just above the spin-gap energy at $T \ll T_c$, with n_h estimated from T_c, from Dai *et al.* [17]. The triangles indicate the incommensurability at 20 meV (upper) and 30 meV (lower) for $La_{2-x}Sr_xCuO_4$ with $x = 0.10$ and 0.16 [40] and $La_{1.875}Ba_{0.125}CuO_4$ [42].

for $YBa_2Cu_3O_{6+x}$ are represented by circles and squares in Figure 6.17. For comparison, the triangles indicate the effective incommensurabilities found at energies of 20 and 30 meV in $La_{2-x}Sr_xCuO_4$ with $x = 0.10$ and 0.16 [40] and in $La_{1.875}Ba_{0.125}CuO_4$ [42]. The trends in the two different cuprate families seem to be similar when one accounts for the dispersion. (Comparable behavior in $YBa_2Cu_3O_{6+x}$ and $La_{2-x}Sr_xCuO_4$ was also noted by Balatsky and Bourges [216].)

6.6.2. Doping Dependence of Energy Scales

The doping dependence of E_r in $YBa_2Cu_3O_{6+x}$ and $Bi_2Sr_2CaCu_2O_{8+\delta}$ has received considerable attention. In optimally-doped and slightly under-doped $YBa_2Cu_3O_{6+x}$, the scattering at E_r (for $T < T_c$) is relatively strong and narrow in $\mathbf{Q}$ and ω. Of course, when the intensity is integrated over $\mathbf{Q}$ and ω one finds that it corresponds to a small fraction of the total expected sum-rule weight [26]; it is also a small fraction of the total spectral weight that is actually measured (which is much reduced from that predicted by the sum rule [90]).

Figure 6.18(a) presents a summary, from Sidis et al. [18], of experimental results for E_r from neutron scattering and for twice the superconducting gap maximum, $2\Delta_m$, from other techniques. For these materials, the resonance energy is found to scale with T_c and fall below $2\Delta_m$. Unfortunately, a major deviation from these trends occurs in $La_{2-x}Sr_xCuO_4$ [see Figure 6.18(b)], where E_r tends to be larger than $2\Delta_m$, and any constant of proportionality between E_r and kT_c is considerably larger than the value of ~ 5 found for $YBa_2Cu_3O_{6+x}$.

As discussed in Section 6.2.2, there may be a more general correlation between the spin-gap energy and T_c. Figure 6.19 shows the variation of the spin-gap energy with T_c for a range of dopings in $YBa_2Cu_3O_{6+x}$ as obtained by Dai et al. [17]. The correlation seen there looks very much like that shown in Figure 6.6 for different cuprate families at optimum doping. For $La_{2-x}Sr_xCuO_4$, a true spin gap is not observed for $x < 0.14$ [221], and this might have a connection with the rapid disappearance of the spin gap in $YBa_2Cu_3O_{6+x}$ for $x < 0.5$ [17].

Neutron Scattering Studies

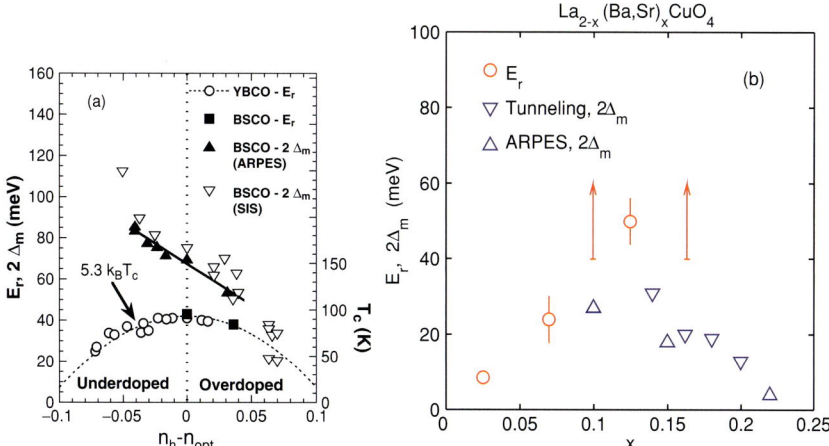

Figure 6.18. (a) Summary of results for the resonance energy E_r from neutron scattering measurements on YBa$_2$Cu$_3$O$_{6+x}$ (open circles) and Bi$_2$Sr$_2$CaCu$_2$O$_{8+\delta}$ (filled squares), and twice the maximum of the superconducting gap, $2\Delta_m$, from angle-resolved photoemission (ARPES, filled triangles) and tunneling (open triangles) measurements on Bi$_2$Sr$_2$CaCu$_2$O$_{8+\delta}$, taken from Sidis et al. [18]. (b) E_r (circles) for La$_{1.875}$Ba$_{0.125}$CuO$_4$ [42] and estimated for La$_{2-x}$Sr$_x$CuO$_4$ from measurements on $x = 0.024$ [129], $x = 0.07$ [217], $x = 0.10$ and 0.16 [40]; $2\Delta_m$ for La$_{2-x}$Sr$_x$CuO$_4$ from tunneling (downward triangles) [218, 219] and ARPES (upward triangles) [220].

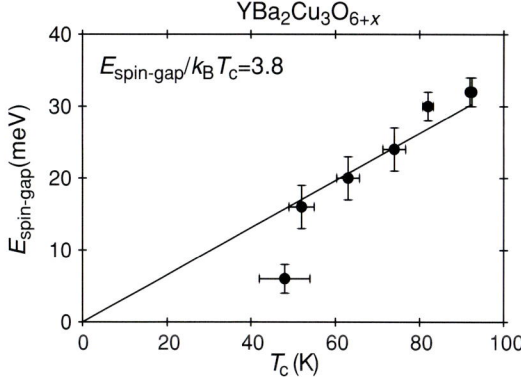

Figure 6.19. Spin-gap energy vs. T_c for YBa$_2$Cu$_3$O$_{6+x}$ from Dai et al. [17].

6.6.3. Temperature-Dependent Effects

A detailed study of the thermal evolution of the magnetic excitations ($E \leq 15$ meV) in La$_{1.86}$Sr$_{0.14}$CuO$_4$ was reported by Aeppli et al. [12]. Fitting the **Q** dependence of the incommensurate scattering with a lorentzian-squared peak shape, they found that κ, the Q-width as a function of both frequency and temperature, can be described fairly well by the formula

$$\kappa^2 = \kappa_0^2 + \frac{1}{a^2}\left[\left(\frac{kT}{E_0}\right)^2 + \left(\frac{\hbar\omega}{E_0}\right)^2\right], \quad (6.12)$$

where $\kappa_0 = 0.034$ Å^{-1}, a is the lattice parameter, and $E_0 = 47$ meV. For $T \geq T_c$, the low-frequency limit of $\chi''(\mathbf{Q}_\delta, \omega)/\omega$ (where $\mathbf{Q}_\delta$ is a peak position) varies with temperature essentially as $1/T^2$. They argued that these behaviors are consistent with proximity to a quantum critical point, and that the type of ordered state that is being approached at low temperature is the stripe-ordered state.

In a study of $La_{2-x}Sr_xCuO_4$ crystals at somewhat higher doping ($x = 0.15, 0.18$, and 0.20), Lee et al. [222] found evidence for a spin pseudogap at $T \geq T_c$. The pseudogap (with a hump above it) was similar in energy to the spin gap that appears at $T < T_c$ and was most distinct in the $x = 0.18$ sample, where the effect is still evident at 80 K but absent at 150 K.

For $YBa_2Cu_3O_{6+x}$, the studies of temperature dependence have largely concentrated on the scattering near E_r. For fully oxygenated $YBa_2Cu_3O_7$, the intensity at E_r appears fairly abruptly at, or slightly below, T_c and grows with decreasing temperature, with essentially no shift in E_r [16, 31]. For underdoped samples, the intensity at E_r begins to grow below temperatures $T^* > T_c$, with the enhancement at T_c decreasing with underdoping [31, 36, 47].

6.7. Effects of Perturbations on Magnetic Correlations

6.7.1. Magnetic Field

An important initial study of the impact of an applied magnetic field on magnetic correlations in a cuprate superconductor was done by Dai et al. [50] on $YBa_2Cu_3O_{6.6}$ ($T_c = 63$ K). They showed that applying a 6.8-T field along the c-axis caused a 30% reduction in the low-temperature intensity of the resonance peak (at 34 meV). The lost weight presumably is shifted to other parts of phase space, but it was not directly detected. (Applying the field parallel to the CuO_2 planes has negligible effect.) In an earlier study on $YBa_2Cu_3O_7$, Bourges et al. [223] applied an 11.5 T field and found that the resonance peak broadened in energy but did not seem to change its peak intensity. The difference in response from $YBa_2Cu_3O_{6.6}$ is likely due to the difference in H_{c2}, which is about five times larger in $YBa_2Cu_3O_7$ [224].

A series of studies on $La_{2-x}Sr_xCuO_4$ samples with various dopings have now been performed [51, 52, 225, 226], and a schematic summary of the results is presented in Figure 6.20. For samples with lower doping ($x = 0.10$ [226] and 0.12 [225]) there is a small elastic,

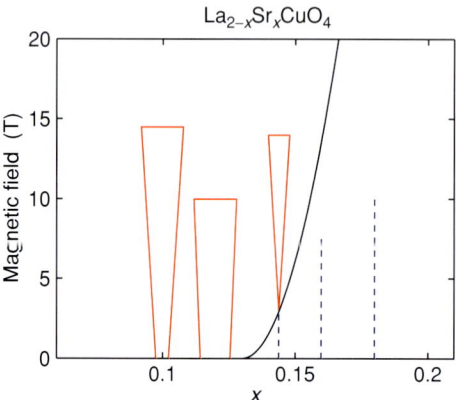

Figure 6.20. Schematic summary of neutron scattering experiments on $La_{2-x}Sr_xCuO_4$ in a magnetic field at $T \ll T_c$. Solid bars indicate observations of elastic, incommensurate peaks; width indicates variation of peak intensity with field. Experiments on $x = 0.10, 0.12, 0.144, 0.163$, and 0.18 from [226], [225], [232], [52], and [51]. The solid curve suggests the shape of a boundary between a state with spin-density-wave order and superconductivity on the left and superconductivity alone on the right, as first proposed by Demler et al. [227].

incommensurate, magnetic peak intensity in zero field that is substantially enhanced by application of a c-axis magnetic field. The growth of the intensity with field is consistent with

$$I \sim (H/H_{c2}) \ln(H_{c2}/H), \qquad (6.13)$$

where H_{c2} is the upper critical field for superconductivity [226]. This behavior was predicted by Demler et al. [227] using a model of coexisting but competing phases of superconductivity and spin-density-wave (SDW) order. In their model, local reduction of the superconducting order parameter by magnetic vortices results in an average increase in the SDW order. (For an alternative approach, in which the competing order is restricted to "halo" regions centered on vortex cores, see, e.g., [228].) Interestingly, the spin–spin correlation length for the induced signal is >400 Å, which is at least 20 times greater than the radius of a vortex core [226]. Very similar results have been obtained on oxygen-doped La_2CuO_4 [229,230]. There is an obvious parallel with the charge-related "checkerboard" pattern observed at vortices in superconducting $Bi_2Sr_2CaCu_2O_{8+\delta}$ by scanning tunneling microscopy [231].

For $La_{2-x}Sr_xCuO_4$ crystals with $x = 0.163$ [52] and 0.18 [51] there is no field-induced static order (at least for the range of fields studied). Instead, the field moves spectral weight into the spin gap. The study on $x = 0.18$ indicated that the increase in weight in the gap is accompanied by a decrease in the intensity peak above the gap [51], the latter result being comparable to the effect seen in $YBa_2Cu_3O_{6.6}$ [50]. For $x = 0.163$, an enhancement of the incommensurate scattering was observed below 10 K for $\hbar\omega = 2.5$ meV.

For an intermediate doping concentration of $x = 0.144$, Khaykovich et al. [232] have recently shown that, although no elastic magnetic peaks are seen at zero field, a static SDW does appear for $H > H_c \sim 3$ T. Such behavior was predicted by the model of competing phases of Demler et al. [227], and a boundary between phases with and without SDW order, based on that model, is indicated by the solid curve in Figure 6.20.

Although evidence for field-induced charge-stripe order in $La_{2-x}Sr_xCuO_4$ has not yet been reported, it seems likely that the SDW order observed is the same as the stripe phase found in $La_{1.875}Ba_{0.125}CuO_4$ [41] and $La_{1.6-x}Nd_{0.4}Sr_xCuO_4$ [173]. Consistent with this scenario, it has been shown that an applied magnetic field has no impact on the Cu magnetic order or the charge order in the stripe-ordered phase of $La_{1.6-x}Nd_{0.4}Sr_xCuO_4$ with $x = 0.15$ [233]; however, the field did effect the ordering of the Nd "spectator" moments.

Returning to $La_{2-x}Sr_xCuO_4$ with $x < 0.13$, it has been argued in the case of $x = 0.10$ that the zero-field elastic magnetic peak intensity observed at low temperature is extrinsic [226]. This issue deserves a short digression. It is certainly true that crystals of lesser quality can yield elastic scattering at or near the expected positions of the incommensurate magnetic peaks; in some cases, this scattering has little temperature dependence. Of course, just because spurious signals can occur does not mean that all signals are spurious. Let us shift our attention for a moment to $x = 0.12$, where the low-temperature, zero-field intensity is somewhat larger [234]. A muon-spin-relaxation (µSR) study [183] on a crystal of good quality has shown that the magnetic order is not uniform in the sample—at low temperature, only $\sim$20% of the muons see a static local hyperfine field. Further relevant information comes from electron diffraction studies. The well-known low-temperature orthorhombic (LTO) phase tends to exhibit twin domains. Horibe, Inoue, and Koyama [235] have taken dark-field images using a Bragg peak forbidden in the LTO structure but allowed in the LTT structure, the phase that pins stripes in $La_{1.875}Ba_{0.125}CuO_4$ and $La_{1.6-x}Nd_{0.4}Sr_xCuO_4$. They find bright lines corresponding to the twin boundaries, indicating that the structure of the twin boundaries is different from the LTO phase but similar to the LTT. (Similar behavior has been studied in $La_{2-x}Ba_xCuO_4$ [236].) The twin boundaries are only a few nanometers wide; however, given that magnetic vortices can pin spin stripes with a substantial correlation length, and we will see next that Zn dopants can also pin spin stripes, it seems likely that LTT-like twin boundaries should be able to pin stripe order with a significant correlation length. Thus, the low-temperature magnetic peaks found in $La_{2-x}Sr_xCuO_4$ with $x = 0.12$ [234] are likely due to stripes pinned at twinned boundaries, giving order in only a small volume fraction, consistent with µSR [183]. Taking into account the fact that stripe order is observed in $La_{1.6-x}Nd_{0.4}Sr_xCuO_4$ for a substantial range of x (but with strongest ordering at $x = 0.12$) [173], it seems reasonable to expect a small volume fraction of stripe order pinned at twin boundaries in $La_{2-x}Sr_xCuO_4$ with $x = 0.10$. Is this order extrinsic? Are twin boundaries extrinsic? This may be a matter of semantics. In any case, I would argue that the low-temperature zero-field peaks measured in good crystals reflect real materials physics of the pure compound.

6.7.2. Zn Substitution

The effects of Zn substitution for Cu are quite similar to those caused by an applied magnetic field. For $La_{2-x}Sr_xCuO_4$ with $x = 0.15$, substituting about 1% or less Zn causes the appearance of excitations within the spin gap of the Zn-free compound [237,238]. Substitution of 1.7% Zn is sufficient to induce weak elastic magnetic peaks. For $x = 0.12$, where weak elastic magnetic peaks are present without Zn, substitution of Zn increases the peak intensity, but also increases the Q-widths of the peaks [179, 239]. Wakimoto et al. [240] have recently found that Zn-substitution into overdoped samples ($x > 0.2$) significantly enhances the low-energy (<10 meV) inelastic magnetic scattering.

In $YBa_2Cu_3O_{6+x}$, Zn substitution causes weight to shift from E_r into the spin gap [241, 242]. While it causes some increase in the Q-width of the scattering at E_r [243], it does not make a significant change in the $\mathbf{Q}$ dependence of the (unresolved) incommensurate scattering at lower energies [242]. Muon-spin rotation studies indicate that Zn-doping reduces the superfluid density proportional to the Zn concentration [244], and this provides another parallel with the properties of the magnetic vortex state.

6.7.3. Li-Doping

An alternative way to dope holes into the CuO_2 planes is to substitute Li^{1+} for Cu^{2+}. In this case, the holes are introduced at the expense of a strong local Coulomb potential that one might expect to localize the holes. Surprisingly, the magnetic phase diagram of $La_2Cu_{1-x}Li_xO_4$ is rather similar to that for $La_{2-x}Sr_xCuO_4$ with $x < 0.06$ [245]. In particular, the long-range Néel order is destroyed with $\sim 0.03\%$ Li. The nature of the magnetic correlations in the paramagnetic phase is different from that in $La_{2-x}Sr_xCuO_4$ in that the inelastic magnetic scattering remains commensurate [246]. Studies of the spin dynamics indicate ω/T scaling at high temperatures, but large deviations from such behavior occur at low temperature [247].

6.8. Electron-Doped Cuprates

Electron-doped cuprates are very interesting because of their similarities and differences from the hole-doped materials; however, considerably less has been done in the way of neutron scattering on the electron-doped materials, due in part to challenges in growing crystals of suitable size and quality. Initial work focused on the systems $Nd_{2-x}Ce_xCuO_4$ and

$Pr_{2-x}Ce_xCuO_4$. A striking difference from hole doping is the fact that the Néel temperature is only gradually reduced by electron-doping. This was first demonstrated in a µSR study of $Nd_{2-x}Ce_xCuO_4$ [248], where it was found that the antiferromagnetic order only disappears at $x \approx 0.14$ where superconductivity first appears. The magnetic order was soon confirmed by neutron diffraction measurements on single crystals of $Pr_{2-x}Ce_xCuO_4$ [249] and $Nd_{2-x}Ce_xCuO_4$ [250].

A complication with these materials is that to obtain the superconducting phase, one must remove a small amount of oxygen from the as-grown samples. The challenge of the reduction process is to obtain a uniform oxygen concentration in the final sample. This is more easily done in powders and thin films than in large crystals. As-grown crystals with x as large as 0.18 are antiferromagnetic [251, 252]. Reducing single crystals can result in superconductivity; however, it is challenging to completely eliminate the antiferromagnetic phase [251]. In trying to get a pure superconducting phase, the reducing conditions can sometimes cause a crystal to undergo partial decomposition, yielding impurity phases such as $(Nd,Ce)_2O_3$ [253, 254].

The effective strength of the spin–spin coupling has been probed through measurements of the spin correlation length as a function of temperature in the paramagnetic phase. The magnitude of the spin stiffness is clearly observed to decrease with doping [249, 251, 252]. Mang et al. [252] have shown that this behavior is consistent with that found in numerical simulations of a randomly site-diluted 2D antiferromagnet. In the model calculations, the superexchange energy is held constant, and the reduction in spin stiffness is due purely to the introduction of a finite concentration of nonmagnetic sites. To get quantitative agreement, it is necessary to allow for the concentration of nonmagnetic sites in the model to be about 20% greater than the Ce concentration in the samples.

Yamada and coworkers [255] were able to prepare crystals of $Nd_{1.85}Ce_{0.15}CuO_4$ with sufficient quality that it was possible to study the low-energy magnetic excitations associated with the superconducting phase. They found commensurate antiferromagnetic fluctuations. In a crystal with $T_c = 25$ K, they found that a spin-gap of approximately 4 meV developed in the superconducting state. Commensurate elastic scattering, with an in-plane correlation length of 150 Å, was also present for temperatures below ~60 K; however, the growth of the elastic intensity did not change on crossing the superconducting T_c.

While the magnetic excitations are commensurate and incompatible with stripe correlations, there are, nevertheless, other measurements that suggest electronic inhomogeneity. Henggeler et al. [256] used the crystal-field excitations of the Pr ions in $Pr_{2-x}Ce_xCuO_4$ as a probe of the local environment. They found evidence for several distinct local environments, and argued that doped regions reached the percolation limit at $x \approx 0.14$, at the phase boundary for superconductivity. Recent NMR studies have also found evidence of electronic inhomogeneity [257, 258].

Motivated by the observation of magnetic-field-induced magnetic superlattice peaks in hole-doped cuprates (Section 6.7.1), there has been a series of experiments looking at the effect on electron-doped cuprates of a field applied along the c axis. An initial study [259] on $Nd_{2-x}Ce_xCuO_4$ with $x = 0.14$ and $T_c \sim 25$ K found that applying a field as large as 10 T had no effect on the intensity of an antiferromagnetic Bragg peak for temperatures down to 15 K. Shortly after that came a report [260] of large field-induced enhancements of antiferromagnetic Bragg intensities, as well as new field-induced peaks of the type $(1/2, 0, 0)$, in a crystal of $Nd_{2-x}Ce_xCuO_4$ with $x = 0.15$ and $T_c = 25$ K. It was soon pointed out that the new $(1/2, 0, 0)$ peaks, as well as most of the effects at antiferromagnetic reflections, could be explained by the magnetic response of the $(Nd,Ce)_2O_3$ impurity phase [253, 254]. There now seems to be a consensus that this is the proper explanation [261, 262]; however, a modest field-induced intensity enhancement has been seen at (1/2, 1/2, 3) that is not explained by the impurity-phase model [261].

In an attempt to clarify the situation, Fujita et al. [263] turned to another electron-doped superconductor, $Pr_{1-x}LaCe_xCuO_4$. This compound also has to be reduced to obtain superconductivity, and reduced crystals exhibit a $(Pr,Ce)_2O_3$ impurity phase; however, the Pr in the impurity phase should not be magnetic. They found a weak field-induced enhancement of an antiferrromagnetic peak intensity for a crystal with $x = 0.11$ ($T_c = 26$ K), but no effect for $x = 0.15$ ($T_c = 16$ K). The induced Cu moment for $x = 0.11$ at a temperature of 3 K and a field of 5 T is $\sim 10^{-4} \mu_B$. Dai and coworkers [264, 265] have studied crystals of $Pr_{0.88}LaCe_{0.12}CuO_{4-\delta}$ in which they have tuned the superconductivity by adjusting δ. They have emphasized the coexistence of the superconductivity with both 3D and quasi-2D antiferromagnetic order [264]. They report a very slight enhancement of the quasi-2D antiferromagnetic signal for a c-axis magnetic field [265].

6.9. Discussion

6.9.1. Summary of Experimental Trends in Hole-Doped Cuprates

There are a number of trends in hole-doped cuprates that one can identify from the results presented in this chapter. To begin with, the undoped parent compounds are Mott–Hubbard (or, more properly, charge-transfer) insulators that exhibit Néel order due to antiferromagnetic superexchange interactions between nearest-neighbor atoms. The magnitude of J is material dependent, varying between roughly 100 and 150 meV.

Doping the CuO_2 planes with holes destroys the Néel order; in fact, the presence of holes seems to be incompatible with long-range antiferromagnetic order. The observed responses to hole doping indicate that some sort of phase separation is common. In some cases, stripe modulations are found, and in others, finite clusters of antiferromagnetic order survive.

In under- and optimally doped cuprate superconductors, the magnetic spectrum has an hour-glass-like shape, with an energy scale comparable to the superexchange energy of the parent insulators. The strength of the magnetic scattering, when integrated over momentum and energy, decreases gradually as one increases the hole concentration from zero to optimal doping. A spin gap appears in the superconducting state (at least for optimal doping), with spectral weight from below the spin gap being pushed above it. The magnitude of the spin gap correlates with T_c.

Underdoped cuprates with a small or negligible spin gap are very sensitive to perturbations. Substituting nonmagnetic Zn for Cu or applying a magnetic field perpendicular to the planes tends to induce elastic incommensurate magnetic peaks at low temperature. For samples with larger spin gaps, the perturbations shift spectral weight from higher energy into the spin gap. Breaking the equivalence between orthogonal Cu–O bonds within a CuO_2 plane can result in charge-stripe order, in addition to the elastic magnetic peaks.

The magnetic correlations within the CuO_2 planes are clearly quite sensitive to hole doping and superconductivity. While their coexistence with a metallic normal state is one of the striking characteristics of the cuprates, their connection to the mechanism of hole-pairing remains a matter of theoretical speculation.

6.9.2. Theoretical Interpretations

The nature and relevance of antiferromagnetic correlations has been a major theme of much of the theoretical work on cuprate superconductors. While some theoretical concepts are discussed in more detail in other chapters of this book, it seems appropriate to briefly review some of them here.

Given that techniques for handling strongly-correlated hole-doped antiferromagnets continue to be in the development stage, some researchers choose to rely on a conventional weak-coupling approach to describing magnetic metals. This might be appropriate if one imagines starting out in the very over-doped regime, where Fermi-liquid theory might be applicable, and then works downward towards optimum doping. The magnetic susceptibility can be calculated in terms of electrons being excited across the Fermi level from filled to empty states. Interactions between quasiparticles due to Coulomb or exchange interactions are assumed to enhance the susceptibility near Q_{AF}, and this is handled using the random-phase approximation (RPA). In the superconducting state, one takes into account the superconducting gap Δ with d-wave symmetry. The gapping of states carves holes into the continuum of electron–hole excitations. The RPA enhancement can then pull resonant excitations down into the region below 2Δ [266–268]. With this approach, it has been possible, with suitable adjustment of the interaction parameter, to calculate dispersing features in χ'' that resemble those measured in the superconducting state of optimally doped $YBa_2Cu_3O_{6+x}$ [269–271].

The RPA approach runs into difficulties when one considers $La_{2-x}Sr_xCuO_4$, $La_{2-x}Ba_xCuO_4$, and underdoped $YBa_2Cu_3O_{6+x}$. It predicts that the magnetic excitations should be highly over damped at energies greater than 2Δ; however, there is no obvious change in the experimental spectra at $E > 2\Delta$ in these materials. Furthermore, the dispersive features in $La_{1.875}Ba_{0.125}CuO_4$ are observed in the normal state. Even if one tries to invoke a d-wave pseudogap, the energy scale is likely to be too small, as indicated by Figure 6.18(b). It is also unclear how one would rationalize, from a Fermi-liquid perspective, the observation that the energy scale of the magnetic excitations is of order J, as superexchange is an effective interaction between local moments in a correlated insulator, and has no direct connection to interactions between quasiparticles [53].

The fact that superexchange seems to remain relevant in the superconducting phase suggests that it may be profitable to approach the problem from the perspective of doped antiferromagnets. The resonating-valence-bond model was one of the first such attempts [1, 272, 273]. The model is based on the assumption that the undoped system is a quantum spin liquid. In such a state, all Cu spins would be paired into singlets in a manner such that the singlet–triplet spectrum is gapless. When a hole is introduced, one singlet is destroyed, yielding a free spinon; all other Cu spins still couple in singlet states. In such a picture, one would expect that the singlet–triplet excitations would dominate the magnetic excitation spectrum measured with neutrons; surprisingly, there has been little effort to make specific theoretical predictions of this spectrum for comparison with experiment. Instead, the analysis has been done in terms of particle–hole excitations of spinons [268]. An alternative treatment has been given in [274]. Again, while these calculations bear some resemblance to the experimental results for $YBa_2Cu_3O_{6+x}$, there are challenges to describing $La_{2-x}Sr_xCuO_4$ and $La_{2-x}Ba_xCuO_4$.

Another alternative is a spiral spin-density wave, as has been proposed several times [275–278]. A spiral state would be compatible with the incommensurate antiferromagnetic excitations at low energy [275, 277], and can also be used to model the full magnetic spectrum [278]. A look at the experimental record shows that a spiral phase cannot be the whole story. In the case of $La_{1.875}Ba_{0.125}CuO_4$ and $La_{1.6-x}Nd_{0.4}Sr_xCuO_4$, where static magnetic order is observed, charge order is also found [279]. When there is charge order present, it follows that the spin-density modulation must have a collinear component in which the magnitude of the local moments is modulated [204]. There could also be a spiral component, but it is not essential. Furthermore, if holes simply cause a local rotation of the spin direction, then it is unclear why the ordering temperature of the Néel phase is so rapidly reduced by a small density of holes.

Given that stripe order is observed in certain cuprates (Section 6.5.2) and that the magnetic excitations of the stripe-ordered phase are consistent with the universal spectrum of good superconductors (Figure 6.3), the simplest picture that is compatible with all of the data is to assume that charge stripes (dynamic ones in the case of the superconducting samples) are a common feature of the cuprates, at least on the underdoped side of the phase diagram. There is certainly plenty of theoretical motivation for stripes [205, 280–283], and a number of calculations based on a stripe model provide a reasonable description of the universal magnetic spectrum [187–190], especially when one allows for stripe fluctuations [284]. The relevance of charge inhomogeneity to the superconducting mechanism is discussed in the chapter by Kivelson and Fradkin [285].

One suprising experimental observation is the minimal amount of damping of the magnetic excitations in underdoped cuprates, especially in the normal state. One would expect the continuum of electron–hole excitations to cause significant damping [90]. Could it be that the antiphase relationship of spin correlations across a charge stripe acts to separate the spin and charge excitations in a manner similar to that in a one-dimensional system [286, 287]? With over doping, there is evidence that regions of conventional electronic excitations become more significant. This is also the regime where magnetic excitations become weak. Could it be that the interaction of conventional electron–hole excitations with stripe-like patches causes a strong damping of the spin excitations? There is clearly plenty of work left to properly understand the cuprates.

Acknowledgments

I am grateful to S. A. Kivelson and M. Hücker for critical comments. My work at Brookhaven is supported by the Office of Science, U.S. Department of Energy, under Contract No. DE-AC02-98CH10886.

Bibliography

1. P. W. Anderson, Science **235**, 1169 (1987).
2. D. Vaknin, S. K. Sinha, D. E. Moncton, D. C. Johnston, J. M. Newsam, C. R. Safinya and J. H. E. King, Phys. Rev. Lett. **58**, 2802 (1987).
3. M. A. Kastner, R. J. Birgeneau, G. Shirane, and Y. Endoh, Rev. Mod. Phys. **70**, 897 (1998).
4. P. Bourges, in *The Gap Symmetry and Fluctuations in High Temperature Superconductors*, J. Bok, G. Deutscher, D. Pavuna, and S. A. Wolf (eds.) (Plenum, New York, 1998), p. 349.
5. L. P. Regnault, P. Bourges, and P. Burlet, in *Neutron Scattering in Layered Copper-Oxide Superconductors*, A. Furrer (ed.) (Kluwer, Dordrecht, 1998), pp. 85–134.
6. S. M. Hayden, in *Neutron Scattering in Layered Copper-Oxide Superconductors*, A. Furrer (ed.) (Kluwer, Dordrecht, 1998), pp. 135–164.
7. T. E. Mason, in *Handbook on the Physics and Chemistry of Rare Earths, Vol. 31: High-Temperature Superconductors—II*, J. K. A. Gschneidner, L. Eyring, and M. B. Maple (ed.) (Elsevier, Amsterdam, 2001), pp. 281–314.

8. H. B. Brom and J. Zaanen, in *Handbook of Magnetic Materials, Vol. 15*, K. H. J. Buschow (ed.) (Elsevier, Amsterdam, 2003), pp. 379–496.
9. G. Shirane, S. M. Shapiro, and J. M. Tranquada, *Neutron Scattering with a Triple-Axis Spectrometer: Basic Techniques* (Cambridge University Press, Cambridge, 2002).
10. G. L. Squires, *Introduction to the Theory of Thermal Neutron Scattering* (Dover, Mineola, NY, 1996).
11. S.-W. Cheong, G. Aeppli, T. E. Mason, H. Mook, S. M. Hayden, P. C. Canfield, Z. Fisk, K. N. Clausen, and J. L. Martinez, Phys. Rev. Lett. **67**, 1791 (1991).
12. G. Aeppli, T. E. Mason, S. M. Hayden, H. A. Mook, and J. Kulda, Science **278**, 1432 (1997).
13. K. Yamada, C. H. Lee, K. Kurahashi, J. Wada, S. Wakimoto, S. Ueki, Y. Kimura, Y. Endoh, S. Hosoya, G. Shirane, et al., Phys. Rev. B **57**, 6165 (1998).
14. J. Rossat-Mignod, L. P. Regnault, C. Vettier, P. Bourges, P. Burlet, J. Bossy, J. Y. Henry, and G. Lapertot, Physica B **180&181**, 383 (1992).
15. H. A. Mook, M. Yethiraj, G. Aeppli, T. E. Mason, and T. Armstrong, Phys. Rev. Lett. **70**, 3490 (1993).
16. H. F. Fong, B. Keimer, D. Reznik, D. L. Milius, and I. A. Aksay, Phys. Rev. B **54**, 6708 (1996).
17. P. Dai, H. A. Mook, R. D. Hunt, and F. Doğan, Phys. Rev. B **63**, 054525 (2001).
18. Y. Sidis, S. Pailhès, B. Keimer, C. Ulrich, and L. P. Regnault, Phys. Status Solidi (b) **241**, 1204 (2004).
19. P. Bourges, Y. Sidis, H. F. Fong, B. Keimer, L. P. Regnault, J. Bossy, A. S. Ivanov, D. L. Milius, and I. A. Aksay, in *High Temperature Superconductivity*, S. E. Barnes, J. Ashkenazi, J. L. Cohn, and F. Zuo (eds.) (American Institute of Physics, Woodbury, NY, 1999), pp. 207–212.
20. H. F. Fong, P. Bourges, Y. Sidis, L. P. Regnault, A. Ivanov, G. D. Gu, N. Koshizuka, and B. Keimer, Nature **398**, 588 (1999).
21. J. Mesot, N. Metoki, M. Bohm, A. Hiess, and K. Kadowaki, Physica C **341**, 2105 (2000).
22. H. He, Y. Sidis, P. Bourges, G. D. Gu, A. Ivanov, N. Koshizuka, B. Liang, C. T. Lin, L. P. Regnault, E. Schoenherr, et al., Phys. Rev. Lett. **86**, 1610 (2001).
23. H. He, P. Bourges, Y. Sidis, C. Ulrich, L. P. Regnault, S. Pailhès, N. S. Berzigiarova, N. N. Kolesnikov, and B. Keimer, Science **295**, 1045 (2002).
24. M. Eschrig and M. R. Norman, Phys. Rev. Lett. **85**, 3261 (2000).
25. A. Abanov, A. V. Chubukov, and J. Schmalian, J. Electron. Spectrosc. Relat. Phenom. **117–118**, 129 (2001).
26. H.-Y. Kee, S. A. Kivelson, and G. Aeppli, Phys. Rev. Lett. **88**, 257002 (2002).
27. S. M. Hayden, H. A. Mook, P. Dai, T. G. Perring, and F. Doğan, Nature **429**, 531 (2004).
28. C. Stock, W. J. L. Buyers, R. A. Cowley, P. S. Clegg, R. Coldea, C. D. Frost, R. Liang, D. Peets, D. Bonn, W. N. Hardy, et al., Phys. Rev. B **71**, 024522 (2005).
29. V. Hinkov, S. Pailhès, P. Bourges, Y. Sidis, A. Ivanov, A. Kulakov, C. T. Lin, D. P. Chen, C. Bernhard, and B. Keimer, Nature **430**, 650 (2004).
30. P. Bourges, H. F. Fong, L. P. Regnault, J. Bossy, C. Vettier, D. L. Milius, I. A. Aksay, and B. Keimer, Phys. Rev. B **56**, R11 439 (1997).
31. P. Dai, H. A. Mook, S. M. Hayden, G. Aeppli, T. G. Perring, R. D. Hunt, and F. Doğan, Science **284**, 1344 (1999).
32. S. M. Hayden, G. Aeppli, T. G. Perring, H. A. Mook, and F. Doğan, Phys. Rev. B **54**, R6905 (1996).
33. D. Reznik, P. Bourges, H. F. Fong, L. P. Regnault, J. Bossy, C. Vettier, D. L. Milius, I. A. Aksay, and B. Keimer, Phys. Rev. B **53**, R14741 (1996).
34. H. A. Mook, P. Dai, S. M. Hayden, G. Aeppli, T. G. Perring, and F. Doğan, Nature **395**, 580 (1998).
35. M. Arai, T. Nishijima, Y. Endoh, T. Egami, S. Tajima, K. Tomimoto, Y. Shiohara, M. Takahashi, A. Garrett, and S. M. Bennington, Phys. Rev. Lett. **83**, 608 (1999).
36. P. Bourges, Y. Sidis, H. F. Fong, L. P. Regnault, J. Bossy, A. Ivanov, and B. Keimer, Science **288**, 1234 (2000).
37. S. Pailhès, Y. Sidis, P. Bourges, V. Hinkov, A. Ivanov, C. Ulrich, L. P. Regnault, and B. Keimer, Phys. Rev. Lett. **93**, 167001 (2004).
38. D. Reznik, P. Bourges, L. Pintschovius, Y. Endoh, Y. Sidis, T. Matsui, and S. Tajima, Phys. Rev. Lett. **93**, 207003 (2004).
39. M. Ito, H. Harashina, Y. Yasui, M. Kanada, S. Iikubo, M. Sato, A. Kobayashi, and K. Kakurai, J. Phys. Soc. Jpn **71**, 265 (2002).
40. N. B. Christensen, D. F. McMorrow, H. M. Rønnow, B. Lake, S. M. Hayden, G. Aeppli, T. G. Perring, M. Mangkorntong, M. Nohara, and H. Tagaki, Phys. Rev. Lett. **93**, 147002 (2004).
41. M. Fujita, H. Goka, K. Yamada, J. M. Tranquada, and L. P. Regnault, Phys. Rev. B **70**, 104517 (2004).
42. J. M. Tranquada, H. Woo, T. G. Perring, H. Goka, G. D. Gu, G. Xu, M. Fujita, and K. Yamada, Nature **429**, 534 (2004).
43. J. M. Tranquada, B. J. Sternlieb, J. D. Axe, Y. Nakamura, and S. Uchida, Nature **375**, 561 (1995).

44. J. M. Tranquada, H. Woo, T. G. Perring, H. Goka, G. D. Gu, G. Xu, M. Fujita, and K. Yamada, J. Phys. Chem. Solids **67**, 511 (2006).
45. T. Barnes and J. Riera, Phys. Rev. B **50**, 6817 (1994).
46. H. A. Mook, P. Dai, F. Doğan, and R. D. Hunt, Nature **404**, 729 (2000).
47. C. Stock, W. J. L. Buyers, R. Liang, D. Peets, Z. Tun, D. Bonn, W. N. Hardy, and R. J. Birgeneau, Phys. Rev. B **69**, 014502 (2004).
48. B. Lake, G. Aeppli, T. E. Mason, A. Schröder, D. F. McMorrow, K. Lefmann, M. Isshiki, M. Nohara, H. Takagi, and S. M. Hayden, Nature **400**, 43 (1999).
49. J. P. Lu, Phys. Rev. Lett. **68**, 125 (1992).
50. P. Dai, H. A. Mook, G. Aeppli, S. M. Hayden, and F. Doğan, Nature **406**, 965 (2000).
51. J. M. Tranquada, C. H. Lee, K. Yamada, Y. S. Lee, L. P. Regnault, and H. M. Rønnow, Phys. Rev. B **69**, 174507 (2004).
52. B. Lake, G. Aeppli, K. N. Clausen, D. F. McMorrow, K. Lefmann, N. E. Hussey, N. Mangkorntong, M. Nohara, H. Takagi, T. E. Mason, et al., Science **291**, 1759 (2001).
53. P. W. Anderson, Adv. Phys. **46**, 3 (1997).
54. J. M. Tranquada, A. H. Moudden, A. I. Goldman, P. Zolliker, D. E. Cox, G. Shirane, S. K. Sinha, D. Vaknin, D. C. Johnston, M. S. Alvarez, et al., Phys. Rev. B **38**, 2477 (1988).
55. Y. S. Lee, R. J. Birgeneau, M. A. Kastner, Y. Endoh, S. Wakimoto, K. Yamada, R. W. Erwin, S.-H. Lee, and G. Shirane, Phys. Rev. B **60**, 3643 (1999).
56. J. M. Tranquada, D. E. Cox, W. Kunnmann, H. Moudden, G. Shirane, M. Suenaga, P. Zolliker, D. Vaknin, S. K. Sinha, M. S. Alvarez, et al., Phys. Rev. Lett. **60**, 156 (1988).
57. T. Freltoft, J. E. Fischer, G. Shirane, D. E. Moncton, S. K. Sinha, D. Vaknin, J. P. Remeikas, A. S. Cooper, and D. Harshman, Phys. Rev. B **36**, 826 (1987).
58. M. A. Kastner, R. J. Birgeneau, T. R. Thurston, P. J. Picone, H. P. Jensen, D. R. Gabbe, M. Sato, K. Fukuda, S. Shamoto, Y. Endoh, et al., Phys. Rev. B **38**, 6636 (1988).
59. T. Thio and A. Aharony, Phys. Rev. Lett. **73**, 894 (1994).
60. A. Zheludev, S. Maslov, I. Zaliznyak, L. P. Regnault, T. Masuda, K. Uchinokura, R. Erwin, and G. Shirane, Phys. Rev. Lett. **81**, 5410 (1998).
61. D. Coffey, T. M. Rice, and F. C. Zhang, Phys. Rev. B **44**, 10112 (1991).
62. J. Stein, O. Entin-Wohlman, and A. Aharony, Phys. Rev. B **53**, 775 (1996).
63. B. O. Wells, Y. S. Lee, M. A. Kastner, R. J. Christianson, R. J. Birgeneau, K. Yamada, Y. Endoh, and G. Shirane, Science **277**, 1067 (1997).
64. B. Keimer, A. Aharony, A. Auerbach, R. J. Birgeneau, A. Cassanho, Y. Endoh, R. W. Erwin, M. A. Kastner, and G. Shirane, Phys. Rev. B **45**, 7430 (1992).
65. K. Yamada, E. Kudo, Y. Endoh, Y. Hidaka, M. Oda, M. Suzuki, and T. Murakami, Solid State Commun. **64**, 753 (1987).
66. T. Yildirim, A. B. Harris, O. Entin-Wohlman, and A. Aharony, Phys. Rev. Lett. **72**, 3710 (1994).
67. T. Yildirim, A. B. Harris, O. Entin-Wohlman, and A. Aharony, Phys. Rev. Lett. **73**, 2919 (1994).
68. R. Coldea, S. M. Hayden, G. Aeppli, T. G. Perring, C. D. Frost, T. E. Mason, S.-W. Cheong, and Z. Fisk, Phys. Rev. Lett. **86**, 5377 (2001).
69. D. Vaknin, S. K. Sinha, C. Stassis, L. L. Miller, and D. C. Johnston, Phys. Rev. B **41**, 1926 (1990).
70. M. Greven, R. J. Birgeneau, Y. Endoh, M. A. Kastner, M. Matsuda, and G. Shirane, Z. Phys. B **96**, 465 (1995).
71. Y. Tokura, S. Koshihara, T. Arima, H. Takagi, S. Ishibashi, T. Ido, and S. Uchida, Phys. Rev. B **41**, 11657 (1990).
72. D. Vaknin, L. L. Miller, and J. L. Zarestky, Phys. Rev. B **56**, 8351 (1997).
73. M. Matsuda, K. Yamada, K. Kakurai, H. Kadowaki, T. R. Thurston, Y. Endoh, Y. Hidaka, R. J. Birgeneau, M. A. Kastner, P. M. Gehring, et al., Phys. Rev. B **42**, 10098 (1990).
74. S. Skanthakumar, J. W. Lynn, J. L. Peng, and Z. Y. Li, Phys. Rev. B **47**, 6173 (1993).
75. J. W. Lynn and S. Skanthakumar, in *Handbood on the Physics and Chemistry of Rare Earths, Vol. 31*, J. K. A. Gschneidner, L. Eyring, and M. B. Maple (eds.) (Elsevier, Amsterdam, 2001), pp. 315–350.
76. P. Bourges, H. Casalta, A. S. Ivanov, and D. Petitgrand, Phys. Rev. Lett. **79**, 4906 (1997).
77. I. W. Sumarlin, J. W. Lynn, T. Chattopadhyay, S. N. Barilo, D. I. Zhigunov, and J. L. Peng, Phys. Rev. B **51**, 5824 (1995).
78. H. Casalta, P. Schleger, E. Brecht, W. Montfrooij, N. H. Andersen, B. Lebech, W. W. Schmahl, H. Fuess, R. Liang, W. N. Hardy, et al., Phys. Rev. B **50**, 9688 (1994).
79. J. Mizuki, Y. Kubo, T. Manako, Y. Shimakawa, H. Igarashi, J. M. Tranquada, Y. Fujii, L. Rebelsky, and G. Shirane, Physica C **156**, 781 (1988).

80. D. Vaknin, E. Caignol, P. K. Davies, J. E. Fischer, D. C. Johnston, and D. P. Goshorn, Phys. Rev. B **39**, 9122 (1989).
81. R. Sachidanandam, T. Yildirim, A. B. Harris, A. Aharony, and O. Entin-Wohlman, Phys. Rev. B **56**, 260 (1997).
82. D. Petitgrand, S. V. Maleyev, P. Bourges, and A. S. Ivanov, Phys. Rev. B **59**, 1079 (1999).
83. P. Burlet, J. Y. Henry, and L. P. Regnault, Physica C **296**, 205 (1998).
84. A. Jánossy, F. Simon, T. Fehér, A. Rockenbauer, L. Korecz, C. Chen, A. J. S. Chowdhury, and J. W. Hodby, Phys. Rev. B **59**, 1176 (1999).
85. H. Kadowaki, M. Nishi, Y. Yamada, H. Takeya, H. Takei, S. M. Shapiro, and G. Shirane, Phys. Rev. B **37**, 7932 (1988).
86. E. Brecht, W. W. Schmahl, H. Fuess, H. Casalta, P. Schleger, B. Lebech, N. H. Andersen, and T. Wolf, Phys. Rev. B **52**, 9601 (1995).
87. N. H. Andersen and G. Uimin, Phys. Rev. B **56**, 10840 (1997).
88. S. Shamoto, M. Sato, J. M. Tranquada, B. J. Sternlieb, and G. Shirane, Phys. Rev. B **48**, 13817 (1993).
89. M. E. Lines, J. Phys. Chem. Solids **31**, 101 (1970).
90. J. Lorenzana, G. Seibold, and R. Coldea, Phys. Rev. B **72**, 224511 (2005).
91. L. Capriotti, A. Läuchli, and A. Paramekanti, Phys. Rev. B **72**, 214433 (2005).
92. C. Yasuda, S. Todo, K. Hukushima, F. Alet, M. Keller, M. Troyer, and H. Takayama, Phys. Rev. Lett. **94**, 217201 (2005).
93. K. M. Kojima, Y. Fudamoto, M. Larkin, G. M. Luke, J. Merrin, B. Nachumi, Y. J. Uemura, N. Motoyama, H. Eisaki, S. Uchida, et al., Phys. Rev. Lett. **78**, 1787 (1997).
94. P. W. Anderson, Phys. Rev. **115**, 2 (1959).
95. T. Oguchi, Phys. Rev. **117**, 117 (1960).
96. R. R. P. Singh, Phys. Rev. B **39**, 9760 (1989).
97. K. B. Lyons, P. A. Fleury, J. P. Remeika, A. S. Cooper, and T. J. Negran, Phys. Rev. B **37**, 2353 (1988).
98. V. J. Emery and G. Reiter, Phys. Rev. B **38**, 4547 (1988).
99. A. B. Van Oosten, R. Broer, and W. C. Nieupoort, Chem. Phys. Lett. **257**, 207 (1996).
100. D. Muñoz, F. Illas, and I. P. R. Moreira, Phys. Rev. Lett. **84**, 1579 (2000).
101. I. A. Zaliznyak, H. Woo, T. G. Perring, C. L. Broholm, C. D. Frost, and H. Takagi, Phys. Rev. Lett. **93**, 087202 (2004).
102. M. Takahashi, J. Phys. C **10**, 1289 (1977).
103. M. Roger and J. M. Delrieu, Phys. Rev. B **39**, 2299 (1989).
104. A. H. MacDonald, S. M. Girvin, and D. Yoshioka, Phys. Rev. B **37**, 9753 (1988).
105. D. K. Morr, Phys. Rev. B **58**, R587 (1998).
106. J. F. Annett, R. M. Martin, A. K. McMahan, and S. Satpathy, Phys. Rev. B **40**, 2620 (1989).
107. J. M. Tranquada, G. Shirane, B. Keimer, S. Shamoto, and M. Sato, Phys. Rev. B **40**, 4503 (1989).
108. J. D. Jorgensen, B. W. Veal, A. P. Paulikas, L. J. Nowicki, G. W. Crabtree, H. Claus, and W. K. Kwok, Phys. Rev. B **41**, 1863 (1990).
109. E. F. Shender, Sov. Phys. JETP **56** (1982).
110. B. Keimer, R. J. Birgeneau, A. Cassanho, Y. Endoh, M. Greven, M. A. Kastner, and G. Shirane, Z. Phys. B **91**, 373 (1993).
111. C. J. Peters, R. J. Birgeneau, M. A. Kastner, H. Yoshizawa, Y. Endoh, J. Tranquada, G. Shirane, Y. Hidaka, M. Oda, M. Suzuki, et al., Phys. Rev. B **37**, 9761 (1988).
112. L. Shekhtman, O. Entin-Wohlman, and A. Aharony, Phys. Rev. Lett. **69**, 836 (1992).
113. C. Vettier, P. Burlet, J. Y. Henry, M. J. Jurgens, G. Lapertot, L. P. Regnault, and J. Rossat-Mignod, Phys. Scripta **T29**, 110 (1989).
114. J. Rossat-Mignod, L. P. Regnault, M. J. Jurgens, C. Vettier, P. Burlet, J. Y. Henry, and G. Lapertot, Physica B **163**, 4 (1990).
115. G. Shirane, Y. Endoh, R. J. Birgeneau, M. A. Kastner, Y. Hidaka, M. Oda, M. Suzuki, and T. Murakami, Phys. Rev. Lett. **59**, 1613 (1987).
116. R. J. Birgeneau, M. Greven, M. A. Kastner, Y. S. Lee, B. O. Wells, Y. Endoh, K. Yamada, and G. Shirane, Phys. Rev. B **59**, 13788 (1999).
117. M. Greven, R. J. Birgeneau, Y. Endoh, M. A. Kastner, B. Keimer, M. Matsuda, G. Shirane, and T. R. Thurston, Phys. Rev. Lett. **72**, 1096 (1994).
118. S. Chakravarty, B. I. Halperin, and D. R. Nelson, Phys. Rev. B **39**, 2344 (1989).
119. P. Hasenfratz and F. Niedermayer, Phys. Lett. B **268**, 231 (1991).
120. S. Chakravarty, B. I. Halperin, and D. R. Nelson, Phys. Rev. Lett. **60**, 1057 (1988).
121. B. J. Suh, F. Borsa, L. L. Miller, M. Corti, D. C. Johnston, and D. R. Torgeson, Phys. Rev. Lett. **75**, 2212 (1995).

122. A. Cuccoli, T. Roscilde, R. Vaia, and P. Verrucchi, Phys. Rev. Lett. **90**, 167205 (2003).
123. Y. J. Kim, R. J. Birgeneau, F. C. Chou, R. W. Erwin, and M. A. Kastner, Phys. Rev. Lett. **86**, 3144 (2001).
124. P. C. Hohenberg and B. I. Halperin, Rev. Mod. Phys. **49**, 435 (1977).
125. C. Niedermayer, C. Bernhard, T. Blasius, A. Golnik, A. Moodenbaugh, and J. I. Budnick, Phys. Rev. Lett. **80**, 3843 (1998).
126. O. P. Vajk, P. K. Mang, M. Greven, P. M. Gehring, and J. W. Lynn, Science **295**, 1691 (2002).
127. M. Matsuda, M. Fujita, K. Yamada, R. J. Birgeneau, Y. Endoh, and G. Shirane, Phys. Rev. B **65**, 134515 (2002).
128. S. Wakimoto, R. J. Birgeneau, M. A. Kastner, Y. S. Lee, R. Erwin, P. M. Gehring, S. H. Lee, M. Fujita, K. Yamada, Y. Endoh, et al., Phys. Rev. B **61**, 3699 (2000).
129. M. Matsuda, M. Fujita, K. Yamada, R. J. Birgeneau, M. A. Kastner, H. Hiraka, Y. Endoh, S. Wakimoto, and G. Shirane, Phys. Rev. B **62**, 9148 (2000).
130. S. Wakimoto, R. J. Birgeneau, Y. S. Lee, and G. Shirane, Phys. Rev. B **63**, 172501 (2001).
131. M. Fujita, K. Yamada, H. Hiraka, P. M. Gehring, S. H. Lee, S. Wakimoto, and G. Shirane, Phys. Rev. B **65**, 064505 (2002).
132. F. C. Chou, N. R. Belk, M. A. Kastner, R. J. Birgeneau, and A. Aharony, Phys. Rev. Lett. **75**, 2204 (1995).
133. S. Wakimoto, S. Ueki, Y. Endoh, and K. Yamada, Phys. Rev. B **62**, 3547 (2000).
134. J. M. Tranquada, in *Neutron Scattering in Layered Copper-Oxide Superconductors*, A. Furrer (ed.) (Kluwer, Dordrecht, The Netherlands, 1998), p. 225.
135. J. D. Jorgensen, B. Dabrowski, S. Pei, D. G. Hinks, L. Soderholm, B. Morosin, J. E. Schirber, E. L. Venturini, and D. S. Ginley, Phys. Rev. B **38**, 11337 (1988).
136. P. C. Hammel, A. P. Reyes, Z. Fisk, M. Takigawa, J. D. Thompson, R. H. Heffner, S.-W. Cheong, and J. E. Schirber, Phys. Rev. B **42**, 6781 (1990).
137. V. P. Gnezdilov, Y. G. Pashkevich, J. M. Tranquada, P. Lemmens, G. Gntherodt, A. V. Yeremenko, S. N. Barilo, S. V. Shiryaev, L. A. Kurnevich, and P. M. Gehring, Phys. Rev. B **69**, 174508 (2004).
138. A. Renault, J. K. Burdett, and J.-P. Pouget, J. Solid State Chem. **71**, 587 (1987).
139. H. Maletta, E. Pörschke, B. Rupp, and P. Meuffels, Z. Phys. B **77**, 181 (1989).
140. R. J. Cava, A. W. Hewat, E. A. Hewat, B. Batlogg, M. Marezio, K. M. Rabe, J. J. Krajewski, J. W. F. Peck, and J. L. W. Rupp, Physica C **165**, 419 (1990).
141. G. Uimin and J. Rossat-Mignod, Physica C **199**, 251 (1992).
142. J. Rossat-Mignod, P. Burlet, M. J. Jurgens, C. Vettier, L. P. Regnault, J. Y. Henry, C. Ayache, L. Forro, H. Noel, M. Potel, et al., J. Phys. (Paris) **49**, C8 (1988).
143. C. Stock, W. J. L. Buyers, Z. Yamani, C. L. Broholm, J.-H. Chung, Z. Tun, R. Liang, D. Bonn, W. N. Hardy, and R. J. Birgeneau, Phys. Rev. B **73**, 100504 (2006).
144. C. Ulrich, S. Kondo, M. Reehuis, H. He, C. Bernhard, C. Niedarmayer, F. Bourée, P. Bourges, M. Ohl, H. M. Rønnow, et al., Phys. Rev. B **65**, 220507 (2002).
145. M. Hücker, Y.-J. Kim, G. D. Gu, J. M. Tranquada, B. D. Gaulin, and J. W. Lynn, Phys. Rev. B **71**, 094510 (2005).
146. N. L. Wang, P. Zheng, T. Feng, G. D. Gu, C. C. Homes, J. M. Tranquada, B. D. Gaulin, and T. Timusk, Phys. Rev. B **67**, 134526 (2003).
147. R. J. Cava, B. Batlogg, T. T. Palstra, J. J. Krajewski, W. F. Peck Jr., A. P. Ramirez, and L. W. Rupp Jr., Phys. Rev. B **43**, 1229 (1991).
148. S. Shinomori, Y. Okimoto, M. Kawasaki, and Y. Tokura, J. Phys. Soc. Jpn **71**, 705 (2002).
149. S.-W. Cheong, H. Y. Hwang, C. H. Chen, B. Batlogg, J. L. W. Rupp, and S. A. Carter, Phys. Rev. B **49**, 7088 (1994).
150. H. Yoshizawa, T. Kakeshita, R. Kajimoto, T. Tanabe, T. Katsufuji, and Y. Tokura, Phys. Rev. B **61**, R854 (2000).
151. J. M. Tranquada, D. J. Buttrey, V. Sachan, and J. E. Lorenzo, Phys. Rev. Lett. **73**, 1003 (1994).
152. V. Sachan, D. J. Buttrey, J. M. Tranquada, J. E. Lorenzo, and G. Shirane, Phys. Rev. B **51**, 12742 (1995).
153. J. M. Tranquada, D. J. Buttrey, and V. Sachan, Phys. Rev. B **54**, 12318 (1996).
154. P. Wochner, J. M. Tranquada, D. J. Buttrey, and V. Sachan, Phys. Rev. B **57**, 1066 (1998).
155. K. Yamada, T. Omata, K. Nakajima, S. Hosoya, T. Sumida, and Y. Endoh, Physica C **191**, 15 (1992).
156. J. M. Tranquada, Y. Kong, J. E. Lorenzo, D. J. Buttrey, D. E. Rice, and V. Sachan, Phys. Rev. B **50**, 6340 (1994).
157. J. M. Tranquada, J. E. Lorenzo, D. J. Buttrey, and V. Sachan, Phys. Rev. B **52**, 3581 (1995).
158. J. Li, Y. Zhu, J. M. Tranquada, K. Yamada, and D. J. Buttrey, Phys. Rev. B **67**, 012404 (2003).
159. Y. Yoshinari, P. C. Hammel, and S.-W. Cheong, Phys. Rev. Lett. **82**, 3536 (1999).
160. I. M. Abu-Shiekah, O. O. Bernal, A. A. Menovsky, H. B. Brom, and J. Zaanen, Phys. Rev. Lett. **83**, 3309 (1999).
161. S.-H. Lee and S.-W. Cheong, Phys. Rev. Lett. **79**, 2514 (1997).

162. K. Ishizaka, T. Arima, Y. Murakami, R. Kajimoto, H. Yoshizawa, N. Nagaosa, and Y. Tokura, Phys. Rev. Lett. **92**, 196404 (2004).
163. P. Bourges, Y. Sidis, M. Braden, K. Nakajima, and J. M. Tranquada, Phys. Rev. Lett. **90**, 147202 (2003).
164. A. T. Boothroyd, D. Prabhakaran, P. G. Freeman, S. J. S. Lister, M. Enderle, A. Hiess, and J. Kulda, Phys. Rev. B **67**, 100407(R) (2003).
165. H. Woo, A. T. Boothroyd, K. Nakajima, T. G. Perring, C. D. Frost, P. G. Freeman, D. Prabhakaran, K. Yamada, and J. M. Tranquada, Phys. Rev. B **72**, 064437 (2005).
166. K. Yamada, M. Arai, Y. Endoh, S. Hosoya, K. Nakajima, T. Perring, and A. Taylor, J. Phys. Soc. Jpn **60**, 1197 (1991).
167. A. T. Boothroyd, P. G. Freeman, D. Prabhakaran, A. Hiess, M. Enderle, J. Kulda, and F. Altorfer, Phys. Rev. Lett. **91**, 257201 (2003).
168. S.-H. Lee, J. M. Tranquada, K. Yamada, D. J. Buttrey, Q. Li, and S.-W. Cheong, Phys. Rev. Lett. **88**, 126401 (2002).
169. T. Katsufuji, T. Tanabe, T. Ishikawa, Y. Fukuda, T. Arima, and Y. Tokura, Phys. Rev. B **54**, R14230 (1996).
170. C. C. Homes, J. M. Tranquada, Q. Li, A. R. Moodenbaugh, and D. J. Buttrey, Phys. Rev. B **67**, 184516 (2003).
171. J. M. Tranquada, J. Phys. Chem. Solids **59**, 2150 (1998).
172. M. Fujita, H. Goka, T. Adachi, Y. Koike, and K. Yamada, Physica C **426–431**, 257 (2005).
173. N. Ichikawa, S. Uchida, J. M. Tranquada, T. Niemöller, P. M. Gehring, S.-H. Lee, and J. R. Schneider, Phys. Rev. Lett. **85**, 1738 (2000).
174. J. D. Axe and M. K. Crawford, J. Low Temp. Phys. **95**, 271 (1994).
175. M. Fujita, H. Goka, K. Yamada, and M. Matsuda, Phys. Rev. Lett. **88**, 167008 (2002).
176. M. Fujita, H. Goka, K. Yamada, and M. Matsuda, Phys. Rev. B **66**, 184503 (2002).
177. H. Kimura, H. Goka, M. Fujita, Y. Noda, K. Yamada, and N. Ikeda, Phys. Rev. B **67**, 140503(R) (2003).
178. B. V. Fine, Phys. Rev. B **70**, 224508 (2004).
179. J. M. Tranquada, N. Ichikawa, K. Kakurai, and S. Uchida, J. Phys. Chem. Solids **60**, 1019 (1999).
180. M. v. Zimmermann, A. Vigliante, T. Niemöller, N. Ichikawa, T. Frello, S. Uchida, N. H. Andersen, J. Madsen, P. Wochner, J. M. Tranquada, et al., Europhys. Lett. **41**, 629 (1998).
181. H. A. Mook, P. Dai, and F. Doğan, Phys. Rev. Lett. **88**, 097004 (2002).
182. S. Sanna, G. Allodi, G. Concas, A. D. Hillier, and R. D. Renzi, Phys. Rev. Lett. **93**, 207001 (2004).
183. A. T. Savici, Y. Fudamoto, I. M. Gat, T. Ito, M. I. Larkin, Y. J. Uemura, G. M. Luke, K. M. Kojima, Y. S. Lee, M. A. Kastner, et al., Phys. Rev. B **66**, 014524 (2002).
184. C. D. Batista, G. Ortiz, and A. V. Balatsky, Phys. Rev. B **64**, 172508 (2001).
185. F. Krüger and S. Scheidl, Phys. Rev. B **67**, 134512 (2003).
186. E. W. Carlson, D. X. Yao, and D. K. Campbell, Phys. Rev. B **70**, 064505 (2004).
187. M. Vojta and T. Ulbricht, Phys. Rev. Lett. **93**, 127002 (2004).
188. G. S. Uhrig, K. P. Schmidt, and M. Grüninger, Phys. Rev. Lett. **93**, 267003 (2004).
189. D. X. Yao, E. W. Carlson, and D. K. Campbell, Phys. Rev. Lett. **97**, 017003 (2006).
190. G. Seibold and J. Lorenzana, Phys. Rev. Lett. **94**, 107006 (2005).
191. M. Ito, Y. Yasui, S. Iikubo, M. Sato, A. Kobayashi, and K. Kakurai, J. Phys. Soc. Jpn **72**, 1627 (2003).
192. E. Fawcett, H. L. Alberts, V. Y. Galkin, D. R. Noakes, and J. V. Yakhmi, Rev. Mod. Phys. **66**, 25 (1994).
193. E. Fawcett, Rev. Mod. Phys. **60**, 209 (1988).
194. A. W. Overhauser and A. Arrott, Phys. Rev. Lett. **4**, 226 (1960).
195. W. M. Lomer, Proc. Phys. Soc. London **80**, 489 (1962).
196. S. K. Sinha, G. R. Kline, C. Stassis, N. Chesser, and N. Wakabayashi, Phys. Rev. B **15**, 1415 (1977).
197. T. Fukuda, Y. Endoh, K. Yamada, M. Takeda, S. Itoh, M. Arai, and T. Otomo, J. Phys. Soc. Jpn **65**, 1418 (1996).
198. E. Kaneshita, M. Ichioka, and K. Machida, J. Phys. Soc. Jpn **70**, 866 (2001).
199. R. S. Fishman and S. H. Liu, Phys. Rev. Lett. **76**, 2398 (1996).
200. S. M. Hayden, R. Doubble, G. Aeppli, T. G. Perring, and E. Fawcett, Phys. Rev. Lett. **84**, 999 (2000).
201. J. B. Staunton, J. Poulter, B. Ginatempo, E. Bruno, and D. D. Johnson, Phys. Rev. Lett. **82**, 3340 (1999).
202. R. Pynn, W. Press, S. M. Shapiro, and S. A. Werner, Phys. Rev. B **13**, 295 (1976).
203. J. M. Tranquada, J. D. Axe, N. Ichikawa, Y. Nakamura, S. Uchida, and B. Nachumi, Phys. Rev. B **54**, 7489 (1996).
204. O. Zachar, S. A. Kivelson, and V. J. Emery, Phys. Rev. B **57**, 1422 (1998).
205. S. A. Kivelson, I. P. Bindloss, E. Fradkin, V. Oganesyan, J. M. Tranquada, A. Kapitulnik, and C. Howald, Rev. Mod. Phys. **75**, 1201 (2003).
206. S. Chakravarty, R. B. Laughlin, D. K. Morr, and C. Nayak, Phys. Rev. B **63**, 094503 (2001).

207. X.-G. Wen and P. A. Lee, Phys. Rev. Lett. **76**, 503 (1996).
208. H. A. Mook, P. C. Dai, S. M. Hayden, A. Hiess, J. W. Lynn, S.-H. Lee, and F. Doğan, Phys. Rev. B **66**, 144513 (2002).
209. C. Stock, W. J. L. Buyers, Z. Tun, R. Liang, D. Peets, D. Bonn, W. N. Hardy, and L. Taillefer, Phys. Rev. B **66**, 024505 (2002).
210. H. A. Mook, P. C. Dai, S. M. Hayden, A. Hiess, S.-H. Lee, and F. Doğan, Phys. Rev. B **69**, 134509 (2004).
211. C. M. Varma, Phys. Rev. B **55**, 14554 (1997).
212. S.-H. Lee, C. F. Majkrzak, S. K. Sinha, C. Stassis, H. Kawano, G. H. Lander, P. J. Brown, H. F. Fong, S.-W. Cheong, H. Matsushita, et al., Phys. Rev. B **60**, 10405 (1999).
213. M. E. Simon and C. M. Varma, Phys. Rev. Lett. **89**, 247003 (2002).
214. B. Fauqué, Y. Sidis, V. Hinkov, S. Pailhès, C. T. Lin, X. Chaud, and P. Bourges, Phys. Rev. Lett. **96**, 197001 (2006).
215. S. Wakimoto, H. Zhang, K. Yamada, I. Swainson, H. Kim, and R. J. Birgeneau, Phys. Rev. Lett. **92**, 217004 (2004).
216. A. V. Balatsky and P. Bourges, Phys. Rev. Lett. **82**, 5337 (1999).
217. H. Hiraka, Y. Endoh, M. Fujita, Y. S. Lee, J. Kulda, A. Ivanov, and R. J. Birgeneau, J. Phys. Soc. Jpn **70**, 853 (2001).
218. T. Nakano, N. Momono, M. Oda, and M. Ido, J. Phys. Soc. Jpn **67**, 2622 (1998).
219. T. Kato, S. Okitsu, and H. Sakata, Phys. Rev. B **72**, 144518 (2005).
220. A. Ino, C. Kim, M. Nakamura, T. Yoshida, T. Mizokawa, Z.-X. Shen, A. Fujimori, T. Kakeshita, H. Eisaki, and S. Uchida, Phys. Rev. B **65**, 094504 (2002).
221. C.-H. Lee, K. Yamada, Y. Endoh, G. Shirane, R. J. Birgeneau, M. A. Kastner, M. Greven, and Y.-J. Kim, J. Phys. Soc. Jpn **69**, 1170 (2000).
222. C. H. Lee, K. Yamada, H. Hiraka, C. R. Venkateswara Rao, and Y. Endoh, Phys. Rev. B **67**, 134521 (2003).
223. P. Bourges, H. Casalta, L. P. Regnault, J. Bossy, P. Burlet, C. Vettier, E. Beaugnon, P. Gautier-Picard, and R. Tournier, Physica B **234–236**, 830 (1997).
224. Y. Ando and K. Segawa, Phys. Rev. Lett. **88**, 167005 (2002).
225. S. Katano, M. Sato, K. Yamada, T. Suzuki, and T. Fukase, Phys. Rev. B **62**, R14677 (2000).
226. B. Lake, H. M. Rønnow, N. B. Christensen, G. Aeppli, K. Lefmann, D. F. McMorrow, P. Vorderwisch, P. Smeibidl, N. Mangkorntong, T. Sasagawa, et al., Nature **415**, 299 (2002).
227. E. Demler, S. Sachdev, and Y. Zhang, Phys. Rev. Lett. **87**, 067202 (2001).
228. S. A. Kivelson, D.-H. Lee, E. Fradkin, and V. Oganesyan, Phys. Rev. B **66**, 144516 (2002).
229. B. Khaykovich, Y. S. Lee, R. W. Erwin, S.-H. Lee, S. Wakimoto, K. J. Thomas, M. A. Kastner, and R. J. Birgeneau, Phys. Rev. B **66**, 014528 (2002).
230. B. Khaykovich, R. J. Birgeneau, F. C. Chou, R. W. Erwin, M. A. Kastner, S.-H. Lee, Y. S. Lee, P. Smeibidl, P. Vorderwisch, and S. Wakimoto, Phys. Rev. B **67**, 054501 (2003).
231. J. E. Hoffman, E. W. Hudson, K. M. Lang, V. Madhavan, H. Eisaki, S. Uchida, and J. C. Davis, Science **295**, 466 (2002).
232. B. Khaykovich, S. Wakimoto, R. J. Birgeneau, M. A. Kastner, Y. S. Lee, P. Smeibidl, P. Vorderwisch, and K. Yamada, Phys. Rev. B **71**, 220508(R) (2005).
233. S. Wakimoto, R. J. Birgeneau, Y. Fujimaki, N. Ichikawa, T. Kasuga, Y. J. Kim, K. M. Kojima, S.-H. Lee, H. Niko, J. M. Tranquada, et al., Phys. Rev. B **67**, 184419 (2003).
234. T. Suzuki, T. Goto, K. Chiba, T. Shinoda, T. Fukase, H. Kimura, K. Yamada, M. Ohashi, and Y. Yamaguchi, Phys. Rev. B **57**, R3229 (1998).
235. Y. Horibe, Y. Inoue, and Y. Koyama, Physica C **282–287**, 1071 (1997).
236. Y. Zhu, A. R. Moodenbaugh, Z. X. Cai, J. Tafto, M. Suenaga, and D. O. Welch, Phys. Rev. Lett. **73**, 3026 (1994).
237. H. Kimura, M. Kofu, Y. Matsumoto, and K. Hirota, Phys. Rev. Lett. **91**, 067002 (2003).
238. M. Kofu, H. Kimura, and K. Hirota, Phys. Rev. B **72**, 064502 (2005).
239. H. Kimura, K. Hirota, H. Matsushita, K. Yamada, Y. Endoh, S.-H. Lee, C. F. Majkrzak, R. Erwin, G. Shirane, M. Greven, et al., Phys. Rev. B **59**, 6517 (1999).
240. S. Wakimoto, R. J. Birgeneau, A. Kagedan, H. Kim, I. Swainson, K. Yamada, and H. Zhang, Phys. Rev. B **72**, 064521 (2005).
241. Y. Sidis, P. Bourges, B. Hennion, L. P. Regnault, R. Villeneuve, G. Collin, and J. F. Marucco, Phys. Rev. B **53**, 6811 (1996).
242. K. Kakurai, S. Shamoto, T. Kiyokura, M. Sato, J. M. Tranquada, and G. Shirane, Phys. Rev. B **48**, 3485 (1993).

243. Y. Sidis, P. Bourges, H. F. Fong, B. Keimer, L. P. Regnault, J. Bossy, A. Ivanov, B. Hennion, P. Gautier-Picard, G. Collin, et al., Phys. Rev. Lett. **84**, 5900 (2000).
244. B. Nachumi, A. Keren, K. Kojima, M. Larkin, G. M. Luke, J. Merrin, O. Tchernyshöv, Y. J. Uemura, N. Ichikawa, M. Goto, et al., Phys. Rev. Lett. **77**, 5421 (1996).
245. T. Sasagawa, P. K. Mang, O. P. Vajk, A. Kapitulnik, and M. Greven, Phys. Rev. B **66**, 184512 (2002).
246. W. Bao, R. J. McQueeney, R. Heffner, J. L. Sarrao, P. Dai, and J. L. Zarestky, Phys. Rev. Lett. **84**, 3978 (2000).
247. W. Bao, Y. Chen, Y. Qiu, and J. L. Sarrao, Phys. Rev. Lett. **91**, 127005 (2003).
248. G. M. Luke et al., Phys. Rev. B **42**, 7981 (1990).
249. T. R. Thurston, M. Matsuda, K. Kakurai, K. Yamada, Y. Endoh, R. J. Birgeneau, P. M. Gehring, Y. Hidaka, M. A. Kastner, T. Murakami, et al., Phys. Rev. Lett. **65**, 263 (1990).
250. I. A. Zobkalo, A. G. Gukasov, S. Y. Kokovin, S. N. Barilo, and D. I. Zhigunov, Solid State Commun. **80**, 921 (1991).
251. M. Matsuda, Y. Endoh, K. Yamada, H. Kojima, I. Tanaka, R. J. Birgeneau, M. A. Kastner, and G. Shirane, Phys. Rev. B **45**, 12548 (1992).
252. P. K. Mang, O. P. Vajk, A. Arvanitaki, J. W. Lynn, and M. Greven, Phys. Rev. Lett. **93**, 027002 (2004).
253. P. K. Mang, S. Larochelle, and M. Greven, Nature **426**, 139 (2003).
254. P. K. Mang, S. Larochelle, A. Mehta, O. P. Vajk, A. S. Erickson, L. Lu, W. L. Buyers, A. F. Marshall, K. Prokes, and M. Greven, Phys. Rev. B **70**, 094507 (2004).
255. K. Yamada, K. Kurahashi, T. Uefuji, M. Fujita, S. Park, S.-H. Lee, and Y. Endoh, Phys. Rev. Lett. **90**, 137004 (2003).
256. W. Henggeler, G. Cuntze, J. Mesot, M. Klauda, G. Saemann-Ischenko, and A. Furrer, Europhys. Lett. **29**, 233 (1995).
257. F. Zamborszky, G. Wu, J. Shinagawa, W. Yu, H. Balci, R. L. Greene, W. G. Clark, and S. E. Brown, Phys. Rev. Lett. **92**, 047003 (2004).
258. O. N. Bakharev, I. M. Abu-Shiekah, H. B. Brom, A. A. Nugroho, I. P. McCulloch, and J. Zaanen, Phys. Rev. Lett. **93**, 037002 (2004).
259. M. Matsuda, S. Katano, T. Uefuji, M. Fujita, and K. Yamada, Phys. Rev. B **66**, 172509 (2002).
260. H. J. Kang, P. Dai, J. W. Lynn, M. Matsuura, J. R. Thompson, S.-C. Zhang, D. N. Argyriou, Y. Onose, and Y. Tokura, Nature **423**, 522 (2003).
261. M. Matsuura, P. Dai, H. J. Kang, J. W. Lynn, D. N. Argyriou, K. Prokes, Y. Onose, and Y. Tokura, Phys. Rev. B **68**, 144503 (2003).
262. M. Matsuura, P. Dai, H. J. Kang, J. W. Lynn, D. N. Argyriou, Y. Onose, and Y. Tokura, Phys. Rev. B **69**, 104510 (2004).
263. M. Fujita, M. Matsuda, S. Katano, and K. Yamada, Phys. Rev. Lett. **93**, 147003 (2005).
264. P. Dai, H. J. Kang, H. A. Mook, M. Matsuura, J. W. Lynn, Y. Kurita, S. Komiya, and Y. Ando, Phys. Rev. B **71**, 100502(R) (2005).
265. H. J. Kang, P. Dai, H. A. Mook, D. N. Argyriou, V. Sikolenko, J. W. Lynn, Y. Kurita, S. Komiya, and Y. Ando, Phys. Rev. B **71**, 214512 (2005).
266. Y.-J. Kao, Q. Si, and K. Levin, Phys. Rev. B **61**, R11898 (2000).
267. M. R. Norman, Phys. Rev. B **61**, 14751 (2000).
268. J. Brinckmann and P. A. Lee, Phys. Rev. Lett. **82**, 2915 (1999).
269. A. V. Chubukov, B. Jankó, and O. Tchernyshyov, Phys. Rev. B **63**, 180507R (2001).
270. I. Eremin and D. Manske, Phys. Rev. Lett. **94**, 067006 (2005).
271. I. Eremin, D. K. Morr, A. V. Chubukov, K. H. Bennemann, and M. R. Norman, Phys. Rev. Lett. **94**, 147001 (2005).
272. P. W. Anderson, P. A. Lee, M. Randeria, T. M. Rice, N. Trivedi, and F. C. Zhang, J. Phys. Condens. Matter **16**, R755 (2004).
273. P. A. Lee, N. Nagaosa, and X.-G. Wen, Rev. Mod. Phys. **78**, 17 (2006).
274. F. Onufrieva and P. Pfeuty, Phys. Rev. B **65**, 054515 (2002).
275. B. I. Shraiman and E. D. Siggia, Phys. Rev. Lett. **62**, 1564 (1989).
276. N. Hasselmann, A. H. Castro Neto, and C. Morais Smith, Phys. Rev. B **69**, 014424 (2004).
277. O. P. Sushkov and V. N. Kotov, Phys. Rev. B **70**, 024503 (2004).
278. P.-A. Lindgård, Phys. Rev. Lett. **95**, 217001 (2005).
279. P. Abbamonte, A. Rusydi, S. Smadici, G. D. Gu, G. A. Sawatzky, and D. L. Feng, Nat. Phys. **1**, 155 (2005).
280. J. Zaanen, O. Y. Osman, H. V. Kruis, Z. Nussinov, and J. Tworzydło, Philos. Mag. B **81**, 1485 (2001).

281. S. Sachdev and N. Read, Int. J. Mod. Phys. B **5**, 219 (1991).
282. K. Machida, Physica C **158**, 192 (1989).
283. C. Castellani, C. Di Castro, and M. Grilli, Phys. Rev. Lett. **75**, 4650 (1995).
284. M. Vojta, T. Vojta, and R. K. Kaul, Phys. Rev. Lett. **97**, 097001 (2006).
285. S. A. Kivelson and E. Fradkin, in *Handbook of High Temperature Superconductivity*, J. R. Schrieffer (ed.) (Springer, New York, 2007), p. 569.
286. S. A. Kivelson and V. J. Emery, Synth. Met. **80**, 151 (1996).
287. H. V. Kruis, I. P. McCulloch, Z. Nussinov, and J. Zaanen, Phys. Rev. B **70**, 075109 (2004).

7

Optical Conductivity and Spatial Inhomogeneity in Cuprate Superconductors

J. Orenstein

We present an overview of the microwave and millimeter wave response of cuprate superconductors, emphasizing two basic types of low-frequency optical conductivity, $\sigma(\omega)$, that these materials exhibit. The first type, exemplified by ultra-pure and stoichiometric $YBa_2Cu_3O_{7-\delta}$ (YBCO) single crystals, is well described by a single component originating from the Drude response of thermal quasiparticles. In other cuprate systems that have been studied $\sigma(\omega)$ has an additional component beyond the quasiparticle contribution, also centered at $\omega = 0$. The existence of this peak has not been widely appreciated because most of its spectral weight lies in the "terahertz gap" between microwave and infrared regimes. After reviewing the evidence for this spectral feature in a wide variety of cuprate compounds, we trace its origin to quenched spatial variation in the superfluid density, ρ_s. We show that the trends in optical conductivity as a function of hole carrier concentration in a series of $Bi_2Sr_2Ca_{1-y}Dy_yCu_2O_{8+\delta}$ (BSCCO) thin films can be understood by adding a component generated by spatial inhomogeneity to the quasiparticle Drude peak. We conclude by discussing the role of optical conductivity measurements in investigating the existence, origin, and importance of inhomogeneity in cuprate superconductors.

7.1. Introduction

7.1.1. Optical Conductivity of Superconductors

The dynamical conductivity, $\sigma(q, \omega)$, is the linear response function that relates current density to electric field. The $q \to 0$ limit of $\sigma(q, \omega)$ is referred to as the optical conductivity, or $\sigma(\omega)$, because it describes the response of the medium to electromagnetic waves with wavelength much longer than the characteristic length scales of condensed electronic systems. The real part of the optical conductivity, $\sigma_1(\omega)$, describes the dissipation of electromagnetic energy in the medium, while the imaginary part, $\sigma_2(\omega)$, describes screening of the applied field.

Measurement of $\sigma(\omega)$ in a superconductor is a powerful method for probing the dynamics of quasiparticle excitations and the size of the energy gap. According to BCS theory, there are three dissipative processes that determine $\sigma_1(\omega)$ in a superconductor: superfluid acceleration, pair creation, and quasiparticle scattering. The first is the work required to accelerate electrons to achieve the Meissner screening current. This contribution appears as a δ-function at zero frequency in $\sigma_1(\omega)$, whose spectral weight is the superfluid density, ρ_s. The latter

---

J. Orenstein • Physics Department, University of California, Berkeley, CA, USA
Materials Science Division, Lawrence Berkeley National Laboratory, Berkeley, CA 94720, USA

two contributions appear above zero frequency. In pair creation, electromagnetic energy is dissipated when a photon excites a pair of quasiparticles out of the BCS vacuum; in quasiparticle scattering the photon promotes a quasiparticle, already excited out of the vacuum due to thermal excitation, to a higher energy state.

7.1.2. Optical Conductivity and the Cuprates

Some special properties of the high-T_c cuprate superconductors make the study of $\sigma(\omega, T)$ particularly valuable in these materials. Because the gap is large, even relatively poor samples are in the clean limit where the scattering rate is smaller than the gap frequency, or $1/\tau \ll \Delta$. As a consequence the spectral weight associated with pair creation is very small and the contribution from quasiparticle scattering can be clearly resolved. Furthermore, $\sigma(\omega)$ associated with quasiparticle scattering can be quite accurately modeled by the Drude response of a dilute gas of weakly interacting particles [1]. This is not true of BCS s-wave superconductors, where the density of states singularity at the gap energy and the coherence factors strongly affect the conductivity spectra [2]. In d-wave superconductors, these influences are considerably weakened because there is a broad range of gap values extending from zero to the maximum gap, Δ_0.

The combination of clean-limit dynamics and d-wave density of states suggests that a simple two-fluid model (TFM) is applicable to $\sigma(\omega, T)$ in the cuprates. Because pair creation has vanishingly small spectral weight, the total conductivity should comprise only two components: the condensate δ-function and a Drude-like peak associated with thermal quasiparticles. The spectral weight of these two components are the normal and superfluid densities, or ρ_n and ρ_s, respectively. The conductivity sum rule requires $\rho_n(T) + \rho_s(T)$ to be independent of T.

The TFM does indeed provide an excellent overall description of the microwave properties of optimally-doped YBa$_2$Cu$_3$O$_{7-\delta}$ (YBCO) single crystals [3], particularly for temperature, T, above a few K. By fitting $\sigma(\omega, T)$ using the TFM, $\rho_s(T)$, $\rho_n(T)$, and $1/\tau(T)$, are determined [4]. However, the microwave properties of all other cuprate systems have turned out to be very different [5–9], showing clear indications of much greater disorder than YBCO. Even more significant is the fact that $\sigma(\omega)$ in these systems is inconsistent with the "dirty d-wave" picture [10], which provides a mean-field description of the effects of disorder on the conductivity.

According to the dirty d-wave picture, the quasiparticle spectrum is sensitive to disorder below a characteristic energy scale, E^* (or temperature scale, $T^* \equiv E^*/k_B$). For $E < E^*$ the density of states approaches a nonzero value, $N(0)$, at the chemical potential, while remaining linear in E above E^*. The existence of states at the chemical potential affects the low T properties of ρ_n and the quasiparticle conductivity. Instead of vanishing linearly with T, as expected for a clean d-wave superconductor, $\rho_n(T) \to \alpha T^* + \mathcal{O}(T^2)$ as $T \to 0$, where α is the temperature coefficient of $\rho_n(T)$ in the clean-limit. The residual quasiparticle spectral weight gives rise to conductivity that approaches a "universal" value $\sigma_d = (v_F/v_\Delta)\sigma_Q/\pi^2$ at low T (where $\sigma_Q \equiv e^2/\hbar d$ and d is the interplanar separation) [11, 12].

The dirty d-wave picture implies a correlation between the low T behaviors of ρ_s and ρ_n that is not observed in measurements on cuprates other than YBCO. In this article, we trace the inconsistency to the mean-field aspect of dirty d-wave, i.e., the assumption that disorder generates a nonzero $N(0)$ that is spatially homogeneous. Instead we argue that the density of states at the chemical potential, $N(0)$, and therefore both ρ_n and ρ_s, vary with position in the medium [13]. This form of inhomogeneity changes qualitatively the nature of the optical properties, essentially by invalidating the selection rule that forbids coupling of the longitudinal excitations of the order parameter to $\sigma(\omega)$. The plausibility of an explanation for anomalies in $\sigma_1(\omega, T)$ based on spatial inhomogeneity has been strengthened by scanning tunneling microscopy (STM) measurements [14, 15]. The STM experiments demonstrate that the local density of states (LDOS) of $Bi_2Sr_2CaCu_2O_{8+\delta}$ (BSCCO) varies in space, with spatial fluctuations that have a minimum wavelength of ~ 50 Å. The variations in LDOS suggest quenched inhomogeneity in local carrier concentration, x, and therefore in the local ρ_s.

The purpose of this article is to describe what can be learned from microwave and millimeter-wave (terahertz) spectra of cuprate superconductors whose $\sigma(\omega, T)$ cannot be described by a mean-field picture of disorder. The presentation is organized as follows. In Section 7.2 we compare the YBCO microwave data with results obtained on other optimally doped cuprates, highlighting qualitative differences in $\sigma(\omega, T)$. Section 7.3 presents a survey of (some previously unpublished) THz spectra, as measured in films of BSCCO with hole concentration varying from under to overdoped. These results help to clarify the nature of the difference in low-frequency conductivity between YBCO crystals and other cuprates. We show that a second Lorentzian peak, in addition to the quasiparticle Drude response, is needed to describe the observed $\sigma(\omega, T)$ in this latter class of materials. The weight of this peak *increases with decreasing* T in proportion to the superfluid density, distinguishing it from the quasiparticle contribution to $\sigma(\omega, T)$. In Section 7.4 we discuss the origin of the additional component, suggesting that the proportionality to ρ_s implies a connection to quantum phase fluctuations of the superconducting order parameter. We argue that the spectral weight displaced from the condensate by quantum phase fluctuations is strongly enhanced in the presence of quenched spatial inhomogeneity in ρ_s. Through a simple model, we show that the spectrum of the displaced spectral weight depends on the spatial correlation of ρ_n and ρ_s variations in the the medium. Specifically, the spectrum shifts from the plasma frequency to near zero frequency as the correlation varies from positive (such that regions of large ρ_s have large ρ_n) to negative (regions with large ρ_s have small ρ_n). In the last part of Section 7.4 we compare the theoretical modeling with the measured $\sigma(\omega, T)$. Finally, in Section 7.5 we discuss the relevance of the low-frequency optical conductivity to the debate over the origin and role of inhomogeneity in the cuprates, and indicate directions for future research.

7.2. Low Frequency Optical Conductivity in the Cuprates

7.2.1. YBCO Single Crystals: Success of the Two-Fluid Model

Detailed microwave studies of $\sigma(\omega, T)$ in high quality YBCO single crystals have been performed by the UBC group. Up to date reviews of this research have been provided by Bonn and Hardy (this volume) and by Maeda et al. [5]. Two sets of measurements exemplify the behavior of these materials. Figure 7.1 shows σ_1 of an optimally doped crystal as a function of T for several representative frequencies that range from 1 to 75 GHz [3]. Figure 7.2 shows data obtained using an experimental technique that enables continuous scan of frequency [17]. Here $\sigma_1(\omega)$ is displayed for temperatures in the range from 1 to 9 K, also for an optimally doped sample.

The main features of the microwave conductivity in YBCO are in excellent agreement with the predictions of the TFM of a d-wave superconductor in the clean limit. For our purposes, the essential feature of the data is the T-dependence of the normal fluid spectral

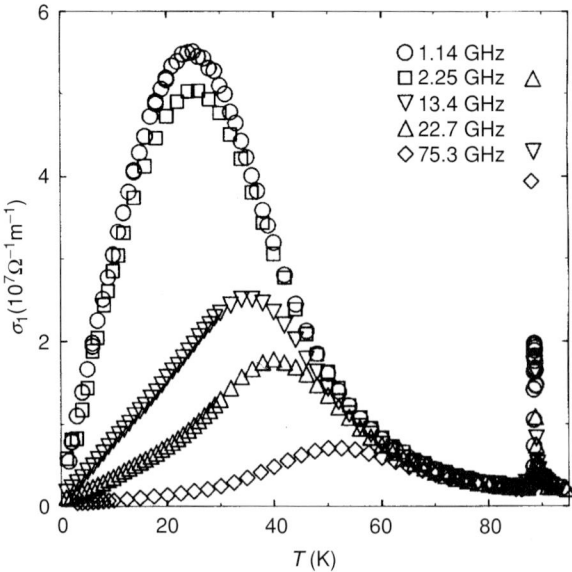

Figure 7.1. σ_1 vs. temperature for various frequencies in the microwave (GHz) range of the spectrum, as measured on a high-purity, optimally doped YBCO single crystal. From [3].

weight, which is depicted in the inset to Figure 7.2. As we will be discussing conductivity weight throughout this article, it is useful to settle on one consistent system of units throughout. The conductivity spectral weight is the integral of $\sigma_1(\omega)$ over ω. In the MKS system we can express spectral weight as the square of a plasma frequency through the relation, $\omega_p^2 = (2/\pi\epsilon_0)\int \sigma_1 d\omega$. Since for typical spectral weights we encounter, ω_p^2 expressed in s^{-2} is cumbersome, we suggest defining one spectral weight unit (SWU) as $10^{30}\,s^{-2}$. A superfluid condensate with 1 SWU corresponds to a London penetration depth of 300 nm. The London length in optimally doped YBCO (parallel to the a-axis) is approximately 150 nm, corresponding to 4 SWU.

According to the dirty d-wave model, the quasiparticle spectral weight that remains uncondensed is constrained by the relation $\rho_n(0) \simeq \alpha T^*$, where T^* is the T below which the increase of $\rho_s(T)$ is no longer linear. As indicated in the Figure 7.2 inset, $\rho_s(T)$ remains linear in this sample down to at least 1.3 K. Therefore, T^* can be no larger than 1 K and, taking $\alpha = 0.02$ SWU/K, dirty d-wave predicts that $\rho_n(0) < 0.02$ SWU. Referring to the solid symbols in the inset, we see that $\rho_n(0)$ calculated by direct numerical integration of the measured $\sigma_1(\omega)$ indeed satisfies this condition, extrapolating to less than 0.01 SWU as $T \to 0$. The maximum residual spectral weight is less than 0.25% of the condensate at $T = 0$.

As the residual spectral weight is so small, the conductivity at all temperatures above a few K satisfies the relation, $\rho_n(T) + \rho_s(T) = \rho_s(0)$. The above expression for the normal fluid spectral weight, together with a Drude conductivity spectrum,

$$\sigma_1(\omega, T) = \frac{\rho_n(T)\tau(T)}{1 + \omega^2\tau^2(T)}, \qquad (7.1)$$

is sufficient to capture all the remarkable features of the quasiparticle conductivity seen in Figure 7.1 (1) the large peak in $\sigma_1(T)$ that occurs at temperatures below T_c, (2) the crossover

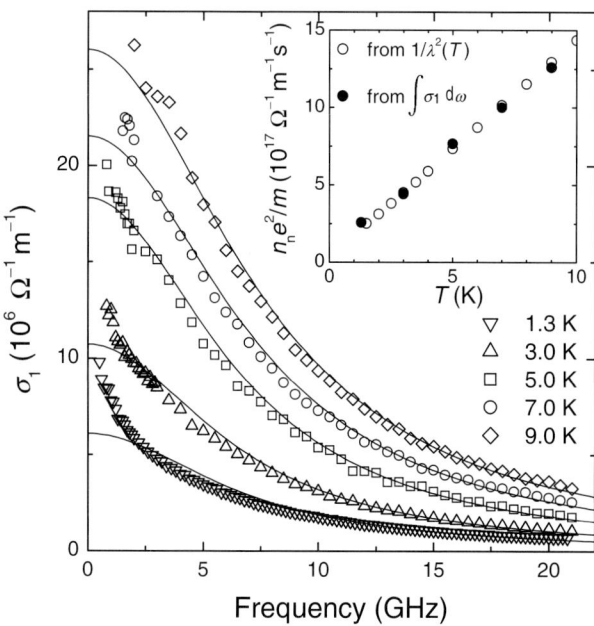

Figure 7.2. *Main panel:* σ_1 vs. frequency for several temperatures in the range 1.3–9.0 K, as measured on a high-purity, optimally doped YBCO single crystal. The solid lines show the best fit to the data obtained using the Drude model for the quasiparticle conductivity. *Inset:* The temperature dependence of the quasiparticle spectral weight as determined from numerical integration of the measured σ_1 and as inferred from the penetration depth. From [17].

of the peak T from ω-dependent to ω-independent behavior with decreasing frequency, and (3) the collapse of $\sigma_1(T)$ at all frequencies to a single curve at high T [4]. According to Eq. (7.1), σ_1 approaches a frequency-independent value, $\rho_n \tau$, in the high T regime where $\omega \tau(T) \ll 1$, thus explaining (3). The increase in σ_1 with decreasing T in this regime indicates that τ increases faster than ρ_n decreases. This is the evidence for the famous "collapse" of the quasiparticle scattering rate upon entering the superconducting state [18, 19], an observation that plays a crucial role in theories of high-T_c superconductivity. Regarding features (1) and (2), Eq. (7.1) predicts that $\sigma_1(T)$ will start to decrease when τ begins to exceed ω^{-1}. This accounts for a frequency-dependent peak T. However, as the measurement frequency is made lower, it will eventually become smaller than the low T limit of $1/\tau$ determined by elastic scattering. In this regime, $\sigma_1(T)$ peaks when the increase of τ can longer overcome the decrease of ρ_n with decreasing T. The frequency scale at which the peak becomes T-independent (~ 10 GHz in YBCO), provides an estimate of the elastic scattering rate. Note that an elastic scattering rate of $2\pi \times 10$ GHz, together with a Fermi velocity of $\sim 2.5 \times 10^7$ cm s^{-1}, corresponds to mean-free-path in the neighborhood of several microns.

7.2.2. The BSCCO System: Failure of the Two-Fluid Description

In the BSCCO family of superconductors, $\sigma(\omega, T)$ has been studied in nearly as much detail as in YBCO crystals [5]. Figure 7.3 shows an example of microwave data on an optimally doped BSCCO crystal [6] presented in the same format as Figure 7.1. The differences are quite striking, especially considering that the BSCCO and YBCO samples have almost the same T_c. In BSCCO, $\sigma_1(T)$ is independent of ω in the same range of frequency where

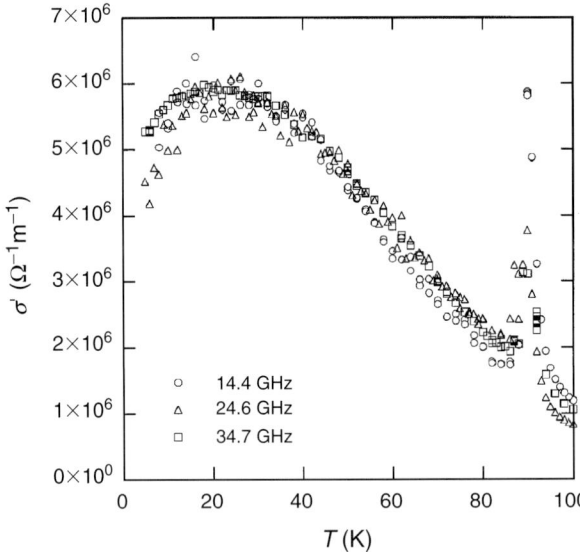

Figure 7.3. σ_1 vs. temperature for various frequencies in the GHz range, as measured in an optimally doped BSCCO sample. Note that the optical conductivity is independent of frequency throughout the same range that it depends strongly on frequency in YBCO. From [6].

$\sigma_1(T)$ in YBCO depends very strongly on ω. This contrast immediately suggests that $1/\tau$ is significantly larger in BSCCO as a result of increased disorder. However, despite the evidence for increased disorder, $\rho_s(T)$ increases linearly with decreasing T (see Figure 7.4) to at least 5 K [6,7].

Based on these results, we can say that $\sigma(\omega, T)$ in BSCCO does not satisfy the conditions for internal consistency required by the dirty d-wave picture. From $\rho_s(T)$ we see that T^* cannot be greater than $\sim$3 K. Using the same value of α as in YBCO, dirty d-wave predicts an upper bound on $\rho_n(0)$ of 0.06 SWU. However, this is much smaller than the uncondensed quasiparticle spectral weight inferred directly from $\int \sigma_1(\omega) d\omega$. As σ_1 is ω-independent up to 35 GHz, the characteristic roll-off frequency of the Drude peak cannot be less than $\sim$100 GHz, placing a lower bound on $\rho_n(0)$ of $\sim$0.45 SWU. Thus, this BSCCO sample is highly ordered based on $\rho_s(T)$, but highly disordered based on $\rho_n(T)$.

THz spectroscopy performed on BSCCO films captures the frequency dependence of σ_1 that takes place above the range of microwave spectroscopy [20, 21]. In Figure 7.5 we show $\sigma_1(\omega, T)$ vs. T for an MBE-grown, slightly sub-optimally doped BSCCO thin film, for a series of representative frequencies from 150 GHz to 800 GHz. The lowest frequency THz data are quite similar to the highest frequency microwave data. Only above 150 GHz does σ_1 begin to change. These data prove that the essential frequency dependence of $\sigma_1(T)$ in BSCCO occurs at frequencies almost two orders of magnitude larger than in YBCO single crystals. Using these data, a more rigorous lower bound on $\rho_n(0)$ of 0.6 SWU is obtained by numerical integration of σ_1 up to 1 THz. *The uncondensed spectral weight in this sample is more than a factor of 60 larger than in the UBC-grown YBCO single crystals, and represents 30% of $\rho_s(0)$.*

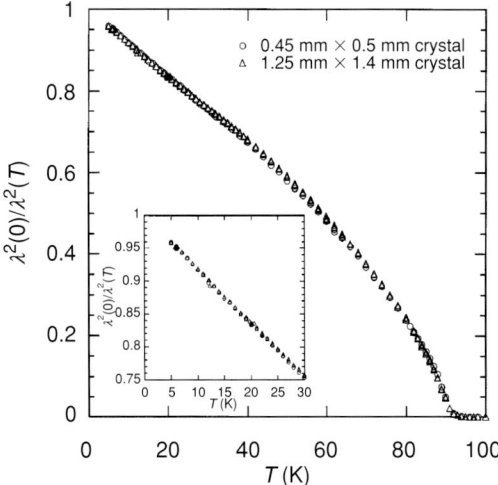

Figure 7.4. The temperature dependence of the inverse square of the penetration depth (proportional to the superfluid density, ρ_s), normalized to its value in the $T \to 0$ limit, for the same sample as Figure 7.3. $\rho_s(T)$ in BSCCO shows the same clean-limit behavior as YBCO, despite its much greater quasiparticle scattering rate. From [6].

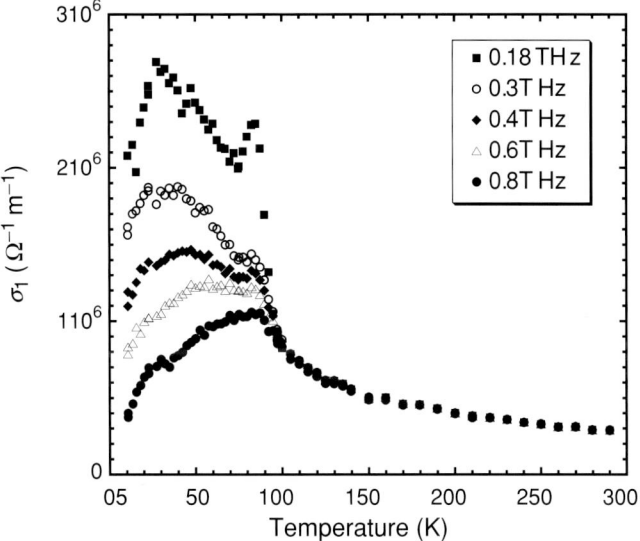

Figure 7.5. σ_1 vs. temperature for various frequencies in the THz range, as measured in a slightly underdoped BSCCO thin film with $T_c = 85$ K.

That $\rho_n(0)$ is comparable to $\rho_s(0)$ may be surprising to many readers. The more familiar picture of the optical conductivity in BSCCO comes from measurements of IR reflectivity, R, on bulk single crystals [22], an example of which is shown in Figure 7.6. Kramers–Kronig analysis of $R(\omega)$ provides an extremely accurate measurement of $\sigma(\omega)$ over a wide range

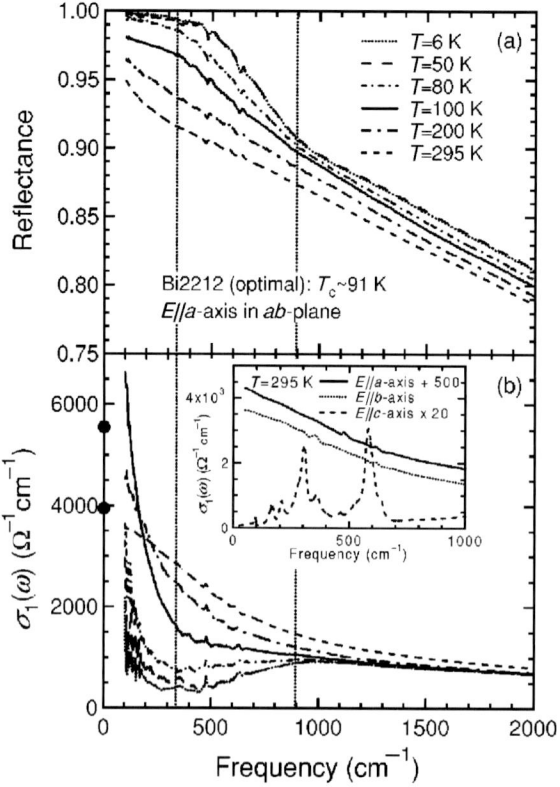

Figure 7.6. *Upper panel:* Reflectance as measured on an optimally doped BSCCO crystal as a function of frequency (in wavenumbers) at the temperatures indicated in the legend. *Lower panel:* Real part of optical conductivity as obtained by Kramers–Kronig transformation of the reflectance. From [22].

of frequency. However, the Kramers–Kronig analysis becomes unreliable when R becomes too close to unity, typically for frequencies below $\sim 100\,\text{cm}^{-1}$, or roughly 3 THz. As a result, IR reflection measurements are not sensitive to the low-frequency component of $\sigma(\omega)$ that is seen in microwave and THz spectroscopy. We note that this component, which approaches $6 \times 10^6\,\Omega^{-1}\,\text{m}^{-1}$ as $\omega \to 0$, is a factor of 30 larger than the value of the conductivity at the low-frequency limit of IR reflectivity measurements.

A plot of σ_1 as a function of frequency from 0.01 to 100 THz, shown in Figure 7.7, presents a more complete picture of the low T optical conductivity of optimally doped BSCCO. The broad-band spectrum was assembled from microwave [6], terahertz [20], and infrared [22] data. For reference, the value of the universal d-wave conductivity is shown as a triangle symbol on the left-hand axis. Considering that the plot is a composite of data obtained from three different techniques (and three different samples), the continuity of $\sigma_1(\omega)$ is remarkable. The composite plot clearly demonstrates that the low T conductivity is dominated by a component centered at $\omega = 0$, with width $\sim 300\,\text{GHz}$.

For comparison, we show (in Figure 7.8) a composite of optical conductivity data as measured in optimally doped YBCO crystals (open symbols). The YBCO spectrum is

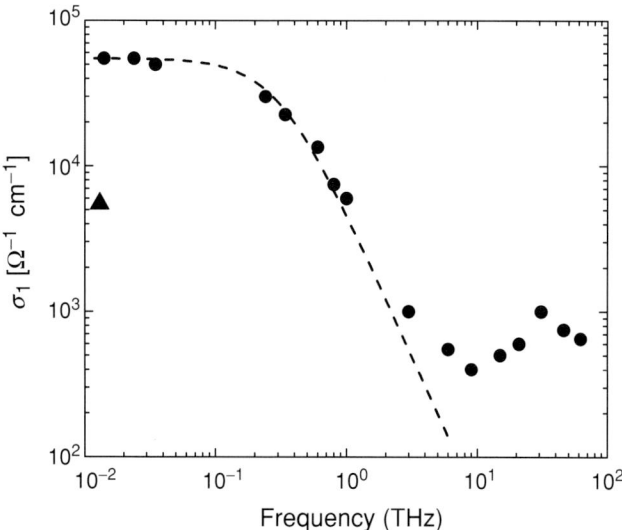

Figure 7.7. Composite plot of σ_1 at $T \leq 10\,\text{K}$ as a function of frequency in THz. The spectrum is assembled from microwave [6], terahertz [20], and infrared [22] data. At low T the conductivity is dominated by a component centered at $\omega = 0$, with width $\simeq 300\,\text{GHz}$, whose magnitude as $\omega \to 0$ is much greater than the "universal d-wave conductivity" (triangle). The dashed line shows a Drude spectrum for scattering rate $(2\pi) \times 3 \times 10^{11}\,\text{s}^{-1}$ and plasma frequency $1.1 \times 10^{15}\,\text{s}^{-1}$, corresponding to 1.2 SWU in the units defined in the text.

assembled from microwave [3, 17] and infrared [23] data. The composite spectrum highlights several interesting features of the optical conductivity in YBCO. The conductivity in the low ω, T limit is comparable to that of optimal BSCCO and is larger than the universal d-wave value by a factor of approximately 20. In YBCO, the uncondensed spectral weight in the microwave region is much smaller than in BSCCO because the characteristic cutoff frequency is $\sim 3\,\text{GHz}$, compared with $\sim 300\,\text{GHz}$ in BSCCO. However, we cannot conclude with certainty that the total uncondensed spectral weight in YBCO is small because, despite intense study of this material, there is a large regime of frequency (0.08–3.0 THz) in which $\sigma(\omega)$ has not been measured. Moreover, a linear extrapolation of the power law decay of $\sigma(\omega)$ in the microwave regime to higher frequency far underestimates the conductivity at the low frequency limit of the infrared data. For purposes of illustration we have indicated (as a dashed line) one possible behavior of the conductivity in the THz regime. If the conductivity did indeed plateau near the universal d-wave value, the uncondensed spectral weight would be ~ 0.2–0.3 SWU, which is still less than 10% of the condensate spectral weight in this material. Of course, the conductivity could, instead, exhibit a deep valley or large peak in the unmeasured frequency range.

7.2.3. Additional Examples

BSCCO is not the only example of a cuprate superconductor in which as much as one-third of the quasiparticle spectral weight remains uncondensed as $T \to 0$. Another example is optimally doped $\text{La}_{1-x}\text{Sr}_x\text{CuO}$ (LSCO), whose $\sigma(\omega)$ in the THz frequency regime has been

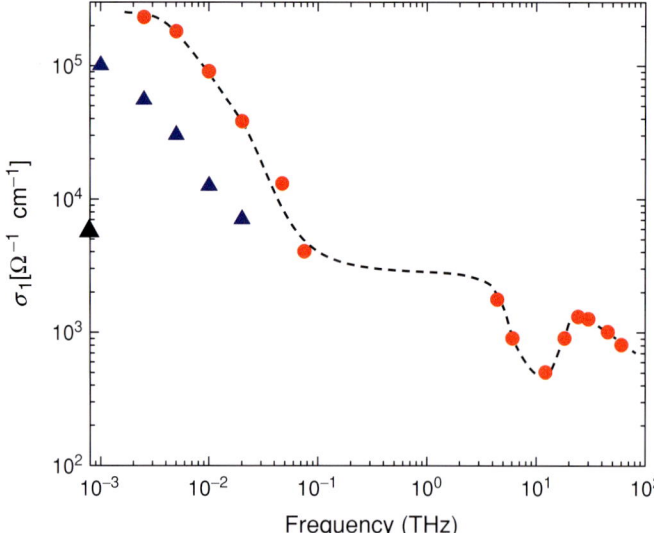

Figure 7.8. Composite plot of σ_1 in optimally doped YBCO, assembled from microwave [3,17] and infrared [23] data. The blue and red symbols correspond to sample temperatures of 1.3 and 9 K, respectively. The line is a guide to the eye, illustrating one of the many possible ways that σ_1 could vary through the THz region where it has not been measured.

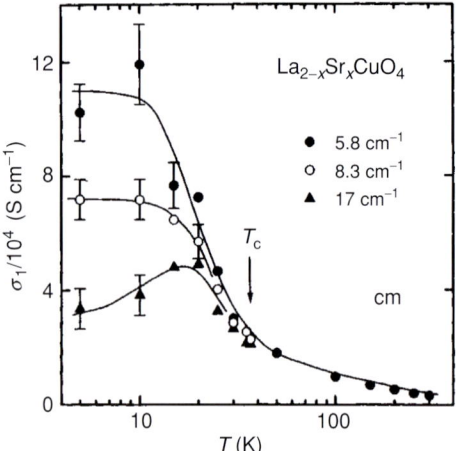

Figure 7.9. σ_1 vs. temperature for various frequencies in the THz range, as measured in an optimally doped thin film of LSCO with $T_c = 35$ K. The frequencies are given in wavenumbers, which can be expressed in Hz using the conversion 1 wavenumber = 30 GHz. From [8].

reported by Pronin et al. [8] (Figure 7.9). These authors analyze their data using a Drude model for the quasiparticle conductivity in which $\rho_n(T) + \rho_s(T)$ is constrained to be constant, but not necessarily equal to $\rho_s(0)$, thus allowing for the possibility of uncondensed quasiparticle spectral weight. From this analysis they conclude that $\rho_n(0)$ is *four times larger* than $\rho_s(0)$. The total spectral weight, ~ 2 SWU, is nearly the same in LSCO as in BSCCO. However, the condensed portion in LSCO is only 0.4 SWU, which corresponds to a London length of 400 nm. Thus we see that the larger London length in LSCO, which is well known from other

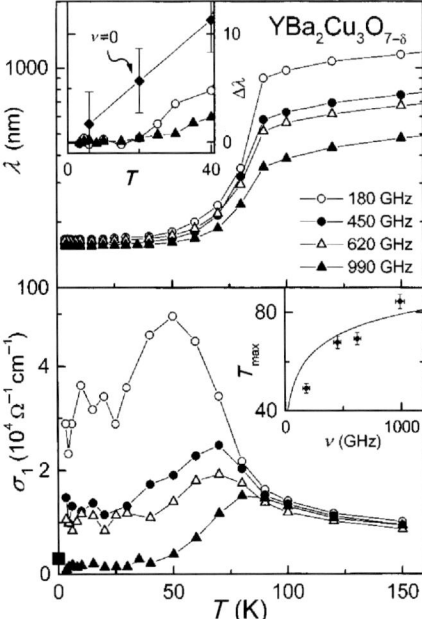

Figure 7.10. *Upper panel:* Dynamical penetration depth (proportional to $(\omega\sigma_2)^{-1/2}$ as a function of temperature for several frequencies in the THz regime. *Upper panel inset:* Penetration depth at low temperature plotted on a linear scale. Diamonds indicate extrapolation to zero frequency. *Lower panel:* σ_1 as a function of temperature for the same frequencies indicated in the upper panel. *Lower panel inset:* Positions of σ_1 peaks in frequency–temperature plane. From [25].

measurements [24], is not a consequence of smaller low-energy spectral weight, but rather results from the fact that only a small fraction of the low-energy spectral weight actually condenses to form the superfluid.

Another example of a system with a large residual spectral weight is YBCO in thin film form. The lower panel of Figure 7.10 shows $\sigma_1(T)$ in the THz regime for an optimally doped YBCO thin film [25]. These data show clearly the presence of uncondensed spectral weight in the $T \to 0$ limit. It is straightforward to estimate that $\rho_n(0) \simeq 1$ SWU, which is approximately one-third of the spectral weight that condenses. These results demonstrate that nearly compete condensation is not a general property of the YBCO system, but rather it is a unique property of ultra-pure, stoichiometric YBCO single crystals.

7.3. Optical Conductivity vs. Hole Concentration in BSCCO

7.3.1. Systematics of the Conductivity Anomaly

In this section we review THz conductivity measurements performed on a set of BSCCO films (which includes the optimally doped film discussed in Section 7.2) with varying hole concentration, x [20, 21, 26, 27]. The trends in $\sigma(\omega, T)$ with x clarify the nature of the uncondensed spectral weight. The data suggest strongly that an additional contribution to $\sigma_1(\omega)$, beyond the Drude response of quasiparticles, is required to understand the optical conductivity of the majority of cuprate samples.

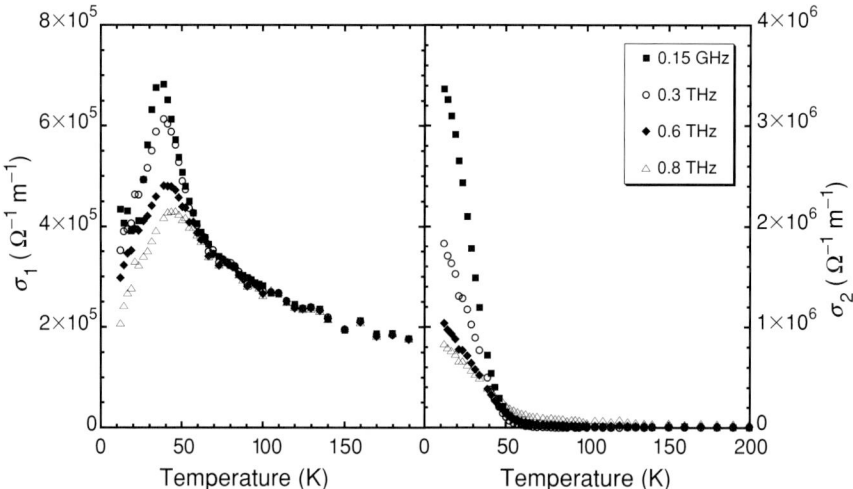

Figure 7.11. *Left panel:* σ_1 and *right panel:* σ_2 at the THz range frequencies indicated in the legend, as measured on an underdoped BSCCO thin film with $T_c = 51$ K. In contrast with optimally doped films, σ_1 peaks near T_c at all frequencies and the residual conductivity in the $T \to 0$ limit is small. The dynamical superfluid density, proportional to σ_2, is essentially linear with from low T to T_c.

The T_cs of the films range from 51 K (underdoped) to 75 K (overdoped). The real and imaginary parts of $\sigma(\omega, T)$ for each sample are presented, in the left and right-hand panels, respectively, in Figures 7.11, 7.12, and 7.13. To illustrate the trend most clearly, we first compare $\sigma(\omega, T)$ in the end members of the series, the most underdoped (Figure 7.11) and the most overdoped (Figure 7.13). In the overdoped sample $\sigma_1(\omega, T)$ is similar to that in the near-optimally doped sample discussed in Section 7.2. The peak in $\sigma_1(T)$ moves systematically to lower T with decreasing ω. However, σ_1 does not approach zero below the peak T; a substantial amount of spectral weight remains uncondensed in the limit that $T \to 0$. The THz conductivity of the most underdoped sample is markedly different, in that $\sigma_1(T)$ peaks just below T_c, independent of frequency. Moreover, spectral weight decreases at low T as expected for a clean d-wave superconductor. The σ_1 spectra for the intermediate sample ($T_c = 71$ K) follows the trend in that the low T dissipation, while clearly visible in the 100 GHz data, is much smaller than in the more heavily doped samples.

We next discuss why the Drude conductivity of superconducting quasiparticles is insufficient to model the $\sigma(\omega, T)$ observed in this set of BSCCO films. There are three basic reasons:

1. *The single component Drude spectrum cannot describe the ω and T dependence of σ.* A robust prediction of any single component model is that $\sigma_1(T)$ for all ω merge as $T \to T_c$ (as seen in the YBCO single crystal data). This prediction does not depend on a specific choice of frequency dependence for the quasiparticle peak, such as Lorentzian. It is seen in all single-component models because the initial increase in $\sigma(T)$ with decreasing T requires $\omega\tau \ll 1$ near T_c. In this limit $\sigma_1 \to \sigma(0, T)$, which is, of course, independent of ω. As is most clear in the spectra in the optimal and overdoped BSCCO films, $\sigma_1(T)$ at different ω fans out immediately below T_c, a feature which is at odds with the single component picture.

Optical Conductivity and Spatial Inhomogeneity in Cuprate Superconductors

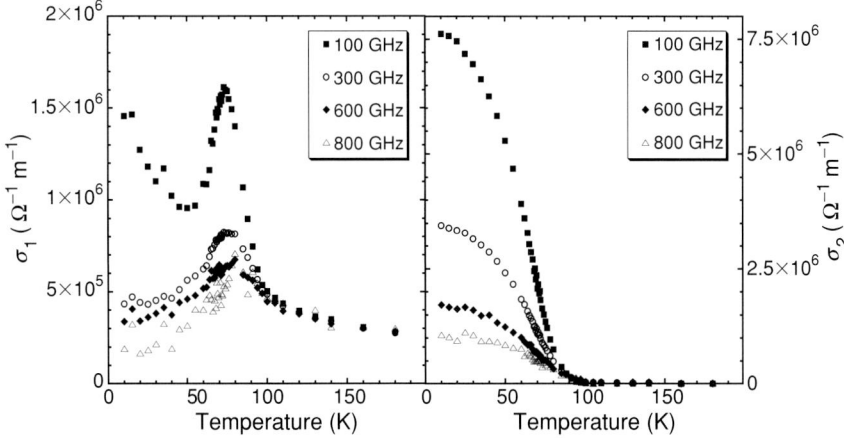

Figure 7.12. *Left panel:* σ_1 and *right panel:* σ_2 at the THz range frequencies indicated in the legend, as measured on an underdoped BSCCO thin film with $T_c = 71$ K. The behavior of σ_1 is intermediate between that of very underdoped and optimal samples in that the residual conductivity is apparent only at the lowest measurement frequency.

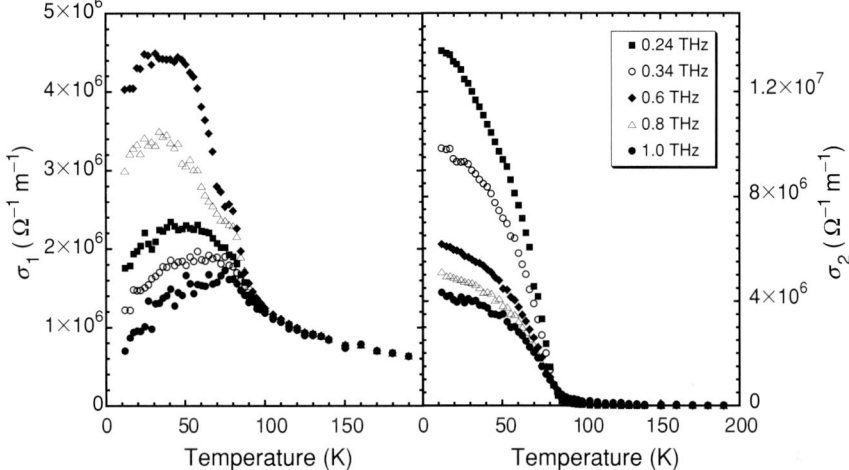

Figure 7.13. *Left panel:* σ_1 and *right panel:* σ_2 at the THz range frequencies indicated in the legend, as measured on an overdoped BSCCO thin film with $T_c = 75$ K. Here the residual conductivity is even larger than in the optimal sample, demonstrating clearly the trend that the spectral weight that remains uncondensed as $T \to 0$ is a monotonically increasing function of the hole concentration.

2. *Forcing a single component fit yields frequency dependent ρ_n and τ.* A useful perspective on "fitting" optical conductivity data is to recognize that the real and imaginary parts of $\sigma(\omega)$ constitute two independent observables at each frequency (although they are nonlocally related through Kramers–Kronig relations). Therefore, if the two free parameters, ρ_n and τ, are allowed to vary with ω, any $\sigma(\omega)$ can be ascribed to a single component. In a sense, this is fitting the data to an infinite number of parameters. The test, of course, is whether $\rho_n(\omega)$ and $\tau(\omega)$ turn out to be physically reasonable. When this fitting procedure is performed on the THz data, $\rho_n(\omega)$ is

found to increase by more than a factor of three from 0.8 to 0.2 THz, implying a large low-frequency renormalization of the quasiparticle mass. This seems highly unlikely in that (1) a single value of m^* is sufficient to describe $\sigma(\omega)$ in the UBC crystals and (2) ARPES yields a single value for the renormalized Fermi velocity, independent of ω and T [28].

3. *In many samples the single component description requires $\rho_n(T)$ to vary as $\rho_n(0) + \alpha T$.* There is no way to understand, in a mean-field picture of disorder, how $\rho_n(T)$ be singular as $T \to 0$ in the presence of a large $N(0)$.

7.3.2. Quantitative Modeling of $\sigma(\omega, T)$

Below, we describe a model for $\sigma_1(\omega, T)$ in BSCCO thin films that allows a quantitative description of the optical conductivity with a minimum of free parameters. The model posits the presence of two components of the conductivity (in addition to the condensate δ-function). The first component, which describes the entire microwave conductivity in YBCO, comes from the quasiparticles that eventually enter the condensate as $T \to 0$. The second component does not vanish in the low T limit, and indeed its spectral weight actually increases with decreasing T. Later, we will associate this component with fluctuations of the condensate phase.

To model the first component we assume the same Drude spectrum for the quasiparticle conductivity that could account for the entire $\sigma(\omega, T)$ in the case of YBCO. The spectral weight of this component is the normal fluid density that ultimately forms the condensate, which is given by $\rho_s(0) - \rho_s(T)$. We note that $\rho_n(T)$ thus defined is not a free parameter, but is determined directly from measurements of $\sigma_2(\omega, T)$. The only free parameter introduced by this component is $\tau(T)$. The measured $\sigma_1(T)$ at any one ω completely determines $\tau(T)$ and hence the predictions of the Drude model for all ω and T in the superconducting state. Figure 7.14 shows the Drude component at several frequencies, together with the data at a measurement frequency of 200 GHz. $\tau(T)$ has been determined from $\sigma_1(T)$ measured at 800 GHz, near the upper limit of the experimental range. Note that the choice of $\tau(T)$ that describes the data at 800 GHz completely fails at 200 GHz.

As a first step in modeling the additional component of the conductivity, one can compute the spectral weight left over after subtracting the Drude conductivity shown in Figure 7.14. When this subtraction is performed, it is found that the spectral weight of the remainder *increases as T is lowered*. Furthermore, the spectral weight increases in fixed proportion to the growth of the superfluid density $\rho_s(T)$. In view of the connection to the superfluid density, we refer hereafter to this contribution to $\sigma(\omega)$ as the "collective" component.

Including the collective contribution yields the following parameterization of the conductivity,

$$\sigma_1(\omega, T) = \frac{\rho_n \tau}{1 + \omega^2 \tau^2} + \frac{\kappa \rho_s / \Gamma_{cm}}{1 + \omega^2 / \Gamma_{cm}^2}. \quad (7.2)$$

In the above formula, the first term is the Drude contribution from condensing quasiparticles and the second term is another Lorentzian peak which models the collective mode. The latter introduces only two additional, T-independent parameters: κ is the fraction of $\rho_s(0)$ that remains uncondensed at $T = 0$, and Γ_{cm} is width of the uncondensed contribution.

The formulation described above provides a remarkably good description of the optical conductivity for all the samples. The comparison of the model with the measured conductivity for the optimal sample is illustrated in Figure 7.15 [20]. The collective mode parameters are $\kappa = 0.30$ and $\Gamma_{cm} = 1.5$ THz. The T-dependence of $1/\tau$ that provides the best fit to the

Optical Conductivity and Spatial Inhomogeneity in Cuprate Superconductors

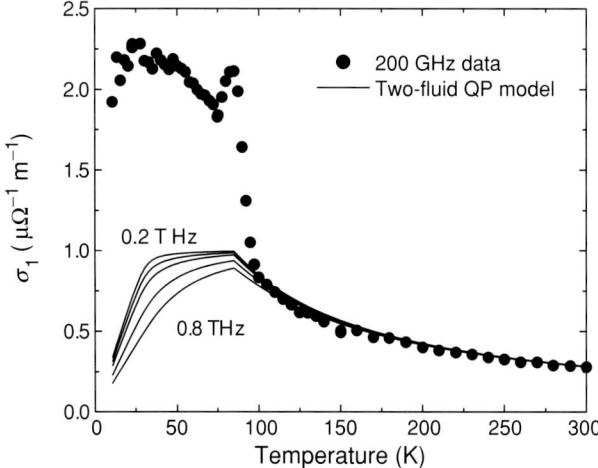

Figure 7.14. Illustration of the difficulty of modeling the measured $\sigma_1(\omega, T)$ with the Drude conductivity of thermal quasiparticles (Eq. (7.1)). *Solid lines:* Plots of Eq. (7.1) at several frequencies from 200 to 800 GHz, with $\rho_n(T)$ extracted from the measured σ_2 and $\tau(T)$ chosen to fit $\sigma_1(T)$ as measured at 800 GHz. *Solid symbols:* $\sigma_1(T)$ at 200 GHz as measured on a optimally doped BSCCO thin film.

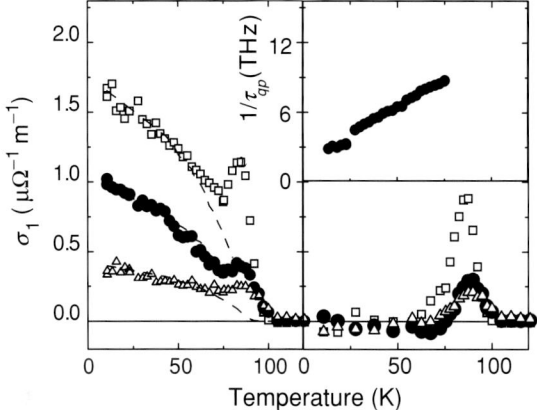

Figure 7.15. Comparison of the 2 + 1 fluid model (Eq. (7.2)) with σ_1 as measured on an optimally doped BSCCO thin film. *Left panel:* Difference between the measured $\sigma_1(T)$ and quasiparticle Drude contribution (first term of Eq. (7.2)), plotted at 0.2, 0.36, and 0.64 THz as squares, circles, and triangles, respectively. The dashed lines are the collective mode conductivity (second term of Eq. (7.2)). *Lower right panel:* The difference between the measured conductivity and the best fit using Eq. (7.2) is plotted on the same scale as the left panel, and can be attributed to thermal phase fluctuations that occur in the neighborhood of T_c. *Upper right panel:* The quasiparticle scattering rate $1/\tau$, as determined from the best fit to Eq. (7.2), plotted vs. temperature with the T scale given by the panel below.

condensing quasiparticle contribution is shown in the upper right hand panel of Figure 7.15. The left-hand panel shows the result of subtracting the quasiparticle contribution, as modeled by the first term of Eq. (7.2) from the measured conductivity. Clearly, what remains increases monotonically with decreasing T over the entire range of measurement frequency. The dashed lines indicate the values of the second term in Eq. (7.2). Finally the lower right-hand panel shows the remainder when both terms of Eq. (7.2) are subtracted from the measured

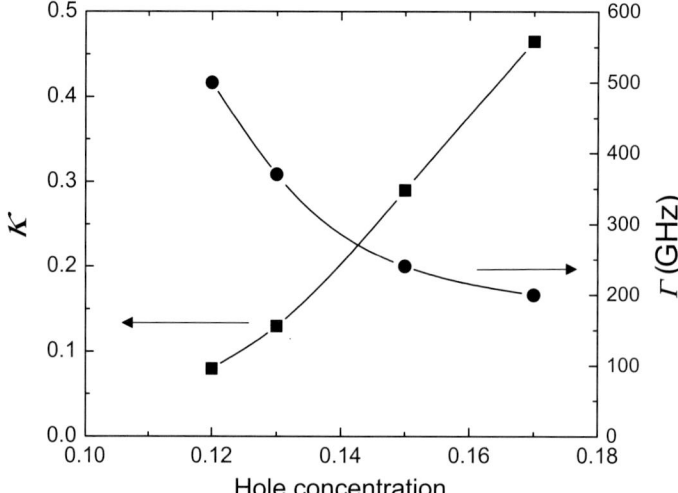

Figure 7.16. The two parameters, κ and Γ, that describe the collective mode conductivity as a function of hole concentration, δ. *Squares:* κ, the spectral weight of the collective mode expressed as a fraction of the condensate spectral weight, increases monotonically with hole concentration. *Circles:* Γ, the width of the collective mode conductivity, decreases with increasing hole concentration.

conductivity, illustrating that the model describes the data very well except for the peak near T_c that results from thermally driven superconducting fluctuations [26].

One of the main benefits of the modeling is that it facilitates comparison of the data from one sample to another. Figure 7.16 shows the variation in κ and Γ_{cm} across the measured range of hole concentration. Note that κ increases rapidly with increasing concentration, reaching almost 0.5 for the overdoped sample. The spectral weight of the collective mode contribution to σ_1 increases even more rapidly than does κ because ρ_s itself increases monotonically with x. Figure 7.17 illustrates that ρ_s increases monotonically over the measured range of hole concentration, despite the fact that T_c decreases beyond optimal doping.

7.4. Collective Mode Contribution to Optical Conductivity

7.4.1. Origin of the Collective Contribution

In Section 7.3 we showed that the optical conductivity of a set of BSCCO films could be modeled by adding to the Drude conductivity of quasiparticles a component whose spectral weight tracks $\rho_s(T)$ as T varies. The proportionality to $\rho_s(T)$ suggests a connection to current fluctuations of the condensate. However, thermal fluctuations are clearly ruled out because they decrease with decreasing T. Likewise, the quantum phase fluctuation conductivity of a spatially homogeneous superconductor, which is $\sim \sigma_Q \sim 3 \times 10^5 \, \Omega^{-1} \, cm^{-1}$ [29], is far too small to account for the observed conductivity.

A calculation performed by Doniach and Inui [30] suggests the conditions under which the conductivity that arises from quantum phase fluctuations can become large. The supercurrent density arising from fluctuations of the order parameter phase is given by $\vec{J}_s = \rho_s \nabla \phi$. This relationship implies that phase-fluctuation currents at wavevector q vanish in the $q \to 0$

Optical Conductivity and Spatial Inhomogeneity in Cuprate Superconductors

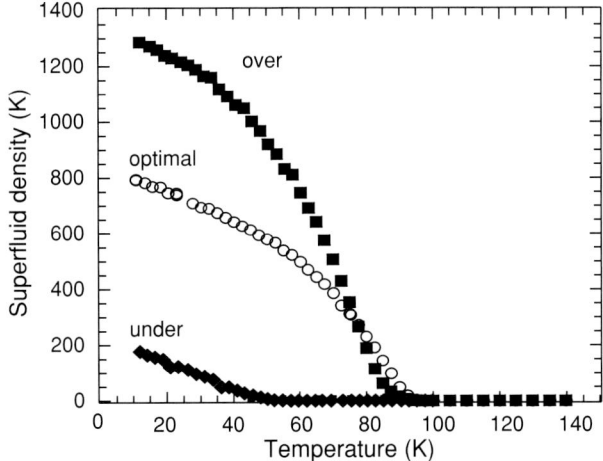

Figure 7.17. Superfluid density as a function of temperature for representative under, optimal, and overdoped BSCCO thin films. The superfluid density per Cu–O bilayer is expressed as an effective phase stiffness temperature through the relation $k_B T_\theta \equiv \hbar \omega G_2(\omega)/G_Q$, where G_2 is the imaginary part of the conductance per layer and G_Q is the conductance quantum.

limit, and therefore make no contribution to the optical conductivity. The lowest order contribution to $\sigma(\omega, q \to 0)$ arises from combined amplitude and phase fluctuations with equal but opposite wavevector, resulting in a $q = 0$ supercurrent. The fact that the current is second order in fluctuations of the order parameter leads to a small value for the conductivity. However, an implication of the Doniach and Inui calculation is that the phase fluctuation conductivity could be enhanced by quenched disorder, as static variations of the order parameter amplitude allow phase fluctuations of all wavevectors to contribute in first order to the optical conductivity.

Another argument supporting the above conclusion is the following: consider a system of clean-limit BCS superconductor islands connected by weak links with coupling J. If the islands were infinite then superfluid density would equal the entire normal state spectral weight. However, in the granular system the phase stiffness is limited by the weak links, leading to a condensate spectral weight that is only $\sim J$. In the limit that J is small, virtually the entire low-frequency spectral weight does not contribute to the superfluid density, i.e., remains uncondensed as $T \to 0$.

Barabash and Stroud [31] demonstrated that the general arguments above apply to disordered superconducting systems by considering a Josephson junction network with quenched variation δJ about the mean coupling $\bar{J}$. They showed that the global phase stiffness J is less than $\bar{J}$ (by $\langle \delta J^2 \rangle / \bar{J}$) because the phase varies more rapidly in regions where the stiffness is below the average value. The connection to optical conductivity is as follows. The spectral weight of the condensate δ-function is proportional to J, which is then less than total spectral weight of condensate current fluctuations, proportional to $\bar{J}$. Thus inhomogeneity of the superfluid density must displace spectral weight $\sim \langle \delta J^2 \rangle / \bar{J}$ from the δ-function to nonzero frequency. The fraction of spectral weight removed from the condensate (the parameter we have previously defined as κ) is therefore to be identified with $\langle \delta J^2 \rangle / \bar{J}^2$. We note that if J varies, for example with T, in such as fashion as the relative variation $\langle \sqrt{\delta J^2} \rangle / \bar{J}$ remains the same, then the fraction of the condensate weight that is displaced will also be constant.

7.4.2. Optical Conductivity in the Presence of Inhomogeneity

We have seen that, in the presence of static inhomogeneity, quantum phase fluctuations contribute substantially to the optical conductivity. To determine whether quenched inhomogeneity in ρ_s is responsible for the anomalies in $\sigma(\omega, T)$ discussed previously, we need to consider where in ω-space the spectral weight removed from the condensate will reappear. Conventional wisdom has it that, in a granular superconductor, the conductivity will appear at the natural oscillation frequency of the order parameter phase, which is the Josephson plasma frequency, ω_s. In optimal cuprates this is a very large frequency, that is $\omega_s/2\pi \sim \sqrt{\text{SWU}} \cdot 150\,\text{THz}$, whereas the anomalous dissipation is found below ~ 1 THz. Inhomogeneity can only explain the anomalous $\sigma(\omega)$ in the cuprates if the displaced spectral weight actually appears at frequencies much smaller than ω_s.

With further reflection it is clear that the shifting of all spectral weight to ω_s is a peculiar feature of a "single-fluid" model, i.e., a superconductor with zero quasiparticle density. Consider first a disordered superconductor in which the superfluid density varies with position as $\langle \rho_s \rangle + \delta\rho_s(\vec{r})$ and $\rho_n = 0$. According to previous arguments, spectral weight $\langle \delta\rho_s^2 \rangle / \langle \rho_s \rangle$ is removed from the condensate as a result of the disorder. In the neighborhood of ω_s this medium is simply a disordered conductor, and will have strong absorption bands related to the Mie absorption of small metallic particles. Therefore it is reasonable that the spectral weight removed from the condensate reappears at ω_s. Next, consider a system with the same distribution of superfluid density, but with nonzero normal fluid density distributed such that $\delta\rho_n(\vec{r}) = -\delta\rho_n(\vec{r})$. Although this system is disordered at the frequency of superconducting correlations, it is optically homogeneous near the Josephson plasma frequency (the fractional variation in optical conductivity near ω_s is $\sim 1/(\omega_p\tau)^2$). Clearly for this system the spectral weight that is shifted from the condensate cannot appear near ω_s.

The difference in the two "Gendanken samples" described above is the number of fluid components (or collective degrees of freedom). The superfluid-only sample has one collective (phase) mode which lies near the plasma frequency. On the other hand, the two-fluid medium has collective modes in which the super and normal components are either co- or counterpropagating. The counterpropagating mode will have $\nabla \cdot \vec{J} \simeq 0$ and therefore have an acoustic rather than plasmonic spectrum. We may expect that this mode will be overdamped due to the dissipation associated with the normal currents and therefore manifest as a Drude-like contribution to the optical conductivity. In Section 7.4.3. we present a model calculation [32] of the optical conductivity of a two-fluid system with quenched spatial inhomogeneity which supports the qualitative predictions discussed above.

7.4.3. Extended Two-Fluid Model

To treat the conductivity in the presence of inhomogeneity we apply the extended two-fluid phenomenology developed by Pethick and Smith [33] and Kadin and Goldman [34]. This approach successfully describes quenched inhomogeneity at the normal–superconductor interface and fluctuating inhomogeneity, as in the Carlson–Goldman oscillations [35]. In the extended two-fluid model the superfluid is accelerated by gradients of the chemical potential as well as electric fields, that is,

$$\dot{\vec{J}}_s = \rho_s \left[\vec{E} - \nabla(\mu_s/e) \right]. \tag{7.3}$$

The chemical potential has the subscript "s" because in a superconductor μ is the energy per electron required to add a pair to the condensate. The corresponding equation for the normal fluid current requires solving the Boltzmann equation for the quasiparticle distribution function. However, for frequencies less than $1/\tau$ the distribution function is the equilibrium distribution shifted by the "quasiparticle chemical potential" or μ_n. If the normal fluid is in local equilibrium with the condensate, $\mu_n = \mu_s$, which differs from "global" equilibrium where $\mu_n = 0$. In the low-frequency regime the constitutive relation for the normal fluid has the simple form:

$$\dot{\vec{J}}_n = \rho_n \left[\vec{E} - \nabla(\mu_n/e)\right] - \vec{J}_n/\tau. \quad (7.4)$$

A closed system of equations requires continuity relations. The total charge of the superconductor separates naturally into a normal component, Q_n, that depends on both the coherence factors and the distribution function,

$$Q_n \equiv \sum_k q_k f_k, \quad (7.5)$$

where $q_k^2 \equiv u_k^2 - v_k^2$, and a superfluid component that depends only on coherence factors,

$$Q_s \equiv \sum_k 2ev_k^2 = 2eN_F\mu_s. \quad (7.6)$$

Under conditions for which μ_n can be defined, the normal fluid charge is given by,

$$Q_n = 2N_F\lambda(\mu_s - \mu_n), \quad (7.7)$$

where N_F is the normal state density of states at the Fermi level. The parameter λ relates the normal fluid charge to the shift of μ_n away from local equilibrium.

In a superconducting medium the normal and superfluid charge are not separately conserved. Interconversion of normal and superconducting charge occurs through two types of processes. In the first process, Q_n changes as quasiparticles recombine or scatter. In the second, the quasiparticle charge changes even if f_k remains constant. In this process, Q_n varies because the quasiparticle excitation spectrum, and consequently the effective charge, adjusts to the local value of μ_s. Continuity equations that include both types of exchange between the two fluids are:

$$\dot{Q}_{n,s} + \nabla \cdot \vec{J}_{n,s} = (-,+)(\frac{Q_n}{\tau_Q} - \lambda \dot{Q}_s), \quad (7.8)$$

where τ_Q is the rate of conversion of normal charge into superfluid charge due to scattering and recombination processes. The above system of equations is closed by $\nabla \cdot \vec{E} = Q/\epsilon_0$.

To see how quenched inhomogeneity affects the optical conductivity, we consider the simplest possible model: a static one-dimensional sinusoidal variation in normal and superfluid density. We assume that the densities of the two components vary as $\rho_{s,n}(x) = \langle \rho_{s,n} \rangle + Re\{\rho_{sq,nq}e^{iqx}\}$, where $\langle \rho \rangle$ is the average density and ρ_q is the amplitude of the inhomogeneity at wavevector q. Because the medium is inhomogeneous, a uniform applied field generates a field at q. Solving the extended two-fluid equations to lowest order in $\rho_{nq,sq}$, we obtain,

$$\frac{E_q}{E_0} = -\frac{\rho_{sq} + \rho_{nq}F}{\omega_s^2 - \omega^2(1-\lambda) + (\omega_n^2 - i\omega\lambda/\tau)F}, \quad (7.9)$$

where,

$$F \equiv \frac{\omega(\omega + i/\tau_Q^*) - v_s^2q^2/(1-\lambda)}{(\omega + i/\tau)(\omega + i/\tau_Q^*) - v_n^2q^2/\lambda}, \quad (7.10)$$

and $\omega_{s,n}^2 \equiv \rho_{s,n}/\epsilon_0 + v_{s,n}^2q^2$, $v_{s,n}^2 \equiv \rho_{s,n}/2N_Fe^2$, and $\tau_Q^* \equiv \tau_Q(1-\lambda)$.

The uniform current density in response to these fields is given by $J_0 = \sigma_0 E_0 + \sigma_q E_{-q}$, where σ_0 is the uniform two-fluid conductivity and σ_q is the conductivity that varies with wavevector q. In the two-fluid model these are given by,

$$\sigma_{0,q} = \frac{i\rho_{s,sq}}{\omega} + \frac{i\rho_{n,nq}}{\omega + i/\tau}. \tag{7.11}$$

The effective conductivity of the medium, $\sigma = \sigma_1 + i\sigma_2$, is the ratio of the uniform current density to the uniform field, so that, $\sigma = \sigma_0 + \sigma_q E_{-q}/E_0$. The second term in this equation is the change in the optical conductivity due to inhomogeneity, or $\Delta\sigma$.

The extra term in the conductivity is particularly simple if $\rho_n = 0$, in which case,

$$\Delta\sigma_2 = -\frac{\rho_{sq}}{\omega}\frac{\rho_{sq}}{\omega_s^2 - \omega^2}. \tag{7.12}$$

Equation (7.12) shows that the inhomogeneity in the superfluid density indeed removes spectral weight ρ_{sq}^2/ρ_s from the condensate δ-function, in agreement with the results of [31]. In the absence of a normal fluid component the spectral weight reappears in a δ-function at the Josephson plasma frequency.

We next assume that $\rho_n \neq 0$, and describe how this assumption affects $\Delta\sigma(\omega)$. We focus on the behavior of $\Delta\sigma$ when the normal fluid density fluctuations are either perfectly correlated or anticorrelated with those of the superfluid density. We take for two-fluid parameters values that are suggested by the terahertz and microwave experiments: $(\rho_s/\epsilon_0)^{1/2} = 1{,}000$ THz, $(\rho_n/\epsilon_0)^{1/2} = 800$ THz, and $\tau^{-1} = \tau_Q^{-1} = 3$ THz. If we make the reasonable approximation of neglecting BCS coherence factors for the quasiparticle states introduced by disorder, then $\lambda = N(0)/N_F$.

We begin with the case where the density fluctuations are perfectly anticorrelated, so that $\rho_{sq} = -\rho_{nq}$ and the total fluid density is uniform throughout the medium. Figure 7.18 shows $\Delta\sigma_1(\omega)$ with $(\rho_{sq}/\epsilon_0)^{1/2} = 100$ THz, for several values of $v_s q \tau$. $\Delta\sigma_1(\omega)$ is positive and centered at $\omega = 0$ rather than ω_s. The spectra depend strongly on $v_s q \tau$. For $v_s q \tau \ll 1$, $\Delta\sigma_1$ has a Drude-like spectrum. As $v_s q \tau$ increases beyond unity, the spectral weight drops and a peak near the Carlson–Goldman frequency $v_s q$ appears in the spectrum.

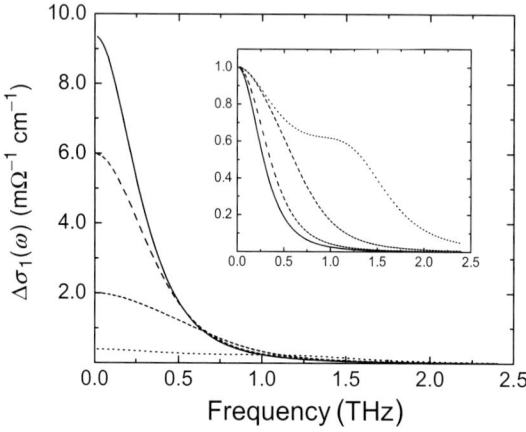

Figure 7.18. $\Delta\sigma_1$ as a function of frequency $(\omega/2\pi)$ for anticorrelated variations in ρ_s and ρ_n. Spectral weight decreases for increasing $v_s q \tau$: 0.25, 0.5, 1.0, and 2.0. The same curves are shown in a normalized plot in the inset.

Optical Conductivity and Spatial Inhomogeneity in Cuprate Superconductors

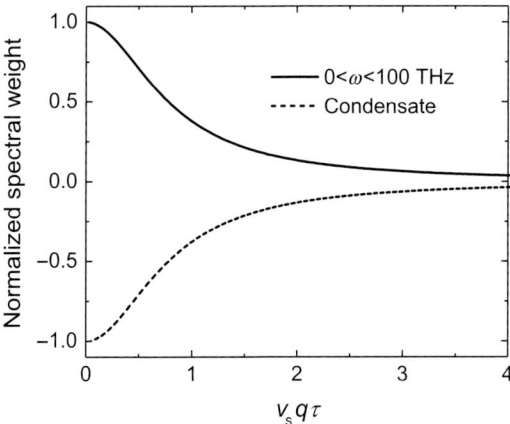

Figure 7.19. Comparison of the spectral weight displaced from the condensate by the quenched variation in ρ_s and the spectral weight that appears in $\Delta\sigma_1$ at low frequency.

The key issue is the fraction of the displaced condensate spectral weight that appears in the low-frequency peak, as opposed to frequencies near ω_s. In Figure 7.19 we compare the reduction in condensate weight with the increase in dissipation at low frequency. The change in condensate spectral weight was evaluated from the $\lim_{\omega \to 0}(\pi/2)\omega\Delta\sigma_2(\omega)$. The low-frequency spectral weight was obtained by numerically computing the integral of $\Delta\sigma_1$ with respect to ω from 0 to 100 THz. Figure 7.19 shows these two quantities, normalized to ρ_{sq}^2/ρ_s, as a function of $v_s q\tau$. They are equal in magnitude but opposite in sign, which shows that *all of the spectral weight removed from the condensate appears at low frequency and none appears at ω_s*. Moreover, the decrease of condensate spectral weight coincides exactly with the prediction of Barabash et al. [31] as $v_s q\tau \to 0$, but vanishes for $v_s q\tau \gg 1$.

The results presented above are a straightforward consequence of the anticorrelation of the density fluctuations. There is no dissipation near ω_s because the response of the medium is homogeneous at high frequencies. The anomalous dissipation appears instead at low frequency where the conductivity has strong spatial variations. σ is smaller in the regions that are superfluid poor and normal fluid rich. E will be larger in such regions, which is precisely equivalent to more rapid order parameter phase variation in regions of low stiffness. Thus the additional low-frequency dissipation arises ultimately from an amplification of E in regions with greater than average ρ_n. Finally, the dynamical inhomogeneity disappears when $q \ll (v_s\tau)^{-1}$ because the normal and superfluid response again become indistinguishable in this regime.

We are now prepared to understand the sharp difference in $\Delta\sigma$ if the super and normal fluid density waves are positively correlated. Figure 7.20 shows $\Delta\sigma_1(\omega)$ calculated with identical parameters as in Figure 7.1, except that $\rho_{sq} = \rho_{nq}$. The change in conductivity is negative, and the spectra are nearly independent of q in contrast to the strongly q dependent and positive change generated when $\rho_{sq} = -\rho_{nq}$. The reduction in conductivity is exactly as expected from the previous arguments: now regions of large E coincide with normal fluid poor regions and the dissipation is attenuated. The spectra are nearly independent of q because the dynamical inhomogeneity does not tend to zero when the normal and superfluid response become indistinguishable. The spectral weight is displaced equally from the condensate and the normal fluid, and weight $2\rho_{sq}^2/\rho_s$ shifts to very high frequency. Finally, although

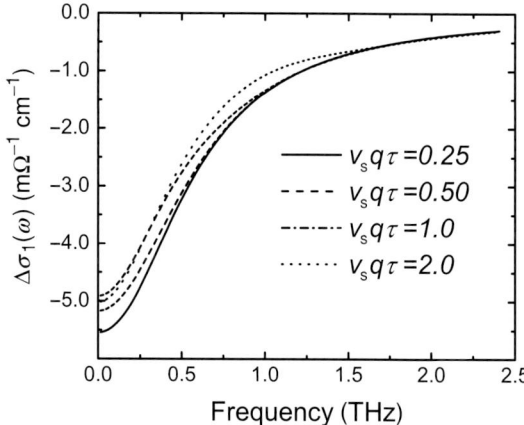

Figure 7.20. $\Delta\sigma_1$ vs. frequency for the same parameters as Figure 7.1, except variations in ρ_s and ρ_n are positively correlated.

the sensitivity of the conductivity spectrum to correlations in ρ_s and ρ_n are demonstrated for a sinusoidal distribution, the essential predictions of the above model are supported by theories that treat random spatial variations in the fluid densities [36, 37].

7.4.4. Comparison of Model and Experiment

In this section, we describe how the behavior of the optical conductivity in disordered cuprate superconductors can be readily understood in the context of the model described above. The three essential features to be explained are:

1. *Many cuprates exhibit uncondensed spectral weight at low T, suggestive of a large density of states at the chemical potential, while $\rho_s(T)$ remains linear to very low T, apparently inconsistent with large $N(0)$.*
   Point (1) is difficult to explain in a mean-field picture of a disordered d-wave superconductor because large $N(0)$ implies that as $T \to 0$ the superfluid density has the asymptotic form, $\rho_s(T) = \rho_s(0) - BT^2$. However, it *is* consistent with an inhomogeneous form of disorder in which regions of clean superconductor coexist with metallic regions that provide the large $N(0)$. Uemura has proposed superconductor/normal metal coexistence in cuprate superconductors on the basis of muon spin resonance data as well as analogies to liquid ^{3}He [38]. Such metallic regions need not have $\rho_s = 0$ as superconductivity could be induced by the proximity effect.
2. *The anomalous spectral weight appears at low frequency and, in a given sample, is proportional to $\rho_s(T)$.*
   Point (2) follows if the local values of $\rho_s(\vec{r})$ and $\rho_n(\vec{r})$ are negatively correlated, such that regions with large ρ_s have small ρ_n and vice versa.
3. *The magnitude of the low-frequency spectral weight grows monotonically as samples evolve from under to overdoped.*
   Point (3) is consistent with the picture of inhomogeneity implied by the LDOS measurements in BSCCO crystals [14, 15]. These experiments reveal a distribution of electronic properties that suggests variation in local hole concentration, x. As samples evolve from under to overdoped the mean of the distribution, $\bar{x}$, shifts toward larger values. Eventually the distribution in x should span the d-wave superconductor

(dSC)/normal metal phase boundary, with coexistence of super and normal patches an inevitable consequence. Note that the model also explains why the uncondensed spectral weight in underdoped samples appears to be small, despite the fact that the fractional variation in the local ρ_s should be just as large as in optimal or overdoped samples. In these samples the "weak links," or regions with smaller ρ_s, are not metallic. If anything their ρ_n is also smaller, so that the fluid densities are positively, rather than negatively correlated. As we saw in the previous section, in this case the order parameter phase fluctuations are not screened and the characteristic absorption appears near the plasma frequency. Indeed, this offers a possible explanation for the anomalous mid-IR absorption observed in underdoped cuprates.

7.5. Summary and Outlook

7.5.1. Summary

We have presented an overview of the microwave and millimeter wave properties of cuprate superconductors, illustrating the two types of low-frequency optical conductivity that these materials exhibit. The first type, exemplified by ultra-pure and stoichiometric YBCO single crystals, is well described by a single component originating from the Drude response of thermal quasiparticles. In other cuprate systems that have been studied $\sigma(\omega)$ has an additional component beyond the quasiparticle contribution, also centered at $\omega = 0$, with a typical width of ~300 GHz. The existence of this peak has not been widely appreciated because its spectral weight lies in the gap between microwave and infrared regimes where most measurements of optical conductivity are performed. However, when microwave, terahertz, and infrared data are assembled in a composite plot, the presence of a Drude-like peak that spans these three ranges is clearly revealed. We also presented an analogous plot of the optical conductivity of ultra-pure single crystalline YBCO, in which no terahertz measurements have been performed. This composite plot reveals that the microwave conductivity does not extrapolate smoothly to the infrared conductivity, hinting at the possible presence of structure in the unmeasured region of the spectrum.

We showed that the extra component of $\sigma(\omega)$ is a collective mode that appears in the $q = 0$ response function when the order parameter amplitude, $|\Psi|$, and the density of states at the chemical potential, $N(0)$, vary with position in the medium. Such variation in microscopic parameters leads to a macroscopic description in terms of spatially varying super and normal fluid densities. A simple model of one-dimensional sinusoidal variation in ρ_s and ρ_n showed that the characteristic frequency of the collective mode depended strongly on the spatial correlation of the fluid density variations. The collective mode spectral weight appears at the low frequencies seen in experiments when the spatial variations are anticorrelated, such that the total fluid density is roughly constant in the medium. Finally, we showed that the frequency, temperature, and doping dependence of the optical conductivity in a series of BSCCO thin films could be understood to arise from the combination of a quasiparticle Drude peak and the collective mode described above.

7.5.2. Outlook and Directions of Future Research

The existence, origin, and ultimate importance of inhomogeneity are central questions facing researchers seeking to understand high-T_c superconductivity in the cuprates (see the article by Kivelson and Fradkin in this volume). The existence of strong quenched variation in the superconducting order parameter remains an area of controversy. It has been argued that the spatial variations seen directly by STM spectroscopy are confined to the surface of the samples that have been studied [39]. Measurements of specific heat, NMR Knight shift, and fluctuations near T_c are claimed to be inconsistent with the large amplitude fluctuations seen in STM. Most relevant to optical conductivity measurements is the claim that the specific heat varies as T^2 at low temperature, with no indication of a component linear in T that could be identified with a nonzero $N(0)$ in the superconducting state. However, this conclusion is contradicted by other measurements which indicate that the linear coefficient of the specific heat, γ, does not vanish even in "high-quality" single crystalline samples [40]. Indeed the quantitative modeling of the low temperature specific heat in the cuprates is complicated, involving possible contributions from normal electrons, d-wave quasiparticles, paramagnetic centers, and phonons, all of which vanish as $T \to 0$ (see the article by Phillips in this volume). When compared to the specific heat, the optical conductivity is a much more direct probe of $N(0)$. Not only are there fewer possible contributors, but $N(0)$ appears directly as a component of the spectral weight that tends to a nonzero value as $T \to 0$.

If we interpret the combination of $\sigma(\omega)$ and STM data as solid evidence for the existence of bulk inhomogeneity in cuprate superconductors, its origin becomes the central question. This question can be framed in terms of intrinsic vs. extrinsic mechanisms, although from certain perspectives the distinction is not always clear. Extrinsic inhomogeneity is that component arising from spatial variation in the defects or chemical dopants that shift the carrier concentration away from the half-filled Mott insulating state. Intrinsic inhomogeneity occurs when CDW and/or SDW states compete with d-wave superconducting order. In the absence of disorder, intrinsic inhomogeneity is expected to vary with time, although potentially at low-frequency scales. However, the presence of chemical or structural disorder could cause quenching of low-frequency variations, generating static spatial inhomogeneity.

The crucial test for intrinsic inhomogeneity is "universality." If the spatial variations arise intrinsically from the physics of competing interactions, then they should appear in all the cuprate families of compounds. As a test for universality, measurements of low-frequency optical conductivity as a function of carrier concentration in more cuprate systems would be quite useful. Of special interest would be the $La_{2-x}Sr_xCuO_4$ system, where quenched stripe-like inhomogeneities are observed by neutron scattering [41] and large values of uncondensed spectral weight have already been reported for optimally doped samples [8].

In connection with universality, characterization of the ultra-pure and highly ordered YBCO system must been given special consideration. In mathematics, a single counter example is sufficient to disprove a conjecture. In the field of high-T_c superconductivity it is difficult to apply such rigorous logic, since experimental "truth" often changes over time. However, it seems clear that the most stringent test for intrinsic inhomogeneity would be to detect it, albeit in fluctuating form, in the cleanest of high-T_c materials. To date, interpretation of optical conductivity in the YBCO crystal system has not required invoking strong spatial inhomogeneity. The microwave conductivity is consistent with weak elastic scattering, while the infrared conductivity appears to involve Holstein-like coupling to a bosonic spectrum with energies in the 40 meV region, associated either with phonons or spin fluctuations [42]. However, the composite broadband spectra presented in Figure 7.8 show that characterization of the optical conductivity in YBCO is not complete. The missing range of the spectrum is the region above carrier scattering rates but below the gap frequency, where intrinsic fluctuations may occur. Future measurements that fill this information gap in YBCO, and extend our knowledge in other cuprate systems, will be of great value in assessing the origin and role of inhomogeneity in high-T_c superconductors.

I would like to thank Jim Eckstein and John Corson for making the study of BSCCO thin films so enjoyable, and to acknowledge valuable discussions with D.-H. Lee, J.C. Davis, S. A. Kivelson, A.J. Millis, and P.A. Lee. This work was supported by NSF-9870258 and DOE-DE-AC03-76SF00098.

Bibliography

1. P. J. Hirschfeld, W. O. Putikka, and D. J. Scalapino, Phys. Rev. Lett. **71**, 3705 (1993).
2. D. Mattis and J. Bardeen, Phys. Rev. **111**, 412 (1958).
3. A. Hosseini, R. Harris, S. Kamal, P. Dosanjh, J. Preston, R. Liang, W. N. Hardy, and D. A. Bonn, Phys. Rev. B **60**, 1349 (1999).
4. D. A. Bonn, R. Liang, T. M. Riseman, D. J. Baar, D. C. Morgan, K. Zhang, P. Dosanjh, T. L. Duty, A. MacFarlane, G. D. Morris, J. H. Brewer, et al., Phys. Rev. B **47**, 11314 (1993).
5. A. Maeda, H. Kitano, and R. Inoue, J. Phys.: Condens. Matter **17**, 143 (2005).
6. S. -F. Lee, D. C. Morgan, R. J. Ormeno, D. M. Broun, R. A. Doyle, J. R. Waldram, and K. Kadowaki, Phys. Rev. Lett. **77**, 735 (1996).
7. T. Jacobs, S. Sridhar, Q. Li, G. D. Gu, and N. Koshizuka, Phys. Rev. Lett. **75**, 4516 (1995).
8. A. V. Pronin, B. P. Gorshunov, A. A. Volkov, H. S. Somal, D. van der Marel, B. J. Feenstra, Y. Jaccard, and J. P. Locquet, JETP Lett. **68**, 6432 (1998).
9. J. Waldram, D. M. Broun, D. C. Morgan, R. Ormeno, and A. Porch, Phys. Rev. B **59**, 1528 (1999).
10. P. Hirschfeld and N. Goldenfeld, Phys. Rev. B **48**, 4219 (1993).
11. P. Lee, Phys. Rev. Lett. **71**, 1887 (1993).
12. A. Durst and P. Lee, Phys. Rev. B **62**, 1270 (2000).
13. A. Ghosal, M. Randeria, and N. Trivedi, Phys. Rev. B **65**, 014501 (2002).
14. S. H. Pan, J. P. O'Neal, R. L. Badzey, C. Chamon, H. Ding, J. R. Engelbrecht, Z. Wang, H. Eisaki, S. Uchida, A. K. Gupta, et al., Nature **413**, 282 (2001).
15. K. M. Lang, V. Madhavan, J. E. Hoffman, E. W. Hudson, H. Eisaki, S. Uchida, and J. C. Davis, Nature **415**, 412 (2002).
16. D. Bonn and W. Hardy, in *Physical Properties of High Temperature Superconductors V*, D. Ginsberg (ed.) (Singapore, World Scientific, 1996), p. 237005.
17. P. J. Turner, R. Harris, S. Kamal, M. E. Hayden, D. M. Broun, D. C. Morgan, A. Hosseini, P. Dosanjh, G. K. Mullins, J. S. Preston, et al., Phys. Rev. Lett. **90**, 237005 (2003).
18. M. C. Nuss, P. M. Mankiewich, M. L. OMalley, E. H. Westerwick, and P. B. Littlewood, Phys. Rev. Lett. **66**, 3305 (1991).
19. D. Bonn, P. Dosanjh, R. Liang, and W. N. Hardy, Phys. Rev. Lett. **68**, 2390 (1992).
20. J. Corson, J. Orenstein, S. Oh, J. O'Donnell, and J. N. Eckstein, Phys. Rev. Lett. **85**, 2569 (2000a).
21. J. Corson, J. Orenstein, J. O. S. Oh, and J. N. Eckstein, Physica B **280**, 212 (2000b).
22. J. Tu, C. Homes, G. Gu, D. Basov, G. Loureiro, R. Cava, and M. Strongin, Phys. Rev. B **66**, 144514 (2002).
23. C. C. Homes, S. V. Dordevic, D. A. Bonn, R. Liang, and W. N. Hardy, Phys. Rev. B **69**, 024514 (2004).
24. T. Shibauchi, H. Kitano, K. Uchinokura, A. Maeda, T. Kimura, and K. Kishio, Phys. Rev. Lett. **72**, 2263 (1994).
25. A. Pimenov, A. Loidl, G. Jakob, and H. Adrian, Phys. Rev. B **59**, 4390 (1999).
26. J. Corson, J. Orenstein, S. Oh, and J. Eckstein, Nature **398**, 221 (1999).
27. J. Corson, Ph.D Thesis University of California, Berkeley (2001).
28. X. J. Zhou, T. Yoshida, A. Lanzara, P. V. Bogdanov, S. A. K. K. M. Shen, W. L. Yang, F. Ronning, T. Sasagawa, T. Kakeshita, T. Noda, et al., Nature **423**, 398 (2003).
29. S. Sachdev, *Quantum Phase Transitions* (Cambridge University Press, New York, 1999).
30. S. Doniach and M. Inui, Phys. Rev. B **41**, 6668 (1990).
31. S. Barabash and D. Stroud, Phys. Rev. B **61**, 14924 (2000).
32. J. Orenstein, Physica C **67**, 144506 (2003).
33. C. Pethick and H. Smith, Ann. Phys. **119**, 133 (1979).
34. A. Kadin and A. Goldman, in *Nonequilibrium Superconductivity*, D. N. Langenberg and A. I. Larkin (ed.) (1096), p. 253.
35. R. Carlson and A. Goldman, Phys. Rev. Lett. **34**, 67 (1976).
36. J. Han, Phys. Rev. B **66**, 054517 (2002).
37. S. Barabash and D. Stroud, Phys. Rev. B **67**, 144506 (2003).

38. Y. Uemura, Solid State Commun. **126**, 23 (2003).
39. J. Loram, J. Tallon, and W. Liang, Phys. Rev. B **69**, 060502 (2004).
40. J. Emerson, D. Wright, B. Woodfield, J. Gordon, R. Fisher, and N. Phillips, Phys. Rev. Lett. **82**, 1546 (1999).
41. J. M. Tranquada, B. Sternlieb, J. Axe, Y. Nakamura, and S. Uchida, Nature **375**, 561 (1995).
42. D. Basov and T. Timusk, Rev. Mod. Phys. **48**, 4219 (2005).

8

What T_c can Teach About Superconductivity

T. H. Geballe and G. Koster

We compare the T_cs found in different families of optimally doped high-T_c cuprates and find, contrary generally accepted lore, that pairing is not exclusively in the CuO_2 layers. Evidence for additional pairing interactions, that occur outside the CuO_2 layers, is found in two different classes of cuprates, namely the charge reservoir and the chain-layer cuprates. The additional pairing in these layers suppresses fluctuations and hence enhances T_c. T_cs higher than 100 K, are found in the cuprates containing charge reservoir layers with cations of Tl, Bi, or Hg that are known to be negative-U ions. Comparisons with other cuprates that have the same sequence of optimally doped CuO_2 layers, but have lower T_cs, show that T_c is increased by factors of two or more upon insertion of the charge reservoir layer(s). The Tl ion has been shown to be an electronic pairing center in the model system (Pb,Tl)Te and data in the literature that suggest it behaves similarly in the cuprates. A number of other puzzling results that are found in the Hg, Tl, and Bi cuprates can be understood in terms of negative-U ion pairing centers in the charge reservoir layers. There is also evidence for additional pairing in the chain-layer cuprates. Superconductivity that originates in the double "zigzag" Cu chains layers that has been recently demonstrated in NMR studies of Pr-247 leads to the suggestion of a linear, charge 1, diamagnetic quasiparticle formed from a charge-transfer exciton and a hole. Other properties of the chain-layer cuprates that are difficult to explain using models in which the pairing is solely confined to the CuO_2 layers can be understood if supplementary pairing in the chain layers is included. Finally, we speculate that these same linear quasiparticles can exist in the two-dimensional CuO_2 layers as well. It is possible that these particles will propagate chiefly in either the x- or y-direction and be appropriate candidates for fluctuating stripes and for d-wave superconductivity.

8.1. Introduction

The transition into the superconducting state reflects the totality of all the underlying microscopic interactions in the system. As such, paying attention to the occurrence of T_c throughout the Periodic Table of the Elements and its magnitude as a function of controlled external parameters such as pressure, composition, strain, dimension, and defects, is the most general approach for gaining an understanding of superconductivity. Discovering superconductors has been a fruitful enterprise for opening new fields of physics ever since Kamerlingh Onnes discovered back in 1911 that Hg loses all of its resistance abruptly just below 4.2 K [1].

T. H. Geballe • Department of Applied Physics and Department of Materials Science, Stanford university, Stanford, CA
G. Koster • Geballe Laboratory for Advanced Materials, Stanford University, Stanford, CA

For more than four decades superconductors were uncommon and poorly understood laboratory curiosities; there was no basis for predicting their occurrence and little connection with normal state properties. Hans Meissner found the barely metallic compound CuS to be superconducting whereas elemental Cu was not [2]. The rare and unpredictable occurrence of superconductivity, and the lack of an underlying microscopic theory, led Enrico Fermi at the University of Chicago around 1950 to encourage two young colleagues, John Hulm and Bernd Matthias, to undertake a search for new superconductors. They soon found a number of new intermetallic alloys and compounds [3] and extended the work of Meissner, who in the 1930s had also found superconducting intermetallic borides [4]. In the same time period a parallel program was carried out in Russia by Alekseevskii and coworkers [5]. Whether there are still more bulk superconductors with novel properties remaining to be discovered is, of course, impossible to predict; but since there are now opportunities for synthesizing entirely new classes of materials and structures beyond equilibrium phases, made possible by advances using thin film deposition and characterization techniques, there is good reason to believe that higher T_cs will be found.

The Periodic Table was a valuable guide for predicting new superconductors particularly when Matthias noted that there is an amazingly simple dependence (known as Matthias' Rule [6]) of the magnitude of T_c upon the average number of valence electrons per atom in elements and also in intermetallic compounds—i.e., T_c is related simply to the electron density [7]. As a consequence of this "rule," superconductivity changed from being rare to being common. As the database increased, refinements were incorporated; superconductivity was found to be favored in specific structures [8], in particular in the A15 structure (also referred to as beta tungsten). This structure had the highest known T_cs up until 1986 as well as other unexpected low temperature instabilities. It contains an unusual arrangement of nonintersecting chains of closely spaced metal transition metal atoms that impose features on the Fermi surface [9]. The discovery of the superconductivity of V_3Si [10] and Nb_3Sn [11] had a great impact, surpassed in impact only by the discovery of cuprate superconductivity by Bednorz and Mueller more than three decades later, and led the way to new concepts in physics and a new high field–high current technology. The discovery of A15 superconductivity in Nb_3Sn illustrates the fact that in searching for new superconductors, even though it may be a high-risk endeavor, can have major consequences well beyond the original scope of the work [12]. In our opinion there is still much to be gained by continuing the research for new superconductors.

Little is known about the limits of superconductivity in the cuprates today. In this chapter we interpret the wide variation in T_c that is found in different cuprate structures in terms of plausible intuitive models that are interesting in their own right, and might be of value in guiding paths to even higher T_cs.

8.2. Cuprate Superconductivity

There is nothing to compare with the impact made by the discovery of High Temperature Superconductivity in the cuprates [13]. After two decades of intensive research there is no accepted theory nor is there consensus as to the superconducting pairing mechanism [14]. The cuprate charge carriers are highly correlated electronic systems that have sometimes been designated as "bad metals" [15] because Fermi-liquid theory is inadequate for treating the normal state properties.

Our aim in this chapter is modest. We start from the insulating side using a simple ionic model because we believe it provides a reliable way of gaining an intuitive understanding. The ionic model has long been used successfully for modeling insulating oxides and for understanding their magnetic properties. It is a limiting case of very strong correlation and thus has credibility as an initial approximation for the nearly insulating cuprate superconductors. In the Born approximation, the large attractive Madelung energy is balanced by the repulsive overlap energy. We have found it useful, following Moyzhes and Suprun et al. [16] to modify the Born's equation by using the high-frequency dielectric constant to account for the electronic polarizability of near neighbor ions and the low-frequency dielectric constant to account for the ionic polarizability of more distant ions. This modification leads to a statistically significant classification of a large number of oxides, and predictions as to their stability, metallicity, and instability to within about 1 eV [17]. The modified Born equation is a simplified representation of the local density approximation (LDA) [18].

8.2.1. Pairing and T_cs in the Cuprates

All the cuprates with high T_cs contain two-dimensional layers of CuO_2 that upon sufficient doping become superconducting fluctuations. T_c is found to increase when the number of CuO_2 layers per unit cell increases from $n = 1$ to 3. This is not surprising because the close spacing of the n layers within the unit cell can be expected by quantum tunneling [19] to stabilize the three-dimensional fluctuations. The decrease in T_c with further increases in n is discussed later. However focusing exclusively on the CuO_2 layers cannot explain some significant variations in T_c that are found in structures that have the same sequences of CuO_2 layers but have different intervening layers, and that is the subject we address here.

The Cu Ion

Before proceeding to discuss the cuprates we recall some facts that make the Cu ion unique. In the vapor phase Cu^{2+} has the highest third ionization potential of the transition metals. This large energy is retained in the condensed state as is evident from the electrode potentials of ions in aqueous solution [20]. Electrode potentials provide rough estimates of the relative ionic energies in crystalline oxides because in both the aqueous and crystalline environments the cations are coordinated by oxygen ions. The standard electrode potential for charge transfer $Cu^{3+} + e^- = Cu^{2+}$, $E(0) = +2.4\,eV$ is very high. It follows that in cuprates the doped holes will reside mainly on oxygen sites (in contrast to other transition metal oxides where the cations are oxidized upon hole doping). On the other hand, the standard electrode potential for the reaction $Cu^{2+} + e^- = Cu^{1+}$ is quite low, $E(0) = -0.15\,eV$, showing that Cu^{2+} can easily coexist with Cu^{1+}. Consequently in the condensed state Cu^{1+} and Cu^{2+} are close in energy which translates in the Hubbard model to a moderate U that splits a half-filled narrow band due to the on-site coulomb (Hubbard U) repulsion. Of course in the crystalline cuprates, the crystal field states, the band, exchange, and correlation energies must be included. But the ionic energies are the largest so we can assume without further calculations that Cu^{3+} (d^8 configuration) does not play a significant role in the dynamics of the cuprates with the consequence that the cuprates are "charge transfer insulators" rather than Mott insulators [21].

In the undoped parent compounds the CuO_2 layers are insulating antiferromagnets containing Cu^{2+} ions. The Cu is in a d^9 state; two of the four electrons needed for charge neutrality in the CuO_2 layer come from other layers such as the La layers in La-214. The overlap of the half-filled Cu $(x^2 - y^2)$ d-levels results in a narrow band that, when U is greater than the band width, splits into the upper and lower (Hubbard) bands [22]. In this chapter we restrict the discussion to the doping of holes in the CuO_2 layers that can be achieved by substitution of a cation with a lower valence (e.g., Sr for La), or by cation valence reduction

Table 8.1. Variation in T_c

CuO_2/c	$n=1$		$n=2$		$n=3$	
	T_c (K)	Separations (Å)	T_c (K)	Separations (Å)	T_c (K)	Separations (Å)
LSCO-214	40	6.6	–	–	–	–
Hg-12$(n-1)n$	98	9.5	127	9.5	134	9.5
Tl-12$(n-1)n$	–	–	103	–	133	–
Tl-22$(n-1)n$	95	11.5	118	11.5	125	11.5
Bi-22$(n-1)n$	38	–	96	–	120	–
Y123 (6 GPa)	–	–	95	7.9	–	–
Y124 (6 GPa)	–	–	105	9.8	–	–

(e.g., Tl^{3+} to Tl^{1+}), or by the addition of negative oxygen ions. Upon hole doping, the charge resides mainly on the oxygen site and the antiferromagnetism is rapidly destroyed. Concentrations >0.05 holes per Cu become superconducting [23]. A dome-shaped curve [24] is found for all the cuprates. It is commonly assumed that the dome shape is universal with the maximum T_c found at an optimum doping concentration $p = 0.16$ holes/Cu. However, it is obvious that the three-dimensional superconducting condensation measured by T_c depends upon coupling between the layers and, as we argue below the coupling is not universal, and therefore we should not expect there to be a single concentration for which T_c becomes optimum. Thus the results of Karppinen that T_c of Bi-2212 occurs for $p = 0.12$ (see below) should not be surprising [25]. However we believe that the comparison of the optimum T_cs in the different families of cuprates, Table 8.1, is meaningful. T_c increases across the rows for $n = 1, 2, 3$ (and decreases for $n > 3$) for reasons that will be discussed later. What concerns us here are the differences in the T_cs of optimally doped cuprates that have the same number, n, of CuO_2 layers in the unit cell (the columns in Table 8.1). In order to account for these strong variations in T_c it is necessary to assume either that the superconductivity of those with the higher T_cs is enhanced, or that the superconductivity in the CuO_2 layers of the cuprates with the lower T_cs is depressed, or, that both effects are present.

Many mechanisms are known to depress T_c, including the competition with other kinds of long-range order, local site disorder, and structural deformation (e.g., layer buckling). Competition with commensurate or nearly commensurate density waves causes large decreases in T_c compared to charge reservoir cuprates. For example, large dip in T_c for $x = 1/8$ in $(La_{1-x}, Sr_x)_2CuO_4$, or its complete destruction in $(Nd_{1-x}, Sr_x)_2CuO_4$ [26], and $(La_{1-x}, Ba_x)_2CuO_4$ [27] at the same 1/8 doping level. However, there has been no evidence for competitive ordering in optimally doped $(La_{1-x}, Sr_x)_2CuO_4$ that can account for the ~ 50 K reduction in T_c. Disorder is also known to depress T_c but again in smaller magnitudes than needed. The T_cs of the Bi-cuprates are somewhat lower than in the corresponding Hg and Tl cuprates (Table 8.1). It is possible that the Bi cuprates are more disordered or that the negative-U Bi ions are somewhat less effective pairing centers. Typically there are excess Bi^{3+} that replace Sr^{2+} next to the apical oxygen. When that disorder is replaced by a less intrusive disorder by substituting Y^{3+} for Ca^{2+} between the CuO_2 layers, Eisaki et al. [28] found that T_c increases from 90 to 96 K.

8.3. Interactions Beyond the CuO_2 Layers

We first present plausible evidence that interactions outside the CuO_2 layers must be taken into account. We examine T_cs in two classes of cuprates (i) The first consists of the charge reservoir layer cuprates from which we infer that there is an electronic pairing mechanism involving the negative-U center ions and (ii) the second has layers consisting of quasi-one-dimensional double-chains of CuO (sometimes described as "zigzag" chains) in which pairing is found. In the next sections we consider structural, compositional, pressure-dependent transport, and Nuclear Magnetic Resonance/Nuclear Quadrapole Resonance (NMR/NQR) data in the literature that collectively provide persuasive evidence for the enhancement hypothesis in both the charge reservoir and chain-layer cuprates.

8.3.1. Pairing Centers in the Charge Reservoir Layer Cuprates

The charge reservoir layer cuprates contain additional layers of the oxides of the ions of Tl, Bi that are well known in aqueous solution and in solids as well to have unstable paramagnetic $6s^1$ configurations that disproportionate to form diamagnetic ions with $6s^0$ and $6s^2$ configurations. Hg exists in solution as a two-center diamagnetic ion that exchanges charges in units of two just as the Tl and Bi ions do.

In solids the ionic picture must of course be modified but, as already noted, the ionic energies found in aqueous solution are large and useful initial approximations for insulating oxides. The three heavy Hg, Tl, and Bi ions in the cuprates that have the highest T_cs (>100 K) are in the so-called charge reservoir layers, separated from the CuO_2 layers by the apical oxygen ions in the alkaline earth-oxide layers. The charge reservoir layers are so named because of their ability to dope the CuO_2 layers. We propose that they have the additional important function of providing negative-U pairing centers and it is more accurate to call them negative-U charge reservoir layers. The T_c of Hg-1223 under pressure reaches the highest recorded $T_c \sim 160$ K [29]. Comparison with other cuprates that do not contain negative-U center ions provides convincing evidence that the negative-U ion layers, directly or indirectly, are responsible for enhancement of T_c. Our interpretation is that the enhancement is due to interactions with the charge reservoir layers as we now show.

In the well-known 214 family of cuprates (based upon La_2CuO_4) T_c reaches a $T_{c,max} \sim 40$ K when the CuO_2 layer is optimally doped with Sr; T_cs above 50 K can be reached in strained epitaxial thin films when the doping occurs with the insertion of oxygen interstitials to form the staged compound La_2CuO_{4+x} [30], and up to 45 K in oxygen doped samples, but there is no evidence for higher T_cs.

What must be addressed is the mechanism by which the T_c of the "optimally" doped 214 compound is increased to >90 K by inserting charge reservoir layers containing ions of Tl, Hg, or Bi and oxygen. As can be seen in the Table 8.1 the 6.6 Å distance between the CuO_2 layers in the 214 compound is increased by another 5 Å by the insertion of TlO layers between them, a change that by itself would be expected to decrease T_c.

In our model the pairing induced by the negative-U ions enhances T_c. We have considered the possibilities that either a given negative-U ion acts as a resonant pair tunneling center, or that clusters of negative-U ions become coherent with CuO_2 layer, or that the negative-U ion layer itself develops an independent 2D order that subsequently becomes coherent with the CuO_2. We are not aware of any experiment that might distinguish them, however theoretical considerations [31] rule out the likelihood that of there being independent two-dimensional order parameters. It is our opinion that the most likely case is the intermediate one where clusters of fluctuating negative-U ions form and then utilize the most favorable sites to become coherent with the adjacent CuO_2 layers. This latter possibility gains support in the theoretical model of a structure consisting of negative-U centers in the barrier of a

Josephson-junction [32]. An interesting prediction of that model is that there will be a strong enhancement of $I_c R$, the product of the critical current and the normal resistance in the c-direction.

8.3.2. Negative-U Center Electronic Pairing in a Model System

Anderson [33] introduced the concept of a negative-U center to explain the failure to observe Electron paramagnetic resonance (EPR) signals in chalcogenide glasses and simultaneously the pinning of the Fermi energy. He noted that the lattice relaxation around a localized electron could overcome the repulsive coulomb (Hubbard U) energy of adding a second electron, and thus result in an effective negative-U. A further analysis by Moyzhes and Suprun showed that in PbTe doped with valence skipping ions the electronic response (i.e., the polarization) of the surrounding medium can over compensate the coulomb repulsion. The relaxation of the polarization charge, set by the high-frequency dielectric constant in PbTe, is large, $\sim$30. In the model (Pb,Tl)Te system experiments support an electronic superconducting pairing mechanism as we now briefly discuss.

PbTe in which $\sim$1% of the +2 Pb is replaced by Tl has long been a good model system for studying negative-U center induced superconductivity [34]. Recent investigations by Matsushita et al. [35] and by Schmalian et al. [36] have illuminated the role of Tl ions. For low concentrations <0.3% Tl acts like a shallow acceptor forming Tl^{1+} as evidenced by the Hall constant that shows one hole is doped into the valence band per added Tl ion. Above $\sim$0.5%, however, the Hall constant becomes more nearly independent of the doping; the Fermi level is pinned as a result of disproportionation reaction: $2Tl^{2+}(6s^1) \rightarrow Tl^{1+}(6s^2) + Tl^{3+}(6s^0)$.

The near degeneracy of the 1+ and 3+ states is suggested by a systematic study of the temperature and field dependence of the resisitivity. The data can be fit by the charge-Kondo model that requires that the two charge states of Tl ions be degenerate within $\sim k_B T_c$. Kondo-like behavior and T_c set in at nearly the same concentration where the Hall effect indicates the Fermi level is pinned. The superconducting transition is driven by the gain in energy when the pairing on the different Tl ions becomes coherent presumably by interacting through the valence band states.

The lesson taken from the above (Pb,Tl)Te investigations is that for pairing to occur in the charge reservoir layers of the cuprates, the negative-U ion configurations of at least some of the centers must be nearly degenerate (within $k_B T$) in energy. If the doping from the charge reservoir layers is achieved by a change of valence of the negative-U ion, the Fermi level is automatically pinned and the condition for degeneracy is assured. A possible approach for discovering new superconductors is to identify structures in which the negative-U ions can be incorporated. Potential negative-U ions are listed by Koster et al. The difficult step is to find those systems where the chemical potential can be adjusted so as to bring the two levels into degeneracy. In terms of the Emery–Kivelson model [37] optimal doping is determined by the intersection of the pair amplitude which decreases as a function of doping and the superfluid density which increases, see Figure 8.1. For the cuprates 214 the optimal density occurs at 0.16 and $T_c = 40$ K. Assuming the optimum is at 0.16, for all cuprates as is frequently done, the doubling of $T_{c,\text{optimum}}$ would require both the stiffness and the pair amplitude to change accordingly. However, the introduction of pairing centers in layers between the CuO_2 layers will mostly suppress fluctuations and consequently will shift the occurrence of $T_{c,\text{optimum}}$ to a lower superfluid density as shown schematically in Figure 8.1. In fact, experimental evidence shows that $T_{c,\text{optimum}}$ for the Bi-2212 does occur near the 1/8 concentration rather than at 0.16, as discussed below.

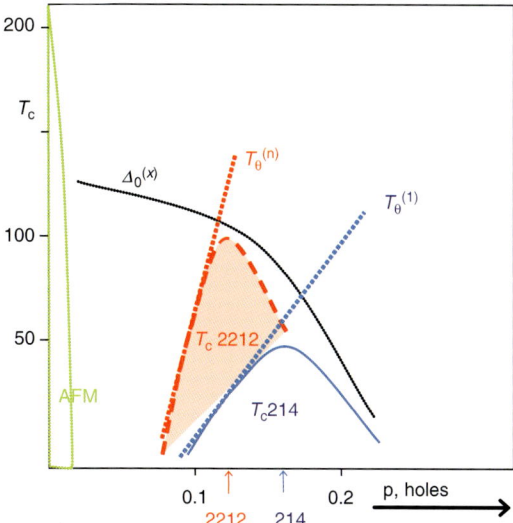

Figure 8.1. A schematic phase diagram (T_c vs. superfluid density p) that illustrates the enhancement of T_c due to the insertion of charge reservoir layers that contain pairing centers. Thin dotted curve, the pairing amplitude (mean field); dotted curves $T_{\theta 1}$ and $T_{\theta 2}$, the phase ordering temperature without and with the charge reservoir layers, respectively; blue solid curve, T_c of 214; red dash-dot curve, suggested T_c curve T_c of 2212. As a consequence of the suppression of fluctuations the model of Kivelson and Fradkin (Figure 8.4 in Kivelson's chapter in this book) would predict that p_{opt} should shift to the left (lower superfluid density). The results of Karppinen et al. [25] are taken as evidence for this shift.

Thallium Cuprates

The T_cs of ceramic samples of the (Cu,Tl)-1223 and Tl-1223 cuprates as prepared are ~100 K and increase monotonically up to 133 K upon annealing temperatures up to 550 °C in vacuum [39]. The 4f 7/2 core level of the Tl ion in the TlO charge reservoir layers in the as prepared samples are rather broad [38] and are centered around the peak of Tl^{3+} found in the Tl_2O_3 reference compound. Terada et al. found upon annealing, along with the 33 degree increase in T_c, the core level peak shift to midway between the peaks found for the +1 and +3 reference states. It is our interpretation that the shift is due to an increased presence of Tl^{1+} (as we have argued above the Tl^{2+} paramagnetic configuration is at higher energy); however, the spectra have not been fully analyzed [39,40].

Mercury Cuprates

The mercury cuprates are interesting for several reasons beyond having the highest known $T_c > 160$ K found in the Hg-1223 compound under pressure. The homologous series $HgBa_2Ca_{n-1}Cu_nO_{2n+2+\delta}$ has been synthesized [41] all the way from $n = 1$ to $n = 7$ with T_c for the optimally doped samples rising from 97 K for $n = 1$ to a maximum at $n = 3$ and then falling at a decreasing rate with further increase in n until $T_c = 80$ K for $n = 7$.

The increase from $n = 1$ to $n = 3$ is common to all the cuprates (Table 8.1) and follows from the coupling of the layers by quantum tunneling [42]. The maximum observed for $n = 3$ and subsequent decrease with higher n can be understood from NMR investigations that show that the CuO_2 layers are not uniformly doped [43]. In $n = 5$ that have been optimally doped by annealing in oxygen the inner layers are antiferromagnetic, $T_N = 60$ K with $\sim 0.35\,\mu_B$/Cu [44] while the outer layers are superconducting with a $T_c = 108$ K. The robust persistence of the superconductivity as evidenced by a negative curvature of the dependence of T_c upon n for $n > 3$ would be difficult to understand if the superconducting interactions were confined to the single outer CuO_2 layers, see Figure 8.2.

Mukada et al. [45] found that for an $n = 5$ underdoped sample, the three inner layers are antiferromagnetic, ($T_N = 290$ K with $0.68\,\mu_B$/Cu) while the two outer layers are both antiferromagnetic ($1\,\mu_B$/Cu) and superconducting, $T_c = 72$ K. In a somewhat comparable structure, but one without negative-U centers, Bozovic et al. [46] found that an isolated, 1 unit cell thick film (two CuO_2 layers) of optimally doped $(La_{1-x}, Sr_x)_2CuO_4$ (sandwiched between antiferromagnetic films of undoped La_2CuO_4 by sharp interfaces) have T_cs of only 30 K. The much higher T_cs ~ 80 K found for the $n > 5$ Hg cuprates that have single layers of doped CuO_2 interfaced on one side with AFM layers and on the other with BaO–HgO layers make it plausible that the latter layers are contributing to the pairing.

The possibility has been raised that the T_cs in the Hg cuprates are unusually high because the layers are flat (O–Cu–O bonds are 180°), rather than buckled that is known to be

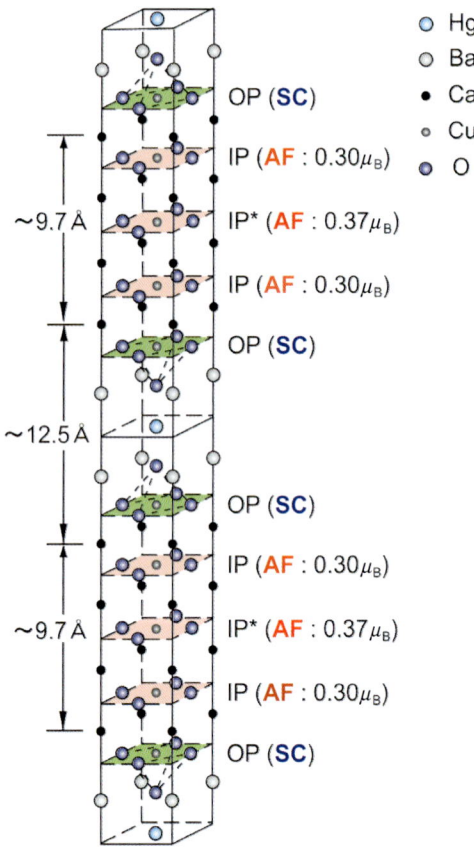

Figure 8.2. The crystal structure of Hg-1245 ($a = 3.850$ Å, $c = 22.126$ Å) [104]. The Outer Planes (OP) undergo the SC transition at $T_c = 108$ K, whereas the three underdoped Inner Planes (IPs) do an AF transition below $T_N \sim 60$ K with the respective Cu moments of $\sim 0.30\,\mu_B$ and $0.37\,\mu_B$ at IP and IP* (after Kotekawa et al.).

detrimental to T_c [47,48]. However, these explanations are at odds with the neutron diffraction data for Hg-1212 where under pressures of ~100 kbar the CuO$_2$ layers become buckled to the same extent found in the 214 cuprate while the apical oxygen moves closer to the planes, and the T_c increases (Figure 8.3). As noted by Jorgensen [49] if the buckling could be prevented T_c should be even higher. While to date only the Hg-1212 diffraction patterns have been studied under pressure it seems likely that the buckling under pressure will occur in all the Hg cuprates because the pressure dependence of T_c scales for the $n = 1, 2,$ and 3 as shown in Figure 8.3.

The reason dT_c/dP is constant= 2.0 K/GPa from low doping levels to optimum doping [50], as shown in Figure 8.4, certainly does not follow from models that assume the changes are due to charge transfer. Such models would reasonably expect to find a steadily decreasing pressure coefficient from low to optimal doping. However, the behavior is not inconsistent

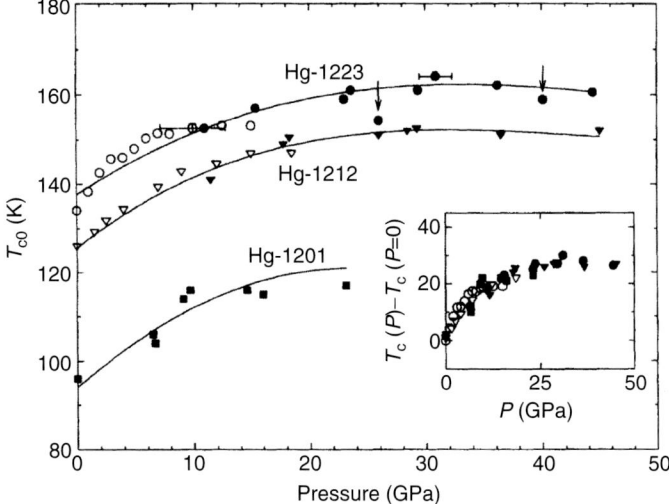

Figure 8.3. T_{c0} vs. P for Hg-1223, Hg-1212, and Hg-1201; Inset $T_{c0}(P) - T_{c0}(P = 0)$ for these three compounds. Open symbols taken from [105] and [106]. Arrows indicate reversibility (from Gao et al. [107]).

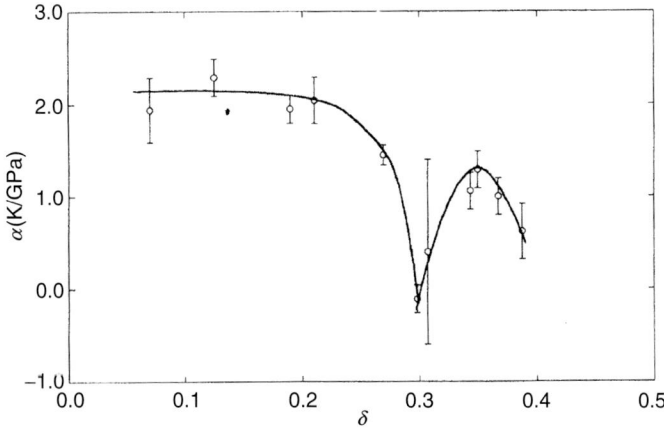

Figure 8.4. Pressure coefficient of T_c as a function of oxygen doping for Hg-1201 (from Cao et al.).

with a negative-U model because pressure should increase the overlap of the pairing centers in the HgO layers with the CuO$_2$ layers. Raman data suggest that the overlap is through the apical oxygen ions [51, 52].

The HgO–BaO layers are highly disordered. There are a large number of oxygen vacancies in the HgO layers. X-ray Absorption Fine Structure (XAFS) measurements of the Hg–Hg distances are of such poor quality that they cannot be modeled [53]. Consequently, the negative U-ion is probably a more complex entity than the idealized two center ion. In the model (Pb,Tl)Te system it is estimated that only a few percent of the Tl negative U ions are pairing centers [35], thus it is not unreasonable that substitution of a substantial concentration of Re or Cu cations substituted for Hg layers has little effect on T_c. Experiments to determine the Hg valency such as has been done for the Tl cuprates (see above) would be helpful. For this purpose better and larger Hg cuprate single crystals are becoming available [54].

The Bismuth Cuprates

The bismuth cuprates have T_cs that are somewhat lower than the corresponding Hg and Tl cuprates (Table 8.1). This suggests that either the Bi negative-U centers are not such effective pairing centers or that their enhancement is counteracted by a competing effect. Disorder is not an unreasonable possibility because there is known to be considerable antisite disorder and excess Bi on the Sr sites [55]. Perhaps even more important is the incommensurate superstructure found in the BiO layers [56, 57] that causes displacements throughout the unit cell including large amplitude waves of CuO buckling along the a-axis. The inhomogeneous images observed by STM [58, 59] are also evidence of disorder although it is not obvious how much of the observed disorder may be due to the surface layer reconstruction because the tunneling must be through the orbitals extending from the surface.

The investigation of Karppinen et al. that utilizes independent electrochemical and spectroscopic means of analysis [25], and finds them to be in agreement, has two significant findings [61]. First, half of the charge introduced by substituting Sr for Y on sites between the two CuO$_2$ layers in Bi-2212 ends up in the nonadjacent BiO layers. The second is that the $T_{c,\text{optimum}}$ for Bi-2212 occurs when the carrier concentration in the CuO$_2$ layers is 0.12 (see Figure 8.5). This is most significant because it is near the same 1/8 concentration where, it is well known from experiments on cuprates that do not contain charge reservoir layers, that T_c is depressed or nonexistent because charge ordering and static stripe formation compete successfully with superconductivity [60] nor, as mentioned above, the Emery–Kivelson model predicts that the intersection of the pairing amplitude the phase fluctuations curves, and $T_{c,\text{optimum}}$ will occur at lower doping levels (as sketched in Figure 8.1).

8.3.3. The Chain-Layer Cuprates

Many investigations have been carried out in the single and double chain cuprates that lead to the conclusion that the chain layers support pairing. We now consider three different chain-layer cuprate structures that are comparable insofar as their layering sequence is concerned, but differ in the structure of the chain layers themselves. The chain layers consist of either single CuO chains, or double ("zigzag") CuO chains, or a combination of alternating single and double layers as shown in Figure 8.6 [62]. In all three structures the quasi-one-dimensional chains are separated from the blocks of $n = 2$ CuO$_2$-(Y,Pr)-CuO$_2$ layers by layers of BaO. Evidence given below suggests that the chains that run in the b-direction interact

What T_c can Teach About Superconductivity

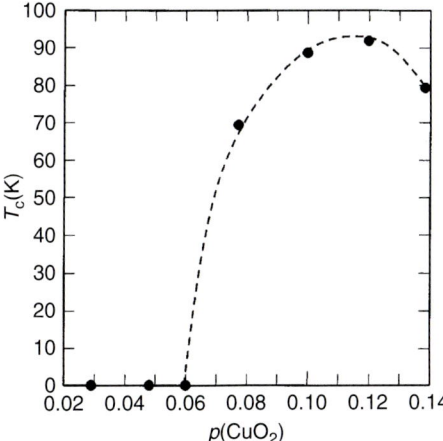

Figure 8.5. The relationship between T_c and the CuO_2-plane hole concentration, $p(CuO_2)$, in the $Bi_2Sr_2(Y_{1-x}Ca_x)Cu_2O_{8-d}$ system. Note that, $p(CuO_2)$ is taken as an average of the values determined for the CuO_2-plane hole concentration by coulometric redox analysis and by $CuL3$ – edge XANES spectroscopy. The actual cation doping level is two times $p(CuO_2)$. The threshold hole concentration for the appearance of superconductivity is seen at $p(CuO_2) = 0.06$ (taken from Karppinen et al. [25]).

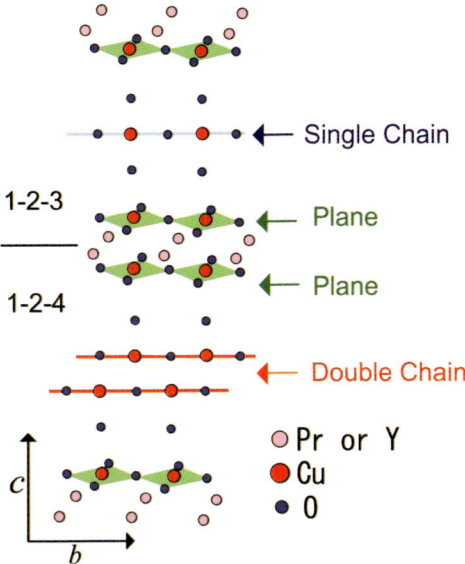

Figure 8.6. Structure of $Pr_2Ba_4Cu_7O_{15-d}$ (Pr247). Pr247 consists of the Pr123 unit ("1-2-3") and the Pr124 unit ("1-2-4"). In addition to two CuO_2 planes, the "1-2-3" ("1-2-4") contains a single chain (a double chain). The Cu atoms in the double chain do not form a "ladder" structure but a "zigzag" chain (taken from Sasaki et al. [62]).

with each other in the a-direction indirectly via the CuO_2 layers. An important difference is that the oxygen ion concentration in the single 123 chain layers is variable whereas in the 124 (both the 124 and 248 notations are used interchangeably in the literature and we do likewise) it is fixed. This permits doping over a wide range in the CuO_2 layers of the 123 cuprates, but not in the 124 cuprates. In the nonstoichiometric 123 cuprates the vacancy diffusion leads to various kinds of short- and long-range ordered structures [63], whereas the 124 cuprates are stoichiometric and the diffusion is very much slower. Doping on the CuO_2 layers of course is possible by cation substitution on the Y site.

Evidence from Nuclear Quadrupole Resonance in the Double Chains

The NQR investigation of Sasaki et al. [62] provides direct evidence that the superconductivity discovered by Matsukawa et al. [64] in Pr-247, originates in the double chains layers. As can be seen in Figure 8.6, the 247 structure is composed of alternating Pr-123 and Pr-124 units. Neither of the units by themselves has been found to be superconducting and, as initially prepared by sintering Pr-247, also is not superconducting. However, it becomes superconducting with zero resistance at $T \sim 10\,K$ when annealed in vacuum at 400 K [64]. The NQR Cu resonances associated with the four different Cu sites in the Pr-247 structure are well resolved allowing them to be followed separately [62]. The CuO_2 layers order antiferromagnetically around 280 K (see Figure 8.7b) as they do in Pr-123 and Pr-124. The relaxation data observed in the Pr-247 samples provide the evidence that the superconductivity resides in the double chain layers, as can be seen in Figure 8.7a. Near T_c the temperature dependence of the nuclear relaxation rate of the double chain Cu nuclei changes markedly due to a gap opening in density of the electronic states. This can hardly be coincidental, and must be due to the super-conductivity.

While $\sim 10\,K$ may not be "high temperature" in comparison with other cuprates it is very high when compared with other comparable 1D systems such as the polymer $(SN)_x$ [65].

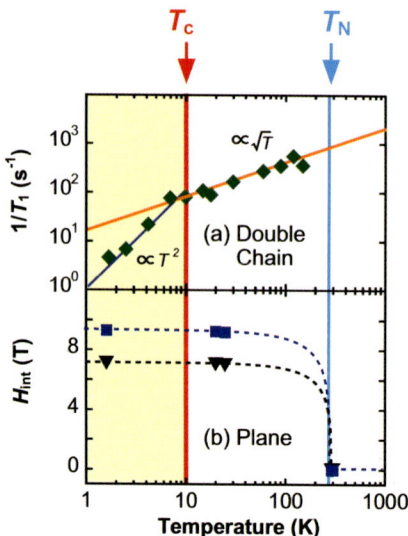

Figure 8.7. (a) Temperature dependence of $1/T_1$ of the double-chain in Pr-247. Above T_c, the T_1 process exhibited a single-exponential time evolution, which yields a unique value of T_1. Below T_c, the T_1 process was reproduced by a biexponential function with two time constants, $T_1 S$ and $T_1 L$, indicating 20% of the chain copper nuclei belong to the superconducting phase. (b) Shows the antiferromagnetic magnetization of the two different copper oxide planes in the 247 unit cell (taken from Sasaki et al. [62]).

The fact that it originates in a cuprate where the CuO_2 layers are insulating and antiferromagnetic [78] is significant; and leads us to suggest the existence of the linear diamagnetic quasiparticles discussed below.

At this time we offer no explanation for how the annealing turns on the superconductivity other than an indirect effect from the reduced single chain. The temperature dependence of the resistivity before and after annealing gives evidence for transport by parallel chain and plane conduction paths. The annealing increases the room temperature resistance presumably due to the single chains becoming insulating. Upon cooling there is a striking increase in the conductivity of the annealed sample that culminates in the superconducting transition. Comparable annealing experiments of the Pr-124 double chain cuprates show no such effects. In order to account for the 1D transport and superconductivity we suggest the formations of a linear diamagnetic bound exciton-hole (eh) quasiparticle (Figure 8.8c) that is discussed below.

Evidence from Anisotropy

The CuO chains running in the b-crystal direction are directly or indirectly responsible for the considerable planar anisotropies observed in dc and optical conductivities in the normal states of Y-123 and Y-124, and in their penetration depths in the superconducting state. Basov et al. [66] found from far infrared data that the planar anisotropies, in agreement with transport data, are large and temperature independent. At room temperature $\sigma_b/\sigma_a = 1.8$ [67] in the Y-123 and in the Y-124 it is even larger ~ 3 [68]. Corresponding penetration depth measurements find rather interestingly that the anisotropy of the superfluid densities is almost the same. If it is assumed that the anisotropy is simply due to the orthorhombicity of the CuO_2 layers, then why is it greater in the 124 when the 124 is less orthorhombic? If it is attributed to a proximity-induced superfluid density on layers and metallic CuO chains [69] then it fails to predict the observed wide temperature range over which the anisotropy is temperature independent. However, the experiments are consistent with models that assume intrinsic pairing occurs in the chain layers.

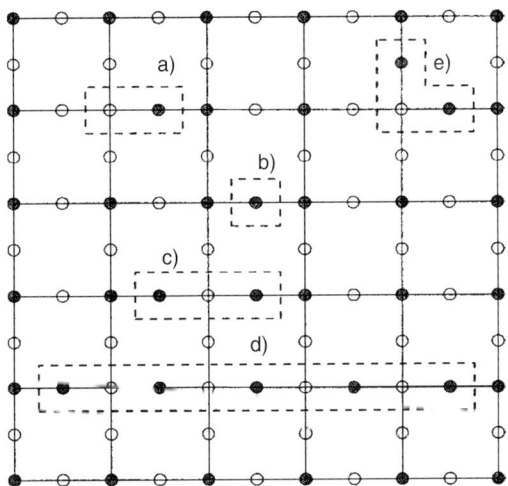

Figure 8.8. Projection of the CuO_2 layer. Open circles, ions with filled shells, either d^{10} or p^6; closed circles, ions, with 1 hole in outer shell (a) charge transfer exciton; (b) doped hole on oxygen; (c) bound exciton-hole; (d) extended bound exciton-hole; (e) high-energy (V_{pp}) configuration of the bound exciton-hole (after Kivelson et al. [92]).

Evidence from Pressure

Superconducting pairing in the double chains is a likely explanation for why the T_cs of Y-124 rise above those in the single chain Y-123 as the pressure is raised. The stoichiometric double chain Y-124O_{15} cuprates are underdoped, $T_c = 80$ K at atmospheric pressure, whereas T_c for the nonstoichiometric optimally doped Y-123$O_{6.93}$ is 93 K. However, T_c for Y-124O_{15} rises to 108 K at 6 GPa [70] exceeding the T_c of optimally doped Y-123 at any pressure, a result that does not follow from any proximity effect theory. The abnormally large increase in T_c of the Y-124 with pressure might in some part be due to additional charge transfer; however charge transfer does not explain the abnormally large anisotropic strain dependence, of T_c of the Y-124 as is evident from the following. Strain dependence in the a-direction (when the chain–chain planar distance is reduced) [71, 72] is much larger than in the c-direction (when the layer spacing is reduced and that should affect charge transfer) suggesting that chain–chain coupling plays a key role. Evidence from Zn doping discussed below indicates that the coupling is made through the CuO_2 layers.

Evidence from Cation Substitution in the $n = 2$ of CuO_2 Layers

If the reasonable assumption is made that substituting Pr for Y between the CuO_2 layers has the same effect in 123 and 124, then there is further evidence for pairing on the double chains. For instance, $(Y_{0.4}Pr_{0.6})$123 is not superconducting while the same composition in the 124 has $T_c \sim 50$ K, a result that is difficult to explain by any proximity effect [73].

Zn is known to dope in the CuO_2 planes and destroys T_c rapidly in all the cuprates [74]. In the Y123 and Y124 cuprates T_c decreases rapidly and roughly at the same rate with Zn substitution. This alone might suggest that all the superconductivity is in the CuO_2 layers, but such an interpretation is not viable in the light of other evidence such as the large dT_c/da mentioned above [75]. Evidence we now cite suggests that the double chains interact with each other by coupling with each other indirectly through the CuO_2 planes. In nonsuperconducting Pr-124 the b-axis (chain direction) shows good metallic conductivity, the a-axis resistivity peaks around 140 K. The planar transport anisotropy is 1,000 at 4 K [76]. The results can be modeled by parallel paths of the conducting chains and the semiconducting CuO_2 planes. Upon doping with Zn the material becomes insulating along the a-axis while the b-axis continues to show metallic behavior [77]. The disappearance of the Fermi level as found in an ARPES investigation [78] is explained by the increased one dimensionality of the double chains. They presumably become more decoupled when the coherence via the CuO_2 layers is destroyed by the Zn, and a competitive 1D instability such as a charge density wave becomes the ground state. The coupling between the chain layers and the CuO_2 layers is likely to be through the apical oxygen for which there is independent evidence. More complete literature references are given in the chapter by Valo and Leskela [79].

8.3.4. Other Chain Layer Compounds

The Ladder Compound

A comparable double "zigzag" CuO chain arrangement to that found in the 124 cuprates is found in $Sr_{14-x}Ca_{x-12}Cu_{24}O_{41}$ (14-12-24-41) which undergoes a broad superconducting transition ($T_c \sim 10$ K) under pressure [80]. The structure contains alternating layers of single chain Sr_2CuO_3 and double chain $SrCuO_2$ layers. The double chains are separated in each

SrCuO$_2$ layer by the rungs of two-leg ladders. There is evidence that the superconductivity originates in the SrCuO$_2$ layers that is due to charge transfer from the single to the double chain layers [81].

A theoretical model predicted the superconductivity in the 14-12-24-41 structure prior to discovery [82] by assuming that the pairing occurs in the two-leg ladders and confirming experimental evidence has been found [83]. In this model the spins of the double chain coppers are assumed to be connected by ferromagnetic superexchange via the oxygen p_x and p_y orbitals. The fact, that the same double chain configuration in the Pr-247 cuprate becomes superconducting rather than ferromagnetic, shows that subtle differences in coupling or doping can result in major changes in the ordering of quasi-one-dimensional systems.

Finite Chain Lengths

Infrared studies show that there is no anisotropy in the single chain cuprates for oxygen concentrations <6.65 per unit cell, or equivalently for chain lengths <15–20 Å for which $T_c \sim 60$ K [84]. The authors show that for higher doping when the chain length fragments exceed 20 Å there is a marked change in properties. The electromagnetic response in the normal state becomes coherent and quasi-one-dimensional. Correspondingly the superfluid density in the b-direction grows rapidly while in the a-direction it remains flat. As pointed out above, proximity effect models have difficulty in accounting for the identical temperature dependences over a wide temperature range in the a- and b-direction. Strain-dependent measurements in the single chain cuprates are ambiguous because the oxygen mobility allows for different oxygen ordering on the chains [85].

8.4. Superconductivity Originating in the CuO$_2$ Layers

We speculate that the linear spinless charge 1 quasiparticles that can explain the superconductivity of the Pr-247 quasi-one-dimensional double chains may equally well exist in the CuO$_2$ layers of all the cuprates. In the limit of negligible oxygen–oxygen near neighbor hopping (t_{pp}) such a quasiparticle model leads naturally to fluctuating stripes and d-wave superconductivity. These considerations lead to the prediction that if a two-dimensional CuO layer (i.e., a structure where the vacant sites in the CuO$_2$ layer are filled with Cu) could be synthesized and properly doped it could have double the number of quasiparticles than presently known cuprates [86].

Low-energy charge-transfer excitations involve the transfer of electrons from the highest-lying oxygen level to the upper Hubbard band. They are estimated to be ≤ 2 eV for La$_2$CuO$_4$ in the antiferromagnetic state at low temperatures [87], and in the same energy range HgBa$_2$CuO$_4$, and likely all the high T_c cuprates. There is also considerable subgap structure. The lowest peak at $\sim$0.4 eV is in reasonable agreement with an ionic estimate of the lowest energy charge-transfer exciton taken to be the gap energy less the screened interaction between the bound charges giving an energy $E_{ex} = E_g - q^2/\varepsilon r$. Here, q is the absolute value of the charges, r is their separation, and ε is the dielectric response. Putting $E_g = 2$ eV and $r = 2$ Å and making the reasonable assumption that for the short distance, $\varepsilon = \varepsilon_\infty = 5$, gives $E_{ex} + 0.5$ eV. Some of the subgap structure may also be due to multimagnon/phonon processes [88].

Upon doping the bands broaden and the gap edge is lowered. In fact RIXS data [89] suggest that the spectral weight of the lowest lying exciton in the undoped compound is transferred to the continuum intensity below the gap. In our model this occurs when the charge transfer exciton combines with the doped hole to form a bound exciton-hole (eh) quasiparticle. The various ionic configurations to be considered in the CuO_2 layers are shown in Figure 8.8.

For reasons given earlier, doped holes mainly reside on the oxygen sites (see Figure 8.8b). In the ionic model the low-energy singlet forms when the hole is attracted to the polarization cloud of the lowest lying charge transfer exciton (Figure 8.8a) resulting in a new quasiparticle that we call an eh (exciton–hole) particle (Figure 8.8c). The eh particle is a linear charge-one spin-zero quasiparticle with an electrostatic energy $E = E_{ex} - [(q^2/\varepsilon r) - (q^2/2\varepsilon r)] = +0.5 - 0.72\,\text{eV} = -0.2\,\text{eV}$. The well-known Zhang–Rice (ZR) singlet is an alternate configuration that places the doped hole in a symmetrical molecular orbital, the oxygen ions surrounding a given Cu ion [90], and is stabilized by exchange energy [91]. However, the eh-singlet is stabilized by Coulombic energy [78] and more importantly is a favorable configuration for stripe formation. A bent configuration rather than linear configuration would also have higher energy due to interaction V_{pp}, between the oxygen ions, (Figure 8.8e) [92].

In the limit $t_{pp} = 0$ the electron dynamics are purely one dimensional as depicted in Figure 8.8c. Cluster calculations, however, suggest that t_{pp}/t_{pd} is in the range of 0.3 [93] raising some question about the validity of one-dimensional transport. The eh particle (Figure 8.8d) will be dressed; in fact the extended version of the eh-particle (Figure 8.8e), in which t_{pp}/t_{pd} should be close to zero, has an even lower coulombic energy [94]. At the higher temperatures, however, entropy will favor the eh particle.

The ionic version of the phase diagram in Figure 8.1, is qualitatively consistent with generally accepted phase diagrams [95] except that we have allowed for enhancement of T_c by negative U charge reservoirs layers.

In the underdoped region below some not-well-defined-temperature T^*, well above T_c, anomalies are observed in various phenomena such as Knight-shifts, spin–lattice relaxation [96], transport, and a reduction of the effective magnetic moments of the charge carriers. These are interpreted as crossover phenomena that we ascribe to the formation of the eh particles that coexist with the paramagnetic doped holes. As the temperature decreases further the concentration of eh particles increases to the extent that the superconducting fluctuations observed by Ong and coworkers [97], occur, still well above T_c. The quasi-one-dimensionality of the eh-particles leads to fluctuating stripes and charge–spin separation [98]. In this model there is no necessity to postulate separate regions of (01) and (10) domains because of the d-wave symmetry that insures opposite phase relation for stripes in the (01) and (10) directions at the Cu crossing points. There would be no corresponding increase in kinetic energy because of the nodes in the $d_{x^2-y^2}$ wave functions at the crossing points. However, there are neutron data that indicate at least in some cases that the spin domains in the two directions are not congruent [99].

It is of interest to consider the properties of a layer in which the number of Cu sites is doubled by filling the vacant sites in the CuO_2 layer so as to form a cubic CuO structure. Such a structure upon doping should have twice the superfluid density. Real space images of naturally occurring monoclinic CuO (known as tenorite) show evidence of spin–charge separation, and anisotropic transport that is consistent with stripe formation [100].

Finally, as in all models the 3d superconducting transition occurs when the temperature is lowered to T_c and the 2D fluctuations condense [101]. While d-wave symmetry is favored in the CuO_2 layers, a small s-wave component must exist in the chain-layer cuprates, a consequence of orthorhombicity, and is also likely because of disorder in all cuprates. Once a small s-component exists there is no restriction as to how large it can grow in the regions between the CuO_2 layers. Hence to first order there is no symmetry restriction preventing the negative-U ions or ion clusters coupling with the CuO_2 layers and enhancing T_c [102, 103].

8.5. Summary

We have considered large amounts of data from the vast number of experiments concerning cuprate superconductors that have been reported over the past decade. Contrary to the commonly made assumption that interactions are confined to the CuO_2 planes we conclude that they are insufficient to explain the striking differences in T_cs that are found. We suggest that the T_cs found in the charge reservoir cuprates are enhanced due to superconducting pairing interactions involving the negative-U ions Hg, Tl, and Bi. A striking example is the doubling of T_c (from ∼45 to >90 K) found when an HgO layer is inserted into the unit cell of the 214 cuprates. Attempts to attribute this difference in T_c to effects that depress the T_cs of the 214 cuprates are ruled out by the pressure dependence experiments that further favor our model and by other considerations as well. The collective sum of the data we have considered and interpreted makes an impressive case for the importance of negative-U pairing centers.

The superconductivity found in the double chain 247 cuprates provides convincing evidence that pairing occurs outside the CuO_2 layers and originates in the one-dimensional chain layers. In order to account for this superconductivity and the normal state properties we hypothesize a linear diamagnetic (eh) quasiparticle that is stabilized by coulombic interactions. We speculate that this (eh) quasiparticle can exist in the CuO_2 layers of all the cuprates and that it offers a consistent basis for understanding the complex phase diagram of the underdoped to optimally doped cuprates.

Acknowledgments

We have profited from many discussions with Steven Kivelson and Boris Moyzhes, and would also like to acknowledge helpful interactions with many other colleagues including (from A to Z) P.W. Anderson, Y. Ando, M.R. Beasley, T. Claeson, M. Greven, J. Mannhart, D. Scalapino, Y. Maeno, S. Sasaki, Z.X. Shen, H. Yamamoto, and Jan Zaanen. The work has been supported by DOE and GK thanks the Netherlands Organization for Scientific Research (NWO, VENI) for support.

Bibliography

1. Early Leiden communicators termed the phenomenon "supraconductivity" (i.e., beyond conductivity), a more apt description than superconductivity.
2. W. Meissner, Z. fur Phys. **58**, 570 (1929); Helium was rare, and its liquefaction required considerable effort. Meissner's use of his limited cryogenic resources to investigate such a far out material as CuS suggests a scientific instinct of the same caliber as the more famous Meissner–Ochsenfeld experiments.
3. J. K. Hulm and B. T. Matthias, Phys. Rev. **87**, 799 (1952).
4. W. Meissner, H. Franz, and H. Westerhoff, Z. fur Phys. **75**, 521 (1932); It is surprising that the superconductivity in MgB_2 was missed throughout the 20th century and not discovered until Akimitsu's group did it; J. Nagamatsu, N. Nakagawa, T. Muranaka, Y. Zenitani, and J. Akimitsu, Nature **410**, 63 (2001); Part of the reason may that the very large differences in the melting points of Mg and B precludes simple synthesis.

5. References to these workers, and the few others engaged in finding new superconductors can be found in B.W. Roberts, J. Phys. Chem. Ref. Data **5**, 581 (1976).
6. B.T. Matthias, Prog. Low. Temp. Phys. **2**, 183 (1957).
7. This is not at all obvious because the chemical energies of the order 1 eV from which the Periodic Table is built are several orders of magnitude greater than the superconducting energy scales.
8. B.T. Matthias, T.H. Geballe, and E. Corenzwit, Rev. Mod. Phys. **35**, 1 (1963); S.V. Vonsovsky, Y.A. Izyumov, and A.P. Zavaritski, *Superconductivity of Transition Metals* (Springer, Berlin Heidelberg, New York, 1982).
9. J. Labbe and J.J. Friedel, J. Phys. Radium **27**, 708 (1960) and M. Weger and I.B. Goldberg, Solid State Physics **28**, 1 (1973); L.R. Testard, Rev. Mod. Phys. **42**, 637 (1975), see also Ref. [5].
10. G.F. Hardy and J.K. Hulm, Phys. Rev. **93**, 1004 (1954).
11. B.T. Matthias, T.H. Geballe, S. Geller, and E. Corenzwit, Phys. Rev. **95**, 1435 (1954).
12. T.H. Geballe, G. Koster, Strongly correlated electron materials; physics and nanoengineering, I. Bozovic, D. Pavuna (eds.), *Proceedings of SPIE* 5932 (SPIE, Bellingham, WA, 2005).
13. This is not to imply that it is one of the great fundamental discoveries (e.g., relativity) but its almost instantaneous impact, where, within a short time high school students the world over were performing experiments that had previously not been possible in even the most advanced laboratories, we believe, is unprecedented. Google cites 283,000 entries for "High T_c cuprate superconductivity" as of January 2006.
14. See for example D.J. Scalapino's chapter in this book.
15. V.J. Emery, S.A. Kivelson, Phys. Rev. Lett. **74**, 3253 (1995).
16. B.Y. Moyzhes and S.G. Suprun, Sov. Phys. Solid State **24**, 309 (1982); also earlier Ref. to L.A. Drabkin, B. Moyzhes, Fiz. Tverd. Tela **550**, (1982)
17. G. Koster, T.H. Geballe, and B. Moyzhes, Phys. Rev. B **66**, 085109 (2002).
18. L.A. Drabkin, B.Y. Mozyhes, and S.G. Suprun, Sov. Phys. Solid State **27**, (1985).
19. S. Chakravarty, A. Sudho, P.W. Andersen, and S. Strong, Science **261**, 351 (1993).
20. F.A. Cotton and G. Wilkenson, *Advanced Inorganic Chemistry*, 5th edition (Wiley, New York, 1988).
21. J. Zaanen, G.A. Sawatzky, and J.W. Allen, Phys. Rev. Lett. **55**, 418 (1985); the fact the cuprates are charge transfer insulators was pointed out by Emery?
22. See for example M. Imada, A. Fujimori, and Y. Tokura, Rev. Mod. Phys. **70**, 1040 (1998).
23. A review of the properties of doped La_2CuO_4 is given by M.A. Kastner, R.J. Birgeneau, G. Shirane, Y. Eudok, Rev. Mod. Phys. **70**, 897 (1998). Y. Ando, A.N. Lavpov etal., Phys. Rev. Lett. **87**, 17001 (2001).
24. Many phase diagrams with roughly similar characteristics are discussed in literature, see for example, J. Orenstein and A. Millis, Science (2002), E.W. Carlson, V.J. Emery and S.A. Kivelson, and D. Orged, Chapter in *The Physics of Conventional and Unconventional Superconductors*, K.H. Benneman and J.B. Ketterson (eds.) (Springer, Berlin Heidelberg New York, 2004)
25. M. Karppinen, M. Kotiranta, T. Nakane et al., Phys. Rev. B **67**, 134522 (2003).
26. J.M. Tranquada, B.J. Sternleib, J.D. Axe et al., Nature **375**, 561 (1995).
27. M. Fujita, H. Goke, K. Yamada, J.M. Tranquada, and L.P. Regnault, Phys. Rev. B **70**, 104517 (2004).
28. H. Eisaki, N. Kaneko, D.L. Feng, A. Damascelli, P.K. Mang, K.M. Shen, Z.X. Shen, and M. Greven, Phys. Rev. B **69**, 064512 (2004).
29. L. Gao, Y.Y. Xue, F. Chen, Q. Xiong, R.L. Meng, d. Ramirez, C.W. Chu, J.H. Eggert, and M.K. Mao, Phys. Rev. B **50**, 4260 (1994).
30. G. Bozovic, I. Logvenov et al., Phys. Rev. Lett. **89**, 103001 (2002).
31. P.W. Anderson, private communication.
32. V. Oganesyan, S. Kivelson, T.H. Geballe, and B. Y. Moyzhes, Phys. Rev. B **65**, 1725041 (2002).
33. P.W. Anderson, Phys. Rev. Lett. **34**, 953 (1975).
34. S.A. Nemov and I.Yu. Ravish, Phys.-Usp. **41**, 735 (1998); V.I. Kardanov, S.A. Nemov, R.V. Parfenev, and D.V. Shamskur, JETP Lett. **35**, 639 (1952).
35. Y. Matsushita, H. Bluhm, T.H. Geballe, and I. Fisher, Phys. Rev. Lett. **94**, 157002 (2005).
36. M. Derzo and J. Schmalian, Phys. Rev. Lett. **94**, 9126 (2005).
37. V.J. Emery and S.A. Kivelson, Nature **374**, 434 (1995).
38. N. Terada et al., IEEE (2001), Figure 8.2: Note that the +1 state has the stronger binding energy.
39. T. Suzuki et al., Phys. Rev. B **40**, 5184 (1989), compare core level spectra of Tl-2223, Tl_2O_3, and Tl_2O.
40. We thank N. Terada for a useful discussion pointing out that the oxygen removal could account for the increase in T_c due to electron transfer directly to the CuO_2 layers thus decreasing the hole concentration if CuO_2 layers were initially overdoped before annealing. However because of the positive pressure coefficient of T_c [Lin et al., Phys. C **175**, 627 (1991)] of Tl-2223, it seems unlikely that the CuO_2 layers before anneal were overdoped. On

the other had the pressure coefficient of Tl-2201 is negative, so that a repeat of the measurements or a Tl-2201 sample would be interesting.

41. J.G. Kuzemskaya, A.L. Kuzemsky, and A.A. Cheglokov, J. Low Temp. Phys. **118**, 147 (2000).
42. S.Chakravarty, H.-Y. Kee, and K. Voelker, Nature **428**, 53 (2004).
43. H. Kotakawa et al., Phys. Rev. B **69**, 014501 (2004).
44. The middle layer is reported to have $0.37\mu_B$ and the adjacent inner layers to have $0.30\mu_B$.
45. H. Mukuda, M. Abe, Y. Araki, H. Kotegawa, Y. Kitaoka, K. Tokiwa, T.Watanabe, A. Iyo, and Y. Tanaka, preprint.
46. I. Bozovic, G. Logvenov, M.A.J. Verhoeven et al., Nature **422**, 873 (2004).
47. O. Chmaissen, J.D. Jorgenson et al., Nature **397**, 45 (1999).
48. S. Kambe and O. Ishii, Physica C **341–348**, 555 (2000).
49. Jorgensen, private communication.
50. Y. Cao, Q. Xiong, Y.Y. Xue, and C.W. Chu, Phys. Rev. B **52**, 6854 (1995).
51. Alternatively, the involvement of apical oxygen in pairing exchange may explain some pressure dependence on T_c Recent pressure dependent Raman data indeed show a clear change of electron–phonon interaction with pressure in the relevant range of Figure 8.3. (Z.X. Shen, private communication.)
52. F. Duc and P. Bordet, *Studies of High Temperature Superconductors*, vol. 30 (Nova Science, Commack, NY, 1999).
53. C.H. Booth, F. Bridges, E.D. Bauer, G.G. Li, J.B. Boyce, T. Claeson, C.W. Chu, and Q. Xiong, Phys. Rev. B **52**, 15745 (1995).
54. L. Lu et al., Cond-mat 0501436 (2005).
55. Y. Petricek, Y. Gao, P. Lee, and P. Coppens, Phys. Rev. B **42**, 387 (1990).
56. M.D. Kirk, J. Nogami, A.A. Baski, D. Mitzi et al., Science **242**, 1673 (1988).
57. J.P. Attfield, A.L.O. Kharlanov, J.A. McAllister, Nature **394**, 126 (1998).
58. A.C. Fang, L. Capriotti, D.J. Scalapino et al., Phys. Rev. Lett. **96**, 017007 (2006).
59. C. Howald, P. Fournier, and A. Kapitulnik, Phys. Rev. B **64**, 100504 (2001); K. McElroy, J. Lee, J.A. Slezak, D.–H. Lee, H. Eisaki, S. Uchida, and J.C. Davis, Science **309**, 1048 (2005).
60. See Basov's chapter, this book.
61. For titration results, the reproducibility is very good. For each point the analysis has been repeated five times. The error in hole concentration due to systematic and nonsystematic errors is ≤ 0.02 M. Karppinen, private communication.
62. S. Sasaki, S. Watanabe, Y. Yamada, F. Ishikawa, S. Kukuda, S. Sekiya, cond-mat/0603067 (2006).
63. D. de Fontaine et al., Nature **343**, 544 (1990).
64. M. Matsukawa, Y. Yamada et al., Physica C **411**, 101 (2004).
65. R.L. Greene and G.B. Street PRL **34**, 577 (1975).
66. D.N. Basov, P. Diany, and D.A. Bonn et al., Phys. Rev. B **74**, 598 (1995).
67. U. Welp, S. Fleischer, W.K. Kok et al., Phys. Rev. B **42**, 1089 (1990).
68. B. Bucher, J. Karpinski, E. Kaldis, and P. Wachter, J. Less Common Metals **20**, 164 (1990).
69. T.Z. Kresin and S.A. Wolf, Phys. Rev. B, 6458 (1992); References to other proximity models are reviewed by D.N. Basov and T. Timusk, Rev. Mod. Phys. **77**, 744 (2005).
70. R.J. Wijngaarden, D. Tristan Jover, R. Griessen, Physica B **265**, 128 (1999).
71. U. Welp et al., J. Supercond. **7**, 159 (1994).
72. C Meingast, F. Gugenberger, O. Kraut, and H. Wuhl, Physica C **235–240**, 1313 (1994).
73. The reported $T_c = 90$ K in zone grown single crystal of Pr-123 is not relevant to the present discussion for reasons given by 67.
74. S. Fujiyama, M. Takigawa, and S Hori, Phys. Rev. Lett. **90**, 147004 (2003).
75. M.H. Julien, Y. Tokunaga, T. Feher et al., cond-mat 0505213 (2005).
76. S. Hori et al., Phys. Rev. B **61**, 6327 (2000) and/or N.E. Hussey et al., Phys. Rev. Lett. **89**, 86601 (2002).
77. K. Nakada, H. Ikuta, S. Hou et al., Physica B **357**, 186 (2001).
78. T. Mizokawa, K. Nakada, C. Kim, Z.X. Zhen et al., Phys. Rev. B **65**, 193101 (2002).
79. *Studies of High Temperature Superconductors*, A.V. Narlikar (ed.), vol. 25 (Nova Science, Commack, New York, 1987).
80. M. Vehara, T. Nagata, J. Akimitsu et al., J. Phys. Soc. Jpn **65**, 2764 (1996).
81. Y. Piskunnov, D. Jerome, P. Auban-Senzier et al., cond-mat (2005) 0505561.
82. E. Dagotto, J. Riera, and D. Scalapino, Phys. Rev. B **45**, 5744 (1992).
83. T. Nagata, M. Yetara, T. Gota et al., Phys. Rev. B **81**, 1091 (1998).
84. X.-Y.S. Lee et al., cond-mat. 0504233 (2005).

85. S. Sadewasser, S. Wang, Y. Schilling et al., Phys. Rev. B **56**, 14168 (1997).
86. T.H. Geballe and B.Y. Moyzhes, Ann. Phys. **13**, 20 (2004).
87. M.A. Kastner, R.J. Birgeneau, G. Shirane, Y. Endoh, Rev. Mod. Phys. **70**, 897 (1998).
88. J. Lorenzana and G. A. Sawatzky, Phys. Rev. Lett. **74**, 1867 (1995); J. Lorenzana and G. A. Sawatzky, Phys. Rev. B **52**, 9576 (1995).
89. Y.J Kim, J.P. Hill, Seiki Komiya, Yoichi Ando et al., Phys. Rev. B **70**, 094524 (2004)—The fact that higher energy charge transfer excitons may exist upon doping is not relevant here.
90. F.C. Zhang and T.M. Rice, Phys. Rev. B **37**, 3759 (1988).
91. Neither the ZR nor the eh particles account for the spin that is added with the doped hole; both tacitly assume charge–spin separation. This can be seen by noting that before doping the layer is AFM with no spin. After doping one of the Cu spins is assigned to the localized singlet leaving the rest of the lattice with a missing spin.
92. S.A. Kivelson, E. Fradkin, and T.H.Geballe, Phys. Rev. B **69**, 144505 (2004).
93. A.K. McMAhan and S. Satpathy, Phys. Rev. B **38**, 6650 (1988).
94. Kivelson, private communication.
95. J. Orenstein and A.J. Millis, Science **288**, 458 (2000).
96. T. Timusk and B. Statt, Rept. Prog. Phys. **62**, 61 (1999).
97. N.P. Ong, Yayu Wang, S. Ono, Yoichi Ando, and S. Uchida, Ann. Phys. **13**, 200310034 (2004).
98. J.M. Tranquada, J.D. Axe, N. Ichikawa et al., Phys. Rev. B **54**, 7489 (1996).
99. J. Tranquada, private communication.
100. X.G. Zheng et al., Phys. Rev. Lett. **85**, 5170 (2000).
101. J. Lawrence and S. Doniach, Proc. Int. Conf. Low Temp. Phys. **12**, 361 (1971).
102. We have not discussed Bi in this chapter but there is every reason to believe that Bi^{+3} and Bi^{+5} with the same nominal $6s^2$ and $6s^0$ configurations as Tl be negative-U pairing centers.
103. Kleiner et al., Phys. Rev. Lett. **76**, 2161 (1996).
104. J. Akimoto, K. Tokiwa, A. Iyo, H. Ihara, K. Kawaguchi, M. Sohma, H. Hayakawa, Y. Gotoh, and Y. Oosawa, Physica C **281**, 237 (1997).
105. C.W. Chu et al., Nature **365**, 323 (1993).
106. L. Gao et al., Philos. Mag. B **68**, 345 (1993).
107. L. Gao, Y.Y. Xue, F. Chen et al., Phys. Rev. B **50**, 4260 (1994).

9

High-T_c Superconductors: Thermodynamic Properties

R. A. Fisher, J. E. Gordon, and N. E. Phillips

Thermodynamic properties, primarily the specific heat, of the high-T_c cuprate superconductors are reviewed. This topic was covered in a number of reviews that appeared in the early years of research on these materials. Here the emphasis is on more recent experimental results, including many measurements in magnetic fields, and on features related to phenomena that have been recognized more recently. Calorimetric evidence bearing on the symmetry of the order parameter, the nature of the transition at T_c, the effects of chemical substitutions, fluctuation effects, the melting of the vortex lattice, the existence of a pseudogap, and effects that appear to be related to stripe formation are discussed. Brief summaries of the different experimental techniques, with evaluations of their strengths and weaknesses, and of the problems and uncertainties that arise in analyses of the data are included.

9.1. Introduction

9.1.1. Scope and Organization of the Review

We review the thermodynamic properties of the high-T_c cuprate superconductors (HTS), primarily the specific heat. Measurements of the specific heat have made a substantial contribution to the understanding of the HTS, but the pace of the research has abated in recent years, making this an appropriate time for a review. It must be recognized, however, that the interpretation of many of the measurements has been limited by the quality of the samples that were available, and recent improvements in sample quality suggest that some of the still open questions will be answered in the future by new measurements on better samples.

There have been a number of reviews of the properties of HTS, e.g., Junod [1]; Atake [2]; Phillips, Fisher, and Gordon [3]; Fisher, Gordon, and Phillips [4]; Greene and Bagley [5]; and Malozemoff [6], references to several of which will be made here for early results. In this review the emphasis is on features that were not known at the time of the earlier reviews, or for which there have been significant advances in the results or their interpretation subsequent to those reviews: evidence for lines of nodes in the energy gap (Section 9.2.2); thermal effects associated with stripes (Section 9.4); fluctuation effects and the nature of the transition at T_c (Section 9.5); vortex-lattice melting (Section 9.6); the pseudogap (Section 9.7). New results on the effects of chemical substitutions, which are relevant to some of the other properties, are also included (Section 9.3).

R. A. Fisher, N. E. Phillips • Department of Chemistry, University of California at Berkeley and Lawrence Berkeley National Laboratory, Berkeley, CA 94720, USA
J. E. Gordon • Physics Department, Amherst College, Amherst, MA 01002, USA

Section 9.1.2 includes an overview of the structures and nomenclature for the HTS, representative values for the important characteristic parameters, and a generic phase diagram. The origin of the substantial uncertainties in the critical fields, as determined by magnetization measurements, and in parameters derived from them, is described briefly in Section 9.1.3, but the reader is referred to an earlier review [6] for the details. The component contributions to the specific heat and the expressions for their temperature and magnetic field dependences, which are used in analyses of experimental data, are outlined in "Specific Heat: Component Contributions; Field and Temperature Dependences; Nomenclature" in Section 9.1.4. Some understanding of the experimental techniques, and their shortcomings and limitations is relevant to the evaluation of the results, and this is taken up in "Specific Heat: Experimental Techniques" in Section 9.1.4. Some more general problems, and limitations associated with the specific-heat data, are mentioned in "Specific Heat: Problems and Uncertainties in Analysis of Data" in Section 9.1.4.

9.1.2. Cuprate Superconductors: Occurrence; Structures; Nomenclature; Phase Diagram; Characteristic Parameters

The HTS are derived from antiferromagnetic Mott insulators by doping, either with holes (the more common case) or electrons. They occur in anisotropic layered perovskite structures, which have strong 2D character. They do not occur in the cubic perovskite, $CaTiO_3$ structure, although there are other superconductors with this structure, $SrTiO_3$ [7], $Ba(Pb,Bi)O_3$ [8], and $(Ba,K)BiO_3$ [9]. The cubic perovskite structure consists of a 3D network of vertex-sharing O octahedra with metal ions at their centers and in interstitial positions. The first cuprate superconductor [10], a La–Ba–Cu–O compound, has a layered perovskite, K_2NiF_4 structure. That structure is illustrated, for $(La_{2-x}Sr_x)CuO_4$ (LSCO), which is hole doped by substitution of Sr on La sites, in Figure 9.1. The apical O atoms are not shared by adjacent octahedra in the c-direction, leaving a quasi-2D structure of CuO_2 sheets with Cu and O in a square-planar configuration in the ab plane. The CuO_2 sheets are separated by square-planar (La,Sr)O sheets that include the apical O atoms of the octahedra. The electron-doped superconductors, $(Nd,Ce)CuO_4$ and $(Nd,Sr,Ce)CuO_4$ have similar structures but with displacements of the apical O atoms to other positions. In other cuprate superconductors some of the apical O atoms are missing (an example is shown in Figure 9.2) but the infinitely extended CuO_2 planes, in which the superconductivity is thought to originate, remain.

The structure of $YBa_2Cu_3O_{6+x}$, $0 \leq x \leq 1$, (YBCO) is shown in Figure 9.2, for the fully oxygenated, $x=1$, structure. The O content on the CuO chains is variable, and hole doping of the antiferromagnetic insulator, with $x=0$, is obtained by increasing x. This structure, and those of other cuprates, can be thought of as comprising "conducting blocks" of CuO_2 planes, terminated by "spacing" or "capping" layers (the BaO planes containing the apical oxygens in YBCO), "insulating layers" between the conducting blocks (the CuO chains in YBCO), and "separating" layers between the CuO_2 planes within a block (the Y layer in YBCO). Particularly for the Tl, Bi, and Hg cuprates, the different structures are often distinguished by a four-number label. The four numbers represent both the relative numbers of the different metal ions and the numbers of planes of each type, in which those ions are contained: The first number is the number of insulating layers between conducting blocks; the second, always two, is the number of capping layers on a conducting block; the fourth is the number of CuO_2 planes in a block, n; the third is the number of separating layers between CuO_2 planes in a block, $n-1$. In this scheme, Bi-2223 is $Bi_2Sr_2Ca_2Cu_3O_{10}$, with two insulating layers of BiO,

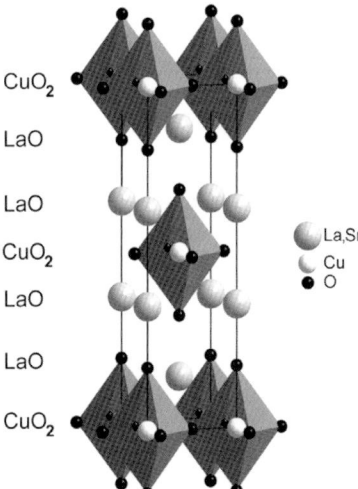

Figure 9.1. The structure of $(La_{2-x}Sr_x)CuO_4$. The atoms in each plane of one unit cell are shown at the left of the plane. For clarity, the O atoms are shown at about one third size relative to the other atoms.

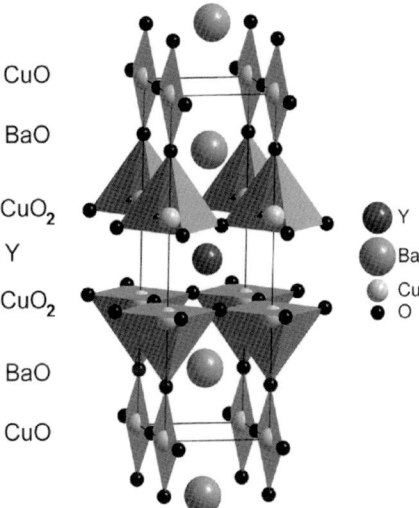

Figure 9.2. The structure of $YBa_2Cu_3O_7$. The atoms in each plane of one unit cell are shown at the left of the plane. Cu is present in the CuO "chains" and the CuO_2 "planes." For clarity, the O atoms are shown at about one third size relative to the other atoms.

two capping layers consisting of Sr and apical oxygens, two layers of Ca separating three CuO_2 planes, the central one of which is without apical O atoms. Table 9.1 gives abbreviations commonly used for some of the cuprate superconductors and maximum values of the critical temperature (T_c) for each series.

Figure 9.3 is a generic phase diagram, temperature vs. doping level, for the HTS. On the hole-doped side, at low temperatures, increasing the concentration of holes brings about a transition from an antiferromagnetic insulator to a metal. Superconductivity occurs in a

Table 9.1. Cuprate Superconductors: Nomenclature; T_c

Abbreviation	Formula	T_c (maximum) (K)
LMCO; M = C, S, B	$(La_{2-x}M_x)CuO_4$; M = Ca, Sr, Ba	$\sim 20, \sim 36, \sim 34$, respectively
YBCO	$YBa_2Cu_3O_{6+x}$; $0 \leq x \leq 1$	95 ($x = 0.89$)
RBCO	$RBa_2Cu_3O_{6+x}$; R = rare earth	~ 90 for most R
Bi-2223	$Bi_2Sr_2Ca_2Cu_3O_6{}^a$	110
BSCCO	Any member of the above series	
Tl-2223	$Tl_2Ba_2Ca_2Cu_3O_{10}{}^a$	125
TBCCO	Any member of the above series	
Hg-1223	$HgBa_2Ca_2Cu_3O_8{}^a$	133
HBCCO	Any member of the above series	

[a] The formula given is for the member of the series with the highest T_c. See the text for an explanation of the 4-digit labeling scheme.

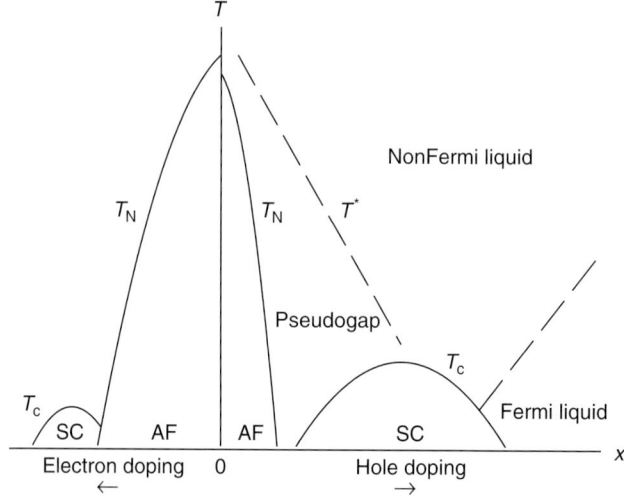

Figure 9.3. Generic phase diagram for the cuprate superconductors, showing antiferromagnetic (AF) and superconducting (SC) regions. See text for further explanation.

limited range of doping, with T_c increasing from zero in the "underdoped" region, peaking in the "optimally" doped region, and going to zero again in the "overdoped" region. The Neel temperature (T_N) decreases with increasing hole concentration; there is a region of reduced density of low-energy excitations, the "pseudogap" region, below the crossover line at the pseudogap temperature (T^*); regions of both Fermi-liquid and nonFermi-liquid behavior have been recognized. The region of "stripe" phases, with spin and charge ordering, which occurs in the underdoped range of hole concentration is not well defined and not shown. Related features occur on the electron-doped side of the phase diagram, which has been less thoroughly studied.

The values of many of the parameters that are important for characterizing the HTS and understanding their properties are generally not well known. The most reliable values are those for YBCO, which has been studied most intensively. Table 9.2 gives selected values for a number of these parameters for optimally doped samples of YBCO. It includes values for the 0-K upper, lower, and thermodynamic critical fields, $B_{c2}(0)$, $B_{c1}(0)$, and $B_c(0)$, respectively, the coherence length (ξ), the penetration depth (λ), and the Ginzberg–Landau parameter

Table 9.2. Characteristic parameters for optimally doped YBCO. The anisotropy ratio, $\gamma = \xi_{ab}/\xi_c = \lambda_c/\lambda_{ab} = B_{c1\perp}/B_{c1\parallel} = B_{c2\perp}/B_{c2\parallel}$, is approximately 3. $B_c(0)$ was calculated from $B_c(0) = [B_{c1}(0)B_{c2}(0)/\ln\kappa]^{1/2}$. (Note: $\parallel$ and $\perp$ refer to the c-axis.)

Parameter	Value	Reference
T_c(K)	95	[11]
γ_n(mJ K^{-2} mol^{-1})	16[a]	See footnote
$B_{c1\parallel}(0)$(G)	180	[12]
$B_{c1\perp}(0)$(G)	530	[12]
$B_{c2\parallel}(0)$(T)	40	[13]
$B_{c2\perp}(0)$(T)	110	[13]
$B_{c\parallel}(0)$(G)	5,800	[12–14]
$B_{c\perp}(0)$(G)	7,100	[12–14]
ξ_c (Å)	8	[14]
ξ_{ab} (Å)	25	[14]
λ_c (Å)	4,500	[12]
λ_{ab} (Å)	1,300	[12]
$\kappa_c = \lambda_c/\xi_c$	560	[12, 14]
$\kappa_{ab} = \lambda_{ab}/\xi_{ab}$	52	[12, 14]

See *Handbook of Superconductivity*, ed. C.P. Poole (Academic, London, 2000) for a comprehensive listing of parameters for the HTS.
[a] An average value of γ_n from: Ref. 3 [16]; Ref. 15 [15]; Ref. 16 [18].

($\kappa = \lambda/\xi$), for directions $\parallel$ c-axis and in the ab plane, i.e., $\perp$ c-axis. (Note that the symbol λ is also used for the electron–phonon interaction parameter, in "Specific Heat: Component Contributions; Field and Temperature Dependences; Nomenclature" in Section 9.1.4.) The values of these parameters are reasonably representative of those reported for the other HTS. The short coherence length has a direct effect on the properties of the vortex cores, including their contribution to the specific heat in the mixed state, and the resulting values of κ make the HTS extreme type-II superconductors.

Although ξ and λ are functions of the temperature, in Table 9.2 and throughout this chapter, we use the symbols ξ and λ for the low-temperature ($T \ll T_c$) values, which are often represented by $\xi(0)$ or ξ_0 and $\lambda(0)$ or λ_0. It is primarily the low-temperature values that are of interest here. At low temperatures, in the clean limit, the BCS, Ginzberg–Landau, and Pippard coherence lengths are essentially the same.

9.1.3. Magnetic Properties; Critical-Field Measurements

Measurements of the magnetic moment (M) as a function of temperature and magnetic field (B) are a primary source of information on B_{c1} and B_{c2}, the closely related quantities ξ and λ, and the anisotropies of all of these parameters. In addition to being of practical importance, knowledge of the values of these parameters is essential to an understanding of the HTS at a microscopic level. The earliest measurements of the magnetization [17, 18] revealed the irreversibility line in the B–T phase diagram, which is associated with the "melting" of the vortex lattice (see Section 9.6), a phenomenon that has not been observed in conventional superconductors. They also gave indications of the time-dependent pinning effects and the difference between measurements on field-cooled and zero-field-cooled samples that complicate measurements of B_{c2}.

There are substantial difficulties in determining both B_{c1} and B_{c2} from measurements of M: In principle, B_{c1} appears as a small change in slope of M vs. B. Its detection would at least require very high precision in the measurements, but there are additional problems of rounding of the feature by pinning, sample-shape effects, and inhomogeneity of the sample. The measurement of the temperature derivative of B_{c2} at T_c and extrapolation to $T = 0$ by the WHH relation [19] does not give reliable values of $B_{c2}(0)$. Consequently, determination of $B_{c2}(T)$ over the temperature range of interest requires measurements in very high fields, obtainable only with pulsed techniques with their attendant problems (see, e.g., [13]). There are also the usual problems with nonequilibrium effects. An early review of the magnetic properties of the HTS [6] describes these problems, and emphasizes the uncertainties in the derived parameters.

9.1.4. Specific-Heat Measurements

Specific Heat: Component Contributions; Field and Temperature Dependences; Nomenclature

There are four contributions to the specific heat (C): the lattice, or phonon, contribution (C_{lat}); the conduction-electron contribution (C_e); a "magnetic" contribution (C_{mag}), associated with paramagnetic centers; a hyperfine contribution (C_{hyp}), associated with nuclear moments:

$$C = C_{lat} + C_e + C_{mag} + C_{hyp}. \qquad (9.1)$$

In general, but with the exception of C_{lat}, which is assumed to be magnetic-field independent, the values depend on B, and the value of B is introduced in parentheses when appropriate. For C_e an additional subscript, s, m, or n, is added, when useful, to distinguish values in the superconducting, mixed, and normal states, respectively.

At low temperatures, and in the harmonic-lattice approximation, C_{lat} can be expanded in a series in odd powers of T,

$$C_{lat} = B_3 T^3 + B_5 T^5 + B_7 T^7 + \cdots. \qquad (9.2)$$

The 0-K Debye temperature, $\Theta_0 = (12\pi^4 R/5B_3)^{1/3}$, is usually calculated using the value of B_3 for 1 g atom because C_{lat} approaches the Dulong–Petit limit for that amount of material, but that value does not have the simple relation to the low-frequency acoustic phonons implied by the Debye model. At high temperatures, in the same approximation, C_{lat} can be expanded in a series in T^{-2}. Dilatation and anharmonic effects produce additional, T-proportional terms in C_{lat}, which are important at high temperatures.

In the normal state, the conduction-electron specific heat is usually approximated by a "linear" (T-proportional) term,

$$C_{en} = \gamma_n T, \qquad (9.3)$$

with γ_n a constant that is proportional to the electron density of states (EDOS) at the Fermi level. However, γ_n includes a factor, $1 + \lambda$, that represents an enhancement of the EDOS by the electron–phonon interaction (λ). In principle, λ, and therefore γ_n, is temperature dependent. With increasing temperature, λ is expected to increase from its 0-K value, go through a shallow maximum, and then drop to low values, with the changes taking place at temperatures related to the relevant phonon frequencies. There is no really satisfying experimental evidence for the associated temperature dependence of γ_n because of the difficulty in separating it at high temperatures from the T-proportional contributions arising from anharmonic effects and dilation in C_{lat}. Although the theoretically predicted T dependence of γ_n is generally believed to be real, it has been ignored in essentially all analyses of specific-heat data on the HTS. This is probably not a serious error at temperatures below $\sim 10\,\text{K}$, but the difference between the high- and low-T values are probably important in relation to measurements that extend over wide temperature intervals, such as those that show evidence of the pseudogap. For example, for YBCO, λ may be of the order of 0.5 at 10 K and below, but less than 10% of that in the vicinity of 100 K and above.

For a "conventional" BCS superconductor, with an isotropic energy gap, C_{es} is an approximately exponential function of temperature

$$C_{\text{es}} = a\gamma_n T_c \exp(-bT_c/T), \tag{9.4}$$

where a and b are weakly temperature dependent parameters. For the HTS, a number of recent measurements give a T^2 term in the low-T limit

$$C_{\text{es}} = \alpha T^2, \tag{9.5}$$

which is associated with line nodes in the energy gap. In addition, even in zero field, most samples show a "residual" linear term (C_{er}), a normal-state-like contribution to C

$$C_{\text{er}} = \gamma_r T. \tag{9.6}$$

The paramagnetic centers that contribute to C_{mag} are associated with chemical impurities, other defects, or, in the case of chemical substitutions on the Cu sites, the moments of the substituent ions or moments they induce on neighboring Cu sites. In zero applied field these moments are ordered by the internal interactions. For sufficiently low concentrations the order develops below 1 K, and in the vicinity of 1 K and above C_{mag} appears as an "upturn" in C/T, which can be represented by empirical expressions of the form

$$C_{\text{mag}}(0) = \Sigma A_n T^{-n}, \tag{9.7}$$

with $n = 2, 3, \ldots$. In applied fields of a few T or more, the paramagnetic centers order under the influence of the applied field, at temperatures above 1 K. If the concentration is sufficiently low the internal interactions are unimportant, and C_{mag} is well represented by a Schottky anomaly. There are examples in which the specific-heat anomaly corresponds, in both temperature and field dependence, to a two-level Schottky anomaly for the moments in the applied field,

$$C_{\text{Sch}}(B) = Rz^2 e^z/(e^z + 1)^2, \tag{9.8}$$

where $z = 2gS\mu_B B/k_B T$, S is the spin, and g is the spectroscopic splitting factor. In other cases the anomaly is broadened by the internal interactions, and appears as a "Schottky-like" anomaly.

The hyperfine contribution arises from the interaction of nuclear magnetic moments with magnetic fields or nuclear quadrupole moments with internal electric field gradients. For the HTS the relevant magnetic field is usually the external applied field, but in principle internal hyperfine fields produced by ordered electronic magnetic moments could contribute. For the temperature range of interest here, only the first-order term in the high-temperature expansion of a Schottky anomaly is important, and C_{hyp} is well approximated as

$$C_{\text{hyp}}(B) = D(B)/T^2. \tag{9.9}$$

In the absence of quadrupole moments and internal hyperfine fields, $D(B) \propto B^2$ with B the applied field, and $D(B)/B^2$ is determined by the nuclear moments. However, the full contribution to C is generally not observed, apparently because the nuclear-spin relaxation time (T_1) in parts of the sample is so long that those nuclei do not contribute. T_1 is expected to be long at low temperatures in the superconducting state, and the fraction of the theoretical value of the coefficient $D(B)$ that is found experimentally tends to be larger in samples with large values of γ_r, for which the "volume fraction of superconductivity" is smaller (see Section 9.2.1).

Specific Heat: Experimental Techniques

The specific heat has been measured by a number of techniques that differ with respect to precision and susceptibility to sources of error. The accuracy also depends on details of the implementation, e.g., the accuracy of the temperature scale. A display of the experimental data often gives a measure of the precision of the measurements. However, some techniques are more prone to errors associated with internal equilibrium times, which may not be obvious in the data. In cases in which this is a possibility, it is important to take the differences in the techniques into account when evaluating experimental data, particularly discrepancies between different results.

The heat-pulse technique is the most precise, and, other factors being equal, the most accurate. Ideally, the temperature of a thermally isolated sample is measured, a pulse of energy is introduced, and the temperature is remeasured after equilibrium is attained. An important advantage of this method is that problems with internal equilibrium times are evident, and can be taken into account. The most serious disadvantage is that large samples are required: measurements have been made only on large polycrystalline samples, which are more likely to be inhomogeneous, than small single crystals.

In the continuous-heating method, the sample is heated continuously with a known power, a correction is made for the "background" heat leak, and the specific heat is calculated from the time derivative of the temperature. The heat capacity is obtained as an essentially continuous function of temperature, but errors associated with internal equilibrium times, which are most likely at low temperatures, may go undetected. There are several ingenious adaptations of this technique to differential measurements. One [20], which gives the difference in the specific heats of a sample and a reference sample with high precision, has been used to separate C_e from C_{lat}. In another [21], which has been used to study vortex-lattice melting in YBCO, the sample and the reference sample are loosely coupled to a temperature controlled bath that is heated continuously. The temperature difference between the sample and the reference gives the latent heat at a first-order transition with high precision.

Most specific-heat measurements on small samples at low temperatures have been made by relaxation calorimetry [22]: The sample is connected to a precisely controlled, constant-temperature bath by a weak thermal link. Starting from the bath temperature, the sample is heated at constant power until equilibrium is obtained at a temperature above the bath temperature, the power is switched off, and the sample returns to the bath temperature. The specific heat is calculated from the time derivative of the temperature and the conductance of the thermal link, which is measured separately. Usually the increase in sample temperature is of the order of $T/20$ or less, and the time dependence of the temperature is a simple exponential, but in measurements by the Geneva group [23], the temperature is approximately doubled, and a point-by-point analysis of the temperature/time data is necessary. The major shortcoming of this technique is error associated with internal relaxation times in the vicinity of a few kelvin and below. The problem is illustrated by data on a YBCO sample with magnetic impurities [24], and was subsequently noted by the same group in connection with measurements on another material [23]. The YBCO data cannot be fitted by the usual expressions for a specific heat that includes contributions from magnetic impurities [25], and was originally interpreted in terms of a concentration of magnetic impurities that increased with increasing field [24]. Below 5 K or so, the region in which magnetic impurities contribute to the specific heat, internal relaxation times tend to increase with decreasing temperature. Depending on the time constant in the measuring system, the results may give underestimates of the concentration of magnetic impurities and errors in the other contributions to the specific heat. The effect of long nuclear spin relaxation time is usually different: The full value of C_{hyp} is rarely observed, even in heat-pulse measurements, but T_1 for the nuclei that do not contribute is evidently so long relative to the time constant of the measurements that the derived values for the other contributions to C are usually not affected. However, in one exceptional case [15] a small effect was noticed.

The physical arrangement for ac calorimetry [26] is essentially the same as for relaxation calorimetry, but the sample is heated continuously by ac power. With an appropriate choice of frequency, which is determined by the sample-to-bath and internal relaxation times, there is an ac response in the sample temperature that is proportional to $1/C$. As applied to measurements on HTS, the method usually gives only relative values of C. However, it can give exceptionally high precision, and has been used very successfully to study transitions in the vortex lattice.

Specific Heat: Problems and Uncertainties in Analysis of Data

The high values of B_{c2} for the HTS preclude low-temperature measurements in the normal state. The clean separation of C_{lat} and C_{en}, and the unambiguous determination of γ_n that are possible for many conventional superconductors, cannot be made. The reported values of γ_n are *estimates*, based on various indirect methods. Because T_c is so high $C_{lat} \gg C_{es}$ at T_c, and the specific-heat anomaly is only a few percent of C, as illustrated in Figure 9.4. The separation of C_{lat} and C_{es}, which is necessary, e.g., for identifying a fluctuation contribution, requires an arbitrary assumption about C_{lat} with a consequent uncertainty in C_{es}.

Sample purity is also a substantial problem. The three-component phase diagrams that govern the preparation of the HTS compounds ensure the existence of competing phases. A number of Y and Ba compounds are troublesome, particularly because YBCO is the most intensively studied HTS. Among them are several forms of $BaCuO_{2+x}$, one that contains paramagnetic centers, which contribute to the low-temperature upturn in C/T, and another that contributes to the residual linear term, $\gamma_r T$. The upturn, which is also illustrated in Figure 9.4, constitutes a significant obstacle to accurate determinations of the more interesting low-temperature contributions to C.

9.2. Low-Temperature Specific Heat

The contributions to the specific heat at low temperatures that are of most interest are $\gamma_r T$ and $C_e(B)$. The residual, zero-field linear term $\gamma_r T$, is largely an impurity effect, but there are theoretical predictions of an intrinsic contribution of this form. $C_e(B)$ gives evidence of the presence of line nodes in the energy gap. The problem of separating these contributions

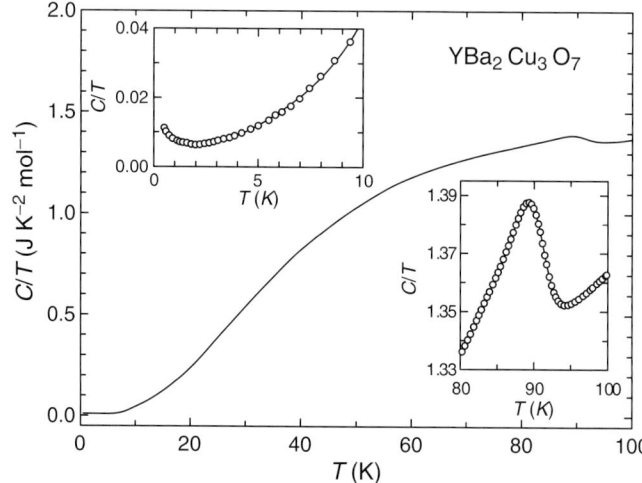

Figure 9.4. The specific heat of Polycrystalline a YBCO sample with a relatively low concentration of paramagnetic impurities. The insets show the low-T "upturn" associated with those impurities and the specific-heat anomaly at T_c. (Figures 9.4–9.9 show the results of measurements on the same sample, Wright et al. [27].)

from C_{lat}, C_{mag}, and C_{hyp} is illustrated in Figure 9.5, which shows all five contributions for a polycrystalline sample of YBCO that has a relatively low concentration of paramagnetic centers. Most measurements on YBCO have been made on samples with substantially larger C_{mag}, and the difficulty of separating $\gamma_r T$ and $C_e(B)$ from the other contributions is correspondingly greater. $C_e(B)$ and $\gamma_r T$ are significant fractions of C only in a narrow window of temperature between regions in which C is dominated by C_{hyp} and C_{mag} or C_{lat}. The separation of the components represented in Figure 9.5 was based on a simultaneous (global) fit to data for eight magnetic fields in the range 0.8–12 K [27]. Figure 9.6 shows the experimental data for the same sample, and gives an impression of the precision with which $C_e(B)$ is determined.

9.2.1. Zero-Field "Linear" Term

At the time of the earliest specific-heat measurements there was a high level of interest in the zero-field linear term because an early version of Anderson's RVB theory [28] predicted a specific-heat contribution of this form. More recently, a BCS to Bose–Einstein crossover model for the pairing in the underdoped, pseudogap region of the phase diagram [29] suggested the existence of a term with approximately this form. This theoretical result makes the linear term in underdoped samples of particular interest. However, impurity-related contributions to γ_r make the identification of any intrinsic effect extremely difficult.

There are a number of possible impurity-related contributions to γ_r: Pair breaking by magnetic scattering centers produces "gapless" superconductivity, with a nonzero γ_r, and accompanying reductions in both T_c and the discontinuity in specific heat at T_c, $\Delta C(T_c)$. Examples of this effect in HTS are described in Sections 9.3.2 and 9.3.3 Because essentially all samples contain paramagnetic centers, apparently associated with chemical impurities or other defects, a pair-breaking contribution to γ_r of this type is a general possibility. For a number of YBCO samples, substantial reductions in the values of $\Delta C(T_c)$ and other parameters that measure the magnitude of the specific-heat anomaly at T_c, are correlated with increases in γ_r, *but*

High-T_c Superconductors: Thermodynamic Properties

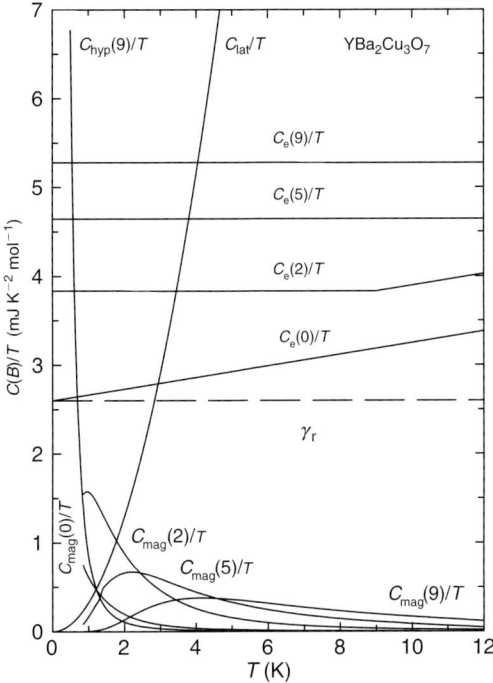

Figure 9.5. The five component contributions to the low-temperature specific heat of a polycrystalline YBCO sample with a relatively low concentration of paramagnetic centers. (Figures 9.4–9.9 show the results of measurements on the same sample, Wright et al. [27].)

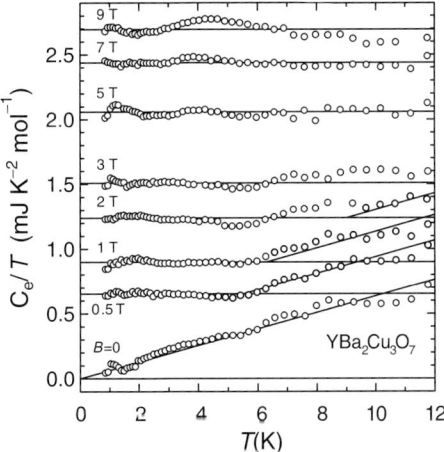

Figure 9.6. The conduction-electron contributions to the specific heat in the superconducting and mixed states. The lines represent $C_e(B)/T$ as derived from the fits; the points are experimental data points from which C_{hyp}, C_{mag}, C_{lat}, and $\gamma_r T$ have been subtracted. (Figures 9.4–9.9 show the results of measurements on the same sample, Wright et al. [27].)

with no reduction in T_c [30]. This suggests a different mechanism for a contribution to γ_r: the suppression of superconductivity by impurities or other defects, and the existence of *normal material* in regions of the order of the coherence volume in size. The short coherence length is consistent with the local suppression of superconductivity by defects on the scale of a lattice parameter. The correlations among these properties were described in terms of the "volume fraction of superconductivity." Measurements of the superfluid density by µSR techniques also suggested a microscopic phase separation into normal and superconducting regions, and the term "Swiss cheese" was suggested [31]. Both the Swiss cheese and pair-breaking mechanisms produce a nonzero γ_r, but the microscopic interpretations, and the effects on T_c, are different. Apparently, both are at work in the HTS. Nonmagnetic scattering centers are also pair breaking for a d-wave order parameter, and, in the case of resonant scattering, their effect on T_c is relatively weak [32]. Resonant scattering by nonmagnetic impurities produces a residual EDOS that can be recognized in the T dependence of the penetration depth [32], and which could be another contribution to γ_r that is not associated with substantial reductions in T_c. Ba-containing impurity phases can also contribute to γ_r (see, e.g., [3] and references therein), a particular problem in the case of YBCO, even in the case of "single crystals."

The difficulty in separating these impurity-related contributions to γ_r from intrinsic effects is illustrated by the differing conclusions reached by the authors of this review in collaborations with other colleagues: The conclusions were based on correlations of the concentrations of paramagnetic centers in the superconducting material and in impurity phases with component contributions to γ_r for a series of YBCO samples. For the original series it was concluded that there was no measurable contribution to γ_r that was not associated with paramagnetic centers [33]; however, with the inclusion of several additional samples, and the omission of two Zn-substituted samples, it was concluded [30] that there was an intrinsic contribution of $2\,\mathrm{mJ\,K^{-2}mol^{-1}}$. While there is certainly a general increase in γ_r with increasing concentrations of paramagnetic centers, more recent information on the nature of paramagnetic centers in YBCO [11] shows that the separation of paramagnetic centers into the two types was an over simplification, and that neither analysis was valid. For one polycrystalline YBCO sample with a particularly low concentration of paramagnetic centers (the sample represented in Figures 9.4–9.9) $\gamma_r \sim 3\,\mathrm{mJ\,K^{-2}\,mol^{-1}}$, for both optimally and slightly overdoped O concentrations [27]. For the same sample slightly underdoped, $\gamma_r \sim 4\,\mathrm{mJ\,K^{-2}\,mol^{-1}}$, but the concentration of spin-1/2 paramagnetic centers is also higher, so the increase in γ_r is not necessarily an intrinsic effect. Detwinning an optimally doped YBCO single crystal reduced γ_r from 3.2 to $2.2\,\mathrm{mJ\,K^{-2}\,mol^{-1}}$, and, in contrast with the results for a polycrystalline sample mentioned above, increasing the O content into the overdoped region, reduced γ_r to $1.2\,\mathrm{mJ\,K^{-2}\,mol^{-1}}$ [34]. For another high quality, overdoped YBCO single crystal, $\gamma_r = 2.4\,\mathrm{mJ\,K^{-2}\,mol^{-1}}$ [15]. For YBCO, the lowest values of γ_r reported are of the order of $2\,\mathrm{mJ\,K^{-2}\,mol^{-1}}$, and the effect of doping level is not clear.

For LSCO the values of γ_r are generally lower than for YBCO, no doubt in part due to the absence of a contribution from the Ba compounds present as impurities in the YBCO samples. Nevertheless, zero values have not been reported. For two optimally doped ($x = 0.15$), polycrystalline (La$_{2-x}$Sr$_x$)CuO$_4$ samples, the values of γ_r, 0.44 and $1.23\,\mathrm{mJ\,K^{-2}\,mol^{-1}}$, are among the lowest reported [35]. For a series of single-crystal samples, $\gamma_r \sim 2\,\mathrm{mJ\,K^{-2}\,mol^{-1}}$, for the *overdoped* and *optimally* doped samples, but rises sharply in the *underdoped* region, to $\sim 4\,\mathrm{mJ\,K^{-2}\,mol^{-1}}$ for $x = 0.069$, with no evidence of an impurity effect that might account for the increase [36].

For BSCCO, C_{lat} is substantially greater than for YBCO or LSCO, and for many samples C_{mag}, the low-temperature upturns in zero field, are also large. These contributions to C

complicate the determination of the linear term. When they are properly taken into account, however, the preponderance of measurements give $\gamma_r = 0$, to within the experimental uncertainty (see, e.g., [3]). As for LSCO, the absence of Ba compounds as impurities is undoubtedly a factor that favors low values of γ_r, but it does not explain why zero values are found for BSCCO and not for LSCO. Most of these measurements have been made on optimally doped samples, but $\gamma_r = 0$ has been reported [37] for one underdoped sample of Bi-2201, with $T_c = 12.5$ K.

There have been relatively few specific-heat measurements on TBCCO. Most of them give high values of γ_r, presumably associated with Ba-containing impurities, but zero values have been reported [38–40].

In general, there is no persuasive evidence for an intrinsic linear term in the HTS. The nonzero values of γ_r are probably associated with impurities or imperfections of one kind or another. The one possible exception, which is of particular interest because it occurs in the underdoped region, is for Bi-2201 [37].

9.2.2. Evidence for Line Nodes in the Energy Gap

It is widely believed that the superconducting-state electron pairing in the cuprates is d-wave, but this conclusion has been questioned (see, e.g., [41–45]). Specific-heat measurements have given evidence of the line nodes in the energy gap that are expected for d-wave pairing, and support that scenario. However, the specific-heat, and other measurements of the EDOS such as the Knight shift and the nuclear-spin relaxation time, do not distinguish between a d-wave order parameter, which changes sign, and extreme anisotropy of the energy gap that might have a different origin. The uncertainty about the origin of the line nodes notwithstanding, the specific-heat results are of interest because the specific heat is a bulk property, while the measurements that give the phase of the order parameter may be affected by abnormalities associated with the surfaces (see, e.g., [43]).

For a "fully gapped" superconductor, i.e., with an essentially isotropic gap, C_{es} is given by the exponential BCS expression, Eq. (9.4). For line nodes in the gap, it is expected on quite general grounds that $C_{es} = \alpha T^2$ in the low-T limit, with α determined by the shape of the node, and $\alpha \sim \gamma_n/T_c$. For a "conventional" type-II superconductor, normal-state-like excitations within the vortex cores give a mixed-state specific heat $C_{em}(B) \propto BT$, where the proportionality to B reflects the number of vortices [46, 47]. For the cuprates, the extremely small value of ξ increases the quantum confinement energy of quasiparticles in the vortex cores, as ξ^{-2}, making this contribution to the EDOS and to C_{em} negligible. However, in the presence of line nodes, a Doppler shift of the quasiparticle spectrum in the outer regions of the vortices gives rise to a $B^{1/2}T$ contribution to the EDOS, and $C_{em}(B) = \beta B^{1/2}T$, where $\beta = k\gamma_n B_{c2}^{-1/2}$ and k is an undetermined parameter of order unity [48]. More generally, in the case of line nodes and very short ξ, there is a crossover between regions of different B and T dependences at the value $z_c \sim B_{c2}^{-1/2}T_c$ of the parameter $z \equiv B^{-1/2}T$. In the limit $B = 0$, $C_{es} = \alpha T^2$; for intermediate B, $C_{em}(B)$ is the sum of a B-independent T^2 term and a T-independent B-proportional term; for high B, $z < z_c$, $C_{em}(B) = AB^{1/2}T$ [49]. These relations are consistent with a scaling relation derived for a d-wave superconductor, $C_{em}(B)/B^{1/2}T = F(B^{-1/2}T)$, where F is an undetermined scaling function [50–52].

The αT^2 and $\beta B^{1/2}T$ terms were first identified in a Stanford/UBC collaboration [53] in data obtained on a single crystal of YBCO. The identification was based on a "global" fit, in which data for six fields, 0–8 T, temperatures from 2.5 to 7 K, were fitted simultaneously, with C_{lat} constrained to be independent of B, and with specific B and T dependences assumed for C_{mag}. Data obtained at LBNL on a polycrystalline YBCO sample, 0.4–10 K, 0–7 T, when fitted without constraints on the B and T dependences of C_{mag}, but otherwise in the same way, gave similar results [54]. However, in both cases fitting the zero-field data alone gave negative values of α, leaving a question about the reality of the T^2 term [55]. The problem is illustrated in Figure 9.5, which emphasizes the narrow range of temperature in which the T^2 term is a substantial fraction of the total. Evidently, the small magnitude of the T^2 term, the interdependence of the six or so parameters required to fit the zero-field data, and the approximate nature of the fitting expression for $C_{mag}(0)$ conspire to give a spurious value of α: using the in-field data to fix the B-independent parameters in C_{lat} and γ_r gives a more nearly correct value. Later measurements on a YBCO sample with a lower concentration of paramagnetic impurities gave somewhat more persuasive evidence of the T^2 term in the zero-field data [27]. The T^2 term for that sample is shown as the $B = 0$ data in Figure 9.6. However, perhaps the most persuasive evidence of the reality of the T^2 term is in LSCO data [35]. The T^2 term has also been reported by other groups in both YBCO [15] and LSCO [56–58]. The substantially different values of α reported are probably more a reflection of various uncertainties in the data and the analyses than systematic trends with, e.g., doping.

The $\beta B^{1/2} T$ term has also been identified in experimental data for both YBCO (see, e.g., [15, 27, 34]) and LSCO (see, e.g., [35, 36, 56]). In one paper the Geneva group presented results for YBCO that contradicted the earlier reports of a $B^{1/2} T$ dependence, which they suggested were spurious results associated with paramagnetic centers [59]. However, in a later paper [15] they reported data on an exceptionally high-quality single crystal that did show an approximate $B^{1/2} T$ dependence, and called attention to an error in the earlier paper that explained the discrepancy. The results for a YBCO sample are illustrated in Figures 9.6 and 9.7: Figure 9.6 shows the determination of the $\gamma(B)$ term in $C_{em}(B)$ and the crossover between the T^2 and $B^{1/2} T$ dependences; Figure 9.7 shows the $B^{1/2}$ dependence of $\gamma(B)$. Since B_{c2} is anisotropic, β depends on the direction of the field, and this has to be taken into account in comparisons among values of β. For a polycrystalline sample the effective mass model can be used to calculate the appropriate average of B_{c2} (see, e.g., [54]). Given the substantial uncertainties in k, γ_n, and the values of B_{c2}, most of the experimental values of β are in satisfactory agreement with theory.

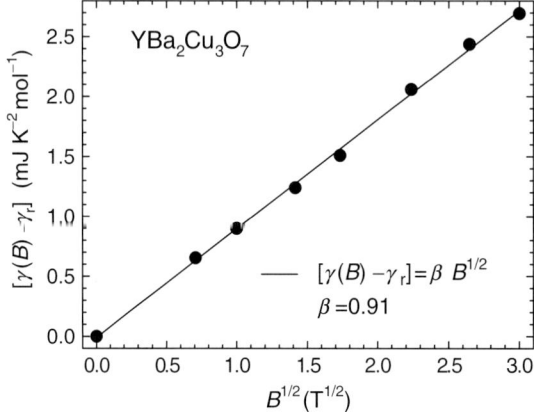

Figure 9.7. The $B^{1/2}$ dependence of $\gamma(B)$. (Figures 9.4–9.9 show the results of measurements on the same sample, Wright et al. [27].)

The suggestion that the $\beta B^{1/2}T$ term is unique to the HTS and not found in conventional superconductors has been questioned on the basis of V_3Si data [60] that show a similar negative curvature of $\gamma(B)$. However, those measurements were made on a zero-field-cooled sample, and at fields near B_{c1}. The observed negative curvature is mainly a consequence of nonequilibrium effects associated with pinning as the flux enters the sample, but some of it may arise from vortex–vortex interactions which are expected to make a contribution to $C_{em}(B)$ for fields near B_{c1} [47]. Since the measurements on the HTS that give the $\beta B^{1/2}T$ term are made on field-cooled samples, and at fields substantially greater than B_{c1}, the V_3Si data are not really relevant. Nevertheless, this suggestion that the usual interpretation of the $B^{1/2}T$ in the HTS may not be valid has been widely cited. Furthermore, the B dependence of $C_{em}(B)$ for conventional type-II superconductors has not been well defined, in part because many of the relevant measurements have been made on A_{15} compounds in which irreversible effects are important, and the measurements have been made in fields not much greater than B_{c1}. In that context, recent measurements on HfV_2 [61], which are free of both of those shortcomings, and which show $\gamma(B) \propto B$, to 14 T, are relevant.

Data that correspond to the $\beta B^{1/2}T$ term necessarily satisfy the predicted scaling law, *in the low-T, high-B, $z = 0$ limit*. Such data have frequently been plotted as $C_{em}(B)/B^{1/2}T$ vs. $z = B^{-1/2}T$ to demonstrate the validity of the scaling law, but in most cases they do not extend beyond the crossover at z_c, and do not constitute a general test of the scaling relation. Data that do extend from close to the $z = 0$ limit to beyond z_c, albeit with limited precision, are shown in Figure 9.8. With a reasonable allowance for the uncertainty in $C_{em}(B)$, they suggest that the scaling relation is valid up to $z \sim 1.5z_c$. Figure 9.9 shows the same data plotted in a different way that illustrates the interpolation of the scaling relation to the other limit, $z = \infty$.

In principle, the dependence of $C_{em}(B)$ on the direction of a field applied in the *ab* plane would determine the location of the gap nodes, which could be compared with expectations for a d-wave order parameter [62]. However, the anisotropy, which may be small, has not been detected in two measurements on YBCO that were designed for the purpose [15, 34]. The negative results may be associated with the orthorhombic structure and twinning.

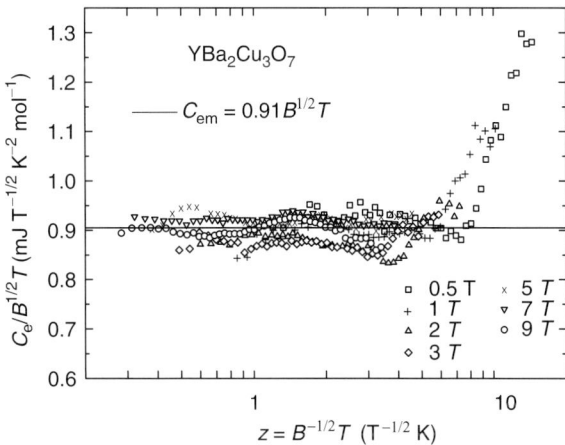

Figure 9.8. A test of the scaling relation. (Figures 9.4–9.9 show the results of measurements on the same sample, Wright et al. [27].)

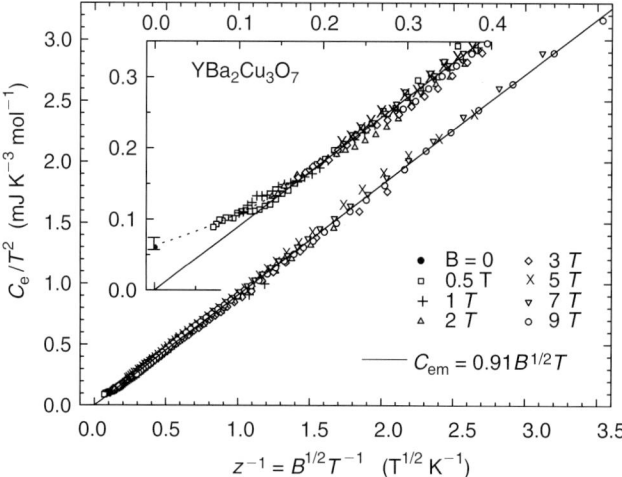

Figure 9.9. A test of the scaling relation that emphasizes the high-z points. (Figures 9.4–9.9 show the results of measurements on the same sample, Wright et al. [27].)

Overall, the specific-heat data are in quite satisfactory agreement with theoretical expectations for line nodes in the energy gap and a short coherence length: The field dependence of the mixed-state specific heat is qualitatively different from that of conventional superconductors, and the superconducting-state specific heat shows a continuum of low-energy excitations. More precise tests of the scaling relation in the $z > z_c$ region, and more conclusive searches for the anisotropy of $C_{em}(B)$ with the field in the ab plane would be of interest.

9.3. Chemical Substitutions

The defect-perovskite structures of the HTS tolerate a variety of elemental substitutions. The effects of these substitutions furnish information about the nature and origin of the superconductivity. The substitutions naturally fall into two groups: substitutions on the La or Y site; substitutions on the Cu sites. There are useful reviews of rare-earth substitutions on the La and Y sites by Markert, Dalichaouch, and Maple [63], Gschneidner, Eyring, and Maple [64], and one dedicated to Pr substitutions by Radousky [65]; and of substitutions on the Cu sites by Green and Bagley [5]. These reviews include references to early work, which are not always repeated here. In the following sections we examine the consequences of elemental substitutions, with emphasis on their effects on magnetic properties and the specific heat. Rare-earth substitutions on the Y and La sites are reviewed in Section 9.3.1, and an overview of the effects of substitutions on the Cu sites is given in Section 9.3.2. Zn substitutions on the Cu sites, which have a particularly dramatic effect in suppressing the superconductivity, have been studied intensively, and some of this work is considered in more detail in Section 9.3.3. Following the identification of stripe phases in LBCO and (La,Nd)SCO there was renewed interest in substitutions for Cu in these materials as an approach to understanding the stripes, and some of the relevant specific-heat measurements are discussed in Section 9.4.

9.3.1. Rare-Earth Substitutions on the Y and La Sites

Lanthanum and all the rare earths except Ce, Pr, and Tb (and radioactive Pm, with a half life of 19 h, which has not been investigated) can be fully substituted for Y in YBCO with retention of the orthorhombic structure, bulk superconductivity with $T_c \sim 90$ K, and similar specific-heat anomalies in the vicinity of T_c. In the case of the larger ions, and in particular La, there is a tendency to substitute for Ba^{2+} because of a close match of ionic radii, and special synthetic procedures are required (see, e.g., [66]). Under the strongly oxidizing conditions normally used for synthesis, Ce and Tb assume their 4+ valence states and different, nonsuperconducting structures are the stable form. However, a laser-ablation procedure [67] was devised that makes possible the synthesis of films with the partial replacement of Y by Ce (30%) or Tb (50%) without phase separation. The partially substituted $(Y_{1-y}Tb_y)$BCO is also a superconductor with $T_c \sim 90$ K. Since magnetic moments usually suppress the transition to a spin-singlet superconducting state, and the electron pairing in the HTS is generally believed to be spin singlet, the lack of an effect of these substitutions was unexpected. Evidently, the interaction of the highly localized 4f electrons with the adjacent CuO_2 planes is too weak to have a significant pair-breaking effect.

When substituted on the Y site in YBCO, the same R^{3+} ions exhibit their usual magnetic moments, crystal electric field (CEF) splitting, and cooperative magnetic ordering. The observed ordering temperatures vary from 0.17 K for HoBCO to 2.25 K for GdBCO. Neutron diffraction shows that Nd, Sm, Gd, Dy, and Er order antiferromagnetically while the others, with no magnetic ordering for $T \geq 0.45$ K, have specific-heat contributions that are related to the CEF splitting. Not surprisingly, the specific heat associated with the antiferromagnetic ordering is well fitted by an anisotropic two-dimensional Ising model. While there is little effect of the substitutions on the superconductivity, there is an indication that the antiferromagnetic ordering involves a weak coupling to the superconducting electron system on the adjacent CuO_2 planes: The Néel temperatures (T_N), which are of a magnitude associated with dipole-dipole coupling, are influenced by oxygen stoichiometry.

Fully substituted PrBCO, with the Pr on the Y sites, is generally thought to be an insulator. Although there are reports of single-crystal PrBCO with 90-K superconductivity [68, 69] it has been suggested that in such cases some of the Pr is on the Ba sites [70]. For the partially substituted $(Y_{1-y}Pr_y)$BCO [65] and $(Y_{1-y}Ce_y)$BCO [67], T_c decreases with increasing y. This is qualitatively similar to what is expected for magnetic pair breaking, but with important differences. Among the mechanisms that have been suggested to explain the behavior of Pr substituted YBCO the three that have received the most attention are (1) hole filling in the CuO_2 planes caused by the presence of 4+ rare-earth ions [65]; (2) hybridization of Pr 4f–O 2p orbitals [71–73]; (3) replacement of Ba by Pr that causes a localization of holes in the CuO_2 planes [69]. However, there is a report of specific heats for $(Y_{1-y}Pr_y)$BCO that show a depression of T_c with no x-ray evidence of Pr in the BaO layers [65]. Similar mechanisms have been suggested to explain the T_c depressions for the Ce substitutions in YBCO [67].

Rare-earth ions Pr, Nd, Sm, Eu, and Gd, which have ionic radii not too different from that of La, can be *partially* substituted on the La site in LMCO with retention of the structure and superconductivity. (Effects of Nd substitution in LSCO that are related to structural transitions and stripe formation, and consequent changes in T_c, are particularly unusual and are covered in Section 9.4.) The effect of these substitutions is strikingly different from that of substitutions for Y in YBCO. T_c for $[(La_{2-y-x}R_y)M_x]CuO_4$ decreases with increasing y and increasing atomic number. The effect on T_c of pair breaking by magnetic moments in conventional superconductors is correlated with the size of the magnetic moment [74]. No such correlation is found for R substitutions in (La,R)MCO, and the effect is much smaller than would be expected on the basis of comparisons with conventional superconductors. There is a correlation with the unit-cell volume and with the Cu–O bond length in the CuO_2 planes, which decrease regularly with increasing rare-earth atomic number, essentially as expected on the basis of the ionic radii, but the correlation is opposite in sign to that expected from the positive pressure dependence of T_c. However, the effects on T_c, from the substitution of R for La, can be understood completely on the basis of variations in carrier concentrations associated with changes in the basicity of the rare-earth ions. An extensive investigation of phase stability, T_c, and carrier concentrations in $[(La_{2-y-x}R_y)Sr_x]CuO_4$ provides the basis for this explanation [75].

The effect of Pr substitution in LMCO is exceptional, as it is in YBCO, but the nature of the exceptional behavior is different. Up to the solubility limit, $y \sim 0.8$, there is essentially no reduction in charge carrier concentration or T_c, which undoubtedly reflects the small mismatch in ionic sizes [71]. This result supports the theory that the suppression of T_c observed in $(Y_{1-y}Pr_y)Ba_2Cu_3O_7$ is caused by hybridization of Pr 4f–O–2p electrons with the conduction band in the CuO_2 planes—because of a coincidental match in energies—that leads to a reduction in charge-carrier concentration [71, 76]. For $[(La_{1.85-y}Pr_y)Sr_{0.15}]CuO_4$ there is no such energy match and no suppression of T_c occurs.

9.3.2. General Effects of Substitutions on the Cu Sites

The substitution of other elements (A) for Cu in both $YBa_2(Cu_{1-y}A_y)_3O_{6+x}$ and $(La_{2-x}M_x)(Cu_{1-y}A_y)O_4$ can have pronounced effects on the superconductivity. These effects on the superconductivity are in marked contrast to those of substitutions for Y in YBCO, which, in general, are small. The larger effects are understandable since the substitutions are on the sites most directly associated with the superconductivity (the CuO_2 planes), or, for YBCO, those that constitute the charge reservoir (the CuO chains), and any disruption of the Cu–O bonding in the planes and/or the oxygen content in the chains would be expected to affect the superconductivity.

For YBCO the interpretation of the effects of substitutions is complicated by the existence of two Cu sites that have different roles in the superconductivity, and there is disagreement in the placement and amount of the substituents incorporated on the lattice (see, e.g., [5]). All of the substituents for Cu have a limited solubility range, but these limits have generally not been precisely defined. In all probability, the synthesis conditions play an important role in the locations and concentrations of the substituted elements. For YBCO, elements with a 2+ valence are generally thought to substitute predominantly on the CuO_2 planes and those with a 3+ valence mainly on the chain sites. However, use of differential anomalous x-ray scattering [77] shows that, *for the samples studied*, Ni and Zn are nearly randomly distributed over chain and plane sites while Fe and Co substitute preferentially on the chains for low y, but for higher y there is some substitution on the planes. It has been reported [78] that for A = Ni, Zn, Fe, Co, and Al the orthorhombic distortion in YBCO decreases with increasing y; the oxygen content changes very little for small y, but exceeds 7 when y is large; and T_c decreases for increasing y.

Measurements of both magnetization and specific heat have shown a number of effects of substitutions on the Cu sites that are common to YBCO and LMCO, and similar for different substituents. Magnetization studies [78–87] on Ti, V, Cr, Mn, Fe, Co, Ni, Zn, Ga, and Al substitutions have shown decreases in the Meissner fraction and T_c. An exception is the substitution of Au for Cu in YBCO that produces a small increase in T_c [88]. The measurements also show that substitution of either magnetic or nonmagnetic ions on the CuO$_2$ planes produces magnetic moments, as is evidenced by the appearance of a significant Curie term in the susceptibility. The nonmagnetic ions, e.g., Zn, produce magnetic moments on the neighboring Cu sites (see Section 9.3.3).

Specific-heat measurements at low temperatures have been reported for YBCO with substitutions of: Cr [89], Fe [90–95], Ni [96], Zn [89, 92, 97–100], Al [101], and Co [102]. Near T_c specific-heat measurements have been made for: Cr [89], Fe [95], Co [102], Ni [96], and Zn [89, 97, 98]. In general, the specific-heat measurements show, in addition to the reduction in T_c, a decrease in $\Delta C(T_c)$, and an increase in the residual, low-T, $\gamma_r T$ term. For any substituted ions on the plane sites, and for magnetic ions on the chain sites, there is an increase in C_{mag}, a low-T upturn in C/T for $B = 0$, and the appearance of Schottky-like anomalies for $B > 0$. Both the upturns and anomalies indicate the presence of paramagnetic centers. These effects, and the reduction in the Meissner fraction, can all be understood as being associated with the paramagnetic centers and other defects introduced by the substitution and a consequent reduction in the "volume fraction of superconductivity" or pair breaking (see Section 9.2.1).

The magnetic properties of overdoped samples of (La$_{2-x}$Sr$_x$)(Cu$_{1-y}$Zn$_y$)O$_4$ have been measured [103] using magnetic susceptibility and neutron scattering. They showed that, as in YBCO, each Zn induces magnetic moments in the CuO$_2$ planes corresponding to $1.2\mu_B$. They also found that for $y = 0$ and $x \geq 0.22$ each additional Sr^{2+} produces a paramagnetic moment of $0.5\mu_B$ that has a similar, but smaller, effect on T_c.

Magnetization and specific-heat measurements have been reported [104] on single-phase (from x-ray analysis), polycrystalline (Bi$_{1.8}$Pb$_{0.2}$)Sr$_2$Ca(Cu$_{1-y}$A$_y$)$_2$O$_{8+x}$ for $y = 0$ ($\gamma_r = 0$) and A = Zn ($y = 0.02, 0.04$) and Co ($y = 0.02, 0.03, 0.04, 0.06, 0.08$) from 2 K to well above T_c. For Co concentrations of 0.06 and 0.08 an impurity phase of Sr$_2$Bi$_2$CuO$_6$ is detected at a level of $\sim$6%. Substitution of Zn for Cu has no effect on the Curie constant above T_c although T_c decreases monotonically. While the addition of Zn decreases T_c, surprisingly, neither γ_r nor $\Delta C(T_c)/T_c$ are affected. In view of the effect of Zn substitution in YBCO and LMCO it is difficult to understand its behavior in (Bi,Pb)SCCO unless Zn is not substituting into the CuO$_2$ planes. Substitution of Co for Cu has a much greater effect than Zn. Above T_c the Curie constant increases (attributed to the paramagnetic Co ions), T_c shifts to lower temperatures, $\Delta C(T_c)/T_c$ is attenuated, and γ_r increases as y increases. For $y \geq 0.04$, $\gamma_r \sim 18$ mJ K^{-2} mol^{-1}, which is taken as γ_n in the normal state since the Curie constant continues to increase as y increases showing that a solubility limit has not been reached. These results are attributed to gapless superconductivity induced by the magnetic Co ion substitution for Cu. They also suggest that no CDW/SDW stripe phases have been destroyed.

The effect of Cr substitutions on C_{mag} is illustrated in Figure 9.10, which shows the 7-T specific-heat anomalies for three samples of YBa$_2$(Cu$_{1-y}$Cr$_y$)$_3$O$_7$, which evolved from the low-temperature upturns in zero field [89]. The samples were all prepared in the same way, with the amounts of Cr introduced corresponding to the nominal values of $y = 0$, 0.0050, and 0.0133, and intended to produce $3y$ moles Cr moments/mole sample. $C_{mag}(7)$ is well represented by $S = 1/2$, $g = 2$ Schottky anomalies for all three samples. However, the numbers of moles of magnetic moments/mole YBCO, as deduced from the fits, are, respectively, $n = 0.0035, 0.0075$, and 0.015. The 0.33% magnetic impurities in the "pure YBCO" sample were not unusual for the time, and illustrate the problems with the preparation of pure samples. The fact that n is substantially less than $3y$ for the other two samples shows that only a

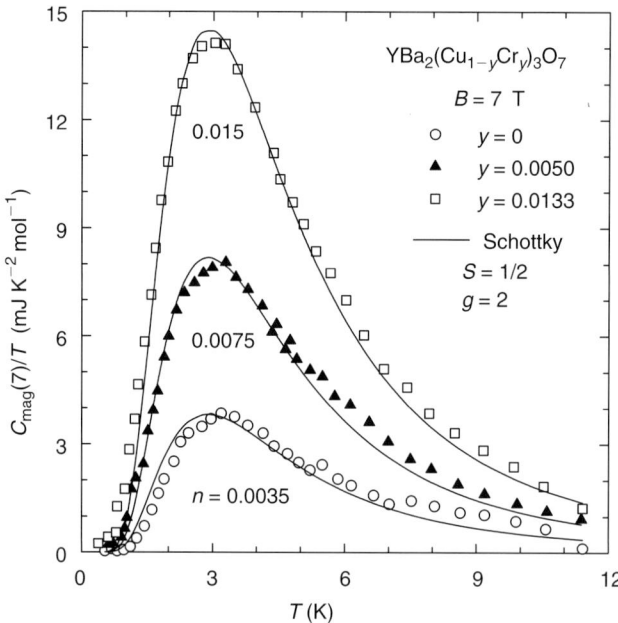

Figure 9.10. $C_{mag}(7)/T$ vs. T for $YBa_2(Cu_{1-y}Cr_y)_3O_7$, where y is the nominal Cr content (see text and Kim et al. [89]). Data for each y were fitted by two-level Schottky functions with spectroscopic splitting factor $g = 2$ and spin $S = 1/2$ to obtain the actual number of moles of paramagnetic centers/mole YBCO (n).

fraction of the Cr appeared in the $YBa_2(Cu_{1-y}Cr_y)_3O_7$ and illustrates both the problems with the preparation of substituted samples and the interpretation of the results.

9.3.3. Effects of Zn Substitution on the Cu Sites

Substitution of 10% nonmagnetic Zn for Cu on the CuO_2 planes depresses the superconductivity to $T_c < 3$ K [80], surprisingly, a much greater effect, ion-for-ion, than that of magnetic ions. However, susceptibility measurements [98] have shown that Zn substitution produces Cu^{2+} moments, and NMR measurements are interpreted as showing the presence of magnetic moments on the four Cu ions adjacent to the Zn ion [105–107]. In addition, magnetic-susceptibility and resistivity data have been interpreted [108] as showing that every Zn ion induces the formation of four $0.32\mu_B$ magnetic moments on the near-neighbor Cu sites, which are coupled ferromagnetically. Presumably the decrease in T_c is produced by pair breaking by potential scattering at the nonmagnetic Zn sites or by magnetic scattering at the magnetic Cu sites. Other studies on $YBa_2(Cu_{1-y}Cr_y)_3O_7$ have shown that for $y > 0.05$ there is incomplete incorporation of the Zn [109, 110], and that ZnO, Y_2BaCuO_5, and $Ba(Cu_{1-y}Zn_y)O_2$ are formed as impurities [109].

Figure 9.11 shows the specific heat for three $YBa_2(Cu_{1-y}Zn_y)_3O_7$ samples, $y = 0$, 0.01, and 0.05, as C/T vs. T [89]. The contributions associated with paramagnetic centers are the low-T upturns in C/T for $B = 0$, which become Schottky-like anomalies in 7 T. The small low-T upturns in the 7-T data are hyperfine contributions. For $B = 0$ the specific heats are fitted with good precision by Eqs. (9.2), (9.6), and (9.7), for C_{lat}, $C_{er} = \gamma_r T$, and $C_{mag}(0)$. Both γ_r and the zero-field, low-T upturns increase as y is increased. This is a case in which the interactions between the paramagnetic centers are sufficiently strong that $C_{mag}(7)$ cannot

High-T_c Superconductors: Thermodynamic Properties

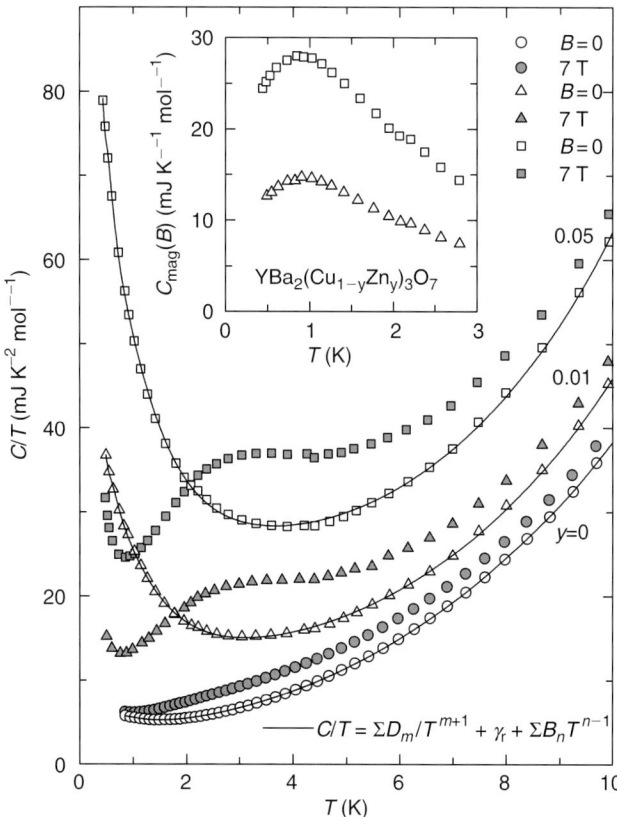

Figure 9.11. Main panel: C/T vs. T for YBa$_2$(Cu$_{1-y}$Zn$_y$)$_3$O$_7$ for $B = 0$ and 7 T (see Kim et al. [89]). The curves are fits to the data using the expression in the figure. In 7 T the $B = 0$ upturns become Schottky-like anomalies and the low-T upturn is the hyperfine component. Inset: C_{mag} vs. T showing the ordering anomalies for the Zn-induced Cu^{2+} moments centered at ~ 1 K.

be fitted with a simple Schottky anomaly, but has the form of a "Schottky-like" anomaly, as shown in the inset. These results, and this interpretation, are typical of data for YBCO samples with paramagnetic centers.

Zero-field specific-heat data on Zn-substituted YBCO by Loram et al. [97], which are similar to those of Kim et al. [89] and also to a third set of data by Roth et al. [100], extend to higher Zn concentrations. They were scanned from Figure 2 of [97] and are plotted together with the data of [89] as C/T vs. T^2 in Figure 9.12. With reasonable allowance for sample-to-sample differences, the two sets of data are in good agreement. The similarity of the data from [89] and [97] is emphasized in Figure 9.13, which shows the values of γ_r and Θ_D obtained by analysis of the two data sets. With increasing y the trend in Θ_D indicates a slight stiffening of the lattice while γ_r increases substantially. A surprising feature of Figure 9.13 is that γ_r increases linearly with y, reaching 40 mJ K^{-2} mol^{-1}, 2.5 times the estimated $\gamma_n = 16$ mJ K^{-2} mol^{-1} for pure YBCO given in Table 9.1, with no indication of saturation. (There are a number of other estimates of γ_n that fall in the range 14–20 mJ K^{-2} mol^{-1} (see, e.g., [3]). The implication is that the pair-breaking effect of the magnetic centers increases the EDOS to a value that exceeds that for the normal state.

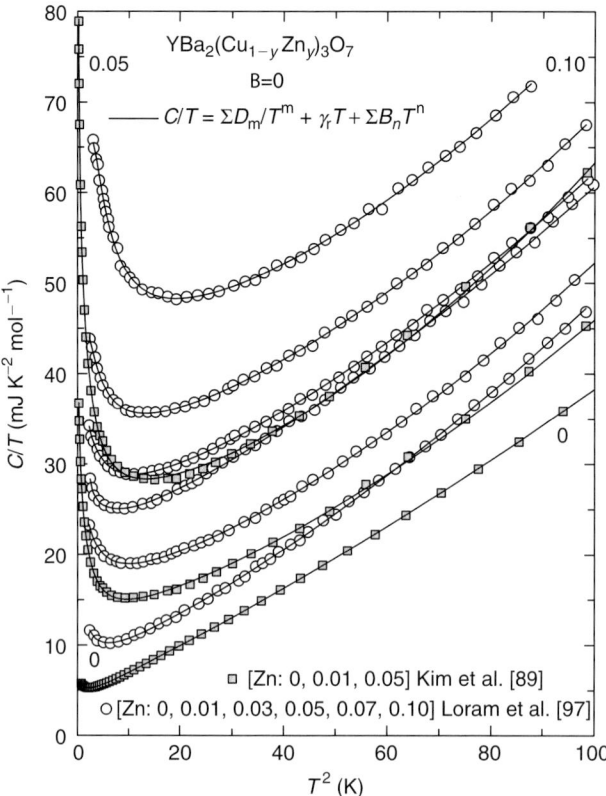

Figure 9.12. The zero-field data are from Figure 9.11 (see Kim et al. [89]), but plotted vs. T^2 and compared with the data of Loram et al. [97]. The two data sets are very similar, particularly for $y = 0.05$, and both are well fitted for all y by the expression given in the figure.

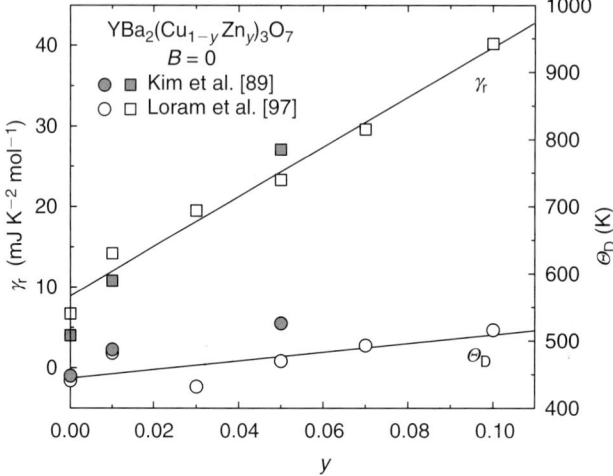

Figure 9.13. The values of γ_r and Θ_D derived from the fits to the data in Figures 9.11 and 9.12, and shown as functions of y. The linear increase in γ_r with y, which shows no sign of saturation, corresponds to an increase in the EDOS (see text for discussion).

Loram et al. [97] propose a different analysis and interpretation of their data that avoids this result, but which is implausible in other ways. In their analysis the low-T upturns and *some* of the increase in the $\gamma_r T$ term are part of an "intrinsic," *but y-proportional*, magnetic contribution. In that interpretation of the data the maximum value of γ_r, at $y = 0.1$, is $\sim 19.5\,\text{mJ}\,\text{K}^{-2}\,\text{mol}^{-1}$, which is close to their estimate of $\gamma_n = 18\,\text{mJ}\,\text{K}^{-2}\,\text{mol}^{-1}$. (That result constitutes one step in reaching their conclusions about the pseudogap, which are discussed in Section 9.7.) However, the similarity to the Kim et al. [89] data, for which the 7-T measurements make an unambiguous separation of the paramagnetic-center and EDOS contributions, the abundance of evidence of paramagnetic centers in Zn-substituted samples, and the regular increase in the linear term that is evident both in Figure 9.12, and in Figure 2 of their paper [97], are persuasive evidence that their interpretation is incorrect, and that γ_r in the Zn substituted samples exceeds the apparent value of γ_n.

The high value of γ_r in Zn substituted samples has to be understood in the context of NMR and NQR data [105–107] that show very little increase in the EDOS for Zn substitutions. The possibility that impurity phases are responsible can be ruled out: Stoichiometric $BaCuO_2$ and $BaCuO_{2+x}$ (where x is small) have very different specific heats [111, 112]. The specific heat of $BaCuO_2$ could be interpreted as linear in T over a limited temperature range, but, under the strongly oxidizing conditions of YBCO synthesis, the most probable form is $BaCuO_{2+x}$ whose specific heats have Schottky-like anomalies and *no* pseudolinear contributions [111, 112]. The $Ba(Cu_{1-y}Zn_y)O_{2+x}$ impurity phase—whose specific heat has not been determined—might have a large, pseudolinear component, but it seems improbable. Even if the impurity phase was $BaCuO_2$ it would need to be present as $\sim 20\%$ of the $y = 0.1$ sample to account for the excess EDOS, which seems unrealistic. In addition, the "intrinsic" magnetic contribution to the specific heat that appears in the interpretation of Loram et al. [97] has no resemblance to that of any of these impurity phases.

The existence of localized holes, or CDW/SDW stripes in YBCO, and their destruction by Zn substitution is perhaps the most plausible explanation of the discrepancy between the maximum value of γ_r and the apparent value of γ_n. There is evidence for stripe phases in YBCO, and it has been shown that substitution of Zn for Cu in LBCO and (La,Nd)SCO destroys localized holes and increases the EDOS (see Section 9.4). It seems reasonable to expect that this increase in the EDOS would also take place for Zn-substituted YBCO. This supposition is supported by Hall-effect measurements [109, 113, 114] on $YBa_2(Cu_{1-y}Zn_y)_3O_7$ that show carrier concentration increases with increasing y at low temperatures. Such an increase in the EDOS with Zn substitutions would be a low-temperature effect occurring only in the temperature region in which the stripes or localized holes exist, and would not be seen in the normal-state NMR and NQR measurements [105–107]. Similar effects on the EDOS of Zn substitution on the Cu sites in LSCO are discussed in Section 9.4.

9.4. Stripes

The "stripe phase" in HTS is a state in which the charges and spins associated with the holes in the CuO_2 planes that promote superconductivity are spatially separated and form "stripes" of ordered charge-density waves (CDW) and spin-density waves (SDW). Stripe-phase formation causes a decrease in the EDOS. It is believed that these collective stripe modes are coupled to the lattice and can occur only for certain geometries. Stripe phases were predicted theoretically and were experimentally verified [115] using neutron-scattering data on single crystals of $(La_{2-x}Ba_x)CuO_4$. In recent years they have been intensively investigated and many theorists believe stripes could hold the key to an understanding of superconductivity in the HTS. For detailed discussions of stripe phases see reviews by Carlson, Emery, Kivelson, and Orgard [116]; Kastner and Birgeneau [117]; Tallon, Benseman, Williams, and Loram [118]; and Wilson [119].

While it is thought that all CuO_2 planes in the cuprates can support the CDW/SDW stripe phases, much of the research has been on the lanthanide-based HTS. Many investigations, by a number of techniques (including specific heat), have concentrated on $(La_{2-x}Ba_x)CuO_4$ and $[(La_{2-y-x}Nd_y)Sr_x]CuO_4$ where stripe phases are unequivocally identified. Recently, investigations have been extended to the other HTS (see, e.g., [116]). For example (1) Neutron-scattering measurements on YBCO have detected both spin and charge fluctuations consistent with dynamic-stripe phases [120–123]. (2) Thermal-conductivity measurements [124] on YBCO, Hg-1201, Hg-1212, and Hg-1223 show the formation of domains of localized holes in the CuO_2 planes near a hole concentration of 1/8, which can be interpreted as a CDW. (3) NMR and NQR measurements [125, 126] also show the presence of domains of localized holes in the CuO_2 planes of LCO and LSCO.

LBCO has a number of lattice structures. On cooling it undergoes a transition from high-temperature tetragonal (HTT), to low-temperature orthorhombic (LTO), to a nonequilibrium coexistence with a low-temperature tetragonal (LTT) phase in the superconducting region of the phase diagram [127–129]. The ratios of the LTT to LTO phases are dependent on x and T. In the region of the mostly LTT phase, for $x = 1/8$, the CDW/SDW phase forms robustly, which leads to a strong suppression of T_c [130,131]. For $x > 1/8$ superconductivity is restored and it is believed that the CDW/SDW phases coexist with "free" superconducting holes in the CuO_2 planes [116–119]. Figure 9.14 is a schematic phase diagram for $(La_{2-x}Ba_x)CuO_4$ and $[(La_{2-y-x}, Nd_y)Sr_x]CuO_4$.

Neutron-scattering and x-ray measurements do not detect an LTT phase in samples of $(La_{2-x}Sr_x)CuO_4$, but there is a weak minimum in T_c at $x \sim 1/8$ [130, 131]. A partial substitution of Nd for La in LSCO, $[(La_{2-y-x}Nd_y)Sr_x]CuO_4$, causes the LTT phase to form with a resulting strong suppression of T_c for $x = 1/8$ [132] (see Figure 9.14). This result provides convincing evidence in support of coupling between the lattice and the CDW/SDW and their formation dependence on the lattice geometry.

The transition of polycrystalline LBCO from the LTO to the LTT phase near 70 K was mapped by specific-heat measurements for various x [133]. Another specific-heat measurement [134] on $(La_{2-x}Ba_x)CuO_4$ with $x = 0.12$ also shows an anomaly centered on 70 K, in agreement with neutron-scattering and x-ray data [127–129] for the LTO to LTT transition. The anomaly was suppressed in a magnetic field of 7 T. Two other field-independent anomalies were observed in the region of 100 K, which have not been detected by other means. For $x = 0.09$ and 0.15, where the amount of LTT phase is diminished [127], no detectable anomaly or field dependence were observed in the 70 K region although those in the 100 K region persisted.

In those same experiments [134] an $(La_{2-x}Sr_x)CuO_4$ sample with $x = 0.09$ was superconducting with $T_c = 30$ K and had a field-dependent anomaly centered on 70 K, which was very similar to that observed for $(La_{2-x}B_x)CuO_4$ with $x = 0.12$. (Samples of $(La_{2-x}Sr_x)CuO_4$ with $x = 0.12$ and 0.15 did not show the anomaly.) Neutron-scattering and x-ray data do not detect any anomalies in this temperature region for any x.

Specific-heat measurements have been reported for polycrystalline samples of $[(La_{1.85-y}Nd_y)Sr_{0.125}]CuO_4$ with $y = 0.4$ and 0.8 [135]. Both samples show y-dependent anomalies spanning the 60–90 K temperature region shown in Figure 9.15. These anomalies are associated with the y-dependent transition from the LTO to the LTT phase [130, 131].

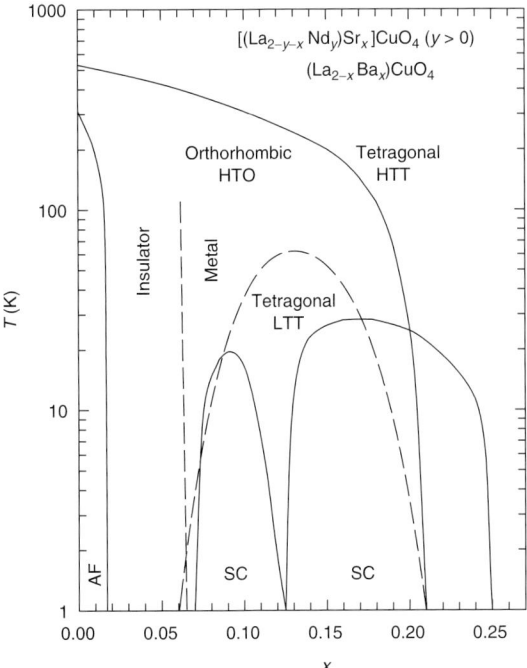

Figure 9.14. Schematic T–x phase diagram for LBCO and (La,Nd)SCO. The occurrence of the LTT structure and the steep decrease toward zero in the T_c–x boundary at $x = 1/8$ are conspicuous features that are absent for the LSCO T_c–x dome, which does, however, have a weak minimum at $x \sim 1/8$. The phase boundaries shown with dashes are not well defined. Conversion of the LTO phase to LTT phase is neither complete nor are they in thermodynamic equilibrium; both phases coexist in the LTT labeled region. The amount of LTT phase maximizes for $x = 1/8$ where superconductivity is sharply diminished.

In the transition from the LTO to LTT structure an intermediate precursor phase has been detected, which is also tetragonal with Pccn symmetry [130, 131]. In Figure 9.15 this transition is observed for $y = 0.4$, but not for $y = 0.8$, suggesting that the Pccn phase does not form for that Nd composition. The entropy (ΔS_{LTT}) and heat content (ΔH_{LTT}) changes for the first-order transitions are given in the figure. Figure 9.16 shows an anomaly near 30 K for $y = 0.4$ and 0.8 and $B = 0$ that is independent of magnetic field to 7 T, which is probably associated with the formation of the CDW/SDW phases. The entropy (ΔS_{CDW}) and heat content (ΔH_{CDW}) changes for the transitions are given in the figure. (A third sample with $y = 0.6$—not shown—gave results very similar to those for $y = 0.8$.) Specific-heat measurements have also been reported for a polycrystalline sample of $[(\text{La}_{1.85-y}\text{Nd}_y)\text{Sr}_{0.15}]\text{CuO}_4$ with $0.12 \leq y \leq 0.7$ that show y-dependent anomalies in the LTO to LTT transition temperature range [136].

Specific heats have been measured [137] for $[(\text{La}_{1.6-x}\text{Nd}_{0.4})\text{Sr}_x](\text{Cu}_{1-y}\text{Zn}_y)\text{O}_4$ from 2 to 300 K. At 15 and 100 K the ratio $[(C - C_{\text{lat}})/15]/[(C - C_{\text{lat}})/100]$ and T_c vs. x for $y = 0$ have significant minima at $x \sim 1/8$, which is interpreted by the authors as showing the formation of the CDW/SDW stripe phases and the suppression of superconductivity. The addition of Zn for a given x reduces T_c, increases γ_r and the EDOS, and, for $x = 1/8$, suppresses the stripe phase.

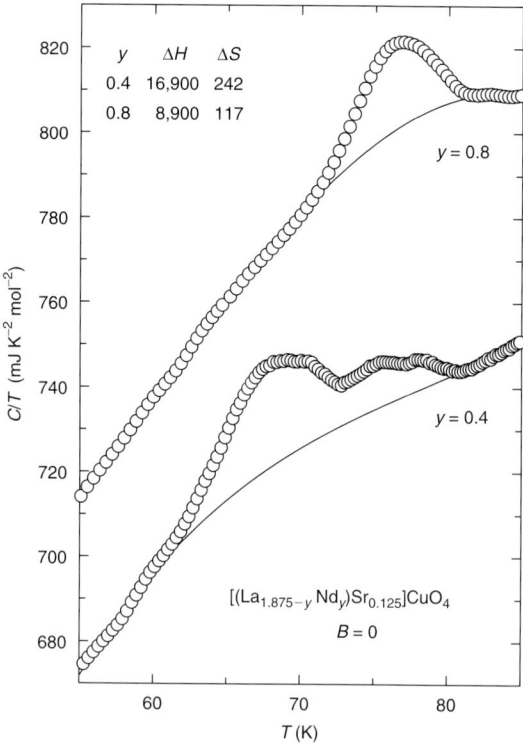

Figure 9.15. The specific heat of (La,Nd)SCO in the vicinity of the LTO to LTT structural transition. The curves, which define a specific-heat anomaly, are polynominal fits to the lattice specific heat above and below the transition. Entropy and heat content changes for the broadened, first-order LTO to LTT transitions are tabulated in the figure. For $y = 0.4$, but not 0.8, there is a "double transition" that is a consequence of the precursor high-temperature phase with Pccn symmetry. (Data are from Wright et al. [135].)

Specific-heat measurements [138] were used to evaluate γ_n for $(La_{2-x}Ba_x)CuO_4$ in the range $0 \leq x \leq 0.24$. The sample for $x = 1/8$ was nonsuperconducting and had a large reduction in γ_n compared to other nonsuperconducting samples with $x \neq 1/8$. This drastic reduction in the EDOS was interpreted as confirmation of the formation of a CDW phase at $x = 1/8$.

A determination of γ_n by an analysis of magnetization and specific-heat measurements was used to assess the effect of substitution of other ions for Cu in the CuO_2 planes for nonsuperconducting $(La_{1.875}Ba_{0.125})(Cu_{1-y}A_y)O_4$ (A = Zn and Ni) [139, 140]. When $y = 0$ the superconductivity is strongly suppressed as a result of the formation of a CDW/SDW stripe phase with $\gamma_r = 3.5$ mJ K^{-2} mol^{-1}. As y is increased, for the Zn-substituted samples, there is a monotonic increase in γ_r for $y \leq 0.06$, to 12 mJ K^{-2} mol^{-1} for $y > 0.06$. An interpretation of this result attributes it to the destruction of the SDW stripes when nonmagnetic Zn ($S = 0$) replaces magnetic Cu ($S = 1/2$) and disrupts the Cu–Cu exchange interaction producing paramagnetic centers. In turn, this causes a destruction of the CDW stripes with the freeing of holes and an increase of the EDOS. On the other hand, substitution of magnetic Ni ($S = 1$) for Cu causes a different behavior. Initially γ_r increases with y with a slope similar to that for the Zn substitution, but for $y = 0.02$ it reaches a constant value of 6.5 mJ K^{-2} mol^{-1}, which is approximately half that of the corresponding Zn-substituted sample. The interpretation [139]

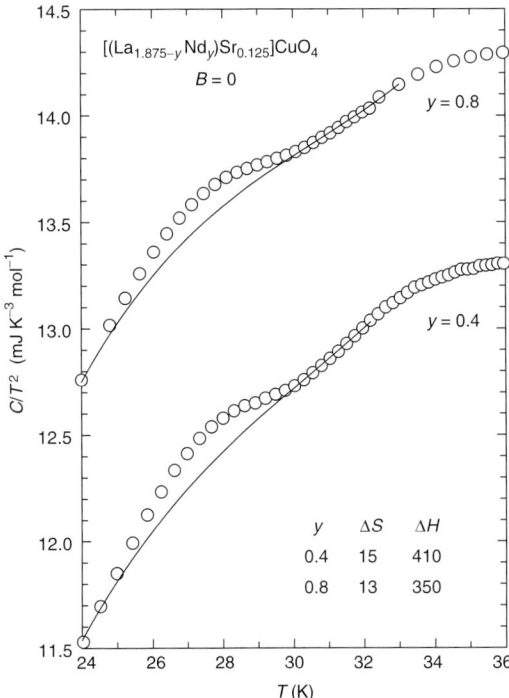

Figure 9.16. Thermal signatures at $T \sim 30\,\mathrm{K}$, which are postulated as showing the formation of the CDW/SDW stripe phases for $x = 1/8$, and $y = 0.4$ and 0.8. The curves were obtained from polynominal fits to the lattice specific heats well above and well below the anomaly. Estimates of entropy and heat-content changes for the transitions are tabulated in the figure data are from Wright et al. [135]).

is that for $x \leq 0.02$, SDW stripes containing Ni ($S = 1$) at their center are distorted (bulge), but are not destroyed. However, they do not connect smoothly with non-Ni containing SDW stripes, which causes spin fluctuations in the boundary region. Holes are freed from the CDW stripes in the neighborhood of these spin fluctuations and γ_r increases. However, above the critical concentration of $y = 0.02$ the hole production ceases and the CDW stripes are "bent" around the Ni sites in the distorted Ni-containing SDW stripes. A "meandering" CDW stripe phase is formed in contrast to the "straight" CDW phase for $y = 0$. This type of charge-ordered phase has been proposed theoretically [141]. In summary, γ_r for $y = 0$ represents an EDOS related to the nonsuperconducting phase that is not part of the ordered CDW/SDW; and, for $y > 0$, as the CDW/SDW stripe phase is gradually suppressed, the freed holes add to this EDOS.

Measurements of the specific heat of $(\mathrm{La}_{1.85}\mathrm{Sr}_{0.15})(\mathrm{Cu}_{1-y}\mathrm{Zn}_y)\mathrm{O}_4$ [142] showed that γ_r increased and T_c and $\Delta C(T_c)$ decreased with increasing y, with superconductivity vanishing for $y = 0.025$. For these samples x-ray analysis showed that the Zn is substituted randomly in the CuO_2 planes. Similar specific-heat measurements [57] were made on polycrystalline samples of $(\mathrm{La}_{2-x}\mathrm{Sr}_x)(\mathrm{Cu}_{1-y}\mathrm{Zn}_y)\mathrm{O}_4$ for $0 \leq x \leq 0.3$ and $0 \leq y \leq 0.05$. The samples were single phase with small γ_r for $y = 0$, attesting the absence of appreciable normal material. For normal $(\mathrm{La}_{2-x}\mathrm{Sr}_x)(\mathrm{Cu}_{1-y}\mathrm{Zn}_y)\mathrm{O}_4$ samples, as x increases from zero there is an increase in γ_n to a maximum at $x \sim 0.1$, a decrease to a minimum (but not zero) at $x \sim 0.13$, an increase to another maximum at $x \sim 0.23$, and a decrease as x increases to 0.3, the limit of the substitution. For a given x in the range $x > 0.16$ there is a monotonic, nonlinear increase in γ_r with increasing y, with a plateau at $y = y_{cr}$, which increases with increasing x. When $x < 0.16$ there is a linear increase of γ_r with increasing y similar to that found for Zn substituted YBCO. It was assumed that γ_n for $x < 0.16$ is the linearly extrapolated value at $y = 0$ of γ_r, which implies that the EDOS increases with increasing Zn concentration. For $x > 0.16$ it was assumed that γ_n is the value for $y > y_{cr}$. Neutron-scattering and x-ray data show no CDW/SDW formations in LSCO, but the minimum at $x \sim 0.13$ for the non-superconducting samples could be interpreted as indicating CDW/SDW fluctuations (see, e.g., [116]). Likewise, for superconducting samples of fixed x, it is plausible that fluctuating CDW/SDW phases [143] are progressively destroyed as y increases with an accompanying increase in the EDOS, which is in addition to the increases associated with pair-breaking by Cu^{2+} moments. The plateau in γ_r at y_{cr} for $x > 0.16$ can be understood to occur when the fluctuating CDW/SDW have been completely suppressed. The value of γ_n for $y = 0$ would be less than γ_r at y_{cr}. This result is similar to that for Zn-substituted YBCO (see Section 9.3.3).

Measurements of magnetization ($2 \leq T \leq 550$ K) and specific-heat ($2 \leq T \leq 7$ K) have been reported [144] for polycrystalline samples of $(La_{2-x}Sr_x)(Cu_{1-y}A_y)O_4$ (A = Zn or Ni) for $0.1 \leq x \leq 0.25$ and $0 \leq y \leq 0.05$. The samples were single phase with small values of γ_r in the superconducting state for $y = 0$. As Zn concentrations increase there are increases in the Curie term above T_c as a result of the formation of localized Cu^{2+} ions adjacent to the Zn, as in YBCO. The specific-heats for the Zn-substituted samples are very similar to those of [57]. However, for Ni substitutions with $y < y_{cr}$ (where y_{cr} is dependent on x) there is no increase in the Curie term, but for $y > y_{cr}$ such a term appears. The authors explain this by postulating that for $y < y_{cr}$ Ni substitutes as nonmagnetic Ni^{3+}, which reduces the hole concentration in the CuO_2 planes. For $y > y_{cr}$ Ni is substituted as magnetic Ni^{2+} with a resulting Curie term. For the Ni-substituted samples the decreases in T_c and increases in γ_r are much less than for the Zn-substituted samples with comparable values of y. Unlike the case of the Zn-substituted samples the values of γ_r increase linearly with y for all x.

9.5. Specific-Heat Anomaly at T_c: Fluctuations; BCS Transition, BEC

As a consequence of the short coherence length, fluctuation effects in the zero-field specific heat near T_c, which have not been observed for conventional bulk superconductors, are readily apparent for the HTS. However, there have been ambiguities as to whether the fluctuations are 2D or 3D, and Gaussian or critical. For the in-field measurements there is also the question of the applicability of several scaling relations. These issues are addressed in Section 9.5.1. There is also another, possibly more fundamental, question related to the specific-heat anomaly at T_c: Does the variation in the shape of the anomaly with doping level reflect a change from a BCS transition to one that is a variant of a Bose–Einstein condensation (BEC)? Evidence relevant to this question is discussed in Section 9.5.2.

9.5.1. Gaussian and Critical Fluctuations:

This section contains references to only a few of the many relevant publications on the fluctuation contribution to the specific heat. More references to early measurements can be found in other reviews: Junod [1]; Phillips, Fisher, and Gordon [3]; and Salamon [145]. References to more recent work can be found in several other papers [146–150].

The discussion of fluctuation effects is in three parts: The specific-heat anomaly at T_c is better defined in YBCO and the RBCO compounds than in other HTS, and in "Fluctuations: Optimally Doped Samples in Zero Field" we focus our attention on measurements on YBCO and DyBCO, near optimal doping ($x \sim 0.92$), and in zero field. We extend the discussion to in-field measurements in "Fluctuations: Optimally Doped Samples in Field" and to over- and under-doped samples more generally in "Fluctuations: Under- and Over-Doped Samples".

Fluctuations: Optimally-Doped Samples in Zero Field

Ginzburg–Landau (GL) theory [151] can successfully explain the mean-field character of the phase transition in a BCS superconductor and can also predict how the transition will be affected when fluctuation effects are small enough to be described by a Gaussian approximation. However, when T is sufficiently close to T_c, that is, when the sample is in the "critical" region, renormalization group (RG) theory [152] rather than GL theory must be used to describe the fluctuations. The temperature regions in which the two kinds of fluctuations are expected to be comparable in size to the mean-field discontinuity $\Delta C_e(T_c)$ are expressed in terms of the Ginzburg criterion [153] and the less-well known Brout criterion [154]. The two criteria can be expressed in terms of the reduced temperature $t \equiv (T/T_c - 1)$. By these criteria the critical region for a 3D superconductor lies within the interval $|t| < t_G \equiv [k_B/\pi \Delta C_e(T_c)\xi^3]^2/32$, where $\xi = (\xi_a \xi_b \xi_c)^{1/3}$ is the GL coherence length of the superconducting order parameter at $T = 0$, and $\Delta C_e(T_c)$ is expressed in units of energy density K^{-1}. Gaussian fluctuations [155] can be expected in the wider reduced interval $1 \gg t_B > |t| > t_G$, where $t_B \sim (t_G)^{1/2}$. The signature of a fluctuation contribution to the specific-heat anomaly near T_c is positive curvature of C_e/T just below T_c coupled with a "tail" just above. There is considerable evidence that superconductivity in the HTS is characterized by strong coupling, which increases $\Delta C_e(T_c)$ and also produces positive curvature below T_c. Therefore, the possibility of strong coupling introduces further uncertainty, in addition to that associated with separating the lattice contribution, into the analysis of the specific-heat data near T_c.

The strong dependence of both t_B and t_G on ξ makes it clear that the coherence length is the determining factor in whether or not fluctuation effects will be measurable. In the case of a typical type-I superconductor, for example Sn, with $\Delta C_e(T_c) \sim 1\,\text{mJ}\,\text{cm}^{-3}\,\text{K}^{-1}$ and $\xi \sim 2{,}000\,\text{Å}$, $t_G \sim 10^{-14}$ and $t_B \sim 10^{-7}$ [156]. The Brout criterion tells us that even a Gaussian fluctuation contribution to C_e would be unobservable unless it was possible to make specific-heat measurements within an interval about T_c of $\sim 1\,\mu\text{K}$ or less. It is not surprising that near T_c the specific-heat anomaly of conventional superconductors is well described by standard mean-field theory. Conversely, in the case of HTS, such as YBCO, with $\Delta C_e(T_c) \sim 30\,\text{mJ}\,\text{cm}^{-3}\,\text{K}^{-1}$, $\xi \sim 10\,\text{Å}$ and $T_c \sim 90\,\text{K}$ [156], t_G and t_B are ~ 0.0015 and 0.04, respectively. The Ginzburg and Brout criteria lead to the expectation that Gaussian fluctuations should be evident over an interval of $\sim 5\,\text{K}$ about T_c, while critical fluctuations should be evident within the much narrower interval of $\sim 0.1\,\text{K}$ about T_c.

A detailed calculation [157, 158] shows that for s-wave pairing the zero-field Gaussian fluctuation contribution, C_f, to C_e can be written:

$$C_f^\pm = A^\pm (k_B/16\pi \xi^d)|t|^{-\alpha}, \tag{9.10}$$

where $\alpha = (4-d)/2$, where $d = $ the dimensionality of the system, and $A^+ = 2$ and $A^- = 2^{d/2}$ for $t > 0$ and $t < 0$, respectively. Early measurements of C_f in the case of YBCO [159] were consistent with $\alpha = 1/2$, as expected for a three-dimensional system, but with neither the predicted values of A^+ and A^- nor their ratio A^+/A^-. Similar results have been obtained on YBCO by a number of other groups (see, e.g., [3] for references to some of the early results). The zero-field C/T data for a single crystal of optimally doped DyBCO [160] are shown in Figure 9.17. The curve through the data is a least-squares fit assuming C to be the sum of a 3D Gaussian fluctuation term, C_f, a mean-field step below T_c of the form $\Delta C_e = hT(1+gt)$, and a quadratic lattice term. While the fit is quite good except close to T_c, the authors' in-field measurements cast doubt on the validity of the assumption that the fluctuation contribution is 3D Gaussian.

In the critical region C_f is predicted to be of the form:

$$C_f^\pm = +(A^\pm/\alpha)|t|^{-\alpha}, \qquad (9.11)$$

where the "critical exponent" α and the coefficients $A^\pm$ are constants [161] that depend upon the "universality class" (3D XY, Ising, etc.). The value of α is close to, but not equal to, zero. For very small values of α, the term $|t|^{-\alpha}$ in Eq. (9.11) can be replaced by $(1 - \alpha \ln|t|)$.

The Brout and Ginzburg criteria are useful guidelines, but are not inviolable rules for distinguishing the regions of Gaussian and critical fluctuations. For example, it has been suggested [162] that for type-II superconductors the boundaries of the critical region exceed the value of t_G as defined above. Moreover, there is disagreement as to the nature of the "crossover" from Gaussian to critical fluctuations. Precisely because there is doubt as to the width of the critical region, as well as to the form of the fluctuation term in any crossover region, it is useful to compare the predictions of Eqs. (9.10) and (9.11) with experimental results. It has been found by a number of groups (see, e.g., the review by Junod et al. [147] for references) that for optimally doped YBCO and DyBCO single crystals the specific heat is well represented by the predictions of the 3D XY model [152, 161], a model that has been found to provide a very good fit to the λ-transition of liquid ^{4}He [163]. In Figure 9.18, the C/T data shown in Figure 9.17 are fitted with the 3D XY model. A mean-field step and a quadratic lattice specific heat are again assumed and the critical exponent α is taken as -0.013, the value obtained from the fit to the ^{4}He data [163]. The fit to the DyBCO data [160] shown in Figure 9.18 yields $A^+/A^- = 1.029 \pm 0.013$, in reasonable agreement with the value of

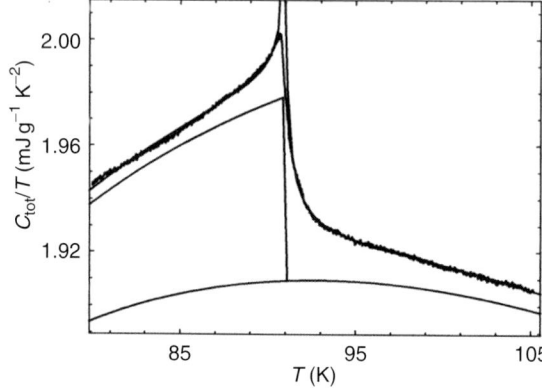

Figure 9.17. Specific-heat data for $DyBaCu_3O_{6+x}$, shown as the "wavy" curve, and a Gaussian-fluctuation fit to the zero-field data (Figure 7 from Garfield et al. [160]). The uppermost solid curve is the sum of the contributions from Gaussian fluctuations, a mean-field step below T_c (see text), and a quadratic representation of the lattice contribution. (Note: C_{tot} in the figure is replaced by C in the text).

High-T_c Superconductors: Thermodynamic Properties

1.058 ± 0.004 found for ^{4}He [163]. Thermal-expansion measurements on optimally doped YBCO [164] gave similar results. This technique has the advantage over specific-heat measurements in that the data analysis does not require correction for a large lattice component of the specific heat.

Fluctuations: Optimally Doped Samples in Field

On the basis of Figures 9.17 and 9.18 alone it would be difficult to choose between the Gaussian and critical representations of the zero-field data. However, in-field specific-heat results on the optimally doped RBCO HTS are better fitted by the critical representation, with the 3D XY model, than by the Gaussian representation. Fairly general arguments [165] were used to show that the field and temperature dependence of C_f can be written:

$$C_f = C_0 - B^{\alpha/2\nu} f(x), \qquad (9.12)$$

where $C_0 = 0$ for $t > 0$ and a constant for $t < 0$, $x = t/B^{1/2\nu}$, α is the specific-heat exponent, and ν is the coherence-length exponent, e.g., $\xi(t) = \xi/t^\nu$. In Eq. (9.12) it is the *scaled temperature*, x, rather than the reduced temperature, t, that appears. A simplified form of Eq. (9.12) has been used [160, 166], namely:

$$[C(0, T) - C(B, T)] B^{\alpha/2\nu} = f(x), \qquad (9.13)$$

where, in the case of the 3D XY model, $\nu = 0.669$, and where it is assumed [163] that $\alpha = -0.013$, the experimental value obtained for ^{4}He. Equation (9.13) has the advantage, in principle, of allowing a test of the scaling prediction without making any assumptions about the lattice specific heat, C_{lat}, except that it is independent of B. In practice, because of uncertainties about the true value of T_c, the authors found it preferable to use fitted values (see Figure 9.18) rather than the measured values of $C(0, T)$ in Eq. (9.11). The results are shown in Figure 9.19. As this figure shows, the agreement is quite good.

The suitability of the 3D XY model for describing the critical behavior of optimally doped RBCO has been verified by a number of groups (see, e.g., [147, 148] for references).

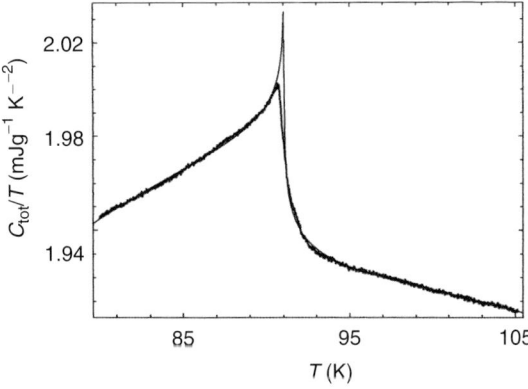

Figure 9.18. Specific heat of DyBaCu$_3$O$_{6+x}$ (Figure 4 from Garfield et al. [160]). The solid curve is from a critical-fluctuation fit to the same data shown in Figure 9.17 plus a quadratic representation of the lattice contribution. (Note: C_{tot} in the figure is replaced by C in the text.)

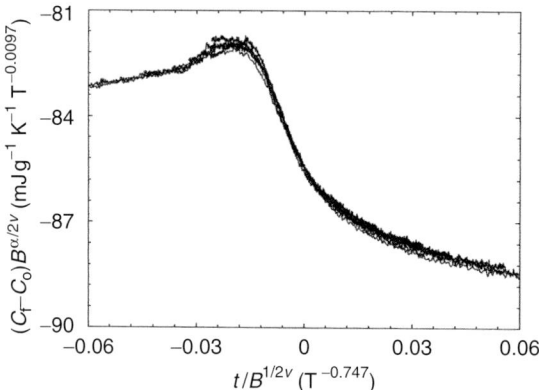

Figure 9.19. 3D XY scaling of the specific heat of $DyBaCu_3O_{6+x}$ for magnetic fields of 1, 2, 4, 6, and 8 T (Figure 6 from Garfield et al. [160]). In the fit it is assumed that the scaling parameters $\alpha = -0.013$ and $\nu = 0.669$. (see text.)

Some authors have tested the 3D XY scaling predictions by examining how the derivatives of C vary with both field and temperature. Others have examined the scaling predictions as they apply to the field dependence of the magnetization. In addition, some groups have tested the applicability of another scaling approach, that of "Lowest-Landau-Level" (LLL) scaling (see, e.g., [167] for references). While there is disagreement among investigators as to the range of B for which one or the other scaling procedure is superior, it has been pointed out [168] that, providing one assumes $B_{c2}(T)$ to vary as $|t|^{4/3}$ rather than as $|t|$, then *both* 3D XY and LLL scaling provide a valid description of the specific heat and magnetization for magnetic fields between 1 and 16 T.

Fluctuations: Under- and Over-Doped Samples

The height of the specific-heat anomaly in the optimally and slightly over-doped RBCO superconductors can be as large as 3–4% of C. In the case of the under-doped superconducting RBCO the height is generally smaller, while the anomalies in the La, Tl, Bi, and Hg HTS rarely exceed $\sim 1.5\%$ of C. Measurements on Pb-stabilized Bi-2212 [169] and on Bi-2212 [170] were reported to be consistent with 2D, rather than 3D, Gaussian fluctuations, since C_f was found to vary as $|t|^{-1}$ rather than as $|t|^{-1/2}$. The specific-heat data for Tl-2201 [171,172], Hg-1201 [173,174], and Hg-1223 [175,176] all show anomalies similar to the anomaly of Bi-2212 that will be discussed below.

9.5.2. BCS to BEC

The Geneva Group [147, 148] has surveyed a series of superconducting anomalies. Figure 9.20a–f (Figure 1 from [147]) is a sequence of plots of $\Delta C(B, T)/T$ vs. T for a number of superconductors. The first plot in the sequence (Figure 9.20a) is for the BCS-like superconductor $Nb_{77}Zr_{23}$ and $\Delta C(B, T) = C_s(B, T) - C_n$. The rest of the plots are for HTS and $\Delta C(B, T) = C(B, T) - C(B_0, T)$ where in all but one plot $B_0 = 14$ T. The advantage of using $\Delta C(B, T)$ rather than $C(B, T)$ as the plot variable is that C_{lat} is virtually eliminated by the subtraction. The sequence of plots (b–f) are arranged in such a way that the specific-heat anomalies about T_c change from being BCS-like (e.g., a sharp change in

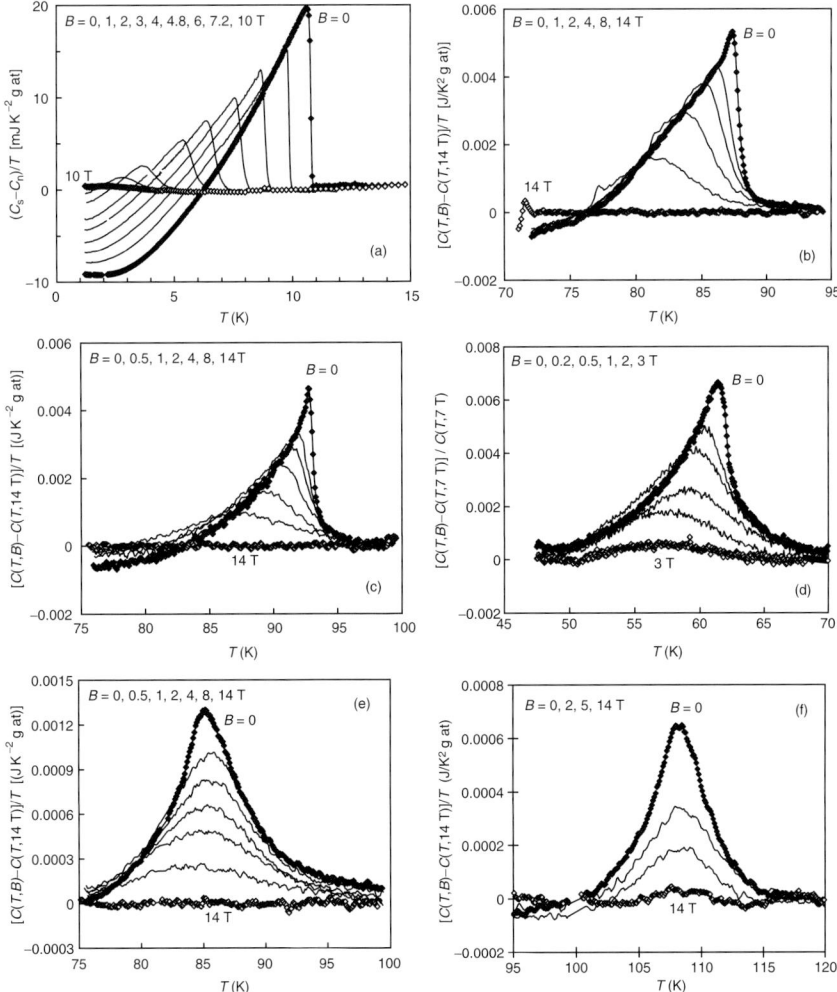

Figure 9.20. Specific heats near T_c for various superconductors showing their evolution with magnetic field (Figure 1a–f from Junod et al. [147]). (a) $Nb_{77}Zr_{23}$, (b) $YBa_2Cu_3O_7$, (c) $YBa_2Cu_3O_{6.92}$, (d) $YBa_2Cu_3O_{6.6}$, (e) $Bi_{2.12}Sr_{1.9}Cu_{1.96}O_8$, (Bi-2212), (f) $Bi_{1.84}Pb_{0.34}Sr_{1.91}Ca_{2.03}Cu_{3.06}O_{10}$(Bi-2223).

$\Delta C(B,T)/T$ at T_c) in the first figure, to being virtually symmetric about T_c in the last. It is also the case that as this symmetry of the specific-heat anomaly increases, the temperature at which the anomaly is a maximum, $T_{\max}(B)$, becomes less and less dependent upon B. In this sequence the over-doped YBCO is more BCS-like than the optimally doped sample, which, in turn is more BCS-like and less symmetric than the under-doped YBCO. The plot for Bi-2212 (Figure 9.20e) is very nearly symmetric and $T_{\max}(B)$ has only a weak T dependence. The final plot in the sequence (Figure 9.20f) is for Bi-2223. Here the anomaly is essentially symmetric about T_c, and $T_{\max}(B)$ is virtually independent of B. The HTS part of the sequence thus proceeds from the BCS-like over-doped YBCO, to 3D-XY-like optimally doped YBCO, to under-doped YBCO, which shows evidence of a pseudogap in the EDOS (see Section 9.7) and has a more symmetric anomaly than the more heavily doped YBCO samples. The last

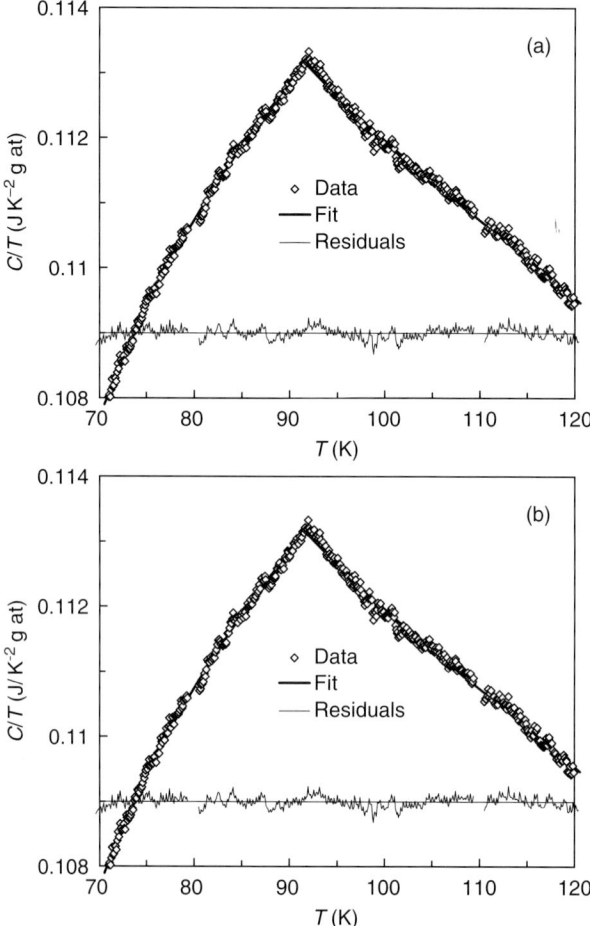

Figure 9.21. Fits of the specific heat of Bi-2212 shown in Figure 9.20e (Figure 3c,d from Junod et al. [147]). (a) Gaussian fluctuation fit with $A^+/A^- = 1$, $\alpha = -0.67$, (b) 3D BEC fit. (See text for details.)

two plots in the sequence, those for the BSCCO compounds, have anomalies that are similar to those predicted for the Bose-Einstein condensation of an ideal Bose gas. Figure 9.21a,b (from Figure 3 in Ref. [147]) demonstrates that the zero-field specific-heat results for the Bi-2212 sample can be represented almost equally well by Eq. (9.11) with $A^+/A^- = 1$, and $\alpha = -0.67$ and by a 3D BEC fit. The authors caution, however, that the fit obtained from Eq. (9.11) deviates from the predictions of the 3D XY model since the exponent α does not satisfy the requirement that $|\alpha| \ll 1$. They conclude that among the HTS there is a crossover from BCS-like to BEC-like behavior with the 3D XY character of optimally doped YBCO representing an intermediate case.

The suggestion that there could exist a continuum of superconducting characteristics that range from BCS-like to BEC-like behavior was proposed a number of years ago [177–179]. It was argued that the BCS ground-state wave function could become BEC-like providing the attractive pairing interactions ($U < 0$) were sufficiently strong. Because the coherence length, ξ, decreases with increasing $|U|$ it is natural to think that this suggestion might be applicable to HTS, for which $\xi \sim 10$ Å. It was suggested [180] that a useful measure of $|U|$ is $(k_F\xi)^{-1}$, where k_F is the k vector at the Fermi level. For weak coupling (BCS), $k_F\xi \gg 1$, whereas $k_F\xi \sim 1$ for optimally doped YBCO [147]. In this picture the crossover from BCS-like to BEC-like behavior occurs in the region $2\pi \geq k_F\xi \geq 1/\pi$, a region where the chemical potential, μ, changes sign from positive (characteristic of Fermions) to negative (characteristic of Bosons). (For a critique of [180], see [181].) A possible $T = 0$ plot of μ vs. $|U|$ contains three regions: the weak-coupled BCS region, the pseudogap (PG) region, and the BEC region [149]. The chemical potential has an essentially constant, positive value in the BCS region. Within the PG region the chemical potential decreases from this constant value to 0 at the boundary between the PG and BEC regions. This boundary corresponds to the value of $|U|$ for which there is no longer a Fermi surface. In the BEC region the chemical potential has $\mu \leq 0$.

In this BCS/BEC scenario the PG region is one where, for temperatures such that $T_c \leq T \leq T^*$, the normal state is neither the usual Fermi-liquid normal state nor a pure Bosonic state. Rather, it is a state comprising two kinds of excitations, Fermions and a collection of tightly bound, incoherent, "preformed" Fermion pairs ("Bosons"). For $T > T^*$ the thermal energy is sufficient to break these Fermion pairs. T^* is a function of doping and increases as the doping decreases, whereas T_c decreases with decreasing doping (except when the doping is close to the optimal value) [149, 182]. In the view taken in [149], the reduced EDOS at the Fermi level in under-doped samples (the pseudogap) is thus correlated with the existence of these incoherent pairs. The Fermion excitations undergo something similar to a BCS transition at T_c. The incoherent Bosonic excitations achieve phase coherence as $T \to 0$ by condensing into a zero-momentum state (they undergo something like a Bose–Einstein condensation). It is not surprising that the shape of the superconducting transition at T_c might reflect the shape of the specific-heat peak associated with BEC, but it is surprising that the shape of the specific-heat anomaly in Bi-2212 appears to resemble so closely the BEC peak. It has been pointed out [150, 183] that the zero-field anomalies in Hg-1223, Tl-2201, and Bi-2212 bear a strong resemblance to this BEC peak. Furthermore, it is calculated how the shape of the BEC peak is affected if the chemical potential of a Bose gas depends upon magnetic field as well as temperature. The calculation is compared with Hg-1223 data [176] and reasonable agreement is found with the observed suppression of the specific-heat anomaly by a magnetic field. It also predicts [150, 183] that $T_{max}(B)$, the temperature for which the specific-heat anomaly has its maximum, will be virtually independent of B, as is, in fact, observed in the 1223, 2201, and 2212 compounds.

There are alternate models for the way in which tightly bound Fermion pairs might be formed, for example, the bipolaron model [150, 184]. For a fuller discussion of the possible relevance of BEC to the HTS see the articles in the book edited by Griffin et al. [185]. The model of tightly bound, preformed Fermion pairs and their relationship to the pseudogap is by no means universally accepted, nor is it consistent with all of the experimental observations of the pseudogap [182, 186, 187]. The high-resolution, thermal-expansion measurements on YBCO [187, 188] have been interpreted as showing that the superconducting transitions of both optimally doped and under-doped YBCO can be accounted for by an anisotropic 3D XY model. In this picture increased anisotropy tends to decrease both T_c and $\Delta C_{es}(T_c)$ and also to shift more of the area under the anomaly above T_c, thereby increasing the symmetry of the anomaly about T_c (see Figure 9.20). In their reply [188] to a comment [189] the authors interpret their data as showing that incoherent Cooper pairs in the HTS begin to form at temperatures well above T_c (but below T^*), and they associate these preformed Cooper pairs with the formation of the pseudogap. As they point out, it is not necessary that the onset of superconductivity at T_c coincide with the onset of pairing. For example, in ^{4}He the superfluidity exists only below T_λ although the Bosons are clearly present at higher temperatures. The anisotropic 3D XY model is regarded [187] as being applicable to the BSCCO HTS as well as to YBCO. The model predicts that in the case of Bi-2212, for example, the specific-heat peak would consist of a broad 2D fluctuation term capped by a very narrow 3D XY divergence at T_c. Specific-heat measurements have been made on microgram-sized crystals of Bi-2223, but so far the divergence has not been observed [190].

There are a number of alternate proposals for explaining the thermodynamic properties of the HTS. Discussion of some of them can be found in recent reviews of the properties of the HTS. (see, e.g., Carlson, Emery, Kivelson, and Orgad [116]; Yeh [191]; Norman and Pépin [192].)

9.6. Vortex-Lattice Melting

9.6.1. Introduction; Early Measurements on YBCO

In the mixed (or vortex) state a magnetic field penetrates a superconductor in an array of "vortices", the Abrikosov "vortex lattice" [193], which can be a "solid" of pinned vortices or a "liquid" in which the vortices are fluid, with the two phases separated by a "melting/freezing" phase boundary, $B_m(T)$. For conventional superconductors the fluid phase is confined to a very narrow region close to $B_{c2}(T)$, the melting has not been observed, and it has not been possible to study the fluid phase. In the HTS, because of the higher values of T_c, lower dimensionality, and smaller coherence lengths, the fluid phase exists over a wide region of the B–T phase diagram, as first shown by magnetization measurements (see Section 9.1.3). $B_m(T)$ is well separated from the fluctuation-dominated crossover to the normal phase at $B_{c2}(T)$, and the melting of the vortex lattice is observable. There is an excellent review of these features of the phase diagram by Blatter, Feigel'man, Geshkenbein, Larkin, and Vinokur [194].

Theory predicts a first-order transition for the melting in "clean" materials [195, 196], but if the material contains a large number of defects, grain boundaries, or twins that cause disorder of the vortex lattice, a second-order transition can occur [167, 197–199]. The physics of the vortex phase is complicated by the importance of disorder, which grows with magnetic field and the concurrent number of vortices. Moderate disorder can change the vortex lattice to a topologically ordered "Bragg Glass" phase in low magnetic fields that retains the ability to melt to a fluid at higher fields [200]. When there is more extensive disorder a glass phase forms for which no thermodynamic transition to a liquid can occur. Furthermore, the nature of the transition may change from first to second order at a critical point $B_{cp}(T)$ on the $B_m(T)$ melting curve. Both types of transition, depending on the degree of pinning in the sample, and other features in the vortex-state phase diagram have been seen experimentally.

A calorimetric measurement of a latent heat (L), $L = T\Delta S$, or a discontinuity in specific heat (ΔC) provides the most direct proof of a first- or second-order transition, but such measurements are difficult because the thermal effects are small ($\sim 1\%$ of C), and they require a very high sensitivity. YBCO has proved to be an ideal substance for observation of the thermal signatures that are associated with the solid-to-fluid transitions. Prior to any calorimetric measurements, however, features that were consistent with the expected transitions had been seen in other measurements on YBCO, resistivity [201–203] and magnetization [204, 205]; and on BSCCO, magnetization [206, 207], muon-spin rotation [208], and neutron-diffraction [209].

Calorimetric measurements were first considered seriously by Schilling and Jeandupeux [21], who recognized that the expected value of L for YBCO should be measurable, and developed special apparatus for the purpose. The technique, a modified differential thermal-analysis method (see "Specific Heat: Experimental Techniques" in Section 9.1.4), allows an unambiguous distinction to be made between a latent heat and other thermal effects associated with higher-order transitions, such as second order. Since the twinned YBCO crystal available for their measurements did not show the signature feature of a first-order transition in the resistivity, their negative result, the absence of a measurable latent heat [21], was not a surprise. However, their measurements proved the effectiveness of the technique, established a new standard of precision for measurements of the specific-heat anomaly at T_c, and set an upper limit to L, *for that crystal*, an order of magnitude below the values expected for the first-order transition. The first calorimetric measurements to show *any* thermal effect at $B_m(T)$ were made by Schilling and Jeandupeux with other colleagues, using that technique on a *different* twinned single crystal of YBCO [210]. They showed clearly resolved "steps" in C/T, i.e., ΔC, as expected for a second-order transition.

The first calorimetric measurements to show the predicted first-order transition from the vortex-solid to the vortex-fluid were made by Schilling et al. [211], again with the differential technique, on an *untwinned* single crystal of YBCO with $T_c = 92$ K. That crystal, made at Argonne National Laboratory, was similar to others that had shown evidence of first-order melting in magnetization and resistivity measurements. The measurements were made with $B \parallel c$-axis, to a maximum $B = 9$ T, and gave $L \approx 0.45 k_B T$ per vortex per superconducting layer. Thermodynamic consistency was verified by comparison with discontinuities in magnetization, ΔM, measured on the same crystal, with the Clapeyron equation. The results for ΔS and ΔM are in satisfactory agreement with a calculation that takes into account the temperature dependence of the Landau parameters [212,213].

Soon after the measurements of ΔC and ΔS by Schilling and colleagues, similar results, obtained with an adiabatic continuous-heating method (see "Specific Heat: Experimental Techniques" in Section 9.1.4), were reported by the Geneva group [214–218]. Roulin et al. [214] reported measurements on a twinned YBCO crystal with $B \parallel c$-axis that showed steps in C for $B \leq 8$ T. A comparison of ΔC and M data with an Ehrenfest relation suggested a reversible thermodynamic second-order transition. On the basis of that result, and several other considerations, including the absence of hysteresis in their measurements and in others, and a comparison with unpublished Grenoble measurements that also showed no latent heat, they concluded that the observed second-order step constitutes the thermal signature of the melting of a vortex solid. However, later measurements, on a twinned crystal with weaker pinning, *did* show a latent heat, $L \approx 0.6 k_B T$, but only for $B \geq 6 T$, where there were no M data for comparison [215,218]. For $B \leq 6 T$, where there was no measurable latent heat, values of ΔM suggested a latent heat an order of magnitude *greater* than that observed for $B \geq 6$ T, showing that, in this case, the transition was not thermodynamically reversible.

9.6.2. Other Measurements on YBCO

The results of specific-heat measurements for $B \leq 9$ T on the sample used by Schilling et al. [211], including measurements with different orientations of the magnetic field, were reported in more detail in later papers [219–221]. For $B \parallel c$-axis, they are shown in Figure 9.22. For $B \geq 0.75 T$, there are sharp peaks in C/T, which give the values of ΔS for a thermodynamic first-order melting transition [211]; for $B \geq 0.25$ T, there are discontinuities in C/T, which give the values ΔC for the transition. The specific heat of the fluid is larger than the solid, as is usually the case, but ΔC was thought to be $\sim$100 times larger than expected for

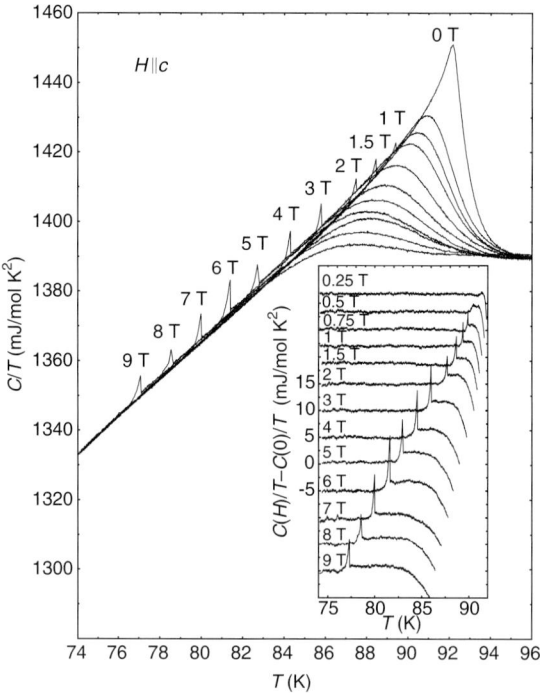

Figure 9.22. Specific heat of an untwinned single crystal of YBCO ($T_c = 92$ K) with $H \parallel c$-axis showing the first thermally detected, first-order, flux-lattice melting (data are from Schilling et al. [219]). The small, sharp anomalies, with magnitudes $\sim 0.8\%$ of C/T, the thermal signatures of the first-order transitions, give ΔS and the latent heat of melting. In the inset, $[C(H) - C(0)]/T$ shows the steps in C/T associated with the discontinuity in specific heat more clearly. The features associated with the transition are a maximum at ~ 6 T, but a transition is still observed for $H = 0.25$ T. (H in the figure, and here in the caption, is B in the text.)

the additional fluctuations for the translational degrees of freedom in the liquid phase. However, the theory [212, 213] that accounts for the values of ΔS and ΔM is also in satisfactory agreement with the values of ΔC.

Measurements on the same sample [211] were repeated and extended to $B = 18$ T, but only for $B \parallel c$-axis, in a collaboration with a group at the National High Magnetic Field Laboratory at Los Alamos National Laboratory [222]. Below 9 T, there was excellent agreement between the two sets of measurements. The latent heat, which had been essentially constant for $B \leq 9$ T, decreased at higher B and went to zero at a critical field B_{cp}, with $12 \leq B_{cp} \leq 13$ T. The results clearly established that the existence of a transition from first-order to second-order melting at B_{cp}. For $B > B_{cp}$ no anomalies were detected in temperature scans at constant B, which was probably because of insufficient instrumental resolution. Sweeps of B, from 0 to 18 T to 0, at fixed T were made to establish the nature of the transition to the fluid phase in both the reversible and irreversible regions of the phase diagram. In the temperature region of the phase diagram where B crossed the melting curve at $B < B_{cp}$ the scans were reversible. At lower temperatures, in the glassy region of the phase diagram, where the crossings were at $B > B_{cp}$, the scans were not reversible, i.e., "jumps" occurred in the T vs. B curves for increasing B, as though flux were increasing in "bundles" instead of smoothly varying. The resulting scatter in the data may also have been caused by one vortex configuration suddenly "jumping" into another as the vortex density was increased. The jumps were not observed for scans where B crossed into the irreversible glassy region from above.

For $B \leq 9\,\text{T}$, measurements were made on the same sample with $B \parallel a$-axis as well as for $B \parallel c$-axis, and with B at intermediate angles of 30, 60, and 75° with the c-axis [219–221]. Both the melting curves, $B_\text{m}(T)$, and the entropies of melting for $B \parallel c$-axis and $B \parallel a$-axis scale with an anisotropy ratio $\gamma = (m_c/m_{ab})^{1/2} = 7.76$, where m_c and m_{ab} are the charge-carrier effective masses. The $B_\text{m}(T)$ curves could be empirically fitted with the scaling relation $B_\text{m}(T,0) = B_\text{m}(T,0)[1 - T/T_c]^n$ for $B \parallel c$-axis, and $B_\text{m}(T) = \gamma B_\text{m}(T,0)[1 - T/T_c]^n$ for $B \parallel a$–axis, where $n = 1.26$, $\gamma = 7.71$, $B_\text{m}(0) = 91.5\,\text{T}$, and $T_c = 91.97\,\text{K}$. The fitted value of n is not very different from the 4/3 for 3D XY scaling. An excellent fit to the data for all angles was obtained using the scaling relationship $B_\text{m}(T,\Theta) = B_\text{m}(T,0)\{\gamma/[\sin^2(\Theta) + \gamma^2 \cos^2(\Theta)]^{1/2}\}$. The value for γ is sample dependent; but the angular scaling rules apply generally.

Junod et al. [223] report on a number of specific-heat measurements made by the Geneva and Grenoble groups. Both twinned and untwinned single crystals of $YBa_2Cu_3O_{6+x}$ ($x = 0.95$–1) were investigated for $0 \leq B \leq 23\,\text{T}$, with $B \parallel c$-axis, and $B \perp c$-axis. They conclude that, depending on flux-pinning and disorder, two processes can occur at the $B_\text{m}(T)$ curve, second-order melting of a vortex glass or first-order melting of a vortex lattice. They identified B_cp, a tricritical point on the $B_\text{m}(T)$ curve at which there is a crossover from vortex-lattice melting (first-order transition) to vortex-glass melting (second-order transition); and $B_\text{end}(x)$, the field at which the steps associated with the vortex-glass melting vanish. For $B < B_\text{cp}$ they found that the data could be fitted with the 3D XY model, to $B_\text{m}(T, x) = f B_\text{m}(0, x)[1 - T/T_c(x)]^{4/3}$, where $f = 1$ or γ for $B \parallel c$-axis or $B \perp c$-axis, respectively. This result is similar to that reported by Schilling et al. [219–221].

The effect on specific heat of the variation of the oxygen content of the chains in $YBa_2Cu_3O_{6+x}$ for $x = 0.94, 0.96$, and 1 was investigated [217] for high-purity, twinned single crystals with $0 \leq B \leq 16\,\text{T}$ for $B \parallel c$-axis. It was reported that $L(x)$, the slope of $B_\text{m}(T, x)$, and $B_\text{cp}(x)$ increase with x; and, the $\gamma(x)$ anisotropy decreases and $B_\text{end}(x)$ vanishes for $x = 1$.

The most complete and detailed vortex-state phase diagram for YBCO is provided by specific-heat measurements, by the ac technique (see "Specific Heat: Experimental Techniques" in Section 9.1.4), and magnetization measurements in magnetic fields to 26 T. The magnetization data showed the relation of the melting line to the irreversibility line, and allowed a correlation to be made between the change in the melting at the critical point and disorder in the solid phase that produces a vortex "glass," for which the melting is second order. The measurements were made by the Grenoble group on a naturally untwinned single crystal with $T_c = 92\,\text{K}$ from Argonne National Laboratory. This crystal is not the one studied by Schilling et al. [211, 219–221], but its properties were similar. The results of the measurements are reported in papers by Bouquet et al. [224,225] and Marcenat et al. [226]. Figure 9.23 shows the evolution of C and ΔS with increasing B, and Figure 9.24 shows the B–T phase diagram for the vortex state. There is a critical point, at $B = B_\text{cp}$, on the melting line, where the melting changes from first order to second order. In this case $B_\text{cp} = 10.5\,\text{T}$. Examples of the specific heats in both regions are shown in Figure 9.23, and the phase diagram derived from them is shown in Figure 9.24. For $B < B_\text{cp}$ the specific heats are characterized by sharp peaks superimposed on steps corresponding, respectively, to the latent heats for a first-order transition and the difference in specific heats between the solid and fluid. The transition is hysteretic, as is frequently observed for a first-order transition. For $B \leq 5\,\text{T}$ the features in C and M are as expected for a thermodynamic first-order melting of a disorder-free lattice. Within

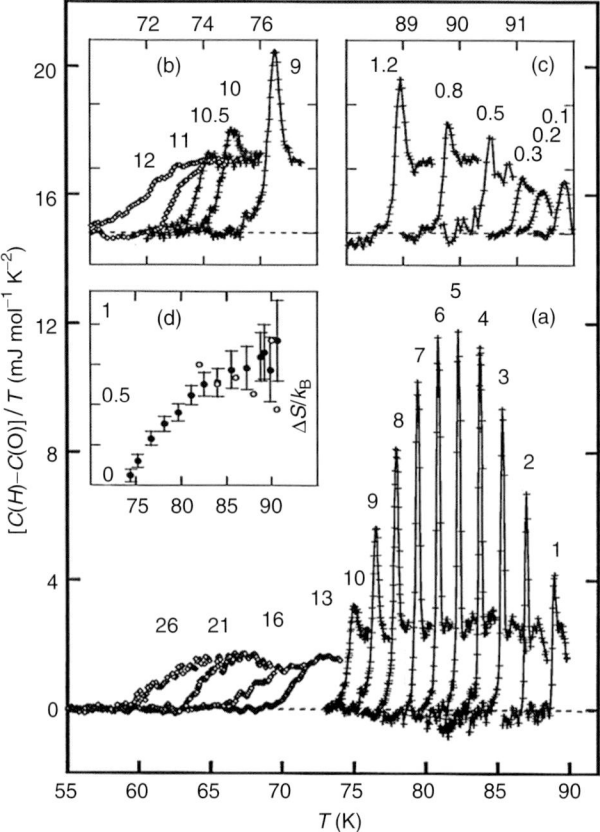

Figure 9.23. ΔS and discontinuities in $[C(H) - C(0)]/T$, at H_m and H_x (constructed from data shown in Figure 2 from Bouquet et al. [225]). (a) Specific-heat data for representative fields from 1 to 26 T. (b) and (c) Specific-heat data on an expanded temperature scale for additional fields in the vicinity of the critical point and for low H, respectively. The data have been shifted vertically to bring them into coincidence at low T. (d) Discontinuities in entropy, ΔS, in units of k_B per vortex per CuO_2 layer, at H_m as measured directly (closed circles) and calculated from ΔM and the Clapeyron equation (open circles). (H in the figure, and here in the caption, is B in the text.)

that range of field the melting obeys the 3D XY scaling relation, $T_m = T_c[1 - (B/B_0)^{3/4}]$ with $T_c = 92.3$ K and $B_0 = 100.3$ T, and the values of ΔS measured directly and those that are derived from ΔM using the Clapeyron equation, $dB_m/dT = -\Delta S/\Delta M$, are in good agreement. Although the transition remains first order for $5 \leq B \leq B_{cp}$, the melting is irreversible, and there are deviations from the 3D XY scaling relation that increase with increasing B. Above B_{cp} the lattice is glass like and transforms irreversibly into a fluid at $B^*(T)$ as shown by magnetization measurements, but with no distinguishable feature in the specific heat. (As noted above, irreversibility was also observed in a similar crystal in this region by Schilling et al. [222] in sweeps of B at constant T.) There is a reversible, thermodynamic, second-order transition from this fluid phase to a second fluid phase at $B_x(T)$. The transition is marked by a step in the specific heat with no hysteresis. The unusual behavior at $B_x(T)$ was predicted theoretically on general duality arguments [227, 228] and by modeling using computer simulations [229]. The transition is unusual in that it is the high-T phase, with unbound vortex loops threading the fluid, for which the order parameter is finite. Theoretical calculations of

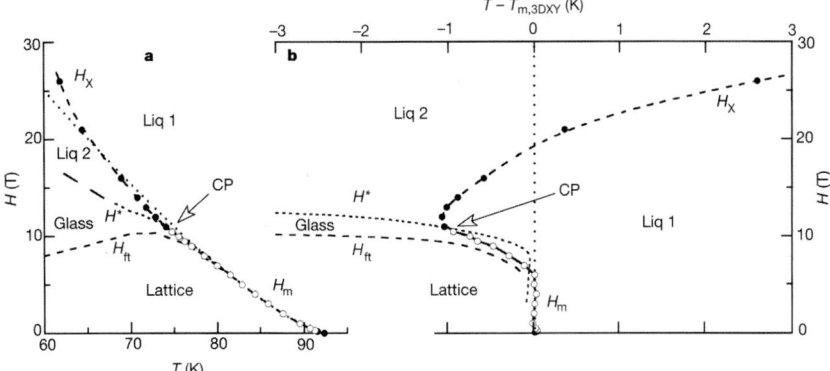

Figure 9.24. Phase diagram for the vortex state in the vicinity of T_c (Figure 9.1 from Bouquet et al. [225]): (a) as H vs. T; (b) on an expanded scale as deviations in T from a 3D XY scaling relation, the dotted line in both (a) and (b), which represents the line of first-order melting for $H \leq 5\,T$. H_m is the curve of first-order melting defined by discontinuities in entropy (open circles). H_x is the phase boundary between the two fluid phases defined by discontinuities in specific heat (filled circles). H_{ft} represents the maxima of the "fishtail" in the magnetization loops and indicates the region in which the lattice-glass transition occurs. H^* is a curve below which irreversibility is important and is defined by the points where the magnetization loops close. CP is the critical point that marks the crossover from first- to second-order melting. (H in the figure, and here in the caption, is B in the text.)

ΔC and ΔS for the first-order transition and B_{cp} [230] are in approximate agreement with the experimental values.

High-resolution magnetic-torque (τ) measurements [231] on an untwinned single crystal of YBCO were used to show that the first-order vortex melting near T_c persists well within the fluctuation region and is detectable to within $T/T_c = 0.995$. A transverse ac magnetic field was used to "shake" the vortices and produce rapid depinning, which dramatically extends the reversible domain in the B–T phase diagram [231, 232]. A theory by Brandt and Mikitik [233] explains the effectiveness of the technique. The same technique was used to measure M as a function of B for different orientations of B to the crystal axes [234]. The torque experienced by an untwinned single crystal of YBCO for $0 < B \leq 7\,T$, below $T_c = 93.3\,K$, was determined as a function of the angle, Θ, between B and the c-axis of the crystal. The measurements allowed the evaluation of $M(B, \Theta)$ and the discontinuity in M, $\Delta M(B, \Theta)$, at $B_m(T, \Theta)$. Thermodynamic relationships were used to show that ΔM is always parallel to M. From the measured $(\partial \tau / \partial B)_T$ it was possible to extract the differences in the reduced specific heat, $\Delta C/T$, between the vortex-fluid and vortex-solid phases, which compared well with the corresponding thermal data. The melting curves $B_m(T, \Theta)$ scale very well according to the scaling rules for anisotropic superconductors up to $T/T_c = 0.99$ (see, e.g., [219, 221]). These measurements were made on a third crystal from Argonne National Laboratory, smaller than, but otherwise similar to, the two mentioned above.

High-resolution dilatometry was used to measure the thermal expansion of a naturally untwinned single-crystal of YBCO [235]. Distinct discontinuities in the thermal expansion are observed at the vortex-lattice melting transition. These results demonstrate that there is coupling between the crystal lattice and vortex lattice. Furthermore, they explain why T_c is so strongly dependent on pressure.

9.6.3. Measurements on Other HTS

Specific-heat measurements have been used to study vortex-lattice melting in a number of RBCO: Specific-heat measurements on a high-purity, twinned, single crystal of DyBa$_2 \times$ Cu$_3$O$_7$ for $0 \leq B \leq 16$ T and $B \parallel c$-axis showed first-order-like transitions for $6 \leq B \leq 16$ T on the $B_m(T)$ melting curve [236]. The measured characteristics for the vortex-lattice melting closely matched those of YBCO. The Geneva group has reported magnetization and specific-heat measurements for EuBa$_2$Cu$_3$O$_7$ and DyBa$_2$Cu$_3$O$_7$ [237] and single-crystal NdBa$_2$Cu$_3$O$_7$ [238] that showed vortex-lattice melting similar to that observed in YBCO. Vortex-lattice melting was observed in specific-heat measurements on an untwinned single crystal of NdBa$_2$Cu$_3$O$_7$ that were essentially the same as YBCO, but with a somewhat larger anisotropy [239]. Vortex-lattice melting in single crystal DyBa$_2$Cu$_3$O$_{6+x}$ was also observed [161] in specific-heat measurements in fields to 8 T.

Steps were observed in $M(B, T)$ near T_c for single crystals of (La$_{2-x}$Sr$_x$)CuO$_4$ ($0.092 \leq x \leq 0.154$), consistent with first-order vortex-lattice melting [240, 241], and in single-crystal (Hg$_{0.8}$Cu$_{0.2}$)Ba$_2$CuO$_4$ [242].

Specific-heat measurements on naturally untwinned, single-crystal Y$_2$Ba$_4$Cu$_8$O$_{16}$ near T_c for $0 \leq B \leq 14$ T and B parallel to either the c- or b-axes, show a broadened step for both orientations that is consistent with a second-order transition from a vortex glass to a fluid [243]. The average anisotropy ratio, $<\gamma> = B_m \parallel b$-axis/$B_m \parallel c$-axis was 9. For both directions $B_m(T)$ was fitted with the empirical relation $B_m(T) = B_m(0)(1 - T/T_c)^n$ with $n = 1.32$ (4/3 to within experimental accuracy) that suggests 3D XY scaling. The values of the other fitted parameters were 78.0 K for T_c, 28 T for $B_m(0) \parallel c$-axis, and 252 T for $B_m(0) \parallel b$-axis. The results for the specific-heat measurements were similar to those obtained with magnetic-torque measurements [244, 245].

Thermal effects associated with vortex-lattice melting in BSCCO, which are expected to be two orders of magnitude smaller than in YBCO, have not been observed. However, in a very impressive series of measurements, vortex-lattice melting in single-crystal Bi$_2$Sr$_2$CaCu$_2$O$_8$ has been extensively studied with magnetization measurements by a group at The Superconductivity Laboratory, Weizmann Institute of Science, and their collaborators [207, 246–255]. (See also their web page, which provides references and very impressive motion pictures of the vortex lattice during solidification/melting, at http://www.weizmann.ac.il/home/fnsup/research.html.) The measurements are made using very sensitive Hall sensor arrays, which can detect the minute differences in magnetization between the fluid and solid phases. Through use of the transverse ac magnetic field technique, pioneered by Willemin et al. [231, 232], they were able to de-pin the vortices and eliminate hysteresis loops in crossing the irreversibility line. The second-order transition becomes a first-order transition when this technique is used, and the tricritical point vanishes. The effects of oxygen content, point defects, and columnar defects (produced by radiation) on the vortex-lattice melting were also studied. Apart from the magnitudes of the thermodynamic effects, the results are similar to those found for RBCO.

9.7. Calorimetric Evidence for the Pseudogap

The experimental evidence for a pseudogap in the normal-state EDOS of under-doped HTS is reviewed by Timusk and Statt [186]. They report that the first evidence came from NMR measurements [256, 257] on two YBa$_2$Cu$_3$O$_{6+x}$ samples, a slightly over-doped sample ($x = 0.97$) and an under-doped sample ($x = 0.64$). The first paper [256] reported marked differences between the two samples in the temperature dependence of the ^{63}Cu spin-relaxation parameter, $1/T_1T$, as the samples were cooled through T_c. The second paper [257] reported that the ^{63}Cu Knight shift for the $x = 0.97$ sample is essentially constant above T_c and then decreases markedly just below T_c, while the data for $x = 0.64$ decrease gradually with T both above and below T_c. The results from both sets of measurements were interpreted as indicating the under-doped sample was characterized by a gap in its EDOS above T_c in addition to the usual gap associated with the superconductivity that opens at T_c. Since 1989, numerous other NMR measurements have been made on YBa$_2$Cu$_3$O$_{6+x}$ as well as on other HTS (see, e.g., [186] for references). Most of these results confirm the existence of a pseudogap in under-doped samples, although there are differences among authors as to the relation between this pseudogap and the superconducting gap.

In addition to NMR, there are a number of other experimental techniques that have been used to study the pseudogap: ARPES, tunneling spectroscopy, transport properties, electric Raman scattering, magnetic nuclear scattering, and specific heats. All of the calorimetric results that provide support for the existence of a pseudogap in the HTS were obtained by Loram and his coworkers. Although this group has reported measurements on several HTS, their first, and most widely quoted results [16], were obtained on YBa$_2$Cu$_3$O$_{6+x}$ with $0 \leq x \leq 0.97$. Moreover, it is only in the case of these measurements that the authors have presented a relatively detailed discussion of the complex experimental and analytic procedures used to obtain their results. Therefore, in the following, we shall concentrate on those measurements.

Loram and his collaborators use a differential calorimeter [20] in which they can measure both $\delta C \equiv C_x - C_r$ and $R_x \equiv C_x/C_r$, where C_r is the heat capacity of a reference sample, C_x is the heat capacity of the sample of interest, and $R_x \approx 1$. It is claimed [20] that with this calorimeter it is possible to measure R_x with a precision of 0.01% and δC with a precision of 1%. The reference sample is chosen to have a lattice specific heat similar to that of the sample of interest. Therefore, a large part of the lattice contribution cancels in δC, which is one of the quantities measured. The electron contribution, the quantity of interest in connection with the pseudogap, is obtained from δC by making corrections for the differences in the magnetic and lattice contributions to the specific heats of the sample and the reference sample. The results for YBCO, for $1.6 \leq T \leq 300$ K, were obtained in two steps: In one paper [97] the specific heats of a series of Zn-substituted YBa$_2$(Cu$_{1-y}$Zn$_y$)$_3$O$_{6+x}$ samples ($0 \leq y \leq 0.10$, $x = 0.97$), including. A fully oxygenated, unsubstituted sample ($y = 0$, $x = 0.97$), were compared with that of a reference sample with $y = 0.07$ and $x = 0.97$. It was concluded that γ_n was independent of both y and temperature, and the electron specific heat of the $y = 0$, $x = 0.97$ sample was determined. Although the qualitative features of the evidence for the pseudogap at $T \geq 100$ K are to a significant degree independent of these results, they are discussed in Section 9.7.1. In a second paper [16] a series of Zn-free YBa$_2$Cu$_3$O$_{6+x}$ samples were compared with the same Zn-substituted reference sample, ($y = 0.07$, $x = 0.97$), to obtain the electron contribution to the specific heat as a function of x and T. These results and the conclusions for the pseudogap and are discussed in Section 9.7.2.

9.7.1. Determination of the Electron Specific Heat of YBa$_2$Cu$_3$O$_{6.97}$

To facilitate comparison with the papers of Loram and colleagues, we use their notation here and in the following section. In that notation γ is used fairly generally for C/T, where C can be the total specific heat or one of the component contributions, and γ is a temperature-dependent quantity. For their fully oxygenated sample of YBa$_2$Cu$_3$O$_{6+x}$, $x = 0.97$, but, for notational convenience, we refer to this as the $x = 1$ sample. It is the variation of

$\gamma_x(T) \equiv C_{ex}/T$, where C_{ex} is C_e for the sample with $(6+x)$ O atoms, with x and T that provides evidence for the pseudogap.

The electron contribution to the specific heat of the Zn-free, fully oxygenated sample was obtained by making corrections to the total specific heat for the lattice and magnetic contributions, which were deduced from the y-dependence of the differences in specific heats of the Zn substituted samples. As noted in Section 9.3.3, the Loram group attributed a part of the electron contribution to the specific heat of the Zn-substituted samples to a magnetic contribution, which they combined with an approximately y-proportional "upturn" to obtain an "intrinsic," but y-dependent, C_{mag} [97]. Our analysis of their data [97] in Section 9.3.3 indicates that as the level of Zn substitution increases γ_r becomes considerably larger than the expected γ_n. Their "raw data" for δC suggest that for low T the T-proportional term in C increases with increasing Zn content (see Figure 4 of [97]) and that for $T \geq 100\,\mathrm{K}$ it decreases with increasing Zn content (see Figures 3 and 4 of [97]). However, these trends in the T-proportional term were attributed to changes in C_{lat} with Zn content. The changes in C_{lat} were said to be in qualitative agreement with changes in the phonon spectrum deduced from neutron-scattering data, but as recognized elsewhere in the same paper, neutron data do not define the phonon spectrum to the accuracy that would make the comparison meaningful. The conclusion that γ_n is independent of Zn content and independent of T is a direct consequence of the conclusions about C_{mag}, γ_r, and C_{lat}, which are not well supported by the experimental data.

9.7.2. Use of the Differential Method to Obtain the Conduction-Electron Specific Heat of $YBa_2Cu_3O_{6+x}$—A Simplified Discussion

The evidence for the pseudogap was obtained by comparing the specific heats of a series of samples of $YBa_2Cu_3O_{6+x}$, with different O contents, x, with that of the reference sample [16]. The total specific heat, C_x, can be written $C_x = C_x' + \gamma_x T$, where C_x' is the non-electronic contribution to C_x. C_x' is essentially the lattice contribution except at quite low temperatures, a temperature region in which a magnetic contribution, C_{mag}, cannot be ignored (see Sections 9.3.1 and 9.7.1). In the case of $x = 0$, C_0' is assumed to have no electronic contribution, i.e., $C_0' = C_0$. While the C_x's are all measured in terms of C_r, it is the C_x's referred to C_0 that are of interest. In particular, we can write:

$$(C_x - C_0) = C_r(R_x - R_0) = (C_x' - C_0) + \gamma_x T, \tag{9.14}$$

and

$$(C_1 - C_0) = C_r(R_1 - R_0) = (C_1' - C_0) + \gamma_1 T. \tag{9.15}$$

The first term on the right side of the second equality in Eqs. (9.14) and (9.15) involves (except at low temperatures) only the differences in phonon terms. It is assumed by the Loram group that these differences are simply related to one another. That is, it is *assumed* that:

$$(C_x' - C_0) = x(C_1' - C_0) = x(C_1 - \gamma_1 T - C_0). \tag{9.16}$$

From Eqs. (9.14), (9.15), and (9.16) it follows that:

$$(C_x - C_0) - (C_1 - C_0) = x(C_1 - C_0) - x\gamma_1 T + \gamma_x T. \tag{9.17}$$

Therefore:

$$\gamma_x = x\gamma_1 + (C_r/T)[(R_x - R_0) - x(R_1 - R_0)]. \tag{9.18}$$

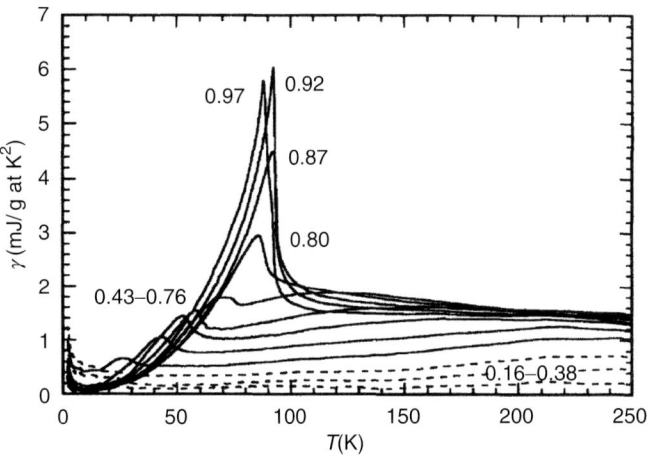

Figure 9.25. Temperature-dependent electronic specific heat coefficient, $\gamma = C_e/T$, for $YBa_2Cu_3O_{6+x}$ (from Loram et al. [16, 258], Figure 4 from [16]) showing the evolution of the pseudogap as a function of x. The dashed lines indicate γ for nonsuperconducting samples. (Note: γ is referred to as γ_x in the text). For a discussion of how γ was obtained, through the use of differential calorimetry, see the text, Loram [20], and Loram et al. [16, 97].

We see from Eq. (9.18) that γ_x differs from γ_1 by a term that depends upon R_0, R_1, and R_x (all known with high precision), and C_r, the reference sample heat capacity that is common to all the measurements. Therefore, the values of $(\gamma_x - x\gamma_1)T/C_r$ can be calculated with considerable precision, and if γ_1 is known the numerical values of γ_x can be determined. Figure 9.25 (from [17]) shows the value of γ_1 derived in [97], as discussed in Section 9.7.1, and the γ_x derived with the use of Eq. (9.18). The curve for γ_1 is essentially constant, at temperatures above $\sim$100 K (except for the high-temperature "tail" of the fluctuation contribution to γ_x—see Section 9.5). The spacing of the γ_x curves relative to γ_1 in this higher-temperature region is due primarily to the first term on the right-hand side of Eq. (9.18), but it is the second term on the right-hand side of Eq. (9.18) that determines the deviations from constancy in γ_x above $\sim$100 K. In particular, it is this second term that is responsible for the positive slopes evident in the γ_x curve for the smaller values of x. These positive slopes correspond to decreases in the normal-state EDOS with decreasing T, and, therefore, to the existence of a pseudogap. The positive slopes are due primarily to the values of $(C_r/T)[(R_x - R_0) - x(R_1 - R_0)]$ and only secondarily to the value of γ_1. That is, the evidence for a pseudogap provided by the calorimetric data does not depend critically upon γ_1, and is, at least qualitatively, not subject to the reservations about the value of γ_1 expressed in Section 9.7.1. However, the values of the EDOS represented by the γ_x at $T \leq 100$ K are subject to the uncertainty about the T independence of γ_1 in that region, and a possible, or even probable, T dependence of γ_1.

The procedure used in [16] to obtain γ_x from their measurements is considerably more complicated than the one we have used to obtain Eq. (9.18). However, we believe the two procedures are essentially equivalent. Both rely on Eq. (9.16), that is, on the assumption that there exists a simple relationship among the lattice specific heats of the various samples of $YBa_2Cu_3O_{6+x}$. This assumption is reasonable and is not inconsistent with neutron-scattering results. Nevertheless, its primary justification lies in the consistency between inferences drawn from Figure 9.25 and determinations of the pseudogap made by other experimental techniques (see, e.g., [186]).

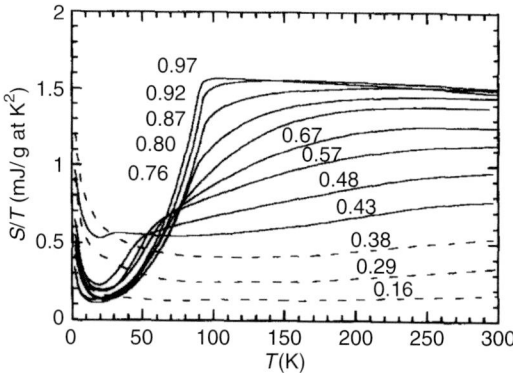

Figure 9.26. Entropy divided by temperature for $YBa_2Cu_3O_{6+x}$ for $0.16 \leq x \leq 0.97$ (from Loram et al. [258], Figure 4). The dashed lines indicate S/T for nonsuperconducting samples.

9.7.3. Other Specific-Heat Results and Their Interpretation

It was pointed out [16] that the changes in the normal-state EDOS will affect entropies both above and below T_c. Figure 9.26 (taken from [258]) is a plot of S/T vs. T (where S is the electronic entropy) for $YBa_2Cu_3O_{6+x}$. The close resemblance of this plot to a plot of magnetic susceptibility vs. T leads the authors to conclude that the normal-state quasiparticles behave like a normal Fermi liquid [259]. A more complete comparison of the entropy and susceptibility data for $YBa_2Cu_3O_{6+x}$ can be found in [260].

Later specific-heat measurements made by Loram and his coworkers on other HTS (see, e.g., [182] for discussions and references) are similar to those reported for $YBa_2Cu_3O_{6+x}$. Many of their conclusions are widely quoted, and, as has been noted, are consistent with other measurements pertaining to the pseudogap. Still, the normal state of the HTS is not well understood. Except for the pioneering work of Loram and his collaborators, there are few accurate determinations of the electronic specific heat above T_c. It is important that their efforts be supplemented by those of other experimental groups. It is unlikely that such measurements can be made without the use of differential calorimetric techniques. The measurements are difficult, but the difficulties involved in interpreting the differential results could be reduced by the use of reference samples that have little or no electronic contribution to C.

Bibliography

1. A. Junod, in: *Physical Properties of High Temperature Superconductors II*, D.M. Ginsberg (ed.) (World Scientific, Singapore, 1990), p. 13.
2. T. Atake, Thermochim. Acta **174**, 291 (1991).
3. N.E. Phillips, R.A. Fisher, and J.E. Gordon, in: *Progress in Low Temperature Physics*, vol. XIII, D.F. Brewer (ed.) (North Holland, Amsterdam, 1992), p. 267.
4. R.A. Fisher, J.E. Gordon, and N.E. Phillips, Annu. Rev. Phys. Chem. **47**, 283 (1996).
5. L.H. Greene and B.G. Bagley, in: *Physical Properties of High Temperature Superconductors II*, D.M. Ginsberg (ed.) (World Scientific, Singapore, 1990), p. 509.
6. A.P. Malozemoff, in: *Physical Properties of High Temperature Superconductors I*, D.M. Ginsberg (ed.) (World Scientific, Singapore, 1989), p. 71.
7. J.F. Schooley, W.R. Hosler, and M.L. Cohen, Phys. Rev. Lett. **12**, 474 (1964).
8. A.W. Sleight, J.L. Gillson, and P.E. Bierstedt, Solid State Commun. **17**, 27 (1975).
9. L.F. Mattheiss, E.M. Gyorgy, and D.W. Johnson, Phys. Rev. B **37**, 3745 (1988).

10. J.G Bednorz and K.A. Müller, Z. Phys. B **64**, 189 (1986).
11. J.P. Emerson, D.A. Wright, B.F. Woodfield, J.E. Gordon, R.A. Fisher, and N.E. Phillips, Phys. Rev. Lett. **82**, 1546 (1999).
12. L. Krushin-Elbaum, A.P. Malozemoff, Y. Yeshurun, D.C. Cronemeyer, and F. Holtzberg, Phys. Rev. B **39**, 2936 (1989).
13. K. Nakao, N. Miura, K. Tatsuhara, H. Takeya, and H. Takei, Phys. Rev. Lett. **63**, 97 (1989).
14. P. Chaudhari, R.T. Collins, P. Freitas, R.J. Gamnino, J. R. Kirtley, R.H. Koch, R.B. Laibowitz, F.K. LeGoues, T.R. McGuire, T. Penney, Z. Schlesinger, A.P. Segmuller, S. Foner, and E.J. McNiff, Jr., Phys. Rev. B **36**, 8903 (1987).
15. Y. Wang, B. Revaz, A. Erb, and A. Junod, Phys. Rev. B **63**, 94508 (2001).
16. J.W. Loram, K.S. Mirza, J.R. Cooper, and W.Y. Liang, Phys. Rev. Lett. **71**, 1740 (1993).
17. J.G. Bednorz, M. Takashige, and K.A. Müller, Europhys. Lett. **3**, 379 (1987).
18. K.A. Müller, M. Takashige, and J.G. Bednorz, Phys. Rev. Lett. **58**, 1143 (1987).
19. N.R. Werthamer, E. Helfand, and P.C. Hohenberg, Phys. Rev **147**, 295 (1996).
20. J.W. Loram, J. Phys. E: Sci. Instrum. **16**, 367 (1983).
21. A. Schilling and O. Jeandupeux, Phys. Rev. B **54**, 9714 (1995).
22. R. Bachmann, F.J. DiSalvo, Jr., T.H. Geballe, R.L. Greene, R.E. Howard, C.N. King, H.C. Kirsch, K.N. Lee, R.E. Schwall, H.-U. Thomas, and R.B. Zubeck, Rev. Sci. Instrum. **43**, 205 (1972).
23. Y. Wang, T. Plackowski, and A. Junod, Physica C **355**, 179 (2001).
24. D. Sanchez, A. Junod, J.-Y. Genoud, T. Graf, and J. Muller, Physica C **200**, 1 (1992).
25. J.P. Emerson, R.A. Fisher, N.E. Phillips and D.A. Wright, Phys. Rev. B **49**, 9256 (1994); R.A. Fisher, unpublished (1993).
26. P.F. Sullivan and G. Seidel, Phys. Rev. **173**, 679 (1968).
27. D.A. Wright, J.P. Emerson, B.F. Woodfield, J.E. Gordon, R.A. Fisher, and N.E. Phillips, Phys. Rev. Lett. **82**, 1550 (1999).
28. P.W. Anderson, G. Baskaran, Z. Zou, and T. Hsu, Phys. Rev. Lett. **58**, 2790 (1987).
29. Q. Chen, I. Kosztin, and K. Levin, Phys. Rev. Lett. **85**, 2801 (2000).
30. N.E. Phillips, J.P. Emerson, R.A. Fisher, J.E. Gordon, B.F. Woodfield, and D.A. Wright, J. Supercond. **7**, 251 (1994).
31. J.Y. Uemura, Solid State Commun. **120**, 347 (2001).
32. R.J. Radtke, K. Levin, H.-B. Schüttler, and M.R. Norman, Phys. Rev. B **48**, 653 (1993); L.S. Borkowski and P.J. Hirschfeld, Phys. Rev. B **49**, 15404 (1994); P.J. Hirschfeld, W.O. Putikka, and D.J. Scalapino, Phys. Rev. B **50**, 10250 (1994).
33. N.E. Phillips, R.A. Fisher, J.E. Gordon, S. Kim, A.M. Stacy, M.K. Crawford, and E.M. McCarron, Phys. Rev. Lett. **65**, 357 (1990).
34. K.A. Moler, D.L. Sisson, J.S. Urbach, M.R. Beasley, A. Kapitulnik, D.J. Baar, R. Liang, and W.N. Hardy, Phys. Rev. B **55**, 3954 (1997).
35. R.A. Fisher, N.E. Phillips, A. Schilling, B. Buffeteau, R. Calemczuk, T.E. Hargreaves, C. Marcenat, K.W. Dennis, R.W. McCallum, and A.S. O'Connor, Phys. Rev. B **61**, 1473 (2000).
36. H.-H. Wen, Z.-Y. Liu, F. Zhou, J. Xiong, W. Ti, J. Xiang, S. Komiya, X. Sun, and Y. Ando, Phys. Rev. B **70**, 214505 (2004).
37. A. Chakraborty, A.J. Epstein, D.L. Cox, E.M. McCarron, and W.E. Farneth, Phys. Rev. B **39**, 12267 (1989).
38. J.S. Urbach, D.B. Mitzi, A. Kapitulnik, J.Y.T. Wei, and D. E. Morris, Phys. Rev. B **39**, 12391 (1989).
39. A. Junod, D. Eckert, G. Triscone, V.Y. Lee, and J. Muller, Physica C **162–164**, 476 (1989).
40. G.Kh. Panova, M.N. Khlopkin, N.A. Chernoplekov, A.V. Suetin, B.I. Savel'ev, A.I. Akimov, L.P. Poluchankma, and A.P. Chernyakova, Supercond. Phys. Chem. Technol. **4**, 60 (1991).
41. K.A. Müller, Nature **377**, 133l (1995).
42. R.A. Klemm, Philos. Mag. **85**, 801 (2005).
43. G.-M. Zhao, Phys. Rev. B **64**, 24503 (2001).
44. B.H. Brandow, Phys. Rev. B **65**, 54503 (2002).
45. B.H. Brandow, Philos. Mag. **83**, 2487 (2003).
46. C. Caroli, P.G. deGennes, and J. Matricon, Phys. Lett. **9**, 307 (1964).
47. K. Maki, Phys. Rev. **139**, A702 (1965).
48. G.E. Volovik, JETP Lett. **58**, 469 (1993).
49. N.B. Kopnin and G.E. Volovik, JETP Lett. **64**, 690 (1996).
50. G.E. Volovik, JETP Lett. **65**, 491 (1997).
51. S.H. Simon and P.A. Lee, Phys. Rev. Lett. **78**, 1548 (1997).

52. G.E. Volovik and N.B. Kopnin, Phys. Rev. Lett. **78**, 5028 (1997).
53. K.A. Moler, D.J. Baar, J.S. Urbach, Ruixing Liang, W.N. Hardy, and A. Kapitulnik, Phys. Rev. Lett. **20**, 2744 (1994).
54. R.A. Fisher, J.E. Gordon, S.F. Reklis, D.A. Wright, J.P. Emerson, B.F. Woodfield, E.M. McCarron, and N.E. Phillips, Physica C **252**, 237 (1995).
55. D.A. Wright, J.P. Emerson, B.F. Woodfield, S.F. Reklis, J.E. Gordon, R.A. Fisher, and N.E. Phillips, J. Low Temp. Phys. **105**, 897 (1996).
56. S.F. Chen, C.F. Chang, H.L. Tsay, H.D. Yang, and J.-Y. Lin, Phys. Rev. B **58**, 14753 (1998).
57. N. Momono, M. Ido, T. Nakano, M. Oda, Y. Okajima, and K. Yamaya, Physica C **233**, 395 (1994).
58. N. Momono and M. Ido, Physica C **264**, 311 (1996).
59. B. Revaz, J.-Y. Genoud, A. Junod, K. Neumaier, A. Erb, and E. Walker, Phys. Rev. Lett. **80**, 3364 (1998).
60. A.P. Ramirez, Phys. Lett. A **211**, 59 (1996).
61. F.R. Drymiotis, J.C. Lashley, T. Kimura, G. Lawes, J.L. Smith, D.J. Thoma, R.A. Fisher, N.E. Phillips, Ya. Mudryk, V.K. Pecharsky, X. Moya, and A. Planes, Phys. Rev. B **72**, 24543 (2005).
62. I. Vekhter, P.J. Hirschfeld, J.P. Carbotte, and W.J. Nicol, Phys. Rev. B **59**, 9023 (1999).
63. J.T. Markert, Y. Dalichaouch, and M.B. Maple, in: *Physical Properties of High Temperature Superconductors I*, D.M. Ginsberg (ed.) (World Scientific, Singapore, 1989), p. 265.
64. K.A. Gschneidner, L. Eyring, and M.B. Maple (eds.), in: *Handbook on the Physics and Chemistry of Rare Earths 30 and 31: Rare Earth High Temperature Superconductivity I and II* (Elsevier, The Netherlands, 2001, 2002).
65. H. Radousky, J. Mater. Res. **7**, 1917 (1992).
66. A. Maeda, T. Yabe, K. Uchinokura, and S. Tanaka, Jpn. J. Appl. Phys. **26**, L1368 (1987).
67. C.R. Fincher and G.B. Blanchet, Phys. Rev. Lett. **67**, 2902 (1991).
68. Z. Zou, J. Ye, K. Oka, and Y. Nishihara, Phys. Rev. Lett. **80**, 1074 (1998).
69. A. Shukla, B. Barbiellini, A. Erb, A. Manuel, T. Buslaps, V. Honkimäki, and P. Suortti, cond-mat/9805225 v2 (1998).
70. V.N. Narozhnyi and S.L. Drechler, Phys. Rev. Lett. **82**, 461 (1999).
71. X.X. Tang, A. Manthiram, and J.B. Goodenough, Physica C **161**, 574 (1989).
72. I.I. Mazin, cond-mat/9903061 v3 (1999).
73. W.E. Pickett and I. I. Mazin, in: *Handbook on the Physics and Chemistry of Rare Earths 30: Rare Earth High Temperature Superconductivity I*, K.A. Gschneidner, L. Eyring, and M.B. Maple (eds.) (Elsevier, The Netherlands, 2000).
74. K. Maki, in: *Superconductivity 2*, R.D. Parks (ed.) (Dekker, New York, NY, 1969), p. 1035.
75. M.A. Subramanian, J. Gopalakrishnan, and A.W. Sleight, J. Solid State Chem. **84**, 413 (1990).
76. G.Y. Guo and W.M. Temmerman, Phys. Rev. B **41**, 6372 (1990).
77. R.S. Howland, T.H. Geballe, S.S. Laderman, A. Fischer-Colbrie, M. Scott, J.M. Tarascon, and P. Barboux, Phys. Rev. B **39**, 9017 (1989).
78. J.M. Tarascon, P. Barboux, P.F. Miceli, L.H. Greene, G.W. Hull, M. Eibschutz, and S.A. Sunshine, Phys. Rev. B **37**, 7458 (1988).
79. G. Xiao, M.Z. Cieplak, A. Gavrin, F.H. Streitz, A. Bakhshai, and C.L. Chien, Solid State Sci. **1**, 323 (1987).
80. G. Xiao, F.H. Streitz, A. Gavrin, Y.W. Du, and C. L. Chien, Phys. Rev. B **35**, 8782 (1987).
81. G. Xiao, M.Z. Cieplak, D. Musser, A. Gavrin, F.H. Streitz, C.L. Chien, J.J. Rhyne, and J.A. Gotaas, Nature (London) **332**, 238 (1988).
82. G. Xiao, M.Z. Cieplak, A. Garvin, F.H. Streitz, A. Bakshai, and C.L. Chien, Phys. Rev. Lett. **60**, 1446 (1988).
83. G. Xiao, F.H. Streitz, A. Garvin, M.Z. Cieplak, C.L. Chien, and A. Bakhshai, J. Appl. Phys. **63**, 4196 (1988).
84. G. Xiao, A. Bakhshai, M.Z. Cieplak, Z. Tesanovic, and C.L. Chien, Phys. Rev. B **39**, 315 (1989).
85. G. Xiao, M.Z. Cieplak, J.Q. Xiao, and C.L. Chien, Phys. Rev. B **42**, 8752 (1990).
86. C.L. Chien, G. Xiao, M.Z. Cieplak, D. Musser, J.J. Rhyne, and J.A. Gotaas, in: *Superconductivity and Its Applications*, H.S. Kwok and D.T. Shaw (eds.) (Elsevier, The Netherlands, 1988), p. 110.
87. M.Z. Cieplak, G. Xiao, A. Bakhshai, and C.L. Chien, Phys. Rev. B **39**, 4222 (1989).
88. M.Z. Cieplak, G. Xiao, C.L. Chien, A. Bakhshai, D. Artymowicz, W. Bryden, J.K. Stalick, and J.J. Rhyne, Phys. Rev. B **42**, 6200 (1990).
89. S. Kim, R.A. Fisher, N.E. Phillips, and J.E. Gordon, Physica C **162–164**, 494 (1989).
90. S. Vilminot, R. Kuentzler, Y. Dossmann, A. Derory, and M. Drillon, Physica C **160**, 575 (1989).
91. I. Felner, Y. Wolfus, G. Hilscher, and N. Pillmayr, Phys. Rev. B **39**, 229 (1989).
92. R. Kuentzler, S. Vilminot, Y. Dossmann, and A. Derory, Physica C **153–155**, 1032 (1988).

93. R. Kuentzler, S. Vilminot, Y. Dossmann, T.L. Wen, and C. His, Physica C **162–164**, 564 (1989).
94. B.D. Dunlap, J.D. Jorgensen, W.K. Kwok, C.W. Kimball, J.L. Matykiewicz, H. Lee, and C.V. Segre, Physica C **153–155**, 1100 (1988).
95. A. Junod, A. Bezinge, D. Eckert, T. Graf, and J. Muller, Physica C **152**, 495 (1988).
96. C.-S. Jee, S. Rahman, A. Kebede, D. Nichols, J. E. Crow, T. Mihalisin, and P. Schlottmann, Bull. Am. Phys. Soc. **33**, 465 (1988).
97. J.W. Loram, K.A. Mirza, and P.F. Freeman, Physica C **171**, 243 (1990).
98. C.-S. Jee, D. Nichols, A. Kebede, S. Rakman, J.E. Crow, A.M. Ponte Goncalves, T. Mihalisin, G.H. Meyer, I. Perez, R.E. Salomon, P. Schlottmann, S.H. Bloom, M.V. Kuric, Y.S. Tao, and R.P. Guertin, Superconductivity **1**, 63 (1988).
99. K. Remschnig, P. Rogl, R. Eibler, G. Hilscher, N. Pillmayr, H. Kirchmayr, and E. Bauer, Physica C **153–155**, 906 (1988).
100. G. Roth, P. Adelmann, R. Ahrends, B. Blank, H. Burkle, F. Gompf, G. Heger, M. Hervieu, M. Nindel, B. Obst, J. Pannetier, B. Raveau, B. Renker, H. Rietschel, B. Rudolf, and H. Wuhl, Physica C **162–164**, 518 (1989).
101. Y. Nakazawa, J. Takeya, and M. Ishikawa, Physica C **174**, 155 (1991).
102. J.W. Loram, K.A. Mirza, P.F. Freeman, and J.J. Tallon, Supercond. Sci. Technol. **4**, S184 (1991).
103. S. Wakimoto, R.J. Birgeneau, A. Kagedan, H. Kim, I. Swainson, K. Yamada, and H. Zhang, Phys. Rev. B **72**, 64521 (2005).
104. M.K. Yu and J.P. Franck, Phys. Rev. B **48**, 13939 (1993).
105. H. Alloul, P. Mendels, H. Casalata, J.F. Marucco, and J. Arabski, Phys. Rev. Lett. **67**, 3140 (1991).
106. R.E. Walstedt, R.F. Bell, L.F. Schneemeyer, J.V. Waszczak, W.W. Warren, R. Dupree, and A. Gencten, Phys. Rev. B **48**, 10646 (1993).
107. A.V. Mahajan, H. Alloul, G. Collins, and J.F. Marucco, cond-mat/9909049 v1 (1999).
108. S. Zagoulaev, P. Monod, and J. Jégoudez, Phys. Rev. B **52**, 10474 (1995).
109. M. Affronte, D. Pavuna, M. Francois, F. Licci, T. Besagni, and S. Cattani, Physica C **162–164**, 1007 (1989).
110. M. Methbod, W. Biberacher, A.G.M. Jansen, P. Wyder, R. Deltour, and P. H. Duvigneaud, Phys. Rev. B **38**, 11813 (1988).
111. J.-Y. Genoud, A. Mirmelstein, G. Triscone, A. Junod, and J. Muller, Phys. Rev. B **52**, 12833 (1995).
112. R.A. Fisher, D.A. Wright, J.P. Emerson, B.F. Woodfield, N.E. Phillips, Z.-R. Wang, and D.C. Johnston, Phys. Rev. B **61**, 538 (2000).
113. H. Zhenuhi, Z. Han, S. Shifang, C. Zuyao, Z. Qirui, and X. Jiansheng, Solid State Commun. **66**, 1215 (1988).
114. M.W. Shafer, T. Penney, B.L. Olson, R.L. Greene, and R.H. Koch, Phys. Rev. B **39**, 2914 (1989).
115. J.M. Tranquada, J.D. Axe, N. Ichikawa, Y. Nakamura, S. Uchida, and B. Nachumi, Phys. Rev. B **54**, 7489 (1996).
116. E.W. Carlson, V.J. Emery, S.A. Kivelson, and D.Orgad, Rev. Mod. Phys. **75**, 1201 (2003).
117. M.A. Kastner and R.J. Birgeneau, Rev. Mod. Phys. **70**, 897 (1998).
118. J.L. Tallon, T. Benseman, G.V.M. Williams, and J.W. Loram, Physica C **415**, 9 (2004).
119. J.A. Wilson, cond-mat/0502666 (2005).
120. H.A. Mook, P. Dai, S.M. Hayden, G. Aeppli, T.G. Perring, and F. Dogan, Nature (London) **395**, 580 (1998).
121. H.A. Mook and F. Dogan, Nature (London) **401**, 145 (1999).
122. H.A. Mook, P. Dai, F. Dogan, and R.D. Hunt, Nature (London) **404**, 729 (2000).
123. L. Pintschovious, Y. Endoh, D. Reznik, H. Hiraka, J.M. Tranquada, W. Reichardt, H. Uchiyama, T. Masui, and S. Tajima, cond-mat/0308357 v1 (2003).
124. J.L. Cohn, C.P. Popoviciu, Q.M. Lin, and C.W. Chu, Phys. Rev. B **59**, 3823 (1999).
125. P.C. Hammel and D. J. Scalapino, Philos. Mag. **74**, 523 (1996).
126. P.C. Hammel, B.W. Statt, R.L. Martin, F.C. Chou, D.C. Johnston, and S.-W. Cheong, Phys. Rev. B **57**, R712 (1998).
127. J.D. Axe, A.H. Moudden, D. Holwein, D.E. Cox, K.M. Mohanty, A.R. Moodenbaugh, and Y. Xu, Phys. Rev. Lett. **62**, 2751 (1989).
128. T. Suzuki and T. Fujita, Physica C **159**, 111 (1989).
129. T. Suzuki and T. Fujita, J. Phys. Soc. Jpn **58**, 1883 (1989).
130. M.K. Crawford, W.E. Farneth, E.M. McCarron, R.L. Harlow, and A.H. Moudden, Science **250**, 1390 (1990).
131. M.K. Crawford, W.E. Farneth, R.L. Harlow, E.M. McCarron, R. Miao, H. Chou, and Q. Huang, in: *Lattice Effects in High-T_c Superconductors*, Y. Bar-Yam, T. Egami, J. Mustre-de Leon, and A. R. Bishop (eds.) (World Scientific, Singapore, 1992), p. 531.

132. M.K. Crawford, R.L. Harlow, E.M. McCarron, W.E. Farneth, J.D. Axe, H.E. Chou, and Q. Huang, Phys. Rev. B **44**, 7749(R) (1991).
133. K. Kumagai, H. Matoba, N. Wada, M. Okaji, and K. Nara, J. Phys. Soc. Jpn **60**, 1448 (1991).
134. D.A. Wright, J.P. Emerson, B.F. Woodfield, R.A. Fisher, N.E. Phillips, M.K. Crawford, and E.M. McCarron, Physica C **185**, 1387 (1991).
135. D.A. Wright, R.A. Fisher, N.E. Phillips, M.K. Crawford, and E.M. McCarron, Physica B **194–196**, 469 (1994).
136. B. Büchner, M. Braden, M. Cramm, W. Schlabitz, W. Schnelle, O. Hoffels, W. Braunisch, R. Müller, G. Heger, and D. Wohlleben, Physica C **185–189**, 903 (1991).
137. J. Takeda, T. Inukai, and M. Sato, J. Phys. Chem. Solids **62**, 181 (2001).
138. Y. Okajima, K. Yamaya, N. Yamada, M. Oda, and M. Ido, Solid State Commun. **74**, 767 (1990).
139. O. Anegawa, Y. Okajima, S. Tanada, and K. Yamaya, Phys. Rev. B **63**, 140506(R) (2001).
140. O. Anegawa, Y. Okajima, S. Tanada, and K. Yamaya, Solid State Commun. **121**, 97 (2002).
141. S.A. Kivelson, E. Fradkin, and V.J. Emery, Nature (London) **393**, 550 (1998).
142. G. Hilscher, N. Pillmayr, R. Eibler, E. Bauer, K. Remschnig, and P. Rogl, Z. Phys. B **72**, 461 (1988).
143. S.-W. Cheong, G. Aeppli, T.E. Mason, H. Mook, S.M. Hayden, P.C. Canfield, Z. Fisk, K.N. Clausen, and J. C. Martinez, Phys. Rev. Lett. **67**, 1791 (1991).
144. T. Nakano, M. Momono, T. Nagata, M. Oda, and M. Ido, Phys. Rev. B **58**, 5831 (1998).
145. M.B. Salamon, in: *Physical Properties of High Temperature Superconductors I*, D.M. Ginsberg (ed.) (World Scientific, Singapore, 1989).
146. A.I. Larkin, and A.A. Varlamov, cond-mat/0109177 (2001).
147. A. Junod, A. Erb, and C. Renner, Physica C **317–318**, 333 (1999).
148. A. Junod, M. Roulin, B. Revaz, and A. Erb, Physica B **280**, 214 (2000).
149. Q. Chen, J. Stajic, S. Tan, and K. Levin, Phys. Rep. **412**, 1 (2005).
150. A.S. Alexandrov, cond-mat/0508769 (2005).
151. M. Tinkham, *Introduction to Superconductivity* (2nd edn) (McGraw-Hill, New York, NY, 1996).
152. N. Goldenfeld, *Lectures on Phase Transitions and the Renormalization Group* (Addison-Wesley, Reading, MA, 1992).
153. V.L. Ginzburg, Sov. Solid State **2**, 61 (1960).
154. R. Brout, Phys. Rev. **118**, 1009 (1960).
155. A. Kapitulnik, M.R. Beasley, C. Castellani, and C. DiCastro, Phys. Rev. B **37**, 537 (1988).
156. G. Mozurkewich, M.B. Salamon, and S.E. Inderhees, Phys. Rev. B **46**, 11914 (1992).
157. J.F. Annett, M. Randeria, and S.R. Renn, Phys. Rev. B **38**, 4660 (1988).
158. S.E. Inderhees, M.B. Salamon, J.P. Rice, and D.M. Ginsberg, Phys. Rev. Lett. **66**, 232 (1991).
159. S.E. Inderhees, M.B. Salamon, N. Goldenfeld, J.P Rice, B.G. Pazol, D.M. Ginsberg, J.Z. Liu, and G.W. Crabtree, Phys. Rev. Lett. **60**, 1178 (1988).
160. N.J. Garfield, M.A. Howson, G. Yang, and S. Abell, Physica C **321**, 1 (1999).
161. T. Schneider and H. Keller, Physica C **207**, 306 (1993).
162. D.S. Fisher, M.A. Fisher, and D.A. Huse, Phys. Rev. B **43**, 130 (1991).
163. J.A. Lipa, D.R. Swanson, J.A. Nissen, T.C.P. Chui, and U.E. Israelsson, Phys. Rev. Lett. **76**, 944 (1996).
164. V. Pasler, P. Schweiss, C. Meingast, B. Obst, H. Wühl, A. I. Rykov, and S. Tajima, Phys. Rev. Lett **81**, 1094 (1998).
165. I.D. Lawrie, Phys. Rev. B **50**, 9456 (1994).
166. M.B. Salamon, J. Shi, N. Overend, and M.A. Howsend, Phys. Rev. B. **47**, 5520 (1993).
167. S.W. Pierson, O.T. Valls, Z. Tesanovic, and M.A. Lindemann, Phys. Rev. B **57**, 622 (1998).
168. M. Roulin, A. Junod, and E. Walker, Physica C **260**, 257 (1996).
169. N. Okazaki, T. Hasegawa, K. Kishio, K. Kitizawa, A. Kishi, Y. Ikeda, M. Takano, K. Oda, H. Kitaguchi, J. Takada, and Y. Miura, Phys. Rev. B **41**, 4296 (1990).
170. W. Schnelle, N. Knauf., J. Bock, E. Preiser, and J. Hudepohl, Physica C **209**, 456 (1993).
171. A. Mirmelstein A. Junod, G. Triscone, K.-Q. Wang, and J. Muller, Physica C **248** 335 (1995).
172. J.W. Radcliffe, J.W. Loran, J.M. Wade, G. Witschek, and J.L. Tallon, J. Low Temp. Phys. **105**, 903 (1996).
173. A. Carrington, C. Marcenat, F. Bouquet, D. Colson, and V. Viallet, Czech. J. Phys. **46**, 3177 (1996).
174. B. Billon, M. Charlambous, O. Riou, J. Chaussy, and D. Polloquin, Phys. Rev. B **56**, 10824 (1997).
175. M. Couach, A.F. Khodar, R. Calemczuk, C. Marcenat, J.-L. Tholence, J.J. Caponi, and M.F. Gorius, Phys. Lett. A **188**, 85 (1994).
176. A. Carrington, C. Marcenat, F. Bouquet, D. Colson, A. Bertinotti, J.F. Marusco, and J. Hammann, Phys. Rev. B **55**, 8674(R) (1997).
177. D.M. Eagles, Phys. Rev. **186**, 456 (1969).

178. A.J. Leggett, J. Phys. (Paris) **41**, C7-19 (1980).
179. P. Nozieres and S. Schmitt-Rink, J. Low Temp. Phys. **59**, 195 (1985).
180. F. Pistolesi and G.C. Strinati, Phys. Rev. B **49**, 6356 (1994).
181. J.J. Rodriguez-Nunez, S. Schafroth, T. Schneider, M.H. Pedersen, and C. Rossel, cond-mat/9405001 (1994).
182. J.W. Loram, J. Luo, J.R. Cooper, W.Y. Liang, and J.L. Tallon, J. Phys. Chem. Solids **62**, 59 (2001).
183. A.S. Alexandrov, W.H. Beere, V.V. Kabanov, and W.Y. Liang, Phys. Rev. Lett. **79**, 1551 (1997).
184. A.S. Alexandrov, Phys. Rev. B **38**, 925 (1988).
185. A. Griffin, D.W. Snoke, and S. Stringari, *Bose–Einstein Condensation*, (eds.) (Cambridge University Press, New York, NY, 1995).
186. T. Timusk and B. Statt, Rep. Prog. Phys. **62**, 61 (1999).
187. C. Meingast, V. Pasler, P. Nagel, A. Rykov, S. Tajima, and P. Olsson, Phys. Rev. Lett. **86**, 1606 (2001).
188. C. Meingast, V. Pasler, P. Nagel, A. Rykov, S. Tajima, and P. Olsson, Phys. Rev. Lett. **89**, 229704 (2002).
189. R.S. Markiewicz, Phys. Rev. Lett. **89**, 229703 (2002).
190. A. Junod, private communication (2005).
191. N.-C. Yeh, cond-mat/0210656 (2002).
192. M.R. Norman and C. Pépin, Rep. Prog. Phys. **66**, 1547 (2003).
193. A.A. Abrikosov, Soviet Phys. JETP **5**, 1174 (1957).
194. G. Blatter, M.V. Feigel'man, V.B. Geshkenbein, A.I. Larkin, and V.M. Vinokur, Rev. Mod. Phys. **66**, 1125 (1994).
195. R.E. Hetzel, A. Sudbø, and D.A. Huse, Phys. Rev. Lett. **69**, 518 (1992).
196. J. Hu and A.H. MacDonald, Phys. Rev. Lett. **71**, 432 (1993).
197. E. Brézin, D.R. Nelson, and A. Thiaville, Phys. Rev. B **31**, 7124 (1985).
198. D.R. Nelson and V.M. Vinokur, Phys. Rev. B **48**, 13060 (1993).
199. A.I. Larkin and V.M. Vinokur, Phys. Rev. Lett. **75**, 4666 (1995).
200. T. Giamarchi and P. Le Doussal, Phys. Rev. B **55**, 6577 (1997).
201. H. Safar, P.L. Gammel, D.A. Huse, D.J. Bishop, J.P. Rice, and D.M. Ginsberg, Phys. Rev Lett. **69**, 284 (1992).
202. W.K. Kwok, S. Fleshler, U. Welp, V.M. Vinokur, J. Downey, G.W. Crabtree, and M.M. Miller, Phys. Rev. Lett. **69**, 3370 (1992).
203. W.K. Kwok, J. Fendrich, S. Fleshler, U. Welp, J. Downey, and G.W. Crabtree, Phys. Rev. Lett. **72**, 1092 (1994).
204. R. Liang, D.A. Bonn, and W.N. Hardy, Phys. Rev. Lett. **76**, 835 (1996).
205. U. Welp, J.A. Fendrich, W.K. Kwok, G.W. Crabtree, and B.W. Veal, Phys. Rev. Lett. **76**, 4809 (1996).
206. H. Pastoriza, M.F. Goffman, A. Arribére, and F. de la Cruz, Phys. Rev. Lett. **72**, 2951 (1994).
207. E. Zeldov, D. Majer, M. Konczykowski, V.B. Geshkenbein, V.M. Vinokur, and H. Shtrikman, Nature (London) **375**, 373 (1995).
208. S.L. Lee, P. Zimmerman, H. Keller, M. Warden, I.M. Savi, R. Schauwecker, D. Zech, R. Cubitt, E.M. Forgan, P.H. Kes, T.W. Li, A.A. Menovsky, and Z. Tarnawski, Phys. Rev. Lett. **71**, 3862 (1993).
209. R. Cubitt, E.M. Forgan, G. Yang, S.L. Lee, D.McK. Paul, H.A. Mook, M. Yethiraj, P.H. Kes, T.W. Li, A.A. Menovsky, Z. Tarnawski, and K. Mortensen, Nature (London) **365**, 407 (1993).
210. A. Schilling, O. Jeandupeux, C. Waelti, H.R. Ott and A. van Otterloo, in: *Proceedings of the 10th Anniversary HTS Workshop on Physics, Materials and Applications*, B. Batlogg et al. (eds.) (World Scientific, Singapore, 1996), p. 349.
211. A. Schilling, R.A. Fisher, N.E. Phillips, U. Welp, D. Dasgupta, W.K. Kwok, and G.W. Crabtree, Nature (London) **382**, 791 (1996).
212. M.J.W. Dodgson, V.B. Geshkenbein, H. Nordborg, and G. Blatter, Phys. Rev. Lett. **80**, 837 (1998).
213. M.J.W. Dodgson, V.B. Geshkenbein, H. Nordborg, and G. Blatter, Phys. Rev. B **57**, 14498 (1998).
214. M. Roulin, A. Junod, and E. Walker, Science **273**, 1210 (1996).
215. M. Roulin, A. Junod, and E. Walker, J. Low Temp. Phys. **105**, 1099 (1996).
216. M. Roulin, A. Junod, E. Walker, and A. Erb, Physica C **282–287**, 1401 (1997).
217. M. Roulin, A. Junod, A. Erb, and E. Walker, Phys. Rev. Lett. **80**, 1722 (1998).
218. A. Junod, M. Roulin, J.-Y. Genoud, B. Revaz, A. Erb, and E. Walker, Physica C **275**, 245 (1997).
219. A. Schilling, R.A. Fisher, N.E. Phillips, U. Welp, W.K. Kwok, and G.W. Crabtree, Phys. Rev. Lett. **78**, 4833 (1997).
220. A. Schilling, R.A. Fisher, N.E. Phillips, U. Welp, W.K. Kwok, and G.W. Crabtree, Physica C **282–287**, 327 (1997).
221. A. Schilling, R.A. Fisher, N.E. Phillips, U. Welp, W.K. Kwok, and G.W. Crabtree, Phys. Rev. B **58**, 11157 (1998).

222. A. Schilling, R.A. Fisher, N.E. Phillips, A. Lacerda, M.F. Hundley, U. Welp, W.K. Kwok, and G.W. Crabtree, unpublished (1999).
223. A. Junod, M. Roulin, J.-Y. Genoud, B. Revaz, E. Walker, A. Erb, C. Marcenat, R. Calemczuk, and F. Bouquet, Physica C **282–287**, 1425 (1997).
224. F. Bouquet, C. Marcenat, R. Calemczuk, U. Welp, W.K. Kwok, G.W. Crabtree, R.A. Fisher, N.E. Phillips, and A. Schilling, in: *Physics and Materials Science of Vortex States, Flux Pinning and Dynamics, NATO Advanced Study Institute, Series E: Applied Sciences 356*, R. Kossowsky, S. Bose, V. Pan, and Z. Durusoy (eds.) (Kluwer, Dordrecht, 1999), p. 743.
225. F. Bouquet, C. Marcenat, E. Steep, R. Calemczuk, W.K. Kwok, U. Welp, G.W. Crabtree, R.A. Fisher, N.E. Phillips, and A. Schilling, Nature (London) **411**, 448 (2001).
226. C. Marcenat, F. Bouquet, R. Calemczuk, U. Welp, W.K. Kwok, G.W. Crabtree, R.A. Fisher, N.E. Phillips, and A. Schilling, Physica C **341–348**, 949 (2000).
227. Z. Tesanović, Phys. Rev. B **51**, 16204 (1995).
228. Z. Tesanović, Phys. Rev. B **59**, 6449 (1999).
229. A.K. Nguyen and A. Sudbø, Phys. Rev. B **60**, 15307 (1999).
230. D. Li and B. Rosenstein, Phys. Rev. Lett. **90**, 167004 (2003).
231. M. Willemin, A. Schilling, H. Keller, C. Rossel, J. Hofer, U. Welp, W.K. Kwok, R.J. Olsson, and G. W. Crabtree, Phys. Rev. Lett. **81**, 4236 (1998).
232. M. Willemin, C. Rossel, J. Hofer, H. Keller, A. Erb, and E. Walker, Phys. Rev. B **58**, 5940(R) (1998).
233. E.H. Brandt and G.P. Mikitik, Phys. Rev. Lett. **89**, 27002 (2002).
234. A. Schilling, M. Willemin, C. Rossel, H. Keller, R.A. Fisher, N.E. Phillips, U. Welp, W.K. Kwok, R.J. Olsson, and G.W. Crabtree, Phys. Rev. B. **61**, 3592 (2000).
235. R. Lortz, C. Meingast, U. Welp, W.K. Kwok, and G.W. Crabtree, Phys. Rev. Lett. **90**, 237002 (2003).
236. B. Revaz, A. Junod, and A. Erb, Phys. Rev. B **58**, 11153 (1998).
237. M. Roulin, B. Revaz, A. Junod, A. Erb, and E. Walker, in: *Physics and Materials Science of Vortex States, Flux Pinning and Dynamics, NATO Advanced Study Institute, Series E: Applied Sciences 356*, R. Kossowsky, S. Bose, V. Pan, and Z. Durusoy (eds.) (Kluwer, Dordrecht, 1999), p. 489.
238. T. Plackowski, Y.Wang, R. Lortz, A. Junod, and Th. Wolf, cond-mat/0502326 (2005).
239. A. Schilling, A., M. Reibelt, and Th. Wolf, unpublished (2000).
240. T, Sasagawa, K. Kishio, Y. Togawa, J. Shimoyama, and K. Kitawaza, Phys. Rev. Lett. **80**, 4297 (1998).
241. T. Sasagawa, Y. Togawa, J. Shimoyama, A. Kapitulnik, K. Kitawaza , and K. Kishio, Phys. Rev. B **61**, 1610 (2000).
242. V. Hardy, A. Daignére, and A. Maignan, Phys. Rev. B **61**, 3610 (2000).
243. O.J. Taylor, A. Carrington, and S. Adachi, in: *Proceedings of the LT24 Conference* (2005), to be published by the American Institute of Physics.
244. D. Zech, C. Rossel, L. Lesne, H. Keller, S.L. Lee, and J. Karpinski, Phys. Rev. B **54**, 12535 (1996).
245. N. Kagawa, T. Ishida, K. Okuda, S. Adachi, and S. Tajima, Physica C **357**, 302 (2001).
246. D. Majer, E. Zeldov, and M. Konczykowski, Phys. Rev. Lett. **75**, 1166 (1995).
247. N. Morozov, E. Zeldov, D. Majer, and M. Konczykowski, Phys. Rev. B **54**, 3784(R) (1996).
248. D.T. Fuchs, E. Zeldov, D. Majer, R.A. Doyle, T. Tamegai, S. Ooi, and M. Konczykowski, Phys. Rev. B **54**, 796(R) (1996).
249. B. Khaykovich, E. Zeldov, D. Majer, T.W. Li, P.H. Kes, and M. Konczykowski, Phys. Rev. Lett. **76**, 2555 (1996).
250. A. Soibel, E. Zeldov, M. Rappaport, Y. Myasoedov, T. Tamegai, S. Ooi, M. Konczykowski, and V.B. Geshkenbein, Nature (London) **406**, 282 (2000).
251. N. Avraham, B. Khaykovich, Y. Myasoedov, M. Rappaport, H. Shtrikman, D.E. Feldman, T. Tamegai, P.H. Kes, M. Li, M. Konczykowski, K. van der Beek, and E. Zeldov, Nature (London) **411**, 451 (2001).
252. M. Menghini, Y. Fasano, F. de la Cruz, S.S. Banerjee, Y. Myasoedov, E. Zeldov, C.J. van der Beek, M. Konczykowski, and T. Tamegai, Phys. Rev. Lett. **90**, 147001 (2003).
253. S.S. Banerjee, A. Soibel, Y. Myasoedov, M. Rappaport, E. Zeldov, M. Menghini, Y. Fasano, F. de la Cruz, C.J. van der Beek, M. Konczykowski , and T. Tamegai, Phys. Rev. Lett. **90**, 87004 (2003).
254. S.S. Banerjee, S. Goldberg, A. Soibel, Y. Myasoedov, M. Rappaport, E. Zeldov, F. de la Cruz, C.J. van der Beek, M. Konczykowski, T. Tamegai , and V.M. Vinokur, Phys. Rev. Lett. **93**, 97002 (2003).
255. M. Menghini, Y. Fasano, F. de la Cruz, S.S. Banerjee, Y. Myasoedov, E. Zeldov, C.J. van der Beek, M. Konczykowski, and T. Tamegai, J. Low Temp. Phys. **135**, 139 (2004).

256. W.W. Warren, R.E. Waldstedt, G.F Brennert, R.J. Cava, R. Tycko, R.F. Bell, and G. Dabbagh, Phys. Rev. Lett. **62**, 1193 (1989).
257. R.E. Walstedt, W.W. Warren, Jr., R.F. Bell, R.J. Cava, G.P. Espinosa, L.F. Schneemeyer, and J. Waszczak, Phys. Rev. B **41**, 9574 (1990).
258. J.W. Loram, K.A. Mirza, J.M. Wade, J.R. Cooper, and W.Y. Liang, Physica C **235–240**, 134 (1994).
259. J.W. Loram, K.A. Mirza, J.R. Cooper, W.Y. Liang, and J.M. Wade, J. Supercond. **7**, 243 (1994).
260. W.Y. Liang, J.W. Loram, K.A. Mirza, N. Athanassopoulou, and J.R. Cooper, Physica C **263**, 277 (1996).

10

Normal State Transport Properties

N. E. Hussey

In this chapter, we summarize the normal state transport properties of high-T_c cuprates with particular emphasis on systematics with doping. Despite some remarkably generic trends, the experimental situation is found to be dogged with complexity, some of which challenge existing paradigms. The leading theoretical proposals, both Fermi-liquid and non-Fermi-liquid based, are examined, and their relative merits and pitfalls reviewed. In the discussion section, we compare transport data with data from other experimental probes and seek ways to merge the two into a coherent description of the normal state.

10.1. Introduction

Just as in conventional superconductors, where the scattering processes that dominate the electrical resistivity provide an important clue to the dominant pairing interaction (via the strength of the electron–phonon coupling), so an understanding of the normal state transport properties of high-T_c cuprates (HTC) is regarded as a key step towards the elucidation of the pairing mechanism for high temperature superconductivity. Whilst this remains the ultimate goal, normal state transport in HTC has emerged as a field in its own right and one of the most challenging (and controversial) topics in modern solid state physics. The ubiquitous T-linear resistivity at optimal doping, extending over a very wide temperature range, the strong T-dependence of the Hall coefficient R_H, the violation of Kohler's rule and the divergence of the resistivity anisotropy are but some of the striking anomalies which have puzzled the community over the past two decades and inspired theorists to develop radical new concepts in many-body theory.

Anderson, for example, boldly extended the ideas of Luttinger liquid theory, with its associated phenomenon of spin–charge separation, into 2D [1], and later Varma and coworkers introduced marginal Fermi-liquid (MFL) phenomenology based on the hypothesis of scale invariance (quantum criticality) [2] and a logarithmically vanishing quasiparticle (qp) weight at the Fermi surface (FS) to explain the emerging physical picture. Such ideas have had a profound influence on the field over the last 20 years and a significant proportion of the experimental data has been compared at some point to the predictions of these non-FL theories. Currently though, these ideas are still only phenomenological in nature and while a full microscopic model for interacting electrons in 2D remains elusive, others consider the abandonment of FL theory to be a little premature. In such a climate, models upholding Landau's qp concept but invoking specific scattering mechanisms [3–6] to explain the observed anomalous behavior still thrive.

N. E. Hussey • H. H. Wills Physics Laboratory, University of Bristol, Tyndall Avenue, Bristol, BS8 1TL, UK

At the time of writing, most theories can be said to remain relevant and it is clearly beyond the scope of this brief article to consider all of these in detail. Instead, I will strive for more modest goals; to summarize the debate, to highlight experimental data that shed light on the main issues and to examine critically (some of) the major theoretical interpretations. In the main, I will focus on more recent experimental papers as most of the early work has been well documented in previous review articles [7, 8]. With so many papers written on the topic, I must apologize at the outset for omitting a vast quantity of high quality work which has helped to shape our understanding of the HTC problem. I have decided, for example, against discussing the effects of superconducting fluctuations on the normal state resistivity as there exists already an excellent review article on this topic [9], whilst my discussion of the thermal transport is somewhat limited by space restrictions.

In Sections 10.2 and 10.3, I outline the systematics of the (zero-field) in- and out-of-plane resistivities, respectively, while in Section 10.4, I review the galvanometric phenomena of HTC, specifically the Hall effect and magnetoresistance (MR). The effects of impurities on the transport behavior are covered in Section 10.5 and the thermal transport properties are briefly reviewed in Section 10.6. Discussion and summary follow in Section 10.7. Throughout, I will discuss experimental data with reference to the predictions of Boltzmann transport theory applied to a quasi-2D FL (for a derivation of the main transport coefficients see the Appendix in [10]). While some readers may feel uncomfortable with this approach, it should be appreciated that the majority of non-FL models, including those of Anderson and Varma, have also used the Boltzmann transport equation in some heuristic form to illustrate their phenomenology.

10.2. Evolution of the In-Plane Resistivity with Doping

10.2.1. Introduction

The in-plane resistivity $\rho_{ab}(T)$ or $\rho_{\parallel}(T)$ of hole-doped (p-type) HTC shows a very systematic evolution with doping that is summarized in Figure 10.1a, where representative data for underdoped (UD), optimally doped (OP) and overdoped (OD) cuprates are shown [11]. (Electron-doped cuprates will be dealt with separately below). In the UD cuprates, $\rho_{ab}(T)$ varies approximately linearly with temperature at high T, but as the temperature is lowered, $\rho_{ab}(T)$ deviates downward from linearity, suggestive of a reduction in scattering or a

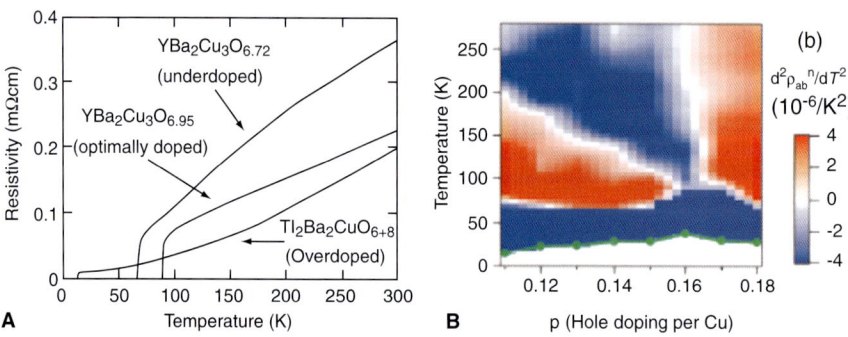

Figure 10.1. (a) Representative $\rho_{ab}(T)$ data for UD, OP and OD cuprates [11]; (b) $d^2\rho_{ab}/dT^2$ for $Bi_2Sr_{2-y}La_yCuO_{6+\delta}$ as a function of hole-doping p [12].

crossover to a higher power T-dependence. OP cuprates on the other hand are characterized by a T-linear resistivity that survives for all $T > T_c$, whilst on the OD side, $\rho_{ab}(T)$ contains a significant supralinear contribution.

Ando et al. recently proposed a novel means of portraying this evolution of $\rho_{ab}(T)$ with doping. The so-called resistivity curvature mapping (RCM) technique maps the second derivative of the resistivity $d^2\rho_{ab}/dT^2$ onto the T–p phase diagram [12]. Figure 10.1b shows a representative RCM for the single-layered cuprate $Bi_2Sr_{2-y}La_yCuO_{6+\delta}$. Similar plots were also obtained for $YBa_2Cu_3O_{7-\delta}$ and $La_{2-x}Sr_xCuO_4$ confirming the remarkable generality of the transport behavior across the phase diagram. S-shaped resistivity on the UD side is manifest by the crossover from sublinear (negative derivative, shaded blue) to linear ($d^2\rho_{ab}/dT^2 = 0$, shaded white) and finally to supralinear (positive derivative, shaded red). The vertical white line around $x = 0.16$ confirms the robust linearity of $\rho_{ab}(T)$ at optimal doping. At higher doping, the RCM is predominantly red indicating the development of a supralinear T-dependence on the OD side. With this overall picture in mind, let us now consider each region of the phase diagram in turn, beginning with optimal doping.

10.2.2. Optimally Doped Cuprates

Despite the large variations in (optimal) T_c and in the crystallography of individual cuprate families, T-linear resistivity is a universal feature at optimal doping, confirming that it is intrinsic to the CuO_2 planes. Moreover, as summarized in Table 10.1, the value of ρ_{ab} at $T = 300\,K$ normalized to a single CuO_2 plane is largely independent of the chemical composition of the charge transfer layers. The values themselves are large when compared with conventional superconductors. Given that the dc resistivity (conductivity) depends on both the normal state plasma frequency Ω_{pn}^2 ($= ne^2/\varepsilon_0 m^*$ in a Drude picture, where n is the carrier density and m^* the effective mass) and the transport scattering rate $1/\tau_{tr} = \Gamma$, it has proved extremely difficult to conclude whether these high values are due to a small coherent spectral weight, i.e., a small number of carriers with long lifetimes, or a large number of heavily damped qp's. This problem is tied inextricably to the interpretation of in-plane optical conductivity $\sigma_{ab}(\omega)$ in HTC. Normally one extracts Ω_{pn} from the optical sum rule by integrating the spectral weight up to some cut-off frequency of order the (renormalized) bandwidth, which in OP cuprates is $\sim$1–1.5 eV. It is still not clear, however, whether one should interpret $\sigma_{ab}(\omega)$ in HTC in terms of a generalized, single component model (with a ω-dependent Γ) or a two-component model comprising a narrow Drude response (with a small constant Γ) and an incoherent Lorentzian mode centered around the mid-infrared.

A simple yet appealing empirical scaling relation was recently established by Homes et al. between the superfluid density ρ_s and the value of the dc conductivity $\sigma_{ab}(0)$ at $T = T_c$, namely $\rho_s = 120\sigma_{ab}(0)T_c$, with $\rho_s = (\Omega_{ps})^2$, the square of superfluid plasma frequency,

Table 10.1. $\rho_{\parallel}(300\,K)$, normalized $\rho_{\parallel}(300\,K)$ and $\rho_\perp/\rho_{\parallel}(T = T_c)$ values for some OP cuprates

Compound	$\rho_{\parallel}(T = 300\,K)\,(\mu\Omega\,cm)$	$\rho_{\parallel}$/layer (300 K) ($\mu\Omega\,cm$)	$\rho_\perp/\rho_{\parallel}(T = T_c)$
$YBa_2Cu_3O_{6.95}$	290 [12]	580	30 [13]
$La_{1.83}Sr_{0.17}CuO_4$	420 [12]	420	300 [14]
$Bi_2Sr_{1.61}La_{0.39}CuO_6$	500 [12]	500	$>10^6$ [15]
$Bi_2Sr_2CaCu_2O_{8+\delta}$	280 [16]	560	10^5 [16]
$Tl_2Ba_2CuO_6$	450 [17]	450	2000 [18]

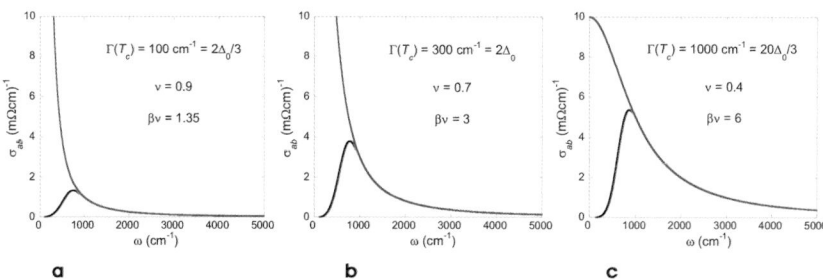

Figure 10.2. Simulated $\sigma_{ab}(\omega)$ above (*blue lines*) and below (*black lines*) T_c for various $\Gamma(T_c)$. For all curves, $2\Delta_0 = 300\,\text{cm}^{-1}$ and the normal state spectral weight is conserved.

expressed in units of cm^{-1} and $\sigma_{ab}(0)$ in $\Omega^{-1}\,\text{cm}^{-1}$ [19]. The beauty of Homes' law is that it allows one to estimate the magnitude of Γ *without* prior knowledge of Ω_{pn}. Writing $\hbar\Gamma = \beta k_B T_c$, $\sigma_{ab}(0) = \varepsilon_0 \Omega_{\text{pn}}^2 \tau$ and converting to SI units, Homes' law reduces to $\beta\nu = \beta(\Omega_{\text{ps}}/\Omega_{\text{pn}})^2 = 24{,}000\pi h/\mu_0 k_B \sim 3$. The key factor is the ratio $\nu = (\Omega_{\text{ps}}/\Omega_{\text{pn}})^2$, the fraction of the total spectral weight at T_c that is condensed into the δ-function at $\omega = 0$. The blue lines in Figure 10.2 represent typical normal state Drude peaks for different (ω-independent) $\Gamma(T_c)$ values. In this illustration, the spectral weight $\int_0^{\omega_c} \sigma(\omega)d\omega$ is assumed to be conserved below a cutoff $\omega_c = 10{,}000\,\text{cm}^{-1}$ corresponding to the effective bandwidth. The black lines are $\sigma_{ab}(\omega)$ below T_c assuming d-wave symmetry and a fixed gap size of $2\Delta_0 = 300\,\text{cm}^{-1}$. According to the Ferrell–Glover–Tinkham sum rule, ρ_s is equal to this "missing" spectral weight, the majority of which comes from $\omega \leq 2\Delta_0$. Thus ν is given by the fraction of the total spectral weight at T_c that is contained within $\omega \leq 2\Delta_0$, which in turn depends on $\Gamma(T_c)$.

Although ν decreases as β increases, the product $\beta\nu$ turns out to be a sensitive function of $\Gamma(T_c)$. Thus Homes' law places considerable constraint on $\Gamma(T_c)$. Indeed, the equality $\beta\nu = 3$ is only found for $\Gamma(T_c) \sim 2\Delta_0 \,(= 4.5 k_B T_c)$, i.e., when the (angle-averaged) scattering rate in OP cuprates at T_c is of order the pairing strength. Given that there is significant anisotropy in Γ, being largest along $(\pi, 0)$ (see below), scattering near these so-called "hot spots" must be even larger than this. It appears then that the resistivity is high in OP HTC because Γ is high [20], and not because the number of mobile carriers is small. One should bear in mind, however, that this analysis relies on the validity of the one-component model, and it is not clear whether such considerations can be applied to UD cuprates, where the case for a two-component picture is much stronger.

After their high T_c, the ubiquitous T-linear resistivity at optimal doping is perhaps the single most striking property of cuprates and it was apparent from the very beginning that such linearity, extending over an anomalously broad temperature range, would be difficult to reconcile within a conventional picture based on electron–phonon (e–ph) coupling [21]. Before discussing possible origins for this T-linear resistivity, however, it is worth perhaps examining the robustness of the linearity itself, or at least its true extent in temperature.

The long-held view that the linear $\rho_{ab}(T)$ observed in OP HTC is the *intrinsic* T-dependence for all $T > 0$ is based primarily on the observation in 1990 of a linear $\rho_{ab}(T)$ in single crystal $Bi_2Sr_2CuO_{6+\delta}$ extending from $T_c = 10$ to 700 K [22]. More recent results, however, question the universality of this result. Figure 10.3a, for example, shows $\rho_{ab}(T)$ data for the La-doped $Bi_2Sr_2CuO_{6+\delta}$ family in zero field (solid lines) and $B = 60T$ (solid symbols) [23]. For OP $Bi_2Sr_{2-y}La_yCuO_{6+\delta}$ ($x = 0.39$), the T-linear $\rho_{ab}(T)$ (highlighted in green and

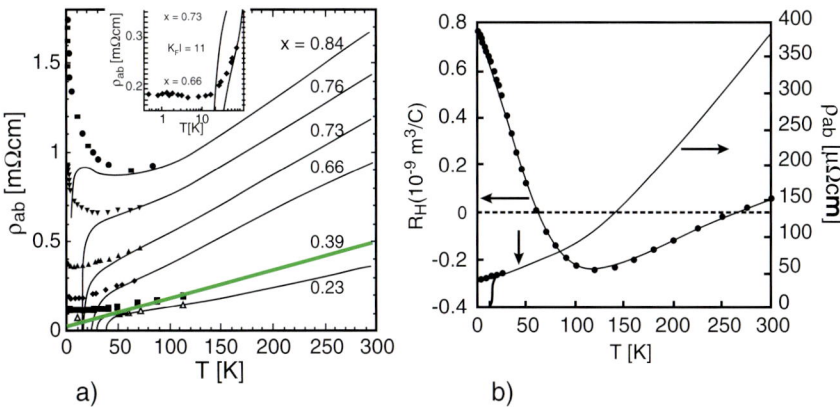

Figure 10.3. (a) $\rho_{ab}(T)$ data for $Bi_2Sr_{2-x}La_xCuO_{6+\delta}$ in 0T (*solid lines*) and 60T (*filled symbols*) [23]. The *green line* is an extrapolation of the zero-field $\rho_{ab}(T)$ to $T = 0$; (b) $\rho_{ab}(T)$ and $R_H(T)$ data for an OP $Pr_{1.83}Ce_{0.17}CuO_4$ thin film [25].

giving a small intercept at $T = 0$ K) *crosses over to a higher power T-dependence* below around 80 K eventually extrapolating to a higher residual value. This behavior is repeated in OP $La_{1.83}Sr_{0.17}CuO_4$ [14] for which $\rho_{ab}(T) \propto T$ over a wide temperature range [21, 24]. In both cases, it is most likely that the T-linear resistivity survives to lower T due to an effective cancellation of two effects; an upward deviation from linearity due to a crossover to the higher power term and a downturn due to the onset of fluctuation effects above T_c. (Indeed, paraconductivity contributions are clearly seen in the RCM plot for $Bi_2Sr_{2-y}La_yCuO_{6+\delta}$ in Figure 10.1b.) A higher power T-dependence as $T \to 0$ can also be inferred from the fact that in some of the cleaner cuprates, such as $YBa_2Cu_3O_{6.95}$ [12] and $Tl_2Ba_2CuO_6$ [17], the slope of the T-linear resistivity extrapolates to a *negative* intercept.

This crossover to a supralinear T-dependence with lowering T in the hole-doped cuprates contrasts markedly with what has been observed in some of the n-type analogs. Figure 10.3b for example shows $\rho_{ab}(T)$ data (along with $R_H(T)$ data) for an OP $Pr_{1.83}Ce_{0.17}CuO_4$ thin film [25]. Here, one can identify a T-linear resistivity surviving to the lowest temperatures (40 mK), which develops upward curvature only at *higher T*. A more recent doping dependent study has revealed that the limiting low-T form of $\rho_{ab}(T)$ could be expressed as $\rho_{ab}(T) = \rho_0 + AT^\beta$ with β tending to unity 1 at a critical doping level $x_c = 0.165$, suggestive of a quantum critical point [26]. Whilst these data are striking evidence for an intrinsic T-linear resistivity in n-type cuprates down to $T = 0$, it should be noted that the T-linear regime coincides with an R_H that is changing sharply with temperature over the same range. Whether such T-linearity persists in the p-type cuprates, where R_H is constant in the low-T limit [27], is not yet clear—it may simply be a question of hitting the right doping level. One important distinction, however, between p- and n-type cuprates is the extent in doping of long-range antiferromagnetic (AFM) order. In n-type cuprates, antiferromagnetism vanishes at $x = 0.15$, close to where this linearity is observed, while in the hope-doped analogs, it is suppressed for $x \leq 0.02$, i.e., well before optimal doping.

With uncertainty surrounding the temperature range of validity of the T-linear resistivity, at least in OP p-type cuprates, one cannot dismiss the possible role of e–ph scattering in HTC. According to the Bloch–Grüneisen formula, the e–ph interaction gives rise to a T-linear scattering rate above $T \sim \Theta_D/4$, Θ_D being the Debye temperature, which typically for HTC ~ 350 K. At elevated temperatures, $1/\tau_{tr}$ is related to the e–ph coupling constant λ_{e-ph} via $\hbar/\tau_{tr} = 2\pi k_B T_c \lambda_{e-ph}$. Taking our cue from Homes' law ($\hbar/\tau_{tr} \sim 4.5 k_B T_c$), and attributing the entire $\rho_{ab}(T)$ to e–ph scattering, one obtains $\lambda_{e-ph} \sim 0.7$. While this is a large value, it is not unfeasibly so. Nevertheless, several other features of the in-plane conductivity appear inconsistent with a picture of dominant e–ph scattering.

Firstly, as illustrated in Figure 10.1b, the T-linear resistivity is observed only in a narrow composition range near optimal doping; the sharp crossover to supralinear resistivity on the OD side being more suggestive of electron correlation effects than phonons. Secondly, the absence of resistivity saturation in OP La$_{2-x}$Sr$_x$CuO$_4$ up to 1000 K argues against a dominant e–ph mechanism [21, 24]. Thirdly, the frequency dependence of $1/\tau_{tr}$, extracted from extended Drude analysis of the in-plane optical conductivity, is inconsistent with an electron–boson scattering response due to phonons [32]. In particular, $\Gamma(\omega)$ does not saturate at frequencies corresponding to typical phonon energies in HTC. Finally, it has proved extremely problematic to explain the quadratic T-dependence of the inverse Hall angle cot $\theta_H(T)$ (see below) in a scenario based solely on e–ph scattering.

Thus it seems that an alternative origin for the T-linear resistivity is required, possibly beyond the realms of FL theory. When discussing deviations from FL behavior, however, one should be careful to distinguish between behavior which is incompatible with predictions for a purely isotropic FS and that which violates the basic assumptions upon which FL theory is founded. In the majority of cuprates, the FS itself displays fourfold cylindrical asymmetry, the Fermi velocity v_F can vary by as much as a factor of 2 around the FS due to the proximity of the saddle point near $(\pi, 0)$ and experimental data from a range of probes show evidence for strong anisotropy in $1/\tau_{tr}$ which may also vary with T. These collective features have been exploited in a variety of models, such as the nearly AFM-FL model [4], nested FL theory [28] and the van Hove scenario [29], to generate a T-linear resistivity down $T = 0$. In each case, however, a careful consideration of *all* scattering around the FS has led to the conclusion that these singular scattering processes must eventually be short-circuited and that $\rho_{ab}(T)$ will always vary as T^2 at the lowest T [30, 31]. Whilst this does not conflict with resistivity data in the hole-doped cuprates, data for Pr$_{2-x}$Ce$_x$CuO$_4$ appear difficult to reconcile within any such FL framework. At this point, we leave further discussion of the T-linear resistivity until Section 10.4.4 where it will be discussed in relation to the magnetotransport data.

10.2.3. Underdoped Cuprates

It is now well established that the UD HTC are characterized by the presence of a pseudogap E_g in the normal-state excitation spectrum [33]. This pseudogap has been observed by a variety of transport, thermodynamic and spectroscopic probes [34] and is found to affect many normal-state properties in an unusual and complex way that can best be interpreted as a reduction in the effective single particle density of states. Its origin remains hotly disputed, however, with the community evenly divided as to whether the pseudogap is a signature of precursor superconductivity (e.g., phase fluctuations) or is an independent state that competes with superconductivity for spectral weight.

As was discussed above, $\rho_{ab}(T)$ of UD cuprates shows marked deviations from the high-T T-linear behavior below a characteristic T^*, well above T_c, the value of which decreases with increasing p in a manner similar to $E_g(p)$. As is shown in Figure 10.4a, where $\rho_{ab}(T)$ for different cuprates with similar p values have been scaled according their values at T_c, there appears to be a close correlation between the form of $\rho_{ab}(T)$ and T/T^* [35]. In the Y-based cuprates, this change of slope was initially interpreted as a "kink" in $\rho_{ab}(T)$

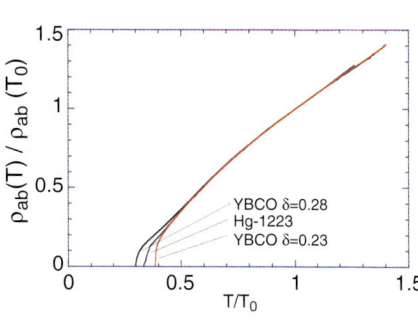 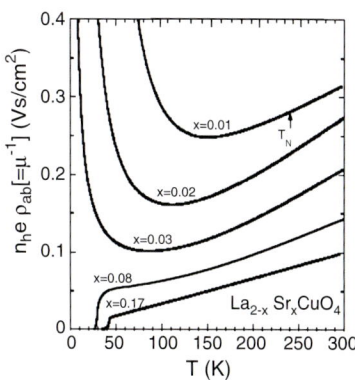

Figure 10.4. (a) Scaled $\rho_{ab}(T)$ plot for HgBa$_2$Ca$_2$Cu$_3$O$_{8+\delta}$ ($T_0 = 370$ K) and YBa$_2$Cu$_3$O$_{7-\delta}$ for $\delta = 0.23$ ($T_0 = 233$ K) and 0.28 ($T_0 = 210$ K) [35]. (b) Inverse mobility of La$_{2-x}$Sr$_x$CuO$_4$ [45].

at $T = T^*$ associated with the removal of the spin scattering channel within the plane in the pseudogap state [36, 37]. Others attribute the reduction in $\rho_{ab}(T)$ to the onset of phase fluctuations [38] or charge ordering phenomena [5]. Plots of the derivative dρ_{ab}/dT showed, however, that $\rho_{ab}(T)$ in fact first deviates from linearity at a much higher T [39]. Moreover, in the vicinity of T^*, there is no additional feature in dρ_{ab}/dT; the change of slope is a very gradual, continuous process with no evidence of a phase transition below T^*. In the more anisotropic cuprates such as La$_{2-x}$Sr$_x$CuO$_4$ [40] and Bi$_2$Sr$_2$CaCu$_2$O$_{8+\delta}$ [41], it has proven difficult to distinguish between deviations from linearity due to genuine pseudogap effects and those due to paraconductivity fluctuations near T_c.

The transport properties of UD p-type cuprates have been studied extensively in recent years by Ando and coworkers, and a number of significant features have emerged. Firstly, $\rho_{ab}(T)$ in "untwinned" lightly doped La$_{2-x}$Sr$_x$CuO$_4$ and YBa$_2$Cu$_3$O$_{7-\delta}$ exhibits marked in-plane aniostropy suggestive of some sort of self-organization of the charge carriers [42], possibly related to dynamical stripe formation. Secondly, transport measurements on UD cuprates in pulsed high magnetic fields have established that $\rho_{ab}(T)$ follows an anomalous $\ln(1/T)$ dependence below T_c, the origin of which is at yet unknown [14, 23]. Significantly, localization sets in for $k_F l$ values one order of magnitude higher than the universal 2D conductivity limit ($k_F l = 1$) [14, 23]. This contrasts markedly with Zn doping which induces a metal/insulator crossover at $k_F l = 1$ in the usual way [43]. Recent high-field measurements [44] on n-type films also suggest that the upturn may be correlated with the onset of spin scattering processes, though this has yet to be confirmed in p-type cuprates.

Finally and perhaps most surprisingly, $\rho_{ab}(T)$ shows metallic behavior (at least at high T) for only a few percent of doped holes [45]. Moreover, as illustrated in Figure 10.4b for La$_{2-x}$Sr$_x$CuO$_4$, the inverse mobility (defined as $\mu^{-1} = 2ex\rho_{ab}/V$ where V is the unit cell volume) is found to be largely insensitive to doping, varying only by a factor of 3 between $x = 0.01$ and 0.17. This variation in mobility becomes even less when one assumes the more usual definition of mobility ($\mu^{-1} = \rho_{ab}/R_H$). Indeed, such analysis gives $\mu^{-1}(x = 0.17) > \mu^{-1}(x = 0.02)$ at $T = 300$ K! Such a surprising result may be reconciled with data from angle-resolved photoemission (ARPES) that show the development in low-doped cuprates of Fermi arcs at (π, π) with relatively long-lived qp's [46]. At optimal doping, the full FS weight is effectively recovered and significant scattering at the hot spots near $(\pi, 0)$ begins to reduce the overall carrier mobility.

At high T, $\rho_{ab}(T)$ of UD cuprates attains very high values of order several mΩcm, well above the expected Mott–Ioffe–Regel (MIR) limit for coherent metallic transport. This mysterious behavior is observed in a number of other strongly correlated systems. Gunnarsson and Calandra attributed the high resistivities in HTC to a marked reduction in the spectral weight due to the strong Coulomb repulsion on the Cu–O sublattice [47]. According to this model, $\rho_{ab}(T)$ eventually saturates at high T once $1/\tau_{tr}$ of the mobile carriers becomes comparable to the bandwidth. Whilst there is evidence for a tendency toward saturation in UD $La_{2-x}Sr_xCuO_4$ [24], optical conductivity data on the same compounds suggest that a fundamental crossover in the charge dynamics occurs well before any saturation threshold is reached [48,49]. Above a certain crossover temperature ($\sim$300 K), spectral weight becomes suppressed at very low ω and is transferred to energies above the bandwidth. This gradual removal of spectral weight at low frequencies effectively induces an additional contribution to the resistivity in UD cuprates; thus the positive slope of $\rho_{ab}(T)$ cannot be simply a continuation of the low T metallic state whose T-dependence is governed solely by the scattering rate.

10.2.4. Overdoped Cuprates

In the OD region of the phase diagram, $\rho_{ab}(T)$ exhibits a T-dependence that can be modeled either as a single power law $\rho_{ab}(T) = \rho_0 + \alpha T^n$ with n varying gradually from $n = 1$ at optimal doping to $n = 2$ for $T_c = 0$ [18], or by a three-component polynomial fit $\rho_{ab}(T) = \rho_0 + \alpha T + AT^2$ with the relative weightings of both contributions changing smoothly with doping. Intriguingly, Mackenzie et al. measured $\rho_{ab}(T)$ of heavily OD $Tl_2Ba_2CuO_{6+\delta}$ down to 0.1 K, suppressing $T_c (= 15$ K$)$ with a large magnetic field, and still found evidence for a finite T-linear term surviving into the $T = 0$ limit [11]. Whether this linear term is intrinsic to the normal state (implying a non-FL response or at the very least, the presence of a very low energy scale in the transport properties) or due to some anomalous form of flux-flow resistivity is unclear at present. In heavily OD, nonsuperconducting $La_{1.7}Sr_{0.3}CuO_4$, the low-T resistivity *in zero-field* is found to be purely quadratic up to $T = 50$ K [50]. This suggests that conventional FL behavior is only achieved as one approaches the nonsuperconducting solubility limit, though significantly quantum oscillations, the classic signature of a FL, have never been observed conclusively in any cuprate. Finally, the observation of T^2 resistivity in $La_{1.7}Sr_{0.3}CuO_4$, coupled with Matthiessen's rule scaling [50], implies that the role of e–ph scattering remains negligible (at least in the transport properties) right out to the heavily OD side of the phase diagram.

10.3. The Out-of-Plane Transport

10.3.1. Introduction

Due to their layered structure, charge transport in HTC is strongly 2D in character and a substantial anisotropy exists between the in- and out-of-plane resistivities. Although the issue of interlayer coherence in low dimensional metals has been explored at length in recent years it is still poorly understood. The key quantity is the interplane transfer integral or hopping rate $t_\perp$. Once $t_\perp$ becomes comparable to other perturbations, such as the temperature or the *intraplane* (intrachain) scattering rate, interlayer hopping is rendered incoherent. When this happens, a number of distinct mechanisms for charge transfer may take over (for an excellent overview of this field, the reader is referred to the article by Cooper and Gray [7]).

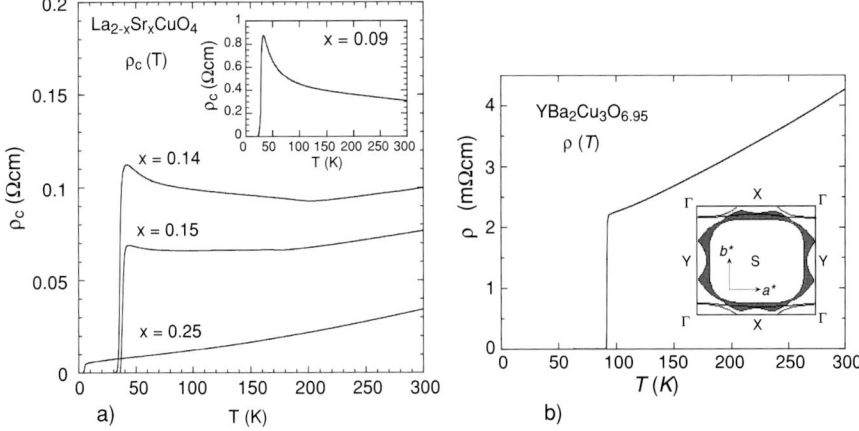

Figure 10.5. (a) $\rho_c(T)$ data for $La_{2-x}Sr_xCuO_4$ at various doping levels [55]. (b) $\rho_c(T)$ of OP $YBa_2Cu_3O_{6.95}$ [13]. Its band structure, projected onto the ab-plane, is shown in the inset.

Illustrative examples of this behavior in oxides are quasi-2D Sr_2RuO_4 [51] and quasi-1D $PrBa_2Cu_4O_8$ [52], both of which develop a highly anisotropic 3D FL ground state with interlayer (chain) resistivities that are metallic and vary as T^2. Above a certain temperature T_{coh} (~30–60 K), $\rho_\perp(T)$ starts to deviate from its limiting low-T quadratic form, reaches a maximum around 100–120 K then becomes thermally activated. In both systems, T_{coh} is found to correlate well with (averaged) values of $t_\perp$ estimated via independent means [53,54]. These text-book examples *in isostructural compounds* make the anomalous behavior of the c-axis resistivity in HTC all the more remarkable.

Figure 10.5a shows the evolution of $\rho_c(T)$ in $La_{2-x}Sr_xCuO_4$ for various x [55]. The doping induced changes are much more pronounced than for $\rho_{ab}(T)$, with $\rho_c(T)$ changing from metallic on the OD side to nonmetallic below $x < 0.14$. Let us now explore the behavior across the phase diagram, beginning as with the previous section with optimal doping.

10.3.2. Optimal Doped Cuprates

In contrast to $\rho_{ab}(T)$, where generic behavior is observed, c-axis transport in HTC at optimal doping is very material specific. As indicated in Table 10.1, the resistive anisotropy ρ_c/ρ_{ab} can vary between 30 in OP $YBa_2Cu_3O_{7-\delta}$ to over 10^6 in $Bi_2Sr_2CuO_6$, though no correlation exists between anisotropy and T_c. Moreover, in OP $Bi_2Sr_{2-y}La_yCuO_6$, $\rho_c(T)$ is insulating down to at least 1 K [56], in $La_{1.83}Sr_{0.17}CuO_4$, it is essentially constant below 200 K (Figure 10.5a), whilst in $YBa_2Cu_3O_{6.95}$, $\rho_c(T)$ remains metallic down to T_c (Figure 10.5b).

According to Boltzmann theory $\rho_c/\rho_{ab} = 1/2 \, (v_F/v_\perp)^2$, where $v_\perp$ is the maximum c-axis velocity component. The experimentally observed ρ_c/ρ_{ab} values in OP cuprates are striking in that they are typically one order of magnitude higher than band structure predictions and vary strongly with T. Within a conventional band picture, this distinct behavior of $\rho_c(T)$ and $\rho_{ab}(T)$ can arise only if the two quantities are controlled by different parts of the FS exhibiting different scattering rates [6]. (The interplane coupling $t_\perp$ can also be significantly renormalized ($t_\perp^* = t_\perp^2/\Delta$ where Δ is a large energy scale such as the charge transfer gap) due, for example, to strong intralayer dephasing [57].) Band structure calculations for $YBa_2Cu_3O_{7-\delta}$ [58], reproduced in Figure 10.5b, predict that $t_\perp$ is highly anisotropic within the CuO$_2$ plane, with minima for momenta parallel to (π, π) and maxima near the zone centers. This implies that c-axis transport could be dominated by carriers near $(\pi, 0)$. If $\rho_{ab}(T)$ is controlled conversely by carriers near (π, π) having a much smaller scattering rate, then the anomalous behavior of $\rho_c(T)$ and $\rho_{ab}(T)$ may be understood. Such a scenario has indeed been inferred from c-axis ellipsometry measurements on Ca-substituted YBa$_2$Cu$_3$O$_{6.95}$ [59]. In non-FL approaches, the ground state inherently drives $t_\perp$ to zero [60] leading to charge confinement in the planes. The insulating character of $\rho_c(T)$ in Bi$_2$Sr$_{2-y}$La$_y$CuO$_6$ in particular is viewed as strong evidence for such a picture, with $t_\perp$ driven to zero by strong electron correlations within the ab-plane.

In OP La$_{2-x}$Sr$_x$CuO$_4$, $\rho_c(T)$ varies linearly with T above 200 K, i.e., above the orthorhombic to tetragonal phase transition. Pressure studies of $\rho_c(T)$ revealed that this T-linear behavior in the tetragonal (HTT) phase is determined predominantly by the effects of thermal expansion along the c-axis [61] and the constant volume $\rho_c(T)$ is in fact essentially constant for all $T < 300$ K, in agreement with optical conductivity data [62]. Moreover for $x = 0.15$, the longitudinal c-axis MR (i.e., with $B||c$) is *larger* than the transverse (orbital) MR at all $T > T_c$ [55, 63] suggestive of incoherent interlayer charge transfer with $\Gamma_{ab} >> t_\perp$.

In YBa$_2$Cu$_3$O$_{6.95}$, ρ_c (300 K) $\sim$ 5mΩcm, i.e., one order of magnitude lower than is typically observed in OP cuprates [13, 64]. As illustrated in Figure 10.5b, however, $\rho_c(T)$ shows significant curvature and a large residual resistivity, in contrast to $\rho_{ab}(T)$ which is linear with a zero or negative intercept. The distinct behavior of the in- and out-of-plane charge dynamics has been confirmed also by optical spectroscopy [65]. While the anisotropic band structure picture described above offers a possible explanation for these discrepancies, anisotropic c-axis MR measurements on YBa$_2$Cu$_3$O$_{6.95}$ show no evidence of an orbital contribution to the c-axis MR for fields applied parallel to the CuO chains (and so only sensitive to the Lorentz force on the in-plane carriers). This suggests that the chains are primarily responsible for coherent c-axis transport in YBa$_2$Cu$_3$O$_{6.95}$ and that those regions of the CuO$_2$ plane states not hybridized with the chains remain effectively 2D, even in the absence of a normal state gap [13]. In short, there is little evidence to suggest that *interplane* transport is coherent in any OP cuprate.

10.3.3. Underdoped Cuprates

In most UD cuprates, such as La$_{2-x}$Sr$_x$CuO$_4$ ($x = 0.09$) in Figure 10.5a, $\rho_c(T)$ shows insulating behavior at all T. This is confirmed by c-axis optical conductivity $\sigma_c(\omega)$ measurements that show c-axis spectral weight gradually becoming suppressed below a certain energy scale with decreasing temperature [66]. A correlation between the c-axis conductivity and the normal state pseudogap has been inferred from a number of studies. In UD YBa$_2$Cu$_3$O$_{6+x}$ for example, the onset of an insulating $\rho_c(T)$ corresponds well to the crossover temperature where $\rho_{ab}(T)$ starts to deviate from linearity [67]. It has been proposed that spin singlet RVB formation blocks the out-of-plane transport, since the singlet pair has to be broken up in order for qp's to propagate from layer to layer, leading to a diverging $\rho_c(T)$ in the UD region at low T [68]. In tandem with the nonmetallic $\rho_c(T)$, an almost isotropic negative MR is commonly observed, suggesting that the spin degrees of freedom are indeed involved in blocking the out-of-plane transport [56, 69]. More recent detailed transport measurements in UD, highly anisotropic Bi$_2$Sr$_{2-y}$La$_y$CuO$_6$ meanwhile have indicated the temperature and doping evolution of $\rho_c(T)$ is a complicated combination of pseudogap effects, charge confinement and localization [15].

One material that bucks the trend in UD cuprates is $YBa_2Cu_4O_8$, a stoichiometric cuprate containing alternating stacks of CuO_2 bilayers and CuO double chain units oriented along the crystallographic b-axis. In $YBa_2Cu_4O_8$ $\rho_c(T)$ displays a crossover from T-independent behavior at high T to *metallic* behavior below 150 K [39]. It has been shown [70], however, that like in $YBa_2Cu_3O_7$, the metallicity of the c-axis in $YBa_2Cu_4O_8$ is due primarily to the emergence of three-dimensionality in the double chain unit, rather than the CuO_2 planes themselves.

10.3.4. Overdoped Cuprates

In heavily OD cuprates, $\rho_c/\rho_{ab}(T)$ eventually becomes T-independent [18,50,71], implying the formation of coherent c-axis motion and the development of a 3D electronic ground state. Such an observation by itself, however, is not proof of interlayer coherence [72], and in OD $Tl_2Ba_2CuO_{6+\delta}$, where $\rho_c\rho_{ab} \sim 1000$ [18], the corresponding l_c values appear too small to be consistent with phase-coherent (Bloch-wave) interlayer transport. Nonetheless, the recent observation [73] of angular MR oscillations (AMRO) in $Tl_2Ba_2CuO_{6+\delta}$ provides strong evidence of coherent c-axis transport on the OD side. AMRO are maxima in ρ_c which occur in a fixed magnetic field whenever $v_\perp$, averaged over many trajectories on the FS, is zero [74]. Figure 10.6a shows polar AMRO data obtained on an OD $Tl_2Ba_2CuO_{6+\delta}$ crystal having $T_c = 20$ K for a variety of azimuthal angles ϕ relative to the Cu–O–Cu bond direction.

Although the observation of polar AMRO by itself does not implicitly require interlayer coherence [75], analysis of the AMRO data in $Tl_2Ba_2CuO_{6+\delta}$ reveals very fine details in the c-axis warping which can only result from coherent hopping. This highly anisotropic $t_\perp(\phi)$, reproduced in Figure 10.6b, is found to contain zeros (nodes) along eight specific symmetry lines of the body-centered-tetragonal Brillouin zone and qualitatively is similar to the projections shown in Figure 10.5b for $YBa_2Cu_3O_{7-\delta}$ [63]. As discussed above, these minima

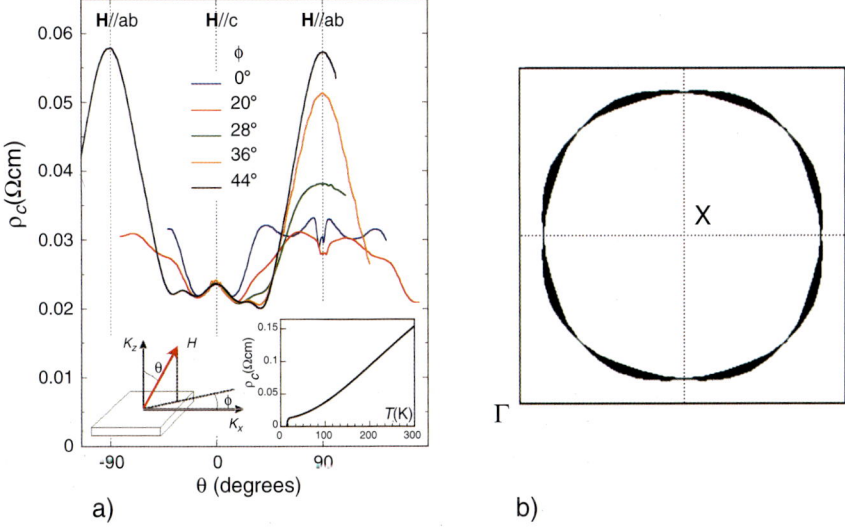

Figure 10.6. (a) Polar AMRO in OD $Tl_2Ba_2CuO_{6+\delta}$ ($T_c\sim20$ K) at $T = 4.2$ K and $H = 45T$ [73]. The azimuthal orientation is given relative to the Cu–O–Cu bond direction (see inset). The other inset shows the zero-field $\rho_c(T)$. (b) Projection of the deduced FS onto the ab-plane.

are consistent with band structure calculations which recognize the role of oxygen bonding and of virtual Cu 4s orbitals in c-axis conduction [76] and form an integral part of FL-based models invoked to explain discrepancies between the in- and out-of-plane charge response in both the normal [6,77] and superconducting states [78]. More significantly, this observation appears to confirm the transition from the highly 2D physics around optimal doping to a more conventional 3D ground state on the OD side.

10.4. The Anomalous Hall Coefficient and Violation of Kohler's Rule

10.4.1. Introduction

Given that they are, for the most part, single band metals, the anomalous T-dependence of the Hall coefficient in HTC is arguably the most striking and least understood of all their normal state transport properties. Unlike the resistivity, the behavior of $R_H(T)$ is not so specific to a particular region of the phase diagram. Thus, in contrast to the previous two sections, we will discuss first the absolute magnitude of $R_H(p)$ across the phase diagram and then consider its T-dependence. At the end of this section, we will critically examine some of the leading models proposed to explain the combined behavior of $\rho_{ab}(T)$ and $\cot\vartheta_H(T)$ and finish by describing the transverse in-plane MR whose behavior is strongly coupled with that of $\cot\vartheta_H(T)$.

10.4.2. Magnitude of R_H

As is shown in Figure 10.7 for $La_{2-x}Sr_xCuO_4$, R_H of hole-doped cuprates varies markedly with both p and T [79]. According to band structure calculations, hybridization of the Cu-3d$_{x^2-y^2}$ and O-2p$_{x,y}$ σ^* orbitals leads to a "large" FS containing $(1-p)$ electrons/Cu ion centered around the X points in the Brillouin zone. In the UD region, however, the carrier density n_H at low T, estimated from the Drude relation $R_H = 1/n_H e$, approaches the "chemical" hole concentration p deduced from the formal valence of Cu^{2+p} [80–82]. The observed scaling of R_H with p thus appears to suggest a violation of the Luttinger sum rule and either the presence of a "small" Fermi pocket containing p holes or a Fermi arc with an active (ungapped) arc length proportional to p. The latter possibility is supported by recent ARPES data on $La_{2-x}Sr_xCuO4$ [46] and $Ca_{2-x}Na_xCuO_2Cl_2$ [83]. It should be noted, however, that this simple Drude relation for R_H is based on the so-called isotropic-l approximation. A striking example of the breakdown of the isotropic-l approximation is La-doped Sr_2RuO_4 [84]. La doping for Sr introduces disorder between the RuO_2 planes, and whilst it has a negligible effect on the de Haas–van Alphen frequencies (and hence the FS volume), it induces both a magnitude and a sign change in the zero-temperature Hall coefficient $R_H(0)$. Similarly, in the cuprates, out-of-plane disorder is believed to induce significant anisotropy in the impurity scattering channel [85], so its effect on $R_H(0)$ may be nontrivial.

Due to the strong T-dependence of $R_H(T)$ and their high T_c values, it is difficult to estimate the true carrier concentration in OP cuprates. Recent high-field Hall measurements on the low-T_c cuprate $Bi_2Sr_{2-y}La_yCuO_6$, however, have allowed the determination of $R_H(0)$ across the entire doping range [27]. A marked drop in $R_H(0)$ is observed near optimal doping that is associated with a marked change in the FS geometry, perhaps due to a quantum phase transition. Again though, the possible role of out-of-plane defects on the low-T Hall coefficient has yet to be fully taken into account.

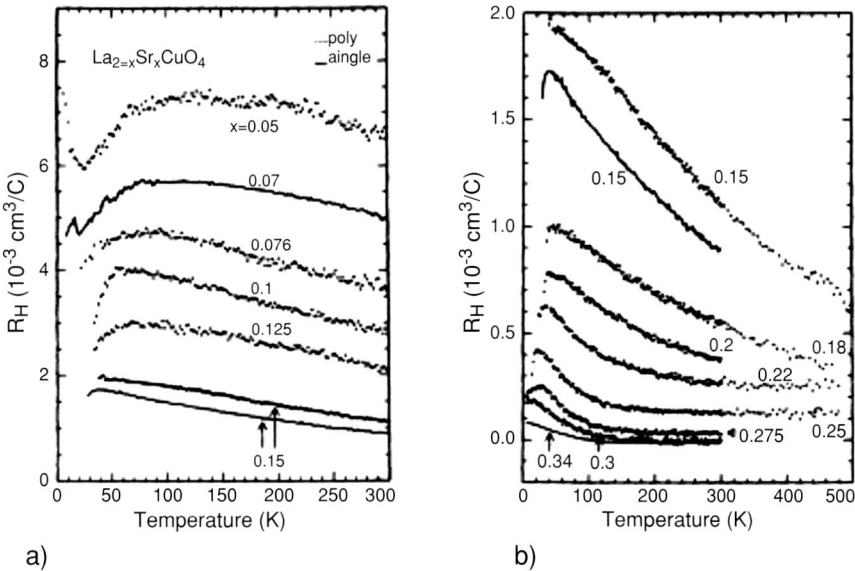

Figure 10.7. $R_H(T)$ for UD (*left panel*) and OD (*right panel*) $La_{2-x}Sr_xCuO_4$ [79].

On increasing the hole concentration beyond optimal doping, R_H shows a rapid decrease such that n_H becomes much larger than p. In the OD normal metal phase realized in $La_{2-x}Sr_xCuO_4$, the magnitude of R_H actually changes sign from positive to negative towards $x = 0.3$ [79]. This crossover is consistent with ARPES measurements which support a fundamental change in the FS topology in $La_{2-x}Sr_xCuO_4$ at high x [86].

At high T, R_H values in HTC agree roughly with standard LDA band structure predictions. Thus there appears to be a crossover from the high-T band-like regime to a low-T anomalous regime. The crossover temperature T_{RH} systematically increases with decreasing hole concentration [79, 87]. A close correlation has been pointed out between T_{RH} and the onset temperature of AFM correlations T_x, defined as a temperature where the uniform susceptibility starts showing a rapid decrease, thereby implying a possible link between the AFM spin correlations and the unusual behavior of $R_H(T)$ [79, 87].

10.4.3. The Inverse Hall Angle $\cot \vartheta_H(T)$

In marked contrast to the T-linear resistivity (at optimal doping), $\cot \vartheta_H(T)$ in HTC typically shows a quadratic T-dependence over a broad temperature range. This is illustrated in Figure 10.8, which shows $\rho_{ab}(T)$ and $\cot \vartheta_H(T)$ data for OP $Tl_2Ba_2CuO_{6+\delta}$ [17]. This implicit "separation of lifetimes" is a classic hallmark of the cuprates, and has led theorists to develop a number of radical ideas beyond conventional FL theory. The prevailing view is that $R_H(T) \sim 1/T$ at optimal doping, and the corresponding inverse Hall angle $\cot \vartheta_H(T) = A + BT^2$. As is often the case, however, the reality is significantly more complicated than that and no single theory has yet to succeed in explaining the emerging empirical picture.

Firstly, close inspection of the data in Figure 10.8 reveals different low-T (extrapolated) intercepts between the T-linear resistivity (small, negative) and quadratic $\cot \vartheta_H(T)$ (large, positive). Given that $Tl_2Ba_2CuO_{6+\delta}$ is believed to be a single band cuprate, this difference is unusual and warrants closer attention. Similar differences in the extrapolated residuals

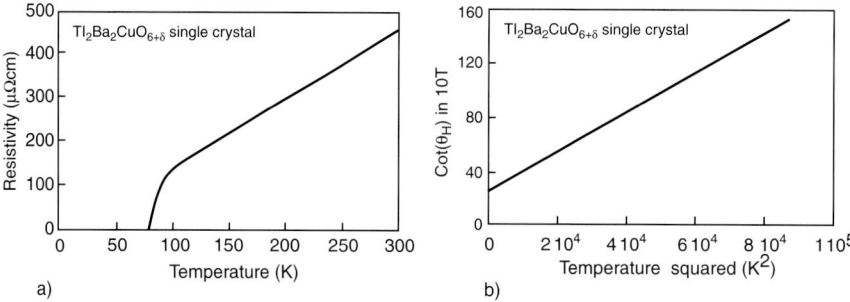

Figure 10.8. (a) $\rho_{ab}(T)$ and (b) $\cot\vartheta_H(T)$ data for OP $Tl_2Ba_3CuO_{6+\delta}$ [17].

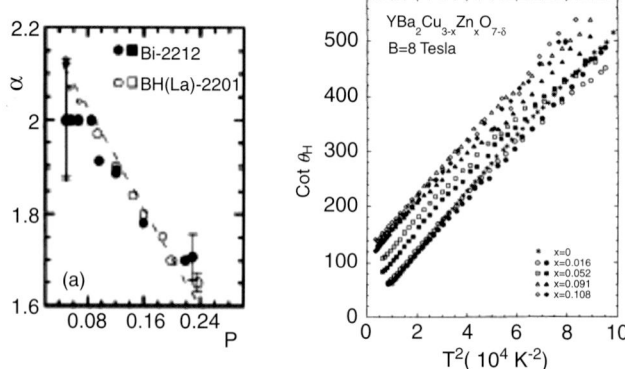

Figure 10.9. (a) p-Dependence of α, the coefficient of $\cot\vartheta_H(T) \sim A + BT^\alpha$ in $Bi_2Sr_2CaCu_2O_8$ and $Bi_2Sr_2CuO_{6+\delta}$ thin films [90]. (b) $\cot\vartheta_H(T)$ for a series of $YBa_2Cu_{3-x}Zn_xO_{7-\delta}$ crystals [88].

are observed in both $La_{2-x}Sr_xCuO_4$ and $Bi_{2-y}La_ySr_2CuO_{6+\delta}$, though it was shown in Section 10.2. that in these compounds, $\rho_{ab}(T)$ deviates upwards from linearity at very low T to give comparable intercepts in both quantities. Only $YBa_2Cu_3O_{7-\delta}$ is exceptional in that the intercept A remains small [88]. These observations fundamentally challenge the robustness of this separation of lifetimes, particularly in the low-T limit.

Secondly, whilst the T^2 dependence of $\cot\vartheta_H$ holds for a wide range of doping in most cuprates, it is not the case for the Bi-based cuprates $Bi_2Sr_2CaCu_2O_8$ and $Bi_2Sr_{2-y}La_yCuO_{6+\delta}$. In these systems, the power exponent of $\cot\vartheta_H$ is closer to 1.75 than 2 [89, 90]. Detailed transport studies of both crystalline and thin film samples of $Bi_2Sr_2CaCu_2O_8$ and $Bi_2Sr_{2-y}La_yCuO_{6+\delta}$ have shown in fact that $\cot\vartheta_H(T) \sim A + BT^\alpha$ with α steadily decreasing from $\alpha \sim 2$ to $\alpha \sim 1.6$–1.7 as one moves from the UD to the OD regime [89, 90]. This variable power law behavior in $\cot\vartheta_H(T)$ (shown in Figure 10.9a) reveals a high level of complexity in the phenomenology of normal state transport in HTC that has yet to be properly addressed.

10.4.4. Theoretical Modeling of $\rho_{ab}T$ and $R_H(T)$ in Cuprates

The origin of the strong T-dependence of R_H and its large magnitude at low doping have been the subjects of intense debate within the community. In a simple Drude picture, the sharp rise in $R_H(T)$ suggests a loss of carriers with decreasing temperature, due perhaps, to the opening of the pseudogap. From this perspective, the nonmonotonic $R_H(T)$ observed in most cuprates at low doping would be interpreted as a repopulation of states. However, this conclusion is not supported by magnetic susceptibility or specific heat data. In a conventional metal, a strong T-dependence of R_H can also arise due to multiple band effects. While this is evident in the electron-doped cuprates (e.g., 25) and other multiple band quasi-2D metals such as Sr_2RuO_4 [91], it is unlikely to be applicable to the majority of hole-doped cuprates where the transport is dominated by a single band.

Attempts to explain the anomalous behavior of $\rho_{ab}(T)$ and $R_H(T)$ in cuprates within a FL scenario have centered around the assumption of a (single) transport scattering rate whose magnitude varies around the in-plane FS. The origin of this approach can be attributed to Ong, who showed that in a 2D metal with an anisotropic $l(\mathbf{k})$, σ_{xy} is determined by the area (curl) swept out by $l(\mathbf{k})$ as it is traced around the FS [92] (Ong referred to this as the "Stokes area"). Thus, anisotropy in $l(\mathbf{k})$, in addition to the local FS curvature, plays a fundamental role in determining both the magnitude and sign of the Hall voltage in 2D metals and $\cot \vartheta_H(T)$ will accordingly be dominated by those regions of the FS where the curvature is greatest and the scattering is *weakest*.

In-plane anisotropy in $1/\tau_{tr}$ has been attributed to anisotropic $e-e$ (Umklapp) scattering [10] as well as to coupling to a singular bosonic mode; be it spin fluctuations [3,4], charge fluctuations [5] or d-wave superconducting fluctuations [6]. Generating a clear "separation of lifetimes" within these single lifetime scenarios, however, has proved very difficult, requiring as it does a very subtle balancing act between different regions in $\mathbf{k}$-space with distinct T-dependences.

Given this reliance on detail, other more exotic models, based on non-FL physics, have gained prominence within the community; most notably the two-lifetime picture of Anderson [1] and the MFL phenomenology of Varma and coworkers [2]. In the two-lifetime approach, scattering processes involving momentum transfer perpendicular and parallel to the FS are governed by independent transport and Hall scattering rates $1/\tau_{tr}$ and $1/\tau_H$ with different T-dependences. The former dominates the resistivity, while the latter is introduced into the (heuristic) Boltzmann equation in the following way:

$$\sigma_{ij}^{(n)} = \frac{e^3}{4\pi^3\hbar} \int v_i \left(\left(-\tau_H [v_k \times B] \frac{\partial}{\partial k} \right)^n v_j \tau_{tr} \left(-\frac{\partial f_0}{\partial \varepsilon} \right) \right) d^3k \qquad (10.1)$$

In conventional FL's of course, τ_{tr} is equal to τ_H. Allowing τ_H to be independent of τ_{tr}, the inverse Hall angle can now be written as $\cot \vartheta_H = \sigma_{xx}/\sigma_{xy} \propto 1/\tau_H$. Thus the different behavior of $\rho_{ab}(T)$ and $\cot \vartheta_H(T)$ reflects the different T dependencies of $1/\tau_{tr}$ and $1/\tau_H$. The enhancement of R_H then comes from the fact that τ_H becomes larger than τ_{tr} at low T. This model received strong support from Hall measurements on Zn-doped $YBa_2Cu_3O_{7-\delta}$ (reproduced in Fig. 10.9b) that showed $\cot \vartheta_H = A + BT^2$ to be robust to Zn doping with B remaining constant and A increasing in proportion to the Zn concentration [88], suggesting that the T^2 (inverse) Hall angle is a well defined fundamental quantity representing $1/\tau_H$. Later measurements on Co-doped $YBa_2Cu_3O_{7-\delta}$ [3] together with MR measurements on $YBa_2Cu_3O_{7-\delta}$ and $La_{2-x}Sr_xCuO_4$ [93] appeared to affirm the robustness of $1/\tau_H$.

The MFL hypothesis argues that optimum T_c lies in proximity to a quantum critical point and as a result, qp weight vanishes logarithmically at the FS with the corresponding imaginary part of the self-energy governed simply by the temperature scale [2]. In contrast to the two-lifetime picture, MFL theory assumes a single T-linear scattering rate but introduces an unconventional expansion in the magnetotransport response. The Hall angle, for example, is given by the *square* of the transport lifetime [85], an idea that has received empirical support from very recent infrared optical Hall angle studies [94]. In order to account for the observed magnetotransport behavior in cuprates, Varma and Abrahams introduced anisotropy into their MFL phenomenology via the (elastic) impurity scattering rate by assuming small angle scattering off impurities located away from the CuO_2 plane [85]. Whilst this hypothesis seems consistent with certain ARPES measurements [95], the legitimacy of the expansion in small scattering angle used in [85] has been subsequently challenged [96, 97]. Moreover, although the predictions of MFL theory appear compatible with the empirical situation in OP cuprates, their applicability to the rest of the cuprate phase diagram is less tangible. In particular, the gradual convergence of the T-dependencies of $\rho_{ab}(T)$ and $\cot\vartheta_H(T)$ in OD cuprates sits uncomfortably with the idea of ϑ_H scaling with the *square* of the transport lifetime.

Evidence for anisotropy in the transport scattering rate in HTC came initially from angular dependent studies of the c-axis MR $\Delta\rho_c/\rho_c$ in OD $Tl_2Ba_2CuO_{6+\delta}$ [98]. On rotating B within the basal plane, $\Delta\rho_c/\rho_c$ was found to exhibit fourfold anisotropy with an amplitude that scaled as B^4 in accordance with Boltzmann transport analysis for an anisotropic quasi-2D FS. The ratio of the amplitude of these fourfold oscillations to the size of the higher order B^4 term in the isotropic MR showed a significant T-dependence from 30 to 150 K (the temperature range of the experiment) that corresponded to an increasing anisotropy of the *in-plane* scattering rate with increasing T. This basal plane anisotropy in the scattering rate was later confirmed by ARPES [95] where ill-defined qp states (in the normal state) at the so-called "antinodal" points at $(\pi, 0)$ coexist with sharp qp peaks along (π, π). The anisotropy of the broadening of the qp self-energy was found to weaken steadily with overdoping [99].

More recent polar AMRO measurements in OD Tl2201 have revealed new details of the form of this anisotropy [100]. By incorporating basal-plane anisotropy in $\omega_c\tau$ into the AMRO analysis, Abdel-Jawad et al. were able to extract the full T- and $\mathbf{k}$-dependence of l_{ab} in 15 K T2201 between 4 and 60 K. The T-dependence of the anisotropy was attributed in full to the scattering rate, and from the resulting fits, the authors were able to conclude that the scattering rate contained two components, an isotropic T^2 scattering rate (presumably due to electron–electron scattering) and an anisotropic T-linear component (of unknown origin) that was maximal along $(\pi, 0)$. Significantly, this form of the scattering rate was able to explain in a self-consistent fashion the T-dependencies of $\rho_{ab}(T)$ and $R_H(T)$ over the same temperature range.

Finally the doping variation of the form of $\cot\vartheta_H(T) = A + BT^\alpha (1.65 \leq \alpha \leq 2)$ [87, 88] has proved difficult to explain within either the two lifetime picture or MFL phenomenology. Indeed, such a gradual weakening of the T-dependence is best understood in pictures based of single anisotropic $\Gamma(\mathbf{k})$ whose anisotropy decreases with increasing doping. As the anisotropy in $\Gamma(\mathbf{k})$ is reduced, other regions of the FS away from the nodes, where $\Gamma(\mathbf{k})$ is larger and the T-dependence is weaker, begin to contribute to σ_{xy}.

The overall experimental situation then tends to support models in which anisotropy in $l(\mathbf{k})$, in conjunction with the FS curvature [92], is responsible for the T-dependence of $R_H(T)$. The origin of the anisotropic scattering term, as revealed by polar AMRO [100], however, is not known at present and intriguingly, none of the current models of anisotropic scattering assume this particular form for $\Gamma(\mathbf{k})$.

10.4.5. In-Plane Magnetoresistance

According to Boltzmann transport theory, the in-plane transverse MR $\Delta\rho_{ab}/\rho_{ab} \propto (\Omega_c\tau_{tr})^2$ where Ω_c is the cyclotron frequency. Thus the product $\Delta\rho_{ab} \cdot \rho_{ab} (= \Delta\rho_{ab}/\rho_{ab} \cdot (\rho_{ab})^2)$ is independent of τ_{tr} and a plot of $\Delta\rho_{ab}/\rho_{ab}$ vs. $(H/\rho_{ab})^2$ is expected to fall on a straight line with a slope that is independent of T (provided the carrier concentration remains constant). This relation, known as Kohler's rule, is obeyed in a large number of standard

Normal State Transport Properties

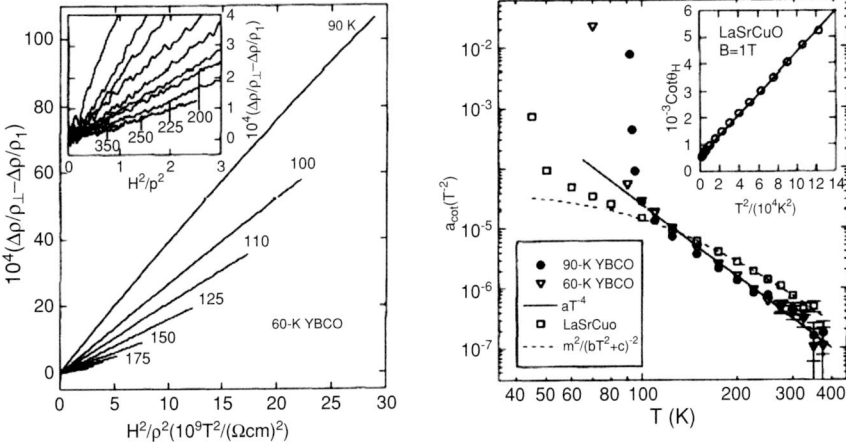

Figure 10.10. (a) Kohler plot for underdoped YBa$_2$Cu$_3$O$_{6.6}$ [93]. (b) Temperature dependence of the orbital part of the MR in YBa$_2$Cu$_3$O$_{6.6}$, optimally doped YBa$_2$Cu$_3$O$_7$ and La$_{1.85}$Sr$_{0.15}$CuO$_4$ [93].

metals. In HTC, however, Kohler's rule is strongly violated. Figure 10.10a shows a typical Kohler plot for UD YBa$_2$Cu$_3$O$_{6.6}$ [93]. Instead of the data collapsing onto a single curve, there is a marked increase in the slope with decreasing temperature. Remarkably, this progression continues up to 350 K (see inset).

This T-dependence of the MR was quantitatively explained by modifying Kohler's rule to incorporate the two scattering lifetime scenario of Anderson [93]. Using (10.1), one obtains a modified Kohler's rule such that $\Delta\rho_{ab}/\rho_{ab} \propto (\Omega_c \tau_H)^2 \propto 1/(A + BT^2)^2$ and consequently $(\Delta\rho_{ab}/\rho_{ab})/\tan^2 \vartheta_H$ becomes constant. This is illustrated in Figure 10.10b for YBa$_2$Cu$_3$O$_{6.6}$, OP YBa$_2$Cu$_3$O$_7$ and La$_{1.85}$Sr$_{0.15}$CuO$_4$ [93]. Similar scaling has also been reported even in OD Tl$_2$Ba$_2$CuO$_{6+\delta}$ [98]. Indeed, only in OD nonsuperconducting La$_{2-x}$Sr$_x$CuO$_4$, where the Hall and transverse rates appear to merge, is conventional Kohler's scaling seemingly recovered [63].

Whilst the two lifetime model of Anderson and coworkers has been successful in reproducing the experimental situation in OP cuprates, it does not appear to be consistent with ARPES results and is yet to explain the evolution of the transport phenomena across the full HTC phase diagram. Within a single (anisotropic) lifetime approach, such marked separation in the T-dependence of the different conductivity coefficients is possible though it requires a very subtle balancing of parameters around the FS. In the "cold spots" model of Ioffe and Millis for example, the phenomenological scattering rate contains two terms, an isotropic FL scattering rate $1/\tau_{FL} \sim T^2$ and a T-independent scattering rate $1/\tau_0$ that is large everywhere except the nodal directions [6]. Whilst this model correctly explains the T-linear resistivity and quadratic Hall angle, the anisotropy required to separate the transport and Hall lifetimes gives an orbital MR that is one order of magnitude too large and has an additional T-dependence [8]. This discrepancy can be removed by the introduction of a "shunt" scattering rate maximum Γ_{max}, compatible with the MIR limit, which acts to reduce the overall effective anisotropy [10]. A modified Kohler's rule of the correct magnitude is then reproduced but again its success relies heavily on a subtle balancing of anisotropies in the elastic and inelastic channels. As discussed above, in the MFL picture small angle impurity scattering is introduced as an additional ingredient to induce anisotropy in the scattering rate. Expansion of the scattering term then leads to a Hall scattering rate $1/\tau_H \sim 1/\tau_{tr}^2$ [85]. It has been argued, however, that the conditions that lead to this separation in lifetimes do not reproduce the violation of Kohler's rule [97]. Thus none of the leading proposals have yet to stand up to close scrutiny with the full complement of experimental data.

Finally, in analyzing the normal state orbital MR, one should be careful to take into account fully the contributions to the orbital MR from paraconductivity terms. In highly anisotropic $Bi_2Sr_2CaCu_2O_{8+\delta}$ [101] for example, apparent Kohler's law violation can be attributed almost entirely to superconducting fluctuations persisting up to room temperature, though it is unlikely that fluctuation effects are omnipresent in cuprates of significantly lower anisotropy. Again we refer the reader to [9] for a review of fluctuation effects in HTC and other superconductors.

10.5. Impurity Studies

The effects of disorder have been widely studied in HTC for a number of reasons, the destruction of superconductivity being the most obvious one. The interplay between the spin and charge degrees of freedom, particularly in the pseudogap regime, can also be investigated by studying the influence of impurities and/or disorder on the transport behavior in UD cuprates. Point defects such as Zn impurities produce a strong suppression of T_c. They are also believed to act as qp scatterers at the unitary limit [43]. Electron irradiation has also proved a powerful tool for studying disorder effects in HTC [102]. Electron irradiation of the correct fluence introduces point defects into the CuO_2 plane and indeed, comparative studies on $YBa_2Cu_3O_7$ show the effects of irradiation and Zn substitution to be very similar [103].

In OP cuprates, Zn substitution lead to upturns in $\rho_{ab}(T)$ at low T once superconductivity has been destroyed (though for $YBa_2Cu_{3-y}Zn_yO_{7-\delta}$, this assumption relies on extrapolation) [42]. Moreover, the onset of localization appears to be correlated with the effective sheet resistance exceeding the universal value $h/4e^2 \sim 6.5\,k\Omega$ per layer. According to some groups [104, 105], the same criterion applies equally to UD $YBa_2Cu_{3-y}Zn_yO_{7-\delta}$ and $La_{2-x}Sr_xCu_{1-y}Zn_yO_4$. This conclusion, however, relies on zero-field data in which the limiting low-T behavior is often obscured by the onset of superconductivity. By applying a large magnetic field to suppress T_c further, Ando and Segawa revealed that the onset of localization in UD $YBa_2Cu_{3-x}Zn_xO_{7-\delta}$ actually begins at significantly *lower* resistivities corresponding to $k_F l \sim 7$ [106]. This lower threshold is similar to that seen in "pure" (i.e., undoped) $La_{2-x}Sr_xCuO_4$ and $Bi_2Sr_{2-y}La_yCuO_{6+\delta}$ [14, 23] where anomalous $\ln(1/T)$ resistivities set in for $k_F l \leq 10$. It has been suggested that this peculiar form of localization is induced by some sort of spin scattering [44], or in the case of $La_{2-x}Sr_xCuO_4$, the tendency towards spin freezing on the UD side [107]. Note that in OD cuprates, T_c is suppressed well before the localization limit has been reached [43], most likely owing to the collapse on the OD side of the superconducting pairing strength.

Results from specific heat [33], NMR [108] and other probes suggest that Zn substitution leads to a gradual disruption of the pseudogap in UD cuprates and a concomitant "filling-in" of the gaps within the spin and charge excitation spectra. How exactly this manifests itself in the transport though is still unclear. Measurements on thin film samples suggested that the anomalous curvature in $\rho_{ab}(T)$, thought to be associated with a steady reduction in the spin-scattering channel below T^*, is completely overridden by Zn substitution [105]. Similar conclusions have also been drawn from an extended Drude analysis of optical measurements on Zn-doped $YBa_2Cu_4O_8$ [109]. This is in marked contrast, however, with anisotropic

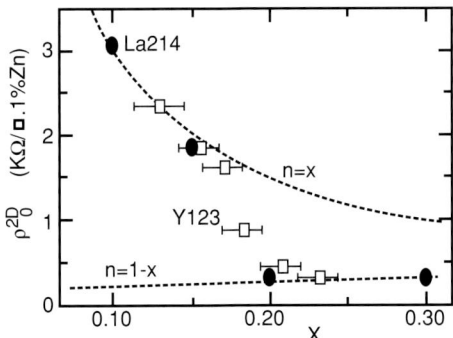

Figure 10.11. Increase in the (2D) residual resistivities in Zn-doped $La_{2-x}Sr_xCuO_4$ and $YBa_2Cu_3O_{7-\delta}$ per 1% of added Zn [43].

transport measurements carried out on $YBa_2Cu_3O_{7-\delta}$ *single crystals* which failed to show any significant effect on either $\rho_{ab}(T)$ or $\rho_c(T)$ [103].

Finally, let us discuss briefly Matthiessen's rule in HTC. In the majority of cases, Matthiessen's rule is found to hold, even for significant levels of disorder. At low doping, the change in residual resistivity ρ_0 is found to be consistent with unitary scattering in the *s*-wave channel, provided the carrier number is taken to be that of the chemical doping p, rather than $1 - p$ to be expected from band structure calculations [42]. At higher doping levels, however, the unitary limit becomes obeyed for $n \sim 1 - p$, as illustrated in Figure 10.11. A phase shift comparable with that expected of unitary scattering ($\delta = \pi/2$) was confirmed by scanning tunneling microscopy (STM) measurements on Zn-doped $Bi_2Sr_2CaCu_2O_8$ [110], suggesting this picture is robust. Measurements by Ong's group on Zn-doped $YBa_2Cu_3O_{7-\delta}$ at optimal doping [94] showed that $\cot \vartheta_H (T) (= A + BT^2)$ is robust also to Zn substitution, the addition of Zn simply increasing A whilst keeping B constant. As discussed above, this was viewed as strong evidence for Anderson's two lifetime picture.

10.6. Thermal Transport

10.6.1. Introduction

Thermal transport properties of HTC, such as thermal conductivity κ and thermoelectric power S, have been less studied relative to the electrical transport coefficients due firstly to the difficulty of performing accurate measurements, and secondly, the problem of interpretating the ensuing results. Their interpretation is compounded of course by the additional contribution to $\kappa(T)$ from heat carrying phonons and the effect of the phonon drag mechanism on $S(T)$. Nevertheless, some notable results have been obtained and the advent of high-field thermal transport measurements has opened up a new vista on the physics of HTC, particularly in the low-doped and highly doped regions where the upper critical fields are sufficiently small to facilitate the exploration of thermal properties in the low-T limit. Particularly striking results in this area include violation of the Wiedemann–Franz (WF) law [111], the remarkable growth of the thermal Hall effect below T_c [112] and the anomalous Nernst effect first reported by Ong's group [113].

10.6.2. Thermoelectric Power

According to the simple Boltzmann picture, the Seebeck coefficient S is governed both by the transport scattering rate (via its energy dependence) and the thermodynamic mass. Separating the effects of these two contributions can prove difficult and it is common practice to apply certain simplifying assumptions that may obscure some of the intrinsic physics especially in relation to the pseudogap.

Whilst the interpretation of $S(T)$ may be difficult, its systematic behavior in the HTC has been well documented. In particular, a remarkable universal correlation was found early on between the room temperature value of S and the doping level [114]. This has been used subsequently to determine the doping concentration in a wide range of materials where determination of the hole concentration is ambiguous. Whilst this relation is not found to hold in $Bi_2Sr_{2-y}La_yCuO_{6+\delta}$ [115], its applicability to other HTC appears robust.

In UD cuprates, $S(T)$ has a large positive value and traces out a broad maximum whose peak temperature decreases with increasing doping. At optimal doping, $S(T)$ remains positive but has a negative linear slope, i.e., $S(T) = \beta - \alpha T$. In some ways, $S(T)$ resembles $R_H(T)$ at optimal doping. As doping increases further, β continues to decrease whilst α remains relatively doping independent. Thus in the most overdoped samples, $S(T)$ is negative at all $T > T_c$. Perhaps understandably, there have been few attempts to explain the phenomenology of $S(T)$ in HTC though a simple picture involving anisotropic (spin-fluctuation dominated) scattering on a large FS has been shown to give a good account of both the form of $S(T)$ and its doping dependence [116].

10.6.3. Thermal Conductivity

The normal state in-plane thermal conductivity κ_{ab} of HTC is dominated by the phonon contribution. Typical estimates of the electronic contribution are of order 10–20% of the total near $T = T_c$. The most striking feature of the thermal conductivity in HTC is the rapid increase in $\kappa_{ab}(T)$ below T_c which peaks typically around $T_c/2$ (for a review, see Uher [7]). The origin of this rapid increase was in dispute for many years, with early reports attributing the enhancement to a rise in the dominant phonon peak [7]. However, later measurements of the thermal Hall effect, which senses only the electronic contribution [112], revealed a 1,000-fold increase in the qp mean-free-path in high-purity $YBa_2Cu_3O_{6.99}$ below T_c [117]. Such a dramatic enhancement of the qp lifetime would more than compensate for the loss of uncondensed carriers below T_c and thus appeared to confirm the electronic origin of the peak in $\kappa_{ab}(T)$ [118]. (While it might be argued that this marked decrease in scattering is a signature that ρ_0 in $YBa_2Cu_3O_{6.99}$ is negligible, as one might expect from extrapolating the T-linear $\rho_{ab}(T)$ down to 0 K, it should be recalled that below T_c the phase space available for scattering is significantly reduced due to the (d-wave) symmetry of the order parameter and confinement of qp's to the nodal regions.)

The WF law, equating the electrical and thermal conductivities in a metal, is often regarded as a key signature of a FL, for which the electrical current transported by the qp's has a one-to-one correspondence with the heat current. The WF law can only be rigorously examined in the low-T region where the electronic (fermionic) and phononic (bosonic) contributions to $\kappa_{ab}(T)$ have well-defined separate T-dependences. The first test of the WF law was carried out on $Pr_{2-x}Ce_xCuO_4$ in the presence of a large magnetic field sufficient to destroy superconductivity [111]. Remarkably, the linear (electronic) coefficient of $\kappa_{ab}(T)$ appeared to vanish in the low-T limit, suggesting a complete breakdown of FL theory in the normal state of a cuprate. It now appears, however, that this suppression of κ_{ab}/T at very low T is an experimental artifact due to decoupling of the electronic and phononic thermal baths [119]. Indeed, other measurements carried out on OP $Bi_2Sr_2CuO_6$ [120] and OD cuprates [50, 121] seem to confirm the WF prediction, at least within the experimental error.

10.6.4. Nernst–Ettinghausen Effect

In 2000, Ong and his coworkers discovered a large Nernst–Ettinghausen effect (hereafter labeled the Nernst effect) that extended to well above T_c in UD $La_{2-x}Sr_xCuO_4$ [113]. The Nernst effect is the transverse voltage generated in the presence of perpendicular magnetic field by a longitudinal thermal gradient. The effect is known to be small in conventional metals due to the Sondheimer cancellation. Whilst a large Nernst signal can appear under certain circumstances (compensated bands [122], the presence of hidden order [123] or interference between itinerant and localized carriers [124]), the overall empirical situation appears to support the presence of short-lived vortices (superconducting phase fluctuations) above T_c but well inside the pseudogap state. The link between the pseudogap and the superconductivity in HTC is not yet clear though since the Nernst signal appears at significantly lower temperatures to T^*. How this impacts on the other normal state transport remains to be seen though claims of significant paraconductivity in $\rho_{ab}(T)$ below T^* in UD cuprates [125] have also implied the presence of short-lived superconducting fluctuations well above T_c.

10.7. Discussion and Summary

If nothing else, the comparison between experiment and theory in Section 10.4.4 should have served to highlight the difficulties inherent in identifying the correct description of the normal state transport properties in HTC. One of the major problems lies with the standard Boltzmann approach itself. Given that essentially all possible manipulations of the Boltzmann equation have been exhausted in the search for the correct phenomenology of HTC, it is tempting to discard this approach altogether in favor of more rigorous treatments, such as the Kubo formulism with its associated current–current correlation function. This would fail, however, to acknowledge just how much insight has been gained from applying the Boltzmann equation, in all its manifestations, to the cuprate problem. Moreover, it is not clear whether the Kubo formulism would be able to capture the low-energy physics of HTC where so much of the action takes place.

The second major problem concerns the fact that transport coefficients are averaged quantities whilst the HTC materials themselves are characterized by intrinsic real-space inhomogeneity and momentum-space anisotropy, both in- and out-of-plane. Developments in the field may therefore come indirectly from other nontransport probes such as ARPES with k-space resolution, or now with the advent of Fourier mapping [126], from real-space probes such as STM. Ultimately though, reconciliation of results from other experimental probes with the dc transport data will remain the imperative. Indeed, it has been one of the chief goals of the ARPES community to understand the transport phenomena of HTC from their k-space resolved measurements of the self-energy. It is equally important to recognize the difference(s), however, in the quantities determined by these probes when making such comparisons; dc and optical conductivities are angle-averaged *bulk* probes of the *two-particle* charge response, whilst ARPES is a angle-resolved *surface* probe of the *single-particle* electronic response.

The imaginary self-energy Im Σ is a measure of the qp relaxation rate or decay of the qp excitation and in the most naïve picture, $2\text{Im } \Sigma = \hbar/\tau_{tr}$. Re Σ, the real part of the self-energy

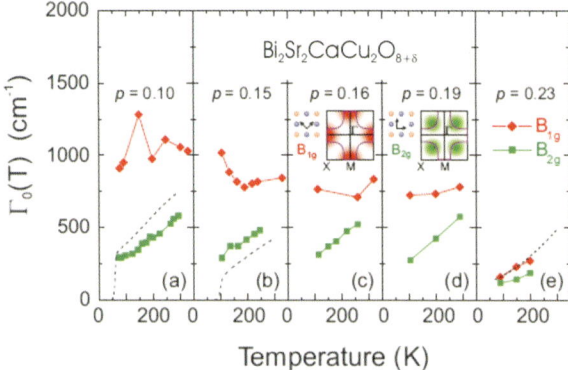

Figure 10.12. Static electron relaxation rates $\Gamma_0(T)$ of $Bi_2Sr_2CaCu_2O_8$ at various doping levels as determined for B_{1g} (*red*) and B_{2g} (*green*) symmetries from Raman scattering. The insets show the regions of the Fermi surface sensitive to the different symmetries indicated. Dashed lines are $\rho_{ab}(T)$ data [129].

is a measure of the renormalization of the single particle energy dispersion and is related to the qp mass enhancement. At the time of writing, there is little consensus within the ARPES community to permit a thorough comparison with the transport properties to be made. The dominant energy dependence of the normal state scattering rate at the nodes, for example, as extracted from the width of momentum distribution curves, is claimed to be either linear [94] or quadratic [127] in energy in OP cuprates or in some cases, to show a kink corresponding to coupling to some bosonic mode [128].

As argued in Section 10.3, the influence of phonons on the normal state transport in HTC is minimal. This contrasts markedly with claims that the ARPES results show strong e–ph coupling in HTC across the entire doping range [128]. One way to reconcile this troubling ambiguity is to recall that ARPES measures the qp lifetime that is affected by *all* scattering events, whilst charge transport is dominated by large-angle scattering. (Note that $\kappa(T)$ is also sensitive to both.) Whilst e–ph scattering involves only small q-transfer at low energies, e–e (Umklapp) and spin-fluctuation scattering are, by their nature, much more effective at perturbing a transport current. Thus the e–ph interaction may be conspicuous to ARPES, but it is effectively transparent to an applied electric current.

Electronic Raman scattering is also a two-particle response function, but unlike dc and optical conductivity, it does have some k-resolution. Regions in k-space can be selected by changing the polarization of the incident and scattered photons and the slope of the spectra at $\omega = 0$ is equal to the k-resolved dc conductivity. Indeed, recent results have again highlighted significant anisotropy in the transport scattering rate, and one that decreases with increasing doping, and for the most part, with increasing T [129]. As shown in Figure 10.12, within the B_{1g} channel (susceptible to the charge response around $(\pi, 0)$), the dc scattering rate $\Gamma_0(T)$ is larger and almost T-independent *in three different families of cuprates*. Conversely, in the B_{2g} channel (associated with the charge response along the nodal directions), $\Gamma_0(T)$ has a linear or supralinear T-dependence, consistent with $\rho_{ab}(T)$ (see figure). This striking difference in the magnitudes *and* T-dependencies in the two regions of k-space is intriguing and deserves further attention.

Clearly, the combination of ARPES, AMRO and Raman paint a consistent picture of strong basal plane anisotropy in the scattering rate (both transport and qp) in HTC intensifying with decreasing doping. As discussed in Section 10.4., such a T-dependent anisotropy will give rise to a T-dependent Hall effect within a quasi-2D FL picture. Given the proximity to the AFM insulator at half-filling, spin fluctuations are the obvious candidate, although theoretical calculations of the scattering rate based on existing susceptibility data are inconclusive as regards both the magnitude and the T-dependence of $\rho_{ab}(T)$. In particular, in the OD regime where the magnetic response is particularly weak, the scattering rate appears to remain high [50]. Indeed, one of the challenges for FL-based models is the absolute magnitude of the scattering. (Recall from Section 10.2 that the scattering rate in cuprates deduced from Homes' law is significantly larger than $k_B T$.) In the non-FL realm, both the two-lifetime picture of Anderson and Varma's MFL theory have had some success in capturing certain aspects of the empirical situation, particularly at optimal doping. There is still significant work to be done, however, particularly with regards to the doping dependence and the inclusion of anisotropy parameters, before those models can claim with any sort of universal applicability.

As mentioned in Section 10.3, the central issue regarding c-axis transport is one of interlayer coherence. In the noninteracting case, the coherent/incoherent crossover is expected to occur once the interlayer hopping rate $t_\perp$ becomes smaller than other energy scales in the system. Given that $\hbar/\tau_{ab} > T$, this suggests that in-plane scattering sets also the scale for interlayer hopping. The experimental data show clearly that interplane transfer in all OP cuprates (*including* $YBa_2Cu_3O_{6.95}$) is incoherent and genuine coherent c-axis motion only appears in the very heavily OD region of the phase diagram. ARPES measurements performed on other oxides suggest that in the 2D state, the loss of interlayer coherence leads to the development of an incoherent *in-plane* response [130]. Hence there may be an interesting correlation between the large basal plane scattering and the crossover to non-FL physics which should be explored further in the cuprates.

The UD regime is without doubt the most unusual region of the cuprate phase diagram. One of the most fundamental problems here for the FL-based models is the violation of the Luttinger sum rule implied from Hall effect measurements, with the carrier density at low T scaling with p rather than $1 - p$. This contrasts markedly with the situation at optimal doping where Luttinger's theorem appears to be restored. Recently, however, new averaging techniques have been exploited in ARPES to reveal a complete FS in UD cuprates with an area that *is* compatible with Luttinger's theorem [131], albeit with a suppressed spectral weight away from the qp nodes. There are many possible origins for the loss of spectral weight at the antinodes. Recent combined studies of ARPES and STM for example have pointed to the possibility of charge ordering between near-nested regions of the FS around $(\pi, 0)$ [84, 132]. Whether this charge ordering is a consequence of strong correlation effects near half-filling, or some form of FS reconstruction below T^* to minimize the ever increasing scattering energies near the antinodes remains to be seen. With regards the transport properties though, the big question that arises is again one of coherence. At what level of spectral weight suppression does the qp cease to exist or cease to contribute to the longitudinal or transverse charge response? This question may only be properly addressed once the correct interpretation of the optical response in HTC is established, i.e., two components (one coherent, one incoherent) vs. a single component with a large frequency-dependent scattering rate. Such questions of course lie at the heart of correlated many-body physics and we look forward to new insight into this fundamental problem in due course.

The fact that the parent HTC compound at half-filling is an AFM Mott insulator rather than a metal is direct evidence that strong electron correlations in the square planar CuO_2 lattice play a central role in the cuprate problem. The introduction of holes into a highly correlated 2D magnetic background gives rise to an exceptional metallic ground state at low doping, characterized by the highly anomalous physics of the UD and OP cuprates (spin/charge freezing, nanoscale inhomogeneity, stripe formation and pseudogap phenomena) and ultimately, d-wave superconductivity. Upon further doping, superconductivity is rapidly destroyed and the ground state that finally emerges has all the hallmarks of a conventional 3D FL. Given this bewildering array of observed phenomena, it seems remarkable that such systematic transport behavior *across all the different cuprate families* is observed at all. This universality, however, does suggest that the key to the transport problem in HTC lies in the physics of a single CuO_4 unit and that the development of a complete coherent description of the normal state transport properties of HTC is still an achievable, albeit daunting, task.

Bibliography

1. P. W. Anderson, *The Theory of Superconductivity in the High-T_c Cuprates* (Princeton University Press, Princeton, 1997).
2. C. M. Varma, P. B. Littlewood, S. Schmitt-Rink, E. Abrahams, and A. E. Ruckenstein, Phys. Rev. Lett. **63**, 1996 (1989).
3. A. Carrington, A. P. Mackenzie, C. T. Lin, and J. R. Cooper, Phys. Rev. Lett. **69**, 2855 (1992).
4. P. Monthoux and D. Pines, Phys. Rev. B **49**, 4261 (1992).
5. C. Castellani, C. di Castro, and M. Grilli, Phys. Rev. Lett. **75**, 4650 (1995).
6. L. B. Ioffe and A. J. Millis, Phys. Rev. B **58**, 11631 (1998).
7. N. P. Ong, *Physical Properties of High Temperature Superconductors*, edited by D. M. Ginsberg (World Scientific, Singapore, 1991), Vol. II, p. 459; Y. Iye, *ibid*, (1993), Vol. III, p. 285; C. Uher, *ibid*, (1993), Vol. III, p. 159; S. L. Cooper and K. E. Gray, *ibid.* (1994), Vol. IV p. 61.
8. H. Takagi and N. E. Hussey, *Proceedings of the International School of Physics "Enrico Fermi"*, 227 (1998).
9. A. I. Larkin and A. A. Varlamov, *The Physics of Superconductors: Vol. I Conventional and High-T_c Superconductors,* edited by K.-H. Bennemann and J.B. Ketterson (Springer-Verlag, Tokyo, 2002).
10. N. E. Hussey, Eur. J. Phys. B **31**, 495 (2003).
11. A. P. Mackenzie, S. R. Julian, D. C. Sinclair, and C. T. Lin, Phys. Rev. B **53**, 5848 (1996).
12. Y. Ando, S. Komiya, K. Segawa, S. Ono, and Y. Kurita, Phys. Rev. Lett. **93**, 267001 (2004).
13. N. E. Hussey, H. Takagi, Y. Iye, S. Tajima, A. I. Rykov, and K. Yoshida, Phys. Rev. B **61**, R6475 (2000).
14. G. S. Boebinger, Y. Ando, A. Passner, T. Kimura, M. Okuya, J. Shimoyama, K. Kishio, K. Tamasaku, N. Ichikawa, and S. Uchida, Phys. Rev. Lett. **77**, 5417 (1996).
15. S. Ono and Y. Ando, Phys. Rev. B **67**, 104512 (2003).
16. M. Giura, R. Fastampa, S. Sarti, and E. Silva, Phys. Rev. B **68**, 134505 (2003).
17. A. W. Tyler and A. P. Mackenzie, Physica **282–287C**, 1185 (1997).
18. T. Manako, Y. Kubo, and Y. Shimakawa, Phys. Rev. B **46**, 11019 (1992).
19. C. C. Homes, S. V. Dordevic, M. Strongin, D. A. Bonn, R. Liang, W. N. Hardy, S. Komiya, Y. Ando, G. Yu, N. Kaneko, X. Zhao, M. Greven, D. N. Basov, and T. Timusk, Nature, **430**, 539 (2004).
20. J. Zaanen, Nature, **430**, 512 (2004).
21. M. Gurvitch and A. T. Fiory, Phys. Rev. Lett. **59**, 1337 (1987).
22. S. Martin, A. T. Fiory, R. M. Fleming, L. F. Schneemeyer, and J. V. Waszczak, Phys. Rev. B **41**, 846 (1990).
23. S. Ono, Y. Ando, T. Murayama, F. F. Balakirev, J. B. Betts, and G. S. Boebinger, Phys. Rev. Lett. **85**, 638 (2000).
24. H. Takagi, B. Batlogg, H. L. Kao, J. Kwo, R. J. Cava, J. J. Krajewski, and W. F. Peck Jr., Phys. Rev. Lett. **69**, 2975 (1992).
25. P. Fournier, P. Mohanty, E. Maiser, S. Darzens, T. Venkatesan, C. J. Lobb, G. Czjzek, R. A. Webb, and R. L. Greene, Phys. Rev. Lett. **81**, 4720 (1998).
26. Y. Dagan, M. M. Qazilbash, C. P. Hill, V. N. Kulkarni, and R. L. Greene, Phys. Rev. Lett. **92**, 167001 (2004).
27. F. Balakirev, J. B. Betts, A. Migliori, S. Ono, Y. Ando, and G. S. Boebinger, Nature **424**, 912 (2003).
28. R. S. Markiewicz, Physica **168C**, 195 (1990).
29. A. Virosztek, and J. Ruvalds, Phys. Rev. B **42**, 4064 (1990).
30. S. Fujimoto, H. Kohno, and K. Yamada, J. Phys. Soc. Jpn. **60**, 2724 (1991).
31. R. Hlubina and T. M. Rice, Phys. Rev. B **51**, 9253 (1995).
32. D. van der Marel, H. J. A. Molegraaf, J. Zaanen, Z. Nussinov, F. Carbone, A. Damascelli, H. Eisaki, M. Greven, P. H. Kes, and M. Li, Nature **425**, 271 (2003).
33. J. R. Cooper and J. W. Loram, J. Phys. **6**, 1 (1996).
34. T. Timusk and B. Statt, Rep. Progr. Phys. **62**, 61 (1999).
35. A. Carrington, D. Colson, Y. Dumont, C. Ayach, A. Bertinotti, and J. F. Marucco, Physica **234C**, 1 (1994).
36. B. Bucher, P. Steiner, J. Karpinski, E. Kaldis, and P. Wachter, Phys. Rev. Lett. **70**, 2012 (1993).
37. T. Ito, K. Takenaka, and S. Uchida, Phys. Rev. Lett. **70**, 3995 (1993).

38. V. J. Emery, S. A. Kivelson, and O. Zachar, Phys. Rev. B **56**, 6120 (1997).
39. N. E. Hussey, K. Nozawa, H. Takagi, S. Adachi, and K. Tanabe, Phys. Rev. B **56**, R11423 (1997).
40. T. Nakano, N. Momono, M. Oda, and M. Ido, J. Phys. Soc. Jpn. **67**, 2622 (1998).
41. T. Watanabe, T. Fujii, and A. Matsuda, Phys. Rev. Lett. **79**, 2113 (1997).
42. Y. Ando, K. Segawa, S. Komiya, and A. N. Lavrov, Phys. Rev. Lett. **88**, 137005 (2002).
43. Y. Fukuzumi, K. Mizuhashi, K. Takenaka, and S. Uchida, Phys. Rev. Lett. **76**, 684 (1996).
44. Y. Dagan, M. C. Barr, W. M. Fisher, R. Beck, T. Dhakal, A. Biswas, and R. L. Greene, Phys. Rev. Lett. **94**, 057005 (2005).
45. Y. Ando, A. N. Lavrov, S. Komiya, K. Segawa, and X. F. Sun, Phys. Rev. Lett. **87**, 017001 (2001).
46. T. Yoshida, X. J. Zhou, T. Sasagawa, W. L. Yang, P. V. Bogdanov, A. Lanzara, Z. Hussain, T. Mizokawa, A. Fujimori, H. Eisaki, Z.-X. Shen, T. Kakeshita, and S. Uchida, Phys. Rev. Lett. **91**, 027001 (2003).
47. M. Calandra and O. Gunnarsson, Europhys. Lett. **61**, 88 (2003).
48. K. Takenaka, J. Nohara, R. Shiozaki, and S. Sugai, Phys. Rev. B **68**, 134501 (2003).
49. N. E. Hussey, K. Takenaka, and H. Takagi, Philos. Mag. **84**, 2847 (2004).
50. S. Nakamae, K. Behnia, S. J. C. Yates, N. Mangkorntong, M. Nohara, H. Takagi, and N. E. Hussey, Phys. Rev. B **68**, 100502(R) (2003).
51. N. E. Hussey, A. P. Mackenzie, J. R. Cooper, Y. Maeno, S. Nishizaki, and T. Fujita, Phys. Rev. B **57**, 5505 (1998).
52. M. N. McBrien, N. E. Hussey, P. J. Meeson, and S. Horii, J. Phys. Soc. Jpn. **71**, 701 (2002).
53. A. P. Mackenzie, S. R. Julian, A. J. Diver, G. J. McMullan, M. P. Ray, G. G. Lonzarich, Y. Maeno, S. Nishizaki, and T. Fujita, Phys. Rev. Lett. **76**, 3786 (1996).
54. N. E. Hussey, M. N. McBrien, L. Balicas, J. M. Brooks, S. Horii, and H. Ikuta, Phys. Rev. Lett. **89**, 086601 (2002).
55. N. E. Hussey, J. R. Cooper, Y. Kodama, and Y. Nishihara, Phys. Rev. B **58**, R611 (1998).
56. Y. Ando, G. S. Boebinger, A. Passner, N. L. Wang, C. Geibel, and F. Steglich, Phys. Rev. Lett. **77**, 2065 (1996).
57. A. J. Leggett, Braz. J. Phys. **22**, 129 (1992).
58. O. K. Anderson, O. Jepsen, A. I. Liechtenstein, and I. I. Mazin, Phys. Rev. B **49**, 4145 (1994).
59. C. Bernhard, D. Munzar, A. Wittlin, W. König, A. Golnik, C. T. Lin, M. Kläser, Th. Wolf, G. Müller-Vogt, and M. Cardona, Phys. Rev. B **59**, R6631 (1999).
60. D. G. Clarke, S. P. Strong, and P. W. Anderson, Phys. Rev. Lett. **74**, 4499 (1995).
61. F. Nakamura, M. Kodama, S. Sakita, Y. Maeno, T. Fujita, H. Takahashi, and N. Mori, Phys. Rev. B **54**, 10061 (1996).
62. S. Uchida, K. Tamasaku, and S. Tajima, Phys. Rev. B **53**, 14558 (1996).
63. T. Kimura, S. Miyasaka, H. Takagi, K. Tamasaku, H. Eisaki, S. Uchida, K. Kitazawa, M. Hiroi, M. Sera, and N. Kobayashi, Phys. Rev. B **53**, 8733 (1996).
64. T. A. Friedmann, M. W. Rabin, J. Giapintzakis, J. P. Rice, and D. M. Ginsberg, Phys. Rev. B **42**, 6217 (1990).
65. S. Tajima, J. Schützmann, S. Miyamoto, I. Terasaki, Y. Sato, and R. Hauff, Phys. Rev. B **55**, 6051 (1997).
66. C. C. Homes, T. Timusk, R. Liang, D. A. Bonn, and W. N. Hardy, Phys. Rev. Lett. **71**, 1645 (1993).
67. K. Takenaka, K. Mizuhashi, H. Takagi, and S. Uchida, Phys. Rev. B **50**, 6534 (1994).
68. N. Nagaosa, Phys. Rev. B **52**, 10561 (1995).
69. Y. F. Yan, P. Matl, J. M. Harris, and N. P. Ong, Phys. Rev. B **52**, 751 (1995).
70. N. E. Hussey, M. Kibune, H. Nakagawa, N. Miura, Y. Iye, H. Takagi, S. Adachi, and K. Tanabe, Phys. Rev. Lett. **80**, 2909 (1998).
71. Y. Nakamura and S. Uchida, Phys. Rev. B **47**, 8369 (1993).
72. N. Kumar and A. M. Jayannavar, Phys. Rev. B **45**, 5001 (1992).
73. N. E. Hussey, M. Abdel-Jawad, A. Carrington, A. P. Mackenzie, and L. Balicas, Nature **425**, 814 (2003).
74. K. Yamaji, J. Phys. Soc. Jpn. **58**, 1520 (1989).
75. R. H. McKenzie and P. Moses, Phys. Rev. Lett. **81**, 4492 (1998).
76. O. K. Andersen, A. I. Liechtenstein, O. Jepsen, and F. Paulsen, J. Phys. Chem. Solids **56**, 1573 (1995).
77. D. van der Marel, Phys. Rev. B **60**, R765 (1999).
78. For a review, see N. E. Hussey, Adv. Phys. **51**, 1685 (2002).
79. H. Y. Hwang, B. Batlogg, H. Takagi, H. L. Kao, R. J. Cava, J. J. Krajewski, and W. F. Peck Jr., Phys. Rev. Lett. **72**, 2636 (1994).
80. N. P. Ong, Z. Z. Wang, J. Clayhold, J. M. Tarascon, L. H. Greene, and W. R. McKinnon, Phys. Rev. B **35**, 8807 (1987).
81. H. Takagi, T. Ido, S. Ishibashi, M. Uota, S. Uchida, and Y. Tokura, Phys. Rev. B **40**, 2254 (1989).
82. Y. Ando, Y. Kurita, S. Komiya, S. Ono, and K. Segawa, Phys. Rev. Lett. **92**, 197001 (2004).
83. N. Kikugawa, A. P. Mackenzie, C. Bergemann, and Y. Maeno, Phys. Rev. B **70**, 174501 (2004).

84. K. M. Shen, F. Ronning, D. H. Lu, F. Baumberger, N. J. C. Ingle, W. S. Lee, W. Meevasana, Y. Kohsaka, M. Azuma, M. Takano, H. Takagi, and Z.-X. Shen, Science **307**, 901 (2005).
85. C. M. Varma and E. Abrahams, Phys. Rev. Lett. **86**, 4652 (2001).
86. A. Ino, C. Kim, M. Nakamura, T. Yoshida, T. Mizokawa, A. Fujimori, Z.-X. Shen, T. Kakeshita, H. Eisaki, and S. Uchida, Phys. Rev. B **65**, 094504 (2002).
87. T. Nishikawa, J. Takeda, and M. Sato, J. Phys. Soc. Jpn. **63**, 1441 (1994).
88. T. R. Chien, Z. Z. Wang, and N. P. Ong, Phys. Rev. Lett. **67**, 2088 (1991).
89. Y. Ando and T. Murayama, Phys. Rev. B **60**, R6991 (1999).
90. Z. Konstantinovic, Z. Z. Li, and H. Raffy, Phys. Rev. B **62**, R11989 (2000).
91. A. P. Mackenzie, N. E. Hussey, A. J. Diver, S. R. Julian, Y. Maeno, S. Nishizaki, and T. Fujita, Phys. Rev. B **54**, 7425 (1996).
92. N. P. Ong, Phys. Rev. B **43**, 193 (1991).
93. J. M. Harris, Y. F. Yan, P. Matl, N. P. Ong, P. W. Anderson, T. Kimura, and K. Kitazawa, Phys. Rev. Lett. **75**, 1391 (1995).
94. M. Grayson, L. B. Rigal, D. C. Schmadel, H. D. Drew, and P.-J. Kung, Phys. Rev. Lett. **89**, 037003 (2002).
95. T. Valla, A. V. Fedorov, P. D. Johnson, Q. Li, G. D. Gu, and N. Koshizuka, Phys. Rev. Lett. **85**, 828 (2000).
96. R. Hlubina, Phys. Rev. B **64**, 132508 (2001).
97. E. Carter and A. J. Schofield, Phys. Rev. B **66**, 241102 (2002).
98. N. E. Hussey, J. R. Cooper, J. M. Wheatley, I. R. Fisher, A. Carrington, A. P. Mackenzie, C. T. Lin, and O. Milat, Phys. Rev. Lett. **76**, 122 (1996).
99. Z. Yosuf, B. O. Wells, T. Valla, A. V. Fedorov, P. D. Johnson, Q. Li, C. Kendziora, Sha Jian, and D. G. Hinks, Phys. Rev. Lett. **88**, 167006 (2002).
100. M. Abdel-Jawad, M. P. Kennett, L. Balicas, A. Carrington, A. P. Mackenzie, R. H. McKenzie, and N. E. Hussey, arXiv:condmat/0609763 (2006).
101. Y. I. Latyshev, O. Laborde, and P. Monceau, Europhys. Lett. **29**, 495 (1995).
102. F. Rullier-Albenque, P. A. Vieillefond, H. Alloul, A. W. Tyler, P. Lejay, and J. F. Marucco, Europhys. Lett. **50**, 81 (2000).
103. A. Legris, F. Rullier-Albenque, E. Radeva, and P. Lejay, J. Phys. (France) I**3**, 1605 (1993).
104. K. Mizuhashi, K. Takenaka, Y. Fukuzumi, and S. Uchida, Phys. Rev. B **52**, R3884 (1995).
105. D. J. C. Walker, A. P. Mackenzie, and J. R. Cooper, Phys. Rev. B **51**, 15653 (1995).
106. K. Segawa and Y. Ando, Phys. Rev. B **59**, R3948 (1999).
107. C. Panagopoulos, unpublished.
108. G.-q. Zheng, T. Odaguchi, T. Mito, Y. Kitaoka, K. Asayama, and Y. Kodama, J. Phys. Soc. Jpn. **62**, 2591 (1993).
109. D. N. Basov, R. Liang, B. Dabrowski, D. A. Bonn, W. N. Hardy, and T. Timusk, Phys. Rev. Lett. **77**, 4090 (1996).
110. S. H. Pan, E. W. Hudson, K. M. Lang, H. Eisaki, S. Uchida, and J. C. Davis, Nature **403**, 746 (2000).
111. R. W. Hill, C. Proust, L. Taillefer, P. Fournier, and R. L. Greene, Nature **414**, 711 (2001).
112. K. Krishana, J. M. Harris, and N. P. Ong, Phys. Rev. Lett. **75**, 3529 (1995).
113. Z. A. Xu, N. P. Ong, Y. Wang, T. Kakeshita, and S. Uchida, Nature **406**, 486 (2000).
114. S. D. Obertelli, J. R. Cooper, and J. L. Tallon, Phys. Rev. B **46**, 14928 (1992).
115. Y. Ando, Y. Hanaki, S. Ono, T. Murayama, K. Segawa, N. Miyamoto, and S. Komiya, Phys. Rev. B **61**, R14956 (2000).
116. J. R. Cooper and A. Carrington, *Advances in Superconductivity V*, edited by Y. Bando and H. Yamauchi (Spinger-Verlag, Tokyo, 1993), p. 95.
117. Y. Zhang, N. P. Ong, P. W. Anderson, D. A. Bonn, R. Liang, and W. N. Hardy, Phys. Rev. Lett. **86**, 890 (2001).
118. R. C. Yu, M. B. Salamon, J. P. Lu, and W. C. Lee, Phys. Rev. Lett. **69**, 1431 (1992).
119. M. F. Smith, J. Paglione, M. B. Walker, and L. Taillefer, Phys. Rev. B **71**, 014506 (2005).
120. R. Bel, K. Behnia, C. Proust, P. van der Linden, D. Maude, and S. I. Vedeneev, Phys. Rev. Lett. **92**, 177003 (2004).
121. C. Proust, E. Boaknin, R. W. Hill, L. Taillefer, and A. P. Mackenzie, Phys. Rev. Lett. **89**, 147003 (2002).
122. R. Bel, K. Behnia, and H. Berger, Phys. Rev. Lett. **91**, 066602 (2003).
123. K. Behnia, R. Bel, Y. Kasahara, Y. Nakajima, H. Jin, H. Aubin, K. Izawa, Y. Matsuda, J. Flouquet, Y. Haga, Y. Ōnuki, and P. Lejay, Phys. Rev. Lett. **94**, 156405 (2005).
124. A. S. Alexandrov and V. N. Zavaritsky, Phys. Rev. Lett. **93**, 217002 (2004).
125. B. Léridon, A. Défossez, J. Dumont, J. Lesueur, and J. P. Contour, Phys. Rev. Lett. **87**, 197007 (2001).
126. K. McElroy, R. W. Simmonds, J. E. Hoffman, D.-H. Lee, J. Orenstein, H. Eisaki, S. Uchida, and J. C. Davis, Nature **422**, 592 (2003).

127. A. A. Kordyuk, S. V. Borisenko, A. Koitzsch, J. Fink, M. Knupfer, B. Büchner, H. Berger, G. Margaritondo, C. T. Lin, B. Keimer, S. Ono, and Y. Ando, Phys. Rev. Lett. **92**, 257006 (2004).
128. A. Lanzara, P. V. Bogdanov, X. J. Zhou, S. A. Kellar, D. L. Feng, E. D. Lu, T. Yoshida, H. Eisaki, A. Fujimori, K. Kishio, J-I. Shimoyama, T. Noda, S. Uchida, Z. Hussain, and Z.-X. Shen, Nature **412**, 510 (2001).
129. R. Hackl, L. Tassini, R. Venturini, C. Hartinger, A. Erb, N. Kikugawa, and T. Fujita, *Advances in Solid State Physics* Vol. 48, edited by B. Kramer (Springer-Verlag, Berlin, Heidelberg, 2005) p. 227.
130. T. Valla, P. D. Johnson, Z. Yusof, B. Wells, Q. Li, S. M. Loureiro, R. J. Cava, M. Mikami, Y. Mori, M. Yoshimura, and T. Sasaki, Nature **417**, 627 (2002).
131. X. J. Zhou, T. Yoshida, D.-H. Lee, W. L. Yang, V. Brouet, F. Zhou, W. X. Ti, J. W. Xiong, Z. X. Zhao, T. Sasagawa, T. Kakeshita, H. Eisaki, S. Uchida, A. Fujimori, Z. Hussain, and Z.-X. Shen, Phys. Rev. Lett. **92**, 187001 (2004).
132. T. Hanaguri, C. Lupien, Y. Kohsaka, D.-H. Lee, M. Azuma, M. Takano, H. Takagi, and J. C. Davis, Nature **430**, 1001 (2004).

11

High-Pressure Effects

J. S. Schilling

Experiments under hydrostatic and uniaxial pressure serve not only as a guide in the synthesis of materials with superior superconducting properties but also allow a quantitative test of theoretical models. In this chapter the pressure dependence of the superconducting properties of elemental, binary, and multiatom superconductors are explored, with an emphasis on those exhibiting relatively high values of the transition temperature T_c. In contrast to the vast majority of superconductors, where T_c decreases under pressure, in the cuprate oxides T_c normally increases. Uniaxial pressure studies give evidence that this increase arises mainly from the reduction in the area A of the CuO_2 planes ($T_c \propto A^{-2}$), rather than in the separation between the planes, thus supporting theoretical models which attribute the superconductivity primarily to intraplanar pairing interactions. More detailed information would be provided by future experiments in which the hydrostatic and uniaxial pressure dependences of several basic parameters, such as T_c, the superconducting gap, the pseudogap, the carrier concentration, and the exchange interaction are determined for a given material over the full range of doping.

11.1. Introduction

Pressure, like temperature, is a basic thermodynamic variable which can be applied in experiment over an enormous range, leading to important contributions in such diverse areas of science and technology as astrophysics, geophysics, condensed matter physics, chemistry, biology, and food processing [1, 2]. The field of superconductivity is no exception. The first high-pressure studies on a superconductor were carried out in 1925 by Sizoo and Onnes [3] and revealed that for Sn and In, as for most superconductors [4], the superconducting transition temperature T_c *decreases* under pressure. As will be seen below, the explanation for this pressure-induced decrease in T_c rivals the isotope effect in its simplicity.

It is no accident that many groups active in the synthesis of novel superconducting materials, particularly the many outstanding scientists who emerged from the groups of the late Bernd Matthias in La Jolla or Werner Buckel in Karlsruhe, routinely use the high-pressure technique as an important diagnostic tool. Why? Because high-pressure experiments can provide valuable assistance in the search for superconductors with higher values of T_c. In contrast to magnetic materials, which owe their enormous technological importance to the fact that their magnetism is stable to temperatures well above ambient, current materials do not become superconducting unless they are artificially cooled to temperatures at least 160 K below ambient, an inconvenient and expensive process in large-scale applications. An overriding goal in technology-oriented superconductivity research is, therefore, to find materials where T_c surpasses room temperature.

J. S. Schilling • Department of Physics, Washington University, CB 1105, One Brookings Dr., St. Louis, Missouri 63130, USA

One way to estimate whether a new superconducting material is capable of higher T_c values is to determine how much T_c changes under variation of the chemical composition and/or the pressure. A large value of $|dT_c/dP|$ gives hope that higher values of T_c are possible. We give three examples. A notably successful application of this strategy was the early high-pressure experiments of Paul Chu's group [5] on the $La_{2-x}Ba_xCuO_4$ cuprate (La-214); the large value of dT_c/dP (+8 K GPa^{-1}) led to the substitution of the smaller Y^{3+} cation for La^{3+} and the discovery of the famous $YBa_2Cu_3O_{7-\delta}$ compound (Y-123), the first superconductor with T_c above the boiling point of liquid N_2 ($\sim$77 K). A second example: in the oxide $La_{2-x}Sr_xCuO_4$ T_c is found to increase if compressed in one direction, but decrease if compressed in another [6]; Locquet et al. [7] used this fact to appropriately strain thin films of this oxide by growing them epitaxially on a substrate, thus doubling the value of T_c from 25 to 49 K. In a third example, the observed increase in T_c under pressure for the Hg-oxides [8] (and for most cuprate oxides for that matter) prompted very high-pressure experiments on $HgBa_2Ca_2Cu_3O_8$ (Hg-1223) whereby T_c increased from 134 K to temperatures near 160 K [9]. In less than 10 years, therefore, the record high value of T_c increased *sevenfold* from 23 K for Nb_3Ge to $\sim$160 K for Hg-1223. A further increase by only a factor of two would place T_c above room temperature! Forty years ago Neil Ashcroft [10] raised the possibility that elemental hydrogen may become a room temperature superconductor, if only sufficient pressure is applied. The metallization of hydrogen and the development of a viable theory for high-T_c cuprates remain two central goals in current condensed matter research.

High-pressure experiments contribute to the field of superconductivity in diverse ways. (1) As mentioned above, a large magnitude of dT_c/dP is a good indicator that higher values of T_c may be possible for a given superconductor at ambient pressure through chemical substitution or epitaxial growth techniques. (2) Some superconducting materials can only be properly synthesized through the simultaneous application of high pressure and high temperature [11]. (3) Many nonsuperconducting materials become superconducting if sufficiently high pressures are applied. As seen in Figure 11.1, there are 29 elemental superconductors at ambient pressure. Under pressure 23 more become superconducting (Li, B, O, Si, P, S, Ca, Sc, Fe, Ge, As, Se, Br, Sr, Y, Sb, Te, I, Cs, Ba, Bi, Ce, and Lu); more than half of these were discovered by Jörg Wittig in the 1960s and 1970s [1, 12]. (4) The basic electronic and lattice properties of a material change with decreasing temperature due to the thermal contraction of the lattice. High-pressure experiments change the lattice parameters directly at any temperature and thus allow one to correct for the thermal contraction effects at ambient pressure, yielding isochores. (5) Determining the dependence of T_c and other superconducting properties on the individual lattice parameters of a single sample allows a clean quantitative test of theoretical models and gives information on the pairing mechanism. For example, if superconductivity in the high-T_c cuprates results primarily from interlayer coupling, one would anticipate a particularly strong change in T_c if uniaxial pressure is applied perpendicular to the layers.

Unfortunately, all high-pressure experiments are not created equal! In superconductivity the pressure dependence of T_c may depend on the pressure medium used and other factors, as illustrated in Figure 11.2 for Pb: $T_c(P)$ using the relatively stiff pressure medium methanol:ethanol lies clearly above that when helium is used [15]. Ideally, the applied pressure should be either purely hydrostatic or purely uniaxial. A purely hydrostatic experiment, however, is only possible over a limited pressure/temperature range since all fluids solidify under pressure, the last one being liquid He which requires 12 GPa to freeze at room temperature. Solid He is very soft, i.e., it can only support very weak shear stresses. Dense He is, therefore, the pressure medium of choice in high-pressure experiments [16, 17]. One practical way to test whether or not a given experimental result is sensitive to shear stress effects is

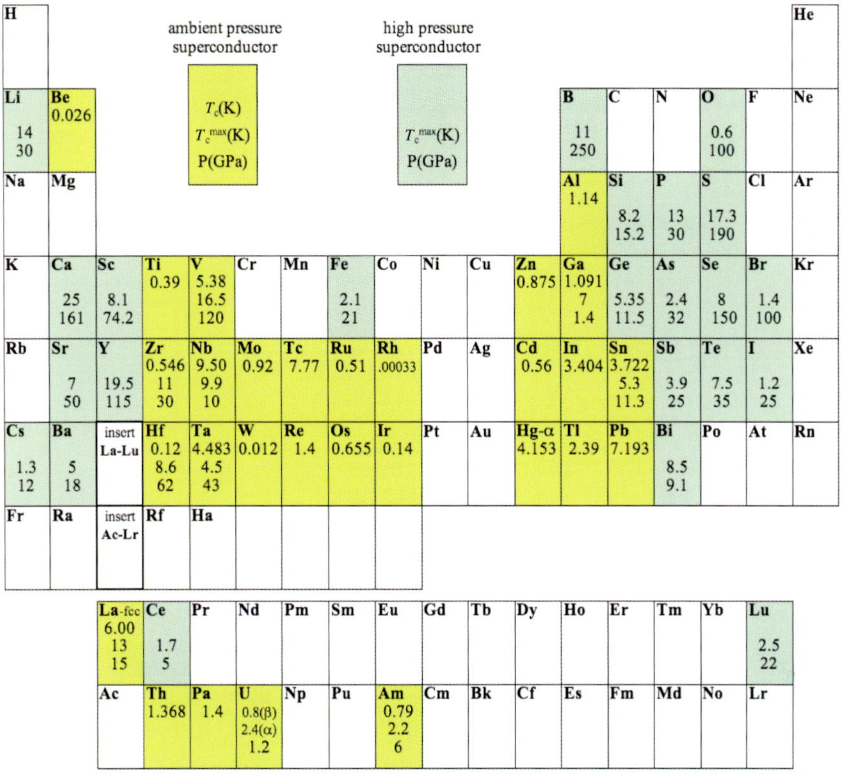

High-Pressure Effects

H																	He
Li	Be 0.026 14 30		ambient pressure superconductor $T_c(K)$ $T_c^{max}(K)$ $P(GPa)$				high pressure superconductor $T_c^{max}(K)$ $P(GPa)$					B 11 250	C	N	O 0.6 100	F	Ne
Na	Mg											Al 1.14	Si 8.2 15.2	P 13 30	S 17.3 190	Cl	Ar
K	Ca 25 161	Sc 8.1 74.2	Ti 0.39	V 5.38 16.5 120	Cr	Mn	Fe 2.1 21	Co	Ni	Cu	Zn 0.875	Ga 1.091 7 1.4	Ge 5.35 11.5	As 2.4 32	Se 8 150	Br 1.4 100	Kr
Rb	Sr 7 50	Y 19.5 115	Zr 0.546 11 30	Nb 9.50 9.9 10	Mo 0.92	Tc 7.77	Ru 0.51	Rh .00033	Pd	Ag	Cd 0.56	In 3.404	Sn 3.722 5.3 11.3	Sb 3.9 25	Te 7.5 35	I 1.2 25	Xe
Cs 1.3 12	Ba 5 18	insert La-Lu	Hf 0.12 8.6 62	Ta 4.483 4.5 43	W 0.012	Re 1.4	Os 0.655	Ir 0.14	Pt	Au	Hg-α 4.153	Tl 2.39	Pb 7.193	Bi 8.5 9.1	Po	At	Rn
Fr	Ra	insert Ac-Lr	Rf	Ha													

	La-fcc 6.00 13 15	Ce 1.7 5	Pr	Nd	Pm	Sm	Eu	Gd	Tb	Dy	Ho	Er	Tm	Yb	Lu 2.5 22
	Ac	Th 1.368	Pa 1.4	U 0.8(β) 2.4(α) 1.2	Np	Pu	Am 0.79 2.2 6	Cm	Bk	Cf	Es	Fm	Md	No	Lr

Figure 11.1. Periodic Table listing 29 elements superconducting at ambient pressure (yellow) and 23 elements which only superconduct under high pressure (green). For each element the upper position gives the value of $T_c(K)$ at ambient pressure; middle position gives maximum value $T_c^{max}(K)$ in a high-pressure experiment at $P(GPa)$ (lower position). In many elements multiple phase transitions occur under pressure. If T_c decreases under pressure, only the ambient pressure value of T_c is given. Sources for T_c values at ambient pressure are given in [13]. Sources for T_c values under high pressure are given in [14].

to carry out the experiment using *two different* pressure media; if the pressure dependence in question remains the same, it is unlikely that shear stress effects play a major role.

This chapter will restrict itself primarily to the final (fifth) benefit of high-pressure investigations as applied to elemental, binary, and multiatom superconductors. The focus will be on those materials with the highest values of T_c since it can be argued that a thorough understanding of such materials will be most likely to lead to further increases in T_c. Attaining the highest values of T_c demands careful optimization of the relevant electronic and lattice (structural) properties. This optimization is most difficult to realize in elemental solids; here the maximum value of T_c has been limited to the temperature range 9–20 K (for Nb at ambient pressure and for Li, B, P, S, Ca, V, Y, Zr, and La under very high pressures, as seen in Figure 11.1). It is not surprising that multiatom systems exhibit higher values of T_c since their structural flexibility allows a higher degree of optimization. The highest values of T_c are exhibited by quasi 2D solids such as MgB_2 and the high-T_c cuprate oxides. The cuprates, however, exhibit great structural and electronic complexity under both ambient and high-pressure conditions, a fact which has greatly hampered attempts to reach a basic understanding of the physical mechanisms responsible for the superconducting state. We will, therefore, begin by

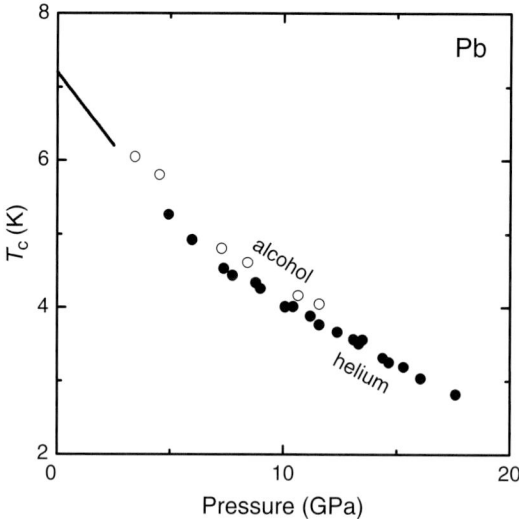

Figure 11.2. Pressure dependence of T_c for Pb from [15] using helium (●) and methanol–ethanol (○) pressure media. Straight line gives initial dependence from [18].

discussing in some detail what we can learn from high-pressure experiments on the relatively simple elemental and binary superconductors before tackling the much more difficult high-T_c oxides.

Rather than attempting to review the results of high-pressure studies on all known superconducting materials, this chapter will attempt to highlight the *new* information high-pressure experiments provide, information not readily available using other techniques. We refer the reader to excellent reviews of the relatively low-T_c heavy Fermion [19–23] and organic [24] superconductors which are not included here.

11.2. Elemental Superconductors

Referring to Figure 11.1, let us begin by considering those superconductors where the pressure dependence of T_c is easily understood, namely, the ten simple s, p metals which are superconducting at ambient pressure: Be, Al, Zn, Ga, Cd, In, Sn, Hg, Tl, and Pb. Under sufficiently high pressures, the number of simple s, p metal superconductors is increased by 14: Li, B, O, Si, P, S, Ge, As, Se, Br, Sb, Te, I, and Bi. The four s, p elements Cs, Ca, Sr, and Ba also become superconducting under pressure, but their superconductivity is likely rooted in the fact that they exhibit strong s → d transfer under pressure and thus effectively become transition metals. The remaining 24 elemental superconductors in Figure 11.1 are either transition metals, rare-earth metals, or actinide metals for all of which the conduction electron character is dominated by d-electrons.

11.2.1. Simple Metals

Nonalkali Metals

The isotope effect played a pivotal role in the development of the BCS theory [25] for conventional phonon-mediated superconductivity. This is due to the fact that isotopic

High-Pressure Effects

substitution primarily affects only a *single* property, the lattice vibration (phonon) spectrum. Considering the BCS expression in Eq. (11.1), changes in the isotopic mass M primarily affect the *prefactor* Θ_D, the Debye temperature, and not the exponent, whence the simple relation $T_c \propto \Theta_D \propto M^{-(1/2)}$. On the other hand, if a superconductor is subjected to high pressures, the *exponent* in Eq. (11.1) is affected since important changes in *both* the lattice vibration *and* the conduction electron states occur. The dependence of T_c on pressure, therefore, may be rather complex, as we shall see below. However, in simple s, p metal superconductors like Sn, In, Zn, Pb, and Al, the pressure-induced stiffening of the lattice vibration spectrum completely dominates over the minor changes in the electronic properties, leading to a *ubiquitous decrease* in T_c with increasing pressure [26], as seen, for example, in Figure 11.2 for Pb. Pressure-induced structural phase transitions in simple metals may prompt T_c to jump to higher (or lower) values [27], but otherwise dT_c/dP exhibits a negative slope. A diamond-anvil cell was first used to study superconductivity in the beautiful experiments by Gubser and Webb [28] on Al in 1975; T_c was found to decrease under ~6 GPa pressure *15-fold* from 1.18 to 0.08 K.

The above discussion can be made more concrete by analyzing the BCS expression [25] for the transition temperature

$$T_c \simeq 1.13 \Theta_D \exp\left\{\frac{-1}{N(E_f)\mathcal{V}_{\text{eff}}}\right\}, \quad (11.1)$$

where $N(E_f)$ is the electronic density of states at the Fermi energy and $\mathcal{V}_{\text{eff}}$ is the effective attractive pairing interaction (for simplicity we set $k_B = \hbar = 1$). Since the s, p electrons in simple metals are normally nearly free, one expects in a 3D system $N(E_f) \propto V^{+2/3}$ so that under pressure $N(E_f)$ decreases even more slowly than the sample volume V. The principal source for the observed decrease in T_c with pressure in simple s, p metals, however, is not the decrease in $N(E_f)$ but the sizeable decrease in the pairing interaction $\mathcal{V}_{\text{eff}}$ itself. The argument can be made more explicit by neglecting the Coulomb repulsion and using McMillan's [29] expression for the electron–phonon coupling parameter λ

$$N(E_f)\mathcal{V}_{\text{eff}} = \lambda = \frac{N(E_f)\langle I^2 \rangle}{M\langle \omega^2 \rangle}, \quad (11.2)$$

where $\langle I^2 \rangle$ is the average square electronic matrix element and $\langle \omega^2 \rangle$ the average square phonon frequency. Making the simplifying assumptions that $\Theta_D \approx \langle \omega \rangle \approx \sqrt{k/M}$, where k is the lattice spring constant, and $M\langle \omega^2 \rangle \approx M\langle \omega \rangle^2 \approx M(k/M) = k$, Eq. (11.1) becomes

$$T_c \approx \sqrt{\frac{k}{M}} \exp\left\{\frac{-k}{N(E_f)\langle I^2 \rangle}\right\}. \quad (11.3)$$

In the isotope effect, M appears explicitly only in the prefactor, so that one obtains the canonical BCS relation $T_c \propto M^{-(1/2)}$. In a high-pressure experiment, the changes in T_c are relatively large since they arise principally from the terms in the exponent. In simple metal superconductors, for example, the quantity in Eq. (11.3) which changes most rapidly under pressure is the spring constant k, the denominator in the exponent being only weakly pressure dependent, as we discuss below. As k increases with pressure, the modest increase of the prefactor $\sqrt{k}$ is overwhelmed by the decrease from the $-k$ in the exponent, leading to the universal rapid decrease in T_c with pressure for simple s, p metal superconductors. For example, Al, Sn, and Pb, where $T_c(0) \simeq 1.14$, 3.73, and 7.19 K, respectively, have the pressure dependences d ln T_c/d$P \simeq -0.25, -0.13$, and -0.051 GPa^{-1} [30, 31]. What is also immediately evident from these data is that |d ln T_c/dP| is largest when $T_c(0)$ is smallest. It can be easily shown that this inverse correlation follows directly from the fact that T_c depends *exponentially* on the solid-state parameters $N(E_f)\mathcal{V}_{eff}$. To show this, take the logarithm of both sides of Eq. (11.1) and then the derivative with respect to pressure to obtain

$$\frac{d \ln T_c}{dP} = \frac{d \ln \Theta_D}{dP} + \left[\ln \frac{\Theta_D}{T_c}\right]\left[\frac{d \ln N(E_f)\mathcal{V}_{eff}}{dP}\right]. \tag{11.4}$$

The first term on the right side of this equation is normally small and can be neglected. The quantity in the left square bracket is positive. The sign of d ln T_c/dP, therefore, is determined by that of [d ln $N(E_f)\mathcal{V}_{eff}$/dP] which is negative in the present case. Since ln[Θ_D/T_c] becomes larger for decreasing T_c (unless Θ_D decreases substantially) the magnitude of d ln T_c/dP would be expected to increase for smaller T_c, as observed. Note that such an inverse correlation would not be obtained were T_c to only depend on some (high) power of the solid-state parameters.

To put this discussion on a more quantitative basis, we consider the McMillan equation [29]

$$T_c \simeq \frac{\langle\omega\rangle}{1.20} \exp\left\{\frac{-1.04(1+\lambda)}{\lambda - \mu^*(1+0.62\lambda)}\right\}, \tag{11.5}$$

which goes beyond weak coupling and connects the value of T_c with fundamental parameters such as the mean phonon frequency $\langle\omega\rangle$, the electron–phonon coupling parameter λ, and the Coulomb repulsion μ^*. Within this framework, it can be shown that the anticipated change in μ^* with pressure is normally very small and can be neglected [32]; here we set μ^* equal to the constant value $\mu^* = 0.1$. However, one should be aware that this assumption for μ^* may not hold in a more rigorous theoretical framework [33] where the electron–electron and electron–phonon coupling effects are treated on the same footing; this framework yielded for the alkali metal Li the estimate $T_c \simeq 0.4$ mK, in contrast to the value $T_c \approx 1$ K from conventional electronic structure calculations [34].

Taking the logarithmic volume derivative of T_c in Eq. (11.5), we obtain the simple relation

$$\frac{d \ln T_c}{d \ln V} = -B\frac{d \ln T_c}{dP} = -\gamma + \Delta\left\{\frac{d \ln \eta}{d \ln V} + 2\gamma\right\}, \tag{11.6}$$

where B is the bulk modulus, $\gamma \equiv -$ d ln $\langle\omega\rangle$ /d ln V the Grüneisen parameter, $\eta \equiv N(E_f)\langle I^2\rangle$ the Hopfield parameter [35], and $\Delta \equiv 1.04\lambda[1 + 0.38\mu^*][\lambda - \mu^*(1 + 0.62\lambda)]^{-2}$. Eq. (11.6) has a simple interpretation. The first term on the right, which comes from the prefactor to the exponent in the above McMillan expression for T_c, is usually very small relative to the second term. The sign of the pressure derivative dT_c/dP, therefore, is determined by the relative magnitude of the two terms in the curly brackets. The first "electronic" term involves the derivative of the Hopfield parameter $\eta \equiv N(E_f)\langle I^2\rangle$ which can be calculated directly in electronic-structure theory [36]. McMillan [29] pointed out that whereas individually $N(E_f)$ and $\langle I^2\rangle$ may fluctuate appreciably, their product $\eta \equiv N(E_f)\langle I^2\rangle$ changes only gradually, i.e., η is a well-behaved "atomic" property. One would thus anticipate that η changes in a relatively well-defined manner under pressure, reflecting the character of the electrons near the Fermi energy [35]. An examination of high-pressure data on simple s, p metal superconductors, in fact, reveals that Eq. (11.6) is obeyed if η increases under pressure at the approximate rate d ln η/d ln $V \approx -1$ [17], a result also obtained from electronic structure calculations [37]. We also note that Chen et al. [32] derived for s, p metals the approximate expression $d \ln \eta / d \ln V = -[d \ln N(E_f)/d \ln V] - 2/3$ which yields for a 3D free-electron gas $d \ln \eta / d \ln V = -4/3 \approx -1$.

Let us now apply Eq. (11.6) to an analysis of dT_c/dP for simple-metal superconductors. The expression in the curly brackets is positive since the lattice term is positive ($2\gamma \approx +3$ to $+5$) and dominates over the negative electronic term $d \ln \eta / d \ln V \approx -1$. Since Δ is always positive and the first term $-\gamma$ is relatively small, the sign of dT_c/dP must be negative. This accounts for the universal *decrease* of T_c with pressure in simple metals due to lattice stiffening.

Let us now consider a specific example in more detail. In Sn T_c decreases under pressure at the rate $dT_c/dP \simeq -0.482 \, \text{K GPa}^{-1}$ which leads to $d \ln T_c/d \ln V \simeq +7.2$ [30]. Inserting $T_c(0) \simeq 3.73$ K, $\langle \omega \rangle \simeq 110$ K [38], and $\mu^* = 0.1$ into the McMillan equation, we obtain $\lambda \simeq 0.69$ from which follows $\Delta \simeq 2.47$. Inserting these values into Eq. (11.6) and setting $d \ln \eta / d \ln V = -1$, we can solve for the Grüneisen parameter to obtain $\gamma \simeq +2.46$, in reasonable agreement with the experimental value $\gamma \approx +2.1$ [30]. Similar results are obtained for other conventional simple metal BCS superconductors [17]. Hodder [39] used the McMillan formula and the measured pressure-dependent phonon spectrum for Pb to estimate $dT_c/dP \simeq -0.36 \, \text{K GPa}^{-1}$, in good agreement with experimental values [18, 26, 30].

From the above it is clear that the observed ubiquitous decrease in T_c with pressure for simple metals results from a weakening of the electron–phonon coupling λ due to the shift of the phonon spectrum to higher frequencies. This weakening of λ is also primarily responsible for the almost universal decrease in the electrical resistivity of simple metals under pressure [40].

Alkali Metals

Alkali metals are widely believed to be simple, nearly free electron metals *par excellence* where each atom donates a single s electron to the conduction band, resulting in a nearly spherical Fermi surface. No alkali metal is known to be superconducting at ambient pressure. In lieu of a structural phase transition, high pressure would not be expected to induce superconductivity in an alkali metal since, as discussed above, pressure weakens the electron–phonon coupling λ. In fact, conventional wisdom tells us that high pressure should enhance the free electron behavior of a metal since compressing a solid normally broadens bands and narrows energy gaps.

It was thus with some trepidation that Lin and Dunn [41] reported in 1986 that above 20 GPa the lightest alkali metal, Li, exhibits both a *positive* resistivity derivative $d\rho/dP$ *and* some type of phase transition near 5 K, perhaps a superconducting transition. The matter attracted little attention until 1997 when Neaton and Ashcroft [42] argued on general grounds that under extreme compression the electronic properties of Li could become quite complex and nonfree-electron-like due to the near overlap of the atomic 1s cores; the anticipated enhancement in the electron–lattice interaction would be expected to lead to low-symmetry crystal structures, possible superconductivity, and an increase in the electrical resistivity. These results corraborated earlier electronic structure calculations [43] which indicated band-narrowing and gap-widening in Li under extreme compression, i.e., drastic deviations from free-electron behavior.

Three years later two groups [44, 45] subjected Li metal to very high pressures and reported superconductivity above 20 GPa, T_c rising to temperatures approaching 20 K at 30 GPa in the resistivity onset [44]. In these three studies on Li, either a solid pressure medium was used [41] or no pressure medium at all [44, 45], the sample coming in direct contact with

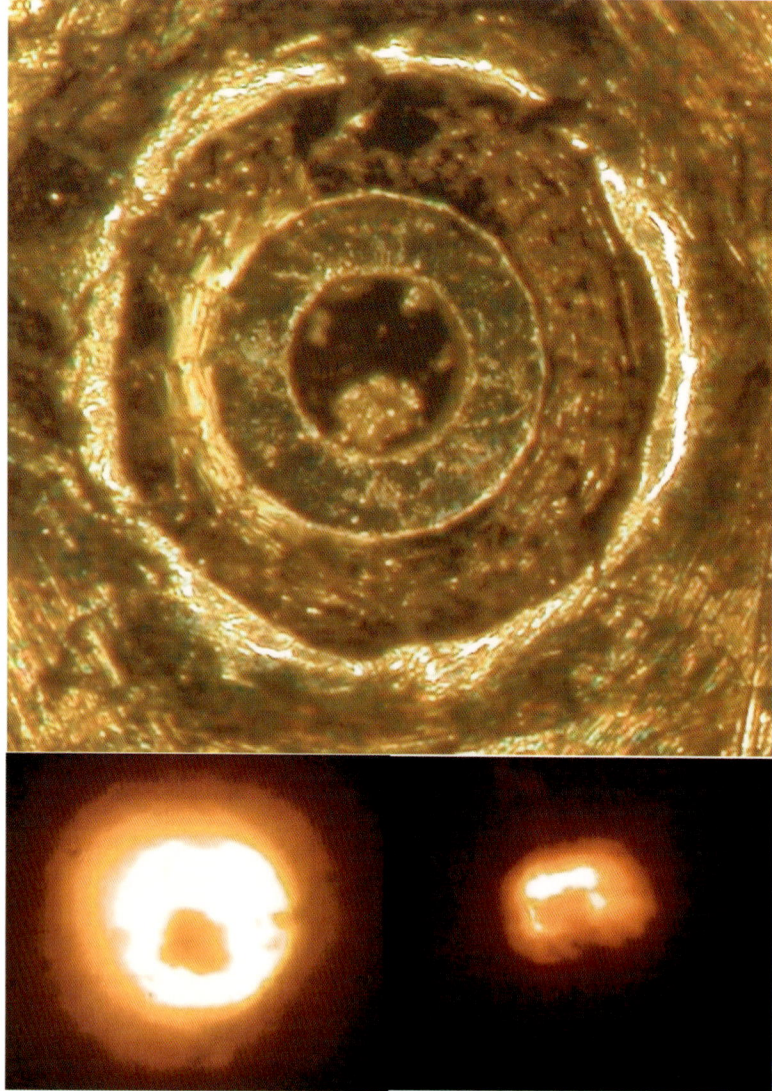

Figure 11.3. (top) Gold-plated rhenium gasket with 250 μm dia hole containing Li sample. (bottom) Transmitted-light photograph of hole containing Li sample at (left) ambient pressure and (right) 30 GPa. Figure taken from [46].

the ultrahard diamond anvils. To determine whether the reported superconductivity might have resulted from shear stresses on the Li sample, a third group [46] surrounded the sample with liquid helium in a diamond-anvil cell, as seen in Figure 11.3, resulting in nearly hydrostatic pressure conditions. These studies confirmed that Li does indeed become superconducting at 5 K for 20 GPa, T_c rising rapidly to 14 K at 30 GPa, as seen in Figure 11.4. In addition, the superconducting phase diagram $T_c(P)$ of Li was accurately mapped out to nearly 70 GPa; several structural phase transitions are indicated at 20, 30, 67, and possibly 55 GPa. The pressure-induced structural transitions in Li have been investigated to 50 GPa in X-ray diffraction studies [47] and to 123 GPa in very recent optical spectroscopic studies [48] at

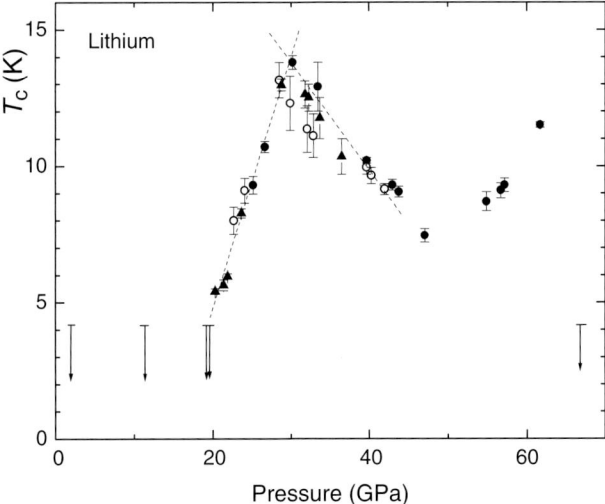

Figure 11.4. Superconducting phase diagram $T_c(P)$ of Li metal under nearly hydrostatic pressure (helium) to 67 GPa from [46]. Dashed lines are guides to eye. Several structural phase transitions are indicated.

variable temperatures; a unifying scheme for the structural transition mechanisms in all alkali metals has been proposed [49]. These two results, (1) that Li becomes superconducting under pressure and (2) that T_c increases rapidly with pressure, are quite remarkable and confirm that at elevated densities the electronic structure of Li deviates markedly from that of a free-electron gas, the anticipated Fermi surface becoming highly nonspherical [50].

Neaton and Ashcroft [51] applied a similar analysis to the next heavier alkali metal, Na, predicting similar results to those for Li, but at higher pressures. To date, no pressure-induced superconductivity has been found above 4 K in Na to 65 GPa or in K to 43.5 GPa (to 35 GPa above 1.5 K) [150], nor in Rb above 0.05 K to 21 GPa [52]. Very recent studies [53] show that the melting temperature of Na actually decreases for pressures above 30 GPa, falling particularly rapidly above 80 GPa in the fcc phase before passing through a minimum near 110 GPa. These results give strong evidence for highly anomalous electronic behavior in Na in the pressure range above 30 GPa and the likelihood of superconductivity, particularly in the fcc phase above 80 GPa. Further s, p metal systems which likely exhibit anomalous electronic behavior include S which becomes metallic for $P \geq 85$ GPa with a superconducting transition temperature as high as 17 K at ~ 200 GPa nonhydrostatic pressure [54] and P where T_c reaches 18 K at 30 GPa [55], as seen in Figure 11.5.

The first alkali metal to become superconducting under high pressure is Cs [56, 57]. Unlike Li and Na, Cs possesses an empty d-band which lies relatively near the Fermi energy. Since it can be shown on general grounds that Cs's half-filled 6s-band moves up under pressure more rapidly than the bottom of the empty 5d-band [58], electrons from Cs's 6s band are transferred into the 5d band (s → d transfer), so that under sufficient pressure Cs becomes, in effect, a transition metal. Nonmagnetic transition metals with their higher electronic density of states are normally superconducting, as seen in Figure 11.1. Wittig has shown that Cs becomes superconducting at temperatures between 0.05 and 1.5 K for quasihydrostatic pressures 11–15 GPa [56, 57], respectively, a pressure range over which a number of structural transitions occur. McMahan [59] has estimated that in Cs the s → d transfer is complete for $P \geq 15$ GPa. Considerably higher values of T_c appear possible at higher pressures, in spite

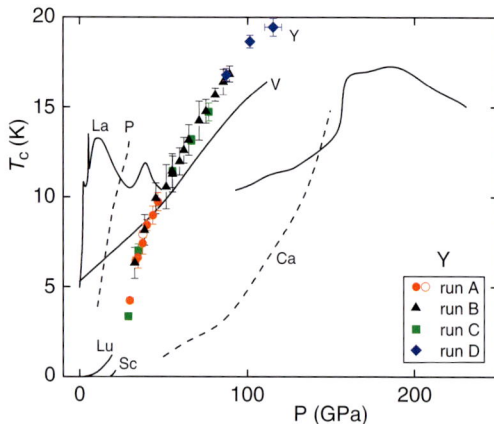

Figure 11.5. Superconducting phase diagrams $T_c(P)$ for nonalkali elements with the highest values of T_c under pressure (Ca [65], La [66], P [67], S [68], V [69], Y [70,71]) as well as for Lu [72] and Sc [73]. For clarity only the three highest-pressure data are shown for Y in run D.

of Cs's 40× higher ionic mass ($M = 133$) compared to Li ($M = 7$). We note that the transition metal superconductor La ($M = 139$) reaches values near $T_c \approx 13$ K at 12 GPa (see Figure 11.5).

Similar scenarios, including superconductivity, would be expected to occur for the next lighter alkali metals, Rb and K, where pressure-induced 5s → 4d and 4s → 3d transfer is estimated to be complete at 53 and 60 GPa, respectively [59]. It thus seems likely that under sufficiently high pressures all alkali metals will become superconducting.

Although the superconducting properties of the alkali metals become highly anomalous under extreme compression, these properties can still be understood within a conventional BCS framework where the electron pairing arises through the electron–phonon interaction, as for the other simple s, d metals, and, in fact, for the transition metal superconductors which we now briefly discuss.

11.2.2. Transition Metals

In transition metals the d-electron character of the conduction band leads to an enhanced density of states $N(E_f)$ which favors superconductivity at higher temperatures than in simple s, p metals. Because of their importance in technological applications, transition metal superconductors have received a great deal of attention, particularly in the 1960s and 1970s. The status of high-pressure experiments on d-band metals and their theoretical interpretation in terms of electron–phonon mediated superconductivity were comprehensively reviewed by Smith [60] and Garland and Bennemann [61], respectively, in the early 1970s. These same analyses were successfully applied to later systematic studies on transition metal alloys [62, 63].

Although in the majority of transition metal superconductors T_c decreases with pressure, in many cases T_c is found to increase. A positive sign of dT_c/dP for d-band superconductors may be understood as arising from a much more rapid increase of the Hopfield parameter under pressure ($d \ln \eta / d \ln V \approx -3$ to -4 [35, 61, 64]) than in s, p-band superconductors ($d \ln \eta / d \ln V \approx -1$). If, in Eq. (11.6), the electronic term $d \ln \eta / d \ln V$ becomes larger in magnitude than the lattice term 2γ, T_c would be expected to *increase* with pressure; this is, in

High-Pressure Effects

fact, observed in V, La, and Zr, for example [60], and is seen in Figure 11.5 for V, La, Y, Lu, and Sc.

Another reason that the pressure dependence $T_c(P)$ may be particularly complex in transition metals is that the number of d electrons in the conduction band increases under pressure due to s → d transfer [58], enhancing the possibility of pressure-induced structural transitions or electronic (Lifshitz) transitions; such effects are likely responsible for the unusually complex $T_c(P)$ dependence of LaAg where $T_c(P)$ to 2.5 GPa passes through two maxima and minima [74]. See the review by Lorenz and Chu [75] for examples of electronic transitions.

It is well known that the number of d electrons, n_d, is a particularly significant quantity in determining the crystal structure and the electronic properties of transition metal [76,77], rare-earth [78], and actinide [77] solids. The pressure-induced superconductivity in the pre-transition elements Cs, Ca, Sr, and Ba is likely the result of s → d electron transfer.

In Figure 11.5 $T_c(P)$ data are compared for the trivalent transition metals La, Y, Sc, and Lu. The very recent nearly hydrostatic [70] and nonhydrostatic [71] studies on Y metal differ substantially from earlier quasihydrostatic work [56, 72] and reveal that T_c increases monotonically from 5 K at 30 GPa to 19.5 K (midpoint) or 20 K (onset) at 115 GPa, the highest value of T_c ever measured in the magnetic susceptibility for an elemental superconductor (see Figure 11.1); remarkably, the dependence of T_c on sample volume is nearly linear over the entire pressure range 33–115 GPa [70, 71]. The initial slope $dT_c/dP \approx +1$ K GPa^{-1} for La is particularly large, possibly due to the anomalously low value of the Grüneisen constant $\gamma \approx 1$ for this metal [61]. Experiments on V metal show that T_c increases slowly, but nearly linearly, with pressure (+0.1 K GPa^{-1}) from 5 to 17 K at 120 GPa [69]. Unlike for s, p metals, the pressure dependence $T_c(P)$ for transition metals follows no universal behavior, reflecting the additional complexity (and potency!) of the electronic properties in a d-electron system.

Can Eq. (11.6) account for the observed pressure dependence of T_c for V? Setting $T_c(0) = 5.3$ K, $\mu^* = 0.1$, and the Debye temperature $\Theta_D = 399$ K [29] in the McMillan equation, where $\langle\omega\rangle = 0.83\Theta_D$, we obtain $\lambda = 0.538$ and thus $\Delta = 3.547$. Inserting now into Eq. (11.6) the volume derivative of the Hopfield parameter (d $\ln\eta$/d $\ln V \simeq -3.3$) calculated for V by Evans et al. [37] and the Grüneisen parameter $\gamma \simeq 1.5$ [79], we obtain d $\ln T_c$/d $\ln V \simeq -2.56$. Using for the bulk modulus $B = 162$ GPa [79], we obtain, finally, $dT_c/dP = -[T_c(0)/B]$d $\ln T_c$/d $\ln V \simeq +0.084$ K GPa^{-1}, in good agreement with the experiments of Ishizuka et al. (0.1 K GPa^{-1}) [69] and the earlier studies of Smith (0.062 K GPa^{-1}) [80].

As seen in Fig. 11.1, the highest values of T_c yet achieved for an elemental superconductor lie in the range 15–25 K for both s-, p-, and d-electron metals under high pressure. It would be expected that higher values of T_c should be possible for binary or pseudobinary compounds where the flexibility afforded by two elements should allow a superior optimization of the parameters. Indeed, binary superconductors reach values of T_c which are more than twice as high as those for elemental superconductors.

11.3. Binary Superconductors

11.3.1. A-15 Compounds

Until the discovery of the cuprate oxides in late 1986, the binary A-15 compounds Nb$_3$Ge ($T_c \simeq 23$ K), Nb$_3$Sn ($T_c \simeq 17.8$ K), and V$_3$Si ($T_c \simeq 16.6$ K) exhibited the highest values of T_c. High-pressure studies on the A-15s were reviewed in 1972 by Smith [81].

Hopfield [35] noted that the near doubling of the value of T_c from Nb to Nb$_3$Sn could be simply understood, using the above relation d ln η/d ln $V \approx -3.5$, as resulting from an enhancement in η by $\sim 60\%$ due to the reduced Nd–Nd separation in Nb$_3$Sn.

In the A-15 compounds the competition between subtle structural transitions and superconductivity has been extensively studied. A case in point are parallel studies by two groups [82, 83] on "nontransforming" V$_3$Si crystals where T_c increases under pressure from 16.6 K to approximately 17.7 K at 3 GPa, whereupon $T_c(P)$ exhibits a break in slope signalling a cubic-to-tetragonal structural transformation predicted by Larsen and Ruoff [84]. Further details and references on A-15 compounds are contained in a recent review by Lorenz and Chu [75]. We will see below that the high-T_c oxides provide numerous examples for the influence of structural defects and transitions on superconductivity, perhaps more than one would like!

Following the discovery of high-T_c superconductivity in the cuprates, two binary compounds, MgB$_2$ and Rb$_3$C$_{60}$, were discovered which have substantially higher transition temperatures than the A-15s. We now consider high-pressure studies on these two compounds.

11.3.2. A Special Case: MgB$_2$

The binary superconductor with the highest known value of the transition temperature, MgB$_2$ with $T_c \approx 40$ K, was discovered in early 2001 [85]; Buzea and Yamashita [86] have reviewed its superconducting properties. MgB$_2$ is a quasi-2D material with strong covalent bonding within the graphite-like B$_2$ layers. Understandably, the compressibility is highly anisotropic, being 64% greater along the c axis than the a axis, with bulk modulus $B = 147.2(7)$ [87]. The anisotropy in the superconducting properties is also appreciable, but less than that observed in the high-T_c oxides [88].

Several studies of the dependence of T_c on pressure for polycrystalline MgB$_2$ were carried out shortly after the discovery of its superconductivity [89–92]. The first studies used either solid (steatite) [89] or fluid (Fluorinert) [90, 91] pressure media and agreed that T_c decreases under pressure, but disagreed widely on the rate of decrease which ranged from -0.35 to -1.9 K GPa^{-1}. The first truly hydrostatic measurement of $T_c(P)$ was carried out to 0.7 GPa using He gas on an isotopically pure (^{11}B) sample [92]; it was found that T_c decreases reversibly under hydrostatic pressure at the rate dT_c/d$P \simeq -1.11 \pm 0.02$ K GPa^{-1}, yielding d ln T_c/d ln $V = B$d ln T_c/d$P \simeq +4.16 \pm 0.08$. This latter result was confirmed subsequently by He-gas studies on MgB$_2$ single crystals to 0.6 GPa as well as parallel diamond-anvil-cell studies in dense He to nearly 30 GPa [93] which are shown in Figure 11.6; the latter are in excellent agreement to 20 GPa with parallel studies in dense He by Goncharov and Struzhkin [94]. On the other hand, diamond-anvil-cell studies on the same samples using methanol:ethanol [95] or Fluorinert [93] pressure media resulted in a substantially more negative slope dT_c/dP, apparently arising from shear stress effects in these frozen pressure media.

Ultrahigh-resolution thermal expansion and specific heat measurements on MgB$_2$ yield through the Ehrenfest relation dT_c/d$P \simeq -1.05 \pm 0.13$ K GPa^{-1}, in excellent agreement with the dependence -1.07 ± 0.03 K GPa^{-1} obtained in He-gas studies, all on the same sample [96]. On cooling through T_c, both the thermal expansion coefficient and the Grüneisen function change from positive to negative, the latter showing a dramatic increase to large positive values at low temperature. These results suggest anomalous coupling between superconducting electrons and low-energy phonons [96].

High-Pressure Effects

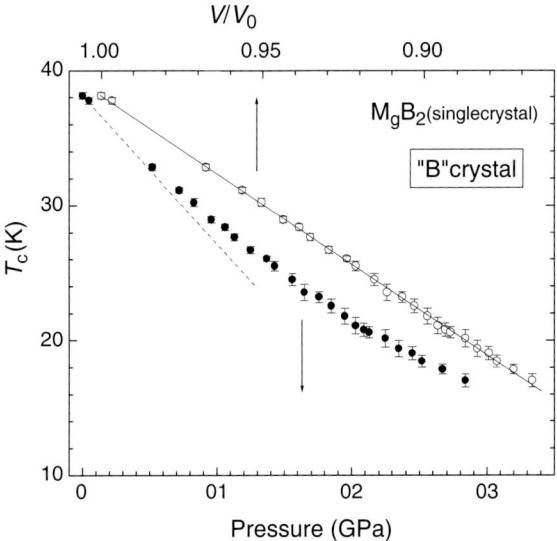

Figure 11.6. Dependence of T_c for a MgB$_2$ single crystal on nearly hydrostatic pressure P and relative volume V/V_o in a He-loaded diamond-anvil cell. Figure reproduced from [93].

We now apply the same analysis carried out above for simple s, p metal superconductors to MgB$_2$ to see whether the measured dependence $dT_c/dP \simeq -1.11 \pm 0.02$ K GPa^{-1} is consistent or not with BCS theory (electron–phonon coupling). Using the average phonon energy from inelastic neutron studies [97] $\langle\omega\rangle = 670$ K, $T_{c0} \simeq 39.25$ K, and $\mu^* = 0.1$, we obtain from the above Eqs. (11.5) and (11.6) $\lambda \simeq 0.90$ and $\Delta \simeq 1.75$. Our estimate of $\lambda \simeq 0.90$ agrees well with those of other authors [98, 99]. Since the pairing electrons in MgB$_2$ are believed to be s, p in character [98, 100–102], we set $d \ln \eta/d \ln V = -1$, a value close to that $d \ln \eta/d \ln V = Bd \ln \eta/dP \approx -0.81$, where $d \ln \eta/dP \approx +0.55$ %/GPa, from first-principles electronic structure calculations by Medvedera et al. [103]. Inserting the above values of $d \ln T_c/d \ln V = +4.16$, $\Delta = 1.75$, and $d \ln \eta/d \ln V = -1$ into Eq. (11.6), we find for the Grüneisen parameter $\gamma = 2.36$, in reasonable agreement with the values $\gamma \approx 2.9$ from Raman spectroscopy studies [48] or $\gamma \approx 2.3$ from ab initio electronic structure calculations on MgB$_2$ [104]. A similar analysis of the data in Figure 11.6 to 30 GPa, based on an analysis by Chen et al. [32], also gives excellent agreement. See [93] for a full discussion and a comprehensive summary of all high-pressure studies on MgB$_2$.

The He-gas $T_c(P)$ data are thus clearly consistent with electron–phonon pairing in MgB$_2$, in agreement with high precision isotope effect experiments [105, 106]. The fact that the B isotope effect is 15 times that for Mg [106] is clear evidence that the superconducting pairing originates within the graphite-like B$_2$ layers.

11.3.3. Doped Fullerenes A$_3$C$_{60}$

A particularly interesting class of superconductors with high values of T_c are the alkali-doped fullerides A$_3$C$_{60}$, where A = K, Rb, Cs [107], each alkali atom donating one s electron to the conduction band. K$_3$C$_{60}$ and Rb$_3$C$_{60}$ have T_c values of 19 and 29.5 K, respectively; evidence has been found for superconductivity in Cs$_3$C$_{60}$ near 40 K [108], but this has yet to be duplicated. The increase in T_c from K_3C_{60} to Rb_3C_{60} to Cs_3C_{60} is mainly related to lattice expansion (negative pressure) effects [109].

T_c for the alkali-doped fullerides is found to *decrease* under the application of hydrostatic pressure [109]. For Rb_3C_{60}, for example, $dT_c/dP \simeq -8.7\,\mathrm{K\,GPa^{-1}}$, as seen in Figure 11.7 [110]. Since the bulk modulus of Rb_3C_{60} is given by $B = 18.3\,\mathrm{GPa}$ [111], one can estimate $d\ln T_c/d\ln V = B(d\ln T_c/dP) \simeq +5.4$, a value intermediate between that for

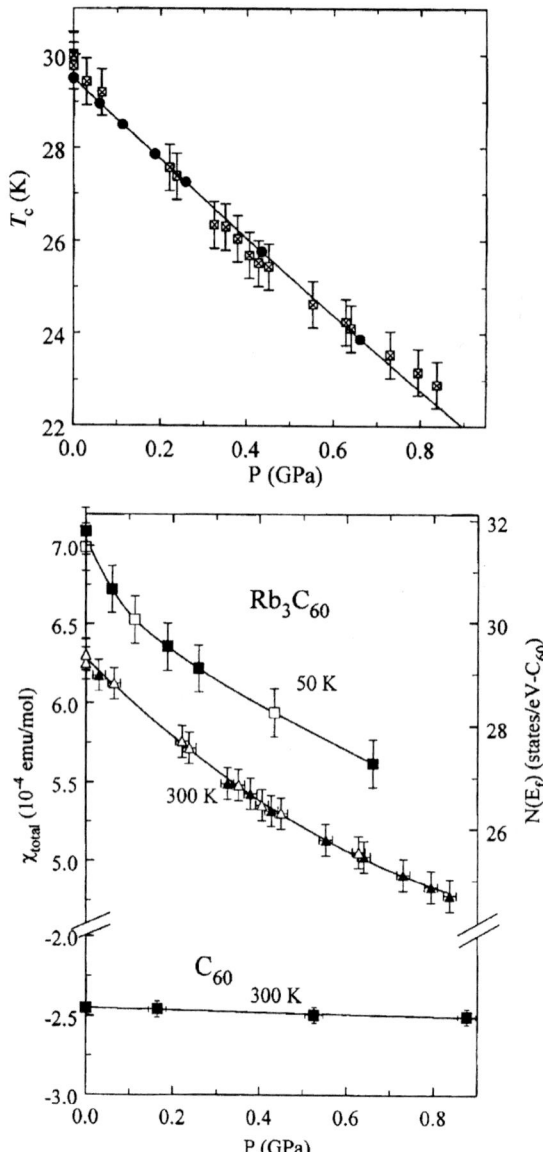

Figure 11.7. Results of hydrostatic pressure studies on Rb_3C_{60} from [110]: (top) superconducting transition temperature T_c; (bottom) magnetic susceptibility and estimated electronic density of states $N(E_f)$ at 50 and 300 K. Data for C_{60} at 300 K are also shown. Both T_c and $N(E_f)$ decrease rapidly with pressure.

High-Pressure Effects

MgB$_2$ (+4.16) and Sn (+7.2). It would thus be reasonable to expect that the reason for the negative value of dT_c/dP for the alkali-doped fullerenes is the same as for MgB$_2$, Sn, and other s, p metal superconductors, namely, lattice stiffening.

To test this hypothesis, let's attempt an analysis of the above data in terms of electron–phonon coupling using the above McMillan equation and its pressure derivative, invoking the *intermolecular* lattice vibrations for Rb$_3$C$_{60}$ which are in the range 15–150 K [112]. Setting the average value $\langle\omega\rangle \approx 80$ K and using $\mu^* = 0.2$ [113], Eq. (11.5) yields a *negative* value for λ, an impossibility, implying that this equation must be invalid for the given set of parameters. Even setting $\langle\omega\rangle \approx 150$ K, the upper limit for intermolecular vibrations, $\lambda \approx 5$ would be required by Eq. (11.5), a value clearly beyond the range of validity of the McMillan equation ($\lambda \leq 1.5$). To proceed, we use the simple expression

$$T_c = \frac{0.26 E_{\text{char}}}{\sqrt{e^{2/\lambda} - 1}}, \tag{11.7}$$

valid for all values of λ [114], where E_{char} is the characteristic lattice-vibration energy. Setting $E_{\text{char}} = \langle\omega\rangle \approx 80$ K and $T_c(0\,\text{K}) = 29.5$ K, Eq. (11.7) yields $\lambda = 5$. Taking the pressure derivative of Eq. (11.7), and using a typical value of the Grüneisen parameter $\gamma \approx +2$, it is easy to show [110] that the above value of dT_c/dP is only possible if $d\ln\eta/d\ln V \approx +10$! This value of $d\ln\eta/d\ln V$ differs grossly in both magnitude and sign from that typically found for conventional simple-metal (-1) or transition-metal (-3.5) superconducting elements, alloys, or compounds [35]. What is likely wrong is the above assumption that the *intermolecular* lattice vibrations are responsible for the superconductivity.

On the other hand, if we assume the characteristic lattice-vibration energy is given by the high-frequency *intramolecular* (on-ball) vibrational modes, where $E_{\text{char}} = \langle\omega\rangle \approx$ 350–2,400 K, then we cannot account for the negative value of dT_c/dP through lattice stiffening since, due to the extreme stiffness of the C$_{60}$ molecule, the average frequency of the on-ball phonons $\langle\omega\rangle$ and the mean square electron–phonon matrix element $\langle I^2\rangle$ are essentially independent of pressure.

So what is responsible for the rapid decrease in T_c under pressure in Rb$_3$C$_{60}$? Perhaps electronic effects are important here, in contrast to simple s, p electron metals. The answer to this question is provided by measurements of the pressure-dependent electronic density of states $N(E_f)$ which is found [110] to decrease sharply under pressure, as seen in Figure 11.7. This decrease is a direct result of the rapid increase in the width of the conduction band as the C$_{60}$ molecules are pressed together.

We are now confronted with a very different situation than in conventional superconductors. Utilizing our knowledge of $N(E_f)(P)$ in the McMillan equation, one can use the pressure-independent value of $\langle\omega\rangle$ as a parameter to obtain the best fit to the experimental $T_c(P)$ data. A detailed analysis [110] reveals that weak-coupling theory can account for the experimental pressure dependences as long as the characteristic energy of the intermediary boson lies between $\langle\omega\rangle \approx 300$ and 800 K, typical energies for the high-frequency on-ball phonons. The reason for the large negative value of dT_c/dP in Rb$_3$C$_{60}$, therefore, is *not* lattice stiffening, but a sharp decrease in the electronic density of states $N(E_f)$ with pressure. The increase in T_c going from K$_3$C$_{60}$ to Rb$_3$C$_{60}$ to Cs$_3$C$_{60}$ is due mainly to the enhancement in the density of states $N(E_f)$ as the progressively larger interstitial alkali cations expand the lattice, increase the separation between neighboring C$_{60}$ molecules, and thus narrow the conduction band.

11.4. Multiatom Superconductors: High-T_c Oxides

As outlined in Section 11.1, the high-pressure technique led directly to the discovery of $YBa_2Cu_3O_{7-\delta}$ (Y-123) [5], one of the most important high-T_c superconductors (HTSC), and generated in $HgBa_2Ca_2Cu_3O_{8+\delta}$ (Hg-1223) the highest transition temperature $T_c \approx 160$ K [9] for the resistivity onset of any known superconductor (see Figure 11.8); very recently Monteverde et al. [115] reportedly bested this value by 3–4 K by applying 23 GPa to a fluorinated Hg-1223 sample.

In this section we will attempt to determine the "intrinsic" dependence of T_c on pressure for hole-doped HTSC and from this intrinsic $T_c(P)$ to identify what, if any, new information is provided regarding the mechanism(s) responsible for, and the appropriate theoretical description of, superconductivity in the high-T_c oxides. No attempt will be made to summarize all available results; we refer the reader to previous reviews covering high-pressure effects in the high-T_c cuprates: Wijngaarden and Griessen in 1989 [116], Schilling and Klotz in 1991 [17], Takahashi and Môri in 1997 [117], Núñez-Regueiro and Acha in 1997 [118], Lorenz and Chu in 2004 [75], and an all-too-short but interesting paper by Wijngaarden et al. in 1999 [119].

We also restrict our consideration here to hole-doped HTSC. As is evident from the above reviews, electron-doped HTSC have received relatively little attention; in the few high-pressure studies carried out, T_c is normally found to decrease with pressure [120]. The fact that electron-doped HTSC must be slightly reduced to induce superconductivity means that oxygen ordering effects will likely play an important role in the pressure dependence of T_c, as discussed below for their hole-doped counterparts. Definitive high-pressure studies on well-characterized electron-doped HTSC which separate "intrinsic" from "oxygen ordering" effects are encouraged.

We begin by showing in Figures 11.8 and 11.9 the pressure dependence of T_c for a number of hole-doped HTSC, including the one-, two- and three-layer Hg-compounds $HgBa_2CuO_{4+\delta}$ (Hg-1201), $HgBa_2CaCu_2O_{6+\delta}$ (Hg-1212), and $HgBa_2Ca_2Cu_3O_{8+\delta}$ (Hg-1223). With the lone exception of $Tl_2Ba_2CuO_{6+y}$ (Tl-1201), $T_c(P)$ is seen to initially *increase* with pressure and pass through a maximum at higher pressures. The nearly ubiquitous initial

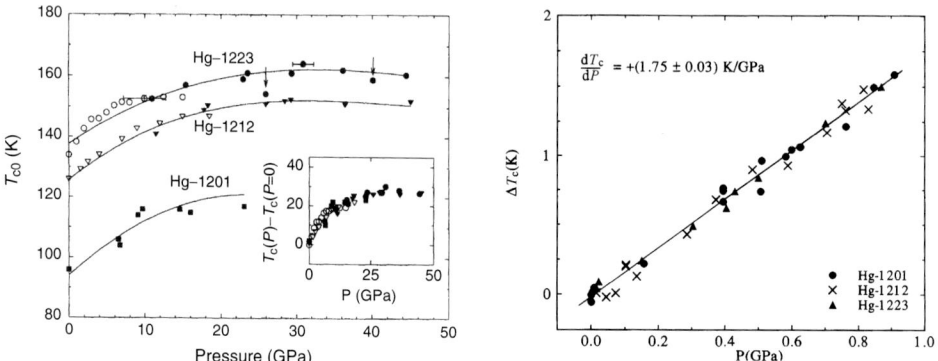

Figure 11.8. Hg-compounds: (left) T_c vs. pressure to 45 GPa from [9]; (right) change in T_c vs. pressure to 0.9 GPa. Figure reproduced from Ref. [8]. The initial pressure dependence dT_c/dP for all three Hg-compounds is identical.

High-Pressure Effects

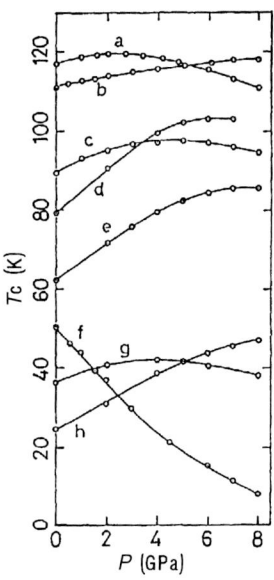

a: $Tl_2Ba_2CaCu_2O_8$
b: $Bi_{1.68}Pb_{0.32}Ca_{1.85}Sr_{1.75}Cu_{2.65}O_z$
c: $Y_{0.9}Ca_{0.1}Ba_2Cu_4O_8$
d: $YBa_2Cu_4O_8$
e: $CaBaLaCu_3O_{7-y}$
f: $Tl_2Ba_2CuO_{6+y}$
g: $La_{1.85}Sr_{0.15}CuO_4$
h: $Nd_{1.3}Ce_{0.2}Sr_{0.5}CuO_4$

Figure 11.9. T_c vs. pressure for several HTSC. Figure reproduced from [122].

increase in T_c with pressure, which was first pointed out by Schirber et al. [121], is a hallmark of hole-doped high-T_c cuprates.

A central question is whether or not the measured pressure dependence $T_c(P)$ in the superconducting cuprates gives evidence for an unconventional (non electron–phonon) pairing mechanism. As seen in Figure 11.5, as for the high-T_c oxides, $T_c(P)$ is known to pass through a maximum for La and S, both of which are believed to superconduct via the standard electron–phonon interaction. As discussed in detail above, for the majority of conventional simple and transition metal superconductors T_c decreases with pressure.

The evident similarity in the $T_c(P)$ dependences for the HTSC systems in Figures 11.8 and 11.9, particularly for the three Hg-compounds, gives strong evidence that the nature of the superconductivity is the same for all. This is not particularly surprising since all HTSC share one common structural element, the CuO_2 planes. The question remains, however, whether the most important interactions for the high-T_c superconductivity take place *within* these planes or *between* them. Uniaxial pressure experiments, in particular, hold promise to shed some light on this question.

The pressure dependences $T_c(P)$ in Figures 11.8 and 11.9 bear some resemblance to the canonical inverted parabolic dependence of $T_c(n)$ for HTSC on the hole carrier concentration n per Cu cation in the CuO_2 sheet

$$T_c(n) = T_c^{\max}[1 - \beta(n - n_{\text{opt}})^2], \qquad (11.8)$$

illustrated in Figure 11.10, where $\beta \simeq 82.6$ and $n_{\text{opt}} \simeq 0.16$ [123, 124]. According to Eq. (11.8) $T_c(n)$ initially increases with n on the underdoped side from 0 K for $n \approx 0.05$ to a maximum value $T_c^{\max}$ at optimal doping $n = n_{\text{opt}}$ before falling back to 0 K for $n \approx 0.27$ on the overdoped side. For underdoped samples one has $dT_c/dn > 0$, for optimally doped $dT_c/dn = 0$, and for overdoped $dT_c/dn < 0$. Since n has been found to initially increase

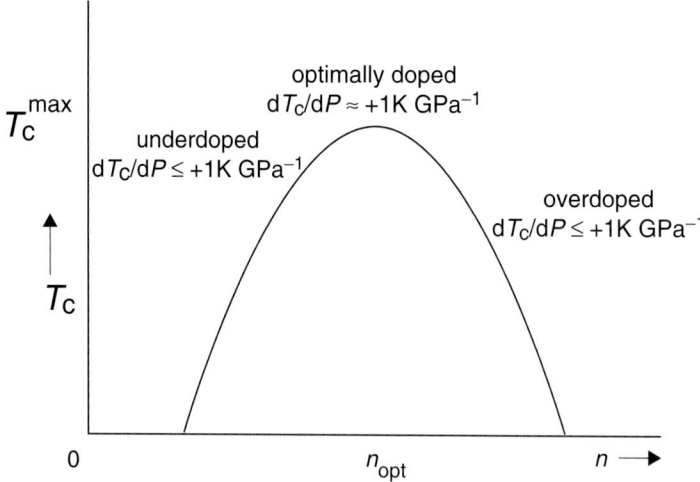

Figure 11.10. Canonical dependence of T_c on carrier concentration n for HTSC according to Eq. (11.8). Typical values for dT_c/dP in the underdoped, optimally doped, and overdoped regions are given.

with pressure ($dn/dP > 0$) in the majority of cuprates studied [17, 75, 117, 125], one might conjecture that with increasing pressure $T_c(P)$ simply traces out the inverted parabolic shape of $T_c = T_c(n)$ in Figure 11.10, yielding the $T_c(P)$ dependences seen in Figures 11.8 and 11.9. In such a "Simple Charge-Transfer Model," where only n is assumed pressure dependent, the pressure derivative is given by

$$\frac{dT_c}{dP} = \left(\frac{dT_c}{dn}\right)\left(\frac{dn}{dP}\right) = -2\beta T_c^{max}(n - n_{opt})\frac{dn}{dP}, \quad (11.9)$$

each system having a particular initial value of n. Within this model, the negative value of dT_c/dP for Tl-2201 in Figure 11.9 would result from the increase of n with pressure and the well-known fact that this compound is overdoped, i.e., $dT_c/dn < 0$.

That "life with the cuprates" is not so simple is seen by the data in Figure 11.11(a) on five Y-123 samples for increasing oxygen content x from $A \rightarrow E$, four underdoped ($A \rightarrow D$), and one nearly optimally doped (E). The measured $T_c(P)$ dependences run contrary to the expectations of the "Simple Charge Transfer Model" for underdoped samples, namely, that the higher the initial value of T_c, the lower the pressure needed to reach $T_c = T_c^{max}$. The data on the Hg-compounds in Figure 11.8, with initial slope $dT_c/dP \simeq +1.75\,\mathrm{K\,GPa^{-1}}$, also violate Eq. (11.9). Since all three Hg-compounds are nearly optimally doped, i.e., $dT_c/dn \simeq 0$, one would expect $dT_c/dP \simeq 0$ from Eq. (11.9). Evidently the "Simple Charge Transfer Model" is too simple! Neumeier and Zimmermann [126] extended this model by hypothesizing that the change in T_c with pressure derives from two contributions (1) an "intrinsic" contribution reflecting pressure-induced changes in T_c resulting solely from the reduction of the lattice parameters (no structural transitions, oxygen ordering effects, nonhydrostatic strains, or changes in the carrier concentration) and (2) the above contribution to dT_c/dP in Eq. (11.9) originating from the normal increase in n under pressure. The pressure derivative in this "Modified Charge-Transfer Model" is thus given by the general expression

$$\frac{dT_c}{dP} = \left(\frac{dT_c}{dP}\right)_{intr} + \left(\frac{dT_c}{dn}\right)\left(\frac{dn}{dP}\right). \quad (11.10)$$

High-Pressure Effects

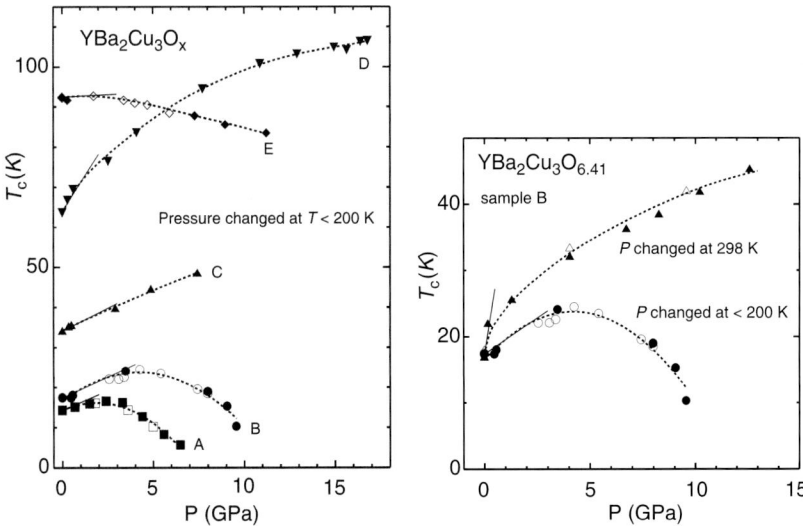

Figure 11.11. Dependence of T_c on pressure from [128] (left) for five $YBa_2Cu_3O_x$ samples $A \rightarrow E$ with increasing oxygen content x; (right) for the underdoped sample B. See text for details.

If one now substitutes Eq. (11.8) in this expression and assumes that β and n_{opt} are independent of pressure [127], one obtains

$$\frac{dT_c}{dP} = \frac{dT_c^{max}}{dP}\left[1 - \beta\left(n - n_{opt}\right)^2\right] - \frac{dn}{dP}\left[\left(2\beta T_c^{max}\right)\left(n - n_{opt}\right)\right]. \quad (11.11)$$

Note that T_c^{max} is the maximum value of T_c when the carrier concentration n alone is varied at constant pressure; T_c^{max} is *not* the maximum value of T_c when the pressure is varied (unless $dn/dP = 0$). Comparing Eqs. (11.10) and (11.11) we see that the intrinsic component of the pressure derivative is given by

$$(dT_c/dP)_{intr} = (dT_c^{max}/dP)\left[1 - \beta\left(n - n_{opt}\right)^2\right]. \quad (11.12)$$

Note that $dT_c/dP = (dT_c/dP)_{intr} \equiv dT_c^{max}/dP$ only for optimally doped samples, where $n = n_{opt}$. If the sample is nearly optimally doped, then we can neglect the term in Eq. (11.11) quadratic in $(n - n_{opt})$, leaving the following expression linear in $(n - n_{opt})$

$$\frac{dT_c}{dP} = \frac{dT_c^{max}}{dP} - \frac{dn}{dP}\left[\left(2\beta T_c^{max}\right)\left(n - n_{opt}\right)\right]. \quad (11.13)$$

A linear dependence of dT_c/dP on $(n - n_{opt})$ was indeed found in a careful high-pressure (He-gas) study [126] on the $Y_{1-y}Ca_yBa_2Cu_3O_x$ compound series where the Ca and O contents were varied to change n near optimal doping; from the slope of this dependence it was determined that $dn/dP \simeq +0.0055$ holes GPa^{-1}. At optimal doping the intrinsic pressure dependence was found to be $dT_c/dP = (dT_c/dP)_{intr} = +0.96$ K GPa^{-1} [126].

We should not be surprised that the "Simple Charge-Transfer Model," which only considers the single charge-transfer contribution, fails to satisfactorily account for the experimental results. We have seen that the $T_c(P)$ dependences for transition metal superconductors can only be understood by taking into account *two* distinct contributions: from both lattice vibrations and electronic properties. One should expect materials as complex as the high-T_c cuprates with their distorted quasi-2D perovskite structures to be a good deal more complex than the transition metals. That this is indeed the case is the reason why it has proven so difficult to reach a basic understanding of HTSC, the results of high-pressure studies being no exception.

Ideally, in a high-pressure experiment we would like to determine the change in the superconducting properties of a given high-T_c oxide under variation of both the intraplanar lattice parameter(s) and the interplanar separation. In actual high-pressure experiments, however, a number of additional effects may occur which considerably complicate the interpretation of the data (1) structural phase transitions, (2) oxygen ordering effects, and (3) effects due to shear stress from nonhydrostatic pressure media.

11.4.1. Nonhydrostatic Pressure Media

As pointed out above, not all high-pressure experiments are created equal. Ideally, a fluid pressure medium is used which transmits hydrostatic pressure to the sample. The use of solid pressure media, or no pressure media at all, may simplify the experimentation, but results in the sample being subjected to varying degrees of nonhydrostatic shear stress which may cause important changes in the superconducting state, in particular in the pressure dependence of the transition temperature $T_c(P)$, as we have seen above for Pb in Figure 11.2. Shear-stress effects on $T_c(P)$ are well known from studies on such diverse superconducting materials as organic metals [129], MgB$_2$ [46], Re metal [130], and Hg [131].

The differing $T_c(P)$ results on the high-T_c cuprates by different groups may arise from differences in samples, in the pressure medium, and/or in the method used to determine T_c. Gao et al. [9] suggested that the fact that their value of $T_c(30\,\text{GPa}) \approx 160\,\text{K}$ for Hg-1223 lies 10–15 K higher than that found by other groups [117, 132] may have its origin in shear stress effects. Klotz et al. [133] carried out two experiments on a single sample of Bi$_2$CaSr$_2$Cu$_2$O$_{8+\delta}$ (Bi-2212) in a diamond-anvil cell, one in helium and the other with no pressure medium whatsoever, and obtained very different $T_c(P)$ dependences. On the other hand, Wijngaarden et al. [119] report that the $T_c(P)$ dependences for YBa$_2$Cu$_4$O$_8$ (Y-124) found by different groups using varying pressure media do not differ widely. Also, a recent purely hydrostatic He-gas experiment to 0.6 GPa on an overdoped Y-123 single crystal agrees within experimental error with the initial pressure dependence $dT_c/dP \approx -1\,\text{K GPa}^{-1}$ found in a parallel diamond-anvil-cell experiment using solid steatite as pressure medium [134]. As discussed in Section 11.1, for quantitative investigations fluid pressure media, particularly helium, are to be preferred over solid media. To test whether or not shear stresses play a role in the pressure-induced changes obtained, it is prudent to carry out the experiment using two different pressure media.

11.4.2. Structural Phase Transitions

As for the A-15 compounds, $T_c(P)$ for HTSC can be a sensitive function of structural instabilities. The initial rate of increase of T_c with pressure for La$_{2-x}$Sr$_x$CuO$_4$ is relatively large at +3.0 K GPa^{-1} [135]; it is even much larger for La$_{2-x}$Ba$_x$CuO$_4$ (+8 K GPa^{-1}) [136]. This led Wu et al. [5] to the discovery of Y-123, as discussed above. The reason for the anomalously large positive value of dT_c/dP for La$_{2-x}$Ba$_x$CuO$_4$ is the existence of a low-temperature-tetragonal (LTT) phase below 60 K which strongly suppresses T_c for x in the range 0.07–0.18, as seen in Figure 11.12. Applying pressure eventually suppresses this LTT

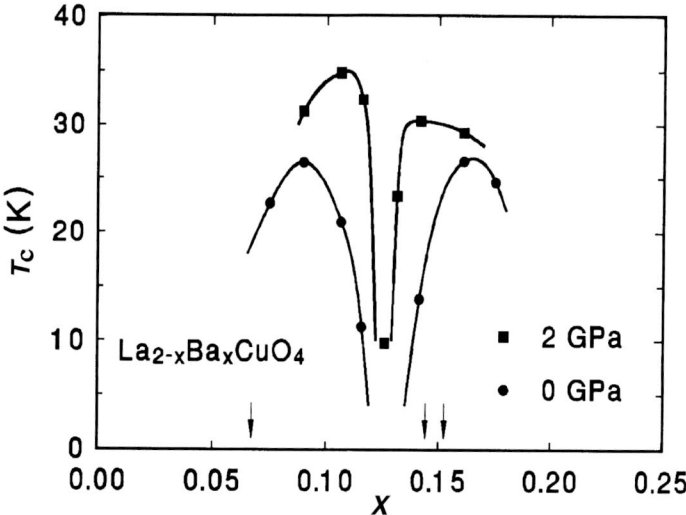

Figure 11.12. T_c vs. Ba content x for $La_{2-x}Ba_xCuO_4$ at 0 and 2 GPa pressure. Figure reproduced from [137]. There is a marked influence of the LTT phase transition on $T_c(x)$ for $x \simeq 0.125$.

phase transition, leading to the anomalously large increase in T_c under pressure seen in the data where dT_c/dP reaches values as large as $+12$ K GPa^{-1} [137]. At 2 GPa the $T_c(x)$ dependence in Figure 11.12 begins to resemble the canonical bell-shaped $T_c(n)$ dependence in Figure 11.10, except in a very narrow range of x centered at $x = 0.125$.

In a further compound system in the same family, $La_{2-x-y}Nd_ySr_xCuO_4$, the doping level or the crystal structure can be independently controlled by varying x or y, respectively [138]. In the phase diagram for $La_{1.48}Nd_{0.4}Sr_{0.12}CuO_4$ in Figure 11.13 it is seen that at ambient pressure the high-temperature-tetragonal (HTT) phase transforms below 500 K into a low-temperature-orthorhombic (LTO1) phase, followed by a phase change below 70 K to the LTT phase [139]. High pressure is seen to suppress the low-temperature phases until above 4 GPa only the HTT phase remains. These phase transitions are seen in Figure 11.13 to have a dramatic effect on the pressure dependence of the superconducting transition temperature $T_c(P)$ which peaks near 5 GPa. Evidently, structural instabilities play an important role in the doped La_2CuO_4 oxide family, making it almost impossible to extract the intrinsic dependence of T_c on pressure from experiment.

11.4.3. Oxygen Ordering Effects

In the majority of HTSC oxygen defects are present with a relatively high mobility, even at ambient temperatures. Many HTSC can thus be readily doped simply by varying the oxygen defect concentration through annealing at controlled oxygen partial pressures at elevated temperatures. The normal and superconducting state properties of HTSC depend not only on the concentration of oxygen defects, but on the relative positions assumed by these defects in the lattice on a local scale. Such oxygen ordering effects were first observed at ambient pressure in strongly underdoped Y-123 samples where the T_c value could be sharply reduced simply by quenching the sample from elevated temperatures into liquid nitrogen [140]. A simple model developed by Veal et al. [140] was able to account for this phenomenon in terms of a reduction

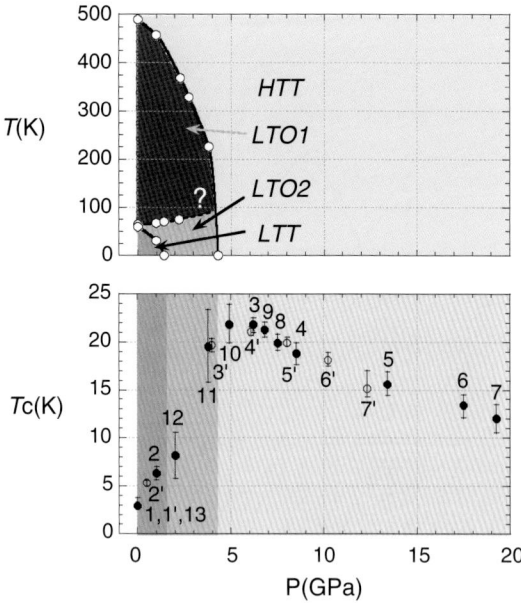

Figure 11.13. Results for $La_{1.48}Nd_{0.4}Sr_{0.12}CuO_4$ from [139]. (top) Pressure–temperature phase diagram showing high-temperature tetragonal (HTT), low-temperature tetragonal (LTT), and low-temperature orthorhombic (LTO1, LTO2) phase regions. (bottom) T_c vs. pressure showing influence of phase transitions.

in the hole-carrier concentration n in the CuO_2 planes due to reduced local order of oxygen defects in the Y-123 chains containing the ambivalent Cu cations. Oxygen ordering effects on T_c are only *observed* if (1) oxygen defects are present, (2) there are vacant sites available which oxygen defects can move into, and (3) the sample is not optimally doped (if optimally doped, $dT_c/dn = 0$ so that small changes in n due to oxygen ordering have no effect on T_c).

A second way to change the oxygen ordering state is through high pressure. The application of high pressure at room temperature prompts the mobile oxygen defects to order locally and thus enhance the hole-carrier concentration n in the CuO_2 planes. Pressure-induced oxygen ordering thus "turbo-charges" the normal enhancement of n with pressure. Significant pressure-induced oxygen ordering effects have been observed for Y-123 by Fietz et al. [141] and others [128], as illustrated in Figure 11.11 (right) for an underdoped sample. Whereas the lower $T_c(P)$ curve in this figure was measured in an experiment carried out completely at temperatures low enough ($T < 200$ K) to prevent the ordering of oxygen defects in the chains as the pressure is changed, the upper curve was obtained for pressure changes at ambient temperature. The difference between the two $T_c(P)$ dependences is substantial indeed! In Y-123 the time-dependent relaxation of T_c following a change in pressure can be best fit using the stretched exponent $\beta \simeq 0.6$ [128]. Phillips [142] has argued that this gives evidence for the importance of the electron–phonon interaction in HTSC and supports his model for defect-induced superconductivity [143].

Pressure-induced oxygen ordering effects in HTSC were first observed in overdoped Tl-2201 samples by Sieburger and Schilling [144] and then extensively studied by Klehe et al. [145, 146], as illustrated in Figure 11.14. If pressure is applied at room temperature, T_c is seen to decrease rapidly, as found earlier by Môri et al. [122] (see Figure 11.9); however,

High-Pressure Effects

if the pressure is released at temperatures low enough (55 K) to freeze in the oxygen defects, T_c does *not* increase back to its initial value, but actually decreases further! The intrinsic pressure derivative for Tl-2201 is thus positive $(dT_c/dP)_{intr} > 0$. As seen in Figure 11.14, if the sample is then annealed at progressively higher temperatures, each for 1 h, T_c relaxes back towards its initial value in a two-step fashion, indicating two distinct relaxation pathways. The low-temperature relaxation stagnates for temperatures near 110 K, but picks up again for temperatures above 180 K where a smaller high-temperature relaxation sets in. For Tl-2201, therefore, the measured pressure dependence of T_c depends on the entire pressure/temperature history of the sample, $T_c = T_c(T, P, \text{time})$. As one would expect, the importance of oxygen ordering effects in Tl-2201 depends strongly on the oxygen defect concentration [144].

Pressure-induced oxygen ordering effects have been observed on numerous other HTSC, including Hg-1201, Nd-123, Gd-123, $TlSr_2CaCu_2O_{7-y}$, $Sr_2CuO_2F_{2+y}$, and superoxygenated La_2CuO_{4+y} [148, 149]. The activation energies for oxygen diffusion in Tl-2201, Y-123, and Hg-1201 were found by Sadewasser et al. [128] to increase with pressure, as expected; the activation volumes obtained allow an estimate of the most probable diffusion pathways for oxygen defects through the respective HTSC lattice. For further discussion of oxygen ordering effects in $La_{2-x}Sr_xCuO_4$ and other HTSC see the recent review by Lorenz and Chu [75].

From the above discussion it is apparent that oxygen ordering effects must be suppressed before the intrinsic pressure dependence $T_c^{intr}(P)$ can be established. There are three known ways to accomplish this (1) carry out the entire experiment at sufficiently low temperatures

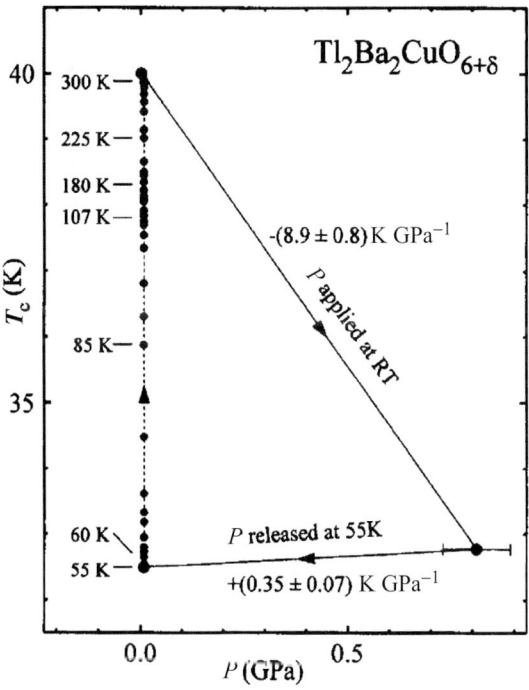

Figure 11.14. T_c vs. pressure for an overdoped $Tl_2Ba_2CuO_{6+\delta}$ single crystal, demonstrating the marked influence of oxygen ordering effects. Pressure is first applied at room temperature but released at 55 K, leaving the sample in a metastable state. T_c relaxes back to its initial value if the sample is annealed at progressively higher temperatures to 300 K. Figure reproduced from [147].

that oxygen ordering effects are frozen out; (2) determine the initial pressure dependence dT_c/dP only on optimally doped samples since at the extremum $T_c(n = n_{opt}) = T_c^{max}$ the additional pressure-induced charge transfer from the oxygen ordering will have no effect; (3) study samples either with no mobile oxygen defects or with the maximum number of oxygen defects so that no empty defect sites are left.

The method (3) above was employed in the beautiful specific heat experiments to 10 GPa by Lortz et al. [134] on a fully oxygenated overdoped $YBa_2Cu_3O_7$ sample; the measurement of such a basic thermodynamic property as the specific heat allows the determination of the pressure dependence not only of the transition temperature $T_c(P)$ but also of the superconducting condensation energy $U_o(P)$, as seen in Figure 11.15. For comparable change in T_c, the observed change in U_o for underdoped YBCO is three times larger, reflecting the presence of superconducting fluctuation or pseudogap effects. In addition, from these results the pressure derivative of the carrier concentration is estimated to be $dn/dP \approx +0.0018$ to $+0.0026$ holes $Cu^{-1}GPa^{-1}$.

Sadewasser et al. [128] applied method (1) above to suppress oxygen ordering effects in an extensive study of Y-123 at different doping levels by maintaining the sample at temperatures below 200 K during the entire experiment in a He-loaded diamond-anvil cell. The results are shown in Figure 11.11 (left). Disappointingly, no simple systematics in $T_c^{intr}(P)$ are evident in these data. The "Modified Charge Transfer Model" as outlined above is unable to account for the data. Y-123 is evidently a VERY complex system, even without oxygen ordering effects. The presence of variably doped chains in Y-123 evidently adds a considerable (and unnecessary!) level of complexity. Y-123 and Y-124 are the only HTSC with CuO chains. To make advances in our understanding of the origins of HTSC, it is essential to study in depth the simplest systems possible. The tetragonal Hg-compounds, which exhibit relatively weak oxygen-ordering effects, appear to be particularly attractive for further detailed studies and comparison with theory.

Very recently Tomita et al. [150–152] have carried out extensive studies of the critical current density J_c across single grain boundaries in bicrystalline Y-123 samples for various oxygen concentrations and grain boundary mismatch angles. In all cases J_c increases markedly with pressure. Interestingly, J_c also exhibits relaxation effects following pressure changes at ambient temperature; the relaxation time is shorter than that for T_c, consistent with the usual picture that oxygen defects have a higher mobility in the grain boundary than in the bulk. That J_c exhibits relaxation effects at all is evidence that some oxygen defect sites in the

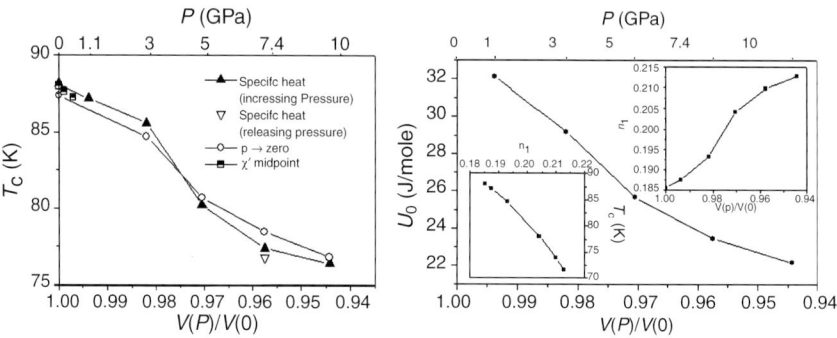

Figure 11.15. Results of specific heat measurements under pressure on an overdoped $YBa_2Cu_3O_7$ crystal. Figure reproduced from [134]. (left) T_c vs. both pressure P and relative volume $V(P)/V(0)$. (right) Superconducting condensation energy vs. pressure P and relative volume $V(P)/V(0)$. Inferred values of hole-carrier concentration n_h are given in insets.

High-Pressure Effects

grain boundary must be vacant, i.e., high-pressure experiments can be used as a probe to test whether the grain boundaries are fully oxygenated ot not [151]. Since it has been shown for Y-123 that J_c increases with oxygen content [152], irrespective of the doping level, further enhancements in J_c should be possible if all vacant sites can be filled with oxygen, for example, by subjecting the sample to a pure oxygen atmosphere at elevated temperatures and pressures.

11.4.4. Intrinsic Pressure Dependence $T_c^{intr}(P)$

In spite of the great complexity of HTSC materials, a number of empirical guidelines have been identified [153] for enhancing the value of T_c (1) vary the carrier concentration n in the CuO$_2$ planes until its optimal value is reached (see Figure 11.10); (2) increase the number of CuO$_2$ planes which lie close together (in a packet) in the oxide structure while maintaining optimal doping—"healthy" one-plane systems, like Tl-2201, have T_c values in the range 90–100 K, two-plane systems in the range 100–120 K, and three-plane systems in the range 120–140 K; (3) try to position defects as far from the CuO$_2$ planes as possible; and (4) since T_c is diminished with increasing buckling angle in the CuO$_2$ planes, develop structures where the CuO$_2$ planes are as flat as possible. Note that according to the above, the system Y-123 with $T_c^{max} \simeq 92$ K is not particularly "healthy."

We now pose the question: what can we learn from high-pressure experiments about how to further enhance the value of T_c? To answer this question, we should carefully select the systems we choose for experimentation, preferably picking "healthy" HTSC systems with relatively high values of T_c. Experimentation on "pathological" low-T_c systems results in numerous factors changing at the same time, making the interpretation difficult if not impossible. The single-layer La-214 oxides are examples of such "pathological" systems, only possessing T_c values in the range 30–40 K, far below the 90–100 K expected according to the above criteria for "healthy" single-layer systems such as Hg-1201 and Tl-2201. It is thus not surprising that in the La-214 oxides T_c increases relatively rapidly with pressure as the structural distortions, which result in considerable buckling in the CuO$_2$ planes, are diminished. The La-214 systems are thus not suitable for further studies aimed at determining $T_c^{intr}(P)$. Similar structural transition effects led to early reports that the rate of increase of the transition temperature in HTSC with pressure, $|d \ln T_c/dP|$, is inversely related to the value of T_c [120,154,155]. A closer examination of the relevant data to exclude systems with structural transitions, however, gave no evidence for such a correlation [17]. For further discussion we will focus on HTSC systems, like the one-, two- or three-layer Tl- or Hg-oxides or the two- or three-layer Bi-oxides, which are free of structural transition issues.

From the measured pressure dependences $T_c(P)$ for these systems, we would like to extract $T_c^{intr}(P)$, the "intrinsic" pressure dependence of T_c for a given fixed carrier concentration n. This separation is extremely difficult for arbitrary doping levels since n generally increases under pressure and T_c is a particularly sensitive function of n, as seen in Figure 11.10. Such a separation has been attempted for Hg-1201 [156] and a Tl-1212 compound [119] under strong simplying assumptions; such studies will only become really quantitative if the pressure dependence of n is determined independently over the entire range of doping and pressure. Fortunately, for one value of n, namely, $n = n_{opt}$, the separation becomes simple, at least for the initial slope dT_c/dP, since at this extremum of $T_c(n)$ we have in Eq. (11.10) $dT_c/dn = 0$ so that we obtain simply $dT_c/dP = (dT_c/dP)_{intr}$. Restricting our attention to optimally doped samples has the great advantage that the initial slope dT_c/dP is free from the influence of changes in the carrier concentration n, and, as a bonus, oxygen ordering effects play no important role since they affect T_c primarily through their influence on n.

For these reasons we now focus our attention on "healthy" optimally doped HTSC systems. The Hg-compounds are of particular interest here since their superconducting and structural properties have been studied on the same samples to high accuracy under purely hydrostatic pressure conditions in dense helium [8], as well as under quasihydrostatic pressures above 40 GPa (see Figure 11.8) [9]. The optimally doped one-, two-, and three-layer Hg-compounds Hg-1201, Hg-1212, and Hg-1223 have, respectively, $T_c(0)$ values of 94, 127, and 134 K, initial pressure derivatives $dT_c/dP = +1.75 \pm 0.05$ K GPa^{-1} for all three, relative pressure derivatives $d\ln T_c/dP \simeq +17.6, +14.2,$ and $+12.9 \times 10^{-3}$ GPa^{-1}, and bulk moduli $B \simeq$ 69.4, 84.0, and 92.6 GPa to 1.4% accuracy [157]. From these values the relative volume derivatives can be accurately determined to be $d\ln T_c/d\ln V = -B(d\ln T_c/dP) \simeq -1.22 \pm 0.05, -1.19 \pm 0.06,$ and -1.20 ± 0.05. It is quite remarkable that the relative pressure derivatives $D\ln T_c/dP$ differ by more than 30%, whereas the relative volume derivatives $d\ln T_c/d\ln V$, which would be expected to be of more direct physical relevance [17], turn out to be *identical* for all three Hg-compounds! This invariance of the relative volume derivative gives strong evidence that the superconducting state, including the pairing mechanism, in the one-, two-, and three-layer Hg-compounds is the same. If one understands the nature of the superconductivity, and the mechanism(s) responsible for it, in the one-layer compound Hg-1201, one understands these basic properties in all three. This conclusion is underscored by the fact that the pressure dependence $T_c(P)$ to 40 GPa is nearly the same for all three Hg-compounds [9], as seen in Figure 11.8.

We now consider the values of the relative volume derivative $d\ln T_c/d\ln V$ for further optimally doped HTSC systems: Y-123 (-1.25 ± 0.06), Tl-2201 (-1.35 ± 0.4), Tl-2212 (-0.9 ± 0.2), Tl-2223 (-1.16 ± 0.3), Bi-2212 (-1.04 ± 0.15), and Bi-2223 (-1.36) [8, 146, 158]; the bulk modulus is known to lesser accuracy for the Tl- and Bi-systems than for Y-123 and the Hg-compounds. It is indeed remarkable that for all these optimally doped HTSC systems the intrinsic relative volume derivative turns out to be nearly the same $d\ln T_c/d\ln V \approx -1.2$, corresponding to the volume dependence

$$T_c \propto V^{-1.2}. \tag{11.14}$$

This is strong evidence that the nature of the superconductivity, and the mechanism(s) responsible for it, are the same in all high-T_c cuprate superconductors. We note that this HTSC volume derivative $d\ln T_c/d\ln V \approx -1.2$ has the opposite sign, and is much weaker in magnitude, than the volume derivatives $d\ln T_c/d\ln V \simeq +7.2$ for Sn and $+4.16$ for MgB$_2$ which we discussed above. This fact by itself does not imply, however, that the electron–phonon interaction plays no role in HTSC. Negative volume derivatives are found in a number of transition metal systems, like La, Y, Lu, Sc, or V (see Figure 11.5) where the superconductivity is believed to be phonon mediated.

We are now in a position to understand why in the optimally doped Hg-compounds Hg-1201, Hg-1212, and Hg-1223, T_c increases with pressure over such a relatively wide pressure range, resulting in the highest values of T_c at 30 GPa for any known superconductor with the same number of CuO$_2$ layers. The very weak increase in the carrier concentration n under pressure measured for Hg-1201 [159], and calculated for Hg-1223 [160], means that a relatively high pressure is required to increase n sufficiently in the $T_c(n)$ phase diagram in Figure 11.10 that the negative slope $(dT_c/dn)(dn/dP)$ becomes equal to the intrinsic positive slope $(dT_c/dP)_{\text{intr}} \simeq +1.75$ K GPa^{-1}, at which point $T_c(P)$ passes through a maximum at $P \approx 30$ GPa. This maximum value $T_c^{\text{max}}(30\,\text{GPa})$, can be estimated from Eq. (11.11) for Hg-1223 by setting $T_c^{\text{max}}(0) = 134$ K and $dT_c^{\text{max}}/dP \simeq +1.75$ K GPa^{-1}, and assuming n_{opt} and T_c^{max} are independent of P and n, respectively. If one now asks what value of dn/dP is required that $T_c(P)$ reaches its maximum value, where $dT_c/dP = 0$, at 30 GPa, out comes $dn/dP \simeq +0.00129$ hole GPa^{-1}. If this value of dn/dP is then inserted in Eq. (11.11), one obtains the estimate $T_c^{\max}(30\,\text{GPa}) \simeq 163$ K which is close to the measured value (see Figure 11.8). This value of dn/dP, which is somewhat smaller than that estimated for Y-123, agrees reasonably well with a calculation by Singh et al. [160]. Thermopower measurements by Chen et al. [159] indicate an even smaller value. If, as suggested by Xiong et al. [161], $\beta = 50$ is substituted for $\beta = 82.6$ in the Tallon formula, $dn/dP \simeq +0.00166$ is obtained, but the estimate $T_c(30\,\text{GPa}) \simeq 163$ K remains the same.

We would now like to explore the question as to the origin of the relatively weak dependence of T_c on sample volume $T_c \propto V^{-1.2}$ in HTSC materials. When hydrostatic pressure is applied to an HTSC, the unit cell is compressed in all three directions. However, with the exception of the La-214 compound family, the compressibility in the direction perpendicular to the CuO$_2$ planes, the c-direction, is in general approximately twice as large as in a direction parallel to the CuO$_2$ planes [17]. The central question at hand is: does the intrinsic increase of T_c with pressure, reflected in $T_c \propto V^{-1.2}$, originate primarily from the reduction in the *separation* between the CuO$_2$ planes or from the reduction in the *area* of these planes? To answer this question, we must turn to uniaxial pressure experiments which have the potential to unravel the information hidden in the hydrostatic pressure studies.

11.4.5. Uniaxial Pressure Results

Uniaxial pressure experiments are technologically very difficult and require high quality single crystals of sufficient size. The partial pressure derivatives along the crystallographic axes dT_c/dP_a, dT_c/dP_b, and dT_c/dP_c can be determined either by applying force directly to the crystal along the respective crystallographic directions [162], or through combined ultrahigh resolution thermal expansion and specific heat measurements using the Ehrenfest relation $dT_c/dP_i = \Delta\alpha_i V_m T_c/\Delta C_p$, where $\Delta\alpha_i$ and ΔC_p are the mean-field jumps of the thermal expansion coefficient and specific heat, respectively, and V_m is the molar volume [163]. Note that the hydrostatic pressure derivative can be written as the sum of the respective partial pressure derivatives $dT_c/dP = dT_c/dP_a + dT_c/dP_b + dT_c/dP_c$. The result of the "Modified Charge Transfer Model" in Eq. (11.11) can be applied by simply replacing dT_c/dP by the respective partial pressure derivative dT_c/dP_i where $i = a, b, c$.

The results of detailed thermal expansion studies by Meingast et al. [127] on crystals from the Y$_{1-y}$Ca$_y$Ba$_2$Cu$_3$O$_x$ compound series are shown in Figure 11.16. At ambient pressure $T_c(n)$ is seen to pass through a maximum at $T_c \simeq 93$ K for $n = n_{\text{opt}} \simeq 0.16$. The partial pressure derivatives generally change from positive to negative as the carrier concentration n increases, reflecting the influence of pressure-induced charge transfer. At optimal doping one has $n = n_{\text{opt}}$ and $dT_c/dn = 0$ so that the partial pressure derivatives give the intrinsic effect directly. In Figure 11.16(c) we see that at optimal doping $dT_c/dP_c \approx 0$; this implies that enhancing the interplanar coupling by pushing the CuO$_2$ planes closer together has no measureable effect on the superconducting state. This, together with the fact that dT_c/dP_c depends linearly on n, in agreement with Eq. (11.13), gives strong evidence that the primary effect of compression in the c direction is to enhance the carrier concentration n.

On the other hand, as seen in Figure 11.16(b), compressing the CuO$_2$ planes themselves enhances T_c at the rate $\sim +1$ K GPa^{-1}, in good agreement with hydrostatic pressure studies on the same compound series [126]. Parallel thermal expansion studies [163] on an optimally doped detwinned YBa$_2$Cu$_3$O$_x$ crystal give the following partial pressure derivatives: $dT_c/dP_a \simeq -1.9$ K GPa^{-1}, $dT_c/dP_b \simeq +2.2$ K GPa^{-1}, and $dT_c/dP_c \simeq 0$ K GPa^{-1}, in

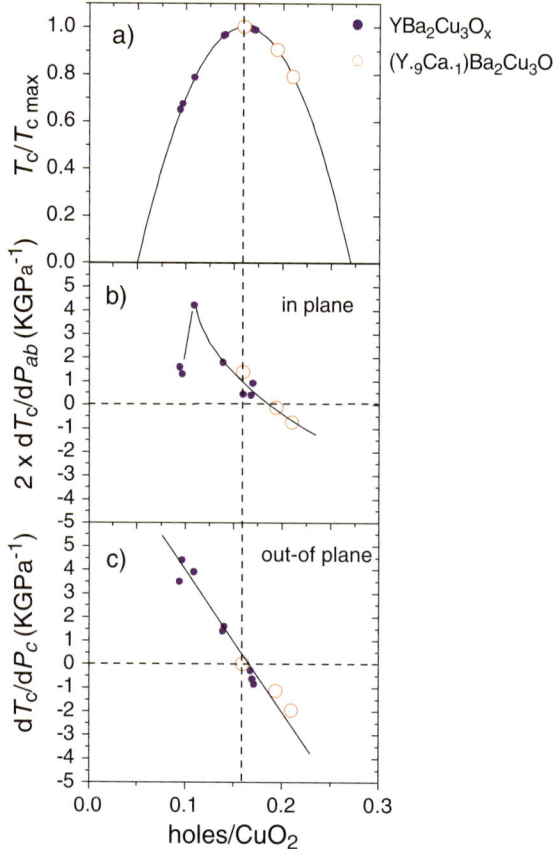

Figure 11.16. Results of ultrahigh resolution thermal expansion experiments on $Y_{1-y}Ca_yBa_2Cu_3O_x$ crystals. Figure reproduced from [127]. (a) T_c vs. hole concentration n. (b) In-plane partial pressure derivative $2(dT_c/dP_{ab})$ vs. n. (c) Out-of-plane partial pressure derivative dT_c/dP_c vs. n.

excellent agreement with later studies by Kund and Andres [164] as well as with direct uniaxial pressure experiments by Welp et al. [162]. All these studies confirm that the intrinsic pressure effect within the CuO_2 planes is large, in contrast to the negligible effect along the c axis perpendicular to these planes. The opposite sign of the partial pressure derivatives in the a and b directions is simply a reflection of the above T_c–optimization rule (4) whereby the CuO_2 planes should be as flat (therefore tetragonal) as possible to maximize T_c; Chen et al. [165] have developed a model which accounts for the dT_c/dP_i anisotropies in terms of anisotropies in both the hole dispersion and the pairing interaction.

In the above experiments, a compression along one axis (unfortunately) leads to an expansion along the other two axes, so that all three change. The partial pressure derivatives, however, can be converted into the partial strain derivatives $dT_c/d\epsilon_a$, $dT_c/d\epsilon_b$, and $dT_c/d\epsilon_c$, if the elastic constants are known to sufficient accuracy. For Y-123 the dominant strain derivative at optimal doping turns out to be $dT_c/d\epsilon_b$, the other two being at least $5\times$ smaller [166]. When pressure is applied to a Y-123 crystal, the intrinsic effect on T_c is predominantly caused by a strain in the CuO_2 plane along the b (chain) direction. On the other hand, in the double-chain system Y-124 compression along the a-direction is dominant [167]. Studies of bond-length systematics in $RBa_2Cu_4O_8$ across a portion of the rare-earth series R both at ambient [168] and high pressure [169] have revealed that T_c correlates well with the Cu–O bond lengths within, rather than perpendicular to, the CuO_2 planes.

The above results underscore the considerable complexity of the two HTSC compounds, Y-123 and Y-124, which contain CuO chains, a superfluous structural element unnecessary for high-T_c superconductivity. Indeed, the HTSC systems with the highest values of T_c have no chains. To make significant progress in our understanding of the basic issues regarding superconductivity in HTSC, one would be well advised to focus on tetragonal systems free from CuO chain structures.

High-resolution thermal expansion experiments have been carried out on a nearly optimally doped Bi-2212 crystal by Meingast et al. [170] with results: $dT_c/dP_i \simeq +1.6, +2.0$, and $-2.8\,\mathrm{K\,GPa^{-1}}$ for $i = a, b, c$. Kierspel et al. [171] obtain somewhat different values for Bi-2212: $dT_c/dP_i \simeq +0.9, +0.9$, and $< 0.4\,\mathrm{K\,GPa^{-1}}$, respectively, yielding a total pressure derivative $dT_c/dP \approx 2\,\mathrm{K\,GPa^{-1}}$, in good agreement with the hydrostatic pressure dependence [172].

Unfortunately, uniaxial pressure results have yet to be published for the tetragonal Hg- and Tl-compound families due to the difficulty in obtaining high-quality crystals of sufficient size. Further experimentation on the Hg cuprates in particular is strongly encouraged since these oxides are blessed with a relatively simple structure and thus offer an excellent opportunity for obtaining definitive results.

11.5. Conclusions and Outlook

Taken together, the above experiments support the picture that the dimensions of the CuO_2 planes, rather than the separation between them, primarily determines the maximum value of T_c in a given HTSC: the closer the planes are to being square and flat, and the smaller their area, the higher the value of T_c. A similar conclusion regarding the relative importance of the in-plane and out-of-plane lattice parameters was reached in a review by Schilling and Klotz [17] in 1991 and in a paper by Wijngaarden et al. [119] in 1999 who commented that "There is quite some evidence that c mainly influences doping, while a mainly influences the intrinsic T_c." The high-pressure experiments carried out to-date thus lend support to those theories where the interactions within the CuO_2 planes are primarily responsible for the superconducting pairing.

From this it follows that the ubiquitous intrinsic volume dependence $T_c \propto V^{-1.2}$ for nearly optimally doped HTSC arises from the compression of the CuO_2 planes, rather than from a reduction in the separation between them. To obtain the intrinsic dependence of T_c on the in-plane lattice parameter a, we evaluate $d\ln T_c/d\ln a = -\kappa_a^{-1}(d\ln T_c/dP)$, where $\kappa_a \equiv -d\ln a/dP$ is the a-axis compressibility. Using the above values of $d\ln T_c/dP$ for the one-, two-, and three-layer Hg-compounds as well as the κ_a-values 4.26, 2.94, and $2.57 \times 10^{-3}\,\mathrm{GPa^{-1}}$ [173], respectively, we obtain $d\ln T_c/d\ln a = -4.1, -4.8$, and -5.0 which translates into the approximate in-plane lattice parameter dependence

$$T_c \propto a^{-\delta}, \text{ where } \delta = 4.5 \pm 0.5. \tag{11.15}$$

This expression implies that at optimal doping the intrinsic T_c is roughly proportional to the inverse square of the area A of the CuO_2 planes, $T_c \propto A^{-2}$. Similar results are obtained for other optimally doped HTSC systems.

This is one of the single most significant results to be distilled from high-pressure experiments on HTSC materials and is information not readily available through other means. Besides giving information on the superconducting mechanism, this dependence points to an additional strategy for further increasing T_c. To the above five "T_c optimization rules," we can now add: "(6) seek out structures which apply maximal compression to the CuO_2 planes without causing them to buckle." According to the above relations, if we were to apply sufficient pressure to an optimally doped Hg-1223 sample to compress its in-plane dimension by about 20%, without adding defects or increasing the number of charge carriers, T_c should increase from 134 to 304 K and we would have the world's first room temperature superconductor!

HTSC systems with the same number of CuO_2 planes generally have different values of $T_c = T_c^{max}$ at optimal doping. It is interesting to inquire whether this difference arises from a variation in the in-plane lattice parameter a, i.e., $T_c^{max} \propto a^{-4.5}$. From Figure 11.17 one can see that no such simple correlation exists. The single-plane material with the highest value of T_c, Hg-1201 with $T_c^{max} \simeq 98$ K, has the *largest* value of a, and the compound in Figure 11.17 with the lowest value of T_c, $La_{1.85}Sr_{0.15}CuO_4$ with $T_c^{max} \simeq 36$ K, has the *smallest* value of a. Perhaps $La_{1.85}Sr_{0.15}CuO_4$ owes its anomalously low value of T_c to an overcompression of its CuO_2 plane, resulting in strong structural distortions and plane buckling, effects which are known to degrade T_c. The a values of the other systems listed differ by only 1.4% which corresponds to ~5 GPa or a change in T_c by only 7–8 K. Raising the compression level from 1.4 to 20% is a worthy goal but constitutes a very difficult challenge for materials scientists.

An important point remaining is to identify what information the above dependence $T_c \propto a^{-4.5}$ gives on the nature of the superconducting state in HTSC. If we assume two electrons are bound in a Cooper pair by electron–phonon, electron–electron, electron–magnon, or other effective interactions, $\mathcal{V}_{eff}$, a BCS-like expression is appropriate for weak interactions

$$T_c \simeq \langle \omega \rangle \exp[-1/\mathcal{V}_{eff} N(E_f)], \quad (11.16)$$

where $\langle \omega \rangle$ is the characteristic energy of the intermediary bosons. Since both $\mathcal{V}_{eff}$ and $N(E_f)$ are in the exponent in Eq. (11.16), it is likely that their pressure dependence is responsible for that of T_c. Early high-pressure measurements of the spin susceptibility of

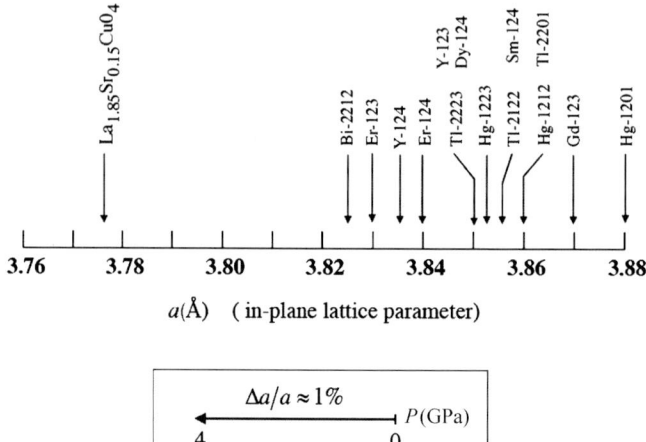

Figure 11.17. Average lattice parameter in the CuO_2 plane for representative HTSC systems at ambient pressure. Figure reproduced from [174].

La$_{1.85}$Sr$_{0.15}$CuO$_4$ [175] and Y-123 [176], and band-structure calculations for Hg-1223 [160] found the changes in $N(E_f)$ under pressure to be less than 0.2, 0.1, and 0.5%/GPa, respectively. For La$_{1.85}$Sr$_{0.15}$CuO$_4$ this change in $N(E_f)$ is too small to account for the rapid increase of T_c under pressure; to make a similar evaluation for Y-123 and Hg-1223, where $d \ln T_c/dP$ is much smaller, the accuracy of the $dN(E_f)/dP$ determination would have to be considerably enhanced.

The question remains: why does $\mathcal{V}_{\text{eff}} N(E_f)$ increase with pressure at a rate such that $T_c \propto a^{-4.5}$? Unfortunately, this $T_c(a)$ dependence alone gives insufficient information to allow one to unequivocally identify the pairing interaction. The intrinsic pressure dependence $dT_c/dP \approx +1$ to 2 K GPa^{-1} for "healthy" HTSC easily falls within the wide range of dependences observed for transition metal superconductors (see discussion above) where electron–phonon pairing is well established. From an analysis of the high-pressure results on HTSC, Neumeier [177] has come to a similar conclusion, namely, electron–phonon coupling is a possible pairing interaction for HTSC.

The above analysis leading to the intrinsic relation $T_c \propto a^{-4.5}$ has, unfortunately, only been carried out on HTSC near optimal doping. This restriction was necessitated by the need to eliminate pressure-induced changes in n. To establish $T_c^{\text{intr}}(P)$ over a wide range of doping, an independent determination of dn/dP over this entire range must first be carried out.

To shed light on the pairing interaction through high-pressure studies, it will be necessary to combine $T_c(P)$ determinations under hydrostatic and uniaxial pressure with simultaneous measurements (preferably on the same crystal) of other important superconducting- and normal-state properties such as the superconducting gap, the pseudo-gap in the underdoped region, superconducting condensation energy, magnetic susceptibility, electrical resistivity, Hall effect, thermoelectric power, etc. Aronson et al. [178] made an early attempt along these lines by carrying out high-pressure Raman scattering studies on antiferromagnetic La$_2$CuO$_4$ and found that the superexchange interaction J increases approximately as $J \propto a^6$. Such studies, if expanded to other HTSC, have the potential to test the viability of spin-fluctuation theories; measurements of the pressure-dependent magnetic susceptibility at elevated temperatures would provide similar information on $J(P)$. Further studies on the La-214 system would seem ill advised since the rampant structural distortions and transitions in this system make a quantitative analysis extremely difficult. It would be of considerable interest to attempt such studies on crystals from the Hg- and Tl-compound families over the full range of doping. Special emphasis should be placed on uniaxial pressure experiments since they can provide the kind of detailed information needed to make real progress.

Acknowledgments

The author would like to acknowledge research support by the National Science Foundation under grant DMR-0404505 and thank J. Hamlin for technical assistance.

Bibliography

1. See, for example: J.S. Schilling, Hyperfine Interact. **128**, 3 (2000).
2. See, for example: J.S. Schilling, J. Phys. Chem. Solids **59**, 553 (1998).
3. G.J. Sizoo and H.K. Onnes, Commun. Phys. Lab. Univ. Leiden, No. 180b (1925).
4. M. Levy and J.L. Olsen, Solid State Commun. **2**, 137 (1964); see also, M. Levy and J.L. Olsen, in *High Pressure Physics*, A. Van Itterbeek (ed.) (North-Holland Publishing Company, Amsterdam, 1965).

5. M.K. Wu, J.R. Ashburgn, C.J. Torng, P.H. Hor, R.L. Meng, L. Gao, Z.J. Huang, Y.Q. Wang, and C.W. Chu, Phys. Rev. Lett. **58**, 908 (1987).
6. F. Gugenberger, C. Meingast, G. Roth, K. Grube, V. Breit, T. Weber, H. Wühl, S. Uchida, and Y. Nakamura, Phys. Rev. B **49**, 13137 (1994).
7. J.-P. Locquet, J. Perret, J. Fompeyrine, E. Mächler, J.W. Seo, and G. Van Tendeloo, Nature **394**, 453 (1998).
8. A.-K. Klehe, A.K. Gangopadhyay, J. Diederichs, J.S. Schilling, Physica C **213**, 266 (1993); A.-K. Klehe, J.S. Schilling, J.L. Wagner, and D.G. Hinks, Physica C **223**, 313 (1994).
9. L. Gao, Y.Y. Xue, F. Chen, Q. Xiong, R.L. Meng, D. Ramirez, C.W. Chu, J.H. Eggert, and H.-K. Mao, Phys. Rev. B **50**, 4260 (1994).
10. N.W. Ashcroft, Phys. Rev. Lett. **21**, 1748 (1968).
11. B.M. Andersson, B. Sundqvist, J. Niska, B. Loberg, and K. Easterling, Physica C **170**, 521 (1990); J. Karpinski, E. Kaldis, E. Jilek, S. Rusiecki, and B. Bucher, Nature **336**, 660 (1988).
12. For a recent review of pressure-induced superconductivity in elemental solids, see: C. Buzea and K. Robbie, Supercond. Sci. Technol. **18**, R1 (2005); K. Shimizu, K. Amaya, and N. Suzuki, J. Phys. Soc. Jpn **74**, 1345 (2005).
13. Source of values of T_c in Fig. 11.1 at ambient pressure: (**Be, Al, Ti, V, Zn, Ga, Zr, Nb, Mo, Tc, Ru, Cd, In, Sn, La, Hf, Ta, W, Re, Os, Ir, Hg-α, Tl, Pb, Th, Pa**) C. Kittel, *Introduction to Solid State Physics*, 7th edition (Wiley, New York, 1996) p. 336; (**Rh**) Ch. Buchal et al., Phys. Rev. Lett. **50**, 64 (1983); (**U-β**) B.T. Matthias et al., Science **151**, 985 (1966). <note that U-α doesn't superconduct above 0.1 K, see, J.C. Ho et al., Phys. Rev. Lett. **17**, 694 (1966)>; (**Am**) J.L. Smith and R.G. Haire, Science **200**, 535 (1978).
14. Source of values of T_c^{max} in Fig. 11.1 at high pressure: (**Li**) Ref. [46]; (**B**) M.I. Eremets et al., Science **293**, 272 (2001); (**O**) K. Shimizu et al., Nature **393**, 767 (1998); (**Si**) K.J. Chang et al., Phys. Rev. Lett. **54**, 2375 (1985); (**P**) from Ref. [67]; (**S**) from Ref. [68]; (**Ca**) T. Yabuuchi et al., J. Phys. Soc. Jpn **75**, 083703 (2006); (**Sc**) J. J. Hamlin, and J.S. Schilling (unpublished); (**V**) from Ref. [69]; (**Fe**) K. Shimizu et al., Nature **412**, 316 (2001); (**Ga**) from Ref. [27]; (**Ge**) J. Wittig, Z. Phys. **195**, 215 (1966); (**As**) A.L. Chen et al., Phys. Rev. B **46**, 5523 (1992); (**Se**) from Ref. [68]; (**Br**) K. Shimizu et al., Rev. High Pressure Sci. Technol. **6**, 498 (1995); (**Sr**) K.J. Dunn and F.P. Bundy, Phys. Rev. B **25**, 194 (1982); (**Y**) from Ref. [70,71]; (**Zr**) Y. Akahama et al., J. Phys. Soc. Jpn **59**, 3843 (1990); (**Nb**) V.V. Struzhkin et al., Phys. Rev. Lett. **79**, 4262 (1997); (**Sn**) J. Wittig, Z. Phys. **195**, 228 (1966); (**Sb**) J. Wittig, Mater. Res. Soc. Symp. Proc. (High Pressure Sci. Technol., Pt. 1) **22**, 17 (1984); (**Te**) Akahama et al., Solid State Commun. **84**, 803 (1992); (**I**) K. Shimizu et al., J. Supercond. **7**, 921 (1994); (**Cs**) from Ref. [57] where data from Ref [56] is replotted; (**Ba**) K.J. Dunn and F.P. Bundy, Phys. Rev. B **25**, 194 (1982); (**La**) from Refs. [66] and [57]; (**Ce**) J. Wittig, Phys. Rev. Lett. **21**, 1250 (1968); (**Lu**) from Ref. [72]; (**Hf**) I.O. Bashkin et al., JETP Lett. **80**, 655 (2004); (**Ta**) see reference for Nb; (**Bi**) M.A. Il'ina et al., Sov. Phys. JETP **34**, 1263 (1972); (**U-α**) T.F. Smith and E.S. Fisher, J. Low Temp. Phys. **12**, 631 (1973); (**Am**) J.-C. Griveau et al., Phys. Rev. Lett. **94**, 097002 (2005).
15. J. Thomasson, C. Ayache, I.L. Spain, and M. Villedieu, J. Appl. Phys. **68**, 5933 (1990).
16. A. Jayaraman, Rev. Mod. Phys. **55**, 65 (1983).
17. J.S. Schilling and S. Klotz, in *Physical Properties of High Temperature Superconductors*, vol. III, D. M. Ginsberg (ed.) (World Scientific, Singapore, 1992), p. 59.
18. In experiments on Pb to 0.6 GPa using dense He as pressure medium, $dT_c/dP \simeq -0.40 \pm 0.03 \, \text{K GPa}^{-1}$, T. Tomita and M. Debessai (private communication).
19. J.D. Thompson and J.M. Lawrence, in *Handbook on the Physics and Chemistry of Rare Earths and Actinides*, vol.19, K.A. Gschneidner, L. Eyring, G. Lander, and G. Choppin, (eds.) (North-Holland Publishing Company, Amsterdam, 1994), p. 385.
20. J.L. Sarrao and J.D. Thompson, in *NATO Science Series, II: Mathematics, Physics and Chemistry* (2003), **110** (*Concepts in Electron Correlation*), p. 345.
21. F. Steglich, J. Phys. Soc. Jpn **74**, 167 (2005)
22. J. Flouquet, "On the Heavy Fermion Road", preprint in: arXiv: cond-mat/0501602.
23. J. Flouquet, G. Knebel, D. Braithwaite, D. Aoki, J.P. Brison, F. Hardy, A. Huxley, S. Raymond, B. Salce, and I. Sheikin, "Magnetism and Superconductivity of Heavy Fermion Matter", preprint in: arXiv: cond-mat/0505713.
24. D. Jerome and H.J. Schulz, Adv. Phys. **51**, 293 (2002).
25. J. Bardeen, L.N. Cooper, and J.R. Schrieffer, Phys. Rev. **108**, 1175 (1957).
26. See, for example: T.F. Smith and C.W. Chu, Phys. Rev. **159**, 353 (1967), and references therein.
27. N.B. Brandt and N.I. Ginzburg, Sci. Am. **224**, 83 (1971).
28. D.U. Gubser and A.W. Webb, Phys. Rev. Lett. **35**, 104 (1075).
29. W.L. McMillan, Phys. Rev. **167**, 331 (1968).
30. A. Eiling and J.S. Schilling, J. Phys. F: Metal Phys. **11**, 623 (1981).

31. D. Gross and J.L. Olsen, Cryogenics, **1**, 91 (1990).
32. X.J. Chen, H. Zhang, and H.-U. Habermeier, Phys. Rev. B **65**, 144514 (2002).
33. C.F. Richardson and N.W. Ashcroft, Phys. Rev. B **55**, 15130 (1997).
34. P.B. Allen and M.L. Cohen, Phys. Rev. **187**, 525 (1969).
35. J.J. Hopfield, Physica (Amsterdam) **55**, 41 (1971).
36. G.D. Gaspari and B.L. Gyorffy, Phys. Rev. Lett. **28**, 801 (1972).
37. R. Evans, V.K. Ratti, and B.L. Gyorffy, J. Phys. F: Metal Phys. **3**, L199 (1973); R. Evans, G.D. Gaspari, and B.L. Gyorffy, J. Phys. F: Metal Phys. **3**, 39 (1973).
38. R.C. Dynes, Solid State Commun. **10**, 615 (1972).
39. R.E. Hodder, Phys. Rev. **180**, 530 (1969).
40. W. Paul, in *High Pressure Physics and Chemistry*, R.S. Bradley (ed.) (Academic Press, New York, 1963), p. 299.
41. T.H. Lin and K.J. Dunn, Phys. Rev. B **33**, 807 (1986).
42. J.B. Neaton and N.W. Ashcroft, Nature **400**, 141 (1999).
43. J.C. Boettger and S.B. Trickey, Phys. Rev. B **32**, 3391 (1985).
44. K. Shimizu, H. Ishikawa, D. Takao, T. Yagi, and K. Amaya, Nature **419**, 597 (2002).
45. V.V. Struzhkin, M.I. Eremets, W. Gan, H.K. Mao, and R.J. Hemley, Science **298**, 1213 (2002).
46. S. Deemyad and J.S. Schilling, Phys. Rev. Lett. **91**, 167001 (2003).
47. M. Hanfland, K. I. Syassen, N.E. Christensen, and D.L. Novikov, Nature **408**, 174 (2000).
48. A.F. Goncharov, V.V. Struzhkin, H.-K. Mao, and R.J. Hemley, Phys. Rev. B **71**, 184114 (2005).
49. H. Katzke and P. Tolédano, Phys. Rev. B **71**, 184101 (2005).
50. A. Rodriguez-Prieto and A. Bergara, in *Proceedings of the Joint 20th AIRAPT—43rd EHPRG Meeting,* June 27–July 1, 2005 in Karlsruhe, Germany.
51. J.B. Neaton and N.W. Ashcroft, Phys. Rev. Lett. **86**, 2830 (2001).
52. K. Ullrich, C. Probst, and J. Wittig, J. Physique Coll. **1**, 6463 (1978); *ibid.*, Bericht Kernforschungsanlage Jülich (Juel-1634) (1980).
53. E. Gregoryanz, O. Degtyareva, M. Somayazulu, R.J. Hemley, and H.-K. Mao, Phys. Rev. Lett. **94**, 185502 (2005).
54. V.V. Struzhkin, R.J. Hemley, H.-K. Mao, and Y.A. Timofeev, Nature **390**, 382 (1997).
55. J. Wittig and B.T. Matthias, Science **160**, 994 (1968).
56. J. Wittig, Phys. Rev. Lett. **24**, 812 (1970).
57. J. Wittig, in *Superconductivity in d- and f-Band Metals,* W. Buckel and W. Weber (eds.) (Kernforschungszentrum, Karlsruhe, 1982) pp. 321–329.
58. M. Ross and A.K. McMahan, Phys. Rev. B **26**, 4088 (1982).
59. A.K. McMahan, Phys. Rev. B **29**, 5982 (1984).
60. T.F. Smith, in *Superconductivity in d- and f-Band Metals,* D.H. Douglass (ed.) (AIP Conf. Proc., New York, 1972), p. 293.
61. J.W. Garland and K.H. Bennemann, in *Superconductivity in d- and f-Band Metals,* D.H. Douglass (ed.) (AIP Conf. Proc., New York, 1972), p. 255.
62. T.F. Smith and R.N. Shelton, J. Phys. F: Metal Phys. **5**, 911 (1975).
63. R.N. Shelton, T.F. Smith, C.C. Koch, and W.E. Gardner, J. Phys. F: Metal Phys. **5**, 1916 (1975).
64. V.K. Ratti, R. Evans, and B.L. Gyorffy, J. Phys. F: Metal Phys. **4**, 371 (1974).
65. S. Okada, K. Shimizu, T.C. Kobayashi, K. Amaya, and S. Endo, J. Phys. Soc. Jpn **65**, 1924 (1996).
66. V.G. Tissen, E.G. Ponyatovskii, M.V. Nefedova, F. Porsch, and W.B. Holzapfel, Phys. Rev. B **53**, 8238 (1996).
67. I. Shirotani, H. Kawamura, K. Tsuji, K. Tsuburaya, O. Shimomura, and K. Tachikawa, Bull. Chem. Soc. Jpn **61**, 211 (1988).
68. E. Gregoryanz, V.V. Struzhkin, R.J. Hemley, M.I. Eremets, H.-K. Mao, and Y.A. Timofeev, Phys. Rev. B **65**, 064504 (2002).
69. M. Ishizuka, M. Iketani, and S. Endo, Phys. Rev. B **61**, R3823 (2000).
70. J.J. Hamlin, V.G. Tissen, and J.S. Schilling, Phys. Rev. B **73**, 094522 (2006).
71. J.J. Hamlin, V.G. Tissen, and J.S. Schilling (unpublished).
72. C. Probst and J. Wittig, in *Handbook on the Physics and Chemistry of Rare Earths,* K.A. Gschneidner, Jr. and L. Eyring (eds.) (North-Holland Publishing Company, Amsterdam, 1978), p. 749.
73. J. Wittig, C. Probst, F.A. Schmidt, and K.A. Gschneidner, Jr., Phys. Rev. Lett. **42**, 469 (1979).
74. J.S. Schilling, S. Methfessel, and R.N. Shelton, Solid State Commun. **24**, 659 (1977).
75. B. Lorenz and C.W. Chu, in *Frontiers in Superconducting Materials,* A. Narlikar (ed.) (Springer, Berlin Heidelberg New York, 2005), p. 459.

76. D.G. Pettifor, J. Phys. C **3**, 367 (1970).
77. H.L. Skriver, Phys. Rev. B **31**, 1909 (1985).
78. J.C. Duthrie and D.G. Pettifor, Phys. Rev. Lett. **38**, 564 (1977).
79. K.A. Gschneidner, Jr., Solid State Phys. **16**, 275 (1964).
80. T.F. Smith, J. Phys. F **2**, 946 (1972).
81. T.F. Smith, J. Low Temp. Phys. **6**, 171 (1972).
82. G. Fasol, J.S. Schilling, and B. Seeber, Phys. Rev. Lett. **41**, 424 (1978).
83. C.W. Chu and V. Diatschenko, Phys. Rev. Lett. **41**, 572 (1978).
84. R.W. Larsen and A.L. Ruoff, J. Appl. Phys. **44**, 1021 (1973).
85. J. Nagamatsu, N. Nakagawa, T. Muranaka, Y. Zenitani, and J. Akimitsu, Nature **410**, 63 (2001).
86. C. Buzea and T. Yamashita, Supercond. Sci. Technol. **14**, R115 (2001).
87. J.D. Jorgensen, D.G. Hinks, and S. Short, Phys. Rev. B **63**, 224522 (2001).
88. See, for example: U. Welp, W.K. Kwok, G.W. Crabtree, K.G. Vandervoort, and J.Z. Liv, Phys. Rev. Lett. **62**, 1908 (1989).
89. M. Monteverde, M. Núñez-Regueiro, N. Rogado, K.A. Regan, M.A. Hayward, T. He, S.M. Loureiro, and R.J. Cava, Science **292**, 75 (2001).
90. B. Lorenz, R.L. Meng, and C.W. Chu, Phys. Rev. B **64,** 012507 (2001).
91. E. Saito, T. Taknenobu, T. Ito, Y. Iwasa, K. Prassides, and T. Arima, J. Phys. Condens. Matter **13,** L267 (2001).
92. T. Tomita, J.J. Hamlin, J.S. Schilling, D.G. Hinks, and J.D. Jorgensen, Phys. Rev. B **64,** 092505 (2001).
93. S. Deemyad, T. Tomita, J.J. Hamlin, B.R. Beckett, J.S. Schilling, D.G. Hinks, J.D. Jorgensen, S. Lee, and S. Tajima, Physica C **385**, 105 (2003).
94. A.F. Goncharov and V.V. Struzhkin, Physica C **385**, 117 (2003).
95. V.G. Tissen, M.V. Nefedova, N.N. Kolesnikov, and M.P. Kulakov, Physica C **363**, 194 (2001).
96. J.J. Neumeier, T. Tomita, M. Debessai, J.S. Schilling, P.W. Barnes, D.G. Hinks, and J.D. Jorgensen, Phys. Rev. B **72**, R220505 (2005).
97. R. Osborn, E.A. Goremychkin, A.I. Kolesnikov, and D.G. Hinks, Phys. Rev. Lett. **87**, 017005 (2001).
98. Y. Kong, O.V. Dolgov, O. Jepsen, and O.K. Andersen, Phys. Rev. B **64**, 020501 (2001).
99. J.M. An and W.E. Pickett, Phys. Rev. Lett. **86**, 4366 (2001).
100. J. Kortus, I.I. Mazin, K.D. Belashchenko, V.P. Antropov, and L.L. Boyer, Phys. Rev. Lett. **86**, 4656 (2001).
101. N.I. Medvedeva, A.L. Ivanovskii, J.E. Medvedeva, and A.J. Freeman, Phys. Rev. B **64**, 020502 (2001).
102. J.B. Neaton and A. Perali, preprint cond-mat/0104098.
103. N.I. Medvedera, A.L. Ivanovskii, J.E. Medvedeva, A.J. Freeman, and D.L. Novokov, Phys. Rev. B **65**, 052501 (2002).
104. D. Roundy, H.J. Choi, H. Sun, S.G. Louie, and M.L. Cohen, post deadline session on MgB_2, March 12, 2001, APS March Meeting in Seattle, Washington.
105. S.L. Bud'ko, G. Lapertot, C. Petrovic, C.E. Cunningham, N. Anderson, and P.C. Canfield, Phys. Rev. Lett. **86**, 1877 (2001).
106. D. Hinks, G.H Claus, and J. Jorgensen, Nature **411**, 457 (2001).
107. K. Tanigaki, T.W. Ebbesen, S. Saito, J. Mizuki, J.S. Tsai, Y. Kubo, and S. Kuroshima, Nature **352**, 222 (1991).
108. T.T.M. Palstra, O. Zhou, Y. Iwasa, P.E. Sulewski, R.M. Fleming, and B.R. Zegarski, Solid State Commun. **93**, 327 (1995).
109. J.S. Schilling, J. Diederichs, and A.K. Gangopadhyay, in *Fullerenes: Recent Advances in the Chemistry and Physics of Fullerenes and Related Materials*, vol. IV, K.M. Kadish and R.S. Ruoff (eds.) (Electrochemical Society, 1997), p. 980.
110. J. Diederichs, A.K. Gangopadhyay, and J.S. Schilling, Phys. Rev. B **54**, R9662 (1996).
111. J. Diederichs, J.S. Schilling, K.W. Herwig, and W.B. Yelon, J. Phys. Chem. Solids **58**, 123 (1997).
112. A.F. Hebard, M.J. Rosseinsky, R.C. Haddon, D.W. Murphy, S.H. Glarum, T.T.M. Palstra, A.P. Ramirez, and A.R. Kortan, Nature (London) **350**, 600 (1991).
113. M. Schlüter, M. Lannoo, M. Needels, G.A. Baraff, and D. Tománek, Phys. Rev. Lett. **68**, 526 (1992).
114. V.H. Crespi and M.L. Cohen, Phys. Rev. B **53**, 56 (1996).
115. M. Monteverde, C. Acha, M. Núñez-Regueiro, D.A. Pavlov, K.A. Lokshin, S.N. Putilin, and E.V. Antipov, Europhys. Lett. **72**, 458 (2005).
116. R.J. Wijngaarden and R. Griessen, in *Studies of High Temperature Superconductors,* vol. **2** (Nova Science, New York, 1989), p. 29.
117. H. Takahashi and N. Môri, in *Studies of High Temperature Superconductors*, vol. 16/17, A.V. Narlikar (ed.) (Nova Science Publishers, Inc., New York, 1995), p. 1.

118. M. Núñez-Regueiro and C. Acha, *Studies of High Temperature Superconductors: Hg-Based High-T_c Superconductors,* vol. 24 (Nova Science, New York, 1997), p. 203.
119. R.J. Wijngaarden, D.T. Jover, and R. Griessen, Physica B **265**, 128 (1999).
120. See, for example: J.T. Markert, J. Beille, J.J. Neumeier, E.A. Early, C.L. Seaman, T. Moran, and M.B. Maple, Phys. Rev. Lett. **64**, 80 (1990).
121. J.E. Schirber, E.L. Venturini, B. Morosin, and D.S. Ginley, Physica C **162**, 745 (1989).
122. N. Môri, C. Murayama, H. Takahashi, H. Kaneko, K. Kawabata, Y. Iye, S. Uchida, H. Takagi, Y. Tokura, Y. Kubo, H. Sasakura, and K. Yamaya, Physica C **185–189**, 40 (1991).
123. M.R. Presland, J.L. Tallon, R.G. Buckley, R.S. Liu, and N.E. Flower, Physica C **176**, 95 (1991).
124. J.L. Tallon, C. Bernhard, H. Shaked, R.L. Hitterman, and J.D. Jorgensen, Phys. Rev. B **51**, 12911 (1995).
125. C. Murayama, Y. Iye, T. Enomoto, N. Môri, Y. Yamada, T. Matsumoto, Y. Kubo, Y. Shimakawa, and T. Manako, Physica C **183**, 277 (1991).
126. J.J. Neumeier and H.A. Zimmermann, Phys. Rev. B **47**, 8385 (1993).
127. C. Meingast, T. Wolf, M. Kläser, and G. Müller-Vogt, J. Low Temp. Phys. **105**, 1391 (1996).
128. S. Sadewasser, J.S. Schilling, A.P. Paulikas, and B.W. Veal, Phys. Rev. B **61**, 741 (2000).
129. J.E. Schirber, D.L. Overmyer, K.D. Carlson, J.M. Williams, A.M. Kini, H.H. Wang, H.A. Charlier, B.J. Love, D.M. Watkins, and G.A. Yaconi, Phys. Rev. B **44**, 4666 (1991), and references therein.
130. C.W. Chu, T.F. Smith, and W.E. Gardner, Phys. Rev. B **1**, 214 (1970).
131. J.E. Schirber and C.A. Swenson, Phys. Rev. **123**, 1115 (1961).
132. D.T. Jover, R.J. Wijngaarden, H. Wilhelm, and R. Griessen, Phys. Rev. B **54**, 4265 (1996).
133. S. Klotz and J.S. Schilling, Physica C **209**, 499 (1993).
134. R. Lortz, A. Junod, D. Jaccard, Y. Wang, C. Meingast, T. Masui, and S. Tajima, J. Phys.: Condens. Matter **17**, 4135 (2005).
135. J.E. Schirber, E.L. Venturini, J.F. Kwak, D.S. Ginley, and B. Morosin, J. Mater. Res. **2**, 421 (1987).
136. C.W. Chu, P.H. Hor, R.L. Meng, L. Gao, Z.J. Huang, and J.Q. Wang, Phys. Rev. Lett. **58**, 405 (1987).
137. M. Ido, N. Yamada, M. Oda, Y. Segawa, N. Momono, A. Onodera, Y. Okajima, K. Yamaya, Physica C **185–189**, 911 (1991).
138. M.K. Crawford, R.L. Harlow, E.M. McCarron, W.E. Farneth, J.D. Axe, H.E. Chou, and Q. Huang, Phys. Rev. B **44**, R7749 (1991).
139. M.K. Crawford, R.L. Harlow, S. Deemyad, V. Tissen, J.S. Schilling, E.M. McCarron, S.W. Tozer, D.E. Cox, N. Ichikawa, S. Uchida, and Q. Huang, Phys. Rev. B **71**, 104513 (2005).
140. B.W. Veal, A.P. Paulikas, H. You, H. Shi, Y. Fang, and J.W. Downey, Phys. Rev. B **42**, 6305 (1990).
141. W.H. Fietz, R. Quenzel, H.A. Ludwig, K. Grube, S.L. Schlachter, F.W. Hornung, T. Wolf, A. Erb, M. Kläser, and G. Müller-Vogt, Physica C **270**, 258 (1996).
142. J.C. Phillips, Inter. J. Mod. Phys. B **15** (24&25), 3153 (2001).
143. J.C. Phillips, Phys. Rev. Lett. **59**, 1856 (1987).
144. R. Sieburger and J.S. Schilling, Physica C **173**, 423 (1991).
145. A.-K. Klehe, C. Looney, J.S. Schilling, H. Takahashi, N. Môri, Y. Shimakawa, Y. Kubo, T. Manako, S. Doyle, and A.M. Hermann, Physica C **257**, 105 (1996).
146. A.-K. Klehe, Ph.D. Thesis, Washington University (1995).
147. C. Looney, J.S. Schilling, S. Doyle, and A.M. Hermann, Physica C **289**, 203 (1997).
148. S. Sadewasser, J.S. Schilling, and A.M. Hermann, Phys. Rev. B **62**, 9155 (2000).
149. See J.S. Schilling, C. Looney, S. Sadewasser, and Y. Wang, Rev. High Pressure Sci. Technol. **7**, 425 (1998), and references therein.
150. T. Tomita, S. Deemyad, J.J. Hamlin, J.S. Schilling, V.G. Tissen, B.W. Veal, L. Chen, and H. Claus, J. Phys.: Condens. Matter **17**, S921 (2005).
151. T. Tomita, J.S. Schilling, L. Chen, B.W. Veal, and H. Claus, Phys. Rev. Lett. **96**, 077001 (2006).
152. T. Tomita, J.S. Schilling, L. Chen, B.W. Veal, and H. Claus (unpublished).
153. J.D. Jorgensen, D.G. Hinks, O. Chmaisem, D.N. Argyriou, J.F. Mitchell, and B. Dabrowski, in *Lecture Notes in Physics—Proceedings of the first Polish–US Conference on High Temperature Superconductivity,* 11–15 September 1995, Wroclaw-Duszniki Zdr., Poland, Springer Berlin Heidelberg New York.
154. R. Griessen, Phys. Rev. B **36**, 5284 (1987).
155. C. Murayama, N. Mori, S. Yomo, H. Takagi, S. Uchida, and Y. Tokura, Nature (London) **339**, 293 (1989).
156. X.D. Ziu, Q. Xiong, L. Gao, Y. Cao, Y.Y. Xue, and C.W. Chu, Physica C **282–287**, 885 (1997).
157. B.A. Hunter, J.D. Jorgensen, J.L. Wagner, D.G. Hinks, P.G. Radaelli, D.G. Hinks, H. Shaked, R.L. Hitterman, and R.B. von Dreele, Physica C **221**, 1 (1994).

158. J.S. Schilling, A.-K. Klehe, and C. Looney, *Proceedings of the Joint XV AIRAPT & XXXIII EHPRG International Conference*, Warsaw, Poland, Sept. 11–15, 1995 (World Scientific Publishing Company, 1996), p. 673.
159. F. Chen, X.D. Qiu, Y. Cao, L. Gao, Y.Y. Xue, and C.W. Chu, *Proceedings of the 10th Anniversary HTS Workshop on Physics, Materials and Applications*, Houston, March 12–16, 1996 (World Scientific, Singapore, 1997), pp. 263–266; F. Chen, Q.M. Lin, Y.Y. Xue, and C.W. Chu, Physica C **282–287**, 1245 (1997).
160. D.J. Singh and W.E. Pickett, Physica C **233**, 237 (1994).
161. Q. Xiong, Y.Y. Xue, Y. Cao, F. Chen, Y.Y. Sun, J. Gibson, and C.W. Chu, Phys. Rev. B **50**, 10346 (1994).
162. U. Welp, M. Grimsditch, S. Fleshler, W. Nessler, J. Downey, and G.W. Crabtree, Phys. Rev. Lett. **69**, 2130 (1992).
163. C. Meingast, O. Kraut, T. Wolf, H. Wühl, A. Erb, and G. Müller-Vogt, Phys. Rev. Lett. **67**, 1634 (1991).
164. M. Kund and K. Andres, Physica C **205**, 32 (1993).
165. X.J. Chen, H.Q. Lin, W.G. Yin, C.D. Gong, and Y.-U. Habermeier, Phys. Rev. B **64**, 212501 (2001).
166. C. Meingast (private communication).
167. C. Meingast, J. Karpinski, E. Jilek, and E. Kaldis, Physica C **209**, 551 (1993).
168. K. Mori, Y. Kawaguchi, T. Ishigaki, S. Katano, S. Funahashi, and Y. Hamaguchi, Physica C **219**, 176 (1994).
169. J. Tang, Y. Okada, Y. Yamada, S. Horii, A. Matsushita, T. Kosaka, and T. Matsumoto, Physica C **282–287**, 1443 (1997).
170. C. Meingast, A. Junod, and E. Walker, Physica C **272**, 106 (1996).
171. H. Kierspel, H. Winkelmann, T. Auweiler, W. Schlabitz, B. Büchner, V.H.M. Duijn, N.T. Hien, A.A. Menovsky, and J.J.M. Franse, Physica C **262**, 177 (1996).
172. R. Sieburger, P. Müller, and J.S. Schilling, Physica C **181**, 335 (1991).
173. B.A. Hunter, J.D. Jorg, J.L. Wagner, D.G. Hinks, P.G. Radaelli, R.L. Hitterman, and R.B. VonDreele, Physica C **221**, 1 (1994).
174. J.S. Schilling, in *Frontiers of High Pressure Research II: Application of High Pressure to Low-Dimensional Novel Electronic Materials*, H.D. Hochheimer et al. (eds.) (Kluwer, Amsterdam, 2001), p. 345.
175. C. Allgeier, J.S. Schilling, H.C. Ku, P. Klavins, and R.N. Shelton, Solid State Commun. **64**, 227 (1987).
176. C. Allgeier, J. Heise, W. Reith, J.S. Schilling, and K. Andres, Physica C **157**, 293 (1989).
177. J.J. Neumeier, Physica C **233** (3&4), 354 (1994).
178. M.C. Aronson, S.B. Dierker, B.S. Dennis, S.-W. Cheong, and Z. Fisk, Phys. Rev. B **44**, 4657 (1991).

12

Superconductivity in Organic Conductors

J. S. Brooks

The physical and electronic structure of organic conductors is first described, and their "conventional" superconducting properties are presented. Two of the most studied organic conductor systems TM_2X, a quasi-one-dimensional system, and ET_2X, a quasi-two-dimensional system, when practical, are treated in parallel for comparison. The generalized phase diagrams for both systems are presented, where it is noted that superconductivity in many cases lies in proximity to an antiferromagnetic or spin density wave ground state. The nature of the band filling and the close relationship between organic conductor structure and the Mott–Hubbard model is reviewed. Examples of organic conductors that exhibit unconventional superconducting behavior, including p-wave or d-wave pairing are then presented, and a comparison is made between organic conductors and cuprate systems. In the summary, the future of organic conductor science is considered.

12.1. Introduction

In 1987 many, or maybe most of us, were drawn away from heavy Fermion materials, organic conductors, semiconductors, quantum Hall systems, and so forth by the lure of high temperature superconductivity (HTSC). Although some balance in representation between these various areas has returned, the chapters in this treatise attest to the formidable (and irresistible) theoretical and experimental challenges that HTSC still presents. Indeed, reflecting on the topic of the present chapter, the question that is often posed is "to what extent are organic conductors similar to the cuprates?" This query, if answered favorably, helps justify the labors of those involved in organic conductor research. Perhaps the best answer is "all of the above": the area of organic, or molecular, materials, embraces most of the phenomena we view as contemporary condensed matter science [1].

The history leading up to the realization of organic superconductors is long, and predominant in this was Little's postulate that superconductivity mediated by organic spine and side chain interactions might occur in organic macromolecules [2,3]. It is interesting to revisit this visionary paper, which refers to even earlier speculations by London [4], and suggests that T_c could be near room temperature, and even relevance to biological systems is discussed. Although this "polaronic" superconductivity has not yet been observed, the nature of superconductivity in organic conductors has been described as unconventional, and the superconducting phases are observed to lie in close proximity to other magnetic and/or insulating ground states which vary with chemistry, pressure, temperature, and magnetic field.

One important impact that the organic conductor community has had on the HTSC (cuprates, perovskites, etc.) community involves the study of anisotropic materials and

---

James S. Brooks • Physics/NHMFL Florida State University 1800 East Paul Dirac Drive Tallahassee, FL 23210 USA brooks@magnet.fsu.edu

Fermiology. After the discovery of superconductivity in quasi-one-dimensional (Q1D) organic conductors by Jerome and coworkers in $(TMTSF)_2PF_6$ [5] and in the quasi-two-dimensional (Q2D) organic conductor $(BEDT-TTF)_4(ReO_4)_2$ by Parkin and coworkers [6], an enormous level of activity followed to characterize the superconducting behavior and normal state electronic structure. Of particular importance was the use of magnetic fields, not only in conventional de Haas van Alphen and Shubnikov-de Haas (dHvA and SdH) studies, but also in the area of angular magnetoresistance (AMRO) where the resistance of a single crystal is monitored as the field direction is changed systematically for a fixed magnetic field and temperature. In Q2D systems with cylindrical-like Fermi surfaces this is often called the "Yamaji" effect [7] first observed by Kartsovnik et al. [8] in β-$(BEDT-TTF)_2IBr_2$ and Kajita et al. [9] in θ-$(BEDT-TTF)_2I_3$. In Q1D systems the effect is often called the Lebed effect as first reported by Osada et al. [10] and Naughton et al. [11] in $(TMTSF)_2ClO_4$. These effects, which only require the existence of anisotropic magnetoresistance, can be used to map out details of the low dimensional Fermi surfaces not readily accessible by standard dHvA or SdH measurements. Hence as someone involved in high magnetic field research, I have seen many cases where the methods developed for organic conductor research have been effectively applied to problems in inorganic correlated electron materials. Two nice examples where the Q2D Fermi surfaces were examined by AMRO are the work by Ohmichi et al. [12] on the perovskite Sr_2RuO_4 and Hussey et al. [13] on $Tl_2Ba_2CuO_{6+\delta}$.

The purpose of this short review is to communicate both the excitement and essential physical aspects of organic superconducting materials to an audience who does not work in this area. (To equitably review nearly 30 years of research herein would be impossible.) There are many excellent books and reviews on this subject, and a selection is discussed in Appendix 1. Many of the remarkable properties of these materials occur in the normal state, or in nonsuperconducting members of the organic conductor family. Although we concentrate here on the superconducting properties, there is much to say about their quantum oscillation behavior, their magnetic field induced phase transitions, and Mott-insulating, Peierls, semiconducting, and charge-ordered behavior. In this review, an attempt is made to present the material in a format where the highly representative quasi-one-dimensional (Q1D) and quasi-two-dimensional (Q2D) organic conductor systems are discussed in parallel for the purpose of comparison. After introducing their general physical properties, the unconventional aspects of their superconducting ground states presented. A comparison of organic and cuprate materials is then made, and in the final section a summary and discussion of future prospects for new physics in these systems is presented. To start, it is important to define the area in brief and general terms.

12.2. Organic Building Blocks and Electronic Structure

Organic normally means a material with carbon and carbon bonds. The building blocks of organic conductors are organic molecules where a central double carbon–carbon bond is often key to its structure. Built out from the center are rings with carbon and chalcogenide (sulfur or selenium) atoms, which are terminated with hydrogen methyl or ethyl groups (Figure 12.1). These large molecules, which are highly planar, can stack like dominoes in linear chains, or in two-dimensional planes, where the chains or planes are separated by smaller inorganic molecules, as shown in Figure 12.2. In the planar staking of the organic molecules, there can be many different motifs as shown in Figure 12.3, where Greek letters are used to describe each symmetry. These assemblies are salts, since in most cases, two of the organic

Superconductivity in Organic Conductors

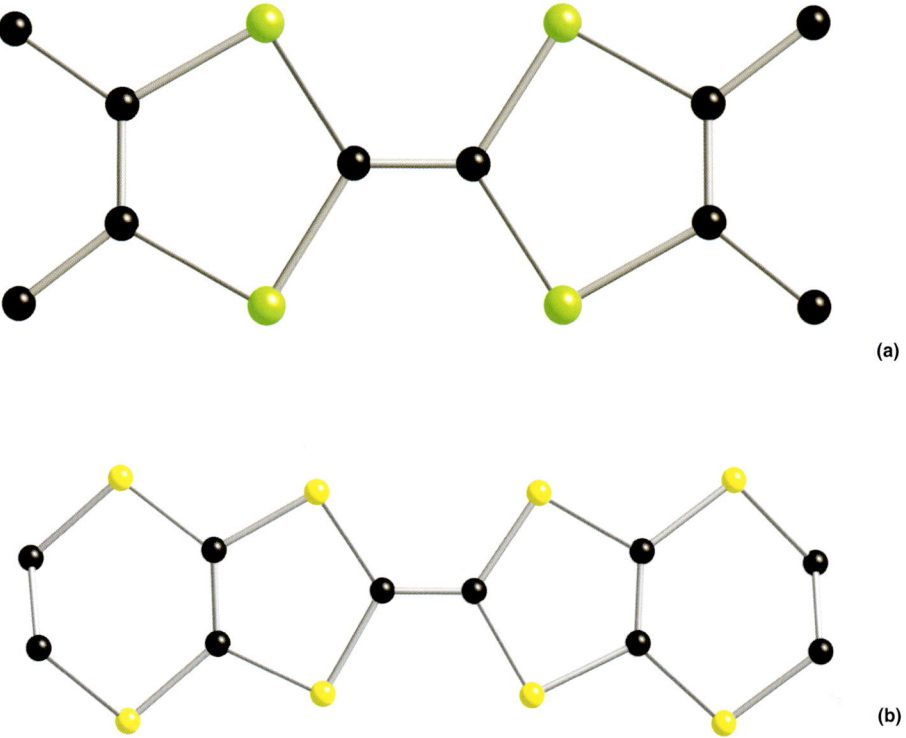

Figure 12.1. Common building blocks of organic conductors (here black=carbon, green=selenium, and yellow=sulfur). (a) Tetramethyltetreselenfulvalne (or TMTSF) represents the donor found in Q1D Bechgaard salt materials. By replacing the selenium atoms with sulfur atoms, one obtains the Fabre salts based on tetramethyltetrathiofulvalene or (TMTTF). (b) Bis(ethylenedithio)-tetrathiafulvalene (or BEDT-TTF or "ET"). By replacing the four central sulfur atoms with selenium atoms, one obtains Bis(ethylenedithio)-tetraselenafulvalene (or BETS). The terminal hydrogens for the methyl and ethyl groups are not shown.

molecules (cations or donors—D) contribute one electron to the inorganic species (anion or acceptor—A). Hence the charge transfer is D_2A, and at room temperature most of these materials are conducting. The electronic structure of these materials to first order is well described by the extended Hückel tight binding (EHTB) model which treats each donor molecule in terms of a molecular orbital that contributes a single band. With either 2 or 4 molecules per unit cell, the D_2A charge transfer leads to a $1/4$ filled band at the Fermi level for undimerized donors.

Representative Fermi surfaces derived from EHTB are shown in Figure 12.4. The TMTSF-type materials, termed the Bechgaard salts, are generally quasi-one-dimensional (Q1D) where the carriers move along the a-axis with bandwidth $\sim$250 meV. Smaller dispersions along the b-axis ($\sim$25 meV) and the c-axis ($\sim$0.03 meV) also allow electronic motion. Although the TMTSF materials are considered Q1D with open orbits, generally their a–b plane will act as a 2D conducting plane due to the significant b-axis bandwidth, and under conditions of nesting Q2D orbits can appear [21]. The BEDT-TTF-type materials are quasi-two-dimensional (Q2D), where the carriers move in the planes on both open and closed orbits with average bandwidth $\sim$25 meV, with a much smaller interplane dispersion of order $1/100\, t_{\text{in-plane}}$ or less. For the Q2D systems, extensive Fermiology studies are in reasonable

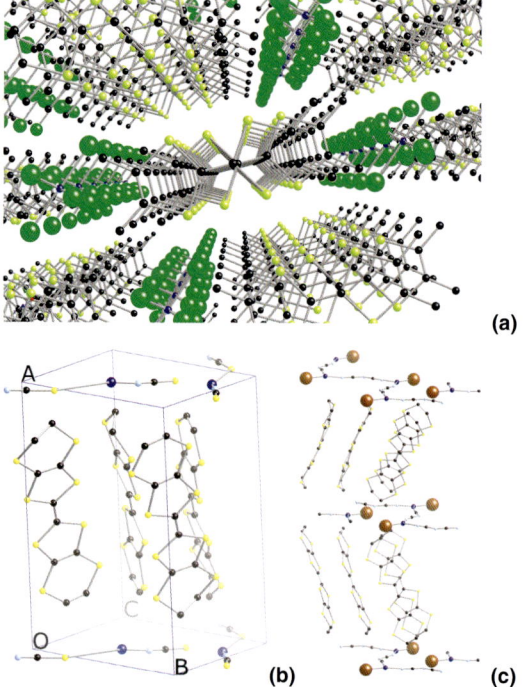

Figure 12.2. Common crystal structures of D_2A organic salts. (a) $(TMTSF)_2PF_6$ [14]. Here the $TMTSF^{+1/2}$ cation molecules stack along the a-direction, and the PF_6^{-1} anions lie between the chains. Conduction is along the a-direction. (b) κ-$(BEDT\text{-}TTF)_2Cu(NCS)_2$ [15]. Here the $ET^{+1/2}$ cation molecules stack in a planar "kappa-phase" configuration with the $Cu(NCS)_2^{-1}$ anions in the interplanar positions. Conduction is in the b–c plane. (c) κ-$(BEDT\text{-}TTF)_2Cu(N[CN]_2)Br$ [16]. This is similar to (b), but the unit cell is doubled along the interplane direction, and the $Cu(N[CN]_2)Br$ anions are connected (polymeric) in between the donor planes.

agreement with the Fermi surface topologies derived from the EHTB calculations [22, 23]. Angular dependent magnetotransport and other magnetoresistance measurements show similar agreement for the Q1D systems [21].

12.3. "Conventional" Properties of Organic Superconductors

Typical behavior for organic superconductors is shown in terms of the temperature dependent resistivity of both Q1D and Q2D materials in Figure 12.5. The immediate message is that molecular conductors are sensitive to the cooling rate due to the details of their complex molecular structure. Hence even in a high quality single crystal with perfect stoichiometry, disorder may be present if, for instance anion ordering is not achieved near 24 K as in the case of $(TMTSF)_2ClO_4$, or ethylene conformation is not complete as in the case of κ-$(BEDT\text{-}TTF)_2Cu([N(CN)_2]Br$ in the range 70–100 K. In both cases, this structural disorder can compromise the superconducting transitions, as well as magnetotransport phenomena associated with the Fermi surface topologies.

How high is T_c in organic conductors? The highest superconducting transition temperature in an organic charge transfer salt at present [26] is 14.2 K in β'-$(BEDT\text{-}TTF)_2ICl_2$. However, this is under unusual conditions where hydrostatic pressure of order 70 kbar (7 GPa)

Superconductivity in Organic Conductors

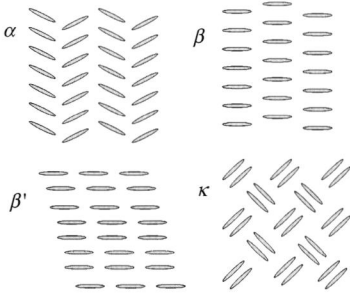

Figure 12.3. Four of the common stacking motifs for the BEDT-TTF donor. The alpha (α) stacking is characteristic of $1/4$ filled systems, and the kappa (κ) stacking, where the donors are dimerized, is characteristic of $1/2$ filling. The donors in Figure 12.2b in the case of κ-(BEDT-TTF)$_2$Cu(NCS)$_2$ for instance, will yield this κ pattern in the b–c plane when viewed along the a-axis. Note that other stacking motifs are variants of those shown: α' and θ are variants of α, λ is a variant of β, and β'' is a variant of β' (see Fig. 12.13 in Ref. [17]). Band fillings will depend on the degree of dimerization, and also on the dihedral angle between donors [18].

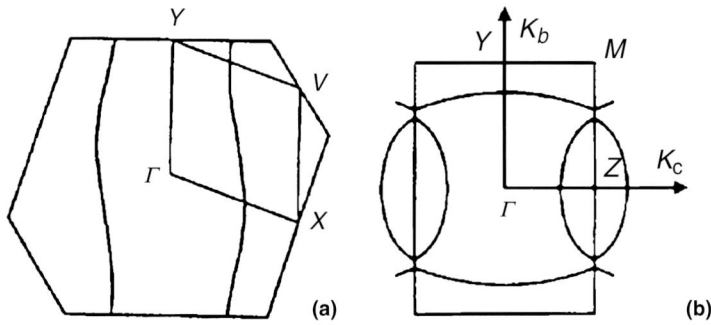

Figure 12.4. Fermi surfaces derived from tight binding calculations. (a) TMTSF$_2$PF$_6$ (after [19]. Here there are two open orbit bands, where the conduction is mainly along the a-axis. (b) κ-(BEDT-TTF)$_2$Cu(NCS)$_2$ (after [20]). Here both closed orbit and open orbit bands exist.

are necessary to stabilize the superconducting state starting from a parent Mott insulator with an antiferromagnetic ground state with $T_N = 22$ K. Here the resistive transition is used to define the onset of T_c. Under ambient conditions the highest T_c in charge transfer salts is 11.6 K obtained inductively from the rf penetration depth [27] in κ-(BEDT-TTF)$_2$Cu([N(CN)$_2$]Br. A slightly higher onset of T_c (>13 K from resistance) is seen in the related material κ-(BEDT-TTF)$_2$Cu([N(CN)$_2$]Cl under a few hundred bar of pressure [28]. (These three compounds represent the highest T_c values in Figure 12.6a.)

In the organic conductor series, the BEDT-TTF Q2D-type materials generally have the higher transition temperatures, whereas the TMTSF and related Q1D materials have transition temperatures usually less than 2 K at ambient pressure and only slightly higher transition temperatures ~3 K under pressure. An overview of the extent and perhaps limitations for superconductivity in these materials may be obtained from "global" descriptions as shown in Figure 12.6 where the transition temperatures for a wide variety of organic conductors are plotted vs. key parameters. In Figure 12.6a, T_c for a variety of organic conductors is plotted vs. the calendar year of discovery. Here the lower T_c of the TMTSF systems were found early on, and progress in raising the T_c occurred with the BEDT-TTF materials, which appears to have peaked with the BEDT-TTF materials with polymeric anions (see Figure 12.2c), an

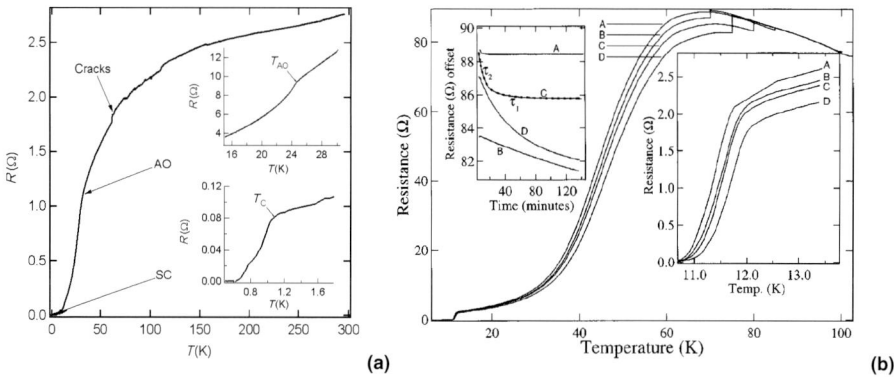

Figure 12.5. Temperature dependent resistance behavior of Q1D and Q2D organic superconductors. (a) TMTSF$_2$ClO$_4$ (courtesy of E.S. Choi). Although the resistance drops monotonically with decreasing temperature, small changes due to strain and cracks generally appear at intermediate temperatures, followed by the signature of anion ordering of the tetrahedral anion ClO$_4$ at $T_{AO} \sim 24$ K. The onset of superconductivity observed below ~ 1.2 K is sensitive to the rate of cooling through the anion ordering transition [24]. (b) κ-(BEDT-TTF)$_2$Cu([N(CN)$_2$]Br. A maximum in the resistance at lower temperatures is characteristic of some of the ET superconductors, where in this case a time dependent ethylene ordering behavior is observed. The transition temperature and also the quasiparticle scattering rate associated with quantum oscillatory behavior is dependent on the degree of ethylene ordering [25].

observation that is further supported in Figure (12.6b. The general description in Figure 12.6b is that a larger unit cell will provide a narrower band width and also soften the lattice, thereby increasing T_c. Strategies such as the use of polymeric anions to increase V_{eff} and therefore T_c appear to support this trend, but as V_{eff} increases, the lattice will eventually become unstable and/or insulating behavior will emerge. So, carbon based superconductors such as A$_3$C$_{60}$ and graphite intercalated systems notwithstanding, at present the organic charge transfer complexes appear to have reached a plateau in T_c at 15 K. (Other, more optimistic diagrams have appeared in the literature, I should admit, as in for instance Figure C4 of [29].) One final mention should be made about the interesting case of the phases of the β-ET$_2$X compounds (see Fig. 12.6 (b)) pursued early on by the Tsukuba and Tokyo groups [30], where the history of the pressure dependence could produce either the "high T$_c$ or low T$_c$" states.

Magnetic fields provide a wealth of information about the superconducting state, where the most apparent effect is the anisotropy of the upper critical field H_{c2}. Experimentally, H_{c2} has been estimated from both dc and ac transport and magnetic measurements. In principle, the onset of diamagnetism is considered the most rigorous determination, since transport will show dissipation as soon as vortices de-pin at the irreversibility line. In Figure 12.7 representative critical fields are shown for both TMTSF$_2$ClO$_4$ [30], κ-(BEDT-TTF)$_2$ X (X = Cu(NCS)$_2$ and Cu[N(CN)$_2$]Br) [31–33]. Because of the small interplane coupling and large unit cell structure, many organic conductors have high anisotropy in their critical fields. For materials with low (or no) anisotropy, a conventional Abrikosov type vortex lattice should be present in a superconducting sample for any field direction. However, when the interplane coherence length is comparable or less than the layer spacing, new effects can arise since Josephson tunneling across layers becomes important. For finite magnetic field perpendicular to the layers the vortex lattice will become pancake-like, where the superconducting component will be confined to the molecular planes. This can lead to novel lock-in effects in tilted fields [34], and also to Josephson plasma resonant effects [35,36]. For fields near or at the in-plane orientation, a Josephson vortex lattice can occur where vortices lie in the layers. It is in this case where

Superconductivity in Organic Conductors

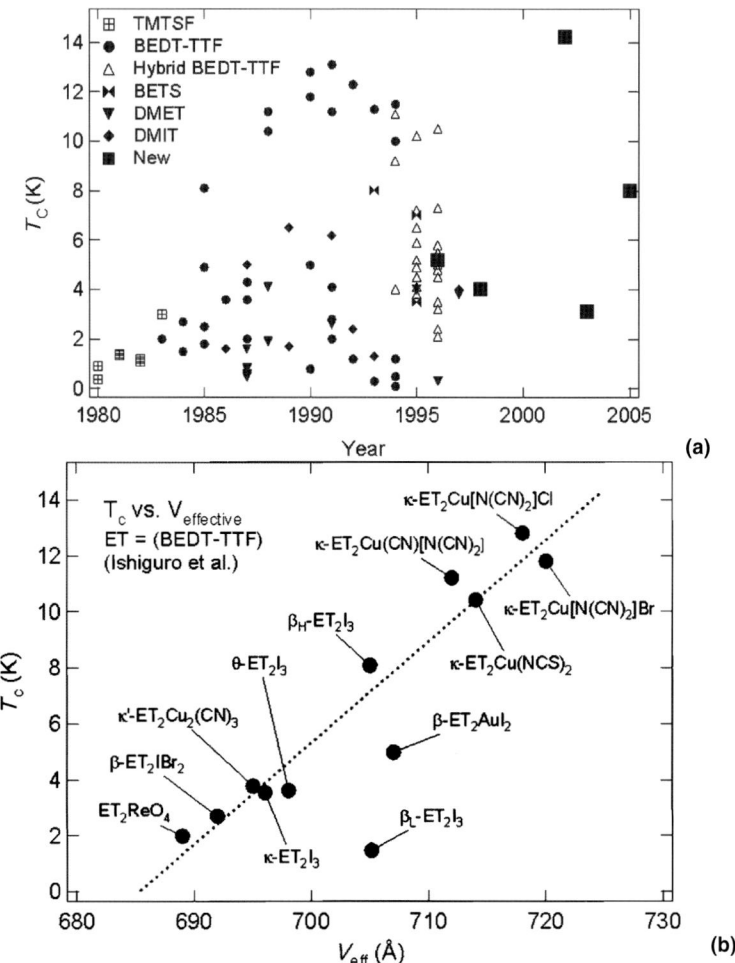

Figure 12.6. (a) T_c vs calendar year for a variety of organic conductors under ambient or pressurized conditions (from tables in [21] and references herein). (b) T_c for selected BEDT-TTF organic conductors vs. the effective unit cell parameter V_{eff} which is the space filling of the donor molecule in the unit cell per conduction electron (redrawn after [21]).

the orbital component of the critical field is most highly suppressed, and where the Zeeman (Pauli limit) field can be approached. It is in the latter case where the possibility of new physics, including p-wave paring [37] or Fulde Ferrell Larkin Ovshinnikov ground states [38] can be potentially accessed, as we will discuss in Section 12.5 below. One of the most beautiful examples of the vortex lock-in for in-plane magnetic fields comes from NMR on κ-(BEDT-TTF)$_2$Cu(SCN)$_2$ [39] where the relaxation rate due to vortex dynamics decreases by orders of magnitude for fields less than ±0.12 of a degree away from B parallel to the layers.

Specific heat studies yield a bulk thermodynamic signature of superconductivity. From the electronic and phonon terms γ and β in $C_p(T) = \gamma T + \beta T^3$ above T_c, the jump in the specific heat ΔC at T_c, and the temperature dependence of $C_p(T)$ below T_c, one can determine the electronic density of states, the Debye temperature θ_D, the BCS coupling parameter

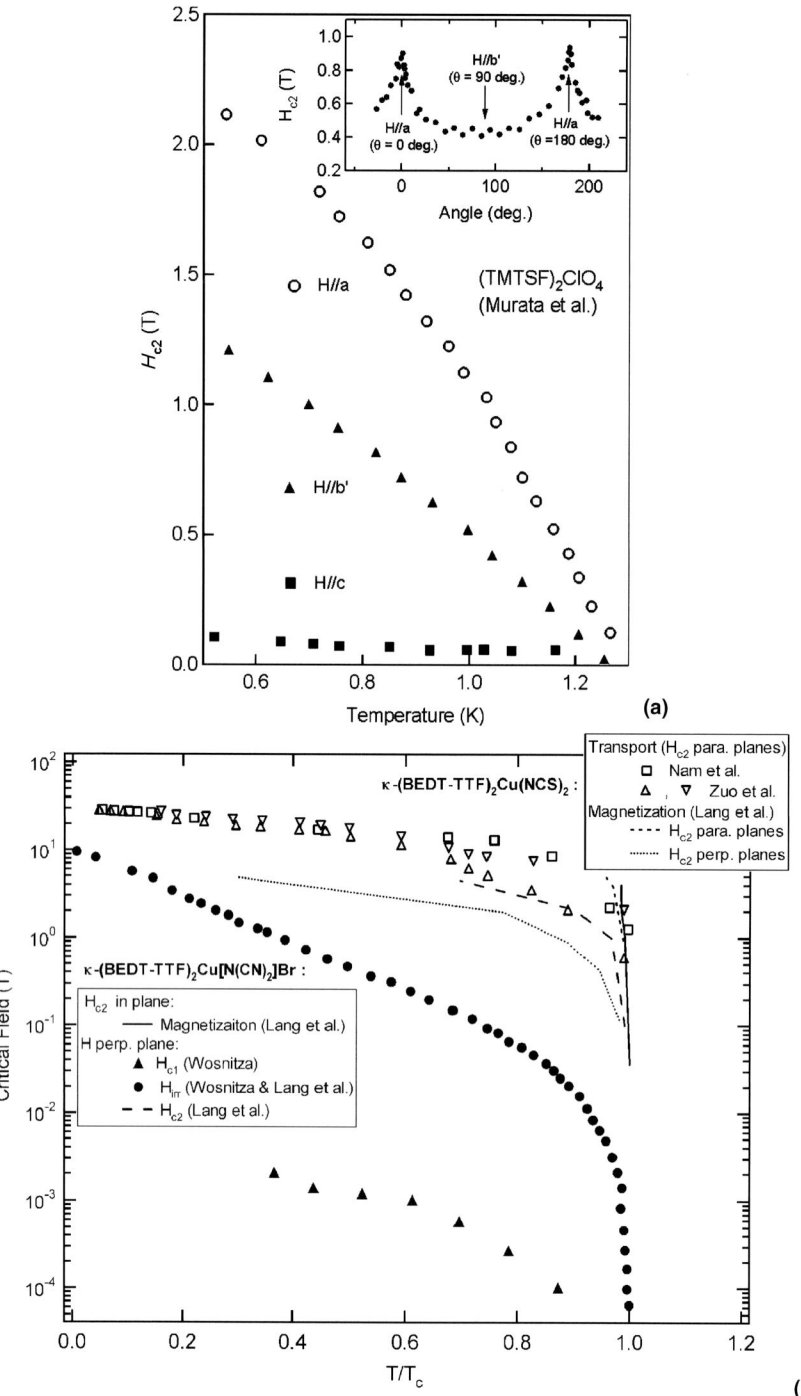

Figure 12.7. Continued

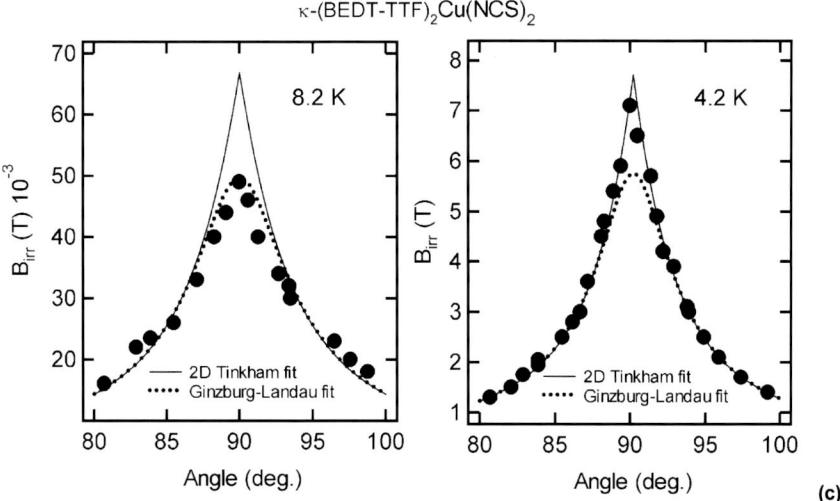

Figure 12.7. Critical field anisotropy for Q1D and Q2D organic superconductors. (a) H_{c2} for $(TMTSF)_2ClO_4$ for the three principal axes directions (re-drawn after [40]). Inset: anisotropy of H_{c2} in the $a-b$ plane. (Lower temperature data for $B//b$ is shown in Figure 12.15 below.) (b) H_{c1} and anisotropic H_{c2} for κ-$(BEDT$-$TTF)_2Cu[N(CN)_2]Br$ and κ-$(BEDT$-$TTF)_2Cu(NCS)_2$ (re-drawn after [31–33]). (c) Crossover behavior of B_{irr} from 3D (at 8.2 K-left panel, where the anisotropic Ginzburg-Landau behavior is seen) to 2D (at 4.2 K-right panel, where the 2D description prevails) for κ-$(BEDT$-$TTF)_2Cu(NCS)_2$ in the vicinity of the $B//b-c$ plane. The sharpness of the cusp at 90° (for $B//b-c$ planes) increases as the ratio of the interplane coherence length to the layer thickness $\xi_\perp(T)/s$ decreases below T_c (re-drawn after [33]).

$\Delta C/\gamma T_c$, and the superconducting gap 2Δ. An advantage of specific heat is that it can be done in the superconducting state without magnetic field, and hence no vortex contributions are present. However, due to the relatively small signature of the superconducting jump at T_c, it is a common practice to perform the measurement at both zero field and for fields above H_{c2} to obtain the difference $\Delta C = C_p(T,0) - C_p(T, H > H_{c2})$ where in principle the normal state electronic and phonon contributions (assumed to be field independent) are subtracted. Since $C_p(T < T_c) \sim \exp(-2\Delta/kT)$ and C_p(normal state) $\sim \gamma T$, ΔC will become negative at lower temperatures, as shown in Figure 12.7b for κ-$(BEDT$-$TTF)_2Cu(NCS)_2$. The results in Figure 12.8 yield a BCS-like weak coupling behavior for $(TMTSF)_2ClO_4$ [24,41,43] and a BCS-like strong coupling behavior for κ-$(BEDT$-$TTF)_2Cu(NCS)_2$ [42]. We note that κ-$(BEDT$-$TTF)_2Cu[N(CN)_2]Br$ (see Section 12.5 below) also shows a strong coupling BCS behavior and based on the specific heat studies [44] is described as fully gapped (no nodes) below T_c.

The pressure dependence of the superconducting transition temperature is another key factor in describing the nature of the superconductivity in these materials. Because they are relatively soft materials, pressure can strongly affect the ground states in several ways (a) For material in an insulating state (charge order, AF insulator, spin density wave, etc.) pressure, including uniaxial pressure, can actually induce the superconducting state. (b) For materials that are already superconducting, pressure generally suppresses T_c to zero very quickly ($\sim$5 kbar) as compared with elemental superconductors, but in some cases pressure and/or uniaxial pressure can increase T_c over some range. A global phase diagram for the Fabre $(TMTTF)_2X$ and Bechgaard $(TMTSF)_2X$ Q1D salts are shown in Figure 12.9a where anions in the range $X = SbF_6$ up to ClO_4 play the role of increasing chemical pressure [45]. In the Fabre

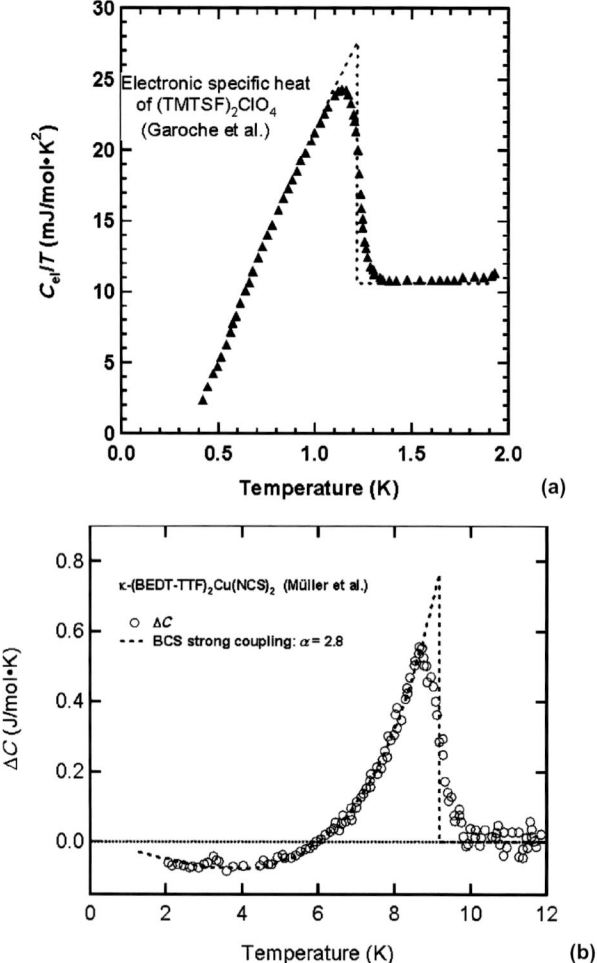

Figure 12.8. Superconducting specific heat signature of (a) $(TMTSF)_2ClO_4$ (redrawn from [41]) and (b) κ-$(BEDT-TTF)_2Cu(NCS)_2$ (redrawn from [42]). In the latter case the difference between the superconducting and normal specific heat (for $B > B_{c2}$) is plotted.

salts sulfur (tetrathio) replaces the selenium (tetraselena) in the Bechgaard salts, leading to a smaller overlap of the molecular orbitals, and hence the Fabre salts have nonmetallic ground states at ambient external pressure. Superconductivity is stabilized above a critical pressure of about 6–8 kbar in $(TMTSF)_2PF_6$. Indeed, this was the first observation of superconductivity in an organic conductor [5]. Subsequently, superconductivity was discovered at ambient pressure (but effective higher relative pressure) in $(TMTSF)_2ClO_4$ [46]. In Figure 12.8, a number of factors also evolve with increasing effective pressure, including the crossover in influence from electron–phonon to electron–electron interactions, and the increase in interchain bandwidth [45]. Recently, a coexistence of the spin density wave and superconducting states has been established, first by Vuletic et al. [47] with later confirmations by other workers [48, 49]. (Pressure induced superconductivity has also been reported in $(TMTTF)_2Br$ above 25 kbar [50] and in $(TMTTF)_2PF_6$ above 50 kbar [51].) As discussed above, $(TMTSF)_2ClO_4$ is

Superconductivity in Organic Conductors 473

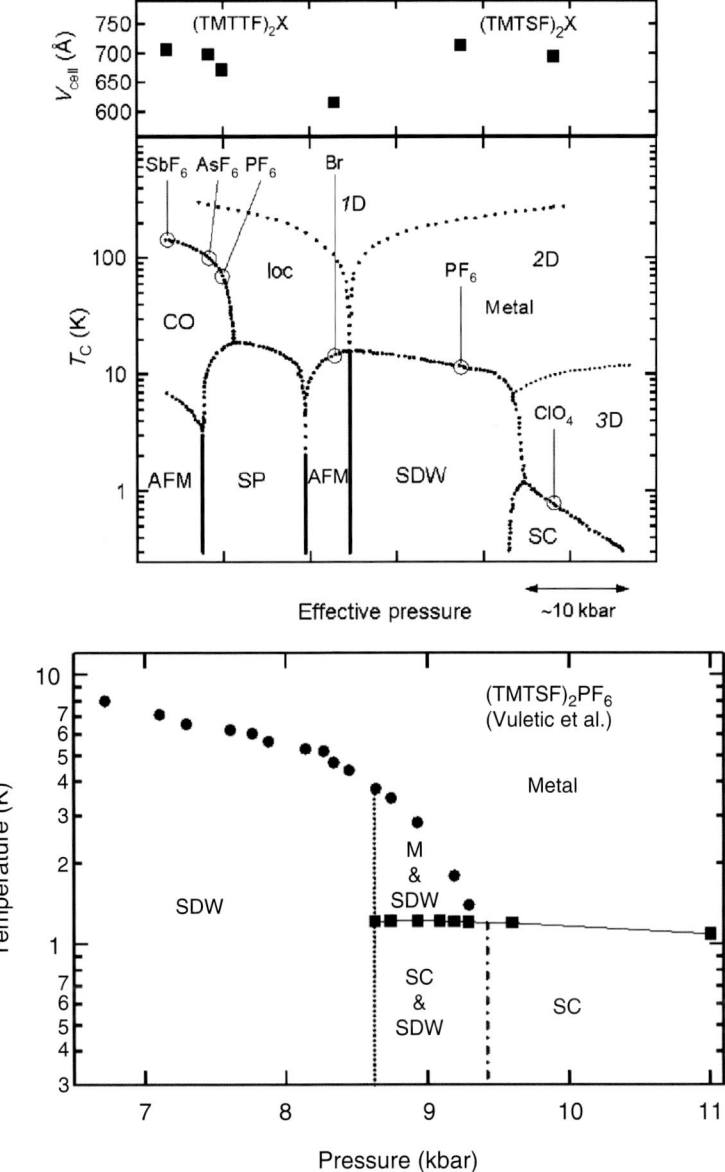

Figure 12.9. Continued

a superconductor under ambient conditions, and like $(TMTSF)_2PF_6$, T_c decreases with increasing pressure [52].

In Figure 12.10 the generalized phase diagram for the quasi-two-dimensional κ-$(BEDT\text{-}TTF)_2X$ series is shown vs. effective chemical pressure and/or the ration of the onsite Coulomb repulsion to the bandwidth U/W (i.e., the well known "Kanoda" diagram redrawn after [39]). Although pressure generally reduces the unit cell volume for a given system, here the unit cell for different components actually monotonically increases with

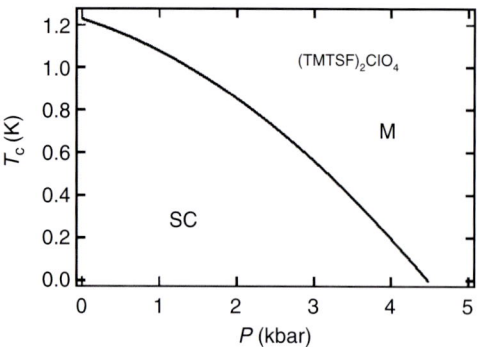

Figure 12.9. Generalized phase diagram for the ground states in the most common Q1D systems (adapted from the famous "Jerome" diagram [45]) Starting with the Fabre (TMTTF) salts at low effective pressure up to the Bechgaard salts at higher effective pressure. For insulating compounds, increasing pressure moves the ground state to the right into the superconducting regime (see text). The ambient pressure unit cell volumes for $(TMTTF)_2(X = SbF_6, AsF_6, PF_6, Br)$ and $(TMTSF)_2(X = PF_6, ClO_4)$ are: 707, 698, 672, 616 [53] and 714.3, 694.3 Å^3 [46], respectively. (b) Coexistance region for $(TMTSF)_2PF_6$ (redrawn after Vuletic et al. [47], see also [48, 49]). (c) *Approximate* pressure dependence of T_c estimated from [52])

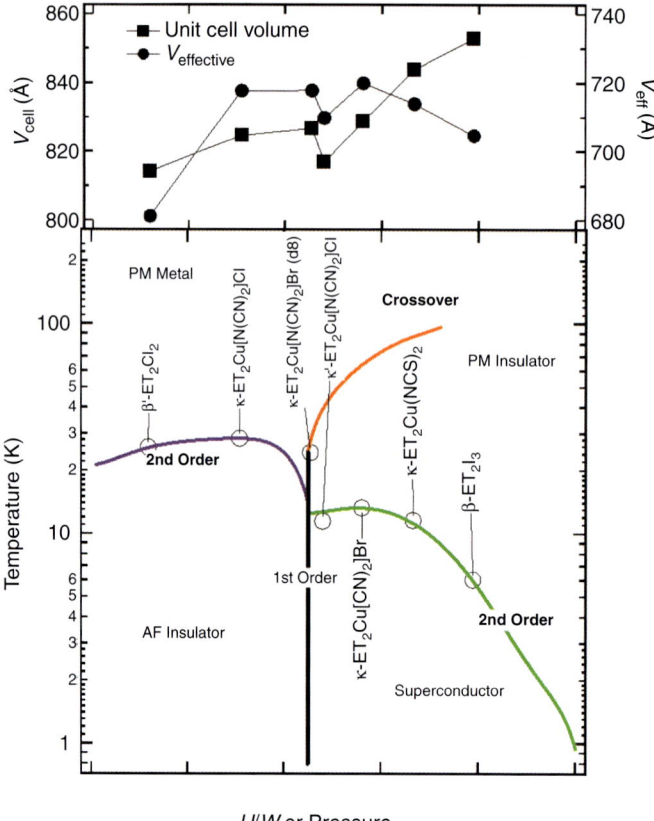

Figure 12.10. Continued

Superconductivity in Organic Conductors

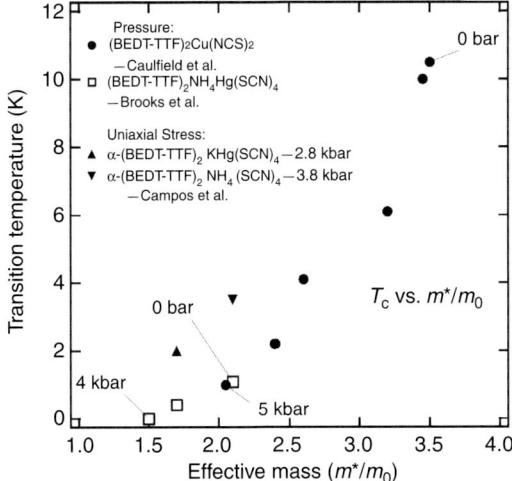

Figure 12.10. (a) Generalized phase diagram of the Q2D systems based on (BEDT-TTF)$_2$X (after [39]). κ'-(BEDT-TTF)$_2$Cu[N(CN)$_2$]Cl data are from [55]. The unit cell volumes and effective cell volumes (with the anion volume subtracted) per carrier are from [21] and [55]. Pressure dependence of transition temperature and effective masses in κ-(BEDT-TTF)$_2$Cu(NCS)$_2$ after [56]. (b) Relationship of T_c to effective mass from pressure studies for α-(BEDT-TTF)$_2$NH$_4$Hg(SCN)$_4$ [57] and κ-(BEDT-TTF)$_2$Cu(NCS)$_2$ [56]. Uniaxial stress induced superconductivity in α-(BEDT-TTF)$_2$KHg(SCN)$_4$ and increased T_c in α-(BEDT-TTF)$_2$NH$_4$Hg(SCN)$_4$ [58]. Filamentary superconductivity in α-(BEDT-TTF)$_2$KHg(SCN)$_4$ was first observed by Ito et al. [59], and subsequently, hydrostatic pressure was found by Andres et al. [60] to induce a complete superconducting transition below 0.1 K.

decreasing U/W. Since the closed orbit quantum oscillation (de Haas van Alphen and Shubnikov de Haas) behavior is accessible in these materials, the effective mass may be determined from the temperature dependence of the oscillatory amplitudes via the Lifshitz–Kosevich analysis [54]. This allows a comparison of T_c and the effective mass m^*/m_e with pressure. The insets of Figure 12.9 indicate the very strong dependence of T_c on pressure, and on the effective mass, indicating the importance of many body interactions to the superconducting state.

12.4. The "Standard Model" for Metallic, Insulating, and Antiferromagnetic Ground States

12.4.1. Band Filling and Its Consequences

In the last section, we saw that there were many factors involving the donor molecule, anion species, the stacking motif, unit cell volume, etc. that control the ground states of these materials. Materials with TMTSF or TMTSF donors are generally 1D due to the linear stacking arrangements, and materials based on the BEDT-TTF structure are generally 2D and planar in their stacking arrangements. The Q1D materials are more susceptible to instabilities such as Fermi surface nesting, whereas in the Q2D systems other factors to be discussed below lead to Mott-insulator type ground states where band instabilities are less likely to occur. How close the competing ground states can be is evident in the κ-(BEDT-TTF)$_2$Cu[N(CN)$_2$]Br and κ-(BEDT-TTF)$_2$Cu[N(CN)$_2$]Cl systems where small changes in pressure [61], deuteration [62], or stacking [55] determine either insulating or superconducting ground states in the vicinity of the AFI-SC boundary in Figure 12.10a. Indeed, in both Q1D and Q2D systems, the highest T_c is seen in close proximity to an insulating AF or SDW state. However, superconductivity does not always occur at an insulating boundary.

The physical, electronic, and magnetic structure of organic materials are unique in several important ways, and a "Standard Model" [63,64] has emerged that describes many aspects of their physical properties. (We note that there is also a "Standard Model" for the formation of field induced spin density waves in Q1D systems as discussed in [52] and references therein.) Since the anion structure is a closed shell, insulating layer, the electronic properties are mostly determined by the remaining partially filled donor orbitals according to the charge transfer D_2A. (In this description we will limit our discussion to the more mainstream materials.) Hence the donors represent the sites in the Hubbard models used to describe these systems. The sites may be insolated donors, or the sites may be dimers or even tetramers of the original donors. Charge transfer yields a single charge and spin $1/2$ per two donor units, and the outcome (ground state) will depend on a number of competing energies (where we have suppressed the crystallographic indices in some cases):

U is the on-site coulomb interaction for double occupancy of a single site.
V is the intersite coulomb interaction.
t is the transfer integral parameter derived from the extended Huckle tight binding model [65,66].
Δ is the splitting of the two HOMO levels that occurs in dimerization (related to the intradimer t).
W is the bandwidth $= 4t$.
J is the intersite spin interaction.

Since the donor stacking can represent ladder, square, or triangular arrangements, the Hubbard model (or extended Hubbard model) is most often used to describe molecular conductors. The general form is [64]

$$H = \sum_{i,j}\sum_{\sigma}(t_{ij}c_{i\sigma}^*c_{j\sigma} + \text{h.c.}) + \sum_i U n_{i\uparrow}n_{i\downarrow} + \sum_{ij}V_{ij}n_in_j,$$

where i and j are the lattice sites, and σ is the spin index for up and down spins. In the model, these parameters, along with pressure, are used in different limits of strength and relative size to describe various families of organic conducting systems. For the $1/2$ filled, or $1/4$ filled systems described below, the extended Hubbard model will typically involve U and t, or V and t, respectively.

Band filling. In the extended Huckel tight binding model, each donor yields one orbital, which in the neutral state is occupied by two electrons (see Figure 12.11a). Because of the D_2A charge transfer, each single donor site will have statistically $1(1/2)$ electrons, and in the solid state the band associated with this orbital will therefore be $3/4$ (or equivalently $1/4$) filled. However, if the donors dimerize, thereby constituting a single site (where statistically there are three electrons in the HOMO levels), then due to the intradonor interaction, the two molecular orbitals will split into a higher energy antibonding level with one electron and a lower energy bonding level with two electrons. Hence in the solid state the singly occupied band will be $1/2$ filled.

Distribution of the Charge. In the absence of Coulomb repulsion (U and $V = 0$) the charge in both filling schemes will be evenly distributed for finite band width W (or transfer

Superconductivity in Organic Conductors

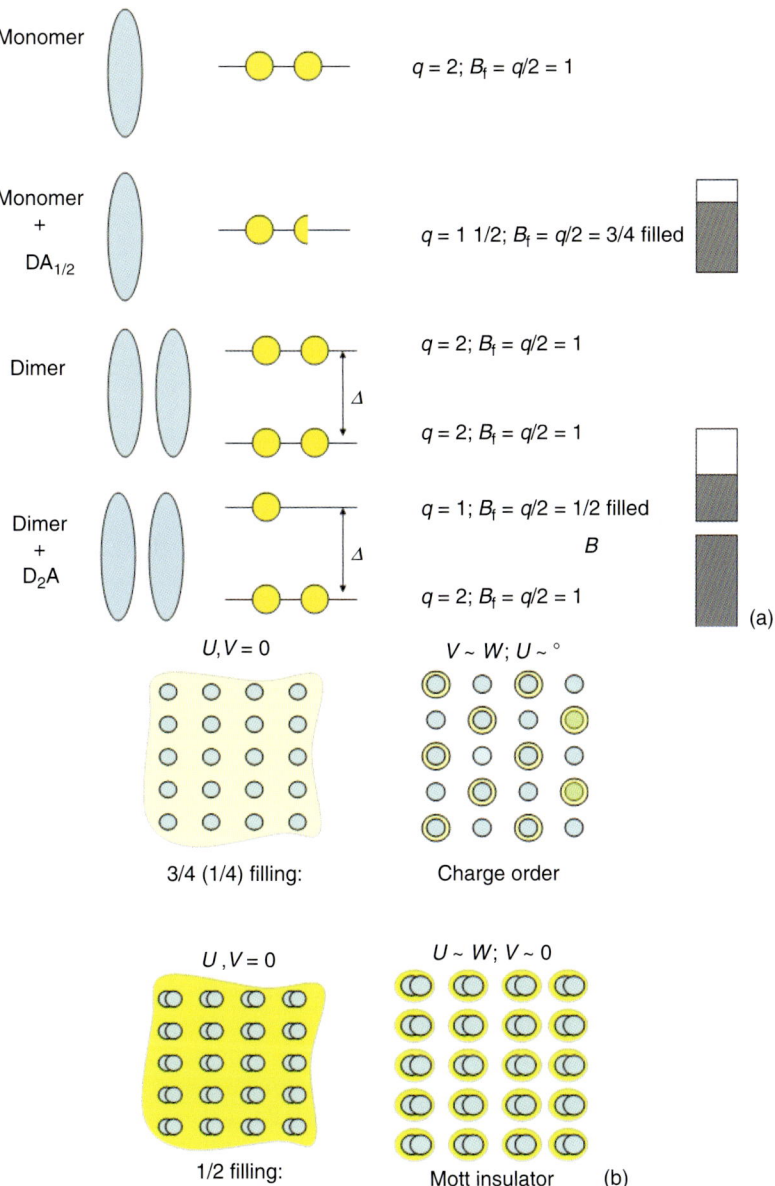

Figure 12.11. (a) Relationship of monomers and dimers to band filling and (b) charge distribution in D_2A charge transfer systems. Note that for the $1/2$ filling, the sites are dimers, not single donors. See text for discussion.

energy t) as shown on the left of Figure 12.11b. However, for $U > W$ and/or $V > W$, then the charges will tend to localize on specific sites. In the case of $3/4$ filling, charges will tend to localize on alternate sites (termed charge order or CO) due to V (where we assume U is infinite and no double occupancy can occur); and in the case of $1/2$ filling, charges will tend to localize on every site (termed the Mott insulator state where we assume that V is negligible).

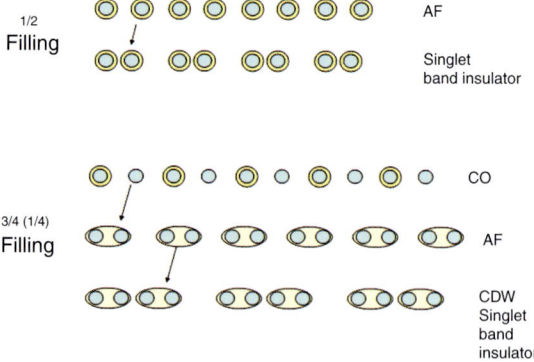

Figure 12.12. Evolution of band insulators. For $1/2$ filling dimerization produces a singlet (spin Peierls) state with $2k_F$ periodicity. For $3/3$ filling, dimerization produces a $2k_F$ AF Mott insulator, and tetramerization produces a $4k_F$ charge density wave singlet state.

Distribution of Spin. In Figure 12.11, the spin on the occupied sites will be $1/2$. Such a configuration leads directly to a spin lattice, which generally results in an antiferromagnetic ground state. However, in some cases the stacking motif of dimer systems can approach a nearly triangular lattice and spin frustration can become important, and even suppress magnetic order [67, 68]. It is also possible to introduce localized spins through the anion substitution, as in the case of λ (or κ)-(BEDT-TSF)$_2$FeX$_4$ (X = Cl, Br, etc.) [69]. This leads to a π–d exchange interaction between the conduction electrons and the localized spins, which in turn can yield very important new phenomena, as described in Section 12.5. (See also Enoki and Miyazaiki [70] for a review of magnetism in TTF-based organic conductors.)

The Special Case of Q1D. The description of the CO and Mott states can be generalized to a linear chain system as shown in Figure 12.12, where a 1D Heisenberg model will be realized. In addition, in the linear Q1D chain systems the 1D instabilities (Fermi surface nesting, i.e., k-space effects) often determine the low temperature ground states. These can be spin Peierls, antiferromagnetic, charge density wave, or spin density wave in form, where additional parameters, such as the degree of inter chain (transverse) coupling and the electron–phonon coupling will have strong influence on the nature of the ground state. In the spin density wave, the spin structure can be incommensurate with the lattice, and is also the result of delocalized electrons, although the spin wave structure itself is static. As shown in Figure 12.9 above, pressure, which increases dimensionally, will cause the commensurate AF state to move toward an incommensurate and/or delocalized SDW ground state, and eventually to a superconducting state. Giamarchi recently has considered Q1D systems within the frame work of Mott Insulators and Luttinger liquids [71]. He points out that dimerization and interchain interactions can greatly complicate any theoretical treatment of the Q1D materials.

In Figure 12.13 another phase diagram is given for materials with $1/4$ filled bands (after [72]). In this case, it is the intersite interaction V, rather than the on site interaction U, that is generally more important. This phase diagram, as discussed in the next section, is most relevant to materials with $α$, $τ$, $θ$, or $β''$ stacking where the dihedral angle between donors does not favor dimerization. Dressel and coworkers [73, 74] have used optical methods to determine the progression of Drude-like metallic behavior toward charge order behavior in a number of organic conductors, as indicated in Figure 12.12 for the alpha-phase materials.

Superconductivity in Organic Conductors

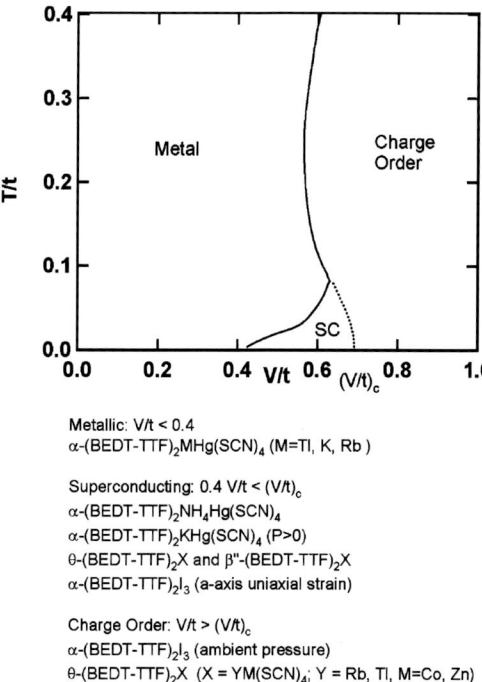

Metallic: V/t < 0.4
α-(BEDT-TTF)$_2$MHg(SCN)$_4$ (M=Tl, K, Rb)

Superconducting: 0.4 V/t < (V/t)$_c$
α-(BEDT-TTF)$_2$NH$_4$Hg(SCN)$_4$
α-(BEDT-TTF)$_2$KHg(SCN)$_4$ (P>0)
θ-(BEDT-TTF)$_2$X and β"-(BEDT-TTF)$_2$X
α-(BEDT-TTF)$_2$I$_3$ (a-axis uniaxial strain)

Charge Order: V/t > (V/t)$_c$
α-(BEDT-TTF)$_2$I$_3$ (ambient pressure)
θ-(BEDT-TTF)$_2$X (X = YM(SCN)$_4$; Y = Rb, Tl, M=Co, Zn)

Figure 12.13. Phase diagram for $1/4$ filled systems (after Ref. [72]). Superconductivity near the SC – CO boundary (dashed line) is predicted to have d_{xy} symmetry [72]. The α-(BEDT-TTF)$_2$MHg(SCN)$_4$ series [73,74], and also materials with θ-(BEDT-TTF)$_2$X and β"-(BEDT-TTF)$_2$X stackings are expected to follow this behavior [72]. Note that for ambient pressure, α-(BEDT-TTF)$_2$I$_3$ has a charge ordered state, but for a-axis uniaxial strain, superconductivity is induced. [75]. See also Seo et al. [76] for a recent review.

12.4.2. Can Superconductivity Emerge From the "Standard Model"?

The phase diagrams given in Figures 12.9 and 12.10 show that superconductivity can lie in close proximity to insulating and/or antiferromagnetic states (although it is noted that charge ordered states [75] or metallic states [77] can also share the boundary with superconductivity). The proximity of AF and SC phases has motivated theoretical work starting from the Mott–Hubbard model, to describe the phase diagrams of these systems. In a seminal paper by Kino and Fukuyama [78] ground states of three different compounds κ-(BEDT-TTF)$_2$X, α-(BEDT-TTF)$_2$I$_3$, and α-(BEDT-TTF)$_2$MHg(SCN)$_4$ were described using the Mott–Hubbard Hamiltonian. By varying the on-site energy U, the dimerization energy (t_{b1}), and the kinetic energy (t_{b4}), a paramagnetic metal, an antiferromagnetic metal, and finally an insulating antiferromagnetic ground state were obtained. As pointed out by McKenzie in his 1997 paper [79], this work provided a frame work for modifying the Hubbard Hamiltonian (i.e., the *extended Hubbard model*) to allow predictions for the superconducting ground state, the pair interaction, and the symmetry. In 1998, Schmalian [80] used a two-band model and the tight binding parameters to treat the problem of κ-(BEDT-TTF)$_2$X. This system is dimerized, half-filled, and the two parameters U and t (the onsite repulsion and interdimer hopping elements, respectively) were used in the model. It was shown that there was a transition, at 13 K, to a superconducting state, mediated by *spin fluctuations, with d-wave pairing*

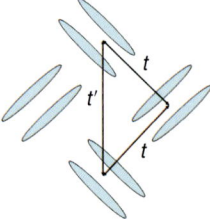

Figure 12.14. The triangular lattice relationship in the kappa stacking system. In $1/2$ filling, each BEDT-TTF dimer has one electron and spin. For $t' > t$, AF order is favored in addition to a Mott insulator, but for $t' = t$, spin frustration can suppress AF order even in the Mott insulating phase.

closely related to $d_{x^2-y^2}$ *symmetry*. Subsequently, Merino and McKenzie [72] considered the $1/4$ filled case of θ and β'' structures. In this case, the extended Hubbard Hamiltonian includes the intersite interaction V and the transfer energies t. U is set to infinity. They find an attractive interaction, mediated by *charge fluctuations, with* d_{xy} *symmetry*. A quantum critical point at $(V/t)_c$ between the metallic-superconducting and charge ordered state is also predicted (see Figure 12.13 above).

The approximation of the kappa-stacking motif shown in Figure 12.3 has also been viewed in terms of a triangular lattice, as can be seen in Figure 12.14. As pointed out by McKenzie [81], the kappa stacking in the dimerized $1/2$ filled case can be viewed as a spin $1/2$ triangular lattice system with two transfer integrals t and t', that is, a Hubbard model on an anisotropic triangular lattice with one electron and spin per site. For relative values of U, t, and t', AF, SDW or SC, PM, and decoupled chain ground states emerge from the model. This has motivated a number of uniaxial stress/strain studies that can vary the in-plane lattice parameters, thereby altering the relative values of t and t'. For instance, as $t \to t'$, SC will be favored since AF order will be reduced due to the onset of spin frustration. The effects of uniaxial pressure on organic conductors has been recently reviewed by Kagoshima and Kondo [82]. Initial studies on κ-(BEDT-TTF)$_2$Cu(NCS)$_2$ have been promising, where T_c showed a nonmonotonic dependence on in-plane stress/strain, but the results were not fully conclusive [83, 84]. There is a tendency for spin frustration when t and t' become equal, an example of which is κ-(BEDT-TTF)$_2$Cu$_2$(CN)$_3$ where $t'/t = 1.06$. Although κ-(BEDT-TTF)$_2$Cu$_2$(CN)$_3$ is a Mott insulator at low temperatures, there is no magnetic order at least to 34 mK, a characteristic of a spin liquid system and under either uniaxial strain or hydrostatic pressure κ-(BEDT-TTF)$_2$Cu$_2$(CN)$_3$ becomes superconducting ($T_c \leq 3.9$ K) [67, 68, 85]. Hence, there is *no* proximity of an AF insulating phase to the superconducting phase, as in the case of κ-(BEDT-TTF)$_2$Cu[N(CN)$_2$]Cl where $t'/t = 0.75$ and the AF phase *does* meet the SC phase (see Figure 12.10 above).

The $1/4$ filled systems can under go charge ordering, but in the event that the transfer integrals t can be modified experimentally, superconductivity can be induced (see Seo et al. [76] for a recent review.), and several examples where the SC phase does not border an AF phase should be mentioned. Tajima et al. [75] have applied a-axis strain to the material α-(BEDT-TTF)$_2$I$_3$ which is insulating below 135 K at ambient pressure due to charge order. Here they find that with increasing a-axis *in-plane* strain, the material becomes superconducting around 3 kbar, and at higher strain becomes a semiconductor. The same group [86] has also induced an insulating to superconducting transition in θ-(DIETS)$_2$[Au(CN)$_4$] under about 10 kbar of uniaxial c-axis strain. However, in this case, *inter-plane* strain is used to move the donor molecules closer together laterally, so the mechanism is slightly different.

Returning finally to the issue of AF/SC proximity, it is important to note that other models besides the Hubbard model that treat the proximity of antiferromagnetic and superconductivity, such as the SO(5) renormalization group approach [87]. Here a rotational symmetry between the AF and SC states is found which is consistent with d-wave symmetry, and also with NMR experiments as discussed in the next section.

12.4.3. But What if it is Really Just Phonons?

Girlando and coworkers [88, 89] have considered the problem of superconductivity in the BEDT-TTF salts from the point of view of phonon interactions. They find that both the electron–lattice phonon coupling and the electron–molecular vibration couplings are necessary for pairing. Starting from the Eliashberg function obtained from the computed phonon and vibrational spectrum, the Allen-McMillan equation is used to compute T_c for κ-(BEDT-TTF)$_2$I$_3$ and β^*-(BEDT-TTF)$_2$I$_3$. They obtain reasonable agreement with the experimental T_c's (3.4 an 8.1 K, respectively) and with the specific heat jumps at T_c. They conclude that "phonons are mainly responsible for the coupling mechanism" in the BEDT-TTF systems, and that many different kinds of "entangled" phonon modes are involved, where the interaction terms ($\lambda = 0.91$ and 0.74 for β^* and κ, respectively) are in the moderate to strong coupling regime. They note, however, that even with an accounting of all of the phonon modes, in for instance the kappa phase, that the T_c values for κ-(BEDT-TTF)$_2$Cu(NCS)$_2$ (10.4 K) and κ-(BEDT-TTF)$_2$Cu[N(CN)$_2$]Br (11.5 K) are too large to be computed from the phonon/vibration spectrum alone. So again, the close proximity of these systems to antiferromagnetic order may indicate that electron–electron interactions also play an important role. It would be interesting to see to what extent the very high sensitivity of T_c to pressure and uniaxial stress or strain (see Figure 12.9b) could be accounted for by changes in the phonon spectrum and tight binding parameters. Our own early work [58, 90] indicated that pressure induced lattice changes alone were not sufficient to account for the changes in T_c and m^*.

12.5. "Unconventional" Properties of Organic Superconductors

12.5.1. Q1D Materials and p-Wave Pairing

Even in the very early days of organic superconductivity, there was discussion [91] of the superconductivity in (TMTSF)$_2$X being unconventional, and perhaps p-wave in character. In a series of beautiful experiments first motivated by the theoretical work of Lebed [37, 92, 93], Naughton, Chaikin, Lee, Brown, and coworkers [94–96] have explored this possibility in great detail, and they have found strong evidence for unconventional superconductivity in both (TMTSF)$_2$ClO$_4$ and (TMTSF)$_2$PF$_6$. A key step was the precise study of the upper critical field of these materials where the magnetic field was aligned precisely parallel to the conducting planes (the a–b' plane direction), thereby excluding Abrikosov vortex states. The results for $B//b'$ are shown in Figure 12.15. Since T_c is of order 1.2 K, the corresponding Pauli limit $H_{c2} \sim 1.84\, T_c \sim 2.2\,\text{T}$ is exceeded in all cases where the field is in the a (not shown) or b' directions. Moreover, there appears to be an upward curvature in H_{c2} at the lowest temperature. In the simplest model, if the pairing is p-wave, then the magnetic field will not break the Cooper pair since the spins are already polarized. To further check this possibility, NMR was used to monitor the Knight shift upon crossing T_c in (TMTSF)$_2$PF$_6$ for $H//b'$. As shown in Figure 12.15, there was no observable Knight shift, a finding consistent with the fact that no singlet pairing occurs in a p-wave transition. For s-wave pairing, the pairs form a

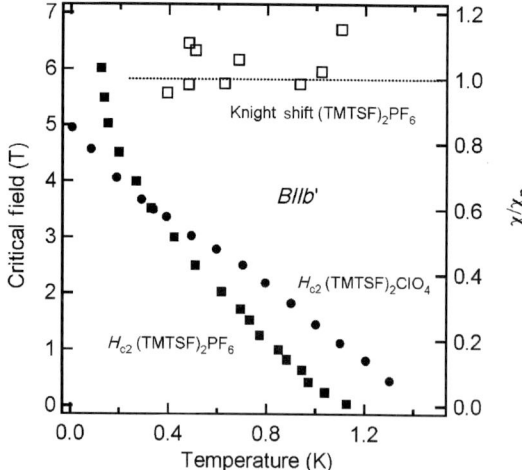

Figure 12.15. Upper critical fields for $(TMTSF)_2PF_6$ (6 kbar, solid squares [95]) determined from transport and $(TMTSF)_2ClO_4$ (solid circles [96]) from both transport and magnetization. The normalized susceptibility from NMR Knight shift measurements (open squares) for $(TMTSF)_2PF_6$ is also shown on the left axis [94]. In all cases, the magnetic field is precisely aligned in the conducting planes in the b'-axis direction.

singlet spin state, the spin susceptibility vanishes, and this would provide a clear Knight shift below T_c, which is not observed experimentally.

12.5.2. Q2D Materials and d-Wave Pairing

The κ-$(BEDT-TTF)_2X$ materials (predominantly $X = Cu(NCS)_2$ and $Cu[N(CN)_2]Br$) show experimental evidence for unconventional superconductivity based on a number of different measurements including NMR, penetration depth, isotope effects, specific heat, etc. However, the pairing mechanism and the symmetry remain controversial, since there are contradictory experiments and interpretations in the literature, as recently discussed by Wosnitza [33] and also Miyagawa et al. [39]. Residing strongly on the conventional side of the argument (see Figure 12.16) are experiments like the careful specific heat work on κ-$(BEDT-TTF)_2Cu[N(CN)_2]Br$ which shows a fully gapped, strong coupling behavior [44] and also measurements of the temperature dependent penetration depth $\lambda(T)$ (determined from the reversible magnetization [97]), which indicate that $\lambda(T)$ is well-described by a conventional BCS description below T_c for both κ-$(BEDT-TTF)_2Cu[N(CN)_2]Br$ and κ-$(BEDT-TTF)_2Cu(NCS)_2$. These results contradict a simple d-wave pairing mechanism where nodes in the gap would not allow manifestations of a fully gapped order parameter. However, equally careful NMR experiments [39,98,99] show no Hebel-Slichter peak and a T^3 temperature dependence for the NMR relaxation rate, both indicating nodes in the gap. All other NMR studies are consistent with these results [100, 101]. A summary of NMR results from [39] is shown in Figure 12.17a. The one caveat concerning NMR experiments is that a finite magnetic field is needed to make the measurements in the superconducting state, although care is taken to minimize possible vortex related effects by careful sample alignment (see inset of Figure 12.17a.). Further support for non-s-wave superconductivity is the temperature dependence of the penetration depth for κ-$(BEDT-TTF)_2X$ which exhibits a $\lambda \sim T^{3/2}$ dependence [102] as shown in Figure 12.17b. This further confirmation that a gap with nodes is present.

Superconductivity in Organic Conductors

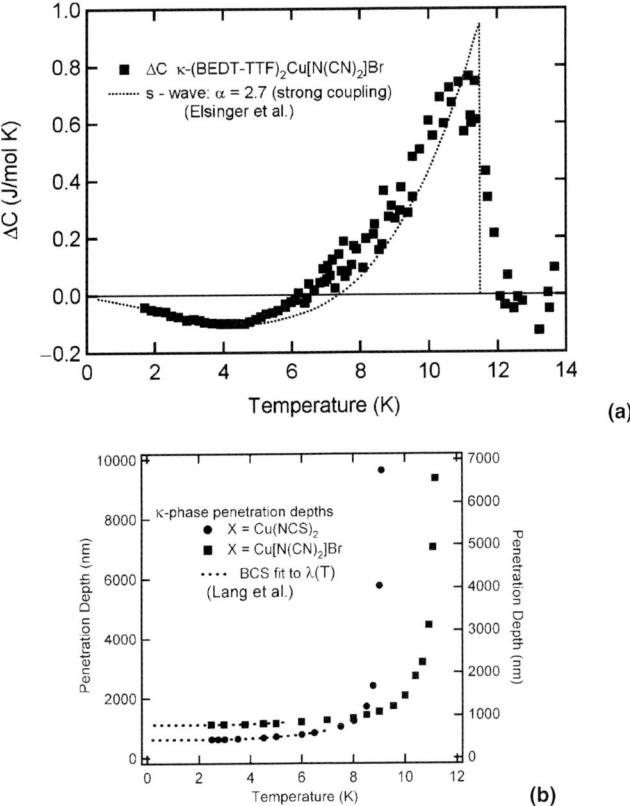

Figure 12.16. Evidence for s-wave superconductivity in κ-(BEDT-TTF)$_2$Cu[N(CN)$_2$]Br. (a) Specific heat (redrawn from Elsinger et al. [44]). Dashed line is the s-wave, strong coupling fit to the data. (b) Temperature penetration depth $\lambda(T)$ for two κ-phase superconductors (re-drawn after Ref. [97]). BCS theory (dashed lines) well describe $\lambda(T)$ for both compounds at low temperature.

One final topic in this section of importance is tunneling. Arai and coworkers [103] performed scanning tunneling microscope (STM) spectroscopy measurements on κ-(BEDT-TTF)$_2$Cu(NCS)$_2$ for $T/T_c \sim 0.15$ normal to the b–c conducting plane, and also along the edges of the sample in the b-c plane. For the former case, they satisfactorily modeled the differential conductance data using an approximate $\Delta = \Delta_0 \cos(2\theta)$ form characteristic of d-wave symmetry where $\Delta_0 = 3$ meV. In the latter case, for the in-plane tunneling at the edges, they observed an angle dependent gap which varied from 5.7 meV in the c-direction to 4 meV near the expected d-wave nodal direction (see Figure 12.18) where they took into account the **k**-dependence of the tunneling matrix. Based on the symmetry of the electronic structure, the STM results are consistent with $d_{x^2-y^2}$ symmetry, where the nodal direction is $\pi/4$ between the k_b and k_c axes.

12.5.3. Magnetic Field Induced Superconductivity and Possible FFLO States

A very important and relatively recent discovery is the observation of a magnetic field induced superconducting state in λ-(BEDT-TSF)$_2$FeCl$_4$ by Uji and coworkers [104]. This novel ground state emerges from the more conventional anisotropic type II superconducting

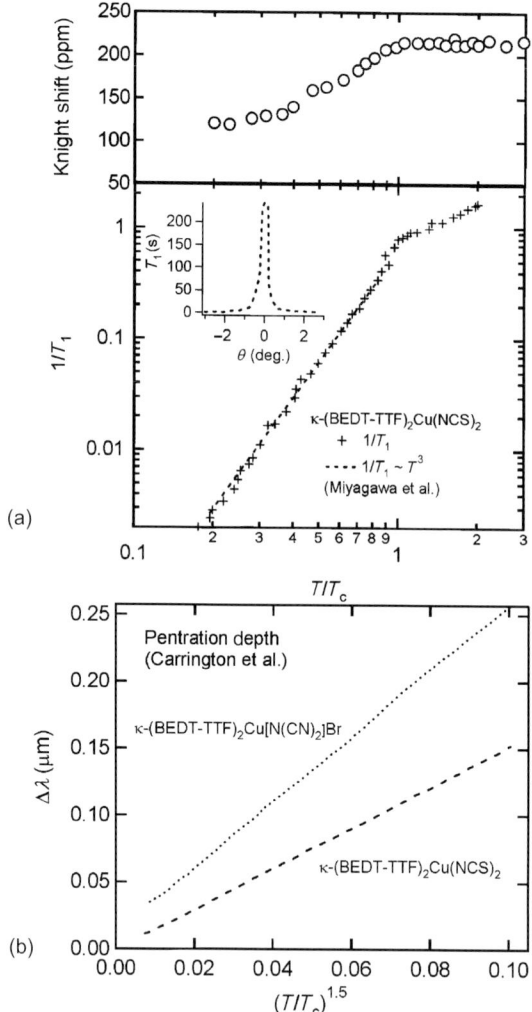

Figure 12.17. Evidence for d-wave paring. (a) NMR data for κ-(BEDT-TTF)$_2$Cu(NCS)$_2$ for $B//a-c$ plane (redrawn from Miyagawa et al. [39]) vs. T/T_c ($T_c \sim 10.4$ K). Main panel: ^{13}C NMR relaxation rate $1/T_1$ exhibits a power law in T below T_c. Inset: angular dependence of ^{1}H NMR relaxation time vs. magnetic field direction at 5 K. $\theta = 0$ represents $B//a$-c plane. Upper panel: Knight shift for ^{13}C NMR data for $\theta = 0$. (b) Penetration depth data for κ-(BEDT-TTF)$_2$Cu[N(CN)$_2$]Br and κ-(BEDT-TTF)$_2$Cu(NCS)$_2$ showing a $T^{3/2}$ dependence on temperature (redrawn after Carrington et al. [102]).

state of the nonmagnetic λ-(BEDT-TSF)$_2$GaCl$_4$ as Fe (with spin 3/2 d electrons) is substituted for nonmagnetic Ga in the alloy sequence λ-(BEDT-TSF)$_2$Fe$_x$Ga$_{1-x}$Cl$_4$ for $0 < x < 1$, as shown in the global phase diagram [105] in Figure 12.19. As the Fe$_x$ concentration increases, the π–d exchange interaction between the donor orbitals and the d-electron spins increases, thereby stabilizing an antiferromagnetic, insulating ground state. In magnetic fields, the AFI state is suppressed, and a paramagnetic metallic state is stabilized. For magnetic fields precisely aligned parallel to the conducting planes, the π–d exchange field J will be cancelled for some value of the external field, and superconductivity can then be stabilized over a range of total *internal* magnetic field ($B = B_{\text{ext}} + |J|$) less than the Pauli limiting field (~ 11 T for

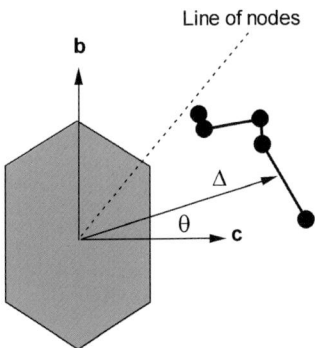

Figure 12.18. Schematic representation of in-plane STM measurements [103] of the superconducting gap in κ-(BEDT-TTF)$_2$Cu(NCS)$_2$ for $T/T_c \sim 0.15$. The gap (solid circles) for $\theta = 0$ is $= 5.7$ meV, and diminishes to 4 meV near the expected d-wave node line.

$x = 0$). By inspection of Figure 12.19, the superconducting phase diagram for $x = 0$ is essentially displaced with increasing x, and for $x = 1$ appears symmetrically for $\sim \pm 14$ T with respect to the external field ($B_{\text{ext}} \sim 32$ T). In Figure 12.20 the critical field phase diagram for λ-(BEDT-TSF)$_2$GaCl$_4$ has been displaced (graphically) by the value exchange field to show the similarity with the field induced superconducting state for λ-(BEDT-TSF)$_2$FeCl$_4$. The exchange field cancellation mechanism, first observed in the Chevrel materials, is due to the Jaccarino–Peter effect [106]. A splitting of the Shubnikov–de Haas oscillation frequencies is observed for magnetic field perpendicular to the conducting planes [107], which is directly proportional to the Fe$_x$ concentration and thereby the exchange field J. More recently, the same behavior has been beautifully demonstrated by Konike et al. [108] in the related compound κ-(BEDT-TSF)$_2$FeBr$_4$.

A closer look at the nature of the H_{c2} phase diagram near the effective Pauli field for the λ-(BEDT-TSF)$_2$Fe$_x$Ga$_{1-x}$Cl$_4$ systems shows an indication of "wings" for $T \to 0$ near or greater than the upper or lower Pauli limiting fields, as indicated in Figure 12.20. Both theoretical [38, 109] and experimental [110, 111] work has suggested the possibility that this behavior is related to the Fulde–Ferrell–Larkin–Ovchinnikov [112, 113] (FFLO) ground state. In the FFLO state, the superconducting order parameter is expected to oscillate with a period λ in real space. This has interesting implications for the pinning of inter-plane Josephson vortices since as a function of in-plane magnetic field, there can be a commensurability relation between the field dependent distribution of the vortices and λ. Here the nodes of the order parameter can provide pinning sites for the vortices. This has been investigated by Uji et al. [114] through current density dependent transport measurements in the Josephson vortex state, where strong evidence for field dependent-commensurate pinning effects are present in the "wings" of the FISC phase diagram. In a second investigation, Tanatar et al. [115] have investigated the thermal conductivity near the upper critical field of λ-(BEDT-TSF)$_2$GaCl$_4$ for samples of different impurity levels, and find that the "wing" structure (see Figure 12.20) is most evident in the more pure samples, which is again consistent with a FFLO ground state.

FFLO-like signatures have also been seen via structure in the susceptibility upon passing through the upper critical field region of κ-(BEDT-TTF)$_2$Cu(SCN)$_2$ for $T < T_c/2$ [116], and in another report of additional structure in the upper critical field via transport measurements by Su et al. [31], also for $T < T_c/2$.

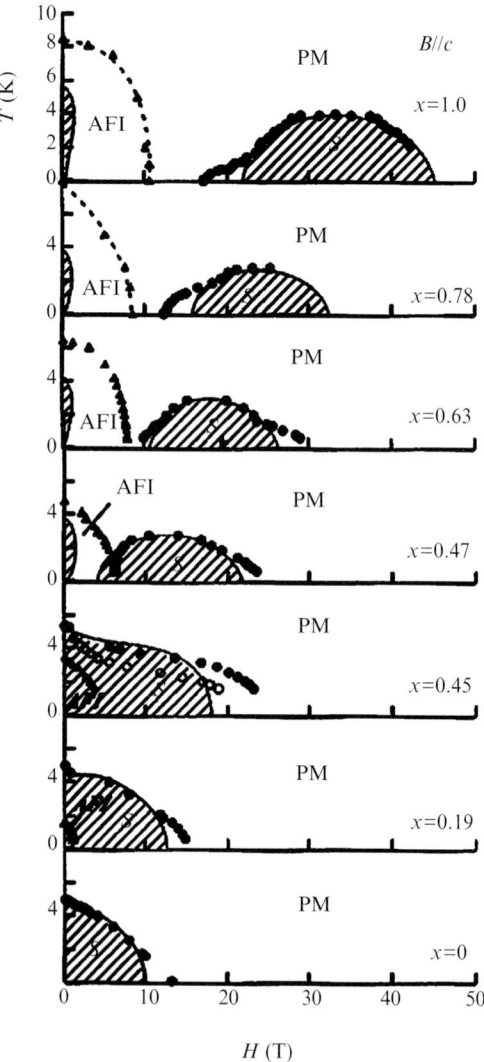

Figure 12.19. Global phase diagram for λ-(BETS)$_2$Fe$_x$Ga$_{1-x}$Cl$_4$ [105] for B//c. For increasing Fe concentration, the field induced superconducting state appears at higher magnetic fields. The shaded areas are the predicted superconducting phases based on the Jaccarino–Peter exchange-field cancellation model [106].

12.6. Comparison of High T_c Superconductors with Organic Conductors

In this section, we summarize the likenesses and differences between organic superconductors and the cuprates, ruthenates, etc.

Similarities: Following McKenzie [79] (and also earlier comparisons [3]), both systems share the following general characteristics:

- They are layered and anisotropic in their physical properties.

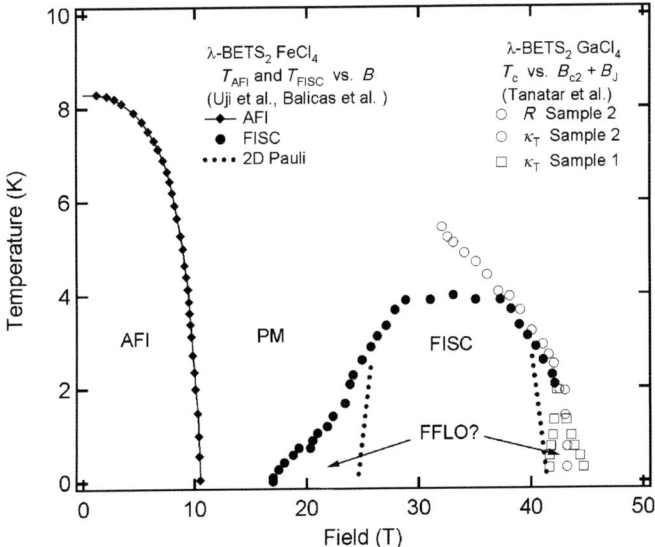

Figure 12.20. Phase diagram for λ-(BEDT-TSF)$_2$FeCl$_4$ [104, 117] and λ-(BEDT-TSF)$_2$GaCl$_4$ [115] displaced graphically by the value of the exchange field. The field is directed parallel to the conducting planes along the c-axis. At low temperatures, "wings" appear in the FISC critical fields (upper and lower) that exceed the Pauli limit, and which may be evidence for the FFLO state [114].

- Antiferromagnetic and superconducting ground states can be in close proximity.
- In the phase diagrams, which bear many strong similarities, pressure (and/or U, V, and t) in organic systems replaces doping in the cuprates.
- Metallic phases can be nonFermi liquid like, $k_F l$ is small, and there is evidence for pseudogap behavior in properties like the optical conductivity [74] and in NMR where $1/T_1$ can be orders of magnitude too large for normal Fermi liquid behavior [39].
- In the Q2D organic systems and the cuprates there is strong evidence for d-wave symmetry.
- From a theoretical standpoint, the ground states of organic systems follow from the extended Mott–Hubbard Hamiltonian, and McKenzie suggested that the formalism of Kino and Fukuyama [78] be extended to include superconductivity. As discussed above in Section 12.5, subsequent work showed that in different cases charge [72] or spin [80] fluctuations could lead to superconductivity with d-wave symmetry.
- That FFLO states may be similar between the BETS systems and the 115 materials [118].
- That comparisons for p-wave symmetry may be made between the Q1D systems and the Ruthenates [119].

Differences. Again, following McKenzie [79], we may again consider the comparison above that *"pressure (and/or U, V, and t) in organic systems replaces doping in the cuprates."* Recently Pratt and Blundell [120] used muon spin resonance/relaxation (μSR) to measure the penetration depth λ of a series of organic conductors. (λ is related to the superfluid stiffness $\rho_s = c^2/\lambda^2$.) A plot of T_c vs. $1/\lambda^2$ for a series of superconductors is known as a "Uemura plot" [121], which yields a scaling relationship for materials of a particular superconducting family. Previously the cuprates had been treated by Uemura et al. [122], and then more

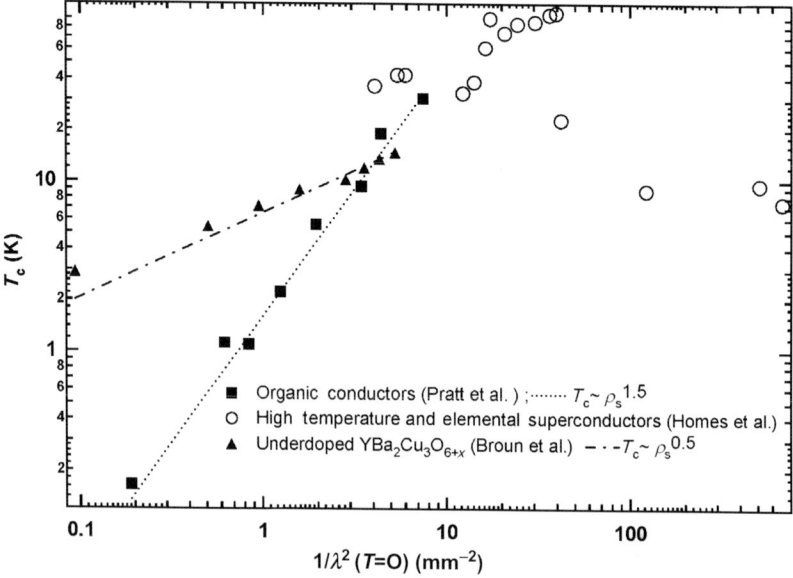

Figure 12.21. The "Uemura plot" of T_c vs. inverse penetration depth $1/\lambda^2$ (or superfluid stiffness ρ_s/c^2) for organic (solid dots) [120], cuprate and elemental superconductors (open circles) [123] (redrawn after [120]), and underdoped YBCO [124].

recently by Homes et al. [123]. In the latter case, a scaling of ρ_s to $\sigma(T_c)T_c$ (the product of T_c with the conductivity at T_c) was found to give a complete scaling of the data. However, for the organic conductors, Pratt and Blundell found that T_c scales as $\rho_s^{3/2}$, and that when plotted on a log–log scale, the two systems (organic vs. cuprates) clearly do not follow the same scaling relationship [120]. The "Uemura plot" for both systems is given in Figure 12.21. A possible argument made to explain the difference is simply that in the organics, the charge transfer (doping if you will) remains fixed, and it is only the details of the molecular structure (stacking motif, unit cell volume, anion and cation species, etc.) that change T_c. In contrast, T_c in the cuprates is generally controlled by the doping. Another obvious difference is that organic conductors are basically low T_c, and in the absence of doping variation other factors and energy scales will be important. However, recently Broun et al. [124] systematically studied a single sample of YBCO as it progressively self doped on the under-doped side of the phase diagram, where T_c is also lower. Here the scaling was found to be $T_c \sim \rho_s^{1/2}$, which is distinctly different from *both* the organic and more optimally doped cuprates.

In the above, it is clear that to make stronger links and comparisons between the two systems, two things are probably necessary (a) figure out how to get to higher T_c in the organic charge transfer systems and break the 15 K barrier shown in Figure 12.6 and (b) figure out how to dope the organic systems to get away from the rigid rules of D_2A charge transfer. Both tasks are formidable.

12.7. Summary and Future Prospects

I think the essence of this chapter, if the reader wants to quickly take away a message about organic conductors, goes something like this.

Q2D. The molecular structures that comprise organic conductors are complex systems, and there are a lot of things going on. However, because of the unique nature of the D_2A charge transfer and the lattice-like stacking of the donor molecules, the Q2D organic materials are essentially Hubbard models with either $1/4$ or $1/2$ band fillings. Starting with the tight binding model and optical data to get the parameters needed to formulate an extended Hubbard model Hamiltonian, theorists can go a long way in developing realistic phase diagrams. It seems a triumph in this respect has been the consideration of the triangular lattice model, and its consequences in spin frustration and in-plane uniaxial pressure induced ground states, including superconductivity. In all of the above, there is a theme that in $1/2$ filled systems AF and SC states are in proximity, and in $1/4$ filled systems CO and SC can be in proximity. The models now can predict transition temperatures, order parameter symmetries, and the nature of the interaction (charge or spin). However, experimentally, careful measurements on Q2D systems fall on both sides of the interaction and symmetry questions: strong coupling (mostly phonon) s-wave (perhaps anisotropic s-wave) represents one camp; and d-wave AF mediated interactions represent the other camp. Hence more experimental work is needed to resolve these issues. Since unconventional parings states are expected to be sensitive to impurities sample quality [125] and also lower temperatures [126] will need to be considered as priorities in future studies to access a fully developed ground state.

Q1D. In the $1/4$ filled Fabre and Bechgaard systems, variations in the inter-chain coupling and 1D instabilities (Peierls, band nesting, etc.), and the expected enhancement of interactions due to the 1D nature make the modeling situation a bit more complex, but there has been progress in using Mott–Hubbard models to describe these systems, as recently discussed by Giamarchi [71]. It is interesting to note that in $(TMTSF)_2PF_6$ the SDW and SC phases coexist at their boundary (see Figure 12.9), whereas in the κ-$(BEDT-TTF)_2X$ systems, the AF insulating and SC phases seem to be more exclusive (see Figure 12.10). In terms of pairing, the work on the Bechgaard salts provides evidence for p-wave pairing with no contradictory experimental results at this time.

Functionalization, Synthesis, and the Future. By synthesizing new materials with supramolecular structures, or hybrid metallic or magnetic characteristics, or modifying the 2:1 charge transfer, these materials can go in entirely new directions distinct from the two basic scenarios described above. An emerging area is to consider biomolecules as components of charge transfer systems, or other novel arrangements of complex and/or functionalized structures. Doping away from the rigid D_2A charge transfer would open a whole new area of comparisons with the cuprates and other perovskite systems. And, going to the other extreme, by synthesizing materials that are less complex, i.e., more one-dimensional, or more two- dimensional, the gap between modeling and experiment could perhaps be closed a bit.

Appendix I. Further Reading in the Area of Organic Conductors

To aid in accessing a deeper (and more accurate!) account of the topics involving organic conductor science, the following references are offered.

In terms of current reviews and proceedings, a superb up-to-date account of the subject is given in the review "Molecular Conductors" edited by Batail in Chemical Reviews, **104** (2004). Here, 32 review papers cover in detail most current aspects of the field in synthesis, physical properties (including techniques), and theory, and many topics not covered in the present chapter including optics, high magnetic fields, magnetic complexes, X-ray studies, etc. are accessible in this comprehensive review. The treatise on Q1D systems by Jerome and Schulz [127] is a classic. The most recent proceedings of the International Symposium on Crystalline Organic Metals, Superconductors and Ferromagnets (ISCOM) also covers most of the topics relevant to molecular materials as well [128]. A slightly older, but very comprehensive review edited by Bernier et al. [129] is also very informative, and covers a wider representation of synthetic organic materials. The Handbook of Organic Conductive Molecules and Polymers, edited by H. Nalwa (Wiley, New York, 1997) is again a very comprehensive account of both crystalline and polymeric materials.

As for books, monographs, and earlier work, one of the most comprehensive books on the subject is by Ishiguro, Yamaji, and Saito [21] which has an advantage in that it was written by an experimentalist, a theorist, and a synthetic chemist. Likewise, Williams and seven coauthors with diverse expertise [29] cover the areas of "synthesis, structure, properties, and theory". In the area of magnetic fields, Wosnitza gives a careful expose of Fermiology [22], as has Singleton [23, 130]. Lang has focused on superconductivity in the Q2D systems [131]. In 1990, Kresin and Little held a conference on Organic Superconductivity that nicely punctuated the field shortly after the discovery of High T_c, where four of the papers discuss comparisons between organic conductors and high T_c oxides [3]. Recently Brooks and Chung have reviewed Q1D systems in high magnetic fields [132]. Further reading on organic superconductors can be found in a recent springer series on super conductivity [133], and Wosnitza has recently completed on updated review [134]. Finally, Powell and Mckenzie have carried out a systematic treatment of electron correlation effects in organic materials, as outlined in a recent review and in a series of previous paper [135].

Acknowledgments

The author and his collaborators would like to acknowledge present and past support from the National Science Foundation (presently NSF DMR 0203532), and also the National High Magnetic Field Laboratory, which is supported by NSF Cooperative Agreement No. DMR-0084173, by the State of Florida, and by the DOE. The author is also grateful to T. Tokumoto for help in preparing some of the figures, and thanks M. Lang, J. Wosntiza, and K. Murata for their important Contribution to particular sections of the manuscripts.

Bibliography

1. P. M. Chaikin. Paul Chaikin often remarks only partially in jest that the organic conductor $(TMTSF)_2PF_6$ exhibits most of the known ground states in condensed matter physics.
2. W. A. Little, Phys. Rev. **134**, A1416 (1964).
3. Z. Kresin and W. A. Little, *Organic Superconductivity* (Plenum, New York, 1990).
4. F. London, *Superfluids* (Wiley, New York, 1950).
5. D. Jerome, A. Mazaud, M. Ribault, et al., J. de Physique Lett. (Paris) **41**, L95 (1980).
6. S. S. P. Parkin, E. M. Engler, R. R. Schumaker, et al., Phys. Rev. Lett. **50**, 270 (1983).
7. K. Yamaji, J. Phys. Soc. Jpn **58**, 1520 (1989).
8. M. V. Kartsovnik, P. A. Kononovich, V. N. Laukhin, et al., Pis'ma Zh. Eksp. Teor. Fiz. **48**, 498 (1988).
9. K. Kajita, Y. Nishio, T. Takahashi, et al., Solid State Commun. **70**, 1181 (1989).
10. T. Osada, A. Kawasumi, S. Kagoshima, et al., Phys. Rev. Lett. **66**, 1525 (1991).
11. M. J. Naughton, O. H. Chung, M. Chaparala, et al., Phys. Rev. Lett. **67**, 3712 (1991).
12. E. Ohmichi, H. Adachi, Y. Mori, et al., Phys. Rev. B **59**, 7263 (1999).
13. N. E. Hussey, M. Abdel-Jawad, A. Carrington, et al., Nature (London) **425**, 814 (2003).

14. N. Thorup, G. Rindorf, H. Soling, et al., Acta Crystallogr. B **37**, 1236 (19811).
15. M. Rahal, D. Chasseau, J. Gaultier, et al., Acta Crystallogr. **B 53**, 159 (1997).
16. U. Geiser, A. M. Kini, H. H. Wang, et al., Acta Crystallogr. C**A 7**, 190 (1991).
17. M. Dressel and N. Drichko, Chem. Rev. **104**, 5689 (2004).
18. H. Mori, N. Sakurai, S. Tanaka, et al., Chem. Mater. **12**, 2984 (2000).
19. P. M. Grant, J. de Physique **44**, C3 (1983).
20. H. Urayama, H. Yamochi, G. Saito, et al., Synth. Met. **27**, A393 (1988).
21. T. Ishiguro, K. Yamaji, and G. Saito, *Organic Superconductors II* (Springer, Berlin, Heidelberg, New York, 1998).
22. J. Wosnitza, *Fermi Surfaces of Low Dimensional Organic Metals and Superconductors* (Springer, Berlin, Heidelberg, New York 1996).
23. J. Singleton, Rep. Prog. Phys. **63**, 1111 (2000).
24. P. Garoche, R. Brusetti, and K. Bechgaard, Phys. Rev. Lett. **49**, 1346 (1982).
25. T. F. Stalcup, J. S. Brooks, and R. Haddon, Phys. Rev. B **60**, 9309 (1999).
26. H. Taniguchi, H. Miyashita, K. Uchiyama, et al., Phys. Soc. Jpn **72**, 468 (2003).
27. A. M. Kini, U. Geiser, H. H. Wang, et al., Inorg. Chem. **29**, 2555 (1990).
28. J. E. Schirber, D. L. Overmyer, K. D. Carlson, et al., Phys. Rev. B **44**, 4666 (1991).
29. J. M. Williams, J. R. Ferraro, R. J. Thorn, et al., *Organic Superconductors (including Fullerenes)* (Prentice-Hall, Englewoods Cliffs, NJ, 1992).
30. $(TMTSF)_2ClO_4$: K. Murata, M. Tokumoto, H. Anzai, et al., Jpn. J. Appl. Phys.Suppl. **26-3**, 1367 (1987); β-ET_2I_3: K. Murata, M. Tokumoto, H. Anzari, et al., J. Phys. Soc. Japan, **54** 1236 (1985), ibid **54** 2084 (1985); M. Tokumoto, H. Anzari, K. Murata, et al., Jpn. J. Appl. Phys., Suppl. **26-3** 1977 (1987).
31. F. Zuo, J. S. Brooks, R. H. McKenzie, et al., Phys. Rev. B **61**, 750 (2000).
32. M.-S. Nam, J. A. Symington, J. Singleton, et al., J. Phys.: Condens. Matter. **11**, L477 (1999).
33. J. Wosnitza, in *Studies of High Temperature Superconductors*, vol.34, by A. V. Narlikar (ed.) (Nova Science, NewYork, 2000). 97
34. P. A. Mansky, P. M. Chaikin, and R. C. Haddon, Phys. Rev. B **50**, 15929 (1994).
35. L. N. Bulaevskii, V. L. Pokrovsky, and M. P. Maley, Phys. Rev. Lett. **76**, 1719 (1996).
36. M. M. Mola, J. T. King, C. P. McRaven, et al., Phys. Rev. B **62**, 5965 (2000).
37. A. G. Lebed and K. Yamaji, Phys. Rev. Lett. **80**, 2967 (1998).
38. M. Houzet, A. Buzdin, L. Bulaevskii, et al., Phys. Rev. Lett. **88**, 227001/4 (2002).
39. K. Miyagawa, K. Kanoda, and A. Kawamoto, Chem. Rev., **104**, 5635 (2004).
40. K. Murata, H. Yoshino, H. O. Yadav, et al., Rev. Sci. Instrum. **68**, 2490 (1997).
41. P. Garoche, R. Brusetti, D. Jerome, et al., J. Physique Lett. **43**, L147 (1982).
42. J. Müller, M. Lang, R. Helfrich, et al., Phys. Rev. B **65**, 140509 (2002).
43. R. Brusetti, P. Garoche, and K. Bechgaard, J. Phys. C: Solid State Phys. **16**, 3535 (1983).
44. H. Elsinger, J. Wosnitza, S. Wanka, et al., Phys. Rev. Lett. **84**, 6098 (2000).
45. D. Jerome, Science **252**, 1509 (1991).
46. K. Bechgaard, K. Carneiro, M. Olsen, et al., Phys. Rev. Lett. **46**, 852 (1981).
47. T. Vuletic, P. Auban-Senzier, C. Pasquier, et al., Eur. Phys. J. B **25**, 319 (2002).
48. A. V. Kornilov, V. M. Pudalov, Y. Kitaoka, et al., Phys. Rev. B **69**, 224404 (2004).
49. I. J. Lee, J. Korean Phys. Soc. **47**, 380 (2005).
50. L. Balicas, K. Behnia, W. Kang, et al., J. Phys. France **4**, 1539 (1994).
51. T. Adachi, E. Ojima, K. Kato, et al., J. Am. Chem. Soc. **122**, 3238 (2000).
52. W. Kang, S. T. Hannahs, and P. M. Chaikin, Phys. Rev. Lett. **70**, 3091 (1993).
53. F. Nad, P. Monceau, T. Nakamura, et al., J. Phys.: Condens. Matter. **17**, L399 (2005).
54. D. Shoenberg, *Magnetic Oscillations in Metals* (Cambridge University Press, Cambridge, 1984).
55. E. B. Yagubskii, N. D. Kushch, A. V. Kazakova, et al., JETP Lett. **82**, 93 (2005).
56. J. Caulfield, W. Lubczynski, F. L. Pratt, et al., J. Phys. Condens. Matter. **6**, 2911 (1994).
57. J. S. Brooks, X. Chen, S. J. Klepper, et al., Phys. Rev. B **52**, 14457 (1995).
58. C. E. Campos, J. S. Brooks, P. J. M. v. Bentum, et al., Phys. Rev. B **52**, R7014 (1995).
59. H. Ito, H. Kaneko, T. Ishiguro, et al., Solid State Commun. **85**, 1005 (1993).
60. D. Andres, M. V. Kartsovnik, W. Biberacher, et al., J. Phys. IV France **12**, 87 (2002).
61. J. M. Williams, A. M. Kini, H. H. Wang, et al., Inorg. Chem. **29**, 3272 (1990).
62. A. Kawamoto, K. Miyagawa, and K. Kanoda, Phys. Rev. B **55**, 14140 (1997).
63. T. Mori, Chem. Rev. **104**, 4947 (2004).
64. H. Seo, C. Hotta, and H. Fukuyama, Chem. Rev. **104**, 5005 (2004).

65. R. Hoffmann, J. Chem. Phys. **39**, 1397 (1963).
66. T. Mori, A. Kobayashi, T. Sasaki, et al., Bull. Phys. Soc. Jpn **57**, 627 (1984).
67. Y. Shimizu, K. Miyagawa, K. Kanoda, et al., Phys. Rev. Lett. **91**, 107001 (2003).
68. Y. Kurosaki, Y. Shimizu, K. Miyagawa, et al., Phys. Rev. Lett. **95**, 177001 (2005).
69. A. Kobayashi, T. Udagawa, H. Tomita, et al., Chem. Lett., 2179 (1993).
70. T. Enoki and A. Miyazaki, Chem. Rev. **104**, 5449 (2004).
71. T. Giamarchi, Chem. Rev. **104**, 5037 (2004).
72. J. Merino and R. H. McKenzie, Phys. Rev. Lett. **87**, 237002 (2001).
73. M. Dressel, N. Drichko, and J. Merino, Phys. B: Condens. Matter. **359–361**, 454 (2005).
74. M. Dressel, N. Drichko, J. Schlueter, et al., Phys. Rev. Lett. **90**, 167002 (2003).
75. N. Tajima, A. Ebina-Tajima, M. Tamura, et al., J. Phys. Soc. Jpn **71**, 1832 (2002).
76. H. Seo, J. Merino, H. Yoshioka, et al., cond-mat/0603008, 21 (2006).
77. H. H. Wang, K. D. Carlson, U. Geiser, et al., Phys. C: Supercond.**166**, 57 (1990).
78. H. Kino and H. Fukuyama, J. Phys. Soc. Jpn **65**, 2158 (1996).
79. R. H. McKenzie, Science **278**, 820 (1997).
80. J. Schmalian, Phys. Rev. Lett. **81**, 4232 (1998).
81. R. H. McKenzie, Comments Condens. Matter Phys. **18**, 309 (1998).
82. S. Kagoshima and R. Kondo, Chem. Rev. **104**, 5593 (2004).
83. E. S. Choi, J. S. Brooks, S. Y. Han, et al., Philos. Mag. B **81**, 399 (2001).
84. T. Mizutani, M. Tokumoto, T. Kinoshita, et al., Synth. Met. **133**, 229 (2003).
85. Y. Shimizu, M. Maesato, G. Saito, et al., Synth. Met. **133**, 225 (2003).
86. N. Tajima, T. Imakubo, R. Kato, et al., J. Phys. Soc. Jpn **72**, 1014 (2003).
87. S. Murakami and N. Nagaosa, J. Phys. Soc. Jpn **69**, 2395 (2000).
88. A. Girlando, M. Masino, A. Brillante, et al., Phys. Rev. B **66**, 100507 (2002).
89. A. Girlando, M. Masino, A. Brillante, et al., in *"New Developments in Superconductivity Research"*, E. R. W. Stevens (ed.) (Nova Science, New York, 2003), p. 15.
90. C. E. Campos, P. S. Sandhu, J. S. Brooks, et al., Phys. Rev. B **53**, 12725 (1996).
91. R. Greene and P. Chaikin, Physica B **126**, 431 (1984).
92. A. G. Lebed, JETP Lett. **43**, 174 (1986).
93. A. G. Lebed and P. Bak, Phys. Rev. Lett. **63**, 1315 (1989).
94. I. J. Lee, S. E. Brown, W. G. Clark, et al., Phys. Rev. Lett. **88**, 017004 (2002).
95. I. J. Lee, M. J. Naughton, G. M. Danner, et al., Phys. Rev. Lett. **78**, 3555 (1997).
96. J. I. Oh and M. J. Naughton, Phys. Rev. Lett. **92**, 067001/4 (2004).
97. M. Lang, N. Toyota, T. Sasaki, et al., Phys. Rev. Lett. **69**, 1443 (1992); M. Lang, N. Toyota, T. Sasaki, et al., Phys. Rev. B **46**, 5822 (1992).
98. K. Kanoda, K. Miyagawa, A. Kawamoto, et al., Phys. Rev. B **54**, 76 (1996).
99. K. Kanoda, Hyperfine Interact. **104**, 235 (1997).
100. H. Mayaffre, P. Wzietek, D. Jérome, et al., Phys. Rev. Lett. **75**, 4122 (1995).
101. S. M. D. Soto, C. P. Slichter, A. M. Kini, et al., Phys. Rev. B **52**, 10364 (1995).
102. A. Carrington, I. J. Bonalde, R. Prozorov, et al., Phys. Rev. Lett. **83**, 4172 (1999).
103. T. Arai, K. Ichimura, K. Nomura, et al., Phys. Rev. B **63**, 104518 (2001).
104. S. Uji, H. Shinagawa, C. Terakura, et al., Nature (London) **410**, 908 (2001).
105. S. Uji, T. Terashima, C. Terakura, et al., J. Phys. Soc. Jpn **72**, 369 (2003).
106. Ø. Fischer, H. W. Meul, M. G. Karkut, et al., Phys. Rev. Lett. **55**, 2972 (1985).
107. S. Uji, C. Terakura, T. Terashima, et al., Phys. Rev. B **65**, 113101 (2002).
108. T. Konoike, S. Uji, T. Terashima, et al., Phys. Rev. B **70**, 094514 (2004).
109. H. Shimahara, J. Phys. Soc. Jpn **71**, 1644 (2002).
110. L. Balicas, V. Barzykin, K. Storr, et al., Phys. Rev. B **70**, 092508 (2004).
111. L. Balicas, J. S. Brooks, K. Storr, et al., in *Physical Phenomena in High Magnetic Fields IV*, Z. Fisk (ed.) (World Scientific, Santa Fe, 2001).
112. P. Fulde and R. A. Ferrell, Phys. Rev. **135**, A550 (1964).
113. A. I. Larkin and Y. N. Ovchinnikov, Sov. Phys. JETP **20**, 762 (1965).
114. S. Uji, T. Terashima, T. Yamaguchi, et al., Phys. Rev. Lett. **97**, 157001 (2006).
115. M. A. Tanatar, T. Ishiguro, H. Tanaka, et al., Phys. Rev. B **66**, 134503 (2002).
116. J. Singleton, J. A. Symington, M.-S. Nam, et al., J. Phys. Condens. Matter. **12**, L641 (2000).
117. L. Balicas, J. S. Brooks, K. Storr, et al., Phys. Rev. Lett. **87**, 067002 (2001).
118. H. A. Radovan, N. A. Fortune, T. P. Murphy, et al., Nature (London) **425**, 51 (2003).

119. Y. Maeno, T. M. Rice, and M. Sigrist, Phys. Today **54**, 42 (2001).
120. F. L. Pratt and S. J. Blundell, Phys. Rev. Lett. **94**, 097006 (2005).
121. Y. J. Uemura, L. P. Le, G. M. Luke, et al., Phys. Rev. Lett. **66**, 2665 (1991).
122. Y. J. Uemura, G. M. Luke, B. J. Sternlieb, et al., Phys. Rev. Lett. **62**, 2317 (1989).
123. C. C. Homes, S. V. Dordevic, M. Strongin, et al., Nature (London) **430**, 539 (2004).
124. D. M. Broun, P. J. Turner, W. A. Huttema, et al., cond-mat/0509223 (2005).
125. M. Pinteric, S. Tomic, M. Prester, et al., Phys. Rev. B **66**, 174521 (2002).
126. K. Izawa, H. Yamaguchi, T. Sasaki, et al., Phys. Rev. Lett. **88**, 027002 (2002).
127. D. Jerome and H. J. Schulz, Adv. Phys. **51**, 293 (2002).
128. ISCOM, J. Low Temp. Phys. **142 3/4** (2006).
129. P. Bernier, S. Lefrant, and G. Bidan, (Elsevier, Amsterdam, 1999).
130. J. Singleton and R. S. Edwards, in *High Magnetic Fields: Science and Technology*, vol. 2 F. Herlach and N. Miura (eds.) (World Scientific, Singapore, 2003).
131. M. Lang, Supercond. Rev. **2**, 1 (1996).
132. J. S. Brooks and O. H. Chung, in *High Magnetic Fields: Science and Technology*, vol. 3 F. Herlach and N. Miura (eds.) (World Scientific, Singapore, 2006).
133. See also in "The physics of Superconductivity", Vol. II, K. H. Bennemann and J.B. Ketterson eds., Springer, (2004) articles by L. Lang and Muller pp 453–554, and M. B. Maple et al., pp. 555–730.
134. J. Wosnitza, J. Low Temp. Phys., in press.
135. B. Powell and McKenzie,Cond-mat/0607078, ibid Cond-mat/0607079, Phys, Rev. Lett. **94**, 047004(2005), J. Phys.; Condens. Matter **16** L367 (2004), Phys. Rev. B **69**, 024519 (2004).

13

Numerical Studies of the 2D Hubbard Model

D. J. Scalapino

Numerical studies of the two-dimensional Hubbard model have shown that it exhibits the basic phenomena seen in the cuprate materials. At half-filling one finds an antiferromagnetic Mott–Hubbard groundstate. When it is doped, a pseudogap appears and at low temperature d-wave pairing and striped states are seen. In addition, there is a delicate balance between these various phases. Here we review evidence for this and then discuss what numerical studies tell us about the structure of the interaction which is responsible for pairing in this model.

13.1. Introduction

A variety of numerical methods have been used to study Hubbard and $t-J$ models and there are a number of excellent reviews [1–8]. The approaches have ranged from Lanczos diagonalization [1, 2, 9–11] of small clusters to density-matrix-renormalization-group studies of n-leg ladders [8, 12–14] and quantum Monte Carlo simulations of two-dimensional lattices [3, 15–24]. In addition, recent cluster generalizations of dynamic mean-field theory [4, 6, 7, 25–33] are providing new insight into the low temperature properties of these models. A significant finding of these numerical studies is that these basic models can exhibit antiferromagnetism, stripes, pseudogap behavior, and $d_{x^2-y^2}$ pairing. In addition, the numerical studies have shown how delicately balanced these models are between nearly degenerate phases. Doping away from half-filling can tip the balance from antiferromagnetism to a striped state in which half-filled domain walls separate π-phase-shifted antiferromagnetic regions. Altering the next-near-neighbor hopping t' or the strength of U can favor $d_{x^2-y^2}$ pairing correlations over stripes. This delicate balance is also seen in the different results obtained using different numerical techniques for the same model. For example, density matrix renormalization group (DMRG) calculations for doped 8-leg $t-J$ ladders find evidence for a striped ground state [12]. However, variational and Green's function Monte Carlo calculations for the doped $t-J$ lattice, pioneered by Sorella and co-workers [23, 24], find groundstates characterized by $d_{x^2-y^2}$ superconducting order with only weak signs of stripes. Similarly, DMRG calculations for doped 6-leg Hubbard ladders [14] find stripes when the ratio of U to the near-neighbor hopping t is greater than 3, while various cluster calculations [27, 30–33] find evidence that antiferromagnetism and $d_{x^2-y^2}$ superconductivity compete in this same parameter regime. These techniques represent present day state-of-the-art numerical approaches. The fact that they can give different results may reflect the influence of different boundary

---

D. J. Scalapino • Department of Physics, University of California, Santa Barbara, CA 93106-9530, USA

conditions or different aspect ratios of the lattices that were studied. The n-leg open boundary conditions in the DMRG calculations can favor stripe formation. Alternately, the cluster lattice sizes and boundary conditions can frustrate stripe formation. It is also possible that these differences reflect subtle numerical biases in the different numerical methods. Nevertheless, these results taken together show that both the striped and the $d_{x^2-y^2}$ superconducting phases are nearly degenerate low energy states of the doped system. Determinantal quantum Monte Carlo calculations [21] as well as various cluster calculations show that the underdoped Hubbard model also exhibits pseudogap phenomena [27–32]. The remarkable similarity of this behavior to the range of phenomena observed in the cuprates provides strong evidence that the Hubbard and t–J models indeed contain a significant amount of the essential physics of the problem [34].

In this chapter, we will focus on the one-band Hubbard model. Section 13.2 provides an overview of the numerical methods which were used to obtain the results that will be discussed. We have selected three methods, determinantal quantum Monte Carlo, the dynamic cluster approximation and the density matrix renormalization group. In principle, these methods provide unbiased approaches which can be extrapolated to the bulk limit or in the case of the DMRG to "infinite length" ladders. This choice has left out many other important techniques such as the zero variance extrapolation of projector Monte Carlo [23, 24], variational cluster approximations [25, 26, 29–32], renormalization group flow techniques [35–37], high temperature series expansions [38–40], and the list undoubtedly goes on. It was motivated by the need to write about methods with which I have direct experience.

In Section 13.3 we review the numerical evidence showing that the Hubbard model can indeed exhibit antiferromagnetic, $d_{x^2-y^2}$ pairing and striped low-lying states as well as pseudogap phenomena. From this we conclude that the Hubbard model provides a basic description of the cuprates, so that the next question is what is the interaction that leads to pairing in this model? From a numerical approach, this question is different from determining whether the groundstate is antiferromagnetic, striped, or superconducting. Here, one would like to understand the structure of the underlying effective interaction that leads to pairing. Although on the surface this might seem like a more difficult question to address numerically, it is in fact easier than determining the nature of the long-range order of the low temperature phase. The phase determination problem involves an extrapolation to an infinite lattice at low, or in the case of antiferromagnetism, to zero temperature. However, the pairing interaction is short ranged and is formed at a temperature above the superconducting transition so that it can be studied on smaller clusters and at higher temperatures.

As discussed in Section 13.4, the effective pairing interaction is given by the irreducible particle–particle vertex Γ^{pp}. Using Monte Carlo results for the one- and two-particle Green's functions, one can determine Γ^{pp} and examine its momentum and Matsubara frequency dependence [41, 42]. One can also determine if it is primarily mediated by a particle–hole magnetic ($S = 1$) exchange, a charge density ($S = 0$) exchange, or by a more complex mechanism. Section 13.5 contains a summary and our conclusions.

13.2. Numerical Techniques

In this chapter we will be reviewing numerical results which have been obtained for the 2D Hubbard model. It would, of course, be simplest if one could say that these results were obtained by "exact" diagonalization on a sequence of $L \times L$ lattices with a "finite-size scaling" analysis used to determine the bulk limit. While one might not know the exact details of how this was done, one understands what it means. Unfortunately the exponential growth of the Hilbert space with lattice size limits this approach and other less familiar and often less transparent methods are required.

In this chapter, we will discuss results obtained using the determinantal quantum Monte Carlo algorithm [16, 43], a dynamic cluster approximation [6, 27], and the density matrix renormalization group [5, 44, 45]. All of these methods have been described in detail in the literature. However, to provide a context for the numerical results discussed in Sections 13.3 and 13.4, we proceed with a brief overview of these techniques.

13.2.1. Determinantal Quantum Monte Carlo

The determinantal quantum Monte Carlo method was introduced in order to numerically calculate finite temperature expectation values:

$$\langle A \rangle = \frac{\mathrm{Tr}\, e^{-\beta H} A}{Z}. \tag{13.1}$$

Here, H includes $-\mu N$ so that $Z = \mathrm{Tr}\, e^{-\beta H}$ is the grand partition function.

For the Hubbard model, the Hamiltonian is separated into a one-body term

$$K = -t \sum_{\langle ij \rangle \sigma} \left(c_{i\sigma}^\dagger c_{j\sigma} + c_{j\sigma}^\dagger c_{i\sigma} \right) - \mu \sum_{i\sigma} n_{i\sigma} \tag{13.2}$$

and an interaction term

$$V = U \sum_i \left(n_{i\uparrow} - \frac{1}{2} \right) \left(n_{i\downarrow} - \frac{1}{2} \right). \tag{13.3}$$

Then, dividing the imaginary time interval $(0, \beta)$ into M segments of width $\Delta\tau$, we have

$$e^{-\beta H} = \left[e^{-\Delta\tau (K+V)} \right]^M \simeq \left[e^{-\Delta\tau K} e^{-\Delta\tau V} \right]^M. \tag{13.4}$$

In the last term, a Trotter breakup has been used to separate the noncommuting operators K and V. This leads to errors of order $tU\Delta\tau^2$ which can be systematically treated by reducing the size of the time slice $\Delta\tau$. Then, on each $\tau_\ell = \ell\Delta\tau$ slice and for each lattice site i, a discrete Hubbard–Stratonovich field [46] $S_i(\tau_\ell) = \pm 1$ is introduced so that the interaction can be written as a one-body term

$$e^{-\Delta\tau U \left(n_{i\uparrow} - \frac{1}{2} \right)\left(n_{i\downarrow} - \frac{1}{2} \right)} = \frac{e^{-\frac{\Delta\tau U}{4}}}{2} \sum_{S_i(\tau_\ell)=\pm 1} e^{-\Delta\tau \lambda S_i(\tau_\ell)\left(n_{i\uparrow} - n_{i\downarrow} \right)} \tag{13.5}$$

with λ set by $\cosh(\Delta\tau\lambda) = \exp(\Delta\tau \frac{U}{2})$. The interacting electron problem has now been replaced by the problem of many electrons coupled to a τ-dependent Hubbard–Stratonovich Ising field $S_i(\tau_\ell)$ which is to be averaged over. This average will be done by Monte Carlo importance sampling.

For an $L \times L$ lattice, it is useful to introduce a one-body $L^2 \times L^2$ lattice Hamiltonian $h_\sigma(S(\tau_\ell))$ for spin σ electrons interacting with the Hubbard–Stratonovich field on the

τ_ℓ-imaginary time slice

$$\sum_{\langle ij \rangle} c_{i\sigma}^\dagger h_\sigma(S(\tau_\ell)) c_{j\sigma} = -t \sum_{\langle ij \rangle} \left(c_{i\sigma}^\dagger c_{j\sigma} + c_{j\sigma}^\dagger c_{i\sigma} \right)$$
$$- \mu \sum_i n_{i\sigma} \pm \lambda \sum_i S_i(\tau_\ell) n_{i\sigma} . \tag{13.6}$$

The plus sign is for spin-up ($\sigma = 1$) and the minus sign is for spin-down ($\sigma = -1$). Then, tracing out the fermion degrees of freedom, one obtains

$$Z = \sum_{\{S\}} \det M_\uparrow(S) \det M_\downarrow(S) . \tag{13.7}$$

The sum is over all configurations of the $S_i(\tau_\ell)$ field and $M_\sigma(S)$ is an $L^2 \times L^2$ matrix which depends upon this field

$$M_\sigma(S) = I + B_M^\sigma B_{M-1}^\sigma \cdots B_1^\sigma . \tag{13.8}$$

I is the unit $L^2 \times L^2$ matrix and $B_\ell^\sigma = e^{-\Delta\tau h_\sigma(S(\tau_\ell))}$ acts as an imaginary time propagator which evolves a state from $(\ell - 1)\Delta\tau$ to $\ell\Delta\tau$.

The expectation value $\langle A \rangle$ becomes

$$\langle A \rangle = \sum_{\{S\}} A(S) \frac{\det M_\uparrow(S) \det M_\downarrow(S)}{Z} \tag{13.9}$$

with $A(S)$ the estimator for the operation A which depends only upon the Hubbard–Stratonovich field. Typically, we are interested in Green's functions. For example, the estimator for the one-electron Green's function is

$$G_{ij\sigma}(\tau_\ell, (S)) = \frac{1}{1 + B_M^\sigma B_{M-1}^\sigma \cdots B_1^\sigma} B_\ell^\sigma B_{\ell-1}^\sigma \cdots B_1^\sigma \tag{13.10}$$

and

$$G_{ij\sigma}(\tau_\ell) = \sum_{\{S\}} G_{ij}(T_\ell, (S)) \frac{\det M_\uparrow(S) \det M_\downarrow(S)}{Z} . \tag{13.11}$$

The calculations of various susceptibilities and multiparticle Green's functions are straightforward since once the Hubbard–Stratonovich transformation is introduced, one has a Wick theorem for the fermion operators. The only thing that one needs to remember is that disconnected diagrams must be retained because they can become connected by the subsequent average over the $S_i(\tau_\ell)$ field.

In computing the product of the B matrices, one must be careful to control round-off errors as the number of products becomes large at low temperatures or at large U where $\Delta\tau$ must be small. In addition, there can be problems inverting the ill-conditioned sum of the unit matrix I and the product of the B matrices needed in Eqs. (13.8) and (13.10). Fortunately, a matrix stabilization procedure [16] overcomes these difficulties.

For the half-filled case ($\mu = 0$), provided there is only a near-neighbor hopping, H is invariant under the particle–hole transformation $c_{i\downarrow} = (-1)^i c_{i\downarrow}^\dagger$. Under this transformation, the last term in Eq. (13.6) for $\sigma = -1$ becomes

$$-\lambda \sum_i S_i(\tau_\ell)(1 - n_{i\downarrow}) \tag{13.12}$$

so that

$$\det M_\downarrow(S) = \prod_{i\ell} e^{\lambda \Delta \tau S_i(\tau_\ell)} \det M_\uparrow(S). \tag{13.13}$$

This means that $\det M_\uparrow(S) \det M_\downarrow(S)$ is positive definite. In this case, the sum over the Hubbard–Stratonovich configurations can be done by Monte Carlo importance sampling with the probability of a particular configuration $\{S(\tau_\ell)\}$ given by

$$P(S) = \frac{\det M_\uparrow(S) \det M_\downarrow(S)}{Z}. \tag{13.14}$$

Given M-independent configurations, selected according to the probability distribution Eq. (13.14), the expectation value of A is

$$\langle A \rangle = \frac{1}{M} \sum_{\{S\}} A(S). \tag{13.15}$$

When the system is doped away from half-filling, the product $\det M_\uparrow(S) \det M_\downarrow(S)$ can become negative. This is the so-called "fermion sign problem." In this case, one must use the absolute value of the product of determinants to have a positive definite probability distribution for the Hubbard–Stratonovich configurations:

$$P_\|(S) = \frac{|\det M_\uparrow(S) \det M_\downarrow(S)|}{\sum_{\{S\}} |\det M_\uparrow(S) \det M_\downarrow(S)|}. \tag{13.16}$$

Then, in order to obtain the correct results for physical observables, one must include the sign of the product of determinants [47]

$$s = \mathrm{Sgn}(\det M_\uparrow(S) \det M_\downarrow(S)) \tag{13.17}$$

in the measurement

$$\langle A \rangle = \frac{\langle As \rangle_\|}{\langle s \rangle_\|}. \tag{13.18}$$

The $\|$ subscript denotes that the average is over configurations generated with the probability distribution $P_\|$ given by Eq. (13.16). If the average sign $\langle s \rangle_\|$ becomes small, there will be large statistical fluctuations in the Monte Carlo results. For example, if $\langle s \rangle_\| = 0.1$, one would have to sample of the order of 10^2 times as many independent configurations in order to obtain the same statistical error as when $\langle s \rangle_\| = 1$. On general grounds, one expects that $\langle s \rangle_\|$ decreases exponentially as the temperature is lowered.

The average sign $\langle s \rangle_\|$ also decreases as U increases and makes it (exponentially) difficult to obtain results at low temperatures for U of order the bandwidth and dopings of interest. Figure 13.1 illustrates just how serious this problem is and why other methods are required.

13.2.2. The Dynamic Cluster Approximation

In the determinantal quantum Monte Carlo approach, one could imagine carrying out calculations on a set of $L \times L$ lattices and then scaling to the bulk thermodynamic limit. The dynamic cluster approximation [6] takes a different approach in which the bulk lattice is replaced by an effective cluster problem embedded in an external bath designed to represent the remaining degrees of freedom. In contrast to numerical studies of finite-sized systems in

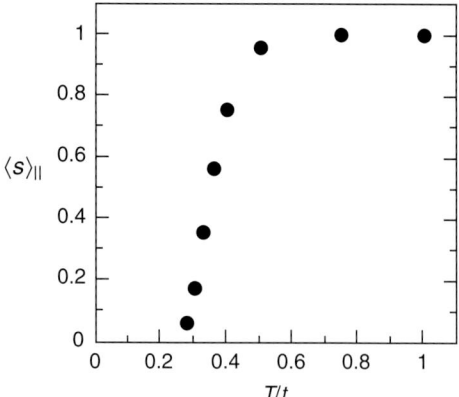

Figure 13.1. The average of the sign of the product of the fermion determinants, Eq. (13.17), that enters in the determinantal Monte Carlo calculations is shown vs. temperature for an 8×8 lattice with $U = 8t$ and $\langle n \rangle = 0.87$. One can understand why calculations for $U = 8t$ with $T < 0.3$ become extremely difficult (Scalapino [19]).

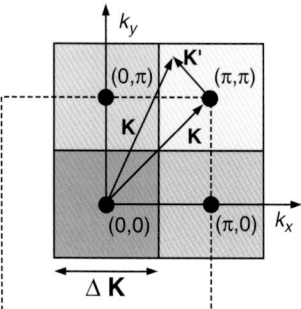

Figure 13.2. In the dynamic cluster approximation the Brillouin zone is divided into N_c cells each represented by a cluster momentum K. Then the self-energy and 4-point vertices are calculated on the cluster using an action determined by the inverse of the coarse-grained cluster-excluded propagator $\mathcal{G}^{-1}(\mathbf{K}, \omega_n)$, Eq. (13.21). This figure illustrates this coarse graining of the Brillouin for $N_c = 4$ (Maier et al. [6]).

which the exact state of an $L \times L$ lattice is determined and then regarded as an approximation to the bulk thermodynamic result, the cluster theories give approximate results for the bulk thermodynamic limit. Then, as the number of cluster sites increases, the bulk thermodynamic result is approached.

In the dynamic cluster approximation, the Brillouin zone is divided into $N_c = L^2$ cells of size $(2\pi/L)^2$. As illustrated in Figure 13.2, each cell is represented by a cluster momentum **K** placed at the center of the cell. Then the self-energy $\Sigma(\mathbf{k}, \omega_n)$ is approximated by a coarse grained self-energy

$$\Sigma(\mathbf{K} + \mathbf{k}', w_n) \simeq \Sigma_c(\mathbf{K}, \omega_n) \tag{13.19}$$

for each $\mathbf{k}'$ within the **K**th cell. The coarse grained Green's function is given by

$$\bar{G}(\mathbf{K}, \omega_n) \cong \frac{N_c}{N} \sum_{\mathbf{k}'} \frac{1}{i\omega_n - (\varepsilon_{\mathbf{K}+\mathbf{k}'} - \mu) - \Sigma_c(\mathbf{K}, \omega)}, \tag{13.20}$$

where the lattice self-energy is replaced by the coarse grained self-energy. Given $\bar{G}$ and Σ_c, one can set up a quantum Monte Carlo algorithm [6, 48] to calculate the cluster Green's

Numerical Studies of the 2D Hubbard Model

function. Here, the bulk lattice properties are encoded by using the cluster-excluded inverse Green's function

$$\mathcal{G}^{-1}(\mathbf{K}, \omega_n) = \bar{G}^{-1}(\mathbf{K}, \omega_n) + \Sigma_c(\mathbf{K}, \omega_n) \tag{13.21}$$

to set up the bilinear part of the cluster action. In Eq. (13.21), the cluster self-energy has been removed from $\mathcal{G}$ to avoid double counting.

Then, the interaction on the cluster

$$\frac{U}{N_c} \sum_{\mathbf{K},\mathbf{K}',\mathbf{Q}} c^\dagger_{\mathbf{K}+\mathbf{Q}\uparrow} c_{\mathbf{K}\uparrow} c^\dagger_{\mathbf{K}'-\mathbf{Q}\downarrow} c_{\mathbf{K}'\downarrow} \tag{13.22}$$

is linearized by introducing a discrete τ-dependent Hubbard–Stratonovich field on each τ-slice and for each $\mathbf{K}$ point. In this way, the cluster problem is transformed into a problem of noninteracting electrons coupled to τ-dependent Hubbard–Stratonovich fields. Integrating out the bilinear fermion field and using importance sampling to sum over the Hubbard–Stratonovich fields one evaluates the cluster Green's function $G_c(\mathbf{K}, \omega_n)$. From this, one evaluates the cluster self-energy

$$\Sigma_c(\mathbf{K}, \omega_n) = \mathcal{G}^{-1}(\mathbf{K}, \omega_n) - G_c^{-1}(\mathbf{K}, \omega_n). \tag{13.23}$$

Then, using this new result for $\Sigma_c(\mathbf{K}, \omega_n)$ in Eq. (13.20), these steps are iterated to convergence.

Measurements of correlation functions and the 4-point vertex are made in the same manner as for the determinantal Monte Carlo. That is, after the Hubbard–Stratonovich field has been introduced, one has a Wick's theorem for decomposing products of various time-ordered operators. However, in this case there is an additional coarse-graining of the Green's function intermediate state legs [6, 42]. Since one is using a determinantal Monte Carlo method, there is also a sign problem for the doped Hubbard model. However, the self-consistent bath and the coarse-graining of the momentum space significantly reduce this problem so that lower temperatures can be reached [49].

13.2.3. The Density Matrix Renormalization Group

The density-matrix-renormalization-group method was introduced by White [44, 45]. Here, as illustrated in Figure 13.3 for a one-dimensional chain of length L, the system under study is separated into four pieces. A block B_ℓ containing $\ell = L/2 - 1$ sites on the left, a reflected $B^R_{\ell'}$ (right interchanged with left) block containing $\ell' = L/2 - 1$ sites on the right, and two additional sites in the middle. The left-hand block B_ℓ and its near-neighbor site are taken to be the system to be studied, while the block $B^R_{\ell'}$ plus its adjacent site are considered to be the "environment." The entire system is diagonalized using a Lanczos or Davidson algorithm to

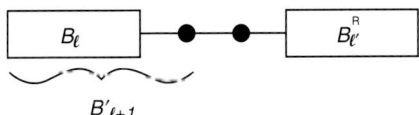

Figure 13.3. The configuration of blocks used for the density matrix renormalization group algorithm (White [45]).

obtain the ground state eigenvalue and eigenvector ψ_o. Then, one constructs a reduced density matrix from ψ_o

$$\rho_{ii'} = \sum_j \psi^*_{oij} \psi_{oi'j} . \qquad (13.24)$$

Here, $\psi_{oij} = \langle i | \langle j | \psi_o \rangle$ with $|i\rangle$ a basis state of the $\ell + 1$ block and $|j\rangle$ a basis state of the $\ell' + 1$ "environment" block. Then the reduced density matrix $\rho_{ii'}$ is diagonalized and m eigenvectors, corresponding to the m largest eigenvalues are kept. The Hamiltonian $H_{\ell+1}$ for the left-hand block and its added site $B'_{\ell+1}$ is now transformed to a reduced basis consisting of the m leading eigenstates of $\rho_{ii'}$. The right-hand environment block $H^R_{\ell'+1}$ is chosen to be a reflection of the system block including the added site. Finally, a superblock of size $L + 2$ is formed using $H_{\ell+1}$, $H^R_{\ell'+1}$, and two new single sites. Open boundary conditions are used. Typically, several hundred eigenstates of the reduced density matrix are kept and thus, although the system has grown by two sites at each iteration, the number of total states remains fixed and one is able to continue to diagonalize the superblock.

This infinite system method suffers because the groundstate wave function ψ_o continues to change as the lattice size increases. This can lead to convergence problems. Therefore, in practice, a related algorithm in which the length L is fixed has been developed. In this case, instead of trying to converge to the infinite system fixed point under iteration, one has a variational convergence to the ground state of a finite system. This finite chain algorithm is similar to the one we have discussed but in this case the total length L is kept fixed and the separation point between the system and the environment is moved back and forth until convergence is achieved [5, 45]. Following this, one can consider scaling L to infinity.

In a sense, the density-matrix-renormalization-group method is a cluster theory. It embeds a numerical renormalization procedure in a larger lattice in which an exact diagonalization is carried out. The division of the chain into the system and the environment is similar in spirit to the embedded cluster and $\mathcal{G}^{-1}$. The use of the reduced density matrix, corresponding to the groundstate, to carry out the basis truncation provides an optimal focus on the low-lying states.

An important aspect of this approach is how rapidly the eigenvalues of the reduced density matrix fall off. This determines how many m states one needs to obtain accurate results. Unfortunately, for the study of n-leg Hubbard ladders, this fall-off becomes significantly slower as n increases and many more states must be kept. In addition, it appears that the behavior of the pairfield–pairfield correlation function is particularly sensitive to the number of states m that are kept. A measure of the errors associated with the truncation in the number m of density matrix eigenstates that are kept is given by the discarded weight

$$W_m = \sum_{i=m+1}^{D} w_i . \qquad (13.25)$$

Here, D is the dimension of the density matrix and w_i is its ith eigenvalue. A useful approach is to increase m and extrapolate quantities to their values as $W_m \to 0$. The error in the groundstate eigenvalue varies as W_m and a typical extrapolation is shown in Figure 13.4. For observables which do not commute with H, the errors vary as $\sqrt{W_m}$.

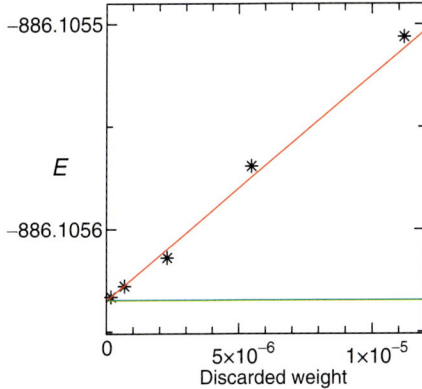

Figure 13.4. DMRG results for the ground state energy of a 2000-site Heisenberg spin-one-half chain vs. the discarded weight W_m. The exact Bethe ansatz energy is shown as the line at the bottom of the figure (S.R. White).

13.3. Properties of the 2D Hubbard Model

As we have discussed, the particle–hole symmetry of the half-filled Hubbard model with a near-neighbor hopping t leads to an absence of the fermion sign problem. In this case, the determinantal Monte Carlo algorithm [43] provides a powerful numerical tool for studying the low temperature properties of this model. In a seminal paper, Hirsch [15] presented numerical evidence that the groundstate of the half-filled two-dimensional Hubbard model with a near-neighbor hopping t and a repulsive on-site interaction $U > 0$ had long-range antiferromagnetic order. In this work, simulations on a set of $L \times L$ lattices were carried out. For each lattice, simulations were run at successively lower temperatures and extrapolated to the $T = 0$ limit. Then, a finite-size scaling analysis was used to extrapolate to the bulk $T = 0$ limit.

This work sets the standard for what one would like to do in numerical studies of the Hubbard model. Unfortunately, the fermion sign problem prevents one from carrying out a similar determinantal Monte Carlo analysis for the doped case. However, various other methods have been developed which provide information on the doped Hubbard model. Here, we will discuss results obtained from a dynamic cluster Monte Carlo algorithm [6]. This method also provides a systematic approach to the bulk limit as the cluster size increases. As noted in Section 13.2, the dynamic cluster Monte Carlo still suffers from the fermion sign problem, although to much less of a degree than the standard determinantal Monte Carlo. Maier et al. [27] have made the important step of studying the doped system on a sequence of different-sized clusters ranging up to 26 sites in size. Furthermore, this work recognized the importance of cluster geometry and developed a Betts'-like [50] grading scheme for determining which cluster geometries are the most useful in determining the finite-size scaling extrapolation.

Following this, we review a density-matrix-renormalization-group (DMRG) study [14] of a doped 6 leg Hubbard ladder in which the authors extrapolate their results to the limit of zero discarded weight and to legs of infinite length. This work provides evidence that static stripes exist in the ground state for large values of U ($U \simeq 12t$) but are absent at weaker coupling ($U \lesssim 3t$). We conclude this section with a discussion of the pseudogap behavior which is observed in the lightly doped Hubbard model when U is of the order of the bandwidth.

13.3.1. The Antiferromagnetic Phase

Determinantal quantum Monte Carlo results for the equal-time magnetization–magnetization correlation function

$$C(\ell) = \langle m^z_{i+\ell} m^z_i \rangle \qquad (13.26)$$

with $m^z_i = (n_{i\uparrow} - n_{i\downarrow})$ are plotted in Figure 13.5. These results are for a half-filled Hubbard model on a 10×10 lattice at a temperature $T = 0.1t$ with $U = 4t$. At this temperature, the antiferromagnetic correlation length exceeds the lattice size and the cluster is essentially in its groundstate. Strong antiferromagnetic correlations are clearly visible in $C(\ell)$.

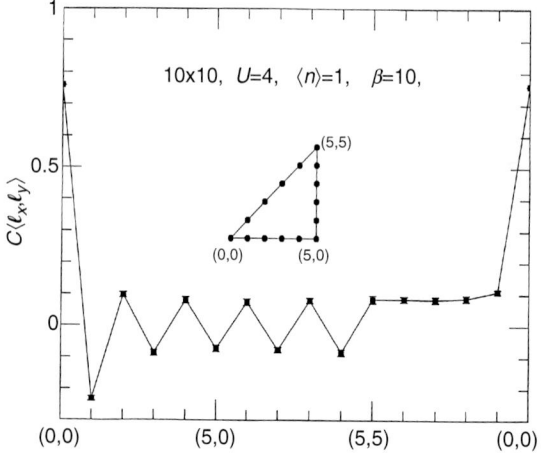

Figure 13.5. The equal-time magnetization–magnetization correlation function $C(\ell_x, \ell_y)$ on a 10×10 lattice with $U = 4t$, $\langle n \rangle = 1$ and $T = 0.1t$. The horizontal axis traces out the triangular path on the lattice shown in the inset. Strong antiferromagnetic correlations are seen (Hirsch [15], White et al. [16]).

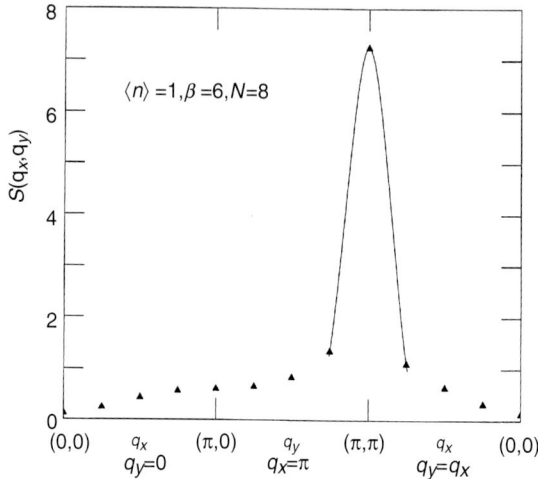

Figure 13.6. $S(q_x, q_y)$ vs. q_x, q_y for $\langle n \rangle = 1$, $U = 4t$ and $T = 0.167t$. The solid line is a fit to guide the eye (Hirsch [15], Moreo et al. [17]).

The magnetic structure factor, shown in Figure 13.6,

$$S(q) = \frac{1}{N} \sum_\ell e^{-i\mathbf{q}\cdot\ell} \langle m^z_{i+\ell} m_i \rangle \tag{13.27}$$

has a peak at $q = (\pi, \pi)$ reflecting the antiferromagnetic correlations. As shown in Figure 13.7, as the temperature is lowered $S(\pi, \pi)$ grows and then saturates when the antiferromagnetic correlations extend across the lattice. If there is long-range antiferromagnetic order in the groundstate, the saturated value of $S(\pi, \pi)$ will scale with the number of lattice sites $N = L \times L$. Furthermore, based upon spin-wave fluctuations [51], one expects that the leading correction will vary as $N^{\frac{1}{2}}$ so that

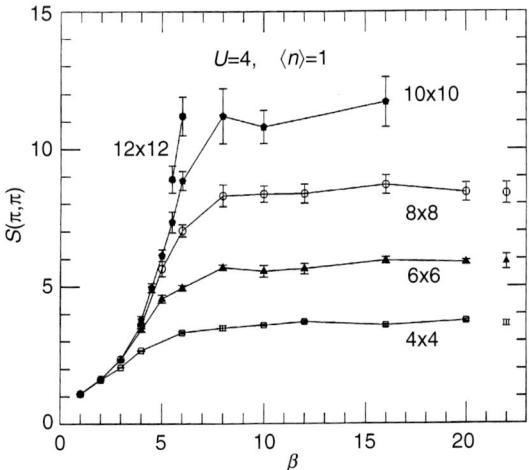

Figure 13.7. The antiferromagnetic structure factor $S(\pi, \pi)$ for $\langle n \rangle = 1$ and $U = 4$ as a function of the inverse temperature β for various lattice sizes. $S(\pi, \pi)$ saturates when the coherence length exceeds the lattice size (Hirsch [15], White et al. [16]).

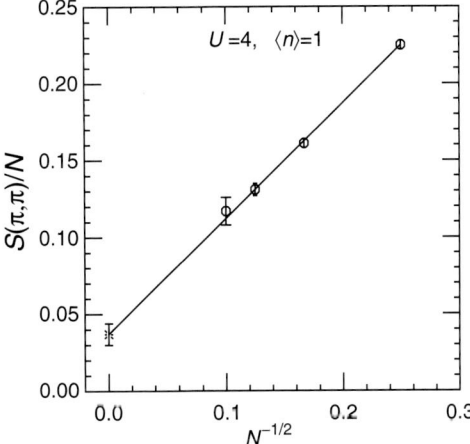

Figure 13.8. The zero-temperature limit of $S(\pi, \pi)/N$ vs. $1/N^{1/2}$. The results extrapolate to a finite value as $N \to \infty$ implying that there is long-range antiferromagnetic order in the groundstate of the infinite lattice (Hirsch [15], White et al. [16]).

$$\lim_{N\to\infty} \frac{S(\pi,\pi)}{N} = \frac{\langle m_x \rangle^2}{3} + \frac{A}{N^{\frac{1}{2}}}. \tag{13.28}$$

Figure 13.8 shows $S(\pi,\pi)/N$ versus $N^{-\frac{1}{2}}$ for $U = 4t$ and one sees that the groundstate has long-range antiferromagnetic order. In his original paper, Hirsch [15] concluded that the groundstate of the half-filled 2D Hubbard model with a near-neighbor hopping t would have long-range antiferromagnetic order for $U > 0$.

13.3.2. $d_{x^2-y^2}$ Pairing

The structure of the pairing correlations in the doped 2D Hubbard model was initially studied using the determinantal Monte Carlo method. The d-wave pairfield susceptibility

$$P_d = \int_0^\beta d\tau \, \langle \Delta_d(\tau) \Delta_d^\dagger(0) \rangle \tag{13.29}$$

with

$$\Delta_d^\dagger = \frac{1}{2\sqrt{N}} \sum_{\ell,\delta} (-1)^\delta c_{\ell\uparrow}^\dagger c_{\ell+\delta\downarrow}^\dagger \tag{13.30}$$

was calculated. Here δ sums over the four near-neighbor sites of ℓ and $(-1)^\delta$ gives the $+-+-$ sign alteration characteristic of d-wave pairing. The doped Hubbard model has a fermion sign problem, so that the Hubbard–Stratonovich fields must be generated according to the probability distribution $P_\parallel(S)$ given by Eq. (13.16).

In this case, it is essential to include the sign factor s in the evaluation of observables. The red circles in Figure 13.9 show results [47] for $P_d(T)$ obtained on a 4×4 lattice with $\langle n \rangle = 0.875$ and $U = 4t$. If the sign s is not included, one obtains the (blue) squares. The neglect of this sign in early work [52] left the false impression that the Hubbard model did not support $d_{x^2-y^2}$ pairing.

As seen, when the sign is included, the d-wave pairfield susceptibility increases as the temperature is lowered. However, over the temperature range accessible to the determinantal

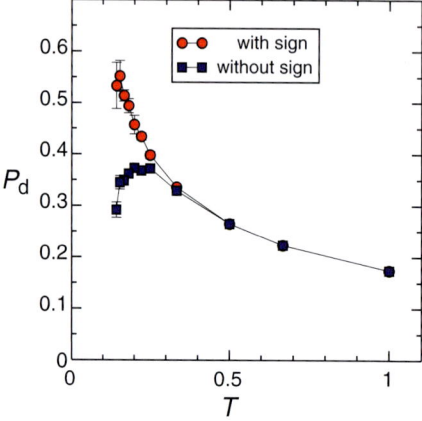

Figure 13.9. The d-wave pairfield susceptibility $P_d(T)$ (red circles) for a 4×4 lattice with $U = 4t$ and $\langle n \rangle = 0.875$ vs. temperature T measured in units of the hopping t. The blue squares show the erroneous result that is found if the fermion sign is ignored (Loh et al. [47]).

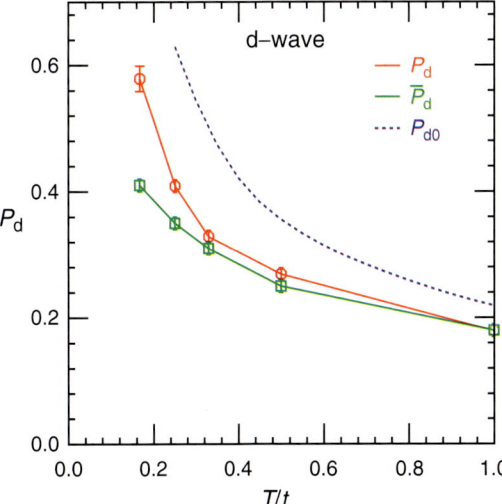

Figure 13.10. The d-wave pair-field susceptibility $P_d(T)$ is shown as the open (red) circles. The open (green) squares show results for the "noninteracting" pair-field susceptibility $\overline{P}_d(T)$ calculated using dressed single-particle Green's functions, Eq. (13.31), while the dashed (blue) curve is the noninteracting susceptibility P_{d0} calculated with the bare Green's functions. (White et al. [16])

Monte Carlo, it remains smaller than the $U = 0$ result P_{d0}, shown as the (blue) dashed line in Figure 13.10. In [16], it was argued that this behavior was due to the renormalization of the single particle spectral weight and that the significant feature to note was that $P_d(T)$ was enhanced over

$$\overline{P}_d(T) = \frac{T}{N} \sum_{pn} G(p) G(-p) (\cos p_x - \cos p_y)^2. \quad (13.31)$$

Here, $G(p)$ is the dressed single particle Green's function determined from the Monte Carlo simulation and $\overline{P}_d$ corresponds to the contribution of a pair of dressed but noninteracting holes. The fact that $\overline{P}_d(T)$ lays below $P_d(T)$ implies that there is an attractive $d_{x^2-y^2}$-pairing interaction between the holes. $\overline{P}_d(T)$ is shown as the (green) curve labeled with open squares in Figure 13.10.

In order to determine what happens at lower temperatures, Maier et al. [27] have determined $P_d(T)$ using a dynamic cluster approximation. In a systematic study, they provided evidence that the doped Hubbard model contained a $d_{x^2-y^2}$ pairing phase. In this work, the authors adapted a cluster selection criteria originally introduced by Betts et al. [50] in a numerical study of the 2D Heisenberg model. For the Heisenberg model, Betts et al. [50] showed that an important selection criteria for a cluster was the completeness of the "allowed neighbor shells" compared to an infinite lattice. They found that a finite-size scaling analysis was greatly improved when clusters with the most complete shells were selected. For a d-wave order parameter, Maier et al. noted that one needs to take into account the nonlocal 4-site plaquette structure of the order parameter in applying this criteria. As illustrated in Figure 13.11, the 4-site cluster encloses just one d-wave plaquette.

Denoting the number of independent near-neighbor plaquettes on a given cluster by Z_d, the 4-site cluster has no near-neighbors so that $Z_d = 0$. In this case the embedding action does not contain any pair field fluctuations and hence T_c is over estimated.

Alternatively, the 8A cluster has space for one more 4-site plaquette ($Z_d = 1$) and the same neighboring plaquette is adjacent to its partner on all four sides. In this case the phase fluctuations are over estimated and T_c is suppressed. For the 16B cluster, one has $Z_d = 2$ while $Z_d = 3$ for the oblique 16A cluster. Thus, one expects that the pairing correlations for the 16B cluster will be suppressed relative to those for the 16A cluster. The number of independent neighboring d-wave plaquettes Z_d for the clusters shown in Figure 13.11 is listed in Table 13.1.

Results for the inverse of the pair field susceptibility vs. T for $U = 4t$ and $\langle n \rangle = 0.9$ are shown in Figure 13.12. As expected, the 4-site cluster results over estimate T_c and the results for the 8A and 18A clusters do not give a positive value for T_c. However, successive $Z_d = 4$ results on larger lattices fall nearly on the same curve. These results suggest that the 2D Hubbard model with $U = 4t$ and $\langle n \rangle = 0.9$ has a $d_{x^2-y^2}$ pairing phase. The dynamic cluster approximation leads to a mean field behavior close to T_c [53]. Values of T_c obtained using a mean field linear fit of the low temperature data for the various clusters are listed in Table 13.1.

Figure 13.11. Cluster sizes and geometries used by Maier et al. [27] in their study of the d-wave pair-field susceptibility. The shaded squares represent independent d-wave plaquettes within the clusters. In small clusters, the number of neighboring d-wave plaquettes Z_d listed in Table 1 is smaller than 4, i.e., than that for an infinite lattice (Maier et al. [27]).

Table 13.1. Number of independent neighboring d-wave plaquettes Z_d and the value T_c obtained from a linear fit of the pair-field susceptibility in Figure 13.12 (Maier et al. [27]).

Cluster	Z_d	T_c/t
4	0 (MF)	0.056
8A	1	−0.006
18A	1	−0.022
12A	2	0.016
16B	2	0.015
16A	3	0.025
20A	4	0.022
24A	4	0.020
26A	4	0.023

Numerical Studies of the 2D Hubbard Model

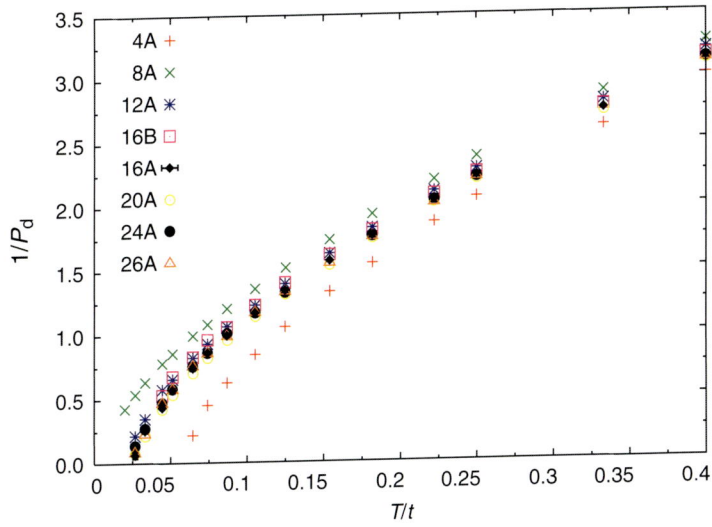

Figure 13.12. The inverse of the d-wave pair-field susceptibility is plotted vs. T/t for various clusters. Here $U = 4t$ and $\langle n \rangle = 0.9$. (Maier et al. [27]).

If $T_c \simeq 0.02t$ and we take $t = 0.2\,\text{eV}$, this gives $T_c \sim 50\,\text{K}$. We believe that T_c will increase with U, reaching a maximum when U is of order the bandwidth. In addition, we expect that the transition temperature is sensitive to the one-electron tight binding parameters. An example which illustrates this is known from density matrix renormalization group calculations for the 2-leg Hubbard ladder [54]. Figure 13.13 shows an average of the rung–rung pairfield correlations

$$\overline{D} = \sum_\ell \langle \Delta(i+\ell) \Delta^\dagger(i) \rangle \tag{13.32}$$

for a 2×16 Hubbard ladder versus the ratio of the rung to leg hopping parameters $t_\perp/t$. Here

$$\Delta^\dagger(i) = c^\dagger_{i1\uparrow} c^\dagger_{i2\downarrow} - c^\dagger_{i1\downarrow} c^\dagger_{i2\uparrow} \tag{13.33}$$

creates a pair on the ith rung of the ladder. The pairing, as measured by $\overline{D}$ exhibits a maximum at a value of $t_\perp/t$ when the minimum of the antibonding band at $k_x = 0$ and the maximum of the bonding band at $k_x = \pi$ approach the fermi surface.

For the half-filled noninteracting system, this would occur when $t_\perp/t = 2$. The doping and the interaction U leads to a reduction of this ratio and to a flattening of the dispersion which further enhances the single particle spectral weight near the fermi energy. If one considers the antibonding band to have $k_y = \pi$ and the bonding band to have $k_y = 0$, then this behavior is similar to increasing the single-particle spectral weight near $(0, \pi)$ and $(\pi, 0)$ in the 2D Hubbard model. One also sees that the largest peak in $\overline{D}$ occurs when U is of the order of bandwidth.

Having argued that the bandstructure plays a key role in determining T_c, it is important to note that this raises a puzzle. State-of-the-art LDA calculations, as Andersen and co-workers have shown [55], can be folded down to give material specific near-neighbor t, next-near-neighbor t', etc. hopping parameters. For the one-band Hubbard model one would then have

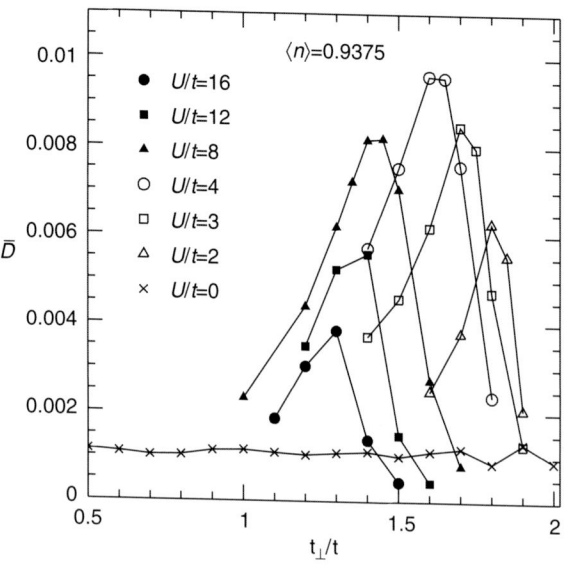

Figure 13.13. $\bar{D}$ vs. $t_\perp/t$ for various values of U/t at a filling $\langle n \rangle = 0.9375$ (Noack et al. [54]).

for the one-electron energy

$$\varepsilon_k = -2t(\cos k_x + \cos k_y) - 4t' \cos k_x \cos k_y - \cdots \qquad (13.34)$$

From an analysis of a large number of hole-doped cuprates, it was found that T_c is correlated with the range of the intralayer hopping [56]. For the one-band Hubbard model that we have discussed, this analysis implies that T_c should increase as t'/t becomes more negative. The opposite trend is seen in both dynamic cluster [57] and density matrix renormalization group calculations [13, 58]. However, a projected fermion calculation [59] finds that t' enters the effective interaction and can lead to an increase in T_c which is consistent with the conclusions of [56]. The resolution of this puzzle represents an important open problem.

13.3.3. Stripes

In a DMRG study of 7r×6 Hubbard ladders with 4r holes, Hager et al. [14] found that the ground state was striped for strong coupling values of U ($U = 12t$). Using a systematic extrapolation they gave evidence that such stripes exist in infinitely long 6-leg ladders. These studies also found that for small values of U ($U = 3t$) there were no stripes in the ground state. This work extended earlier work [60] on a 7×6 system with four holes which found that a well-defined stripe formed for $U/t \sim 8$ to 12. The absence of stripes for weak coupling is consistent with the fact that weak coupling renormalization group studies of the Hubbard model find no evidence of a stripe instability [36, 37].

Using the DMRG technique, the ground state expectation values of the hole density

$$h(x) = \sum_{y=1}^{6} (1 - \langle n(x, y) \rangle) \qquad (13.35)$$

Numerical Studies of the 2D Hubbard Model

and the staggered spin density

$$s(x) = \sum_{y=1}^{6} (-1)^{x+y} \langle n_\uparrow(x,y) - n_\downarrow(x,y) \rangle \tag{13.36}$$

were evaluated for $7r \times 6$ ladders with $4r$ holes. Periodic boundary conditions were used for the 6-site direction and open boundaries were used in the leg direction. Results for the hole $h(x)$ and spin $s(x)$ densities for a 21×6 ladder doped with 12 holes are shown in Figure 13.14. One sees from the modulation of the hole density along the leg direction that three stripes have formed. These stripes, each associated with four holes, run around the 6-site cylinder. In earlier t–J studies [12], a preferred stripe density of half-filling was found and we believe that the 2/3 filling seen in Figure 13.14 is a consequence of the restriction to 6 legs. Just as in the t–J ladder calculations, the staggered spin density undergoes a π-phase shift where the hole density is maximal. The finite staggered spin density is an artifact of the DMRG procedure in which no spin symmetry was imposed. This, along with the open boundary conditions which break the translational symmetry, allows the charge and spin density structures to appear in the $h(x)$ and $s(x)$ expectation values.

While stripe-like structures are seen in Figure 13.14 for both $U/t = 12$ and $U/t = 3$, the amplitude of the charge density modulations for $U/t = 3$ are both smaller and less regular than the $U/t = 12$ results. As discussed in Section 13.2, DMRG results for operators which are nondiagonal in the energy basis are expected to deviate from their exact values by the square root of the discarded weight $\sqrt{W_m}$ as the number of basis states is increased. Thus, to determine whether there are stripes in the ground state of an infinite ladder, Hager et al. [14] extrapolated their results for a set of $7r \times 6$ ladders to $W_m \to 0$ and then took $R = 7r \to \infty$.

They did this for the wave-vector power spectrum of the charge density

$$H^2 = \sum_{k_x} |H(k_x)|^2 \tag{13.37}$$

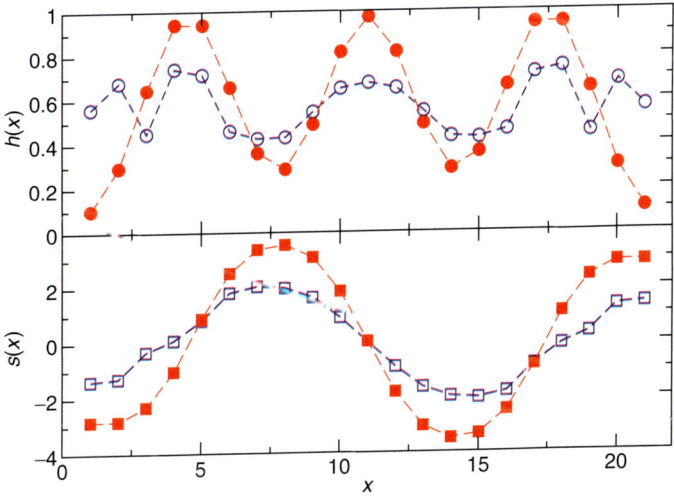

Figure 13.14. The hole $h(x)$ (circle) and staggered spin $s(x)$ (square) densities in the leg x-direction are plotted for a 21×6 ladder with 12 holes for $U = 12t$ (solid symbols) and $U = 3t$ (open symbols) (Hager et al. [14]).

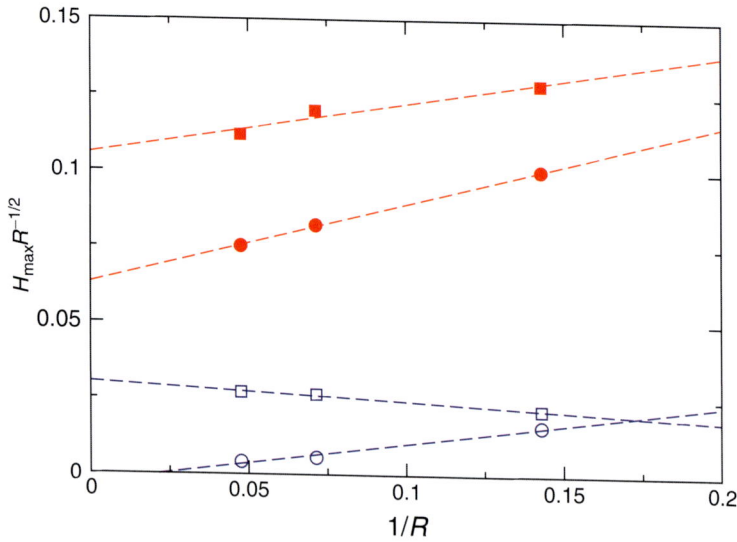

Figure 13.15. The amplitude of the power spectrum component $|H(k_x^*)|/\sqrt{R}$ for the hole density modulation. Results for a fixed number ($6000 \leq m \leq 8000$) of density-matrix eigenstates (squares) and results extrapolated to the limit $W_m \to 0$ (circles) are shown as a function of the inverse ladder length $1/R$ for $U = 12t$ (solid symbols) and $U = 3t$ (open symbols). The dashed lines are linear fits (Hager et al. [14]).

with

$$H(k_x) = \sqrt{\frac{2}{R+1}} \sum_x \sin(k_x x) \langle h(x) \rangle . \quad (13.38)$$

For a ladder with a periodic array of stripes separated by 7 sites, the maximum contribution to H^2 is associated with the wave vector

$$\frac{k_x^*}{\pi} = \frac{2r+1}{R+1} \to \frac{2}{7} \quad (13.39)$$

and

$$H_{\max} = |H(k_x^*)| \propto \sqrt{R}\, h_0 \quad (13.40)$$

as R goes to infinity. In Figure 13.15, the amplitude $H_{\max}(k_x^*)/\sqrt{R}$ is plotted for $U/t = 12$ and $U/t = 3$ vs. the inverse of the ladder length R^{-1}. The solid squares show the results when a fixed number ($6000 \leq m \leq 8000$) of density-matrix eigenstates are retained. The solid circles are the extrapolated $W_m \to 0$ ($m \to \infty$) results for $U/t = 12$. Similar results are shown using open symbols for $U/t = 3$. When the $W_m \to 0$ results are then extrapolated to $R \to \infty$, one sees clear evidence for stripes when $U/t = 12$ and an absence of stripes for $U/t = 3$. Note the importance of the $W_m \to 0$ extrapolation in determining the absence of stripes for $U/t = 3$.

13.3.4. The Pseudogap

Besides the antiferromagnetic, d-wave pairing, and striped phases, the cuprates exhibit a normal state pseudogap below a characteristic temperature T^* when they are underdoped. This pseudogap manifests itself in a variety of ways [61]. There is a decrease in the Knight

Numerical Studies of the 2D Hubbard Model

shift, reflecting a decrease in the magnetic susceptibility [62]. This was interpreted in terms of the opening of a pseudogap in the spin degrees of freedom. Observations of a similar suppression in the tunneling density of states [63], the c-axis optical conductivity [64] and the specific heat [65] made it clear that there was a pseudogap in both the spin and charge degrees of freedom. ARPES studies show that in the hole-doped materials, a pseudogap opens near the $(\pi, 0)$ antinodal regions while in the electron-doped materials, at the lowest dopings, it opens along the nodal direction near $(\frac{\pi}{2}, \frac{\pi}{2})$ [66]. The pseudogap appears in the underdoped region of the phase diagram and weakens as optimal doping is approached. If the Hubbard model is to contain the essential physics of the cuprates, it should exhibit the pseudogap phenomenon.

Before looking for evidence of pseudogap behavior in the doped Hubbard model, it is useful to first look at the structure of the single particle spectral weight for the half-filled Hubbard model. An important paper on this was that of Preuss et al. [20]. Here, determinantal Monte Carlo calculations of the finite temperature single particle Green's function $G(k, \tau)$ were carried out on an 8×8 periodic lattice. The spectral weight

$$A(k, \omega) = -\frac{1}{\pi} \text{Im}\, G(k, i\omega_n \to \omega + i\delta) \tag{13.41}$$

was then determined using a numerical maximum entropy continuation. Results for the half-filled case with $U = 8t$ and $T = 0.1t$ are shown in Figure 13.16a. Here, $A(k, \omega)$ is plotted vs. ω for various k values in the Brillouin zone. Figure 13.16b summarizes these results using a standard "band structure" ω vs. k plot in which the dark areas signify a large spectral weight. This work and related studies [67] showed that when U was of order the bandwidth or larger, there were four bands consisting of two incoherent upper and lower Hubbard bands and two quasiparticle-like, narrow bands nearer $\omega = 0$. The inner bands were found to have a dispersion set by $J \cong 4t^2/U$ while the outer, upper, and lower Hubbard bands, appear as an essentially dispersionless incoherent background.

The left-hand part of Figure 13.17 shows the single particle density of states for the half-filled case. Here, when the temperature is small compared to the exchange energy $J \sim 4t^2/U$, one clearly sees the broad upper and lower Hubbard bands and the narrow inner bands. When the system is hole-doped, the chemical potential moves down into the narrow coherent band that lays below $\omega = 0$ for the half-filled case and at the same time the upper coherent band loses weight and disappears as shown on the right hand side of Figure 13.17. This is also seen in Figure 13.18 which shows $A(k, \omega)$ for $\langle n \rangle = 0.95$ from [20]. Here, one sees that the dispersing band below $\omega = 0$ in the insulator and the band that the holes are doped into as the system becomes metallic are quite similar. At the same time, the narrow dispersing band that lays just above $\omega = 0$ in the insulating state has lost most of its spectral weight.

For the doped system, the fermion sign problem limited the temperature to $T = t/3$ for the determinantal data shown in Figure 13.18, but later similar determinantal Quantum Monte Carlo runs at $T = 0.25t$ and a filling of 0.95 found evidence for the formation of a pseudogap near $(\pi, 0)$ [21]. In this work, the spin susceptibility was shown to have a large spectral weight at well-defined spin excitations for the doping and temperature range in which the pseudogap appeared. There was no pseudogap in the overdoped $\langle n \rangle \lesssim 0.8$ regions where the spectral weight of the spin susceptibility became broad and featureless.

Dynamic cluster Monte Carlo calculations [28] with $U = 6t$ and $\langle n \rangle = 0.95$ find that the magnetic spin susceptibility exhibits a clear decrease below a temperature $T \cong 0.1t$, as shown in the inset of Figure 13.19 and simulations at $T = 0.06t$ gave the results for $A(k, \omega)$ shown in Figure 13.19. Here, a pseudogap has opened for $k = (\pi, 0)$, while the nodal region with $k = (\frac{\pi}{2}, \frac{\pi}{2})$ remains gapless. In addition, a variety of other cluster calculations [7, 29,

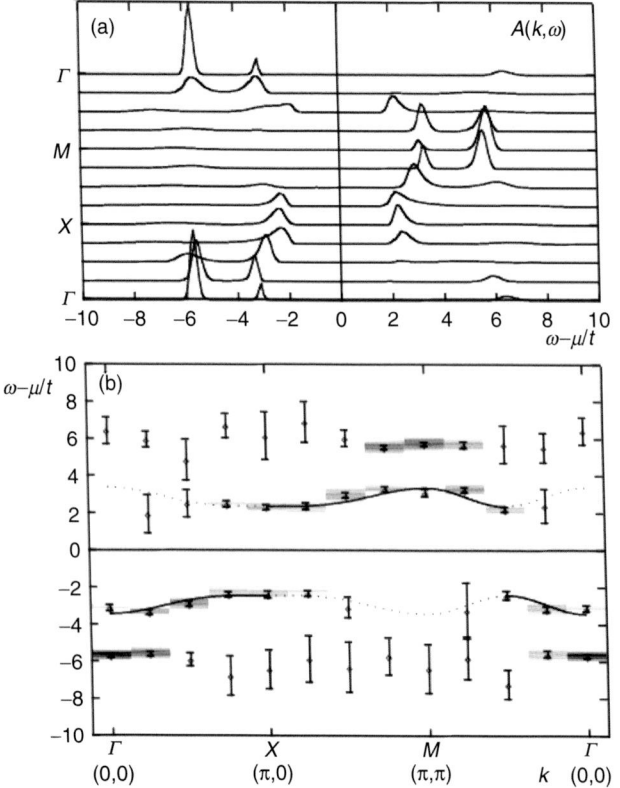

Figure 13.16. Single-particle spectral weight $A(k, \omega)$ for an 8×8 Hubbard model at half-filling $\langle n \rangle = 1$ with $U = 8t$ and $T = 0.1t$. (a) $A(k, \omega)$ vs. ω for different values of k and (b) ω vs. k plotted as a "band-structure" where sizable structure in $A(k, \omega)$ is represented by the strongly shaded regions and peaks by error bars (Preuss et al. [20].)

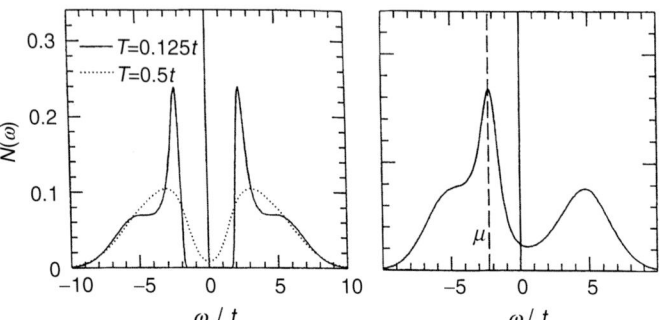

Figure 13.17. On the left, the single particle density of states $N(\omega)$ vs. ω for $U = 8t$ and $\langle n \rangle = 1$. On the right, $N(\omega)$ for the hole doped $\langle n \rangle = 0.875$, $U = 8t$ case at $T = 0.33t$ (Scalapino [68]).

30, 32] have found pseudogap behavior in both hole- and electron-doped Hubbard models and studied its dependence on the next near-neighbor hopping t'. The t' dependence as well as the doping dependence are consistent with renormalization-group calculations which show the importance of umklapp scattering processes [37] and the short-range antiferromagnetic spin correlations.

Numerical Studies of the 2D Hubbard Model

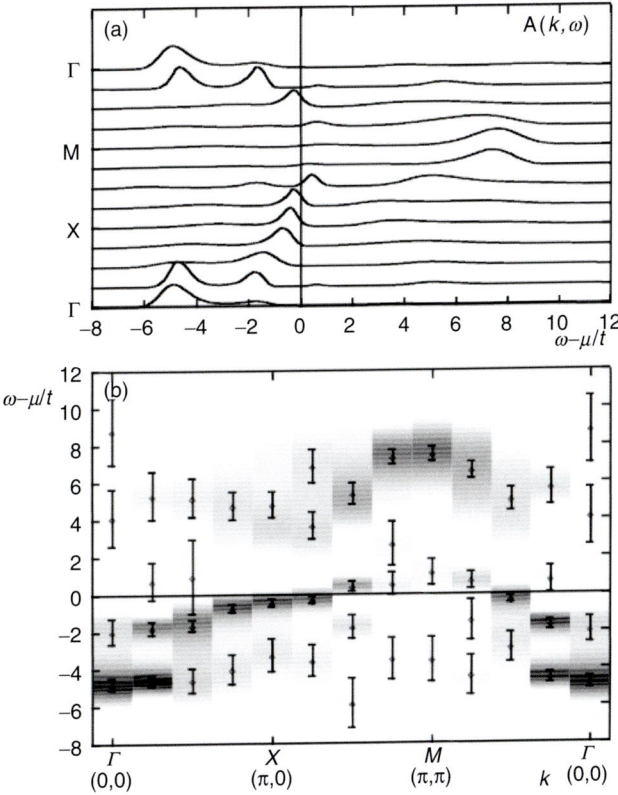

Figure 13.18. The single-particle spectral weight $A(k, \omega)$ for the hole doped $\langle n \rangle = 0.95$ system at a temperature $T = 0.33t$. This plot is similar to Figure 13.16 and shows what happens as holes are doped into the Mott–Hubbard insulator (Preuss et al. [20]).

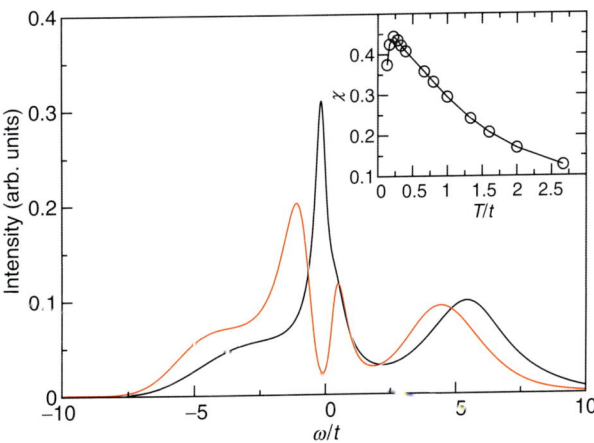

Figure 13.19. The single particle spectral weight $A(k, \omega)$ vs. ω for the antinodal $k = (\pi, 0)$ (red curve) and nodal $k = (\pi/2, \pi/2)$ (black curve) momenta of an underdoped $\langle n \rangle = 0.95$, $U = 8t$ Hubbard model at $T = 0.11t$. The inset shows the temperature dependence of the magnetic susceptibility (M. Jarrell).

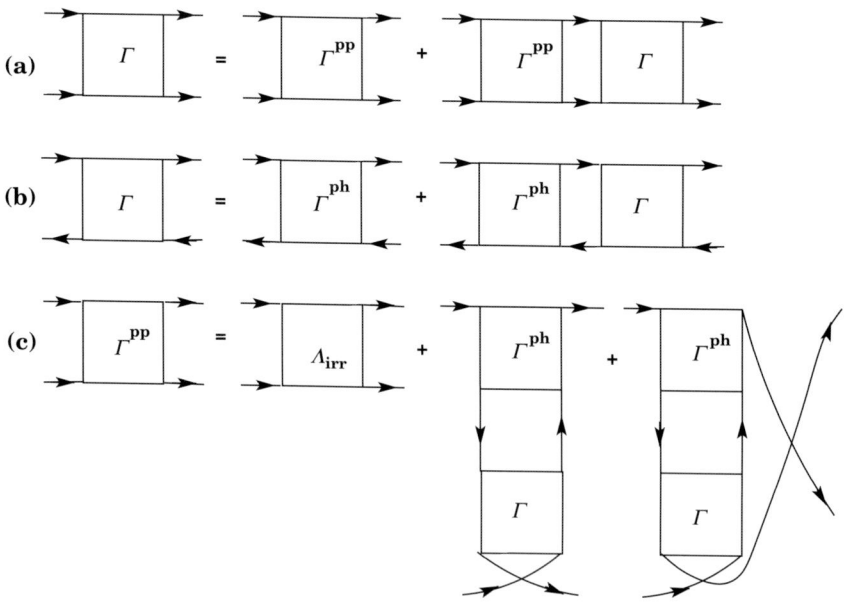

Figure 13.20. Bethe–Salpeter equations for (a) the particle–particle and (b) the particle–hole channels showing the relationship between the full vertex, the particle–particle irreducible vertex Γ^{pp}, and the particle–hole irreducible vertex Γ^{ph}, respectively. (c) Decomposition of the irreducible particle–particle vertex Γ^{pp} into a fully irreducible two-fermion vertex Λ_{irr} plus contributions from the particle–hole channels (Maier et al. [42]).

In the next section, we turn to a discussion of the effective pairing interaction. Specifically, the structure of the two-particle irreducible vertex and its associated d-wave eigenfunction are analyzed.

13.4. The Structure of the Effective Pairing Interaction

As discussed in Section 13.3, determinantal quantum Monte Carlo studies of the doped two-dimensional Hubbard model find that $d_{x^2-y^2}$ pairing correlations develop as the temperature is lowered and a dynamic cluster quantum Monte Carlo calculation on Betts' clusters finds evidence for a finite temperature d-wave superconducting phase. Here we discuss how one can use numerical techniques to determine the structure of the interaction responsible for the pairing. The basic idea is to numerically calculate the 4-point vertex Γ and the single particle propagator G (solid lines) shown in Figure 13.20. Then, using the particle–particle Bethe–Salpeter equation (Figure 13.20a), one can extract the two-particle irreducible vertex Γ^{pp} which is the pairing interaction. As we will discuss, the 4-point vertex Γ also contains information on the particle–hole magnetic ($S = 1$) and charge ($S = 0$) channels. Thus, it provides a natural framework for understanding the relationship of the pairing interaction to these other channels.

Using quantum Monte Carlo simulations, one can calculate both the one- and two-fermion Green's functions

$$G(x_2, x_1) = -\langle T c_\sigma(x_2) c_\sigma^\dagger(x_1) \rangle \tag{13.42}$$

and

$$G_2(x_4, x_3, x_2, x_1) = -\langle T c_{\sigma_4}(x_4) c_{\sigma_3}(x_3) c_{\sigma_2}^\dagger(x_2) c_{\sigma_1}^\dagger(x_1) \rangle. \tag{13.43}$$

Here, $c^\dagger_\sigma(x_\ell)$ creates an electron with spin σ at site x_ℓ and imaginary time τ_ℓ. T is the usual τ-ordering operator and we have suppressed the σ indices. Fourier transforming on both the space and imaginary time variables, one obtains $G(p)$ and

$$G_2(p_4, p_3, p_2, p_1) = -G(p_1)G(p_2)(\delta_{p_1,p_4}\delta_{p_2,p_3} - \delta_{p_1,p_3}\delta_{p_2,p_4}) + \frac{T}{N}\delta_{p_1+p_2, p_3+p_4} G(p_4) G(p_3) \Gamma(p_4, p_3; p_2, p_1) G(p_2) G(p_1) \quad (13.44)$$

with $p = (\mathbf{p}, i\omega_n)$. Then, using the Monte Carlo results for G and G_2, one can determine the 4-point vertex Γ from Eq. (13.44).

Given Γ and G, one can solve the Bethe–Salpeter equations shown in Figure 13.20a and b to obtain the irreducible particle–particle and particle–hole vertices Γ^{pp} and Γ^{ph}. For example, in the zero center of mass and energy channel, the particle–particle Bethe–Salpeter equation shown in Figure 13.20a gives

$$\Gamma(p'|p) = \Gamma^{\text{pp}}(p'|p) - \frac{T}{N}\sum_k \Gamma^{\text{pp}}(p'|k) G_\uparrow(k) G_\downarrow(-k) \Gamma(k|p) \quad (13.45)$$

with $\Gamma(p'|p) = \Gamma(p', -p'; p, -p)$. Given Γ and G, one can then determine the irreducible particle–particle vertex Γ^{pp}. This procedure is essentially the opposite of what one does in the traditional diagrammatic approach. There, one introduces an approximation for the irreducible vertex Γ^{pp} and solves Eq. (13.45) for Γ. Here, we use Monte Carlo results for Γ and G and solve Eq. (13.45) for Γ^{pp}. The Monte Carlo results for Γ satisfy crossing symmetry and $\Gamma^{\text{pp}}(p'|p)$ obtained from Eq. (13.45) is the effective particle–particle interaction. There is no approximation except for the fact that a finite lattice is used and one has the usual statistical Monte Carlo errors (and the small systematic finite $\Delta\tau$ errors which can be eliminated by extrapolating $\Delta\tau \to 0$).

The dominant pairing response, at low temperatures, is found to occur in the even frequency $d_{x^2-y^2}$ channel. Since this channel is even in both the relative frequency and momentum, it must be a spin singlet. Note that there are also spin singlet pairing channels which are odd in the relative frequency and momentum. However, the pairing instability in the doped Hubbard model comes from the even frequency and even momentum part of the irreducible particle–particle vertex

$$\Gamma^{\text{pp}}_e(p'|p) = \frac{1}{2}\left[\Gamma^{\text{pp}}(p'|p) + \Gamma^{\text{pp}}(-p'|p)\right]. \quad (13.46)$$

Determinantal quantum Monte Carlo results [41] for $\Gamma^{\text{pp}}_e(p'|p)$ obtained from an 8×8 lattice with $U = 4t$ and $\langle n \rangle = 0.87$ are shown in Figure 13.21. Here, $\Gamma^{\text{pp}}_e(p'|p)$ is plotted for various temperatures as a function of $\mathbf{q} = \mathbf{p}' - \mathbf{p}$ with $\mathbf{p} = (\pi, 0)$ and $\omega_n = \omega_{n'} = \pi T$. One sees that as the temperature is lowered, Γ^{pp}_e peaks at large momentum transfers. The size of the effective pairing interaction Γ^{pp}_e also depends upon the energy transfer $\omega_m = \omega_{n'} - \omega_n$, and falls off with ω_m on a scale set by the characteristic spin-fluctuation energy.

To obtain a more intuitive picture of the way in which the local repulsive $Un_{i\uparrow}n_{i\downarrow}$ Hubbard interaction can lead to an effective attractive pairing interaction in the singlet channel, it is useful to construct the real space Fourier transform

$$\Gamma^{\text{pp}}_e(\mathbf{R}) = \frac{1}{N^2}\sum_{\mathbf{p},\mathbf{p}'} e^{i(\mathbf{p}'-\mathbf{p})\cdot\mathbf{R}} \Gamma^{\text{pp}}_e(\mathbf{p}', i\pi T; \mathbf{p}, i\pi T). \quad (13.47)$$

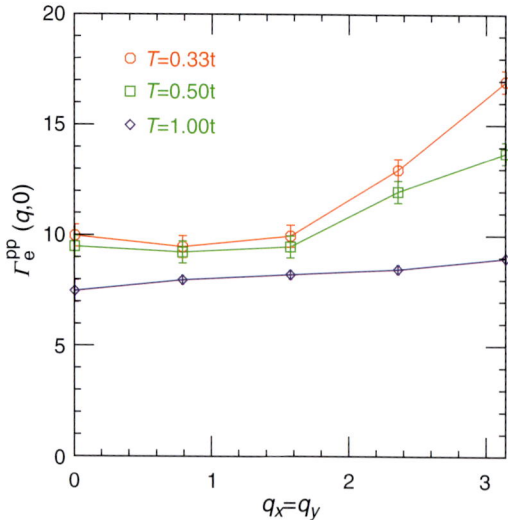

Figure 13.21. The even irreducible particle–particle vertex $\Gamma_e^{pp}(\mathbf{q}, \omega_m = 0)$ for $\mathbf{q} = \mathbf{p}' - \mathbf{p}$ and $\mathbf{p} = (\pi, 0)$ vs. momentum transfer $\mathbf{q}$ along the $(1, 1)$ direction. Here $U = 4t$ and $\langle n \rangle = 0.875$. As the temperature decreases below the temperature where spin–spin correlations develop, the strength of the interaction is enhanced at large momentum transfers (Bulut et al. [41]).

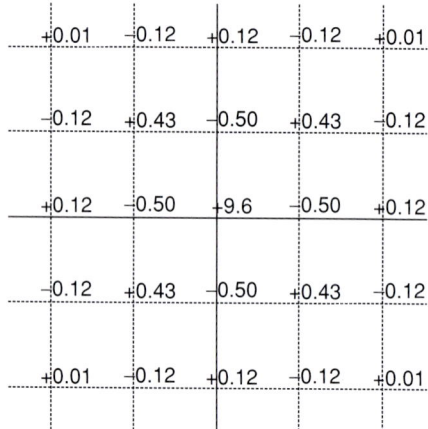

Figure 13.22. The real-space structure of $\Gamma_e^{pp}(\mathbf{R})$ at a temperature $T = 0.25t$ for $U = 4t$ and $\langle n \rangle = 0.87$. When the singlet electron pair is separated by one lattice spacing, $\mathbf{R} = \mathbf{x}$ or $\mathbf{y}$, the interaction is attractive, while it is strongly repulsive when $\mathbf{R} = 0$ and the pair occupies the same site (Bulut et al. [41]).

Values for $\Gamma_e^{pp}(\mathbf{R})$ are shown in Figure 13.22, with the distance $\mathbf{R}$ between the two fermions measured from the central point. If two fermions occupy the same site, spin-up and spin-down, $\Gamma_e^{pp}(\mathbf{R} = 0) \simeq 9.6t$. That is, the effective pairing interaction is even more repulsive than the bare $U = 4t$ onsite Coulomb interaction. However, if two fermions in a singlet state are on near-neighbor sites, the effective interaction $\Gamma_e^{pp}(\mathbf{R} = \hat{\mathbf{x}} \text{ or } \hat{\mathbf{y}}) \simeq -0.5t$ is attractive.

In order to determine the structure of the pairing correlations which are produced by Γ_e^{pp}, we turn to the homogenous Bethe–Salpeter equation

Numerical Studies of the 2D Hubbard Model

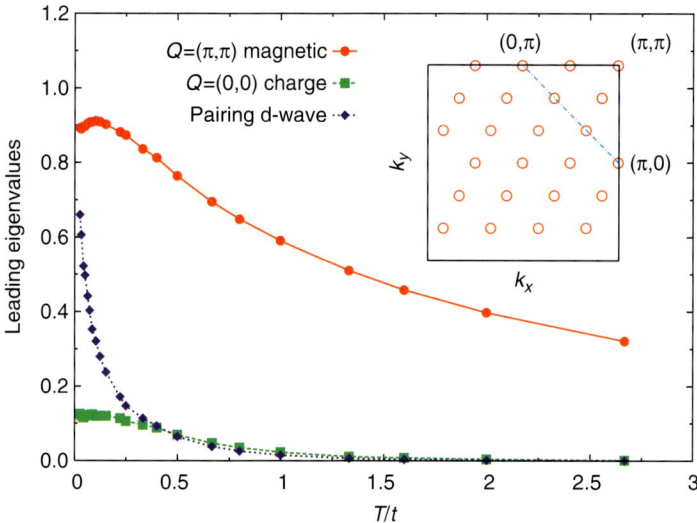

Figure 13.23. Leading eigenvalues of the Bethe–Salpeter equations in various channels for $U/t = 4$ and a site occupation $\langle n \rangle = 0.85$. The $\mathbf{Q} = (\pi, \pi)$, $\omega_m = 0$, $S = 1$ magnetic eigenvalue is seen to peak at low temperatures. The leading eigenvalue in the even singlet $\mathbf{Q} = (0, 0)$, $\omega_m = 0$ particle–particle channel has $d_{x^2-y^2}$ symmetry and increases toward 1 at low temperatures. The largest charge density eigenvalue occurs in the $\mathbf{Q} = (0, 0)$, $\omega_m = 0$ channel and saturates at a small value. The inset shows the distribution of k-points for the 24-site cluster (Maier et al. [42]).

$$-\frac{T}{N} \sum_{p'} \Gamma_e^{pp}(p|p') \, G_\uparrow(p') \, G_\downarrow(-p') \phi_\alpha(p') = \lambda_\alpha \phi_\alpha(p) \,. \tag{13.48}$$

The temperature dependence of the leading eigenvalue in the particle–particle channel is plotted vs. the temperature in Figure 13.23. When this eigenvalue reaches 1, it signals an instability into a superconducting phase. Here, $U = 4t$ with $\langle n \rangle = 0.85$ and we are showing results obtained using the dynamic cluster approximation [42] for the 24-site k-cluster discussed in Section 13.3. The distribution of $\mathbf{k}$ points for the 24-site cluster is shown in the inset of Figure 13.23. Similar results for $T \geq 0.25t$ have been obtained using the determinantal Monte Carlo algorithm on an 8×8 lattice [41].

The eigenfunction corresponding to the leading particle–particle eigenvalue is a singlet and its K dependence, plotted in the inset of Figure 13.24, shows that it has $d_{x^2-y^2}$ symmetry. The frequency dependence of this eigenfunction at the antinodal point $K = (\pi, 0)$ is shown in the main part of Figure 13.24. Here, $\phi((\pi, 0), \omega_n)$ has been normalized so that at $\omega_n = \pi T$ its value is 1. It is even in ω_n as it must be for a d-wave singlet to satisfy the Pauli principle. Also shown in this figure is the ω_m-dependence of the $Q = (\pi, \pi)$ spin susceptibility $\chi(Q, \omega_m)$ normalized by $(\chi(Q, 0) + \chi(Q, 2\pi T))/2$ for comparison with $\phi((\pi, 0), \omega_n)$. The boson Matsubara frequency dependence, $\omega_m = 2m\pi T$, of the susceptibility is seen to interlace with the fermion, $\omega_n = (2n + 1)\pi T$, dependence of the eigenfunction. The momentum and frequency dependence of $\phi_{d_{x^2-y^2}}(K, \omega)$ reflects the structure of the pairing interaction Γ_e^{pp}. The numerical results show that Γ_e^{pp} is an increasing function of momentum transfer and is characterized by a similar energy scale to that which enters the spin susceptibility $\chi(Q, \omega_m)$.

In a similar way, one can use Γ and G to solve for the irreducible particle–hole vertex Γ^{ph} shown in Figure 13.20b. The homogenous Bethe–Salpeter equation for the channel with

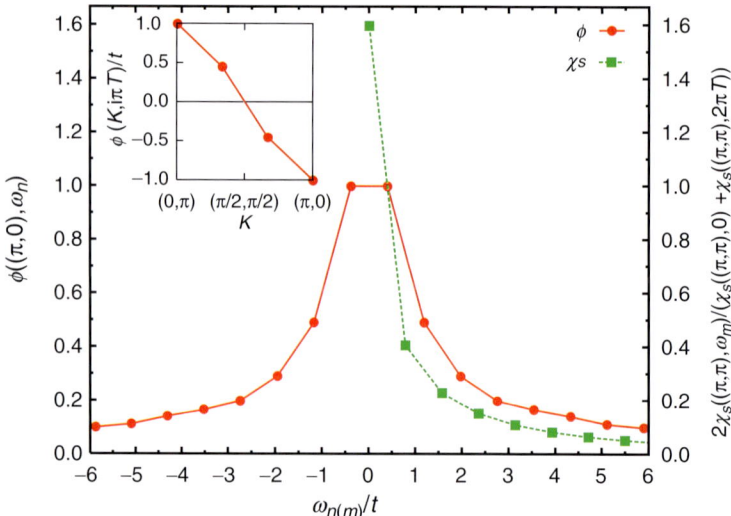

Figure 13.24. The Matsubara frequency dependence of the eigenfunction $\phi_{d_{x^2-y^2}}(\mathbf{K}, \omega_n)$ of the leading particle–particle eigenvalue of Figure 13.23 for $\mathbf{K} = (\pi, 0)$ normalized to $\phi(\mathbf{K}, \pi T)$ (red). Here, $\omega_n = (2n+1)\pi T$ with $T = 0.125t$. For comparison, the Matsubara frequency dependence of the normalized magnetic spin susceptibility $2\chi(\mathbf{Q}, \omega_m)/[\chi(\mathbf{Q}, 0) + \chi(\mathbf{Q}, 2\pi T)]$ for $\mathbf{Q} = (\pi, \pi)$ vs. $\omega_m = 2m\pi T$ is also shown (green). In the inset, the momentum dependence of the eigenfunction $\phi_{d_{x^2-y^2}}(\mathbf{K}, \pi T)$ normalized to $\phi_{d_{x^2-y^2}}((0, \pi), \pi T)$ shows its $d_{x^2-y^2}$ symmetry. Here, $\omega_n = \pi T$ and the momentum values correspond to values of $\mathbf{K}$ which lay along the dashed line shown in the inset of Figure 13.23. (Maier et al. [42]).

center-of-mass momentum Q, Matsubara frequency $\omega_m = 0$ and z-component of spin $S_z = 0$ is

$$-\frac{T}{N}\sum_{k'} \Gamma^{\text{ph}}(k+Q, k; k'+Q, k') G_{\uparrow}(k'+q) G_{\downarrow}(k') \phi_{Q\alpha}(k') = \lambda_\alpha(Q) \phi_{Q\alpha}(k). \quad (13.49)$$

The leading eigenvalue in the particle–hole channel occurs for $Q = (\pi, \pi)$ for the 24-site k-cluster and carries spin 1. Earlier determinantal quantum Monte Carlo studies [17] on 8×8 lattices show that for this doping the peak response is, in fact, slightly shifted from (π, π), but the 24-site k-cluster used in the dynamic cluster calculation lacks the resolution to show this. As seen in Figure 13.23, for this doping, the antiferromagnetic eigenvalue initially grows as the temperature is reduced, peaking at low temperatures. The largest eigenvalue in the $S = 0$ charge density channel occurs for $Q = (0, 0)$ and $\omega_m = 0$. Its temperature dependence is also plotted in Figure 13.23.

Returning to the question of the structure of the irreducible particle–particle vertex Γ_e^{pp}, we have seen that Γ_e^{pp} peaks at large momentum transfers and has a frequency dependence reflected in $\Phi_{d_{x^2-y^2}}(K, \omega_n)$ which is similar to the spin susceptibility. However, we would like to understand one further aspect. Is the dominant contribution to the $d_{x^2-y^2}$ pairing interaction associated with an $S = 1$ particle–hole channel? Alternatively, for example, one could have a charge density $S = 0$ channel or a more complicated multiparticle–hole exchange process such as that suggested by the spin-bag picture [69].

In order to address this, we will make use of the representation of Γ^{pp} shown diagrammatically in Figure 13.20c. Here, Γ^{pp} is decomposed into a fully irreducible vertex Λ_{irr} plus contribution from particle–hole exchange channels. Because of the spin rotation invariance of the Hubbard model, one can separate the particle–hole channels into a charge density $S = 0$ contribution and a spin $S = 1$ magnetic part. For the even frequency and even momentum (singlet pairing) part of the irreducible particle–particle vertex, Eq. (13.46), one has

$$\Gamma_e^{pp}(p'|p) = \Lambda_{\text{irr}}(p'|p) + \frac{1}{2} \Phi_d(p', p) + \frac{3}{2} \Phi_m(p', p). \quad (13.50)$$

The subscripts d and m denote the charge density ($S = 0$) and magnetic ($S = 1$) particle–hole channels, respectively, with

$$\Phi_{d/m}(p', p) = \frac{1}{2} \Big[\Gamma_{d/m}(p' - p; p, -p') - \Gamma_{d/m}^{ph}(p' - p; p, -p') \\ + \Gamma_{d/m}(p' + p; -p, -p') - \Gamma_{d/m}^{ph}(p' + p; -p, -p') \Big]. \quad (13.51)$$

Here, on the right-hand side, the center of mass and relative wave vectors and frequencies in these channels are labeled by the first, second, and third arguments, respectively. Results for the irreducible particle–particle interaction Γ_e^{pp} obtained from the 24-site dynamic cluster approximation are shown in Figure 13.25. As we have seen, when the temperature is lowered,

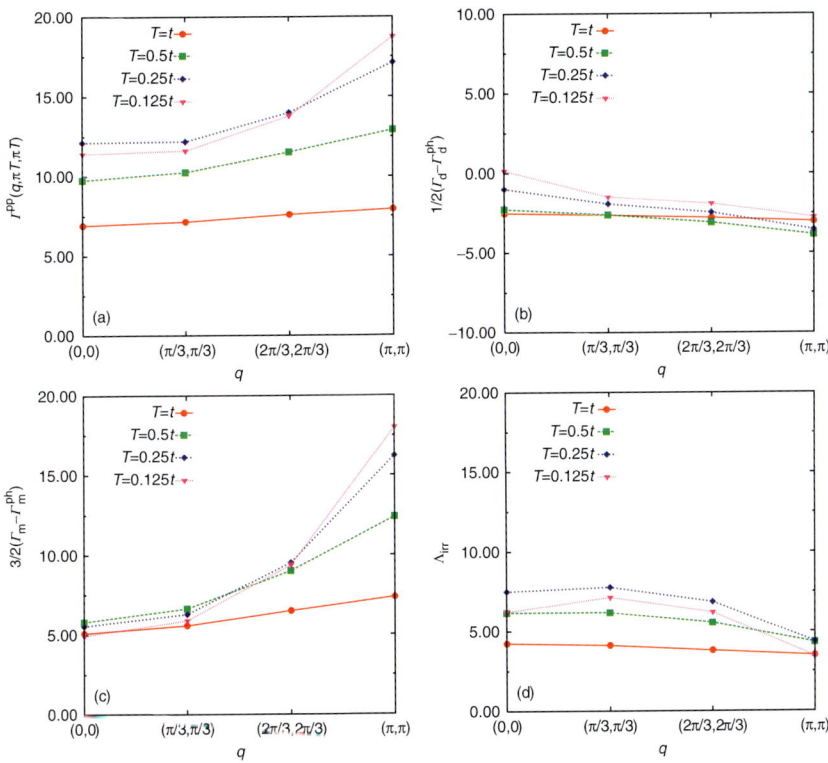

Figure 13.25. (a) The irreducible particle–particle vertex Γ_e^{pp} vs. $\mathbf{q} = \mathbf{K} - \mathbf{K}'$ for various temperatures with $\omega_n = \omega_{n'} = \pi T$. Here, $\mathbf{K} = (\pi, 0)$ and $\mathbf{K}'$ moves along the momentum values of the 24-site cluster which lay on the dashed line shown in the inset of Figure 13.23. Note that the interaction increases with the momentum transfer as expected for a d-wave pairing interaction. (b) The $\mathbf{q}$-dependence of the fully irreducible two-fermion vertex Λ_{irr}. (c) The $\mathbf{q}$-dependence of the charge density ($S = 0$) channel $\frac{1}{2}\Phi_d$ for the same set of temperatures. (d) The $\mathbf{q}$-dependence of the magnetic ($S = 1$) channel $\frac{3}{2}\Phi_m$ (Maier et al. [42]).

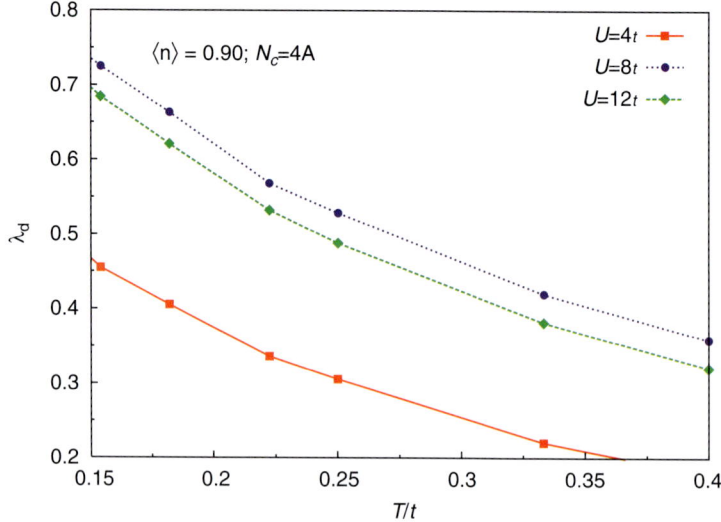

Figure 13.26. The $d_{x^2-y^2}$-wave eigenvalue λ_d vs. temperature T/t for $\langle n \rangle = 0.85$ with $U = 4t$ (red), $U = 8t$ (blue) and $U = 12t$ (green). These results were obtained for a 4-site k-cluster (Maier et al. [70]).

Γ_e^{pp} increases as the momentum transfer $\mathbf{q} = \mathbf{p}' - \mathbf{p}$ increases. Using the results for Γ^{ph}, Γ, and G one can calculate the contributions Φ_d from the $S = 0$ charge density and Φ_m for the $S = 1$ magnetic channels. Subtracting these from Γ_e^{pp} gives Λ_{irr} and results for each of these contributions are shown in Figure 13.25. The dominant $d_{x^2-y^2}$ pairing contribution to Γ_e^{pp} clearly comes from the $S = 1$ channel.

At larger values of U, 4-site k-cluster calculations [70] of the temperature dependence of the $d_{x^2-y^2}$ eigenvalue for $\langle n \rangle = 0.85$ and $U = 4t, 8t$, and $12t$ are shown in Figure 13.26. Over the temperature range [71] shown in Figure 13.26, the $d_{x^2-y^2}$ eigenvalue is largest for $U = 8t$. This is consistent with the 2-leg ladder result shown in Figure 13.13 and the expectation that the maximum transition temperature occurs for U of order the bandwidth. The $d_{x^2-y^2}$ eigenfunction $\phi_{d_{x^2-y^2}}(K, \omega_n)$ has the expected d-wave K dependence and its Matsubara frequency dependence for $U = 4t$ and $8t$ are shown in Figure 13.26. Here, as before, we also show the ω_m dependence of the spin susceptibility $\chi(Q, \omega_m)$. As U increases, both $\phi_{d_{x^2-y^2}}(K, \omega_n)$ and $\chi(Q, \omega_m)$ fall off more rapidly, reflecting the reduction in the frequency scale set by $J \sim 4t^2/U$.

13.5. Conclusions

The numerical studies of the Hubbard model that we have reviewed show that it exhibits the basic properties that are observed in the cuprate materials: antiferromagnetism, $d_{x^2-y^2}$-pairing, stripes, and pseudogap phenomena. Numerical methods have also been used to study the structure of the interaction responsible for pairing in the Hubbard model. As discussed in Section 13.4, this can be done by directly calculating the irreducible particle–particle vertex Γ^{pp} or by studying the momentum and frequency dependence of the gap function $\phi_{d_{x^2-y^2}}(K, \omega)$. The decomposition of Γ^{pp} showed that the dominant pairing interaction arose from a spin-one particle–hole exchange. The strength of Γ^{pp} was found to increase with momentum transfer leading to $d_{x^2-y^2}$-pairing. Alternately, the $(\cos K_x - \cos K_y)$ momentum dependence of the gap function $\phi_{d_{x^2-y^2}}(K, \omega)$ and the similarity of its ω_n dependence to that of the $Q = (\pi, \pi)$ spin susceptibility leads to the same conclusion: the pairing interaction in the doped Hubbard model is repulsive on site, attractive between near-neighbor sites and retarded on a time scale set by the inverse of the spin-fluctuation spectrum. It is important to recognize that this spectrum includes a particle–hole continuum.

Now, one can ask whether this interaction is actually the mechanism responsible for pairing in the high T_c cuprate materials and how one would know this from experiments? As far as the momentum dependence of the interaction is concerned, ARPES studies [66] of the k-dependence of the gap along with a variety of transport [72] and phase dependent studies [73, 74] provide strong evidence for the nodal d-wave character of the gap. While it is known that the chains in YBCO lead to an admixture of s-wave [75–77] and the momentum regions probed are primarily along the fermi surface, there is good reason to believe from the observed k-dependence of the gap that the pairing interaction is indeed repulsive on site and attractive for singlets formed between near-neighbor sites. It will be interesting to compare calculations for an orthorhombic Hubbard model with experiments [76, 77], to see if the observed k-dependence of the gap can provide additional help in identifying the pairing mechanism.

Another characteristic of the interaction is its frequency dependence. Here, less is known but it seems likely that the frequency dependence of the gap and renormalization parameter will provide important insight into the mechanism. As one knows, it was the frequency dependence of the gap for the traditional low T_c superconductors that provided the ultimate fingerprint identifying the phonon exchange pairing interaction, although at the time few doubted that this was the mechanism. In the high T_c case, the initial hope was that the d-wave momentum dependence of the gap would provide a sufficiently precise fingerprint. However, this has not been the case. For example, the exchange of B_{1g} phonons is known to favor d-wave pairing [55, 78], although its overall contribution to T_c is small within the standard theory. A two-band Cu–O model, in which fluctuations in circulating currents provide a d-wave pairing mechanism, has also been proposed [79]. Even within the framework of the Hubbard model there are different views regarding the dynamics. In the "Plain-Vanilla-RVB" picture [80], it has been suggested that the dynamics is set by an energy scale associate with the Mott–Hubbard gap [81]. However, our numerical results support a picture in which the dominant contributions come from particle–hole excitations within the relatively narrow band that the doped holes enter giving an energy scale of several times J. While the spectrum of these excitations extends down to zero energy, the main strength is associated with a broad spin-fluctuation continuum [82, 83]. Thus it seems likely that the dynamics will again be important in identifying the mechanism.

In addition to the traditional electron tunneling [84] and infrared conductivity [85] measurements, ARPES experiments provide an important tool for probing the frequency dependence of the renormalization parameter and the gap. Advances in the energy and momentum resolution of both ARPES [66] and neutron scattering [86] along with material preparation techniques that allow ARPES and neutron scattering to be done on the same material are opening new opportunities. Various RPA-BCS approximations have been used to model both the ARPES [87, 88] and neutron scattering data [89, 90]. One would clearly like to extend the numerical Hubbard model studies so that they can be used in making such experimental comparisons.

Finally, in addition to the frequency and momentum dependence of the interaction, there is the question of its strength. The estimate for the transition temperature in Section 13.3 with $U = 4t$ was relatively small. As discussed, we believe that for larger values of U (of order the bandwidth) and a more optimal bandstructure, T_c will increase. Beyond this, the actual Cu–O structure has additional exchange paths and it is known that $t-J-U$ Hubbard ladders can exhibit stronger pairing correlations [91]. Nevertheless, the question of the strength of the pairing interaction remains. It is not that several times J is not a wide spectral range compared to the phonon scale of the traditional low temperature superconductors or that the system isn't strongly coupled with U of order the bandwidth. Rather it is that the strong coupling has created a delicately balanced system [92]. As discussed in Section 13.3, different numerical methods on different lattices find evidence in one case for d-wave pairing and in another for stripes. Thus small changes in local parameters may alter the nature of the correlations and there is a question regarding the role of inhomogeneity in the cuprates. An interesting theory of "dynamic inhomogeneity-induced pairing" is discussed in another chapter of this treatise [93]. In this approach, pairing from repulsive interactions appears as a mesoscopic effect and the phenomena of high temperature superconductivity is viewed as arising from the existence of mesoscale structures [93, 94]. Recent STM measurements of impurities and inhomogeneities in BSCCO are providing important new information on the question of the local modulation of the pairing and its strength [95–97].

Thus, two decades after Bednorz' and Müller's [98] discovery of the high T_c cuprates the question of the pairing mechanism remains open. However, it is clear that the desire to understand these materials has driven dramatic advances in the experimental energy and momentum resolution of ARPES and neutron scattering and the energy and spatial resolution of STM. It was also largely responsible for the development of a variety of numerical techniques which are providing new insights into the electronic properties of a wide class of strongly correlated materials.

Acknowledgments

I would like to acknowledge past graduate students N. Bulut, J.M. Byers, M. Jarrell, E. Loh, R. Melko, R.M. Noack, S. Quinlan, and R. Scalettar and postdocs F. Assaad, N.E. Bickers, L. Capriotti, E. Dagotto, T. Dahm, J. Freericks, M.E. Flatte, R. Fye, J. Hirsch, M. Imada, P. Monthoux, A. Moreo, M. Salkola, H.B. Schuttler, and S.R. White who have been willing to teach me new things for so long. I would also like to acknowledge the pleasure I have had discussing and working on this problem with A.V. Balatsky, W. Hanke, P. Hirschfeld, S.A. Kivelson, T. Maier, and D. Poilblanc. Finally I want to thank my long time collaborator R.L. Sugar for his insights and encouragement. This work was supported by NSF Grant DMR02-11166 and the Department of Energy under FG02-03ER46048.

Bibliography

1. E. Dagotto, *Rev. Mod. Phys.* **66**, 763 (1994).
2. J. Jaklic and P. Prelovsek, *Adv. Phys.* **49**, 1 (1999).
3. N. Bulut, *Adv. Phys.* **51**, 1587 (2002).
4. A. Georges, G. Kotliar, W. Krauth, and M.J. Rozenberg, *Rev. Mod. Phys.* **68**, 13 (1996).
5. R.M. Noack and S.R. Manmana, *AIP Conf. Proc.* **789**, 93 (2005), cond-mat/0510321.
6. T. Maier, M. Jarrell, T. Pruschke, and M. Hettler, *Rev. Mod. Phys.* **77**, 1027 (2005).

7. A.-M.S. Tremblay, B. Kyung, and D. Senechal, cond-mat/0511334.
8. E. Dagotto and T.M. Rice, *Science* **271**, 618 (1996).
9. D. Poilblanc, *Phys. Rev. B* **48**, 3368 (1993); *Phys. Rev. B* **49**, 1477 (1994).
10. F. Becca, A. Parola, and S. Sorella, *Phys. Rev. B* **61**, R16287 (2000).
11. P.W. Leung, *Phys. Rev. B* **62**, R6112 (2000).
12. S.R. White and D.J. Scalapino, *Phys. Rev. Lett.* **81**, 3227 (1998).
13. S.R. White and D.J. Scalapino, *Phys. Rev. B* **60**, R753 (1999).
14. G. Hager, G. Wellein, E. Jeckelmann, and H. Fehske, *Phys. Rev. B* **71**, 75108 (2005).
15. J.E. Hirsch, *Phys. Rev. B* **31**, 4403 (1985).
16. S.R. White, D.J. Scalapino, R.L. Sugar, E.Y. Loh, J.E. Gubernatis and R.T. Scalettar, *Phys. Rev. B* **40**, 506 (1989); S.R. White, D.J. Scalapino, R.L. Sugar, N.E. Bickers and R.T. Scaletter, *Phys. Rev. B* **39**, 839 (1989).
17. A. Moreo, D.J. Scalapino, R.L. Sugar, S.R. White, and N.E. Bickers, *Phys. Rev. B* **41**, 2313 (1990).
18. A. Moreo, *Phys. Rev. B* **48**, 3380 (1993).
19. D.J. Scalapino, *Modern Perspectives in Many-Body Physics* (Sixth Physics Summer School, Australian National University), M.P. Das and J. Mahanty (eds.) (World Scientific, Singapore, 1994), pp. 199–225.
20. R. Preuss, W. Hanke, and W. von der Linden, *Phys. Rev. Lett.* **75**, 1344 (1995).
21. R. Preuss, W. Hanke, C. Grober, and H.G. Evertz, *Phys. Rev. Lett.* **79**, 1122 (1997).
22. D. Duffy, A. Nazarenko, S. Haas, A. Moreo, J. Riera, and E. Dagotto, *Phys. Rev. B* **56**, 5597 (1997).
23. S. Sorella, G. Martins, R. Becca, C. Gazza, L. Capriotti, A. Parola, and E. Dagotto, *Phys. Rev. Lett.* **88**, 117002 (2002).
24. V.I. Anisimov, M.A. Korotin, I.A. Nekrasov, Z.V. Pehelkina, and S. Sorella, *Phys. Rev. B* **66**, 100502 (2002).
25. M. Potthoff, *Eur. Phys. J.* **B32**, 429436 (2003).
26. C. Dahnken, M. Aichhorn, W. Hanke, E. Arrigoni, and M. Potthoff, *Phys. Rev. B* **70**, 245110 (2004).
27. T.A. Maier, M. Jarrell, T.C. Schulthess, P.R.C. Kent, and J.B. White, *Phys. Rev. Lett.* **95**, 237001 (2005).
28. C. Huscroft, M. Jarrell, T. Maier, S. Moukouri, and A.N. Tahvildarzadeh, *Phys. Rev. Lett.* **86**, 139 (2001); A. Macridin, M. Jarrell, T. Maier, and P.R.C. Kent, cond-mat/0509166.
29. B. Kyung, V. Hankevych, A.-M. Dare, and A.-M.S. Tremblay, *Phys. Rev. Lett.* **93**, 147004 (2004); B. Kyung, S.S. Kancharla, D. Senechal, A.-M. Tremblay, M. Civelli, and G. Kotliar, cond-mat/0502565.
30. C. Dahnken, M. Potthoff, E. Arrigoni, and W. Hanke, cond-mat/0504618.
31. S.S. Kancharla, M. Civelli, M. Capone, D. Senechal, G. Kotliar, and A.-M.S. Tremblay, cond-mat/0508205.
32. M. Aichhorn, E. Arrigoni, M. Potthoff, and W. Hanke, cond-mat/0511460.
33. A.I. Lichtenstein and M.I. Katsnelson, cond-mat/9911320.
34. P.W. Anderson, *Science* **235**, 1196 (1987).
35. M. Salmhofer, *Commun. Math. Phys.* **194**, 249 (1998).
36. C.J. Halboth and W. Metzner, *Phys. Rev. B* **61**, 7364 (2000).
37. C. Honerkamp, M. Salmhofer, N. Furukawa, and T.M. Rice, *Phys. Rev. B* **63**, 35109 (2001).
38. W.O. Putikka and M.U. Luchini, *Phys. Rev. B* **62**, 1684 (2000).
39. L.P. Pryadko, S.A. Kivelson, and O. Zachar, *Phys. Rev. Lett.* **92**, 67002 (2004).
40. T. Koretsune and M. Ogata, cond-mat/0505618.
41. N. Bulut, D.J. Scalapino, and S.R. White, *Phys. Rev. B* **47**, R6157 (1993); N. Bulut, D.J. Scalapino, and S.R. White, *Phys. Rev. B* **50**, 9623 (1994).
42. T.A. Maier, M. Jarrell, and D.J. Scalapino, *Phys. Rev. Lett.* **96**, 47005 (2006).
43. R. Blankenbecler, D.J. Scalapino, and R.L. Sugar, *Phys. Rev. D* **24**, 2278 (1981).
44. S.R. White, *Phys. Rev. Lett.* **69**, 2863 (1992).
45. S.R. White, *Phys. Rev. B* **48**, 10345 (1993).
46. J. Hirsch, *Phys. Rev. Lett.* **51**, 1900 (1983).
47. E.Y. Loh, J.E. Gubernatis, R.T. Scalettar, S.R. White, D.J. Scalapino, and R.L. Sugar, *Phys. Rev. B* **41**, 9301 (1990).
48. J.E. Hirsch and R.M. Fey, *Phys. Rev. Lett.* **56**, 2521 (1986).
49. M. Jarrell et al., *Phys. Rev. B* **64**, 195130 (2001).
50. D. Betts, H. Lin, and J. Flynn, *Can. J. Phys.* **77**, 353 (1999).
51. D.A. Huse, *Phys. Rev. B* **37**, 2380 (1988).
52. J.E. Hirsch and H.Q. Lin, *Phys. Rev. B* **37**, 5070 (1988).
53. In the infinite cluster limit, one expects that P_d will follow the low temperature Kosterlitz–Thouless behavior $P_d^{-1} \sim A \exp(-2B/(T-T_c)^{0.5})$.
54. R.M. Noack, N. Bulut, D.J. Scalapino, and M.G. Zacher, *Phys. Rev. B* **56**, 7162 (1997).
55. O.K. Andersen, *J. Phys. Chem. Solids* **56**, 1573 (1995); *J. Low Temp. Phys.* **105**, 285 (1996); *Phys. Rev. B* **62**, R16219 (2000).

56. E. Pavarini, I. Dasgupta, T. Saha-Dasgupta, O. Jepsen, and O.K. Andersen, *Phys. Rev. Lett.* **87**, 47003 (2001).
57. T. Maier, M. Jarrell, T. Pruschke, and J. Keller, *Phys. Rev. Lett.* **85**, 1524 (2000).
58. G.B. Martins, J.C. Xavier, L. Arrachea, and E. Dagotto, *Phys. Rev. B* **64**, 180513 (2001).
59. P. Prelovsek and A. Ramsak, *Phys. Rev. B* **65**, 174529 (2002) cond-mat/0502044.
60. S.R. White and D.J. Scalapino, *Phys. Rev. Lett.* **91**, 136403 (2003).
61. T. Timusk and B. Statt, *Rep. Prog. Phys.* **62**, 61 (1999).
62. H. Alloul, T. Ohno, and P. Mendels, *Phys. Rev. Lett.* **63**, 1700 (1989).
63. Ch. Renner, B. Revaz, J.-Y. Genoud, K. Kadowaki, and O. Fischer, *Phys. Rev. Lett.* **80**, 149 (1998).
64. C.C. Homes, T. Timusk, R. Liang, D.A. Bonn, and W.N. Hardy, *Phys. Rev. Lett.* **71**, 1645 (1993).
65. J.W. Loram, K.A. Mirza, J.R. Cooper, W.Y. Liang, and J.M. Wade, *J. Supercond.* **7**, 243 (1994).
66. A. Damascelli, Z. Hussain, and Z.X. Shen, *Rev. Mod. Phys.* **75**, 474 (2004).
67. A. Moreo, S. Haas, A.W. Sandirk, and E. Dagotto, *Phys. Rev. B* **51**, 12045 (1995).
68. D.J. Scalapino, *Tr. J. Phys.* **20**, 560 (1996).
69. A. Kampf and J.R. Schrieffer, *Phys. Rev. B* **41**, 6399 (1990).
70. T.A. Maier, M. Jarrell, and D.J. Scalapino, *Phys. Rev. B* **74**, 94513 (2006).
71. The eigenvalue for $U = 4t$ on the 4-site cluster lays above the result obtained for the 24-site cluster. As discussed in Section 13.3, this is expected because the 4-site cluster encloses just one d-wave plaquette and the embedding action does not contain pairfield fluctuations.
72. M. Chiao, R. Hill, C. Lupien, L. Taillefer, P. Lambert, R. Gagon, and R. Fournier, *Phys. Rev. B* **62**, 3554 (2000).
73. D.J. Van Harlinger, *Rev. Mod. Phys.* **67**, 515 (1995).
74. C.C. Tsuei and J.R. Kirtley, *Rev. Mod. Phys.* **72**, 969 (2000).
75. K.A. Kouznetsov et al., *Phys. Rev. Lett.* **79**, 3050 (1997).
76. H.J.H. Smilde, A.A. Golubov, Ariando, G. Rijnders, J.M. Dekkers, S. Harkema, D.H.A. Blank, H. Rogalla, and H. Hilgenkamp, *Phys. Rev. Lett.* **95**, 257001 (2005).
77. J.R. Kirtley, C.C. Tsuei, Ariando, C.J.M. Verwijs, S. Harkema, and H. Hilgenkamp, *Nat. Phys.* **2**, 190 (2006).
78. T.P. Devereaux, T. Cuk, Z.-X. Shen, and N. Nagaosa, *Phys. Rev. Lett.* **93**, 117004 (2004).
79. C.M. Varma, *Phys. Rev. B* **55**, 14554 (1997).
80. P.W. Anderson, P.A. Lee, M. Randeria, T.M. Rice, N. Trivedi, and F.C. Zhang, *J. Phys. Condens. Matter* **16**, R755 (2004).
81. P.W. Anderson, cond-mat/0512471.
82. D.J. Scalapino, *Phys. Rep.* **250**, 330 (1995).
83. A.V. Chubukov, D. Pines, and J. Schmalian, chapter in *The Physics of Conventional and Unconventional Superconductors*, K.H. Bennemann and J.B. Kettersen (eds.), (Springer, NewYork, 2002).
84. B. Barbiellini, O. Fischer, A.D. Kent, D.B. Mitzi, and A. Kapitulnik, *Physics B* **194**, 1689 (1994).
85. J.P. Carbotte, E. Schachinger, and D. Basou, *Nature (London)* **401**, 354 (1999).
86. S. Pailhes, C. Ulrich, R. Fauque, V. Hinkov, Y. Sidis, A. Ivanov, C.T. Lin, B. Keimer, and P. Bourges, cond-mat/0512634.
87. M. Eschrig and M.R. Norman, *Phys. Rev. B* **67**, 144503 (2003).
88. T. Dahm, P.J. Hirschfeld, D.J. Scalapino, and L. Zhu, *Phys. Rev. B* **72**, 214512 (2005).
89. N. Bulut and D.J. Scalapino, *Phys. Rev. B* **53**, 5149 (1996).
90. J. Brinckmann and P.A. Lee, *Phys. Rev. B* **65**, 14502 (2002).
91. S. Daul, D.J. Scalapino, and S.R. White, *Phys. Rev. Lett.* **84**, 4188 (2000).
92. D.J. Scalapino, *Phys. Canada* **56**, 267 (2000).
93. S.A. Kivelson and E. Fradkin, cond-mat/0507459.
94. V.J. Emery and S.A. Kivelson, *Phys. Rev. Lett.* **74**, 3253 (1995); E. Arrigoni, E. Fradkin, and S.A. Kivelson, *Phys. Rev. B* **69**, 214519 (2004).
95. A. Yazdani, C.M. Howald, C.P. Lutz, A. Kapitulnik, and D.M. Eigler, *Phys. Rev. Lett.* **83**, 176 (1999).
96. E.W. Hudson, S.H. Pan, A.K. Gupta, K.-W. Ng, and J.C. Davis, *Science* **285**, 88 (1999).
97. S.H. Pan, E.W. Hudson, K.M. Lang, H. Eisaki, S. Uchida, and J.C. Davis, *Nature* **403**, 746 (2000).
98. J.G. Bednorz and K.A. Müller, *Z. Phys. B: Condens. Matter* **64**, 189 (1986).

14

t–J Model and the Gauge Theory Description of Underdoped Cuprates

Patrick A. Lee

We review the effort to understand the physics of high temperature superconductors from the point of view of doping a Mott insulator. We begin with a discussion of the basic electronic structure of the cuprates, emphasizing the physics of strong correlation and establishing the model of a doped Mott insulator as a starting point. We review the analytic treatment of the t–J model, with the goal of putting the RVB idea on a more formal footing. The slave-boson formalism is introduced to enforce the constraint of no double occupation. The implementation of the local constraint leads naturally to gauge theories. We follow the historical order and first review the $U(1)$ formulation of the gauge theory. Some inadequacies of this formulation for underdoping are discussed, leading to the $SU(2)$ formulation and its extension to nonzero hole doping. Then we digress with a discussion of the role of gauge theory in describing the spin liquid phase of the undoped Mott insulator. We emphasize the difference between the high energy gauge group in the formulation of the problem vs. the low energy gauge group which is an emergent phenomenon. We emphasize that d-wave superconductivity can be considered as evolving from a stable $U(1)$ spin liquid. We apply these ideas to the high T_c cuprates, and discuss their implications for the vortex structure and the phase diagram. A possible test of the topological structure of the pseudogap phase is discussed.

14.1. Introduction

The discovery of high T_c superconductivity in cuprates was a major milestone in condensed matter physics. Not only was the transition temperature raised, but the fact that superconductivity was discovered in an unexpected class of materials, the transition metal oxide, made it clear that some new physics must be at work. By now it is clear that superconductivity emerges by doping a Mott insulator. The understanding of the strong correlation physics which is central to the Mott insulator and its doped state is a challenge that has to be met. Experimentalists and materials scientists have made great strides in studying these materials and it is clear that the cuprates embody a wealth of new phenomena not encountered previously, and that superconductivity is only one part of a fascinating phase diagram which must be understood in its entirety.

The present chapter reviews the theoretical work over the past two decades which addresses the strong correlation problem head-on. We review the basic electronic structures of the cuprates and argue that the important physics can be modeled by a single band Hubbard model, or its strong coupling limit, the t–J model. The strategy is to first see whether this

Patrick A. Lee • Department of Physics, Massachusetts Institute of Technology, Cambridge, MA 02139, USA

deceptively simple model has enough physics to describe the high T_c phenomena. The most unusual part of the phase diagram is the underdoped region, where the proximity to the Mott insulator is most strongly felt. After a brief review of the phenomenology, we explain the RVB picture proposed by Anderson [1]. To a large degree, our work is an effort to put these ideas on a firm theoretical footing. There are really not too many analytic tools available to address these strongly correlated problems which involve the imposition of the constraint of no double occupation in the case of the t–J model. The best tool available to us is the slave-boson formulation. It turns out that the mean field theory already gives a good qualitative account of the phase diagram, including the existence of d-wave superconductivity and the pseudogap regime. The imposition of the constraint invariably leads to gauge theories. The original $U(1)$ theory and a more recent $SU(2)$ version are then reviewed. These theories are treated with various levels of approximations. The goal is to make contact with experiments and to predict new measurements as much as possible. Since the problem of slave particles coupled to gauge fields remains one of strong coupling without a well-controlled expansion parameter, our conclusions remain largely qualitative in nature and confrontation with experiments is essential for future progress. Meanwhile, important progress has been made in the past decade toward understanding the spin liquid state which forms the backbone of the RVB theory. The relation of the existence of the spin liquid state with the deconfined phase of gauge theories is now firmly established. Fractionalized particles and gauge fields are emergent properties which describe the low energy physics. After describing these advances we return to the high T_c problem and see what insight can be gained in light of the recent progress. From this more general vantage point, one challenging aspect of the high T_c problem is that the ground states, which range from the antiferromagnetic to the superconductor to the Fermi liquid as the doping increases, are all phases where the gauge field is confining. Consequently, unusual phenomena associated with exotic phases of matter are not manifest at zero temperature. While the confined phase is difficult to treat theoretically, this observation by itself is in agreement with experiment because the low temperature properties of the Néel and superconducting states are quite conventional, as far as we can tell. But this also means that possible novel phases only reveal themselves as cross-over phenomena at finite temperatures. One such scenario is described, where the novel phase is assumed to be the algebraic spin liquid, which emerged as a promising candidate in the $SU(2)$ slave-boson theory. An experimental test of this hypothesis is then proposed.

A fuller treatment of the gauge theory approach is available in the review article by Lee, Nagaosa, and Wen [2]. References to other reviews and a more complete reference list can also be found there.

14.2. Basic Electronic Structure of the Cuprates

It is generally agreed that the physics of high T_c superconductivity is that of the copper oxygen layer, as shown in Figure 14.1. In the parent compound such as La_2CuO_4, the formal valence of Cu is 2+, which means that its electronic state is in the d^9 configuration. The copper is surrounded by six oxygens in an octahedral environment (the apical oxygen lying above and below Cu are not shown in Figure 14.1). The distortion from a perfect octahedron due to the shift of the apical oxygens splits the e_g orbitals so that the highest partially occupied d orbital is x^2–y^2. The lobes of this orbital point directly to the p orbital of the neighboring oxygen, forming a strong covalent bond with a large hopping integral t_{pd}. As we shall see, the strength of this covalent bonding is responsible for the unusually high energy scale for the exchange

t–J Model and the Gauge Theory Description of Underdoped Cuprates

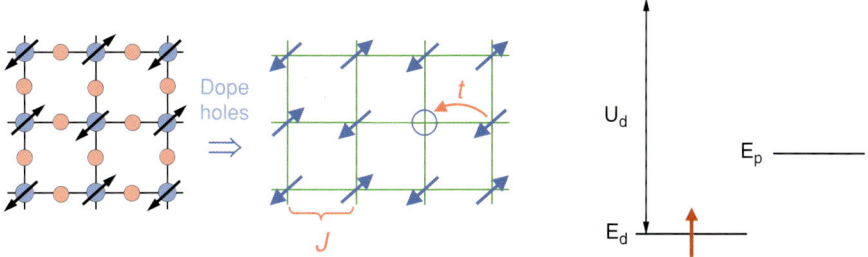

Figure 14.1. The two-dimensional copper–oxygen layer (left) is simplified to the one-band model (right). Bottom figure shows the copper d and oxygen p orbitals in the hole picture. A single hole with $S = 1/2$ occupies the copper d orbital in the insulator.

interaction. Thus the electronic state of the cuprates can be described by the so-called three-band model, where in each unit cell we have the Cu $d_{x^2-y^2}$ orbital and two oxygen p orbitals [3,4]. The Cu orbital is singly occupied while the p orbitals are doubly occupied, but these are admixed by t_{pd}. In addition, admixtures between the oxygen orbitals may be included. These tight-binding parameters may be obtained by fits to band structure calculations [5,6]. However, the largest energy in the problem is the correlation energy for doubly occupying the copper orbital. To describe these correlation energies, it is more convenient to go to the hole picture. The Cu d^9 configuration is represented by energy level E_d occupied by a single hole with $S = \frac{1}{2}$. The oxygen p orbital is empty of holes and lies at energy E_p which is higher than E_d. The energy to doubly occupy E_d (leading to a d^8 configuration) is U_d, which is very large and can be considered infinity. The lowest energy excitation is the charge transfer excitation where the hole hops from d to p with amplitude $-t_{pd}$. If $E_p - E_d$ is sufficiently large compared with t_{pd}, the hole will form a local moment on Cu. This is referred to as a charge transfer insulator in the scheme of Zaanen et al. [7]. Essentially, $E_p - E_d$ plays the role of the Hubbard U in the one-band model of the Mott insulator. Experimentally an energy gap of 2.0 eV is observed and interpreted as the charge transfer excitation (see [8]).

Just as in the one-band Mott–Hubbard insulator, where virtual hopping to doubly occupied states leads to an exchange interaction $J\mathbf{S}_1 \cdot \mathbf{S}_2$ where $J = 4t^2/U$, in the charge-transfer insulator, the local moments on nearest neighbor Cu prefer antiferromagnetic alignment because both spins can virtually hop to the E_p orbital. Ignoring the U_p for doubly occupying the p orbital with holes, the exchange integral is given by

$$J = \frac{t_{pd}^4}{(E_p - E_d)^3}. \qquad (14.1)$$

The relatively small size of the charge transfer gap means that we are not deep in the insulating phase and the exchange term is expected to be large. Indeed, experimentally the insulator is found to be in an antiferromagnetic ground state. By fitting Raman scattering to two magnon excitations [9], the exchange energy is found to be $J = 0.13$ eV. This is one of the largest exchange energies known. (It is even larger in the ladder compounds which involve the same Cu–O bonding.) This value of J is confirmed by fitting spin wave energy to theory, where an additional ring exchange terms is found [10].

By substituting divalent Sr for trivalent La, the electron count on the Cu–O layer can be changed in a process called doping. For example, in $La_{2-x}Sr_xCuO_4$, x holes per Cu is added to the layer. As seen in Figure 14.1, due to the large U_d, the hole will reside on the oxygen p orbital. The hole can hop via t_{pd} and due to translational symmetry, the holes are mobile and form a metal, unless localization due to disorder or some other phase transition intervenes. The full description of the hole hopping in the three-band model is complicated. On the other hand, there is strong evidence that the low energy physics (on a scale small compared with t_{pd} and $E_p - E_d$) can be understood in terms of an effective one-band model, and we shall follow this route in this review. The essential insight is that the doped hole resonates on the four oxygen sites surrounding a Cu and the spin of the doped hole combines with the spin on the Cu to form a spin singlet. This is known as the Zhang–Rice singlet [11]. This state is split off by an energy of order $t_{pd}^2/(E_p - E_d)$ because the singlet gains energy by virtual hopping. On the other hand, the Zhang–Rice singlet can hop from site to site. Since the hopping is a two step process, the effective hopping integral t is also of order $t_{pd}^2/(E_p - E_d)$. Since t is the same parametrically as the binding energy of the singlet, the justification of this point of view relies on a large numerical factor for the binding energy which is obtained by studying small clusters.

By focusing on the low lying singlet, the hole doped three-band model simplifies to a one-band tight binding model on the square lattice, with an effective nearest neighbor hopping integral t given earlier and with E_p–E_d playing a role analogous to U. In the large E_p–E_d limit this maps onto the t–J model

$$H = P \left(-\sum_{\langle ij \rangle, \sigma} t_{ij} c_{i\sigma}^\dagger c_{j\sigma} + J \sum_{\langle ij \rangle} \left(S_i \cdot S_j - \frac{1}{2} n_i n_j \right) \right) P, \qquad (14.2)$$

where the projection operator P restricts the Hilbert space to one which excludes double occupation of any site. J is given by $4t^2/U$ and we can see that it is the same functional form as that of the three-band model described earlier. It is also possible to dope with electrons rather than holes. The typical electron doped system is $Nd_{2-x}Ce_xCuO_{4+\delta}$ (NCCO). The added electron corresponds to removal of a hole from the copper site in the hole picture (Figure 14.1), i.e., the Cu ion is in the d^{10} configuration. This vacancy can hop with t_{eff} and the mapping to the one-band model is more direct than the hole doped case. Note that in the full three-band model the object which is hopping is the Zhang–Rice singlet for hole doping and the Cu d^{10} configuration for electron doping. These have rather different spatial structure and are physically quite distinct. For example, the strength of their coupling to lattice distortions may be quite different. When mapped to the one-band model, the nearest neighbor hopping t has the same parametric dependence, but could have a different numerical constant. As we shall see, the value of t derived from cluster calculations turn out to be surprisingly similar for electron and hole doping. For a bipartate lattice, the t–J model with nearest neighbor t has particle–hole symmetry because the sign of t (but not that of next nearest neighbor hopping t') can be absorbed by changing the sign of the orbital on one sublattice. Experimentally the phase diagram exhibits strong particle–hole asymmetry. On the electron doped side, the antiferromagnetic insulator survives up to much higher doping concentration (up to $x \approx 0.2$) and the superconducting transition temperature is quite low (about 30 K). Many of the properties of the superconductor resemble that of the overdoped region of the hole doped side and the pseudogap phenomenon, which is so prominent in the underdoped region, is not observed with electron doping. It is as though the greater stability of the antiferromagnet has covered up any anomalous regime that might exist otherwise.

A promising possibility is that particle–hole asymmetry may be accounted for by including further neighbor hopping t'. This point of view has been tested extensively by Hybertson et al. [12] who used ab initio local density functional theory to generate input parameters for the three-band Hubbard model and then solve the spectra exactly on finite clusters. The results are compared with the low energy spectra of the one-band Hubbard model and the t–t'–J model. They found an excellent overlap of the low lying wavefunctions for both the one-band Hubbard and the t–t'–J model and were able to extract effective parameters. They found J to be 128 ± 5 meV, in excellent agreement with experimental values. Furthermore they found $t \approx 0.41$ eV and 0.44 eV for electron and hole doping, respectively. The near particle–hole symmetry in t is surprising because the underlying electronic states are very different in the two cases, as already discussed. Based on their results, the commonly used parameter J/t for the t–J model is $1/3$. They also found a significant next nearest neighbor t' term, again almost the same for electron and hole doping.

More recently, Andersen et al. [13] have pointed out that in addition to the three-band model, an additional Cu 4s orbital has a strong influence on further neighbor hopping t' and t'' where t' is the hopping across the diagonal and t'' is hopping to the next-nearest neighbor along a straight line. Recently Pavarini et al. [14] emphasized the importance of the apical oxygen in modulating the energy of the Cu 4s orbital and found a sensitive dependence of t'/t on the apical oxygen distance. They also pointed out an empirical correlation between optimal T_c and t'/t. Thus t' may play an important role in determining T_c and in explaining the difference between electron and hole doping. However, in view of the fact that on-site repulsion is the largest energy scale in the problem, it would make sense to begin our modeling of the cuprates with the t–J model and ask to what extent the phase diagram can be accounted for. As we shall see, even this is not a simple task and will constitute the major thrust of this review.

14.3. Phenomenology of the Underdoped Cuprates

The essence of the problem of doping into a Mott insulator is readily seen from Figure 14.1. When a vacancy is introduced into an antiferromagnetic spin background, it would like to hop with amplitude t to lower its kinetic energy. However, after one hop its neighboring spin finds itself in a ferromagnetic environment, at an energy cost of $\frac{3}{2}J$ if the spins are treated as classical $S = \frac{1}{2}$. It is clear that the holes are very effective in destroying the antiferromagnetic background. This is particularly so when $t \gg J$ when the hole is strongly delocalized. The basic physics is the competition between the exchange J and the kinetic energy which is of order t per hole or xt per unit area. When $xt \gg J$ we expect the kinetic energy to win and the system should be a Fermi liquid metal with weak residual antiferromagnetic correlation. When $xt \leq J$, however, the outcome is much less clear because the system would like to maintain the antiferromagnetic correlation while allowing the hole to move as freely as possible. Experimentally we know that Néel order is destroyed with 3% hole doping, after which d-wave superconducting state emerges as the ground state up to 30% doping. Exactly how and why superconductivity emerges as the best compromise is the centerpiece of the high T_c puzzle but we already see that the simple competition between J and xt sets the correct scale $x = J/t = \frac{1}{3}$ for the appearance of nontrivial ground states. We shall focus our attention on the so-called underdoped region, where this competition rages most fiercely. Indeed it is known experimentally that the "normal" state above the superconducting T_c behaves differently from any other metallic state that we have known about up to now. Essentially an energy gap appears in some properties and not others. This region of the phase diagram is referred to as the pseudogap region and is well documented experimentally.

As seen in Figure 14.2 Knight shift measurement in the YBCO 124 compound shows that while the spin susceptibility χ_s is almost temperature independent between 700 and

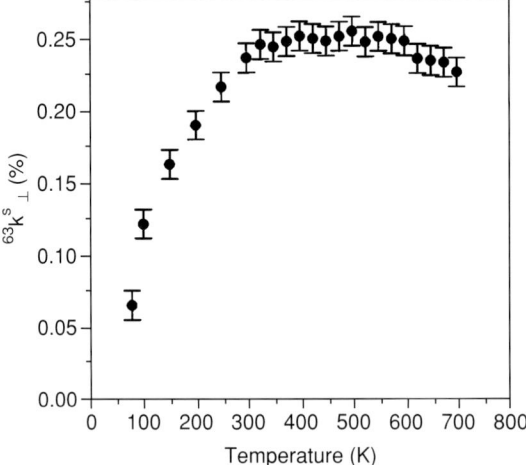

Figure 14.2. The Knight shift for $YB_2Cu_4O_8$. It is an underdoped material with $T_c = 79$ K. From Curro et al. [15].

300 K, as in an ordinary metal, it decreases below 300 K and by the time the T_c of 80 K is reached, the system has lost 80% of the spin susceptibility [15]. Similar phenomena have been seen in YBCO and LSCO, making this a universal property of the cuprates.

A second indication of the pseudogap comes from the linear T coefficient of the specific heat, which shows a marked decrease below room temperature. Furthermore, the specific heat jump at T_c is greatly reduced with decreasing doping [16]. It is apparent that the spins are forming into singlets and the spin entropy is gradually lost. On the other hand, as shown in Figure 14.3 the frequency dependent conductivity behaves very differently depending on whether the electric field is in the ab plane (σ_{ab}) or perpendicular to it (σ_c).

At low frequencies (below 500 cm^{-1}) σ_{ab} shows a typical Drude-like behavior for a metal with a width which decreases with temperature, but an area (spectral weight) which is independent of temperature [17]. Thus there is no sign of the pseudogap in the spectral weight. This is surprising because in other examples where an energy gap appears in a metal, such as the onset of charge or spin density waves, there is a redistribution of the spectral weight from the Drude part to higher frequencies. An important observation concerning the spectral weight is that the integrated area under the Drude peak is found to be linear in x [18–21]. In the superconducting state this weight collapses to form the delta function peak, with the result that the superfluid density n_s/m is also linear in x. It is as though only the doped holes contribute to charge transport in the plane. In contrast, angle-resolved photoemission shows a Fermi surface at optimal doping very similar to that predicted by band theory, with an area corresponding to $(1-x)$ electrons. With underdoping, this Fermi surface is partially gapped in an unusual manner which we shall next discuss.

In contrast to the metallic behavior of σ_{ab}, it was discovered by Homes et al. (1993) that below 300 K $\sigma_c(\omega)$ is gradually reduced for frequencies below 500 cm^{-1} and a deep hole is carved out of $\sigma_c(\omega)$ by the time T_c is reached. This is clearly seen in the lower panel of Figure 14.3.

Finally, angle-resolved photoemission shows that an energy gap (in the form of a pulling back of the leading edge of the electronic spectrum from the Fermi energy) is observed near momentum $(0, \pi)$. However, the lineshape is extremely broad and completely incoherent. The onset of superconductivity is marked by the appearance of a small coherent peak at this

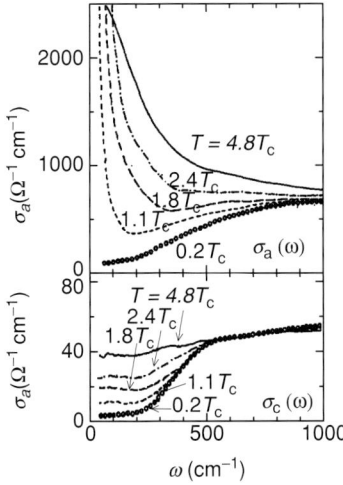

Figure 14.3. The frequency dependent conductivity with electric field parallel to the plane ($\sigma_a(\omega)$ top figure) and perpendicular to the plane ($\sigma_c(\omega)$ bottom figure) in an underdoped YBCO crystal. From Uchida [23].

gap edge. The size of the pull back of the leading edge is the same as the energy gap of the superconducting state as measured by the location of the coherence peak. This gap energy increases with decreasing doping, while the superconducting T_c decreases. This trend is also seen in tunneling data.

It is possible to map out the Fermi surface by tracking the momentum of the minimum excitation energy in the superconducting state for each momentum direction. Along the Fermi surface the energy gap does exactly what is expected for a d-wave superconductor. It is maximal near $(0, \pi)$ and vanishes along the line connection $(0, 0)$ and (π, π) where the excitation is often referred to as nodal quasiparticles. Above T_c the gapless region expands to cover a finite region near the nodal point, beyond which the pseudogap gradually opens as one moves toward $(0, \pi)$. This unusual behavior is sometimes referred to as the Fermi arc [24–26]. It is worth noting that unlike the antinodal direction [near $(0, \pi)$] the lineshape is relatively sharp along the nodal direction even above T_c. From the width in momentum space, a lifetime which is linear in temperature has been extracted for a sample near optimal doping [27] and recently this width is found to decrease rapidly below T_c [28]. A narrow lineshape in the nodal direction has also been observed in LSCO [29] and in Na doped $Ca_2CuO_2Cl_2$ [30]. So the notion of relatively well-defined nodal excitations in the normal state is most likely a universal feature.

As mentioned earlier, the onset of superconductivity is marked by the appearance of a sharp coherence peak near $(0, \pi)$. The spectral weight of this peak is small and gets even smaller with decreasing doping. Note that this behavior is totally different from conventional superconductors. There the quasiparticles are well defined in the normal state and according to BCS theory, the sharp peak pulls back from the Fermi energy and opens an energy gap in the superconducting state.

Yet another indication that the superconducting transition is different from BCS theory comes from the measurement of the change in kinetic energy through the transition. In conventional BCS theory, pairing between quasiparticles leads to a gain in the attractive potential energy at the expense of increasing the kinetic energy, since Fermi distribution is smeared by the creation of the energy gap. By carefully monitoring the optical spectral weight above and below T_c, it was found that while optimally doped samples behave as expected for BCS superconductors, underdoped samples exhibit the opposite behavior in that the kinetic energy is lowered by the onset of superconductivity [31–34].

In the literature, the pseudogap behavior is often associated with anomalous behavior of the nuclear spin relaxation rate $1/T_1$. In normal metals the nuclear spin relaxes by exciting low energy particle–hole excitations, leading to the Koringa behavior, i.e., $1/T_1T$ is temperature independent. In high T_c materials, it is rather $1/T_1$ which is temperature independent, and the enhanced relaxation (relative to Koringa) as the temperature is reduced is ascribed to antiferromagnetic spin fluctuations. It was found that in underdoped YBCO, the nuclear spin relaxation rate at the copper site reaches a peak at a temperature T_i^* and decreases rapidly below this temperature [35–37]. The resistivity also shows a decrease below T_i^*. In some literature T_i^* is referred to as the pseudogap scale. However, we note that T_i^* is lower than the energy scale we have been discussing so far, especially compared with that for the uniform spin susceptibility and the c-axis conductivity. Furthermore, the gap in $1/T_1$ is not universally observed in cuprates. It is not seen in LSCO. In $YBa_2Cu_4O_8$, which is naturally underdoped, the gap in $1/T_1T$ is wiped out by 1% Zn doping, while the Knight shift remains unaffected [38]. It is known from neutron scattering that the low lying spin excitations near (π, π) are sensitive to disorder. Since $1/T_1$ at the copper site is dominated by these fluctuations, it is reasonable that $1/T_1$ is sensitive as well. In contrast, the gap-like behavior we described thus far in a variety of physical properties is universally observed across different families of cuprates (wherever data exist) and are robust. Thus we prefer not to consider T_i^* as the pseudogap temperature scale.

14.4. Introduction to RVB and a Simple Explanation of the Pseudogap

We explained in the last section that the Néel spin order is incompatible with hole hopping. The question is whether there is another arrangement of the spin which achieves a better compromise between exchange energy and the kinetic energy of the hole. For $S = \frac{1}{2}$ it appears possible to take advantage of the special stability of the singlet state. The ground state of two spins S coupled with antiferromagnetic Heisenberg exchange is a spin singlet with energy $-S(S+1)J$. Compared with the classical large spin limit, we see that quantum mechanics provides an additional stability in the term unity in $(S+1)$ and this contribution is strongest for $S = \frac{1}{2}$. Let us consider a one-dimensional spin chain. A Néel ground state with $S_z = \pm\frac{1}{2}$ gives an energy of $-\frac{1}{4}J$ per site. On the other hand, a simple trial wavefunction of singlet dimers already gives a lower energy of $-\frac{3}{8}J$ per site. This trial wavefunction breaks translational symmetry and the exact ground state can be considered to be a linear superposition of singlet pairs which are not limited to nearest neighbors, resulting in a ground state energy of $0.443J$. In a square and cubic lattice the Néel energy is $-\frac{1}{2}J$ and $-\frac{3}{4}J$ per site, respectively, while the dimer variational energy stays at $-\frac{3}{8}J$. It is clear that in a 3D cubic lattice, the Néel state is a far superior starting point, and in two dimensions the singlet state may present a serious competition. Historically, the notion of a linear superposition of spin singlet pairs spanning different ranges, called the resonating valence bond (RVB), was introduced by Anderson [39] and Fazekas and Anderson [40] as a possible ground state for the $S = \frac{1}{2}$ antiferromagnetic Heisenberg model on a triangular lattice. The triangular lattice is of special interest because an Ising-like ordering of the spins is frustrated. Subsequently, it was decided that the ground state forms a $\sqrt{3} \times \sqrt{3}$ superlattice where the moments lie on the same plane and form 120° angles between neighboring sites [41]. Up to now there is no known spin Hamiltonian with full $SU(2)$ spin rotational symmetry outside of one dimension which is known to have an RVB ground state. However, there have been suggestions that ring exchange or proximity to the Mott insulator may stabilize such a state [42–45]. There are also examples which either violate spin rotation or which permit charge fluctuations.

The Néel state has long-range order of the staggered magnetization and an infinite degeneracy of ground states leading to Goldstone modes which are magnons. In contrast, the RVB state is a unique singlet ground state with either short range or power law decay of antiferromagnetic order. This state of affairs is sometimes referred to as a spin liquid. Soon after the discovery of high T_c superconductors, Anderson [1] revived the RVB idea and proposed that with the introduction of holes the Néel state is destroyed and the spins form a superposition of singlets. The vacancy can hop in the background of what he envisioned as a liquid of singlets and a better compromise between the hole kinetic energy and the spin exchange energy may be achieved. Many elaborations of this idea followed, but here we argue that the basic physical picture described above gives a simple account of the pseudogap phenomenon. The singlet formation explains the decrease of the uniform spin susceptibility and the reduction of the specific heat γ. The vacancies are responsible for transport in the plane. The conductivity spectral weight in the ab plane is given by the hole concentration x and is unaffected by the singlet formation. On the other hand, for c-axis conductivity, an electron is transported between planes. Since an electron carries spin $\frac{1}{2}$, it is necessary to break a singlet. This explains the gap formation in $\sigma_c(\omega)$ and the energy scale of this gap should be correlated with that of the uniform susceptibility. In photoemission, an electron leaves the solid and reaches the detector, the pull back of the leading edge simply reflects the energy cost to break a singlet. The lowering of the kinetic energy below the onset of superconductivity may also be explained qualitatively in this picture, because superconductivity is driven by the phase coherence of holes which lower the kinetic energy, while the cost of smearing out the Fermi surface by the creation of the gap has already been paid by the creation of the spin gap at higher temperatures.

A second concept associated with the RVB idea is the notion of spinons and holons, and spin charge separations. Anderson postulated that the spin excitations in an RVB state are $S = \frac{1}{2}$ fermions which he called spinons. This is in contrast with excitations in a Néel state which are $S = 1$ magnons or $S = 0$ gapped singlet excitations.

Initially the spinons are suggested to form a Fermi surface, with Fermi volume equal to that of $1-x$ fermions. Later it was proposed that the Fermi surface is gapped to form d-wave type structure, with maximum gap near $(0, \pi)$. This $\mathbf{k}$ dependence of the energy gap is needed to explain the momentum dependence observed in photoemission.

The concept of spinons is a familiar one in one-dimensional spin chains where they are well understood to be domain walls. In two dimensions the concept is a novel one which does not involve domain walls. Instead, a rough physical picture is as follows. If we assume a background of short-range singlet bonds, forming the so-called short-range RVB state, a cartoon of the spinon is shown in Figure 14.4. If the singlet bonds are "liquid," two $S = \frac{1}{2}$ formed by breaking a single bond can drift apart, with the liquid of singlet bonds filling in the space between them. They behave as free particles and are called spinons. The concept of holons follows naturally [46] as the vacancy left over by removing a spinon. A holon carries charge e but no spin.

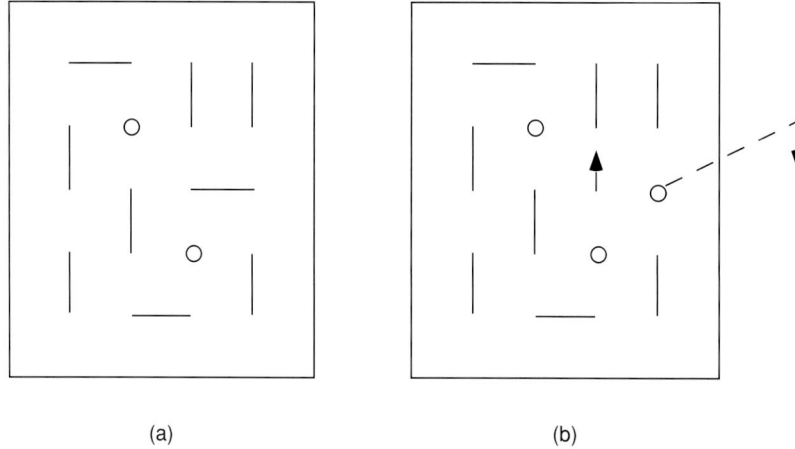

Figure 14.4. A cartoon representation of the RVB liquid or singlets. Solid bond represents a spin singlet configuration and circle represents a vacancy. In (b) an electron is removed from the plane in photoemission or c-axis conductivity experiment. This necessitates the breaking of a singlet.

While the RVB idea is qualitative successful in describing the phenomenology, many conceptual questions remain. The remainder of this review discusses effort to put these ideas on a more formal footing.

14.5. Slave-Boson Formulation of *t–J* Model and Mean Field Theory

We begin with the *t–J* model given by Eq. (14.2). The effect of the strong Coulomb repulsion is represented by the fact that double occupation is forbidden. This is written as the inequality

$$\sum_\sigma c^\dagger_{i\sigma} c_{i\sigma} \leq 1, \tag{14.3}$$

which is very difficult to handle. A powerful method to treat this constraint is the use of projected wavefunctions. One writes down trial wavefunctions of the form

$$\Psi = P_G |\psi_0\rangle, \tag{14.4}$$

where $P_G = \prod_i (n_{i\uparrow} n_{i\downarrow})$ is called the Gutzwiller projection operator and ψ_0 is typically a mean field trial wavefunction. The projection is treated numerically using quantum Monte Carlo methods. It has been found that the choice of the d-wave BCS wavefunction for ψ_0 gives an excellent account of the ground state properties, as well as other properties such as the quasiparticle's spectral weight [47–49]. A useful analytic method is called the Gutzwiller approximation, which imposes the constraint only approximately by treating the available configuration for the hopping and exchange operators in a statistical basis [50]. The Gutzwiller approximation is closely related to the slave-boson mean field theory, which we discuss below. In the slave-boson method [51, 52] the electron operator is represented as

$$c^\dagger_{i\sigma} = f^\dagger_{i\sigma} b_i, \tag{14.5}$$

where $f^\dagger_{i\sigma}$ is the fermion operators, while b is the slave-boson operators. This representation together with the constraint reproduces all the algebra of the

$$f_{i\uparrow}^\dagger f_{i\uparrow} + f_{i\downarrow}^\dagger f_{i\downarrow} + b_i^\dagger b_i = 1 \tag{14.6}$$

electron (fermion) operators. This constraint can be enforced with a Lagrangian multiplier λ_i. Note that Eq. (14.5) is not an operator identity and the R.H.S. does not satisfy the fermion commutation relation. Rather, the requirement is that both sides have the correct matrix elements in the reduced Hilbert space with no doubly occupied states. For example, the Heisenberg exchange term is written in terms of $f_{i\sigma}^\dagger$, $f_{i\sigma}$, only [53]

$$\mathbf{S_i} \cdot \mathbf{S_j} = -\frac{1}{4} f_{i\sigma}^\dagger f_{j\sigma} f_{j\beta}^\dagger f_{i\beta} - \frac{1}{4}\left(f_{i\uparrow}^\dagger f_{j\downarrow}^\dagger - f_{i\downarrow}^\dagger f_{j\uparrow}^\dagger\right)\left(f_{j\downarrow} f_{i\uparrow} - f_{j\uparrow} f_{i\downarrow}\right) + \frac{1}{4}\left(f_{i\alpha}^\dagger f_{i\alpha}\right). \tag{14.7}$$

We write

$$n_i n_j = \left(1 - b_i^\dagger b_i\right)\left(1 - b_j^\dagger b_j\right). \tag{14.8}$$

Then $\mathbf{S_i} \cdot \mathbf{S_j} - \frac{1}{4} n_i n_j$ can be written in terms of the first two terms of Eq. (14.7) plus quadratic terms, provided we ignore the nearest-neighbor hole–hole interaction $\frac{1}{4} b_i^\dagger b_i b_j^\dagger b_j$. We then decouple the exchange term in both the particle–hole and particle–particle channels via the Hubbard–Stratonovich (HS) transformation.

Then the partition function is written in the form

$$Z = \int \mathrm{D}f \mathrm{D}f^\dagger \mathrm{D}b \, \mathrm{D}\lambda \, \mathrm{D}\chi \, \mathrm{D}\Delta \exp\left(-\int_0^\beta \mathrm{d}\tau L_1\right), \tag{14.9}$$

where

$$L_1 = \tilde{J} \sum_{\langle ij \rangle} \left(|\chi_{ij}|^2 + |\Delta_{ij}|^2\right) + \sum_{i\sigma} f_{i\sigma}^\dagger (\partial_\tau - i\lambda_i) f_{i\sigma} - \tilde{J}\left[\sum_{\langle ij \rangle} \chi_{ij}^* \left(\sum_\sigma f_{i\sigma}^\dagger f_{j\sigma}\right) + \text{c.c.}\right]$$

$$+ \tilde{J}\left[\sum_{\langle ij \rangle} \Delta_{ij} \left(f_{i\uparrow}^\dagger f_{j\downarrow}^\dagger - f_{i\downarrow}^\dagger f_{j\uparrow}^\dagger\right) + \text{c.c.}\right] + \sum_i b_i^*(\partial_\tau - i\lambda_i + \mu_B) b_i - \sum_{ij} t_{ij} b_i b_j^* f_{i\sigma}^\dagger f_{j\sigma}$$

$$\tag{14.10}$$

with χ_{ij} representing fermion hopping and Δ_{ij} representing fermion pairing corresponding to the two ways of representing the exchange interaction in terms of the fermion operators. From Eqs. (14.7) and (14.10) it is concluded that $\tilde{J} = J/4$ but in practice the choice of $\tilde{J}_{ij}$ is not so trivial, namely one would like to study the saddle point approximation (SPA) and the Gaussian fluctuation around it, and requires SPA to reproduce the mean field theory. The latter requirement is satisfied when only one HS variable is relevant, but not for the multicomponent HS variables [54, 55]. In the latter case, it is better to chose the parameters in the Lagrangian to reproduce the mean field theory. In the present case, $\tilde{J} = 3J/8$ reproduces the mean field self-consistent equation which is obtained by the Feynman variational principle [56].

We note that L_1 in Eq. (14.10) is invariant under a local $U(1)$ transformation

$$\begin{aligned} f_i &\to e^{i\varphi_i} f_i \\ b_i &\to e^{i\varphi_i} b_i \\ \chi_{ij} &\to e^{-i\varphi_i} \chi_{ij} e^{i\varphi_j} \\ \Delta_{ij} &\to e^{i\varphi_i} \Delta_{ij} e^{i\varphi_j} \\ \lambda_i &\to \lambda_i + \partial_\tau \varphi_i, \end{aligned} \tag{14.11}$$

which is called $U(1)$ gauge transformation. Due to such a $U(1)$ gauge invariance, the fluctuations λ_i and the phase of χ_{ij} have the dynamics of $U(1)$ gauge field.

Now we describe the various mean field theory corresponding to the saddle point solution to the functional integral. The mean field conditions are

$$\chi_{ij} = \sum_\sigma \langle f_{i\sigma}^\dagger f_{j\sigma} \rangle, \tag{14.12}$$

$$\Delta_{ij} = \langle f_{i\uparrow} f_{j\downarrow} - f_{i\downarrow} f_{j\uparrow} \rangle. \tag{14.13}$$

Let us first consider the t–J model in the undoped case, i.e., the half-filled case. There are no bosons in this case, and the theory is purely that of fermions. The original one, i.e., uniform RVB state, proposed by Baskaran et al. [53] is given by

$$\chi_{ij} = \chi = \text{real} \tag{14.14}$$

for all the bond and $\Delta_{ij} = 0$. The fermion spectrum is that of the tight binding model

$$H_{\text{uRVB}} = -\sum_{\mathbf{k}\sigma} 2\tilde{J}\chi \left(\cos k_x + \cos k_y \right) f_{\mathbf{k}\sigma}^\dagger f_{\mathbf{k}\sigma}, \tag{14.15}$$

with the saddle point value to the Lagrange multiplier $\lambda_i = 0$. The so called "spinon Fermi surface" is large, i.e., it is given by the condition $k_x \pm k_y = \pm\pi$ with a diverging density of states (van Hove singularity) at the Fermi energy. Soon after, many authors found lower energy states than the uniform RVB state. One can easily understand that lower energy states exist because the Fermi surface is perfectly nested with the nesting wavevector $\vec{Q} = (\pi, \pi)$ and the various instabilities with $\vec{Q}$ are expected. Of particular importance are the d-wave state and the staggered flux state. The d-wave state is described by $\chi_{ij} = \chi_0$ for nearest neighbors, and $\Delta_{ij} = \Delta_0$ for $\mathbf{j} = \mathbf{i} + \hat{x}$ and $-\Delta_0$ for $\mathbf{j} = \mathbf{i} + \hat{y}$. The eigenvalues are the well-known BCS spectrum

$$E_\mathbf{k} = \sqrt{(\varepsilon_\mathbf{k} - \mu)^2 + \Delta_\mathbf{k}^2}, \tag{14.16}$$

where

$$\varepsilon_\mathbf{k} = -2\chi_0(\cos k_x + \cos k_y), \tag{14.17}$$

$$\Delta_\mathbf{k} = 2\Delta_0(\cos k_x - \cos k_y). \tag{14.18}$$

A variety of mean-field wavefunctions were soon discovered which give identical energy and dispersion. Notable among these is the staggered flux state [57]. In this state the hopping χ_{ij} is complex, $\chi_{ij} = \chi_0 \exp{(i(-1))^{i_x+j_y} \Phi_0}$, and the phase is arranged in such a way that it describes free fermion hopping on a lattice with a fictitious flux $\pm 4\Phi_0$ threading alternative plaquettes. Remarkably, the eigenvalues of this problem are identical to that of the d-wave superconductor given by Eq. (14.16), with

$$\tan \Phi_0 = \frac{\Delta_0}{\chi_0}. \tag{14.19}$$

The case $\Phi_0 = \pi/4$ called the π-flux phase, is special in that it does not break the lattice translation symmetry. The key feature is that the energy gap vanishes at the nodal points located at $\left(\pm\frac{\pi}{2}, \pm\frac{\pi}{2}\right)$. Around the nodal points the dispersion rises linearly, forming a cone which resembles the massless Dirac spectrum. For the π-flux state the dispersion around the node is isotropic. For Φ_0 less than $\pi/4$ the gap is smaller and the Dirac cone becomes progressively anisotropic.

The reason various mean-field theories have the same energy was explained by [58] and [59] as being due to a certain $SU(2)$ symmetry. It corresponds to the following particle–hole transformation

$$f_{i\uparrow}^\dagger \to \alpha_i f_{i\uparrow}^\dagger + \beta_i f_{i\downarrow}$$
$$f_{i\downarrow} \to -\beta_i^* f_{i\uparrow}^\dagger + \alpha_i^* f_{i\downarrow}. \qquad (14.20)$$

Note that the spin quantum number is conserved. It describes the physical idea that adding a spin-up fermion or removing a spin-down fermion are the same state after projection to the subspace of singly occupied fermions. Let us write

$$\Phi_{i\uparrow} = \begin{pmatrix} f_{i\uparrow} \\ f_{i\downarrow}^\dagger \end{pmatrix}, \Phi_{i\downarrow} = \begin{pmatrix} f_{i\downarrow} \\ -f_{i\uparrow}^\dagger \end{pmatrix}. \qquad (14.21)$$

Then Eq. (14.10) can be written in the more compact form

$$L_1 = \frac{\tilde{J}}{2} \sum_{\langle ij \rangle} \text{Tr}\left(U_{ij}^\dagger U_{ij}\right) + \frac{\tilde{J}}{2} \sum_{\langle ij \rangle, \sigma} \text{Tr}\left(\Phi_{i\sigma}^\dagger U_{ij} \Phi_{j\sigma} + \text{c.c.}\right) + \sum_{i,\sigma} f_{i\sigma}^\dagger (\partial_\tau - i\lambda_i) f_{i\sigma}$$
$$+ \sum_i b_i^*(\partial_\tau - i\lambda_i + \mu_B) b_i - \sum_{ij,\sigma} t_{ij} b_i b_j^* f_{i\sigma}^\dagger f_{j\sigma}, \qquad (14.22)$$

where

$$U_{ij} = \begin{pmatrix} -\chi_{ij}^* & \Delta_{ij} \\ \Delta_{ij}^* & \chi_{ij} \end{pmatrix}. \qquad (14.23)$$

At half filling $b = \mu_B = 0$ and the mean field solution corresponds to $\lambda_i = 0$. The Lagrangian is invariant under

$$\Phi_{i\sigma} \to W_i \Phi_{i\sigma}, \qquad (14.24)$$
$$U_{ij} \to W_i U_{ij} W_j^\dagger, \qquad (14.25)$$

where W_i is an $SU(2)$ matrix. To give an explicit example, the π-flux and d-RVB states are represented as

$$U_{ij}^{\pi\text{-flux}} = -\chi\left(\tau^3 - i(-1)^{i_x+j_y}\right) \qquad (14.26)$$

and

$$U_{i,i+\mu}^d = -\chi\left(\tau^3 + \eta_\mu \tau^1\right), \qquad (14.27)$$

respectively, where τ^i are the Pauli matrices and $\eta_x = -\eta_y = 1$. These two are related by

$$U_{ij}^{SF} = W_i^\dagger U_{ij}^d W_j, \qquad (14.28)$$

where

$$W_j = \exp\left[i(-1)^{j_x+j_y}\frac{\Pi}{4}\tau^1\right]. \qquad (14.29)$$

Therefore the $SU(2)$ transformation of the fermion variable

$$\Phi'_\mathbf{i} = W_\mathbf{i} \Phi_\mathbf{i} \tag{14.30}$$

relates the π-flux and d-RVB states. Here some remarks are in order. First it should be noted that at half filling we are discussing the Mott insulating state and its spin dynamics. The charge transport is completely suppressed by the constraint Eq. (14.6). Implementation of the constraint will be discussed in the next section where the mean field theory is elaborated into gauge theory. Secondly, it is now established that the ground state of the two-dimensional antiferromagnetic Heisenberg model shows the antiferromagnetic long-range ordering (AFLRO). This corresponds to the third (and most naive) way of decoupling the exchange interaction in the spin channel. However, even with the AFLRO, the singlet formation represented by $\chi_{\mathbf{ij}}$ and $\Delta_{\mathbf{ij}}$ dominates and AFLRO occurs on top of it. This view has been stressed by Hsu [60, 61] generalizing the π-flux state to include the AFRLO, and is in accord with the energetics of the projected wavefunctions which find that the best trial mean-field state requires both flux and AFRLO. An alternative route to reach the AF ground state is to start with the π-flux mean field state and include gauge fluctuations. The phenomenon of confinement in lattice gauge theory will also lead to AF order.

Now we turn to the doped case, i.e., $x \neq 0$. Then the behavior of the bosons are crucial for the charge dynamics. At the mean field level, the bosons are free and condensed at T_{BE}. In three-dimensional system, T_{BE} is finite while $T_{BE} = 0$ for purely two-dimensional system. If we assume weak three-dimensional hopping between layers, we obtain finite T_{BE} roughly proportional to the boson density x. This materializes the original idea by Anderson that RVB turns into the real superconductivity via the Bose condensation of holons. Kotliar and Liu [62] found the d-wave superconductivity in the slave-boson mean field theory presented above, and the schematic phase diagram is given in Figure 14.5. There are five phases classified by the order parameters χ, Δ, and $b = \langle b_i \rangle$ for the Bose condensation. In the incoherent state at high temperature, all the order parameters are zero. In the uniform RVB state (IV in Figure 14.5), only χ is finite. In the spin gap state (II), Δ and χ are nonzero while $b = 0$. This corresponds to the spin single "superconductivity" with incoherent charge motion, and can be viewed as the precursor phase of the superconductivity. This state has been interpreted as

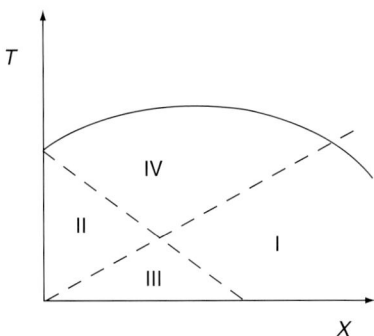

Figure 14.5. Schematic phase diagram of the $U(1)$ mean field theory. The solid line denotes the onset of the uniform RVB state ($\chi \neq 0$). The dashed line denotes the onset of fermion pairing ($\Delta \neq 0$) and the dotted line denotes mean field Bose condensation ($b \neq 0$). The four regions are (I) Fermi liquid $\chi \neq 0, b \neq 0$; (II) spin gap $\chi \neq 0, \Delta \neq 0$; (III) d-wave superconductor $\chi \neq 0, \Delta \neq 0, b \neq 0$; and (IV) strange metal $\chi \neq 0$ (from Lee and Nagaosa [67]).

the pseudogap phase [63]. We note that at the mean field level, the $SU(2)$ symmetry is broken by the nonzero μ_B in Eq. (14.22) and the d-wave pairing state is chosen because it has lower energy than the staggered flux state. We shall return to this point later. In the Fermi liquid state (I), both χ and b are nonzero while $\Delta = 0$. This state is similar to the slave-boson description of heavy fermion state. Lastly when all the order parameter is nonzero, we obtain the d-wave superconducting state (III). This mean field theory, in spite of its simplicity, captures rather well the experimental features as described earlier.

14.6. U(1) Gauge Theory of the URVB State

The mean field theory only enforces the constraint on the average. Furthermore, the fermions and bosons introduce redundancy in representing the original electron, which results in an extra gauge degree of freedom. To include these effects we need to consider fluctuations around the mean field saddle points, which immediately become gauge theories, as first pointed out by Baskaran and Anderson [64]. Here, we review the early work on the $U(1)$ gauge theory, which treats gauge fluctuations on the Gaussian level [65–68]. The theory can be worked out in some detail, leading to a nontrivial recipe for obtaining physical response functions in terms of the fermion and boson ones, called the Ioffe–Larkin composition rule. It highlights the importance of calculating gauge invariant quantities and the fact that the fermion and bosons only enter as useful intermediate steps. The Gaussian $U(1)$ gauge theory was mainly designed for the high temperature region of the optimally doped cuprate, i.e., the so-called strange metal phase in Figure 14.5. We will describe its failure in the underdoped region, which leads to the $SU(2)$ formulation of the next two sections. The Gaussian theory also misses the confinement physics which is important for the ground state.

As discussed earlier, the phenomenology of the optimally doped Mott insulator is required to describe the two seemingly contradicting features, i.e., the doped insulator with small hole carrier concentration and the electrons forming the large Fermi surface. The former is supported by various transport and optical properties, representatively the Drude weight proportional to x, while the latter by the angle resolved photoemission spectra (ARPES) in the normal state of optimal doped samples. In the conventional single-particle picture, the reduction of the 1st Brillouin zone due to the antiferromagnetic long-range ordering (AFLRO) distinguishes these two. Namely small hole pockets with area x are formed in the reduced 1st BZ in the AFLRO state, while the large metallic Fermi surface of area $1-x$ appears otherwise. The challenge for the theory of the optimally doped case is that aspects of the doped insulator appear in some experiments even with the large Fermi surface. Also it is noted that the ARPES shows that there is no sharp peak corresponding to the quasiparticle in the normal state, especially at the antinodal region near $\mathbf{k} = (\pi, 0)$. The fermi surface is defined by a rather broad peak dispersing near the Fermi energy. These strongly suggests that the normal state of high temperature superconductors is not described in terms of the usual Landau Fermi liquid picture.

A promising theoretical framework to describe this dilemma is the slave-boson formalism introduced above. It has the two species of particles, i.e., fermions and bosons, due to the strong correlation, and the electron is "fractionalized" into these two particles. However, one must be mindful that the fermions and bosons are not gauge invariant and they are strongly coupled to the gauge field. Under the gauge transformation [Eq. (14.11)] the Green's functions for fermions and bosons $G_F(\mathbf{i}, \mathbf{j}; \tau) = -\left\langle T_\tau f_{\mathbf{i}\sigma}(\tau) f_{\mathbf{j}\sigma}^\dagger \right\rangle$ and $G_B(\mathbf{i}, \mathbf{j}; \tau) = -\left\langle T_\tau b_\mathbf{i}(\tau) b_\mathbf{j}^\dagger \right\rangle$

transforms as

$$G_F(\mathbf{i}, \mathbf{j}; \tau) \to e^{i(\varphi_i - \varphi_j)} G_F(\mathbf{i}, \mathbf{j}; \tau)$$
$$G_B(\mathbf{i}, \mathbf{j}; \tau) \to e^{i(\varphi_i - \varphi_j)} G_B(\mathbf{i}, \mathbf{j}; \tau). \quad (14.31)$$

Nevertheless, the fermions and bosons are useful in the intermediate step of the theory to calculate the physical (gauge invariant) quantities.

The question is often asked, whether the fermions and bosons are real particles. Strictly speaking, the physical electron operator is also not gauge invariant under the electromagnetic field gauge transformation. Yet, due to the small size of the coupling constant e, we can mentally turn off the coupling for the electromagnetic field and have no trouble thinking of the electron as real. In our case, for the fermion and boson to emerge as useful concepts, we require that on some short distance scale (or finite temperature) the confinement effects due to the compactness of the gauge fields are not important and the problem can be treated as noncompact as is done below. Even granting this, we find that the fermions and bosons are not close to being free particles, but are coupled to the gauge field with coupling constant of order unity. (This coupling has been reduced from infinity to unity by screening.) Thus in the following we regard the fermions and bosons as intermediate steps in the theory and focus on the calculation of physical (gauge invariant) quantities. The notion of spinons as emergent low energy excitations will be discussed further in a later section.

At the mean field level, the constraint was replaced by the averaged one $<Q_i> = 1$. This average is controlled by the saddle point value of the Lagrange multiplier field $<\lambda_i> = \lambda$. Originally λ_i is the functional integral variable and is a function of (imaginary) time. When this integration is done exactly, the constraint is imposed. Therefore we have to go beyond the mean field theory and take into account the fluctuation around it. In other words, the local gauge symmetry is restored by the gauge fields which transform as

$$a_{ij} \to a_{ij} + \varphi_i - \varphi_j$$
$$a_0(\mathbf{i}) \to a_0(\mathbf{i}) + \frac{\partial \varphi_i(\tau)}{\partial \tau}. \quad (14.32)$$

The fields satisfying this condition are already in the Lagrangian Eq. (14.10). Namely the phase of the HS variable χ_{ij} and the fluctuation part of the Lagrange multiplier λ_i are a_{ij} and $a_0(\mathbf{i})$, respectively.

Let us study this $U(1)$ gauge theory for the uRVB state in the phase diagram Figure 14.5. This state is expected to describe the normal state of the optimally doped cuprates, where the $SU(2)$ particle–hole symmetry described by Eq. (14.21) is not so important. Here we neglect Δ field, and consider χ and λ field. There are amplitude and phase fluctuations of χ field, but the former one is massive and does not play important roles in the low energy limit. Therefore the relevant Lagrangian to start with is

$$L_1 = \sum_{\mathbf{i},\sigma} f_{\mathbf{i}\sigma}^* \left(\frac{\partial}{\partial \tau} - \mu_F + i a_0(\mathbf{r_i}) \right) f_{\mathbf{i}\sigma} + \sum_{\mathbf{i}} b_{\mathbf{i}}^* \left(\frac{\partial}{\partial \tau} - \mu_B + i a_0(\mathbf{r_i}) \right) b_{\mathbf{i}}$$
$$- \tilde{J} \chi \sum_{\langle \mathbf{ij} \rangle \sigma} \left(e^{i a_{ij}} f_{\mathbf{i}\sigma}^* f_{\mathbf{j}\sigma} + \text{h.c.} \right) - t\eta \sum_{\langle \mathbf{ij} \rangle} \left(e^{i a_{ij}} b_{\mathbf{i}}^* b_{\mathbf{j}} + \text{h.c.} \right), \quad (14.33)$$

where η is the saddle point value of another HS variable to decouple the hopping term. We can take $\eta = \chi$ using Eq. (14.12). Equation (14.33) takes the form of a lattice gauge theory. The spatial component of the gauge fields are a_{ij} defined on (ij) link while the time component $a_0(\mathbf{r}_i)$ is defined on the lattice site $\mathbf{r}_i$. Note the a_{ij} appears as a phase variable, i.e., the Lagrangian is invariant under the transformation $a_{ij} \to a_{ij} + 2\pi$, which identifies this theory as a compact $U(1)$ gauge theory. The gauge fields are coupled to fermions and bosons hopping on the lattice. The fermions and bosons are referred to as matter fields in the field theory literature. We also note that the usual Maxwell term (familiar in electrodynamics), $\frac{1}{g}f_{\mu\nu}^2$ where $f_{\mu\nu} = \partial_\mu a_\nu - \partial_\nu a_\mu$, which controls the gauge fluctuations and describes their dynamics, is absent in Eq. (14.33). In other words, the coupling constant g is infinite. This is because the gauge field represents the constraint; by integrating over the gauge field we obtain the original problem with the constraint.

In the mean field theory the gauge fluctuations are completely ignored. One consequence is that the entropy is grossly overestimated, since extra degrees of freedom have been introduced. This was shown explicitly by Hlubina et al. [69], who compared the entropy of the mean field theory with high temperature expansion and found that it is too large by a factor of two. They also found that by including gauge fluctuations in the RPA approximation, the agreement is improved considerably. In the RPA approximation we exchange the order of the integration between the gauge field (a_{ij}, a_0) and the matter fields (fermions and bosons). Namely the matter fields are integrated over first, and we obtain the effective action for the gauge field.

$$e^{-S_{\text{eff.}}(a)} = \int Df^*Df\, Db^*Db\, e^{-\int_0^\beta L_1}. \tag{14.34}$$

However, this integration cannot be done exactly, and an approximation is introduced here. The most standard one is the Gaussian approximation or RPA, where the effective action is obtained by perturbation theory up to the quadratic order in a. For this purpose we introduce here the continuum approximation to the Lagrangian L_1 in Eq. (14.33).

$$L = \int d^2 r \Bigg[\sum_\sigma f_\sigma^*(\mathbf{r}) \left(\frac{\partial}{\partial \tau} - \mu_F + i a_0(\mathbf{r}) \right) f_\sigma(\mathbf{r}) + b^*(\mathbf{r}) \left(\frac{\partial}{\partial \tau} - \mu_B + i a_0(\mathbf{r}) \right) b(\mathbf{r})$$
$$- \frac{1}{2 m_F} \sum_{\sigma, j=x,y} f_\sigma^*(\mathbf{r}) \left(\frac{\partial}{\partial x_j} + i a_j \right)^2 f_\sigma(\mathbf{r}) - \frac{1}{2 m_B} \sum_{j=x,y} b^*(\mathbf{r}) \left(\frac{\partial}{\partial x_j} + i a_j \right)^2 b(\mathbf{r}) \Bigg], \tag{14.35}$$

where the vector field $\mathbf{a}$ is introduced by $a_{ij} = (\mathbf{r}_i - \mathbf{r}_j) \cdot \mathbf{a}[(\mathbf{r}_i + \mathbf{r}_j)/2]$. Note $1/m_F \approx \tilde{J}\chi$ and $1/m_F \approx t\chi$. The coupling between the matter fields and gauge field is given by

$$L_{\text{int}} = \int d^2 r \left(j_\mu^F + j_\mu^B \right) a_\mu, \tag{14.36}$$

where j_μ^F (j_μ^B) is the fermion (boson) current density.

Note that integration over a_0 recovers the constraint Eq. (14.6) and integration over the vector potential $\mathbf{a}$ yields the constraint

$$\mathbf{j}_F + \mathbf{j}_B = 0, \tag{14.37}$$

i.e., the fermion and boson can move only by exchanging places. Thus the Gaussian approximation apparently enforces the local constraint exactly. We must caution that this is true only in the continuum limit, and an important lattice effect related to the π periodicity of the phase variable, i.e., the *compactness* of the gauge field, has been ignored. These latter effects lead to instantons and confinement, as will be discussed later. Thus it is not surprising that the "exact" treatment of Lee [70] yields the same Ioffe–Larkin composition rule which is derived based on the Gaussian theory as we next discuss.

We now proceed to reverse the order of integration. We integrate out the fermion and boson fields to obtain an effective action for a_μ. We then consider the coupling of the fermions and bosons to the gauge fluctuations which are controlled by the effective action. To avoid double counting, it may be useful to consider this procedure in the renormalization group sense, i.e., we integrate out the high energy fermion and boson fields to produce an effective action of the gauge field which in turn modifies the low energy matter field. This way we convert the initial problem of infinite coupling to one of finite coupling. The coupling is of order unity but may be formally organized as a $1/N$ expansion by artificially introducing N species of fermions. Alternatively, we can think of this as an RPA approximation, i.e., a sum of fermion and boson bubbles. The effective action for a_μ is given by the following

$$S_{\text{eff}}^{\text{RPA}}(a) = \left(\Pi_{\mu\nu}^{F}(q) + \Pi_{\mu\nu}^{B}(q)\right) a_\mu(q) a_\nu(-q), \tag{14.38}$$

where $q = (\mathbf{q}, \omega_n)$ is a three-dimensional vector. The current–current correlation function $\Pi_{\mu\nu}^{F}(q)$ $\left(\Pi_{\mu\nu}^{B}(q)\right)$ of the fermions (bosons) is given by $\Pi_{\mu\nu}^{\alpha}(q) = \langle j_\mu^\alpha(q) j_\nu^\alpha(-q)\rangle$ with $\alpha = F, B$. Taking the transverse gauge by imposing the gauge fixing condition $\nabla \bullet a = 0$ the scalar ($\mu = 0$) and vector parts of the gauge field dynamics are decoupled. The scalar part $\Pi_{00}^\alpha(q)$ corresponds to the density–density response function and does not show any singular behavior in the low energy/momentum limit. On the other hand, the transverse current–current response function shows singular behavior for small $\mathbf{q}$ and ω. Explicitly the fermion correlation function is given by

$$\Pi_T^F(q) = i\omega \sigma_{F1}^T(\mathbf{q},\omega) - \chi_F \mathbf{q}^2, \tag{14.39}$$

where $\chi_F = 1/(24\pi m_F)$ is the fermion Landau diamagnetic susceptibility. The first term describes the dissipation and the static limit of σ_{F1}^T (real part of the fermion conductivity) for $\omega < \gamma_\mathbf{q}$, $\sigma_{F1}^T(\mathbf{q}, \omega) = \rho_F/(m_F \gamma_\mathbf{q})$ where ρ_F is the fermion density and

$$\begin{aligned}\gamma_\mathbf{q} &= \tau_{\text{tr}}^{-1} \text{ for } |\mathbf{q}| < (v_F \tau_{\text{tr}})^{-1} \\ &= v_F |\mathbf{q}|/2 \text{ for } |\mathbf{q}| > (v_F \tau_{\text{tr}})^{-1},\end{aligned} \tag{14.40}$$

where τ_{tr} is the transport lifetime due to the scatterings by the disorder and/or the gauge field. It turns out that $\Pi^B \ll \Pi^F$ and the propagator of the transverse gauge field is given by

$$\langle a_\alpha(q) a_\beta(-q)\rangle = \left(\delta_{\alpha\beta} - q_\alpha q_\beta/|\mathbf{q}|^2\right) D_T(q), \tag{14.41}$$

$$D_T(q) = \left[\Pi_T^F(q) + \Pi_T^B(q)\right]^{-1} \cong \left[i\omega\sigma(\mathbf{q}) - \chi_d \mathbf{q}^2\right]^{-1}. \tag{14.42}$$

Here

$$\begin{aligned}\sigma(\mathbf{q}) &\cong k_0/|\mathbf{q}| \text{ for } |\mathbf{q}|\ell > 1 \\ &\cong k_0 \ell \text{ for } |\mathbf{q}|\ell < 1,\end{aligned} \tag{14.43}$$

where ℓ is the fermion mean free path and k_0 is of the order k_F of the fermions.

This gauge field is coupled to the fermions and bosons and leads to their inelastic scatterings. By estimating the lowest order self-energies of the fermion and boson propagators, it is found that these are diverging at any finite temperature. It is because of the singular behavior of $D_T(q)$ for small $|\mathbf{q}|$ and σ. This kind of singularity was first noted by Reizer [71] for the problem of electrons coupled to a transverse electromagnetic field, even though related effects such as nonFermi liquid corrections for the specific heat have been noted earlier by Holstein et al. [72]. As an example, we consider the conductivity of fermions and bosons. (Note that these are still not "physical" because one must combine these to obtain the physical conductivity as discussed in the next section.) The integral for the (inverse of) transport life-time τ_{tr} contains the factor $1 - \cos\theta$ where θ is the angle between the initial and final momentum for the scattering. This factor scales with $|\mathbf{q}|^2$ for small $\mathbf{q}$, and gets rid of the divergence. The explicit estimate gives

$$\frac{1}{\tau_{tr}^F} \cong \xi_\mathbf{k}^{4/3} \text{ for } \xi_\mathbf{k} > kT$$
$$\cong T^{4/3} \text{ for } \xi_\mathbf{k} < kT \tag{14.44}$$

for the fermions while

$$\frac{1}{\tau_{tr}^B} \cong \frac{kT}{m_B \chi_d} \tag{14.45}$$

for bosons. These results are interpreted as the scattering by the fluctuating gauge flux whose propagator is given by the loop representing the particle–hole propagator for the two-particle current–current correlation function.

Now some words on the physical meaning of the gauge field are in order. For simplicity let us consider the three sites, and that the electron is moving around these. The quantum mechanical amplitude for this process is

$$P_{123} = \langle \chi_{12}\chi_{23}\chi_{31} \rangle = \left\langle f_{1\alpha}^\dagger f_{2\alpha} f_{2\beta}^\dagger f_{3\beta} f_{3\gamma}^\dagger f_{1\gamma} \right\rangle. \tag{14.46}$$

One can prove that

$$(P_{123} - P_{132})/(4i) = \mathbf{S}_1 \cdot (\mathbf{S}_2 \times \mathbf{S}_3) \tag{14.47}$$

and the right-hand side of the above equation corresponds to the solid angle subtended by the three vectors $\mathbf{S}_1, \mathbf{S}_2, \mathbf{S}_3$, and is called spin chirality [73]. The left-hand side of Eq. (14.47) is proportional $\sin\phi$, where ϕ is the flux of the gauge field as seen by the fermions. Therefore the gauge field fluctuation is regarded as that of the spin chirality. A possible way to measure the chirality fluctuation using resonant Raman scattering has been discussed by Shastry and Shraiman [74].

In order to discuss the physical properties of the total system, we have to combine the information obtained for fermions and bosons. This has been first discussed by Ioffe and Larkin [65]. Let us start with the physical conductivity σ, which is given by

$$\sigma^{-1} = \sigma_F^{-1} + \sigma_B^{-1} \tag{14.48}$$

in terms of the conductivities of fermions (σ_F) and bosons (σ_B). This formula corresponds to the sequential circuit (not parallel) of the two resistance, and is intuitively understood from the fact that both fermions and bosons have to move subject to the constraint. This formula can be derived in terms of the shift of the gauge field $\mathbf{a}$, and resultant backflow effect. In the presence of the external electric field $\mathbf{E}$, the gauge field $\mathbf{a}$ and hence the internal electric field $\mathbf{e}$ is induced. Let us assume that the external electric field $\mathbf{E}$ is coupled to the fermions. Then the effective electric field seen by the fermions is

$$\mathbf{e}_F = \mathbf{E} + \mathbf{e} \tag{14.49}$$

while that for the boson is

$$\mathbf{e}_B = \mathbf{e}. \tag{14.50}$$

The fermion current $\mathbf{j}_F$ and boson current $\mathbf{j}_B$ are induced, respectively, as

$$\mathbf{j}_F = \sigma_F \mathbf{e}_F, \quad \mathbf{j}_B = \sigma_B \mathbf{e}_B. \tag{14.51}$$

The constraint $\mathbf{j}_F + \mathbf{j}_B = 0$ given by Eq. (14.37) leads to the relation

$$\mathbf{e} = -\frac{\sigma_F}{\sigma_F + \sigma_B}\mathbf{E}. \tag{14.52}$$

The physical current $\mathbf{j}$ given by

$$\mathbf{j} = \mathbf{j}_F = -\mathbf{j}_B = \frac{\sigma_F \sigma_B}{\sigma_F + \sigma_B}\mathbf{E} \tag{14.53}$$

leading to the expression for the physical conductivity σ in Eq. (14.48). It is also noted here that the same result is obtained if instead we couple the e.m. field to bosons. In this case the internal electric field $\mathbf{e}$ is different, but $\mathbf{e}_F$ and $\mathbf{e}_B$ remain unchanged. Therefore it is not a physical question which particle is charged, i.e., fermion or boson. Note that $\sigma_F \gg \sigma_B$ in the uRVB state, we conclude that $\sigma \cong \sigma_B = x \tau_{tr}^B / m_B$ which is inversely proportional to the temperature T. Furthermore the Drude weight of the optical conductivity is determined by x/m_B as is observed experimentally. It remains true that the superfluidity density ρ_S in the superconducting state is given by the missing oscillator strength below the gap, this also means that $\rho_S \propto x$.

The Ioffe–Larkin rule can be extended to various other physical quantities. For example, the thermopower $S = S_B + S_F$ and the electronic thermal conductivity the $\kappa = \kappa_B + \kappa_F$ are sum of the bosonic and fermionic contributions [67].

Compared with the two-particle correlation functions discussed above, the single particle Green's function is more complicated. At the mean field level, the electron Green's function is given by the product of those of fermions and boson in the $(\mathbf{r}, \tau)$ space. Therefore in the momentum–frequency space, it is given by the convolution. The spectral function is composed of the two contributions, one is the quasiparticle peak with the weight $\sim x$ while the other is the incoherent background. Even the former one is broadened due to the momentum distribution of the noncondensed bosons, i.e., there is no quasiparticle peak in the strict sense.

Combined with the discussion on the transport properties and the electron Green's function, the present uniform RVB state in the $U(1)$ formulation offers an explanation on the dichotomy between the doped Mott insulator and the metal with large Fermi surface. In particular, the conclusion that the conductivity is dominated by the boson conductivity $\sigma \approx \sigma_B \approx x \tau_{tr}^B / m_B \approx xtT$ explains the linear T resistivity which has been taken as a sign of nonFermi liquid behavior from the beginning of high T_c research. However, we must caution that this conclusion was reached for $T > T_{BE}^{(0)}$ while in the experiment the linear T behavior persists to much lower temperature near optimal doping. It is possible that gauge fluctuations suppress the effective Bose condensation. Lee et al. [75] attempted to include the effect of strong gauge fluctuations on the boson conductivity by assuming a quasi-static gauge fluctuation and treating the problem by quantum Monte Carlo. The picture is that the boson tends to make self-retracing paths to cancel out the effect of the gauge field [76]. They indeed find that the boson conductivity remains linear in T down to much lower temperature than $T_{BE}^{(0)}$.

14.7. *SU*(2) Slave-Boson Theory of Doped Mott Insulators

The $U(1)$ gauge theory described up to now encounters a number of difficulties when applied to the underdoped region. First, it is known from neutron scattering that spin correlation at (π, π) is enhanced as the doping is reduced. This happens at the same time the spin gap is formed in the pseudogap regime. In the $U(1)$ mean field theory the pseudogap is explained by fermion pairing and it is not clear how to enhance the spin correlation except by introducing phenomenological RPA interactions [56]. Secondly, the $U(1)$ theory predicts that the hc/e vortex is more stable than the $hc/2e$ superconducting vortex in the underdoped limit [77, 78]. This is because one can wind the boson phase by 2π around the vortex, while keeping the fermion gap intact inside the core. An $hc/2e$ vortex would necessarily destroy the fermion pairing inside the core. This contradicts the observation of a pseudogap inside the vortex core by STM [79, 80]. Finally, on a formal level, the $U(1)$ mean field theory ignores the $SU(2)$ symmetry at $x = 0$, which states that many apparently different mean field states are degenerate. Slightly away from $x = 0$, the degeneracy is slightly broken, but the $U(1)$ mean field state picks out only the most stable one (d-wave pairing) and completely ignores the other states which are nearby in energy. This motivates us to develop an $SU(2)$ theory for finite doping which connects smoothly to $x = 0$ and includes the myriad low lying states already at the mean field level [81].

14.7.1. $SU(2)$ Slave-Boson Mean-Field Theory at Finite Doping

The generalized $SU(2)$ slave-boson theory involves two $SU(2)$ doublets ψ_i and $h_i = \begin{pmatrix} b_{1i} \\ b_{2i} \end{pmatrix}$. Here b_{1i} and b_{2i} are two spin-0 boson field and $\psi_i = \begin{pmatrix} f_{i\uparrow} \\ f_{i\downarrow}^\dagger \end{pmatrix}$. The additional boson fields allow us to form $SU(2)$ singlet to represent the electron operator c_i:

$$c_{\uparrow i} = \tfrac{1}{\sqrt{2}} h_i^\dagger \psi_i = \tfrac{1}{\sqrt{2}} \left(b_{1i}^\dagger f_{\uparrow i} + b_{2i}^\dagger f_{\downarrow i}^\dagger \right)$$
$$c_{\downarrow i} = \tfrac{1}{\sqrt{2}} h_i^\dagger \bar{\psi}_i = \tfrac{1}{\sqrt{2}} \left(b_{1i}^\dagger f_{\downarrow i} - b_{2i}^\dagger f_{\uparrow i}^\dagger \right), \quad (14.54)$$

where $\bar{\psi} = i\tau^2 \psi^*$ which is also an $SU(2)$ doublet. The t–J Hamiltonian can now be written in terms of our fermion–boson fields. The Hilbert space of the fermion–boson system is larger than that of the t–J model. However, the local $SU(2)$ singlets satisfying $\left(\psi_i^\dagger \tau \psi_i + h_i^\dagger \tau h_i \right) |\text{phys}\rangle = 0$ form a subspace that is identical to the Hilbert space of the t–J model. On a given site, there are only three states that satisfy the above constraint. They are $f_\uparrow^\dagger |0\rangle$, $f_\downarrow^\dagger |0\rangle$, and $\tfrac{1}{\sqrt{2}} \left(b_1^\dagger + b_2^\dagger f_\uparrow^\dagger f_\downarrow^\dagger \right) |0\rangle$ corresponding to a spin up and down electron, and a vacancy, respectively. Furthermore, the fermion–boson Hamiltonian H_{tJ}, as a $SU(2)$ singlet operator, acts within the subspace, and has same matrix elements as the t–J Hamiltonian.

We note that just as in Eq. (14.8), our treatment of the $\tfrac{1}{4} n_i n_j$ term introduces a nearest neighbor boson attraction term which we shall ignore from now on. Now the partition function Z is given by

$$Z = \int D\psi\, D\bar{\psi}\, Dh\, Da_0^1 Da_0^2 Da_0^3 DU \exp\left(-\int_0^\beta d\tau\, L_2 \right)$$

with the Lagrangian taking the form

$$L_2 = \tilde{J} \sum_{\langle ij \rangle} \mathrm{Tr}\left[U_{ij}^\dagger U_{ij} \right] + \tilde{J} \sum_{\langle ij \rangle} \left(\psi_i^\dagger U_{ij} \psi_j + \text{c.c} \right) + \sum_i \psi_i^\dagger \left(\partial_\tau - ia_{0i}^\ell \tau^\ell \right) \psi_i$$
$$+ \sum_i h_i^\dagger \left(\partial_\tau - ia_{0i}^\ell \tau^\ell + \mu \right) h_i - \frac{1}{2} \sum_{\langle ij \rangle} t_{ij} \left(\psi_i^\dagger h_i h_j^\dagger \psi_j + \text{c.c} \right). \quad (14.55)$$

Following the standard approach with the choice $\tilde{J} = \frac{3}{8}J$, we obtain the following mean-field Hamiltonian for the fermion–boson system

$$H_{\text{mean}} = \sum_{\langle ij \rangle} \frac{3}{8} J \left[\frac{1}{2} \text{Tr}\left(U_{ij}^\dagger U_{ij} \right) + \left(\psi_i^\dagger U_{ij} \psi_j + \text{h.c.} \right) \right] - \frac{1}{2} \sum_{\langle ij \rangle} t \left(h_i^\dagger U_{ij} h_j + \text{h.c.} \right)$$

$$- \mu \sum_i h_i^\dagger h_i + \sum_i a_0^l \left(\psi_i^\dagger \tau^l \psi_i + h_i^\dagger \tau^l h_i \right). \quad (14.56)$$

The value of the chemical potential μ is chosen such that the total boson density (which is also the density of the holes in the t–J model) is

$$\left\langle h_i^\dagger h_i \right\rangle = \left\langle b_{1i}^\dagger b_{1i} + b_{2i}^\dagger b_{2i} \right\rangle = x.$$

The values of $a_0^l(\mathbf{i})$ are chosen such that

$$\left\langle \psi_i^\dagger \tau^l \psi_i + h_i^\dagger \tau^l h_i \right\rangle = 0.$$

For $l = 3$ we have

$$\left\langle f_{i\alpha}^\dagger f_{i\alpha} + b_{1i}^\dagger b_{1i} - b_{2i}^\dagger b_{2i} \right\rangle = 1. \quad (14.57)$$

We see that unlike the $U(1)$ slave-boson theory, the density of the fermion $\left\langle f_{i\alpha}^\dagger f_{i\alpha} \right\rangle$ is not necessarily equal to $1-x$. This is because a vacancy in the t–J model may be represented by an empty site with a b_1 boson, or a doubly occupied site with a b_2 boson.

To obtain the mean-field phase diagram, we have searched the minima of the mean-field free energy for the mean-field ansatz with translation, lattice, and spin rotation symmetries. We find a phase diagram with six different phases (see Figure 14.6) [81].

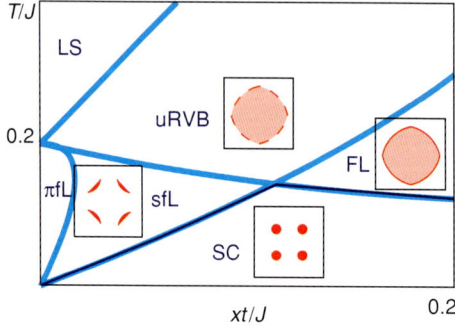

Figure 14.6. $SU(2)$ mean-field phase diagram for $t/J = 1$. The phase diagram for $t/J = 2$ is quantitatively very similar to the $t/J = 1$ phase diagram, when plotted in terms of the scaled variable xt/J, except the π fL phase disappears at a lower scaled doping concentration. We also plotted the Fermi surface, the Fermi arcs, or the Fermi points in some phases (from Wen and Lee [81]).

(1) The d-wave superconducting (SC) phase is described by the following mean-field ansatz

$$U_{i,i+\hat{x}} = -\chi\tau^3 + \Delta\tau^1$$
$$U_{i,i+\hat{y}} = -\chi\tau^3 - \Delta\tau^1$$
$$a_0^3 \neq 0, \quad a_0^{1,2} = 0$$
$$\langle b_1 \rangle \neq 0, \quad \langle b_2 \rangle = 0.$$
(14.58)

Notice that the boson condenses in the SC phase despite the fact that in our mean-field theory the interactions between the bosons are ignored.

(2) The Fermi liquid (FL) phase is similar to the SC phase except that there is no fermion pairing ($\Delta = 0$).

(3) Staggered flux liquid (sfL) phase:

$$U_{i,i+\hat{x}} = -\tau^3\chi - i(-)^i\Delta$$
$$U_{i,i+\hat{y}} = -\tau^3\chi + i(-)^i\Delta$$
$$a_0^l = 0, \quad \langle b_{1,2} \rangle = 0.$$
(14.59)

The U matrix is the same as that of the staggered flux phase in the $U(1)$ slave-boson theory, which breaks transition symmetry. Here the breaking of translational invariance is a gauge artifact. In fact, a site dependent $SU(2)$ gauge transformation $W_i = e^{-i\pi\tau^1/4}e^{-i\pi(i_x+i_y)(\tau^1/2+1)}$ maps the sfL ansatz to the d-wave pairing ansatz:

$$U_{i,i+\hat{x}} = -\chi\tau^3 + \Delta\tau^1$$
$$U_{i,i+\hat{y}} = -\chi\tau^3 - \Delta\tau^1$$
$$a_0^l = 0, \quad \langle b_{1,2} \rangle = 0,$$
(14.60)

which is explicitly translation invariant. However, the staggered flux representation of Eq. (14.59) is more convenient because the gauge symmetry is immediately apparent. Since this U matrix commutes with τ^3, it is clearly invariant under τ^3 rotation, but not τ^1 and τ^2, and the gauge symmetry has been broken from $SU(2)$ down to $U(1)$. For this reason we shall refer to this state as the staggered flux liquid (sfL). We emphasize that this $U(1)$ gauge field is distinct from the one discussed in the earlier section in connection with the uniform RVB state.

In the sfL phase, a_0^3 and we have $\langle f_{\alpha i}^\dagger f_{\alpha i} \rangle = 1$ and $\langle b_1^\dagger b_1 \rangle = \langle b_2^\dagger b_2 \rangle = \frac{x}{2}$.

(4) The π-flux liquid (πfL) phase is the same as the sfL phase except here $\chi = \Delta$.

(5) The uniform RVB (uRVB) phase is described by Eq. (14.60) with $\Delta = 0$.

(6) A localized spin (LS) phase has $U_{ij} = 0$ and $a_{0i}^l = 0$, where the fermions cannot hop.

Note that the topology of the phase diagram is similar to that of $U(1)$ mean field theory shown in Figure 14.5. The uRVB, sfL, πfL, and LS phases contain no boson condensation and correspond to unusual metallic states. As temperature is lowered, the uRVB phase changes into the sfL or πfL phases. A gap is opened at the Fermi surface near $(\pi, 0)$ which reduces the low energy spin excitations. Thus the sfL and πfL phases correspond to the pseudogap phase.

The FL phase contains boson condensation. In this case the electron Green's function $\langle c^\dagger c \rangle = \langle (\psi^\dagger h)(h^\dagger \psi) \rangle$ is proportional to the fermion Green's function $\langle \psi^\dagger \psi \rangle$. Thus the electron spectral function contain δ-function peak in the FL phase. Therefore, the low energy excitations in the FL phase are described by electron-like quasiparticles and the FL phase corresponds to a Fermi liquid phase of electrons.

The SC phase contains both the boson and the fermion-pair condensations and corresponds to a d-wave superconducting state of the electrons. Just like the $U(1)$ slave-boson theory, the superfluid density is given by

$$\rho_s = \frac{\rho_s^b \rho_s^f}{\rho_s^b + \rho_s^f},$$

where ρ_s^b and ρ_s^f are the superfluid density of the bosons and the condensed fermion-pairs, respectively. We see that in the low doping limit, $\rho_s \sim x$ and one needs the condensation of both the bosons and the fermion-pairs to get a superconducting state.

We would like to point out that the different mean-field phases contain different gapless gauge fluctuations at classical level, i.e., the gauge groups for gapless gauge fluctuations are different in different mean-field phases. The uRVB and the πfL phases have trivial $SU(2)$ flux and the gapless gauge fluctuations are $SU(2)$ gauge fluctuations. In the sfL phase, the ansatz Eq. (14.60) breaks the $SU(2)$ gauge structure to a $U(1)$ gauge structure. In this case the gapless gauge fluctuations are $U(1)$ gauge fluctuations. In the SC and FL phases, $\langle b_a \rangle \neq 0$. Since b_a transform as a $SU(2)$ doublet, there is no pure $SU(2)$ gauge transformation that leave mean-field ansatz (U_{ij}, a_0^l, b_a) invariant. As a result, the $SU(2)$ gauge structure is completely broken and there is no low energy gauge fluctuations.

14.7.2. Effect of Gauge Fluctuations: Enhanced (π, π) spin Fluctuations in Pseudogap Phase

As mentioned earlier, the pseudogap phase has a very puzzling property which seems hard to explain. As the doping is lowered, it was found experimentally that both the pseudogap and the antiferromagnetic (AF) spin correlation in the normal state increase. Naively, one expects the pseudogap and the AF correlations to work against each other. That is the larger the pseudogap, the lower the single particle density of states, the fewer the low energy spin excitations, and the weaker the AF correlations. It turns out that the gapless $U(1)$ gauge fluctuations present in the sfL phase play a key role in resolving the above puzzle [82, 83]. Due to the $U(1)$ gauge fluctuations, the AF spin fluctuations in the sfL phase are enhanced despite the presence of the pseudogap.

To see how the $U(1)$ gauge fluctuation in the sfL phase enhance the AF spin fluctuations, we map the lattice effective theory for the sfL state onto a continuum theory. In the low doping limit, the low energy excitations consist of nodal fermions centered at $\left(\pm\frac{\pi}{2}, \pm\frac{\pi}{2}\right)$ and bosons which are coupled to a $U(1)$ gauge field. At half filling the bosons are absent and this problem can be treated as a $\frac{1}{N}$ expansion, where N is the number of independent four-component nodal fermions. Rantner and Wen [83] showed that coupling to gauge fields leads to a singularity in the (π, π) spin fluctuation spectrum, in that the spectral fluctuation Im $\chi(q, \omega)$ is proportional to $(\omega^2 - q^2)^{1/2-\alpha}$. They found α to be $32/(3\pi^2 N)$. This singularity may explain why the neutron scattering detects enhanced staggered spin correlations in the pseudogap regime.

Rantner and Wen [84] also calculated the electron Green's function by combining the fermion and boson Green's function in a gauge invariant way. In the single hole limit, they find that the nodal quasiparticle at $\left(\frac{\pi}{2}, \frac{\pi}{2}\right)$ is destroyed and replaced by a very broad spectrum. At finite doping, by introducing binding between fermions and bosons due to gauge fluctuations in a phenomenological way, Wen and Lee [81, 85] produced electron spectra which display the Fermi arcs seen by ARPES experiment. Essentially, the nodal fermions are shifted towards (0, 0) and stretched out into an arc, while the gap at the antinodal points around $(0, \pi)$ remain intact.

The coupled problem of nodal fermions with a $U(1)$ gauge field leads to an interesting state where the staggered spin correlation functions have power law decay. This has been called the algebraic spin liquid (ASL). We stress that this is a phase of matter, and not a critical point in a phase transition. This state is unusual in that while the fermions are low energy excitations, they cannot be treated as free quasiparticles and response functions contain branch cuts rather than holes. As such, the ASL is reminiscent of the Luttinger liquid in one dimension. However, in this discussion, so far the compactness of the gauge field has been ignored in the $\frac{1}{N}$ expansion. In a compact gauge theory, instantons can appear which lead to confinement. Fortunately, it has recently been shown that instantons are irrelevant for sufficiently large N [86]. Thus confinement can be avoided and the large N expansion is internally consistent.

In the next section we shall further explore the properties of the $U(1)$ spin liquid upon doping. We approach the problem from the low temperature limit and work our way up in temperature. This regime is conveniently described by a nonlinear σ-model effective theory.

14.7.3. σ-Model Effective Theory and New Collective Modes in the Superconducting State

Here we attempt to reduce the large number of degrees of freedom in the partition function in Eq. (14.55) to the few which dominate the low energy physics. We shall ignore the amplitude fluctuations in the fermionic degree of freedom which are gapped on the scale of J. The bosons tend to Bose condense. We shall ignore the amplitude fluctuation and assume that its phase is slowly varying on the fermionic scale, which is given by $\xi = \varepsilon_F/\Delta$ in space. In this case we can have an effective field theory σ-model) description where the local boson phases are the slow variables and the fermionic degrees of freedom are assumed to follow them. We begin by picking a mean field representation $U_{ij}^{(0)}$. The choice of the staggered flux state U_{ij}^{SF} given by Eq. (14.60) is most convenient because U_{ij}^{SF} commutes with τ^3, making explicit the residual $U(1)$ gauge symmetry which corresponds to a τ^3 rotation. Thus we choose $U_{ij}^{(0)} = U_{ij}^{(SF)} e^{i a_{ij}^3 \tau^3}$ and replace the integral over U_{ij} by an integral over the gauge field a_{ij}^3. It should be noted that any $U_{ij}^{(0)}$ which are related by $SU(2)$ gauge transformation will give the same result. At the mean field level, the bosons form a band with minima at Q_0. Writing $h = \tilde{h} e^{i Q_0 \cdot r}$, we expect $\tilde{h}$ to be slowly varying in space and time. We transform to the radial gauge, i.e., we write

$$\tilde{h}_\mathbf{i} = g_\mathbf{i} \begin{pmatrix} b_\mathbf{i} \\ 0 \end{pmatrix}, \tag{14.61}$$

where $h_\mathbf{i}$ can be taken as real and positive and $g_\mathbf{i}$ is an $SU(2)$ matrix parametrized by

$$g_\mathbf{i} = \begin{pmatrix} z_{\mathbf{i}1} & -z_{\mathbf{i}2}^* \\ z_{\mathbf{i}2} & z_{\mathbf{i}1}^* \end{pmatrix}, \tag{14.62}$$

where

$$z_{\mathbf{i}1} = e^{i\alpha_\mathbf{i}} e^{-i\frac{\phi_\mathbf{i}}{2}} \cos\frac{\theta_\mathbf{i}}{2} \tag{14.63}$$

and
$$z_{i2} = e^{i\alpha_i} e^{i\frac{\phi_i}{2}} \sin\frac{\theta_i}{2}. \tag{14.64}$$

We ignore the boson amplitude fluctuation and replace b_i by a constant b_0.

An important feature of Eq. (14.56) is that L_2 is invariant under the $SU(2)$ gauge transformation

$$\tilde{h}_i = g_i^\dagger h_i, \tag{14.65}$$

$$\tilde{\psi}_i = g_i^\dagger \psi_i, \tag{14.66}$$

$$\tilde{U}_{ij} = g_i^\dagger U_{ij}^{(0)} g_j, \tag{14.67}$$

and

$$\tilde{a}_{0i}^\ell \tau^\ell = g^\dagger a_{0i}^\ell \tau^\ell g - g(\partial_\tau g^\dagger). \tag{14.68}$$

Starting from Eq. (14.55) and making the above gauge transformation, the partition function is integrated over g_i instead of h_i and the Lagrangian takes the form

$$L_2' = \frac{\tilde{J}}{2} \sum_{\langle ij \rangle} \text{Tr}\left(\tilde{U}_{ij}^\dagger \tilde{U}_{ij}\right) + \tilde{J} \sum_{\langle ij \rangle} \psi_i^\dagger \tilde{U}_{ij} \psi_j + \text{c.c.} + \sum_i \psi_i^\dagger \left(\partial_\tau - i a_{0i}^\ell \tau^\ell\right) \psi_i$$
$$+ \sum_i \left(-i a_{0i}^3 + \mu_B\right) b_0^2 - \sum_{ij,\sigma} \tilde{t}_{ij} b_0^2 f_{j\sigma}^\dagger f_{i\sigma}. \tag{14.69}$$

We have removed the tilde from $\tilde{\psi}_{i\sigma}$, $\tilde{f}_{i\sigma}$, $\tilde{a}_0^\ell$ because these are integration variables and $\tilde{t}_{ij} = t_{ij}/2$. Note that g_i appears only in $\tilde{U}_{ij}$. For every configuration $\{g_i(\tau), a_{ij}^3(\tau)\}$ we can, in principle, integrate out the fermions and a_0^ℓ to obtain an energy functional. This will constitute the σ-model description. In practice, we can make the slowly varying g_i approximation and solve the local mean field equation for a_{0i}^ℓ. This is the approach taken by Lee et al. [85].

The σ-model depends on $\{g_i(\tau), a_{ij}^3(\tau)\}$, i.e., it is characterized by $\alpha_i, \theta_i, \phi_i$, and the gauge field a_{ij}^3. α_i is the familiar overall phase of the electron operator which becomes half of the pairing phase in the superconducting state. To help visualize the remaining dependence of freedom, it is useful to introduce the local quantization axis

$$\mathbf{I}_i = z_i^\dagger \tau z_i = (\sin\theta_i \cos\phi_i, \sin\theta_i \sin\phi_i, \cos\theta_i). \tag{14.70}$$

Note that $\mathbf{I}_i$ is independent of the overall phase α_i, which is the phase of the physical electron operator. Then different orientations of $\mathbf{I}$ represent different mean field states in the $U(1)$ mean field theory. This is shown in Figure 14.7. For example, $\mathbf{I}$ pointing to the north pole corresponds to $g_i = \mathbf{I}$ and the staggered flux state. This state has $a_0^3 \neq 0$, $a_0^1 = a_0^2 = 0$ and has small Fermi pockets. It also has orbital staggered currents around the plaquettes. $\mathbf{I}$ pointing to the south pole corresponds to the degenerate staggered flux state whose staggered pattern is shifted by one unit cell. On the other hand, when $\mathbf{I}$ is in the equator, it corresponds to a d-wave superconductor. Note that the angle ϕ is a gauge degree of freedom and states with different ϕ anywhere along the equator are gauge equivalent. A general orientation of $\mathbf{I}$ corresponds to some combination of d-SC and s-flux.

t–J Model and the Gauge Theory Description of Underdoped Cuprates

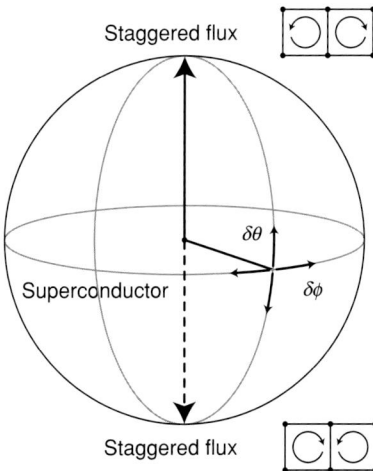

Figure 14.7. The quantization axis **I** in the $SU(2)$ gauge theory. The north and south poles correspond to the staggered flux phases with shifted orbital current patterns. All points on the equators are equivalent and correspond to the d-wave superconductor. In the superconducting state one particular direction is chosen on the equator. There are two important collective modes. The θ modes correspond to fluctuations in the polar angle $\delta\theta$ and the ϕ gauge mode to a spatially varying fluctuation in $\delta\phi$.

At zero doping, all orientations of **I** are energetically the same. This symmetry is broken by doping, and the **I** vector has a small preference to lie on the equator. At low temperature, there is a phase transition to a state where **I** lies on the equator, i.e., the d-SC ground state. It is possible to carry out a small expansion about this state and work out explicitly the collective modes [87]. In an ordinary superconductor, there is a single complex order parameter Δ and we expect an amplitude mode and a phase mode. For a charged superconductor the phase mode is pushed up to the plasma frequency and one is left with the amplitude mode only. In the gauge theory we have in addition to Δ_{ij} the order parameter χ_{ij}. Thus it is natural to expect additional collective modes. From Figure 14.7 we see that two modes are of special interest corresponding to small θ and ϕ fluctuations. Physically the θ mode corresponds to local fluctuations of the s-flux states which generate local orbital current fluctuations. These currents generate a small magnetic field (estimated to be $\sim 10\,\text{G}$) which couples to neutrons. Lee and Nagaosa [87] predict a peak in the neutron scattering cross-section at (π, π), at energy just below $2\Delta_0$, where Δ_0 is the maximum d-wave gap. This is *in addition* to the resonance mode seen experimentally which is purely spin fluctuation in origin. The orbital origin of this mode can be distinguished from the spin fluctuation by its distinct form factor [61, 88].

The ϕ mode is more subtle because ϕ is the phase of a Higgs field, i.e., it is part of the gauge degree of freedom. It turns out to correspond to a relative oscillation of the *amplitudes* of χ_{ij} and Δ_{ij} and is again most prominent at (π, π). Since $|\chi_{ij}|$ couples to the bond density fluctuation, inelastic Raman scattering is the tool of choice to study this mode, once the technology reaches the requisite 10 meV energy resolution. Due to the special nature of the buckled layers in LSCO, this mode couples to photons and may show up as a transfer of spectral weight from a buckling phonon to a higher frequency peak. Such a peak was reported experimentally [89], but it is apparently not unique to LSCO as the theory would predict, and hence its interpretation remains unclear at this point.

From Figure 14.7 it is clear that the σ-model representation of the $SU(2)$ gauge theory is a useful way of parameterizing the myriad $U(1)$ mean field states which become almost degenerate for small doping. The low temperature d-SC phase is the ordered phase of the σ-model, while in the high temperature limit we expect the **I** vector to be disordered in space and time, to the point where the σ-mode approach fails and one crosses over to the $SU(2)$ mean field description. The disordered phase of the σ-model then corresponds to the pseudo-gap phase. How does this phase transition take place? It turns out that the destruction of superconducting order proceeds via the usual route of BKT proliferation of vortices. To see how this comes about in the σ-model description, we have to first understand the structure of vortices.

14.7.4. Vortex Structure

The σ-model picture leads to a natural model for a low energy $hc/2e$ vortex [90]. It takes advantage of the existence of two kinds of bosons b_1 and b_2 with opposite gauge charges but the same coupling to electromagnetic fields. Far away from the vortex core, $|b_1| = |b_2|$ and b_1 has constant phase while b_2 winds its phase by 2π around the vortex. As the core is approached $|b_2|$ must vanish in order to avoid a divergent kinetic energy, as shown in Figure 14.8 (top). The quantization axis **I** provides a nice way to visualize this structure [Figure 14.8 (bottom)]. It smoothly rotates to the north pole at the vortex core, indicating that at this level of approximation, the core consists of the staggered flux state. The azimuthal angle winds by 2π as we go around the vortex. It is important to remember that **I** parameterizes only the internal gauge degrees of freedom θ and ϕ and the winding of ϕ by 2π is different from the usual winding of the overall phase α by π in an $hc/2e$ vortex. To better understand the phase winding we write down the following continuum model for the phase θ_1, θ_2 of b_1 and b_2, valid far away from the core.

$$D = \int d^2x \frac{K}{2} \left[(\nabla\theta_1 - \mathbf{a} - \mathbf{A}) + (\nabla\theta_2 + \mathbf{a} - \mathbf{A})^2 \right] + \cdots, \quad (14.71)$$

where **a** stands for the continuum version of a_{ij}^3 in the last section and **A** is the electromagnetic field (e/c has been set to be unity). We now see that the $hc/2e$ vortex must contain a half integer vortex of the **a** gauge flux with an opposite sign. Then θ_1 sees zero flux while θ_2 sees 2π flux, consistent with the windings chosen in Figure 14.8. This vortex structure has low energy for small x because the fermion degrees of freedom remain gapped in the core and one does not pay the fermionic energy of order J as in the $U(1)$ gauge theory. Physically, the above description takes advantage of the states with almost degenerate energies (in this case

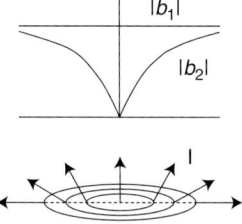

Figure 14.8. Structure of the superconducting vortex. Top: b_1 is constant while b_2 vanishes at the center and its phase winds by 2π. Bottom: The isospin quantization axis points to the north pole at the center and rotates toward the equatorial plane as one moves out radially. The pattern is rotationally symmetric around the $\hat{z}$ axis.

the staggered flux state) which is guaranteed by the $SU(2)$ symmetry near half filling. There is direct evidence from STM tunneling that the energy gap is preserved in the core [79, 80]. This is in contrast to theoretical expectations for conventional d-wave vortex cores, where a large resonance is expected to fill in the gap in the tunneling spectra [91].

We can clearly reverse the roles of b_1 and b_2 to produce another vortex configuration which is degenerate in energy. In this case **I** in Figure 14.8 points to the south pole. These configurations are sometimes referred to merons (half of a hedgehog) and the two halves can tunnel to each other via the appearance of instantons in space–time. The time scale of the tunneling event is difficult to estimate, but should be considerably less than J. Depending on the time scale, the orbital current of the staggered flux state in the core generates a physical staggered magnetic field which may be experimentally observable by NMR (almost static), μSR (intermediate time scale), and neutron (short time scale). The experiment must be performed in a large magnetic field so that a significant fraction of the area consists of vortices and the signal of the staggered field should be proportional to H. A μSR experiment on underdoped YBCO has detected such a field dependent signal with a local field of ± 18 G [92]. However, μSR is not able to determine whether the field has an orbital or spin origin and this experiment is only suggestive, but by no means definitive, proof of orbital currents in the vortex core. In principle, neutron scattering is a more definitive probe, because one can use the form factor to distinguish between orbital and spin effects. However, due to the small expected intensity, neutron scattering has so far not yielded any definite results.

As discussed earlier, we expect enhanced (π, π) fluctuations to be associated with the staggered flux liquid phase. Indeed, the s-flux liquid state is our route to Néel order and if gauge fluctuations are large, we may expect to have quasistatic Néel order inside the vortex core. Experimentally, there are reports of enhanced spin fluctuations in the vortex core by NMR experiments [93–95]. There are also reports of static incommensurate spin order forming a halo around the vortex in the LSCO family [96–99]. One possibility is that these halos are the condensation of pre-existing soft incommensurate modes known to exist in LSCO, driven by quasistatic Néel order inside the core. We emphasize the s-flux liquid state is our way of producing antiferromagnetic order starting from microscopies and hence is fully consistent with the appearance of static or dynamical antiferromagnetism in the vortex core. Our hope is that gauge fluctuations (including instanton effects) are sufficiently reduced in doped systems to permit a glimpse of the staggered orbital current. The detection of such currently fluctuations will be a strong confirmation of our approach.

Finally, we note that orbital current does not show up directly in STM experiments, which are sensitive to the local density of states. However, Kishine et al. [100] have considered the possibility of interference between Wannier orbitals on neighboring lattice sites, which could lead to modulations of STM signals *between* lattice positions. STM experiments have detected 4×4 modulated patterns in the vortex core region and also in certain underdoped regions. Such patterns appear to require density modulations which are in addition to our vortex model.

14.7.5. Phase Diagram

We can now construct a phase diagram of the underdoped cuprates starting from the d-wave superconductor ground state at low temperatures. The vortex structure allows us to unify the σ-model picture with the conventional picture of the destruction of superconducting order in two dimensions, i.e., the Berezinskii–Kosterlitz–Thouless (BKT) transition via the unbinding of vortices. The σ-model contains in addition to the pairing phase 2α, the phases θ and ϕ. However, we saw in the last section that a particular configuration of θ and ϕ is favored inside the vortex core. The $SU(2)$ gauge theory provides a mechanism for cheap vortex core energy which is necessary for a BKT description. If the core energy is too large, the system will behave like a superconductor on any reasonable length scale above T_{BKT}, which is not in accord with experiment. On the other hand, if the core energy is small compared with T_c, vortices will proliferate rapidly. They overlap and lose their identity. There is now strong experimental evidence based on the Nernst effect that vortices survive over a considerable temperature range above T_c [101–103]. Taken as a whole, these experiments require the vortex core energy to be cheap, but not too cheap, i.e., of the order of T_c. Honerkamp and Lee [104] have attempted a microscopic modeling of the proliferation of vortices. They assume an s-flux core and estimate the energy from projected wavefunction calculations. They indeed found that there is a large range of temperature above the BKT transition where vortices grow in number but still maintain their identity. This forms a region in the phase diagram which may be called the Nernst region shown in Figure 14.9. The corresponding picture of the **I**

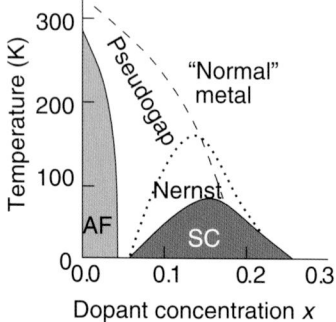

Figure 14.9. Schematic phase diagram showing the phase fluctuation regime where the Nernst effect is large. Note that this regime is a small part of the pseudogap region for small doping.

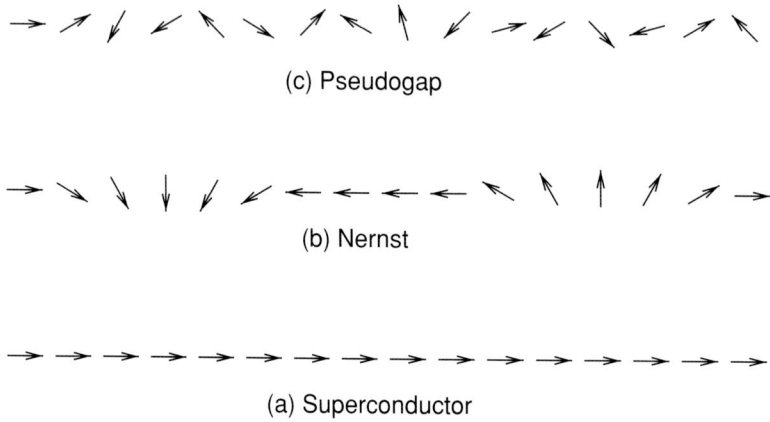

Figure 14.10. Schematic picture of the quantization axis **I** in different parts of the phase diagram shown in Figure 14.9. (a) In the superconducting phase **I** is ordered in the $x-y$ plane. (b) In the Nernst phase, **I** points to the north or south pole inside the vortex core. (c) The pseudogap corresponds to a completely disordered arrangement of **I**. (**I** is a three-dimensional vector and only a two-dimensional projection is shown.)

vector fluctuation is shown in Figure 14.10. Above the Nernst region the **I** vector is strongly fluctuating and is almost isotropic. This is the strongly disordered phase of the σ-model. The vortices have lost their identity and indeed the σ-model description which assumes well-defined phases of b_1 and b_2 begin to break down. Nevertheless, the energy gap associated with the fermions remains. This is the pseudogap part of the phase diagram in Figure 14.9. In the $SU(2)$ gauge theory this is understood as the algebraic spin liquid (ASL) discussed earlier. There is no order parameter in the usual sense associated with this phase, as all fluctuations including staggered orbital currents and d-wave pairing become short range. Is there a way to characterize this state of affairs other than the term spin liquid? This question is addressed later, but first we digress to discuss recent advances in understanding the spin liquid state and its relation to deconfinement in gauge theories.

14.8. Spin Liquids, Deconfinement, and the Emergence of Gauge Fields and Fractionalized Particles

The spin liquid state was first proposed by Anderson [39] and Fazekas and Anderson [40] as the ground state for the spin-$\frac{1}{2}$ AF Heisenberg model on a triangular lattice. One important consequence of the spin liquid state is that the low energy excitations are spin-$\frac{1}{2}$ particles, called spinons. This is in contrast to the Néel ordered AF, in which case the low energy excitations are magnons which carry $S = 1$. The spinons are considered an example of fractionalized quasiparticles. Doping holes into a spin liquid leads naturally to a superconductor, and this idea forms the backbone of Anderson's RVB theory. However, the ground states of the AF Heisenberg model on the triangular lattice is now known to be Néel ordered, with neighboring spins forming 120° angles. Furthermore, up to now no example of Heisenberg models on any lattice has been firmly established as having a spin liquid ground state, either theoretically of experimentally. (Some promising possible examples will be mentioned later.) Thus the spin liquid concept and, by implication, the RVB idea have been met with skepticism by the community. Yet much progress has been made in the past several years in characterizing the spin liquid state and in the understanding of the phenomenon of fractionalization. Here we summarize what is known and also attempt to clear up some common misconceptions.

By now we have many theoretically well-established examples of ground states which do not show Néel order and which support fractionalized excitations. For example, Kitaev [105] has found an exactly soluble model on the honeycomb lattice and Wen [106] on the square lattice. These models support fermionic excitations and Z_2 gauge fields. The interaction involves multispin interactions which break spin rotation symmetry and thus these are not Heisenberg models per se. Earlier, a dimer model on a triangular lattice, admitted not a spin Hamiltonian, has been shown to support Z_2 gauge fields and fractionalized spinons [107]. Other examples where strong arguments for fractionalization with varying degrees of rigor can be made include certain spin-$\frac{1}{2}$ models on the Kagome lattice [108] and bosonic models which in principle can be realized by Josephson junction arrays [109]. Recently, a model of multiorbital exciton condensations has been shown to support $U(1)$ gauge photons and either bosonic or fermionic fractionalized particles depending on the coupling constants [110].

There has also been progress on more realistic spin models using less reliable methods. Exact diagonalization of small clusters of the $S = \frac{1}{2}$ Heisenberg model on the Kagome lattice and on the triangular lattice with ring exchange [42] suggests the existence of a disordered ground state. The latter model has recently been studied by Motrunich [43] using projected trial wavefunction methods. Also, it has been proposed that the Hubbard model on the triangular lattice may have a spin liquid ground state just on the insulating side of the Mott transition based on numerical [44] and analytic [45] considerations. These latter works were motivated by experiments on the organic compound κ-(BED-TTF)$_2$Cu$_2$(CN)$_3$ which appears to be a promising candidate for the spin liquid state [111, 112].

In all these examples, the emergence of fractionalized particles at low energies is invariably accompanied by the emergence of a gauge field. The gauge structure may be Z_2 or $U(1)$, but they must be in the deconfined phase. Thus the phenomenon of fractionalization is identified with deconfinement of certain gauge structures. We have seen that gauge fields emerge very naturally in the slave particle representation of the spin operator. When **S** is written as $f_\alpha^\dagger \sigma_{\alpha\beta} f_\beta$, we see that **S** is invariant under the gauge transformation $f_\alpha \to f_\alpha e^{i\theta}$, i.e., there is a redundancy in the representation which forces one to introduce a $U(1)$ gauge field in this case. However, there are many different ways to represent the spin operators. One can write them either in terms of slave bosons or fermions and these are originally introduced as formed devices. How can they emerge as real particles? These are reasonable questions but they also lead to a number of misconceptions which we address below.

(1) What is the distinction between high energy and low energy gauge groups. As we mentioned earlier, the spin operator can be decoupled into fermions or bosons. Furthermore, the gauge field associated with this decoupling is not unique. It can be $U(1)$, $SU(2)$ as we have seen earlier, or Z_2 [113]. These different formulations of the theory are distinct on the lattice scale, but in principle they are all exact and they are all equivalent to the same Heisenberg model. We refer to the different formulations as the high energy gauge group. On the other hand, fractionalization and gauge fields are emergent quantities which describe the low energy physics. We refer to these as the low energy gauge structures. In principle a $U(1)$ low energy gauge field can emerge from a $SU(2)$ or Z_2 high energy gauge structure. It is a matter of convenience which formulation one chooses to begin with, in that the emergent gauge field may correspond closely to the mean-field decoupling of one formulation and not another. To show that fractionalization occurs, one has to show that a deconfined phase of the low energy gauge structure exists.

(2) If the initial gauge coupling is infinite, how can a deconfined state emerge? This question was raised by Nayak [114] and addressed by several subsequent comments [115, 116]. The gauge field is introduced as a way of enforcing constraint. There is not Maxwellian restoring force, i.e., the coupling constant is infinite. In pure compact gauge theory this leads to confinement. In the presence of matter fields such as fermions and bosons, the situation is more complicated and examples of deconfinement with infinite coupling are cited in [115]. A clear example is given in the recent work which studies the Bose condensation of excitons with multiple band indices $a = 1, \ldots, N$ [110]. On the short distance scale the exciton field χ_{ab} made up of band a electrons and band b holes. The exciton field can be represented by a product of bosonic fields $\psi_a^\dagger \psi_b$. In a world-line picture of the excitons, the Bose particle is always part of an exciton, i.e., it is never unbound on the lattice scale. This is enforced by an infinite gauge coupling constant. Nevertheless, excitons scatter with each other and exchange partners, thus rapidly losing their identity. However, the band index is conserved and the world line of an individual particle, which carries a fixed band index, emerges as a low energy excitation which lives much longer than an individual exciton χ_{ab}. It is also clear that new particles are not free but coupled

to each other through a web of other exciton world lines. It can be shown that this web of exciton world lines is equivalent to fluctuating $U(1)$ gauge fields. In the large N limit, the effective gauge coupling is small despite the initial infinite coupling, essentially due to screening. Thus a Coulomb phase of fractionalized particles and gauge photons emerge as low energy excitations. This example also shows that the emergent particles may be fermion or boson, depending on the coupling constant. The statistics and spin of the emergent particles are governed by dynamics of the microscopic Hamiltonian, and are independent of the method of decoupling. This again reinforces the distinction between high energy and low energy gauge groups.

A related objection that is often made is that the field ψ_a (and the slave fermion or boson fields used to decouple the spin operators) are not gauge invariant and therefore cannot be real. Well, the world we live in is the Coulomb phase of the compact QED, and the electron operator is also not gauge invariant under an electromagnetic gauge transformation. What it means is that electrons are always accompanied by electromagnetic fields and do not live in isolation. Since the gauge fluctuations are so weak, we get accustomed to fix the gauge and think of electrons as real objects. Strictly speaking, electrons are real only in the sense that gauge invariant quantities can be easily understood in terms of them. The same is true with fractionalized particles, provided we are in the deconfined phase and the coupling to the gauge field is relatively weak. If the coupling is of order unity, the resulting ground state can be complicated. Nevertheless, the emergent particles and gauge fields are the appropriate starting points to address the following question.

(3) What is the ultimate fate of the emergent fractionalized particles? It turns out that depending on the situation, the fractionalized particles may or may not behave as quasiparticles, i.e., with poles in the spectral function, in the low field limit. A number of examples of fractionalization is based on the Z_2 gauge group, which is discrete and has a well understood deconfined phase even in $2 + 1$ dimensions. In the deconfined phase, excitations of the gauge fields (called visons) are gapped and therefore dilute. It is possible to fix the gauge and the matter fields become well-defined quasiparticles which may be either gapped or gapless. In the case where fractionalization is based on a low energy $U(1)$ gauge group, if the matter fields are gapped, deconfinement can occur only in $3 + 1$ dimensions, in which case the emergent fractionalized particles are also well defined quasiparticles. This is the case for the exciton condensate described above. The algebraic spin liquid (ASL) discussed in the last section turns out to be a special example. There the emergent particle is a gapless nodal fermion which is coupled to a $U(1)$ gauge field. As a result of the coupling, the low energy excitations are not well-defined $S = \frac{1}{2}$ quasiparticles (spinons) and the spin excitations take on branch cuts as discussed earlier. The case of a gauge field coupled to an emergent Fermi sea of fermions and nonrelativistic bosons encountered in the $U(1)$ gauge theory is even more difficult to control and is less understood at present.

14.9. Application of Gauge Theory to the High T_c Superconductivity Problem

Now we summarize how the gauge theory concepts we have described may be applied to the high T_c problem. The central observation is that high T_c superconductivity emerges upon doping a Mott insulator. The antiferromagnetic order of the Mott insulator disappears

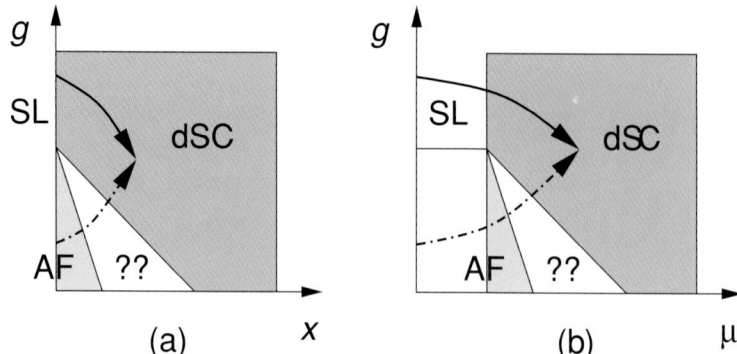

Figure 14.11. (a) Schematic zero temperature phase diagram showing the route between the antiferromagnetic Mott insulator and the d-wave superconductor. The vertical axis is labeled by a parameter g which may be taken as a measure of the frustration in the interaction between the spins in the Mott insulator. AF represents the antiferromagnetically ordered state. SL is a spin liquid insulator that could potentially be reached by increasing the frustration. The path taken by the cuprate materials as a function of doping x is shown in a thick dashed-dot line. The question marks represent regions where the physics is not clear at present. Doping the spin liquid naturally leads to the dSC state. The idea behind the spin liquid approach is to regard the superconducting system at nonzero x as resulting from doping the spin liquid as shown in the solid line, though this is not the path actually taken by the material. (b) Same as in Figure 14.11(a) but as a function of chemical potential rather than hole doping. Here the AF includes an insulating and a lightly doped (light shaded) regions.

rather rapidly and is replaced by the superconducting ground state. The "normal" state above the superconducting transition temperature exhibits many unusual properties which we refer to as pseudogap behavior. How does one describe the simultaneous suppression of Néel order and the emergence of the pseudogap and the superconductor from the Mott insulator? The approach we take is to first understand the nature of a possible nonmagnetic Mott state at zero doping, the spin liquid state, which naturally becomes a singlet superconductor when doped. This is the central idea behind the RVB proposal [1] and is summarized in Figure 14.11. The idea is that doping effectively frustrates the Néel order so that the system is pushed across the transition where the Néel order is lost. In the real system the loss of Néel order may proceed through complicated states, such as incommensurate charge and spin order, stripes or inhomogeneous charge segregation [117]. However, in this direct approach the connection with superconductivity is not at all clear. Instead it is conceptually useful to arrive at the superconducting state via a different path, starting from a spin liquid state. Recently, Senthil and Lee [118] have elaborated upon this point of view which we summarize below.

14.9.1. Spin Liquid, Quantum Critical Point, and the Pseudogap

It is instructive to consider the phase diagram as a function of the chemical potential rather than the hole doping as shown in Figure 14.11b. Consider any spin liquid Mott state that when doped leads to a d-wave superconductor. As a function of chemical potential, there will then be a zero temperature phase transition where the holes first enter the system. For concreteness we will simply refer to this as the Mott transition. The associated quantum critical fixed point will control the physics in a finite nonzero range of parameters. The various crossovers expected near such transitions are well known and are shown in Figure 14.12.

Sufficiently close to this zero temperature critical point many aspects of the physics will be universal. The regime in which such universal behavior is observed will be limited by

t–J Model and the Gauge Theory Description of Underdoped Cuprates

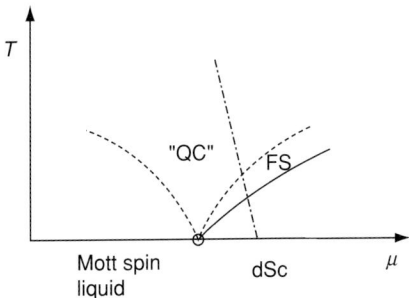

Figure 14.12. Schematic phase diagram for a doping induced Mott transition between a spin liquid insulator and a d-wave superconductor. The bold dot-dashed line is the path taken by a system at hole density x that has a superconducting ground state. The region marked FS represents the fluctuation regime of the superconducting transition. The region marked QC is the quantum critical region associated with the Mott critical point. It is separated from the Mott spin liquid state by a cross-over line (dashed). The QC region may be identified with the high temperature pseudogap phase in the experiments.

"cut-offs" determined by microscopic parameters. In particular we may expect that the cutoff scale is provided by an energy of a fraction of J (the exchange energy for the spins in the Mott insulator). We note that this corresponds to a reasonably high temperature scale.

Now consider an underdoped cuprate material at fixed doping x. Upon increasing the temperature this will follow a path in Figure 14.12 that is shown schematically. The properties of the system along this path may be usefully discussed in terms of the various crossover regimes. In particular it is clear that the "normal" state above the superconducting transition is to be understood directly as the finite temperature "quantum critical" region associated with the Mott transition. Empirically this region corresponds to the pseudogap regime. Thus our assertion is that the pseudogap regime is controlled by the unstable zero temperature fixed point associated with the (Mott) transition to a Mott insulator.

What are the candidates for the spin liquid phase? There have been several proposals in the literature. One proposal is the dimer phase [119]. Strictly speaking, this is a valence bond solid and not a spin liquid: it is a singlet state which breaks translational symmetry. It has been shown by Read and Sachdev [120] that within the large N Schwinger boson approach the dimer phase emerges upon disordering the Néel state. Sachdev and collaborators have shown that doping the dimer state produces a d-wave superconductor [121]. However, such a superconductor also inherits the dimer order and has a full gap to spin excitations, at least for low doping. As we have seen in this review, there are strong empirical evidence for gapless nodal quasiparticles in the superconducting state. In our view, it is more natural to start with translation invariant spin liquid states which produce d-wave superconductors with nodal quasiparticles when doped.

We have seen that the spin liquid states are rather exotic beasts in that their excitations are conveniently described in terms of fractionalized spin $\frac{1}{2}$ "spinon" degrees of freedom. We discussed that spin liquids are characterized by their low energy gauge group. Among spin liquids with nodal fermionic spinons, two versions, the Z_2 and the $U(1)$ spin liquids have bee proposed. The Z_2 gauge theory was advocated by Senthil and Fisher [113]. It can be considered as growing out of the fermion pairing phase of the $U(1)$ mean field phase diagram shown in Figure 14.5. The pairing of fermions $\Delta_{ij} = \langle f_{i\uparrow} f_{j\downarrow} - f_{i\downarrow} f_{j\uparrow} \rangle$ breaks the $U(1)$ gauge symmetry down to Z_2, i.e., only $f \rightarrow -f$ remains unbroken. One feature of this theory is that in the superconducting state hc/e vortices tend to have lower energy than $hc/2e$

vortices, particularly at low doping. We that $hc/2e$ vortices involve suppression of the pairing amplitude $|\Delta_{ij}|$ at the center and cost a large energy of order J. On the other hand, one can form an hc/e vortice by winding the boson phase by 2π, leaving the fermion pairing intact inside the core. Another way of describing this from the point of view of Z_2 gauge theory is that the $hc/2e$ vortex necessarily involves the presence of a Z_2 gauge flux (called a vison by Senthil and Fisher) in its core. The finite energy cost of the Z_2 flux dominates in the low doping limit and raises the energy of the $hc/2e$ vortices. Experimental proposals were made [122] to provide for a critical test of such a theory by detecting the vison excitation or by indirectly looking for signatures of stable hc/e vortices. To date, all such experiments have yielded negative results and provided fairly tight bounds on the vison energy [123].

We are then left with the $U(1)$ spin liquid (also called the algebraic spin liquid ASL) as the final candidate. The mean field basis of this state is the staggered flux liquid state of the $SU(2)$ mean field phase diagram (Figure 14.6). The low energy theory of this state consists of fermions with massless Dirac spectra (nodal quasiparticles) interacting with a $U(1)$ gauge field. Note that this $U(1)$ gauge field refers to the low energy gauge group and is not to be confused with the high energy gauge group. This state has enhanced (π, π) spin fluctuations but no long-range Néel order, and the ground states becomes a d-wave superconductor when doped with holes. As we have seen, a low energy $hc/2e$ vortex can be constructed, thus overcoming a key difficulty of the Z_2 gauge theory. Furthermore, an objection in the literature about the stability of the $U(1)$ spin liquid has been overcome, at least for sufficiently large N [86]. It has also been argued by Senthil and Lee [118] that even if the physical spin $\frac{1}{2}$ case does not possess a stable $U(1)$ liquid phase, it can exist as a critical state separating the Néel phase from a Z_2 spin liquid and may still have the desired property of dominating the physics of the pseudogap and the superconducting states. An example of deconfinement appearing at the critical point between two ordered phases is recently pointed out by Senthil et al. [124].

14.9.2. Signature of the Spin Liquid

If the pseudogap region is controlled by the $U(1)$ spin liquid fixed point, is it possible to characterize this region in a certain precise way? The spin liquid is a deconfined state, meaning that instantons are irrelevant. Then the $U(1)$ gauge flux is a conserved quantity. Unfortunately, it is not clear how to couple to this gauge flux using conventional probes. We note that the flux associated with the $\mathbf{a}^3$ gauge field is *different* from the $U(1)$ gauge flux considered in Section 14.6., which had the meaning of spin chirality.

In the superconducting state the gauge flux is localized in the vortex core and fluctuations between $\pm$ half integer vortices are possible via instantons, because the instanton action is finite. The superconductor is in a confined phase as far as the $U(1)$ gauge field is concerned. As the temperature is raised toward the pseudogap phase this gauge field leaks out of the vortex cores and begins to fluctuate more and more homogeneously. The observation that the gauge flux is associated with the vortex core led Senthil and Lee [124] to propose a way to generate and detect the existence of conserved gauge fluxes. Their proposed experiment involves creating a disc in the pseudogap phase with two concentric rings of superconductors, i.e., if we denote T_c of the outer ring, inner ring and the disc as T_{c1}, T_{c2}, T_{c3} then $T_{c3} < T_{c2} < T_{c1}$.

Now consider the following set of operations on such a sample.

(i) First cool in a magnetic field to a temperature T_{in} such that $T_{c2} < T_{in} < T_{c1}$. The outer ring will then go superconducting while the rest of the sample stays normal. In the presence of the field the outer ring will condense into a state in which there is a net vorticity on going around the ring. We will be interested in the case where this net vorticity is an odd multiple of the basic $hc/2e$ vortex. If as assumed the physical penetration depth is much bigger than the thickness ΔR_o then the physical magnetic flux enclosed by the ring will not be quantized.

(ii) Now consider turning off the external magnetic field. The vortex present in the outer superconducting ring will stay (manifested as a small circulating persistent current) and will give rise to a small magnetic field. As explained above if the vorticity is odd, then it must be associated with a flux of the internal gauge field that is $\pm\pi$. This internal gauge flux must essentially all be in the inner "normal" region of the sample with very small penetration into the outer superconducting ring. It will spread out essentially evenly over the full inner region. We have thus managed to create a configuration with a nonzero internal gauge flux in the nonsuperconducting state.

(iii) How do we detect the presence of this internal gauge flux? For that imagine now cooling the sample further to a temperature T_{fin} such that $T_{c3} < T_{\text{fin}} < T_{c2}$. Then the inner ring will also go superconducting. This is to be understood as the condensation of the two boson species $b_{1,2}$. But this condensation occurs in the presence of some internal gauge flux. When the bosons $b_{1,2}$ condense in the inner ring, they will do so in a manner that quantizes the internal gauge flux enclosed by this inner ring into an integer multiple of π. If as assumed the inner radius is a substantial fraction of the outer radius then the net internal gauge flux will prefer the quantized values $\pm\pi$ rather than be zero. However, configurations of the inner ring that enclose quantized internal gauge flux of $\pm\pi$ also necessarily contain a physical vortex that is an odd multiple of $hc/2e$. With the thickness of the inner ring being smaller than the physical penetration depth, most of the physical magnetic flux will escape. There will still be a small residual physical flux due to the current in the inner ring associated with the induced vortex. This residual physical magnetic flux can then be detected.

Note that the sign of the induced physical flux is independent of the sign of the initial magnetic field. Furthermore the effect obtains only if the initial vorticity in the outer ring is odd. If on the other hand the initial vorticity is even the associated internal gauge flux is zero, and there will be no induced physical flux when the inner ring goes superconducting.

14.10. Summary and Outlook

In this review we have summarized a large body of work which views high temperature superconductivity as the problem of doping of a Mott insulator. We have argued that the $t - J$ model, supplemented by t' terms, contains the essence of the physics. Superconductivity with d-wave pairing emerges as a natural candidate for the ground state. The driving force is the exchange interaction J and the temperature scale for superconductivity is set by xt. These simple observations answer the question: what is unique about the cuprate. The answer is that the exchange constant J in the cuprates is among the highest known and the two dimensional $S = \frac{1}{2}$ system is also unique in that the effect of quantum fluctuations is maximized. Already at the mean field level, the phase diagram includes the d-wave superconductor and the pseudogap state and captures the essential features of the experiment. Variational Monte Carlo calculations using projected wavefunctions further improve these comparisons. Further progress on analytic theory hinges on the treatment of the constraint of no double occupation. The redundancy in the representations used to enforce the constraint naturally leads to various gauge theories. We argue that with doping, the gauge theory may be in a deconfined phase, in which case the slave-boson and fermion degrees of freedom, which were introduced as mathematical devices, take on a physical meaning in that they are sensible starting points to describe physical phenomena. However, even in the deconfined phase, the coupling to gauge fluctuations is still of order unity and approximation schemes (such as large N expansion) are needed to calculate physical properties such as spin correlation and electron spectral function. These results qualitatively capture the physics of the pseudogap phase, but certainly not at a quantitative level. Nevertheless, our picture of the vortex structure and how they proliferate gives us a reasonable account of the phase diagram and the onset of T_c.

One direction of future research is to refine the treatment of the low energy effective model, i.e., fermions and bosons coupled to gauge fields, and attempt more detailed comparison with experiments such as photoemission lineshapes, etc. On the other hand, it is worthwhile to step back and take a broader perspective. What is really new and striking about the high temperature superconductors is the strange "normal" metallic state for underdoped samples. The carrier density is small and the Fermi surface is broken up by the appearance of a pseudogap near $(0, \pi)$ and $(\pi, 0)$, leaving a "Fermi arc" near the nodal points. All this happens without doubling of the unit cell via breaking translation or spin rotation symmetry. How this state comes into being in a lightly doped Mott insulator is the crux of the problem. We can distinguish between two classes of answers. The first, perhaps the more conventional one, postulates the existence of a symmetry-breaking state which gaps the Fermi surface, and further assumes that thermal fluctuation prevents this state from ordering. A natural candidate for the state is the superconducting state itself. However, it now appears that phase fluctuations of a superconductor can explain the pseudogap phenomenon only over a relatively narrow temperature range, which we called the Nernst regime. Alternatively, a variety of competing states which have nothing to do with superconductivity have been proposed, often on a phenomenological level, to produce the pseudogap. We shall refer to this class of theory as "thermal" explanation of the pseudogap.

A second class of answer, which we may dub the "quantum" explanation, proposes that the pseudogap is connected with a fundamentally new quantum state. Thus, despite its appearance at high temperatures, it is argued that it is a high frequency phenomenon which is best understood quantum mechanically. The gauge theory reviewed here belongs to this class, and views the pseudogap state as derived from a new state of matter, the quantum spin liquid state. The spin liquid state is connected to the Néel state at half filling by confinement. At the same time, with doping a d-wave superconducting ground state is naturally produced. We argue that rather than following the route taken by the cuprate in the laboratory of evolving directly from the antiferromagnet to the superconductor, it is better conceptually to start from the spin liquid state and consider how AF and superconductivity develop from it. In this view the pseudogap is the closest we can get to obtaining a glimpse of the spin liquid which up to now is unstable in the square lattice $t–J$ model.

Is there a "smoking gun" signature to prove or disprove the validity of this line of theory? Our approach is to make specific predictions as much as possible in the hope of stimulating experimental work. This is the reason we make special emphasis on the staggered flux liquid with its orbital current fluctuations, because it is a unique signature which may be experimentally detectable. Our predictions range from new collective modes in the superconducting state, to quasi-static order in the vortex core. Unfortunately the physical manifestation of the orbital current is a very weak magnetic field, which is difficult to detect, and to date we have not found experimental verification. Besides orbital current, we also propose an experiment involving flux generation in a special geometry. This experiment addresses the fundamental issue of the quantum spin liquid as the origin of the pseudogap phase.

In the past several years signification progress has been made in our understanding of the concept of fractionalization and the characterization of the spin liquid state. In physics we are familiar with the binding of particles to form more complicated entities, electrons and nuclei form atoms, atoms form molecules and solids, etc. Fractionalization is the reverse process, where particles which carry a fraction of the quantum number of the original particles emerge as low energy excitations. What we have learned is that gauge field invariably accompany these particles, and the fractionalized particles and gauge fields should be considered as *emergent* phenomena since there is no sign of them in the original short distance starting point. As time progresses we will see more concrete examples of this phenomena, and hopefully real experiments as well. Progress in this direction will help establish the point of view of high temperature superconductivity which is summarized in this review.

Acknowledgments

We would like to thank X.-G. Wen, Naoto Nagaosa, and T. Senthil for their collaboration and many helpful discussions. We acknowledge support by NSF grant number DMR-0517222.

Bibliography

1. P.W. Anderson, Science **235**, 1196 (1987).
2. P.A. Lee, N. Nagaosa, and X.G. Wen, Rev. Mod. Phys. **78**, 17 (2006).
3. V.J. Emery, Phys. Rev. Lett. **58**, 3759 (1987).
4. C.M. Varma, S. Schmitt-Rink, and E. Abrahams, Solid State Commun. **62**, 681 (1987).
5. L.F. Mattheiss, Phys. Rev. Lett. **58**, 1028 (1987).
6. J. Yu, A. Freeman, and J.-H. Xu, Phys. Rev. Lett. **58**, 1035 (1987).
7. J. Zaanen, G. Sawatzky, and J. Allen, Phys. Rev. Lett. **55**, 418 (1985).
8. M.A. Kastner, R. Birgeneau, G. Shirane, and Y. Endoh, Rev. Mod. Phys. **70**, 897 (1998).
9. P.E. Sulewsky, P. Fleury, K. Lyons, S. Cheong, and Z. Fisk, Phys. Rev. B **41**, 225 (1990).
10. R. Coldea, S. Hayden, G. Aeppli, T. Perrig, C. Frost, T. Mason, S. Cheong, and Z. Fisk, Phys. Rev. Lett. **86**, 5377 (2001).
11. F.C. Zhang and T. Rice, Phys. Rev. B **37**, 3759 (1988).
12. M.S. Hybertson, E. Stechel, M Schuter, and D. Jennison, Phys. Rev. B. **41**, 11068 (1990).
13. O.K. Andersen et al., J. Low Temp. Phys. **105**, 285 (1996).
14. E. Pavarini, I. Dasgupta, T. Saha-Dasgupta, O. Jasperson, and O. Andersen, Phys. Rev. Lett. **87**, 047003 (2001).
15. N.J. Curro, T. Imai, C. Slichter, and B. Dabrowski, Phys. Rev. B **56**, 877 (1997).
16. J.W. Loram, K. Mirza, J. Cooper, and W. Liang, Phys. Rev. Lett. **71**, 1740 (1993); J.W. Loram, J. Luo, W. Liang, and J. Tallon, J. Phys. Chem. Solids **62**, 59 (2001).
17. A.F. Santander-Syro, R.P.S.M. Lobo, N. Bontemps, Z. Konstantinovic, Z. Li, and H. Raffy, Phys. Rev. Lett. **88**, 097005 (2002).
18. J. Orenstein, G. Thomas, A. Millis, S. Cooper, D. Rapkine, T. Timusk, L. Schneemeyer, and J. Waszczak, Phys. Rev. B **42**, 6342 (1990).
19. S.L. Cooper, D. Reznik, A. Kotz, M.A. Karlow, R. Liu, M.V. Klein, W.C. Lee, J. Giapintzakis, D.M. Ginsberg, B.W. Veal, et al., Phys. Rev. B **47**, 8233 (1993).
20. S. Uchida, T. Ido, H. Takagi, T. Arima, Y. Tokura, and S Tajima, Phys. Rev. B **43**, 7942 (1991).
21. W.J. Padilla, Y.S. Lee, M. Dumm, G. Blumberg, S. Ono, K. Segawa, S. Komiya, Y Ando, and D. Basov, Phys. Rev. B **72**, 060511 (2005).
22. C.C. Homes, T. Timusk, R. Liang, D. Bonn, and W. Hardy, Phys. Rev. Lett. **71**, 4210 (1993).
23. S. Uchida, Physica C **282**, 12 (1997).
24. A.C. Loeser, Z.-X. Shen, D. Dessau, D. Marshall, C. Park, P. Fournier, and A. Kapitulnik, Science **273**, 325 (1996).

25. D.S. Marshall, D.S. Dessau, A.G. Loeser, C.-H. Park, A.Y. Matsuura, J.N. Eckstein, I. Bozovic, P. Fournier, A. Kapitulnik, W.E. Spicer, et al., Phys. Rev. Lett. **76**, 4841 (1996).
26. H. Ding, T. Yokoya, J. Campuzano, T.T. Takahashi, M. Randeria, M. Norman, T. Mochiku, and J. Giapintzakis, Nature **382**, 51 (1996).
27. T. Valla, A. Fedorov, P. Johnson, B. Wells, S. Hulbert, Q. Li, G. Gu, and N. Koshizuka, Science **285**, 2110 (1999).
28. T. Valla et al., cond-mat/0512685.
29. T. Yoshida et al., Phys. Rev. Lett. **91**, 027001 (2003).
30. F. Ronning, T. Sasagawa, Y. Kohsaka, K.M. Shen, A. Damascelli, C. Kim, T. Yoshida, N.P. Armitage, D.H. Lu, D.L. Feng, et al., Phys. Rev. B **67**, 165101 (2003).
31. H.J.A. Molegraaf, C. Pressura, D. van der Marel, P.H. Kes, and M. Li, Science **295**, 2239 (2002).
32. A.V. Boris, N.N. Kovaleva, O.V. Dolgov, T. Holden, C.T. Lin, B. Keimer, and C. Berkhard, Science **304**, 708 (2004).
33. A.B. Kuzmenko, H.J.A. Molegraaf, F. Carbone, and D. van der Marel, Phys. Rev. B **72**, 144503 (2005).
34. A.F. Santander-Syro and N. Bontemps, cond-mat/0503768; G. Deutscher, A.F. Santander-Syro, and N. Bontemps, Phys. Rev. B **72**, 092504 (2005).
35. W.W. Warren, R. Walstedt, G. Brennert, R. Cava, R. Tyeko, R. Bell, and G. Dabbaph, Phys. Rev. Lett. **62**, 1193 (1989).
36. H. Yasuoka, T. Imai, and T. Shimizu, in *Strong Correlation and Superconductivity*, H. Fukuyama, S. Maekawa, and A.P. Malozemoff (eds.) (Springer, Berlin, Heidelberg, NewYork, 1989), p. 254.
37. M. Takigawa, A.P. Reyes, P.C. Hammel, J.D. Thompson. R.H. Heffner, Z. Fisk, and K.C. Ott, Phys. Rev. B **43**, 247 (1991).
38. G.-Q. Zheng, T. Odaguchi, T. Mito, Y. Kitaoka, K. Asayama, and Y. Kodama, J. Phys. Soc. Jpn **62**, 2591 (2003).
39. P.W. Anderson, Mat. Res. Bull. **8**, 153 (1973).
40. P. Fazekas and P. Anderson, Philos. Mag. **30**, 432 (1974).
41. D.A. Huse and V. Elser, Phys. Rev. Lett. **60**, 2531 (1988).
42. G. Misguich and C. Lhuillier, in *Frustrated Spin Systems*, H.T. Diep (ed.) (World Scientific, Singapore, 2004).
43. O.I. Motrunich, Phys. Rev. B **72**, 045105 (2005).
44. M. Imada, T. Mizusaki, and S. Watanabe, cond-mat/0307022; H. Morita, S. Watanabe, and M. Imada, J. Phys. Soc. Jpn **71**, 2109 (2002).
45. S.-S. Lee and P.A. Lee, Phys. Rev. Lett. **95**, 036403 (2005).
46. S.A. Kivelson, D.S. Rokhsar, and J.P. Sethna, Phys. Rev. B **35**, 8865 (1987).
47. C. Gros, Ann. Phys. **189**, 53 (1988).
48. A. Paramekanti, M. Randeria, and N. Trivedi, Phys. Rev. Lett. **87**, 217002 (2001).
49. P.W. Anderson, M. Randeria, T. Rice, N. Trivedi, and F. Zhang, J. Phys. Condens. Matter **16**, R755 (2004).
50. F.C. Zhang, C. Gros, T.M. Rice, and H. Shiba, Supercond. Sci. Technol. **1**, 36 (1988).
51. S.E. Barnes, J. Phys. F **6**, 1375 (1976).
52. P. Coleman, Phys. Rev. B **29**, 3035 (1984).
53. G. Baskaran, Z. Zou, and P.W. Anderson, Solid State Commun. **63**, 973 (1987).
54. J.W. Negele and H. Orland, *Quantum Many-Particle Systems* (Addison-Wesley, Reading, MA, 1987).
55. M. Ubbens and P.A. Lee, Phys. Rev. B **46**, 8434 (1992).
56. J. Brinckmann and P.A. Lee, Phys. Rev. B **65**, 014502 (2001).
57. I. Affleck and J.B. Marston, Phys. Rev. B **37**, 3774 (1988).
58. I. Affleck, Z. Zou, T. Hsu, and P.W. Anderson, Phys. Rev. B **38**, 745 (1988).
59. E. Dagotto, E. Fradkin, and A. Moreo, Phys. Rev. B **38**, 2926 (1988).
60. T.C. Hsu, Phys. Rev. B **41**, 11379 (1990).
61. T. Hsu, J.B. Marston, and I. Affleck, Phys. Rev. B **43**, 2866 (1991).
62. G. Kotliar and J. Liu, Phys. Rev. B **38**, 5142 (1988).
63. H. Fukuyama, Prog. Theor. Phys. Suppl. **108**, 287 (1992).
64. G. Baskaran and P.W. Anderson, Phys. Rev. B **37**, 580 (1988).
65. L. Ioffe and A. Larkin, Phys. Rev. B **39**, 8988 (1989).
66. N. Nagaosa and P.A. Lee, Phys. Rev. Lett. **64**, 2450 (1990).
67. P.A. Lee and N. Nagaosa, Phys. Rev. B **46**, 5621 (1992).
68. L.B. Ioffe and G. Kotliar, Phys. Rev. B **42**, 10348 (1990).
69. R. Hlubina, W. Putikka, T.M. Rice, and D. Khveshchenko, Phys. Rev. B **46**, 11224 (1992).
70. D.-H. Lee, Phys. Rev. Lett. **84**, 2694 (2000).

71. M. Reizer, Phys. Rev. B **39**, 1602 (1989).
72. T. Holstein, R. Norton, and P. Pincus, Phys. Rev. B **8**, 2649 (1973).
73. X.-G. Wen, F. Wilczek, and A. Zee, Phys. Rev. B **39**, 11413 (1989).
74. B.S. Shastry and B.I. Shraiman, Phys. Rev. Lett. **65**, 1068 (1990).
75. D.K.K. Lee, D. Kim, and P.A. Lee, Phys. Rev. Lett. **76**, 4801 (1996).
76. N. Nagaosa and P.A. Lee, Phys. Rev. B **43**, 1233 (1991).
77. S. Sachdev, Phys. Rev. B **45**, 389 (1992).
78. N. Nagaosa and P.A. Lee, Phys. Rev. B **45**, 966 (1992).
79. I. Maggio-Aprile, C. Renner, A. Erb, E. Walker, and O. Fischer, Phys. Rev. Lett. **75**, 2754 (1995).
80. S.-H. Pan, E. Hudson, A. Gupta, K.-W. Ng, H. Eisaki, S. Uchida, and J. David, Phys. Rev. Lett. **85**, 1536 (2000).
81. X.-G. Wen and P.A. Lee, Phys. Rev. Lett. **76**, 503 (1996).
82. D.H. Kim and P.A. Lee, Ann. Phys. **272**, 130 (1999).
83. W. Rantner and X.-G. Wen, Phys. Rev. B **66**, 144501 (2002).
84. W. Rantner and X.-G. Wen, cond-mat/ 0105540.
85. P.A. Lee, N. Nagaosa, T.-K. Ng, and X.-G. Wen, Phys. Rev. B **57**, 6003 (1998).
86. M. Hermele, T. Senthil, M.P.A. Fisher, P.A. Lee, N. Nagaaosa, and X.-G. Wen, Phys. Rev. B **70**, 214437 (2004).
87. P.A. Lee and N. Nagaosa, Phys. Rev. B **68**, 024516 (2003).
88. S. Chakravarty, H. Kee, and C. Nayak, Int. J. Mod. Phys. B **16**, 3140 (2002).
89. A.B. Kuzmenko, N. Tombros, H. Molegraaf, M. Greuninger, D. van der Marel, and S. Uchida, Phys. Rev. Lett. **91**, 037004 (2003).
90. P.A. Lee and X.-G. Wen, Phys. Rev. B **63**, 224517 (2001).
91. Y. Wang and A. MacDonald, Phys. Rev. B **52**, 3876 (1995).
92. R.I. Miller, R.F. Kiefel, J.H. Brewer, J.E. Sonier, J. Chakhalian, S. Dunsiger, G.D. Morris, A.N. Price, D.A. Bonn, W.H. Hardy, et al., Phys. Rev. Lett. **88**, 137002 (2002).
93. N.J. Curro, C. Milling, J. Haase, and C. Slichter, Phys. Rev. B **62**, 3473 (2000).
94. V.F. Mitrovic, E. Sigmund, H. Bachman, M. Eschrig, W. Halperin, A. Reyes, P. Kuhns, and W. Moulton, Nature **413**, 505 (2001); V.F. Mitrovic, E. Sigmund, W. Halperin, A. Reyes, P. Kuhns, and W. Moulton, Phys. Rev. B **67**, 220503 (2003).
95. K. Kakuyanagi, K. Kumagai, and Y. Matsuda, Phys. Rev. B **65**, 060503 (2002).
96. S.M. Kitano, K. Yamada, T. Suzuki, and T. Fukase, Phys. Rev. B **62**, 14677 (2000).
97. B. Lake, G. Aeppli, K. Clausen, D. McMorrow, K. Lefmann, N. Hussey, N. Mangkorntong, M. Nohara, H. Takagi, T. Mason, et al., Science **291**, 1759 (2001).
98. B. Lake, H. Ronnow, N. Christensen, G. Aeppli, K. Lefmann, D. McMorrow, P. Vorderwisch, P. Smeibidl, M. Mangkorntong, T. Sasagawa, et al., Nature **415**, 299 (2002).
99. B. Khaykovich, Y.S. Lee, R. Erwin, S.-H. Lee, S. Wakimoto, K. Thomas, M. Kastner, and R. Birgeneau, Phys. Rev. B **66**, 014528 (2002).
100. J. Kishine, P.A. Lee, and X.-G. Wen, Phys. Rev. B **65**, 064526 (2002).
101. Y. Wang, N.P. Ong, Z. Xu, T. Kakeshita, S. Uchida, D. Bonn, R. Liang, and W. Hardy, Phys. Rev. Lett. **88**, 257003 (2002).
102. Y. Wang, S. Ono, Y. Onose, G. Gu, Y. Ando, Y. Tokura, S. Uchida, and N.P. Ong, Science **299**, 86 (2003).
103. Y. Wang, Z. Yu, T. Kakeshita, S. Uchida, S. Ono, Y. Ando, and N.P. Ong, Phys. Rev. B **64**, 224519 (2001).
104. C. Honerkamp and P.A. Lee, Phys. Rev. Lett. **39**, 1201 (2004).
105. A.Y. Kitaev, Ann. Phys. (N.Y.) **303**, 2 (2003).
106. X.-G. Wen, Phys. Rev. Lett. **90**, 016803 (2003).
107. R. Moessner and S.L. Sondhi, Phys. Rev. Lett. **86**, 1881 (2001).
108. L. Balents, M.P.A. Fisher, and S.M. Girvin, Phys. Rev. B **65**, 224412 (2002).
109. T. Senthil and O. Motrunich, Phys. Rev. B **66**, 205104 (2002).
110. S.S. Lee and P.A. Lee, Phys. Rev. B **72**, 235104 (2005).
111. Y. Shimizu, K. Miyagawa, K. Kanoda, M. Maesato, and G. Saito, Phys. Rev. Lett. **91**, 107001 (2003).
112. A. Kawamoto, Y. Honma, and K. Kumagai, Phys. Rev. B **70**, 060510 (2004).
113. T. Senthil and M.P.A. Fisher, Phys. Rev. B **62**, 7850 (2000).
114. C. Nayak, Phys. Rev. Lett. **85**, 178 (2000).
115. I. Ichinose and T. Matsui, Phys. Rev. Lett. **86**, 942 (2001).
116. M. Oshikawa, Phys. Rev. Lett. **91**, 109901 (2003).
117. E.W. Carlson, V. Emery, S. Kivelson, and D. Orgad, in *Physics of Conventional and Unconventional Superconductors*, K.H. Bennemann and J.B. Ketterson (eds.) (Springer, Berlin, Heidelberg, NewYork, 2003).

118. T. Senthil and P.A. Lee, Phys. Rev. B **71**, 174515 (2005).
119. S. Sachdev, Rev. Mod. Phys. **75**, 913 (2003).
120. N. Read and S. Sachdev, Phys. Rev. B **42**, 4568 (1990).
121. M. Vojta and S. Sachdev, Phys. Rev. Lett. **83**, 3916 (1999).
122. T. Senthil and M.P.A. Fisher, Phys. Rev. Lett. **86**, 292 (2001).
123. D.A. Bonn, J. Wynn, B.W. Gardner, R. Liang, W. Hardy, J. Kirteley, and K. Moler, Nature **414**, 887 (2001).
124. T. Senthil, A. Vishwanath, L. Balents, S. Sachdev, and M.P.A. Fisher, Science **303**, 1490 (2004).

15

How Optimal Inhomogeneity Produces High Temperature Superconductivity

Steven A. Kivelson and Eduardo Fradkin

Before Vic Emery's untimely death, we had the privilege of working closely with him on the role of Coulomb frustrated phase separation in doped Mott insulators, and on the consequences of the resulting local electronic structures on the "mechanism" of high temperature superconductivity. In the present paper, we discuss the resulting perspective on superconductivity in the cuprates, and on the more general theoretical issue of what sorts of systems can support high temperature superconductivity. We discuss some of the general, qualitative aspects of the experimental lore which we think should constrain any theory of the mechanism, and show how they are accounted for within the context of our theory.

The focus of this paper is a "dynamic inhomogeneity-induced pairing" mechanism of high temperature superconductivity (HTC) in which the pairing of electrons originates directly from strong repulsive interactions.[1] Repulsive interactions can be shown, by exact solution, to lead to a form of local superconductivity on certain mesoscale structures, but the strength of this pairing tendency decreases as the size of the structures increases above an optimal size. Moreover, the same physics responsible for pairing within a structure provides the driving force for the Coulomb frustrated phase separation that leads to the formation of mesoscale electronic structures in many highly correlated materials. From this perspective, the formation of mesoscale structures (such as "stripes") in the cuprate superconductors may not be a problem for the mechanism of superconductivity but rather a part of the mechanism itself. This mechanism is not based, as is the BCS mechanism [1], on the pairing of preexisting well-defined and essentially free quasiparticles. Rather, it is based on the physics of strong correlations and low dimensionality. In this approach, coherence and quasiparticles are *emergent phenomena* at low energy, not an assumed property of the "high energy physics" from which this state derives.

The existence of strong local pairing does not guarantee a large critical temperature, since in a system of electronically isolated structures, the phase ordering (condensation) temperature is suppressed by phase fluctuations, often to $T = 0$. Thus, the highest possible superconducting transition temperature is obtained at an intermediate degree of inhomogeneity. A corollary of this is that the optimal T_c always occurs at a point of crossover from a pairing dominated regime when the system is too homogeneous, to a phase ordering regime with a pseudo-gap when the system is too granular.

---

[1] By "dynamic inhomogeneity" we mean inhomogeneity, whether static or fluctuating, which is generated dynamically by the strongly interacting degrees of freedom.

Steven A. Kivelson • Department of Physics, Stanford University, Stanford CA 93105, USA
Department of Physics and Astronomy, University of California Los Angeles, Los Angeles, CA 90095-1547, USA
Eduardo Fradkin • Department of Physics, University of Illinois at Urbana-Champaign, 1110 West Green Street, Urbana, IL 61801-3080, USA

Coulomb frustrated phase separation leads to mesoscale electronic structures as a generic feature of highly correlated electronic systems. (By "mesoscale" we mean on length scales longer than but of order of the superconducting coherence length, ξ_0.) Usually this tendency leads to dominant charge density wave (CDW) and spin density wave (SDW) order, or possibly to more exotic electronic liquid crystalline phases, which can coexist with but tend to compete with superconductivity. However, we argue that one feature that is special about the cuprate high temperature superconductors is that the intrinsic electronic inhomogeneity is strong enough to produce high temperature pairing, but strongly fluctuating enough that it does not entirely kill phase coherence.[2]

In Section 15.1, we discuss the reasons that HTC is difficult, and hence why there are so few high temperature superconductors. In Section 15.2 we discuss the inhomogeneity induced pairing mechanism of HTC. Section 15.3 reports the latest theoretical development in this area—a solved model, the "striped Hubbard model," for which a well-controlled theoretical treatment is possible, and many of the qualitative points made in the other sections can be illustrated explicitly. Then, in Section 15.4 we briefly discuss the ways in which incipient charge order, especially due to Coulomb frustrated phase separation, can lead to the sort of local (slowly fluctuating) electronic inhomogeneities required for the proposed mechanism, as well as to a host of interesting "competing ordered" phases; a much more complete discussion of these aspects of the problem, with an extensive review of the experimental evidence in the cuprates, is contained in [3]. Sections 15.5, which discusses the relative merits of the weak and strong coupling perspectives, and 15.6, which examines what is so special about the cuprates, deal explicitly with HTC in the cuprates, as opposed to the more abstract issues treated in the first part. These sections can be viewed as a set of commentaries, rather than a coherent exegesis. In Section 15.7, we highlight some of the salient conclusions. Finally, in Appendix A we give a theoretical definition of HTC.

With the exception of Section 15.3, the discussion in this paper is entirely qualitative and descriptive. For all but the most recent developments, a more detailed and technical discussion can be found in a review article [4], which also includes extensive references to the original literature.

15.1. Why High Temperature Superconductivity is Difficult

Before 1986, all but a few lonely voices proclaimed that superconductivity with transition temperatures much above 20 K was impossible. Since the experimental discovery of high temperature superconductivity in the cuprates, scores of different theoretical arguments have been presented demonstrating that any number of simple model Hamiltonians are superconducting below a temperature which is "high" in the sense that it is equal to a number of order one times a microscopic electronic energy scale. These calculations, however, are typically uncontrolled, in the sense that they cannot be justified either as exact solutions of the stated model, or as asymptotic expansions in powers of a small parameter—they rely on physical intuition rather than systematic solution in any traditional sense of the word.

It seems to us that the answer cannot be so simple. The arguments (some of which are reviewed later and in [4]) made before 1986 were not ill-considered, even if they may have

---

[2] That the building blocks of an appropriate theory of strongly correlated systems should involve various self-organized mesoscale structures, rather than simple weakly interacting quasiparticles, is genetically related to the point of view articulated by P. W. Anderson in his famous monograph, *More is different* [2]. He, however, may deny paternity.

been accepted somewhat too uncritically—in materials that are basically good metals (Fermi liquids) there are, indeed, serious reasons to suspect that high temperature superconductivity is implausible. Moreover, even now, that we have learned to expand our horizons to include "bad metals" (i.e., resistively challenged materials which are not well described by Fermi liquid theory), the number of high temperature superconducting materials remains extremely small; maybe it is only the cuprates that can legitimately be called high temperature superconductors, or the class may include some subset of alkali doped C_{60}, $Ba_{1-x}K_xBiO_3$, $(TMTSF)_2ClO_4$, BEDT, MgB_2, and $Na_{0.3}CoO_{2y}H_2O$.

In Figure 15.1, we show the distribution of superconducting transition temperatures among over 500 superconducting materials, as tabulated by Geballe and White [5] in 1979. The definition of what constitutes a distinct "material" is somewhat arbitrary (e.g., at what point, as one varies the concentration of two constituents of an alloy, does it become a new material). However, what is clear from the figure is that materials with transition temperatures above 15 K are, already, extremely rare exceptions. Indeed, for reasons which, as far as we know are still not clear, all the materials known prior to 1979 with T_c in excess of 18 K are alloys of Nb with the A15 crystal structure. We have added to the figure (blue hatched bars)

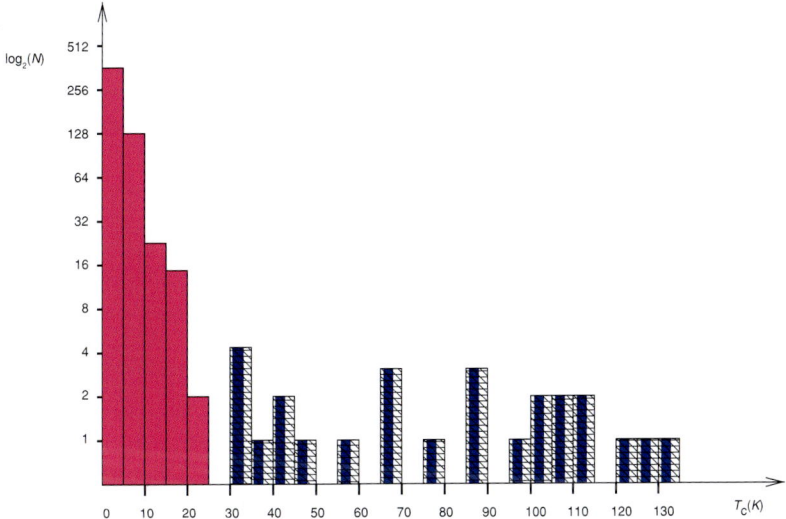

Figure 15.1. Distribution of superconducting transition temperatures. The solid magenta bars represent the number of materials, N, whose transition temperatures are tabulated in Figure VI.2 from [5], which includes over 500 superconducting materials known prior to 1979. Note that the numbers are shown on a log scale. We have added to the figure (the blue hatched bars) superconductors discovered since 1979 with transition temperatures in excess of 20 K. Since all the cuprate superconductors contain nearly square Cu–O planes, which are thought to be the central structure responsible for HTC, one might think of them all as one superconducting material. However, there are also notable differences between different cuprates, including the fact that some are n-type and some p-type, they have different numbers of proximate Cu–O planes, they can have different elements making up the charge reservoir layer, etc. There were 26 distinct crystal structures for cuprate superconductors tabulated in the 1994 monograph by Shaked et al. [6], so we have taken this as our definition of "distinct" materials. In each case, we have reported the highest transition temperature among different materials with the same crystal structure, restricting ourselves, however, to data at atmospheric pressure in bulk materials. C_{60} can be doped with different metal ions or mixtures of metal ions, but they all have more or less the same crystal structure and charge density, so we have counted this as one material (with a maximum $T_c = 31$ K in Rb_2CsC_{60}). One point is for BaKBiO ($T_c = 31$ K). We have also added one point for MgB_2 ($T_c = 39$ K). All of the organic superconductors and $Na_{0.3}CoO_{2y}H_2O$ have T_c less than our arbitrary cutoff, and so have not been included.

some of the new superconductors with T_c in excess of 18 K that have been discovered since this figure was made, using arbitrary definitions of our own. (See caption of Figure 15.1.)

The paucity of materials that exhibit high temperature superconductivity suggests that there must be a number of fairly stringent conditions on the character of the interactions that give rise to HTC. Many theories of high temperature superconductivity give no indication of why this should be the case—applying the stated (uncontrolled) analysis used in these approaches to a wide variety of Hubbard-like models on different lattices would suggest the existence of a high temperature superconducting phase in all of them.

There are several reasons why high temperature superconductivity is hard to attain [7] and why we should be pleasantly surprised that it occurs at all, rather than being shocked that it does not lurk in every third new material. At the crudest level, the dominant interaction between electrons is the strongly repulsive Coulomb interaction—for electrons to pair at all must involve subtle many-body effects which will therefore tend to be rather delicate. In BCS theory, it is the fact that the Coulomb interaction, μ, is well screened (short-ranged), and that the phonon-induced attraction, λ, is highly retarded, that combine to make a net effective attraction, $\lambda_{\text{eff}} = \lambda - \mu^*$, between electrons at low energy. (This important point is stressed, for instance, in the classic treatise on the subject, [1].) Since the downward renormalization of the Coulomb repulsion, $\mu^* = \mu[1 + \mu \log(E_F/\omega_0)]^{-1}$, is only logarithmic, it is effective only when the scale of retardation, ω_0, is very small compared to the Fermi energy, E_F. Moreover, since there are all sorts of polaronic and structural instabilities which occur if λ is large compared to one, λ_{eff} can never be much larger than one. Combined, these considerations imply that superconductivity in normal metals must satisfy the hierarchy of energy scales, $E_F \gg \omega_0 \gg T_c \sim \omega_0 \exp(-1/\lambda_{\text{eff}})$.

Another important issue is that superconductivity has two distinct features: the electrons must pair and the pairs must condense. Rather than approaching the problem from the normal state, if we try to understand the physics of T_c by asking what sorts of fluctuations destroy the superconducting order as the system is heated from $T = 0$, we find that T_c is roughly determined by the lower of the two characteristic energy scales corresponding to these two features [8]. The energy scale which characterizes pair formation is the maximum gap, Δ_0. The energy scale, T_θ, of bose condensation (or more precisely, the temperature above which phase fluctuations destroy the order) is proportional to the superfluid density, $T_\theta \propto \rho_s(T=0)/m^*$. In good metals, T_θ is enormous. As is correctly captured by mean field theory, T_c is determined entirely by the pairing scale. However, strong interactions tend to localize electrons, either collectively (through formation of charge or spin density wave states) or through small polaron formation. Thus, as the strength of the interactions increases, Δ_0 can increase, but correspondingly T_θ will decrease. Eventually, in the strong interaction limit, T_c is set by T_θ, and so decreases as the strength of the pairing increases.

The opposing tendencies of Δ_0 and T_θ mean that there is generally an optimal T_c, i.e., one that does not grow without bound as the interaction strength is varied. This also suggests that, within a class of model systems, or even possibly in a class of materials, the optimal T_c will always occur at a point of crossover from a pairing dominated transition to a phase ordering transition.

15.2. Dynamic Inhomogeneity-Induced Pairing Mechanism of HTC

In order to obtain high temperature superconductivity, we would like to eliminate the middle man. Rather than relying on a weak induced attraction, the pairing should arise directly from the *strong short-range repulsion between electrons*. It might not be a priori obvious that any such mechanism exists, but we have by now demonstrated, by controlled solution of several model problems, that it does. Clearly any such pairing mechanism must be highly collective (since the pairwise interaction is repulsive), and must be "kinetic energy driven" in the sense that the energy cost of pairing two mutually repelling electrons must be more than compensated by the gain in some sort of energy of motion.[3]

One of the main reasons we have reached the conclusion that mesoscale inhomogeneities are essential to the mechanism of high temperature superconductivity is that all the model systems in which pairing from repulsive interactions has been clearly established share this feature. This observation may reflect our limited model solving abilities rather than a characteristic of nature. However, enormous effort has been devoted to numerical searches for superconductivity in various uniform Hubbard and t–J related models, with results that are, at least, ambiguous. (For a review, see [4].) It seems to us that if superconductivity with characteristic energy and length scales of order the microscopic scales in the problem were indeed a robust feature of these models, that unambiguous evidence of it would have been found by now.

15.2.1. Pairing in Hubbard Clusters

The properties of the Hubbard model on various clusters has been studied [13–18] extensively, both numerically and analytically. A finite cluster cannot be a superconductor, but there are two local indicators of superconductivity that can be investigated: existence of a spin-gap and pair binding. If we wish to think of a Hubbard cluster as being a superconducting grain, then we certainly expect it to have a spin-gap. Even if we think of it as a grain of a d-wave superconductor, since nodal quasiparticles only occur at discrete points (sets of measure 0) in k-space, and since k is effectively quantized in a small grain, we expect there to be a true spin-gap in almost all cases. Pair-binding is less obvious—on small superconducting grains, the energy to add one quasiparticle can be less (by the charging energy) than the energy to add a pair. However, especially in models (such as the Hubbard model) in which the long-range Coulomb interaction is neglected, pair-binding is also a reasonable indicator of local superconductivity.

What is found in the cited studies is that many, but certainly not all, small Hubbard clusters exhibit spin-gaps and pair-binding in an appropriate range of strength of (repulsive) Hubbard interaction, U, and electron concentration. This effect is typically strongest at half-filling (one electron per site). It occurs most strongly for intermediate values of U/t, and the pair-binding is lost when U/t gets either very large or very small [19]. Finally, there is a general tendency for the magnitude of both the pair-binding and the spin-gap to decrease as the size of the cluster increases, suggesting that this is intrinsically an effect associated with mesoscale structure.

Among the Hubbard clusters that have been found to exhibit this locally superconducting behavior are [20] the $4n$ membered Hubbard ring, with n from 1 to 250, the cube, the truncated tetrahedron, and various pieces of the 2D square lattice on a torus. Closely related studies [21, 22] have been carried out on clusters that are effectively infinite in one direction but are mesoscale transverse to it. These clusters include Hubbard ladders with up to eight

---

[3] This latter statement is intuitively compelling, but cannot be made completely precise since, by the time one is dealing with effective Hamiltonians, it is never completely clear how each remaining interaction is related to the microscopic kinetic energy of the constituent electrons. Note, the attractiveness of a kinetic energy driven mechanism has been emphasized by several other authors, including [9–12].

legs, and the circumference 4 Hubbard cylinders. In these "fat" 1D systems, the size of the spin-gap, and with it the magnitude of the pair-binding energy, tend to decrease exponentially with the transverse size of the clusters.[4]

The physics of spin-gap formation is at the core of this problem. It is inherited from the properties of the cluster at half-filling where, at least for large U/t, the system can better be thought of as a grain of a Mott insulator. The spin-gap is then associated with the quantum disordering of the electron spins. In the limit of infinite cluster size there is no spin-gap since (except, perhaps, on special, highly frustrating lattices) the spin rotation symmetry spontaneously broken, and there are gapless spin-waves. For instance, if one considers a ladder of width L to be a finite size version of the square lattice quantum antiferromagnet, whose interacting spin-waves one treats in the continuum limit, then one can derive an expression for the spin-gap [23], $\Delta_s \sim 3.347\ J\ \exp(-0.682L/a)\left[1 + O(L/a)\right]$, which agrees quantitatively with the results of numerical simulations [24–26]. Again, this argument makes clear that the spin-gap is a mesoscale effect, which tends to decrease rapidly with the size of the cluster.

The remaining question is why does the spin-gap survive away from half filling, and why does the existence of a spin-gap (in many, but not all cases) lead to pair-binding? There are two distinct intuitive arguments that rationalize this observation.

The first is based [14] on the notion of a local form of spin-charge separation [27]. If we add one hole to each of two half-filled Hubbard clusters, we must make on each cluster an excitation carrying spin 1/2 and charge e. If we add two electrons to a single cluster, they can form a spin singlet, in which case we need to make excitations carrying only charge $2e$. If we can approximate the excitations as holons (charge e spin 0) and spinons (charge 0 and spin 1/2), then by adding two electrons to one cluster we save twice the spinon creation energy. Even if this description is invalid (due to confinement) at long length scales, in some circumstances, it may give us a good handle on the local energetics.

The second line of argument is similar to those that lead to phase separation in doped antiferromagnets [28], or the spin-bag ideas of pairing [29]. Under some circumstances the state of the system at half-filling is anomalously stable, since the system can take maximal advantage of Umklapp scattering. A large spin-gap is a measure of this anomalous stability. When adding two electrons to two identical clusters, we have the choice of adding one electron to each cluster, in which case the particularly favorable correlations are disturbed on both clusters, or we can add both to one cluster (even if they have a direct repulsion between them), since in that case only one cluster is disturbed. Thus, paradoxically, it could be the strength of the insulating correlations in the half-filled cluster that give rise to superconductivity when the system is lightly doped.

15.2.2. Spin-Gap Proximity Effect

The arguments in the previous section are general and intuitive, but supported mainly by anecdotal evidence. (In a few cases, the origin of the pair-binding can be understood analytically for small U/t on the basis of perturbation theory [14, 15], but here the effects are weak and the strong correlation physics, which is so central in the actual materials, is only present in ghostly form.) In the case of "fat" 1D systems, various ladders or sets of coupled

---

[4] Tsunetsugu, M. Troyer and T. M. Rice [22] studied arrays of two-leg $t-J$ ladders as a way to understand the physics of the translationally invariant 2D system. Although the model they studied nominally corresponds to the period 2 case we discuss later, and some of their discussion prefigures the present analysis, the questions asked by these authors were quite different. In particular they did not consider the mechanism of superconductivity in inhomogeneous 2D systems which we discuss here.

ladders, we have sufficient theoretical understanding of the problem that we can analyze in some detail the conditions under which superconducting correlations emerge directly from the repulsive interactions.

In a single band one-dimensional electron gas (1DEG) with short-ranged repulsive interactions, superconductivity is suppressed relative to noninteracting electrons—there is no tendency toward a spin-gap (rather, there is quasi-long-range antiferromagnetic order) and the superconducting susceptibility is not even logarithmically divergent as $T \to 0$. Technically speaking, the low-energy physics is governed by the Luttinger liquid fixed point (gapless, bosonic modes with spin–charge separation) with the charge Luttinger exponent, $K_c < 1$. However, in multiband 1D systems, under many circumstances [30–32], the low-energy physics is governed by a strong-coupling Luther–Emery fixed point [33], with a spin-gap, Δ_s, and with a charge Luttinger exponent in the range $0 < K_c < 2$. This fixed point exhibits incipient superconductivity in the sense that the singlet superconducting susceptibility diverges for $T \ll \Delta_s$ so long as $K_c > 1/2$,

$$\chi_{SC} \sim \Delta_s / T^{2-K_c^{-1}}. \tag{15.1}$$

To complicate matters, it also exhibits incipient CDW order in the sense that the CDW susceptibility diverges at wave number $Q = 2k_F$ for $T \ll \Delta_s$ so long as $K_c < 2$,

$$\chi_{CDW}(Q) \sim \Delta_s / T^{2-K_c}. \tag{15.2}$$

Why are the multiband cases so different from the single band case? In particular, since spin-gap formation is the 1D version of singlet pairing, what is it that causes pairing to be a common feature of multiband systems and not of the single band problem? The new physics comes from interband pair scattering, and has been explained intuitively by Emery, Zachar, and one of us [11] as "the spin-gap proximity effect."

Consider coupling two distinct 1D systems. From the weak coupling perspective, one can think of these as being two bands arising from the existence of more than one atom per unit cell. From a strong coupling perspective, one could think of these as two chemically distinct chains in close physical proximity to one another. Assuming that the two systems have distinct values of the Fermi wave vector, k_F and k_F', low-energy processes in which an odd number of electrons are scattered from one system to the other are forbidden by momentum conservation. Coupling of CDW fluctuations, which are singular at different values Q and Q', are negligible (i.e., it is an irrelevant interaction). However, scattering of electron pairs with zero center of mass momentum from one system to the other is, under many circumstances, peturbatively relevant. It is the renormalization of these interband pair-scattering terms, and their feedback on the other interactions in the system, that can drive the system to the Luther–Emery fixed point.

The physical origin of this effect is simply understood. The electrons can gain zero-point energy by delocalizing between the two bands. In order to take advantage of this, however, the electrons need to pair, which may cost some energy. When the energy gained by delocalizing between the two bands exceeds the energy cost of pairing, the system is driven to a spin-gap phase. In this sense, the physics is very analogous to the ordinary proximity effect in superconductivity. Here, a normal metal, even one with residual repulsive interactions between electrons, is brought in contact with a superconductor. In order for the electrons to be delocalized over the combined system, the electrons in the metal must pair. In this case, even though this costs energy, the gain in zero point "kinetic energy" always makes the proximity effect favorable. In this sense this is a kinetic energy driven mechanism. As is well known, the result is that superconductivity is induced in the normal metal over a distance which diverges as $T \to 0$.

The spin-gap proximity effect is not quite so robust—it occurs only if a certain exponent inequality is satisfied. If one of the two subsystems already has a spin-gap, then the price (pairing) only needs to be paid in the other, so the exponent inequality is easier to satisfy. It is an interesting, and still largely unexplored issue, what local "chemistry" does or does not give rise to a Luther–Emery liquid with a large spin-gap in a variety of multicomponent 1D systems. We do know that the two-leg ladder in both weak and strong coupling has a robust Luther–Emery phase. We also know, as mentioned above, that the spin-gap of the half-filled $2N$ leg ladder in strong coupling decreases exponentially with N. Similar behavior is seen in weak coupling, where the spin-gap in the entire Luther–Emery phase can be shown [34] to decay exponentially with N. Together, these two observations reinforce our belief that pairing directly from repulsion is a mesoscopic effect, which disappears rapidly if the relevant dimensions of the system in question get too large.

15.3. Superconductivity in a Striped Hubbard Model: A Case Study

In this section, we present a theoretically well-controlled solution of an explicit model in which high temperature superconductivity arises directly from the repulsive interactions and the existence of mesoscale structures.[5] In collaboration with Arrigoni, we discuss this model in some detail in [36].

The model has modulated interactions in one direction, so that it breaks into an array of weakly coupled two-leg ladders (hence the name "striped Hubbard model"). Perhaps one can view this as a caricature of the spontaneous symmetry breaking that occurs in stripe phases in real materials, but there are troubles with this identification. Primarily, we would like this model to be viewed as a solvable model in which the basic mechanism of mesoscale inhomogeneity-induced pairing can be studied.

Because the solution of the ladder problem is so well characterized, it is possible to treat the coupled ladder problem reliably so long as the coupling between ladders is sufficiently weak. Within this model, we establish the occurrence of superconductivity directly from the repulsive interactions, document the important role of competing (CDW) order in the phase diagram, and analyze the circumstances under which the optimal T_c is obtained. A very schematic representation of the resulting phase diagram is shown in Figure 15.2.[6]

The striped Hubbard model (sketched in Figure 15.3) is

$$H = -\sum_{<\vec{r},\vec{r}'>,\sigma} t_{\vec{r},\vec{r}'}[c^\dagger_{\vec{r},\sigma} c_{\vec{r}',\sigma} + \text{h.c.}] + \sum_{\vec{r},\sigma}[\epsilon_{\vec{r}} c^\dagger_{\vec{r},\sigma} c_{\vec{r},\sigma} + (U/2) c^\dagger_{\vec{r},\sigma} c^\dagger_{\vec{r},-\sigma} c_{\vec{r},-\sigma} c_{\vec{r},\sigma}],$$

where $<\vec{r},\vec{r}'>$ designates nearest-neighbor sites, $c^\dagger_{\vec{r},\sigma}$ creates an electron on site $\vec{r}$ with spin polarization $\sigma = \pm 1$ and satisfies canonical anticommutation relations, and $U > 0$ is the

---

[5] The same sort of physics was studied in weak coupling in [35].

[6] In the schematic phase diagram of Figure 15.2, we have illustrated qualitatively several important effects discussed in the text: (a) at low x, T_c grows linearly with x; (b) for somewhat larger values of x, one can use the low-temperature form of the susceptibility of the spin-gap phase to estimate T_c; (c) although for larger values of x non universal effects are important, as $x \to x_c$ the spin-gap vanishes and so does T_c. We have simplified the figure by taking the T_c curves for the periods 2 and 4 stripes to coincide, so as to highlight the main difference, i.e., the critical x shifts to larger values as the period increases. In fact, however, the entire curve should be somewhat different in the two cases.

How Optimal Inhomogeneity Produces High Temperature Superconductivity 577

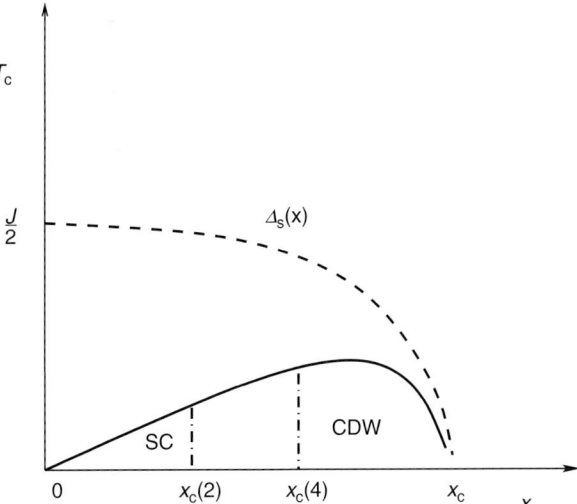

Figure 15.2. Schematic phase diagram for a period 2 and a period 4 striped Hubbard model, at fixed (and small) δt. The broken line is the spin-gap $\Delta_s(x)$ as a function of doping x, which labels the horizontal axis; $x_c(2)$ and $x_c(4)$ indicate the SC-CDW quantum phase transition for the period 2 and period 4 cases. These, most likely, are first-order transitions. For $x \gtrsim x_c$ the isolated ladders do not have a spin-gap; in this regime the physics is different involving low-energy spin fluctuations.

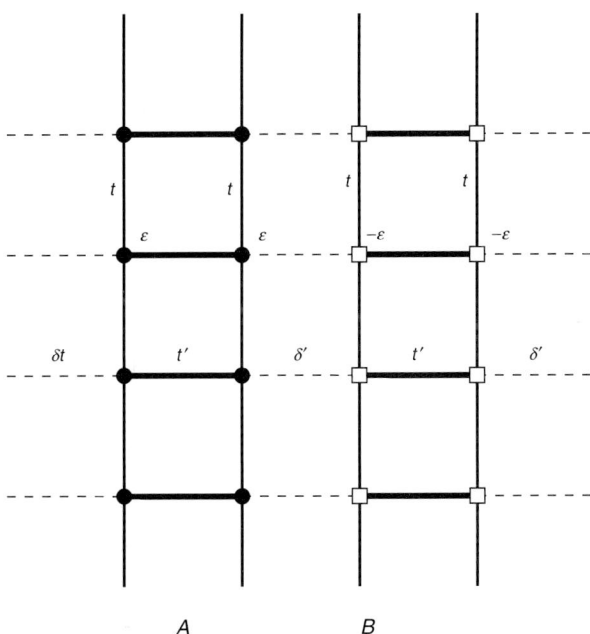

Figure 15.3. Schematic representation of the striped Hubbard model analyzed in this paper. See the surrounding text for details; here A and B are the two types of ladders discussed in the text.

repulsion between two electrons on the same site. In the limit of strong repulsions, $U \gg t_{\vec{r},\vec{r}'}$, this model reduces approximately to the corresponding $t-J$ model, which operates in the subspace without doubly occupied sites, but with an exchange coupling, $J_{\vec{r},\vec{r}'} = 4|t_{\vec{r},\vec{r}'}|^2/U$ between neighboring spins. Our results only depend on the low-energy physics of the ladder and, thus, apply equally to the $t-J$ and Hubbard models.

In the translationally invariant Hubbard model, $t_{\vec{r},\vec{r}'} = t$ and $\epsilon_{\vec{r}} = 0$. The striped version of this model is still translationally invariant along the stripe direction (which we take to be the y axis), so $t_{\vec{r},\vec{r}+\hat{y}} = t$. However, perpendicular to the stripes the hopping matrix takes on alternately large and small values: $t_{\vec{r},\vec{r}+\hat{x}} = t'$ for r_x = even, and $t_{\vec{r},\vec{r}+\hat{x}} = \delta t \ll t' \sim t$ for r_x = odd. This defines a "period 2 striped Hubbard model," as shown in Figure 15.3. For the "period 4 striped Hubbard model," we include a modulated site energy, $\epsilon_{\vec{r}} = \pm \epsilon$ on alternate ladders with $\epsilon \gg \delta t$. Ladders with site energy ϵ will be called A ladders and ladders with site energy $-\epsilon$ will be called B ladders.

15.3.1. Zeroth-Order Solution: Isolated two-Leg Ladders

For $\delta t = 0$, the model breaks up into a series of disconnected two-leg ladders. Considerable analytic and numerical effort has gone into studying the properties of two-leg $t-J$ and Hubbard ladders, and much is known about them. For $x = 0$, the undoped two-leg ladder has a unique, fully gapped ground state. In the large U limit, the magnitude of the spin-gap of the undoped [24, 37] ladder is approximately $\Delta_s \approx J/2$. Then, for a substantial range of x ($0 < x < x_c$), the ladder exhibits a Luther–Emery phase, with a spin-gap that drops smoothly[7] with increasing x, and vanishes at a critical value of the doping, $x = x_c$. (This particular Luther–Emery liquid is known [30, 37–41] to have "d-wave-like" superconducting correlations, in the sense that the pair-field operator has opposite signs along the edge of the ladder (y-direction) and on the rungs (x-direction).) For $x > x_c$, there remain uncertainties concerning the exact character of the possible gapless phases.

For the purposes of the present paper, we will confine ourselves to the range of parameters where both A and B type ladders are in the Luther–Emery phase. The low-energy physics (at all energies less than Δ_s) of the two-leg ladder in the Luther–Emery phase is contained in the effective (free) bosonized Hamiltonian for the collective charge degrees of freedom

$$H = \int dy \left\{ \frac{v_c}{2} \left[K(\partial_y \theta)^2 + \frac{1}{K}(\partial_y \phi)^2 \right] + \cdots \right\}, \quad (15.3)$$

where ϕ is the CDW phase and θ is the superconducting phase. These two fields are dual to each other, and so satisfy the canonical equal-time commutation relations, $[\phi(y'), \partial_y \theta(y)] = i\delta(y - y')$. Specifically, the component of the charge density operator at the wave-vector $P = 2\pi x$ of the incipient CDW order is

$$\hat{\rho}_P(y) \propto \sqrt{\Delta_s} \exp[iPy + i\sqrt{2\pi}\phi(y)], \quad (15.4)$$

while the singlet pair creation operator,

$$\hat{\Phi}(y) \propto \sqrt{\Delta_s} \exp[i\sqrt{2\pi}\theta]. \quad (15.5)$$

---

[7] For a restricted range of x, the authors of [37] show numerical evidence indicating that the spin-gap decreases smoothly with increasing x. We are not aware of any published studies that carefully trace the spin-gap as a function of x, and in particular ones that accurately determine the critical doping, x_c, at which it vanishes.

This effective Hamiltonian is general and physical; the precise x dependence of the spin-gap, Δ_s, the charge Luttinger exponent, K, the charge velocity, v_c, and the chemical potential, $\mu(x)$, depends on details such as the values of U/t and t'/t. For certain cases [37–39] Δ_s, K, v_c, and v_s have been accurately computed in Monte Carlo studies, and these studies could be straightforwardly extended to other values of the parameters.[8]

The ellipsis in Eq. (15.3) represent cosine potentials, which we will not explicitly exhibit here, that produce the Mott gap Δ_M at $x = 0$. Because of these terms, for $x \to 0$ the elementary excitations are charge $2e$ solitons that can either be viewed as spinless Fermions or hard-core bosons, with a dispersion relation $E(k) \simeq \Delta_M + \tilde{t}k^2$. One consequence of this is that [39, 42] $K \to 2$ and $v_c \to 2\pi\tilde{t}x$ as $x \to 0$. A second consequence is that the renormalized harmonic theory, which retains only the explicitly exhibited terms in Eq. (15.3), is valid in a range of energies which is small in proportion to the effective Fermi energy, $\tilde{E}_F^{(1D)} = 2\pi\tilde{t}x^2$. (An estimate of $\tilde{t} \approx t/2$ can be obtained from the DMRG study of the t–J ladder with $J/t = 1/3$ in [38].)

For larger x, the numerical studies [38, 39, 43] generally find that both K and Δ_s drop monotonically with increasing x. By the time $x = x_1 \approx 0.1$, K is close to 1, and by $x = x_c \approx 0.3$, Δ_s has dropped to values that are indistinguishable from 0, and $K \approx 0.5$. Thus, over most of the Luther–Emery phase, both the SC and the CDW susceptibilities are divergent. However, the SC susceptibility is the more divergent only at rather small values of $x < x_1$.

Before leaving the single-ladder problem, it is worth mentioning a useful intuitive caricature of its electronic properties. We picture a singlet pair of electrons on neighboring sites as being a hard-core bosonic "dimer." The undoped ladder can be thought of as a Mott insulating state of these dimers, with one dimer per rung of the ladder, i.e., a "valence bond crystal" with lattice spacing one. To remove one electron from the system, we need to destroy one dimer and remove one electron, leaving behind a single electron with spin 1/2 and charge e. However, when we remove a second electron from the system, we have the choice of either breaking another dimer, thus producing two quasiparticles with the quantum numbers of an electron, or of removing the unpaired electron left behind by the first removal, thus producing a new boson—a missing dimer—with charge $2e$ and spin 0. The persistence of the spin-gap upon doping the ladder can thus be interpreted as implying that the energy needed to break a dimer (of order Δ_s) is sufficiently large that one charge $2e$ boson costs less than two charge e quasiparticles. At finite x, the missing dimers can be treated as a dilute gas of hardcore bosons. That the elementary excitations of the undoped ladder can be constructed in this simple manner reflects the fact that this is a confining phase [44–47], not a spin liquid.[9]

15.3.2. Weak Inter-Ladder Interactions

We now address the effect of a small, but nonzero coupling (i.e., single-particle hopping) between ladders, $\delta t > 0$. Because of the spin-gap, δt is an irrelevant perturbation in the renormalization group sense, and so does not directly affect the thermodynamic state of the system. However, second-order processes result in various induced interactions between neighboring ladders. These consist of marginal forward scattering interactions, which are negligible for

---

[8] Note that the normalization convention on the fields used in the present paper differs from that of White and coworkers [39], so that our K is the same as their $2K_{c,+}$.

[9] In a confining phase, all finite energy excitations have quantum numbers equal to those of an integer number of electrons and holes; a deconfining phase supports excitations with "fractional" quantum numbers such as those of a "spinon": spin 1/2 and charge 0.

small δt, and potentially relevant Josephson tunneling and back-scattering density–density interactions.

The important (possibly relevant) low-energy pieces of these latter interactions are most naturally expressed in terms of the bosonic collective variables defined above:

$$H' = -\sum_j \int dy \left\{ \mathcal{J} \cos[\sqrt{2\pi}(\theta_j - \theta_{j+1})] + \mathcal{V} \cos[(P_j - P_{j+1})y + \sqrt{2\pi}(\phi_j - \phi_{j+1})] \right\}, \tag{15.6}$$

where $P_j = 2\pi x_j$, with x_j the concentration of doped holes on ladder j, and ϕ_j and θ_j are the charge field and its dual on each ladder. Here, again, the form of the low-energy interactions between two Luther–Emery liquids is entirely determined by symmetry considerations, but the magnitude of the Josephson coupling $\mathcal{J}$ and the induced interaction between CDW's, $\mathcal{V}$, must be computed from microscopics; they are renormalized parameters which result from "integrating" out the high-energy degrees of freedom with energies between the bandwidth $W \sim 4t$ and the renormalized cutoff, Δ_s, or with wavelengths between a and $\xi_s \equiv v_s/\Delta_s$ where v_s is the spin-wave velocity.

So long as x is not too near x_c, the spin-gap is large, $\Delta_s \sim J$. In this case, the spin physics really occurs on a microscopic scale, and hence the coupling constants are not qualitatively changed in this first stage or renormalization. In this case, a rough estimate of $\mathcal{J}$ and $\mathcal{V}$ can be made from second-order perturbation theory:

$$\mathcal{J} \approx \mathcal{V} \sim \frac{(\delta t)^2}{J}. \tag{15.7}$$

As $x \to x_c$, and hence $\Delta_s \to 0$, the problem becomes more subtle, as discussed in [36].

15.3.3. Renormalization-Group Analysis and Inter-Ladder Mean Field Theory

The effect of these interchain couplings can be deduced from an analysis of the lowest-order perturbative renormalization group equations in powers of the couplings $\mathcal{V}$ and $\mathcal{J}$. However, *equivalent* results are obtained from inter-ladder mean-field theory [43, 48], which is conceptually simpler. These equations are the analogue of the BCS gap equations applied to this model, and are expected to give a quantitatively accurate estimate of T_c for small $\delta t/\Delta_s$ for precisely the same reason. A discussion of the accuracy of *interchain* mean-field theory is given in the Appendix of [36]. In the present two-dimensional system, T_c should be interpreted as the onset of quasi-long range order, i.e., as a Kosterlitz–Thouless transition.

To implement this mean-field theory, we need to compute the expectation value $M_j(h_j) = \langle \cos[\sqrt{2\pi}\theta_j] \rangle$ of the pair creation operator on an isolated ladder, where the expectation value is taken with respect to the mean-field Hamiltonian

$$H_{\text{MF}} = H_j - h_j \int dy \cos[\sqrt{2\pi}\theta_j] \tag{15.8}$$

in which H_j is the effective Hamiltonian in Eq. (15.3) with parameters appropriate to ladder j, and h_j represents the mean-field due to the neighboring ladders, and so satisfies the self-consistency condition,

$$h_j = \mathcal{J}[M_{j+1} + M_{j-1}]. \tag{15.9}$$

The expression for the mean-field transition temperature can be expressed in terms of the corresponding susceptibility, $\tilde{\chi}_{\text{SC}}^{(j)} = \partial M_j(h)/\partial h|_{h=0}$, which is related to the superconducting susceptibility in Eq. (15.1) by a proportionality constant which depends on the expectation value of the spin-fields. In the case in which all the ladders are equivalent, this yields the implicit relation $2\mathcal{J}\tilde{\chi}_{SC}(T_c) = 1$. For an alternating array of A and B type ladders, the expression for the superconducting T_c is easily seen to be

$$(2\mathcal{J})^2 \tilde{\chi}_{SC}^{(A)}(T_c) \tilde{\chi}_{SC}^{(B)}(T_c) = 1. \tag{15.10}$$

Notice that in the case in which the A and B type ladders are identical Eq. (15.10) reduces properly to the expression for equivalent ladders. The expression for χ_{SC} from Eq. (15.1) can be used to invert Eq. (15.10) to obtain the estimate for T_c

$$T_c \sim \Delta_s \left(\frac{\mathcal{J}}{\tilde{W}}\right)^\alpha ; \alpha = \frac{2 K_A K_B}{[4 K_A K_B - K_A - K_B]} \tag{15.11}$$

where $\mathcal{J}$ is the effective coupling given in Eq. (15.7), and $\tilde{W}$ is a high energy cutoff which, so long as x is not too close to x_c, is also of order J. Although T_c is small for small $\mathcal{J}$, it is only power law small. In fact typically $\alpha \sim 1$. A perturbative renormalization-group treatment for small $\mathcal{J}$ yields the same power law dependence as Eq. (15.11), suggesting that this expression is asymptotically exact for $\mathcal{J} \ll \tilde{W}$.

The mean-field equations for the CDW order are obtained similarly. The expression for the transition temperature for CDW order with wave-vector P is

$$(2\mathcal{V})^2 \tilde{\chi}_{CDW}^{(A)}(P, T_c) \tilde{\chi}_{CDW}^{(B)}(P, T_c) = 1, \tag{15.12}$$

where the notation is the obvious extension of that used in the superconducting case. The best ordering vector is that which maximizes T_c. For $P = P_A$, $\chi_{CDW}^{(A)}(P_A, T)$ diverges with decreasing temperature as in Eq. (15.2), but $\chi_{CDW}^{(B)}(P_A, T)$ saturates to a finite, low temperature value when $T \sim v_c |P_A - P_B|$. Thus, even if $\chi_{CDW}^{(A)}(P_A, T)$ diverges more strongly with decreasing temperature than $\chi_{SC}^{(A)}$, there are two divergent susceptibilities in the expression for the superconducting T_c, and only one for the CDW T_c. So long as the exponent inequalities

$$2 > K_A^{-1} + K_B^{-1} - K_A; \quad 2 > K_A^{-1} + K_B^{-1} - K_B \tag{15.13}$$

are satisfied, the superconducting instability wins out.

15.3.4. The $x \to 0$ Limit

Since $K \to 2$ as $x \to 0$, there is necessarily a regime of small x in which the superconducting susceptibility on the isolated ladder is more divergent than the CDW susceptibility. Here, in the presence of weak inter-ladder coupling, even the period 2 striped Hubbard model (i.e., with $\epsilon = 0$) is superconducting. However, care must be taken in this limit, since, as mentioned above, the range of energies over which H in Eq. (15.3) is applicable vanishes in proportion to x^2. Fortunately, a complementary treatment of the problem, which takes into account the additional terms, the ellipsis in Eq. (15.3), can be employed in this limit. The small x problem can be mapped onto a problem of dilute, hard-core charge $2e$ bosons (with concentration x per rung) with an anisotropic dispersion, $E(\vec{k}) = \tilde{t}k_y^2 - \mathcal{J}\cos[2k_x]$. (The 2 reflects the ladder periodicity.) Consequently, for small x

$$T_c \approx 2\pi \sqrt{2\mathcal{J}\tilde{t}} \, x F(x) \sim |\delta t| x, \tag{15.14}$$

where $F(x) \sim 1/\ln(1/x)$ is never far from 1, and the logarithm reflects [49] the fact $d = 2$ is the marginal dimension for Bose condensation. (This result is not substantially different for the period 4 striped Hubbard model, so long as ϵ is not too large.) There is a complicated issue of order of limits when both δt and x are small; roughly, we expect that T_c will be determined by whichever expression, Eq. (15.10) or Eq. (15.14), gives the higher T_c, but with the understanding that χ_{SC} must be computed taking into account the terms represented by the ellipsis in Eq. (15.3) which cause the susceptibility to vanish as $x \to 0$.

15.3.5. Relation to Superconductivity in the Cuprates

The striped Hubbard model realizes the idea that the pairing scale, in this case the spin-gap, can be inherited from a parent Mott insulating state. Moreover, like the underdoped cuprates, the gap scale is a decreasing function of increasing x, while the actual superconducting transition occurs at a T_c typically much smaller than $\Delta_s/2$, and is determined by the phase ordering temperature rather than the pairing scale. Hence, for x not too close to x_c, this model exhibits a pseudogap regime for temperatures between T_c and $T^* \sim \Delta_s/2$, reminiscent of that seen in underdoped cuprates. However, T_c is always bounded from above by Δ_s and so tends to zero as $x \to x_c$. The model also exhibits a competition between SC and CDW order, which is somewhat akin to the competition with fully developed stripe order and SC that occurs in certain cuprates.[10]

However, as mentioned earlier, the model cannot be thought of as a literal model of superconductivity in the cuprates. First, most of the cuprates have, at most, local fluctuating charge stripe order (see [3] for an extensive discussion of the present status of this issue), and even where such order occurs, it occurs through spontaneous symmetry breaking. Moreover, the striped Hubbard model possesses a large spin-gap, and so does not contain any of the physics of low energy incommensurate spin-fluctuations which are the principle experimental signatures to date of stripe correlations in the cuprates. Last, although the superconducting state is "d-wave-like" in the sense that the order parameter changes sign under rotations by $\pi/2$, since the striped Hamiltonian explicitly breaks this symmetry, there is no precise symmetry distinction between d-wave and s-wave superconductivity. Indeed, the superconducting state is not even truly adiabatically connected to the superconducting state observed in the cuprates, because the existence of a spin-gap implies the absence of gapless "nodal" quasiparticles in the superconducting state.[11]

There is a strong tendency in our contentious field to set up straw men which can easily be toppled by (purposely?) misinterpreting carefully caveated statements. We therefore reiterate that the striped Hubbard model is a solvable (and, we believe, fascinating) case study—not a "realistic" model of superconductivity in the striped phase of the cuprates.

15.4. Why There is Mesoscale Structure in Doped Mott Insulators

The cuprate high temperature superconductors are strongly correlated electronic systems, in which the short-range repulsions between the electrons are larger than the bandwidth. They are doped descendants of a strongly correlated (Mott) insulating "parent compound" which is antiferromagnetically ordered. While HTC is, seemingly, uniquely a property of the cuprates, many other aspects of the strong correlation physics are features of a much broader

---

[10] For $x > x_c$ the low-energy physics is dominated by spin fluctuations and by single-particle (electron) tunneling. Low T_c superconductivity can occur in this regime by conventional BCS-like mechanisms.

[11] However, simplified models of this type can have 2D anisotropic superconducting phases both with and without low-energy nodal quasiparticles; see, e.g., [50].

class of strongly correlated materials including various manganites, nickelates, cobaltates, and ruthenates. Magnetism, and various forms of charge order (to be discussed later) are among the clearest signatures of the strong correlation physics.

Of great fundamental importance is the failure of the Fermi liquid description of the "normal" state at room temperature and above. This fact was clear already at the time of the discovery of high temperature superconductivity and it has been a leit motif of much of the research done since then [51, 52]. A directly related and associated fact is that these doped Mott insulators are "bad metals" [53]: above the superconducting T_c they exhibit a metallic T dependence of the conductivity, the famous linear resistivity, while at the same time there appears to be no evidence of well-defined quasiparticles (in the sense of Landau), and the resistivity passes the Ioffe-Regel limit without taking any notice of it. It may often be the case that well-defined quasiparticles develop as emergent phenomena at low T and energy; those who treat the normal state as a Fermi liquid, despite the evidence to the contrary, are, in the immortal words of Landau [54], "Enemies of the working class."

Whether their ground states exhibit long-range magnetic order or not, most models of undoped Mott insulators share an intrinsic tendency towards *electronic phase separation* [28,55], an effect which was found quite early on in analytic studies and numerical simulations of models of strongly correlated systems. The physics behind electronic phase separation is quite simple, and is related to the mechanism of pair-binding in clusters, discussed above. The addition of a single hole induces a "defect" in the correlations of the Mott insulator. The energy associated with the subsequent addition of holes is less if they clump together, since this disrupts the favorable correlations of the insulating state to a lesser extent. Thus, even though all the microscopic interactions are repulsive, there are effective attractive forces between the doped holes.

On the other hand, since the undoped systems are insulators, the long-range piece of the repulsive Coulomb interactions between the charges is poorly screened. This gives rise to *Coulomb-frustrated phase separation*—states which have as their constituents mesoscopic puddles of charges whose size and shape [56] are determined by the competition between the short-range tendency to phase separation and the Coulomb interaction. Electronic phases with self-organized mixtures of high- and low-density regions have been called [56,57] "electronic microemulsions." In a precise sense, the mesoscale structure defines the set of relevant degrees of freedom responsible for the low-energy physics of strongly correlated systems.

At sufficiently small T, depending on how large the effective mass of a puddle, they can remain mobile (a puddle fluid), or can freeze into a variety of possible charge ordered states. (In the presence of quenched disorder, they can also be pinned.) Among the possible charge ordered states are a variety of *electronic liquid crystal phases* which exhibit a varying degree of charge inhomogeneity and spatial anisotropy [58]. As far as the mechanism of HTC is concerned, the existence of local structures on length scales greater than or of order of the superconducting coherence length, ξ_0, is what is important, not the manner in which the structures themselves order or not. However, it is much easier experimentally to identify the states of broken spatial symmetry that arise from Coulomb frustrated phase separation. Thus, both because of their intrinsic interest, and as a way of gaining insight into the nature of the structures produced by Coulomb frustrated phase separation, there has been considerable interest in studying these phases.[12]

Since electronic liquid crystalline phases are in some ways ordered and in some ways fluid, they are more subtle to identify in experiments than typical CDWs. Elsewhere, we have

---

[12] A related proposal is developed in [59, 60].

discussed the evidence in the cuprates [3], of the existence of such ordered phases, especially smectic (stripe ordered) and Ising nematic phases. In many respects, electronic liquid crystal phases are similar to the analogous phases of complex classical fluids [61]. However, while in classical liquid crystals, the rich phase diagram originates form the microscopic anisotropic structure of complex molecules (e.g., nematogens, chiral molecules, viruses, "molecular bananas," etc.), electronic liquid crystals are the quantum ground states of systems of point particles (holes); the role of the complex molecules is played by the self-organized structures produced by Coulomb-frustrated phase separation. It cannot get more politically correct than this: complex "soft quantum matter" from self-assembling nanostructures!

15.5. Weak Coupling Vs. Strong Coupling Perspectives

Much of the commonly adopted theoretical analysis of the mechanism of high temperature superconductivity is, at core, the same as the BCS/Eliashberg theory, but (possibly) with a different collective excitation (spin-wave, phonon, exciton, director wave, ...) playing the role of the "glue." However, an essential feature of BCS theory is that the normal state is a good Fermi liquid [1], with well-defined quasiparticles at all energies small compared to the retardation scale (the frequency of the collective mode). It is, of course, possible to simply evaluate the same class of diagrams that are sanctified by Eliashberg theory, even when whatever peaks there are in the single particle spectral function are too broad to be classified as quasiparticles; however, in this case, there is no known justification for summing this particular class of diagrams (which sum the leading logarithms in a Fermi liquid). Whether or not one is comfortable with this sort of uncontrolled extrapolation of the (beautifully well controlled) weak coupling theory is a matter of personal taste. A distinguishing feature of these theories is that, for them, the strongly correlated nature of the cuprates is an inconvenient side issue. Indeed in all these theories, if the single particle spectral function, $A(k, \omega)$, (often taken phenomenologically from experiment) were replaced by a Fermi liquid $A(k, \omega)$, with well-defined quasiparticles, the resulting calculated T_c would actually increase!

In contrast, a smaller but highly visible set of theories start from the viewpoint that the strong correlation physics is central to the physics of high temperature superconductivity. In this case, the mechanism is not based on pairing of well-defined quasiparticles. Theories based on proximity to quantum critical points are of this sort. In these theories, the same physics (quantum critical fluctuations) that is supposed to be responsible for the pairing is also presumed to be responsible for the non-Fermi liquid character of the normal state, so it does not make sense to ask what would happen were the normal state replaced by a Fermi liquid. Of course, theories based on a fractionalized normal-state, with spin–charge separation, also fall in this category. The ideas we have discussed, in which mesoscale (and/or mesotime) inhomogeneity plays a crucial role in the pairing, share some features with both of these other non-Fermi liquid-based approaches. Since in the cleanest versions of our mechanism, coherence between different clusters occurs with the advent of superconducting order, these ideas provide a very concrete implementation of a mechanism of superconductivity in which the normal state has no coherently propagating quasiparticles.

It may be possible to discriminate between the strong correlation and the more BCSish approaches experimentally. In the strong correlation approaches, it would be unexpected to find a material with a high superconducting transition temperature and well-defined quasiparticles in the normal state. This finds some support in the observation that, with increasing doping in the overdoped regime, as the single-particle spectral function becomes more Fermi liquid like, T_c drops rapidly. From the more BCSish viewpoint, one would be unsurprised to find some materials, even materials in which T_c is optimized, in which the normal state is well described by Fermi liquid theory, and the single-particle spectral function exhibits well-defined quasiparticles.

In this context, it is important not to over-interpret ARPES evidence for or against the existence of quasiparticles. On the one hand, it is possible for quenched disorder, especially at the sample surface, to broaden what would have been a sharp peak in $A(k, \omega)$, making it too broad to be clearly identified as a quasiparticle—so long as this broadening is due to strictly elastic scattering process, a quasiparticle description remains valid despite the negative evidence from ARPES. Probably, this can be checked with STM by looking for Friedel oscillations with random phases, but long distance power-law fall-off associated with the introduction of a known scatterer at a point in space. On the other hand, the spectral function of the one-dimensional Luttinger liquid, even with moderately strong interactions, possesses a reasonably clear Fermi-liquid-like peak, although the elementary excitations of the system have no overlap with a single electron [62]. Thus, one should be cautious about concluding, without rather detailed theoretical analysis, that any particular observed spectral function is or is not exhibiting quasiparticle behavior.

15.6. What is so Special About the Cuprates?

Until now, the issues we have discussed were mostly abstract, based on an analysis of the behavior of model Hamiltonians. Ultimately, however, we are interested in understanding the superconductivity in the cuprates. Moreover, since it is the one place where we all agree that a new phenomenon called high temperature superconductivity occurs, we would like to gain intuition about what is essential for high temperature superconductivity more generally, by analyzing what is essential to its occurrence in the cuprates.

15.6.1. Is Charge Order, Or Fluctuating Charge Order, Ubiquitous?

We have argued that some form of mesoscale spatial structure is essential to the mechanism of pairing. This structure could be static or slowly fluctuating, so long as the fluctuation frequency is less than the pairing scale. For this statement to be true, it is necessary that any material which exhibits high temperature superconductivity should also exhibit the requisite inhomogeneities. Since in the cuprates, T_c is not terribly sensitive to out of plane disorder, but, if anything, it increases as materials get cleaner, it seems implausible to us that the inhomogeneities in question can be directly linked to any sort of chemical inhomogeneity. This sort of inhomogeneity is certainly present in some materials—for instance, it is well documented [63–65] in STM studies on $Bi_2Sr_2CaCu_2O_{8+\delta}$, and may play a role in the superconductivity in that material.[13] However, more plausibly, in our opinion, the inhomogeneities in question are primarily associated with slow fluctuations of a proximate charge ordered state, of which the best documented example is the stripe phase [67].

Stripe order has been clearly documented in cuprates with reduced or vanishing T_c [68]. Clearly, where the stripe order is fully developed, the inhomogeneity is too strong—the superfluid density is highly suppressed and with it, T_c [69]. However, fluctuating stripe order has been clearly seen in numerous materials with moderately high T_cs, as discussed in depth

---

[13] For comparison, it is interesting to note that similar STM evidence of stripes has been found in the manganates [66].

in a recent review article of ours [3]. It remains an open issue whether such fluctuating order is universal in materials with high transition temperatures. In this regard, it is most important to study the evidence [70] of stripe fluctuations in YBa_2Cu_3O, the material in which the greatest degree of chemical homogeneity has seemingly been achieved. While the evidence for stripe-like fluctuations in this material is not unambiguous, the magnetic structure seen with neutrons is extremely reminiscent of that seen in stripe-ordered $La_{2-x}Ba_xCuO_4$, and is in many ways suggestive of the existence of some remnant tendency to striping. (See [3, 71–73].)

15.6.2. Does the "Stuff" Between the Cu–O Planes Matter?

One structural feature of the cuprates which has a much discussed systematic relation with T_c is the variation with the number of Cu–O planes stacked together between each "charge reservoir layer." For instance, in the sequence of materials $HgBa_2Ca_mCu_nO_y$, $T_c(n) = 98, 128, 135, 125$, and 108 K for $n = 1, 2, 3, 4, 5$, respectively. The peak in T_c at $n = 3$ is seen in all families of high temperature superconductors in which n can be systematically varied. There are many ideas concerning what this variation means. It is important to note that for $n > 2$, the different layers are not all equivalent, and so there is every reason to expect different doping levels on the different layers [74–76].

In the present context, three aspects of the layer number systematics seem suggestive. In the first place, this is a clear example of a situation in which there is an optimal inhomogeneity for superconductivity—apparently, $n = 3$ is in some way an optimal scale for superconductivity. Second, where phase fluctuations play a substantial role in determining T_c, it is clear that interplane couplings will suppress phase fluctuations and hence increase T_c. For instance [77], for the classical cubic lattice XY model on a slab n layers thick, the transition temperatures (computed by Monte Carlo) are $T_c(1) = 0.89J$, $T_c(2) = 1.38J$, $T_c(\infty) = 2.38J$. Finally, the $n = 3$ problem may reflect still more directly the way in which inhomogeneity can enhance T_c—where one has underdoped layers in good contact with overdoped layers, the combined system can inherit the high pairing scale from the underdoped layers and the large phase stiffness (superfluid density) from the overdoped layers [75].

Different "families" of high temperature superconductors are defined by subtle differences in the crystal structure and in the chemical character of the "charge reservoir layers" that lie between the Cu–O planes. There are substantial differences between the optimal T_cs in different families. For instance, double layer YBCO has an optimal $T_c \approx 92$ K, while double layer Tl 2212 has $T_c = 118\,K$ and double layer Hg 2212 has $T_c = 128\,K$. The differences are still more extreme if we compare the single layer cuprates, where the optimal T_c in the 214 family is $T_c = 42$ K for Stage IV O-doped LCO, while it is $T_c = 94\,K$ in Hg 1221. Thus, the variation of T_c with family is stronger still than its variation with n, as has been stressed by Leggett [74], Chakravarty et al. [76], and Geballe and Moyzhes [78]. Relatively little thought has been given to this striking observation, possibly because it makes one reflect uncomfortably about the importance of the solid state chemistry. One exception is the appealing idea of Geballe and Moyzhes [78], which is discussed in the article by Geballe elsewhere in this volume [79]. It is clear to us that this is an issue worth considerably more attention than it has so far received.

While it may well be true that interlayer tunneling [76] and/or electronic interactions in the charge reservoir layers in some way enhances the pairing, there is another possible explanation for the strong dependence of T_c on the three-dimensional structure of the materials. This is illustrated in the schematic phase diagram in Figure 15.4. We suppose, as indicated by the dashed-dotted line, that the pairing scale, i.e., the superconducting gap magnitude $\Delta_0(x)$,

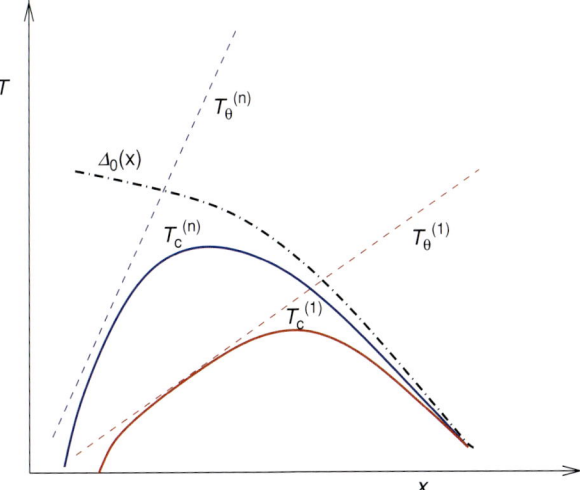

Figure 15.4. Schematic phase diagram for a high temperature superconductor for a single layer and a multilayer cuprate (n layers) as a function of doping x. The rationale for this figure is discussed in the text. The dashed lines are the putative classical phase-ordering temperature (were all other fluctuations suppressed), and the dashed-dotted curve is the pairing scale or mean-field transition temperature. The solid lines are the transition temperatures.

is a monotonically falling function of doping, x. Were fluctuations negligible, the material would order at a mean-field transition temperature $\sim \Delta_0/2$. However, in the underdoped regime, the small superfluid density implies [8, 80, 81] a large, fluctuation-induced reduction of T_c to a phase ordering temperature, $T_\theta \sim Ax$, as shown by the dashed lines in Figure 15.4.

Since pairing involves short-distance physics (on the scale of ξ_0), we take as a working hypothesis that it is largely a single plane property, so $\Delta_0(x)$ is largely insensitive to structures outside of the Cu–O plane.[14] However, since the phase ordering involves long-wave-length fluctuations (at length scales large compared to ξ_0), it is reasonable to expect the proportionality constant, A, to depend on the number of layers, n, and the electronic structure of the charge reservoir layer. Specifically, from the Monte Carlo calculations on the classical XY model mentioned earlier [77], we know that it is reasonable for A to vary by 50% or so with n. Since the pair tunneling amplitude through the charge reservoir layer can clearly depend on its electronic structure [82], it is likewise possible that A depends on "family."

The two different T_θ lines in the figure are thus supposed to represent materials with different three-dimensional structures.[15] The actual superconducting transition, T_c, is bounded above by T_θ and $\Delta_0/2$, as shown schematically by the solid curves in the figure. (In drawing the figure, we have assumed that quantum fluctuations will drive $T_c \to 0$ at a critical $x_c > 0$.) A consequence of this scheme is that in comparing the properties of "optimally doped" materials, those with a higher $T_c(x_{opt})$ should (unsurprisingly) have a larger gap, $\Delta_0(x_{opt})$, and

---

[14] This is certainly an oversimplification. For instance, in $La_{2-x}Sr_xCuO_4$ the gap at all doping levels is much smaller than in $YBa_2Cu_3O_{6+y}$, at optimal doping.

[15] For graphical simplicity, we have assumed that in all cases, T_θ, which is proportional to the low-frequency Drude weight, is linear in x, but the same qualitative physics is obtained if a more complex x dependence is assumed; what is important is that T_θ vanishes as x gets small (approaching the Mott insulator) and increases monotonically with increasing x.

a *smaller* value of the optimal doping, x_{opt}. (This latter correlation, which as far as we know has never been tested, is a slightly nontrivial prediction.)

15.6.3. What About Phonons?

There are phonons in the cuprates—they are seen in neutron scattering and thermal conductivity. They show up clearly in the optical absorption spectrum, so they must involve charge motion. There is evidence in support of the obvious fact that they affect the electron dynamics obtained from an analysis of the ARPES spectra, and the Raman spectra [83]. Despite the moral injunction against mentioning the "P word" in certain company, it is respectable—even desirable—to think about the relevance of phonons for high temperature superconductivity.

Two obvious facts argue against the usual role for phonons in the mechanism. First, there is the d-wave character of the superconductivity: most phonons are pair-breaking in the d-wave channel [84]. Second, the isotope effect is nearly zero at optimal doping; it is, of course, possible to have zero isotope effect even in the context of a conventional phonon-mediated BCS mechanism from a competition between the isotope dependence of the prefactor and μ^*. However, were this to occur precisely where T_c is maximum would smack of a joke by a malicious deity.

In underdoped cuprates, there is often an appreciable isotope effect, one that can be larger than those observed in simple metallic superconductors and which can apparently diverge as $x \to 1/8$ in some cases [85, 86]. However, the fact that this isotope effect occurs where T_c is suppressed, and in particular its singular doping dependence near $x = 1/8$, suggests that the isotope effect is indirect as far as superconductivity is concerned, and is probably better thought of as an isotope-dependent enhancement of the tendency to stripe order. In the underdoped regime, where the inhomogeneity is more than optimal, if replacing O^{17} with O^{18} tends to further stabilize the charge order, it will consequently tend to suppress the superconducting T_c.

15.6.4. What About Magnetism?

The empirical evidence suggests that antiferromagnetic correlations are an important feature of the electronic correlations in the cuprates, even when doped. Exactly what role this plays in the mechanism of HTC is much debated. It seems clear, by now, that whatever antiferromagnetism survives in the optimally doped superconductor is very short-ranged, so exchange of well-defined magnon like elementary excitations cannot be the mechanism of HTC. In addition, as shown by Schrieffer [87], excitations that too closely resemble Goldstone modes decouple from the electrons, and so are particularly ineffective for inducing pairing. However, short-range magnetic correlations can [88], and in our opinion are likely do play a role in the mechanism of HTC. These are the principle correlations responsible for the pair binding on Hubbard clusters.

In other strongly correlated systems, such as the manganites and nickelates, there is ample magnetism, but no superconductivity. Any mechanism that involves magnetism must rationalize why these other materials are not superconducting. In our view, there are several features that are responsible for this. The higher spin (spin 1/2 in the cuprates, spin 1 in the nickelates and spin 3/2 in the manganates) means that the magnetism is less quantum mechanical, and less easily quantum disordered in the presence of weak inhomogeneity. In addition, the presence of stronger electron phonon coupling and of other orbital degrees of freedom increases the tendency of these other materials to condense into other (nonsuperconducting) ordered ground states. In particular, the strong electron–phonon coupling in many standard perovskites, much enhances their tendency to form insulating "classical" charge-ordered states relative to the cuprates.

15.6.5. Must We Consider Cu–O Chemistry and the Three-Band Model?

It is a standard assumption in this field that the 2D Hubbard model, i.e., without any additional interactions or other embellishments, is "the Standard Model of Strongly Correlated Systems."[16] There are other "simple models," such as the Emery or three-band model [89, 90], which are more complicated (and hence "uglier") but which may be, in some ways, more "realistic." It is unclear to us whether the microscopic differences between the Emery and Hubbard models are essential to the mechanism of HTC, or unimportant. However, one thing that we have realized recently is that the Emery model, by virtue of its greater complexity, can be studied in various limits where certain aspects of the physics can be seen more simply and with better mathematical control than in the Hubbard model. For instance, the Emery model has an even stronger tendency to electronic phase separation than its simpler cousin, the Hubbard model. In addition, we have recently shown [91] the Emery model supports charge (Ising) nematic long range order, and probably other electron liquid crystal phases. (See also [92, 93].) Hence, competing interactions over microscopic length scales can (and do) give rise to relevant mesoscale structures.

15.6.6. Is d-Wave Crucial?

The answer to this question depends on what one means by "d-wave." If by d-wave one means a precise symmetry under rotations by $\pi/2$ this is clearly not essential as many materials, notably YBa_2Cu_3O, are orthorhombic. In the particular case of $YBa_2Cu_4O_8$ the anisotropy is so large that the ratio of the superfluid densities in the a and b directions is as large as $\rho_s^a/\rho_s^b \sim 7$; this material is essentially quasi one-dimensional [94]. (At the very least, this means that there must be order one s–d mixing.) On the other hand, even in this case, the *sign* of the order parameter alternates as seen clearly in corner junction [95] and tri-crystal [96] experiments. So far, all the existing experimental evidence in the cuprates is consistent with "d-wave like" superconductivity, in this sense.

What is less evident is how essential are the nodal quasiparticles. The experimental evidence in most cuprates [83, 97–99] is consistent with the existence of nodal excitations in the superconducting state,[17] while they are either manifestly absent or poorly defined above T_c, in the pseudogap regime [83, 99, 100]. One of the puzzles of this problem, and one that makes it interesting, is why there are nodal quasiparticles below T_c even though they do not exist in the "normal." In the BCS mechanism, or in any other weak coupling approach, the quasiparticles of the superconducting state are a "left-over" of the states of the parent normal (Fermi liquid) state. While it is clear that as the interactions become stronger the *symmetry* of the superconducting state may be "protected," it is not obvious that the quasiparticles themselves should be. From the perspective of a strong coupling approach, such as the one advocated here which does not assume a state with well-defined quasiparticles in the parent state, the nodal quasiparticles are an emergent phenomenon, and one can perfectly conceive a d-wave state with or

---

[16] The enshrinement of this simple model as a sort of "Theory of Everything" is peculiar in a field that stresses the fundamental importance of "emergent" and the misleading assertions of the "fundamental."

[17] In fact, even in the superconducting state the nodal quasiparticles in high temperature superconductors are never as well-defined as in conventional metals, e.g., even at temperatures as low as 5 K, the energy width of a nodal quasiparticle is at least comparable to its energy.

without nodal quasiparticles. In fact, the transition between a node-less and nodal d-wave-like state was studied in [50, 101], where it was found to be a mean field (Lifshitz) transition with relatively little effect on T_c.

15.6.7. Is Electron Fractionalization Relevant?

The discovery of high temperature superconductivity and the realization that the underlying physics of these systems is inconsistent with the venerable Landau Theory of the Fermi Liquid, launched an all-out effort to develop a "new" theory of strongly correlated systems. Many interesting and novel phases of matter were (and are) proposed, some of which were hoped to contain the fundamental (pardon our language) correlations responsible for high temperature superconductivity and in particular for the high values of T_c. Thus, in addition to the conventional Néel antiferromagnetic state, other nonmagnetic ground states have been proposed, such as spin liquids with and without time-reversal symmetry breaking, as well as valence bond crystals which break translation and rotation invariance to various degrees [46, 47, 102–104]. However, perhaps following the "Bell Labs Rule" (a New Jersey version of Occam's Razor) that of all possible theories the most boring one (the one with the standard answer) is the one most likely to be correct, it has turned out that the ground states of simple models of undoped strongly correlated systems are typically antiferromagnets with long range Néel order [105, 106].

A number of interesting theories of spin liquid states, with [107, 108] and without [27, 109–113] time-reversal symmetry breaking, have been proposed over the years. Electron fractionalization and deconfinement are a defining feature of all these spin liquid phases. However, while recent advances in this subject [114] have put some of these proposals on firmer theoretical footing (by proving that they are the ground states of reasonably local Hamiltonians), most simple models of strongly correlated systems do not seem to naturally have these phases [4, 45, 103]. Moreover, in apparent accordance with the Bell Labs Rule, there is no compelling experimental evidence (yet) in support of their relevance, at least in the cuprates. Typically, the simple spin models thus far explored, even those with significant ring exchange interactions, have either spin ordered phases or valence bond ordered phases, and confinement on relatively short length scales, although there are known counterexamples [115]. We should note, however, that it is also possible to have phases with extremely long confinement length scales, e.g., the Cantor Deconfinement phases of [116], which for all practical purposes can do the job just as well.

As noted in Section 15.2.1, both the spin liquid scenario and the mechanism explored here have in common the existence of a high energy pairing scale associated with spin-gap formation.

15.7. Coda: High Temperature Superconductivity is Delicate But Robust

By whatever measure one might devise, the set of materials which exhibit high temperature superconductivity is a very small subset of electronically active materials. However, within the cuprates, materials that share the basic motif of Cu–O planes, high temperature superconductivity is robust in the sense that the transition temperature is not wildly sensitive to many sorts of chemical substitutions, structural differences, and degrees of quenched disorder.[18] It seems reasonable to us to expect that any theory of high temperature superconductivity should be able to answer the question: why is high temperature superconductivity so rare?

Part of the answer is clearly the role of competing order. At weak coupling, the only instability of a Fermi liquid is the Cooper instability, so low temperature superconductivity should be (and is) reasonably generic. At strong coupling, many sorts of ordered states can be stabilized, including spin and charge density wave states, and more exotic states such as orbital antiferromagnetism [117] (dDW), which, in general, compete with superconductivity. Thus, precisely in those materials in which the couplings are strong enough that they could produce high T_c, other ordered phases occur which can quench the superconductivity substantially.

In our view, another feature is the necessity of an optimal degree (and character) of inhomogeneity—self-organized or otherwise. If the system is too homogeneous, then a high pairing scale is unattainable. If the system is too inhomogeneous, the coherence scale is strongly suppressed, and with it T_c. Obtaining a high T_c requires a rather delicate balance between these two extremes.

There are several other special features of the cuprates which likely also are essential. It seems to us that the fact that the cuprates are doped Mott insulators (with local moments), and that the insulating state in question is highly quantum mechanical (spin 1/2) are likely to be essential features of the physics, although the fact that the undoped system has a Néel ordered ground-state is probably not crucial. It is clear to us that overly strong electron–phonon coupling would produce too strong a tendency toward charge ordering [118], and hence would be destructive of high temperature superconductivity. From this point of view, the relatively *weakness* of the electron–phonon coupling in the cuprates in comparison with other perovskites (e.g., the nickelates and the manganates) is one of the important features of the cuprates that makes them high temperature superconductors. On the other hand, it seems to us likely that the tendency toward self-organized inhomogeneity found in theoretical studies of the Hubbard and related models is too weak to provide the necessary mesoscale inhomogeneity. In this sense, the electron–phonon coupling in the cuprates likely plays an important role in producing high temperature superconductivity—not that phonons serve as the glue but that they help with the self-organization of the necessary inhomogeneities.

Acknowledgments

We would like to acknowledge the contributions of our many collaborators on several aspects of this many-faceted subject, especially Victor Emery, Sudip Chakravarty, Oron Zachar, John Tranquada, Ted Geballe, Aharon Kapitulnik, Vadim Oganesyan, Erica Carlson, Dror Orgad, and Enrico Arrigoni. SAK would particularly like to acknowledge formative discussions with J.R. Schrieffer (and who could know better) on the mechanism of superconductivity, and in particular on the critical role of retardation for obtaining an effective attraction. We are also grateful to him for giving us an opportunity to present our prejudices unhindered by the pernicious influence of referees and other savage beasts. This work was supported, in part, by the National Science Foundation through the grants NSF DMR 04-42537 at the University of Illinois (EF), NSF DMR-04-21960 at Stanford University (SAK), and by the Department of Energy through the grant DE-FG03-00ER45798 at UCLA (SAK), and DEFG02-91ER45439 at the Frederick Seitz Materials Research Laboratory of the University of Illinois (EF).

---

[18] T_c is sensitive to some changes, such as Cu substitution in the Cu–O planes and some features of the interplane arrangements and chemistry, so this statement has some exceptions.

Appendix A: What Defines "High Temperature Superconductivity"

The term "high temperature superconductivity" is rather vague, since of course the question arises, high compared to what? From Figure 15.1, it is clear that, from a material science viewpoint, high temperature superconductivity means T_c larger than 20 K. However, as an abstract issue in theory, it is less clear what is meant.

What we would like to find are models that are "physical," although not necessarily "realistic," and which have superconducting transition temperatures that are the of order of a microscopic energy scale. By "physical," we mean that the model must satisfy certain sets of constraints, such as having electrons with spin-$1/2$ which are fermions with dominantly repulsive bare interactions. Of course, in some sense, the closer a model is to reflecting the essential solid state chemistry of a particular material of interest, the more clearly physical it is, but for the purposes of understanding the mechanism, we would prefer to study as simple a model as possible, rather than one that has extraneous bells and whistles that happen to be part of the electronic structure of one material or another.

Alas, upon reflection, this rough definition of what constitutes high temperature superconductivity ceases to make any sense. Presumably, in any model in which the strength of the various interactions are all comparable to each other, if the model is superconducting at all, T_c must be equal to a number of order 1 times a microscopic scale. It then becomes a question of how big the number of order 1 must be to be considered high. (For the negative U Hubbard model with $U = -4t$ the superconducting transition temperature has been estimated [119] from quantum Monte Carlo to be $T_c = 0.14t$. Putting aside the "unphysical" nature of the microscopically attractive interactions in this model, it is not clear whether one should or should not classify this as "high temperature superconductivity.")

We [36] have therefore proposed a different purely theoretical definition of HTC. In all cases we know of in which T_c can be computed reliably (other than by Monte Carlo or related numerical methods), there is a small parameter, $\lambda \ll 1$, which is exploited in the calculation. In BCS theory, λ is the dimensionless electron–phonon coupling, and T_c depends exponentially on $1/\lambda$. If we agree that we can trust BCS theory when $\lambda < 1/5$ (to choose a number arbitrarily), this means that on the basis of this theory, we can claim to have a good understanding of the mechanism of superconductivity only so long as T_c is at least two orders of magnitude smaller than the typical microscopic scale. In contrast, mechanisms we wish to associate with high temperature superconductivity should have a much weaker dependence on the small parameter, $T_c \propto \lambda^\alpha$, where the smaller α the better. For such a mechanism, say with $\alpha \sim 1$, if we accept the same criterion for the range of λ for which the theory is trustworthy, we have a valid theoretical understanding of the superconductivity even when T_c is fully $1/5$ of a microscopic scale.

Bibliography

1. J. R. Schrieffer, *Theory of Superconductivity* (Addison-Wesley, Redwood City, CA, 1964).
2. P. W. Anderson, Science **177**, 393 (1972).
3. S. A. Kivelson, E. Fradkin, V. Oganesyan, I. Bindloss, J. Tranquada, A. Kapitulnik, and C. Howald, Rev. Mod. Phys. **75**, 1201 (2003), arXiv:cond-mat/02010683.
4. E. W. Carlson, V. J. Emery, S. A. Kivelson, and D. Orgad, in *The Physics of Conventional and Unconventional Superconductors*, K. H. Bennemann and J. B. Ketterson (eds.) (Springer, Berlin, Heidelberg, New York, 2004), arXiv:cond-mat/0206217.
5. R. M. White and T. H. Geballe, *Long Range Order in Solids* (Academic, New York, 1979).

6. H. Shaked, et al., *Crystal Structures of the High T_c Superconducting Copper-Oxides* (Elsevier, Amsterdam, 1994).
7. S. Chakravarty, Science **266**, 386 (1994).
8. V. J. Emery and S. A. Kivelson, Nature **374**, 434 (1995).
9. S. Chakravarty, A. Sudbø, P. W. Anderson, and S. Strong, Science **261**, 337 (1993).
10. J. E. Hirsch, Physica C **199**, 305 (1992).
11. V. J. Emery, S. A. Kivelson, and O. Zachar, Phys. Rev. B **56**, 6120 (1997), arXiv:cond-mat/9610094.
12. S. Chakravarty, H.-Y. Kee, and E. Abrahams, Phys. Rev. B **67**, 100504 (2003).
13. M. Boninsegni and E. Manousakis, Phys. Rev. B **47**, 11897 (1993).
14. S. Chakravarty, M. Gelfand, and S. A. Kivelson, Science **254**, 970 (1991).
15. S. R. White, S. Chakravarty, M. Gelfand, and S. A. Kivelson, Phys. Rev. B **45**, 5062 (1992).
16. S. Trugman and D. J. Scalapino, Philos. Mag. B **74**, 607 (1996), arXiv:cond-mat/9604008.
17. R. M. Fye, D. J. Scalapino, and R. T. Scalettar, Phys. Rev. B **46**, 8667 (1992).
18. S. Chakravarty and S. A. Kivelson, Phys. Rev. B **64**, 64511 (2001), arXiv:cond-mat/0012305.
19. S. Chakravarty, L. Chayes, and S. A. Kivelson, Lett. Math. Phys. **23**, 265 (1991).
20. S. Chakravarty and S. Kivelson, Eur. Phys. Lett. **16**, 751 (1991).
21. S. R. White, R. M. Noack, and D. J. Scalapino, Phys. Rev. Lett. **73**, 886 (1994), arXiv:cond-mat/9403042.
22. H. Tsunetsugu, M. Troyer, and T. M. Rice, Phys. Rev. B **51**, 16456 (1995), arXiv:cond-mat/9401050.
23. S. Chakravarty, Phys. Rev. Lett. **77**, 4446 (1996), arXiv:cond-mat/9608124.
24. S. R. White, R. M. Noack, and D. J. Scalapino, Phys. Rev. Lett. **73**, 886 (1994), arXiv:cond-mat/9409065.
25. M. Greven, R. J. Birgeneau, and U. J. Wiese, Phys. Rev. Lett. **77**, 1865 (1996), arXiv:cond-mat/9605068.
26. O. F. Syljuåsen, S. Chakravarty, and M. Greven, Phys. Rev. Lett. **78**, 4115 (1997), arXiv:cond-mat/9701197; and references therein.
27. S. A. Kivelson, D. Rokhsar, and J. P. Sethna, Phys. Rev. B **35**, 865 (1987).
28. V. J. Emery, S. A. Kivelson, and H. Q. Lin, Phys. Rev. Lett. **64**, 475 (1990).
29. J. R. Schrieffer, X.-G. Wen, and S.-C. Zhang, Phys. Rev. Lett. **60**, 944 (1988).
30. L. Balents and M. P. A. Fisher, Phys. Rev. B **53**, 12133 (1996), arXiv:cond-mat/9503045.
31. H. H. Lin, L. Balents, and M. P. A. Fisher, Phys. Rev. B **58**, 1794 (1998), arXiv:cond-mat/9801285.
32. V. J. Emery, S. A. Kivelson, and O. Zachar, Phys. Rev. B **59**, 15641 (1999), arXiv:cond-mat/9810155.
33. A. Luther and V. J. Emery, Phys. Rev. Lett. **33**, 589 (1974).
34. H. H. Lin, L. Balents, and M. P. A. Fisher, Phys. Rev. B **56**, 6569 (1997), arXiv:cond-mat/9703055.
35. I. Martin, D. Podolsky, and S. A. Kivelson, Phys. Rev. B **72** (2005), unpublished; arXiv:cond-mat/0501659.
36. E. Arrigoni, E. Fradkin, and S. A. Kivelson, Phys. Rev. B **69**, 214519 (2004), arXiv:cond-mat/0309572.
37. R. M. Noack, S. R. White, and D. J. Scalapino, Phys. Rev. B **56**, 7162 (1997), arXiv:cond-mat/9601047.
38. T. Siller, M. Troyer, T. M. Rice, and S. R. White, Phys. Rev. B **63**, 195106 (2001), arXiv:cond-mat/0006080.
39. S. White, I. Affleck, and D. J. Scalapino, Phys. Rev. B **65**, 165122 (2002), arXiv:cond-mat/0111320.
40. E. Dagotto and T. M. Rice, Science **271**, 618 (1996), arXiv:cond-mat/9509181.
41. C. Wu, W. V. Liu, and E. Fradkin, Phys. Rev. B **68**, 115104 (2003), arXiv:cond-mat/0206248.
42. H. J. Schulz, Phys. Rev. B **59**, 2471 (1999), arXiv:cond-mat/9807328.
43. E. W. Carlson, D. Orgad, S. A. Kivelson, and V. J. Emery, Phys. Rev. B **62**, 3422 (2000), arXiv:cond-mat/0001058.
44. E. Fradkin and S. A. Kivelson, Mod. Phys. Lett. B **4**, 225 (1990).
45. E. Fradkin, *Field Theories of Condensed Matter Systems* (Addison-Wesley, Redwood City, 1991), chapter 6.
46. N. Read and S. Sachdev, Nucl. Phys. B **316**, 609 (1989).
47. N. Read and S. Sachdev, Phys. Rev. Lett. **62**, 1694 (1989).
48. D. J. Scalapino, Y. Imry, and P. Pincus, Phys. Rev. B **11**, 2042 (1975).
49. D. S. Fisher and P. C. Hohenberg, Phys. Rev. B **37**, 4936 (1988).
50. M. Granath, V. Oganesyan, S. A. Kivelson, E. Fradkin, and V. J. Emery, Phys. Rev. Lett. **87**, 167011 (2001), arXiv:cond-mat/0010350.
51. P. W. Anderson, in *"Frontiers and Borderlines in Many Particle Physics,"* Proceedings of the Enrico Fermi International School of Physics, Varenna (North Holand, Amsterdam, the Netherlamds, 1987).
52. P. W. Anderson, Science **235**, 1169 (1987).
53. V. J. Emery and S. A. Kivelson, Phys. Rev. Lett. **74**, 3253 (1995).
54. I. M. Khalatnikov, private communication.
55. V. J. Emery and S. A. Kivelson, Physica C **209**, 597 (1993).
56. R. Jamei, S. Kivelson, and B. Spivak, Phys. Rev. Lett. **94**, 056805 (2005), arXiv:cond-mat/0408066.
57. B. Spivak and S. A. Kivelson, Phys. Rev. B **70**, 155114 (2004), arXiv:cond-mat/0310712.

58. S. A. Kivelson, E. Fradkin, and V. J. Emery, Nature **393**, 550 (1998), arXiv:cond-mat/9707327.
59. E. Demler, W. Hanke, and S.-C. Zhang, Rev. Mod. Phys. **76**, 909 (2004).
60. H.-D. Chen, S. Capponi, F. Alet, and S.-C. Zhang, Phys. Rev. B **70**, 024516 (2004).
61. P. G. de Gennes and J. Prost, *The Physics of Liquid Crystals* (Oxford Science Publications/Clarendon Press, Oxford, UK, 1993).
62. D. Orgad, S. A. Kivelson, E. W.Carlson, V. J. Emery, X. J. Zhou, and Z. X. Shen, Phys. Rev. Lett. **86**, 4362 (2001).
63. C. Howald, R. Fournier, and A. Kapitulnik, Phys. Rev. Lett. **64**, 100504 (2001), arXiv:cond-mat/0101251.
64. K. M. Lang, V. Madhavan, J. E. Hoffman, E. W. Hudson, H. Eisaki, S. Uchida, and J. C. Davis, Nature **415**, 412 (2002), arXiv:cond-mat/0112232.
65. T. Cren, D.Roditchev, W. Sacks, J. Klein, J. B. Moussy, C. Deville-Cavellin, and M. Lagues, Phys. Rev. Lett. **84**, 147 (2000).
66. C. Renner, G. Aeppli, B.-G. Kim, Y.-A. Soh, and S.-W. Cheong, Nature **416**, 518 (2002).
67. J. M. Tranquada, B. J. Sternlieb, J. D. Axe, Y. Nakamura, and S. Uchida, Nature **375**, 561 (1995).
68. J. M. Tranquada, J. D. Axe, N. Ichikawa, A. R. Moodenbaugh, Y. Nakamura, and S. Uchida, Phys. Rev. Lett. **78**, 338 (1997).
69. N. Ichikawa, S. Uchida, J. M. Tranquada, T. Niemöller, P. M. Gehring, S.-H. Lee, and J. R. Schneider, Phys. Rev. Lett. **85**, 1738 (2000).
70. H. A. Mook, P. Dai, S. M. Hayden, G. Aeppli, T. G. Perring, and F. Doan, Nature **395**, 580 (1998).
71. J. M. Tranquada, H. Woo, T. G. Perring, H. Goka, G. D. Gu, G. Xu, M. Fujita, and K. Yamada, Nature **429**, 534 (2004), arXiv:cond-mat/0401621.
72. N. B. Christensen, D. F. McMorrow, H. M. Rønnow, B. Lake, S. M. Hayden, G. Aeppli, T. G. Perring, M. Mangkorntong, M. Nohara, and H. Takagi, Phys. Rev. Lett. **93**, 147002 (2004), arXiv:cond-mat/0403439.
73. V. Hinkov, S. Pailhes, P. Bourges, Y. Sidis, A. Ivanov, A. Kulakov, C. Lin, D. Chen, C. Bernhard, and B. Keimer, Nature **430** (2004), arXiv:cond-mat/0408379.
74. A. J. Leggett, Phys. Rev. Lett. **83**, 392 (1999).
75. S.A.Kivelson, Physica B **318**, 61 (2002).
76. S. Chakravarty, H.-Y. Kee, and K. Voelker, Nature **428**, 53 (2004), arXiv:cond-mat/0309209.
77. E. W. Carlson, S. A. Kivelson, V. J. Emery, and E. Manousakis, Phys. Rev. Lett. **83**, 612 (1999).
78. T. H. Geballe and B. Y. Moyzhes, Annalen der Physik **13**, 20 (2004).
79. T. H. Geballe (2005), chapter in this volume.
80. Y. Wang, Z. A. Xu, T. Kakeshita, S. Uchida, S. Ono, Y. Ando, and N. P. Ong, Phys. Rev. B **64**, 224519 (2001), arXiv:cond-mat/0108242.
81. Y. Wang, N. P. Ong, Z. A. Xu, T. Kakeshita, S. Uchida, D. A. Bonn, R. Liang, and W. N. Hardy, Phys. Rev. Lett. **88**, 257003 (2002), arXiv:cond-mat/0205299.
82. V. Oganesyan, S. Kivelson, T. Geballe, and B. Moyzhes, Phys. Rev. B **65**, 172504 (2002).
83. A. Damascelli, Z.-X. Shen, and Z. Hussain, Rev. Mod. Phys. **75**, 473 (2003), arXiv:cond-mat/0208504.
84. N. Bulut and D. J. Scalapino, Phys. Rev. B **54**, 14971 (1996).
85. M. K. Crawford, W. E. Farneth, E. M. M. III, R. L. Harlow, and A. H. Moudden, Science **250**, 1390 (1990).
86. M. K. Crawford, R. L. Harlow, E. M. McCarron, W. E. Farneth, J. D. Axe, H. Chou, and Q. Huang, Phys. Rev. B **44**, 7749 (1991).
87. J. Schrieffer, J. Low Temp. Phys. **99**, 397 (1995).
88. D. J. Scalapino, E. L. Jr., and J. E. Hirsch, Phys. Rev. B **34**, 8190 (1986).
89. V. J. Emery, Phys. Rev. Lett. **58**, 2794 (1987).
90. C. M. Varma, S. Schmitt-Rink, and E. Abrahams, Solid State Commun. **62**, 550 (1987).
91. S. A. Kivelson, E. Fradkin, and T. H. Geballe, Phys. Rev. B **69**, 144505 (2004), arXiv:cond-mat/0302163.
92. J. Lorenzana and G. Seibold, Phys. Rev. Lett. **89**, 136401 (2002), arxiv:cond-mat/0207206.
93. V. I. Anisimov, M. A. Korotin, A. S. Mylnikova, A. V. Kozhevnikov, D. M. Korotin, and J. Lorenzana, Phys. Rev. B **70**, 172501 (2004), arXiv:cond-mat/0402162.
94. D. Broun (2001), Cambridge University PhD Thesis, unpublished.
95. D. A. Wollman, D. J. V. Harlingen, W. C. Lee, D. M. Ginsberg, and A. J. Leggett, Phys. Rev. Lett. **71**, 2134 (1993).
96. C. C. Tsuei, J. R. Kirtley, C. C. Chi, L. S. Yu-Jahnes, A. Gupta, T. Shaw, J. Z. Sun, and M. B. Ketchen, Phys. Rev. Lett. **73**, 593 (1994).
97. R. I. Miller, R. F. Kiefl, J. H. Brewer, J. E. Sonier, J. Chakhalian, S. Dunsiger, G. D. Morris, A. N. Price, D. A. Bonn, W. H. Hardy, et al., Phys. Rev. Lett. **88**, 137002 (2002), arXiv:cond-mat/0111550.

98. J. E. Hoffman, E. W. Hudson, K. M. Lang, V. Madhavan, H. Eisaki, S. Uchida, and J. C. Davis, Science **295**, 466 (2002), arXiv:cond-mat/0201348.
99. J. C. Campuzano, M. R. Norman, and M. Randeria, in *Physics of Conventional and Unconventional Superconductors*, K. H. Bennemann and J. B. Ketterson (eds.) (Springer, Berlin Heiderberg, New York, 2003), arXiv:cond-mat/0209476.
100. M. Vershinin, S. Misra, S. Ono, Y. Abe, Y. Ando, and A. Yazdani, Science **303**, 1005 (2004), arXiv:cond-mat/0402320.
101. S. Sachdev, Physica A **313**, 252 (2002), and references therein.
102. D. Rokhsar and S. A. Kivelson, Phys. Rev. Lett. **61**, 2376 (1988).
103. S. Sachdev, Rev. Mod. Phys. **75**, 913 (2003), arXiv:cond-mat/0211005.
104. S. Sachdev, in *Quantum magnetism*, U. Schollwöck, J. Richter, D. J. J. Farnell, and R. A. Bishop (eds.) (Springer, Berlin Heiderberg, New York, 2004), arXiv:cond-mat/0401041.
105. S. Chakravarty, B. I. Halperin, and D. R. Nelson, Phys. Rev. B **39**, 2344 (1989).
106. Y. Endoh, K. Yamada, R. J. Birgeneau, D. R. Gabbe, H. P. Jenssen, M. A. Kastner, C. J. Peters, P. J. Picone, T. R. Thurston, J. M. Tranquada, et al., Phys. Rev. B **37**, 7443 (1988).
107. V. Kalmeyer and R. B. Laughlin, Phys. Rev. Lett. **59**, 2095 (1988).
108. X. G. Wen, F. Wilczek, and A. Zee, Phys. Rev. B **39**, 11413 (1989).
109. N. Read and S. Sachdev, Phys. Rev. B **42**, 4568 (1990).
110. C. Mudry and E. Fradkin, Phys. Rev. B **49**, 5200 (1994), arXiv:cond-mat/9309021.
111. T. Senthil and M. P. A. Fisher, Phys. Rev. B **62**, 7850 (2000), arXiv:cond-mat/9910224.
112. P. A. Lee, N. Nagaosa, and X.-G. Wen, Rev. Mod. Phys. **78**, 17 (2006), arXiv:cond-mat/0410445.
113. J. Zaanen, O. Y. Osman, H. V. Kruis, Z. Nussinov, and J. Tworzydlo, Philos. Mag. **81**, 1485 (2001).
114. R. Moessner and S. L. Sondhi, Phys. Rev. Lett. **86**, 1881 (2001), arXiv:cond-mat/0007378.
115. K. S. Raman, R. Moessner, and S. L. Sondhi, Phys. Rev. B **72**, 064413 (2005), arXiv:cond-mat/cond-mat/0502146.
116. E. Fradkin, D. A. Huse, R. Moessner, V. Oganesyan, and S. L. Sondhi, Phys. Rev. B **69**, 224415 (2005), arXiv:cond-mat/0311353.
117. S. Chakravarty, R. B. Laughlin, D. K. Morr, and C. Nayak, Phys. Rev. B **63**, 094503 (2001).
118. J. Zaanen and P. B. Littlewood, Phys. Rev. B **50**, 7222 (1994).
119. R. T. Scalettar, E. Y. Loh, J. E. Gubernatis, A. Moreo, S. R. White, D. J. Scalapino, R. L. Sugar, and E. Dagotto, Phys. Rev. Lett. **62**, 1407 (1989).

16

Superconducting States on the Border of Itinerant Electron Magnetism

Emma Pugh, Siddharth Saxena, and Gilbert Lonzarich

16.1. Introduction

In recent years there has been growing evidence for the existence of novel metallic states on the border of long-range magnetic order in certain materials. At sufficiently low temperatures these unfamiliar metallic states can be unstable and lead to the formation of unconventional forms of superconductivity that cannot be fully explained in terms of the Bardeen–Cooper–Schrieffer (BCS) model [1] in its usual form. As this superconductivity is observed in a narrow regime on the border of magnetism, it is thought that the pairing of carriers might arise from a spin–spin or magnetic interaction rather than solely from the traditional electron–phonon interaction of the BCS model.

This chapter begins by considering the simplest deviations that are observed from the standard low-temperature theory of metals on the border of long-range ferromagnetic order in metals where superconductivity is not observed. It then goes on to discuss the border of antiferromagnetism where superconducting instabilities are prevalent. We consider, in particular, the role of the effective dimensionality of the material and of the combined effect of magnetic and density instabilities that have led to a clarification of the nature of Cooper pair formation in some heavy-fermion superconductors. Until relatively recently superconductivity on the border of itinerant ferromagnetism was elusive. The search and discovery of this new and important class of superconductors is discussed. The chapter concludes by looking forward to possible future advances in this area of research. Throughout, examples are demonstrated by showing data from experiments and open questions to our understanding are highlighted.

16.2. Uncharted Territory: The New Frontier

By the precise control of matter it is possible to observe in detail the crossover region as one goes from one state of matter to another at low temperature. Figure 16.1 shows a possible phase diagram of a magnetic metal when it is subjected to applied pressure. In the past much research effort has focused on the nature of magnetic ordering in materials (left-hand side of Figure 16.1) and on the conventional metallic state (right-hand side of Figure 16.1). For low temperatures the right-hand side of the figure is well described by Landau's Fermi-liquid theory and the left-hand side is described in terms of an exchange split Fermi surface together

---

E. Pugh, S. S. Saxena, and G. G. Lonzarich • Cavendish Laboratory, University of Cambridge, J.J. Thomson Avenue, Cambridge CB3 0HE, UK

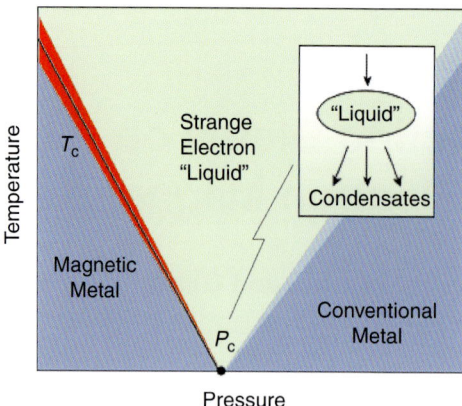

Figure 16.1. The temperature–pressure phase diagram of a ferromagnetic or antiferromagnetic metal. The crossover region between the magnetic metal and the conventional metal can be occupied by a strange quantum "liquid" which defies conventional description. In sufficiently pure samples and at sufficiently low temperatures, the electron liquid can condense into other quantum phases. These include exotic forms of superconductivity that can differ from that of the traditional Bardeen–Cooper–Schrieffer (BCS) model. (Figure is taken from reference [100].)

with collective excitations corresponding to spin waves. The crossover region between the magnetic and conventional metal can be occupied by a strange metallic state or quantum "liquid" which defies conventional description and a quantum critical point can occur in this region at low temperature. This liquid state, normally viewed as an intermediate-temperature phenomenon, can extend down to the lowest temperatures explored near the quantum critical point. In sufficiently pure samples and at sufficiently low temperatures, the electron liquid can condense into other quantum phases. These include exotic forms of superconductivity. (For recent reviews see, e.g., [2, 3]).

16.3. Logarithmic Fermi Liquid

The mildest form of quantum liquid that deviates from the Fermi-liquid theory is the marginal or logarithmic Fermi liquid state. A review of several underlying models that yield such a state can be found in a 1991 book by Baym and Pethick [4]. For example, in the logarithmic Fermi liquid that is expected to arise where the Curie temperature tends toward absolute in an isotropic system, one expects a breakdown of the standard Fermi-liquid character and instead observe a $T \ln(1/T)$ dependence of the heat capacity, a $T^{5/3}$ temperature dependence of the resistivity, an inverse susceptibility varying as $T^{4/3}$ and a linear quasi-particle relaxation rate. (For discussions and references to the earlier literature, see [5, 6]. A renormalisation group approach to this problem is given in [7, 8].) Experimental evidence for such behaviour is mounting. Examples can be found in cubic metals which are on the border of ferromagnetism and which have a nearly continuous ferromagnetic transition when the Curie temperature is suppressed to zero with increasing pressure. Recent re-examinations of this problem have been carried out for example in Ni_3Al [9] and $ZrZn_2$ [10].

The observation of the logarithmic Fermi liquid appears to be sensitive to the exact form of the ferromagnetic transition as pressure is applied. If the transition switches from second order to first order then this state may not be clearly seen. However, this does not mean that the only other possibility is a return to Fermi-liquid behaviour. It is thought that behaviour which cannot be explained by either the Fermi-liquid or logarithmic Fermi-liquid models have been observed, such as in MnSi.

16.4. The Puzzle of MnSi

The d-electron metal MnSi has proved a puzzle to our understanding of systems on the border of magnetism. It has a cubic structure that orders into a long-wave spin-spiral state at ambient pressure with a Curie temperature of around 30 K. If clean samples of MnSi are subjected to hydrostatic pressure, the Curie temperature is pushed towards absolute zero at around a critical pressure of 1.5 GPa [11–14]. The Curie temperature begins by decreasing in a second order way and then switches over to first order [12, 15, 16]. Above this later critical pressure MnSi shows anomalous normal state properties with clear deviations from the Fermi liquid temperature dependence of the resistivity. In place of the Fermi liquid T^2 dependence one observes an unexpected $T^{3/2}$ dependence. Figure 16.2 shows the temperature–pressure phase diagram for MnSi. As the magnetic transition with pressure is first order close to the critical pressure, models predict that if cooled to sufficiently low temperature the T^2 Fermi-liquid behaviour should be recovered. This has not yet been observed in MnSi and remains at the robust $T^{3/2}$ value down to the lowest temperatures currently measured and at least up to 2.5 GPa

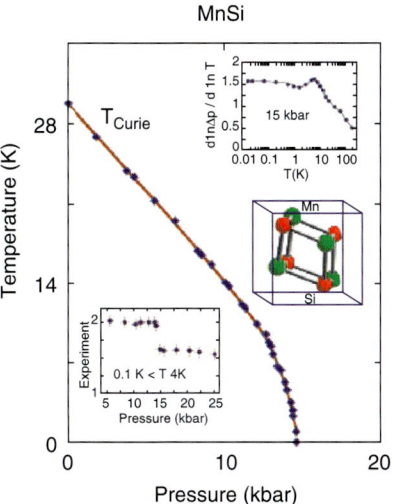

Figure 16.2. Temperature–pressure phase diagram of MnSi [11–17]. The magnetic transition is continuous or second order at low pressures but is discontinuous or first order near the critical pressure $P_c = 15$ kbar where the Curie temperature vanishes. Since the transition from the magnetic to the non-magnetic state is discontinuous at P_c, the logarithmic (marginal) Fermi liquid discussed in the text is not expected to be observed in this material. Thus the exponent characterising the temperature dependence of the resistivity is predicted to tend to the Fermi-liquid value of 2 both above and below P_c for temperatures below about 10 K. The recovery of a T^2 regime at much lower temperature is not ruled out but would not significantly alter the discrepancy between theory and experiment. As shown in the inset on the lower left, this is not observed. The resistivity exponent drops suddenly and unexpectedly at P_c to a value of around 3/2. The resistivity exponents are determined from the logarithmic derivative of $\Delta\rho = \Delta\rho - \rho_0$ where ρ_0 is the extrapolated residual resistivity, which is $0.2\,\mu\Omega$ cm for the samples investigated (top inset on right). Analysis of another logarithmic derivative, $d(\ln(T\, d\rho/dT))/d(\ln T)$, that does not depend on the way ρ_0 is extrapolated gives a similar result. The lower inset on the right gives the B20 structure of MnSi (10 kbar = 1 GPa).

[14, 17]. It is possible that a new ground state is forming in this material near and above the critical pressure. There is so far no evidence for the formation of a superconducting condensate in MnSi even in extremely pure samples with mean free paths of the order of 10,000 Å.

16.5. Superconductivity on the Border of Magnetism

For conventional superconductors that can be described by the BCS theory, the presence of magnetic interactions tends to destroy superconductivity [18]. However, in the heavy-fermion compounds and a number of other systems, superconductivity appears to arise in the presence of strong magnetic fluctuations. One possibility is that the magnetic interactions, rather than destroying superconductivity as in the conventional superconductors, could instead act as "glue" and actually bind electrons together to form pairs. It is important to discern between two basic classes of magnetic superconductor. In some systems, such as the Chevrel phase superconductors, different electrons are responsible for the magnetism and superconductivity. Examples of Chevrel phase materials that exhibit antiferromagnetic and superconducting states are $GdMo_6Se_8$ and $ErMo_6Se_8$ [19, 20]. There are some materials, however, where it is believed that the magnetic order and superconductivity arise from the same electrons. Examples include, in particular, the heavy-electron compounds in which 4f electrons in Ce or 5f electrons in U form narrow quasi-particle bands characterised by quasi-particle effective masses two to three orders of magnitude greater than that of bare electrons (see, e.g., [21] for a recent discussion).

The magnetic interaction model for Cooper pair formation and superconductivity has a long history that has been reviewed recently, e.g., in [22, 23]. The results of model calculations show that magnetic pairing can be exceedingly sensitive to the details of the lattice, electronic and magnetic structures. In the simplest models considered, spin-triplet pairing is favoured on the border of ferromagnetism, whereas anisotropic spin-singlet pairing is most likely to arise on the border of antiferromagnetism. In the latter case the superconductivity dome is expected to peak near the critical pressure or critical carrier concentration when the Néel temperature tends towards absolute zero.

An experimental example of this behaviour is shown in Figure 16.3 for $CePd_2Si_2$ [24–29]. A somewhat more complex, but related, case is that of $CeRh_2Si_2$ [30, 31]. In $CePd_2Si_2$ the superconducting state arises out of an unconventional normal metallic state that is characterised by a quasi-linear temperature dependence of the resistivity. A similar behaviour has also been seen more recently in the parent compound $YbRh_2Si_2$ [32] and its doped relatives in which a quasi-linear resistivity is found to extend over a wide temperature range. This behaviour is not cut off by any superconducting transition down to low millikelvin temperatures. The puzzling absence of superconductivity in this and related Yb analogues of superconducting Ce compounds will be discussed in a later section.

16.6. Three Dimensional vs. Quasi-Two-Dimensional Structures

The magnetic interaction model has proved to be a useful tool in directing experimental research and can be used to consider materials in which one might expect an elevated superconducting transition temperature. In particular elementary analysis suggests that the dimensionality of the materials is likely to be important [26, 28, 33–36]. In tetragonal materials on the border of antiferromagnetism the robustness of magnetic pairing increases gradually with

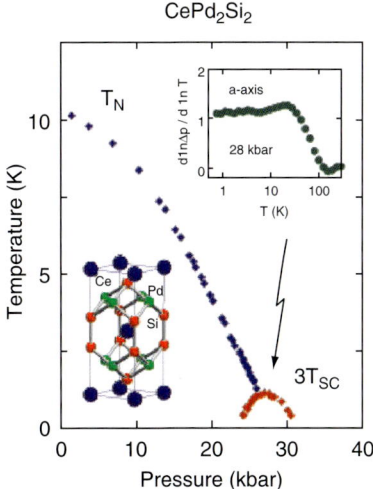

Figure 16.3. Temperature–pressure phase diagram of $CePd_2Si_2$ [24–29]. The antiferromagnetic ordering temperature is suppressed to zero around 28 kbar with the application of pressure. In this region the temperature exponent of the resistivity is well below the Fermi-liquid value of 2 and is close to unity (upper inset). Around this region superconductivity is observed. T_{sc} values are scaled by a factor of 3 for clarity. The bottom inset shows the tetragonal crystal structure of $CePd_2Si_2$. Another example is $CeRh_2Si_2$ [30, 31] (10 kbar = 1 GPa).

the strength of the magnetic correlations and a more anisotropic lattice. Therefore, all else being equal one might expect materials with a quasi-two-dimensional lattice to have elevated superconducting transition temperatures compared to their three dimensional cousins. Some experimental evidence exists for this behaviour. $CeIn_3$ has a simple cubic lattice and has been shown to superconduct in a narrow window of applied pressure with a superconducting transition temperature maximum of around 0.2 K [31, 37, 38]. If this simple cubic lattice is stretched along one axis by the inclusion of non-magnetic layers to form the quasi-two-dimensional tetragonal compound $CeMIn_5$, where M is Co, Rh or Ir, the maximum superconducting transition temperature increases by nearly an order of magnitude to around 2 K and exists over a much wider pressure range [31, 39–41]. Another interesting effect of the dimensionality is the change in the non-Fermi liquid temperature exponents of the resistivity. For example in $CeIn_3$ the temperature exponent of the resistivity has a $T^{3/2}$ dependence, whereas in the tetragonal $CeMIn_5$ systems it has a quasi-linear dependence. Figure 16.4 shows the superconducting regions for $CeIn_3$ and $CeRhIn_5$.

16.7. Density Mediated Superconductivity

There is increasing evidence that other forms of instabilities may be responsible for superconductivity in some materials. An example of particular relevance to the heavy-fermion compounds concerns the border of density instabilities.

Often when there is a volume change in a material it is accompanied by a change in the crystal structure. However this is not always the case. An example is the α–γ or valence instability transition in some Ce compounds. The change of valence and corresponding change in volume can occur as a first-order transition line in a way reminiscent of the familiar liquid–gas transition. The volume collapse can be as high as 15%, e.g., in the α–γ transition of

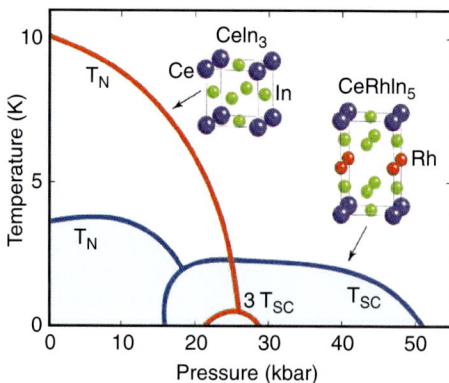

Figure 16.4. Effect of dimensionality on the superconducting transition temperature as illustrated by the superconducting regions for cubic CeIn$_3$ and quasi-two-dimensional tetragonal CeRhIn$_5$ [31, 37–41]. CeIn$_3$ superconducts over a narrow window of applied pressure with a superconducting transition temperature maximum of around 0.2 K. If this simple cubic lattice is stretched along one axis by the inclusion of non-magnetic layers to form the quasi-two-dimensional tetragonal compound CeRhIn$_5$, the maximum superconducting transition temperature increases by nearly an order of magnitude to around 2 K and exists over a much wider pressure range. (Figure is taken from reference [22].)

elemental cerium. It is thought that superconductivity can occur if the first-order transition line has a critical end point at suitably low temperatures.

A simple example of this behaviour may have been found in CeNi$_2$Ge$_2$ [27,42–44]. This system can be considered as the high-pressure analogue of CePd$_2$Si$_2$. However, in CeNi$_2$Ge$_2$ a second superconducting dome was found which is disconnected from the antiferromagnetic boundary. This suggested that more than one mechanism for pairing could act in close proximity. Recent work on the heavy-fermion superconductor CeCu$_2$Si$_2$ [45] and CeCuGe$_2$ [46] has helped shed further light on what is happening in these systems.

CeCu$_2$Si$_2$ has an antiferromagnetic quantum critical point and superconductivity near ambient pressure. The superconducting transition temperature vs. pressure plot however does not exhibit a simple dome structure, but instead remains constant at around 0.7 K and then increases to a maximum value of around 2.2 K at around 4 GPa [47]. Recent studies of super-conductivity in which disorder is introduced via Ge doping have split this large coalescing single superconducting region into two separate superconducting domes [48]. One dome is on the border of antiferromagnetism and the other near an α–γ density transition [48, 49]. The presence of two distinct domes suggests two quantum critical points occur in close proximity. The critical end point for the α–γ transition is believed to occur at low temperatures (10–20 K) and so density fluctuations may be sufficiently important to cause pairing. The superconductivity in this system has been modelled in terms of the effects of magnetic and valence fluctuations [50,51]. Figure 16.5 shows the phase diagram of the CeCu$_2$Si$_2$/CeCu$_2$Ge$_2$ system.

16.8. The Search for Superconductivity on the Border of Itinerant Ferromagnetism

At first sight simple forms of the magnetic interaction model suggest that ferromagnetic pairing looks more promising to produce superconductivity than in the antiferromagnetic case

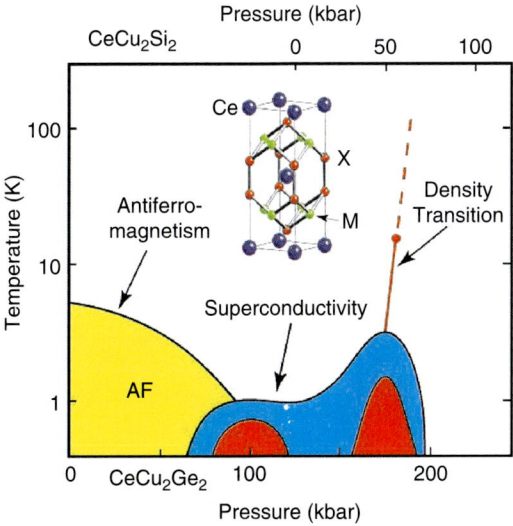

Figure 16.5. Phase diagram of the $CeCu_2Si_2/CeCu_2Ge_2$ system [27, 42–49]. $CeCu_2Si_2$ has an antiferromagnetic quantum critical point and superconductivity near ambient pressure with a large superconducting transition region which increases to a maximum some distance from the border of antiferromagnetism (blue region). By the application of hydrostatic pressure or doping $CeCu_2Si_2$ with Ge, the superconducting region can be split into two separate superconducting domes, one on the border of antiferromagnetism and the other near an α–γ density transition. The presence of two distinct domes suggests two quantum critical points occur in close proximity in this system. The first-order density transition line in the $CeCu_2Si_2/CeCu_2Ge_2$ system is believed to have a critical end point at sufficiently low temperature to permit pairing. (Figure is taken from reference [22].)

(see, e.g., [22] for a recent review). For a long time the absence of superconductivity on the border of ferromagnetism was a puzzle, particularly as an ever increasing number of examples on the border of antiferromagnetism were being shown. Suggestions for the relative scarcity of superconductivity on the border of ferromagnetism have now been made.

In the spin-triplet case, which might occur on the border of ferromagnetism, only the longitudinal component of the spin fluctuations contribute to pairing, whereas the transverse fluctuations reduce the superconducting transition temperature by increasing the quasi-particle self-interaction energy arising from the exchange of magnetic fluctuations which is ultimately pair breaking. In the spin-singlet case on the border of antiferromagnetism however, all three components of the magnetic interaction contribute to pairing. This has the effect that all else being equal, the attraction between quasi-particles, can be on average three times smaller in magnitude for the spin-triplet state than the spin-singlet state for quantum spin-1/2 particles. This comparison involves a number of qualifications and subtleties that have been reviewed in [52, 53].

In ferromagnetic systems with strong magnetic anisotropy however, the effect of the transverse fluctuations on the self-interaction energy is reduced while the beneficial longitudinal fluctuations that result in pairing can remain strong [52]. Therefore, materials on the border of ferromagnetic order, which possess strong spin–orbit coupling and have strong spin anisotropy, may be considered as promising candidates for ferromagnetically mediated superconductivity.

Bearing this in mind the ambient pressure ferromagnetic material UGe_2 was studied with hydrostatic pressure to suppress the ferromagnetic ordering temperature. This proved

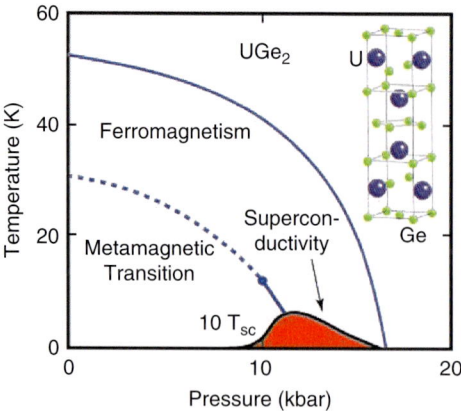

Figure 16.6. Temperature–pressure phase diagram of UGe$_2$ [54–62] showing the co-existence of itinerant-electron ferromagnetism and superconductivity. A strongly first-order ferromagnetic transition occurs along the solid line, and a metamagnetic transition is seen along the dashed line. The red area denotes a superconducting region, T_{sc}. (T_{sc} values are scaled by a factor of 10 for clarity.) More recent examples include URhGe [69, 70] and UIr [73]. (Figure is taken from reference [22].)

to be very fruitful and in 1999 produced the first example of the co-existence of superconductivity and itinerant-electron ferromagnetism [54–59] in this new class of superconductors. Figure 16.6 shows the temperature–pressure phase diagram for UGe$_2$. At ambient pressure UGe$_2$ orders ferromagnetically with a Curie temperature of 52 K. As pressure is applied the Curie temperature decreases. Superconductivity is observed in a narrow window of pressure (1.0–1.5 GPa) and temperatures below 1 K. Specific heat [60] and NMR [61, 62] measurements showed the superconductivity to be bulk in nature. Neutron-diffraction experiments showed the coexistence of superconductivity and ferromagnetism [57]. The temperature–pressure phase diagram of UGe$_2$ is somewhat more complicated than originally expected. A strongly first-order ferromagnetic transition occurs along the solid line in Figure 16.6, and a weaker more nearly continuous ferromagnetic or more precisely metamagnetic transition is seen along the dashed line. The superconducting transition observed in this system is centred near the end point of the metamagnetic transition. The expected superconductivity dome around the end of the upper ferromagnetic transition line may be absent because this transition is found to be strongly first order. Firstly, the magnetic pairing potential is expected to be reduced compared to a continuous transition and secondly, the system may be accompanied by strong magnetic heterogeneities which may be pair breaking. Spin-triplet pairing has been proposed as a possible form of the superconducting state in UGe$_2$ [63–65]. Alternative mechanisms based on a spin-singlet order parameter have also been proposed [66–68]. Because the exchange splitting near the superconducting dome is expected to be much larger than the superconducting energy gap in an itinerant electron ferromagnet such as UGe$_2$, spin-triplet pairing is the most natural choice.

Since the discovery of superconductivity in ferromagnetic UGe$_2$ physicists have looked for other examples that may show similar behaviour. URhGe [69, 70] has been shown to be a ferromagnetic superconductor similar to UGe$_2$ but superconducts at ambient pressure. The possibility of superconductivity on the border of the elemental ferromagnets was reconsidered shortly before the discovery of superconductivity in UGe$_2$ [54, and references therein] and was subsequently discovered in iron [71, 72] (Figure 16.7). Iron orders ferromagnetically at ambient pressure below 1,042 K. At an applied pressure of 14 GPa it undergoes a structural

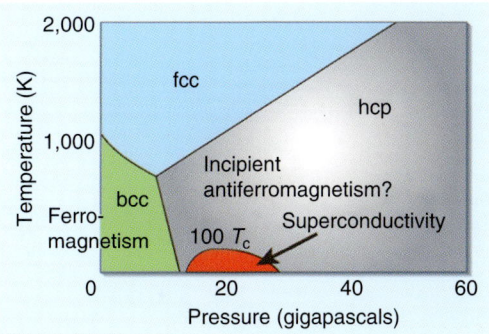

Figure 16.7. The temperature–pressure phase diagram of iron [54, 71, 72]. The low-pressure and low-temperature phase of iron has a body-centred cubic (bcc) structure and is strongly ferromagnetic. The higher-pressure phase has a hexagonal close-packed (hcp) structure and energy band calculations predict that it might be weakly or nearly antiferromagnetic at low temperature [93]. Below 2 K and at pressures above 10 GPa (100 kbar) the hcp phase of iron becomes superconducting. (Here the superconducting transition temperature, T_c, has been enlarged by a factor of 100.) At higher temperatures iron exists in the face centred cubic (fcc) structure. (Figure is taken from reference [94].)

phase transition from the body centred cubic α-phase to the hexagonal close packed ε-phase. Recently, superconductivity has also been observed on the border of ferromagnetism in another U system, UIr that crystallises in a somewhat unusual lattice structure not possessing inversion symmetry [73].

However, despite the difficulties associated with ferromagnetically mediated superconductivity described above, the relative scarcity of superconductivity on the border of metallic ferromagnetism has still been seen by some as surprising [22]. One possibility of this scarcity is because real materials can have very complex band structures, the details of which have not been fully included in theoretical models of magnetically mediated superconductivity. The Lindard function is the building block for developing the magnetic interaction in several such theories, e.g., [53], and has oscillatory structure that would not be found, for example, in jellium, i.e., the starting model of early theoretical work on nearly ferromagnetic systems. These oscillations lead to peaks and troughs in the magnetic interaction that, analyses show, tend to frustrate the triplet state [53, 74].

Although the observation of superconductivity on the border of ferromagnetism is still relatively rare, ferromagnetic quantum critical points are proving to be a rich arena for observing new quantum states of matter as illustrated in Figure 16.8.

16.9. Why Don't All Nearly Magnetic Materials Show Superconductivity?

There are a large and growing number of examples of superconductivity on the border of antiferromagnetism and ferromagnetism, however there are also a significant number of other examples where superconductivity is not seen, not only on the border of ferromagnetism but also antiferromagnetism. Notable cases are found among d-electron metals and most especially among the Yb hole-analogues of the Ce band heavy fermion superconductors. Superconductivity has recently been discovered in the ytterbium system YbC_6 [75, 76] at ambient pressure. However, in this system ytterbium appears to be non-magnetic under these conditions.

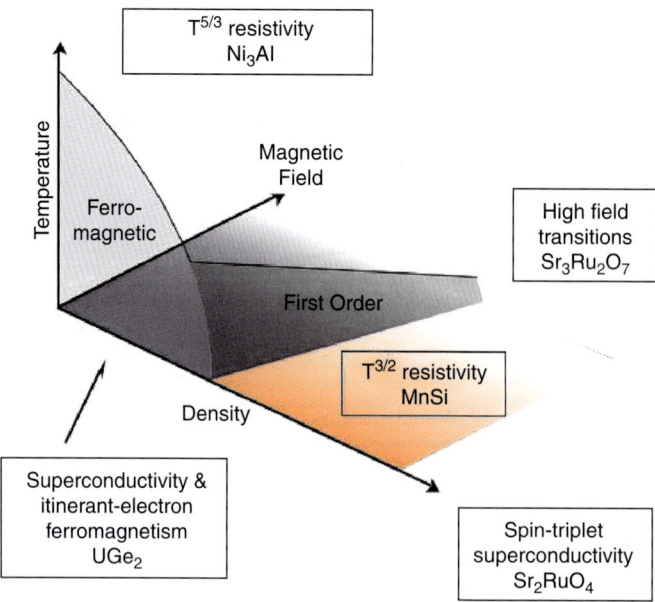

Figure 16.8. Examples of new quantum states on the border of ferromagnetism. When the ferromagnetic ordering temperature is suppressed to low temperature by the application of pressure a wide variety of new quantum states can occur particularly when magnetic field is used as an additional quantum tuning parameter. For example, UGe_2 [54–62] shows the co-existence of itinerant electron ferromagnetism and superconductivity, Sr_2RuO_4 [95, 96] shows p-wave superconductivity, Ni_3Al [9] exhibits logarithmic (marginal) Fermi-liquid behaviour, $Sr_3Ru_2O_7$ shows anomalous high field states [97, 98] and MnSi [11–17] appears to have new transport behaviour.

The lack of observed superconductivity in these types of system leads us to ask why this is the case. There are a number of possible reasons. Some of the effects which are detrimental to superconductivity within the framework of the magnetic interaction model are listed below:

1. *Effects of quenched disorder.* The mean free path must be greater than the superconducting coherence length for the anticipated anisotropic superconductivity to occur. For example, this requires residual resistivities of the order of or below $1\,\mu\Omega\,cm$ in d-metal superconductors such as ε-Fe [72] and Sr_2RuO_4 [77].
2. *Measurement range.* The superconducting transition could be outside of the range of temperature and pressure so far studied or the pressure resolution is too coarse. For example, the superconducting dome in UIr [73] at the magnetic quantum critical point is only a few kilobars wide and this could readily be missed.
3. The presence of first-order transitions which may suppress the strength of the magnetic interaction and introduce magnetic heterogeneities.
4. The effects of competing quasi-particle interactions. Co-existence and competition between ferromagnetic and antiferromagnetic fluctuations, for example, can be detrimental to magnetically mediated superconductivity [53, 78].
5. Frustrated magnetic interactions in, e.g., triangular lattices [79].
6. Multiplicity of energy bands near the Fermi level and effects of orbital fluctuations [80].

The effects of these complications are not fully understood, but some analyses suggest that they can be detrimental to magnetic pairing. The missing superconductivity, for example, in d-electron metals such as V_2O_3 and NiS_2 on the border of metallic antiferromagnetism (see [81] for a review of these and related materials) may be due to the high degree of frozen-in disorder in the samples that have been reported thus far as well as due to the multiplicity of bands that are thought to be important near the Fermi level. On the other hand, the absence of superconductivity in the Yb analogues of the heavy electron Ce compounds may be due to the fact that the characteristic spin–spin (intersite) interaction parameter is typically lower in the former than in the latter. This leads to lower Néel temperatures and to more local behaviours of the magnetic response in the Yb than in the corresponding Ce systems. In turn this leads to a lower predicted superconducting transition temperature, making the Cooper pairs more sensitive to the pair breaking effects of impurities.

16.10. From Weak to Strong Coupling

In the early models of magnetic pairing based on the Hubbard model the effective spin–spin interaction was evaluated in perturbation theory in terms of the dimensionless coupling parameter U/W, where U is the intra-atomic Coulomb energy and W is the electronic band width. The predictions of these models were largely restricted to the weak coupling limit ($U \ll W$). An approach appropriate to the opposite limit ($U \gg W$) was subsequently introduced and extensively studied (for recent reviews, see, e.g., [23, 82]). More recently, numerical techniques that bridge these limits have been developed and a typical prediction of one of these, the so-called Cellular Dynamical Mean Field Theory, is shown in Figure 16.9. Although the different methods involve subtle and important distinctions, it nevertheless seems that in

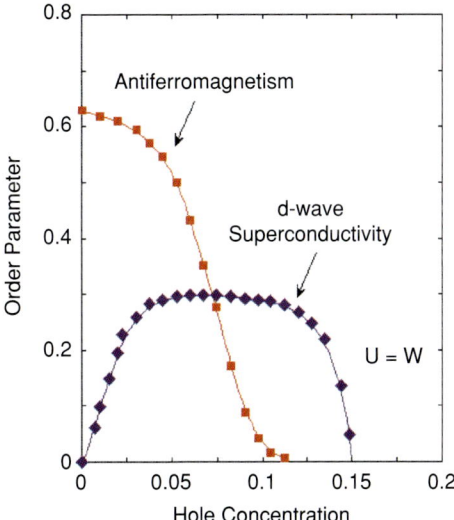

Figure 16.9. Predicted d-wave superconductivity on the border of antiferromagnetism in the Hubbard model. The calculations are based on the cellular dynamical mean field theory [99] for a single nearly half filled band in a square lattice and at absolute zero. The hole concentration is measured from the half filling of the band, U is the Coulomb interaction for two electrons in the same atomic site and W is here defined as $4t$ (the width of the occupied portion of the band at half filling), where t is the hopping matrix element for the band. The calculations are for intermediate coupling ($U = W$). A detailed description of the model and scaled order parameters is given in [99].

all of these approaches superconductivity on the border of magnetism may be thought of as arising from an effective induced spin–spin interaction.

16.11. Superconductivity Without Inversion Symmetry

Recently superconductivity has been shown in materials in which space and time inversion symmetry is broken (for recent comments see [83]). Examples include CePt$_3$Si [84], UIr [73] and CeRhSi$_3$ [101]. A Cooper pair in crystals like CePt$_3$Si is likely therefore to be a mixture of spin-triplet and spin-singlet states. Although magnetically mediated pairing is thought to be relevant in such systems, most theoretical interpretation to date has been for parity-conserving materials. The presence of these new superconducting states which do not possess inversion symmetry are initiating the development and analysis of new theoretical models (e.g., [85–87]).

16.12. Quantum Tuning

In order to push forward our understanding of unconventional forms of superconductivity and novel forms of quantum order such as those discussed above, we have relied on a series of low noise measurements on high purity samples. These can be exacting experiments, particularly if meaningful temperature exponents of the resistivity are to be found. This research has particularly benefited from recent developments in high pressure methods. Pressure has become as important a control parameter as temperature and magnetic field in probing the many-body states of matter. These three parameters (temperature, pressure, magnetic field) together make very powerful tools to the modern condensed matter physicist and allow for precise control and tuning of systems as is demonstrated in Figure 16.8 for materials on the border of ferromagnetism. With the availability of good quality samples and with these techniques we have a rich arena for pushing forward our understanding of these systems and to discover new phenomena.

Pressure has the advantage over other control parameters in that it only normally alters the lattice spacing within the material and the evolution into new quantum states can be studied on the same sample. Doping studies, although of great importance, introduce new elements into a material or vary the quantity of existing elements therein in a random way and can make theoretical analysis of the results difficult. They can also suppress certain physical states, including anisotropic superconductivity when the mean free path is below the superconducting coherence length as stated above.

The use of high pressure in science research is expanding and it has now become a multidisciplinary tool and has applications in physics, materials sciences, chemistry, earth sciences, engineering, astronomy and biological sciences. High pressure methods are discussed elsewhere in this book so we will not go into detail in this chapter. However, we will consider briefly the possible impact of future developments in this area of research. For studies on unconventional superconductivity and novel quantum order, resistivity measurements are particularly useful as if done with sufficient sensitivity they can show deviations from Fermi-liquid theory by seeing deviations from the T^2 temperature term of the resistivity. These measurements should ideally be carried out under hydrostatic pressure conditions. This places certain constraints on the pressure environment, and together with difficulties in attaching four wires to tiny samples in pressure cells has meant that for a long time the main work horse for these

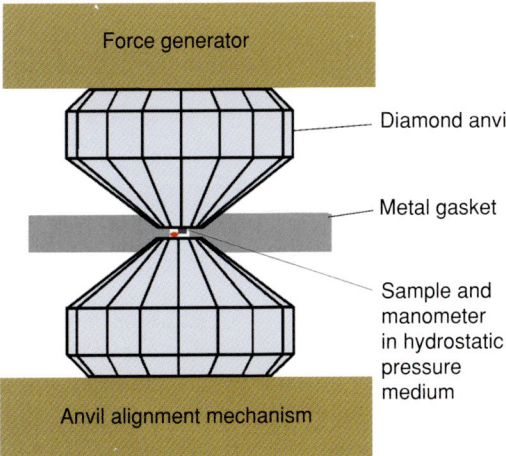

Figure 16.10. Diamond anvil high pressure cell. The diamond anvils are accurately positioned so that they are centred and aligned parallel to each other. The samples, pressure gauge and hydrostatic pressure medium are contained within a small hole (<0.5 mm diameter) in the centre of a preindented region of the gasket. As force is applied the diamonds press on the gasket causing the hole in the centre of the gasket to collapse. The collapse of this hole pressurises the hydrostatic medium and hence the sample contained within the hole.

type of measurements was the large piston cylinder clamp cells which make use of hydrostatic liquid pressure medium. These cells however are limited to a pressure of around 3 GPa. To achieve higher pressures, cells which make use of opposing anvils (e.g., diamond, moissanite, sapphire) need to be used (Figure 16.10). The difficulty with these however is that the sample size is considerably smaller ($250 \times 250 \times 40 \, \mu$m or less).

Recent technological advances promise to change high pressure research into a routine technique which can be utilised by a wide range of researchers. Future high pressure developments are likely to involve new advances in growth of single crystal large synthetic diamond, nanolithography of conducting wires and diamond chemical vapour deposition as well as new micromanipulation methods. Figure 16.11 shows a possible future direction for resistivity measurements in the diamond anvil cell. "Designer anvils" are produced by lithographically depositing metal wires on to diamond anvils and then covering them with a protective, electrically insulating cap of diamond deposited on the top by chemical vapour deposition of diamond (e.g., [88] and references therein). Increasing the size of diamonds available for research is also likely to be hugely advantageous as it means that greater pressure homogeneities can be achieved on smaller samples or larger samples can be used which gives a larger signal to noise in measurements and makes the experiments easier to set up. Using chemical vapour deposition of synthetic diamond it is possible to produce large single crystal colourless diamonds (up to 10 carats). Further studies on the use of alternatives to diamond as an anvil material, e.g., moissanite [89], is also likely to be beneficial.

The exploitation of high pressure methods for the investigation of novel forms of quantum order and unconventional superconductivity has gone in tandem with developments of other quantum control parameters, in particular low temperature methods. This has been part of ongoing research programmes which have led to advances in cryogenic methods such that the high pressure experiments can be performed at millikelvin temperatures in a routine manner with a fast turn around time. The recent advances in cryogen free systems are making

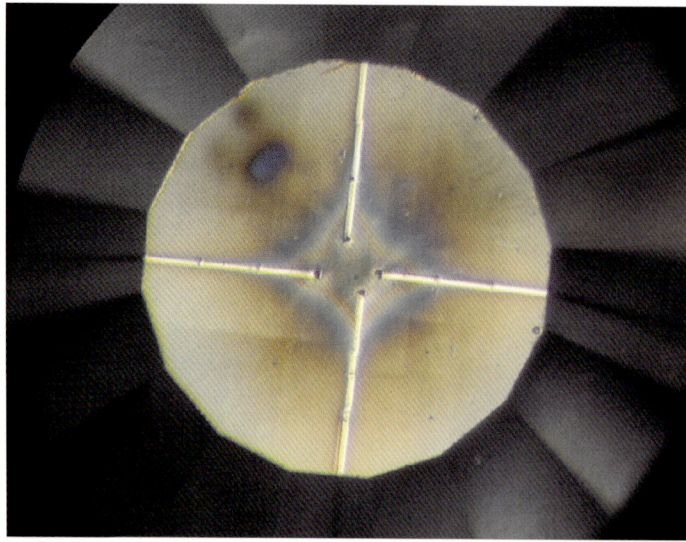

Figure 16.11. Possible future of high pressure resistivity measurements. Tungsten wires are vapour deposited on the surface of a diamond anvil. A protective cap of diamond will be deposited over the tungsten wires by chemical vapour deposition. (Figure courtesy of Dr. Konstantin Kamenev of the Centre for Science at Extreme Conditions, University of Edinburgh, UK).

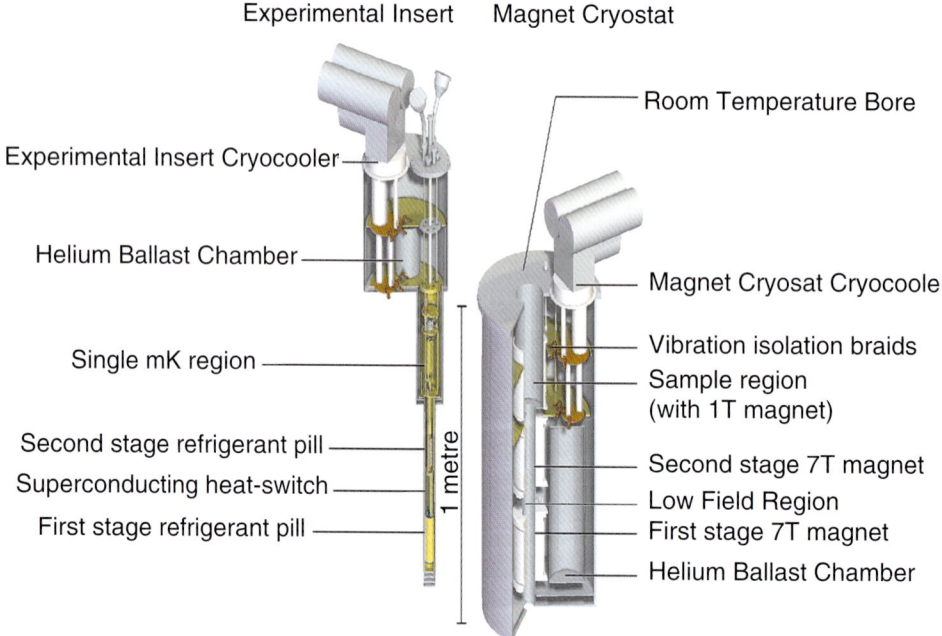

Figure 16.12. Next generation cryogen free cooler to operate continuously from room temperature to 1 mK in development at the Cavendish Laboratory in collaboration with Cambridge Magnetic Refrigeration Ltd. The system is based around low vibration pulse tube refrigerators to cool from room temperature to 4 K and a two-stage adiabatic demagnetisation refrigerator to cool to 1 mK. For ultra-low noise operation the apparatus is equipped with a passive helium ballast to maintain base temperature for 1–2 days with the pulse tube turned off.

low temperatures more accessible to a wider number of researchers. They also remove problems usually associated with liquid cryogens. These coupled with adiabatic demagnetisation systems mean that it is possible to get to temperatures normally as low as 1 mK in a turn-key fashion without the use of cryogenic liquids (liquid helium or nitrogen) (Figure 16.12).

16.13. Concluding Remarks

The magnetic interaction model can be used to consider the make up of systems that are likely to give elevated superconducting transition temperatures. As already discussed in this chapter, suitable materials on the border of antiferromagnetism can produce superconducting Cooper pairs which all else being equal are likely to be a more robust than that formed by materials on the border of ferromagnetic ordering. Also mentioned are the improvements to the superconducting transition temperature when one goes from a cubic three dimensional structure to an anisotropic quasi-two-dimensional tetragonal structure. The model also predicts that a single band having a relatively high energy scale is likely to be beneficial to producing a higher superconducting transition temperature provided the other desirable features are also present (see, e.g., [52]). To increase the energy scale one might look for suitable materials among the 5f metals, e.g., $PuCoGa_5$ [90–92], which have larger bands than 4f metals. The effective Fermi temperature and hence the superconducting transition temperature is found to increase by an order of magnitude in $PuCoGa_5$ compared with the 4f analogue $CeCoIn_5$. The next step would be to go to d-metal systems that generally have a larger bandwidth compared to the f-metal systems without losing other requirements for superconductivity.

Interestingly, the cuprate high temperature superconductors possess many of these features. The parent compounds are all antiferromagnetic insulators. They crystallise in an anisotropic tetragonal structure. The superconductivity is thought to be caused by a single dominant copper quasi-two-dimensional d-band. The magnetic interaction model, both in the weak and strong coupling limits (for recent reviews see, e.g., [22, 23] and references cited therein), predicts very robust spin-singlet d-wave pairing not too far from half filling of this d-band. This is found to be the case for the cuprates. A density interaction model (section 16.7) may also be relevant to these systems.

This chapter has given an overview of some of the interesting and unusual physics that occurs between a magnetically ordered state and a conventional metal at low temperatures. It has been shown that the new quantum states that emerge cannot be explained by the conventional low temperature theories of matter, but instead we must turn to new theoretical models. Advances in experimental methods, particularly in the areas of material preparation, high-pressure and cryogenics, promise to lead to further understanding and in particular to the discovery of still more exotic phenomena on the border of quantum phase transitions in general.

Acknowledgements

We would like to thank in particular P. Coleman, F. M. Grosche, S. R. Julian, P. Monthoux, and D. Pines for recent discussions on this and related topics. We acknowledge support from The Royal Society, EPSRC and The Newton Trust of the United Kingdom.

Bibliography

1. J. Bardeen, L. N. Cooper, and J. R. Schrieffer, Phys. Rev. **106**, 162 (1957).
2. E. Pugh, Phil. Trans. R. Soc. Lond. A **361**, 2715 (2003).
3. P. Coleman and A. J. Schofield, Nature **433**, 226 (2005).
4. G. Baym and C. Pethick, *Landau–Fermi Liquid Theory* (Wiley, New York, 1991), Chap. 3, and references therein.
5. T. Moriya, *Spin Fluctuations in Itinerant Electron Magnetism* (Springer, Berlin, Germany, 1985), and references therein.
6. G. G. Lonzarich, *Electron*, edited by M. Springford (Cambridge University Press, Cambridge, UK, 1997), Chap. 6, and references therein.
7. J. A. Hertz, Phys. Rev. B **14**, 1165 (1976), and references therein.
8. A. J. Millis, Phys. Rev. B **48**, 7183 (1993), and references therein.
9. P. G. Niklowitz, PhD Thesis, University of Cambridge, UK (2003); P. G. Niklowitz, F. Beckers, G. G. Lonzarich, G. Knebel, B. Salce, J. Thomasson, N. Bernhoeft, D. Braithwaite, and J. Flouquet, Phys. Rev. B **72**, 024424 (2005).
10. R. Smith, PhD Thesis, University of Cambridge, UK (2006).
11. J. D. Thompson, Z. Fisk, and G. G. Lonzarich, Physica B **161**, 317 (1989).
12. C. Pfleiderer, G. J. McMullan, S. R. Julian, and G. G. Lonzarich, Phys. Rev. B **55**, 8330 (1997).
13. C. Thessieu, C. Pfleiderer, A. N. Stepanov, and J. Flouquet, J. Phys.: Condens. Matter **9**, 6677 (1997).
14. C. Pfleiderer, S. R. Julian, and G. G. Lonzarich, Nature **414**, 427 (2001).
15. A. E. Petrova, V. Krasnorussky, J. Sarrao, and S. M. Stishov, Phys Rev. B **73**, 052409 (2004).
16. D. Belitz, T. R. Kirkpatrick, and J. Rollbühler, Phys. Rev. Lett. **94**, 247205 (2005).
17. N. Doiron-Leyraud, I. R. Walker, L. Taillefer, M. J. Steiner, S. R. Julian, and G. G. Lonzarich, Nature **425**, 595 (2003).
18. N. F. Berk and J. R. Schrieffer, Phys. Rev. Lett. **17**, 433 (1966).
19. R. W. McCallum, D. C. Johnston, R. N. Shelton, and M. B. Maple, *Proc. 2nd Rochester Conf. on Superconductivity in d and f Band Metals*, edited by D. H. Douglas (Plenum, New York, 1976), p. 625.
20. R. W. McCallum, D. C. Johnston, R. N. Shelton, W. A. Fertig, and M. B. Maple, Solid State Commun. **24**, 501 (1977).
21. P. Thalmeier, G. Zwicknagl, O. Stockert, G. Sparn, and F. Steglich, *Superconductivity in Heavy Fermion Compounds* (Frontiers in Superconducting Materials), edited by A. Narlikar) (Springer, Berlin, Germany, 2004).
22. P. Monthoux, D. Pines, and G. G. Lonzarich, to be published
23. P. W. Anderson, P. A. Lee, M. Randeria, T. M. Rice, N. Trivedi, and F. C. Zhang, J. Phys.: Condens. Matter **16**, R755 (2004).
24. N. D. Mathur, PhD thesis, University of Cambridge, UK (1995); F. M. Grosche, S. R. Julian, N. D. Mathur and G. G. Lonzarich, Physica B **224**, 50 (1996).
25. S. R. Julian, C. Pfleiderer, F. M. Grosche, N. D. Mathur, G. J. McMullan, A. J. Diver, I. R. Walker, and G. G. Lonzarich, J. Phys.: Condens. Matter **8**, 9675 (1996).
26. N. D. Mathur, F. M. Grosche, S. R. Julian, I. R. Walker, D. M. Freye, R. K. W. Haselwimmer, and G. G. Lonzarich, Nature **394**, 39 (1998).
27. F. M. Grosche, P. Agarwal, S. R. Julian, N. J. Wilson, R. K. W. Haselwimmer, S. J. S. Lister, N. D. Mathur, F. V. Carter, S. S. Saxena, and G. G. Lonzarich, J. Phys.: Condens. Matter **12**, L533 (2000).
28. A. Demuer, A. T. Holmes, D. Jaccard, J. Phys.: Condens. Matter **14**, L529 (2002).
29. G. Knebel, D. Braithwaite, P. C. Canfield, G. Lapertot, and J. Flouquet, High Press. Res. **22**, 167 (2002).
30. R. Movshovich, T. Graf, D. Mandrus, J. D. Thompson, J. L. Smith, and Z. Fisk, Phys. Rev. B **53**, 8241 (1996).
31. Y. Onuki, R. Settai, K. Sugiyama, T. Takeuchi, T. C. Kobayashi, Y. Haga, E. Yamamoto, J. Phys. Soc. Jpn **73**, 769 (2004).
32. J. Custers, P. Gegenwart, H. Wilhelm, K. Neumaier, Y. Tokiwa, O. Trovarelli, C. Geibel, F. Steglich, C. Pepin, and P. Coleman, Nature **424**, 524 (2003).
33. R. Arita, K. Kuroki, and H. Aoki, J. Phys. Soc. Jpn **69**, 1181 (2000).
34. P. Monthoux and G. G. Lonzarich, Phys. Rev. B **63**, 054529 (2001).
35. P. Monthoux and G. G. Lonzarich, Phys. Rev. B **66**, 224504 (2002).
36. H. Fukazawa and K. Yamada, J. Phys. Soc. Jpn **72**, 2449 (2003).
37. I. R. Walker, F. M. Grosche, D. M. Freye, and G. G. Lonzarich, Physica C **282**, 303 (1997).
38. F. M. Grosche, I. R. Walker, S. R. Julian, N. D. Mathur, N. D. Freye, M. J. Steiner, and G. G. Lonzarich, J. Phys.: Condens. Matter **13**, 2845 (2001).

39. H. Hegger, C. Petrovic, E. G. Moshopoulou, M. F. Hundley, J. L. Sarrao, Z. Fisk, and J. D. Thompson, Phys. Rev. Lett. **84**, 4986 (2000).
40. C. Petrovic, R. Movshovich, M. Jaime, P. G. Pagliuso, M. F. Hundley, J. L. Sarrao, Z. Fisk, and J. D. Thompson, Europhys. Lett. **53**, 354 (2001).
41. C. Petrovic, P. G. Pagliuso, M. F. Hundley, R. Movshovich, J. L. Sarrao, J. D. Thompson, Z. Fisk, and P. Monthoux, J. Phys.: Condens. Matter **13**, L337 (2001).
42. F. M. Grosche, S. J. S. Lister, F. V. Carter, S. S. Saxena, R. K. W. Haselwimmer, N. D. Mathur, S. R. Julian, and G. G. Lonzarich, Physica B **239**, 62 (1997).
43. P. Gegenwart, F. Kromer, M. Lang, G. Sparn, C. Geibel, and F. Steglich, Phys. Rev. Lett. **82**, 1293 (1999).
44. D. Braithwaite, T. Fukuhara, A. Demuer, I. Sheikin, S. Kambe, J.-P. Brison, K. Maezawa, T. Naka, and J. Flouquet, J. Phys.: Condens. Matter **12**, 1339 (2000).
45. F. Steglich, J. Aarts, C. D. Bredl, W. Lieke, D. Meschede, W. Franz, and H. Schäfer, Phys. Rev. Lett. **43**, 1892 (1979).
46. D. Jaccard, K. Behnia, and J. Sierro, Phys. Lett. **163A**, 475 (1992).
47. B. Bellarbi, A. Benoit, D. Jaccard, J. M. Mignot, and H. F. Braun, Phys. Rev. B **30**, 1182 (1984).
48. H. Q. Yuan, F. M. Grosche, M. Deppe, C. Geibel, G. Sparn, and F. Steglich, Science **302**, 2104 (2003).
49. A. T. Holmes, D. Jaccard, and K. Miyake, Phys. Rev. B **69**, 024508 (2004).
50. Y. Onishi and K. Miyake, J. Phys. Soc. Jpn **69**, 3955 (2000).
51. P. Monthoux and G. G. Lonzarich, Phys. Rev. B **69**, 064517 (2004).
52. P. Monthoux and G. G. Lonzarich, Phys. Rev. B **59**, 14598 (1999).
53. P. Monthoux and G. G. Lonzarich, Phys. Rev. B **71**, 054504 (2005).
54. E. Pugh, PhD Thesis, University of Cambridge, UK (1999).
55. S. S. Saxena, P. Agarwal, K. Ahilan, F. M. Grosche, R. K. W. Haselwimmer, M. J. Steiner, E. Pugh, I. R. Walker, S. R. Julian, P. Monthoux, G. G. Lonzarich, A. Huxley, I. Sheikin, D. Braithwaite, and J. Flouquet, Nature **406**, 587 (2000).
56. P. Coleman, Nature **406**, 580 (2000).
57. A. Huxley, I. Sheikin, E. Ressouche, N. Kernavanois, D. Braithwaite, R. Calemczuk, and J. Flouquet, Phys. Rev. B **63**, 144519 (2001).
58. N. Tateiwa, T. C. Kobayashi, K. Hanazono, K. Amaya, Y. Haga, R. Settai, and Y. Onuki, J. Phys.: Condens. Matter **13**, L17 (2001).
59. P. Coleman, Nature **410**, 320 (2001).
60. N. Tateiwa, T. C. Kobayashi, K. Amaya, Y. Haga, R. Settai, and Y. Onuki, Phys. Rev. B **69**, 180513 (2004).
61. H. Kotegawa, S. Kawasaki, A Harada, Y. Kawasaki, K. Okamoto, G. Q. Zheng, Y. Kitaoka, E. Yamamoto, Y. Haga, Y. Onuki, K. M. Itoh, and E. E. Haller, J. Phys.: Condens. Matter **15**(28), S2043 (2003).
62. H. Kotegawa, A. Harada, S. Kawasaki, Y. Kawasaki, Y. Kitaoka, Y. Haga, E. Yamamoto, Y. Onuki, K. M. Itoh, E. E. Haller, and H. Harima, J. Phys. Soc. Jpn **74**, 705 (2005).
63. K. G. Sandeman, G. G. Lonzarich, and A. J. Schofield, Phys. Rev. Lett. **90**, 167005 (2003).
64. A. B. Shick, V. Janis, V. Drchal, and W. E. Pickett, Phys. Rev. B **70**, 134506 (2004).
65. H. Kaneyasu and K. Yamada, J. Phys. Soc. Jpn **74**, 527 (2005).
66. K. B. Blagoev, J. R. Engelbrecht, and K. S. Bedell, Phys. Rev. Lett. **82**, 133 (1999).
67. H. Suhl, Phys. Rev. Lett. **87**, 167007 (2001).
68. A. A. Abrikosov, J. Phys.: Condens. Matter **13**, L943 (2001).
69. D. Aoki, A. Huxley, E. Ressouche, D. Braithwaite, J. Flouquet, J.-P. Brison, E. Lhotel, and C. Paulsen, Nature **413**, 613 (2001).
70. F. Levy, I. Sheikin, B. Grenier, and A. D. Huxley, Science **309**, 1343 (2005).
71. K. Shimizu, T. Kimura, S. Furomoto, K. Takeda, K. Kontani, Y. Onuki, and K. Amaya, Nature **406**, 316 (2001).
72. D. Jaccard, A. T. Holmes, G. Behr, Y. Inada, and Y. Onuki, Phys. Lett. A **299**, 282 (2002).
73. T. Akazawa, H. Hidaka, T. Fujiwara, T. C. Kobayashi, E. Yamamoto, Y. Haga, R. Settai, and Y. Onuki, J. Phys.: Condens. Matter **16**, L29 (2004).
74. C. Honerkamp and T. M. Rice, J. Low Temp. Phys. **131**, 159 (2003).
75. T. E. Weller, M. Ellerby, S. S. Saxena, R. P. Smith, and N. T. Skipper, Nat. Phys. **1**, 39 (2005).
76. G. Csanyi, P. B. Littlewood, A. H. Nevidomskyy, C. J. Pickard, and B. D. Simons, Nat. Phys. **1**, 42 (2005).
77. A. P. Mackenzie, R. K. W. Haselwimmer, A. W. Tyler, G. G. Lonzarich, Y. Mori, S. Nishizaki, and Y. Maeno, Phys. Rev. Lett. **80**, 161 (1998).
78. P. McHale and P. Monthoux, Phys. Rev. B **67**, 214512 (2003).
79. T. Koretsune and N. Ogata, Physica B: Condens. Matter **359**, 545 (2005).
80. T. Takimoto, T. Hotta, and K. Ueda, Phys. Rev. B **69**, 104504 (2004).

81. N. Mott, *The Metal–Insulator Transition* (Taylor & Francis, London, UK 1990).
82. P. A. Lee, N. Nagaosa, and X.-G. Wen, Cond-mat/0410445, v1 (2004).
83. S. S. Saxena and P. Monthoux, Nature **427**, 799 (2004).
84. E. Bauer, G. Hilscher, H. Michor, C. Paul, E. W. Scheidt, A. Gribanov, Y. Seropegin, H. Noel, M. Sigrist, and P. Rogl, Phys. Rev. Lett. **92**, 027003 (2004).
85. L. P. Gor'kov and E. I. Rashba, Phys. Rev. Lett. **87**, 037004 (2001).
86. P. Frigeri, D. F. Agterberg, A. Koga, and M. Sigrist, Phys. Rev. Lett. **92**, 097001 (2004).
87. K. V. Samokhin, E. S. Zijlstra, and S. K. Bose, Phys. Rev. B. **69**, 094514 (2004).
88. S. T. Weir, J. Akella, C. A. Ruddle, Y. K. Vohra, and S. A. Catledge, Appl. Phys. Lett. **77**, 3400 (2000).
89. J. A. Xu and H. K. Mao, Science **290**, 183 (2000).
90. J. L. Sarrao, L. A. Morales, J. D. Thompson, B. L. Scott, G. R. Stewart, F. Wastin, J. Rebizant, P. Boulet, E. Colineau, and G. H. Lander, Nature **420**, 297 (2002).
91. E. D. Bauer, J. D. Thompson, J. L. Sarrao, L. A. Morales, F. Wastin, J. Rebizant, J. C. Griveau, P. Javorsky, P. Boulet, E. Colineau, G. H. Lander, and G. R. Stewart, Phys. Rev. Lett. **93**, 147005 (2004).
92. N. J. Curro, T. Caldwell, E. D. Bauer, L. A. Morales, M. J. Graf, Y. Bang, A. V. Balatsky, J. D. Thompson, and J. L. Sarrao, Nature **434**, 622 (2005).
93. G. Steinle-Neumann, R. E. Cohen, and L. Stixrude, J. Phys.: Condens. Matter **16**, S1109 (2004), and references therein.
94. S. S. Saxena and P. B. Littlewood, Nature **412**, 290 (2001).
95. Y. Maeno, H. Hashimoto, K. Yoshida, S. NishiZaki, T. Fujita, J. G. Bednorz, and F. Lichtenberg, Nature **372**, 532 (1994).
96. A. P. Mackenzie and S. A. Grigera, J. Low. Temp. Phys. **135**, 39 (2004).
97. S. A. Grigera, R. S. Perry, A. J. Schofield, M. Chiao, S. R. Julian, G. G. Lonzarich, S. I. Ikeda, Y. Maeno, A. J. Millis, and A. P. Mackenzie, Science **294**, 329 (2001).
98. A. G. Green, S. A. Grigera, R. A. Borzi, A. P. Mackenzie, R. S. Perry, and B. D. Simons, Phys. Rev. Lett. **95**, 086402 (2005).
99. M. Capone and G. Kotliar, Cond-mat/0603227 v1 9th March 2006.
100. G. G. Lonzarich, Nat. Phys. **1**, 11 (2005).
101. N. Kimura, K. Ito, K. Saitoh, Y. Umeda, and H. Aoki, Phys. Rev. Lett. **95**, 247004 (2005).

Index

^{16}O & ^{18}O, 2, 13, 114, 216, 253
3D XY model, 374–376, 378–380, 383–386

A-15 compounds, 326, 359, 437–438, 446, 571
A_3C_{60}, 439, 468
acceptor, 330, 465
acoustic phonons, 12, 350
actinides, 458
activation energy, 11, 253
adiabatic demagnetization, 610
algebraic spin liquid (ASL), 551, 557, 559, 562
alloy, 484, 571
Aluminum (Al), 65, 158–159, 224, 266–267, 362–363, 430–431, 458
Andreev band, 31
Andreev bound state, 26
Andreev reflection, 8, 21, 23, 25–26, 28–29, 31, 59
Andreev scattering, 19, 58–59, 66
Andreev state, 29
angle resolved photoemission spectroscopy (ARPES), 2, 47, 51, 60, 71–72, 87–88, 90–92, 98, 106, 111, 116, 118–119, 124, 126, 128–131, 135, 137, 200, 283, 312, 338, 387, 405, 410–411, 414–415, 419–421, 513, 523–524, 541, 551, 585, 588
angular dependent magnetoresistance oscillations (AMRO), 409, 414, 420, 464
anion, 465–466, 468, 475–476, 478, 488
anisotropic superconductivity, 606, 608
anisotropy (resistivity), 11, 407, 399
anisotropy ratio, 154, 349, 383, 386
anomalies, 16, 65, 116, 301, 316, 340, 361, 363–365, 367–369, 376, 378–379, 382, 399
antibonding, 47, 96–97, 105, 118–119, 476, 509
antiferromagnet, 244, 248, 272, 287, 530, 564, 574

antiferromagnetic correlations, 44, 135, 257, 289, 504–505, 588
antiferromagnetic ordering, 106, 266, 361, 601
antiferromagnetism, 258, 264, 328, 403, 495–496, 522, 555, 588, 591, 597, 600, 602–603, 605, 607, 611
antinodal, 95, 106, 111, 118–120, 123–124, 129–135, 137–138, 414, 513, 515, 519, 533, 541, 551
apical oxygen, 95, 101, 136–137, 220, 328–329, 333–334, 338, 343, 528, 531
atomic force microscopy (AFM), 2, 332, 344, 403, 411, 420–421

backward scattering, 155
bad metals, 326, 571, 583
BaKBiO, 571
band filling (1/2, 1/4), 465, 467, 76–80, 89
band structure, 45–46, 48–49, 69, 71, 89, 92, 96–98, 118, 122–124, 129, 153, 178, 223, 407–408, 410–411, 417, 513, 529
band theory, 125, 274, 276, 532
band-filling, 463, 467, 475–477, 489
Bardeen-Cooper-Schrieffer (BCS), 15, 45, 47–48, 50–52, 73, 87, 99, 118, 145, 152, 154–155, 167–168, 175–176, 178, 180–181, 202, 215, 223–224, 240, 299–300, 315, 318, 349, 351, 354, 357, 372–373, 376, 378–379, 430–431, 433, 436, 439, 469, 471, 482–483, 533–534, 536, 538, 569, 572, 580, 584, 588–589, 592, 597–598, 600
Barium (Ba), 100, 251, 274, 346, 348, 353, 356–357, 361, 364, 367, 428, 430, 437, 447, 458
BEDT-TTF (ET), 464–469, 471–473, 475, 479–485, 489

Berezinski–Kosterlitz–Thouless model (BKT) 26, 48, 207
Beryllium (Be), 116–117, 430, 458
beta (β)-NMR, 161
Bi-2201, 2–3, 70, 98, 112–113, 357
Bi-2212, 47–48, 50–54, 57–58, 60–63, 70–74, 328, 330, 334, 376–380, 412, 446, 452, 455
$Bi_2Sr_2CaCu_2O_{8+\delta}$, 38–39, 43, 52, 56–58, 70, 92, 112, 114, 174, 181, 188–193, 195–197, 199–200, 205–206, 208, 259, 264, 282–283, 285, 301, 386, 401, 405, 412, 416–417, 420, 585
bilayers, 23–25, 33, 38, 46, 92, 98, 118–119, 173, 188, 269, 274, 315
binary compounds, 427, 429–430, 437–438
binding energy, 2, 88, 94–95, 114–115, 117, 124–126, 129, 342, 530
bipartate lattice, 530
bipolaron model, 379
bipolarons, 1–2, 5, 7, 11–13, 17
BiSCCO, 39
bismutates, 3
Bismuth (Bi), 37, 57, 70, 325, 328–329, 334, 341, 344, 346, 363, 376, 412, 428, 430, 458, 451–452
Boltzmann (transport), 317, 400, 407, 413–414, 418–419
bond length, 362, 454–455
bond-buckling, 122–124
bonds, 9, 137, 265, 271, 362, 409, 455, 464, 528, 534–536, 538, 553, 561, 579, 590
border (borderlines), 22, 152, 480, 597–608, 611
Born (approximation), 103, 125–127, 155–156, 194, 196–198, 327
Bose, 64, 107, 354, 372, 378–379, 395–396, 540, 546, 551, 558, 572, 581, 614
Bose-Einstein Condensation (BEC), 372, 376, 378–379
bosonic, 2, 48, 75, 99, 116, 119, 134, 322, 379, 413, 418, 420, 546, 557–558, 575, 579–580
bosons, 113, 116, 118, 379–380, 456, 538, 540–546, 549–551, 554, 558–559, 563–564, 579, 581
Bragg, 259, 261, 266–267, 272, 274, 280, 285, 287, 380
breathing modes, 122, 136
Brillouin zone, 2–3, 7, 95, 101, 111–112, 118–123, 130–131, 133, 135, 137, 153, 259, 268–269, 409–410, 500, 513, 541
Brout criterion, 373

C_{60}, 440–441, 571
$Ca_2CuO_2Cl_2$, 124–128, 266, 533
carrier doping, 146
cation, 35, 152, 180, 191–192, 327, 335–336, 338, 428, 443, 466, 488
c-axis transport, 39, 177–178, 201, 407–409, 421
$CeCoIn_5$, 611
$CeCu_2Ge_2$, 602–603
$CeIn_3$, 601–602
cellular dynamical mean field theory, 607
$CeNi_2Ge_2$, 602
$CePd_2Si_2$, 600–602
$CePt_3Si$, 608
$CeRh_2Si_2$, 600–601
$CeRhIn_5$, 601–602
Cerium (Ce), 287–288, 346, 350, 352–355, 361, 373–374, 388–389, 428, 458, 600–602, 605, 607,
CeS, 351, 353, 357, 379
charge density wave (CDW), 279, 338, 363, 367–372, 478, 570, 575–576, 578–582, 591
charge Kondo model, 330
charge localization, 276
charge order, 132, 135, 144, 250, 275–277, 279, 285, 290, 334, 348, 405, 421, 471, 477–478, 480, 570, 583, 585, 588, 591
charge reservoir, 325, 328–331, 334, 341, 362, 571, 586–587
charge stripes, 274–276, 281, 290
charge transfer salt, 466
charge transfer, 97, 129, 136, 288, 325, 327, 333, 337–340, 342, 344, 401, 406–408, 444–445, 450, 453, 465–468, 476–477, 488–489, 529
charge-Kondo model, 330
checkerboard, 277, 285
chemical potential, 8, 111, 126–131, 300, 316–317, 320–321, 330, 379, 513, 548, 560, 579
chemical pressure, 471, 473
chemical substitutions, 345, 351, 360, 590
chemical vapor deposition of diamond, 609
Chevrel materials, 485, 600
Chromium (Cr), 65, 264, 279, 362–364, 387–389
Clapeyron equation, 381, 384
clean limit, 28, 147, 178, 187, 300–301, 305, 315, 349
clusters, 1, 9, 11, 17, 274, 288, 329, 341, 495–496, 503, 507–509, 516, 530–531, 557, 573–574, 583–584, 588
Cobalt (Co), 38, 362–363, 477–479, 489, 601

Index

coexistence, 11, 17, 21, 52, 61, 67, 75, 129, 274, 277, 288, 320–321, 327, 340, 368–369, 414, 472, 489, 570, 604, 606
coherence length, 9, 12, 21, 24–25, 28, 35, 37, 62, 69, 145–146, 151, 160, 202–203, 348–349, 356, 360, 372–373, 375, 378, 380, 468, 471, 506, 570, 583, 606, 608
coherence peak, 72, 74, 155, 180–181, 239, 533
coherent c-axis transport, 408–409, 421
coherent hopping, 177, 409
collective excitations, 98, 112, 598
collective modes, 199, 312–314, 321, 584
commensurate effects, 245, 259–260, 273–274, 286–287, 328, 478, 485
competing order (phases, states), 19–20, 62, 151, 274, 280, 285, 353, 564, 570, 591
condensate, 58, 147, 149, 300–302, 307, 312, 314–319, 559, 600
conductance, 19, 22–24, 26–27, 35, 45–47, 49–50, 52–53, 61, 64–67, 70–72, 74–75, 315, 352, 483
conduction band, 128, 362, 433, 436–437, 439, 441
conductivity tensor, 146–147, 157
confinement, 178, 357, 408, 418, 540–543, 551, 558, 564, 574, 590
conventional superconductors, 7, 19–20, 42, 44, 53, 56, 63, 71–73, 87, 99, 116, 152, 180, 183, 202, 239, 349, 353, 359–360, 362, 373, 380, 399, 401, 441, 463, 533
Cooper pairs, 19, 20, 25, 29, 87, 99, 135, 202, 379, 456, 481, 597, 600, 607–608, 611
Copper (Cu), 1–2, 4–5, 13–14, 34–35, 55, 57, 68, 70–72, 95–97, 100–101, 104–105, 122–123, 136, 180, 185–186, 195, 215–217, 220, 225–228, 230–231, 233–234, 237, 239–244, 246–251, 253–254, 258–259, 265–269, 271–272, 274, 276–277, 280, 285–286, 288–289, 315, 325–328, 331–332, 334–336, 339–340, 344, 346–347, 351, 360, 362–364, 367, 370, 406, 409–410, 443, 448, 450, 455, 466, 468, 482, 523–524, 528–531, 571, 586–587, 589–591
correlation length, 151, 230–231, 234, 244, 271–272, 274, 285–287, 504
Coulomb frustrated phase, 569–570, 583–584
Coulomb interactions, 65, 98, 136–137, 583
Coulomb pseudopotential, 19, 72–73
Coulomb repulsion, 8, 12, 96–97, 137, 267, 277, 330, 406, 431–432, 473, 476, 536, 572

coupling constants, 5, 6, 8, 98, 103, 106, 109, 110, 127, 221, 244, 404, 542–543, 557–559, 580
coupling parameters, 431–432, 469, 607
critical exponents, 203, 205–206
critical pressure, 472, 599–600
crystal field, 95–96, 123, 253, 327
CuO chains, 55, 70, 72, 146, 167, 172–173, 180, 191, 197–198, 216, 261–262, 266, 268, 334, 337–338, 346, 362, 408, 450, 455
CuO planes, 34, 188, 216, 571, 586–587, 590–591
CuO_2 layers, 101, 178, 191, 207, 257, 266, 268, 277, 325, 327–342, 452
CuO_2 plane, 1, 4–6, 35, 71, 95–97, 101, 122, 154, 199, 258–259, 270, 276–277, 288, 401, 408, 414, 416, 454, 456
CuO_2 planes, 35, 38, 57, 95, 97, 101, 103, 118, 151, 167, 177, 185, 192, 216, 224, 226, 236, 250, 253, 258, 264–267, 271, 274, 280, 284, 286, 288, 335, 338, 341, 346–347, 361–364, 367–368, 370–372, 401, 409, 427, 443, 448, 451, 453–456
cuprate superconductors, 1, 11, 17–20, 38–39, 42–43, 45, 50–51, 58, 64, 67, 69, 71–74, 87–88, 95, 99, 124, 138, 150, 160, 257, 259, 288–289, 299–301, 320–322, 327, 341, 345–348, 452, 569, 571
Curie (term, temperature), 363, 372, 598–599, 604

Debye temperature, 350, 403, 431, 437, 469
degeneracy, 6, 103, 107, 330, 535, 547
d-electrons, 430
delocalization, 2, 194, 478, 531, 575
demagnetization, 175, 221, 610
density matrix renormalization group (DMRG) approach, 495–496, 503, 510–511, 579
density mediated superconductivity, 601
density of states (DOS), 19–20, 23–24, 26, 45–46, 51, 60–61, 64, 70–73, 75, 109–110, 117–118, 123, 154–155, 178, 180, 198, 221, 223–224, 300–301, 317, 320–321, 350, 404, 431, 435–436, 440–441, 469, 513–514, 538, 550, 555
density wave, 338, 463, 471–472, 478, 570, 572, 591
de-pinning, 385–386, 468
deuteration, 475
diagonal stripe order, 274–275
diamagnetic susceptibility, 544
diamond anvil cell, 609

dichotomy (doping), 106, 115, 546
dichotomy (nodal), 130–134,
diffraction, 3, 91, 157, 215, 245, 257, 264–267, 275, 277, 279, 285, 287, 333, 361, 434
diffuse scattering, 5, 272, 274
diffusion, 24, 65, 336, 449
dimensionless coupling constant, 103, 106
dimer, 478, 480, 534, 557, 561, 579
dimerization, 467, 476, 478–479
dirty limit, 24, 28, 147
disorder, 22, 28, 36, 72, 106, 147, 151–152, 171, 180, 185, 189, 192, 199, 206, 273, 300–301, 304, 312, 315–316, 318, 320, 322, 328, 334, 340, 380, 383, 410, 416–417, 466, 530, 534, 544, 583, 585, 591, 602, 606–607
dispersion, 2–3, 29, 31, 44, 50, 71–72, 87, 89, 94–95, 101, 104–107, 109–116, 118–120, 125–130, 152–153, 259–262, 267–271, 278, 281–282, 420, 454, 465, 509, 513, 538, 579, 581
distortion, 8, 17, 110–111, 253, 266, 362, 528
donor, 465–467, 469, 475–476, 480, 484, 489
doped Mott insulator, 151, 527, 541, 546, 564
double chain cuprates, 334, 337
double chain, 334–339, 341, 409
double resonance, 220, 242, 245–246
Drude behavior, 147, 154–156, 160, 166–167, 182–184, 196–199, 299–304, 307–310, 312–314, 321, 401–402, 404, 410, 412, 416, 532, 541, 546, 587
d-wave character, 523, 588
d-wave order parameter, 22, 34, 37, 87, 177, 356–357, 359, 507
d-wave superconductivity, 39, 51, 173, 325, 339, 421, 527–528, 540, 607
d-wave symmetry, 26, 60, 99, 135, 176, 289, 340, 402, 418, 481, 483, 487
$d_{x^2-y^2}$ symmetry, 19–20, 22, 42–44, 46–47, 49–50, 53–60, 63, 66–67, 69–71, 75, 96, 109, 135–137, 151–153, 155–156, 239, 340, 480, 483, 495–496, 506–508, 516–517, 519–520, 522–523, 529
d_{xy} symmetry, 479–480
$DyBaCu_3O_{6+x}$ (DyBCO), 373–376
dynamic inhomogeneity, 569
dynamic susceptibility, 270, 278
dynamical coexistence, 17
dynamical mean field theory, 495, 607
Dysprosium (Dy), 361, 580
Dzyaloshinsky–Moriya (DM), 253, 265, 271

effective dimensionality, 597
effective mass, 1, 105, 111, 116, 147, 358, 401, 475, 583
effective volume, 158–159
Einstein modes, 107–109
elastic constants, 454
electrical transport, 152, 154–155, 399, 433, 417, 457
electrodynamics, 145–147, 151, 162, 166–167, 171, 178, 209, 543
electron correlations, 100, 102, 151, 155, 404, 408, 458, 421, 490
electron density of states (EDOS), 350, 356–357, 365–367, 369–372, 377, 379, 386–387, 389–390
electron doped systems, 2, 32, 38, 43, 61, 68, 175, 530
electron paramagnetic resonance (EPR), 1, 4–6, 9–14, 17, 252–253, 330
electron spin resonance (ESR), 161, 204, 215–216, 250, 252
electron-electron correlations, 100, 151, 155
electron-electron interactions, 88, 99–100, 102, 116, 138, 184, 223, 472, 481
electron-electron pairing, 74
electron-electron scattering, 184, 414
electronic correlations, 17, 135, 279, 588
electronic structure, 67, 69, 87–88, 90, 95–96, 98, 129, 132, 137, 151, 432–433, 435, 439, 463–465, 483, 527–528, 587, 592
electron-phonon coupling, 72, 75, 99, 100–103, 109, 110–112, 114–115, 119, 122–124, 127–128, 135–138, 141, 399, 431–433, 436, 441, 457, 478, 591–592
electron-phonon interaction, 72–73, 85, 88–89, 99–102, 106–107, 109, 116, 124, 136–137, 143, 349–350, 436, 443, 448, 452, 597
elemental superconductors, 428, 430, 437, 471, 488
Eliashberg, 103, 106–108, 116–117, 128, 138, 141, 481, 584
emergent phenomena, 565, 569, 583
energy bands, 8, 125, 130, 605
energy dispersion, 110, 114–115, 126, 420
energy distribution curve (EDC), 91, 93–94, 110, 113–116, 119–121, 125, 130, 132
entropy, 116–117, 340, 369–371, 384–385, 390, 513, 532, 543
$ErMo_6Se_8$, 600
ethylene conformation, 466
exact diagonalization, 496, 502, 557

Index

exact solution, 111, 569
excitation spectrum, 151–152, 154, 257–258, 261, 263, 289, 317, 404
excluded volume techniques, 158–159, 161, 164
extended X-ray absorption fine structure (EXAFS), 1–2, 4–5

Fabry–Perot resonator, 166
Fano function, 101–102
Fermi arc, 131–132, 410, 533, 564
Fermi level, 25, 50, 70, 94–95, 114–115, 120, 129, 131, 133–134, 155, 289, 317, 330, 338, 350, 379, 465, 606–607
Fermi liquid (theory), 64, 87, 97, 103, 112, 154–155, 177, 194–195, 348, 390, 487, 528, 531, 540–541, 549–550, 571, 583–585, 589–591, 598–599, 611
Fermi-surface nesting, 279
Ferromagnetic ordering, 603, 606, 611
Ferromagnetism, 597–598, 600, 602–606, 608
field induced superconductivity, 483
first order transitions, 352, 369, 380–383, 385–386, 577, 601–602, 606
fluctuation effects, 178, 202, 205, 208, 345, 372–373, 403, 416
flux quantum (Φ_0), 40–41, 43, 55, 62, 160, 203, 538, 539
flux-lattice melting, 382
form factor, 123, 136, 231, 234, 244, 267, 270, 280, 553, 555
forward scattering, 137, 579
fractionalization, 557–559, 565, 590
frustration, 62, 266, 478, 480, 489, 535, 560, 566, 569–570, 583, 606
Fulde–Ferrell–Larkin–Ovchinnikov (FFLO) state, 483, 485, 487

Gallium (Ga), 36, 362, 430, 458, 484
gapless superconductivity, 354, 363
gauge fields, 528, 538, 541–546, 549–552, 557–559, 562–565
gauge theory, 527–528, 540–543, 546, 551, 553–554, 556–559, 561–564
Gaussian (distributions, fluctuations), 48, 51, 92, 127, 203, 205, 218, 233–234, 372–376, 378, 537, 541, 543
$GdMo_6Se_8$, 600
Ginzburg Criterion, 373
Ginzburg-Landau (GL), 202, 348, 349, 373
grain boundary, 21–22, 32–36, 38, 43, 53, 55, 67, 450–451
graphite, 468

Green's functions, 107, 495–496, 498, 500–501, 507, 513, 516, 541, 546, 549–550, 614
g-value, 250

half-filling, 267, 420–421, 495, 499, 511, 514, 539–540, 550, 555, 564, 573–574, 607, 611
half-flux quantum, 43–44, 55, 57
Hall angle, 404, 411, 413, 415
Hall coefficient, 106, 131, 399, 410
Hall constant, 330
Hall effect, 330, 400, 417–418, 420–421, 457
Hall voltage, 413
H_c, 66, 285
H_{c1}, 62, 160, 471
H_{c2}, 62, 203, 284–285, 468, 471, 481, 485
heat capacity, 206, 352, 387, 389, 598
heavy Fermion materials, 39, 225, 430, 458, 463, 541, 597, 605, 600–602, 607, 612
heavy Fermion superconductors, 39, 605
Hebel-Slichter peak, 221, 223, 254, 482
Heisenberg exchange, 534, 537
Heisenberg Hamiltonian, 267, 270
Heisenberg model, 249, 268, 478, 507, 534, 540, 557–558
HfV_2, 359
Hg-2212, 586
$HgBa_2Ca_mCu_nOy$, 586
HgO, 332, 334, 341
high magnetic fields, 63, 251, 405, 489–490, 492–493
high pressure (effects, methods), 99, 180, 428–429, 433, 435, 437, 447–448, 452, 455, 457–459, 461–462, 608–610
high T_c (superconductors), 44, 50, 59–60, 64–65, 75, 99–100, 106, 135, 146, 149, 165–166, 190, 210, 214–215, 217–218, 223, 253, 322, 339, 342, 345, 393, 402, 410, 422, 461, 468, 486, 490, 523–524, 527–528, 531, 534–535, 546, 559, 591, 593
high temperature superconductivity (HTS), 1–2, 6, 9, 15, 17, 20–22, 32–35, 37–38, 64, 210, 345–351, 353–354, 356–357, 359–361, 367–368, 372–373, 375–380, 386–387, 390, 395, 462
high temperature tetragonal (HTT) symmetry, 368, 408, 447–448
highest occupied molecular orbital (HOMO), 476
hole density, 247, 274–275, 281, 510–512, 561
hole-doped cuprates, 1, 17, 43, 176, 257, 287–288, 403–404, 410, 413, 510
Holmium (Ho), 14, 458

hopping integrals, 8, 15–17
Hubbard ladder, 503, 509
Hubbard model, 97, 267–268, 327, 463, 476, 479–481, 489, 495–497, 501, 503–504, 506–510, 513–517, 521–523, 527, 531, 558, 570, 573, 576–578, 581–582, 589, 592, 607
Hubbard U, 2, 327, 330, 529
hybridization, 96, 267, 361–362, 410
hydrostatic pressure, 427–428, 434–435, 437–440, 446, 452–453, 455, 457, 466, 475, 480, 599, 603, 608–609
hyperfine (coupling, fields), 225, 229–231, 233, 244, 252, 285, 350–352, 364–365, 457, 492
hysteresis, 39, 59, 381, 383–384, 386

impurities (magnetic), 155, 353–354, 358, 363
impurities (non-magnetic), 70, 146, 155, 356
impurity effects, 57, 185, 199, 201, 265, 353, 356
impurity phases, 194, 287–288, 356, 363, 367
impurity scattering, 67, 69, 147, 156, 177, 187, 194, 410, 414
impurity states, 64, 70–71
incommensurate effects, 133, 215, 244–245, 259–263, 272–273, 275–279, 280–286, 288–289, 334, 478, 555, 560, 582
inelastic neutron scattering, 1–4, 11, 119, 261, 276
infrared active, 136
infrared conductivity, 57, 321–322, 523
infrared reflectivity, 101, 159, 162
infrared spectroscopy, 100
inhomogeneity, 20, 179, 181, 187, 199, 218, 221, 276–278, 287, 290, 299, 301, 315–322, 350, 419, 421, 524, 569–570, 583–586, 588, 591
insulators, 35, 44, 71, 96, 105–106, 126, 128, 130, 151–152, 257, 261, 274, 280, 289, 346–347, 361, 405, 420–421, 467, 471, 477–478, 480, 513, 515, 527–531, 535, 541, 546, 559–561, 563–564, 574, 583, 587, 613
interchain coupling, 489
intermediate coupling, 127, 607
intermetallic compounds, 326
invariance, 57, 399, 452, 498, 521, 537–539, 541–542, 549–550, 552, 558–559, 561, 574, 578, 590
ion, 4, 13, 35, 37, 67, 161, 252, 267, 276, 280, 325, 327, 329–331, 334–335, 340–341, 363–364, 410, 530
ionization potential, 327
Iron (Fe), 250, 362–363, 428, 458, 484, 486, 492, 604–606
irradiation, 34–35, 67, 185, 416

isotope effect, 1–2, 8–9, 12–17, 99–100, 114, 427, 430–431, 439, 482, 588
isotope shift, 14, 16
itinerant electron, 597, 604, 606, 612

Jaccarino-Peter effect, 485–486
Jahn–Teller (effects), 1–2, 5–7, 9, 11, 13, 15, 17, 253
jellium, 605
Josephson behavior, 21, 32, 37
Josephson contacts, 39
Josephson coupling, 30–31, 39, 56–57, 175–176, 580
Josephson critical current, 40, 43, 56
Josephson current, 29–31, 37, 52–53, 57, 59, 74
Josephson effect, 21–22, 25, 27–28, 38
Josephson fluxons, 43
Josephson junction, 27, 30, 39–40, 60, 315, 557
Josephson phenomenology, 22
Josephson plasma frequency, 316, 318
Josephson plasma resonance, 316, 468
Josephson semifluxons, 40
Josephson strength, 52
Josephson supercurrent, 56
Josephson tunneling, 56, 468, 580
Josephson vortex, 40, 43, 468, 485
Josephson weak-link, 21

K_3C_{60}, 439–441
Kagome lattice, 557
k_Fl, 405, 416, 487
Knight shift, 215, 217, 222–229, 231–232, 235–238, 245–247, 249, 253, 322, 357, 387, 481–482, 484, 512, 531–532, 534
Kohler's rule, 399, 410, 414–416
Kramers–Kronig, 108–109, 116, 148, 150, 157, 159, 166, 305–306, 311

$La_2CuO_{4+\delta}$, 1, 37, 95, 97, 124–125, 128–129, 180, 241, 244, 248–249, 254, 257, 264–269, 271–273, 277, 285, 329, 332, 339, 342, 447, 449, 457, 528
$La_{2-x}Sr_xCuO_4$ (LSCO), 2–5, 9, 11–12, 14, 37, 94, 98, 100–101, 104, 112–115, 117, 122–123, 129–134, 215, 220, 240, 243–247, 249–250, 253, 264, 307–308, 346, 356–358, 361, 367–369, 372, 532–534, 553, 555
ladders, 261, 278, 335, 338–339, 476, 495–496, 502–503, 509–512, 522, 524, 529, 573–581
Lanthanum (La), 13, 100, 241–242, 247, 250–251, 274, 327, 346, 360–362,

367–370, 376, 395–396, 410, 427, 429, 436–437, 443, 452, 458, 529
large N expansion, 551, 564
lattice specific heat, 353, 370, 374–375, 387
lattice vibrations, 100, 103, 118, 431, 441, 481
Lead (Pb), 35, 42–43, 45–46, 65, 67–68, 119, 138, 164, 175, 194, 325, 330, 334, 346, 363, 428, 430–431, 433, 446, 458, 581
line nodes, 351, 353, 357, 360
linear resistivity, 99, 399, 401–404, 411, 415, 583, 600
Lithium (Li), 79–82, 84–85, 139–140, 142, 144, 209–214, 286, 292, 294–295, 323, 343, 395–396, 422, 424–425, 428–430, 432–436, 458, 565–566
local density approximation (LDA), 97, 126, 136–137, 178, 327, 411, 509
local susceptibility, 258, 263, 278
localization, 9, 105–106, 276, 361, 405, 408, 416, 530
localized electrons, 103, 106, 330, 478
localized holes, 5, 367–368
localized spins, 478, 549
localized states, 37, 53
logarithmic Fermi liquid, 598
London (model, penetration depth), 1, 15, 17, 85, 139, 141, 144–146, 149, 160, 163, 214, 255, 295, 302, 308, 349, 392–396, 460–461, 463, 490, 492–493, 526, 613
long-range antiferromagnetic order, 272, 274, 288, 403, 503, 505–506, 540–541, 575
Long-range Coulomb interaction, 105, 573
long-range ferromagnetic order, 597
Lorentzian line shape, 10, 95, 127, 199, 301, 310, 312, 401
low temperature superconductivity, 591
low temperature tetragonal (LTT) symmetry, 4–5, 248, 277–278, 281, 285, 368–370, 446–448
lowest-Landau-Level (LLL), 376
low-frequency conductivity, 301
low-temperature orthorhombic (LTO) symmetry, 5, 285, 368–370
Luttinger (liquid), 399, 410, 421, 478, 551, 575, 579, 585

Mach–Zehnder interferometer, 166, 168
macroscopic quantum tunneling, 34, 39, 59
magnetic correlations, 234, 257–258, 263–264, 280–281, 284, 286, 288, 588, 601
magnetic excitations, 117, 132–133, 257–259, 263–264, 278–281, 283, 287, 289–290
magnetic field dependence, 62, 67–68, 346

magnetic incommensurability, 275, 280–282
magnetic interaction model, 600, 602, 606, 611
magnetic ordering, 279, 361, 597
magnetic polaron, 6
magnetic properties, 158, 327, 349–350, 360, 363
magnetic structure, 265–266, 280, 476, 505, 586
magnetic susceptibility, 2, 158, 227, 236, 243, 289, 363, 390, 413, 437, 440, 457, 513, 515
magnetic transition, 273, 599
magnetism, 176, 279, 427, 458, 478, 583, 588, 595, 597, 599–600, 608, 612
magnetization, 55, 163, 218, 225, 228, 236, 336, 346, 349, 362–363, 370, 372, 376, 380–381, 383–386, 482, 504, 535
magnetoresistance, 400, 414, 464, 466
magnetotransport, 404, 413–414, 466
magnons, 19, 72, 74, 106, 127, 456, 529, 588
Manganese (Mn), 216, 251, 253, 279, 362
marginal Fermi liquid (MFL), 64, 87, 112–113, 399, 413–415, 421, 599
Matsubara frequency, 496, 519, 520, 522
Matthiessen's rule, 406, 417
mean free path, 12, 37, 145, 185, 198, 544, 606, 608
mean-field (theory, transitions, etc.), 15, 61–62, 180, 202–204, 235, 300–301, 312, 320, 331, 373–374, 453, 495, 508, 551–552, 554, 528, 536–543, 546–547, 549, 547–550, 558, 561–563, 572, 580–581, 587, 590, 607
Meissner screening, 39, 66, 69, 228, 236, 299, 326
Mercury (Hg), 325, 328–329, 331–334, 341, 346, 376, 430, 446, 455, 458, 586
metamagnetic transition, 604
MgB_2, 98, 341, 429, 438–439, 441, 446, 452, 460, 571
microwave (cavity perturbation), 157–159, 163, 165, 167–168, 172, 174, 180, 188, 195, 203–204
microwave conductivity, 146, 179, 182, 184, 195, 198–202, 205, 301, 312, 321–322
mid-IR absorption, 321
Migdal-Eliashberg theory, 103, 106, 116, 128, 137–138
millimeter-waves, 209, 301, 321
Millis, Monien, and Pines theory (MMP), 215, 229–230, 234, 244
MnSi, 599–600, 606
mobility, 2, 339, 405, 447, 450
momentum distribution curve (MDC), 94–95, 110–111, 113–115, 119–120, 130, 132–133

momentum transfer, 122, 132–133, 137, 271, 413, 517–523
Monte Carlo, 111, 126, 128, 137, 495–497, 499–501, 503–504, 506–507, 513, 516–517, 519–520, 536, 546, 563, 579, 586–587, 592
Mott (insulator, transition), 2, 17, 69, 99, 106, 124, 128, 141, 151–152, 288, 322, 327, 346, 406, 421, 463, 467, 477–480, 487, 489, 495, 515, 523, 527–529, 531, 535, 540–541, 546, 558–561, 563–564, 569, 574, 579, 582–583, 587, 591, 613
multiband effects, 96, 575
muon spin resonance (μSR), 1, 13, 15, 29, 100, 160, 162, 167–168, 170, 174–175, 204, 241–243, 247–248, 253, 272–274, 277, 285–287, 320, 327, 329, 334, 346–348, 356, 410, 428, 430, 437, 458, 487, 529, 555
mutual inductance, 161–162, 168

Na–CCOC, 127, 129, 131–132, 134–135
Nb_3Sn, 326, 437–438
NCCO, 2–3, 38, 161, 530
Néel (order, state, structure), 264, 286, 528, 531, 534–535, 555, 557, 560–562, 564, 590–591
Néel temperature, 188, 257, 287, 272–273, 348, 361, 600, 607
Nernst (effect, state), 417, 419, 556–557, 564
Nernst-Ettinghausen effect, 419
nesting, 132–133, 135, 279, 465, 475, 478, 489, 538
neutron (scattering, diffraction), 1–5, 11, 14, 73, 75, 101, 114, 117–119, 132, 137, 183, 215, 227, 244–245, 248, 250, 257–261, 264, 266, 271–272, 275–276, 279–280, 282–284, 286–287, 289, 290–291, 294, 322, 333, 340, 361, 363, 388, 439, 523–524, 534, 546, 550, 553, 555, 586, 588
Ni_3Al, 598, 606
Niobium (Nb), 35–36, 38, 42–44, 55–56, 194, 250, 429, 438, 458, 571
NiS_2, 607
nodal direction, 3, 94–96, 111–114, 116, 123–124, 129–131, 134, 137, 151, 415, 420, 483, 513, 533
nodal dispersion, 112, 114–115
nodal Fermions, 550–551
nodal points, 120, 153, 414, 519, 533, 538, 551, 564
nodal quasiparticles, 59, 129–130, 183–184, 196, 201, 533, 550, 561–562, 573, 582, 589–590
non-adiabatic effect, 124

non-Fermi liquid (NFL), 348, 399, 487, 544, 546, 584, 601
non-hydrostatic pressure, 435, 437, 444, 446
non-local effects, 146, 151, 170, 311, 507
nuclear magnetic resonance (NMR), 2, 4, 64, 160–161, 181, 215–221, 223–228, 230–232, 234, 236, 240, 242–243, 245–250, 252–255, 272, 287, 322, 325, 329, 331, 364, 367–368, 386–387, 416, 469, 481–482, 484, 487, 555, 604
nuclear quadrupole resonance (NQR), 2, 4–5, 13–14, 215, 240, 242, 247, 249–251, 254, 329, 336, 351, 367–368
numerical methods, 495–496, 522, 524, 592

one-band Hubbard model, 496, 509–510, 531
one-dimensional (1D) systems, 28, 70, 91, 157, 290, 302, 317, 321, 336–338, 340–341, 475, 478, 489, 501, 534–535, 574–576, 579, 585, 589
optical conductivity, 106, 274, 276, 299–301, 304–307, 309, 311–312, 314–318, 320–322, 401, 404, 406, 408, 420, 487, 513, 546
optical phonons, 99, 113, 118
optical properties, 301, 541
optical spectroscopy, 100, 408
optimally and slightly over-doped systems, 356, 376
optimally doped systems, 2, 14–15, 38, 42–45, 47, 49–50, 52–53, 55, 57–59, 61–62, 67–71, 74, 87, 98–100, 112, 114, 120–122, 124, 159, 168, 174, 176, 180, 184–185, 188, 190, 192, 200, 203–206, 208, 220, 228, 230, 237, 239–240, 243–246, 248, 257, 261–264, 279, 282, 288–289, 301–304, 306–310, 313–314, 322, 325, 328–329, 330–334, 338, 341, 348–349, 356–357, 373–375, 377–379, 399–407, 410–411, 415, 417–418, 421, 443–445, 448, 450–457, 488, 513, 532–534, 541–542, 546, 587–588
orbital currents, 280, 555, 557
ordered charge, 258
organic conductors, 463–469, 478, 480, 486–490
organic systems, 487–488
orthorhombic symmetry, 8, 55–56, 147, 152, 167, 178, 188, 265–266, 272–274, 285, 337, 359, 361–362, 368, 408, 448, 523, 589
over-doped systems, 2, 35–36, 38, 43, 52–53, 58–60, 63, 68–69, 71, 74, 98, 109, 119, 132, 137–138, 172–173, 253, 281, 286, 289, 301,

310–311, 314–315, 320–321, 342, 348, 356, 363, 373, 376–377, 386, 400, 406, 409, 418, 443–444, 446, 448–450, 513, 530, 584, 586
oxides, 65, 76, 85, 95, 336, 428, 446–447, 451, 527
oxygenated materials, 284, 346, 387–388, 450–451

pairing interactions, 7, 67, 74, 136, 399, 431, 454, 457, 496, 507, 516–524
pairing mechanism, 1, 16, 87, 118, 124, 135, 151, 175, 263, 326, 329–330, 399, 428, 443, 452, 482, 523–524, 569–570, 572–573
pairing states, 184, 236, 541
pairing theory, 152
parabolic bands, 107
parent insulators, 257–258, 264, 288
PbTe, 330
Peierls transition, 464, 478, 489
penetration depth, 1, 15–17, 38, 40–41, 51, 59, 62, 100, 145–147, 149–151, 154, 156–158, 160–164, 166, 169–172, 177–179, 185, 187–188, 195, 202–203, 302–303, 305, 309, 337, 348, 356, 467, 482–484, 487–488, 563
perovskite structure, 346
perovskites, 463, 589, 591
perturbation theory, 543, 574, 580, 607
phase separation, 5, 244, 273–274, 288, 356, 361, 569–570, 574, 583–584, 589
phase transitions, 323, 394, 429, 431, 434–435, 446–448, 464, 611
phonon interaction, 73, 85, 88, 99–102, 106, 109, 124, 136–137, 343, 349–350, 436, 443, 448, 452, 597
phonon modes, 100, 117–118, 120, 122, 135, 137–138, 481
phonon spectrum, 388, 431, 433, 481
photoemission spectroscopy, 60, 74, 87–88, 92, 122
photoemission, 1–3, 48, 51, 57, 60, 74, 87–95, 102, 109, 118, 121–122, 125, 127, 129, 180, 280, 283, 405, 532, 535–536, 541, 564
Pippard coherence length, 349
piston cylinder, 609
plasma frequency, 160, 301–302, 307, 316, 318, 321, 401, 553
polarons, 1–2, 4–9, 17, 99, 105–106, 110–111, 116, 124, 126–129, 133–135, 572
powders, 158, 160, 162–163, 175, 179, 225, 236, 264–265, 274, 287

$Pr_{2-x}Ce_xCuO_4$, 68, 287, 404, 418
Praseodymium (Pr), 100, 266, 287–288, 334, 336, 338, 360–362
pressure cell, 608–609
pressure effect, 454
pressure media, 428–430, 433, 438, 446, 458, 609
propagator, 107, 498, 500, 516, 544–545
proximity effects, 19, 21–25, 36–38, 46, 50, 59, 69, 73, 99, 152, 157, 173, 192, 283, 320, 338–339, 343, 404, 413, 420, 463, 476, 479–481, 487, 489, 528, 535, 574–576, 584, 602–603
pseudogap behavior, 20, 60, 487, 495, 503, 513–514, 534, 560
pseudogap, 1, 11–12, 19–20, 51–53, 58, 60–63, 67, 75, 87, 98, 138, 227, 235, 253, 280, 284, 289, 345, 348, 351, 354, 367, 377, 379, 386–390, 404–405, 408, 412, 416, 418–419, 421, 427, 450, 457, 487, 495–496, 503, 512–514, 522, 527–528, 530–535, 541, 547, 549–550, 554, 556–557, 560–564, 569, 582, 589
puddle fluid, 583
p-wave pairing, 463, 469, 481, 487, 489, 606

quantum critical behavior, 14, 59, 112, 284, 403, 413, 480, 560–561, 584, 598, 602–603, 605–606
quantum critical point (QCP), 59, 284, 403, 413, 480, 560, 598, 602–603, 606
quantum Monte Carlo, 111, 126, 137, 495–497, 499–500, 504, 513, 516–517, 520, 536, 546, 592
quantum-renormalization, 266, 268
quasioptics, 165–166, 168
quasiparticle interactions, 606
quenched disorder, 273, 315, 583, 585, 606
quenched systems, 273, 299, 301, 315–317, 319, 321–322, 583, 585, 590, 606

Raman (scattering, spectra), 11, 100–101, 103–104, 122, 136, 141, 268, 334, 343, 387, 420, 439, 457, 529, 545, 553, 588, 595
Rb_2CsC_{60}, 571
RBCO ($RBa_2Cu_3O_X$, R = Y, Gd, La), 348, 373, 375–376, 386
reflectance, 101, 306
relaxation energy, 103, 110
relaxation rate, 137, 197, 223–224, 239, 241, 244, 247, 336, 419, 469, 482, 484, 534, 598

renormalization group (RG) approach, 8, 15, 73, 89, 107, 111–112, 114, 118–120, 122, 124, 126, 178, 271, 312, 373, 394, 420, 481, 495–497, 501–502, 507, 509–510, 523, 544, 572, 575, 579–580
resistive transition, 467
resistivity, 11, 74, 99, 106, 137, 150, 154, 178, 181, 184, 337–338, 364, 380–381, 399–404, 406–408, 410–411, 413, 415, 417, 433, 442, 457, 466, 534, 546, 583, 598–601, 608–610
resonant cavity, 157, 163, 165, 194,
resonators, 145, 162, 165
retardation, 523, 572, 584, 591
Ruderman-Kittel-Kasuya-Yosida (RKKY) interaction, 220

scaling (theory, parameters), 21–22, 51, 62–63, 75, 197–198, 203–205, 207, 215, 234–236, 243, 249, 254, 264, 272, 286, 357, 359–360, 372, 375–376, 383–386, 401, 406, 410, 414–415, 421, 487–488, 497, 499, 502–503, 507
scanning SQUID microscope, 43–44
scanning tunneling microscopy, (STM), 32, 45, 47–49, 57, 61–64, 70–72, 91, 132, 135, 154, 157, 285, 301, 322, 334, 417, 419, 421, 483, 485, 524, 547, 555, 585
scanning tunneling spectroscopy (STS), 196–197, 200
Schottky anomalies, 351, 363–365
screening effects, 9, 66, 69, 136–137, 147, 151, 161, 163, 180–182, 187, 192, 194, 228, 274, 299, 542, 559
screening length, 9
second order transitions, 110, 117, 164, 267, 315, 380–386, 579–580, 598–599
self-consistent Born approximation, 103, 125–127
self-consistent T-matrix approximation (SCTMA), 154
self-energy, 47, 51, 73, 89, 94–95, 99, 106–109, 114, 116–118, 122, 137, 156, 197, 413–414, 419, 500–501
short-range antiferromagnetic order, 44, 274, 514
short-range effects, 69, 274, 514, 535, 573, 582–583, 588
Shubnikov-de Haas, 464
Sodium (Na), 435, 533
spectral weight unit (SWU), 302, 304, 307–309, 316
spin echo double resonance (SEDOR), 220–221
spin gap, 71, 215, 228, 232, 235, 258, 260–264, 271, 281–283, 284–288, 535, 540, 547, 573–580, 582, 590

spin lattice relaxation time (T_1), 8, 16–17, 215, 218–224, 231–232, 234, 239, 241, 243–244, 247–250, 336, 352–353, 358, 484, 487, 534
spin liquid, 289, 480, 527–528, 535, 551, 557–562, 564–565, 579, 590
spin polarization, 576
spin pseudogap, 227, 284
spin stripes, 258, 280–281, 286
spin structure, 478
spin susceptibility, 184, 215, 217, 222, 227, 229, 232–234, 237–238, 245, 258, 456, 482, 513, 519–520, 522–523, 531–532, 534–535
spin waves, 257, 267–268, 278, 574, 598
spin-charge separation, 60, 178, 340, 399, 421, 535, 574–575, 584
spin-density-wave (SDW), 273, 279, 284–285, 289, 363, 367–372, 476, 478, 480, 489, 570
spin-flip, 39, 253, 280
spin-fluctuations, 60, 65, 75, 136, 178, 184–185, 228, 230–231, 239, 245, 250–252, 266–267, 276–277, 322, 371, 413, 418, 420, 457, 479, 517, 523, 534, 550, 553, 555, 562, 577, 582, 603, 612
spin-glass phase, 215–216, 250–252, 273, 280
spinons, 64, 126, 289, 535, 538, 542, 557, 559, 561, 574, 579
spin-triplet, 237, 600, 603–604, 606, 608
spin-wave, 257, 266–271, 276, 278–279, 478, 505, 529, 574, 580, 584, 598
$Sr_2CuO_2Cl_2$, 124–126, 266–267, 271–272
Sr_2RuO_4, 125, 127, 407, 410, 413, 464, 606
$Sr_3Ru_2O_7$, 606
Sr-doped compounds, 215, 248
stacking, 265–268, 467, 475–476, 478, 480, 488–489
staggered flux phase, 549
staggered (states, phases), 280, 511, 535, 538, 541, 549–555, 557, 562, 564
stiffness (phase), 145–146, 172, 207, 271, 287, 315, 319, 330, 441, 487–488, 586
stoichiometry, 180, 229, 271, 299, 309, 321, 336, 338, 361, 367, 409, 466
stripe order, 13–14, 71, 215, 248, 250, 257–258, 261, 274–279, 281, 284–286, 288, 290, 582, 584–586, 588
striped Hubbard model, 570, 576–578, 581–582
stripes, 1, 4–5, 7, 9, 11, 13, 69, 71–72, 75, 132, 151, 248–249, 255, 258, 274–277, 279–281, 285–286, 290, 325, 339–340, 345, 360, 367–368, 370–371, 495, 503, 510–512, 522, 524, 560, 569, 576, 578, 585
strong coupling limit, 527

Index

strong coupling, 19, 65, 73, 110–11, 127–128, 134–135, 238, 373, 471, 481–483, 489, 510, 524, 527–528, 570, 575–576, 589, 591, 611
strongly correlated systems, 87, 289, 342, 406, 524, 528, 570, 582–584, 588–590
structural transitions, 361, 370, 434–435, 437–438, 444, 451
structure factor, 258, 266, 269–270, 272, 274, 280, 505
superconducting fluctuations, 146, 181, 208, 314, 327, 340, 400, 413, 416, 419
superconducting gap, 8–9, 19–20, 26, 44, 48, 51–53, 57, 60–62, 65, 70–72, 74–75, 87, 89, 91, 98–99, 111, 118–119, 122–123, 147–148, 153, 180, 184, 282–283, 289, 387, 427, 457, 471, 485, 586
superconducting phase, 26, 38, 44, 59, 264, 273–274, 281, 287, 289, 336, 419, 434–436, 480, 485, 516, 519, 556, 572, 578
superconducting transition temperature (T_c), 1, 8, 16, 44, 87, 221, 258, 427, 435, 440, 447, 466, 471, 530, 560, 569, 584, 592, 600–603, 607, 611
superconductivity on the border of antiferromagnetism, 605, 607
superconductivity on the border of ferromagnetism, 603, 605
superconductivity on the border of magnetism, 600, 608
superconductivity without inversion symmetry, 608
superexchange, 257, 260–261, 266, 268, 270, 276, 287–289, 339, 457
superfluid density, 145, 149–150, 152, 154, 167, 169–170, 172–173, 177–179, 194, 200, 203, 205, 286, 299, 301, 305, 310, 312, 315–318, 320, 330–331, 337, 339–340, 356, 401, 532, 550, 572, 585–587
superfluid stiffness, 487–488
superlattice, 4, 258, 267, 273–275, 277, 287, 535
superstructure, 119, 334
surface impedance, 147, 150, 159, 164, 188, 192, 196
surface resistance, 147, 163, 165, 172, 179–181, 185–187, 189, 192–193
surface states, 31, 90, 116
s-wave pairing, 55, 67–68, 87, 135, 239–240, 373, 481
s-wave superconductivity, 30, 56, 65, 70, 108, 146–147, 160, 192–193, 300, 482–483, 582
synchrotron, 87, 90, 92, 118, 139

terahertz (THz), 147, 162, 299, 183, 190, 301, 304–313, 316, 318–319, 321
tetragonal symmetry, 7, 57, 105, 152, 174, 177–178, 248, 264–266, 270, 272–274, 368–369, 408, 448, 450, 454–455, 600–602, 611
t_G, 242, 373–374
Thallium (Tl), 180, 325, 328–331, 334, 341, 344, 346, 376, 430, 451–452, 458, 586
thermal conductivity, 154–155, 183–184, 194–196, 198, 417–418, 485, 546, 588
thermal contraction, 428
thermal expansion, 35–36, 159, 161, 163–164, 385, 408, 438, 453–455
thermal fluctuation, 564
thermal transport, 183, 185, 400, 417
thermodynamic critical field, 66
thermodynamic properties, 203, 345, 380
thermopower (thermoelectric) effects, 417–418, 453, 457, 546
thin films, 21, 32, 35, 38, 43, 58, 149, 157–158, 161, 163, 165–166, 168, 175–176, 179, 183, 186–187, 199, 203, 205, 207, 287, 299, 312, 315, 321, 323, 329, 412, 428
three-dimensional (3D) systems, 56, 88, 96–97, 136, 202, 224, 264, 272, 279–280, 288, 327–328, 340, 346, 372–376, 378–380, 383–386, 407, 409–410, 421, 431, 433, 436, 471, 534, 540, 544, 556, 586–587, 600–601, 611
tight binding models, 50, 105, 122, 129, 153, 465, 467, 476, 479, 481, 489, 509, 529–530, 538
tight-binding band, 122
time reversal symmetry, 19–20, 28, 30, 67, 75, 98, 280, 590,
t–J model, 63, 78, 97–98, 101, 111, 125–128, 256, 293, 495–496, 511, 524, 527–528, 530–531, 536, 538, 547–548, 563–564, 573–574, 578–579
Tl-1223, 331
Tl-2201, 46–47, 50, 343, 376, 379, 444, 448–449, 451–452
Tl-2212, 452, 586
Tl-2223, 342, 348, 452
T-matrix, 154, 183, 196–197, 199
TMTSF, 464–467, 471–475, 481–482, 489–491, 571
TMTTF, 465, 471–472, 474
transfer energy, 97
transfer integral, 406, 476, 480
transition metals, 264, 326–327, 342, 430, 435–437, 441, 443, 445–446, 457, 527

transition-metal-oxide, 264, 327, 527
transmission effects, 24, 26, 34, 44, 48, 53, 57, 67, 157, 162, 164–165, 168
transmission electron microscopy, 57
transport lifetime, 150, 201, 413–414, 544–545
transverse relaxation time (T_{2g}), 96, 215, 218–219, 232–236, 240, 242–243, 245, 248–249
tunneling processes, 6, 19–20, 22–23, 28, 32, 34–35, 37, 39, 44–52, 54, 56–76, 91, 98, 101, 116, 119, 132, 283, 285, 301, 327, 329, 331, 334, 387, 417, 468, 483, 513, 523, 533, 555, 580, 582, 586–587
tunneling spectroscopy, 19, 44, 51–52, 57–58, 63, 70, 72, 75–76, 387
twinned boundaries, 285–286
twinned crystals, 55, 167–168, 204, 260, 286, 381, 383, 386
two-band (model), 479, 523
two-component (model), 1, 7–8, 15, 17, 215, 243, 401–402
two-dimensional (2D) systems, 24, 50, 94, 154, 174–175, 207, 249, 268–272, 274, 276, 287, 329, 340, 346, 372–373, 376, 380, 399, 405–406, 408, 410, 413, 417, 421, 429, 465, 471, 475, 495–496, 503, 506–509, 563, 573–574, 582, 589

UGe$_2$, 603–604, 606
UIr, 604–606, 608
ultra-pure YBCO, 299, 309, 321–322
Umklapp (scattering), 184–185, 413, 420, 514, 574
unconventional pairing, 19–20, 39, 45, 48
unconventional superconductivity, 20, 28, 39–41, 138, 154, 156, 463, 481–482, 608–609
unconventional superconductors, 20, 28, 39, 40–41, 139, 154, 156, 342, 463, 481–482, 526, 567, 592, 595, 608–609
underdoped systems, 5, 12, 15, 36, 51–52, 54, 58, 60–61, 63, 67–68, 70–72, 74, 87, 98, 101, 106, 111–113, 116–117, 127, 129, 131–132, 137–138, 159, 173, 178–180, 200, 208, 232, 253, 259–260, 280, 282, 284, 288–290, 305, 310–311, 321, 332, 338, 340–341, 348, 354, 356–357, 373, 376–377, 379, 386–387, 400, 404, 408, 415, 443–445, 447–448, 450, 457, 488, 496, 512–513, 515, 527–528, 530–534, 541, 546–547, 555, 561, 564, 582, 586–588
undoped LCO, 215

uniaxial pressure, 427–428, 443, 453–455, 457, 471, 480, 489
uniaxial strain, 479–480
uniaxial stress, 475, 480–481
uniform RVB (URVB), 541–542, 546, 548–550
universal magnetic excitation spectrum, 257, 263
universality, 202–203, 206. 322, 374, 402, 422
unpolarized neutrons, 280
untwinned crystals, 55–56, 167, 381–383, 385–386, 405
URhGe, 604
URu$_2$Si$_2$, 48

V$_2$O$_3$, 607
V$_3$Si, 326, 359, 437–438
valence band, 95, 126, 128–129, 131, 330
valence bond - resonating (RVB), 354, 408, 527–528, 534–536, 538–540, 546, 549, 557, 560
valence bond - systems, 561, 579, 590
van Hove singularity, 50, 123, 404, 538
Vanadium (V), 254, 429, 436–438, 458
variational Monte Carlo, 563
vertex (corrections), 103, 107–109, 122–124, 136–138, 155, 194–195, 496, 501, 516–522
vibronic states, 1–2, 6, 17
VO, 439
vortex dynamics, 469
vortex glass, 383, 386
vortex lattice, 63, 345, 349, 352–353, 380, 383, 385–386, 468
vortex lock-in, 469
vortex pinning, 62
vortex solid, 381
vortex-lattice melting, 345, 352, 380, 383, 385–386
vortices, 20, 40, 43, 62–64, 160, 174, 207, 221, 263, 285, 357, 380, 385–386, 419, 468, 485, 554–557, 561–562
VUV lasers, 90, 92

weak coupling, 22, 32, 50–51, 89, 99, 110–111, 127, 152, 167–168, 174–176, 240, 262, 289, 361, 379, 432, 441, 471, 510, 575–576, 584, 589, 591, 607
Werthamer Helfand Hohenberg (WHH) expression, 350
wipeout (intensity), 248–249, 251

X-ray absorption near edge structure (XANES), 1, 4, 13–14, 335
X-ray absorption, 334

Index

Y-123, 70, 337–338, 428, 442, 444, 446–455, 457
Y-124, 337–338, 446, 450, 454–455
$YBa_2Cu_3O_{6+x}$ (YBCO), 3, 146, 149, 156–157, 159–161, 167–168, 170, 172, 174–175, 177–178, 180–181, 183, 185, 187–188, 195, 198, 200, 202–208, 253, 258–262, 264, 266–268, 271–274, 277, 279–284, 286, 289, 346, 348, 383, 386–390, 408
$YBa_2Cu_3O_7$, 45, 55–56, 58, 62, 67, 98, 100, 171, 185, 187, 254, 266, 284, 299–300, 347, 355, 358, 401, 405, 407, 409, 412–413, 415–417, 420, 428, 442, 450
$YBa_2Cu_4O_8$, 146, 180, 409, 416, 446, 534, 589
YbC_6, 605
YBCO ($YBa_2Cu_3O_{6+x}$), 3, 14, 32–38, 42–46, 49–50, 53–56, 58–60, 62–63, 66–69, 101–102, 105, 167, 169, 172, 178, 215–216, 226, 228, 230, 236–237, 239–241, 264, 299–310, 312, 321–322, 346, 348–349, 351–359, 361–365, 367–368, 372–375, 377–383, 385–387, 450, 488, 523, 531–534, 555, 586
YBCO ($YBa_2Cu_3O_{6+x}$), above T_c, 226
$YbRh_2Si2$, 600
Ytterbium, 605
yttria-stabilized zirconia, 33, 168, 182, 185

zero bias effects, 19, 45–46, 50, 52, 58, 64–65, 67, 74–75
zero energy states, 20, 26, 31, 64, 66, 70, 75, 155, 523
zero-point energy, 271–272, 575
zero-point fluctuations, 266–267
Zhang–Rice (ZR) singlet, 97, 137, 225, 340, 344, 530
Zinc (Zn) (substitution, doping), 35, 70, 149, 168–169, 184, 185–187, 195, 272, 286, 288, 338, 360, 362–365, 367, 369–372, 388, 405, 413, 416–417, 430–431, 458, 534
Zirconium (Zr), 429, 437, 458
$ZrZn_2$, 598

Printed in Singapore